CHEMISTRY OF HAZARDOUS MATERIALS

CHEMISTRY OF HAZARDOUS MATERIALS

5th Edition

Eugene Meyer

Pearson

Boston   Columbus   Indianapolis   New York   San Francisco   Upper Saddle River
Amsterdam   Cape Town   Dubai   London   Madrid   Milan   Munich   Paris   Montreal   Toronto
Delhi   Mexico City   Sao Paulo   Sydney   Hong Kong   Seoul   Singapore   Taipei   Tokyo

Library of Congress Cataloging-in-Publication Data

Meyer, Eugene
  Chemistry of hazardous materials / Eugene Meyer. — 5th ed.
    p. cm.
  Includes bibliographical references and index.
  ISBN-13: 978-0-13-504159-8 (alk. paper)
  ISBN-10: 0-13-504159-7 (alk. paper)
  1. Hazardous substances—Fires and fire prevention—Textbooks.
  2. Hazardous substances—Textbooks.   I. Title.
  TH9446.H38M48 2010
  628.9'2—dc22

                                        2009017693

Publisher: Julie Levin Alexander
Publisher's Assistant: Regina Bruno
Executive Editor: Marlene McHugh Pratt
Senior Acquisitions Editor: Stephen Smith
Associate Editor: Monica Moosang
Development Editor: Alexis Breen Ferraro, Triple SSS Press Media Development
Editorial Assistant: Heather Luciano
Director of Marketing: Karen Allman
Executive Marketing Manager: Katrin Beacom
Marketing Specialist: Michael Sirinides
Marketing Assistant: Lauren Castellano
Managing Production Editor: Patrick Walsh
Production Liaison: Julie Li
Production Editor: Karen Fortgang, bookworks publishing services
Manufacturing Manager: Ilene Sanford
Manufacturing Buyer: Pat Brown
Senior Design Coordinator: Christopher Weigand
Cover Designer: Bruce Kenselaar
Cover Photo: Mike Greenlar/Syracuse Newspapers/The Image Works
Composition: Aptara®, Inc.
Printing and Binding: Courier Kendalville
Cover Printer: Phoenix Color Corporation

10 9 8 7 6 5 4 3

www.pearsonhighered.com

ISBN: 13: 978-0-13-504159-8
ISBN: 10: 0-13-504159-7

ABOUT THE COVER

On March 12, 2007, CSX Transportation Train No. Q39010 was traveling from Buffalo, New York, to Selkirk, New York, when 29 of its tankcars derailed near Oneida. In its entirety, the train consisted of 79 cars, 20 of which contained liquefied petroleum gas (LPG), a flammable gas; one contained toluene, a flammable liquid; and one contained ferric chloride, a corrosive liquid. During the derailment, four pressurized tanks containing LPG were punctured, which resulted in the immediate release of their contents into the environment. As shown on the cover, the LPG ignited and burned. Approximately 500 gal (1.9 m^3) of toluene and 17,000 gal (64 m^3) of ferric chloride was also released from each of two nonpressurized tankcars. Emergency responders from the Oneida Fire Department responded to this hazardous materials mishap.*

*Railroad Accident Report No. DCA-07-MR-009, "Derailment of CSX Transportation Train No. Q39010, Oneida, New York, March 12, 2007," NTSB/RAB-08/05 (National Transportation Safety Board, Washington, DC, September 30, 2008).

CONTENTS

Chapter 2: Features of Matter and Energy 42

Chapter 3: Flammable Gases and Flammable Liquids 86

Chapter 4: Chemical Forms of Matter 126

Chapter 7: Chemistry of Some Common Elements 249

Chapter 8: Chemistry of Some Corrosive Materials 298

Chapter 9: Chemistry of Some Water-Reactive Substances 344

Chapter 11: Chemistry of Some Oxidizers 467

Chapter 13: Chemistry of Some Hazardous Organic Compounds: Part II 589

Chapter 15: Chemistry of Some Explosives 702

PREFACE

This is not simply another chemistry textbook. It is a chemistry textbook that meets the specific needs of emergency responders, who require not only a basic knowledge of chemistry but also an understanding of the laws and regulations that affect their work at emergency and disaster scenes.

In writing the fifth edition of *Chemistry of Hazardous Materials*, my principal aim was to balance these two points: introduce responders to the fundamental principles of chemistry while simultaneously showing them how to meet certain occupational objectives by recognizing information on tags, signs, markings, labels, placards, and shipping papers. Although some responders are first introduced to chemistry in either high school or college courses, I have written this book at a level that assumes little or no previous exposure to the subject. As responders proceed from one chapter to the next, their comprehension of chemistry should expand, so they will understand why and how the myriad of occupational, transportation, and environmental regulations affect their daily work.

I have also written this book as an introduction to the subject; thus, despite its size, *Chemistry of Hazardous Materials* is intentionally noncomprehensive in scope. I have reorganized the material presented in earlier editions to facilitate understanding and show responders how to grasp the essential principles more easily and apply them to actual situations. This is accomplished in part by the inclusion within each chapter of several solved exercises. These exercises have been designed to demonstrate how responders may use the material introduced in the chapters to predict what happens when hazardous materials are encountered. I have also included end-of-chapter review exercises to provide opportunities for them to determine just how well they are learning before their instructor tests them on the subject.

When discussing the properties of hazardous substances, I have listed the names of each substance under its formula in every equation so students may comprehend the relevant chemical change more readily. I have also listed the basic description of each hazardous material that appears on shipping papers when it is transported.

ACKNOWLEDGMENTS

It is a pleasure to acknowledge the help of the following people and to thank them for their indispensable roles in reviewing selected chapters for this fifth edition:

Diane W. Boykin, Ph.D.
Instructor
Delaware Technical & Community College
Newark, Delaware

Jeff Dean, PSTI-IV
Assistant Chief
Georgia Fire Academy
Forsyth, Georgia

Ed Dwyer
Division Chief
Estero Fire Rescue
Estero, Florida

William D. Hicks, Jr.
Visiting Assistant Professor
Fire and Safety Engineering Technology Program
Eastern Kentucky University
Richmond, Kentucky

John J. Golder
Senior Special Agent
Bureau of Alcohol Tobacco, Firearms and Explosives
Wilmington, North Carolina

Matthew W. Knott
Illinois Fire Service Institute
University of Illinois
Chicago, Illinois
Rockford Fire Department
Rockford, Illinois

Pano Koukopoulos
Adjunct Instructor
Western Connecticut State University
Woodbury Fire Department
Connecticut Fire Academy
Windsor Locks, Connecticut

Steve Malley
Department Chair
Public Safety Professions
Weatherford College
Weatherford, Texas

Richard Moss
Senior Project Manager
Radiological/Decommissioning Services
Energy Solutions
Bethel, Connecticut

Jerry A. Nulliner
Division Chief
Fishers Fire Department
Fishers, Indiana

Tim H. Peebles, AAS, NREMT-P
EMS Coordinator
Hall County Fire Services
Gainesville, Georgia

Christopher Suprun
Director of Education, Consurgo, LLC
Cedar Hill, Texas
Forest Hill Fire Department
Forest Hill, Texas

Tim Swaim
Fire Marshal
Kernersville Fire Department
Kernersville, North Carolina

I am also deeply grateful to all those whose individual talents guided the production of this edition from a manuscript into a book, especially Alexis Breen Ferraro, Developmental Editor, Triple SSS Press Media Development; Monica Moosang, Associate Editor, Brady Publishing/Pearson Education; Stephen Smith, Editor, Brady Publishing/Pearson Education; Julie Li, Production Liaison, Patrick Walsh, Managing Production Editor, Karen Fortgang of bookworks publishing services; Barbara Liguori, copyeditor; and Ravi Bhatt of Aptara Corporation, New Delhi, India.

It is also a pleasure to thank my daughter, Holly, for her many dedicated efforts that helped to make this fifth edition a reality. Holly has read every word of every chapter, solved every exercise, checked arithmetic, contacted parties for figures and permissions, pointed out where clarification was required, searched for errors and typos, encouraged me to swear at editors when the need arose, and helped me in dozens of other ways. Without her intellectual dedication to this project, I could never had prepared the manuscript in a timely fashion. I am enormously grateful for her hard work.

Finally, as with the previous editions, I extend a special thanks to my wife, Phyllis, the woman with whom I share my life. Phyllis has waited in the wings while I completed this and countless other assignments. Her unconditional love, never-ending encouragement, and eternal patience always serve to influence and motivate me. To her, I dedicate this fifth edition.

Eugene Meyer

Eugene Meyer earned a Ph.D. in chemistry from Florida State University, Tallahassee, Florida, in 1964. He completed a postdoctoral fellowship at the Instituut voor Kernphysisch Onderzoek, Amsterdam, The Netherlands, in 1965. He served as Professor of Chemistry with tenure at Lewis University, Lockport, Illinois, from 1965 to 1979. He joined the technical staff of the U.S. Environmental Protection Agency, Region 5, in 1979, serving as Regional Expert in the Chemistry of Hazardous Waste and Chief of the Technical Programs Section of the Hazardous Waste Division. In 1982, he became President of Meyer Environmental Consultants, Inc., where during the course of his work he consulted with attorneys at the U.S. Department of Justice and private law firms or served in the capacity of an expert witness on more than 200 legal matters concerned with the transportation, treatment, storage, and disposal of hazardous substances. He is now retired and lives with his wife in Las Vegas, Nevada.

The following grid outlines the course requirements of the Hazardous Materials Chemistry course developed as part of the FESHE model curriculum. For your convenience, we have indicated specific chapters where these requirements are located in the text.

Course Requirements	1	2	3	4	5	6	7	8	9	10	11	12	13	14	15	16
Identify the common elements by their atomic symbols on the periodic table and demonstrate an understanding of why the table is organized into columns and groups.				X												
Differentiate among elements, compounds and mixtures, and give examples of each.				X												
Explain the difference between ionic and covalent bonding and be able to predict when each will occur.				X												
Identify, name, and understand the basic chemistry involved with common hydrocarbon derivatives.												X	X			
Comprehend the basic chemical and physical properties of gases, liquids and solids, and predict the behavior of a substance under adverse conditions.			X													
Identify, name, and understand the basic chemistry and hazards involved with the nine U.S. Department of Transportation hazard classes and their divisions.						X	X	X	X	X	X	X	X		X	X
Analyze facility occupancy, transportation documents, shape and size of containers, and material safety data sheets (MSDS) to recognize the physical state and potential hazards of reactivity related to firefighter health and safety.	X		X			X										
Demonstrate the ability to utilize guidebooks to determine an initial course of action for emergency responders.	X					X										

New to This Edition

Students will benefit from the revised presentation of chapter material, a new four-color design, and a variety of new content in this edition. Changes to this edition include:

- Correlation guide to FESHE curriculum learning objectives
- Updated color photos and illustrations throughout the text
- Restructured individual chapters that are segregated into topic areas for more effective student learning
- Separation of Chapter 12 into two chapters: Chapter 12, Chemistry of Some Organic Compounds: Part I; and Chapter 13, Chemistry of Some Organic Compounds: Part II
- More specific information on rail and tankcars
- An added discussion of the need for awareness of secondary explosions
- Expanded coverage of terrorist threats relative to biological agents, chemical agents, and radioactive materials
- Updated information on acceptable practices for handling and storage of hazardous materials
- New end-of-chapter exercises
- Expanded coverage of warning labels and material safety data sheets used by emergency responders
- New lists of hazardous materials
- Additional student review material and instructor resources available at www.myfirekit.com. MyFireKit offers a wealth of tools to reinforce student learning and comprehension, including quizzes, audio glossaries, scenarios, links, videos and animations, and more. MyFireKit also includes a full suite of supplements just for instructors, including Instructor Resource Manuals, Test Banks, and PowerPoint slides.

Emergency incidents caused by natural phenomena have continually been mitigated by firefighters, police officers, and other emergency responders. Wildfires, flash floods, earthquakes, erupting volcanoes, droughts, and typhoons are examples of such incidents. These natural events are not entirely predictable and they cannot yet be controlled or prevented.

In the 1940s, new types of emergency incidents began to surface. They were linked with the behavior of certain chemical, petroleum, and nuclear products collectively called hazardous materials. These types of emergency incidents most often occur when hazardous materials are misused or involved in unintended fires and mishaps.

Hazardous materials are needed for the operations of our technology-based society. They are used in connection with the production and manufacture of a host of products. Among the most common of these products are plastics, rubber, leather, paper, textiles, paints, fertilizers, pesticides, solvents, detergents, fuels, medicines, automobiles, televisions, building materials, electronics, and sporting equipment. Their use makes it possible for us to experience the trappings of the "good life." To eliminate hazardous materials from our society would not only be impractical, it would be undesirable. Instead, to reduce the loss of lives and property and protect the environment, we must learn to live safely in their presence and effectively control the emergency-related incidents they produce.

On September 11, 2001, the United States fell victim to a third type of emergency incident. On 9-11, as it is now called, Islamic extremists hijacked and deliberately flew two commercial planes as missiles into the towers of the World Trade Center in New York. Other terrorists flew a third plane into the Pentagon. A fourth hijacked plane did not reach its intended target but crash-landed in rural Pennsylvania. In connection with these reprehensible acts of violence, 2976 innocent people died with 19 hijackers, and at least another 857 people were injured. To execute these horrific acts aircraft carrying a hazardous material—aviation fuel—were used as a weapon of mass destruction.

Terrorists had earlier chosen the World Trade Center as the site of an attack in 1993. On this occasion, they detonated more than a half-ton of explosives in the parking garage of the building. This action killed six and injured approximately 1000 people. Hence, the 9-11 terrorist incidents were not the first deviant actions undertaken by extremists against the United States. Nonetheless, the 9-11 incidents are generally cited as initiating the war on terror in which the United States and other countries are now engaged.

During the aftermath of 9-11 emergency-response personnel widely acknowledged the evolution of a new component of their workload: the necessity to conduct rescue and recovery work at incidents where hazardous materials are employed en masse for destructive purposes. Although emergency responders continue to serve the nation by saving lives, property, and the environment, they must now also help secure and preserve our homeland and the public welfare.

The 9-11 incidents also spawned an extremist ideology that promotes international terrorism. Islamic extremists have expanded the war on terror into virtually every civilized country. The more notorious terrorist incidents headed by Islamic terrorists occurred in 2004, when bombings of Madrid's commuter trains killed 191 people; in 2005, when bombings in London's underground transit system killed 52 people; and in 2008, when bombings in Mumbai's financial district killed 173 people. In these instances, an explosive was the hazardous material chosen to cause injury and death.

The nature of hazardous materials did not change on 9-11, nor did the manner by which we respond to incidents in which they are involved; but the 9-11 terrorist acts did provide a heightened sense of the unorthodox ways by which individuals or groups may use hazardous materials to intentionally kill massive numbers of people, cause substantial property damage, and affect economic stability. Today, the actions of the so-called suicide bombers have become virtually commonplace. Suicide bombers are individuals who detonate explosives strapped to their bodies, usually within densely populated areas. The 9-11 incident also alerted emergency-response personnel to anticipate the presence of biological, chemical, and nuclear warfare agents within our communities.

We begin our study in this first chapter by broadly examining the general features of all hazardous materials. In particular, we note how different federal statutes regulate certain aspects relating to "hazardous chemicals," "hazardous materials," "hazardous substances," "hazardous wastes," and "dangerous goods." In combination, these expressions provide direction in understanding the meaning of the word *hazardous*. We also observe how these statutes aim to eliminate or reduce the risks associated with their usage, storage, and transportation. Finally, we learn that certain interest groups stand ready to help responders when hazardous materials are involved in emergencies.

- Relate the field of chemistry to the profession of the emergency responder.
- Identify the two main branches of chemistry.

1.1 Why Must Emergency Responders Study Chemistry?

Odds are you have never considered what life would be like without **chemistry**. If you had, it would soon be apparent that chemistry affects everything we do. There is not a single instance during which we are not affected by a chemical substance or a chemical process.

Chemistry is regarded as a *natural science*; that is, it is a subject concerned with studying natural phenomena. Specifically, the science of chemistry is the study of substances, their composition, properties, and the changes they undergo. Chemistry concerns itself not only with substances that occur naturally but also with synthetic substances.

By understanding the interplay of various substances, chemists have successfully contributed to combating the scourges of hunger, disease, and human deprivation throughout the world. By using established methods, chemists aspire to further improve the quality of our lives in the future. For instance, during the 21st century, we anticipate that chemistry will radically alter the methods used to treat sicknesses and disease.

Chemistry is broadly divided into two main branches: **organic chemistry** and **inorganic chemistry**. At one time, chemists presumed that certain substances could exist in the world only if they had originated in living things—plants and animals. These substances included sugars, alcohols, waxes, fats, and oils, all of which were called organic substances, since their origin was believed to require the vital force of life itself.

In 1828, this hypothesis was disproved when Friedrich Wöhler prepared a substance called urea, a known constituent of urine. Wöhler successfully prepared urea from substances having no apparent connection to plants or animals. This achievement caused chemists to abandon the idea that a vital force was required for the production of certain substances. Despite abandonment of the hypothesis, however, the name of this branch of chemistry has been historically retained.

Organic chemistry is now recognized as the study of substances that contain carbon in their chemical composition. Although some organic substances are actually found in living systems, many others have been synthesized that have no known natural counterpart. The study of organic substances that affect the life process is now regarded as a subdivision of organic chemistry called **biochemistry**. By contrast, the study of those substances that do not contain carbon in their composition is known as inorganic chemistry. Inorganic substances include aluminum, iron, sulfur, oxygen, table salt, and many others.

To be an effective emergency responder, is it necessary to study organic and inorganic chemistry? It would be misleading to give the impression that the academic pursuit of chemistry is essential for achievement of successful careers as firefighters, police officers, and other emergency responders. Nonetheless, the study of hazardous materials utilizes the fundamental concepts common to organic and inorganic chemistry courses. Thus, when they understand basic chemistry, modern-day emergency responders are equipped with the technical background essential for the performance of duty.

chemistry
- The natural science concerned with the properties, composition, and reactions of substances

organic chemsitry
- The chemistry of the substances that contain carbon in their chemical composition

inorganic chemsitry
- The chemistry of those substances that do not contain carbon in their composition

biochemistry
- The study of the structures, functions, and reactions of the substances found in living organisms

- Name the general types of "hazardous materials" as this term is used within the emergency-response profession.

1.2 General Characteristics of Hazardous Materials

Emergency responders generally regard a hazardous material as any substance or combination of substances that is potentially damaging to health, well-being, or the environment. Based on this viewpoint, there are seven general classes of hazardous materials. We note here the broad features of these classes, and in Chapter 6, we review the manner by which the U.S. Department of Transportation specifically classifies hazardous materials for shipment from place to place. Although the latter system is much more inclusive than the traditional approach followed by emergency-response professionals, the general approach is useful here as an introduction to the characteristics that are likely to cause a substance or mixture of substances to be regarded as a hazardous material.

The following are the seven classes of hazardous materials traditionally noted by emergency responders:

- *Flammable materials.* These are solid, liquid, vaporous, or gaseous materials that ignite easily and burn rapidly when exposed to an ignition source. Examples of such flammable materials include the commercial solvents toluene and ethanol, flour dust, finely dispersed powders of aluminum and certain other metals, gasoline, and natural gas.

- *Spontaneously ignitable materials.* These are solid or liquid materials that ignite spontaneously without the need of an ignition source. Spontaneous ignition can be initiated by the concentration of heat within a material resulting from oxidation or microbiological action. In other instances, the cause can be traced to the production of a flammable gas or vapor at the surface of a material when it reacts with atmospheric water vapor. Examples of spontaneously ignitable materials are white phosphorus (Section 7.5-A) and aluminum alkyl compounds (Section 9.4).

- *Explosives.* These chemical substances detonate. Examples are dynamite and TNT. The cause of explosive detonations is typically an initiating mechanism such as shock or the localized concentration of heat.

- *Oxidizers.* These substances evolve or generate oxygen at ambient temperatures when exposed to heat. An example of an oxidizer is ammonium nitrate.

- *Corrosive materials.* These are solid or liquid materials that "burn" or otherwise damage skin tissue at the site of contact. An example of a liquid corrosive material is battery acid.

- *Toxic materials.* These substances cause adverse health effects or death in individuals exposed to relatively small doses. An example of a toxic material of primary concern to firefighters is carbon monoxide (Section 10.9).

- *Radioactive materials.* These materials emit radiation, exposure to which can cause adverse health effects or death. An example of a radioactive material is uranium.

National Fire Protection Association
■ The voluntary membership organization that promotes and improves fire protection and prevention and establishes safeguards against loss of life and property by fire (abbreviated NFPA)

NFPA classes of fire
■ The five types of fire, each differentiated by the nature of its fuel

PERFORMANCE GOALS FOR SECTION 1.3:

- Name the types of fuels that constitute NFPA class A, class B, class C, class D, and class K fires.

1.3 NFPA Classes of Fire

The **National Fire Protection Association** (NFPA),* defines five general **classes of fire**: class A, class B, class C, class D, and class K. We now briefly examine the nature of each NFPA class.

*National Fire Protection Association, 1 Batterymarch Park, Quincy, Massachusetts 02269-9101.

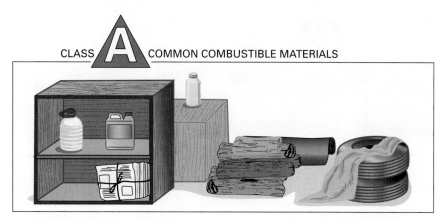

CLASS **A** COMMON COMBUSTIBLE MATERIALS

FIGURE 1.1 The combustion of wood, cotton, plastics, or rubber results in an NFPA class A fire.

1.3-A NFPA CLASS A FIRE

An NFPA class A fire results from the burning of ordinary cellulosic materials such as wood and paper as well as similar natural and synthetic materials like rubber and plastics. The solid residue of a class A fire consists of ashes or embers. Examples of materials that burn to produce class A fires are the combustible items in Figure 1.1. They are typically extinguished through the effective use of water.

1.3-B NFPA CLASS B FIRE

An NFPA class B fire results when flammable gases and flammable or combustible liquids burn. Several examples of hazardous materials that burn as class B fires are illustrated in Figure 1.2. Firefighters generally use carbon dioxide, dry-chemical extinguishers, or foam to extinguish them. The use of water is typically ineffective.

1.3-C NFPA CLASS C FIRE

An NFPA class C fire results from the combustion of materials occurring in or originating from energized electrical circuits, wiring, motors, and other equipment like the types shown in Figure 1.3. The use of carbon dioxide or dry-chemical extinguishers is recommended for extinguishing these class C fires. The use of water is not recommended.

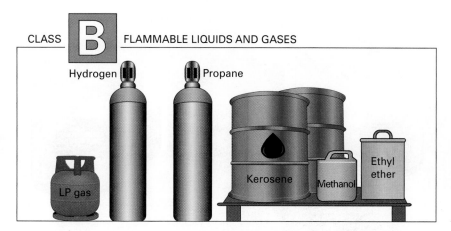

CLASS **B** FLAMMABLE LIQUIDS AND GASES

Hydrogen Propane

Ethyl ether

Kerosene

Methanol

LP gas

FIGURE 1.2 The combustion of a flammable gas or flammable liquid results in an NFPA class B fire.

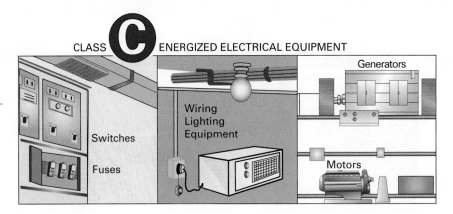

FIGURE 1.3 The combustion processes that occur within energized electrical equipment are denoted as NFPA class C fires.

CLASS **C** ENERGIZED ELECTRICAL EQUIPMENT

Generators

Wiring
Lighting
Equipment

Switches

Fuses

Motors

1.3-D NFPA CLASS D FIRE

An NFPA class D fire results from the combustion of certain metals, some examples of which are shown in Figure 1.4. They include titanium, magnesium, aluminum, and sodium. Class D fires are often extinguished with special types of fire extinguishers, such as graphite-based dry powder or sodium chloride. The use of water is suitable for extinguishing some class D fires, but only when it can be applied in a deluging volume.

1.3-E NFPA CLASS K FIRE

An NFPA class K fire results from the combustion of certain cooking media such as vegetable and animal fats and oils. The "K" in class K refers to "kitchen." This NFPA class may be considered a subclass of an NFPA class B fire. Some examples are shown in Figure 1.5. An NFPA class K fire is often extinguished with special fire extinguishers, such as handheld portable wet-chemical extinguishers consisting predominantly of water solutions of either potassium acetate or potassium carbonate. Professionals do not recommend the sole use of water for extinguishing class K fires.

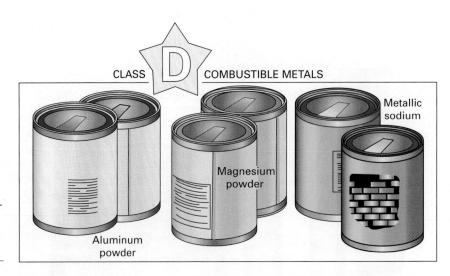

CLASS **D** COMBUSTIBLE METALS

Metallic
sodium

Magnesium
powder

Aluminum
powder

FIGURE 1.4 The combustion processes involving certain chemically reactive metals are denoted as NFPA class D fires.

CLASS **K** KITCHEN FIRES

High — Off   High — Off

PERFORMANCE GOALS FOR SECTION 1.4:

- Identify the information the Consumer Product Safety Commission (CPSC) requires manufacturers to list on labels affixed to containers of regulated consumer products.
- Identify the specific hazardous substances CPSC has banned in all consumer products.
- Identify the information the U.S. Environmental Protection Agency (EPA) requires to be posted on containers of approved pesticides.
- Describe how EPA regulates the manufacture, distribution, and use of chemical substances.
- Identify the specific hazardous substances EPA bans in all products.
- Name the federal laws that provide the CPSC, Food and Drug Administration (FDA), and EPA with the legal authority to require manufacturers to identify the potential hazards of their products to consumers.

1.4 Hazardous Substances Within Consumer Products

There are more than a dozen federal statutes or laws that form the legal basis for regulating hazardous materials in some way. The federal statutes listed in Table 1.1 are especially useful for these purposes, since they obligate the manufacturers, packers, and distributors of consumer products to identify the names of the constituent hazardous substances, their potential hazards, and the precautions to be exercised during their use. This information is conveyed on the **labels** that are affixed to the exterior surfaces of all product containers.

Although it is probably true that most individuals simply scan the label information on consumer product containers, it is important to take the time to read it. The statutes specifically require the posting of this information to protect those who use the products. Notwithstanding this intended purpose, the label information is equally useful for protecting emergency responders when they encounter the products while engaged in work activities.

1.4-A FEDERAL HAZARDOUS SUBSTANCES ACT

When Congress passed the **Federal Hazardous Substances Act**, it directed the U.S. Consumer Product Safety Commission (CPSC) to protect the public against unsuspecting exposure

label
■ The written, printed, and graphic matter affixed to containers of hazardous materials to rapidly communicate hazard information.

Federal Hazardous Substances Act
■ The federal statute that empowers the Consumer Product Safety Commission to protect the public against unsuspecting exposure to hazardous substances contained within household products

TABLE 1.1 — Federal Laws That Affect the Contents and Labeling of Household and Other Products

FEDERAL STATUTE	RESPONSIBLE AGENCY	AUTHORIZATIONS
Federal Hazardous Substances Act	Consumer Product Safety Commission	Establishes requirements for identifying certain hazardous substances found in household products
Federal Food, Drug, and Cosmetic Act	U.S. Food and Drug Administration	Ensures that foods are safe to eat, drugs are safe and effective, and packaging and labeling of foods, drugs, and cosmetics are truthful and informative
Federal Insecticide, Fungicide, and Rodenticide Act	U.S. Environmental Protection Agency	Regulates the manufacture, use, and disposal of pesticides
Toxic Substances Control Act	U.S. Environmental Protection Agency	Requires premarket evaluation of all new chemical products other than food additives, drugs, pesticides, alcohol, and tobacco; regulates existing hazards not covered by other federal laws dealing with toxic substances
Federal Alcohol Administration Act	Alcohol and Tobacco Tax and Trade Bureau	Includes ensuring that distilled spirits (beverage alcohols), wines, and malt beverages intended for nonindustrial use are safe for consumption, and their packaging and labeling are truthful and informative

hazardous substance
■ For purposes of CPSC, any substance or mixture of substances that (i) is toxic, (ii) is corrosive, (iii) is an irritant, (iv) is a strong sensitizer, (v) is flammable or combustible, or (vi) generates pressure through decomposition, heat, or other means if such substance or mixture of substances may cause substantial personal injury or substantial illness during or as a proximate result of any customary or reasonable foreseeable handling or use, including reasonably foreseeable ingestion by children; any radioactive substance; any toy or other article intended for use by children that presents an electrical, mechanical, or thermal hazard; any solder that has a lead content in excess of 0.2%

to hazardous substances contained within consumer products such as household cleaners. For the purposes of CPSC, a **hazardous substance** is defined at 16 C.F.R. §1500.3* as follows:

> any substance or mixture of substances which (i) is toxic, (ii) is corrosive, (iii) is an irritant, (iv) is a strong sensitizer, (v) is flammable or combustible, or (vi) generates pressure through decomposition, heat or other means, if such substance or mixture of substances may cause substantial personal injury or substantial illness during or as a proximate result of any customary or reasonable foreseeable handling or use, including reasonably foreseeable ingestion by children; any radioactive substance . . .; any toy or other article intended for use by children which . . . presents an electrical, mechanical, or thermal hazard; any solder which has a lead content in excess of 0.2 percent

The application of this expression does not apply to pesticides, foods, drugs, and cosmetics, since these consumer products are regulated under other federal statutes.

One way by which CPSC fulfills its congressional mandate is to require appropriate labeling on the containers of household products that contain hazardous substances. At 16 C.F.R. §1500.121,* CPSC requires the following information to be clearly and conspicuously stated on the labels affixed to containers of consumer products having hazardous substances as components:**

- The name and place of business of the manufacturer, packer, distributor, or seller
- The common or usual name or the chemical name of the hazardous substance or of each component that contributes substantially to its hazard
- Initial signal words such as DANGER, WARNING, or CAUTION

*Relevant citations to federal regulations are noted throughout this book. A reference to 16 C.F.R. §1500.121 means Title 16 of the *Code of Federal Regulations*, Part 1500, Section 121. The regulatory parts are also divided into subparts, which are denoted with capital letters A, B, C, and so on. Readers may access the regulations electronically on the Internet.
**Consumers may report hazards caused by consumer products, or product-related injuries, to CPSC by using the toll-free hotline (800) 638-CPSC.

- An affirmative statement of the principal hazard or hazards, such as FLAMMABLE, COMBUSTIBLE, VAPOR HARMFUL, CAUSES BURNS, POISON, SKIN AND EYE IRRITANT, or similar wording that is descriptive of the hazard
- Instructions, when necessary or appropriate, for first-aid treatment
- Instructions for handling and storage of packages that require special care during handling or storage
- The statement KEEP OUT OF REACH OF CHILDREN

The signal word CAUTION appears on products that are least harmful to the user, WARNING appears when a product is more harmful than one bearing a CAUTION label, and DANGER appears on products that are poisonous or corrosive.

When products are stored within self-pressurized containers, CPSC requires the manufacturer at 16 C.F.R. §1500.130 to include the following statements or their equivalents on product labels:

**WARNING—CONTENTS UNDER PRESSURE
DO NOT PUNCTURE OR INCINERATE CONTAINER
DO NOT EXPOSE TO HEAT OR STORE AT TEMPERATURES ABOVE 120 °F (49 °C)
KEEP OUT OF REACH OF CHILDREN**

When the misuse of a product is likely to adversely affect one's health, CPSC requires the manufacturer to include advisory warnings and first-aid actions on product labels. For example, the statements provided in Figure 1.6 on a container of an aerosol oven cleaner provide the user

DANGER: Keep from heat flame; Do not puncture, incinerate or store above 120 °F. KEEP OUT OF REACH OF CHILDREN. CONTAINS PETROLEUM DISTILLATES. HARMFUL OR FATAL IF SWALLOWED. If swallowed, do not induce vomiting. Call a physician immediately. Spray only in well-ventilated area. Avoid prolonged breathing of vapors. In case of eye contact, flush with copious amounts of water.

FIGURE 1.6 The Federal Hazardous Substances Act provides the Consumer Product Safety Commission with the authority to compel manufacturers to label their products with the specific information denoted at 16 C.F.R. §§1500.3 and 1500.130. The label on this aerosol oven cleaner advises the consumer to beware of its flammable and health hazards. To provide confidentiality, the name of the manufacturer, packer, distributor, or seller and the tradename of the product have been intentionally omitted.

with several follow-up actions to be implemented when the product is swallowed or splashed into the eyes.

SOLVED EXERCISE 1.1

A multipurpose lighter is also known as a grill lighter, fireplace lighter, utility lighter, microtorch, or gas match. It is a handheld self-igniting flame-producing consumer product that operates on fuel and is used to ignite items like candles, fuel for fireplaces, charcoal or gas-fired grills, campfires, camp stoves, lanterns, fuel-fired appliances, and pilot lights. Why are multipurpose lighters containing butane or petroleum distillate fuel subject to the CPSC's labeling requirements?

Solution: Butane and petroleum distillate are two flammable substances that generate pressure and may cause substantial personal injury or illness as a proximate result of customary or foreseeable use. For this reason, they are hazardous substances within the meaning provided in CPSC's regulations at 16 C.F.R. §1500.3. Multipurpose lighters that contain fuel when sold to consumers are subject to the CPSC's labeling provisions because they contain a hazardous substance that is intended or packaged in a form suitable for use in the common household.

SOLVED EXERCISE 1.2

DANGER!
EXTREMELY FLAMMABLE LIQUID AND VAPOR
MAY CAUSE FLASH FIRE
CONTENTS UNDER PRESSURE
MAY CAUSE EYE AND SKIN IRRITATION
VAPOR OVEREXPOSURE MAY CAUSE RESPIRATORY TRACT IRRITATION
AND CENTRAL NERVOUS SYSTEM DEPRESSION

Contains: Dimethyl ether; acetone; nonvolatile components; pentane; cyclohexane; and propane

Precautions: Keep away from heat, sparks, open flames, pilot lights and sources of static discharge. Avoid breathing of vapors, mists, or spray. Do not use in confined areas or areas with little or no air movement. Avoid eye and skin contact. Intentional misuse by deliberately concentrating and inhaling the contents can be harmful or fatal. Do not puncture or incinerate. Do not store at temperatures above 120 °F (49 °C).

KEEP OUT OF REACH OF CHILDREN

Suggested First Aid: Eye Contact: Immediately flush eyes with plenty of water. If irritation persists, seek medical advice.

Skin Contact: Wash skin with plenty of soap and water. If irritation persists, seek medical advice. **Inhalation:** Remove to fresh air. If not breathing, give artificial respiration. If breathing is difficult, seek medical advice. **Ingestion:** Do not induce vomiting. Drink two glasses of water. Seek medical advice.

> **Directions for Use:** Shake can before using. Select spray pattern width by turning spray tip to the desired setting. Spray both surfaces to be bonded. Allow adhesive to dry a minimum of 1 minute. Make bond while still tacky. Clean spray tip with turpentine. Allow to dry 2 to 5 minutes.

A manufacturer affixes the preceding label to aerosol cans containing a spray adhesive product. Why has the manufacturer included these specific statements on the label?

Solution: To forewarn the consumer of the presence of hazardous substances in a product, the CPSC is authorized to *compel* its manufacturers at 16 C.F.R. §§1500.3 and 1500.130 to label the product with the following minimum information: advisory warnings; initial precautionary statements indicating the principal hazards associated with use of the product; measures the user should take to reduce or eliminate the risk of a hazard during storage and handling; the common names of the hazardous ingredients; and first-aid instructions. By providing this information on the cans of this product, the manufacturer has complied with the obligations specified by the CPSC. (To provide confidentiality, the name and place of business of the manufacturer, packer, distributor, or seller and the tradename of the consumer product have been intentionally omitted.)

CPSC also fulfills its congressional mandate by banning the use of a hazardous substance in household products under the following condition:

> when the substance possesses such a degree or nature of hazard that adequate cautionary labeling cannot be written and the public health and safety can be served only by keeping such articles out of interstate commerce

At 16 C.F.R. §1500.17, CPSC bans the following hazardous substances as constituents of consumer products intended for use in the United States:

- Carbon tetrachloride (formerly used in fire extinguishers)
- Extremely flammable water repellants
- Certain fireworks devices intended to produce audible effects
- Liquid drain cleaners containing 10% or more by weight potassium hydroxide
- Products containing soluble metallic cyanides
- Paint having a lead content above 0.06% by dry weight (other than artist's paint)
- Self-pressurized products containing vinyl chloride as an ingredient
- General-use garments, tape joint compounds, and artificial fireplace embers containing asbestos

1.4-B FEDERAL FOOD, DRUG, AND COSMETIC ACT

Although foods, drugs, and cosmetics are not regarded as hazardous substances, their use may put people at risk even when they are used in ordinary ways. Under unique circumstances, the use of foods, drugs, and cosmetics may cause sickness, allergic or other undesirable side effects, and death in vulnerable individuals. For this reason, the production and manufacture of foods, drugs, and cosmetics are closely monitored before these products are marketed to the public.

Chemical substances are routinely added to foods to serve unique purposes. They are available commercially as food-grade materials. They function as leavening agents, flavoring agents, preservatives, coloring agents, or antimicrobial agents; others act as stabilizers, thickeners, or emulsifiers; and some improve the nutritional value of foods.

The **Federal Food, Drug, and Cosmetic Act** directs the U.S. Food and Drug Administration (FDA) to ensure that foods are safe to eat as sold in the marketplace,* drugs are safe and effective, and the labeling information for foods, drugs, and cosmetics is truthful and informative. This statute requires those who manufacture, package, and distribute foods, drugs, and cosmetics to strictly comply with the following:

■ Label all containers with the information specified at 21 C.F.R. §§101.4 and 101.9 for foods; 21 C.F.R. §§201.2 and 201.10 and 21 C.F.R. §§201.56 and 201.57 for over-the-counter (OTC) and prescription drugs, respectively; and 21 C.F.R. §§701.3 and 740.10 for cosmetics. Labels on OTC drug packaging, for example, cite the brand or generic name; the quantity of a single dose; the frequency of its administration; the duration of its administration; the time, route, or method of its administration; directions concerning the preparation of the drug for use; warnings, precautions, and overdose information; the active ingredient and its amount; and the inactive ingredients and their amounts.

■ Monitor and assess the health risks potentially caused by exposure to chemical substances in foods, drugs, and cosmetics

■ Prevent the direct use of substances that are regarded as hazardous in other federal statutes within prepared foods, drugs, and cosmetics

■ Limit the ways by which hazardous substances may unknowingly enter the food chain

■ Establish tolerance limits for those hazardous substances that pose a risk to health when they become components of foods as unavoidable residues from environmental or industrial sources

■ Assist in the investigation of public health threats such as product tampering, bioterrorism, food-borne illnesses, and contaminated blood supplies

■ Warn consumers to limit their exposure to certain foods altogether. For example, FDA requires individuals who market swordfish and tuna to issue specific warning advisories to pregnant women. One such advisory notes that the mercury content of swordfish and tuna may be harmful to the women's unborn children. This advisory is based on scientific studies that confirm the potential for the development of neurological problems in a developing fetus when a mother consumes fish containing minutely low levels of mercury.

1.4-C FEDERAL INSECTICIDE, FUNGICIDE, AND RODENTICIDE ACT

In 1947, Congress passed the **Federal Insecticide, Fungicide, and Rodenticide Act**, often called FIFRA, and in 1970, it provided the U.S. Environmental Protection Agency (EPA) with the authority to administer the statute. EPA has used the law to regulate the manufacture, use, and disposal of pesticides for agricultural, forestry, household, and other activities. For the purposes of FIFRA, a **pesticide** is defined in part at 40 C.F.R. §152.3 as follows:

> any substance or mixture of substances intended for preventing, destroying, repelling, or mitigating any pest, or any substance or mixture of substances intended for use as a plant regulator, defoliant, or drying agent

FIFRA provides EPA with the authority to require manufacturers to obtain an establishment number, to register their pesticides, and to classify them for general or restricted use. EPA assigns a number to each pesticide approved for use when the prevailing information indicates that use of the pesticide will not pose an unreasonable adverse impact on public health and the environment.

FIFRA also provides EPA with the authority to cancel a pesticide registration if unreasonable adverse effects to the environment and public health develop when a product is used according to widespread and commonly recognized practice.

---

*Currently, the U.S. Department of Agriculture is responsible for assessing the safety of meat, poultry, and processed eggs destined for human consumption, whereas the FDA is responsible for assuring the safety of nearly all other foods, including fish and fresh produce.

FIFRA also provides EPA with the authority to require informative and accurate labeling on pesticide containers. At 40 C.F.R. §156.10, EPA requires manufacturers and distributors to label pesticide containers with the following information, prominently placed with such conspicuousness and in such terms as to render it likely to be read and understood by the ordinary individual:

- The product name, brand, or trademark under which the product is sold
- The name and address of the producer, registrant, or person for whom produced
- The net weight or measure of contents
- The EPA registration number for the product
- The EPA number for the producing establishment
- An ingredient statement, giving the name and percentage by weight of each active ingredient and the total percentage by weight of all inert ingredients
- Appropriate directions for use in terms that, if followed, will adequately protect human health and the environment and can be easily read and understood by the average person likely to use or supervise the use of the pesticide
- The following statement placed immediately below the statement of use:

IT IS A VIOLATION OF FEDERAL LAW TO USE THIS PRODUCT IN A MANNER INCONSISTENT WITH ITS LABELING

- Specific directions concerning the storage and disposal of the pesticide and its container
- Appropriate hazard and precautionary statements applicable to human and domestic animal hazards and environmental hazards, including physical hazards, acute toxicity hazards, and certain toxicity statements relating to ecological hazards
- A statement of the use classification under which it is registered (general or restricted)

SOLVED EXERCISE 1.3

A portion of the labeling on aerosol cans of Insect Zappo, an insecticide, follows:

INSECT ZAPPO

KILLS FAST—KEEPS ON KILLING UP TO FOUR MONTHS
KEEP OUT OF REACH OF CHILDREN

ACTIVE INGREDIENTS:

Tralomethrin (1R,4S) 3[(1′RS)(1′,2′,2′,2′-tetrabromomethyl)]-2,2-dimethyl-cyclopropanecarboxylic acid(S)-alpha-cyano-3-phenoxybenzyl ester	0.03%
d-*trans*-Allethrin	0.05%
Inert ingredients	99.92%

HOW TO USE:

1. Shake well before each use.
2. Hold can upright. Aim opening away from people. Spray in short bursts to prevent excessive wetting and waste of product.

3. Do not allow the spray to contact food. Thoroughly wash dishes, utensils, and counter-tops with soap and water if they are sprayed with this product. Do not allow people or pets to contact treated surfaces until spray has dried. Remove pets and cover aquariums before spraying.

INDOORS:

1. Contact as many insects as possible with the spray in addition to thoroughly treating all parts of the room where these pests may hide.
2. Pay special attention to cracks, dark corners, hidden surfaces under sinks, behind stoves and refrigerators, food storage areas, and wherever these pests are suspected. The treatment continues to work for up to four months.
3. To kill insects and prevent them from entering your home, spray around door and window frames, inside and outside. Spray upward to treat active insect nests. Do not remove insect nests until spray has dried. Repeat as necessary.

OUTDOORS:

1. Spray outside surfaces of screens, doors, window frames, or wherever these pests may enter the home.
2. Also spray surfaces around light fixtures on porches, in garages, and other places where they alight or congregate. Repeat as necessary.

PRECAUTIONARY STATEMENTS:

Hazards to Humans and Domestic Animals: Harmful if absorbed through the skin. Avoid contact with skin, eyes, or clothing. Wash thoroughly with soap and water after handling. Avoid contamination of food and drinking water. All food preparation surfaces and utensils should be covered during treatment or thoroughly washed before use. Remove pets and cover aquariums before spraying.

STATEMENT OF PRACTICAL TREATMENT: If on skin, wash with plenty of soap and water. Get medical attention if irritation persists. If swallowed, do not induce vomiting because of aspiration hazard. Get medical attention. If in eyes, flush with plenty of water. Get medical attention.

ENVIRONMENTAL HAZARDS. This product is toxic to fish. Do not apply directly to water.

PHYSICAL AND CHEMICAL HAZARDS. Contents under pressure. Do not use or store near heat or open flame. Do not puncture or incinerate container. Exposure to temperature above 130 °F may cause bursting.

Why has the pesticide manufacturer included these label statements on the cans of this product?

Solution: Using the authority of FIFRA, EPA *compels* manufacturers and distributors at 40 C.F.R. §156.10 to label their pesticide products with the following minimum information: the name, brand, or trademark; directions for use that are "necessary for effecting the purpose for which the product is intended" and adequate to protect public health and the environment; an ingredient statement, giving the name and percentage of each active ingredient and the percentage of all inert ingredients; a statement of the use classification under which it is registered; the name and address of the manufacturer; and appropriate warning or caution statements. By providing this information on the pesticide containers, the manufacturer or distributor has complied with the regulatory requirements specified by EPA.

1.4-D TOXIC SUBSTANCES CONTROL ACT

In 1976, Congress passed the **Toxic Substances Control Act** (TSCA). This law provides EPA with the broad authority to regulate the manufacture, distribution, use, and disposal of chemical substances. To accomplish the mandate of this law, EPA uses its authority in two ways. First, it obtains from industry production and test data on commercial chemical products whose manufacture, processing, distribution, use, or disposal could pose an unreasonable risk of injury to public health and the environment. Then, it evaluates this information and, when warranted, imposes limitations or prohibitions on their manufacture and use.

Eight product categories are exempt from TSCA's authority: pesticides, tobacco, nuclear materials, firearms and ammunition, food, food additives, and cosmetics. These product categories are regulated under other federal laws. For example, health and environmental issues related to pesticides and drugs are covered by the Federal Insecticide, Fungicide, and Rodenticide Act and the Federal Food, Drug, and Cosmetic Act, respectively.

Using the authority of TSCA, EPA selects a control action to prevent unreasonable risks to public health and the environment from exposure to chemical products. The nature of the control action varies with the severity of the risks. In some instances, EPA requires the manufacturer to post a hazard warning on product labels, although in others it bans outright the manufacture or severely restricts the use of a product.

In the TSCA regulations, chemical substances are classified as either "existing chemicals" or "new chemicals." Existing chemicals are listed in a document known as EPA's Chemical Substance Inventory, which is available in most libraries. There are approximately 75,000 existing chemicals. When manufacturers or importers intend to use an existing chemical in a product, they are required to review the inventory to determine whether there are restrictions on its manufacture or use.

Substances not listed in the inventory are regarded as new chemicals. As used here, the word *new* implies that the substance was not already present in commerce at the time the inventory was first prepared or later supplemented. TSCA requires anyone who manufactures or imports a product containing a new chemical to submit a premanufacture notice to EPA before initiating the activity.

In Subpart E of 40 C.F.R. Part 721, EPA has published a list of chemical substances whose manufacture, processing, distribution, use, or disposal could pose an unreasonable risk of injury to public health and the environment. At 40 C.F.R. §721.72, EPA obligates manufacturers who intend to distribute in commerce a product that contains one of these listed substances to label each product container with the following information:

- The name and address of the manufacturer
- A statement regarding the health hazards resulting from exposure to the substance and precautionary measures that minimize or reduce the hazards
- The identity by which the substance is commonly recognized
- A statement of environmental hazards linked with the substance
- A statement of exposure hazards linked with the substance

These labeling requirements apply not only to those products manufactured within the United States but also to products containing listed substances that are imported into the United States for distribution in commerce.

The manufacture or use of some substances has been banned or severely restricted by EPA using the authority of TSCA. The properties of these hazardous substances are noted in future chapters in this book. They include polychlorinated biphenyls (PCBs) (Section 12.17), chlorofluorocarbons (CFCs) (Section 12.16), 2,3,7,8-tetrachlorodibenzo-*p*-dioxin (TCDD) (Section 13.4-A), asbestos (Section 10.18), acrylamide grout, hexavalent chromium (Section 11.13), and nitrosating agents used in metalworking operations.

Toxic Substances Control Act
■ The federal statute that empowers EPA to regulate all aspects of the manufacture of toxic substances other than pesticides and drugs, including the obtaining of production and test data from industry on those toxic substances whose manufacture, processing, distribution, use, or disposal could present an unreasonable risk of injury or damage to the environment (abbreviated TSCA)

Polychlorinated biphenyls (PCBs) are chemical substances that were formerly used as electrical transformer and capacitor fluids. When EPA evaluated scientific and other data obtained from industry, it concluded that exposure to PCBs could cause cancer, birth defects, liver damage, acne, impotence, and death. EPA then banned their manufacture, processing, and distribution; restricted their future use; and regulated the manner by which they are treated, stored, or disposed of in the United States. Which federal statute provided EPA with the legal authority to take this action?

Solution: EPA first used the authority of TSCA to obtain the data that were evaluated as to whether exposure to PCBs could present an unreasonable risk of injury or damage to public health and the environment. On reaching a positive conclusion, EPA then used the authority of TSCA again to prohibit the manufacture, processing, and distribution of PCBs, restrict their future use, and regulate the manner by which they are treated, stored, or disposed of within the United States.

PERFORMANCE GOALS FOR SECTION 1.5:

- Name the six criteria air pollutants and distinguish between the nature of the information provided by their "primary" and "secondary" standards.
- Describe the general nature of the hazardous air pollutants identified by EPA whose airborne emissions are regulated by EPA through national emission standards.
- Describe the general nature of the toxic pollutants identified by EPA whose discharge into waterways is regulated by EPA.
- Describe the general nature of the hazardous wastes whose treatment, storage, and disposal are regulated by EPA.
- Describe the general nature of the hazardous substances whose improper treatment, storage, and disposal at facilities mandates EPA to clean up and restore the natural resources at these sites.
- Name the federal laws that provide EPA with the legal authority to regulate emissions and other discharges of criteria air pollutants, hazardous air pollutants, toxic pollutants, hazardous substances, and hazardous wastes.

Clean Air Act
■ The federal statute that empowers EPA to establish national ambient air quality standards for specified air contaminants

criteria air pollutants
■ The six most commonly encountered air pollutants in the United States for which EPA has developed emission standards based on specific health criteria, namely, carbon monoxide, nitrogen dioxide, ozone, lead, sulfur dioxide, and particulate matter

1.5 Hazardous Constituents of Pollutants and Wastes

Using the authority of the federal statutes in Table 1.2, EPA has been charged with the responsibility of assuring protection of human health and the environment from exposure to the hazardous substances contained in polluted air, water, and chemical wastes. We note the important components of each statute separately.

1.5-A CLEAN AIR ACT

The modern approach to regulating the quality of the air we breathe was initiated with passage in 1970 of the **Clean Air Act**. To assure Americans of their right to breathe clean air, Congress empowered EPA to establish national ambient air quality standards for a number of air contaminants called **criteria air pollutants**. To date, EPA has established standards for six criteria air pollutants: sulfur dioxide, particulate matter, carbon monoxide, ground-level ozone, nitrogen dioxide, and lead.

TABLE 1.2	Some Federal Environmental Statutes
STATUTE	**EPA'S RESPONSIBILITY**
Clean Air Act	Establishes national ambient air quality standards for specified air contaminants; devises and promulgates technology-based emission reduction standards
Federal Water Pollution Prevention and Control Act	Restores and maintains the chemical, physical, and biological integrity of the nation's waters
Resource Conservation and Recovery Act	Regulates the generation, transportation, treatment, storage, and disposal of hazardous wastes within a cradle-to-grave framework
Comprehensive Environmental Response, Compensation, and Liability Act	Leads efforts to clean up sites at which hazardous substances were improperly disposed of in the past
Superfund Amendments and Reauthorization Act	Considers the standards and requirements found in other environmental laws and regulations and includes citizen participation when considering a remedial-response action

Each national ambient air quality standard is composed of two parts:

- *Primary standard.* This standard identifies a contaminant concentration in the ambient air that protects human health. When EPA sets the primary standard, it considers not only healthy individuals but the health of "sensitive" populations such as children, people with asthma, and the elderly.
- *Secondary standard.* This standard identifies a contaminant concentration that protects public welfare. In setting the standard, EPA accounts for such subjects as reduced air visibility and damage to animals and vegetation.

The contaminant concentrations provided in primary and secondary standards are generally annual arithmetic means. When they are exceeded, EPA is compelled to take action to protect public health and welfare. For example, much of the nitrogen dioxide in the ambient air is linked with its presence in vehicular exhaust. Concerned with reducing the concentration of this toxic gas, EPA approached automobile manufacturers to determine whether they could construct and operate automobiles in a fashion that would reduce the amount of nitrogen dioxide within the exhaust. In response, the manufacturers built new automobiles in which they installed catalytic converters. The converters facilitate the change of nitrogen dioxide into nontoxic nitrogen and thereby reduce the amount of nitrogen dioxide discharged into the air when cars and trucks are driven.

The Clean Air Act also provides EPA with the authority to establish national emission standards for airborne substances that cause or are suspected to cause serious health problems. These substances are called **hazardous air pollutants**, or **air toxics**. They are listed at 40 C.F.R. §61.01. The standards are called **National Emission Standards for Hazardous Air Pollutants** (NESHAPs). To date, EPA has published NESHAPs for 189 hazardous air pollutants.

The NESHAP regarding airborne asbestos (Section 10.18) is familiar to emergency-response personnel who provide standby services during demolition work involving asbestos-containing materials. It is published at 40 C.F.R. §61.145. The NESHAP requires the persons intending to conduct asbestos-related activities to notify EPA of the proposed action, follow specified procedures to protect public health and the environment, and adopt specific work procedures to prevent the release of asbestos fibers into the air.

hazardous air pollutant (air toxic)
■ A substance designated by EPA pursuant to enactment of the Clean Air Act and listed at 40 C.F.R. §61.01

National Emission Standards for Hazardous Air Pollutants
■ For purposes of EPA regulations, the standards that relate to the air emission of hazardous air pollutants that cause an increase in fatalities or a serious irreversible or incapacitating illness

The Clean Air Act also provides EPA with the authority to regulate the prevention of accidental, potentially catastrophic releases of airborne hazardous substances. This program obligates the owners and operators of facilities using large quantities of toxic or flammable substances to identify potential sources from which they could be released into the air. It also places responsibility on owners and operators to prevent and minimize the risks associated with these potential catastrophes. Companies that make or use large quantities of toxic or flammable substances are obligated to compile *risk management plans*, which provide key information to emergency planners, fire departments, and local residents for responding to emergencies involving them.

Finally, the Clean Air Act established the U.S. Chemical Safety and Hazard Investigation Board. Modeled after the National Transportation Safety Board, the Chemical and Hazard Investigation Board conducts investigations and reports on findings regarding the causes of major chemical accidents. These findings are then used to recommend actions to improve the safety of operations involved in the production, transportation, industrial handling, use, and disposal of chemical substances.

1.5-B FEDERAL WATER POLLUTION PREVENTION AND CONTROL ACT

Federal Water Pollution Prevention and Control Act
■ The federal statute that empowers EPA to restore and maintain the chemical, physical, and biological integrity of the waters of the United States

Clean Water Act
■ The federal statute that empowers EPA to restore and maintain the chemical, physical, and biological integrity of the waters of the United States

toxic pollutant
■ For purposes of the FWPPCA, any substance designated by EPA and listed at 40 C.F.R. §401.15

The federal water pollution control effort began seriously with the passage of the **Federal Water Pollution Prevention and Control Act** (FWPPCA) in 1948, but a comprehensive regulatory program for controlling water pollution discharges was first established with the creation of EPA in 1970 and passage of amendments to the FWPPCA in 1972, 1977, and 1987. FWPPCA and its combination of amendments is commonly called the **Clean Water Act** (CWA).

The primary regulatory approach of the CWA now consists of the issuance of industry-by-industry technology-based effluent limitations that apply to the discharge of toxic pollutants into all surface waters within the United States and its territories. Thus, EPA may permit the discharge of an aqueous waste into a nearby river but only when the toxic pollutant concentrations in the waste are equal to or less than established limits.

A **toxic pollutant** is defined in the statute as a substance that

> after discharge and upon exposure, ingestion, inhalation or assimilation into any organism . . . . , will, on the basis of information . . . , cause death, disease, behavioral abnormalities, cancer, genetic mutations, physiological malfunctions (including malfunctions in reproduction) or physical deformations, in organisms or their offspring.

EPA lists toxic pollutants at 40 C.F.R. §401.15. To date, EPA has used the authority of the Clean Water Act to regulate the discharge of 189 toxic pollutants. In addition, it regulates the conventional water pollutants designated as *biochemical oxygen demand,* total suspended solids, pH, fecal coliform, and oil and grease. Biochemical oxygen demand (BOD) is a standardized means of estimating the degree of contamination of a water supply, particularly one contaminated by sewage or industrial wastes.

Using the authority of the Clean Water Act, EPA regulates not only the direct discharges of toxic pollutants into waters of the United States but also indirect discharges, such as those directed into storm drains or sewers that have the potential for ultimate entry into navigable waters. The U.S. Coast Guard has jurisdiction over discharges into marine waters, whereas EPA has jurisdiction over discharges into fresh waters. When either agency receives notification of a discharge, it responds by locating the point source from which the toxic pollutant is entering the waterway and takes appropriate measures to stop the discharge and correct the damage to natural resources.

hazardous substance
■ For purposes of the FWPPCA, a substance designated by EPA and listed at 40 C.F.R. §117.3

The Clean Water Act also authorizes EPA to prescribe quantities of oil and any **hazardous substance** the discharge of which in any quantity into or upon the navigable waters of the United States affects natural resources belonging to or exclusively managed by the United States and presents an imminent and substantial danger to the public health or welfare including fish, shellfish, wildlife, shorelines, and beaches. The discharge of hazardous substances into or upon the navigable waters of the United States subjects their owners to certain reporting and response requirements. A listing of these hazardous substances is provided at 40 C.F.R. §117.3.

1.5-C RESOURCE CONSERVATION AND RECOVERY ACT

In 1976, Congress enacted the **Resource Conservation and Recovery Act** (RCRA), and in 1979, EPA promulgated regulations that aim to ensure protection of public health and the environment by facilities that generate, transport, treat, store, or dispose of hazardous wastes. The statute specifically defines the term *hazardous waste* as a material that "because of its quantity, concentration, or physical, chemical, or infectious characteristics may

- cause, or significantly contribute to an increase in mortality or an increase in serious irreversible, or incapacitating, reversible illness; or
- pose a substantial present or potential hazard to human health or the environment when improperly treated, stored, or disposed of, or otherwise managed."

EPA identifies hazardous wastes in two ways. First, EPA *lists* a series of wastes at 40 C.F.R. §§261.31, 261.32, 261.33(e), and 261.33(f). Second, EPA provides criteria for *characterizing* a hazardous waste as ignitable, corrosive, reactive, or toxic. The nature of these defining criteria is identified at 40 C.F.R. §§261.21, 261.22, 261.23, and 261.24. We revisit the RCRA characteristics later, in Sections 3.2, 5.9, 8.14, and 10.1-B.

1.5-D COMPREHENSIVE ENVIRONMENTAL RESPONSE, COMPENSATION, AND LIABILITY ACT

In 1980, Congress enacted the **Comprehensive Environmental Response, Compensation, and Liability Act** (CERCLA) to correct for the past mistakes resulting from the improper disposal of hazardous substances. The law is commonly known by its acronym CERCLA, although outside the legal profession, it is also known colloquially as the *Superfund law*.

CERCLA addresses the prevention and cleanup of hazardous substances released at a site and the restoration and replacement of the natural resources that were damaged or lost because of the release. As defined in the statute, a **hazardous substance** is any of the following:

- Any substance designated as a hazardous substance under the Federal Water Pollution Prevention and Control Act
- Any substance specifically designated as a hazardous substance pursuant to the act
- Any hazardous waste listed in the RCRA regulations or possessing the characteristic of ignitability, corrosivity, reactivity, or toxicity
- Any substance designated as a toxic pollutant pursuant to the Federal Water Pollution Prevention and Control Act
- Any substance designated as a hazardous air pollutant pursuant to the Clean Air Act
- Any imminently hazardous substance designated pursuant to the Toxic Substances Control Act

Hazardous substances pursuant to CERCLA are listed at 40 C.F.R. §302.4.

EPA has used the legal authority of CERCLA to correct the problems of abandoned or uncontrolled sites that threaten public health or the environment. The agency initially accomplished this task through the use of a corporate tax imposed on the chemical and petroleum industries. The tax monies were deposited into a federal trust fund called the *superfund*. When those responsible could not be located, or when responsible parties were unwilling or unable to pay for the damages they had caused, EPA used superfund monies to investigate and remediate the damages.

CERCLA also gives EPA the authority to require those responsible for contaminating sites to investigate and conduct remedial activities or to reimburse EPA for doing the work. To date, EPA has used the authority of CERCLA to clean up 321 sites at which hazardous substances were improperly treated, stored, or disposed of and to recover the cleanup costs from the parties responsible for the contamination. The hazardous substances that initially contaminated another 712 sites have also been removed, but these sites still require additional remediation.

In 1986, Congress enacted the **Superfund Amendments and Reauthorization Act**, commonly called SARA. This statute requires EPA when selecting a remedial action, to consider the standards and requirements identified in other environmental laws and regulations and to include

Resource Conservation and Recovery Act
■ The federal statute that empowers EPA to regulate the treatment, storage, and disposal of hazardous wastes (abbreviated RCRA)

Comprehensive Environmental Response, Compensation, and Liability Act
■ The federal statute that empowers EPA to identify and clean up sites at which hazardous substances were released into the environment (abbreviated CERCLA); called colloquially the superfund law

hazardous substance
■ For purposes of CERCLA, a substance designated by EPA and listed at 40 C.F.R. §302.4

Superfund Amendments and Reauthorization Act
■ The federal statute that requires EPA when selecting a remedial action, to consider the standards and requirements in other environmental laws and regulations and to include citizen participation in the decision-making process (abbreviated SARA)

citizen participation in the decision-making process. SARA also establishes an assistance program for training and educating workers engaged in activities such as emergency-response efforts.

The term **extremely hazardous substance** is defined in SARA as follows:

> substances which could cause serious irreversible health effects from accidental releases" and are "most likely to induce serious acute reactions following short-term exposure

When a concentration equal to or exceeding a published reportable quantity of an extremely hazardous substance is released into the environment, local and state emergency-response authorities are notified as soon as practical. Citizen exposure to the extremely hazardous substance may cause serious irreversible health effects. The extremely hazardous substances and their reportable quantities are listed in Appendices A and B following 40 C.F.R. §355.

PERFORMANCE GOALS FOR SECTION 1.6:

- Describe how OSHA works to reduce or eliminate the incidence of occupational illness and injury caused by employee exposure to hazardous chemicals in the workplace.
- Identify the 16 categories of information OSHA requires manufacturers to provide on an MSDS.

1.6 Hazardous Chemicals Within the Workplace

The **Occupational Safety and Health Act** is a federal statute that aims to protect employees in the workplace from occupational illnesses and injuries caused by exposure to hazardous chemicals. When Congress passed this act, it empowered the U.S. Occupational Safety and Health Administration (OSHA) to reduce or eliminate the incidence of chemically induced occupational illnesses and injuries by regulating certain conditions in the workplace.

When hazardous substances are present in the workplace, OSHA requires employers to limit employee exposure to certain permissible limits averaged over the 8-hr workday. OSHA establishes these limits, but on occasion, the National Institute for Occupational Safety and Health (NIOSH) recommends a revision of the limits based on its own research efforts. It is the responsibility of OSHA and NIOSH to ensure that employers provide each working individual with a safe and healthful working environment. The agencies accomplish this task by setting threshold limit values (Section 10.6-D) and permissible exposure limits (Section 10.6-F) for so-called hazardous chemicals, and by establishing requirements for worker training, safe work practices, labeling, ventilation, proper storage, and worker-protective equipment.

OSHA defines a **hazardous chemical** at 29 C.F.R. §1910.1200(c) as "any chemical which is a physical hazard or a health hazard." These chemicals are listed in Subpart Z of 29 C.F.R. §1910.

In 1983, OSHA first enacted a standard that sets minimum requirements to which employers must adhere for communicating information to workers; it is often referred to as the **right-to-know law**. Briefly, its intent is to assure workers of their right to know about the hazards associated with substances to which they are being exposed in their places of employment.

The right-to-know law requires chemical manufacturers and importers to assess the hazards of the chemical substances and products they produce or import and to transmit the hazard information to their users. This information must be sufficiently comprehensive to allow user-employers to develop appropriate employee protection programs and to give employees the information they require to protect themselves against the potential risks associated with the substances and products. This regulation applies to all chemical substances known to be present within the workplace, substances to which employees could be exposed under normal working conditions, and substances to which employees could be exposed in a foreseeable emergency.

---

extremely hazardous substance
- For purposes of SARA, a substance listed in Appendices A and B following 40 C.F.R. §355; a substance that can cause serious irreversible health effects from accidental releases and is most likely to induce serious acute reactions following short-term exposure

Occupational Safety and Health Act
- The federal statute that empowers the Occupational Safety and Health Administration (OSHA) to protect employees from occupational illnesses and injuries caused by exposure to hazardous chemicals in the workplace

hazardous chemical
- For purposes of OSHA regulations, a chemical that is a physical hazard or a health hazard and is listed at 29 C.F.R. §1910, Subpart Z

right-to-know law
- For purposes of OSHA regulations, the federal statute that addresses workers' right to know about the hazards associated with substances to which they are being exposed in the workplace

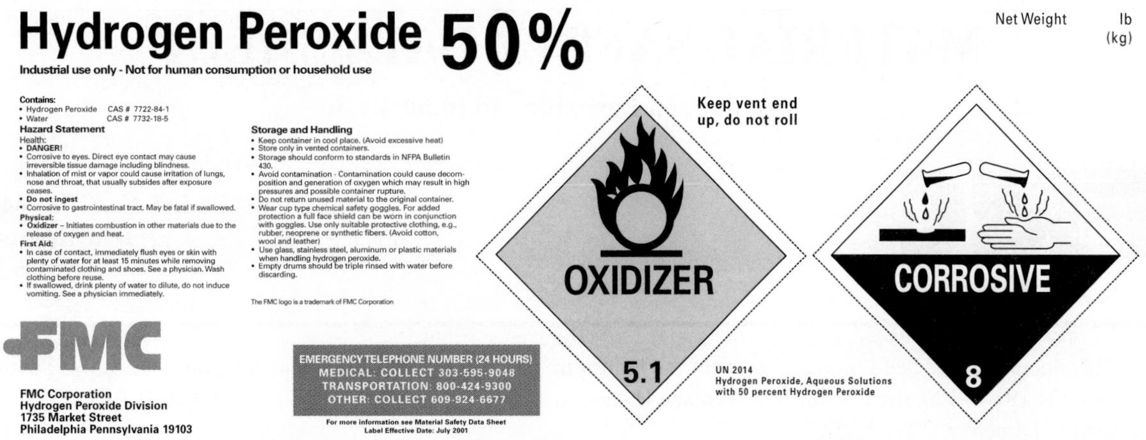

FIGURE 1.7 The label affixed to containers of 50% hydrogen peroxide provides a hazard statement, storage and handling information, and the name and address of the manufacturer. The U.S. Department of Transportation requires shippers and carriers to affix labels to containers of hydrogen peroxide designated for shipment. The nature of these labels is described in Chapter 6. (*Courtesy of FMC Corporation, Hydrogen Peroxide Division, Philadelphia, Pennsylvania.*)

1.6-A WARNING LABELS

OSHA requires every manufacturer and importer of a chemical substance to ensure that each container of a hazardous substance is labeled, tagged, and marked with the identity of the product, appropriate hazard warnings, storage and handling information, and the name and address of the manufacturer, importer, or other responsible party. Symbols, pictures, and/or words are used on the labels to present the hazard-warning messages.

Figure 1.7 is an example of a typical warning label for a chemical product. FMC Corporation affixes this label to its containers of 50% hydrogen peroxide.

1.6-B MATERIAL SAFETY DATA SHEETS

A **material safety data sheet** (MSDS) is a technical bulletin containing detailed information about a hazardous substance. OSHA requires manufacturers to prepare an MSDS for each hazardous substance it sells in the marketplace. OSHA requires manufacturers and distributors to provide a copy of the relevant MSDS when each sample or order of a hazardous substance is shipped by the manufacturer or other carrier to a location for the first time.

To ensure that the MSDS serves as a comprehensive source of written information on a product, OSHA requires that it address the following 16 categories of information:

- Chemical product and company identification
- Chemical composition and information on ingredients
- Hazards identification
- First-aid measures
- Firefighting measures
- Accidental release measures
- Handling and storage information
- Exposure controls/personal protection
- Physical and chemical properties
- Stability and reactivity
- Toxicological properties
- Ecological information
- Disposal considerations
- Transport information
- Regulatory information
- Other information deemed important

material safety data sheet
■ A technical bulletin prepared by manufacturers and marketers in accordance with the provisions of 29 C.F.R. §1910.1200 to provide workers with detailed information about the properties of a commercial product that contains hazardous substances (abbreviated MSDS)

An example of an MSDS is provided in Figure 1.8. It consists of nine pages of information on the chemical product hydrogen peroxide. Note that the MSDS is a compilation of information relating to each of the 16 categories previously cited. We note the properties of hydrogen peroxide in Section 11.5.

MATERIAL SAFETY DATA SHEET

Hydrogen Peroxide (40 to 60%)

MSDS Ref. No.: 7722-84-1-4
Date Approved:
Revision No.:

This document has been prepared to meet the requirements of the U.S. OSHA Hazard Communication Standard, 29 CFR 1910.1200; the Canada's Workplace Hazardous Materials Information System (WHMIS), and the EC Directive, 2001/58/EC.

1. PRODUCT AND COMPANY IDENTIFICATION

PRODUCT NAME: Hydrogen Peroxide (40 to 60%)

ALTERNATE PRODUCT NAME(S): Durox® Reg. & LR 50%, Oxypure® 50%, Semiconductor Reg & Seg 50%, Standard 50%, Technical 50%, Chlorate Grade 50%, Super D® 50%

GENERAL USE: Durox® 50% Reg. and LR - meets the Food Chemical Codex requirements for aseptic packaging and other food related applications.

Oxypure® 50% - certified by NSF to meet NSF/ANSI Standard 60 requirements for drinking water treatment.

Standard 50% - most suitable for industrial bleaching, processing, pollution abatement and general oxidation reactions.

Semiconductor Reg. & Seg. 50% - conforms to ACS and Semi Specs., for wafer etching and cleaning, and applications requiring low residues.

Super D® 50% - meets US Pharmacopoeia specifications for 3% topical solutions when diluted with proper quality water. While manufactured to the USP standards or purity and to FMC's demanding ISO 9002 quality standards, FMC does not claim that its Hydrogen Peroxide is manufactured in accordance with all pharmaceutical cGMP conditions.

Technical 50% - essentially free of inorganic metals, suitable for chemical synthesis.

Chlorate Grade 50% - specially formulated for use in chlorate manufacture or processing.

SynergOx™ - combination of a proprietary catalyst and 50% hydrogen peroxide, at the point of use, for environmental applications.

FIGURE 1.8 The material safety data sheet for water solutions containing 40% to 60% hydrogen peroxide. Technical information regarding this chemical product is presented in Section 11.5. (*Courtesy of FMC Corporation, Hydrogen Peroxide Division, Philadelphia, Pennsylvania.*)

MANUFACTURER	EMERGENCY TELEPHONE NUMBERS
FMC CORPORATION Hydrogen Peroxide Division 1735 Market Street Philadelphia, PA 19103 (215) 299-6000 (General Information) FMC of Canada Ltd. Hydrogen Peroxide Division PG Pulp Mill Road Prince George, BC V2N2S6 (250) 561-4200 (General Information)	(800) 424-9300 (CHEMTREC - U.S.) (613) 996-6666 (CANUTEC) (303) 595-9048 (Medical - U.S. - Call Collect) (281) 474-8750 (Plant: Pasadena, TX, US - Call Collect) (250) 561-4221 (Plant: Prince George, BC, Canada - Call Collect)

2. HAZARDS IDENTIFICATION

EMERGENCY OVERVIEW:

- Clear, colorless, odorless liquid
- Oxidizer.
- Contact with combustibles may cause fire.
- Decomposes yielding oxygen that supports combustion of organic matters and can cause overpressure if confined.
- Corrosive to eyes, nose, throat, lungs and gastrointestinal tract.

POTENTIAL HEALTH EFFECTS: Corrosive to eyes, skin, nose, throat and lungs. May cause irreversible tissue damage to the eyes including blindness.

3. COMPOSITION / INFORMATION ON INGREDIENTS

Chemical Name	CAS#	Wt.%	EC No.	EC Class
Hydrogen Peroxide	7722-84-1	40 - 60	231-765-0	C, R34
Water	7732-18-5	40 - 60	231-791-2	Not classified as hazardous

4. FIRST AID MEASURES

EYES: Immediately flush with water for at least 15 minutes, lifting the upper and lower eyelids intermittently. See a medical doctor or ophthalmologist immediately.

SKIN: Immediately flush with plenty of water while removing contaminated clothing and/or shoes, and thoroughly wash with soap and water. See a medical doctor immediately.

INGESTION: Rinse mouth with water. Dilute by giving 1 or 2 glasses of water. Do not induce vomiting. Never give anything by mouth to an unconscious person. See a medical doctor immediately.

INHALATION: Remove to fresh air. If breathing difficulty or discomfort occurs and persists, contact a medical doctor.

FIGURE 1.8 (*continued*)

NOTES TO MEDICAL DOCTOR: Hydrogen peroxide at these concentrations is a strong oxidant. Direct contact with the eye is likely to cause corneal damage especially if not washed immediately. Careful ophthalmologic evaluation is recommended and the possibility of local corticosteroid therapy should be considered. Because of the likelihood of corrosive effects on the gastrointestinal tract after ingestion, and the unlikelihood of systemic effects, attempts at evacuating the stomach via emesis induction or gastric lavage should be avoided. There is a remote possibility, however, that a nasogastric or orogastric tube may be required for the reduction of severe distension due to gas formation.

5. FIREFIGHTING MEASURES

EXTINGUISHING MEDIA: Flood with water.

FIRE / EXPLOSION HAZARDS: Product is non-combustible. On decomposition releases oxygen, which may intensify fire.

FIREFIGHTING PROCEDURES: Any tank or container surrounded by fire should be flooded with water for cooling. Wear full protective clothing and self-contained breathing apparatus.

FLAMMABLE LIMITS: Non-combustible

SENSITIVITY TO IMPACT: No data available

SENSITIVITY TO STATIC DISCHARGE: No data available

6. ACCIDENTAL RELEASE MEASURES

RELEASE NOTES: Dilute with a large volume of water and hold in a pond or diked area until hydrogen peroxide decomposes. Hydrogen peroxide may be decomposed by adding sodium metabisulfite or sodium sulfite after diluting to about 5%. Dispose according to methods outlined for waste disposal.

Combustible materials exposed to hydrogen peroxide should be immediately submerged in or rinsed with large amounts of water to ensure that all hydrogen peroxide is removed. Residual hydrogen peroxide that is allowed to dry (upon evaporation hydrogen peroxide can concentrate) on organic materials such as paper, fabrics, cotton, leather, wood or other combustibles can cause the material to ignite and result in a fire.

7. HANDLING AND STORAGE

HANDLING: Wear chemical splash-type monogoggles and full-face shield, impervious clothing, such as rubber, PVC, etc., and rubber or neoprene gloves and shoes. Avoid cotton, wool and leather. Avoid excessive heat and contamination. Contamination may cause decomposition and generation of oxygen gas which could result in high pressures and possible container rupture. Hydrogen peroxide should be stored only in vented containers and transferred only in a prescribed manner (see FMC Technical Bulletins). Never return unused hydrogen peroxide to original container, empty drums should be triple rinsed with water before discarding. Utensils used for handling hydrogen peroxide should only be made of glass, stainless steel, aluminum or plastic.

STORAGE: Store drums in cool areas out of direct sunlight and away from combustibles. For bulk storage refer to FMC Technical Bulletins.

COMMENTS: VENTILATION: Provide mechanical general and/or local exhaust ventilation to prevent release of vapor or mist into the work environment.

FIGURE 1.8 (*continued*)

8. EXPOSURE CONTROLS / PERSONAL PROTECTION

EXPOSURE LIMITS

Chemical Name	ACGIH	OSHA	Supplier
Hydrogen Peroxide	1 ppm (TWA)	1 ppm (PEL)	

ENGINEERING CONTROLS: Ventilation should be provided to minimize the release of hydrogen peroxide vapors and mists into the work environment. Spills should be minimized or confined immediately to prevent release into the work area. Remove contaminated clothing immediately and wash before reuse.

PERSONAL PROTECTIVE EQUIPMENT

EYES AND FACE: Use chemical splash-type monogoggles and a full-face shield made of polycarbonate, acetate, polycarbonate/acetate, PETG or thermoplastic.

RESPIRATORY: If concentrations in excess of 10 ppm are expected, use NIOSH/DHHS approved self-contained breathing apparatus (SCBA), or other approved atmospheric-supplied respirator (ASR) equipment (e.g., a full-face airline respirator (ALR)). DO NOT use any form of air-purifying respirator (APR) or filtering facepiece (AKA dust mask), especially those containing oxidizable sorbants such as activated carbon.

PROTECTIVE CLOTHING: For body protection wear impervious clothing such as an approved splash protective suit made of SBR Rubber, PVC (PVC Outershell w/Polyester Substrate), Gore-Tex (Polyester trilaminate w/Gore-Tex), or a specialized HAZMAT Splash or Protective Suite (Level A, B, or C). For foot protection, wear approved boots made of NBR, PVC, Polyurethane, or neoprene. Overboots made of Latex or PVC, as well as firefighter boots or specialized HAZMAT boots are also permitted. DO NOT wear any form of boot or overboots made of nylon or nylon blends. DO NOT use cotton, wool or leather, as these materials react RAPIDLY with higher concentrations of hydrogen peroxide. Completely submerge hydrogen peroxide contaminated clothing or other materials in water prior to drying. Residual hydrogen peroxide, if allowed to dry on materials such as paper, fabrics, cotton, leather, wood or other combustibles can cause the material to ignite and result in a fire.

GLOVES: For hand protection, wear approved gloves made of nitrile, PVC, or neoprene. DO NOT use cotton, wool or leather, for these materials react RAPIDLY with higher concentrations of hydrogen peroxide. Thoroughly rinse the outside of gloves with water prior to removal. Inspect regularly for leaks.

9. PHYSICAL AND CHEMICAL PROPERTIES

ODOR:	Odorless
APPEARANCE:	Clear, colorless liquid
AUTOIGNITION TEMPERATURE:	Non-combustible
BOILING POINT:	110°C (229°F) (40%); 114°C (237°F) (50%)
COEFFICIENT OF OIL / WATER:	Not available
DENSITY / WEIGHT PER VOLUME:	Not available

FIGURE 1.8 (*continued*)

EVAPORATION RATE:	Above 1 (Butyl Acetate = 1)
FLASH POINT:	Non-combustible
FREEZING POINT:	$-41.4°C$ ($-42.5°F$) (40%); $-52°C$ ($-62°F$) (50%)
ODOR THRESHOLD:	Not available
OXIDIZING PROPERTIES:	Strong oxidizer
PERCENT VOLATILE:	100%
pH:	(as is) 1.0 to 3.0
SOLUBILITY IN WATER:	(in H_2O % by wt) 100%
SPECIFIC GRAVITY:	(H_2O = 1)1.15 @ 20°C/4°C (40%); 1.19 @ 20°C/4°C(50%)
VAPOR DENSITY:	Not available(Air = 1)
VAPOR PRESSURE:	22 mmHg @ 30°C (40%); 18.3 mmHg @ 30°C (50%)

COMMENTS:
pH (1% solution): 5.0-6.0

---

10. STABILITY AND REACTIVITY

CONDITIONS TO AVOID:	Excessive heat or contamination could cause product to become unstable.
STABILITY:	Stable (heat and contamination could cause decomposition)
POLYMERIZATION:	Will not occur
INCOMPATIBLE MATERIALS:	Reducing agents, wood, paper and other combustibles, iron and other heavy metals, copper alloys and caustic.
HAZARDOUS DECOMPOSITION PRODUCTS:	Oxygen, which supports combustion.

COMMENTS: Materials to Avoid: Dirt, organics, cyanides and combustibles such as wood, paper, oils, etc.

---

11. TOXICOLOGICAL INFORMATION

EYE EFFECTS: 70% hydrogen peroxide: Severe irritant (corrosive) (rabbit) [FMC Study Number: ICG/T-79.027]

SKIN EFFECTS: 50% hydrogen peroxide: Severe irritant (corrosive) (rabbit) FMC Study Number: I89-1079]

DERMAL LD_{50}: 70% hydrogen peroxide: > 6.5 g/kg (rabbit) [FMC Study Number: ICG/T-79.027]

ORAL LD_{50}: 50% hydrogen peroxide: > 225 mg/kg (rat) [FMC Study Number I86-914]

INHALATION LC_{50}: 50% hydrogen peroxide: > 0.17 mg/L (rat) [FMC Study Number: I89-1080]

TARGET ORGANS: Eye, skin, nose, throat, lungs

FIGURE 1.8 (continued)

ACUTE EFFECTS FROM OVEREXPOSURE: Severe irritant/corrosive to eyes, skin and gastrointestinal tract. May cause irreversible tissue damage to die eyes including blindness. Inhalation of mist or vapors may be severely irritating to nose, throat and lungs.

CHRONIC EFFECTS FROM OVEREXPOSURE: The International Agency for Research on Cancer (IARC) has concluded that there is inadequate evidence for carcinogenicity of hydrogen peroxide in humans, but limited evidence in experimental animals (Group 3 - not classifiable as to its carcinogenicity to humans). The American Conference of Governmental Industrial Hygienists (ACGIH) has concluded that hydrogen peroxide is a Confirmed Animal Carcinogen with Unknown Relevance to Humans'(A3).

CARCINOGENICITY:

Chemical Name	IARC	NTP	OSHA	Other
Hydrogen Peroxide	Listed	Not listed	Not listed	(ACGIH) Listed (A3, Animal Carcinogen)

12. ECOLOGICAL INFORMATION

ECOTOXICOLOGICAL INFORMATION: Channel catfish 96-hour $LC_{50} = 37.4$ mg/L
Fathead minnow 96-hour $LC_{50} = 16.4$ mg/L
Daphnia magna 24-hour $EC_{50} = 7.7$ mg/L
Daphnia pulex 48-hour $LC_{50} = 2.4$ mg/L
Freshwater snail 96-hour $LC_{50} = 17.7$ mg/L
For more information refer to ECETOC "Joint Assessment of Commodity Chemicals No. 22, Hydrogen Peroxide." ISSN-0773-6339, January 1993

CHEMICAL FATE INFORMATION: Hydrogen peroxide in the aquatic environment is subject to various reduction or oxidation processes and decomposes into water and oxygen. Hydrogen peroxide half-life in freshwater ranged from 8 hours to 20 days, in air from 10 to 20 hours and in soils from minutes to hours depending upon microbiological activity and metal contaminants.

13. DISPOSAL CONSIDERATIONS

DISPOSAL METHOD: An acceptable method of disposal is to dilute with a large amount of water and allow the hydrogen peroxide to decompose followed by discharge into a suitable treatment system in accordance with all regulatory agencies. The appropriate regulatory agencies should be contacted prior to disposal.

14. TRANSPORT INFORMATION

U.S. DEPARTMENT OF TRANSPORTATION (DOT)

PROPER SHIPPING NAME: Hydrogen peroxide, aqueous solutions with more than 40% but not more than 60% hydrogen peroxide.

PRIMARY HAZARD CLASS / DIVISION: 5.1 (Oxidizer)

UN/NA NUMBER: UN 2014

FIGURE 1.8 (*continued*)

PACKING GROUP:	II
LABEL(S):	Oxidizer, Corrosive
PLACARD(S):	5.1 (Oxidizer)
ADDITIONAL INFORMATION:	DOT Marking: Hydrogen Peroxide, aqueous solution with more than 40%, but not more man 60% Hydrogen Peroxide, UN 2014
	Hazardous Substance/RQ: Not applicable
	49 STCC Number: 4918775
	DOT Spec: stainless steel/high purity aluminum cargo tanks and rail cars. UN Spec: HDPE drums. Contact FMC for specific details.

INTERNATIONAL MARITIME DANGEROUS GOODS (IMDG)

PROPER SHIPPING NAME:	Hydrogen peroxide, aqueous solutions with not less than 20%, but not more than 60% hydrogen peroxide.

INTERNATIONAL CIVIL AVIATION ORGANIZATION (ICAO)/ INTERNATIONAL AIR TRANSPORT ASSOCIATION (IATA)

PROPER SHIPPING NAME:	Hydrogen peroxide (40 - 60%) is forbidden on Passenger and Cargo Aircraft, as well as Cargo Only Aircraft.

OTHER INFORMATION:

Protect from physical damage. Keep drums in upright position. Drums should not be stacked in transit. Do not store drum on wooden pallets.

15. REGULATORY INFORMATION

UNITED STATES

SARA TITLE III (SUPERFUND AMENDMENTS AND REAUTHORIZATION ACT)
SECTION 302 EXTREMELY HAZARDOUS SUBSTANCES (40 CFR 355, APPENDIX A):
Hydrogen Peroxide > 52%, RQ: 1000 lbs. Planning Threshold: 10,000 lbs.

SECTION 311 HAZARD CATEGORIES (40 CFR 370):
Fire Hazard, Immediate (Acute) Health Hazard

SECTION 312 THRESHOLD PLANNING QUANTITY (40 CFR 370):
The Threshold Planning Quantity (TPQ) for this product, if treated as a mixture, is 10,000 lbs; however, this product contains the following ingredients with a TPQ of less than 10,000 lbs.: None, (conc. <52%) (hydrogen peroxide, 1000 lbs. when conc. is >52%)

SECTION 313 REPORTABLE INGREDIENTS (40 CFR 372):
Not listed

FIGURE 1.8 (*continued*)

CERCLA (COMPREHENSIVE ENVIRONMENTAL RESPONSE, COMPENSATION, AND LIABILITY ACT)
   CERCLA DESIGNATION & REPORTABLE QUANTITIES (RQ) (40 CFR 302.4):
      Unlisted (Hydrogen Peroxide); RQ = 100 lbs.; Ignitability, Corrosivity

TSCA (TOXIC SUBSTANCES CONTROL ACT)
   TSCA INVENTORY STATUS (40 CFR 710):
      Listed

RESOURCE CONSERVATION AND RECOVERY ACT (RCRA)
   RCRA IDENTIFICATION OF HAZARDOUS WASTE (40 CFR 261):
      Waste Number: D001,D002

CANADA

WHMIS (WORKPLACE HAZARDOUS MATERIALS INFORMATION SYSTEM):
   Product Identification Number: 2014
   Hazard Classification / Division: Class C (Oxidizer), Class D, Div. 2, Subdiv. B (Toxic), Class E (Corrosive)
   Ingredient Disclosure List: Listed

EU EINECS NUMBERS:
   008-003-00-9 (hydrogen peroxide)

INTERNATIONAL LISTINGS
   Hydrogen peroxide:
   China: Listed
   Japan (ENCS): (1)-419
   Korea: KE-20204
   Philippines (PICCS): Listed

---

16. OTHER INFORMATION

HAZARD, RISK AND SAFETY PHRASE DESCRIPTIONS:

Hydrogen Peroxide:

EC Symbols:	C	(Corrosive)
EC Risk Phrases:	R34	(Causes burns)
EC Safety Phrases:	S1/2	(Keep locked up and out of reach of children.)
	S3	(Keep in a cool place.)
	S28	(After contact with skin, wash immediately with plenty of water and soap.)
	S36/39	(Wear suitable protective clothing. Wear eye / face protection.)
	S45	(In case of accident or if you feel unwell, seek medical advice immediately - show the label where possible.)

FIGURE 1.8 *(continued)*

HMIS

Health	3
Flammability	0
Physical Hazard	1
Personal Protection (PPE)	H

Protection = H (Safety goggles, gloves, apron, the use of a supplied air or SCBA respirator is required in lieu of a vapor cartridge respirator)

HMIS = Hazardous Materials Identification System

Degree of Hazard Code:
4 = Severe
3 = Serious
2 = Moderate
1 = Slight
0 = Minimal

NFPA

Health	3
Flammability	0
Reactivity	1
Special	0X

SPECIAL = OX (Oxidizer)

NFPA = National Fire Protection Association

Degree of Hazard Code:
4 = Extreme
3 = High
2 = Moderate
1 = Slight
0 = Insignificant

REVISION SUMMARY:

Changes in information are as follows:

New Format, as well as text changes and/or updates to one or more Sections of this MSDS.

Durox, Oxypure, Super D, SynergOx and FMC Logo - FMC Trademarks

© 2004 FMC Corporation. All Rights Reserved.

NOTE: NFPA Reactivity is 3 - when greater than 52%

FMC Corporation believes that the information and recommendations contained herein (including data and statements) are accurate as of the date hereof. NO WARRANTY OF FITNESS FOR ANY PARTICULAR PURPOSE, WARRANTY OF MERCHANTABILITY, OR ANY OTHER WARRANTY, EXPRESSED OR IMPLIED, IS MADE CONCERNING THE INFORMATION PROVIDED HEREIN. The information provided herein relates only to the specific product designated and may not be applicable where such product is used in combination with any other materials or in any process. It is a violation of Federal law to use this product in a manner inconsistent with its labeling. Further, since the conditions and methods of use are beyond the control of FMC Corporation, FMC Corporation expressly disclaims any and all liability as to any results obtained or arising from any use of the product or reliance on such information.

FIGURE 1.8 *(continued)*

1.7 Hazardous Materials in Transit

Using the authority of the **Federal Hazardous Materials Transportation Law**, as amended by the Homeland Security Act of 2002, the U. S. Department of Transportation (DOT), regulates shippers and carriers who offer or accept hazardous materials for intrastate, interstate, or international transportation through certain marking, labeling, placarding, and packaging requirements. The information conveyed by these warning labels, placards, and shipping papers is especially critical for emergency responders. We discuss the details of these requirements throughout Chapter 6.

Federal Hazardous Materials Transportation Law
■ The federal statute that empowers DOT to regulate certain activities of shippers and carriers who offer or accept hazardous materials for intrastate, interstate, or international transportation

1.8 Global Harmonization

Over the years, countries throughout the world have implemented methods of defining relevant hazards and classifying and labeling chemical substances to forewarn consumers and users of the hazards associated with them. In many instances, these procedures, classification schemes, and labels vary significantly from country to country. Even within the United States, there are four federal authorities that have jurisdiction over the issue of warning consumers about product hazards: OSHA, CPSC, EPA, and DOT. Although there are certain similarities among these agencies' methods of classification and labeling, inconsistent and conflicting requirements have tended to create confusion.

To circumvent this use of multiple systems for defining and classifying hazards and communicating hazard information, the United Nations designed the **Globally Harmonized System of Classification and Labeling of Chemical Substances** (GHS). To the emergency responder, the implementation of a globally harmonized system provides a basis for a greater consistency in the classification and labeling of chemical substances, thus providing an enhancement for the safe handling and use of substances when they are transported and used in the workplace and consumer-use settings.

In 2008, OSHA adopted the GHS to communicate hazard information by means of warning labels and information on material safety data sheets. The international system uses a series of pictograms and specific wording for various chemical hazards. Some examples of the pictograms are provided in Figure 1.9.

DOT is also adopting the GHS by implementing procedures for common domestic and international shipping papers for use in transporting hazardous materials, as well as common requirements for marking and labeling packaging, and placarding transport containers. We review the nature of these procedures in Chapter 6.

Globally Harmonized System of Classification and Labeling of Chemicals (abbreviated GHS)
■ The United Nations system for defining and classifying hazards and communicating hazard information on labels and material safety data sheets in a common international fashion

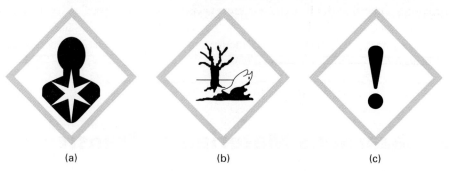

(a)   (b)   (c)

FIGURE 1.9 As nations proceed to implement a globally harmonized system for communicating hazard information, we may encounter pictograms like these affixed to product containers. The pictogram in (a) denotes that the product contains a carcinogen, respiratory sensitizer, reproductive or target-organ toxicant, or mutagen. The pictogram in (b) denotes that the product contains a substance that could have a negative impact on the environment on improper disposal. The pictogram in (c) denotes that the product contains an irritant or dermal sensitizer.

PERFORMANCE GOALS FOR SECTION 1.9:

- Discuss the intent of the Emergency Planning and Community Right-to-Know Act, especially as it bears on the emergency-response profession.

1.9 Hazardous Substances Within Communities

Emergency Planning and Community Right-to-Know Act
- The federal statute that empowers EPA to require a facility handling chemical substances to either submit copies of MSDSs for certain chemical substances present at the facility in amounts greater than specified threshold quantities *or* to provide a list of such substances along with their acute health hazards, chronic health hazards, fire hazards, sudden-release-of-pressure hazards, and reactivity hazards (abbreviated EPCRA)

In 1986, to better prepare communities with information concerning the presence of chemical substances within their boundaries, Congress enacted the **Emergency Planning and Community Right-to-Know Act** (EPCRA). The statute requires any facility handling chemical substances either to submit copies of MSDSs for certain chemical substances present at the facility in amounts greater than specified threshold quantities or to provide a list of such substances along with each substance's acute health hazard, chronic health hazard, fire hazard, sudden-release-of-pressure hazard, and reactivity hazard. The relevant substances include 77 toxic substances, 63 flammable substances, and certain explosive substances. *Threshold quantities*, that is, minimum amounts required for reporting, have been established for toxic substances ranging from 500 to 20,000 lb. For all listed flammable substances, the threshold quantity is 10,000 lb, and for the explosive substances, the threshold quantity is established at 5000 lb.

EPCRA provides EPA with the authority to require submission of the list of hazardous substances or copies of relevant MSDSs to the local emergency-planning committee, the state emergency-response commission, and the local fire department having jurisdiction over the facility within 60 days of receipt of the hazardous substances. From a review of this information, the three groups can readily determine the substances that are present within any community above threshold quantities. When utilized effectively, this information serves to improve the quality of accident-planning exercises, preparedness, mitigation, response, and recovery capabilities of emergency-response teams. It also provides citizens with sufficient knowledge to prepare for incidents that could involve hazardous substances within their local communities.

Following 9-11, EPA acknowledged that it would be imprudent to publish data concerning amounts of hazardous substances present at individual chemical companies. Although this information is useful to communities preparing for an accident or natural disaster, radical organizations could also find the information useful when preparing for terrorist actions. Whereas EPA once withheld this information, the agency again made it available in 2005.

PERFORMANCE GOALS FOR SECTION 1.10:

- Describe the procedure used by NFPA for rapidly conveying information regarding the potential hazards associated with exposure to a given hazardous substance.
- Identify the colors and the significance of each NFPA number denoting a relative rating for health, flammability, and chemical reactivity.

1.10 NFPA System of Identifying Potential Hazards

At the scene of an emergency, how may the potential hazards associated with a given hazardous material be regardly identified? The answer to this question is based on recognizing certain markings that are posted on bulk tanks, exterior building walls, pipelines, small containers, and other locations at which hazardous materials are stored or used.

The NFPA introduced a procedure that provides a means for rapidly identifying the relative degree of three hazards associated with a given hazardous material: its health, flammability, and chemical reactivity hazards. NFPA implements this procedure by assigning one of five numbers, 0 through 4, to each property for a given hazardous material. The numbers in Table 1.3 identify the relative degree of hazard that corresponds to a relevant property. The number 0 signifies that the hazardous material at issue does not possess the relevant hazard, whereas the number 4 denotes that it possesses the highest degree of that hazard.

NFPA hazard diamond
■ A diamond-shaped figure divided into four quadrants, each of which is generally color-coded for each of a substance's three hazards (blue for health hazard, red for fire hazard, and yellow for chemical reactivity hazard), and in which is marked a number ranging from 0 to 4 that designates the degree of the relevant hazard

SOLVED EXERCISE 1.5

What information does NFPA recommend marking on the exterior surfaces of flammable liquid storage tanks?

Solution: When they respond to an emergency situation within a tank farm, firefighters require readily accessible and accurate information concerning the contents of the storage tanks. Accordingly, the information marked on the them should be easily visible and legible and should include the names of the liquids stored within them, their general hazards, and special firefighting precautions of which the firefighters should be aware.

To convey hazard information, NFPA recommends the use of number codes that are marked within the top three quadrants of a diamond-shaped symbol painted on, or otherwise affixed to, the exterior surfaces of storage tanks. Beginning with the left-hand quadrant and proceeding clockwise, the number codes denote the relative degrees of health, flammability, and reactivity hazards of the substance stored within each tank. Firefighters compare each number with the appropriate information listed in Table 1.3. Special codes are provided in the bottom quadrant, such as a "W" to caution against the application of water. This combination of information helps firefighters determine which actions to take and not to take when they first encounter a fire within a storage tank.

Each number is then displayed in the appropriate top three quadrants of the diamond-shaped diagram in Figure 1.10 called the **NFPA hazard diamond**. Each quadrant of the diamond is color-coded, beginning with the left-hand quadrant and proceeding clockwise: blue for the health hazard, red for the fire hazard, and yellow for the chemical reactivity hazard.

TABLE 1.3 — The NFPA Codes for Health, Flammability, and Reactivity

Identification of Health		Identification of Flammability		Identification of Reactivity (Stability)	
Hazard Color Code: BLUE		**Hazard Color Code: RED**		**Hazard Color Code: YELLOW**	
SIGNAL	**TYPE OF POSSIBLE INJURY**	**SIGNAL**	**SUSCEPTIBILITY OF MATERIALS TO BURNING**	**SIGNAL**	**SUSCEPTIBILITY TO RELEASE OF ENERGY**
4	Materials that on very short exposure could cause death or major residual injury even though prompt medical treatment was given	4	Materials that will rapidly or completely vaporize at atmospheric pressure and normal ambient temperature, or that are readily dispersed in air, and will burn readily	4	Materials that in themselves are readily capable of detonation or of explosive decomposition or reaction at normal temperature and pressures
3	Materials that on short exposure could cause serious temporary or residual injury even though prompt medical treatment was given	3	Liquids and solids that can be ignited under almost all ambient temperature conditions	3	Materials that in themselves are capable of detonation or explosive reaction but require a strong initiating source or that must be heated under confinement before initiation or that react explosively with water
2	Materials that on intense or continued exposure could cause temporary incapacitation or possible residual injury unless prompt medical treatment was given	2	Materials that must be moderately heated or exposed to relatively high ambient temperature before ignition can occur	2	Materials that in themselves are normally unstable and readily undergo violent chemical change but do not detonate. Also, materials that may react violently with water or may form potentially explosive mixtures with water
1	Materials that on exposure would cause irritation but only minor residual injury even if no treatment was given	1	Materials that must be preheated before ignition can occur	1	Materials that in themselves are normally stable but which can become unstable at elevated temperatures and pressures or which may react with water with some release of energy but not violently
0	Materials that on exposure under fire conditions would offer no hazard beyond that of ordinary combustible material	0	Materials that will not burn	0	Materials that in themselves are normally stable, even under fire exposure conditions, and which are not reactive with water

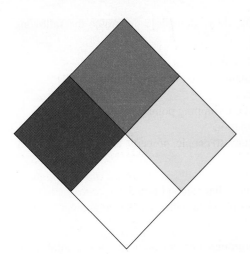

FIGURE 1.10 Four potential hazards of a substance may be rapidly and simultaneously identified by the use of a color-coded numeral system on this NFPA hazard diamond. Beginning with the left-hand quadrant and proceeding clockwise, the quadrants are color-coded as follows: blue for the health hazard; red for the fire hazard; and yellow for the chemical reactivity hazard. A number from 0 to 4 is entered in each of the top three quadrants consistent with the information compiled in Table 1.3 to identify the severity of the health, fire, and chemical reactivity hazards. Each of several symbols, such as W and/or OXY, may also be entered in the bottom white quadrant to provide additional hazard information.

The NFPA also uses three symbols, each of which is displayed in the bottom quadrant of the diamond, when warranted:

- The letter "W" with a line drawn through its center (W) to caution against the application of water
- The letters "OX" or "OXY" to indicate the presence of an oxidizer
- A radiation hazard symbol resembling a three-bladed propeller, or trefoil (Section 16.6), which indicates the presence of a radioactive material

Sometimes, expressions like ACID, ALK, or CORR, are also displayed in the bottom quadrant to indicate the presence of an acid, alkaline material, or corrosive material, respectively.

As we study the properties of individual hazardous materials beginning in Chapter 7, the appropriate NFPA hazard diamond will be displayed in the page margin near the point at which a discussion of each hazardous material first begins.

PERFORMANCE GOALS FOR SECTION 1.11:

- Identify the services that CHEMTREC provides when emergency responders are called to incidents in which hazardous materials are involved.

1.11 CHEMTREC

The Chemical Transportation Emergency Center (**CHEMTREC**) serves as a state-of-the-art communications center that deals with transportation emergencies by reinforcing the effectiveness of specialized emergency-response groups and enhancing hazardous materials transportation security. During an average year, CHEMTREC receives more than 10,000 calls relating to transportation emergencies involving the release of hazardous materials.*

CHEMTREC
■ Acronym for Chemical Transportation Emergency Center; a public service hotline for firefighters, law enforcement, and other emergency responders for obtaining information and assistance for incidents involving hazardous materials

*Within the United States, telephone (800)424-9300. For calls originating outside the United States, telephone collect (703)527-3887. Within Canada, telephone Le Centre canadien d'urgence transport du Ministère des transports (the Canadian Transport Emergency Centre of the Department of Transport), or CANUTEC, at (613)996-6666. Within Mexico, telephone the Secretaría de Comunicaciones y Transportes (Secretariat of Communications and Transportation of Mexico), or SCT, at 52-5-684-1275. These numbers have been widely circulated in the professional literature that is distributed to emergency-service personnel, shippers and carriers, and members of the chemical industry, and they have been further circulated in bulletins of governmental agencies, trade associations, and similar groups.

During each phone conversation, the person on duty attempts to determine the following from the caller:

- The nature of the emergency and where and when it occurred
- The hazardous material involved and its DOT identification number (Section 6.1-B)
- The type and condition of the tanks or containers
- The identity of the shipper or manufacturer and the shipping point
- The identity of the carrier, consignee, and destination
- The general nature and extent of the injuries, if any, to people, property, and the environment
- The prevailing local weather conditions
- The composition of the surrounding area
- The name of the caller and a means by which the caller can be located
- A means by which telephone contact can be reestablished with the caller or another responsible person at the scene

With the caller remaining on the line, the communicator draws on the best available information concerning the hazardous materials that are involved in the emergency-response effort. As immediate first steps in controlling the emergency, specific hazard information and directions on what to do and what not to do are provided to the caller. CHEMTREC obtains this information by searching among the nearly 2.8 million MSDSs in its data base. To preclude unfounded personal speculation regarding the specific features of an emergency, the communicator is under instructions to abide strictly by information previously prepared for use.

Having advised the caller, the communicator proceeds immediately to notify the shipper by telephone. The known particulars of the emergency incident having thus been relayed, the responsibility for further guidance to the emergency-response crew, including dispatching personnel to the scene, passes to the shipper.

Although proceeding to the second stage of assistance becomes more difficult when the shipper is unknown, communicators are armed with other resources on which they can rely. For instance, CHEMTREC can contact the U.S. Department of Energy in incidents involving the release of radioactive materials.

The notification of a transportation emergency involving hazardous materials to CHEMTREC does not constitute compliance with the RCRA, CERCLA, or DOT reporting requirements noted in Sections 1.12 and 6.10.

PERFORMANCE GOALS FOR SECTION 1.12:

- Identify the general function of the National Response Center.

1.12 National Response Center

National Response Center
■ The sole national point of contact for reporting hazardous substance releases into the environment within the United States and its territories

The **National Response Center** (NRC), first began operating when EPA required the services of a federal group to whom responsible parties could report spills, discharges, fires, explosions, and other incidents in which oil and hazardous substances were released into waters of the United States. The existence of this center was first mandated by FWPPCA and has been operated by the U.S. Coast Guard since 1974.

The responsibilities of the NRC were broadened subsequent to enactment of federal transportation and environmental regulations. Today, the NRC serves as the sole federal point of contact for reporting oil, hazardous materials, hazardous substances, biological, etiological (infectious), and radiological spills throughout the United States and its territories. The NRC serves as the link to the **National Response Team** (NRT), a group of individuals from 16 federal agencies having varying interests and expertise in emergency-response matters. When the situation warrants, the NRT is dispatched to the scene of a release incident to provide assistance.

Federal law requires that the NRC must be notified *immediately* whenever anyone who releases into the environment a reportable quantity (Section 6.1-D) of a hazardous substance (including oil when water is or may be affected) or a material identified as a marine pollutant (Section 6.1-E). Reporting a release incident initially involves telephoning NRC at (800) 424-8802 or, in the District of Columbia, at (202)426-2675. The notification should minimally include such information as the date, time, nature of the release incident, the quantity and type of material involved in the release, and the extent of injuries, if any. A follow-up written report to the relevant federal agency is also required.

The NRC maintains a communication link with CHEMTREC personnel, who provide the caller with emergency-response information. The NRC also informs the Chemical Safety and Hazard Investigation Board, which subsequently investigates the nature of the physical evidence found at the accident scene to determine the cause of the accident. EPA dispatches an on-scene coordinator who ensures that the subsequent cleanup follows the relevant environmental laws.

Additional information regarding the reporting of a release of a hazardous substance is noted in Section 6.10.

National Response Team
A federal interagency group chaired by EPA and the U.S. Coast Guard responsible for distributing technical, financial, and operational information about hazardous substance releases and oil spills

Classification of Hazardous Materials

1.1 Assign each of the following to one or more of the seven classes of hazardous materials denoted in Section 1.2:

(a) Carbon monoxide

(b) Fireworks

(c) Battery acid

NFPA Classes of Fire

1.2 Identify the NFPA class of fire associated with the ignition and burning of each of the following:

(a) Peanut oil

(b) Rubber gloves

(c) Magnesium turnings

(d) Energized electrical circuit

Hazardous Substances Within Consumer Products

1.3 Xylene is a flammable liquid that is harmful or fatal if swallowed, and an eye and skin irritant. Why are spot removers containing xylene subject to CPSC's labeling requirements?

1.4 The following label is affixed to the exterior surface of an aerosol can containing an air deodorizer:

WARNING

FLAMMABLE, EYE IRRITANT, CONTENTS UNDER PRESSURE

DIRECTIONS FOR USE: To deodorize the air, spray whenever odors are a problem or occur. Allow a few seconds for superfine mist to bloom. To deodorize fabrics, hold can 14 to 15 in. from surface and spray evenly in a sweeping motion to obtain maximum penetration.

CAUTION: May cause severe eye irritation. Prolonged or repeated contact may dry or defat the skin and lead to inflammation (i.e., dermatitis). Inhalation may be irritating to the eyes, nose, and respiratory tract. Nausea, vomiting, and stomach upset may occur. Misuse or intentional misuse may be harmful or fatal. May affect the nervous system, eyes, or skin upon prolonged or repeated overexposure.

FIRST-AID INSTRUCTIONS: EYES—Immediately flush with plenty of water for at least 15 minutes. Get prompt medical attention. **SKIN**—Wash with soap and plenty of water.

INHALATION—Remove to fresh air. If symptoms develop, seek immediate medical attention. If not breathing, give artificial respiration. **INGESTION**—In the unlikely event of swallowing, get immediate medical attention. Do NOT induce vomiting unless directed by medical personnel. Rinse mouth with water and give another cupful to drink. Never give anything by mouth to an unconscious person.

KEEP AWAY FROM HEAT, SPARKS, AND FLAMES. Do not puncture or incinerate container. Do not expose to heat or store at temperatures above 120 °F (49 °C), as container may burst. Avoid prolonged exposure to sunlight. Use with adequate ventilation.

KEEP OUT OF REACH OF CHILDREN.

INGREDIENTS: Acetone, fragrance, and liquefied petroleum gas.

(To provide confidentiality, the name and place of business of the manufacturer, packer, distributor, or seller and the tradename of the deodorizer have been intentionally omitted.) Which federal law provides the CPSC with the legal authority to require manufacturers to list the information on this label?

1.5 The following label is affixed to the exterior surface of an aerosol can containing a product that kills spiders, flies, and mosquitoes:

KEEP OUT OF REACH OF CHILDREN

ACTIVE INGREDIENTS:

Chloropyrifos [O,O-diethyl O-3,5,6-trichloro-2-pyridinyl phosphorothioate]	0.25%
d-*trans*-Allethrin	0.05%
Inert ingredients	99.70%

HAZARDS TO HUMANS AND DOMESTIC ANIMALS: Avoid contact with eyes, skin, and clothing. Avoid breathing vapors or spray mist. Keep away from food, feedstuffs, and domestic water supplies. Wash thoroughly after handling.

STATEMENT OF PRACTICAL TREATMENT: If swallowed, call a physician or Poison Control Center. Drink 1 or 2 glasses of water and induce vomiting by touching back of throat with finger. Do not induce vomiting or give anything by mouth to an unconscious person. In case of skin contact, remove contaminated clothing and immediately wash skin with plenty of water. If in eyes, flush eyes with plenty of water. Note to physician: Chloropyrifos is a cholinesterase inhibitor. Treat symptomatically. Atropine only by injection is an antidote.

ENVIRONMENTAL HAZARDS: This pesticide is toxic to fish, birds, and other wildlife. Do not apply directly to water. This product is highly toxic to bees exposed to direct treatment or residues on plants.

PHYSICAL OR CHEMICAL HAZARDS: Contents under pressure. Do not use or store near heat or open flame. Do not puncture or incinerate. Exposure to temperatures above 150 °F may cause bursting.

(To provide confidentiality, the name and place of business of the manufacturer, packer, and or distributor and the tradename of the pesticide have been intentionally omitted.) When containers of this product are encountered by first-on-the-scene responders during a warehouse fire, what precautions should they exercise to safeguard lives, property, and the environment?

1.6 At 21 C.F.R. §172.804, FDA obligates the posting of the following potential health hazard on the labels of diet-drink containers and certain other food containers:

PHENYLKETONURICS: CONTAINS PHENYLALANINE

Phenylketonurics are individuals who inherit a biochemical disorder whereby they are unable to breakdown phenylalanine, a protein component. The phenylalanine concentrates within the body and damages the brain. The posted information serves to forewarn phenylketonurics that they should avoid the diet drink or consume it only in moderation. Which federal law provides the FDA with the legal authority to require manufacturers to list this label information?

Hazardous Constituents of Pollutants and Wastes

1.7 EPA has taken appropriate action to ensure that the ozone concentration in the air does not exceed 0.12 parts per million (ppm) within a single hour. Which federal statute provides EPA with the legal authority to take such measures?

Hazardous Substances in the Workplace

1.8 An adhesive caulk used by carpenters is characterized as a thick white or tan paste with a mild sweet odor. Canisters of the caulk bear the following warning statements:

WARNING

May cause eye, skin, nose, throat and respiratory tract irritation. Harmful if swallowed or absorbed through the skin. Presents little or no hazard (if spilled) and/or no unusual hazard if involved in a fire.

POTENTIAL HEALTH EFFECTS

EYE CONTACT: May cause eye irritation.

SKIN CONTACT: May cause allergic skin reaction or sensitization. Prolonged or repeated contact with skin may cause irritation.

INHALATION: Harmful if swallowed.

CHRONIC HAZARDS: Repeated or prolonged exposure may cause skin, respiratory, kidney, cardiovascular, and liver damage.

How does the manufacturer of this adhesive use this warning information when designing the material safety data sheet for this product?

Hazardous Substances Within Communities

1.9 Why are the MSDSs for chemical products manufactured and stored within communities often compiled for review at local fire departments?

NFPA Hazard Diamonds

1.10 First-on-the-scene responders observe the numbers 4, 2, and 2 encoded within the top three quadrants, proceeding clockwise from the far left, respectively, of an NFPA hazard diamond affixed to the front of a burning chemical storage shed. How are these numbers useful to the emergency responders?

1.11 For each of the following pairs of NFPA hazard diamonds, identify the appropriate code that conveys the greater degree of the following hazards:

(a) **Fire**

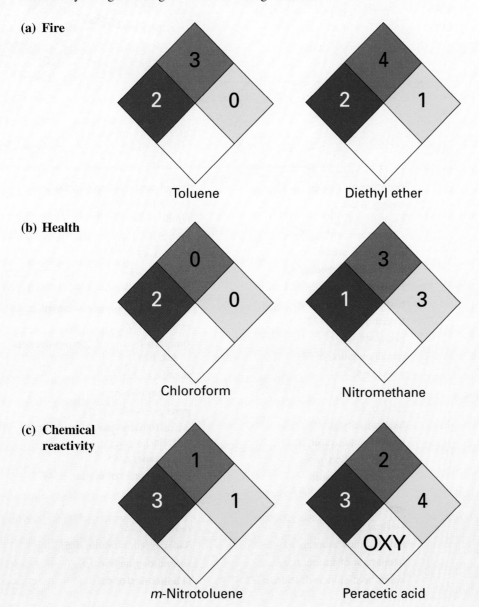

Toluene   Diethyl ether

(b) **Health**

Chloroform   Nitromethane

(c) **Chemical reactivity**

m-Nitrotoluene   Peracetic acid

A t first glance, the mere number of hazardous materials is likely to overwhelm the average nonscientist. How is it possible to learn the individual properties of so many substances and recall them under the disordered conditions that often prevail when lives and property are in jeopardy?

Fortunately, we can associate the hazardous properties of many substances with their state of matter. We learn, for instance, that all gases possess certain common properties; on studying the chemistry of gases, we learn to identify these common properties and then turn our attention later to the features that cause individual gases to be regarded as unique substances.

In this chapter, we learn about some of the general properties of matter and energy and how they influence certain phenomena such as the spread of fire. Also, as we review the properties of matter and energy, we learn how they relate to the issues in fire science. Specifically, in this chapter we learn how the modes of heat transfer contribute to the propagation of fire, the reason liquid water often effectively extinguishes a fire, and why a gas confined within a cylinder is likely to rupture the cylinder when excessively heated.

PERFORMANCE GOALS FOR SECTION 2.1:

- Describe the general features of matter.
- Distinguish between the mass and weight of an object.
- Identify the properties that characterize the three physical states of matter.

2.1 Matter Defined

Each day, we encounter air, water, metals, stone, dirt, animals, and plants. These are the materials of which the world is made. Scientist refer to them as the different kinds of **matter**. When we search for common features among its forms, we note that all matter possesses mass and occupies space. In other words, matter is distinguishable from empty space by its presence in it.

The concept of **mass** is closely related to the concept of **weight**. In our universe, every form of matter is attracted to all other forms by the force we call **gravity**. On Earth, the weight of matter is a measure of the force with which gravity pulls it toward Earth's center. As we leave Earth's surface, the gravitational pull decreases until it becomes virtually insignificant. The *weight* of matter accordingly reduces to zero, yet the matter still possesses the same *mass* as it did on the surface of Earth. For this reason, the expressions "has a mass of" and "weighs" are essentially equivalent on Earth's surface.

matter
- Anything that possesses mass and occupies space

mass
- The quantity of matter possessed by an object regardless of its location

weight
- The force of gravity acting on the mass of a particular object

gravity
- The force of attraction between two bodies

As usually experienced, matter exists in three different forms or states of aggregation: solid, liquid, and gas.* These are called the **physical states of matter**.

Because matter occupies space, a given form of matter is also associated with a definite volume or capacity. Space should not be confused with air, since air is itself a form of matter. **Volume** refers to the actual amount of space that a given form of matter occupies.

2.1-A SOLIDS

A **solid** is the form of matter existing in a rigid state independent of the size and shape of its container. Consider this book. It retains its shape regardless of its position in space and does not need to be placed into a container to retain that shape. Left to itself, it will never spontaneously assume a shape different from what it has now.

Solids also occupy a definite volume at a given temperature and pressure. We can squeeze most solids with all our might or heat or cool them, but their total volume change is relatively insignificant.

Certain solid plastics can behave unlike most solids. Because of their inherent elasticity, they can assume different shapes and volumes when squeezed, stretched, or otherwise manipulated. Solid foams can also behave uncharacteristically, since they contain encapsulated air.

2.1-B LIQUIDS

A **liquid** is a form of matter that does not possess a characteristic shape; rather, its shape depends on the shape of the container it occupies. Consider water within a glass. The liquid water takes the shape of the glass up to the level that it occupies. If we pour the water into a cup, the water takes the shape of the cup; or if we pour it into a bowl, the water takes the shape of the bowl. Of course, sufficient space must always be available for the water within the container; otherwise, the liquid overflows. But assuming that space is available, any liquid assumes whatever shape its container possesses.

Like solids, liquids occupy a specific volume at a given temperature and pressure. They tend to maintain a relatively fixed volume when they are exposed to a change in either of these conditions. Liquids and solids are often considered incompressible, since the application of pressure barely changes their volumes. When heated, liquids do expand, more than solids do, but not nearly to the degree that gases do. We revisit the expansion of heated liquids in Section 2.11.

2.1-C GASES

A **gas**, or **vapor**, is another form of matter that does not possess a characteristic shape and assumes the shape of its container. If a gas or mixture of gases, such as air, is put into a balloon, it assumes the shape of the balloon; or if it is transferred into a tire, it assumes the shape of the tire.

Gases also lack a characteristic volume. When confined to a container with nonrigid, flexible walls, the volume that a confined gas occupies depends on its temperature and pressure. When confined to a balloon, for instance, the gas's volume expands and contracts depending on the prevailing temperature and pressure. When confined to a container with rigid walls, however, the volume of the gas is forced to remain constant. This property of gases can cause rigid containers to explode, a topic we note later in Section 2.12.

*Other physical states of matter aside from the solid, liquid, and gas are known. A fourth state, called *plasma*, exists only at very high temperatures, whereas a fifth state, called the *Bose–Einstein condensate*, exists only at very low temperatures. Although we will not encounter materials in the fourth and fifth states during our study of hazardous materials, it is interesting to note that the Sun, other stars, and other forms of intergalactic matter exist in the plasma state. It is the predominant state of matter within the universe.

The properties of solids, liquids, and gases noted in this section can now be used to formally define the three states of matter:

- Matter in the solid state possesses a definite volume and a definite shape.
- Matter in the liquid state possesses a definite volume but lacks a definite shape.
- Matter in the gaseous state possesses neither a definite volume nor a definite shape.

PERFORMANCE GOALS FOR SECTION 2.2:

- Identify the common English and metric units used to measure length, mass, and volume.
- Learn the equivalent English and metric units of measurement for length, mass, and volume.

2.2 Units of Measurement

The necessity to measure, and to measure with accuracy, is essential to any kind of scientific or technological endeavor. To *measure* means to find the number of units in a sample of something. For instance, when we measure the distance from one point to another along a wall, we generally determine how many feet, yards, or meters are between the two points. The foot, yard, and meter are examples of the common units of length.

Two systems of units have survived the test of time: the English system and the metric system. The English system is still used in the United States (and Liberia and Myanmar), but the British stopped using it in 1959. This system comprises an array of units that have no obvious interrelationship, such as inches (in.), feet (ft), yards (yd), miles (mi), ounces (oz), pounds (lb), tons (tn), pints (pt), quarts (qt), and gallons (gal), the combination of which are still called the **English units of measurement**.

The majority of the world's population and the worldwide scientific community use a system called the **metric system of measurement**. This system has itself been modified so that today, we actually use the **SI system of units**, an abbreviation of its official French name, *Le Système International d'Unités*. The SI system encourages the use of certain fundamental units from which all other measurements are constructed. These fundamental units are called the **SI base units**. Examples of SI base units are the meter and kilogram, for length and mass, respectively.

Certain prefixes are used in the SI system to denote multiples and fractions of the units of measurement. Each prefix is a fraction or multiple of the number 10. For example, when we wish to refer to 1000 meters, we use the word *kilometer*. The prefix *kilo* means 1000 times the meter, the SI base unit.

The four prefixes listed in Table 2.1 are commonly used in studying the chemistry of hazardous materials. These particular prefixes should be committed to memory. They are used to measure certain properties of matter, particularly its length, mass, and volume. It is appropriate to discuss each type of measurement separately.

English units of measurement
- Units based on the yard, pound, and quart, and their fractions and multiples

metric system of measurement
- The decimal system of measurement based upon the meter, kilogram, and cubic meter

SI system of units
- The scientific standard of measurement that employs a set of units describing length, mass, time and other attributes of matter

SI base units
- The accepted units of measurement adopted by the International Bureau of Weights and Measures (Bureau International des Poids et Mesures, Sèvres, France) such as the meter for length, kilogram for mass, and kelvin for temperature

meter
- The SI unit of length in the metric system of measurement

2.2-A LENGTH

Today, the **meter** is defined as the length of the path traveled by light in vacuum in 1/299,792,458 of a second. In ordinary practice, we measure length in the metric system with a metric ruler. By so doing, we discover that one meter (m) is slightly longer than a yard; specifically, one meter equals 39.37 inches (in.).

$$1\,m \ = \ 39.37\,in.$$

TABLE 2.1	Common Prefixes Used in the Metric System		
PREFIX	**SYMBOL**	**MEANING**
kilo-	k	One thousand times the SI base unit[a]
deci-	d	One-tenth of the SI base unit
centi-	c	One-hundredth of the SI base unit
milli-	m	One-thousandth of the SI base unit
micro-	μ	One-millionth of the SI base unit

[a]See text for an exception when using the SI base unit of mass.

One meter is equivalent to 100 centimeters (cm) and to 1000 millimeters (mm).

$$1\,m = 100 \text{ cm} = 1000 \text{ mm}$$

For very large lengths, we use the kilometer (km); once again, 1000 meters is equivalent to 1 kilometer.

$$1\,km = 1000 \text{ m}$$

For very small lengths, we use the micron (μm). One micron is one-millionth of a meter.

$$1\,\mu m = 0.000001 \text{ m}$$

The relationship between the inch and centimeter is shown in Figure 2.1. One inch equals 2.54 centimeters.

$$1\,in. = 2.54 \text{ cm}$$

2.2-B MASS

The SI unit of mass is the *kilogram*. A bit of attention needs to be given to constructing the multiples and fractions of mass measurements, since this unit of mass is the only one that already contains a prefix (kilo-). The names of the various multiples and fractions of the unit of mass are constructed by attaching the appropriate prefix to the word **gram**, not to kilogram. In other words, the gram is used as though it were the SI unit of mass, even though it actually is not.

One kilogram is equivalent to 2.2 pounds. One gram is approximately the mass of a peanut.

Three common metric units of mass are the milligram, microgram, and kilogram. One one-thousandth of a gram is a *milligram* (mg); one one-millionth of a gram is a *microgram* (μg); and 1000 grams is a kilogram (kg).

$$1\,mg = 0.001\,g$$

$$1\,\mu g = 0.000001\,g$$

$$1\,kg = 1000\,g$$

gram
■ One one-thousandth of the mass of 1 kilogram

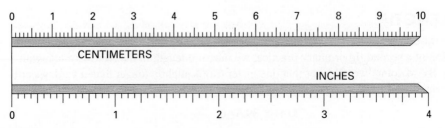

FIGURE 2.1 The relationship between the inch and the centimeter. Note that 1 inch (in.) equals 2.54 centimeters (cm).

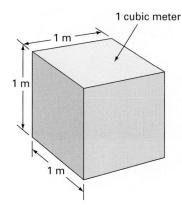

1 cubic meter

FIGURE 2.2 Because this cube measures 1 meter on each edge, its volume is 1 cubic meter (m^3), the approved SI unit of volume.

One milligram is approximately the mass of a grain of sand, whereas one microgram is approximately the mass of a fleck of dust.

For relatively large mass measurements, use is made of the metric ton, or **tonne (t)**.

$$1\,t = 1000\,kg$$

To verbally distinguish the ton and tonne, the latter is sometimes pronounced "tunny."

tonne
■ The metric unit of mass equivalent to 1000 kilograms

2.2-C VOLUME

The approved SI unit of volume is the *cubic meter* (m^3). This unit is derived directly from the SI unit of length and is not a SI base unit itself. We can easily derive the cubic meter by considering the cube in Figure 2.2, which measures 1 meter on each edge. The volume of this cube is determined by taking the product of its length, width, and height.

$$\text{Volume} = 1\,m \times 1\,m \times 1\,m = 1\,m^3$$

Using simple arithmetic, the volume of this cube is determined to be 1 cubic meter. Since it is possible to derive the cubic meter directly from a previously defined unit, it is unnecessary to define some other unit as the SI unit of volume.

Although the cubic meter is the approved unit of volume in the SI system, another unit has been used for many years to measure volume. This unit is the **liter (L)**. One quart is slightly less than one liter.

$$946\,mL = 1\,qt$$

liter
■ The volume occupied by a cube measuring 10 cm to an edge

Chemists continue to use the liter for measuring volume because its size is so convenient for laboratory-scale measurements. The cubic meter is comparatively too large.

One liter is equivalent to one one-thousandth of a cubic meter.

$$1\,L = 0.001\,m^3$$

If we construct a cube measuring 1 meter on each edge and then divide the resulting volume into 1000 equally sized cubes, the volume of each small cube is one liter. Imagine further dividing each liter cube into another 1000 equally sized cubes. These subdivisions are illustrated in Figure 2.3. Because the prefix milli- means one one-thousandth of a unit, the volume of each of the new cubes is one *milliliter* (mL).

$$1\,L = 1000\,mL$$

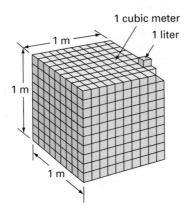
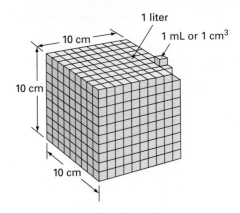

FIGURE 2.3 The cube on the left measures 1 m on each edge and has been divided into 1000 equally sized cubes. The volume occupied by each of the smaller cubes is 0.001 m³, or 1 liter (L). The cube on the right measures 10 cm on each edge and has been subdivided into 1000 equally sized cubes. The volume of the larger cube is I L, and the volume of each smaller cube is 1 milliliter (mL), or 1 cubic centimeter (cm³).

Formerly, the milliliter was known as a cubic centimeter (cm³). Although the use of the cubic centimeter has essentially been phased out in chemistry, it is still used in the medical field.

2.2-D GENERAL USE OF THE METRIC SYSTEM IN THE UNITED STATES

In the international community, the United States stands alone as the sole industrialized nation that has not officially adopted the use of the metric system. The United States's nonuse of the metric system was contributing to unsuccessful competition in some international markets. To address this concern, Congress passed the *Metric Conversion Act* in 1975. Notwithstanding its passage, the act called only for voluntary compliance and accomplished little change in the manner by which Americans measure the properties of matter.

Then, in 1988, Congress passed the **Omnibus Trade and Competitiveness Act**, which mandated the following:

Omnibus Trade and Competitiveness Act
■ The federal statute whose aim is to enhance the competitiveness of American industry throughout international markets based in part on the required use of the metric system

- Identified the metric system as the preferred system for U.S. trade and commerce
- Required use of the metric system by all U.S. government agencies
- Required use of metric units for all federally funded construction projects costing over $10 million
- Required use of metric units for all federally assisted highway construction projects that began on or after October 1, 1996

This act obligated building and highway construction companies to begin using the metric system. In this fashion, the metric system has slowly crept into the United States and taken us one step closer toward common worldwide usage.

PERFORMANCE GOALS FOR SECTION 2.3:

- Use the factor-unit method for converting between English and metric units of measurement.

2.3 Converting Between Units of the Same Kind

Suppose we wish to convert between grams and kilograms, pounds and grams, or liters and gallons. How do we accomplish this task? Problems like these are best solved by using a simple procedure called the **factor-unit method**. Briefly, it consists of the following key steps:

- Identify the desired unit.
- Choose the proper **conversion factor(s)**.
- Multiply the given measurement by the conversion factor(s), being certain to multiply, divide, and cancel equal units.

Choosing the proper conversion factor is a crucial step to obtaining the correct answer to a problem. The conversion factor is a fraction numerically equal to 1 that equates the quantity in the numerator to the quantity in the denominator. For example, we know there are 1000 m in 1 km. Either of the following fractions serves as a correct conversion factor:

$$\frac{1000 \text{ m}}{1 \text{ km}} \quad \text{or} \quad \frac{1 \text{ km}}{1000 \text{ m}}$$

The factors are respectively read as follows: 1000 meters per kilometer; one kilometer per 1000 meters.

Suppose we know that the height of the Sears Tower in Chicago has been measured from ground level to the top as 443.2 m. What is the height of the building in kilometers? Using the factor-unit method, we multiply the given measurement, 443.2 m, by a conversion factor that relates meters to kilometers.

$$443.2 \text{ m} \times \frac{1 \text{ km}}{1000 \text{ m}} = 0.4432 \text{ km}$$

Note that the "m" in 443.2 m cancels with the "m" in 1000 m. Then, dividing 443.2 km by 1000 gives 0.4432 km.

Individuals who are practiced in using the metric system simply move the decimal point from left to right, or vice versa, as the need arises.

Some of the more frequently used metric units and their equivalent English counterparts are given in Table 2.2. These relationships permit us to write conversion factors like the following:

$$\frac{1 \text{ m}}{39.37 \text{ in.}} \qquad \frac{2.54 \text{ cm}}{1 \text{ in.}} \qquad \frac{1 \text{ kg}}{2.2 \text{ lb}} \qquad \frac{946 \text{ mL}}{1 \text{ qt}}$$

factor-unit method
■ A procedure for changing a quantity expressed in one unit to a quantity of another unit by multiplying, dividing, and arithmetically canceling numbers and units

conversion factor
■ A fraction equal to 1 in which the magnitude of one unit is related in the numerator to the magnitude of another unit of the same type in the denominator

TABLE 2.2	Common SI (Metric) Units and Their English Equivalents
METRIC UNIT	**ENGLISH UNIT**
1 m	39.37 in.
2.54 cm	1 in.
1 kg	2.2 lb
454 g	1 lb
946 mL	1 qt

The standard length of fire hose in the United States is 50 ft. Express the equivalent length in meters.

Solution: We first need one or more conversion factors to convert 50 ft into its equivalent length in meters. Since one meter is 39.37 in., and one foot is 12 in., we can construct the following conversion factors:

$$\frac{1 \text{ m}}{39.37 \text{ in.}} \qquad \frac{12 \text{ in.}}{1 \text{ ft}}$$

Through use of the factor-unit method, we then convert 50 ft into meters as follows:

$$50 \text{ ft} \times \frac{12 \text{ in.}}{1 \text{ ft}} \times \frac{1 \text{ m.}}{3.9.37 \text{ in.}} = 15 \text{ m}$$

A firefighter must possess the physical agility to easily carry a 60-lb "bundle" from the base to the top of a 148-ft ladder. Express the bundle mass in kilograms and the length of the ladder in meters.

Solution: According to the information in Table 2.2, 1 kg = 2.2 lb, and 1 m = 39.37 in. By the factor-unit method, a 60-lb bundle is equivalent to a 27-kg bundle.

$$60 \text{ lb} \times \frac{1 \text{ kg}}{2.2 \text{ lb}} = 27 \text{ kg}$$

Furthermore, a 148-ft ladder is equivalent to a 45-m ladder.

$$148 \text{ ft} \times \frac{12 \text{ in.}}{1 \text{ ft}} \times \frac{1 \text{ m}}{39.37 \text{ in.}} = 45 \text{ m}$$

How many football fields fit into a distance of 1 km?

Solution: We know that the length of a football field is 100 yd. We also know that 1 km = 1000 m, and 39.37 in. = 1 m. By the factor-unit method, 1 km is equivalent to 1094 yd.

$$100 \text{ m} \times \frac{39.37 \text{ in.}}{1 \text{ m}} \times \frac{1 \text{ ft}}{12 \text{ in.}} \times \frac{1 \text{ yd}}{3 \text{ ft}} = 1094 \text{ yd}$$

This means that approximately 11 football fields fit into a distance of 1 km.

$$\frac{1094 \text{ yd}}{100 \text{ yd}} = 11$$

2.4 Concentration

Scientists often measure the amount of a substance as a stated unit in a given mass or volume of a mixture. This value is called the substance's **concentration**. A variety of units are used by professionals to measure concentration. In the medical field, for instance, physicians often express the concentration of a substance in *milligrams per deciliter* (mg/dL). Your doctor may have recommended that you choose a diet that aims to provide a total blood cholesterol concentration of less than 150 mg/dL to reduce the onset of heart ailments.

Concentration is sometimes expressed as a percentage. The word **percentage** means parts per hundred. To calculate the percentage of an item in a total, we divide the number of items by the total number and multiply the product by 100. The percentage is expressed as a number followed by the percent sign (%). In chemistry, the concentration of a substance in a mixture is expressed as either a percentage by mass or percentage by volume. In the former instance, we use mass units of the same type, and in the latter case, we use volume units of the same type. For example, suppose we have 100 g of a solution in which 5 g of table salt is dissolved. The concentration of table salt in the solution is expressed as 5% by mass. If we have 100 mL of a solution in which 5 mL of a substance is dissolved, the concentration of the substance in the solution is expressed as 5% by volume.

In this textbook, we shall also encounter hazardous materials in the air or water whose concentrations are expressed in **parts per million (ppm)** or **parts per billion (ppb)**. A concentration of 1 ppm means 1 part of a substance measured as an arbitrary unit in a million parts of the same unit. For example, depending on the units used, 1 ppm could be one molecule in a million molecules, one gram in a million grams, or one ounce in a million ounces. A concentration of 1 ppm is minute. It corresponds approximately to 2 or 3 grains of sand in an Olympic-size swimming pool, or to 1 cent in $10,000.

Similarly, one part per billion means 1 part of a substance measured as a unit in a billion parts of the same unit. Parts per million are related to parts per billion as follows:

$$1 \text{ ppm} = 1000 \text{ ppb}$$

This means that 1 ppb is one one-thousandth of 1 ppm. One part per billion is equivalent to one cent in $10 million.

concentration
- The relative amount of a minority constituent of a mixture or solution to the majority constituent, expressed in such units as grams per liter, pounds per cubic foot, parts per million, and percentage by mass and volume

percentage
- The concentration of a mixture expressed as parts of a unit in 100 of the same units (symbol %)

parts per million
- The concentration of a mixture in which a certain number of units are contained within a million of the same units (abbreviated ppm)

parts per billion
- The concentration of a mixture in which a certain number of units are contained within a billion of the same units (abbreviated ppb)

- Calculate the density of a substance from its mass and volume.
- Establish how the density of a substance is related to its specific gravity or vapor density.
- Memorize the density of water when expressed in lb/gal, lb/ft^3, and g/mL.
- Use the specific gravity of a flammable liquid that is immiscible with water to evaluate its behavior when the liquid and water are mixed.
- Use the vapor density of a gas or vapor to evaluate its behavior when the gas or vapor is released from confinement.

2.5 Density of Matter

density
■ The property of a substance that measures its compactness; the mass of a substance divided by the volume it occupies

If the volume of a given mass of a substance is known, we can readily use division to compute its mass per unit volume. This ratio is a property of substances known as the **density**.

$$\text{Density} = \frac{\text{mass}}{\text{volume}}$$

For instance, one milliliter of liquid mercury has a mass of 13.6 g at 68 °F (20 °C). Hence, the density of mercury is 13.6 g/mL at 68 °F (20 °C).

In the English system of units, the densities of solids and liquids are usually expressed in pounds per gallon (lb/gal) or pounds per cubic foot (lb/ft^3), whereas the densities of gases or vapors are expressed in pounds per cubic foot. In the metric system, the densities of solids and liquids are normally expressed in grams per milliliter (g/mL) or kilograms per cubic meter (kg/m^3), whereas the densities of gases and vapors are expressed in either grams per cubic meter (g/m^3) or kilograms per cubic meter (kg/m^3).

Scientists measure densities in the laboratory by determining the volume occupied by a given mass of a substance. Suppose we weigh exactly 50 lb of water on a scale.

$$\text{Mass} = 50.00 \text{ lb}$$

Then, we transfer this water into containers of known capacity. When this simple exercise is performed, we discover that 50.00 lb of water completely fills six 1-gal containers to the brim at 68 °F (20 °C).

$$\text{Volume} = 6.00 \text{ gal}$$

We then compute the density of water to be 8.33 lb/gal as follows:

$$\text{Density of water} = \frac{50.00 \text{ lb}}{6.00 \text{ gal}} = 8.33 \text{ lb/gal}$$

Scientists establish the densities of other substances by conducting similar exercises. Some results are provided in Table 2.3.

We can also convert a density in one unit into its equivalent in other units. For example, we can determine that a density of 8.33 lb/gal is equivalent to 62.3 lb/ft^3. We calculate this latter value using the conversion factor 1 ft^3 = 7.48 gal, or 7.48 gal/ft^3.

$$8.33 \text{ lb/gal} \times 7.48 \text{ gal/ft}^3 = 6.23 \text{ lb/ft}^3$$

When the density of water is determined using metric units, the obvious result at 20 °C is 1.00 g/mL, or 1000 kg/m^3. This result is obtained because water was once selected as the reference standard for establishing the units of mass and volume in the metric system. One gram was defined as the mass of water that occupied a volume of one milliliter at 20 °C (68 °F).

SOLVED EXERCISE 2.4

The density of dry air in kilograms per cubic meter (kg/m^3) varies with temperature as follows: 1.29 (0 °C); 1.25 (10 °C); 1.21 (20 °C); and 1.16 (30 °C). What do these data illustrate about the nature of dry air during building fires?

Solution: These data denote that the density of dry air decreases as the temperature of the air increases. In other words, a given volume of dry air gets lighter in mass as it becomes hotter. This lighter, hotter air is buoyant; that is, it rises upward. During building fires, the lighter, hotter air concentrates near the undersides of ceilings and roofs, whereas the denser, colder air concentrates near the floors.

TABLE 2.3 | Densities of Some Common Liquids and Solids

SUBSTANCE	DENSITY AT 68 °F (20 °C)	
	g/mL	lb/gal
Acetone	0.790	6.58
Aluminum	2.7	22.5
Benzene	0.879	7.32
Carbon disulfide	1.274	10.6
Carbon tetrachloride	1.595	13.3
Chloroform	1.489	12.4
Diethyl ether	0.715	5.96
Ethanol, 180 proof	0.828	6.90
Gasoline	0.66–0.69	5.5–5.8
Kerosene	0.82	6.8
Lead	11.34	94.5
Mercury	13.6	113
Silver	10.5	87.5
Sulfur	2.07	17.3
Trichloroethylene	1.460	12.2
Water[a]	1.00	8.3

[a]At 39 °F (4 °C).

All substances possess densities that vary with the prevailing temperature and possess a maximum density at a unique temperature. Figure 2.4 shows that the maximum density of water is 1.0000 g/mL at 39.1 °F (3.97 °C). The density of water decreases at temperatures less than and greater than 39.1 °F (3.97 °C).

2.5-A SPECIFIC GRAVITY

Often, the mass of a liquid and solid substance is compared with the mass of an equal volume of water. This comparison yields a dimensionless number called its **specific gravity**. For example, if 1 gal of sulfuric acid is weighed, we find that the scale reads 15.33 lb. This means that the density of sulfuric acid is 15.33 lb/gal. We can compare the density of sulfuric acid with the density of water to compute the specific gravity of sulfuric acid as follows:

specific gravity
■ The mass of a given volume of matter compared with the mass of an equal volume of water

$$\text{Specific gravity} = \frac{15.33 \; \text{lb/gal}}{8.33 \; \text{lb/gal}} = 1.84$$

This computation informs us that any volume of sulfuric acid is 1.84 times heavier than an equal volume of water.

The specific gravities of liquids are directly determined through the use of a device called a *hydrometer*, a cylindrical glass stem containing a bulb weighted with shot so that the stem

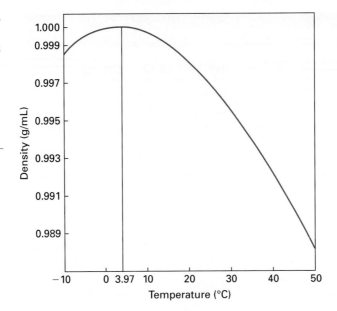

FIGURE 2.4 The density of water at atmospheric pressure near its freezing point as a function of temperature. The density of liquid water is at its maximum at 39.1 °F (3.97 °C).

floats upright in a liquid. A graduated scale is printed on the hydrometer's stem. A sample of the liquid is generally poured into a tall cylinder or jar as in Figure 2.5, and the hydrometer is lowered into the sample until it floats. The specific gravity of the liquid is read from the graduated scale at the point where the surface of the liquid touches the stem.

The specific gravities of some commercially important liquids are provided in Table 2.4. Since the specific gravity is determined by comparing the mass of a liquid or solid with the mass of an equal volume of water, the specific gravity of water is 1.0.

Many substances are soluble in water to some extent, but there are also other substances that are relatively insoluble in water. When a liquid substance does not appreciably dissolve in water, we say that the substance and water are mutually **immiscible**. Oil and water are examples of two mutually immiscible liquids. When combined, mutually immiscible liquids coexist as two separate and distinct phases, one on top of the other.

immiscible

■ Incapable of combining or mixing so as to form a single phase

FIGURE 2.5 A hydrometer is an instrument that is often used to determine the specific gravity of a liquid. It is a sealed graduated tube, weighted at one end with shot. When immersed into a liquid contained within a tall cylinder or jar, the hydrometer floats upright at a unique depth. The hydrometer operates on the principle that an object floats "high" in a liquid of relatively high specific gravity (greater than 1.0) and "low" in a liquid of relatively low specific gravity (less than 1.0). Commercial hydrometers are generally calibrated at 68 °F (20 °C). The specific gravity is read from the graduated scale at the point where the surface of the liquid touches the stem. The specific gravity of the liquid in this cylinder is 0.68.

TABLE 2.4	Properties of Some Common Liquids		
LIQUID	SPECIFIC GRAVITY [68 °F (20 °C)]	MISCIBLE/IMMISCIBLE WITH WATER	FLAMMABLE/ NONFLAMMABLE
Acetone	0.79	Miscible	Flammable
Carbon disulfide	1.26	Immiscible	Flammable
Chlorobenzene	1.11	Immiscible	Flammable
Cyclopentanone	0.95	Immiscible	Flammable
2-Ethylhexanol	0.83	Immiscible	Flammable
Heptane	0.68	Immiscible	Flammable
Hydrochloric acid	1.18	Miscible	Nonflammable
Methyl acetate	0.97	Miscible	Flammable
Methyl ethyl ketone	0.81	Miscible	Flammable
Sulfuric acid	1.84	Miscible	Nonflammable
Trichlorofluoromethane	1.49	Immiscible	Nonflammable

When a liquid is immiscible with water, its specific gravity tells us how the liquid behaves when it is mixed with water as follows:

- An immiscible liquid *floats* on water when its specific gravity is less than 1.0.
- An immiscible liquid *sinks* below water when its specific gravity is greater than 1.0.

All grades of fuel oils have specific gravities of less than 1.0. Consequently, when an oil tanker ruptures at sea, the oil that spills from the tanker floats on the water and forms an oil slick.

Knowledge of the specific gravity of a flammable liquid tells us whether water functions effectively when it is applied to a fire as a fire extinguisher. The usefulness of this information is demonstrated in Figure 2.6 by the following two situations:

- In the first situation, a burning liquid immiscible with water and having a specific gravity of less than 1.0 is confined within a drum or tank. When water is added, it sinks below the surface of the liquid, where it is incapable of extinguishing the fire. If excess water is added, the burning liquid overflows the container or tank before the water does. Consequently, the application of water could potentially worsen the situation by spreading the fire to adjoining areas.
- In the second situation, a burning liquid immiscible with water and having a specific gravity of greater than 1.0 is confined within a drum or tank. When water is added, it floats on the heavier liquid and smothers it. In this instance, the fire is effectively extinguished.

2.5-B VAPOR DENSITY

Air is a mixture of several gases. Its density is 0.08 lb/ft^3, or 1.29 g/L, at 32 °F (0 °C) at atmospheric pressure. Often, the mass of a given volume of a substance is compared with the mass of an equal volume of air. The comparison yields the **vapor density** of that substance. The vapor density is a dimensionless property of all gases.

vapor density
- The mass of a vapor or gas compared with the mass of an equal volume of another gas or vapor, generally air

FIGURE 2.6 On the left, water is added from a hose to a drum of burning benzene. Water and benzene are immiscible liquids. Because the specific gravity of benzene is 0.879, the water settles *below* the benzene and does not ordinarily extinguish the fire. On the right, water is added from a hose to a drum of burning carbon disulfide. Water and carbon disulfide are also immiscible liquids. Because the specific gravity of carbon disulfide is 1.274, the water floats *on* the carbon disulfide and smothers the fire.

Suppose we compare the masses of equal volumes of oxygen and air. A liter of oxygen weighs 1.43 g. This information can be used to compute the vapor density of oxygen relative to air as follows:

$$\text{Vapor density of oxygen} = \frac{1.43 \text{ g/L}}{1.29 \text{ g/L}} = 1.11$$

The value tells us that an arbitrary volume of oxygen is 1.11 times heavier in mass than an equal volume of air. The vapor densities of the gases and vapors in Table 2.5 were calculated from their densities in an identical fashion.

An awareness of the vapor density of a gas or vapor is useful for identifying the initial location of these substances when they are released into the air. The information can be summarized as follows:

- When the vapor density of a gas or vapor is greater than 1.0, an arbitrary volume of the gas or vapor is heavier than the same volume of air. When the gas or vapor is released from its container, it tends to concentrate in low spots.
- When the vapor density of a gas or vapor is less than 1.0, an arbitrary volume of the gas or vapor is lighter than the same volume of air. When it is released, the gas or vapor naturally rises into the air.

A leak of a gaseous substance whose vapor density is greater than 1.0 constitutes a pronounced risk to health and safety. The gas displaces the air and concentrates in low spots before ultimately dissipating into the air. When the substance is poisonous, its presence poses a pronounced risk of inhalation toxicity, and when the substance is flammable, its presence poses a pronounced risk of fire and explosion. Consider gasoline vapor, for example. A given volume of gasoline vapor is heavier than an equal volume of air. When gasoline evaporates, its vapor accumulates in low spots such as the bilges in boats, where exposure to an ignition source could ignite the vapor.

In contrast, a gaseous substance whose vapor density is less than 1.0 rises in the atmosphere before it ultimately dissipates. In the open air, a gaseous substance whose vapor density is less than 1.0 dissipates rapidly. However, when this substance is released within a room, it first accumulates near its ceiling, where, if the substance is flammable, the localized accumulation poses a pronounced risk of fire and explosion.

TABLE 2.5	Vapor Densities of Some Common Gases or Vapors

SUBSTANCE	VAPOR DENSITY (AIR = 1)
Acetylene	0.899
Ammonia	0.589
Carbon dioxide	1.52
Carbon monoxide	0.969
Chlorine	2.49
Fluorine	1.7
Hydrogen	0.069
Hydrogen chloride	1.26
Hydrogen cyanide	0.938
Hydrogen sulfide	1.18
Methane	0.553
Nitrogen	0.969
Oxygen	1.11
Ozone	1.66
Propane	1.52
Sulfur dioxide	2.22

All gases and vapors are totally miscible with air. This means that given adequate time, they become completely mixed with the components of the air. We say that gas and vapors are *infinitely miscible* with air.

SOLVED EXERCISE 2.5

Carbon monoxide is a poisonous gas emitted into the atmosphere from the exhaust pipes of operating motor vehicles. When first emitted from an exhaust pipe, does carbon monoxide tend to concentrate in the lower or upper regions of a confined garage?

Solution: Table 2.5 reveals that carbon monoxide has a vapor density of 0.969, indicating that an arbitrary volume of carbon monoxide is only slightly lighter than an equal volume of air. In addition, the gases emitted from exhaust pipes of operating motor vehicles are hotter than the air into which they are released. From these two facts, we can conclude that the carbon monoxide emitted from a vehicle's exhaust pipe is likely to rise toward the ceiling of the garage. Because the exhaust is hot and the densities of carbon monoxide and air are so similar, the carbon monoxide will not concentrate at the ceiling for long. Instead, it is likely to disperse rapidly within the confined space of the garage and mix with the air.

- Define energy and identify the different forms in which it is manifested.
- Identify the common units in the English and metric systems used to measure energy.
- Describe how mass and energy are interrelated.

2.6 Energy

energy
- The property of matter that enables it to do work

Energy is defined as the capacity to do work; thus, energy is proportional to work. In a broad sense, when an effort is expanded to accomplish some act, the activity is referred to as work. Scientists describe work in terms of a force applied over a distance—like a push or pull of matter that results in moving it.

Energy exists in a variety of forms, although these forms are often abstract: radiant energy (light, heat, and X rays), thermal energy (heat), acoustical energy (sound), mechanical energy, chemical energy, electrical energy, and atomic energy. All forms of matter possess *chemical energy*, which is the energy stored in its chemical makeup. Each hazardous material possesses chemical energy that can constitute the basis for its hazardous nature. Dynamite, for example, possesses a large amount of chemical energy. When dynamite detonates, some of its chemical energy is released into the environment as light, heat, and sound.

Energy is measured in various units. In the United States, energy units like the British thermal unit (Btu) and the calorie (cal) are commonly encountered. The SI unit of energy is the *joule*. One joule (J) is the energy possessed by a two-kilogram mass that moves through a distance of one meter at a velocity of one meter per second. We revisit these units in Section 2.9.

law of conservation of mass and energy
- The observation that the total amount of mass and energy in the universe is constant

Mass and energy are closely interrelated. Neither mass nor energy can be created or destroyed. Certain phenomena require the conversion of matter to energy, and vice versa, but regardless of the nature of the transformation, the total amount of mass and energy in the universe remains constant. This observation is an important law of nature known as the **law of conservation of mass and energy**.

- Describe the concept of temperature.
- Describe the differences between temperature readings on the Fahrenheit, Celsius, Kelvin, and Rankine scales.
- When provided with a reading on one temperature scale, convert it to its equivalent on the other temperature scales.

2.7 Temperature and Its Measurement

temperature
- The condition of a body that determines the transfer of heat to or from other bodies by comparison with established values for water, which is used as the standard

Temperature is the property of matter associated with its degree of hotness or coldness. Hot matter is associated with high temperatures, and cold matter is associated with low temperatures. To say that something is hot or cold merely points out a relative condition, whereas a temperature reading is a measure of the hotness or coldness of the matter. The temperature is an indication of a condition of matter, just as the measurement of its size is an indication of how large or small it is.

The temperature of many substances is most commonly determined using the simple mercury-in-glass column thermometer. This is a sealed glass capillary tube that has been partially filled with mercury and then calibrated according to a prescribed procedure. When the thermometer is inserted

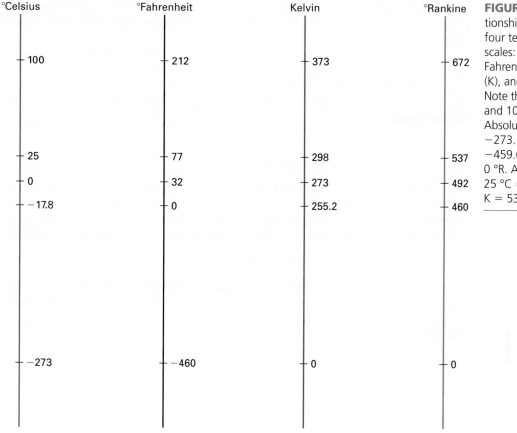

FIGURE 2.7 The relationships among the four temperature scales: Celsius (°C), Fahrenheit (°F), Kelvin (K), and Rankine (°R) Note that 0 °C = 32 °F, and 100 °C = 212 °F. Absolute zero is −273.15 °C, −459.69 °F, 0 K, or 0 °R. Also note that 25 °C = 77 °F = 298 K = 537 °R.

into a substance, the temperature of the glass and mercury rises or falls—causing the mercury to expand or contract, respectively—until it reaches the temperature of the substance itself. The height of the calibrated column of mercury corresponds to the temperature of this substance.

Temperature is measured on a number of scales, four of which are illustrated in Figure 2.7 and discussed in the next two sections.

2.7-A THE FAHRENHEIT AND CELSIUS TEMPERATURE SCALES

Suppose an unmarked capillary tube containing mercury is placed in the steam that evolves above boiling water at sea level. The mercury level rises within the tube and then remains stationary. This height can now be etched on the glass and serves as the first reference point, called the *steam point*. If the same tube is placed in ice water, the mercury level drops and ultimately becomes stationary. This height of the mercury column serves as the second reference point, called the *ice point*. These calibrations are assigned temperature units in at least two common ways:

- On the **Fahrenheit scale**, the ice and steam points are assigned the values of 32 degrees and 212 degrees, respectively. We recognize them as the freezing point (or melting point) and boiling point of water. There are 180 equally spaced divisions between these two reference points, and each division is called one degree Fahrenheit (°F). The capillary tube calibrated in this fashion is called a *Fahrenheit thermometer*.
- On the **Celsius scale**, the melting and boiling points are assigned the values of 0 degrees and 100 degrees, respectively. There are 100 equally spaced divisions between these two reference points and each division is called one degree Celsius (°C). This tube is called a *Celsius thermometer*.

Fahrenheit scale
■ The temperature scale on which water freezes at 32 °F and boils at 212 °F at 1 atm

Celsius scale
■ The temperature scale on which water freezes at 0 °C and boils at 100 °C at 1 atm

Thus, the Fahrenheit and Celsius thermometers are alike insofar as the freezing and boiling points of water are used to define their reference points, but they differ in the numbers that are assigned to these points.

Because the same temperature interval is divided into 180 degrees on the Fahrenheit scale and 100 degrees on the Celsius scale, the temperature range corresponding to one Celsius degree is 180/100, or 9/5, as great as the temperature range corresponding to one Fahrenheit degree. This factor can be used to relate these two temperature scales to each other by formulas like the following:*

$$t(°F) = 9/5\,t(°C) + 32$$
$$t(°C) = 5/9[t(°F) - 32]$$

When using either formula, it is essential to perform the arithmetic exactly as the formula is written. To determine a Fahrenheit reading, nine-fifths of a Celsius reading is taken first, and 32 degrees is added to the product. Suppose we desire to convert 46 °C to its equivalent temperature reading on the Fahrenheit scale. We take nine-fifths of 46, which equals 83, and add 32:

$$t(°F) = (9/5 \times 46) + 32 = 83 + 32 = 115\ °F$$

To determine a Celsius reading, 32 is subtracted from the Fahrenheit reading, and then five-ninths of this difference is taken. Suppose we desire to convert 115 °F to its equivalent temperature on the Celsius scale. We subtract 32 from 115, which equals 83, and then take five-ninths of this number:

$$t(°C) = 5/9 \times (115 - 32) = 5/9 \times 83 = 46\ °C$$

SOLVED EXERCISE 2.6

When exposed to an ignition source, polystyrene begins to burn in the temperature range from 345 to 360 °C. It self-ignites in air (without exposure to an ignition source) at 490 °C.

(a) Will coaxial television cable made of polystyrene ignite at a fire scene whose maximum temperature is 555 °F?
(b) Will the cable ignite without exposure to an ignition source when the temperature is further elevated to 950 °F?

Solution:

(a) The minimum temperature at which polystyrene begins to burn when exposed to an ignition source is 345 °C, or 653 °F.

$$(9/5 \times 345) + 32 = 653\ °F$$

Consequently, when exposed to an ignition source at 555 °F, the polystyrene will not burn.
(b) Polystyrene autoignites in air at 490 °C, or 914 °F.

$$(9/5 \times 490) + 32 = 914\ °F$$

Consequently, when the temperature of the fire scene rises to at least 914 °F, polystyrene will self-ignite.

---

*In this textbook, the lowercase t is used to denote a temperature on the Fahrenheit and Celsius scales, whereas the capital T is used to denote a temperature on the Kelvin and Rankine scales.

Subdivision of the Fahrenheit and Celsius scales can be continued above and below the two reference points. When a thermometer is read below the zero mark, the divisions are read as minus degrees, or as degrees Fahrenheit or Celsius *below zero*. When a temperature reading below zero on either scale is converted to a reading on the other scale, it is important to account for the negative signs algebraically.

2.7-B THE KELVIN AND RANKINE TEMPERATURE SCALES

No upper temperature limit appears to exist. The highest known value is 3.6 million degrees, the average temperature of the Sun's corona, or outer atmosphere.

Conversely, scientific theory and experiment establish the limit below which it is impossible to cool matter. This coldest temperature is $-273.15\ °C$ ($-459.67\ °F$), which is called **absolute zero**. Although this temperature cannot be attained experimentally, scientists at the University of Helsinki and at the Massachusetts Institute of Technology have successfully achieved temperatures of 0.000000001 K and 0.000000005 K, or 0.1×10^{-9} K and 0.5×10^{-9} K, respectively.

Two additional temperature scales are defined by reference to this lowest attainable temperature as zero: the Kelvin and Rankine temperatures scales. The **Kelvin temperature** is defined as follows:

$$T(K) = t(°C) + 273.15\ °C$$

Thus, a Kelvin temperature is obtained by merely adding 273.15 to any Celsius temperature reading. The unit of temperature on the Kelvin scale is called the **kelvin**, without a degree symbol.

The **Rankine temperature** is defined as follows:

$$T(°R) = t(°F) + 459.67\ °F$$

A Rankine temperature is thus obtained by adding 459.67 degrees to the Fahrenheit temperature reading. The unit of temperature on the Rankine scale is the degree Rankine (°R).

Since 0 K (0 °R) represents the lowest attainable temperature, negative numbers are never encountered on either the Kelvin or the Rankine temperature scale.

absolute zero
▪ The lowest temperature attainable by a substance, equal to $-459.67\ °F$ ($-273.15\ °C$)

Kelvin temperature
▪ The absolute temperature obtained by adding 273.15 °C to a Celsius temperature

kelvin
▪ The unit of temperature on the Kelvin scale

Rankine temperature
▪ The absolute temperature obtained by adding 459.67 °F to a Fahrenheit temperature

SOLVED EXERCISE 2.7

Metallic mercury loses all resistance to the flow of electricity when it is cooled to 9 K. Under such conditions, it is said to be a *superconductor*. At which Fahrenheit temperature does mercury become a superconductor?

Solution: A temperature in kelvins is obtained by adding 273.15 °C to a Celsius temperature.

$$K = °C + 273.15\ °C$$

Consequently, 9 K is equivalent to $-264.15\ °C$.

$$9\ K = -264.15\ °C + 273.15\ °C$$

A temperature on the Fahrenheit scale is obtained by adding 32 to 9/5 of a Celsius temperature. Consequently, $-264.15\ °C$ is equivalent to $-443.47\ °F$.

$$[9/5 \times (-264.15)] + 32 = -443.47\ °F$$

- Describe the concept of pressure.
- Distinguish between the gauge pressure and absolute pressure of a confined gas or vapor.
- Identify the common English and metric units used to measure pressure.
- Identify the equivalent units in which the standard atmospheric pressure is most commonly measured.
- Convert a pressure reading in one unit into its equivalent in another unit.
- Describe the concept of vapor pressure and the general relation of the vapor pressure of a flammable liquid to its fire and health hazards.
- Describe the concept of blood pressure, and indicate generally how a blood pressure reading may be used to identify the severity of certain health problems.

pressure
- A force applied to a unit area

psi
- The unit of pressure expressed as pounds per square inch (psi$_a$ means pounds per square inch absolute; psi$_g$ means pounds per square inch gauge)

newton
- The unit of force in the SI system

pascal (newton per square meter)
- The SI unit of pressure

atmospheric pressure
- The force exerted on matter by the mass of the overlying air

atmosphere
- The gaseous layer that surrounds Earth

standard atmospheric pressure
- The average value of the atmospheric pressure at sea level

barometer
- An instrument used for determining the atmospheric pressure, usually expressed as the height of a column of mercury

gauge pressure
- The amount of pressure by which a confined gas exceeds atmospheric pressure

absolute pressure
- The pressure exerted on a material; for confined gases, the sum of the gauge pressure and the atmospheric pressure

2.8 Pressure and Its Measurement

Pressure is the force exerted over a specified area; hence, pressure is often expressed as a unit of force per unit of area such as pounds per square inch **(psi)**. In the SI system, the unit of force is the **newton** (N); it is the force that accelerates a one-kilogram body one meter per second each second. The SI unit of pressure is the **newton per square meter** (N/m^2), which is called a **pascal** (Pa). A unit of pressure commonly encountered is the *kilopascal* (kPa).

The force resulting from the mass of the overlying air produces **atmospheric pressure**. Earth's gravity gives the atmosphere an average downward force of 14.7 psi at sea level; that is, on average, 14.7 lb of air bears down on each square inch of Earth's surface. The average pressure exerted by the atmosphere at sea level supports a column of mercury 760 mm high (760 mmHg); this pressure is also called 1 **atmosphere** (atm). For the purpose of converting a pressure that is expressed in one unit to its equivalent in another unit, each of the following applies:

$$1\,\text{atm} = 760\,\text{mmHg} = 760\,\text{torr} = 101.3\,\text{kPa} = 29.9\,\text{in.Hg} = 14.7\,\text{psi}$$

Each of these values is referred to as **standard atmospheric pressure**.

Atmospheric pressure readings are normally recorded on an instrument called a **barometer**. A simple barometer can easily be constructed by filling a glass tube, longer than 760 mm and closed at one end, with mercury. When the tube is inverted, open-side down, in a dish of mercury, the liquid flows out of the submerged open bottom until the level in the tube reaches an average height of 760 mm, or 1 atm. When the atmospheric pressure decreases below the average value of 1 atm, the mercury level on the barometer is said to "fall"; when it increases above 1 atm, the mercury level is said to "rise." Such changes in the atmospheric pressure are closely monitored and used by meteorologists to forewarn of inclement weather conditions.

A pressure reading is also used for monitoring the amount of a gas or vapor confined within a cylinder or tank. For such purposes, the affixed pressure gauge shown in Figure 2.8 is read. This reading provides the pressure exerted by the contents of the vessel. As portions are removed, the amount remaining in the vessel decreases as the gauge reading approaches zero. Because atmospheric pressure is continuously exerted on all forms of matter on Earth, a pressure gauge measures the amount of pressure by which the gas or vapor exceeds atmospheric pressure. The reading is called the **gauge pressure**. The units of gauge pressure are expressed with the inclusion of a subscript "g," as in psi$_g$.

The true pressure of the contents within a containment vessel, called its **absolute pressure**, is the sum of the gauge pressure and the atmospheric pressure. The absolute pressure is measured with respect to a value of zero, whereas the gauge pressure is measured with respect

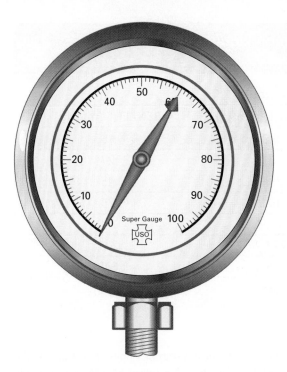

FIGURE 2.8 When the needle points to zero on a pressure gauge, the gauge pressure is zero. When the pressure gauge is connected to a gas cylinder, the cylinder is empty when the gauge reads zero. A gauge pressure of zero corresponds to an actual or absolute pressure of 14.7 psi, or 101.3 kPa.

to the prevailing pressure of the atmosphere; thus, in the English system of units, the absolute pressure on average is expressed as follows:

$$P(\text{psi}_a) = P(\text{psi}_g) + 14.7 \text{ psi}$$

The units of absolute pressure are sometimes noted by inclusion of the subscript "a" with the unit, as shown in the previous expression. Thus, a gauge pressure of 19.4 psi_g corresponds to an average absolute pressure of 34.1 psi_a.

$$19.4 \,\text{psi}_g + 14.7 \,\text{psi} = 34.1 \,\text{psi}_a$$

We sometimes refer to the temperature and pressure at *room conditions*. For our purposes, room conditions means a temperature of 68 °F (20 °C) and a pressure of 14.7 psi (101.3 kPa).

2.8-A VAPOR PRESSURE

When a liquid is confined within a closed container at a given temperature, it evaporates into the headspace above the liquid until equilibrium between the liquid and its vapor is attained. This equilibrium is characterized by two opposing changes that occur simultaneously: the rate at which the liquid evaporates and the rate at which the vapor condenses. At equilibrium, these rates are equal. The pressure exerted by the vapor in equilibrium with its liquid is called the liquid's **vapor pressure**.

Although the vapor pressure can be expressed in any unit of pressure, chemists normally measure it in millimeters of mercury (mmHg) or kilopascals (kPa) at the prevailing temperature. In this book, the vapor pressure is exclusively expressed in mmHg at 68 °F (20 °C). When expressed in this fashion, the vapor pressure is the height of a column of mercury (measured on a meterstick) that the liquid's vapor supports at a given temperature.

The vapor pressure of liquids is a characteristic property. As their temperature increases, more and more of the substances vaporize into the headspace of the container in which each is confined. As Table 2.6 shows for three representative liquids, this causes the substances's vapor pressure to increase accordingly.

Liquids with low boiling points have relatively high vapor pressures, because they evaporate readily. Conversely, liquids with high boiling points have relatively low vapor pressures, since they evaporate more slowly.

vapor pressure
■ The pressure exerted within a confinement vessel by the vapor of a substance in equilibrium with its liquid; a measure of a substance's propensity to evaporate

TABLE 2.6 — Vapor Pressures of Some Common Liquids as a Function of Increasing Temperature

| Temperature | | Liquid | | |
°F	°C	WATER (mmHg)	ETHANOL (mmHg)	BENZENE (mmHg)
14	−10	2.1	5.6	15
32	0	4.6	12.2	27
50	10	9.2	23.6	45
68	20	17.5	43.9	75
86	30	31.8	78.8	118
122	50	92.5	222.2	271
167	75	289.1	666.1	643
212	100	760	1693.3	1360

Generally speaking, the vapor of a substance is its most hazardous physical form. It is the vapor of a flammable liquid that burns, and it is a liquid's vapor that causes adverse health effects when inhaled. Clearly, the vapor pressure of a flammable or toxic substance has a direct impact on its potential fire and health hazards, respectively.

When a flammable or toxic liquid has a relatively low vapor pressure, little vapor evolves at the prevailing temperature. It can neither pose a significant fire and explosion hazard nor an inhalation toxicity hazard. However, a flammable liquid with a comparatively high vapor pressure poses a fire and explosion hazard, and a poisonous liquid with a comparatively high vapor pressure poses an inhalation toxicity hazard.

A flammable or poisonous liquid with a relatively high vapor pressure presents unique transportation problems. During the course of its transportation, a liquid confined within a tank or other vessel can increase in temperature by absorbing heat from its surroundings. As the temperature of the liquid increases, a considerable volume of vapor is produced within the tank. This vapor enters the headspace above the liquid and exerts pressure on the walls of the confining vessel. To retain its integrity, the vessel must be constructed to withstand this internal pressure during the time that it is used to transport the liquid.

Even when a container or transport tank containing a flammable or toxic liquid has been essentially emptied of its contents, the vessel may still constitute a fire or health hazard. These hazards are caused by the presence of residual vapor that may ignite when exposed to an ignition source or pose a health-related problem when inhaled. An "empty" container or tank should always be regarded as potentially lethal because it most likely contains residual vapor.

SOLVED EXERCISE 2.8

The respective vapor pressures of iodine and mercury are 0.3 and 0.0012 mmHg at 68 °F (20 °C). Which substance vaporizes more readily at this temperature?

Solution: Iodine vaporizes more readily than mercury at 68 °F (20 °C), since its vapor pressure exceeds the vapor pressure of mercury. (Notwithstanding this fact, sufficient mercury may vaporize at 68 °F (20 °C) to pose a health threat when inhaled.)

Methanol, ethanol, and isopropanol are flammable liquids whose vapor pressures at 68 °F (20 °C) are 98, 50, and 33 mmHg, respectively. Using only these data, determine which liquid poses the greatest risk of fire and explosion at this temperature.

Solution: The magnitude of the vapor pressure for methanol exceeds the vapor pressures of ethanol and isopropanol. This means that at 68 °F (20 °C), methanol produces a comparatively larger volume of vapor than either ethanol or isopropanol. Because it is the vapor of a substance that burns, prudence dictates that considerable caution should be exercised to prevent its ignition.

2.8-B BLOOD PRESSURE

When dispatched to an emergency scene, paramedics often measure an individual's **blood pressure**. This is the constant force per unit area that is exerted on the walls of the arteries as blood is pumped to the tissues of the body. Blood pressure is measured using a cufflike device called a *sphygmomanometer*, a type of which is shown in Figure 2.9. Blood pressure can also be measured electronically through use of a device that provides digital readings.

When an individual's blood pressure is measured, two numbers are obtained: the **systolic**, which records the pressure as the heart beats; and the **diastolic**, which indicates the pressure as the heart relaxes between beats. A typical "normal" blood pressure reading is 120 mmHg/80 mmHg, referred to as "120 over 80." When a paramedic relays an individual's blood pressure reading to a physician, the critical nature of an illness can be estimated by noting the deviation from the norm.

Blood pressure readings of approximately 140 over 90 are considered abnormal. Emergency-response personnel and other individuals who experience such readings are afflicted with the disease known as **hypertension**, or **high blood pressure**. Such readings signify that the blood vessels are being subjected to more strain than they should experience. An individual's risk of experiencing high blood pressure is increased by being overweight and inactive and consuming high levels of salt in the diet. Although the use of medication can lower an individual's blood pressure,

blood pressure
■ The constant pressure on the walls of the arteries

systolic
■ The top number of a blood pressure reading, typically less than 130 mmHg; the pressure exerted within an individual's arteries as the heart is beating

diastolic
■ The bottom number of a blood pressure reading, typically less than 85 mmHg; the pressure exerted within an individual's arteries as the heart relaxes between beats

high blood pressure (hypertension)
■ The condition in which blood is pumped through the body at a pressure that exceeds approximately 140 over 90

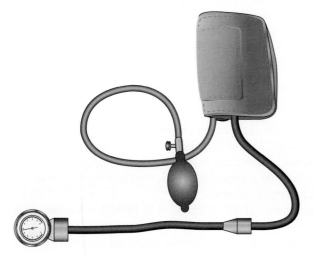

FIGURE 2.9 A manual or electronic sphygmomanometer is used to measure an individual's blood pressure. In this manual device, a flexible cuff is placed around a patient's upper arm over the brachial artery and inflated to about 200 mmHg. The pressure squeezes the artery shut and prevents blood flow. When the pressure is relieved, blood spurts through the artery at the systolic pressure, which creates a turbulent tapping sound. The pressure registered at the moment the first sound is detected is the systolic blood pressure. As the pressure in the cuff is further lowered, the sounds in the artery continue to be detected until the pressure decreases enough to allow diastolic blood flow. At this point, the blood flow becomes smooth, no sounds are heard, and the device records a diastolic blood pressure.

regular exercise, weight control, and lifestyle changes are equally important. Hypertension can damage not only the blood vessels but also the organs that supply blood. In more severe instances, it causes a traumatic health problem like a stroke.

PERFORMANCE GOALS FOR SECTION 2.9:

- Define the concept of heat.
- Describe the mechanisms by which heat is transmitted from one material to another or from spot to spot.
- Identify the common English and metric units used to measure heat.
- Convert a heat measurement in one unit into its equivalent in another unit.
- Describe the manner by which convection and radiation affect the spread of a freely burning fire.
- Describe the manner by which the human body responds to the presence of excessive heat.
- Describe the reactions likely to be experienced by individuals exposed to excessive heat.
- Describe the nature of a first-, second-, and third-degree burn.

heat
■ The form of energy transferred from one body to another because of a temperature difference between them; energy arising from atomic or molecular motion

endothermic process
■ Referring to a process that absorbs heat from its surroundings

exothermic process
■ Referring to a process that emits heat into its surroundings

British thermal unit
■ The amount of heat required to raise the temperature of one pound of water one degree Fahrenheit (abbreviated Btu)

calorie
■ The amount of heat required to raise the temperature of one gram of water one degree Celsius

heat of reaction
■ The energy that is absorbed or evolved when a chemical reaction occurs

heat of combustion
■ The heat emitted into the surroundings when a compound is burned

2.9 Heat and Its Transmission and Impact on the Human Body

Heat is the form of energy associated with the motion of atoms or molecules (Sections 4.4 and 4.6), small particles of which all matter is composed. Heat is manifested when changes in any of the following occur:

- A material's temperature
- The physical state of a substance
- The chemical identity of a substance

Heat is either absorbed or emitted as these processes occur. A process that results in the absorption of heat is called an **endothermic process**, whereas a process that results in the release of heat to the surroundings is called an **exothermic process**.

In Section 2.6, it was noted that energy is measured in British thermal units and calories. One **British thermal unit** represents the heat that must be supplied to raise the temperature of 1 pound of water 1 degree Fahrenheit, specified at the temperature of water's maximum density, 39.1 °F (3.97 °C). One **calorie** is defined as the amount of heat required to raise the temperature of 1 gram of water 1 degree Celsius, from 14.5 to 15.5 °C. Two hundred fifty-two (252) calories is equivalent to one British thermal unit, and 1 calorie is equivalent to 4.184 joules.

$$1 \, \text{Btu} = 252 \, \text{cal}$$
$$1 \, \text{cal} = 4.184 \, \text{J}$$

When chemical reactions occur, substances change their chemical identities. One substance transforms into one or more other substances. The thermal energy accompanying the chemical reaction is called its **heat of reaction**; when a substance burns, the evolved energy is called the **heat of combustion**. Heat is evolved during a variety of combustion processes such as the burning of gasoline, wood, natural gas, and other materials.

The heats of combustion for some common fuels are provided in Table 2.7. You are most likely familiar with the names of these fuels. When they burn, the fuels provide energy and serve as sources of heat, power, and light. On the negative side, however, they can also pose a risk of

TABLE 2.7	Heats of Combustion of Some Common Fuels	
FUEL	**Btu/lb**	**kJ/g**
Natural gas (methane)	23,800	55.5
Propane	21,600	50.3
Acetylene	21,400	49.9
Butane	21,200	49.5
Aviation fuel	20,420	47.6
Motor gasoline	20,100	46.9
Kerosene	19,800	46.2
Diesel fuel	19,400	45.2
Medium heating fuel	18,900	44.1
Wood (dry)	8,600	20.0

fire and explosion. When these materials burn in bulk, they usually ignite secondary fires involving nearby materials.

The proper control of the heat emitted during combustion is an extremely important factor in fire control. Fires continue as self-sustained phenomena only when sufficient heat is released during combustion to substitute for the input energy initially provided from an ignition source. Conversely, many fires cannot be extinguished until some means is undertaken to reduce or eliminate this heat.

Heat is always transferred from warm materials to cooler ones. If several materials near one another have different temperatures, those that are warm become cooler, and those that are cool become warmer, until they achieve a common temperature. Heat is transmitted from one material to another or from one spot to another spot by three independent modes: conduction, convection, and radiation. The nature of each mode of heat transmission is an important factor associated with understanding how fire spreads.

2.9-A CONDUCTION

The handle of a metal spoon that has been inserted into hot coffee becomes hot itself. This transfer of heat between two or more materials in contact—in this case, from the coffee to the spoon—is called **conduction**.

Every material conducts heat to some extent. Metals, such as silver, copper, iron, and aluminum, conduct heat most efficiently; nonmetals, such as glass and air, are not good conductors of heat. Materials that are not good conductors are good insulators, because they delay the transfer of heat.

conduction
- The mechanism by which heat is transferred to the parts of a stationary material or from one material to another with which it is in contact

2.9-B CONVECTION

Heat can also be transferred from spot to spot or even between two or more substances by the natural mixing of their component parts. This happens when cold milk or cream disperses throughout hot coffee without the use of a spoon, or when the air in a room gets warmer when it is hot outside the room. This transmission of heat within a substance or between substances by means of natural mixing is known as **convection**.

The circulation of warm air that exits a heat vent into a room is caused by convection. The popular expression "heat rises" actually means "hot air rises." As warm air issues from a heat vent into a room, it rises toward the ceiling. The surrounding cool air descends to replace the

convection
- The mechanism by which heat is transferred by the movement of the heated material itself from spot to spot

warm air that rose, is heated at the heat vent, and then rises toward the ceiling. In this fashion, warm and cool air exchange to generate air currents.

The origin of the convection phenomenon is associated with the impact that gravity has on matter. In zero gravity, convection can occur only if it is artificially induced. When a substance burns, convection moves its combustion products away from the flame to dissipate into the surrounding atmosphere; but where zero gravity exists, as in a space ship orbiting Earth, the combustion products remain in the immediate area of the combustion zone, where they quickly extinguish the flame.

2.9-C RADIATION

Imagine a 200-W lightbulb hanging from a ceiling. When the light is turned on, heat can be felt when we hold our hands *around* the bulb. This transfer of heat from the bulb to our hands cannot be caused by conduction, since the air between the bulb and our hands is a poor conductor of heat. It also cannot be caused by convection, since hot air currents rise upward. The heat from the bulb is transmitted by a third means called **radiation**.

radiation
■ The mechanism by which heat is transferred between two materials not in contact

Unlike conduction and convection, radiation occurs even when there is no material contact between two objects. Heat from the Sun radiates through space and warms Earth and other celestial bodies.

All matter radiates heat at elevated temperatures; that is, it exhibits the phenomenon called *incandescence.* When the temperature of a heated object exceeds approximately 932 °F (500 °C), the radiation becomes visible. Burning flames, glowing coals, and molten metal are examples of matter hot enough to radiate visible light. Hot objects may also radiate energy at lower temperatures, but the energy emitted is generally in the infrared region of the electromagnetic spectrum, which cannot be observed with the naked eye.

2.9-D SPREAD OF FIRE

The conduction process does not significantly contribute to the sustenance and spread of a normal fire. Figure 2.10 illustrates that convection and radiation are the modes of heat trans-

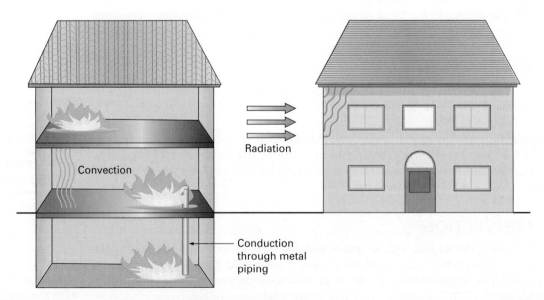

FIGURE 2.10 Conduction, convection, and radiation contribute to the spread of fire. In the building on the left, metal piping first transmits heat by conduction from an unattended basement fire to the first floor. Heat is then transmitted by convention to the second floor and throughout the remainder of the building by convection. Heat is transmitted toward the nearby adjacent building on the right by radiation.

mission primarily responsible for sustaining and spreading freely burning fires in the following ways:

- Convection affects the spread of fire by means of the natural movement of hot and cold air. As hot air rises, heat is transmitted to adjoining materials and initiates their ignition. Cool, fresh air simultaneously flows inward into the fire to replace the hot air, thereby providing the fire with a supply of oxygen.
- Radiation affects the spread of fire by transmitting heat, primarily in a lateral direction, from the immediate fire scene to nearby materials. The transmission of radiant heat initiates the combustion of these materials.

2.9-E ILL EFFECTS RESULTING FROM EXPOSURE TO HEAT

The routine work activities of a firefighter require exposure to high-temperature environments. The heat is transmitted from these hot environments directly to the body. The human organism reacts to its presence and attempts to reduce the body's temperature by the following mechanisms, both of which function in combination:

- *The body circulates blood to the skin.*   The body releases excess heat to the environment through the skin. When the muscles are being used for physical labor, as during firefighting activities, less blood flows to the skin, which can hinder the body's capacity to release its excess heat.
- *The body perspires.*   The body attempts to maintain a stable internal temperature ranging from 96.5 to 99 °F (35.8 to 37.2 °C) by producing perspiration. The skin releases as much as 3 gal/d (11 L/d) of sweat during hot weather, but perspiring is an effective means of cooling only when the prevailing relative humidity is low enough to permit evaporation and if the fluids and salts lost during the process are adequately replaced. Evaporative cooling is limited in the case of individuals—such as firefighters—who are obligated to wear protective gear that hinders the release of perspiration to the atmosphere.

Individuals store the heat they are incapable of eliminating. They first sense a rise in body temperature and heart rate. As additional heat is stored, they experience a lethargic or sick feeling, irritability, an inability to effectively concentrate, and difficulty with focusing on a task. To overcome these sensations, they often respond by fainting.

Overheated individuals may also experience one or more of the following reactions:

- *Heat rash.*   When perspiration is not readily removed from the surface of the skin by evaporation, small itchy, red spots appear on the surface of the skin. In severe cases, this heat rash is so uncomfortable that it impedes an individual's performance.
- *Heat cramps.*   These painful, involuntary spasms of the muscles are often caused by the failure of individuals to replace the minerals lost in perspiration, despite consuming water after exposure to a heated environment. Heat cramps frequently develop in individuals who have exerted muscular activity for an extended period of time.
- *Heat exhaustion.*   Heat exhaustion generally produces certain common symptoms such as weakness, fatigue, giddiness, nausea, and headache. The skin becomes clammy and moist, the complexion pale or flushed, and the body temperature increases slightly higher than normal. Heat exhaustion results when an individual fails to drink adequate volumes of fluids or to replace the salts lost in perspiration following exposure to a heated environment.
- *Heatstroke, or hyperthermia.*   Heatstroke is caused by the failure of the body to satisfactorily regulate its internal temperature on exposure to excessive heat. The body temperature of this individual can rapidly rise to 106 °F (41 °C), or higher. The common symptoms associated with heatstroke are mental confusion, delirium, loss of consciousness, convulsions, and coma. The skin generally becomes red or bluish. Individuals who experience heatstroke must be promptly treated or they will die.

Individuals who have been exposed to excessive heat should be quickly provided with an adequate volume of fluids, especially commercially available products known to be rich in mineral salts. Minerals, including sodium and potassium, are essential for proper cardiac and other bodily functions. As individuals perspire, these minerals are lost from the body. For this reason, individuals who have been exposed to excessive heat should also be assessed for cardiac dysfunction. While they are awaiting medical assistance, individuals experiencing heatstroke should be cooled by soaking their clothing with cool water.

SOLVED EXERCISE 2.10

Gatorade is a drink that is rich in minerals, especially sodium. Why are overheated athletes and firefighters encouraged by physicians to drink Gatorade?

Solution: Overheated athletes and firefighters are encouraged to drink Gatorade to replace the minerals lost during perspiration. By replacing the minerals required by the body for muscular activities, including proper cardiac function, athletes and firefighters mitigate the risk of experiencing muscular cramps.

The skin and its surrounding areas may be damaged or destroyed when a person contacts very hot materials, including superheated air. The extent of the damage and destruction is designated as a first-, second-, or third-degree burn, or a superficial, partial-thickness, or full-thickness burn, respectively. These burn types are illustrated in Figure 2.11. They are associated with the following features:

first-degree burn
■ A superficial injury of the skin caused by exposure to heat

second-degree burn
■ An injury of the skin to a limited depth caused by exposure to heat; a partial-thickness burn

■ During a **first-degree burn**, the surface of an inner layer of the skin, called the *dermis*, is superficially damaged. The afflicted area visually appears red, resembling a bad sunburn, but it is entirely intact. The first-degree burn is typically accompanied by minor swelling, but it generally is the cause of relatively little pain.

■ During a **second-degree burn**, the dermis is gravely damaged, but to a limited depth, such that the surrounding skin swells and blisters. A second-degree burn is intensely red in appearance. The person afflicted with a second-degree burn generally experiences severe pain at the burn site. Although the skin heals with the passage of time, scars may remain at this site throughout the person's lifetime.

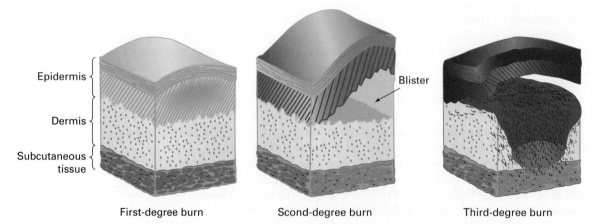

Epidermis

Dermis

Subcutaneous tissue

Blister

First-degree burn   Scond-degree burn   Third-degree burn

FIGURE 2.11 Thermal burns are classified according to the degree of damage to skin and body tissues.

■ During a **third-degree burn**, the damage or destruction of the skin typically involves the subsurface muscular structures, fat, and nerves. When the muscular structure receives a third-degree burn, it may be physically charred. If the nerves remain intact, a third-degree burn is associated with considerable pain; but when the nerves are destroyed, whether partially or wholly, the afflicted person experiences little or no pain.

Firefighters are at risk of experiencing a third-degree burn whenever they inhale superheated air. In the most egregious instances, their lungs become seared. This immediately hampers the ability to breathe and reduces the likelihood of their survival.

Care should be provided to individuals who have been burned by running cool tap water on the damaged skin for a period of 5 minutes to an hour. The cool water absorbs the residual heat and reduces the possibility of further damage to the skin. Cool water also acts as an anesthetic to reduce pain. Ice should *never* be applied to burns, however, since the extreme cold may further damage burnt skin.

Small first- and second-degree burns should be loosely covered with a dry, sterile gauze dressing until medical attention is provided. Larger burn areas can be covered with a clean sheet for protection. Petroleum jelly, antiseptic preparations, ointments, and bandages should *not* ordinarily be applied, since this practice traps heat and restricts air circulation. To reduce the likelihood of infection, the blisters should *not* be broken or cut.

The healing period for first- and second-degree burns varies, but it is typically no more than 3 weeks. A person who has received a third-degree burn should receive specialized medical assistance as quickly as feasible. Ultimately, the corrective measures associated with a full-thickness burn may include grafting and reconstructive surgery. The healing period for a third-degree burn is lengthy but generally no more than 1 year. An individual with a serious third-degree burn can also be at risk of death. When 70% or more of the skin on the body has been burned, the person has only a 1-in-2 chance of survival.

PERFORMANCE GOALS FOR SECTION 2.10:

- Describe the concepts of heat capacity and specific heat.
- Calculate the amount of heat evolved or absorbed by a substance when provided with its heat capacity and the magnitude of a temperature change.
- Identify the latent heat of vaporization and the latent heat of fusion of water.
- Describe the relationship between the vapor pressure of a liquid and its boiling point.
- Describe the relationships between melting and freezing, and between boiling and condensation.
- Calculate the amount of heat involved in each step of a process involving the change of 1 lb of water into 1 lb of steam over the temperature range from −20 °F to 300 °F (−29 °C to 149 °C), and relate the significance of these calculations to the use of water as a fire extinguisher.

2.10 Calculation of Heat

Two or more substances differ from one another by the quantity of heat required to produce a given temperature change in the same mass of each substance. This quantity of heat is called the **heat capacity**. The heat capacities of some common liquids are provided in Table 2.8. For water, the magnitude of the heat capacity varies as a function of its temperature, as indicated. The units of heat capacity in the English and metric systems are the British thermal unit per pound

TABLE 2.8	Heat Capacities of Some Common Liquids	
LIQUID	**HEAT CAPACITY (cal/g °C or Btu/lb °F)**	
Acetone	0.506	
Benzene	0.406	
Carbon tetrachloride	0.201	
Chloroform	0.234	
Diethyl ether	0.547	
Ethanol	0.581	
Methanol	0.600	
Turpentine	0.411	
Water [32 to 212 °F (0 to 100 °C)]	1.00	
Water [<32 °F and > 212 °F (<0 °C and >100 °C)]	0.5	

per degree Fahrenheit (Btu/lb °F) and the calorie per gram per degree Celsius (cal/g °C), respectively. The magnitude of these units is equal:

$$1\,\text{Btu/lb}\,°\text{F} = 1\,\text{cal/g}\,°\text{C}$$

The ratio of the heat capacity of a substance to the heat capacity of water at the same temperature is a dimensionless number called the **specific heat** of the substance. The specific heats of most substances are less than 1.

When heat is transferred to a substance, it may cause a change in its physical state or its chemical identity, or it may merely cause the substance to assume a different temperature. The heat that is transferred to a substance causing a change in its temperature, but no change in its physical state or chemical identity, is calculated by the following formula, in which Q is the heat, m is the mass of the substance, C is its heat capacity, and Δt (delta t) is the difference between the final and initial temperatures of the substance:

$$Q = m \times C \times \Delta t$$

For example, given that copper has a heat capacity of 0.093 cal/g °C, what amount of heat is transferred to 100 g of copper when it is heated from 30 to 100 °C? First, the difference in temperature, Δt, is 100 °C − 30 °C, or 70 °C. Then, the heat transferred is calculated to be 651 cal by multiplication as follows:

$$Q = 100\,\text{g} \times 0.093\,\text{cal/g}\,°\text{C} \times 70\,°\text{C} = 651\,\text{cal}$$

As noted, a change in temperature can also result in a change in the physical state of matter. If liquid water is exposed to the cold, its temperature falls until 32 °F (0 °C) is attained. Then, the temperature remains fixed until the entire liquid mass freezes to solid ice. This latter temperature is called the **freezing point** of water. The quantity of heat per unit mass required to change a liquid substance to a solid at its freezing point is called the **latent heat of fusion** of the substance. For example, the conversion of liquid water to solid ice liberates 144 Btu/lb, or 80 cal/g. This is the latent heat of fusion of water.

We can similarly heat a mass of liquid water to 212 °F (100 °C). Then, the temperature remains fixed while the liquid water changes into steam. This is the temperature at which the

specific heat
■ The ratio of the heat capacity of a substance to the heat capacity of water at the same temperature

freezing point
■ The temperature at which the liquid and solid states of a substance coexist at 1 atm (101.3 kPa)

latent heat of fusion
■ The amount of heat required to convert a unit mass of a solid substance into a liquid

vapor pressure of water equals the atmospheric pressure of 1 atm; it is called the **boiling point** of water. The amount of heat that must be supplied to a material at its boiling point to convert it from a liquid to a vapor is called its **latent heat of vaporization**. The heat of vaporization of water is 970 Btu/lb, or 540 cal/g. This is a relatively high value compared with the heat of vaporization of other substances.

Two phenomena are closely associated with freezing and boiling, respectively:

- *Melting* is the reverse of freezing. As ice melts, 144 Btu/lb, or 80 cal/g, is absorbed from the surroundings.
- *Condensation* is the reverse of boiling. When steam condenses, 970 Btu/lb, or 540 cal/g, is liberated into the surroundings. This explains why exposure of the skin to steam at 212 °F (100 °C) causes a more severe burn than exposure to liquid water at the same temperature.

The amount of heat associated with the conversion of 1 pound of water into 1 pound of steam over the temperature range from −20 to 300 °F (−29 to 149 °C) is illustrated in Figure 2.12. The relevant quantities of heat in Btu can be calculated at each juncture as follows:

- Ice undergoes a change in temperature from −20 to 32 °F (−29 to 0 °C)

$$Q = 1\,\text{lb} \times 0.5\,\text{Btu/lb °F} \times 52\,\text{°F} = 26\,\text{Btu}$$

- Ice melts at 32 °F (0 °C)

$$Q = 1\,\text{lb} \times 144\,\text{Btu/lb} = 144\,\text{Btu}$$

- Liquid water undergoes a change in temperature from 32 to 212 °F (0 to 100 °C)

$$Q = 1\,\text{lb} \times 1.0\,\text{Btu/lb °F} \times 180\,\text{°F} = 180\,\text{Btu}$$

boiling point
- The temperature at which the vapor pressure of a substance equals the average atmospheric pressure

latent heat of vaporization
- The amount of heat required to convert a unit mass of a liquid substance into a gas

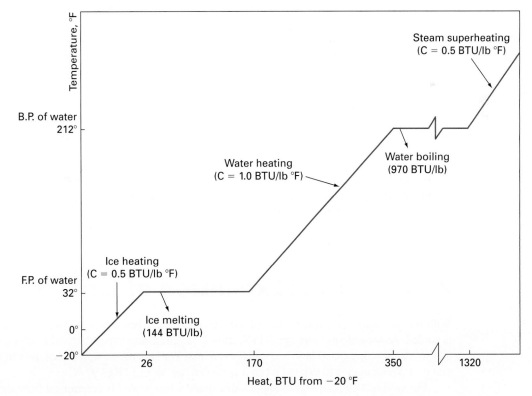

FIGURE 2.12 The heat absorbed in Btu when one pound of pure water is heated at atmospheric pressure. B.P. and F.P. refer to the boiling and freezing points of water, respectively.

- Liquid water evaporates into steam at 212 °F (100 °C)

$$Q = 1 \text{ lb} \times 970 \text{ Btu/lb} = 970 \text{ Btu}$$

- Steam is superheated from 212 to 300 °F (100 to 149 °C)

$$Q = 1 \text{ lb} \times 0.5 \text{ Btu/lb °F} \times 88 \text{ °F} = 44 \text{ Btu}$$

These calculations clearly demonstrate that the largest quantity of heat is associated with the evaporation of water.

Water is commonly employed as a fire extinguisher, since it is nonflammable, plentiful, and inexpensive; but the primary reason water effectively extinguishes fire is associated with its ability to remove heat from the burning material. As water is applied to a fire, it absorbs heat and ultimately vaporizes, each pound carrying 970 Btu of heat into the atmosphere and away from the burning material. Simultaneously, the temperature of the burning material decreases until it ultimately drops to the ambient temperature of the surroundings. A temperature reduction is normally sufficient to extinguish most fires.

SOLVED EXERCISE 2.11

When wood and high-density polyethylene burn, the average heat released into the environment is 8600 Btu/lb and 20,050 Btu/lb, respectively. Compare the potential impact that the burning of products made from equal amounts of these materials has on an ongoing fire.

Solution: The heat of combustion of high-density polyethylene is approximately 2.4 times greater than the heat of combustion of wood. When these two materials burn, energy in the form of heat and light is released into the immediate environment. Because substantially more heat is available from burning polyethylene compared with burning an equivalent amount of wood, the presence of polyethylene products at a fire scene increases the risk of spreading fire in the immediate vicinity. Flammable wooden materials in the same setting may not even ignite, since a lesser amount of heat is released when the same amount of wood burns.

PERFORMANCE GOALS FOR SECTION 2.11:

- Calculate the change in volume of a liquid when it undergoes a temperature change.
- Describe the nature of the potential hazard associated with the expansion of a heated liquid confined within a sealed storage vessel

2.11 Thermal Expansion of Liquids

With few exceptions, the dimensions of a material expand when it is heated and contract when it is cooled. As we note in more detail in Section 2.12, gases and vapors expand more than liquids and solids; liquids expand less; and solids even less. Water is the common exception to this general rule: Water begins to expand when it is cooled below 39.1 °F (3.97 °C).

The volume to which liquids and solids expand when heated is determined from the following formula:

$$V_2 = V_1 + (V_1 \times \alpha \times \Delta t)$$

We first determine the product of V_1, α, and Δt and then add the product to V_1. V_1 and V_2 are the initial and final volumes of an arbitrary liquid, respectively; Δt is the difference between its initial and final temperature; and the Greek letter alpha (α) is a measure of the change in volume per unit of original volume per degree change in temperature. The last term is called the **coefficient of volume expansion**, the units of which are reciprocal degrees Fahrenheit ($°F^{-1}$) and reciprocal degrees Celsius ($°C^{-1}$). Numerical values of this coefficient are given in Table 2.9 for some common liquids.

Imagine a tank or metal drum filled to its brim with a liquid. When the liquid expands, it must overflow the brim. If the tank or container is sealed so that the contents cannot overflow, it must either expand to compensate for the increase in liquid volume or rupture. To prevent its rupture, a tank or container should *never* be completely filled with a liquid product. The amount by which a tank or container of liquid falls short of being completely filled is called its **outage**, or *ullage*, and is generally expressed in percentage by volume.

The manufacturers of liquid chemical products normally acknowledge that an outage must be provided in containers. Nonetheless, the allowance could be inadequate to compensate for the volume increase experienced by liquid materials during a fire. Consider a 55-gal steel drum filled to the brim with flammable benzene liquid at 68 °F (20 °C) that is subsequently heated during a fire to 170 °F (77 °C). As the liquid benzene is heated, it expands to occupy a volume of 59 gal at the new temperature:

$$V_2 = 55 \text{ gal} + \left(55 \text{ gal} \times \frac{0.00071}{°F} \times 102 \, °F \right) = 59 \text{ gal}$$

coefficient of volume expansion
- The change in the volume of a liquid per degree change in temperature

outage
- The percentage by volume by which a container of liquid falls short of being filled to the brim

TABLE 2.9	Coefficients of Volume Expansion for Some Common Liquids	
	COEFFICIENT OF VOLUME EXPANSION	
LIQUID	α ($°F^{-1}$)	α ($°C^{-1}$)
Acetic acid	0.00059	0.00107
Acetone	0.00085	0.00153
Benzene	0.00071	0.00128
Carbon tetrachloride	0.00069	0.00124
Diethyl ether	0.00098	0.00176
Ethanol	0.00062	0.00112
Gasoline	0.00080[a]	0.00144
Glycerine	0.00028	0.00051
Methanol	0.00072	0.00130
Pentane	0.00093	0.00168
Toluene	0.00063	0.00113
Water	0.00012	0.00022

[a]The coefficient of volume expansion for gasoline varies from 0.00055 to 0.00090 $°F^{-1}$. The coefficient listed in the table is a representative value.

This increase in volume is 4 gal, or 7.3% by volume. To relieve the internal pressure accompanying this volume increase, the drum ruptures and releases the benzene into the environment. When the benzene vapor is exposed to an ignition source, it is likely to burst into flame.

When liquids are transferred into containers, rail tankcars, and other cargo tanks for transportation, DOT requires the presence of a headspace above the liquid so the vessels are not completely filled to the brim. DOT requires this space to prevent leakage from, or permanent distortion of, the transport vessel by accounting for the expansion of its contents due to a temperature rise likely to be encountered during transportation. Sufficient outage must also be provided within containers of 110 gal or less so that they are not completely filled with liquid at 130 °F (55 °C). In rail tankcars, an adequate headspace must be provided within the shell when the *dome* of the tankcar does not itself provide sufficient outage. DOT's requirements for outage and filling limits for nonbulk and bulk containers are specified at 49 C.F.R. §§173.24a(d) and 173.24b(a), respectively.

SOLVED EXERCISE 2.12

A manufacturer stores 54 gal of liquid methyl alcohol within each of six 55-gal steel drums. The six drums are assembled next to one another along an interior wall within a refrigerated vault whose average temperature is routinely maintained at 20 °F (7 °C). However, the cooling system fails for a period of 6 days during summer weather, causing the temperature within the vault to rise to 90 °F (32 °C). Are these drums likely to rupture at this increased temperature?

Solution: We need to determine the volume of the liquid at 90 °F (32 °C). To do so, we use the following formula:

$$V_2 = V_1 + V_1 \alpha \Delta t$$
$$= V_1(1 + \alpha \Delta t)$$

$$V_1 = 54 \text{ gal}; \Delta t = 90 \text{ °F} - 20 \text{ °F} = 70 \text{ °F}; \alpha = 0.00072 \text{ °F}^{-1} \text{ (from Table 2.10)}$$

$$V_2 = 54 \text{ gal}[1 + (0.00072 \text{ °F}^{-1} \times 70 \text{ °F})]$$
$$= 54 \text{ gal} \times 1.05$$
$$= 57 \text{ gal}$$

Although the manufacturer provided a 1-gal headspace in each drum, 57 gal of liquid cannot be accommodated within a 55-gal drum at a temperature of 90 °F (32 °C). To relieve the pressure generated by the extra volume at this elevated temperature, the sealed drums are very likely to rupture.

PERFORMANCE GOALS FOR SECTION 2.12:

- State Boyle's law, Charles's law, and the combined gas law.
- Use Boyle's law, Charles's law, and the combined gas law to calculate the volume occupied by a confined gas when it is exposed to different pressure and temperature conditions.
- Describe the potential hazard associated with the expansion of a heated gas or vapor that is confined within a storage vessel.

2.12 General Properties of the Gaseous State

Of the three states of matter, only the gaseous state is capable of being described in comparatively simple terms. This description relates the volume of a gas to its temperature and pressure. We shall independently examine the laws of nature that apply to the gaseous state of matter: Boyle's law, Charles's law, and the combined gas law.

2.12-A BOYLE'S LAW

Robert Boyle, an Irish physicist, demonstrated experimentally that the volume of a confined gas varies inversely with its absolute pressure when the temperature remains fixed. The constant-temperature experiments performed by Boyle demonstrated that the mathematical product of the volume and absolute pressure of a gas is always constant. This observation is commonly called **Boyle's law**. It is expressed mathematically as follows:

$$V_1 \times P_1 = V_2 \times P_2$$

Here, V_1 and P_1 are the initial volume and initial absolute pressure, respectively; V_2 and P_2 are the final volume and final absolute pressure, respectively.

Boyle's law is used to determine the volume that a gas occupies when it remains at a fixed temperature but undergoes a change in pressure. Suppose the pressure of 40 cubic feet (40 ft^3) of an arbitrary gas is changed from 14.7 psi to 450 psi at a fixed temperature. What is the new volume assumed by the gas? Common sense tells us that the new volume must be less than 40 ft^3, since the application of any pressure squeezes the gas into a smaller volume. Using Boyle's law, we can readily compute the new volume as follows:

$$V_2 = 40 \text{ ft}^3 \times \frac{14.7 \text{ psi}}{450 \text{ psi}} = 1.3 \text{ ft}^3$$

> **Boyle's law**
> ▪ The observation that at constant temperature the volume of a confined gas is inversely proportional to its absolute pressure

2.12-B CHARLES'S LAW

Jacques Charles and Joseph Gay-Lussac, two French scientists, independently demonstrated experimentally that the volume of a confined gas increases proportionately to the increase in its absolute temperature when the pressure remains fixed. This statement is now known as **Charles's law**. It is expressed mathematically as follows:

$$V_1 \times T_2 = V_2 \times T_1$$

Here, V_1 and T_1 are the initial volume and initial absolute temperature, respectively; V_2 and T_2 are the final volume and final absolute temperature, respectively.

Suppose an arbitrary gas at atmospheric pressure occupies a volume of 300 mL at 0 °C; what volume does it occupy at 100 °C and the same pressure? Common sense tells us that a heated gas always occupies a larger volume. We must first convert the temperatures in degrees Celsius into absolute temperatures in kelvins. In Section 2.7, we learned that the conversion to absolute temperature is accomplished by adding 273.15 °C to the Celsius temperature readings. On the Kelvin scale, the temperature readings are as follows:

$$0 \,°C = 273.15 \text{ K}$$
$$100 \,°C = 373.15 \text{ K}$$

Now, we can use Charles's law to compute the new volume.

$$V_2 = 300 \text{ mL} \times \frac{373.15 \text{ K}}{273.15 \text{ K}} = 410 \text{ mL}$$

> **Charles' law**
> ▪ The observation that at constant pressure the volume of a confined gas is directly proportional to its absolute temperature

2.12-C COMBINED GAS LAW

Boyle's and Charles's laws can also be combined into the following mathematical expression:

$$V_1 \times P_1 \times T_2 = V_2 \times P_2 \times T_1$$

combined gas law
■ The observation that the volume of a confined gas is directly proportional to its absolute temperature and inversely proportional to its absolute pressure

This is called the **combined gas law**. It is used to calculate the volume of a gas at a new temperature and pressure.

The volume of a gas can be forced to remain fixed, as when a gas is confined within a steel cylinder or other storage vessel. Under this circumstance, $V_1 = V_2$, and the combined gas law assumes the following form:

$$P_1 \times T_2 = P_2 \times T_1$$

The heating of a gas that is initially enclosed within a sealed metal tube at 1000 psi$_a$ at 70 °F (21 °C) is illustrated in Figure 2.13. When the gas is heated, its pressure increases accordingly. When the gas is heated to 570 °F, the pressure of the gas nearly doubles; when it is heated to 1070 °F, it nearly triples. These temperatures are akin to those routinely encountered during building fires. To relieve the strain on their walls, the second and third tubes are likely to rupture.

SOLVED EXERCISE 2.13

A welder purchases a gas cylinder of flammable hydrogen gas and chains it to a wall within a workshop. After periodic usage, the gauge pressure reads 235 psi$_g$ when the temperature is 65 °F. What is the gauge pressure reading when the temperature of the cylinder contents becomes 350 °F during a fire?

Solution: Because a steel cylinder is a constant-volume container, $P_1 \times T_2 = P_2 \times T_1$, where these symbols refer to the *absolute* initial and final pressures and temperatures, respectively.

$$P_1 = 235 \text{ psi}_g + 14.7 \text{ psi} = 250 \text{ psi}_a$$
$$T_1 = 65\,°\text{F} + 459.67\,°\text{F} = 525\,°\text{R}$$
$$T_2 = 350\,°\text{F} + 459.67\,°\text{F} = 810\,°\text{R}$$
$$P_2 = 250 \text{ psi}_a \times \frac{810\,°\text{R}}{525\,°\text{R}} = 386 \text{ psi}_a$$

Then, we determine the gauge pressure by subtracting 14.7 psi from the absolute pressure and obtain 371 psi$_g$.

$$P_2 = 386 \text{ psi}_a - 14.7 \text{ psi} = 371 \text{ psi}_g$$

The absorption of heat causes the internal pressure of the cylinder contents to increase from 235 psi$_g$ to 371 psi$_g$.

PERFORMANCE GOALS FOR SECTION 2.13:

■ Describe the general nature of a cryogen.
■ Describe the general hazards associated with exposure to cryogens.

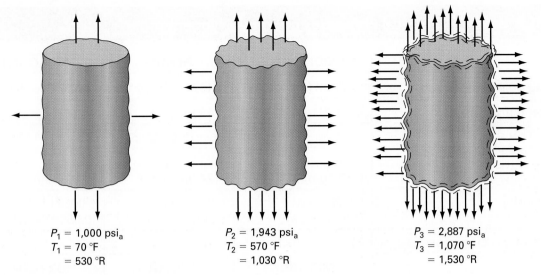

$P_1 = 1{,}000\ \text{psi}_\text{a}$	$P_2 = 1{,}943\ \text{psi}_\text{a}$	$P_3 = 2{,}887\ \text{psi}_\text{a}$
$T_1 = 70\ °\text{F}$	$T_2 = 570\ °\text{F}$	$T_3 = 1{,}070\ °\text{F}$
$= 530\ °\text{R}$	$= 1{,}030\ °\text{R}$	$= 1{,}530\ °\text{R}$

FIGURE 2.13 The effect of applied heat on a gas confined within a constant-volume container such as a sealed metal tube. An increase in the temperature of the gas of 500 °F and 1000 °F causes the internal pressure to nearly double and triple, respectively, which usually causes the containers to rupture.

2.13 General Hazards Resulting from Exposure to Cryogens

The study of matter at temperatures less than approximately −130 °F (−90 °C) is called **cryogenics**. When substances have been cooled to these very low temperatures, they are called **cryogens**, or **cryogenic liquids**.

A gas or vapor can eventually be reduced to a liquid by compressing it and/or lowering its temperature. This liquid can then be reduced to the solid state of matter by lowering its temperature and/or increasing its pressure further. Nonetheless, there is a temperature above which an increased pressure alone cannot cause a gas or vapor to condense; this is called the **critical temperature** of that substance.

Consider gaseous carbon dioxide. When it is cooled to a minimum temperature of 88 °F (31 °C) and compressed by sufficient pressure, carbon dioxide liquefies. But when its temperature is elevated above 88 °F (31 °C), applied pressure is unable to liquefy carbon dioxide. The pressure required to liquefy a gas or vapor maintained at its critical temperature is called the **critical pressure**. The critical pressure of carbon dioxide is 1070 psi_g (7373 kPa). Consequently, when carbon dioxide gas is confined within a vessel and maintained at a temperature equal to or less than 88 °F (31 °C), it liquefies only when compressed by a pressure equal to or greater than 1070 psi_g (7374 kPa).

The critical temperatures, critical pressures, and other physical properties of carbon dioxide and several other common cryogens are listed in Table 2.10.

To retain a gas or vapor as a cryogen, it must be stored in a specially designed and insulated vessel fitted with pressure-regulating valves to control its internal pressure. DOT requires shippers and carriers to transport cryogenic liquids within specialized delivery units or *cryogenic vessels* like the cargo tanks shown in Figure 2.14. DOT regulates the design specifications for these transport vehicles, including the nature of their temperature and pressure control systems, venting mechanisms, valves, and piping.

Tanks approved by DOT for the transportation of cryogenic liquids sometimes utilize a liquid-nitrogen refrigeration system. The cold liquefied nitrogen is dispersed through internal coils

cryogenics
■ The study of the properties of matter at extremely cold temperatures

cryogen (cryogenic liquid)
■ A refrigerated liquefied compressed gas that generally has a boiling point colder than −130 °F (−90 °C) at 14.7 psi_a (101.3 kPa)

critical temperature
■ The temperature above which the vapor of a liquid cannot be condensed by pressure alone

critical pressure
■ The minimum pressure that causes a gas to liquefy at its critical temperature

TABLE 2.10 | Physical Properties of Some Cryogens

	CARBON DIOXIDE	HELIUM	HYDROGEN
Boiling point	−108 °F (−78 °C)	−452 °F (−269 °C)	−423 °F (−252.8 °C)
Critical temperature	88 °F (31.1 °C)	−450°F (−268 °C)	−450 °F (−268 °C)
Critical pressure	1070 psi_g (7374 kPa)	34 psi_g (0.230 kPa)	183 psi_g (1256 kPa)
Liquid density	47.6 lb/ft³ (762 kg/m³)[a]	7.801 lb/ft³ (124.98 kg/m³)[b]	4.43 lb/ft³ (70.96 kg/m³)[b]
Gas density	0.1444 lb/ft³ (1.8333 kg/m³)[a]	0.0103 lb/ft³ (0.165 kg/m³)[a]	0.00521 lb/ft³ (0.08342 kg/m³)[a]
Liquid-to-gas expansion ratio	790	780	865
	METHANE	NITROGEN	OXYGEN
Boiling point	−258 °F (−161 °C)	−321 °F (−196 °C)	−297 °F (−183 °C)
Critical temperature	−116 °F (−82 °C)	−232 °F (−147 °C)	−181 °F (−118 °C)
Critical pressure	673 psi_g (4638 kPa)	492 psi_g (3390 kPa)	736 psi_g (5072 kPa)
Liquid density	26.57 lb/ft³ (425.61 kg/m³)[b]	50.7 lb/ft³ (808.5 kg/m³)[b]	71.23 lb/ft³ (1141 kg/m³)[b]
Gas density	0.04235 lb/ft³ (0.6784 kg/m³)[c]	0.072 lb/ft³ (1.153 kg/m³)[a]	0.083279 lb/ft³ (1.326 kg/m³)[a]
Liquid-to-gas expansion ratio	650	696	861

[a] At 70 °F (21.1 °C) and 1 atm.
[b] At boiling point and 1 atm.
[c] At 60 °F (15.6 °C) and 1 atm.

around the transport tank to cool the cryogen. These specialized tanks used for the transportation of cryogens are designated as DOT Specification 113 and 204 tankcars.

2.13-A EXPANSION OF CRYOGENS DURING VAPORIZATION

When it becomes impossible to adequately maintain the temperature and pressure conditions at which a cryogen is confined, the substance ultimately vaporizes. This event is likely to occur catastrophically, since the difference in volume between the liquid and its gas or vapor is always substantial.

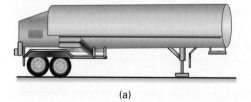
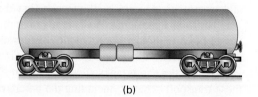

(a)   (b)

FIGURE 2.14 The structural silhouettes of two of several means by which cryogens are transported in bulk: (a) a tank truck and (b) a low-pressure rail tankcar.

Consider cryogenic methane confined within a storage vessel whose pressure-reducing equipment malfunctions. When the entire bulk of the liquid vaporizes, it assumes a volume approximately 650 times its initial volume. This results in a buildup of internal pressure within the storage vessel. This increased pressure exerts force on the walls of the vessel, which most likely will cause the vessel to rupture. As the vessel bursts, the entire mass of methane is released into the environment at once. Because methane is a flammable substance, this situation poses a pronounced risk of fire and explosion.

Fortunately, the storage vessels for cryogens are equipped with *pressure-relief valves* from which the vapor escapes. When the storage vessel has been properly designed, the gas or vapor escapes so slowly that the vessel does not rupture. Nonetheless, when the pressure-reducing equipment fails, adequate ventilation must be provided so that the escaping gas does not reach a concentration that can pose an unreasonable risk to health and safety.

2.13-B IMPACT OF CRYOGENS ON OTHER MATTER

Cryogens are so cold they can affect matter in a number of ways. In particular, when in contact with cryogenic liquids, many materials solidify, become brittle, and easily snap or break. For example, rubber items freeze when in contact with a cryogenic liquid and break into dozens of small pieces when they are dropped on a hard surface.

When bulk quantities of cryogens like hydrogen and helium vaporize into the atmosphere, they cause water vapor and other components of the air to solidify. When cryogens are transferred from one closed vessel into another, the solidification of an air component constitutes a major hazard, because the solid may block venting valves and the passageways within tubing. As noted in Section 2.13-A, this prevents the release of internal pressure and an increased likelihood of vessel failure.

When cryogenic oxygen vaporizes, it causes gases with a boiling point below -297 °F (-183 °C) to condense. These gases include atmospheric oxygen itself. A leak of cryogenic oxygen from its containment vessel generates an environment enriched in oxygen. Although oxygen does not burn, it supports combustion (Section 7.1-C). The presence of an oxygen-enriched environment may cause explosions and other violent chemical reactions.

2.13-C ILL EFFECTS RESULTING FROM EXPOSURE TO CRYOGENS

Exposure to a cryogen may potentially cause several ill effects. In particular, the exposure may cause the development of very serious skin burns. Depending on the length of exposure and the depth to which the skin tissue has been affected, these skin burns resemble first-, second-, and third-degree thermal burns in physical appearance. Dermatologists take advantage of this characteristic when they remove superficial skin growths such as warts. Liquid carbon dioxide or nitrogen is applied to freeze the growths. The skin surrounding them becomes swollen and red and may blister. The skin growths subsequently drop away, leaving healthy new skin.

To avoid the emergence of serious burns, individuals should exercise caution when they handle cryogenic liquids. A face shield or visor should be used to protect the eyes, and loosely fitted gloves and boots should be worn to protect the hands and feet. If a mishap occurs during which a cryogen accidentally flows inside gloves or boots, these items should be removed immediately to minimize the length of time during which the skin and liquid remain in contact.

Living tissue may solidify when it is exposed to cryogenic liquids for an extended period. The solidification causes a local arrest in the circulation of blood. Widespread cellular damage may occur within the solidified tissue, causing it to become vulnerable to bacterial infection. When solidification lasts for hours, physicians may decide to surgically amputate the affected tissue to protect against the onset of gangrene.

The extreme coldness of a cryogenic liquid or solid can also cause bonding between a body part and the uninsulated vessels or pipes containing the cryogen. The bond may be so firm that the flesh rips or tears when a separation attempt is made. Bonding also occurs when a piece of dry ice (solid carbon dioxide) is inserted in jest into the mouth. The dry ice bonds to the tongue. To prevent subsequent damage to the mouth and its organs, persons should refrain from this type of behavior.

Normal body temperature should be restored to tissues that have been exposed to cryogens as soon as practical. The rapid restoration of normal body temperature minimizes the potential for further tissue damage. This is best accomplished by flushing the affected area with large volumes of tepid—not hot—water. However, to avoid aggravating the injury, affected body parts should never be rubbed or massaged.

Individuals generally experience discomfort when they breathe the cold gas or vapor of a cryogen. Prolonged breathing of a very cold gas or vapor can cause serious lung disease.

Finally, caution must be exercised as a cryogenic liquid vaporizes, since its gas or vapor displaces a corresponding amount of air. In a confined or unventilated area, the resulting atmosphere may contain a reduced concentration of oxygen that is incapable of supporting life. Individuals who breathe this atmosphere can suffocate or experience other ill effects.

SOLVED EXERCISE 2.14

Why do experts encourage the use of a water fog at an accident scene involving a spill of a flammable cryogenic liquid but discourage the use of streams of water?

Solution: The discharge of a water fog at an accident scene involving the spill of a cryogen helps dissipate the gas or vapor produced as the cryogen evaporates. When the cryogen is flammable, the use of a water fog accomplishes each of the following:

- A water fog dilutes the flammable vapor within the air, thereby rendering the atmosphere less flammable.
- A water fog quenches or prevents the propagation of an incipient or developed flame front.

Experts discourage the use of streams of water at this type of accident scene because streams of water do not yield similar results.

States of Matter

2.1 In which state of matter does each of the following materials exist at 68 °F (20 °C) and 101.3 kPa:

(a) Steel   (b) Mercury   (c) Silver

(d) Nitrogen   (e) Ammonia   (f) Battery acid

Conversions Between Units of Measurement

2.2 The members of a firefighting brigade in Cincinnati, Ohio, roast a 25-lb turkey in an aluminum roasting pan that measures 12 in. × 16 in. × 3.75 in. A similar group in Montréal, Quebec, roasts a 11.33-kg turkey in an identical pan. What are the measurements in SI units of the roasting pan used by the Montréal brigade?

2.3 The manufacturers of light-duty trucks warn that drivers could readily lose control of their vehicle when towing a trailer that weighs more than 1000 lb. Is the driver of such a truck likely to lose control of the vehicle when pulling a 1500-kg trailer?

Density of Matter

2.4 At 68 °F (20 °C), 5 gal of a flammable liquid weighs approximately 36 lb.

(a) Use Table 2.3 to identify this liquid.
(b) What is the approximate specific gravity of the liquid?

2.5 A piece of metal weighs 13.21 g. When inserted into a cylinder containing exactly 50.00 mL of water, the metal and water occupy a volume of 54.89 mL at 68 °F (20 °C). Use Table 2.3 to identify the metal.

Conversion Between Temperature Readings

2.6 The highest temperature ever recorded on Earth's surface is 136 °F, which was measured at El ´Azizia, Libya, in 1922.

(a) What is the equivalent Celsius temperature?
(b) What is the equivalent Kelvin temperature?

2.7 The lowest temperature ever recorded on Earth's surface is -128 °F, which was measured at the Vostok Ice Station in Antarctica in 1989.

(a) What is the equivalent Celsius temperature?
(b) What is the equivalent Kelvin temperature?

2.8 The four reaction zones of color and temperature of a candle flame are illustrated in Figure 2.15. Incandescent soot particles burning in the "yellow zone," sometimes called the "luminous zone," provide most of the light from a burning candle.

(a) Identify the coolest and hottest zones of the flame.
(b) What is the temperature in degrees Fahrenheit of the blue zone?
(c) What is the temperature in degrees Rankine of the yellow zone?

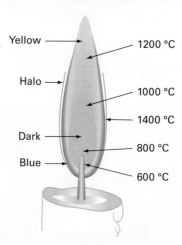

FIGURE 2.15 The zones of color and temperature of a candle flame.

Yellow —————— 1200 °C

Halo —————— 1000 °C

————— 1400 °C

Dark —————— 800 °C

Blue —————— 600 °C

Pressure Readings and Their Significance

2.9 The OSHA regulation at 29 C.F.R. §1910.158(a)(3)(iii) stipulates that employers must provide their personnel in fire brigades, industrial fire departments, and private or contractually arranged fire departments with fire hose of such a length that friction loss resulting from the flow of water through the hose will still provide a dynamic pressure at the nozzle within the approximate range of 210 kPa to 860 kPa. What pressure range at the nozzle does OSHA require in the fire hose when the pressure is expressed in pounds per square inch (psi)?

2.10 The OSHA regulation at 29 C.F.R. §1910.157(f)(6) stipulates that employers must hydrostatically test carbon dioxide hose assemblies having a shutoff nozzle at 300 psi. If an employer hydrostatically tests such a hose assembly at 2100 kPa, is the company in compliance with this OSHA regulation?

Transmission of Heat

2.11 When fire begins on the fourth floor of a high-rise building, why are firefighters more concerned about its immediate spread to upper floors than to lower floors?

2.12 Why do firefighters often ventilate a burning building by opening a large hole at the highest point on its roof?

Significance of Heat

2.13 The human body is approximately 60% water by mass. How is the body capable of maintaining a relatively steady internal temperature of 98.6 °F (37 °C) regardless of the surrounding temperature to which it is exposed?

2.14 Why is more pain usually experienced from a second-degree burn than from a first- or third-degree burn?

General Properties of the Gaseous State

2.15 Exactly 4.00 L of air is inserted into a balloon indoors at 75.2 °F (24 °C). When the balloon is taken outside, where the temperature is −9.4 °F (−23 °C), what will its volume in liters become? (Assume the atmospheric pressure remains constant.)

2.16 The U.S. Consumer Product Safety Commission (Section 1.4) requires manufacturers to provide the following hazard-warning statements on aerosol cans whose contents have been pressurized:

> **CONTENTS UNDER PRESSURE**
> **DO NOT INCINERATE CONTAINER**
> **DO NOT EXPOSURE TO HEAT OR**
> **STORE AT TEMPERATURES ABOVE**
> **120 °F**

   (a) What is the most likely reason CPSC requires manufacturers to provide these hazard-warning statements to consumers?
   (b) What is the scientific basis for these hazard-warning statements?

2.17 The air within a tire on a fire truck is compressed at 32.0 psi_g at 68 °F (20 °C). Following use of the fire truck, the new air pressure in the tire is measured as 36.0 psi_g. What is the new temperature in degrees Celsius of the air within the tire?

Cryogenic Liquids

2.18 Use the data in Table 2.11 to determine whether carbon dioxide exists as a compressed gas or cryogenic liquid at 87 °F (27 °C) and 500 psi_g (3446 kPa).

2.19 Nitrogen is neither flammable nor toxic. When emergency-response crews respond to an incident involving a leak or spill of cryogenic nitrogen, what is the most likely reason they are encouraged to use self-contained breathing apparatus (SCBA) and wear total-encapsulating suits?

3 CHAPTER

Flammable Gases and Flammable Liquids

Large inventories of flammable gases and flammable liquids are often stored within industrial settings in connection with their use during the manufacturing and processing of commercial products. Some flammable gases and flammable liquids are also located within residential environments, albeit in lesser amounts. Given the commonplace nature of these hazardous materials, thousands of emergency incidents involving them occur annually throughout the nation.

Although we note the hazardous properties of individual flammable gases and flammable liquids elsewhere within this book, we note their common features in this chapter. We also examine the safety practices that OSHA and NFPA mandate when employers store flammable gases and flammable liquids in the workplace. Finally, we examine the procedures implemented by emergency responders when an incident involves the release of flammable gases and flammable liquids into the environment.

- Describe the concept of the following terms as they apply to a flammable liquid: lower and upper explosive limit, flammable range, flashpoint, fire point, and autoignition point.
- Describe the nature of a flammable liquid and combustible liquid as used by OSHA and the NFPA.
- Determine a liquid's OSHA/NFPA class from its flashpoint and boiling point.

3.1 Liquid Flammability

Even when exposed to an ignition source, there are unique minimum and maximum vapor concentrations below and above which a flammable liquid does not burn. These minimum and maximum concentrations are called the **lower explosive limit** and **upper explosive limit**, respectively.

For example, the flammable liquid xylene has a lower explosive limit and upper explosive limit in air of 1.1% and 7.0% by volume, respectively. The first limit tells us that a mixture having a concentration of less than 1.1% xylene vapor by volume in air will not burn. We say that it is too "lean" in xylene vapor. Similarly, the second limit tells us that a mixture containing more than 7.0% xylene by volume will not burn. We say that it is too "rich" in xylene vapor.

The **flammable range**, sometimes called the **explosive range**, of flammable gases and the vapors of flammable liquids is the set of concentrations of gases or vapors that lie between their upper and lower explosive limits. Thus, the flammable range of xylene vapor is 1.1% to 7.0% by volume.

A flammable gas or vapor burns only when the concentration of the gas or vapor is exposed to an ignition source *and* when its concentration in the air lies within its flammable range. Thus, all concentrations of xylene vapor between 1.1% and 7.0% by volume ignite when exposed to an ignition source. Explosive limits are always measured in ambient air, that is, 21% oxygen and 78% nitrogen by volume. Lower and upper explosive limits are irrelevant when this composition differs.

A gas or vapor often escapes from its confinement vessel at hatches, apertures, vents, and other openings. If the gas or vapor is flammable, it either ignites or disperses on release into the air. Immediate corrective action must be taken when the leaking gas or vapor is identified during an emergency-response action. Generally, this means by stopping or sealing the leak.

Three temperatures are used to describe the ability with which a flammable liquid burns: the flashpoint, fire point, and autoignition point. These temperatures are also characteristic properties of certain flammable gases and solids.

- *Flashpoint.* The **flashpoint** of a flammable liquid is the minimum temperature at which it gives off sufficient vapor to form an ignitable mixture with air across the surface of the liquid or within a test vessel. One procedure for measuring the flashpoint of a flammable liquid uses the specialized apparatus in Figure 3.1. The flashpoint of xylene has thus been determined to be 84 °F (29 °C).
- *Fire Point.* Above the flashpoint of a substance is a temperature at which self-sustained combustion occurs. This temperature is called the **fire point**. At its fire point, a flammable liquid gives off sufficient vapor so that continued combustion is maintained. The fire points of many liquids are approximately 30 to 50 °F (17 to 27 °C) higher than their flashpoints; for example, the fire point of xylene vapor has been measured to be 111 °F (44 °C), almost 30 °F (17 °C) higher than its flashpoint.
- *Autoignition Point.* As the temperature of the confined vapor of a flammable liquid further increases, a minimum temperature is attained at which self-sustained combustion occurs even in the absence of an ignition source. This is called the **autoignition point**. For example, the autoignition point of xylene is 924 °F (496 °C).

lower explosive limit
- The concentration of a gas or vapor in air below which a flame will not propagate when exposed to an ignition source (abbreviated LEL)

upper explosive limit
- The concentration of a gas or vapor in the air above which a flame will not propagate when exposed to an ignition source (abbreviated UEL)

flammable range
- The set of concentrations of gases or vapors that lie between a flammable substance's lower and upper explosive limits in air

flashpoint
- The minimum temperature at which the vapor of a liquid or solid ignites when exposed to sparks, flames, or other ignition sources

fire point
- The lowest temperature at which a flammable liquid emits vapors at a rate sufficient to support continuous combustion

autoignition point
- The minimum temperature of a liquid necessary to initiate the self-sustained combustion of its vapor in the absence of an ignition source

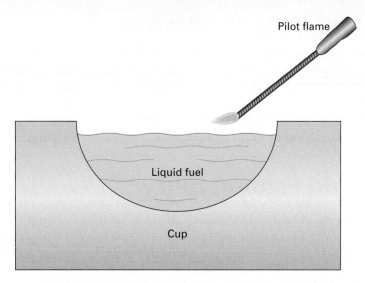

FIGURE 3.1 The test cup of a Pensky-Martens closed-cup flashpoint apparatus. In theory, an analyst determines the flashpoint of a flammable or combustible liquid by heating a sample of the liquid in the test cup while simultaneously applying a pilot flame within the vapor space above the liquid. When a flash of fire is momentarily observed, the analyst records the temperature of the liquid. This temperature is the flashpoint of the liquid.

SOLVED EXERCISE 3.1

The flashpoints of *tert*-butyl mercaptan and phenyl acetate are −15 °F and 80 °C, respectively. Other factors being equal, which liquid poses the greater risk of fire and explosion?

Solution: When two liquids have flashpoints that are widely different, the liquid having the lower flashpoint generally poses the greater risk of fire and explosion. To compare them, their flashpoints must be expressed on the same temperature scale. We calculate that a temperature of −15 °F is equivalent to −26 °C:

$$t(°C) = 5/9 \times \left[-15 - 32\right] = -26 \, °C$$

Because the flashpoint of *tert*-butyl mercaptan is substantially less than the flashpoint of phenyl acetate, *tert*-butyl mercaptan poses the greater risk of fire and explosion.

SOLVED EXERCISE 3.2

Use the following limited data to determine whether liquid aziridine or benzene poses the greater risk of fire and explosion, all other factors being equal:

	Aziridine	*Benzene*
Flashpoint	12 °F (−11 °C)	12 °F (−11 °C)
Vapor pressure at 68 °F (20 °C)	160.0 mmHg	75 mmHg

Solution: First, the data clearly reveal that *both* liquids pose a risk of fire and explosion. Because both have the same flashpoint, the liquid with the higher vapor pressure produces more vapor at a given temperature. It is this vapor that burns when exposed to an ignition source. Consequently, aziridine poses the greater risk of fire and explosion.

When a shipper intends to transport a flammable liquid having a flashpoint equal to or less than 61 °C on water, the Department of Transportation (DOT) requires the shipper to list the liquid's flashpoint on the accompanying waybill in degrees Celsius. What is the most likely reason DOT requires the flashpoint to be communicated in degrees Celsius rather than in degrees Fahrenheit?

Solution: Water shipments are generally bound for international destinations at which degrees Celsius have been used to measure temperature for decades. (As we noted in Section 2.2, the United States, Liberia, and Myanmar are the only nations that still use the English system of measurement, including the use of degrees Fahrenheit to measure temperature.) By requiring the posting of degrees Celsius on a waybill (and not degrees Fahrenheit), most emergency responders are provided with flashpoint information in the format with which they are most familiar. Confusion is also avoided by requiring a single means of communicating flashpoint information.

The flashpoints, fire points, and autoignition points of the most commonly used liquids are compiled in the scientific literature. Emergency responders should identify an easily accessible means (e.g., the Internet) to identify these values prior to their need. Emergency responders should always assume that the ignition of a flammable liquid is imminent when the temperature of the surroundings exceeds the liquid's flashpoint. In a comparison of the risk of fire and explosion for multiple flammable liquids, the greatest risk is potentially posed by the liquid that possesses the lowest flashpoint, ignoring all other factors.

3.1-A OSHA/NFPA DEFINITION OF A FLAMMABLE LIQUID AND A COMBUSTIBLE LIQUID

flammable liquid
■ For purposes of OSHA regulations and NFPA standards, any liquid having a flashpoint below 100 °F (37.8 °C), unless it is a liquid mixture having 99% or more of the volume of its components with flashpoints of 100 °F (37.8 °C) or greater

In the system used by both OSHA and NFPA a liquid is called a **flammable liquid**, or a *class I flammable liquid*, if its flashpoint is below 100 °F (37.8 °C), unless it is a liquid mixture having 99% or more of the volume of its components with flashpoints of 100 °F (37.8 °C) or greater. OSHA and NFPA recognize the following three classes of flammable liquids:

■ A *class IA flammable liquid* is a liquid with a flashpoint below 73 °F (22.8 °C) and a boiling point below 100 °F (37.8 °C). An example of a class IA liquid is *n*-pentane, since its flashpoint and boiling point are −56 °F (−49 °C) and 97 °F (36 °C), respectively.
■ A *class IB flammable liquid* is a liquid with a flashpoint below 73 °F (22.8 °C) and a boiling point at or above 100 °F (37.8 °C). An example of a class IB liquid is acetone (Section 13.5-C), since its flashpoint and boiling point are 0 °F (−18 °C) and 133 °F (56 °C), respectively.
■ A *class IC flammable liquid* is a liquid with a flashpoint at or above 73 °F (22.8 °C) and below 100 °F (37.8 °C). An example of a class IC liquid is turpentine, since its flashpoint lies in the range 95 to 102 °F (35 to 39 °C).

combustible liquid
■ For purposes of OSHA
regulations and NFPA
standards a liquid hav-
ing a flashpoint at or
above 100 °F (37.8 °C)

OSHA and NFPA also recognize a **combustible liquid** as any liquid having a flashpoint at or above 100 °F (37.8 °C). There are three types of combustible liquids:

- A *class II combustible liquid* is a liquid with a flashpoint at or above 100 °F (37.8 °C) but below 140 °F (60 °C), other than a liquid mixture having 99% or more of the volume of its components with flashpoints equal to or greater than 200 °F (93.3 °C). An example of a class II combustible liquid is acetic acid (Section 8.12), since its flashpoint is 109 °F (43 °C).
- A *class IIIA combustible liquid* is a liquid with a flashpoint at or above 140 °F (60 °C) but below 200 °F (93.3 °C). An example of a class IIIA liquid is creosote oil (Section 7.7-C), since its flashpoint ranges from 165 to 185 °F (74 to 85 °C).
- A *class IIIB combustible liquid* is a liquid having a flashpoint at or above 200 °F (93.3 °C). An example of a class IIIB liquid is ethylene glycol (Section 13.2-G), a common antifreeze agent, since its flashpoint is 232 °F (111 °C).

The flowchart in Figure 3.2 is useful for classifying flammable and combustible liquids.

SOLVED EXERCISE 3.4

Cyclohexanone is a liquid with a boiling point and a flashpoint of 313 °F (156 °C) and 146 °F (63 °C), respectively. Determine whether OSHA and NFPA classify cyclohexanone as a flammable or combustible liquid.

Solution: OSHA and NFPA classify a liquid with a flashpoint at or above 140 °F (60 °C) but below 200 °F (93.3 °C) as a class IIIA combustible liquid. Because cyclohexanone has a flashpoint of 146 °F (63 °C), OSHA and NFPA classify it as a class IIIA combustible liquid.

When responding to an emergency involving the release of a flammable gas or vapor, the first-on-the-scene responder often determines the concentration of the substance in the air with a portable combustible gas monitor or sensor like the type shown in Figure 3.3. Using this device, emergency-response crews can readily estimate the likelihood that a spark, flame, or other ignition source will ignite the flammable gas or vapor. When the identity of an ignition source is evident, the responders must eliminate it.

A reading on a combustible gas monitor or sensor must always be used cautiously at scenes where a flammable gas or vapor is released. Because wind currents and other factors interplay, readings must be taken frequently, as the meter response can change from one moment to the next. These scenes are always unsafe for emergency responders even under the best of conditions. When a combustible gas monitor or sensor is unavailable for their use, they should respond to the emergency only from a distance.

3.1-B "EMPTIED" TANKS

When a flammable liquid is stored within a container or tank, its vapor evolves into the headspace above it and mixes with the confined air. When the liquid is removed, some vapor always remains within the container or storage tank. This ostensibly emptied tank or container can pose a risk to health and safety under the following circumstances:

- It is especially hazardous to conduct cutting or welding operations on a tank in which a flammable liquid had been stored, unless specific actions to completely discharge the liquid's vapor from the tank were previously implemented.

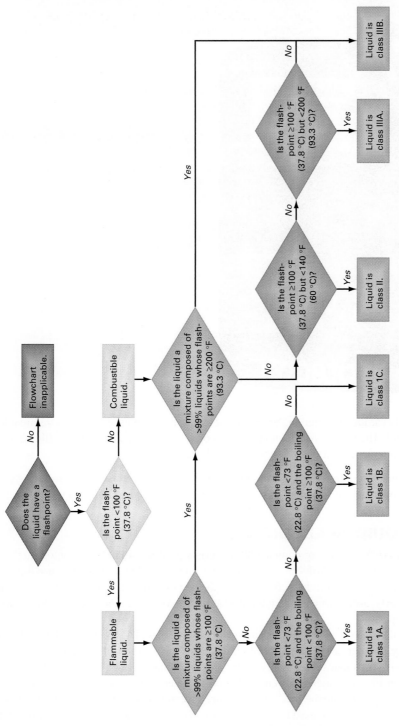

FIGURE 3.2 A flowchart from which the OSHA/NFPA classes of flammable and combustible liquids may be readily established based on the flashpoints and boiling points of unique substances and their liquid mixtures.

FIGURE 3.3 This personal confined-space entry monitor is called the Exotector®. Before an air–gas mixture reaches a concentration at which it will ignite, the Exotector® triggers an audible and visual alarm. The threshold for the alarm is generally set at 10% of the lower explosive limit of the flammable gas. *(Courtesy of GfG Instrumentation, Inc., Ann Arbor, Michigan.)*

■ It is also hazardous to enter an "emptied" tank that had been used to store a flammable or toxic liquid. The inhalation of a nontoxic gas or vapor may cause dizziness, illness, or suffocation, but the inhalation of a toxic gas can be fatal.

For this combination of reasons, precautions should always be exercised when handling or working in or around a tank or container in which a flammable or toxic liquid was once stored.

3.1-C LIQUID ACCELERANTS

accelerant
■ A flammable or combustible material used to initiate or promote fire

Accelerants are flammable or combustible materials used to initiate and promote fire in either a beneficial or detrimental fashion. On the positive side, firefighters sometimes use liquid accelerants to put a controlled burn into effect; but on the negative side, arsonists intent on committing a crime formerly used them to intentionally initiate a fire or accelerate the rate at which it destroyed property.

Because they have low flashpoints and readily ignite, class IA, IB, and IC flammable liquids were formerly popular accelerants used by arsonists. The ignition of relatively small amounts initiated massive fires that subsequently destroyed entire buildings.

After such fires investigators were called on to determine whether an arsonist had used a flammable liquid to initiate the fire. Typically, fire investigators first searched for an area where an accelerant may have been used to initiate the fire. Then, by collecting samples of fire debris in the immediate scene and submitting them to a chemical laboratory for analysis, they linked the cause of the fire with an arsonist's use of a flammable liquid.

Contemporary arsonists have learned to avoid the use of flammable liquids to initiate fires. They now choose to use accelerants that fire investigators cannot easily link with their crimes. These are often combustible materials such as items made of Styrofoam, wicker, or cardboard. Massive fires also result when large amounts of these materials ignite, but when samples of fire debris are subsequently collected and submitted to chemical analysis, the data cannot be easily linked with the accelerant.

- Describe the concept of ignitability as it is used by EPA in RCRA regulations.
- Determine whether a chemical waste exhibits the RCRA characteristic of ignitability.

3.2 RCRA Characteristic of Ignitability

As previously noted in Section 1.5-C, the Resource Conservation and Recovery Act (RCRA) gives EPA the authority to regulate the treatment, storage, and disposal of certain materials called hazardous wastes. Although EPA denotes some hazardous wastes by their chemical names, the nature of other hazardous wastes is established by determining whether they exhibit certain characteristics, one of which is relevant here.

A waste exhibits the RCRA characteristic of **ignitability** if it is any one of the following:

- A *flammable gas*, as this term is defined in the DOT regulations (see the glossary)
- A liquid that possesses a flashpoint equal to or less than 140 °F (60 °C), other than an aqueous solution (with water as the solvent) containing less than 24% alcohol by volume, when tested through use of a Pensky-Martens closed-cup tester like that shown in Figure 3.1
- A material other than a liquid that is capable "of causing fire by friction, absorption of moisture, or spontaneous chemical changes and, when ignited, burns so vigorously and persistently as to create a hazard"
- A *flammable solid*, as this term is defined in the DOT regulations (see the glossary)
- An *oxidizer*, as this term is defined in the DOT regulations (see the glossary)

A waste exhibiting the RCRA characteristic of ignitability is a hazardous waste and its treatment, storage, and disposal are subject to EPA's regulations. A waste exhibiting the characteristic of ignitability is assigned the hazardous waste number D001.

ignitability
- For purposes of RCRA regulations, a characteristic of any of the following wastes: a liquid, a sample of which has a flashpoint equal to or less than 140 °F (60 °C) when determined through use of a Pensky-Martens closed-cup tester; a solid capable of causing fire through friction, absorption of moisture, or spontaneous chemical changes and that burns vigorously and persistently; an oxidizer; or an ignitable compressed gas (40 C.F.R. §261.21)

SOLVED EXERCISE 3.5

Three terms are routinely used to describe the potential for a liquid to burn: flammable, combustible, and ignitable. What are the differences among them?

Solution: Although all three terms are commonly used to denote the relative ease with which a liquid burns, these terms have specific meanings in regulations enacted pursuant to federal laws as follows:

- OSHA (and NFPA) define a *flammable liquid* as a liquid having a flashpoint below 100 °F (37.8 °C), unless it is a liquid mixture having 99% or more of the volume of its components with flashpoints of 100 °F (37.8 °C) or greater.
- OSHA defines a *combustible liquid* as any liquid having a flashpoint at or above 100 °F (37.8 °C).
- In the RCRA regulations, EPA characterizes an *ignitable liquid* as a liquid waste, other than an aqueous alcoholic solution containing less than 24% alcohol by volume, that possesses a flashpoint equal to or less than 140 °F (60 °C).

cylinder
■ For purposes of DOT regulations, a pressure vessel having a circular cross section and designed to safely store and transport non–liquefied and liquefied compressed gases at pressures greater than 40 psi$_a$ (275.6 kPa)

tank
■ For purposes of DOT regulations, a container consisting of a shell and heads that form a pressure-tight vessel having openings designed to accept pressure-tight fittings or closures but excludes any appurtenances, reinforcements, fittings, or closures

compressed gas
■ A gas confined within a vessel under pressure

liquefied compressed gas
■ For purposes of DOT regulations, a gas that when packaged under pressure for transportation is partially liquid at temperatures above −58 °F (−50 °C)

stationary tank
■ Packaging designed primarily to be installed in a stationary position and not intended for loading, unloading, or attachment to a transport vehicle

nesting
■ The method of securing gas cylinders upright in a tight mass using a contiguous three-point contact system wherein the cylinders within a group have a minimum of three points of contact with other cylinders, walls, or bracing

PERFORMANCE GOALS FOR SECTION 3.3:

■ Distinguish between a compressed gas and a liquefied compressed gas.
■ Describe generally the design features of a cylinder intended for the storage and transportation of a gas under pressure.
■ Describe the nature of the information that DOT requires the shipper or carrier to mark on gas cylinders.
■ Identify the nature of the information DOT requires manufacturers and owners to mark on rail tankcars used to transport substances under pressure, including compressed gases and liquefied compressed gases.
■ Describe the general safety practices that should be implemented when handling, storing, and transporting compressed gases and liquefied compressed gases.

3.3 Storing and Transporting Compressed Gases

During storage and shipment, a gas confined under pressure within a steel **cylinder** or **tank** is called a **compressed gas**. Examples are hydrogen, oxygen, and nitrogen. When a confined gas liquefies under the application of moderate pressure, it is called a **liquefied compressed gas**. Examples are methane and propane.

A compressed gas or liquid may be either nonflammable or flammable. Oxygen and nitrogen are nonflammable gases, whereas hydrogen, methane, and propane are flammable gases. The following discussions are pertinent to both types.

3.3-A COMPRESSED GASES IN STORAGE

Bulk volumes of compressed gases are ordinarily stored in spherical **stationary tanks**, whereas nonbulk volumes are stored in the steel cylinders that were used to transport them from their place of purchase. In the latter case, gas cylinders are likely to be stored at one or more locations within cabinets. For example, at hospitals and clinics, cylinders of medical gases are generally encountered within specially identified cabinets. The storage of gas cylinders within such cabinets provides a means for them to be safely segregated from other medical supplies.

Employers should exercise precautions to avoid damaging gas cylinders during their storage, transportation, and use. Experts recommend maintaining them in an upright position and strapping, chaining, bracing, or otherwise securing them in a fixed position like that shown in Figure 3.4. OSHA requires employers to post the following warning sign within the workplace to reduce or eliminate the physical hazards associated with the use of gas cylinders:

> **WARNING**
>
> **CYLINDERS MUST BE CHAINED AT ALL TIMES**

Once the cylinders have been secured, their valve-protection hoods should be kept in place. The hoods should be removed only when use of the gas is required.

When multiple gas cylinders are stored in near proximity, experts recommend that they be nested. **Nesting** refers to a method of storage in which the gas cylinders are situated upright in a tight-mass group using a contiguous three-point contact system. Each cylinder within a group has a minimum of three points of contact with other cylinders, walls, or bracing.

When it is necessary to move cylinders from place to place, they should first be secured or restrained by straps or chains to a cylinder cart having wheels. The cylinders should never be rolled or dragged, as such actions could seriously damage them. Dropping them could result in the catastrophic release of their contents.

Gas cylinders should always be stored in dedicated areas where they may be properly managed and protected. They should never be stored in public hallways. NFPA stipulates the following limitations on the number of gas cylinders allowed in the same storage area: three 10 in. × 50 in. (25 cm × 127 cm) cylinders containing flammable gases or oxygen; and three 4 in. × 15 in. (10 cm × 38 cm) cylinders containing toxic gases. NFPA also recommends storing cylinders containing oxygen and flammable gases in entirely different locations.

SOLVED EXERCISE 3.6

Why do experts recommend turning the valve on a gas cylinder to the closed position when the cylinder is essentially empty?

Solution: A residual amount of gas remains within any cylinder that has been essentially emptied. If the valve is turned to the open position, air may enter the tank and mix with the contents. This situation poses the risk of fire and explosion when the gas is flammable and exposed to an ignition source.

Cylinders containing flammable gases require special attention during storage. In particular, they should be properly grounded to prevent the build-up of static electricity. They should not be stored in the same area as cylinders containing oxidizers like oxygen and chlorine, and precaution should be exercised to ensure the absence of all ignition sources in the storage area. Smoking and the use of open flames and hot surfaces, cutting and welding, and all other potential sources of static, electrical, and mechanical sparks should be eliminated.

3.3-B CYLINDERS OF COMPRESSED GASES IN TRANSPORT

Nonbulk volumes of compressed gases and liquefied compressed gases are generally shipped in cylinders. To ensure that they do not rupture while transported, DOT regulates a variety of features pertaining to their design and construction.

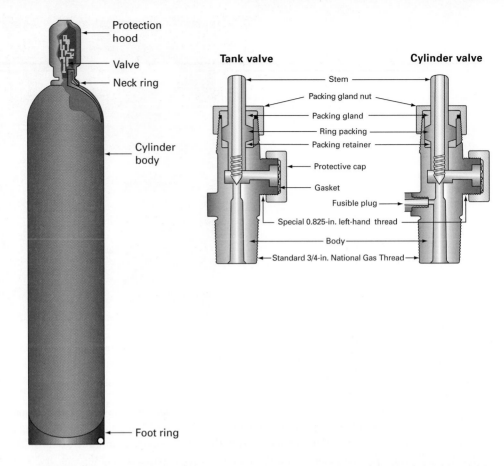

FIGURE 3.5 The design features of a DOT-approved steel cylinder used for containerization and transport of certain compressed gases. Located at the top of the cylinder is a single discharge valve in which a fusible plug is located within the valve body just below the seat. The plug is constructed from a specially forged bronze alloy that melts at a temperature ranging from 158 to 163 °F (70 to 73 °C).

fusible plug
■ A piece of low-melting metal inserted as a pressure-relief device in cylinders, cargo tanks, stationary tanks, and railroad tankcars used to store or transport a compressed gas or flammable liquid

service pressure
■ For purposes of DOT regulations, the pressure designated in units of psi$_g$ at 70 °F (21 °C) for each specific type of authorized compressed gas cylinder

A typical gas cylinder has the design features illustrated in Figure 3.5. An important DOT requirement is that nearly all cylinder types must be equipped with a **fusible plug** that melts when a cylinder is exposed to a temperature ranging from 158 to 163 °F (70 to 73 °C). The melting allows the contents to be slowly released into the environment, which prevents the cylinder from rupturing.

DOT regulates the design and construction of gas cylinders so they will maintain their integrity while their contents are under pressure and exposed to the range of temperatures likely to be encountered during transportation. DOT assigns a unique pressure at 70 °F (21 °C) to each type of gas cylinder. It is called the cylinder's **service pressure**. DOT requires the service pressure to be marked on the exterior surface of the cylinder in units of psi$_g$. For instance, when the service pressure of a gas cylinder is 1800 psi$_g$ (1792 kPa), it is marked on the cylinder as DOT-3E1800. The DOT-3E indicates a type of cylinder that has been constructed in compliance with DOT's specifications for a 3E cylinder.

DOT requires manufacturers to determine the integrity of each type of gas cylinder when the temperature of the contents is increased from 70 °F (21 °C) to 130 °F (54 °C). DOT approves the cylinder for use only when the test results reveal that the pressure of its contents did not exceed five-fourths of the cylinder's service pressure.

DOT also regulates the amount of gas that manufacturers put into a cylinder. At 49 C.F.R. §§173.301 through 173.306, DOT identifies specific filling limits for compressed gases contained within cylinders.

DOT also requires gas cylinders containing nonbulk volumes of compressed gases to be marked and labeled prior to transportation. We visit these marking and labeling requirements in Chapter 6. Once they have been correctly marked and labeled, properly designed and constructed cylinders are frequently loaded onboard a truck or rail boxcar in which they are securely strapped or chained in a fixed position during transportation. Less commonly, large numbers of gas cylinders are secured within tube cars like that illustrated in Figure 3.6 for transportation by railroad.

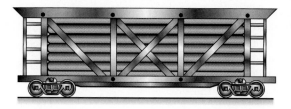

FIGURE 3.6 Large numbers of compressed gas cylinders are sometimes stacked within high-pressure tube cars designed for their transportation by railroad.

DOT-3A2000 is marked on a steel cylinder whose gauge pressure reads 1750 psi at 70 °F (21 °C). Is this cylinder likely to rupture during a fire if the temperature of its contents rises to 300 °F (149 °C)?

Solution: To ensure the integrity of a gas cylinder at elevated temperatures, DOT stipulates that the actual pressure of its contents cannot exceed five-fourths of the service pressure when the cylinder is subjected to a temperature of 130 °F (54 °C). The service pressure of a cylinder marked DOT-3A2000 is 2000 psi. If the gas manufacturer complied with the relevant DOT regulation, the integrity of the cylinder should be maintained as long as the pressure of the contents does not exceed 5/4 × 2000 psi, or 2500 psi, at a temperature of 130 °F (54 °C).

However, an internal pressure greater than 2500 psi is generated when the temperature of the gas cylinder exceeds 130 °F (54 °C). Consequently, a cylinder marked DOT-3A2000 is likely to rupture when heated to 300 °F (149 °C).

3.3-C COMPRESSED GASES DURING TRANSPORT IN BULK

Bulk volumes of compressed gases are transported in a variety of ways. They are sometimes contained within jumbo-sized tubes like those illustrated in Figure 3.7 and transported by public highway. More commonly, however, bulk volumes of compressed gases are shipped either within cargo tanks by public highway or tankcars by railroad. As illustrated in Figure 3.8, these **transport vehicles** are distinctively constructed with hemispherical heads and ends. They are used to ship not only compressed gases but all pressurized substances.

transport vehicle
■ A motor vehicle or rail car used for the transportation of cargo by any mode

DOT authorizes the bulk shipments of compressed gases only when their transport vessels have been constructed according to specified requirements, and the shipper or carrier has complied with allowable maximum filling limits, testing, maintenance, marking, placarding, and other relevant requirements. DOT's marking and placarding requirements are especially relevant to the training of emergency responders. The placarding regulations are noted in Chapter 6.

There are two types of marking requirements that apply to the shipment by railroad of *all* hazardous materials, including compressed gases. One applies to the tankcar manufacturer and the other to its owner.

Marking Requirements of Tankcar Manufacturers

DOT requires rail tankcar manufacturers to legibly mark certain information on the exterior surfaces of the tankcars. Included among this information is the tankcar's specification number. An example

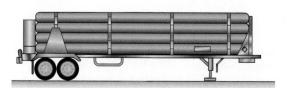

FIGURE 3.7 Bulk volumes of compressed gases and liquefied compressed gases are occasionally transported by public highway in these jumbo-sized tubes, which have been securely strapped side-by-side to the body of the van.

Chapter 3 Flammable Gases and Flammable Liquids **97**

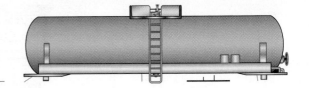

of a specification number for a rail tankcar is DOT-105A500W. The DOT-105A denotes a unique type of tankcar that has been constructed in compliance with DOT's specifications for an A tankcar; the 500 denotes the maximum pressure in psi_g under which the contents may be safely confined; and the W denotes that the tank was built using fusion-welded construction methods.

DOT also requires manufacturers to mark on the tankcar its maximum weight in pounds and kilograms to which it may be filled, its weight when empty, and the month and year of the tankcar's construction. DOT requires this information to be imprinted on the left side of the tankcar as an observer faces the tankcar, generally below the owner's reporting mark (see the following section).

Marking Requirements of Tankcar Owners

DOT also obligates the owner to legibly imprint certain information on the exterior surface of both sides and both ends of a tankcar used to ship a hazardous material. This includes a **reporting mark**, which is a sequence of several identification letters followed by several digits. The letters are assigned by the American Association of Railroads[*] to freight carriers who operate within North America, whereas the digits unique identify a specific tankcar.

When a compressed gas is shipped, DOT generally requires the name of the gas to be imprinted on the right side of a rail tankcar as an observer faces the tankcar. In Section 6.6-B, we note that DOT requires the names of certain other substances aside from compressed gases to be imprinted on the surface of a tankcar, yet there are also many other hazardous substances for which this imprinting requirement does not apply.

Notwithstanding this limitation, when emergency responders encounter a reporting mark on the exterior surface of a tankcar during a transportation mishap, they can rapidly determine the tankcar's owner by consulting a computer compilation.[†] Then, in combination with the information on the waybill that accompanies the shipment, they may easily identify the tankcar's contents.

More commonly, however, emergency responders provide the reporting mark (letters and the digits) to CHEMTREC personnel, who contact the tankcar's owner directly to determine the tank's contents and an appropriate action. CHEMTREC provides this information to the emergency responders.

DOT also requires tankcar owners to test the integrity of their vessels and safety valves and to imprint the test results of these tests on both sides and both ends of the tank.

This combination of marking requirements is illustrated by the following example:

reporting mark
■ A sequence of two to six identification letters assigned by the American Association of Railroads to rail carriers who operate in North America

DOT-105A500W		
UTLX901794		
LD LMT	181700 LB	82400 KG
LT WT	81300 LB	36900 KG
NEW	03 95	

VINYL CHLORIDE	
SAFETY VALVE	75 LB
TESTED	75 LB
TEST DUE	05 96
TANK	100 LB
TESTED	100 LB
TEST DUE	05 96

[*]The American Association of Railroads (AAR) is an industry trade group whose members include representatives of the major freight railroads of North America (Canada, United States, and Mexico).
[†]For example, the reporting marks of shippers can be obtained by accessing the following site on the Internet: http://en.wikipedia.org/wiki/List_of_AAR_reporting_marks:_A.

UTLX901794 is the reporting mark of Union Tank Car Company, which leases its rail tankcars to shippers of hazardous materials. In this example, the railcar contains the flammable and poisonous gas vinyl chloride.

PERFORMANCE GOALS FOR SECTION 3.4:

- Identify the common hazards associated with the presence of compressed gases in cylinders, stationary storage tanks, and transport vessels.

3.4 General Hazards of Compressed Gases

We now examine the general hazards of compressed gases when their cylinders, stationary storage tanks, and transport vessels are involved in emergency-response efforts.

3.4-A COMPRESSED GAS CYLINDERS AND STATIONARY STORAGE TANKS

When a gas cylinder has been constructed and tested for compliance with DOT's specifications, it will not ordinarily rupture under the normal conditions of transport. Even when it is slowly heated to temperatures above ambient conditions, a small quantity of the contents slowly escapes into the environment through pressure-relief devices or fusible plugs, each of which prevents the cylinder from bursting.

However, when *rapidly* heated to temperatures exceeding approximately 130 °F (54 °C), a gas cylinder can readily burst and discharge its contents into the immediate environment. At such temperatures, the cylinder has clearly exceeded its authorized service pressure. The heated contents cannot discharge rapidly enough through the pressure-relief valves or fusible plugs to adequately reduce the internal pressure. If unsecured, the rupturing cylinder can travel as an active airborne missile and jettison from spot to spot, perhaps for hundreds of yards, until virtually all the contents have been released.

When a cylinder or tank containing a flammable gas initially ruptures, the concentration of the released gas is usually within its flammable range. As it mixes with the air the gas ignites explosively, typically engulfing the entire area in flames. Even when the contents are nonflammable, the immediate environment in which the rupture occurs is an extremely fearsome one. The immense force that is exerted on the surroundings is sufficient to shatter windows and cause other types of physical destruction within the immediate area. This force may also rupture eardrums and disembowel or otherwise injure those who are struck by flying debris.

A stationary storage tank may also be used for confining a compressed gas. When the tank is situated inside a building, sensing devices electronically linked to exhaust fans are often installed to provide maximum protection for workers against potential exposure to a gas that may inadvertently be released into the working environment. Once activated, the fans evacuate escaping gases from the enclosure into the outside atmosphere at very high speeds, thereby minimizing the possibility of their concentrating indoors. When a tank is used to store gases with vapor densities greater than 1.0, these sensing devices and fans are positioned near the floors of the buildings in which the tank is located, but when a tank is used to store gases with vapor densities less than 1.0, the sensing devices and fans are positioned near the ceilings.

SOLVED EXERCISE 3.8

A gas cylinder contains compressed nitrogen, a nonflammable gas, at 2000 psi$_g$ and 65 °F. During a fire, the cylinder is engulfed in flames, and the temperature of the cylinder and its contents rapidly rises to 350 °F.

    (a) What is the gauge pressure on the cylinder at 350 °F?

    (b) Is the cylinder likely to withstand this elevated temperature?

Solution:

    (a) The pressure of the nitrogen confined within the cylinder may be calculated using the combined gas law previously noted in Section 2.12-C. Because a steel cylinder is essentially a constant-volume container, $V_1 = V_2$.

First, the initial absolute pressure of the nitrogen at 65 °F in the cylinder is calculated as follows:

$$2000 \text{ psi} + 14.7 \text{ psi} = 2015 \text{ psi}_a.$$

This is P_1.

Then, the initial and final Fahrenheit temperatures are converted into their equivalent readings on the Rankine scale by adding 460 °F to each temperature:

$$T(°R) = 65 °F + 460 °F = 525 °R$$
$$T(°R) = 350 °F + 460 °F = 810 °R$$

These temperatures are T_1 and T_2, respectively.

The absolute pressure of nitrogen at 350 °F, P_2, is then be calculated as follows:

$$2015 \text{ psi}_a \times \frac{810 \text{ °R}}{525 \text{ °R}} = 3109 \text{ psi}_a$$

The gauge pressure is then calculated by subtracting 14.7 psi from the absolute pressure:

$$3109 \text{ psi} - 14.7 \text{ psi} = 3094 \text{ psi}_g$$

    (b) The internal strain on the walls of the steel cylinder has increased from 2000 psi at 65 °F to 3094 psi at 350 °F. To relieve this excessive strain, the cylinder is likely to rupture and expel the contents into the immediate environment. Although nitrogen does not burn, its release from a ruptured cylinder generates an immense force on the surroundings that may cause considerable damage.

SOLVED EXERCISE 3.9

Methyl formate is a liquid that boils at 90 °F (32 °C) and has a flashpoint of −2 °F (−19 °C). Because it has a pleasant odor, methyl formate is often a component of commercial air fumigants. A company that manufactures air fumigants has constructed an indoor storage room for flammable liquids from fire-resistant materials and equipped it with a sprinkling system. What is the maximum number of 55-gal drums containing methyl formate that OSHA and NFPA allow to be stored per pile within the room?

Solution: From the boiling point and flashpoint data, we determine that methyl formate is a class IA liquid. We also determine from Table 3.3 that when the storage is protected, the maximum number of 55-gal drums of methyl formate that OSHA and NFPA allow to be stored per pile within an indoor storage room on either the ground or upper floors is 50. (When the storage is unprotected, the maximum number of 55-gal drums of methyl formate that OSHA and NFPA allow to be stored per pile in an indoor storage room is only 12.)

3.4-B COMPRESSED GAS TRANSPORT VESSELS

As noted in Section 3.4-A, nonbulk quantities of compressed gases and liquefied compressed gases are routinely transported in cylinders, whereas bulk quantities are generally transported within **cargo tanks** and rail tankcars.

A transport vessel containing a liquefied compressed gas is particularly dangerous. The liquefied compressed gas exists within the vessel as two phases: a liquid that settles to the bottom of the vessel and a gas that coexists in the headspace above it. It is this liquid that drips from valves, fittings, or openings. Its vapor is especially hazardous if the gas is flammable or toxic.

PERFORMANCE GOALS FOR SECTION 3.5:

- Use Figures 3.9, 3.10, 3.11 and 3.12 to describe the response actions to be executed when a compressed gas has been released into the environment.

3.5 Responding to Incidents Involving the Release of Flammable Gases

Countless occasions involving the release of compressed gases from cylinders, storage tanks, and transport vessels have occurred. For first-on-the-scene responders, it is always prudent to acknowledge immediately that these vessels are likely to rupture when they are exposed to intense heat. In fact, emergency responders are often called to scenes where cylinders and tanks have already ruptured.

Specific procedures to be implemented during emergency-response actions involving compressed gases in cylinders and tanks are provided in Figures 3.9, 3.10, 3.11, and 3.12. Later, in Section 10.15, we give special attention to emergency-response actions involving toxic compressed gases.

As an emergency-response scene involving a flammable gas is being assessed, certain actions should be quickly executed. The following list, although not intended to be exhaustive, is illustrative:

- If the response action involves the storage of a compressed gas within a stationary tank, personnel should use the information posted on an NFPA hazard diamond to ascertain the degree of hazard relating to fire, health, and chemical reactivity. It is especially important to determine whether the tank contents are flammable or toxic.
- When there is a fire in an area where cylinders or tanks containing compressed gases are located, unmanned monitors should be situated to cool them with direct streams of water—but only when the monitors may be placed without risk to personnel.
- When a small leak from a cylinder is discovered, a water fog should be used to disperse the gas or vapor.
- When personnel decide to move a leaking cylinder containing a flammable compressed gas, a portable combustible gas monitor or similar device may be used to ascertain the concentration of the gas within the immediate area. No attempt to move a leaking cylinder should be made

cargo tank
- For purposes of DOT regulations, bulk packaging that (a) is a tank intended primarily for the carriage of liquids or gases and includes appurtenances, reinforcements, fittings, and closures; (b) is permanently attached to or forms a part of a motor vehicle, or is not permanently attached to a motor vehicle but which, by reason of its size, construction, or attachment to a motor vehicle, is loaded or unloaded without being removed from the motor vehicle; (c) is not fabricated under a specification for cylinders, intermediate bulk containers, multiunit tankcar tanks, portable cars, or tankcars

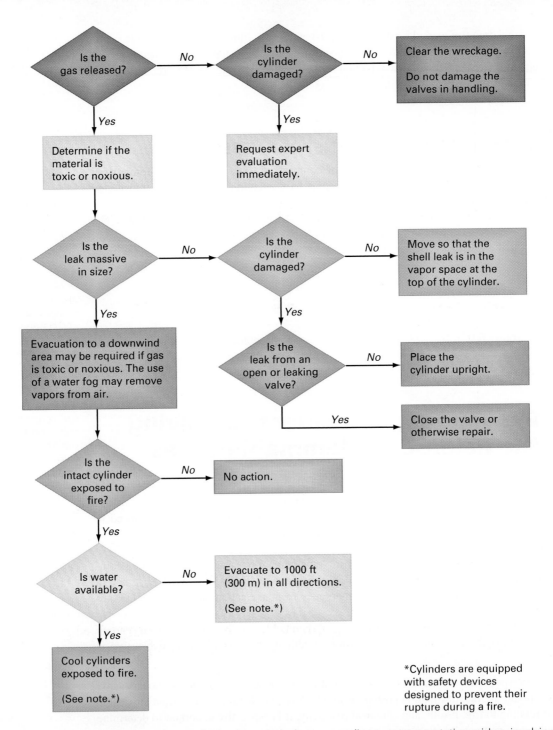

FIGURE 3.9 The recommended procedures to be implemented when responding to a transportation mishap involving the release of a nonflammable compressed gas from its packaging. *Adapted with permission of the American Society for Testing and Materials, from a figure in ASTM STP 825,* A Guide to the Safe Handling of Hazardous Materials Accidents, *second edition. Copyright © 1990, American Society for Testing and Materials.*

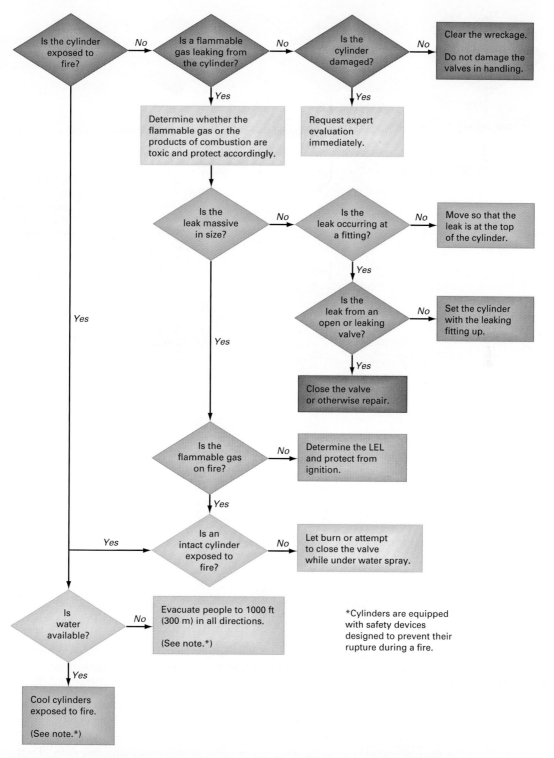

FIGURE 3.10 The recommended procedures to be implemented when responding to a transportation mishap involving the release of a flammable compressed gas from nonbulk packaging. *Adapted with permission of the American Society for Testing and Materials, from a figure in ASTM STP 825,* A Guide to the Safe Handling of Hazardous Materials Accidents, *second edition. Copyright © 1990, American Society for Testing and Materials.*

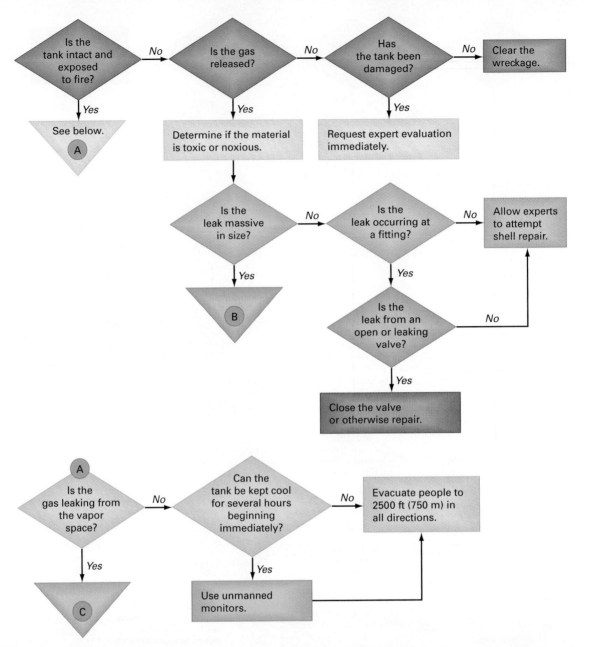

FIGURE 3.11 The recommended procedures to be implemented when responding to a transportation mishap involving the release of a nonflammable compressed gas from bulk packaging such as a cargo tank. *Adapted with permission of the American Society for Testing and Materials, from a figure in ASTM STP 825,* A Guide to the Safe Handling of Hazardous Materials Accidents, *second edition. Copyright © 1990, American Society for Testing and Materials.*

if the concentration of an escaping gas or vapor is within its flammable range. Furthermore, no attempt to stop the escape of the gas from its cylinder should be made unless the concentration can be reduced to less than its lower explosive limit. This may be accomplished by dispersing the gas or vapor with a water fog.

■ A slowly leaking cylinder containing a flammable compressed gas may be moved into an open isolated area, but *only* when the gas has not ignited and when the operation can be safely conducted. When a leaking cylinder is moved, it is prudent to wear a total-encapsulating suit and

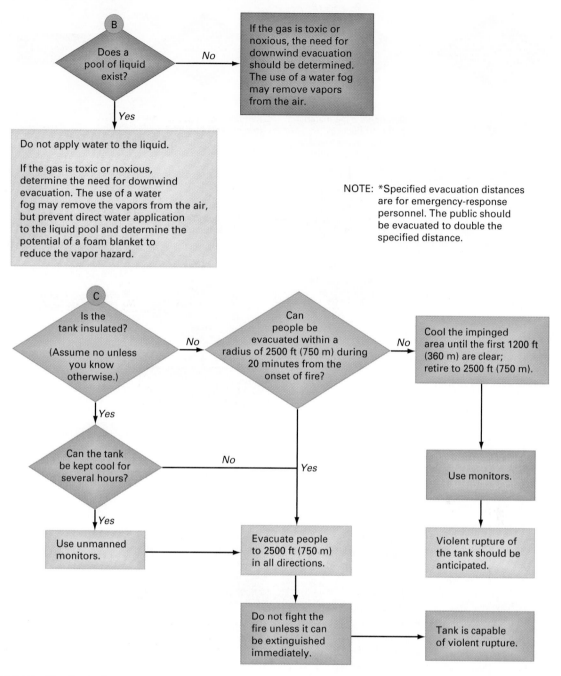

FIGURE 3.11 *(Continued)*

use self-contained breathing apparatus. The latter action is warranted, since the exposure to flammable compressed gases may cause responders to experience dizziness, illness, or suffocation.

■ When a cylinder is moved into an open area, it should be rotated so that the point of leakage becomes the uppermost part of the cylinder. Although emergency responders may elect to allow the gas to slowly escape into the air, the leak may often be stopped by closing the valve or tightening the packing gland nut.

■ Experience indicates that vessels generally rupture at a tank's head or end, near the seam where the head or end was welded to the tank's body. For this reason, emergency responders

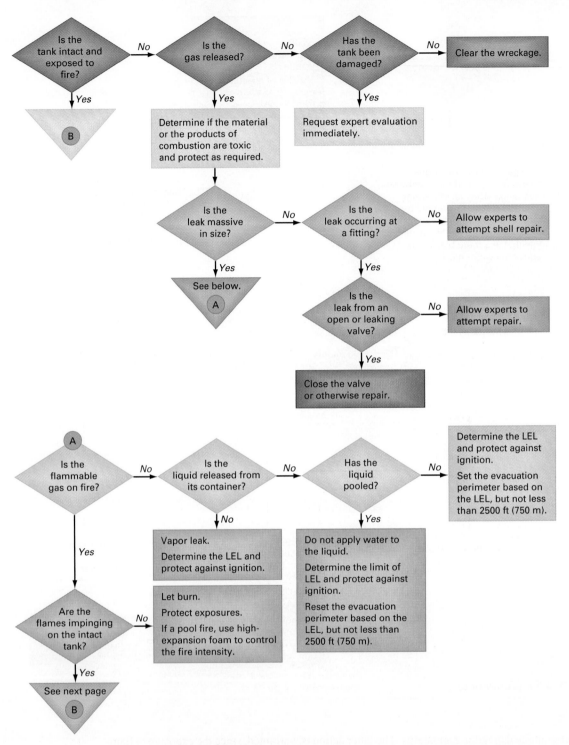

FIGURE 3.12 The recommended procedures to be implemented when responding to a transportation mishap involving the release of a flammable compressed gas from bulk packaging such as a cargo tank. *Adapted with permission of the American Society for Testing and Materials, from a figure in ASTM STP 825,* A Guide to the Safe Handling of Hazardous Materials Accidents, *second edition. Copyright © 1990, American Society for Testing and Materials.*

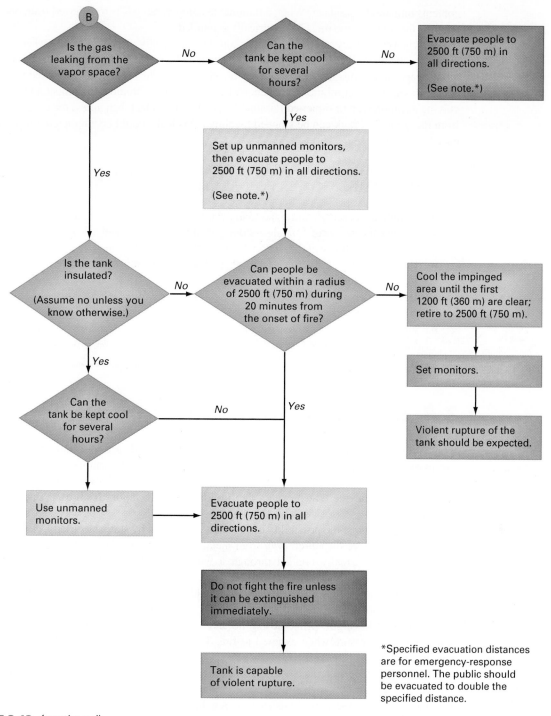

FIGURE 3.12 *(continued)*

should *never* approach the head or end of a cylinder or tank containing a flammable compressed gas during an ongoing fire.

■ Because a compressed gas exerts pressure equally on all internal points of a cylinder or tank in which it is confined, an overturned, nonleaking rail tankcar may be safely uprighted, when practical, without first unloading its contents. This action should be taken *only* by experienced personnel. Sometimes, to lighten a load, responders decide to first transfer a portion of the contents into another tankcar. When a flammable compressed gas is transferred from one tank to another, the transfer line must be electrically grounded.

■ When emergency-response personnel must be physically present in an area where a cylinder or tank containing a flammable compressed or liquefied gas has been identified, they should wear total-encapsulating suits, use self-contained breathing apparatus, and work downwind from it.

■ Based on the combined experience of fire personnel, attempts to combat a major fire involving a compressed or liquefied flammable gas should not be taken *unless* the escape of the gas from its cylinder or tank can be stopped. Sealing a gas leak should be attempted only by experienced personnel.

PERFORMANCE GOALS FOR SECTION 3.6:

■ Identify the practices for safely storing and handling flammable liquids.
■ Identify the practices for safely storing flammable liquids within containers, portable tanks, and stationary tanks within the workplace.
■ Distinguish among the three types of stationary tanks: atmospheric tanks, low-pressure tanks, and pressure vessels.
■ Describe generally the multiple types of vessels in which bulk volumes of flammable liquids are stored.
■ Identify OSHA/NFPA's minimum design and construction standards for bulk storage tanks that are intended to hold flammable liquids.

3.6 Storing Flammable Liquids

At manufacturing and production facilities, flammable liquids are routinely stored for future use in safety cans, bottles, metal cans, metal drums, barrels, portable tanks, or stationary tanks. When they are not in use, these containers and tanks should always be closed by means of a lid or other device that prevents the liquids and their vapors from escaping at ordinary temperatures.

To protect against the possibility of fires that are ignited by static electricity, class I flammable liquids should be dispensed *only* from metal containers or tanks that have been positively grounded and electrically bonded through a wire attached to both containers or through a common ground system.

safety can
■ For purposes of OSHA regulations, an approved, closed container of not more than 5-gal (18.9-L) capacity, having a flash-arresting screen, spring-closing lid and spout cover, and so designed that it will safely relieve internal pressure when subjected to fire exposure

As a safety precaution, flammable liquids should be transferred into a **safety can** from which they are then dispensed as needed. A safety can is an approved container of not more than 5-gal (19-L) capacity, having a spring-closing lid and spout cover and so designed that it will safely relieve internal pressure when exposed to fire. Two examples of OSHA/NFPA-approved safety cans are shown in Figure 3.13.

To minimize or eliminate employee exposure to flammable liquids, OSHA and NFPA regulate the manner by which employers store flammable liquids within the workplace. We next examine certain aspects of these storage regulations that are most relevant to fire-service personnel.*

When employers choose to store flammable liquids within an inside room, they must comply with the size and construction allowances on the volume of the flammable liquid listed in Table 3.1. The storage room must be constructed in compliance with relevant NFPA standards

---

*These regulations do not apply to the following circumstances: liquids stored in bulk plants, service stations, processing plants, refineries, or distilleries; a liquid used in vehicle tanks, portable engines, or stationary engines; when the liquid is a paint, oil, or varnish used for maintenance; a liquid beverage stored in containers having a capacity of less than 1 gal (3.8 L); and liquid medicines, beverages, foodstuffs, cosmetics, and other common consumer items.

FIGURE 3.13 These safety cans have been constructed with leak-tight gasketed lids, dual-density flame arresters, and other requirements identified at 29 C.F.R. §1910.106 and NFPA Code 30, *Flammable and Combustible Liquids Code.* To prevent their rupture in the event of a fire, they slowly vent their confined vapors between 3 and 5 psi$_g$. The OSHA regulation at 29 C.F.R. §1910.144 requires safety cans and other portable containers of flammable liquids having a flashpoint at or below 80 °F (27 °C) to be painted red with some additional clearly visible identification either in the form of a yellow band around the can or the name of the contents conspicuously stenciled or painted on the can in yellow. (*Courtesy of Justrite Manufacturing Company, Des Plaines, Illinois.*)

TABLE 3.1	**Storage of Flammable Liquids Within Inside Rooms**[a]				
		Maximum Size		**Total Allowable Volumes**	
FIRE-PROTECTION SYSTEM NEEDED[b]	**FIRE RESISTANCE**[c]	**ft²**	**m²**	**gal/ft²/ floor area**	**L/m²/ floor area**
Yes	2 hr	500	46.5	10	410
No	2 hr	500	46.5	5	200
Yes	1 hr	150	13.9	4	160
No	1 hr	150	13.9	2	80

[a]29 C.F.R. §1910.106(d), Table H-13.
[b]The fire-protection system must be sprinkler, water spray, carbon dioxide, or a similarly approved system.
[c]The fire-resistance rating of a construction material is determined by exposing it to standardized gas burners within a furnace. The rating is expressed in units of time. Materials having a fire-resistance rating of 1 hr and 2 hr have met the furnace test without failure for 1 hour and 2 hours, respectively.

FIGURE 3.14 This cabinet is specifically marked to identify its sole utilization for the storage of flammable liquids. It is constructed from 18-gauge sheet iron, the doors are equipped with a latching device, and a 2-in. liquid-tight sill is provided. (*Courtesy of A & A Sheet Metal Products, Inc., SE-Cur-All® Products, LaPorte, Indiana*)

that include the use of liquid-tight sills or ramps at least 4 in. (10 cm) in height, approved self-closing doors, and a liquid-tight seal between the walls and floor.

When employers choose to store flammable liquids in the workplace, OSHA and NFPA require that they provide either an inside room used solely for storage purposes or a storage cabinet similar to the one illustrated in Figure 3.14. A flammable liquid storage cabinet approved by OSHA and NFPA is designed and constructed to prevent ignition of the contents by exposure to external fires or other sources of ignition. The manufacturer usually equips it with vents. Although they are available, these vents should remain plugged during use unless local ordinances require otherwise. The cabinet should also be conspicuously stenciled, imprinted, or otherwise marked as follows:

FLAMMABLE—KEEP FIRE AWAY

When employers use a cabinet for the storage of flammable liquids, the OSHA regulation at 29 C.F.R. §1910.106(d)(3)(i) requires that they store no more than 60 gal (230 L) of class I or class II liquids and no more than 120 gal (454 L) of class III liquids. Furthermore, they may use no more than three storage cabinets, separated from one another by a distance of at least 100 ft (30 m) when they are located within a single storage area.

Flammable liquids may also be stored outside a flammable liquid storage room or storage cabinet but only in limited amounts. The OSHA regulation at 29 C.F.R. §1910.106(e)(2)(ii)(b) stipulates that the volume of flammable liquid that may be located *outside* of an inside flammable liquid storage room or storage cabinet located within any one fire area of a building cannot exceed the following:

- 25 gal (95 L) of class IA liquids within containers
- 120 gal (454 L) of class IB, IC, II, or III liquids within containers
- 660 gal (2500 L) of class IB, IC, II, or III liquids within a single portable tank

TABLE 3.2 — Maximum Allowable Sizes of Containers and Portable Tanks[a]

CONTAINER TYPE	Flammable Liquids						Combustible Liquids			
	Class IA		Class IB		Class IC		Class II		Class III	
		L		L	gal	L	gal	L	gal	L
Glass or approved plastic	1 pt	0.5	1 qt	0.95	1	4	1	4	1	4
Metal (not DOT approved)	1 gal	4	5 gal	18.9	5	18.9	5	18.9	5	18.9
Safety can	2 gal	7.5	5 gal	18.9	5	18.9	5	18.9	5	18.9
Metal drum (DOT approved)	60 gal	230	60 gal	230	60	230	60	230	60	230
Portable tank	660 gal	2500	660 gal	2500	660	2500	660	2500	660	2500

[a]29 C.F.R. §1910.106(d), Table H-12.

OSHA and NFPA also restrict the volumes of flammable liquids that employers may store within different types of containers. These restrictions are listed in Table 3.2.

3.6-A STORAGE WITHIN CONTAINERS

OSHA and NFPA impose restrictions on the maximum volume of flammable liquids that employers may store in the workplace within containers. This allowable maximum volume varies with the flashpoint of the liquid, the manner in which the liquid is stored, and whether the storage occurs indoors or outdoors.

When employers choose to store flammable liquids indoors within containers, they must comply with the restrictions in Table 3.3. When two or more classes of flammable or combustible materials are stored in containers in a single pile or group, the maximum volume allowed in the pile is the smallest of their separate maximum volumes. OSHA and NFPA also require that no container be situated more than 12 ft (4 m) from an aisle. The main aisles must be at least 3 ft (0.9 m) wide, and the side aisles must be at least 4 ft (1.2 m) wide.

When employers choose to store flammable liquids outdoors within containers, they must comply with the restrictions in Table 3.4. When two or more classes of flammable or combustible materials are stored within containers in a single pile or group, the maximum allowable volume in the pile is the smallest of their separate maximum volumes. OSHA and NFPA also require the availability of a 12-ft- (4-m-) wide access way for the approach of fire control apparatus.

The distances in Table 3.4 apply to a facility that is "protected for exposure," meaning that all adjacent structures on the facility are adequately provided with fire-protection systems. When these structures are not protected for exposure, the distances in the fourth column of Table 3.4 must be doubled. When the total volume of flammable liquids does not exceed 50% of the maximum per pile, the distances in the fourth and fifth columns may be reduced by half, but in any event, the distance must never be less than 3 ft (0.9 m).

TABLE 3.3 Indoor Container Storage[a]

CLASS OF LIQUID	STORAGE LEVEL	Protected Storage, Maximum per Pile		Unprotected Storage, Maximum per Pile	
		gal	L	gal	L
IA	Ground and upper floors	2750 (50)[b]	10,400	660 (12)	2500
	Basement	Not permitted	Not permitted	Not permitted	Not permitted
IB	Ground and upper floors	5500 (100)	21,000	1375 (25)	5200
	Basement	Not permitted	Not permitted	Not permitted	Not permitted
IC	Ground and upper floors	16,500 (300)	62,400	4125 (75)	15,600
	Basement	Not permitted	Not permitted	Not permitted	Not permitted
II	Ground and upper floors	16,500 (300)	62,400	4125 (75)	15,600
	Basement	5500 (100)	20,800	Not permitted	Not permitted
III	Ground and upper floors	55,000 (1000)	208,000	13,750 (250)	52,000
	Basement	8250 (450)	31,200	Not permitted	Not permitted

[a]29 C.F.R. §1910.106(d), Table H-14.
[b]The numbers in parentheses below the listed gallons refer to the number of 55-gal drums that are equivalent to the indicated volumes.

SOLVED EXERCISE 3.10

Ethanol and acetone are flammable liquids that boil at 174 °F (79 °C) and 133 °F (56 °C) and have flashpoints of 54 °F (12 °C) and 0 °F (−18 °C), respectively. What is the maximum volume of ethanol that OSHA allows to be stored in containers outside a flammable liquid storage cabinet where 25 gal (95 L) of acetone is already stored?

Solution: We determine from the boiling point and flashpoint data that ethanol and acetone are class IB liquids. The OSHA regulation at 29 C.F.R. §1910.106(e)(2)(ii)(b) stipulates that no more than a total of 120 gal (460 L) of class IB flammable liquids may be stored outside a storage cabinet within any one fire area of a building. Because 25 gal (95 L) of acetone is already stored outside the cabinet, OSHA permits 95 gal (360 L) of ethanol to be stored in the same area in compliance with the regulation.

TABLE 3.4 | Outdoor Container Storage[a]

CLASS OF LIQUID	Maximum per Pile		Distance Between Piles		Distance to the Property Line that May Be Stored Upon		Distance to Street, Alley, or Other Public Way	
	gal	L	ft	m	ft	m	ft	m
IA	1100	4200	5	1.5	20	6	10	3
IB	2200	8300	5	1.5	20	6	10	3
IC	4400	16,600	5	1.5	20	6	10	3
II	8800	33,300	5	1.5	10	3	5	1.5
III	22,000	83,000	5	1.5	10	3	5	1.5

[a]29 C.F.R. §1910.106(d), Table H-17.

SOLVED EXERCISE 3.11

A lubricating oil boils at 680 °F (360 °C) and has a flashpoint greater than 300 °F (>149 °C). What is the maximum volume in gallons of this lubricating oil that OSHA and NFPA allow to be stored outdoors within containers?

Solution: We determine from the flashpoint data that the lubricating oil is a class IIIB liquid. From Table 3.4 we determine that a maximum of 22,000 gal of a class III liquid (i.e., any class IIIA and IIIB liquid) may be stored outdoors within containers, as long as the structures adjacent to the storage area are provided with fire-protection systems.

SOLVED EXERCISE 3.12

A major fire at a manufacturing plant within a densely populated industrial park threatens to engulf a stationary tank that is located on adjacent property. The captain of the firefighting team is informed that the tank contains Cellosolve acetate, a class II liquid whose lower explosive limit is 1.75% by volume and whose flashpoint is 131 °F (55 °C). Aside from extinguishing the major fire, what action should the responding team undertake?

Solution: When emergency-response personnel arrive at an emergency incident involving the potential release of a flammable liquid, it is critical that they undertake immediate action to prevent the occurrence of a BLEVE. They may do this by discharging cooling water from unmanned monitors upon the storage tank containing Cellosolve acetate. The water should be directed at the uppermost area of the tank and the actual sites where flame contact with the shell of the tank occurs.

TABLE 3.5 | Indoor Portable Tank Storage[a]

CLASS OF LIQUID	STORAGE LEVEL	Protected Storage, Maximum per Pile		Unprotected Storage, Maximum per Pile	
		gal	L	gal	L
IA	Ground and upper floors	Not permitted	Not permitted	Not permitted	Not permitted
	Basement	Not permitted	Not permitted	Not permitted	Not permitted
IB	Ground and upper floors	20,000	75,700	2000	7570
	Basement	Not permitted	Not permitted	Not permitted	Not permitted
IC	Ground and upper floors	40,000	151,000	5500	20,800
	Basement	Not permitted	Not permitted	Not permitted	Not permitted
II	Ground and upper floors	40,000	151,000	5500	20,800
	Basement	20,000	75,700	Not permitted	Not permitted
III	Ground and upper floors	60,000	227,000	22,000	83,000
	Basement	20,000	75,700	Not permitted	Not permitted

[a]29 C.F.R. §1910.106(d), Table H-15.

3.6-B STORAGE WITHIN PORTABLE TANKS

For the purpose of this section, a **portable tank** is a closed container that is not intended for fixed installation and possesses a liquid capacity of 60 gal (227 L) or more but less than 1000 lb (454 kg). It is designed with mountings to facilitate handling by mechanical means. In common practice, such tanks are called *totes*.

Portable tanks that are approved for the storage of flammable liquids have one or more devices installed within their tops with sufficient emergency-venting capability to limit the internal pressure under fire conditions to 10 psi_g, or 30% of the bursting pressure, whichever is greater.

When employers choose to store flammable liquids indoors within portable tanks, they must comply with the restrictions in Table 3.5. When one or more classes of flammable or combustible materials are stored within a single pile or group of portable tanks, the maximum volume allowed is the smallest of the separate maximum volumes. All piles are required to be separated by at least 4 ft (1.2 m).

When employers choose to store flammable liquids outdoors within portable tanks, they must comply with the restrictions in Table 3.6. When one or more classes of flammable or combustible materials are stored in a single pile or group of portable tanks, the maximum volume allowed is the smallest of the separate maximum volumes.

The distances in Table 3.6 apply to a property that is protected for exposure. When the property is not protected for exposure, the distances in the fourth column of Table 3.6 must be doubled. When the total volume of flammable liquids does not exceed 50% of the maximum per

portable tank
■ For purposes of DOT regulations, a closed container that is not fixed in position, has a liquid capacity of 60 gal (227 L) or more but less than 1000 lb (454 kg), and is designed to be loaded into, or on, or temporarily attached to a transport vehicle and equipped with skids, mounting, or accessories to facilitate handling of the tank by mechanical means

TABLE 3.6 | Outdoor Portable Tank Storage[a]

CLASS OF LIQUID	Maximum Per Pile		Distance Between Piles		Distance to the Property Line That May Be Built Upon		Distance to the Street, Alley, or Other Public Way	
	gal	L	ft	m	ft	m	ft	m
IA	2200	8300	5	1.5	20	6	10	3
IB	4400	16,600	5	1.5	20	6	10	3
IC	8800	33,300	5	1.5	20	6	10	3
II	17,600	66,600	5	1.5	10	3	5	1.5
III	44,000	166,000	5	1.5	10	3	5	1.5

[a]29 C.F.R. §1910.106(d), Table H-17.

pile, the distances in the fourth and fifth columns may be reduced by one half, but in any event, the distance must never be less than 3 ft (0.9 m).

3.6-C STORAGE WITHIN STATIONARY TANKS

Flammable liquids are also stored within stationary tanks, of which there are three common types:

- An **atmospheric tank** is a storage tank designed to operate at pressures ranging from atmospheric pressure through 0.5 psi$_g$ (3 kPa).
- A **low-pressure tank** is a storage tank designed to operate at pressures at or above 0.5 psi$_g$ (3 kPa), but not more than 15 psi$_g$ (100 kPa).
- A **pressure vessel** or **pressure tank** is a storage tank or other vessel designed to operate at pressures above 15 psi$_g$ (100 kPa).

The structural silhouettes of the most common types of stationary tanks are illustrated in Figure 3.15.

Multiple stationary tanks of varying sizes and shapes are often situated at adjacent positions on the same piece of property. This collection of storage tanks is called a **tank farm**. Water monitors and other cooling devices are generally available within a tank farm to use for rapidly extinguishing fires and cooling the surrounding tanks.

When they are encountered, one or more stationary tanks are routinely situated within **secondary containment** areas designed to provide a degree of protection in the event of tank failure. The liquid generated from overfilling, maintenance activities, and piping failures collects within these areas.

NFPA's Code 30 provides guidelines for situating stationary tanks used for the storage of flammable liquids. These guidelines include restrictions on the distances from property lines, public ways, and major buildings on the same property.[*] NFPA stipulates that the storage tanks be situated within a secondary containment area having a capacity equal to at least the capacity of the largest tank within the area. In certain instances, diversion curbs or grading must be provided to protect the adjoining property. The released liquids then drain from the area in which the tank is located into a remote impounding area.

*NFPA No. 30, *Flammable and Combustible Liquids Code* (Quincy, Massachusetts: National Fire Protection Association).

atmospheric tank
■ A tank designed to store a flammable liquid at pressures ranging from atmospheric |pressure through 0.5 psi$_g$ (3 kPa)

low-pressure tank
■ A tank designed to store flammable liquids at pressures above 0.5 psi$_g$ (3 kPa) but not more than 15 psi$_g$ (100 kPa)

pressure vessel (pressure tank)
■ A closed vessel that has been designed to store flammable liquids at pressures above 15 psi$_g$ (100 kPa)

tank farm
■ Property on which multiple stationary storage tanks have been situated

secondary containment
■ A safeguarding method (such as diking around a primary containment vessel) used to prevent the unplanned release of a hazardous material into the environment

FIGURE 3.15
Structural silhouettes of the commonly encountered stationary tanks used to store flammable liquids. Although the capacity of these tanks varies, the maximum capacity of the covered floating-roof tank is 20,000,000 gal (75,680 m³). It is often used to store crude petroleum at oil refineries.

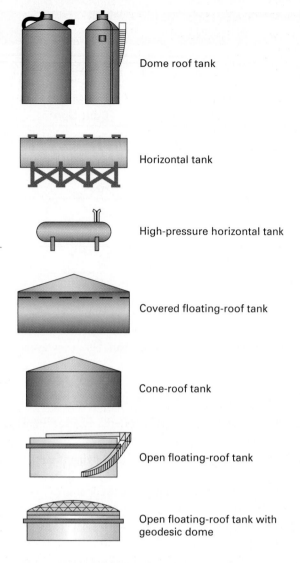

Dome roof tank

Horizontal tank

High-pressure horizontal tank

Covered floating-roof tank

Cone-roof tank

Open floating-roof tank

Open floating-roof tank with geodesic dome

OSHA and NFPA also address the design, construction, fabrication, and installation features that pertain to these tanks. When tanks are used to store flammable liquids, the following principles apply:

- An atmospheric tank should never be used to store a flammable liquid at a temperature equal to or above its boiling point.
- The operating pressure of a low-pressure or pressure vessel should never exceed the design pressure of the vessel.
- The distance between any two *adjacent* outside aboveground tanks storing flammable liquids cannot be less than one-sixth the sum of their diameters. When the diameter of one tank is less than half the diameter of the adjacent tank, the distance between the two tanks cannot be less than half the diameter of the smaller tank.
- The distance between any two outside aboveground tanks storing flammable liquids cannot be less than 3 ft (0.9 m).
- When tanks are connected in three or more rows or in an irregular pattern, greater spacing or other means must be provided so that firefighting crews may access the innermost tanks.

- Where crude petroleum tanks are located at production facilities in isolated areas and have capacities not exceeding 126,000 gal (477 m^3, or three thousand 42-gal barrels), the distance between such tanks cannot be less than 3 ft (0.9 m).
- The minimum separation between a liquefied petroleum gas container and an aboveground atmospheric tank storing flammable liquids must be 20 ft (6 m).

PERFORMANCE GOALS FOR SECTION 3.7:

- Identify the general ways by which flammable liquids are transported.
- Distinguish between nonpressure and pressure rail tankcars.
- Distinguish between thermally insulated and thermally protected rail tankcars.

3.7 Transporting Flammable Liquids

When they are transported, flammable liquids are often confined within bulk transport vessels such as portable tanks, cargo tanks, rail tankcars, or multiunit tankcars. DOT authorizes their shipment only when the carrier has loaded them into transport vessels constructed according to certain DOT prescribed specifications and complied with allowable maximum filling limits, testing, maintenance, marking, placarding, and other relevant requirements.

3.7-A TRANSPORT BY MOTOR CARRIER

Two types of motor trucks are commonly used to transport bulk volumes of hazardous materials: cargo vans and a cargo tank/semitrailer combination. The cargo vans are used to transport flammable liquids contained within steel drums and other approved containers.

3.7-B TRANSPORT BY RAIL TANKCAR

Rail tankcars can be divided into the following two types:

   - *Nonpressure tanks.*   These are DOT Specification-103, -104, -111, and -115 tankcars. Prior to use, the integrity of the tanks is tested by applying internal pressures ranging from 60 to 100 psi (410 to 690 kPa). Although the outlets and external heater lines are located on the bottom of these tankcars, all other fittings, valves and pressure-relief devices are located topside and externally within the dome. Approximately three-fourths of the rail tankcars used to transport flammable liquids are nonpressure tanks.

   - *Pressure tanks.*   These are DOT Specification-105, -109, -112, -114, and -120 tankcars. They are distinctively constructed with hemispherical heads and ends. The integrity of these tanks is tested by applying internal pressures ranging from 100 to 600 psi (690 to 4100 kPa). Although the DOT Specification-114 tankcar has a bottom unloading valve, the fittings, valves, and pressure-relief devices on the DOT Specification-105, -109, -112, and -120 tankcars are located topside and externally within the dome. Approximately one-fourth of the rail tankcars used to transport flammable liquids are pressure tankcars.

   The rail tankcars in use today were typically constructed so their contents are protected during transit in either of the following ways:

- *Thermally insulated tankcars* are constructed so that external heat is transferred to the contents very slowly under the *normal conditions* of transport.
- *Thermally protected tankcars* are constructed so that external heat cannot be transferred to the contents by either conduction or radiation under *abnormal conditions*, such as a transportation mishap.

DOT requires tankcars used for the shipment of a flammable gas and certain flammable liquids to be equipped with thermal protection, either by spraying an insulating material directly on their shells or enclosing them within a jacket that surrounds the shells.

PERFORMANCE GOALS FOR SECTION 3.8:

- Use Figures 3.16 and 3.17 to describe the response actions to be implemented by emergency responders when encountering the release of a flammable liquid into the environment.
- Identify the conditions that cause a BLEVE to occur.
- Describe the actions that emergency responders should implement to prevent the occurrence of a BLEVE.

3.8 Responding to Incidents Involving the Release of a Flammable Liquid

Many mishaps involving the release of a flammable liquid occur annually. They occur not only when the liquid is transported but also when it is in storage at manufacturing and processing industries. Most commonly, these mishaps originate with the leak or spill of the flammable liquid from its containment vessel. The liquid then vaporizes and its vapor reaches a concentration within its flammable range as it mixes with the air. Subsequently, the mixture ignites when exposed to an ignition source.

Specific procedures to be implemented during emergency-response actions involving flammable and combustible liquids are provided in Figures 3.16 and 3.17, respectively. Many are analogous to those procedures associated with a response action involving the release of a flammable gas.

3.8-A BOILING-LIQUID, EXPANDING-VAPOR EXPLOSIONS

The presence of a flammable liquid in an intact transport vessel gives rise to a potentially dangerous incident when an ongoing fire impinges on the vessel for sufficient duration. During a rail tankcar mishap, for example, the fire associated with one burning tankcar may impinge on the exterior metal surface of a second tankcar that contains a flammable liquid. Then, either of the following events is likely to occur:

- When the fire impinges on the metal at an area on the second tank that is associated with the vapor space above the liquid, the heat produced by the fire is primarily absorbed by the metal. This causes the metal to lose its tensile strength and results in a weakening of the shell. The second tankcar then ruptures at the point of the impingement.
- When the fire impinges on the metal at an area on the second tankcar that is associated with the presence of liquid, more and more liquid vaporizes and accumulates within the vapor space. At first, the liquid absorbs the heat produced by the fire, but as excessive vapor is produced, the internal pressure increases within the vessel. This ultimately causes the tankcar to rupture.

In both instances, a colossal mass of vapor is discharged into the atmosphere that ignites in a mushrooming fireball. The vapor is generally ignited within 10 minutes.

This phenomenon is called a **boiling-liquid expanding-vapor explosion (BLEVE)** (ble-vē or blē-vēe). BLEVEs produce fireballs having radii that vary with the volume of the flammable liquid being transported. The relative size of the fireball in Figure 3.18 may be estimated by visually comparing it with the standing water tower in the foreground.

BLEVE (boiling-liquid expanding-vapor explosion)
- The phenomenon in which the rapid build-up of internal pressure within a container or tank is relieved by its rapid and violent rupture, accompanied by the release of a vapor or gas to the atmosphere and propulsion of the container or its pieces

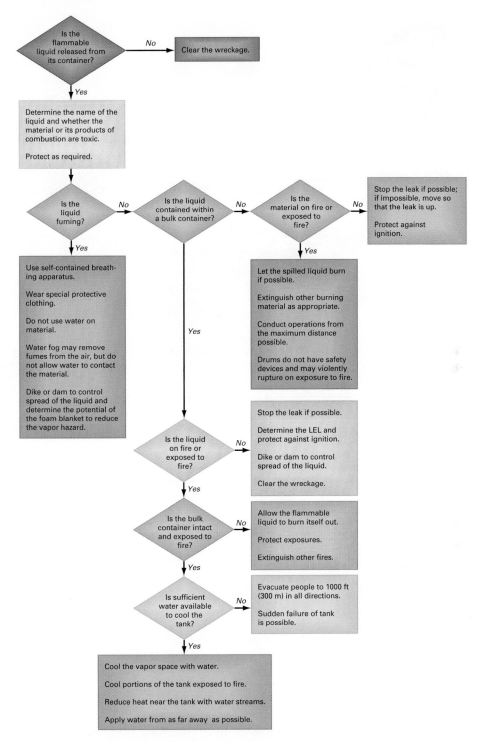

FIGURE 3.16 The recommended procedures to be implemented when responding to a transportation mishap involving the release of a flammable liquid from its packaging. *Adapted with permission of the American Society for Testing and Materials, from a figure in ASTM STP 825,* A Guide to the Safe Handling of Hazardous Materials Accidents, *second edition. Copyright © 1990, American Society for Testing and Materials.*

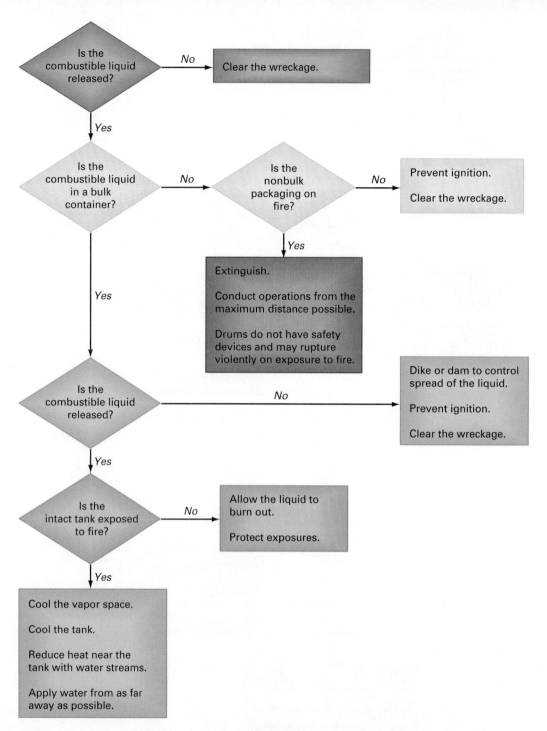

FIGURE 3.17 The recommended procedures to be implemented when responding to a transportation mishap involving the release of a combustible liquid from its packaging. *Adapted with permission of the American Society for Testing and Materials, from a figure in ASTM STP 825,* A Guide to the Safe Handling of Hazardous Materials Accidents, *second edition. Copyright © 1990, American Society for Testing and Materials.*

FIGURE 3.18 A boiling-liquid, expanding-vapor explosion (BLEVE) results when a flammable liquid is rapidly heated to a temperature significantly above its boiling point. Under this condition, the venting mechanism of a tankcar is generally incapable of releasing the buildup of internal pressure that accompanies the expansion of the vapor. Structural failure of the tankcar occurs, resulting in its rupture. During the massive explosion that follows, the vapor is released as a fireball.

When first-on-the-scene responders arrive at a mishap involving the potential release of a flammable liquid, it is critical that they take immediate action to prevent the occurrence of a BLEVE. They may do this by situating unmanned monitors so cooling water is directed on the vessel affected by the fire. The water should be directed at the vessel's main vulnerable points, the uppermost area where the vapor phase is present, and the actual site of flame contact on the shell of the vessel. Experts recommend discharging the cooling water at a rate of approximately 500 gal/min (1.9 m^3/min). Unless the vessel is effectively cooled, failure in its metal structure is likely to occur within 10 to 30 minutes after the initial contact of the fire.

The danger to property, firefighters, and bystanders from the occurrence of a BLEVE cannot be overexaggerated. The fragments of the rupturing vessel often travel like missiles for hundreds of feet in all directions, potentially causing death or injury to anyone in their path. Nearby structures are destroyed by the secondary fires that originate from radiation and rocketing shells. Following the occurrence of a BLEVE, the main assistance firefighters may provide is to extinguish these secondary fires.

3.8-B GENERAL RESPONSE PROCEDURES

The nature of an emergency-response action involving flammable liquids does not always approach the magnitude of an operation concerned with preventing a BLEVE. As they assess the nature of the emergency scene, the first-on-the-scene responders should proceed to undertake the following actions as deemed necessary and appropriate:

- As quickly as possible, the chemical or product name of the flammable liquid should be determined, either by reviewing transportation manifests, labels, markings, and placards, or by speaking with plant or transportation personnel. When the mishap involves a bulk volume of a flammable liquid, attention should also be focused on other hazardous materials that are in the immediate vicinity.

■ When the emergency incident is associated with a transportation mishap, CHEMTREC and, if warranted, the National Response Center should be contacted.

■ If the response action involves the storage of a flammable liquid within a stationary tank, personnel should use the information posted on an NFPA hazard diamond to ascertain the degree of hazard relating to fire, health, and chemical reactivity.

■ Once the chemical nature and approximate volume of the flammable liquid and availability of extinguishers are known, an appropriate fire extinguisher may be selected. The use of either water or carbon dioxide is typically effective when extinguishing a fire involving a relatively small volume of a flammable liquid, whereas the use of an aqueous-film-forming foam (Section 5.11-B) is warranted when extinguishing a class B fire involving a bulk volume of a flammable liquid.

■ When a small spill or leak of flammable liquid from a container or tank is discovered, a water fog should be used to disperse the vapor.

■ Water, fog, foam, or carbon dioxide should be applied for extinguishing or cooling purposes when a container or tank containing a flammable liquid is situated in an area near an ongoing fire, or when such containers or tanks are exposed directly to flames.

■ Quick attention should be given to cooling steel drums used to store flammable liquids, as when they are heated, the drums are particularly susceptible to rupturing. Although they may have tiny pinholes through which accumulated vapor slowly releases, they are never outfitted with the safety-relief devices or rupture disks found on most compressed gas cylinders and rail tankcars. To prevent the drums from rupturing, they should be cooled with direct streams of water.

■ When a liquid's vapor has not yet ignited, or when its fire has been momentarily extinguished, the concentration of the flammable vapor in the air should be measured. If the reading establishes that the vapor concentration is within the flammable range, emergency responders should vacate the area until the use of a water fog within the area reduces the concentration to less than the lower explosive limit. Emergency-response personnel should cautiously attempt to seal or otherwise stop a leak on a stationary tank or transport vessel only when the vapor is not burning and the vapor concentration in the air is less than the lower explosive limit. Sometimes, this task involves nothing more than tightening a valve. If it is determined that the source of the leak cannot be sealed, it may be necessary to transfer the liquid into a new storage vessel. When transfer of the liquid is impractical, the tank or transport vessel should be cooled with unmanned water monitors, but no further attempt to extinguish the fire should be made.

■ The use of water for cooling a leaking storage tank often generates a volume of water so large that the secondary containment system is unable to hold it. This water is contaminated with the flammable liquid that has leaked from the tank. When the water is likely to overtop the secondary containment system, action should be taken to preclude its entrance into nearby sewers or bodies of water or flow to areas where other flammable or combustible materials are stored. This may be accomplished by pumping the mixture into a portable tank or another temporary storage system.

■ When emergency responders encounter a massive leak of a flammable liquid, a foam blanket should be applied to confine the flammable vapor. To the extent feasible, pools of the liquid should be dammed or diked to prevent widespread exposure and contamination.

■ During mishaps associated with the transportation of a flammable liquid, upright cabs and railcars should be disconnected and moved from overturned tankcars whenever practical.

■ When the vapor of a flammable liquid has been released into the atmosphere, it is prudent to wear total-encapsulating suits, use self-contained breathing apparatus, and perform as many response actions as practical from a downwind location. Even when a flammable liquid is nontoxic, the use of respiratory protection equipment is warranted, since inhalation of its vapor may cause dizziness, illness, or suffocation.

■ An overturned *pressure* transport tank may be safely righted without first unloading its liquid contents; however, an overturned *nonpressure* transport tank should not be uprighted until the bulk of the tank's liquid contents has first been transferred into a different tankcar. This transfer should be attempted only by experienced personnel.

Flammability and Ignitability

3.1 Use the following data to identify the OSHA/NFPA class of flammable or combustible liquid to which each substance belongs:

Substance	Boiling point	Flashpoint
(a) benzene	176 °F (80 °C)	12 °F (−11.1 °C)
(b) chlorobenzene	270 °F (132 °C)	84 °F (29 °C)
(c) pentane	97 °F (36.1 °C)	−40 °F (−40 °C)

3.2 The lower and upper explosive limits of liquid *n*-butyl acetate are 1.4% and 7.6% by volume, respectively. When exposed to an ignition source, will *n*-butyl acetate vapor ignite at each of the following concentrations in air by volume:

(a) 0.1%
(b) 1%
(c) 10%

3.3 *n*-Hexane is a liquid constituent of many commercially available solvent mixtures. Some of its physical properties are listed here.

Melting point	−141 °F (−96 °C)
Boiling point	154 °F (68 °C)
Specific gravity at 68 °F (20 °C)	0.66
Vapor density (air = 1)	3.0
Vapor pressure at 68 °F (20 °C)	124 mmHg
Flashpoint	−7 °F (−21.7 °C)
Autoignition point	437 °F (225 °C)
Lower explosive limit	1.1% by volume
Upper explosive limit	7.5% by volume
Solubility in water	Slightly soluble (0.014 g/100 mL)

(a) In which range of temperature does *n*-hexane exist as a liquid?
(b) Identify the NFPA class of flammable liquid to which *n*-hexane belongs.
(c) If steel drums containing *n*-hexane rupture, is *n*-hexane vapor likely to ignite when the surrounding temperature reaches 482 °F (250 °C) in the absence of an ignition source?
(d) If steel drums containing *n*-hexane rupture and produce a vapor concentration ranging from 20% to 45% by volume in air, is the vapor likely to ignite if exposed to an ignition source at 320 °F (160 °C)?
(e) If steel drums containing *n*-hexane rupture, will its vapor initially accumulate in low areas or diffuse upward?
(f) Describe the extent to which water may be used effectively to extinguish a fire involving *n*-hexane inside a drum.

Storing and Transporting Compressed Gases

3.4 DOT-3E1800 is marked on a steel cylinder when its gauge pressure is 1610 psi at 70 °F (21 °C). Is this cylinder likely to rupture during a fire if the temperature of its contents rises to 300 °F (149 °C)?

General Hazards of Compressed Gases

3.5 Why is the rupture of a steel cylinder containing compressed helium gas gravely hazardous, despite the fact that helium is a nonflammable gas?

Responding to Incidents Involving the Release of Compressed Gases

3.6 When a leaking cylinder is moved from a burning building into an open area, why do experts recommend that it be rotated so the point of leakage becomes the uppermost part of the cylinder?

3.7 On arrival at a transportation mishap, the first-on-the-scene responders observe an ongoing fire and learn from the driver of an overturned cargo tank that the consignment is compressed liquefied butane. They also observe a ruptured discharge line from which releasing vapors have ignited.

(a) What immediate action should the team take on arrival at this scene?
(b) What is the approximate distance from this scene to which unauthorized personnel should be evacuated?

Storing Flammable Liquids

3.8 Methyl acetate is a liquid that boils at 135 °F (57 °C) and has a flashpoint of 15 °F (−9 °C). It is a constituent of paint removers. A company that manufactures paint removers has constructed an indoor storage room for flammable liquids from fire-resistant materials and equipped it with a sprinkling system. What is the maximum number of 55-gal drums containing methyl acetate that may be stored per pile in the room in compliance with the OSHA and NFPA regulations?

3.9 Adjoining property and waterways can be protected from the hazards posed by the presence of a flammable or combustible liquid in a stationary storage tank by constructing a means to drain a spilled or leaked liquid from the tank into a remote impounding area. For such purposes, NFPA No. 30, *Flammable and Combustible Liquids Code*, advocates each of the following for a 100,000-gal (378-m^3) storage tank:

- A slope of not less than 1% away from the tank for at least 50 ft (15 m) toward the impounding area
- A capacity of the impounding area not less than the capacity of the tank
- A route of drainage so that if ignited, the burning liquid within the drainage system will not seriously expose tanks or adjoining property
- The location of the impounding area not closer than 50 ft (15 m) from any property line that is or may be built on, or from any tank

These conditions are illustrated below:

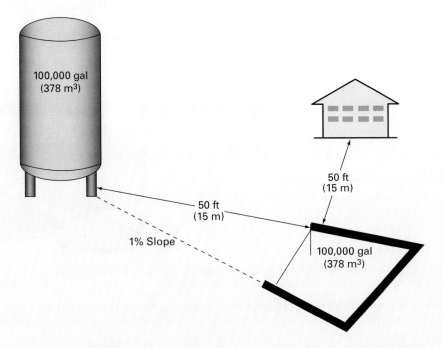

If a processing company stores 50,000 gal of *n*-amyl alcohol, a class IC liquid, in a tank located between two buildings on the same property, each separated from the tank's remote 45,000-gal impounding area by a distance of 50 m, has the company complied with this NFPA recommendation?

Responding to Incidents Involving the Release of Flammable Liquids

3.10. Arriving at the scene of a railroad transportation mishap, the first-on-the-scene responders note that a fire in a ruptured tankcar is impinging on the exterior surface of an adjacent, intact tankcar. No information is immediately available concerning the chemical nature of the contents in either tankcar, but the following reporting marks are imprinted on each side and each end of the two tankcars: MDSB124665 and MDSB124666. To best protect public health, safety, and the environment, what is the preferred action for these responders to take?

Chemical Forms of Matter

Commercial chemical products like paint, pesticide formulations, ammunition, gasoline, and other petroleum fuels are hazardous materials that consist of mixtures of distinctly different components, each separable from the others. Paint, for example, is a mixture of some resin or other film-forming compound, a solvent or thinner, and organic or inorganic pigments. Each chemical product is characterized in a similar fashion as a uniform or nonuniform blend of different components, each of which may be separated from one another.

When any mixture has been separated into its components, the individual components are unique forms of matter. In this chapter, we identify them as either elements or compounds possessing a characteristic set of physical and chemical properties. We also observe how chemists describe the structural nature of these substances in terms of atoms, molecules, and ions. Finally, we learn how chemists name these substances and write their chemical formulas.

PERFORMANCE GOALS FOR SECTION 4.1:

- Distinguish between elements and compounds.
- Learn the symbols of the elements listed in Table 4.2.

4.1 Elements and Compounds

A material that has been separated from all other materials is called a *pure substance* or, more simply, a **substance**. Examples include aluminum metal, copper metal, oxygen, distilled water, and table sugar. No matter how they originated or what procedures are used to purify them, samples of the same substance are indistinguishable from one another. For example, all samples of distilled water are alike and indistinguishable from all other samples; that is, water is the same substance whether it has been distilled from rainwater, well water, seawater, or any other source.

A substance is characterized as a material having a fixed composition, usually expressed in terms of percentage by mass. Distilled water is a pure substance consisting of 11.2% hydrogen and 88.8% oxygen by mass. By contrast, coal is not a pure substance; its carbon content alone may vary anywhere from 35% to 90% by mass. Such materials that are not pure substances are called **mixtures**.

Suppose we take a pure substance like limestone (calcium carbonate) and heat it, as shown in Figure 4.1. The limestone ultimately crumbles into a white powder. Careful examination shows that carbon dioxide evolved from the limestone while it was heated. Substances like limestone that can be broken down into two or more simpler substances are called **compound substances** or, simply, **compounds**. There are more than 50 million different compounds.

The substances that cannot be decomposed into simpler forms of matter are called **elements**. The elements are the fundamental substances of which all compounds are composed. Although there are only 118 known elements, there are billions of known compounds. Of the 118 elements, only 88 are present in detectable amounts on Earth, and many are very rare. Table 4.1 shows that only 10 elements make up approximately 99% by mass of Earth, including its crust, the atmosphere, the surface layer, and the bodies of water. Oxygen is the most abundant element on Earth and is found in the free state in the atmosphere, as well as in combined form with other elements in numerous minerals and ores.

Each element has a specific *name* and **chemical symbol**. The names and symbols of some common elements are listed in Table 4.2. The symbols of the elements consist of either one or two letters with the first letter capitalized. The selection of the symbols of many elements was originally based on their Latin names: Pb for *plumbum* (lead), Ag for *argentum* (silver), and Sn for *stannum* (tin). It is best to memorize the symbols of the elements listed in Table 4.2, since they are frequently encountered in the study of hazardous materials.

The elements beyond uranium (atomic number 92) do not exist on Earth, but scientists have prepared some of them in nuclear laboratories. The following names for the elements having atomic numbers 104 through 110 have been accepted by the *International Union of Pure and Applied Chemistry* (IUPAC): rutherfordium (104), symbol Rf; dubnium (105), symbol Db; seaborgium (106), symbol Sg; bohrium (107), symbol Bh; hassium (108), symbol Hs;

substance
- A homogeneous material having a constant, fixed chemical composition; an element or compound

mixture
- Matter consisting of two or more substances in varying proportions that are not chemically combined

compound (compound substance)
- A substance composed of two or more elements in chemical combination

element
- A substance composed of only one kind of atom

chemical symbol
- A notation using one or two letters to represent an element

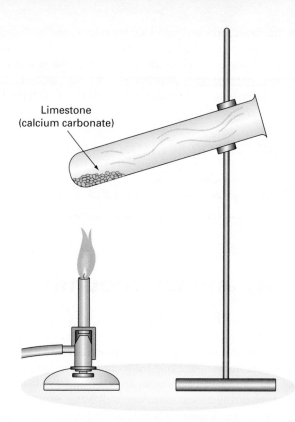

FIGURE 4.1 The heating of limestone (calcium carbonate) in a test tube. Carbon dioxide evolves as the limestone chemically decomposes. This simple experiment establishes that calcium carbonate is a compound and not an element.

TABLE 4.1	Natural Abundance of the Elements (Earth's Crust, Oceans, and the Atmosphere)		
Oxygen	49.5%	Chlorine	0.19%
Silicon	25.7%	Phosphorus	0.12%
Aluminum	7.5%	Manganese	0.09%
Iron	4.7%	Carbon	0.08%
Calcium	3.4%	Sulfur	0.06%
Sodium	2.6%	Barium	0.04%
Potassium	2.4%	Chromium	0.033%
Magnesium	1.9%	Nitrogen	0.030%
Hydrogen	1.9%	Fluorine	0.027%
Titanium	0.58%	Zirconium	0.023%
		All others	<0.1%

TABLE 4.2 | Symbols of the Most Common Elements

ELEMENT	SYMBOL	ELEMENT	SYMBOL
Aluminum	Al	Lithium	Li
Antimony	Sb	Magnesium	Mg
Argon	Ar	Manganese	Mn
Arsenic	As	Mercury	Hg
Barium	Ba	Neon	Ne
Bismuth	Bi	Nickel	Ni
Boron	B	Nitrogen	N
Bromine	Br	Oxygen	O
Cadmium	Cd	Phosphorus	P
Calcium	Ca	Platinum	Pt
Carbon	C	Potassium	K
Chlorine	Cl	Radium	Ra
Chromium	Cr	Silicon	Si
Cobalt	Co	Silver	Ag
Copper	Cu	Sodium	Na
Fluorine	F	Strontium	Sr
Gold	Au	Sulfur	S
Helium	He	Tin	Sn
Hydrogen	H	Titanium	Ti
Iodine	I	Tungsten	W
Iron	Fe	Uranium	U
Lead	Pb	Zinc	Zn

meitnerium (109), symbol Mt; darmstadtium (110), symbol Ds, and roentgenium (111), symbol Rg. The elements having atomic numbers 112 through 118 have been discovered but have not yet been named.*†

A complete listing of all known elements is provided in the appendix at the back of this book.

*In 2004, an international research team composed of scientists from the Joint Institute for Nuclear Research (JINR), Dubna, Russia, and the Lawrence Livermore National Laboratory (LLNL), Berkeley, California, announced the discovery of elements having atomic numbers 113, 114, 115, and 116. In 2006, JINR and LLNL announced that teams from both laboratories had produced three atoms of the element having an atomic number 118. This element has not yet been named.

†Although they are not yet officially named, the elements having atomic numbers 112 through 118 have respectively been given the following temporary names: ununbium, ununtrium, ununquadium, ununpentium, ununhexium, ununseptium, and ununoctium.

4.2 Metals, Nonmetals, and Metalloids

metal
- An element that conducts heat and electricity well and has high physical strength and the properties of ductility and malleability

Each element can be classified as a metal, nonmetal, or metalloid. **Metals** are elements that usually have the following properties:

- Conduct heat and electricity well
- Melt and boil at relatively high temperatures
- Have relatively high densities
- Are normally malleable (able to be hammered into sheets) and ductile (able to be drawn into a wire)
- Display a brilliant luster

Examples of metals are iron, platinum, and silver. Almost all metals are solids at room temperature [68 °F (20 °C)]. Mercury is the only common metal that is liquid at room temperature. No metal is gaseous at room temperature.

nonmetal
- An element that does not conduct heat and electricity well, has low physical strength, and is neither ductile nor malleable

Elements that do not have the physical properties of metals are called **nonmetals**. These elements have the following general properties:

- Melt and boil at relatively low temperatures
- Do not have a luster
- Are less dense than metals
- Poorly conduct heat and electricity

At room temperature, most nonmetals generally exist as either solids or gases. Bromine is the only liquid nonmetal at room temperature. Oxygen, fluorine, and nitrogen are examples of gaseous nonmetals, whereas carbon, sulfur, and phosphorus are examples of solid nonmetals.

Not all metals and nonmetals exhibit the general properties noted here. There are exceptions; for instance, carbon conducts heat and electricity very well, and one form of carbon (diamond) melts at about 6872 °F (~3800 °C), a relatively high temperature.

metalloid (semimetal)
- A metal that exhibits the general physical properties of both metals and nonmetals

Several elements have properties resembling those of both metals and nonmetals. They are called **metalloids**, or **semimetals**. The metalloids are boron, silicon, germanium, arsenic, antimony, tellurium, and polonium. For example, silicon and germanium have the luster associated with metals, but they do not conduct heat and electricity.

PERFORMANCE GOALS FOR SECTION 4.3:

- Classify the changes undergone by a substance as physical changes and chemical changes.
- Identify the representative physical and chemical changes of a substance.
- Identify the nature of the physical properties and chemical properties of a substance.

4.3 Chemical and Physical Changes

chemical bond
- The force by which the atoms of one element become attached to, or associated with, other atoms in a compound

chemical change (chemical reaction)
- A change that results in an alteration of the chemical identity of a substance

The constant composition associated with a given substance is maintained by internal linkages among its units. These linkages are called **chemical bonds**. When a particular process occurs that makes or breaks these bonds, we say that a **chemical change**, or a **chemical reaction**,

FIGURE 4.2 An example of a physical and chemical change in the components of gasoline. On the left, the gasoline evaporates; that is, its constituents change their physical state from the liquid to the vapor. This constitutes a physical change. On the right, a spill of gasoline ignites and burns, becoming carbon dioxide and water vapor. This constitutes a chemical change.

has occurred. Combustion and corrosion are common examples of chemical changes associated with some hazardous materials.

Let's briefly consider the nature of the combustion process. When something burns, it combines with oxygen. The resulting products of combustion are compounds containing oxygen, called *oxides*. For instance, many commercial gasolines are mixtures of several substances, including one called octane. When octane burns completely, it becomes carbon dioxide gas and water vapor. Carbon dioxide is an oxide of carbon, as its name implies, whereas water is an oxide of hydrogen. Carbon dioxide and water vapor are unlike octane or other gasoline components. They have different properties and different compositions. This conversion of octane to carbon dioxide and water is typical of a chemical change.

By contrast, substances can undergo changes during which their compositions remain the same. Such changes are called **physical changes**. Let's consider octane again. Some of its physical and chemical changes are illustrated in Figure 4.2. When exposed to the ambient environment, liquid octane evaporates, but its chemical composition remains unchanged. Such alterations in the physical state of a substance, such as from a liquid to a vapor, are considered physical changes. Other examples of physical changes are melting, freezing, boiling, crushing, and pulverizing.

The types of behavior that a substance exhibits when undergoing chemical changes are called its **chemical properties**. The characteristics that do not involve changes in the chemical identity of a substance are called its **physical properties**. All substances can be distinguished from one another by these properties, in much the same way as certain features—fingerprints or DNA, for example—distinguish one human being from another. The study of hazardous materials is concerned to a great extent with learning the chemical and physical properties of appropriate substances, some examples of which are listed in Table 4.3.

physical change
▪ A change that does not result in an alteration of the chemical identity of a substance, like vaporization or sublimation

chemical property
▪ A type of behavior a substance exhibits when it undergoes a chemical change

physical property
▪ A phenomenon that a substance exhibits when it undergoes a physical change, like melting and boiling

PERFORMANCE GOALS FOR SECTION 4.4:

- Identify the names of the common particles present in an atom.
- Identify the common properties of the common particles present in an atom.
- Determine the number of protons and electrons in an atom from an element's atomic number.
- Describe the general structure of an atom.
- Describe the nature of atomic orbitals.
- Distinguish between the atomic number and atomic weight of an element.
- Given the atomic mass of each stable isotope of an element and its natural abundance, calculate the atomic weight of the element.

TABLE 4.3 | Characteristics of Some Substances

SUBSTANCE	PHYSICAL PROPERTIES	CHEMICAL PROPERTIES
Oxygen, an element	Odorless, colorless gas; does not conduct heat or electricity; density = 1.43 g/L; becomes liquid at −297 °F (−183 °C)	Combines readily with many elements (a chemical reaction called oxidation)
Phosphorus, an element	White or red solid; does not conduct heat or electricity; density = 1.82 g/mL (white) and 2.34 g/mL (red)	Readily combines with oxygen, chlorine, and fluorine; white form spontaneously ignites in dry air
Carbon dioxide, a compound	Odorless and colorless gas; solidifies at −83 °F (−67 °C) under pressure, forming dry ice; soluble in water under pressure	Does not burn; reacts with water-soluble metal compounds, forming metallic carbonates
Hydrogen chloride, a compound	Strong-smelling, colorless gas; density = 1.20 g/mL; soluble in water, forming hydrochloric acid	Reacts with many minerals, forming water-soluble products; reacts with ammonia, forming ammonium chloride

4.4 Some Basic Features of Atoms

atom
■The smallest particle of an element that can be identified with that element

electron
■A fundamental particle that bears a charge of −1

proton
■A fundamental particle of which all atoms are composed, bearing a charge of +1

neutron
A fundamental particle of which matter other than protium is composed, bearing no charge

If a small piece of an element, say aluminum, could be hypothetically divided and subdivided into smaller and smaller pieces, until subdivision was no longer possible, the result would be one particle of aluminum. This smallest particle of the element that is still representative of the element is called an **atom**, from the Greek word *atomos*, meaning "indivisible."

An atom is infinitesimally small, yet it is also composed of even smaller particles known as electrons, protons, and neutrons. **Electrons** are negatively charged particles that are responsible for the chemical reactivity of a given element. **Protons** are positively charged particles. **Neutrons** are neutral particles. Electrons and protons bear the same magnitude of charge but of opposite signs. For convenience, the electron has a charge of −1, the proton of +1, and the neutron of 0.

Protons are relatively heavy particles; they are 1836 times more massive than electrons. Neutrons are slightly more massive than protons. The fundamental characteristics of electrons, protons, and neutrons are summarized in Table 4.4.

TABLE 4.4 | Some Basic Atomic Particles

PARTICLE	PROTON	ELECTRON	NEUTRON
Symbol	P$^+$	e$^-$	n
Relative charge	+1	−1	0
Relative mass	1	About 0[a]	1

[a]The mass of an electron is 1/1836 the mass of the proton.

The protons and neutrons of an atom reside in a central area called the **atomic nucleus**. Electrons reside primarily in designated regions of space surrounding the nucleus, called **atomic orbitals**. There are several types of atomic orbitals; some are close to the nucleus, whereas others are relatively remote from it. Scientists have learned that only a prescribed number of electrons reside in a given type of atomic orbital. Two electrons are always close to the nucleus, in an atom's innermost atomic orbital (with the exception of a hydrogen atom, which possesses only one electron). Most atoms have additional electrons in farther atomic orbitals.

The number of protons in an atom is called the **atomic number**. Atomic numbers are often used in the study of chemistry to determine the number of electrons possessed by neutral atoms of an element. An atom of hydrogen has one electron, helium has two, lithium has three, and so forth. Carbon is an element composed only of carbon atoms, and all carbon atoms exhibit nearly the same physical and chemical properties. Some atoms may have slightly different masses owing to different numbers of neutrons in their nuclei, but these carbon atoms all act the same when they undergo chemical changes. Similarly, oxygen is an element composed of oxygen atoms, and all oxygen atoms possess nearly the same properties. But carbon and oxygen atoms are not alike, since their atoms possess different numbers of electrons, protons, and neutrons.

Although neutral atoms of the same element have an identical number of protons and electrons, they may differ by the number of neutrons in their nuclei. Atoms of the same element having different numbers of neutrons are called the **isotopes** of that element. We note later, in Chapter 16, that some isotopes are radioactive.

Over the past two centuries, scientists have determined the relative masses of many atoms of the known elements. These relative masses are called **atomic weights**. The atomic weights of the elements are not absolute masses but, rather, masses measured relative to the mass of another atom that has been selected as a reference standard. Because atomic weights are relative parameters, they are unitless numbers. A specific isotope of carbon, called carbon-12, has been selected as the atom whose mass serves as the reference standard. The atoms of this carbon isotope consists of six electrons, six protons, and six neutrons. It is assigned a mass of exactly 12.

The atomic weight of a given element is a number that tells us how the mass of an *average* atom of that element compares with the mass of the carbon-12 atom. Scientists establish the atomic weight of an element by first determining the atomic mass and natural abundance of each of its stable isotopes. Suppose an element has three stable isotopes, X-*a*, X-*b*, and X-*c*, whose atomic masses and natural abundances are known. The atomic weight of the element is then determined as follows:

$$\text{Atomic weight} = \frac{(\text{mass}_{X-a} \times \%_{X-a}) + (\text{mass}_{X-b} \times \%_{X-b}) + (\text{mass}_{X-c} \times \%_{X-c})}{100}$$

The atomic weight of any element is obtained from the relative mass of its naturally occurring isotopes, weighted according to their natural abundances.

Use of the carbon-12 standard results in an atomic weight for natural carbon of 12.011. This slight difference results from averaging the masses of its natural isotopic forms. The atomic numbers and atomic weights of the elements are listed on the page following the Hazardous Materials Index at the back of this book.

atomic nucleus
- The region at the center of the atom composed of protons and neutrons

atomic orbital
- The region in the space around an atomic nucleus in which electrons are most likely to be found

atomic number
The number of protons in an atomic nucleus; the number of electrons an atom

isotope
- Any of a group of nuclei having the same number of protons but different number of neutrons

atomic weight
- The mass of an atom compared with the mass of the carbon-12 isotope, which is assigned a mass of exactly 12

PERFORMANCE GOALS FOR SECTION 4.5:

- Explain the basis for the arrangement of the elements on the periodic table.
- Associate the chemical reactivity of an element with the electrons in the outermost atomic orbitals of its atoms.
- Use the periodic table to distinguish between a family and a period.
- Note the location of the principal families on the periodic table: the alkali metals, alkaline earth metals, chalcogens, halogens, and noble gases.

4.5 The Periodic Classification of the Elements

periodic law
■ The observation that the chemical properties of the elements are periodic functions of their atomic numbers

periodic table
■ A compilation of the known elements into periods and families by increasing atomic number, so that elements with similar chemical properties are in the same column (or family)

During the last half of the 19th century, several scientists first noted that the chemical properties of any given element were similar to the chemical properties of certain other elements. For example, they noted that sodium metal reacts explosively with water and burns spontaneously in the air. When these two chemical properties of sodium were compared with the properties of other elements, they found that potassium also explodes on contact with water and burns spontaneously in air. These scientists summarized this observation in the **periodic law**: The properties of the elements vary periodically with their atomic numbers. The term *periodic* reflects this repetition of chemical properties.

Suppose we list each element in a square and then arrange the squares by order of increasing atomic number. This means that the total number of electrons possessed by each element increases in this arrangement, one at a time, as we move from one square to the next. Then, let's further arrange them into columns of elements that possess similar properties. Of course, we would need to know a great deal of chemistry to accomplish this feat. A similar exercise was first performed more than a hundred and thirty years ago, when many elements known today had not yet been discovered.

Such an arrangement of the chemical elements into a chart designed to represent the periodic law is called a **periodic table**. Although a number of versions are in existence, the periodic table in Figure 4.3 provides ample information for emergency responders. The elements positioned

Periodic Table of the Elements

1A	2A	3B	4B	5B	6B	7B	8B	8B	8B	1B	2B	3A	4A	5A	6A	7A	8A
1 **H** 1.00794 Hydrogen																	2 **He** 4.002602 Helium
3 **Li** 6.341 Lithium	4 **Be** 9.012182 Beryllium											5 **B** 10.811 Boron	6 **C** 12.0107 Carbon	7 **N** 14.0067 Nitrogen	8 **O** 15.9994 Oxygen	9 **F** 18.9984032 Fluorine	10 **Ne** 20.1797 Neon
11 **Na** 22.989789 Sodium	12 **Mg** 24.3050 Magnesium											13 **Al** 26.9815386 Aluminum	14 **Si** 28.0855 Silicon	15 **P** 30.973762 Phosphorus	16 **S** 32.065 Sulfur	17 **Cl** 35.453 Chlorine	18 **Ar** 38.948 Argon
19 **K** 39.0983 Potassium	20 **Ca** 40.078 Calcium	21 **Sc** 44.955912 Scandium	22 **Ti** 47.867 Titanium	23 **V** 50.9415 Vanadium	24 **Cr** 51.9961 Chromium	25 **Mn** 54.938045 Manganese	26 **Fe** 55.845 Iron	27 **Co** 58.933195 Cobalt	28 **Ni** 58.6934 Nickel	29 **Cu** 63.546 Copper	30 **Zn** 65.38 Zinc	31 **Ga** 69.723 Gallium	32 **Ge** 72.64 Germanium	33 **As** 74.92160 Arsenic	34 **Se** 78.96 Selenium	35 **Br** 79.904 Bromine	36 **Kr** 83.798 Krypton
37 **Rb** 85.4678 Rubidium	38 **Sr** 87.62 Strontium	39 **Y** 88.90585 Yttrium	40 **Zr** 91.224 Zirconium	41 **Nb** 92.90638 Niobium	42 **Mo** 95.96 Molybdenum	43 **Tc** [98] Technetium	44 **Ru** 101.07 Ruthenium	45 **Rh** 102.90550 Rhodium	46 **Pd** 106.42 Palladium	47 **Ag** 107.8682 Silver	48 **Cd** 112.411 Cadmium	49 **In** 114.818 Indium	50 **Sn** 118.710 Tin	51 **Sb** 121.760 Antimony	52 **Te** 127.60 Tellurium	53 **I** 126.90447 Iodine	54 **Xe** 131.293 Xenon
55 **Cs** 132.90545 Cesium	56 **Ba** 137.327 Barium	57-71 **La-Lu** Lanthanides	72 **Hf** 178.49 Hafnium	73 **Ta** 180.94788 Tantalum	74 **W** 183.84 Tungsten	75 **Re** 186.207 Rhenium	76 **Os** 190.23 Osmium	77 **Ir** 192.217 Iridium	78 **Pt** 195.084 Platinum	79 **Au** 196.966569 Gold	80 **Hg** 200.59 Mercury	81 **Tl** 204.3833 Thallium	82 **Pb** 207.2 Lead	83 **Bi** 208.98040 Bismuth	84 **Po** [209] Polonium	85 **At** [210] Astatine	86 **Rn** [222] Radon
87 **Fr** [223] Francium	88 **Ra** [226] Radium	89-103 **Ac-Lr** Actinides	104 **Rf** [267] Rutherfordium	105 **Db** [268] Dubnium	106 **Sg** [271] Seaborgium	107 **Bh** [272] Bohrium	108 **Hs** [270] Hassium	109 **Mt** [276] Meitnerium	110 **Ds** [281] Darmstadtium	111 **Rg** [280] Roentgenium	112 **Uub** [285] Ununbium	113 **Uut** [284] Ununtrium	114 **Uuq** [289] Ununquadium	115 **Uup** [288] Ununpentium	116 **Uuh** [293] Ununhexium	117 **Uus** [294] Ununseptium	118 **Uuo** [294] Ununoctium

Lanthanides	57 **La** 138.90547 Lanthanum	58 **Ce** 140.116 Cerium	59 **Pr** 140.90765 Praseodymium	60 **Nd** 144.242 Neodymium	61 **Pm** [145] Promethium	62 **Sm** 150.36 Samarium	63 **Eu** 151.964 Europium	64 **Gd** 157.25 Gadolinium	65 **Tb** 158.92535 Terbium	66 **Dy** 162.500 Dysprosium	67 **Ho** 164.93032 Holmium	68 **Er** 167.259 Erbium	69 **Tm** 168.93421 Thulium	70 **Yb** 173.054 Ytterbium	71 **Lu** 174.9668 Lutetium
Actinides	89 **Ac** [227] Actinium	90 **Th** 232.03806 Thorium	91 **Pa** 231.03588 Protactinium	92 **U** 238.02891 Uranium	93 **Np** [237] Neptunium	94 **Pu** [244] Plutonium	95 **Am** [243] Americium	96 **Cm** [247] Curium	97 **Bk** [247] Berkelium	98 **Cf** [251] Californium	99 **Es** [252] Einsteinium	100 **Fm** [257] Fermium	101 **Md** [258] Mendelevium	102 **No** [259] Nobelium	103 **Lr** [262] Lawrencium

Alkali metal	Alkaline earth	Basic metal	Halogen	Noble gas	Nonmetal	Rare earth	Metalloid	Transition metal

FIGURE 4.3 A modern version of the periodic table of the elements. [*Courtesy of Todd Helmenstine*, About Chemistry (*2008*).]

within the same column of the periodic table are called a **family of elements**. Each family is identified by a number and a capital letter at the top of the column, such as 1A and 2A. Thus, for example, helium, neon, argon, krypton, xenon, and radon belong to the same family, identified by 8A.

Elements in the same row of a periodic table are said to belong to the same **period**. The periods are numbered on the far left of the table from 1 to 7. There is one period of 2 elements, two periods of 8 elements each, two more of 18 elements each, one period of 32 elements, and a final period that presently has 28 elements.

The periodic table of the elements is one of the most powerful icons in science: a single table that consolidates an array of valuable information. Some version of the table hangs on the wall of nearly every chemistry laboratory throughout the world. Simply by glancing at it, we can observe immediately the atomic number of any element. We can also readily distinguish among those elements that are metals, nonmetals, and metalloids. The group of salmon-colored squares contain the symbols of the elements that are metalloids. They separate the metals from the nonmetals. Generally, the metals are the elements that fall to the left of the group, and the nonmetals are the elements that fall to the right of it.

The usefulness of a periodic table consists in the manner by which it displays the periodicity, or repetition, in the properties of the elements at regular intervals. In particular, when we observe the elements as members of the same family, we know they possess similar chemical properties. Five families deserve special recognition in this regard. They are identified by unique names:

- Group 1A is called the **alkali metal family**; its members are lithium, sodium, potassium, rubidium, cesium, and francium. As noted earlier, each of these metals reacts explosively with water and ignites on exposure to the air. In Figure 4.3, the squares for the alkali metals are colored pink.

- Group 2A is called the **alkaline earth family**; its members are beryllium, magnesium, calcium, strontium, barium, and radium. These elements are also chemically reactive, but not nearly as reactive as the alkali metals. They cause water to decompose, but the rate of decomposition is slow at ambient temperatures. They ignite in the air, but only after they have been heated or exposed to an ignition source. In Figure 4.3, the squares for the alkaline earth metals are colored dark blue.

- Group 6A is called the **chalcogen family**; its members are oxygen, sulfur, selenium, tellurium, and polonium, each of which is a moderately active substance. In Figure 4.3, the squares for the chalcogens are not uniquely colored.

- Group 7A is called the **halogen family**; its members are fluorine, chlorine, bromine, iodine, and astatine. In Figure 4.3, the squares for the halogens are colored light blue. These elements are nonmetals, each of which is especially reactive.

- Group 8A is called the **noble gas family**; its members are helium, argon, krypton, xenon, and radon. In Figure 4.3, the squares for the noble gases are colored light green. Chemists originally thought that these gases were all inert to chemical combination and called them *inert gases*. Although some of them, such as krypton, are now known to form compounds, the noble gases uniquely stand out as a group of elements lacking the chemical reactivity observed for all other elements.

family of elements
- The group of elements listed in a vertical column of the periodic table, all of whose members possess similar chemical properties

period
- A horizontal row on the periodic table

alkali metal family
- The elements of Group 1A on the periodic table: lithium, sodium, potassium, rubidium, cesium, and francium

alkaline earth family
- The elements in Group 2A on the periodic table: beryllium, magnesium, calcium, strontium, barium, and radium

chalcogen family
- The elements in Group 6A on the periodic table: oxygen, sulfur, selenium, tellurium, and polonium

halogen family
- The elements in Group 7A on the periodic table: fluorine, chlorine, bromine, iodine, and astatine

noble gas family
- The elements in Group 8A on the periodic table: helium, neon, argon, krypton, xenon, and radon

SOLVED EXERCISE 4.1

Because compounds of thallium are highly toxic, they are sometimes commercially available in rodenticides, products that kill rodents. Using the periodic table, answer the following questions:

(a) What is the symbol for thallium?
(b) What are the symbols for the elements immediately adjacent to thallium on the periodic table?

(c) Is thallium a metal, nonmetal, or metalloid?

(d) Provide the symbols of all the elements in the family of which thallium is a member.

(e) Identify the group number of the family of which thallium is a member.

Solution:

(a) Thallium is located in the sixth period. Its chemical symbol is Tl.

(b) The elements immediately adjacent to thallium on the periodic table are mercury and lead, whose symbols are Hg and Pb, respectively.

(c) Thallium is a metal because it is located to the left of the salmon-colored metalloids on the periodic table.

(d) Boron, aluminum, gallium, indium, and thallium are members of the same family of elements.

(e) Thallium is a member of the family of elements denoted as 3A.

PERFORMANCE GOALS FOR SECTION 4.6:

- Describe the molecule as a basic unit of matter.
- Demonstrate how ions are formed from atoms.

4.6 Molecules and Ions

Although the smallest representative particle of an element is the atom, not all uncombined elements exist as single atoms. In fact, only six elements actually exist as single atoms. These are the noble gases. We say that these elements are *monatomic*.

Other gases or liquids at room conditions consist of units containing pairs of like atoms. These units are called **molecules**. For example, hydrogen, oxygen, nitrogen, and chlorine are gaseous elements as we generally encounter them, each of which is composed of a molecule having two atoms. These molecules are said to be *diatomic* and are symbolized by the notations H_2, O_2, N_2, and Cl_2, respectively. They are illustrated in Figure 4.4.

molecule
∎ The smallest neutral unit of some elements and compounds, composed of two or more atoms

ion
∎ An atom (or group of atoms bound together) with a net electric charge, due to the loss or gain of electrons

The smallest particle of many compounds is the molecule. Molecules of compounds contain atoms of two or more elements. For example, the water molecule consists of two hydrogen atoms and one oxygen atom; a molecule of methane consists of one carbon atom and four hydrogen atoms. Chemists denote these molecules as H_2O and CH_4, respectively.

Not all compounds occur naturally as molecules. Many occur as aggregates of oppositely charged atoms or groups of atoms called **ions**. Atoms become charged by gaining or losing some of their electrons. In general, metal atoms *lose* electrons, whereas nonmetal atoms *gain* electrons.

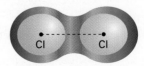

FIGURE 4.4 Four diatomic molecules (not to scale): hydrogen, oxygen, nitrogen, and chlorine. Chemists denote them as H_2, O_2, N_2, and Cl_2, respectively.

Atoms of metals that lose their electrons become positively charged; atoms of nonmetals that gain electrons become negatively charged.

Let's consider the difference between the sodium atom and the sodium ion. By examining Figure 4.3, we learn that the atomic number of sodium is 11. Thus, the neutral sodium atom has 11 electrons and 11 protons. If a sodium atom loses one electron, it then has only 10 left, although it still retains its 11 protons. By losing the electron, the sodium atom becomes a sodium ion. Its net charge is $+1$; that is, $+11 + (-10) = +1$.

Consider magnesium as a second example. The atomic number of magnesium is 12. Thus, the neutral magnesium atom has 12 electrons and 12 protons. When a magnesium atom loses two electrons, it becomes a magnesium ion. The magnesium ion possesses 10 electrons and 12 protons, and its net charge is $+2$.

These two ionization processes for metals can be represented as follows:

$$Na \longrightarrow Na^+ + e^-$$
$$Mg \longrightarrow Mg^{2+} + 2e^-$$

Here, e^- represents an electron; writing e^- to the right of the arrow means an electron is lost from the atom of the metal whose symbol appears to the left of the arrow. Na^+ and Mg^{2+} denote the sodium ion and the magnesium ion, respectively.

As noted earlier, the atoms of a nonmetal tend to *gain* electrons. Let's consider fluorine. The atomic number of fluorine is 9. The neutral fluorine atom possesses nine electrons and nine protons. If the fluorine atom somehow attains another electron, it will then have 10 electrons, but still only 9 protons. Its net charge is -1, that is, $+9 + (-10) = -1$. Because fluorine exists in the form of diatomic molecules, we represent the ionization of fluorine using F_2 to symbolize the fluorine molecule on the left of the arrow. This process can be designated as follows:

$$F_2 + 2e^- \longrightarrow 2F^-$$

Here, a 2 is written in front of e^- and F^- so that an equal number of electrons exist on each side of the arrow.

Metal atoms that have lost one or more electrons to become positively charged ions retain the name of the metal. As noted earlier, Na^+ and Mg^{2+} are called the sodium ion and the magnesium ion, respectively. Atoms of nonmetals that have gained one or more electrons to become negatively charged ions are named by modifying the name of the nonmetal so it ends in *-ide*; F^- is thus named the fluoride ion.

Frequently, two or more nonmetal atoms unite to form a **polyatomic ion**. For instance, one sulfur atom and four oxygen atoms unite to form an ion with a net charge of -2; it is called the *sulfate ion* and is symbolized as SO_4^{2-}. We observe other examples of polyatomic ions throughout this book.

polyatomic ion
■ An ion having more than one constituent atom

PERFORMANCE GOALS FOR SECTION 4.7:

- Describe the nature of a chemical bond.
- Associate the octet rule with the formation of chemical bonds.

4.7 The Nature of Chemical Bonding

As we noted earlier, when compounds form, the atoms of one element become attached to, or associated with, atoms of other elements by forces called chemical bonds. But how do these chemical bonds form? Chemists have pondered the answer to this question for centuries.

A clue to understanding the manner by which chemical bonds form can be deduced from the observation that the noble gases are relatively inert substances. This implies that the electrons in the atoms of the noble gases are specially arranged. Today, we know that the electronic stability of the noble gases is associated with the number of electrons in their outmost atomic orbitals as follows:

- An atom of helium has two electrons, both located in the innermost atomic orbital.
- The atoms of the noble gases other than helium have eight electrons in their outermost atomic orbitals, those orbitals farthest from their nuclei.

Scientists discovered long ago that molecules and ions often form when their constituent atoms acquire the electronic stability of the noble gases. Because there are only a prescribed number of electrons in an atom, they must somehow interact with the electrons from other atoms to acquire the electronic structure of the noble gases. For the elements in the first and second periods, this process occurs as follows:

- Hydrogen, lithium, and beryllium atoms attain the electronic structure of the helium atom. This means that electronic stability is achieved when these atoms have two electrons.
- Atoms of the elements other than hydrogen, helium, lithium, and beryllium attain the electronic structure of the noble gas nearest in atomic number. For the atoms of these elements, electronic stability is achieved by attaining a total of eight electrons in their outermost atomic orbitals. The observation that atoms other than hydrogen, helium, lithium, and beryllium achieve electronic stability by attaining a total of eight electrons is called the **octet rule**.

octet rule
■ The tendency for certain nonmetallic atoms to achieve electronic stability by gaining or sharing a total of eight electrons

There are two mechanisms by which atoms are capable of achieving these electronic structures, called *ionic* and *covalent* bonding. Each represents an extreme situation; in actuality, some degree of each description exists in all substances. The mechanisms are discussed independently in Sections 4.9 and 4.10.

PERFORMANCE GOALS FOR SECTION 4.8:

- Distinguish between the bonding (valence) and nonbonding electrons in an atom.
- Describe the use of a Lewis symbol.
- Illustrate how the Lewis symbols can readily be determined using the periodic table and Table 4.5.

4.8 Lewis Symbols

valence electrons
■ An atom's electrons that participate in chemical bonding to other atoms

nonbonding electrons
■ An atom's electrons that do not participate in bonding with other atoms

Lewis symbol
■ The symbol of an element together with a number of dots representing its bonding electrons

Although an atom of any element possesses a definite number of electrons, only some of them are involved in chemical bonding. These electrons are called the atom's **valence electrons**. The valence electrons are those electrons in an atom's outermost atomic orbital. Electrons that do not participate in bonding are called **nonbonding electrons**.

We display valence electrons by means of a **Lewis symbol**, named after an American chemist, Gilbert N. Lewis. A Lewis symbol consists of the symbol of an element together with a certain number· of dots that represent the atom's valence electrons. For example, Na· is the Lewis symbol for the sodium atom. The sodium atom has only one electron that participates in chemical bonding.

Table 4.5 lists the Lewis symbols of some representative elements important to the study of hazardous materials. Note that a simple way exists for determining the number of valence electrons for any element of an A family: The number that identifies the A family on the periodic table is also the number of valence electrons possessed by elements in that family. For instance, the halogens are located in the family identified as Group 7A, and we note from Table 4.5 that each halogen has *seven* valence electrons.

| TABLE 4.5 | Lewis Symbols of Some Representative Elements | | | | | | | |

FAMILY	1A	2A	3A	4A	5A	6A	7A	
	Li·	·Be·	·B·	·C·	·N·	·O·	:F·	H·
	Na·	·Mg·	·Al·	·Si·	·P·	·S·	:Cl·	
	K·	·Ca·	·Ga·	·Ge·	·As·	·Se·	·Br·	
	Rb·	·Sr·				·Te·	·I·	

- Describe the nature of an ionic bond.
- Describe how an ionic bond forms when metallic atoms and nonmetallic atoms unite.

4.9 Ionic Bonding

Electrons can be *transferred* from an atom of one element to an atom of another element resulting in the formation of positive and negative ions. This phenomenon generally occurs between the atoms of metals and nonmetals. The ions that form are attracted to each other by virtue of their opposite charges. Chemists call this electrostatic force of attraction an **ionic bond**. The formation of the ionic bond is based on a fundamental law of nature by which forms of matter with like charges ($+/+$ or $-/-$) repel each other, whereas forms of matter with unlike charges ($+/-$) attract.

Many atoms of the elements transfer or accept just the number of electrons that gives them eight electrons in their outermost atomic orbitals. By transferring only this number of electrons, these atoms attain the electronic stability of the nearest noble gas to them in atomic number.

For illustration, consider the ionic bonding in sodium fluoride. From Table 4.5 we learn that the Lewis symbols of sodium and fluorine are Na· and :F·, respectively. When these two elements combine at the atomic level, an atom of sodium transfers its single electron to a fluorine atom. By transferring one electron, the sodium atom electronically resembles neon. By accepting it, the fluorine atom electronically resembles neon. The atoms become charged; that is, they become ions. The process can be written schematically as follows:

$$\text{Na·} + \text{:F·} \longrightarrow \text{Na}^+ \ \text{:F:}$$

The attraction between these oppositely charged ions constitutes the ionic bond that binds the two ions together.

ionic bond
- The electrostatic force of attraction between oppositely charged ions

- Describe the nature of a covalent bond.
- Describe how a covalent bond forms when nonmetallic atoms unite.
- Describe the nature of a Lewis structure.
- Identify the nature of single, double, and triple covalent bonds.
- Use Lewis symbols to draw the Lewis structure of a compound.

4.10 Covalent Bonding

Electrons can also be *shared* by atoms of identical or different elements to form molecules of elements or compounds. This sharing of electrons is usually between nonmetal atoms. Atoms of the same nonmetal bond to one another forming molecules of the element; atoms of different nonmetals bond to one another forming molecules of a compound.

The atoms of nonmetals acquire their electronic stability by sharing electrons in a manner that permits them to resemble atoms of the noble gases nearest to them in atomic number. For the atoms in the first and second periods other than hydrogen, helium, lithium, and beryllium, this means sharing a total of eight electrons in their outermost atomic orbital. This includes the atom's valence electrons plus those it shares with another atom. Hydrogen atoms share only two electrons. One shared pair of electrons between any two atoms is called a **covalent bond**.

covalent bond
■ A shared pair of electrons between two atoms

Let's observe how two atoms of hydrogen combine to form a hydrogen molecule. H· is the Lewis symbol for hydrogen. To achieve the electronic structure of helium, the nearest noble gas, one hydrogen atom shares its only electron with the single electron from a second hydrogen atom. We represent this process as follows:

$$\text{H·} + \text{H·} \longrightarrow \text{H:H}$$

The pair of electrons shared between the two hydrogen atoms is a covalent bond. This manner of representing the hydrogen molecule (H:H) is called its **Lewis structure**.

Lewis structure
■ A means of displaying the bonding between the atoms of a molecule by using dashes to represent a shared pair of electrons

Two chlorine atoms combine to form a chlorine molecule. :Cl· is the Lewis symbol for chlorine. To achieve the electronic structure of argon, the nearest noble gas, each chlorine atom shares its unpaired electron. This formation of the chlorine molecule from two chlorine atoms can be represented as follows:

$$\text{:Cl·} + \text{:Cl·} \longrightarrow \text{:Cl:Cl:}$$

Let's consider next the combination of hydrogen and chlorine atoms. A hydrogen atom and a chlorine atom can share an electron pair to form a molecule of the substance called *hydrogen chloride*. The formation of a hydrogen chloride molecule from a hydrogen atom and a chlorine atom can be represented as follows:

$$\text{H·} + \text{:Cl·} \longrightarrow \text{H:Cl:}$$

The hydrogen atom shares an electron pair, so its electronic structure resembles that of helium, whereas the chlorine atom shares an electron pair, so that its electronic structure resembles that of argon.

An atom can also form several covalent bonds with other atoms by simply sharing more than one pair of electrons. For instance, consider the formation of the methane molecule. This molecule consists of one carbon atom and four hydrogen atoms. The Lewis symbols for carbon and hydrogen atoms are ·Ċ· and H·, respectively. In the methane molecule, the carbon atom shares each of its four electrons with the electrons from four hydrogen atoms, as represented here:

$$\text{·Ċ·} + 4\text{H·} \longrightarrow \begin{matrix} & \text{H} & \\ \text{H:} & \text{Ċ:} & \text{H} \\ & \text{H} & \end{matrix}$$

The sharing of electrons in the methane molecule results in an electronic arrangement like the neon atom for the carbon atom and an electronic arrangement like the helium atom for each of the four hydrogen atoms.

Sometimes, two nonmetallic atoms share more than one pair of electrons. This behavior is particularly characteristic of carbon atoms and results in the formation of *multiple bonds*. Two

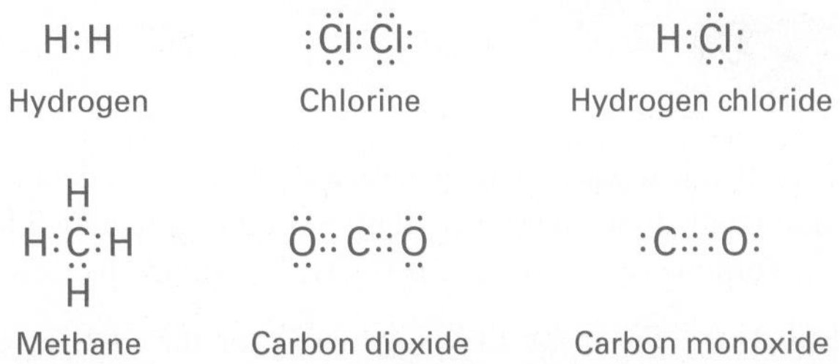

$$H:H \qquad :\overset{..}{C}l:\overset{..}{C}l: \qquad H:\overset{..}{C}l:$$

Hydrogen   Chlorine   Hydrogen chloride

$$\overset{\displaystyle H}{\underset{\displaystyle H}{H:\overset{..}{C}:H}} \qquad \overset{..}{O}::C::\overset{..}{O} \qquad :C:::O:$$

Methane   Carbon dioxide   Carbon monoxide

FIGURE 4.5 The Lewis structures of some simple molecules.

types of multiple bonds exist: double and triple bonds. A **double bond** consists of the sharing of two pairs of electrons (::), whereas a **triple bond** consists of the sharing of three pairs of electrons (:::).

The carbon dioxide molecule consists of one atom of carbon and two atoms of oxygen. The Lewis symbols for carbon and oxygen atoms are $\cdot\overset{..}{C}\cdot$ and $\cdot\overset{..}{O}\cdot$, respectively. The formation of a carbon dioxide molecule from one carbon atom and two oxygen atoms can be represented as follows:

$$\cdot\overset{..}{O}\cdot + \cdot\overset{..}{C}\cdot + \cdot\overset{..}{O}\cdot \longrightarrow O::C::O$$

Notice the existence of two pairs of electrons on each side of the carbon atom in the carbon dioxide molecule. Each of these shared pairs of electrons is a double bond. By sharing electrons in this fashion, the carbon atom and the two oxygen atoms all achieve the electronic arrangement of neon.

The formation of the carbon monoxide molecule from a carbon atom and an oxygen atom can be represented as follows:

$$\cdot\overset{..}{C}\cdot + \cdot\overset{..}{O}\cdot \longrightarrow :C:::O:$$

The three shared pairs of electrons between the carbon and oxygen atoms constitute a triple bond. Once again, the carbon and oxygen atoms achieve the electronic stability of the neon atom by sharing eight electrons between them. The Lewis structures of several other molecules are illustrated in Figure 4.5.

For the sake of simplicity, Lewis structures are usually written by drawing a long dash to represent the shared pair of electrons. The dots representing all other electrons are omitted. In this notation, the compounds previously noted can be represented as follows:

double bond
- A covalent bond composed of four electrons shared between two atoms

triple bond
- A covalent bond composed of six shared electrons between two atoms

$$\text{H}-\text{H} \qquad \text{Cl}-\text{Cl} \qquad \text{H}-\text{Cl} \qquad \overset{\displaystyle H}{\underset{\displaystyle H}{\text{H}-\overset{|}{\underset{|}{\text{C}}}-\text{H}}} \qquad \text{C}\equiv\text{O} \qquad \text{O}=\text{C}=\text{O}$$

Hydrogen   Chlorine   Hydrogen chloride   Methane   Carbon monoxide   Carbon dioxide

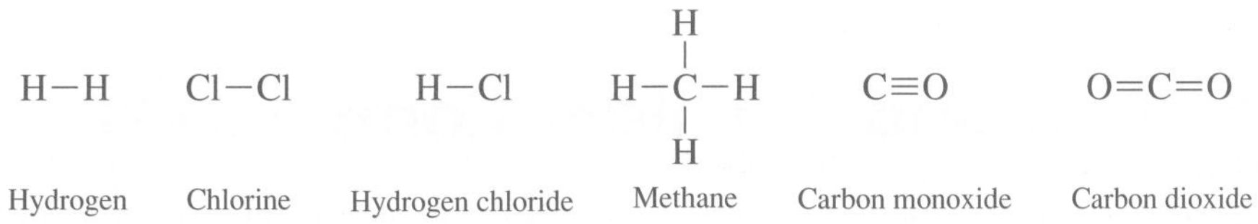

SOLVED EXERCISE 4.2

The chemical warfare agent called *phosgene* was responsible for approximately 80% of the casualties associated with exposures to chemical substances during World War I. It was called a *choking agent*, since the troops who inhaled phosgene subsequently experienced bronchial constriction and were unable to breathe properly. The chemical formula of phosgene is $COCl_2$. What is its Lewis structure?

Solution: The Lewis symbols for the carbon, oxygen, and chlorine atoms are provided in Table 4.5. The Lewis structure of a phosgene molecule can be represented by examining its formation from one carbon atom, one oxygen atom, and two chlorine atoms as follows:

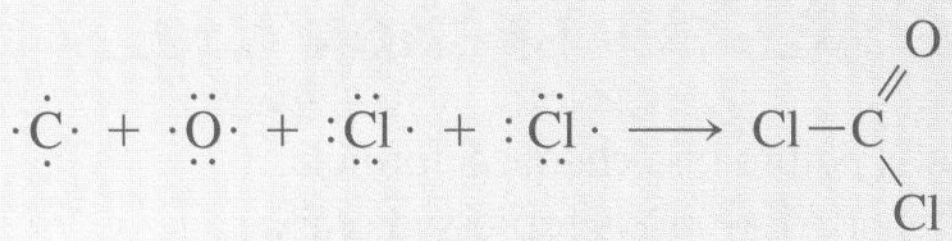

$$\cdot\overset{.}{C}\cdot + \cdot\overset{..}{O}\cdot + :\overset{..}{C}l\cdot + :\overset{..}{C}l\cdot \longrightarrow \text{Cl}-\overset{\displaystyle O}{\overset{\|}{\underset{\displaystyle \diagdown}{C}}}{\underset{\displaystyle \text{Cl}}{}}$$

Chapter 4 Chemical Forms of Matter   141

The chemical warfare agent called *diphosgene* is denoted as a *delay-action casualty agent*, since individuals first experience ill effects approximately 3 hours following its inhalation. The chemical formula of diphosgene is $C_2O_2Cl_4$. What is its Lewis structure?

Solution: From the Lewis symbols for the carbon, oxygen, and chlorine atoms provided in Table 4.5 the Lewis structure of a diphosgene molecule can be represented by examining its formation from two carbon atoms, two oxygen atoms, and four chlorine atoms. It is helpful first to recognize that an oxygen atom bonds between the two carbon atoms. Then, the process can be denoted as follows:

$$\cdot \ddot{C} \cdot + \cdot \ddot{C} \cdot + \cdot \ddot{O} \cdot + \cdot \ddot{O} \cdot + \, : \! \ddot{C} \! \dot{l} \cdot + \, : \! \ddot{C} \! \dot{l} \cdot + \, : \! \ddot{C} \! \dot{l} \cdot + \, : \! \ddot{C} \! \dot{l} \cdot \longrightarrow$$

(Chemists refer to diphosgene as trichloromethyl chloroformate.)

PERFORMANCE GOALS FOR SECTION 4.11:

- Distinguish between the nature of ionic compounds and covalent compounds.
- Contrast the general properties of ionic compounds and covalent compounds.

4.11 Ionic and Covalent Compounds

ionic compound
■ A compound whose atoms are bonded together mainly by ionic bonds

covalent compound
■ A substance whose constituent atoms are bonded together mainly by covalent bonds

Chemical compounds are classified into two groups based on the predominant nature of the bonding between their atoms. Chemical compounds with mainly ionic bonds are called **ionic compounds**. Compounds whose atoms are mainly bonded together by covalent bonds are called **covalent compounds**.

Ionic and covalent compounds are associated with a number of general properties summarized in Table 4.6. Note from this table that ionic compounds usually boil and melt at much higher temperatures than covalent compounds. Most ionic compounds dissolve in water and will conduct electricity when melted or dissolved in solution, whereas the opposite is true of covalent compounds. These properties are generalizations that reflect the characteristics of only the majority of ionic and covalent compounds.

PERFORMANCE GOALS FOR SECTION 4.12:

- Describe the general nature of a chemical formula.
- Indicate how a chemical formula is expressed orally.

TABLE 4.6	Contrasting Properties of Ionic and Covalent Compounds	
IONIC COMPOUNDS	**COVALENT COMPOUNDS**	
High melting points	Low melting points	
High boiling points	Low boiling points	
High solubility in water	Low solubility in water	
Nonflammable	Flammable	
Molten substances and their water solutions conduct an electric current	Molten substances do not conduct an electric current	
Exist predominantly as solids at room temperature	Exist as gases, liquids, and solids at room temperature	

4.12 The Chemical Formula

Chemists condense information regarding the chemical composition of substances by writing a **chemical formula** for each of them. For substances consisting of molecules, the chemical formula indicates the kinds and numbers of atoms present in each molecule. For instance, we observed earlier in this chapter that hydrogen and oxygen are symbolized as H_2 and O_2, respectively; these expressions are the chemical formulas of elemental hydrogen and oxygen. The subscript 2 means that each molecule contains two atoms. These chemical formulas are read verbally as "H-two" and "O-two," respectively.

We also observed that the chemical formulas for hydrogen chloride and methane are HCl and CH_4, respectively. These formulas represent the composition of the individual molecules. One molecule of hydrogen chloride consists of one hydrogen atom and one chlorine atom, whereas a molecule of methane consists of one carbon atom and four hydrogen atoms. We verbally read these chemical formulas as "HCl" and "CH-four," respectively. Figure 4.6 further illustrates the chemical formulas of two substances based on their composition.

However, substances are not always composed of molecules. For instance, sodium fluoride, an ionic compound, is composed of ions held together by an ionic bond. Experimental evidence indicates that the components of ionic compounds are not molecules but, rather, aggregates of

chemical formula
■A method of expressing the number of atoms or ions of specific types in a molecule or unit of matter through the use of chemical symbols for the elements and numbers or proportions of each kind of atom

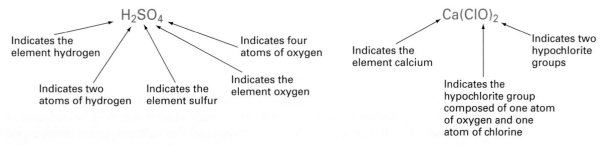

FIGURE 4.6 An explanation of the chemical formulas H_2SO_4 and $Ca(ClO)_2$. The formula H_2SO_4 refers to sulfuric acid, a corrosive liquid used to produce rayon, certain explosives, dyes, other acids, and detergents. The formula $Ca(ClO)_2$ refers to calcium hypochlorite, a constituent of some solid bleaching and disinfecting agents.

positive and negative ions. When we analyze a sample of sodium fluoride, we learn that it is composed of an equal number of sodium ions (Na^+) and fluoride ions (F^-). Chemists write its chemical formula as $Na^+\,F^-$, representing the lowest number of formula units. In common practice, the ionic charges are omitted and we simply write NaF.

In an emergency involving a hazardous material, knowing the names or chemical formulas of its constituents can be vital to the responders. When the constituents have been identified, a prompt and appropriate response can be directed toward handling the hazardous material while minimizing the risk to personnel. During emergency-response actions, the names and formulas of commercial chemical commodities can be determined in several ways. For instance, when the emergency occurs within a manufacturing plant or other workplace, the formulas are provided on MSDSs (Section 1.6-B). During a transportation mishap, the specific names of the hazardous material being transported can readily be obtained from a review of the information provided on the accompanying shipping paper. Sometimes, when a commodity is transported in bulk, its name is stenciled on the side of the transport vehicle. In Chapter 6, we review other ways to rapidly determine the nature of the hazardous materials involved in transportation mishaps.

PERFORMANCE GOALS FOR SECTION 4.13:

- Use Table 4.7 to learn the names and symbols of the common ions.
- Write the chemical formula of an ionic compound when provided its name.
- Name an ionic compound when provided its chemical formula.

4.13 Writing Chemical Formulas and Naming Ionic Compounds

The chemical formulas of ionic compounds are obtained from the symbols of the ions that make up a given substance. The most common ions are listed in Table 4.7. The names and symbols of these ions should be memorized. There are several simple rules you can follow.

4.13-A POSITIVE IONS

The following rules are useful for naming positive ions:

- Alkali metals, alkaline earth metals, aluminum, zinc, and hydrogen form monatomic positive ions, each taking the name of the element from which it was derived. For the hydrogen ion and each alkali metal ion, the net charge is $+1$, and for each alkaline earth metal ion, the net charge is $+2$.

Na^+	sodium ion	Ba^{2+}	barium ion
Al^{3+}	aluminum ion	H^+	hydrogen ion

- If a metal forms more than one positive ion, the ion is named in either of two ways:
  a. The ion takes the English name of the metal from which it is derived, immediately followed by a Roman numeral written in parentheses that indicates the net charge on the ion. This system of naming ionic compounds is called the *Stock system*.

Cu^+	copper(I) ion	Sn^{2+}	tin(II) ion
Cu^{2+}	copper(II) ion	Sn^{4+}	tin(IV) ion

TABLE 4.7 Some Common Ions

POSITIVE IONS	NEGATIVE IONS
Ammonium, NH_4^+	Acetate, $C_2H_3O_2^-$
Copper(I) or cuprous, Cu^+	Bromide, Br^-
Hydrogen, H^+	Chloride, Cl^-
Silver, Ag^+	Chlorate, ClO_3^-
Sodium, Na^+	Chlorite, ClO_2^-
Potassium, K^+	Cyanide, CN^-
	Fluoride, F^-
Barium, Ba^{2+}	Hydrogen carbonate, or bicarbonate, HCO_3^-
Cadmium, Cd^{2+}	Hydrogen sulfate, or bisulfate, HSO_4^-
Calcium, Ca^{2+}	Hydrogen sulfite, or bisulfite, HSO_3^-
Cobalt(II) or cobaltous, Co^{2+}	Hydroxide, OH^-
Copper(II) or cupric, Cu^{2+}	Hypochlorite, ClO^-
Iron(II) or ferrous, Fe^{2+}	Iodide, I^-
Lead(II) or plumbous, Pb^{2+}	Nitrate, NO_3^-
Magnesium, Mg^{2+}	Nitrite, NO_2^-
Manganese(II) or manganous, Mn^{2+}	Perchlorate, ClO_4^-
Mercury(I) or mercurous, Hg_2^{2+}	Permanganate, MnO_4^-
Nickel, Ni^{2+}	
Strontium, Sr^{2+}	
Tin(II) or stannous, Sn^{2+}	Carbonate, CO_3^{2-}
Zinc, Zn^{2+}	Oxide, O^{2-}
	Peroxide, O_2^{2-}
Chromium(III) or chromic, Cr^{3+}	Sulfate, SO_4^{2-}
Iron(III) or ferric, Fe^{3+}	Sulfide, S^{2-}
	Sulfite, SO_3^{2-}
Tin(IV) or stannic, Sn^{4+}	
	Phosphate, PO_4^{3-}

b. The ion takes the Latin name of the metal from which it was derived, together with one of the suffixes -ous or -ic representing the lower and higher net ionic charge, respectively. We call this system of naming ionic compounds the *older system*.

Cu^+	cuprous ion	Sn^{2+}	stannous ion
Cu^{2+}	cupric ion	Sn^{4+}	stannic ion

- There are only two common polyatomic positive ions:

NH_4^+	ammonium ion	Hg_2^{2+}	mercury(I) or mercurous ion

4.13-B NEGATIVE IONS

There are also rules are useful for naming negative ions:

- Monatomic negative ions are named by adding the suffix *-ide* to the stem of their names.

F^-	fluoride ion	Cl^-	chloride ion
O^{2-}	oxide ion	S^{2-}	sulfide ion

- Two polyatomic negative ions are named with the suffix *-ide*.

CN^-	cyanide ion	OH^-	hydroxide ion

- The polyatomic negative ions containing carbon are named uniquely. Two polyatomic negative ions, the carbonate and acetate ions, are commonly encountered.

CO_3^{2-}	carbonate ion	$C_2H_3O_2^-$	acetate ion

- When a nonmetal forms two different negative ions containing oxygen, the suffixes *-ite* and *-ate* are used to name the ion with the lesser and greater number of oxygen atoms, respectively.

NO_2^-	nitrite ion	SO_3^{2-}	sulfite ion
NO_3^-	nitrate ion	SO_4^{2-}	sulfate ion

- When a nonmetal forms more than two different negative ions containing oxygen, the prefixes *hypo-* (meaning lower than usual) and *per-* (meaning higher than usual) are used together with the suffixes *-ite* and *-ate* to name the ions having the lowest and highest number of oxygen atoms, respectively, as follows: *hypo-* _____ *-ite*, _____ *-ite*, _____ *-ate*, and *per-* _____ *-ate*, where the dashed line represents the stem of the identifying nonmetal. For example the negative ions containing one chlorine atom and one to four oxygen atoms are named as follows:

ClO^-	ClO_2^-	ClO_3^-	ClO_4^-
Hypochlorite ion	Chlorite ion	Chlorate ion	Perchlorate ion

- Negative ions containing hydrogen and oxygen atoms (other than the hydroxide and acetate ions) are named *hydrogen* _____, or alternatively, *bi* _____, where the line represents the stem of the identifying nonmetal.

HCO_3^-	hydrogen carbonate, or bicarbonate	HSO_4^-	hydrogen sulfate, or bisulfate

When we know the chemical formulas of substances, we name the associated compounds by simply listing the names of the positive and negative ions in that order. For instance, NaCl is sodium chloride, MgS is magnesium sulfide, and $BaCO_3$ is barium carbonate.

Frequently, we encounter a chemical formula containing parentheses, such as $Fe(ClO_4)_3$. To name such a formula, it is sometimes best to rewrite it and indicate the net charges on the constituent ions. In this case, we have $Fe^{3+}(ClO_4^-)_3$. This shows us that the +3 charge on the iron(III) ion balances the −3 charge on three perchlorate ions. The total positive and negative charges of the ions are equal (+3 and −3), since the compound itself is uncharged, or neutral. The compound is named iron(III) perchlorate by the Stock system and ferric perchlorate by the older system.

Before proceeding further, note the names of the following chemical formulas:

$Al_2(SO_3)_3$	aluminum sulfite	$(NH_4)_2CO_3$	ammonium carbonate
$Ba(CN)_2$	barium cyanide	$Cd_3(PO_4)_2$	cadmium phosphate
$Ca(MnO_4)_2$	calcium permanganate	$Ni(NO_3)_2$	nickel nitrate

Suppose next that we are interested in writing a chemical formula when we know the name of a substance. It is again best to follow two simple rules:

- If the net charge of the positive and negative ions is equal (but opposite in sign), the formula of a substance is obtained by simply listing the symbol of the positive ion first, then the symbol of the negative ion. For instance, NaF, CaS, ZnO, KOH, $NaClO_3$, $BaSO_4$, and $AlPO_4$ all have component ions with numerically equal charges. Their names are sodium fluoride, calcium sulfide, zinc oxide, potassium hydroxide, sodium chlorate, barium sulfate and aluminum phosphate, respectively.
- If the net charges of the positive and negative ions are not numerically equal, first write the symbol of the positive and negative ions together with their respective charges. Then, simply cross the numbers representing the charges (but not the signs) to the opposite ion, as shown in the following examples:

Potassium sulfate	$K_2^+ \quad SO_4^{2-}$,	or K_2SO_4	(1 is not written)
Aluminum oxide	$Al^{3+}{}_2 \quad O^{2-}{}_3$,	or Al_2O_3	
Barium iodide	$Ba^{2+} \quad I^-{}_2$,	or BaI_2	(1 is not written)
Ammonium sulfate	$NH_4^+{}_2 \quad SO_4^{2-}$,	or $(NH_4)_2SO_4$	(1 is not written)

One special case should be noted for using this rule. When the charges are multiples of one another, such as +4 and −2, the lowest multiple is used. For instance, the chemical formula of tin(IV) oxide is SnO_2, not Sn_2O_4. In the study of hazardous materials, this exception arises only for compounds containing the tin(IV) ion.

SOLVED EXERCISE 4.4

The use of a portable chemical extinguisher containing a water solution of potassium acetate is recommended for extinguishing class K fires (Section 1.3-E). What is the chemical formula of potassium acetate?

Solution: By consulting Table 4.7, we determine that the potassium and acetate ions are K^+ and $C_2H_3O_2^-$, respectively. Because their charges are equal in magnitude, the chemical formula of potassium acetate is written as $KC_2H_3O_2$.

Provide acceptable names for the compounds whose chemical formulas are **(a)** $MgCO_3$, **(b)** $FeSO_4$, and **(c)** $Cd_3(PO_4)_2$.

Solution: To answer this question, we use the information in Table 4.7.

(a) Mg^{2+} and CO_3^{2-} are the symbols for the magnesium and carbonate ions, respectively. Because their charges are equal in magnitude, $MgCO_3$ is the formula for the compound whose name is magnesium carbonate.

(b) Iron has two ions: ferrous, or iron(II) (Fe^{2+}), and ferric, or iron(II) (Fe^{3+}); SO_4^{2-} is the symbol for the sulfate ion. Because there is no subscript adjacent to the Fe in $FeSO_4$, the symbol represents the ferrous ion. Acceptable names for the compound whose formula is $FeSO_4$ are ferrous sulfate and iron(II) sulfate.

(c) Cd^{2+} and PO_4^{3-} are the symbols for the cadmium and phosphate ions, respectively. $Cd_3(PO_4)_2$ is the formula for the compound whose name is cadmium phosphate.

PERFORMANCE GOALS FOR SECTION 4.14:

- Learn the common names and formulas of the substances in Table 4.8.
- Using Table 4.9, learn the names of the prefixes that identify the number of atoms of a given type in a covalent compound.
- Write the chemical formula of a simple covalent compound when provided its name.
- Provide the name of a simple covalent compound when provided its chemical formula.

4.14 Some Chemical Formulas and Names of Covalent Compounds

It is generally difficult to write the chemical formula of a covalent compound. In this book, we shall learn many formulas of covalent compounds as the occasions present themselves. Nevertheless, a great many covalent names have acquired common names that are still used as part of the chemical language. Some examples are listed in Table 4.8. These names and formulas should be memorized.

A number of covalent compounds contain only two elements, and these substances can be named in a simple fashion. We name the element that is written first in a chemical formula, preceded by a Greek prefix indicating the number of atoms of the element in the compound; then, we name the other element, preceded also by its relevant prefix to indicate the number of its atoms and modifying its ending to *-ide*. Table 4.9 lists the Greek prefixes corresponding to the numbers that are used to name these simple covalent compounds.

We have encountered the chemical formulas of several simple covalent compounds containing only two elements: HCl, CO, and CO_2. The compounds represented by these formulas are named hydrogen chloride, carbon *mono*xide, and carbon *di*oxide, respectively. Other examples of naming covalent compounds are provided in Table 4.9.

TABLE 4.8 — Common Names for Some Simple Covalent Compounds

FORMULA OF COMPOUND	COMMON NAME
C_2H_2	Acetylene
NH_3	Ammonia
C_6H_6	Benzene
C_2H_6	Ethane
N_2H_4	Hydrazine
H_2S	Hydrogen sulfide
CH_4	Methane
PH_3	Phosphine
C_3H_8	Propane
H_2O	Water

TABLE 4.9 — Greek Prefixes Used in Naming Simple Covalent Compounds

PREFIX		COMPOUND
Mono-	1	Carbon monoxide, CO[a,b]
Di-	2	Carbon dioxide, CO_2
Tri-	3	Phosphorus trichloride, PCl_3
Tetra-	4	Carbon tetrachloride, CCl_4
Penta-	5	Phosphorus pentachloride, PCl_5
Hexa-	6	Sulfur hexafluoride, SF_6
Hepta-	7	Dichlorine heptoxide, Cl_2O_7[a]
Octa-	8	Dichlorine octoxide, Cl_2O_8
Ennea-[c]	9	Tetraiodine enneaoxide, I_4O_9
Deca-	10	Tetraphosphorus decoxide, P_4O_{10}[a]

[a]When two vowels appear next to one another, as oo in monooxide and ao in heptaoxide, the vowel from the Greek prefix is dropped for the sake of euphony.

[b]The prefix mono- is generally used only to avoid ambiguity.

[c]The Latin prefix nona- is also used to denote nine atoms. Thus, I_4O_9 may also be correctly named tetraiodine nonoxide.

Dinitrogen tetroxide has been used to oxidize rocket fuels. What is its chemical formula?

Solution: From Table 4.9, we see that the Greek prefixes *di-* and *tetra-* refer to two and four, respectively. This means that the dinitrogen tetroxide molecule contains two nitrogen atoms and four oxygen atoms. Accordingly, the chemical formula of dinitrogen tetroxide is N_2O_4.

When animal manure decomposes, the compounds having the chemical formulas NH_3 and H_2S are generated. What are the names of these substances?

Solution: Using Table 4.8, we determine that the names of the substances having the formulas NH_3 and H_2S are ammonia and hydrogen sulfide, respectively.

The gaseous substance having the chemical formula SF_4 is used to produce water and oil repellants. What is the name of this substance?

Solution: Using Table 4.9, we determine that each molecule of SF_4 is composed of one atom of sulfur and four atoms of fluorine. Consequently, the name of the substance is sulfur tetrafluoride.

PERFORMANCE GOALS FOR SECTION 4.15:

- Write the names and chemical formulas of the common binary acids.
- Learn the names and chemical formulas of the oxyacids listed in Table 4.10.
- Relate the names of an oxyacid with the name of its corresponding polyatomic ion.

4.15 Naming Acids

We learn in Chapter 8 that acids are compounds that produce hydrogen ions when dissolved in water. This means that all acids must contain hydrogen in their chemical structures.

4.15-A BINARY ACIDS

Binary acids contain hydrogen and one other nonmetal. The chemical formula of a binary acid is HX or H_2X, where X is the symbol of the nonmetal other than hydrogen. These acids are named by placing the prefix *hydro-* before the stem of the name of an identifying nonmetal, attaching the suffix *-ic* to this stem, and adding the word *acid* to the name. There are two important binary acids whose chemical formulas should be noted:

binary acid
■ An acid composed of hydrogen and a nonmetal

    HF   hydrofluoric acid   HCl   hydrochloric acid

The names of these binary acids are valid only when the acids are dissolved in water. As single substances, they are named as compounds of hydrogen: hydrogen fluoride and hydrogen chloride.

4.15-B OXYACIDS

Another important group of acids are the **oxyacids**. In addition to hydrogen and the identifying nonmetal, these acids contain one or more oxygen atoms. The chemical formula of the oxyacids is H_mXO_n, where X is the symbol of the identifying nonmetal, and *m* and *n* are numbers. There are only 10 oxyacids important to the study of hazardous materials. Their names and chemical formulas are listed in Table 4.10. They should be memorized.

oxyacid
■ An acid composed of hydrogen, oxygen, and a nonmetal

The names of oxyacids correlate directly with the names of their associated polyatomic ions:

■ Polyatomic ions whose names end in *-ate* are derived from acids whose names end in *-ic*; for example, the nit*rate* ion is derived from nit*ric* acid, and the perchlor*ate* ion is derived from perchlor*ic* acid.
■ Polyatomic ions whose names end in *-ite* are derived from acids whose names end in *-ous*; for example, the nit*rite* ion is derived from nit*rous* acid, and the hypochlor*ite* ion is derived from hypochlor*ous* acid.

TABLE 4.10	Names and Chemical Formulas of Some Oxyacids
NAME	**CHEMICAL FORMULA**
Sulfurous acid	H_2SO_3
Sulfuric acid	H_2SO_4
Nitrous acid	HNO_2
Nitric acid	HNO_3
Phosphorous acid	H_3PO_3
Phosphoric acid	H_3PO_4
Hypochlorous acid	$HClO$
Chlorous acid	$HClO_2$
Chloric acid	$HClO_3$
Perchloric acid	$HClO_4$

PERFORMANCE GOALS FOR SECTION 4.16:

- Describe how the molecular weight or formula weight of a substance is determined.
- Describe the concept of the mole.
- Calculate the number of moles of a substance when its amount is provided.
- Identify the Avogadro number.
- Use the Avogadro number to calculate how many atoms, molecules, or formula units are present in a given amount.

4.16 Molecular Weights, Formula Weights, and the Mole

molecular weight
■ The sum of the atomic weights of all the atoms in a molecular formula

The relative weight of a compound that occurs as molecules is called the **molecular weight** and is the sum of the atomic weights of each atom that are part of the molecule. Consider the water molecule. Its molecular weight is determined as follows:

$$2 \text{ hydrogen atoms} = 2 \times 1.008 = 2.016$$
$$1 \text{ oxygen atom} = 1 \times 15.999 = \underline{15.999}$$
$$\text{Molecular weight of } H_2O = 18.015$$

formula weight
■The sum of the atomic weights of all the atoms in a chemical formula

The relative weight of a compound that occurs as formula units is called the **formula weight**. It is the sum of the atomic weights of all atoms that make up one formula unit. Consider sodium fluoride. Its formula weight is determined as follows (to five significant figures):

$$1 \text{ sodium ion} = 22.990$$
$$1 \text{ flouride ion} = \underline{18.998}$$
$$\text{Formula weight of NaF} = 41.988$$

mole
■ An Avogadro number of particles; the amount of a substance whose mass equals its molecular weight or formula weight, as relevant

Chemists also frequently make use of a unit quantity called the **mole**. The mole is the approved SI unit for the amount of any substance. As a unit, it is abbreviated as *mol* (without the *e*). One mole of atoms, molecules, or formula units represents the amount of substance that has a mass in grams equal to its atomic weight, molecular weight, or formula weight, respectively. Thus, 1 mol of carbon is an amount of carbon that has a mass of 12.011 g. Also, 1 mol of water is an amount of water that has a mass of 18.105 g.

Because the concept of the mole applies to any type of particle, it is important to identify just what the particle is. For instance, 1 mol of hydrogen *atoms* has a mass of 1.0079 g, but 1 mol of hydrogen *molecules* has a mass of 2.0158 g.

Avogadro number
■The number of atoms, molecules, or other units contained in a mole of a specified substance, 6.02×10^{23}

Finally, 1 mol of atoms, molecules, or formula units contains a definite number (6.022×10^{23}) of these units. This number is called the **Avogadro number**, in honor of the scientist who first suggested its existence. The exponential notation refers to a number in which the decimal point has been moved 23 places to the right. We can also write this number without using an exponent as 602,200,000,000,000,000,000,000. The mole is analogous to units like the dozen or ream, which mean 12 and 500 items, respectively. One mol of hydrogen atoms is 6.022×10^{23} H atoms; 1 mol of hydrogen molecules is 6.022×10^{23} H_2 molecules, and 1 mol of sodium fluoride units is 6.022×10^{23} NaF units.

The Avogadro number is not just huge—it is so enormous that it is difficult to appreciate how large it is. Imagine that you are counting jelly beans at the rate of 3 beans per second. At this rate, it would take 6×10^{15} years to count 1 mol of jelly beans! This is a million times older than the age of Earth. The enormous magnitude of this number reflects the minute dimensions of atoms and molecules on the scale of "ordinary" measurements.

When hydrazine burns, 148.6 kcal/mol of heat evolves to the environment. How many kcal/g of heat evolves?

Solution: From Table 4.8, we see that the chemical formula of hydrazine is N_2H_4. Using this formula, we calculate the molecular weight of hydrazine to be 32.045:

$$
\begin{array}{rl}
\text{2 nitrogen atoms, } 2 \times 14.0067 = & 28.013 \\
\text{4 hydrogen atoms, } 4 \times 1.008 = & \underline{4.032} \\
\text{Molecular weight } = & 32.045
\end{array}
$$

One mole of hydrazine is an amount equal to 32.045 g. When hydrazine burns, 4.5 kcal/g of heat evolves:

$$
\frac{148.6 \text{ kcal/mol}}{32.045 \text{ g/mol}} = 4.5 \text{ kcal/g}
$$

Elements and Compounds

4.1 There are 16 elements known to be essential for healthy plant growth: carbon, hydrogen, oxygen, nitrogen, phosphorus, potassium, calcium, magnesium, sulfur, boron, copper, iron, manganese, zinc, molybdenum, and chlorine. What are the symbols of these elements?

4.2 For the proper functioning and survival of the human organism, trace amounts of 14 metals and metalloids are essential in the diet. Their symbols are Ca, K, Na, Mg, Fe, Zn, Cu, Sn, V, Cr, Mn, Mo, Co, and Ni. What are the names of these elements?

Physical and Chemical Changes

4.3 Classify each of the following phenomena as a physical or chemical change:

(a) The detonation of trinitrotoluene (TNT) produces carbon monoxide, water, nitrogen, and carbon.
(b) When exposed to an ignition source, hexane vapor ignites at -7 °F (-22 °C).
(c) Battery acid is neutralized when mixed with sodium bicarbonate.

Atomic Weights of the Elements

4.4 Iodine has only one naturally occurring isotope, iodine-127. Its atomic mass is 126.9044. What is the atomic weight of iodine?

The Periodic Table

4.5 Consider the element that has an atomic number of 114.

(a) Identify the family of which this element is a member.
(b) Identify the period of which this element is a member.
(c) Identify the element that occupies the position immediately above it on the periodic table.

Atoms, Moles, and Ions

4.6 How many valence electrons do atoms of each of the following elements possess:

(a) sodium	(b) carbon	(c) silicon
(d) aluminum	(e) bromine	(f) oxygen

4.7 Write the Lewis symbol for the following atoms:

(a) nitrogen	(b) calcium	(c) boron
(d) phosphorus	(e) potassium	(f) sulfur

4.8 Determine the net charge of each of the following ions:

(a) sodium (b) oxide (c) hydrogen
(d) chloride (e) magnesium (f) sulfide

Chemical Bonding

4.9 Anaerobic bacteria initiate the decomposition of some forms of matter in the absence of oxygen. The phenomenon is often associated with the generation of hydrogen sulfide, a gas possessing the offensive odor of rotten eggs. What is the Lewis structure for the hydrogen sulfide molecule?

4.10 Dentists frequently use a 2% sodium fluoride solution in water to prevent tooth decay. Illustrate the manner by which a unit of sodium fluoride forms from atoms of the elements sodium and fluorine.

4.11 Fluorite is a mineral that is commonly used in making opalescent glass, as well as enameling cooking utensils. The principal chemical constituent of fluorite is calcium fluoride. Describe the nature of the chemical bonds in a unit of calcium fluoride and show that the chemical formula of calcium fluoride is CaF_2.

Chemical Formulas

4.12 Write the chemical formula of the ionic compound containing each of the following pairs of ions:

(a) Na^+ and ClO_3^- (b) Cu^{2+} and O^{2-} (c) NH_4^+ and SO_4^{2-}
(d) Cd^{2+} and OH^- (e) Pb^{2+} and NO_2^{2-} (f) Mg^{2+} and ClO_4^-

4.13 Name the substance having each of the following chemical formulas:

(a) O_2 (b) FeS (c) KOH
(d) $NiBr_2$ (e) $HgSO_4$ (f) $Al(OH)_3$
(g) $Cd(ClO_4)_2$ (h) Na_2S (i) MgI_2
(j) $Ca(MnO_4)_2$ (k) $Ba(CN)_2$ (l) H_2O_2
(m) H_2SO_4 (n) $Mg(ClO_3)_2$ (o) NiF_2
(p) $Ba(NO_2)_2$ (q) NBr_3 (r) $Pb(C_2H_3O_2)_2$
(s) $SrSO_3$ (t) AgCl (u) CO_2
(v) $Sr(ClO)_2$ (w) MnO (x) HCl (aqueous)

4.14 Give the chemical formula for each of the following substances:

(a) nitrogen (b) chlorine (c) nitric acid
(d) ammonia (e) hydrogen sulfide (f) silver nitrate
(g) barium permanganate (h) cadmium phosphate (i) nickel sulfate
(j) magnesium peroxide (k) stannic cyanide (l) zinc chloride
(m) mercuric nitrate (n) carbon monoxide (o) chromous nitrate
(p) magnesium bisulfite (q) ammonium bromide (r) lead perchlorate
(s) acetylene (t) hydrogen chloride (u) magnesium sulfate
(v) phosphoric acid (w) cobalt(II) carbonate (x) aluminum sulfate
(y) strontium iodide (z) magnesium ammonium phosphate

Molecular Weights, Formula Weights, and Moles

4.15 α-Chloroacetophenone is a very potent lacrimator that law enforcement officers discharge as tear gas during riots and other forms of civil unrest. Its chemical formula is C_8H_7OCl. What is the molecular weight of α-chloroacetophenone?

4.16 Argon is a monatomic noble gas that can be isolated by liquefying the components of air.

(a) How many moles of argon are contained in 119.98 g of argon?

(b) How many argon molecules are contained in 119.98 g of argon?

4.17 An average U.S. penny weighs 3.0 g. Assuming the penny is composed of 100% copper by mass, determine the number of copper atoms in an average penny.

4.18 When propane burns in the air, 526.3 kcal/mol of heat is released to the environment.

(a) How many kcal/g of heat evolves?

(b) How many Btu/lb of heat evolves?

KEY TERMS

O nce we have properly named substances and written their formulas, we can begin to examine how they interact with one another. Although the chemical reactions of individual hazardous materials will be examined throughout most of the remainder of this textbook, we note in this chapter some features that are common to all chemical reactions. This includes writing and balancing chemical equations and learning how certain factors affect reaction rates.

In the study of hazardous materials, combustion is a chemical reaction of major concern, especially to fire-service personnel. Consequently, we note what occurs when a substance burns, as well as how other substances effectively function to extinguish fires.

PERFORMANCE GOALS FOR SECTION 5.1:

- Describe the nature of a chemical change.
- Demonstrate a chemical change by means of an equation.

5.1 The Chemical Reaction

A substance that has undergone a chemical change is no longer the original substance. In other words, it has become one or more new substances. To describe a chemical change is to indicate that a substance "reacted" in a particular manner. For instance, we say that "dynamite exploded," "hydrogen burned," "acid corroded metal," or similar expressions. Each of these statements relates to a chemical change that we generally call a *chemical reaction*.

In chemistry, it is commonplace to summarize the result of any given reaction in the form of a **chemical equation**. An equation is a shorthand method for expressing a reaction in terms of written chemical formulas. For instance, if we wish to describe the combustion of elemental carbon, we may write the equation as follows:

chemical equation
■ The symbolic representation of a chemical process that respectively lists the chemical formulas of its reactants and products on each side of an arrow

reactants
■ Two or more substances that enter into a chemical reaction

products
■ A substance produced as a result of a chemical change

$$C + O_2 \longrightarrow CO_2$$

The chemical formulas of the substances that enter the chemical reaction are written on the left of the arrow; they are called **reactants**. The formulas of the substances formed as a consequence of the reaction are written on the right of the arrow; they are called **products**. The arrow itself is read as "yields," "produces," "forms," or "gives." Consequently, one way to read this equation is "carbon plus oxygen yields carbon dioxide."

Chemical equations are the basic language of chemistry. They should contain as much information as possible concerning the specific chemical change under consideration. We list not only the formulas of the substances reacting and forming but sometimes also their physical states under the temperature and pressure conditions of the reaction. Physical states are indicated in parentheses next to the chemical formulas of the reactants and products with the italicized letters *s*, *l*, and *g*, which symbolize solid, liquid, and gas, respectively. Examples of the symbols used in equations are provided in Table 5.1. Using the relevant symbols, we denote the combustion of elemental carbon by the following equation:

$$C(s) + O_2(g) \longrightarrow CO_2(g)$$

Before we can write a chemical equation, we need to know first the chemical formulas of the reactants and products. To write the more complete form of an equation, we must also know the physical state of the reactants and products. The physical states of reactants and products are determined from chemical reference books or through laboratory experimentation.

TABLE 5.1   Symbols Used in Chemical Equations

SYMBOL	MEANING	EXAMPLES
$(g)^a$	Gaseous reactant or product	$H_2(g)$, $CO_2(g)$
(l)	Liquid reactant or product	$H_2O(l)$, $Br_2(l)$
$(s)^b$	Solid reactant or product	$Fe(s)$, $S_8(s)$
$(aq)^c$	Reactant or product dissolved in water	$NaCl(aq)$, $KNO_3(aq)$
$(conc)^d$	Reactant undiluted with water	$HCl(conc)$

[a]An arrow pointing upward ($\uparrow$) is also used when the gas is a product of a reaction.
[b]An arrow pointing downward ($\downarrow$) is also used when a solid precipitates from solution.
[c]Meaning *aqueous*.
[d]Meaning *concentrated*.

PERFORMANCE GOALS FOR SECTION 5.2:

- Describe the process of balancing an equation.

5.2 Balancing Simple Equations

Not only must an equation summarize what occurs qualitatively during a chemical change, but it must also account for other more fundamental observations. In particular, each equation must be written so that the chemical change at issue adheres to the **law of conservation of mass and energy** (Section 2.6). This law states that there is no apparent change in mass during "ordinary" chemical reactions. The term "ordinary" as used here means chemical reactions occurring at the molecular or ionic level, as opposed to the reactions of atomic nuclei.

> **law of conservation of mass and energy**
> ■ The observation that the total amount of mass and energy in the universe is constant

The conservation of mass and energy requires that during a given chemical change the atoms of any element are neither created nor destroyed. This means that the number of atoms before and after a reaction remains the same. When chemical equations are written, there must be an equal number of atoms for each element on both sides of the arrow. Such an equation is then said to be **balanced**. The simple equation written in Section 5.1 is balanced, since it has one carbon atom and two oxygen atoms on each side of the arrow.

> **balanced**
> ■ The feature of a chemical equation that has an equal number of like atoms on each side of the arrow

Not all equations are directly balanced after we write the chemical formulas of its reactants and products; in fact, they are often unbalanced at this point. We must select a proper coefficient to place *in front of* the appropriate formula so that the equation then becomes balanced. The correct formula of a substance must never be changed when balancing an equation. Furthermore, coefficients are never written in the middle of a formula, such as H_23O.

Most simple equations can be balanced by inspection. Although there are no absolute rules for balancing equations by inspection, the following points are useful when first learning this process:

- Write the correct formula for each reactant and product and separate the reactants from the products with an arrow.
- If known, write the physical state of the reactants and products in parentheses after each formula.
- Choose the formula of the substance containing the greatest number of atoms of an arbitrary element. Insert a number in front of one or both formulas so that the number of atoms for this particular element is balanced.

- If polyatomic ions appear in an equation, balance them as single units only when they retain their identity on both sides of the arrow.
- Balance any remaining atoms or ions.

Let's consider an example. Suppose we wish to write the balanced chemical equation for the reaction that occurs when methane burns in air to form carbon dioxide and water vapor. This is an ordinary combustion reaction involving the chemical combination of methane and atmospheric oxygen. First, we write the chemical formulas of the reactants and products. The formula of methane is CH_4 (from Table 4.8), oxygen is O_2, carbon dioxide is CO_2, and water is H_2O. Under the reaction conditions, each is a gas or vapor. Hence, we initially write the following:

$$CH_4(g) + O_2(g) \longrightarrow CO_2(g) + H_2O(g)$$

Next, we note that this is an unbalanced equation because there are more hydrogen atoms in CH_4 than in H_2O. We balance the hydrogen atoms by inserting a 2 in front of the formula for water, as follows:

$$CH_4(g) + O_2(g) \longrightarrow CO_2(g) + 2H_2O(g)$$

Because there are no ions in this equation, we next balance the number of the oxygen atoms by inserting a 2 in front of O_2, as follows:

$$CH_4(g) + 2O_2(g) \longrightarrow CO_2(g) + 2H_2O(g)$$

When performing such exercises it is usually best to make one final check: there are one carbon atom, four hydrogen atoms, and four oxygen atoms on each side of the arrow.

PERFORMANCE GOALS FOR SECTION 5.3:

- Identify the different types of simple chemical reactions.
- Write equations for each type.

5.3 Types of Chemical Reactions

By now, one point should be apparent: an equation denoting the chemical reaction of a hazardous material cannot be written when the products of the reaction are unknown. Although identifying the reaction products is not always a simple feat, we can frequently determine them by knowing the reaction type. The reactions in which we are interested can be classified as one of four types reviewed in the next section with illustrative examples.

5.3-A COMBINATION (OR SYNTHESIS) REACTIONS

combination reaction (synthesis reaction)
■ The type of chemical reaction involving two substances that react to form a single product

In a **combination reaction**, or **synthesis reaction**, two or more simpler substances combine to form a more complex substance. Some examples of such reactions are illustrated by the following equations:

$$2H_2(g) \quad + \quad O_2(g) \quad \longrightarrow \quad 2H_2O(l)$$
Hydrogen   Oxygen   Water

$$2Na(s) \quad + \quad Cl_2(g) \quad \longrightarrow \quad 2NaCl(s)$$
Sodium   Chlorine   Sodium chloride

$$C(s) \quad + \quad O_2(g) \quad \longrightarrow \quad CO_2(g)$$
Carbon   Oxygen   Carbon dioxide

decomposition reaction
■ The type of chemical reaction involving the breakup of one substance into two or more elementary substances

5.3-B DECOMPOSITION REACTIONS

In a **decomposition reaction**, a relatively complex substance is broken down into several simpler substances. Some examples of decomposition reactions are illustrated by the following equations:

$$2H_2O(l) \longrightarrow 2H_2(g) + O_2(g)$$

Water   Hydrogen   Oxygen

$$Na_2CO_3(s) \longrightarrow Na_2O(s) + CO_2(g)$$

Sodium carbonate   Sodium oxide   Carbon dioxide

$$(NH_4)_2CO_3(s) \longrightarrow 2NH_3(s) + CO_2(g) + H_2O(g)$$

Ammonium carbonate   Ammonia   Carbon dioxide   Water

5.3-C SINGLE REPLACEMENT (OR SINGLE DISPLACEMENT) REACTIONS

In a **single replacement reaction**, or **single displacement reaction**, an element and a compound react so that the free element replaces an element in the compound. Some examples of replacement reactions are illustrated by the following equations:

single replacement reaction (single displacement reaction)
■ The type of chemical reaction in which one element replaces another within a compound

$$Mg(s) + 2HCl(aq) \longrightarrow MgCl_2(aq) + H_2(g)$$

Magnesium   Hydrochloric acid   Magnesium chloride   Hydrogen

$$Cu(s) + 2AgNO_3(aq) \longrightarrow 2Ag(s) + Cu(NO_3)_2(aq)$$

Copper   Silver nitrate   Silver   Copper(II) nitrate

$$2KI(aq) + Cl_2(g) \longrightarrow 2KCl(aq) + I_2(s)$$

Potassium iodide   Chlorine   Potassium chloride   Iodine

5.3-D DOUBLE REPLACEMENT (OR DOUBLE DISPLACEMENT) REACTIONS

In a **double replacement reaction** or **double displacement reaction**, there is an exchange of the positively charged ions in two compounds. Some examples of this type of chemical reaction are illustrated by the following equations:

double replacement reaction (double displacement reaction)
■ The type of chemical reaction in which two different compounds exchange their ions to form two new compounds

$$2NaCN(s) + H_2SO_4(aq) \longrightarrow Na_2SO_4(aq) + 2HCN(g)$$

Sodium cyanide   Sulfuric acid   Sodium sulfate   Hydrogen cyanide

$$PbS(s) + 2HCl(aq) \longrightarrow PbCl_2(aq) + H_2S(aq)$$

Lead(II) sulfide   Hydrochloric acid   Lead(II) chloride   Hydrogen sulfide

$$Na_2SO_4(aq) + BaCl_2(g) \longrightarrow BaSO_4(s) + 2NaCl(aq)$$

Sodium sulfate   Barium chloride   Barium sulfate   Sodium chloride

SOLVED EXERCISE 5.1

Potassium bicarbonate is a useful dry-chemical fire extinguisher. When heated, it produces potassium carbonate, water, and carbon dioxide. Write the balanced chemical equation for this reaction.

Solution: First, it is necessary to write the chemical formulas for the reactant and products associated with this chemical reaction. Following the methods learned in Chapter 3, the chemical formulas of the four relevant compounds are written as follows:

potassium bicarbonate, $KHCO_3(s)$ potassium carbonate, $K_2CO_3(s)$
water, $H_2O(g)$ carbon dioxide, $CO_2(g)$

The physical states of these substances are denoted parenthetically. They are known from experience.

These formulas can now be used to write an unbalanced equation representing the chemical reaction.

$$KHCO_3(s) \longrightarrow K_2CO_3(s) + H_2O(g) + CO_2(g)$$

Finally, we must balance the equation. Because oxygen atoms are more abundant in this equation than any other type of atom, we begin by balancing the number of oxygen atoms. Initially, there are three atoms of oxygen on the left side of the arrow but six on the right side $(3 + 1 + 2)$. Balance oxygen by inserting a 2 in front of the formula for potassium bicarbonate. The 2 also balances the number of potassium atoms on each side of the arrow. The equation now looks as follows:

$$2KHCO_3(s) \longrightarrow K_2CO_3(s) + H_2O(g) + CO_2(g)$$

Performing a final check on the number of atoms, we see that there are two atoms of potassium, two atoms of hydrogen, two atoms of carbon, and six atoms of oxygen on each side of the arrow. Hence, this equation is now balanced.

SOLVED EXERCISE 5.2

Hydrogen cyanide, a highly toxic gas, is produced when hydrochloric acid is mixed with barium cyanide. Identify this type of chemical reaction, and write a balanced equation illustrating the reaction.

Solution: The names of three substances are mentioned in this exercise: hydrogen cyanide, hydrochloric acid, and barium cyanide. Their chemical formulas are HCN, HCl, and $Ba(CN)_2$, respectively. Using them, we write the following partial equation:

$$Ba(CN)_2(s) + HCl(aq) \longrightarrow \underline{\hspace{1cm}} + HCN(g)$$

It is apparent that the nature of the reaction involves an exchange of ions. Consequently, this reaction is an example of a double replacement or double displacement reaction.

Knowing the reaction type, we can now identify the remaining compound as barium chloride. The chemical formula of barium chloride is $BaCl_2$. The unbalanced equation now looks as follows:

$$Ba(CN)_2(s) + HCl(aq) \longrightarrow BaCl_2(aq) + HCN(g)$$

In this form, there are unequal numbers of chloride and cyanide ions on each side of the arrow. To overcome this obstacle, we write a 2 in front of the formulas for hydrochloric acid and hydrogen cyanide as follows:

$$Ba(CN)_2(s) + 2HCl(aq) \longrightarrow BaCl_2(aq) + 2HCN(g)$$

Equal numbers of atoms are now present on each side of the arrow: one barium ion, two cyanide ions, two hydrogen ions, and two chloride ions. This equation is balanced.

PERFORMANCE GOALS FOR SECTION 5.4:

- Describe oxidation and reduction and the phenomenon of an oxidation–reduction reaction.
- For a given oxidation-reduction reaction, identify the oxidizing agent, the reducing agent, the substance oxidized, and the substance reduced.

oxidation–reduction reaction (redox reaction)
■ A chemical reaction between one or more oxidizing and reducing agents; a redox reaction

5.4 Oxidation–Reduction Reactions

Chemists also classify a chemical process in terms of whether it represents an **oxidation–reduction reaction**, frequently called a **redox reaction**. Combination, decomposition, and simple replacement reactions involve oxidation–reduction processes, whereas double replacement reactions do not. Although we study redox reactions in more depth in Chapter 11, a basic under-standing is now required, since we will encounter them frequently.

5.4-A OXIDATION

Oxidation is any of the following processes:

Elements and compounds *oxidize* when they gain oxygen atoms. When a compound is oxidized, each type of atom within the compound combines with oxygen. For example, carbon, hydrogen, and methane oxidize by combining with oxygen.

$$C(s) \;+\; O_2(g) \longrightarrow CO_2(g)$$
Carbon   Oxygen   Carbon dioxide

$$2H_2(g) \;+\; O_2(g) \longrightarrow 2H_2O(l)$$
Hydrogen   Oxygen   Water

$$CH_4(g) \;+\; 2O_2(g) \longrightarrow CO_2(g) \;+\; 2H_2O(g)$$
Methane   Oxygen   Carbon dioxide   Water

Compounds also oxidize when they lose hydrogen atoms. When methanol decomposes, for instance, formaldehyde and hydrogen form.

$$CH_3OH(g) \longrightarrow HCHO(g) \;+\; H_2(g)$$
Methanol   Formaldehyde   Hydrogen

Because methanol loses hydrogen atoms, it is said to be *oxidized*.

An element or ion oxidizes when it becomes less affiliated with its electrons. For ionic sub-stances, this is accomplished by the *loss* of one or more electrons.

$$Na(s) \longrightarrow Na^+(aq) + e^-$$
$$Mg(s) \longrightarrow Mg^{2+}(aq) + 2e^-$$
$$Cu(s) \longrightarrow Cu^{2+}(aq) + 2e^-$$
$$Fe^{2+}(aq) \longrightarrow Fe^{3+}(aq) + e^-$$
$$2Cl^-(aq) \longrightarrow Cl_2(g) + 2e^-$$

In the first three examples, neutral atoms of sodium, magnesium, and copper, respectively, lose either one or two electrons as indicated and become positively charged ions; in the fourth exam-ple, the iron(II) ion loses an electron and becomes the iron(III) ion; and in the fifth example, each of two chloride ions loses an electron to form a neutral molecule of chlorine.

5.4-B REDUCTION

Oxidation is always associated with the accompanying process called **reduction**. Any one of the following processes constitutes reduction:

Compounds *reduce* when they lose oxygen atoms. For example, when sodium perchlorate is heated, it loses oxygen atoms.

$$NaClO_4(s) \longrightarrow NaCl(s) \;+\; 2O_2(g)$$
Sodium perchlorate   Sodium chloride   Oxygen

Therefore, sodium perchlorate is *reduced*.

Compounds also reduce when they gain hydrogen atoms. For example, the organic com-pound ethene combines with hydrogen to become ethane.

$$C_2H_4(g) \;+\; H_2(g) \longrightarrow C_2H_6(g)$$
Ethene   Hydrogen   Ethane

Because it gains hydrogen atoms, ethene is *reduced*.

Substances reduce when they become more affiliated with electrons. For ionic systems, re-duction is associated with the *gain* of electrons.

oxidation
■ A chemical process during which a sub-stance unites with oxygen or another oxidizing agent

reduction
■ A chemical process during which oxygen or another oxidizing agent is removed from a compound; a chemical process that is some-times accompanied by the gain of hydrogen by a species; a chemical process that is some-times accompanied by the gain of electrons

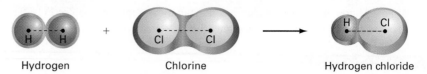

Hydrogen   Chlorine   Hydrogen chloride

FIGURE 5.1 When the oxidation–reduction phenomenon occurs between covalently bonded substances, electrons are not completely transferred from one reactant to the other. In the hydrogen and chlorine molecules shown here to the left of the arrow, the electron pairs are mutually shared between the two like atoms. But in the hydrogen chloride molecule shown to the right of the arrow, the electronic distribution is asymmetric about the center of the molecule. This partial loss and gain of electron density is typical of the oxidation–reduction reactions involving covalently bonded substances.

$$Cl_2(g) + 2e^- \longrightarrow 2Cl^-(aq)$$
$$S_8(s) + 16e^- \longrightarrow 8S^{2-}(aq)$$
$$Fe^{3+}(aq) + e^- \longrightarrow Fe^{2+}(aq)$$
$$Fe^{2+}(aq) + 2e^- \longrightarrow Fe(s)$$

In the first two examples, neutral elements gain electrons and form negative ions; in the third example, the iron(III) ion gains an electron and becomes the iron(II) ion; and in the final example, the iron(II) ion gains two electrons and becomes an atom of elemental iron. The molecules and ions on the left of these arrows are said to be reduced.

Oxidation and reduction also occur in covalent systems, but here, a total transference of electrons does not occur. For instance, consider the chemical reaction represented by the combination of hydrogen and chlorine.

$$H_2(g) \quad + \quad Cl_2(g) \quad \longrightarrow \quad 2HCl(g)$$
Hydrogen   Chlorine   Hydrogen chloride

In the hydrogen and chlorine molecules, the electron pairs in the covalent bonds are shared equally by their respective atoms. In the hydrogen chloride molecule, however, the chlorine atom shares the pair of bonding electrons to a greater degree than the hydrogen atom. This unequal sharing of the electron pair is illustrated in Figure 5.1. It causes an unsymmetrical electron distribution in the molecule of hydrogen chloride. This unsymmetrical distribution of electrons is typical of oxidation in covalent systems. Hydrogen has been oxidized and chlorine has been reduced.

In either ionic or covalent systems, the substances oxidized are called **reducing agents**, and the substances reduced are called **oxidizing agents**, or **oxidizers**. These names result from the effect the agent has on other substances. In the combination of hydrogen and chlorine, chlorine is the oxidizing agent, and hydrogen is the reducing agent.

Consider another example. Decades ago, cameras used flashbulbs to generate a brilliant blaze to lighten a darkened scene. The brilliance was associated with a chemical reaction in which metallic magnesium burned to form magnesium oxide.

$$2Mg(s) \quad + \quad O_2(g) \quad \longrightarrow \quad 2MgO(s)$$
Magnesium   Oxygen   Magnesium oxide

During this reaction, a magnesium atom loses two electrons to become a magnesium ion. It also combines with oxygen. For both reasons, magnesium is *oxidized*. Each atom of the oxygen molecule gains two electrons, and each becomes an oxide ion. The oxygen is *reduced*. Magnesium is the *reducing agent*, and oxygen is the *oxidizing agent*, or *oxidizer*.

reducing agent
■ A substance that takes oxygen from another substance; the substance oxidized during an oxidation–reduction reaction

oxidizing agent
■ An oxygen-rich substance capable of readily yielding some of its oxygen; the substance reduced during an oxidation–reduction reaction

oxidizer
■ The substance reduced in an oxidation–reduction reaction

PERFORMANCE GOALS FOR SECTION 5.5:

- Describe the rate of reaction.
- Identify the seven factors that affect the rate at which a chemical reaction occurs.
- Describe the impact of each factor on the rate of reaction.

5.5 Factors Affecting the Rate of Reaction

Each chemical reaction occurs at a definite speed called its **rate of reaction**. Sometimes the rate of reaction is referenced to a correlating chemical phenomenon, as in the use of terms such as the *rate of combustion*, *rate of corrosion*, or *rate of explosion*. Chemists establish these rates of reaction by experimentally noting the change in concentration of a reactant or product over time.

The speed at which a given substance undergoes a chemical change is often associated with its hazardous nature. This is clearly illustrated by the detonation of nitroglycerin. Several grams of nitroglycerin will completely decompose within a millionth of a second. Fortunately, not all chemical reactions occur as rapidly, or we would have even greater problems when responding to emergencies involving hazardous materials.

The rate of reaction depends on at least seven factors, each of which will be discussed independently in the sections that follow. When appropriate, the influence of each factor is noted as it bears on the rate of combustion.

rate of reaction
■ The speed at which a chemical transformation occurs; the amount of a product formed, or reactant consumed, per unit of time

5.5-A NATURE OF THE MATERIAL

When exposed to air, some substances do not burn at all. Examples are water, carbon dioxide, nitrogen, and the noble gases. Other substances, like hydrogen, magnesium, and sulfur, do not begin burning in air until they are first exposed to a spark, flame, or other source of ignition. Still other substances burn spontaneously in air, even without exposure to an ignition source. An example is elemental phosphorus, which bursts into flame on exposure to the air. These rates of combustion vary from zero to some finite value. It is their individual chemical nature that causes some substances not to burn at all, others to burn only when kindled, and still others to burn spontaneously.

5.5-B SUBDIVISION OF THE REACTANTS

Wooden logs do not burn spontaneously. Initially, they must first be kindled, perhaps by the heat generated from the burning of smaller pieces of wood. By contrast, when the dust from the same type of wood is dispersed or suspended in air and exposed to an ignition source within a confined area, it is likely to ignite and burn with explosive violence throughout its entire mass. The phenomenon is referred to as a **dust explosion**.

Sawdust, coal dust, grain dust, flour dust, and cotton lint are examples of combustible materials that can be dispersed in the air. Their rapid combustion reactions are well-acknowledged phenomena. When they are confined within a space or enclosure, even the static electricity generated as one particle circulates among the others can serve as a source of their ignition.

Why does the dust of a combustible material burn explosively, whereas bulk quantities of the same materials must be kindled before they burn? The answer to this question is associated with particle size. As the particle size of these combustible materials is reduced, the greater is the total exposed surface area of a given mass. Because more molecules become available at the particle surface to react, the likelihood of a dust explosion is accentuated.

More molecules are also available to react in configurations that provide an increased surface area. Figure 5.2 shows that a flammable liquid burns fastest in the vessel that allows it to assume the greatest surface area. In general, whenever the surface area of a given substance is increased, the substance reacts at an increased rate, because its molecules are not internally bound to one another and are more free to react with the molecules of neighboring substances.

dust explosion
■ The combustion of dust particles suspended within a confined space

5.5-C STATE OF AGGREGATION

The rate at which a substance reacts is also affected by its physical state of matter. This is particularly true for the rate of combustion. Generally, reactions involving gases occur much faster than reactions

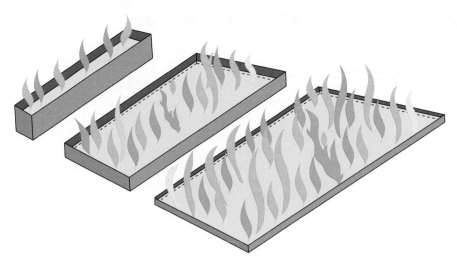

FIGURE 5.2 An equivalent amount of a flammable liquid has been added to three containers of progressively increasing size. When an ignition source is passed along their surfaces at the same instant, the liquid burns at the fastest rate within the container that provides the largest surface area.

involving liquids or solids. These differences in reaction rates are affected by the nature of the gaseous, liquid, and solid states of matter. Molecules of gaseous substances are relatively far apart and exert small attractive forces on one another. This allows diffusion to occur very rapidly. But in the liquid and solid states of matter, the particles are in contact and held tightly together. This hinders the likelihood that they will encounter other particles. Accordingly, their reaction rates are reduced.

5.5-D CONCENTRATION OF REACTANTS

Before a chemical reaction occurs, the particles that make up the structure of the reactants must contact each other. However, particle contact does not signify that a reaction will necessarily occur. The probability that particles will contact one another increases as the number of particles in a given volume increases. In other words, if different concentrations of the same reactants are put into two vessels, the rate of the reaction is generally faster in the vessel containing the greater concentration of reactants.

Imagine, such as in Figure 5.3, that we have four containers of equal volume holding different numbers of two hypothetical molecules, A and B. What is the relative number of collisions between unlike molecules? (We ignore the collisions between like molecules, such as A contacting A, or B contacting B, since these collisions do not cause a chemical reaction.)

Let's consider each container separately. The first container holds one molecule of A and one molecule of B, whereas the second container holds two molecules of A and one molecule of B. It follows that the likelihood that an A molecule will collide with a B molecule is twice as great in the second container as compared with the first one. In the third container, which holds two molecules of A and two molecules of B, the probability of collision between unlike molecules is increased to four times the probability of collision in the first container and two times

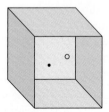 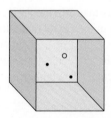 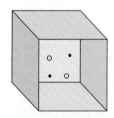 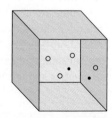

FIGURE 5.3 The probability of a chemical reaction increases as the number of reactant molecules confined within their container increases. The molecules of reactants A and B are designated here as dots and open circles, respectively.

that of the second one. Finally, in the fourth container, which holds two molecules of A and four molecules of B, the probability of collision between unlike molecules is increased to eight times that in the first container. This shows that an increase in the concentration of the reactants causes the rate of a given reaction to increase.

We also noted in Section 3.1 that the concentration of the reactants has an important bearing on the rate at which combustion occurs. Unless the concentration of flammable gases or the vapors of flammable liquids is within the flammable range of a given material, the material does not burn.

The concentration of atmospheric oxygen also affects the rate of combustion. In clean air at sea level, the concentration of oxygen is about 21% by volume. The majority of the remaining 79% by volume is nitrogen gas. Atmospheric nitrogen serves as a diluent during the combustion of materials in the air and retards their combustion rates. Combustion always occurs with an increased intensity within an atmosphere of pure oxygen. Some substances that are ordinarily stable in the air can burn spontaneously in an atmosphere of pure oxygen.

5.5-E ACTIVATION ENERGY

Although some substances can react spontaneously on contact, most must be supplied with a minimum amount of energy before they chemically react. Consider pieces of combustible solids lying exposed to atmospheric oxygen. They do not normally burst into flame, whereas they readily ignite when exposed to an energy source. This minimum amount of energy that must be supplied to initiate a chemical reaction is called the **activation energy**.

Before the combustion of a solid substance can occur, heat must ordinarily be supplied. The heat raises the temperature of the substance to its **kindling point**, the temperature at which the burning is sufficiently rapid to proceed without the need for additional heat from an external source. In exothermic reactions, the activation energy does not need to be continuously supplied as the reaction proceeds, since it is replaced by the energy released in the process. Once combustion has been initiated, it continues until either the material or the oxygen is exhausted.

An amount of activation energy must also be supplied to initiate endothermic reactions. In this case, the chemical phenomenon ceases if the source of energy is removed. We must supply not only the activation energy but enough additional energy to replace the energy absorbed by the reactants. Consider **electrolysis**, the process of decomposing a substance through use of an electric current. Electrolysis is an endothermic phenomenon. When we disconnect the current, the electrolysis stops.

activation energy
■ The minimum energy that must be supplied to a group of reactants before a chemical reaction is initiated

kindling point
■ The temperature at which the combustion of a solid can proceed without further addition of heat from an external source

electrolysis
■ The passage of an electric current through a material, generally resulting in the formation of ions and electrons

5.5-F TEMPERATURE

Heating a mixture of reactants causes its particles to move more rapidly, and this motion increases the probability that the particles will collide. As the speed of the particles increases, the temperature rises accordingly. Thus, a reaction rate increases with a rise in temperature, because more molecules become sufficiently activated at higher temperatures.

According to a classic rule in chemistry, the rate of a chemical reaction doubles for each rise of 18 °F (10 °C). Thus, twice as many molecules are activated when the temperature of the system is increased by 18 °F (10 °C); four times as many are activated when the temperature is increased 36 °F (20 °C); and eight times as many for a 54 °F (30 °C) increase in temperature.

5.5-G CATALYSIS

A **catalyst** is a substance that affects the rate of a reaction and appears to remain unchanged. A familiar example of a catalyzed reaction is the rusting of iron in the presence of atmospheric water vapor. Iron exposed to moisture corrodes much faster than iron stored in a dry atmosphere. An additional example of a catalyzed reaction is the combustion of hydrogen in the presence of platinum. A mixture of hydrogen and oxygen can be kept in a vessel for years without reacting to any noticeable degree; however, if a small amount of platinum is introduced into this vessel, the mixture reacts explosively. At the end of the reaction, hydrogen and oxygen form water but the platinum remains unchanged.

catalyst
■ A substance that increases or decreases the rate of a chemical reaction without itself being permanently altered

During a given reaction, the catalyst itself can undergo a chemical change but then react again and return to its original condition. In the end, its overall chemical identity is unchanged. But a catalyst does not always undergo a chemical change. It can also be altered physically by the absorption of particles one or more reactants on its surface.

SOLVED EXERCISE 5.3

Propane is frequently used as a fuel for heating and cooking food in areas of the country where natural gas is not delivered by pipeline to residences. When propane burns completely, carbon dioxide and water vapor are produced as combustion products. Write a balanced equation representing the complete combustion of propane.

Solution: First, we require the chemical formulas of the reactants and products. The chemical formula of propane was provided in Table 4.8 as C_3H_8. The chemical formulas of elemental oxygen, carbon dioxide, and water vapor are O_2, CO_2, and H_2O, respectively. The unbalanced equation representing the combustion of propane is then written as follows:

$$C_3H_8(g) + O_2(g) \longrightarrow CO_2(g) + H_2O(g)$$

As written, there are three atoms of carbon, eight atoms of hydrogen, and two atoms of oxygen on the left side of the arrow but only one atom of carbon, two atoms of hydrogen, and three atoms of oxygen on the right side of the arrow. To balance this equation, we write a 3 in front of the formula for carbon dioxide. This provides three atoms of carbon on each side of the arrow, but the number of hydrogen and oxygen atoms still remains unbalanced. Hence, we insert a 4 in front of the formula for water. Now, there are eight hydrogen atoms on each side of the arrow, but the number of oxygen atoms remains unbalanced: 2 on the left side and 10 on the right side of the arrow. To overcome this problem, we insert a 5 in front of the formula for oxygen, and the equation is balanced in the following final form:

$$C_3H_8(g) + 5O_2(g) \longrightarrow 3CO_2(g) + 4H_2O(g)$$

PERFORMANCE GOALS FOR SECTION 5.6:

- Describe the ordinary combustion of a substance in air.
- Distinguish between complete and incomplete combustion.
- Describe the nature of chemical energy and the heat of combustion.
- Identify the common commercial fuels.

5.6 The Combustion Process

combustion
■ The rapid oxidation of a material in the presence of oxygen or air

Combustion is a chemical reaction that releases energy to the surroundings as heat and light. The light is ordinarily visible to the naked eye and observable as flames. The flames appear during a combustion reaction when minute particles of the fuel or combustion products are heated to incandescence. When combustion is referred to in everyday practice, we say that something is "burning" or "on fire."

FIGURE 5.4 These Emergency responders use a thermal-imaging camera to locate objects without hindrances from external influences such as thick smoke or darkness. The camera is also used to locate points on storage and transport vessels from which escaping vapors are burning with nearly imperceptible flames. *(Courtesy of Dräger Safety, Inc., Pittsburgh, Pennsylvania.)*

Combustion is not always accompanied by the presence of visible flames. Hydrogen and methanol are examples of substances that burn with nearly imperceptible flames. The energy released to the surroundings is still heat and light, but the light is outside the visible range. To locate the flames of these burning substances as they leak from storage and transport vessels, experts recommend using a thermal-imaging camera, as shown in Figure 5.4, during execution of an emergency-response action.

Two or more substances chemically unite during combustion. Because one of them is typically oxygen, we often hear that a substance *oxidizes* when it burns. There are also combustion reactions that involve the union of one substance with a substance that is not oxygen. Elemental phosphorus, for example, burns in an oxygen environment, but it also burns in a chlorine environment. Although oxygen is not involved in the latter process, we still indicate that the phosphorus *oxidizes* when it combines with chlorine.

Combustion manifests itself as fire when an oxidation occurs relatively fast. For this reason, combustion is often described as *rapid* oxidation. Combustion is also a self-sustaining reaction; that is, unless the process is intentionally extinguished, combustion continues until the concentration of the substance falls below a minimum value.

The substance that burns is called the **fuel**. When a fuel burns, its atoms are never destroyed. Instead, they unite with the atoms of other substances to form one or more new substances. Because a substance typically unites with atmospheric oxygen when it burns, the products of the combustion are often metallic or nonmetallic oxides. Oxygen itself does not burn; it is said to *support* combustion.

fuel
■ A material that burns; any of several products whose combustion serves as a source of energy

complete combustion
■ Combustion process involving complete oxidation

incomplete (partial) combustion
■ Combustion process involving incomplete oxidation

Many flammable and combustible substances contain carbon in their molecular framework. When they burn in air, the constituent carbon atoms become carbon mon*oxide* or carbon di*oxide*. When carbon dioxide forms, the burning is said to constitute **complete combustion**; but when carbon monoxide forms, we regard the process as **incomplete** or **partial combustion**. We represent the incomplete and complete combustion of carbon as follows:

$$2C(s) \quad + \quad O_2(g) \quad \longrightarrow \quad 2CO(g)$$
$$\text{Carbon} \qquad \text{Oxygen} \qquad \text{Carbon monoxide}$$

$$C(s) \quad + \quad O_2(g) \quad \longrightarrow \quad CO_2(g)$$
$$\text{Carbon} \qquad \text{Oxygen} \qquad \text{Carbon dioxide}$$

Incomplete combustion can be a relatively complicated phenomenon. It often occurs when items smolder or burn slowly without a visible flame. During the incomplete combustion of carbon-containing fuels, some carbon atoms do not even unite with oxygen. When this occurs, elemental carbon is produced, which we observe as the tiny particulates of smoke.

fossil fuel
■ An energy source like coal, crude oil, and natural gas, each of which was formed by the decomposition of prehistoric organisms

Carbon-containing fuels are called **fossil fuels**. They were originally formed in nature by the decomposition of prehistoric organisms. They include coal, natural gas, and crude oil. Since the onset of industrialization prodigious amounts of carbon dioxide have been produced by burning fossil fuels for energy. They have usually been discharged into the atmosphere, where the carbon dioxide has accumulated for lengthy periods.

In Section 2.6 we noted that chemical energy is present in each substance. This energy is composed of two types:

■ Some energy is present with a substance by virtue of the motion of its atoms and molecules.
■ Energy is also stored in a substance as the result of its unique structure. This energy is associated with the strength of its chemical bonds. During any chemical reaction, some bonds break, new ones form, and any excess energy is either absorbed or released into the environment. This is the source of the energy released during combustion.

heat of combustion
■ The heat emitted into the surroundings when a compound is burned, yielding carbon dioxide and water vapor

When combustion occurs, heat evolves to the surroundings. The amount of heat varies with the nature of the substance that burns and is called the **heat of combustion**. This value is generally noted in an energy unit per gram, pound, or mole of substance burned, for example, Btu/lb, kcal/g, kJ/kg, or kJ/mol.

Materials that burn and evolve an amount of heat greater than approximately 5000 Btu/lb (~11,600 kJ/kg) are useful commercial and industrial fuels. Some common fuels and their heats of combustion are listed in Table 5.2. These limited data indicate that hydrogen outranks other

TABLE 5.2	Heat Values of Some Common Fuels	
FUEL	**Btu/lb**	**kJ/kg**
Hydrogen	61,600	143,000
Methane (natural gas)	24,100	56,000
Acetylene	21,600	50,200
Propane	21,500	50,000
Diesel fuel	20,700	48,000
Ethanol	12,900	30,000
Methanol	9900	23,000
Wood	8500	19,700

fuels from the viewpoint of the amount of energy liberated per unit of mass. We note the use of hydrogen as a fuel in Section 7.2-A.

Let's consider the combustion of methane as an example. Chemists sometimes denote the complete combustion of methane by an equation like the following:

$$CH_4(g) \ + \ 2O_2(g) \ \longrightarrow \ CO_2(g) \ + \ 2H_2O(g) \ + \ 24{,}100 \text{ Btu/lb (56,000 kJ/kg)}$$

Methane   Oxygen   Carbon dioxide   Water

Methane is the fuel and oxygen is the oxidizing agent. When methane burns, the constituent carbon and hydrogen atoms in each molecule become molecules of carbon dioxide and water vapor, both of which are routinely discharged into the atmosphere. Chemical bonds are broken and new bonds are formed. The amount of heat released during the combustion of methane is 24,100 Btu/lb, or 56,000 kJ/kg.

PERFORMANCE GOALS FOR SECTION 5.7:

- Identify the conditions necessary for the occurrence of spontaneous combustion.

5.7 Spontaneous Combustion

Some substances undergo oxidation so slowly that their fires are initially imperceptible. When the oxidation occurs within a confined space where the air circulates poorly, these substances often absorb their own heat of reaction. The continued absorption ultimately heats these substances to their autoignition points, at which they self-ignite. Because a source of ignition other than the heat of reaction is absent, the process is called **spontaneous combustion**.

There are flammable and combustible animal and vegetable oils that oxidize slowly. An example is linseed oil, which was formerly a component of commercial paints. As demonstrated in Figure 5.5, animal and vegetable oils should be regarded as potential sources of spontaneous combustion. The improper storage or disposal of rags dampened with linseed oil is a well-acknowledged

spontaneous combustion
■ Rapid oxidation initiated by the accumulation of heat in a material undergoing slow oxidation

FIGURE 5.5 Oily rags and improperly stored linseed oil, varnishes, lacquers, and other oil-based paint products are among the materials likely to undergo spontaneous combustion. Considerable loss of property is associated annually with major fires caused by spontaneous combustion.

circumstance in which spontaneous combustion is probable. When such rags are negligently tossed into a pile in a broom closet, for instance, the heat cannot readily dissipate to the surroundings. Instead, the rags absorb the heat; before long, the oil bursts into flame. Because the availability of oxygen is limited, a thick, black smoke usually accompanies these slow-burning fires.

Spontaneous combustion also occurs within stacks of undried agricultural products, such as damp hay and grass. Microorganisms proliferate in these moist materials, in which their physiological activities generate heat. This microbiological activity is known as **thermogenesis**. It is supplemented by chemical oxidation until the temperature rises to approximately 160 °F (71 °C), at which point the microorganisms can no longer survive. Biological activity ceases, but chemical oxidation continues. When heat evolves faster than it can dissipate, the internal temperature increases and the product self-ignites. Since the burning product is damp, evaporating water can evolve from the fire as billows of white smoke.

thermogenesis
■ The physiological activity of microorganisms that results in the slow liberation of heat within a material and when absorbed by the material, can cause its spontaneous combustion

PERFORMANCE GOALS FOR SECTION 5.8:

- Describe the nature of the greenhouse effect and its potential connection with global warming.
- Identify the principal greenhouse gas of environmental concern.
- Identify the impact that global warming has on the workload of firefighters.
- Describe the purpose of the Kyoto protocol.

5.8 The Greenhouse Effect

Our planet is continuously warmed by the radiant energy that enters our atmosphere from the Sun. Earth's surface absorbs this energy, some of which is then emitted back into outer space. Over the past 500,000 years, the rates at which this radiation was absorbed and emitted were balanced on average and produced an average surface temperature that fluctuated between 66 °F (19 °C) and 80 °F (27 °C).

Today, however, the average temperature exceeds the high point of this range. The absorbed energy is now hindered from effectively radiating back into outer space, which has caused scientists to fear that the average temperature of Earth's surface may continue to rise. This interference is caused in part by an elevated concentration of carbon dioxide in the atmosphere compared with its concentration in the past.

The atmospheric carbon dioxide level has slowly increased since industrialization first began during the mid-18th century. The increase has been caused by various processes involving the burning of carbon-bearing substances, as occurs whenever manufacturing plants are powered and motor vehicles are operated. Approximately half the generated carbon dioxide is absorbed from the atmosphere by plants and the oceans, whereas the other half persists within the atmosphere.

Prior to industrialization, the average concentration of carbon dioxide in Earth's atmosphere was only 280 parts per million (ppm), an estimate derived from analyzing samples of air trapped in Antarctic ice. With the onset of industrialization, the carbon dioxide level rose only slowly. It was not until the early 1900s that the average carbon dioxide concentration had risen to 315 ppm. However, by the mid-1970s, it had risen to 330 ppm, and by the mid-1990s, to 360 ppm. The current concentration is 380 ppm, which is higher than any level attained during the past 650,000 years. If this unprecedented trend continues, the average carbon dioxide concentration could reach 500 ppm by the mid-2000s.

Why is the accumulation of carbon dioxide within the atmosphere so problematic? The answer to this question is associated with carbon dioxide's comparatively unique capability of absorbing the energy that would otherwise pass into space. After it absorbs energy from the sun, carbon dioxide then reemits it to Earth's surface instead of into outer space. This process has upset the conventional balance between the rates of radiation absorption and emission, which in turn, has caused our planet to warm. Using computer modeling, researchers predict that in this century the atmospheric accumulation of carbon dioxide could cause Earth to warm roughly 3 to 9 °F (2 to 5 °C).

The increase in global warming has been compared to the manner in which a greenhouse operates. The phenomenon is accordingly called the **greenhouse effect**. The glass in a greenhouse traps radiant energy and reduces the rate at which it dissipates to the outside environment, thereby keeping the enclosure warmer than its outside surroundings. By analogy, the carbon dioxide and other greenhouse gases in the atmosphere trap energy in the atmosphere, which results in the warming of our planet.

As we note in Section 5.8-A, the evidence for global warming as a phenomenon is overwhelming and undebatable, but not all scientists agree about its cause. Some scientists attribute global warming to cyclic variations in solar activities rather than the increased atmospheric concentration of carbon dioxide. Notwithstanding these conflicting positions among scientists, the majority of the scientific community now support the theory that our planet is warming owing to increased levels of certain gases in our atmosphere, especially carbon dioxide.

Each gas that contributes to global warming—such as carbon dioxide—is called a **greenhouse gas**. Carbon dioxide is the main culprit among the greenhouse gases in Earth's atmosphere because it is produced in prodigious amounts when humans burn fossil fuels. Figure 5.6 notes the primary global sources that emit carbon dioxide into the atmosphere. The largest amount (24%) is associated with operation of the world's fossil-fuel-fired power plants.

greenhouse effect
■ The warming of Earth's atmosphere caused by the decrease in the rate of escape of energy from Earth's surface into outer space compared with the rate at which the Sun's radiant energy enters the atmosphere and is absorbed by Earth

greenhouse gas
■ A substance that absorbs atmospheric energy and by so doing, affects the rate of escape of energy from Earth's surface into outer space

5.8-A THE ADVERSE IMPACT OF GLOBAL WARMING

Many scientists believe we have been experiencing the global impact of the greenhouse effect for some time by pointing to melting polar glaciers and rising sea levels. They also note the comparatively excessive number of extreme weather events including shorter and drier winters, earlier springs, hurricanes, heat waves, deepening droughts in Australia, longer and more severe wildfire seasons in the western United States, rising sea levels that caused more floods in Latin America and elsewhere, increasingly stronger typhoons in Asia, and floods that first became common during the mid- to late-1990s and have further increased during this century. In 2007,

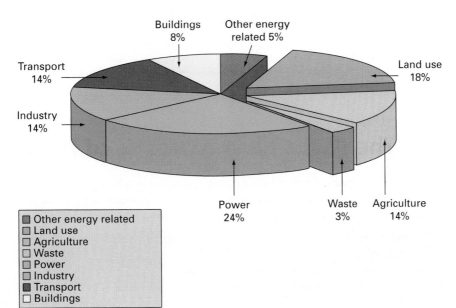

FIGURE 5.6 Global sources of carbon dioxide emissions: 65% originates from energy-related emissions, and 35% originates from non-energy-related emissions. *[Nicholas H. Stern*, Economics of Climate Change, *(New York: Cambridge University Press, 2007).]*

they also noted that Canada's fabled Northwest Passage, the shortest sea route from Europe to the Pacific, was free of obstructing ice for the first time in recorded history.

Until methods are implemented to reduce the atmospheric concentration of greenhouse gases, the predominance of these weather events is likely to continue—particularly in certain areas of the world. It appears plausible that warmer springs and longer summers will be experienced. Emergency responders will be affected by this climate change, since they will be called on more frequently than during any period in the past to mitigate wildfires and other disturbances caused by these warmer climates.

5.8-B ENVIRONMENTAL REGULATIONS INVOLVING CARBON DIOXIDE

In 1997, an international attempt to curb the impact of the greenhouse effect was first spearheaded by the United Nations. The intent was to implement a policy with the aim of producing a worldwide reduction in greenhouse-gas emissions among its industrialized members. This international effort is the subject matter of the Kyoto Protocol to the United Nations Framework Convention on Climate Change or, simply, the **Kyoto protocol**. The protocol required 35 industrialized nations to reduce their greenhouse-gas emissions by the year 2012, when the protocol expires, by 7% below the 1990 level. Developing countries such as Brazil, China, and India are exempt from its obligations.

The United States has not ratified the Kyoto protocol. Because most greenhouse gases generated within our borders are directly linked to numerous industrial processes, to curb the emission of greenhouse gases would simultaneously hamper the economic growth of many American businesses. Rather than agreeing to set specific goals for greenhouse-gas-emission reductions, the United States elected to develop its own program by correlating them with economic output and energy efficiency. The program includes the search for and implementation of environmentally friendly technologies that curb the growth of greenhouse-gas emissions.

Reducing greenhouse gas emissions continues to be an ongoing and challenging international effort. Following the meeting in Kyoto, for instance, representatives of the world's industrialized nations have met repeatedly to discuss procedures for reducing the impact of global warming. In 2007, trade ministers met at a United Nations conference in Bali, Indonesia, but the United States continued to resist a binding timetable for capping greenhouse-gas emissions. Because unchecked global warming remains a contested political subject, its consequences will most likely continue to be experienced worldwide for decades.

PERFORMANCE GOALS FOR SECTION 5.9:

- Describe the concept of reactivity as it is used by EPA in RCRA regulations.
- Determine whether a chemical waste exhibits the RCRA characteristic of reactivity.

5.9 The RCRA Characteristic of Reactivity

The RCRA regulations (Section 1.5-C) denote a material as a hazardous waste if it exhibits one or more characteristics, one of which is reactivity. A material exhibits the characteristic of **reactivity** when it possesses one or more of the following properties:

- It readily undergoes violent change without detonating.
- It reacts violently with water.
- It forms potentially explosive mixtures with water.
- It is a cyanide- or sulfide-bearing substance that when exposed to acid or alkaline solutions generates toxic gases, vapors, or fumes in a quantity sufficient to present a danger to human health or the environment.

Kyoto protocol
■ The international policy aimed at reducing the worldwide impact of global warming through curtailment of greenhouse gas emissions into the atmosphere

reactivity
■ For purposes of RCRA regulations, the characteristic of a waste that exhibits any of the following properties (40 C.F.R. §261.23): readily undergoes a violent change without detonating; reacts violently with water; forms potentially explosive mixtures with water; generates toxic gases, vapors, or fumes in a quantity sufficient to present a danger to human health or the environment; is capable of detonation or an explosive reaction if subjected to a strong initiating source or if heated under confinement; is readily capable of detonation or explosive decomposition or reaction at standard temperature and pressure; is an explosive material in hazard class 1.1, 1.2, or 1.3; or is a forbidden explosive

- It is capable of detonation or an explosive reaction if it is subjected to a strong initiating source or if it is heated under confinement.
- It is readily capable of detonation, explosive decomposition, or reaction at standard temperature and pressure.
- It is an explosive material in hazard class 1.1, 1.2, or 1.3 (Section 6.1-B).
- It is a forbidden explosive (Section 15.4).

A waste that exhibits the characteristic of reactivity is a hazardous waste and is subject to EPA's treatment, storage, and disposal regulations. When a waste exhibits the characteristic of reactivity, it is assigned the hazardous waste number D003.

PERFORMANCE GOALS FOR SECTION 5.10:

- Describe the combustion process by using the fire triangle and the fire tetrahedron.
- Describe the nature of a free radical.
- Demonstrate the importance of free radicals in the combustion process.

5.10 The Fire Triangle and Fire Tetrahedron

Decades ago, the combustion process was visually represented by using the **fire triangle** in Figure 5.7. Each leg of this triangle was assigned a factor that is required for combustion. The intent of this model was to display the process by noting the interplay between fuel, oxygen, and heat. Although scientists initially believed that combustion could be described by the use of only these three factors, it soon became apparent that the model could not adequately account for everything scientists knew about the combustion process.

Scientists also discovered that a fourth factor—the presence of free radicals—is important during combustion. This means that a four-sided geometric figure is needed to visually depict the four components. For this purpose, a **fire tetrahedron** was generated. Each face of the fire tetrahedron in Figure 5.8 is assigned an element of the combustion process. The message conveyed by this figure is that combustion occurs when adequate fuel, oxygen, heat, and free radicals are simultaneously present.

A **free radical** is a molecule that has an unpaired electron. It forms when a covalent bond is broken. Consider the methane molecule. When one of its four chemical bonds is broken or split, we represent the resulting structure as follows:

$$
\begin{array}{c}
H \\
| \\
H-C\cdot \\
| \\
H
\end{array}
$$

This species is called the *methyl free radical*. The dot represents the unpaired electron.

fire triangle
- A triangle each leg of which is used to designate three of the four components of a fire (fuel, oxygen, and heat)

fire tetrahedron
- A four-sided pyramid each face of which denotes an essential fire component (fuel, oxygen, heat, and free radicals)

free radical
- A reactive species resembling a molecule but having an unpaired electron

FIGURE 5.7 The fire triangle. The self-sustenance of an ordinary fire was once associated with only three components: fuel, oxygen, and heat. Today, however, this concept has been broadened into the fire tehrahedron shown in Figure 5.8, which includes a fourth component, free radicals.

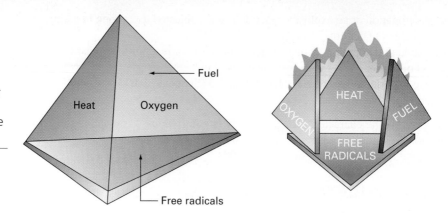

FIGURE 5.8 Two versions of the fire tetrahedron, each of which associates the self-sustenance of an ordinary fire with four components: fuel, oxygen, heat, and free radicals.

Once produced, a free radical rapidly seeks out an atom, a molecule, or another free radical. Because it possesses an unpaired electron, a free radical is an extraordinarily reactive, short-lived species. Nonetheless, during its transient existence, a free radical undergoes many kinds of chemical reactions, which occur in the following four ways:

- An electron is donated to a stable molecule or atom.
- An electron is removed from a stable molecule or atom.
- A group of atoms is removed from a molecule.
- A free radical adds to a molecule.

To illustrate the importance of free radicals during combustion, we need to examine the burning of methane in more detail. Chemists now view the combustion of methane as a series of individual reactions, each of which involves free radical reactions. As the combustion proceeds, the products of one reaction activate additional molecules, which then initiate other reactions. This stepwise description of the manner by which reactants are converted into products is called the **mechanism** of the reaction.

In all, the combustion of methane can be summarized by means of a mechanism consisting of the following steps:

$$CH_4(g) \longrightarrow CH_3{\cdot}(g) \longrightarrow HCHO(g) \longrightarrow$$
Methane   Methyl radical   Formaldehyde

$$HCO{\cdot}(g) \longrightarrow CO(g) \longrightarrow CO_2(g)$$
Formyl radical   Carbon monoxide   Carbon dioxide

This mechanism is an example of a **chain reaction** that is, a series of steps, each of which generates a reactive substance that brings about another step. Let's examine each step more closely.

5.10-A INITIATION

The first mechanistic step involves the production of a methyl free radical and a hydrogen atom, which occurs when a methane molecule first absorbs activation energy from an ignition source.

$$CH_4(g) \longrightarrow CH_3{\cdot}(g) \longrightarrow H{\cdot}(g)$$
Methane   Methyl radical   Hydrogen atom

It is called the **initiation** step of the mechanism.

mechanism
■ The step-by-step description (initiation, propagation, and termination) of the chemical process by which reactants are converted into products

chain reaction
■ A multistep reaction in which a reactive intermediate formed in one step reacts during a subsequent step to generate the species needed in the first step

initiation
■ The first mechanistic step of a chemical reaction during which reactive chemical species like free radicals are produced when heat is applied to a reactant

5.10-B PROPAGATION

The hydrogen atom then reacts with molecular oxygen, which results in the formation of a hydroxyl radical ($\cdot$OH) and an oxygen atom ($\cdot\ddot{O}\cdot$).

$$H\cdot(g) \quad + \quad O_2(g) \quad \longrightarrow \quad \cdot OH(g) \quad + \quad \cdot\ddot{O}\cdot(g)$$

Hydrogen atom   Oxygen   Hydroxyl radical   Oxygen atom

This reaction is an example of a **propagation**, since one reactive species has reacted and generated new ones.

    During the propagation steps that occur during the combustion of methane, new reactive intermediates, atoms, and molecules are produced by reactions involving methyl radicals, hydroxyl radicals, hydrogen atoms, and oxygen atoms. Examples of these propagation steps are represented by a group of equations like the following:

propagation
- The second mechanistic step of a chemical reaction during which reactant species repetitively form new species by means of multiple reactions

$$CH_4(g) \quad + \quad \cdot OH(g) \quad \longrightarrow \quad H_2O(g) \quad + \quad CH_3\cdot(g)$$
Methane   Hydroxyl radical   Water   Methyl radical

$$CH_4(g) \quad + \quad H\cdot(g) \quad \longrightarrow \quad CH_3\cdot(g) \quad + \quad H_2(g)$$
Methane   Hydrogen atom   Methyl radical   Hydrogen

$$CH_3\cdot(g) \quad + \quad \cdot\ddot{O}\cdot(g) \quad \longrightarrow \quad \underset{\substack{| \\ H}}{\overset{\substack{O \\ \|}}{H-C(g)}} \quad + \quad H\cdot(g)$$
Methyl radical   Oxygen atom   Formaldehyde   Hydrogen atom

$$\underset{\substack{| \\ H}}{\overset{\substack{O \\ \|}}{H-C(g)}} \quad + \quad CH_3\cdot(g) \quad \longrightarrow \quad \underset{\substack{| \\ H}}{\overset{\substack{O \\ \|}}{\cdot C(g)}} \quad + \quad CH_4(g)$$
Formaldehyde   Methyl radical   Formyl radical   Methane

$$\underset{\substack{| \\ H}}{\overset{\substack{O \\ \|}}{H-C(g)}} \quad + \quad \cdot OH(g) \quad \longrightarrow \quad \underset{\substack{| \\ H}}{\overset{\substack{O \\ \|}}{\cdot C(g)}} \quad + \quad H_2O(g)$$
Formaldehyde   Hydroxyl radical   Formyl radical   Water

$$\underset{\substack{| \\ H}}{\overset{\substack{O \\ \|}}{H-C(g)}} \quad + \quad H\cdot(g) \quad \longrightarrow \quad \underset{\substack{| \\ H}}{\overset{\substack{O \\ \|}}{\cdot C(g)}} \quad + \quad H_2(g)$$
Formaldehyde   Hydrogen atom   Formyl radical   Hydrogen

$$\underset{\substack{| \\ H}}{\overset{\substack{O \\ \|}}{H-C(g)}} \quad + \quad \cdot\ddot{O}\cdot(g) \quad \longrightarrow \quad \underset{\substack{| \\ H}}{\overset{\substack{O \\ \|}}{\cdot C(g)}} \quad + \quad \cdot OH(g)$$
Formaldehyde   Oxygen atom   Formyl radical   Hydroxyl radical

$$\underset{\substack{| \\ H}}{\overset{\substack{O \\ \|}}{\cdot C(g)}} \quad \longrightarrow \quad CO(g) \quad + \quad H\cdot(g)$$
Formyl radical   Carbon monoxide   Hydrogen atom

$$CO(g) \quad + \quad \cdot OH(g) \quad \longrightarrow \quad CO_2(g) \quad + \quad H\cdot(g)$$
Carbon monoxide   Hydroxyl radical   Carbon dioxide   Hydrogen atom

The combination of the reactions illustrated by these equations allows combustion to continue until the supply of the reactants or the reactive intermediates is exhausted.

5.10-C TERMINATION

termination
■ The third mechanistic step of a chemical reaction during which certain reactive species formed during the propagation step unite to produce the product

Finally, the free radicals and atoms combine in **termination** steps similar to the reactions denoted by the following equations:

$$CH_3{\cdot}(g) \;+\; H{\cdot}(g) \longrightarrow CH_4(g)$$

Methyl radical   Hydrogen atom   Methane

$$CH_3{\cdot}(g) \;+\; CH_3{\cdot}(g) \longrightarrow CH_3{-}CH_3(g)$$

Methyl radicals   Ethane

In the termination steps of the mechanism, two reactive intermediates combine. Because new reactive species are not generated, the combustion process slows or ceases.

SOLVED EXERCISE 5.4

During the combustion of methane, formaldehyde is produced during the propagation step of the mechanism. Because it is an intermediate, chemists indicate that formaldehyde is produced by the *partial oxidation* of methane. Although only a trace of formaldehyde is produced during the normal burning of methane, the reaction can be controlled so as to commercially manufacture formaldehyde.

(a) Write the Lewis formula for formaldehyde.
(b) Write the balanced equation illustrating the production of formaldehyde by the partial oxidation of methane.

Solution:

(a) The chemical formula of formaldehyde is $HCHO$, as first provided in Section 5.4-A. The Lewis structure is

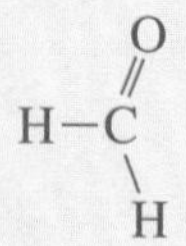

$$
\begin{array}{c}
\quad\quad O \\
\quad\quad \parallel \\
H{-}C \\
\quad\quad \backslash \\
\quad\quad H
\end{array}
$$

(b) We begin by writing an unbalanced partial equation:

$$CH_4(g) + O_2(g) \longrightarrow HCHO(g) + \underline{\quad\quad}$$

After some consideration, it becomes apparent that the remaining substance is water. When included, the production of formaldehyde by the partial oxidation of methane is depicted as follows:

$$CH_4(g) + O_2(g) \longrightarrow HCHO(g) + H_2O(g)$$

- Discuss how water effectively functions as a fire extinguisher.
- Identify the conditions during which the use of water as a fire extinguisher is inadvisable.
- Identify the nature of an emergency-response action in which experts recommend the use of a film-forming foam to extinguish a fire.

5.11 Water as a Fire Extinguisher

Water is generally effective for extinguishing only NFPA class A fires. Nonetheless, it can also be used to extinguish NFPA class B fires in circumstances like the following:

- Water is effective as a **fire extinguisher** when water is immiscible with a burning liquid and floats on its surface. The water extinguishes the fire by absorbing heat and preventing the escape of vapor from the liquid into the atmosphere.
- Water is effective as a fire extinguisher when it dilutes the burning fuel. If a flammable liquid is soluble in water, it becomes nonflammable when a sufficient amount of water has been mixed with it. Water extinguishes such fires primarily by reducing the amount of available fuel for self-sustenance.

fire extinguisher
■ A substance that promotes the extinguishment of fire or the propagation of flame

Water is often selected as a fire extinguisher because it is usually available in relatively large quantities. Yet, water is not a good all-purpose fire extinguisher for the following reasons:

- The application of water to fires often causes considerable damage; in fact, the damage resulting from the use of water frequently surpasses that caused directly by the fire.
- The fluidity of water causes it to be highly inefficient as a fire extinguisher. Most water that is applied to a fire drains from the scene into adjacent areas, where its exposure to the heat generated by the fire may be so limited it does not rapidly vaporize.
- Many parts of the world are cold enough for water to freeze when it is discharged. Although antifreeze agents can be added to keep the water liquid at temperatures as low as $-60\,°F$ ($-51\,°C$), these water solutions are frequently corrosive to metals and require the use of special equipment for mixing and discharging them. Furthermore, when insufficient antifreeze and water are applied to a fire, the water evaporates, allowing the antifreeze to burn.
- Water is often more dense than the burning liquids that constitute the matter of NFPA class B fires. When water is applied to burning liquids that are confined within containers or tanks, it sinks below their surface, where it is incapable of absorbing the evolved heat.
- Water ruins delicate electronic circuitry; thus, it is not recommended as an extinguisher of NFPA class C fires. Water containing dissolved mineral salts conducts electricity, which can put firefighters at an unreasonable risk of being electrocuted.
- Water often reacts with the burning metals that constitute NFPA class D fires and may aid in sustaining their combustion rather than extinguishing their fires.

saturated
■ The description of the air when it contains the maximum amount of water it can hold at a given temperature

Water also acts as a fire extinguisher when the moisture content of the air is relatively high. Fires burn less vigorously than when the water content of the air is low. Although the air always contains some water vapor, there is a limit to the amount of water that it can contain at a given temperature. When the air attains this limit, we say it is **saturated**. A measurement of the **relative humidity** indicates the percentage of water vapor present in the air compared with the amount the air would hold if it were 100% saturated at a given temperature. Weather forecasters warn that grassland fires are most likely to occur when the grass is dry, the wind velocity is high, and the relative humidity is low.

relative humidity
■ A measurement in percent at a given temperature of the amount of water vapor present in the air compared with saturation at 100%

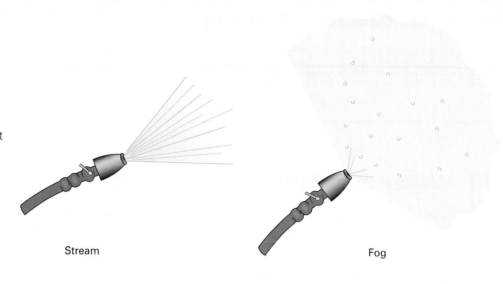

FIGURE 5.9 A comparison between two ways by which firefighters apply water when extinguishing a fire. On the left, the water is discharged from a hose as a solid stream. On the right, it is discharged from a hose as a fog. In the latter case, a larger surface area is presented by the droplets of water. These droplets absorb heat from a fire more effectively than a solid stream.

Stream

Fog

5.11-A DISCHARGING WATER AS A FIRE EXTINGUISHER

The efficiency of water as a fire extinguisher is markedly improved by using a low- or high-velocity fog-nozzle to discharge it as a mist, spray, or fog. As illustrated in Figure 5.9, the increased surface area of the fog particles causes water to vaporize more rapidly. Increasing the rate at which the water vaporizes removes heat faster from the fire scene.

Water spray patterns can range from 10° to 90°. They are used to control a burning fire and to provide exposure protection for personnel, equipment, buildings, storage containers, tanks, and other items.

5.11-B AQUEOUS-FILM-FORMING FOAM

aqueous-film-forming foam
■ A foam prepared by mixing a commercially available concentrate with water, intended for use as an extinguisher of fires involving flammable, water-insoluble liquids (abbreviated AFFF)

surfactant
■ A substance that reduces the interfacial tension between two immiscible liquids

The efficiency of water as a fire extinguisher is also improved when the water has been sealed into a gel or foam called an **aqueous-film-forming foam** (AFFF). Several AFFF concentrates are commercially available for producing foam on demand. One component of the concentrate is a **surfactant**, a compound capable of increasing the wetting and penetrating ability of water. One surfactant is perfluorooctanoic acid, whose properties are noted in Section 13.6-C.

An AFFF concentrate can be mixed with fresh, sea, or brackish water in specialized equipment. Firefighters typically use an aqueous-film-forming foam to extinguish NFPA class B fires involving a bulk volume of a flammable liquid that is water-insoluble, like gasoline, but the use of aqueous-film-forming foam is also recommended for extinguishing some stubborn NFPA class A fires in upholstery, bedding, paper, hay, and brush.

AFFF is an effective fire extinguisher because it functions as a blanketing agent over the burning fuel. Because it produces a barrier between the fuel and the atmosphere, AFFF prevents the fuel from contacting the atmospheric oxygen needed for combustion. The foam also cools the fuel and serves as a vapor-sealing film over the surface of the fuel.

Foam concentrate is often stored in a bladder-tank system. Water pressure squeezes the bladder, which then provides the concentrate at the same pressure. Although little maintenance of the system is normally needed, firefighters soon learn that considerable skill must be developed to effectively use AFFF for extinguishing a major fire. They must apply the foam across the surface of a burning fuel rather than plunging it into the fuel. When it is applied improperly, the foam cannot isolate the burning fuel from atmospheric oxygen, and the fire continues to burn.

There has been some apprehension about using certain aqueous-film-forming foams, since their use can release hydrofluoric acid (Section 8.10) and other fluorides. Their use can also lead to groundwater contamination and failure of the wastewater treatment systems to which they are discharged.

5.11-C ALCOHOL-RESISTANT AQUEOUS-FILM-FORMING FOAM

A versatile and popular fire extinguisher is **alcohol-resistant aqueous-film-forming foam** (AR-AFFF). Several AR-AFFF concentrates are available commercially. When they are mixed with water and properly dispensed, the resulting foams are used to extinguish NFPA class B fires involving both water-soluble and water-insoluble flammable liquids. Fire extinguishers are also available in which the AR-AFFF concentrate is already mixed with water for immediate use, but in this instance, they are primarily useful for extinguishing relatively small fires. Examples of these fire extinguishers are illustrated in Figure 5.10.

alcohol-resistant aqueous film-forming foam
▪ A foam prepared by mixing a commercially available concentrate with water, intended for use as a fire extinguisher on flammable liquid fires (abbreviated AR-AFFF)

5.11-D PROTEIN FOAM

Another example of a fire-extinguishing foam is **protein foam**. This foam is prepared from a natural protein material such as soybeans, fish meal, horn-and-hoof meal, or feather meal. It typically contains from 3% to 6% by weight of a protein concentrate as well as a stabilizing agent to provide permanence. Protein foam has a lubricating nature; thus, it is often used on airport runways to assist disabled aircraft during landing.

Several types of protein foam are commercially available: *regular protein foam* (P); *fluoroprotein foam* (FP); *alcohol-resistant fluoroprotein foam* (AR-FP); *film-forming fluoroprotein*

protein foam
▪ A film-forming foam prepared from natural protein materials and used to extinguish fires

FIGURE 5.10 These portable AR-AFFF fire extinguishers are useful when combating relatively small fires within laboratory, research, or industrial settings. Their operation is simple: remove the locking pin, aim the horn at the base of the fire, squeeze the lever to discharge the foam, and release the lever to stop the discharge. *(Courtesy of Badger Fire Protection, Charlottesville, Virginia.)*

foam (FFFP); and *alcohol-resistant film fluoroprotein foam* (AR-FFFP). Each is used to bring fires of various flammable materials under control.

5.11-E PYROCOOL FIRE-EXTINGUISHING FOAM

It is sometimes possible to circumvent the environmental problems associated with the use of ordinary firefighting foams by using products such as *Pyrocool FEF*. This is a special foaming agent consisting of a blend of surface-active agents that reduce the interfacial tensions between two liquids or between a liquid and a solid, thus permitting the foam to flow freely and rapidly over a liquid or solid compared with the rate at which water flows. There are two significant features of Pyrocool FEF that distinguish it from other fire-extinguishing foams:

- On application to a burning surface, a slippery foam is produced that spreads rapidly and coats the surface. This allows the heat of the fire to transfer rapidly to the water so the fire is extinguished rapidly.
- Pyrocool FEF contains additives that absorb the ultraviolet waves emitted from a fire and then reemits them at a lower energy. Thus, application of the foam effectively serves to lower the available heat to sustain the fire.

  The use of Pyrocool FEF has been alleged to be environmentally benign. Following its application to a fire, no hazardous substances are produced to pollute the nearby environment. In this sense, its application can differ significantly from the use of aqueous-film-forming foam.

PERFORMANCE GOALS FOR SECTION 5.12:

- Discuss the manner by which carbon dioxide effectively functions as a fire extinguisher.
- Identify the nature of the emergency-response action in which the use of carbon dioxide in a total-flooding situation is recommended.

5.12 Carbon Dioxide as a Fire Extinguisher

phase diagram
- A plot of the temperature versus pressure indicating the conditions at which a substance exists as a gas, liquid, and solid

dry ice
- Solid carbon dioxide

sublimation
- A physical change in which a substance passes from the solid directly to the gaseous state of matter, without first liquefying

At ordinary temperatures and pressures, carbon dioxide is encountered as a gas. Because it is nonflammable, carbon dioxide is often used to extinguish NFPA class B, NFPA class C, and even some NFPA class A fires. Its use is never recommended on NFPA class D fires. The effectiveness of carbon dioxide as a fire extinguisher is linked with its vapor density. Since carbon dioxide has a vapor density of 1.52, it is about one and a half times more dense than air. Thus, when it is applied to a fire, carbon dioxide prevents contact between the burning material and atmospheric oxygen.

Figure 5.11 demonstrates the pressure–temperature relationship for carbon dioxide. This graph is called the **phase diagram** of carbon dioxide, since it provides the temperature and pressure conditions at which the substance exists as a gas, liquid, and solid. This phase diagram illustrates that at normal room conditions, carbon dioxide exists either as a solid or vapor but not as a liquid. This is a relatively rare property. Whereas most solids liquefy before they vaporize, solid carbon dioxide, called **dry ice**, changes directly from the solid state of matter into its vapor at normal atmospheric conditions. The physical transformation of a substance directly from its solid state to its gaseous state without becoming liquid is called **sublimation**.

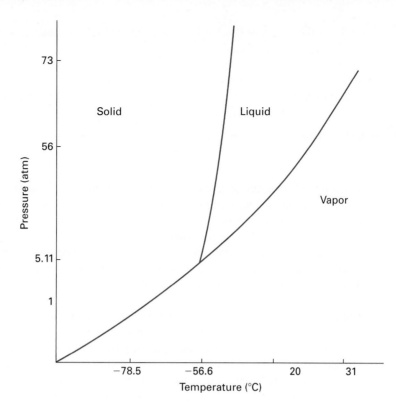

FIGURE 5.11 The phase diagram of carbon dioxide (not to scale). At 1 atm, carbon dioxide exists as either its vapor or solid called dry ice. Liquid carbon dioxide exists only at pressures equal to or greater than 5.11 atm (517 kPa) and temperatures between −69.9 °F (−56.6 °C) and approximately 88 °F (31 °C).

SOLVED EXERCISE 5.5

What is the intended message of the following dry ice pictographs?

DO NOT TOUCH
NO TOQUE
(a)

DO NOT EAT
NO CONSUME
(b)

COULD CAUSE SUFFOCATION
PUEDE CAUSAR SOFOCO
(c)

DO NOT PLACE IN AIR-TIGHT CONTAINERS
NO COLOQUE DENTRO DE RECEPTACULOS
HERMETICOS
(d)

Solution: Dry ice is extremely cold relative to room temperature and should be handled with the specific precautions noted for cryogens in Section 2.13. Each pictograph successively conveys the following message:

(a) Do not allow dry ice to contact the skin, since it can bond so firmly to the tissue that the flesh is ripped or torn when separation is attempted.

(b) Do not attempt to consume dry ice, since it can bond so firmly to the tongue or cheeks that the flesh is ripped or torn when separation is attempted.

(c) The sublimation of dry ice in a confined area is simultaneously associated with a displacement of air. Because carbon dioxide does not support life, the inhalation of an atmosphere in which carbon dioxide has displaced the air may cause suffocation and death.

(d) Dry ice sublimates into a substantially larger volume as the gas. When it is confined within a storage vessel, internal pressure develops as sublimation occurs. This can cause the container to rupture.

When intended for use as a fire-extinguishing agent, carbon dioxide is often encountered as a compressed gas within portable extinguishers like those shown in Figure 5.12. Portable handheld extinguishers containing compressed carbon dioxide in capacities from 2 to 25 lb (0.9 to 10 kg) are available commercially.

Carbon dioxide is also discharged as a gas from storage tanks in which it is confined as a liquid under pressure. Discharged as the gas, the carbon dioxide is much colder than the surrounding atmosphere. It functions not only by smothering the burning materials but also by cooling them.

Unlike water, carbon dioxide does not ordinarily damage the fire area. It is considered a "clean" fire extinguisher. Because it is a gas at normal temperature and pressure conditions, carbon dioxide dissipates into the surrounding atmosphere after use without leaving a residue.

Carbon dioxide is generally applied to fires that occur inside buildings where the atmosphere is calm. Even under this circumstance, however, firefighters often need to apply carbon dioxide repeatedly to fires that have been initially extinguished. The heat generated by the hot

FIGURE 5.12 The use of these portable carbon dioxide fire extinguishers quickly delivers smothering action to flames and suffocates fires supported by atmospheric oxygen. Their operation is simple: remove the locking pin, aim the horn at the base of the fire, squeeze the lever to discharge carbon dioxide, and release the lever to stop the discharge. *(Courtesy of Walter Kidde, Mebane, North Carolina.)*

material causes the carbon dioxide gas to dissipate into the atmosphere. Unless it has cooled sufficiently, the material reignites on contacting atmospheric oxygen.

5.12-A CARBON DIOXIDE IN TOTAL-FLOODING SYSTEMS

Carbon dioxide is often stored in industrial plants for potential use as a fire extinguisher in a total-flooding situation. The term *total flooding* means that the affected area is literally flooded on demand with carbon dioxide. The following two systems involving the use of carbon dioxide for total flooding are commercially available:

- In the *high-pressure system*, liquid carbon dioxide is stored under a pressure of approximately 850 psi_a (5900 kPa) within steel cylinders.
- In the *low-pressure system*, liquid carbon dioxide is stored in tanks maintained at 0 °F (−18 °C) by means of refrigeration under a pressure of approximately 300 psi_a (2100 kPa).

Each system discharges carbon dioxide as the gas even though the substance is stored as the liquid. This phenomenon is illustrated for the low-pressure system in Figure 5.13.

The principal danger posed by the use of carbon dioxide during total flooding is the threat of asphyxiation. When discharged into a confined area, the carbon dioxide plume replaces the ground-level air. Because individuals who inhale carbon dioxide within the treated area could lose consciousness, it is critical to first evacuate everyone from the area *before* the carbon dioxide is discharged.

5.12-B CARBON DIOXIDE PRODUCTION BY CHEMICAL ACTION

Some fire extinguishers produce carbon dioxide by chemical action with other substances. The once-popular soda–acid fire extinguisher used a chemical reaction to produce carbon dioxide and water. "Soda" is the common name of sodium bicarbonate. Separate solutions of soda and

FIGURE 5.13 Low-pressure carbon dioxide is discharged into as evacuated newspaper pressroom to serve as a method of fire protection. *(Courtesy of Chemetron Fire Systems, Matteson, Illinois.)*

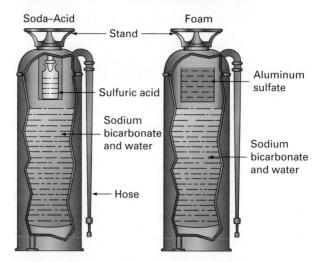

Soda–Acid — Stand — Foam

Sulfuric acid

Aluminum
sulfate

Sodium
bicarbonate
and water

Sodium
bicarbonate
and water

Hose

FIGURE 5.14 On the left is the soda–acid fire extinguisher, now relatively obsolete. The extinguisher held sodium bicarbonate and sulfuric acid solutions. When it was inverted, the two solutions mixed and chemically reacted. The carbon dioxide that was produced forced the solution mixture from the nozzle, which the user directed at a fire, On the right is a modification of the soda–acid fire extinguisher, which uses solutions of sodium bicarbonate and aluminum sulfate to produce the carbon dioxide.

sulfuric acid were arranged so that when the assembly was inverted, the two solutions mixed together. The resulting chemical reaction produced carbon dioxide.

$$2NaHCO_3(aq) \quad + \quad H_2SO_4(aq) \quad \longrightarrow \quad Na_2SO_4(aq) \quad + \quad 2H_2O(l) \quad + \quad 2CO_2(g)$$

Sodium bicarbonate   Sulfuric acid   Sodium sulfate   Water   Carbon dioxide

The carbon dioxide produced in the soda–acid fire extinguisher generated a pressure that forced water through the nozzle. It was the water, not the carbon dioxide, that extinguished the fire. The standard 2½-gal soda–acid fire extinguisher shown in Figure 5.14 provided 2½ gallons of water. Because sodium bicarbonate solutions slowly deteriorate over time, soda-acid extinguishers required periodic recharging with fresh solutions.

Carbon dioxide can also be contained as bubbles within a **chemical foam**. This is another example of a film-forming foam. Here, sodium bicarbonate and aluminum sulfate are stored as separate water solutions or powders. When they are mixed, these substances react to produce carbon dioxide.

chemical foam
■ A fire-extinguishing foam made from sodium bicarbonate and aluminum sulfate solutions, which react to produce a foam consisting of aluminum hydroxide and encapsulated bubbles of carbon dioxide

$$6NaHCO_3(aq) \quad + \quad Al_2(SO_4)_3(aq) \quad \longrightarrow \quad 3Na_2SO_4(aq) \quad + \quad 2Al(OH)_3(s) \quad + \quad 6CO_2(g)$$

Sodium bicarbonate   Aluminum sulfate   Sodium sulfate   Aluminum hydroxide   Carbon dioxide

Sometimes an extract of licorice root is mixed with these solutions to help the bubbles from collapsing. In combination with aluminum hydroxide, a tough coating is formed around each bubble of carbon dioxide. The entire mass emerges as a foam and acts as a wet blanket on a fire. This prevents the burning material from contacting atmospheric oxygen.

Some large petroleum refineries and/or their outlying tank farms employ a two-solution wet-foam system for smothering fires that occur in or around their tanks. Solutions of aluminum sulfate and sodium bicarbonate are stored within separate tanks equipped with foam-mixing chambers, usually located at the top ring or roof of these tanks, where they are combined to produce the chemical foam. When spread on the surface of a burning petroleum product, the foam starves the fire by preventing its contact with atmospheric oxygen.

- Discuss the manner by which the halons effectively function as fire extinguishers.
- Describe why the use of the halons as fire extinguishers was phased out in the United States.

5.13 Halons as Fire Extinguishers

The **halons** are nonflammable organic compounds whose molecules possess one or two carbon atoms and from four to six halogen atoms. They are halogenated derivatives of methane and ethane, that is, compounds in which one or more halogen atoms have been substituted for the hydrogen atoms in the molecules of methane (CH_4) and ethane (C_2H_6).

Halon 1301 and Halon 1211 are two representative halon products. They are two halogenated derivatives of methane having the chemical formulas CF_3Br and CF_2ClBr, respectively. In the first formula, the four hydrogen atoms in the methane molecule have been substituted with three fluorine atoms and one bromine atom. In the second, the four hydrogen atoms have been substituted with one chlorine atom, one bromine atom, and two fluorine atoms. Both compounds were once very popular fire extinguishers. Their vapors were discharged at room temperature as total-flooding and streaming systems.

The word *halon* was first devised by the U.S. Army Corps of Engineers as an abbreviation for halogenated hydrocarbon, a class of organic compounds we visit again in Sections 12.15 and 12.16. The commercial halon products are named by using the word Halon and a three- or four-digit number. A halon whose formula is $C_aF_bCl_cBr_d$ is named commercially as "Halon abcd," but when d is 0, it is dropped.

The halons formerly used as fire extinguishers are listed in Table 5.3. The most broadly used were Halon 1301 and Halon 1211. Both were more effective at extinguishing NFPA class A, class B, and class C fires than either water or carbon dioxide.

In chemistry, the halon products are named as halogenated derivatives of methane or ethane. The halogen atoms are named by replacing the *-ine* suffix on the name of the halogen with *-o*. The number of halogen atoms is indicated by the use of *mono-*, *di-*, *tri-*, and *tetra-*, as relevant, for 1, 2, 3, and 4, respectively, and when multiple halogens are part of the same substance, they are named alphabetically. The prefix *mono* is often avoided when the number of halogen atoms is clearly one. Thus, the chemical names of the halon products having the chemical formulas CF_3Br and CF_2ClBr are bromotrifluoromethane and bromochlorodifluoromethane, respectively.

halon
- A fire suppressant primarily composed of a substance whose molecules have one or two carbon atoms and four to six fluorine, chlorine, or bromine atoms

Bromotrifluoromethane
(Halon 1301)

Bromochlorodifluoromethane
(Halon 1211)

Collectively, the halons possess two properties that cause them to be effective fire extinguishers:

- Their vapors are much denser than air. This means that when a halon is discharged on a fire, its vapor smothers it.
- The halons decompose at the elevated temperatures experienced during fires, producing free radicals and atoms. These reactive species scavenge the combustion intermediates

TABLE 5.3   Some Physical Properties of the Common Halon Agents

HALON AGENT	CHEMICAL COMPONENT	CHEMICAL FORMULA	BOILING POINT	MELTING POINT	SPECIFIC GRAVITY OF LIQUID AT 68 °F (20 °C)
Halon 1011	Bromochloromethane	H–C(H)(Cl)–Br	151 °F (66 °C)	–124 °F (–87 °C)	1.93
Halon 1202	Dibromodifluoromethane	Br–C(Br)(F)–F	76 °F (24.5 °C)	–223 °F (–142 °C)	2.28
Halon 2402	1,2-Dibromo-1,1,2,2-tetrafluoroethane	F–C(Br)(F)–C(Br)(F)–F	117 °F (47 °C)	–167 °F (–111 °C)	2.17
Halon 1301	Bromotrifluoromethane	F–C(Br)(F)–F	–72 °F (–58 °C)	–270 °F (–168 °C)	1.57
Halon 122	Dichlorodifluoromethane	F–C(Cl)(Cl)–F	–22 °F (–30 °C)	–252 °F (–158 °C)	1.31
Halon 1211	Bromochlorodifluoromethane	F–C(Cl)(Br)–F	25 °F (–3.9 °C)	–257 °F (–161 °C)	1.83
Halon 242	1,2-Dichloro-1,1,2,2-tetrafluoroethane	F–C(Cl)(F)–C(Cl)(F)–F	39 °F (3.9 °C)	–137 °F (–94 °C)	1.44

produced during NFPA class B fires. For instance, when Halon 1301 thermally decomposes, trifluoromethyl radicals and bromine atoms are produced.

$$CF_3Br(g) \longrightarrow F{-}\underset{\displaystyle F}{\overset{\displaystyle F}{C}}{\cdot}(g) \ + \ Br{\cdot}(g)$$

These bromine atoms have a far greater affinity for reacting with combustion intermediates than the combustion intermediates have for reacting with one another.

The halons were once popular fire extinguishers for the following reasons:

■ They effectively extinguished NFPA class A, class B, and class C fires.
■ They dissipated into the atmosphere and did not leave a residue following their use, thereby avoiding secondary property damage. This was a highly desirable feature when assessing the protection of delicate and expensive electronic equipment, computer circuitry, and aircraft interiors.
■ They do not conduct electricity.
■ When used with proper exposure controls, they posed only a minimum health risk to firefighters and other personnel. This was a highly desirable feature when considering their use within enclosed areas like aircraft interiors.

The halons were suitable for extinguishing fires within vaults, museums, libraries, and hospitals. They were also used in military and civilian planes to extinguish fires in engines, cargo bays, cockpits, passenger compartments, and washrooms. They were once widely recommended for use in total-flooding systems (such as stationary fire-suppression systems in large computer facilities) and streaming systems (such as handheld portable fire extinguishers).

Given these features, why are they not widely used today? The answer is linked with their ability to deplete the ozone that occupies a stratospheric layer lying at the edge of Earth's atmosphere. Ozone is a form of elemental oxygen having three atoms of oxygen per molecule instead of the usual two. As we examine in more detail in Section 7.1-M, the ozone that exists as a blanket around Earth protects life from overexposure to harmful ultraviolet radiation.

When most substances are released into the atmosphere, they ultimately undergo chemical reactions that convert them into relatively innocuous substances. However, the halons are very stable compounds and do not readily undergo conversion reactions. When released into the environment, their vapors migrate upward and react with the stratospheric ozone. The combination of countless reactions between ozone and compounds like the halons has resulted in depletion of the stratospheric ozone. Scientists fear that this depletion has allowed excessive ultraviolet radiation to penetrate to the surface of our planet, where it has contributed to the incidence of skin cancer and cataracts and caused untold harm in other ways.

SOLVED EXERCISE 5.6

Identify the general risks of which workers and emergency-response personnel should be aware during an unintended fire when areas within a workplace are automatically flooded with carbon dioxide from a fixed fire-extinguishing system.

Solution: Carbon dioxide is a gas with a density one and a half times that of air. Consequently, when discharged from a fixed fire-extinguishing system, it first settles near the floor by displacing the air. Flooding an area with carbon dioxide extinguishes fires by starving them of atmospheric oxygen. However, nearby workers and response personnel are also exposed to an atmosphere having limited oxygen. Breathing inadequate oxygen can cause neurological damage, and in severe cases, it causes death. To protect workers and response personnel from encountering this situation, OSHA requires employers to provide a predischarge employee alarm to alert employees *before* carbon dioxide is released from a fixed fire-extinguishing system.

The import and manufacture of halon products as fire extinguishers have been banned in the developed countries since January 1994. In the United States, EPA permits the supplies of halon products that existed before the ban to be recycled and reused, but the ban has nonetheless substantially reduced the use of halons as fire extinguishers.

International attempts are now being implemented to curb the use of the halons and other substances that contribute to depletion of stratospheric ozone. Most likely, generations will pass before this threat to the ozone layer entirely abates.

PERFORMANCE GOALS FOR 5.14:

- Discuss how dry alkali metal bicarbonates, monoammonium phosphate, and graphite effectively function as fire extinguishers.

5.14 Dry-Chemical Fire Extinguishers

dry-chemical fire extinguisher
- An extinguisher containing any of several solid substances that function by removing one or more of the fire tetrahedron components from the fire scene

The following chemical substances are generally used in the solid state as **dry-chemical fire extinguishers**: sodium bicarbonate, potassium bicarbonate, sodium chloride, potassium chloride, and monoammonium phosphate. Some dry-chemical fire extinguishers effectively suppress NFPA class B and class C fires; others are useful on NFPA class D fires; and some extinguish NFPA class A, class B, and class C fires. They are not clean fire extinguishers, since messy residues remain following their use.

5.14-A SODIUM BICARBONATE AND POTASSIUM BICARBONATE

Fire extinguishers containing sodium bicarbonate and potassium bicarbonate are sometimes referred to as "regular" dry-chemical fire extinguishers. Potassium bicarbonate is usually referred to as the commercial product *Purple K*. Although ineffective on NFPA class A fires, both sodium and potassium bicarbonates effectively suppress and extinguish NFPA class B flammable liquid fires and are often used during initial tactics when fighting a fire involving a flammable compressed gas.

5.14-B ALKALI METAL BICARBONATES

The effectiveness of the alkali metal bicarbonates as fire extinguishers is related in part to their ability to produce carbon dioxide when heated.

$$2NaHCO_3(s) \longrightarrow Na_2CO_3(s) + CO_2(g) + H_2O(g)$$

Sodium bicarbonate　　　　Sodium carbonate　　Carbon dioxide　　Water

Each alkali metal bicarbonate requires heat from the fire to decompose. The carbon dioxide that is produced smothers the fire. In addition, sodium atoms are produced when these compounds are exposed to high temperatures. They react with other reactive chemical species to produce water.

For example, during fires involving petrochemical fuels, among the reactive species produced are hydroxyl radicals (·OH). When alkali metal bicarbonates are used to extinguish the fires, sodium atoms react with these hydroxyl radicals by the following stepwise process:

$$Na·(g) \quad + \quad ·OH(g) \quad \longrightarrow \quad NaOH(g)$$

Sodium atoms Hydroxyl radicals   Sodium hydroxide

$$NaOH(g) \quad + \quad ·H(g) \quad \longrightarrow \quad H_2O(l) \quad + \quad Na·(g)$$

Sodium hydroxide Hydrogen atoms   Water molecules   Sodium atoms

The overall reaction produces water.

$$H·(g) \quad + \quad ·OH(g) \quad \longrightarrow \quad H_2O(l)$$

Hydrogen atoms   Hydroxyl radicals   Water molecules

The sodium atoms serve to catalyze the production of water and remove the reactive free radicals.

5.14-C ABC FIRE EXTINGUISHER

Another dry-chemical fire extinguisher is monoammonium phosphate, also known as ammonium dihydrogen phosphate. This substance is employed in the multipurpose **ABC fire extinguisher** shown in Figure 5.15. The reason for its effectiveness is twofold:

■ When applied to a burning material, monoammonium phosphate decomposes endothermically as follows:

$$4NH_4H_2PO_4(s) \quad \longrightarrow \quad P_4O_{10}(s) \quad + \quad 4NH_3(g) \quad + \quad 6H_2O(g)$$

Monoammonium phosphate   Tetraphosphorus decoxide   Ammonia   Water

Because it absorbs heat as it decomposes, monoammonium phosphate cools the combustible material below the minimum temperature at which it can burn.

ABC fire extinguisher
■ A multipurpose extinguisher containing monoammonium phosphate, which functions primarily by removing heat and free radicals from the scene of an NFPA class A, B, or C fire

FIGURE 5.15 Easily portable, this ABC dry-chemical fire extinguisher containing monoammonium phosphate is recommended for use on fires within the home and small manufacturing and process facilities. The term "ABC" implies that its use effectively suppresses most NFPA class A, class B, and class C fires. The user holds the unit upright, pulls the ring pin, and aims the hose at the base of the fire while squeezing the lever and sweeping the spray from side to side. (*Courtesy of Walter Kidde, Mebane, North Carolina.*)

■ The ammonia generated by the thermal decomposition of monoammonium phosphate reacts with certain atoms and free radicals. For example, ammonia reacts with hydroxyl radicals produced when petroleum fuels burn. The reaction effectively removes the free radicals.

PERFORMANCE GOALS FOR SECTION 5.15:

■ Discuss the manner by which dry powder effectively functions as a fire extinguisher on NFPA class D fires.

5.15 Dry-Powder Fire Extinguishers

dry-powder fire extinguisher
■ An extinguisher usually containing graphite powder that functions by removing heat from the scene of an NFPA class D fire

Commercial chemical products known as **dry-powder fire extinguishers** have uniquely been manufactured to effectively suppress NFPA class D fires. Most are composed primarily of graphite. The graphite-based dry powders extinguish the fires of combustible metals such as magnesium and aluminum (Sections 9.3-A and 9.3-D).

Although graphite-based fire extinguishers are successful for suppressing NFPA class D fires, caution must be exercised when they are used. Applied to a combustible metal fire, graphite produces water-reactive metallic carbides like magnesium carbide and aluminum carbide. On exposure to water, the metallic carbides generate flammable gases including methane and acetylene. For this reason the water reactivity of metallic carbides must be considered when removing the residues following the use of graphite-based dry powders on combustible metal fires. We revisit this issue in Chapter 9.

PERFORMANCE GOALS FOR SECTION 5.16:

■ Discuss the manner by which potassium acetate effectively functions as a fire extinguisher on NFPA class K fires.

5.16 NFPA Class K Fire Extinguishers

wet-chemical fire extinguisher
■ An extinguisher consisting of a substance dissolved in water that is capable of forming a soapy foam blanket when applied to burning oil

saponification
■ The conversion of triglycerides into soaps

As we first noted in Section 1.3-E, a mixture of either potassium acetate or potassium carbonate with water is recommended for use as a fire extinguisher on kitchen fires. It is an example of a **wet-chemical fire extinguisher**. A commercially available dispenser of this mixture is shown in Figure 5.16.

Kitchen fires are NFPA class K fires, which generally involve the burning of fats, oils, and grease within cooking appliances. The addition of potassium acetate or potassium carbonate converts these fats, oils, and greases into noncombustible potassium soaps and glycerol. This type of chemical reaction is called **saponification**. Fats and oils contain complex organic compounds called *triglycerides*. When they saponify, glycerol and soap are produced by means of two chemical reactions that occur simultaneously. Using potassium carbonate as an example, potassium hydroxide is first produced.

$$K_2CO_3(aq) \quad + \quad 2H_2O(l) \quad \longrightarrow \quad 2KOH(aq) \quad + \quad CO_2(g) \quad + \quad H_2O(g)$$

Potassium carbonate   Water   Potassium hydroxide   Carbon dioxide   Water

FIGURE 5.16 The use of this wet-chemical fire extinguisher is recommended on class K fires. The user holds the unit upright, pulls the ring pin, and aims the hose at the base of the fire while squeezing the lever and discharging the contents on the burning surface. *(Courtesy of Walter Kidde, Mebane, North Carolina.)*

Then, the triglycerides react with the potassium hydroxide to form soaps. Although the latter chemical reactions are relatively complex, we can simplify their nature as follows:

$$\text{triglycerides} \ + \ \underset{\text{Potassium hydroxide}}{\text{KOH}(aq)} \ \longrightarrow \ \underset{\text{Glycerol}}{\begin{array}{c} \text{CH}_2-\text{OH} \\ | \\ \text{CH}-\text{OH}(l) \\ | \\ \text{CH}_2-\text{OH} \end{array}} \ + \ \text{soaps}$$

We revisit the saponification of triglycerides in Section 13.13.

The potassium soaps extinguish NFPA class K fires because they do not burn. Notwithstanding their success, wet-chemical fire extinguishers are messy and are not considered "clean" fire extinguishers.

Writing Chemical Equations

5.1 One type of portable oxygen generator is constructed of separate compartments, one containing iron powder and the other containing sodium chlorate. When the compartment contents are mixed and heated, oxygen is generated as follows:

$$Fe(s) + NaClO_3(s) \longrightarrow FeO(s) + NaCl(s) + O_2(g)$$

Show that this equation is balanced.

5.2 To assist in protecting occupants during an automobile collision, air bags rapidly inflate with nitrogen produced when sodium azide (NaN_3) reacts with iron(III) oxide as follows:

$$NaN_3(s) + Fe_2O_3(s) \longrightarrow Fe(s) + Na_2O(s) + N_2(g)$$

Balance this equation.

5.3 When a bulk quantity of ammonium nitrate fertilizer is strongly heated, various gases are produced at explosive rates. Two examples of the decomposition of ammonium nitrate are shown by the following unbalanced equations:

$$NH_4NO_3(s) \longrightarrow N_2(g) + NO_2(g) + H_2O(g)$$
$$NH_4NO_3(s) \longrightarrow N_2(g) + O_2(g) + H_2O(g)$$

How do these equations appear when they are balanced?

Balancing Equations

5.4 Convert each word equation that follows into a formula equation and balance it:

(a) Magnesium metal + oxygen $\longrightarrow$ magnesium oxide
(b) Calcium carbonate $\longrightarrow$ calcium oxide + carbon dioxide
(c) Nickel chlorate $\longrightarrow$ nickel(II) chloride + oxygen
(d) Sulfur tetrachloride + water $\longrightarrow$ sulfurous acid + hydrogen chloride
(e) Iron(III) oxide + carbon monoxide $\longrightarrow$ metallic iron + carbon dioxide
(f) Cadmium hydroxide + hydrochloric acid $\longrightarrow$ cadmium chloride + water
(g) Phosphoric acid + calcium chloride $\longrightarrow$ calcium phosphate + hydrochloric acid

5.5 Balance each of the following equations:

(a) $Na(s) + H_2O(l) \longrightarrow NaOH(aq) + H_2(g)$
(b) $CdO(s) + H_2SO_4(aq) \longrightarrow CdSO_4(aq) + H_2O(l)$
(c) $Fe(s) + O_2(g) + H_2O(g) \longrightarrow Fe(OH)_3(s)$
(d) $CS_2(l) + Cl_2(g) \longrightarrow CCl_4(l) + S_2Cl_2(l)$
(e) $As_2O_3(s) + H_2S(g) \longrightarrow As_2S_3(s) + H_2O(l)$
(f) $K_2Cr_2O_7(s) \longrightarrow Cr_2O_3(s) + K_2O(s) + O_2(g)$
(g) $Fe(OH)_2(s) + H_2O(g) + O_2(g) \longrightarrow Fe(OH)_3(s)$
(h) $PCl_3(l) + H_2O(l) \longrightarrow H_3PO_3(aq) + HCl(g)$

5.6 Supply the name of the reactant or product that completes each of the following word equations:

(a) Aluminum + bromine $\longrightarrow$ _____

(b) Hydrogen + _____ $\longrightarrow$ hydrogen fluoride

(c) Potassium oxide + _____ $\longrightarrow$ potassium hydroxide

(d) _____ + hydrochloric acid $\longrightarrow$ iron(III) chloride + hydrogen

(e) _____ + calcium chloride $\longrightarrow$ lead(II) chloride + calcium acetate

(f) Barium hypochlorite $\longrightarrow$ barium chloride + _____

(g) Manganese chloride + _____ $\longrightarrow$ manganese hydroxide + sodium chloride

(h) Copper(II) sulfite $\longrightarrow$ _____ + sulfur dioxide

Factors Affecting Reaction Rates

5.7 Among the various factors that influence reaction rates, which is involved in each of the following observations:

(a) When heated in a flame and then placed in oxygen, iron wire ignites, whereas when an iron rod having a diameter of 0.4 in. (1 cm) is also heated to the same temperature and placed in oxygen, it does not ignite.

(b) When heated in a flame and then placed in oxygen, iron wire ignites, whereas when heated in a flame and then exposed to the air, it does not ignite.

(c) Gasoline ignites more easily in the combustion chambers of an automobile when the ambient surroundings are warm than when they are cold.

(d) A stack of wooden logs lying exposed to the air does not burn until a single log in the stack has first been kindled.

Combustion Phenomena

5.8 Charcoal briquettes consist almost entirely of elemental carbon. When they burn in the open air, carbon monoxide and carbon dioxide are produced. Write balanced chemical equations illustrating the production of each combustion product.

5.9 Sooty plumes generally billow from petroleum tank fires. What does the presence of these sooty plumes generally indicate about the nature of the combustion phenomena?

Global Warming and the Greenhouse Effect

5.10 In what major way has global warming affected firefighting in the western United States?

Chemistry of Fire Extinguishers

5.11 The OSHA regulation set forth at 29 C.F.R. §1910.162(b)(5) stipulates that employers must provide a predischarge employee alarm system within the workplace to alert employees if a concentration of 4% or greater of carbon dioxide is released from a fixed fire-extinguishing system. What is the most likely reason OSHA enacted this regulation?

Use of the DOT Hazardous Materials Regulations by Emergency Responders

The U.S. Department of Transportation (DOT) regulates hazardous material shipments to protect transportation personnel and equipment from exposure to hazardous materials and to provide emergency responders with communication information in the event of a transportation mishap. These regulations have been promulgated so they conform with generally accepted international standards. Consequently, their proper implementation potentially aids emergency-response personnel, the public, and transportation workers worldwide.

The material presented in this chapter aims to demonstrate how emergency responders may use certain aspects of the hazardous materials regulations when they encounter a spill or leak of a hazardous material during a transportation mishap. Because the regulations are subject to change from time to time, the parties affected by them should always consult 49 C.F.R. Parts 171 through 179 for the latest revisions and amendments.

6.1 DOT Hazardous Materials and Their Proper Shipping Names

A **hazardous material** is uniquely defined by DOT as follows:

> a designated substance or material that has been determined by the Secretary of Transportation to be capable of posing an unreasonable risk to health, safety, and property when transported in **commerce**.

The hazardous materials have designated names that are listed in the **Hazardous Materials Table** at 49 C.F.R. §172.101. The table contains more than 16,000 entries. Many hazardous materials are identified in the table by their chemical names, but some are identified generically with expressions like "flammable liquids, toxic, n.o.s." (not otherwise specified) and "elevated-temperature liquid, flammable, n.o.s." or with technical names like "tear gas substances, liquid, n.o.s." and "organophosphorus pesticides, solid, toxic." DOT regulates their transportation in certain instances only when they are shipped either domestically or internationally (but not both), or by means of aircraft or watercraft.

Although there are 10 columnar entries in the Hazardous Materials Table, only the entries in columns 1 through 7 are important to emergency responders. Table 6.1 is an excerpt of the Hazardous Materials Table that lists the information in the first seven columns only.

6.1-A PROPER SHIPPING NAMES

The **proper shipping name** of a hazardous material appears in column 2 of the Hazardous Materials Table. The proper shipping name is the name that appears in plain, nonitalic type, *not* the name sometimes listed in italics or any other name used to describe the material. DOT permits shippers to select a proper shipping name from several general descriptions in the event that the correct technical name of a material is neither listed in the table nor entirely accurate. The general descriptions often use the letters *n.o.s.*, *n.o.i.* (not otherwise indexed), or *n.o.i.b.n.* (not otherwise indexed by name).

Shippers describe a particular hazardous material by selecting the name from the Hazardous Materials Table that most accurately describes the material to be transported. Suppose a shipper desires to transport ten 55-gal drums containing industrial heating oil. DOT requires the shipper to identify this commodity as "Fuel Oil" on a shipping paper accompanying the shipment, since the Hazardous Materials Table provides only this name as the closest description of the commodity. An arbitrary selection of "Oil," "Heating Oil," or any other similar term does not comply with the intent of these regulations. DOT does allow the shipper to include additional information parenthetically, but only *in addition to* the proper shipping name. In this example, the commodity may be correctly noted as "Fuel Oil (Fuel Oil No. 2)" when this parenthetical entry constitutes the fuel's closest commercial description. On occasion, the word "forbidden" appears in column 3 of the Hazardous Materials Table. The entry indicates that the hazardous material is a **forbidden material**, that is, one that DOT prohibits from being transported by any mode. Most forbidden materials are explosive materials, readily combustible solids, self-heating materials, or self-reactive materials (see the glossary).

hazardous material
- A substance or material in any form that because of its quantity, concentration, chemical, corrosive, flammable, reactive, toxic, infectious, or radioactive characteristics, either separately or in combination with any other substance or substances, constitutes a present or potential threat to human health, safety, welfare, or the environment when improperly stored, treated, transported, disposed or used, or otherwise managed

commerce
- Trade, traffic, transportation, importation, exportation, and similar activities engaged in by shippers and carriers

Hazardous Materials Table
- The compilation of hazardous materials whose transportation is regulated by DOT at 49 C.F.R. §172.101

proper shipping name
- For purposes of DOT regulations, the name in Roman print assigned to a hazardous material listed in the Hazardous Materials Table published at 49 C.F.R. §172.101

forbidden material
- For purposes of DOT regulations, a hazardous material listed in the Hazardous Materials Table at 49 C.F.R. §172.101 with the word "forbidden" in column 3, signifying that the material may not be transported by any mode

TABLE 6.1 | Hazardous Materials Table[a]

SYM-BOLS (1)	HAZARDOUS MATERIALS DESCRIPTIONS AND PROPER SHIPPING NAMES (2)	HAZARD CLASS OR DIVISION (3)	IDENTI-FICATION NUMBER (4)	PACKING GROUP (5)	LABEL CODES (6)	SPECIAL PROVI-SIONS (7)
	Acetone	3	UN1090	II	3	
	Acetonitrile	3	UN1648	II	3	
	Acetyl benzoyl peroxide, solid, or with more than 40 percent in solution	Forbidden				
	Acrolein, stabilized	6.1	UN1092	I	6.1, 3	1
	Alcoholic beverages	3	UN3065	II	3	
		3	UN3065	III	3	
	Aluminum alkyls	4.2	UN3051	I	4.2, 4.3	
I	Ammonia, anhydrous	2.3	UN1005		2.3, 8	
D	Ammonia, anhydrous	2.2	UN1005		2.2	
I	Ammonia solution, *relative density less than 0.880 at 15 °C in water, with more than 50% ammonia*	2.3	UN3318		2.3, 8	4
D	Ammonia solution, *relative density less than 0.880 at 15 °C in water, with more than 50% ammonia*	2.2	UN3318		2.2	
	Ammonia solution, *relative density less than 0.880 at 15 °C in water, with more than 35% but not more than 50% ammonia*	2.2	UN2073		2.2	
	Ammonia solution, *relative density between 0.880 and 0.957 at 15 °C in water, with more than 10% but not more than 35% ammonia*	8	UN2672	III	8	
	Ammunition, *incendiary liquid or gel, with burster, expelling charge, or propelling charge*	1.3J	UN2047	II	1.3J	

[a]Excerpted from the DOT Hazardous Materials Table, 49 C.F.R. §172.101. In this excerpt, column 7 contains only the four codes immediately relevant to emergency-response personnel. Columns 8 through 10 in the complete table have been redacted in their entireties.

SYM-BOLS (1)	HAZARDOUS MATERIALS DESCRIPTIONS AND PROPER SHIPPING NAMES (2)	HAZARD CLASS OR DIVISION (3)	IDENTI-FICATION NUMBER (4)	PACKING GROUP (5)	LABEL CODES (6)	SPECIAL PROVI-SIONS (7)
D	Asbestos	9	NA2212	III	9	
	Barium cyanide	6.1	UN1565	I	6.1	
	Battery fluid, acid	8	UN2796	II	8	
A	Calcium oxide	8	UN1901	III	8	
	Calcium, pyrophoric or Calcium alloys, pyrophoric	4.2	UN1855	I	4.2	
I	Carbon, activated	4.2	UN1362	III	4.2	
I	Carbon, animal or vegetable origin	4.2	UN1361	II	4.2	
		4.2	UN1361	III	4.2	
	Caustic soda, (etc.) see Sodium hydroxide, etc.					
D	Charcoal briquettes, shell screenings, wood, etc.	4.2	NA1361	III	4.2	
	Chlorine	2.3	UN1017		2.3, 8	2
	Chlorine trifluoride	2.3	UN1749		2.3, 5.1, 8	2
	Chlorobenzene	3	UN1134	III	3	
	Coal tar distillates, flammable	3	UN1136	II	3	
		3	UN1136	III	3	
	Cyclohexane	3	UN1145	II	3	
	Dichloromethane	6.1	UN1593	III	6.1	
	Dinitrogen tetroxide	2.3	UN1067		2.3, 5.1, 8	1
	Elevated-temperature liquid, flammable, n.o.s., with flashpoint above 37.8 °C at or above its flashpoint	9	UN3257	III	9	
	Environmentally hazardous substances, liquid, n.o.s.	9	UN3082	III	9	
	Ethyl acetate	3	UN1173	II	3	
	Ethyl mercaptan	3	UN2363	I	3	
	Ethyl methyl ketone or Methyl ethyl ketone	3	UN1193	II	3	

(continued)

TABLE 6.1 Hazardous Materials Table (continued)

SYM-BOLS (1)	HAZARDOUS MATERIALS DESCRIPTIONS AND PROPER SHIPPING NAMES (2)	HAZARD CLASS OR DIVISION (3)	IDENTI-FICATION NUMBER (4)	PACKING GROUP (5)	LABEL CODES (6)	SPECIAL PROVI-SIONS (7)
	Ethylbenzene	3	UN1175	II	3	
	Ethylene	2.1	UN1962		2.1	
	Explosive, blasting, type A	1.1D	UN0081	II	1.1D	
	Fireworks	1.1G	UN0333	II	1.1G	
	Fireworks	1.2G	UN0334	II	1.2G	
	Fireworks	1.3G	UN0335	II	1.3G	
	Fireworks	1.4G	UN0336	II	1.4G	
	Fireworks	1.4S	UN0337	II	1.4S	
G	Flammable liquids, n.o.s.	3	UN1993	I	3	
		3	UN1993	II	3	
		3	UN1993	III	3	
G	Flammable liquids, toxic, n.o.s.	3	UN1992	I	3, 6.1	
		3	UN1992	II	3, 6.1	
		3	UN1992	III	3, 6.1	
G	Flammable solids, toxic, organic, n.o.s.	4.1	UN2926	II	4.1, 6.1	
	Fluorine, compressed	2.3	UN1045		2.3, 5.1, 8	1
	Fuel, aviation, turbine engine	3	UN1863	I	3	
		3	UN1863	II	3	
		3	UN1863	III	3	
D	Fuel oil (No. 1, 2, 4, 5 or 6)	3	NA1993	III	3	
	Gasoline includes gasoline mixed with ethyl alcohol, with not more than 10% alcohol	3	UN1203	II	3	
D, G	Hazardous waste, liquid, n.o.s.	9	NA3082	III	9	
D, G	Hazardous waste, solid, n.o.s.	9	NA3077	III	9	
	Hydrazine, anhydrous	8	UN2029	I	8, 3, 6.1	

SYM-BOLS (1)	HAZARDOUS MATERIALS DESCRIPTIONS AND PROPER SHIPPING NAMES (2)	HAZARD CLASS OR DIVISION (3)	IDENTI-FICATION NUMBER (4)	PACKING GROUP (5)	LABEL CODES (6)	SPECIAL PROVI-SIONS (7)
	Hydrofluoric acid, anhydrous see Hydrogen fluoride, anhydrous					
	Hydrogen, compressed	2.1	UN1049		2.1	
	Hydrogen fluoride, anhydrous	8	UN1052	I	8, 6.1	3
	Hydrogen, refrigerated liquid *(cryogenic liquid)*	2.1	UN1966		2.1	
	Hydrogen sulfide	2.3	UN1053		2.3, 2.1	1
G	Infectious substances, affecting humans	6.2	UN2814		6.2	
	Lime, unslaked, see Calcium oxide					
	Liquefied petroleum gas *see* Petroleum gases, liquefied					
	Lithium aluminum hydride	4.3	UN1410	I	4.3	
	Lithium hydride	4.3	UN1414	I	4.3	
	London purple	6.1	UN1621	II	6.1	
	Lye *see* Sodium hydroxide, solutions					
	Mercury oxide	6.1	UN1641	II	6.1	
	Methyl bromide	2.3	UN1062		2.3	3
	Methyl ethyl ketone *see* Ethyl methyl ketone					
	Methyl isobutyl ketone	3	UN1245	II	3	
	Methyl isocyanate	6.1	UN2480	I	6.1, 3	1
	Methylene chloride see Dichloromethane					
	Nickel carbonyl	6.1	UN1259	I	6.1, 3	1
+	Nitric acid, red fuming	8	UN2032	I	8, 5.1, 6.1	2
	Nitrogen, compressed	2.2	UN1066		2.2	
	Nitrogen dioxide *see* Dinitrogen tetroxide					

(continued)

TABLE 6.1 | Hazardous Materials Table (*continued*)

SYM-BOLS (1)	HAZARDOUS MATERIALS DESCRIPTIONS AND PROPER SHIPPING NAMES (2)	HAZARD CLASS OR DIVISION (3)	IDENTI-FICATION NUMBER (4)	PACKING GROUP (5)	LABEL CODES (6)	SPECIAL PROVI-SIONS (7)
	Nitrous oxide	2.2	UN1070		2.2, 5.1	
	Organic peroxide type A, liquid or solid	Forbidden				
G	Organic peroxide type B, solid	5.2	UN3102	II	5.2, 1	
G	Organic peroxide type B, solid, temperature-controlled	5.2	UN3112	II	5.2, 1	
G	Organic peroxide type D, liquid	5.2	UN3105	II	5.2	
G	Organic peroxide type D, liquid, temperature-controlled	5.2	UN3115	II	5.2	
	Organophosphorus pesticides, solid, toxic	6.1	UN2783	I	6.1	
		6.1	UN2783	II	6.1	
		6.1	UN2783	III	6.1	
	Oxygen, compressed	2.2	UN1072		2.2, 5.1	
+	Oxygen generator, chemical spent	9	NA3356	III	9	
	Oxygen, refrigerated liquid *(cryogenic liquid)*	2.2	UN1073		2.2, 5.1	
	Paint-related material *including paint thinning, drying, removing, or reducing compound*	3	UN1263	I	3	
		3	UN1263	II	3	
		3	UN1263	III	3	
	Petroleum gases, liquefied *or* Liquefied petroleum gas	2.1	UN1075		2.1	
	Phosgene	2.3	UN1076		2.3, 8	1
	Phosphoric acid, solution	8	UN1805	III	8	
	Phosphorus, amorphous	4.1	UN1338	III	4.1	
	Phosphorus, white dry *or* Phosphorus, white, under water *or* Phosphorus, white, in solution, *or* Phosphorus,	4.2	UN1381	I	4.2, 6.1	

SYM-BOLS (1)	HAZARDOUS MATERIALS DESCRIPTIONS AND PROPER SHIPPING NAMES (2)	HAZARD CLASS OR DIVISION (3)	IDENTI-FICATION NUMBER (4)	PACKING GROUP (5)	LABEL CODES (6)	SPECIAL PROVI-SIONS (7)
	yellow dry *or* Phosphorus, yellow, under water *or* Phosphorus, yellow in solution					
	Phosphorus, white, molten	4.2	UN2247	I	4.2, 6.1	
	Phthalic anhydride *with more than 0.05% maleic anhydride*	8	UN2214	III	8	
	Radioactive material, excepted package—empty packaging	7	UN2908		Empty	
	Radioactive material, low specific activity (LSA-III), *nonfissile or fissile excepted*	7	UN3322		7	
	Radioactive material, Type A package, *non-special form, nonfissile or fissile excepted*	7	UN2915		7	
	Radioactive material, uranium hexafluoride, fissile	7	UN2977		7, 8	
	Selenium disulfide	6.1	UN2657	II	6.1	
	Sodium	4.3	UN1428	I	4.3	
	Sodium hydroxide, solid	8	UN1823	II	8	
	Sodium hydroxide solution	8	UN1824	II	8	
		8	UN1824	III	8	
	Sodium peroxide	5.1	UN1504	I	5.1	
	Stannic chloride, anhydrous	8	UN1827	II	8	
D	Sulfur	9	NA1350	III	9	
I	Sulfur	4.1	UN1350	III	4.1	
D	Sulfur, molten	9	NA2448	III	9	
I	Sulfur, molten	4.1	UN2448	III	4.1	
	Sulfuryl fluoride	2.3	UN2191		2.3	4
G	Tear gas substances, liquid, n.o.s.	6.1	UN1693	I	6.1	
		6.1	UN1693	II	6.1	

(*continued*)

| **TABLE 6.1** | Hazardous Materials Table (*continued*) |

SYM-BOLS (1)	HAZARDOUS MATERIALS DESCRIPTIONS AND PROPER SHIPPING NAMES (2)	HAZARD CLASS OR DIVISION (3)	IDENTI-FICATION NUMBER (4)	PACKING GROUP (5)	LABEL CODES (6)	SPECIAL PROVI-SIONS (7)
	1,1,2,2-Tetrachloroethane	6.1	UN1702	II	6.1	
	Toluene	3	UN1294	II	3	
	1,1,1-Trichloroethane	6.1	UN2831	III	6.1	
	Trinitrotoluene and Trinitrobenzene mixtures *or* TNT and trinitrobenzene mixtures *or* TNT and hexanitrostilbene mixtures *or* Trinitrotoluene and hexanitrostilbene mixtures	1.1D	UN0388	II	1.1D	
	Trinitrotoluene, *or* TNT, *dry or wetted with less than 30% water by mass*	1.1D	UN0209	II	1.1D	
	Trinitrotoluene, *wetted with not less than 30% water by mass*	4.1	UN1356	I	4.1	
	Trinitrotoluene mixtures containing Trinitrobenzene and Hexanitrostilbene *or* TNT mixtures containing trinitrobenzene and hexanitrostilbene mixtures	1.1D	UN0389	II	1.1D	
	Xylenes	3	UN1307	II	3	
		3	UN1307	III	3	

6.1-B HAZARD CLASSES AND DIVISIONS

hazard class/division (division number; primary hazard class)
■ For purposes of DOT regulations, any of the nine categories of hazard assigned to a hazardous material because it complies with defining criteria

Each hazardous material is associated with a specific **hazard class** that is characterized by a unique defining criterion. Each of nine hazard classes is symbolized by a number 1 through 9. The defining criterion of each hazard class is provided in the glossary at the end of this book.

Several hazard classes are further divided into divisions. Each division is symbolized by two numbers: the hazard class number and a digit separated by a period, like 1.5. Table 6.2 lists the DOT hazard class numbers and their divisions, if any, that DOT assigns to hazardous materials as well as the names of the classes and divisions associated with them.

There is also an additional unnumbered hazard class that is represented as **other regulated materials** (ORM). **ORM-D** refers to the hazard class of materials whose domestic transportation is regulated as "other regulated materials." These materials pose a limited hazard during trans-

TABLE 6.2		Hazard Classes and Divisions of Hazardous Materials[a]
HAZARD CLASS NUMBER	**DIVISION NUMBER (IF ANY)**	**NAME OF CLASS OR DIVISION**
1	1.1	Explosives (with a mass explosion hazard)
1	1.2	Explosives (with a projection hazard)
1	1.3	Explosives (with predominantly a fire hazard)
1	1.4	Explosives (with no significant blast hazard)
1	1.5	Very insensitive explosives; blasting agents
1	1.6	Extremely insensitive detonating substances
2	2.1	Flammable gas
2	2.2	Nonflammable compressed gas
2	2.3	Poison gas
3		Flammable and combustible liquid
4	4.1	Flammable solid
4	4.2	Spontaneously combustible material
4	4.3	Dangerous-when-wet material
5	5.1	Oxidizer
5	5.2	Organic peroxide
6	6.1	Poisonous materials
6	6.2	Infectious substance
7		Radioactive material
8		Corrosive material
9		Miscellaneous hazardous material
		Other regulated material (ORM-D)

[a]40 C.F.R. §173.2.

other regulated material
■ For purposes of DOT regulations, the classification of a hazardous material that does not conform to the definition of any of the nine hazard classes but is nonetheless considered to possess a dangerous property that could pose a risk to transportation personnel and equipment during its transit (abbreviated ORM)

ORM-D
■ For purposes of DOT regulations, any "other-regulated material" whose transportation is subject to certain regulations when the material is transported domestically

subsidiary hazard
■ For purposes of DOT regulations, any hazard of a material (other than its primary hazard) whose hazard class numbers are listed in column 6 of the Hazardous Materials Table

portation owing to their forms, quantities, and packaging. They are generally consumer commodities like drugs, medicines, hair spray, and other items used for personal care in the home.

Although many hazardous materials are characterized by a single hazard or division class, some hazardous materials possess additional hazards. In such instances, DOT refers to their major hazard class as the **primary hazard class** or **division number** of the designated hazardous materials. All other hazards are called their **subsidiary hazards**.

The primary and subsidiary hazard classes or divisions of a hazardous material are noted in columns 3 and 6 of the Hazardous Materials Table. Column 6 is headed "Label Codes." We note in Section 6.4 that these codes also identify the specific warning labels shippers affix to hazardous material packaging.

Let's consider an example. As noted in column 3 of Table 6.1, compressed oxygen—a non-flammable gas—has a primary division number of 2.2. However, oxygen also supports combustion, which is the defining criterion of an oxidizer. Hence, compressed oxygen has a subsidiary division number of 5.1. Both the primary and subsidiary division numbers are listed in column 6 of Table 6.1.

6.1-C IDENTIFICATION NUMBERS

identification number
■ For purposes of DOT regulations, the UN or NA number listed in column 4 of the Hazardous Materials Table

DOT uses an **identification number** to identify each hazardous material. It is a three- or four-digit number preceded by either the prefix UN (United Nations) or NA (North America). The identification numbers preceded by UN are now used when transporting hazardous materials domestically *and* internationally, whereas identification numbers preceded by NA are used when transporting certain hazardous materials domestically and within Canada *only*. The identification number of a hazardous material is listed in column 4 of the Hazardous Materials Table.

Emergency Response Guidebook
■ A DOT publication that provides emergency actions for implementation at hazardous materials transportation mishaps

An identification number is assigned to a hazardous material to rapidly convey important information concerning the consignment to emergency-response teams. When the identification number is known, team members may quickly identify the physical and chemical nature of a hazardous material and respond effectively to the incident in which the material is involved by referring to a guidebook published by DOT, the **Emergency Response Guidebook** (*ERG*). In this guidebook, each identification number is referenced to a guide number that provides specific instructions relating to appropriate response-actions for a given listing. We examine the use of this guidebook in Section 6.8.

6.1-D PACKING GROUPS

packing group
■ For purposes of DOT regulations, a grouping according to the degree of danger presented by hazardous materials (Packing Group I indicates great danger; Packing Group II, medium danger; and Packing Group III, minor danger)

In most instances, DOT assigns a **packing group** to each hazardous material. This term represents a grouping of hazardous materials according to the collective degree of hazard that is posed by exposure to them. There are only three packing groups. Packing Group I applies to those hazardous materials that are associated with a great degree of danger; Packing Group II, a medium degree of danger; and Packing Group III, a minor degree of hazard.

The packing group of a hazardous material appears as the Roman numeral I, II or III in column 5 of the Hazardous Materials Table. The packing group affects the nature of the packaging shippers and carriers select when they transport a hazardous material. DOT does not assign packing groups to hazard classes 2, 7, and ORM-D, since these hazardous materials require their own specially approved packaging.

DOT permits certain hazardous materials within a single hazard class or division number to be shipped in packaging associated with each of the three packing groups. For example, shippers or carriers may transport flammable liquids (hazard class 3) in Packing Group I, II, or III in accordance with the following flashpoint and boiling point data:

packaging
■ For purposes of DOT regulations, the receptacle and any other components or materials necessary for the receptacle to perform its containment function and to ensure conformance with prescribed minimum requirements

Packing Group I	Boiling point equal to or less than 95 °F (35 °C)
Packing Group II	Flashpoint less than 73 °F (23 °C); boiling point greater than 95 °F (35 °C)
Packing Group III	Flashpoint equal to or greater than 73 °F (23 °C) but equal to or less than 140 °F (60.5 °C); boiling point greater than 95 °F (35 °C)

6.1-E PACKAGING

Packaging is defined by DOT as follows:

package
■ For purposes of DOT regulations, the means of containment used to transport a hazardous material from one location to another

> the receptacle and any other components or materials necessary for the receptacle to perform its containment function and to ensure compliance with packing requirements

It represents the container used to transport a hazardous material from one location to another. The packaging together with its contents is called the **package**.

FIGURE 6.1 Acetone, hexane, methylene chloride, and methanol are liquid hazardous materials that are often encountered in 1-gal (4-L) bottles. When they are transported, these tightly closed glass bottles are over-packed in sturdy cardboard boxes to prevent breakage. (*Courtesy of Mallinckrodt Baker, Inc., Phillipsburg, New Jersey.*)

nonbulk packaging
■ For purposes of DOT regulations, packaging that has (a) a maximum capacity of 119 gal (450 L) or 1ess as a liquid receptacle; (b) a maximum net mass of 882 lb (400 kg) or less and a maximum capacity of 119 gal (450 L) or less as a solid receptacle; *or* (c) a water capacity of 1000 lb (454 kg) or less for a gas receptacle

bulk packaging
■ For purposes of DOT regulations, packaging other than watercraft including a transport vehicle or freight container in which materials are loaded with no intermediate form of containment and have (a) a maximum capacity of greater than 119 gal (450 L) for a liquid receptacle; (b) a maximum net mass greater than 882 lb (400 kg) for a solid receptacle; *or* (c) a water capacity greater than 1000 lb (454 kg) as a gas receptacle

There are two types of performance-oriented packaging for containerization of hazardous materials:

■ **Nonbulk packaging**, such as bottles, drums, cans, cardboard boxes, fiberboard boxes, carboys, cylinders, and bags, sometimes as inner packages that have been overpacked to provide an additional measure of safety during transit.

■ **Bulk packaging**, such as tankcars, cargo tanks, portable tanks, and freight containers.

As illustrated in Figure 6.1, 1-gal (4-L) bottles containing liquid hazardous materials are commonly encountered in laboratories and research facilities. The capacity of this form of non-bulk packaging is very convenient for handling. When they are transported, these bottles are overpacked to eliminate or reduce the likelihood of breakage. Nonetheless, the same liquid hazardous material may be transported in containers of varying capacities. Examples of nonbulk and bulk liquid containers of an identical type are illustrated in Figure 6.2. Each consists of over-packed glass bottles, but only the largest of them is a bulk container.

6.1-F SPECIAL PROVISIONS

DOT identifies certain special provisions that apply to the transportation of a hazardous material. Each provision is identified by a code. We are primarily concerned here only with the provisions that apply directly to the needs of emergency-response personnel.

FIGURE 6.2 These returnable stainless-steel containers are used to overpack bottles of liquids having varying capacities that increase as we proceed from the left to the right of the page. Each of the first four containers is a nonbulk container; the remaining container at the far right is a bulk container. (*Courtesy of Mallinckrodt Baker, Inc., Phillipsburg, New Jersey.*)

When shippers and carriers transport a gaseous or liquid hazardous material *that poses an inhalation hazard*, DOT requires them to establish its **hazard zone**. For gases, the hazard zone is any of the designations Zone A, Zone B, Zone C, or Zone D; and for liquids, it is either Zone A or Zone B. The greatest degree of health hazard is posed by exposure to hazardous materials that are assigned to Zone A.

The code for the hazard zone of a hazardous material that poses an inhalation hazard appears in column 7 of the Hazardous Materials Table. When DOT lists any of the codes 1, 2, 3, or 4 in column 7, shippers and carriers associate them with Zone A, Zone B, Zone C, and Zone D, respectively.

hazard zone
▪ For purposes of DOT regulations, the four hazard levels assigned to poison gases and either of the two levels assigned to liquids that are poisonous by inhalation [49 C.F.R. §§173.116(a) and 173.133(a)]

PERFORMANCE GOALS FOR SECTION 6.2:

- ▪ Illustrate how a shipper of a hazardous material uses the Hazardous Materials Table to prepare a shipping description.
- ▪ Identify the specific information emergency responders may obtain from shipping descriptions.
- ▪ Illustrate the significance to emergency responders of the terms *reportable quantity*, *marine pollutant*, and *hazardous substance*.

shipping paper
▪ The shipping order, bill of lading, air bill, dangerous-cargo manifest, or similar shipping document issued by a shipper or carrier as required by DOT at 49 C.F.R. §§172.202, 172.203, and 172.204

6.2 The Shipping Paper

With few exceptions, DOT requires a hazardous material to be properly described for transportation on a **shipping paper**. There are several forms of a shipping paper, such as a shipping order, bill of lading, manifest, railroad waybill, or similar document. Most resemble the example shown in Figure 6.3, a portion of which we shall use as a template throughout the remainder of

Non-Negotiable
BILL OF LADING

No.

_____OF_____

Date Received:	Dispatch/Pro No:	Driver:	Truck No:	Trailer No:

SHIPPER	CONSIGNEE	CHARGES
Name	Name	
Street Address	Street Address	
City, State	City, State	
Notify/Contact Phone	Notify/Contact Phone	

PARTICULARS FURNISHED BY SHIPPER

Units	HM	Shipping Description (Identification Number, Proper Shipping Name, Primary Hazard Class or Division, Subsidiary Hazard Class or Division, and Packing Group)	Weight (lb)

	Placards Required:	Emergency Telephone:	ERG No.

For hazardous materials transported by vessel: Shipper declares that the packing/loading of freight containers and/or transport vehicles containing hazardous materials has been carried out in accordance with the provisions of 49 C.F.R. §176.27(c).
Signature: _____

This is to certify that the above-named materials are properly classified, described, packaged, marked and labeled, and are in proper condition for transportation according to the applicable regulations of the Department of Transportation (49 C.F.R. §172.204).
Signature: _____

LIABILITY FOR LOSS, DAMAGE, ETC. TO GOODS

Shipper's attention is directed to Section 11 on the reverse side of this Bill of Lading. All goods shall have an agreed release valuation of $0.10 per pound unless Shipper declares a higher value and Carrier accepts that valuation in the space below. For water carriage, see Section 5 on the reverse side of this Bill of Lading.
Shipper's initials:_____ Higher value: $_____ per pound   Carrier's acceptance:_____

ORIGIN	DESTINATION	
time in :	time in :	Received in good order, count and condition unless otherwise noted above.
time out:	time out:	
date :	date :	

SHIPPER _____   **CONSIGNEE** _____

Authorized Signature   Date   Authorized Signature   Date   Authorized Signature   Date

FIGURE 6.3 One form of a shipping paper used in commerce.

this book. For emergency responders, the information entered in the section noted as " Shipping Description" is especially important.

6.2-A SHIPPING DESCRIPTIONS OF HAZARDOUS MATERIALS

shipping description
■ For purposes of DOT regulations, the following minimum information for a hazardous material: proper shipping name (including the technical name, when required); hazard class or division; identification number; packing group; total quantity by mass, volume, or as otherwise appropriate; and the number and types of packages

basic description
■ For purposes of DOT regulations, the sequence of the DOT identification number, proper shipping name, hazard code, subsidiary hazard code, and packing group

DOT requires shippers to provide the **shipping description** of each hazardous material in a given consignment on the accompanying shipping paper. The shipping description of a hazardous material is a composite listing of the following components in the sequence noted:*

- Identification number
- Proper shipping name (including the technical name in parentheses, when applicable)
- Primary hazard class or division
- Subsidiary hazard class or division in parentheses, if any
- Packing group (when applicable)

This combination of components is called the **basic description** of a hazardous material. DOT requires it to be entered on a shipping paper in English.

DOT also requires the total number of packaging units and the total amount of each hazardous material in a consignment to be communicated in the shipping description. The total number of units is indicated by such expressions as 4 boxes, 17 cylinders, 5 drums, 3 glass bottles, 1 tank truck, and so forth. The total amount is typically provided as a gross aggregate mass or volume. Abbreviations are used to express the common units of measurement.

DOT lists six symbols in column 1 of the Hazardous Materials Table. When a G is listed, a technical name must be parenthetically included with this hazardous material's basic description. When a + symbol is listed, the proper shipping name, hazard class, and packing group may not be altered for any reason. When a D or I is listed, the proper shipping name is appropriate for describing the hazardous material in domestic and international transportation, respectively. Finally, when an A or W is listed, DOT respectively regulates the transportation of the given hazardous material by aircraft and watercraft only.

When a consignment comprises hazardous materials and other materials whose transportation is not regulated by DOT, the proper shipping names of the hazardous materials must be entered *first* on the relevant shipping paper, or a contrasting color or an X in the "HM" column can be entered to identify the entries that are hazardous materials. For example, gasoline, nitrogen, and a flammable solid are listed on the following portion of a shipping paper *before* the materials that are not subject to the DOT hazardous materials regulations. The hazardous materials are specifically identified by an X in the column headed "HM."

---

*Beginning January 1, 2007, the DOT regulations described the basic description of a hazardous material as its identification number, proper shipping name, primary hazard class or division number, subsidiary hazards (if any), and packing group (when applicable) in this sequence. The use of this sequence is now mandatory when preparing the shipping descriptions of hazardous materials for *international* transportation. However, when hazardous materials are shipped domestically, transporters may either use this sequence or continue to use the sequence that became effective on December 31, 2006. The latter sequence listed the identification number after the hazard class or division number.

On January 1, 2013, the use of the former sequence becomes mandatory for domestic *and* international shipments. In this book we consistently use the sequence published at 49 C.F.R. 172.202(b): identification number, proper shipping name, primary hazard class or division number, subsidiary hazards (if any), and packing group (when applicable).

Units	HM	Shipping Description (Identification Number, Proper Shipping Name, Primary Hazard Class or Division, Subsidiary Hazard Class or Division, and Packing Group)	Weight (lb)
10 drums	X	UN1203, Gasoline, 3, PGII	4500
40 cylinders	X	UN1066, Nitrogen, compressed, 2.2	800
1 drum	X	UN2926, Flammable solids, toxic, organic, n.o.s. (Cloth/Paper containing 2,4-dinitrophenol), 4.1, (6.1), PGII	452
4 boxes		Advertising materials, paper, n.o.i.	60
1 roll		Paper, printing, newsprint	690
12 sets		Carbon paper	22
		EMERGENCY CONTACT: (000) 000-0000	

This example also illustrates that the shipper must provide an emergency-contact telephone number on a shipping paper for each hazardous material within the consignment. In this instance, there are multiple hazardous materials listed on the same paper, and the single telephone number applies to all of them. During a transportation mishap, emergency responders secure the shipping paper and provide this number to CHEMTREC.

When providing the shipping description of a hazardous material, shippers must be aware of other DOT regulatory requirements that are germane to identifying the given consignment by a selected mode of transportation. Examples of these regulations are provided in Table 6.3. They illustrate that emergency-response personnel obtain valuable information about the nature of a unique hazardous materials consignment from the information that has been entered on shipping papers.

Let's consider another example. Suppose a fuel company desires to transport a tank truck that is loaded with aviation gasoline from a petroleum storage facility to an airport terminal. DOT obligates the company to select the particular entry in the Hazardous Materials Table that most accurately describes the material to be transported. In this instance, the correct selection is "Fuel, aviation, turbine engine." This combination of information is properly entered on a shipping paper as follows:

Units	HM	Shipping Description (Identification Number, Proper Shipping Name, Primary Hazard Class or Division, Subsidiary Hazard Class or Division, and Packing Group)	Volume (gal)
1 tank truck	X	UN1863, Fuel, aviation, turbine engine, 3, PGI	1500

When shippers and carriers transport aviation fuel, DOT requires them to enter only this description of the commodity on a shipping paper. The selection of any other name constitutes an illegal entry.

TABLE 6.3 | Information Required in Shipping Descriptions[a,b]

EXAMPLE (SELECT PHRASES)[c]	RELEVANT REGULATION
Transportation of a Hazardous Material by All Modes	
RQ, UN1052, Hydrogen fluoride, anhydrous, 8, (6.1), PGI (Poison - Inhalation Hazard, Zone C) *or* UN1052, Hydrogen fluoride, anhydrous, 8, (6.1), PGI, RQ (Poison - Inhalation Hazard, Zone C)	When a shipper or carrier transports a hazardous substance in an amount that exceeds its reportable quantity in a single package, the letters RQ are included before or after the shipping description.
RQ, UN1428, Sodium, 4.3, PGI (Dangerous When Wet)	When a hazardous material, by chemical interaction with water, is liable to become spontaneously flammable or give off flammable gases in dangerous quantities, the words "Dangerous When Wet" are included with the shipping description.
UN1831, Sulfuric acid, fuming, 8, (6.1), PGI (Poison - Inhalation Hazard, Zone B)	When a hazardous material possesses multiple hazards, its subsidiary hazard class or division numbers are entered parenthetically immediately following the primary hazard class.
UN1760, Corrosive liquids, n.o.s. (Valeric acid), 8, PGI	When a hazardous material is described with an n.o.s. entry or other generic description in the Hazardous Materials Table, the shipping description includes the name of the substance in parentheses.
UN1992, Flammable liquids, toxic, n.o.s. (contains xylene and methanol), 3, (6.1), PGI	When a mixture or solution of two or more hazardous materials described by an n.o.s. entry in the Hazardous Materials Table is to be transported, the technical names of at least two components most predominantly contributing to the hazards of the mixture or solution are parenthetically entered in the shipping description.
UN2672, Ammonia solutions (contains 12.8% ammonia in water), 8, PGIII	When a concentration range is a component of a proper shipping name, the actual concentration of the hazardous material, if it is within the indicated range, is used in lieu of the range.
UN2789, Acetic acid glacial (contains more than 80% acid by mass), 8, (3), PGII *or* UN2789, Acetic acid glacial, 8, (3), PGII (containing more than 80% acid by mass)	Technical and chemical group names may include an appropriate modifier like "contains" or "containing," which is entered parenthetically between the proper shipping name and the hazard class *or* following the shipping description.

[a]49 C.F.R. §§172.202 and 172.203.

[b]Not provided in this table is the unique information that DOT requires in the shipping descriptions of organic peroxides, explosives, and radioactive materials. This information is discussed independently in Sections 13.14-A, 15.4, and 16.10, respectively.

[c]The component of the shipping description that is applicable to the regulation is provided in blue print.

EXAMPLE (SELECT PHRASES)	RELEVANT REGULATION
RQ, UN2029, Hydrazine anhydrous, 8, (3, 6.1), UN2029, PGI (Poison) or RQ, UN2029, Hydrazine anhydrous, 8, (3, 6.1), PGI (Toxic)	Notwithstanding the hazard class to which a hazardous material is assigned, if a liquid or solid material in a package meets the definition of a Division 6.1, PG I or II, and the fact that it is a poison is not disclosed in its shipping name or class entry, either of the words "Poison" or "Toxic" is entered in the shipping description.
UN2783, Organophosphorus pesticides, solid, toxic (Ciodrin), 6.1, PGII or UN2783, Organophosphorus pesticides, solid, toxic, 6.1, PGI (Ciodrin)	Notwithstanding the hazard class to which a hazardous material is assigned, if the technical name of the compound or principal constituent that causes a material to meet the definition of a Division 6.1, PG I or II, is not included in the proper shipping name for the material, the technical name is entered in parentheses in the shipping description.
RQ, UN2199, Phosphine, 2.3, (2.1) (Poison - Inhalation Hazard, Zone A) or RQ, UN2199, Phosphine, 2.3 (2.1) (Toxic - Inhalation Hazard, Zone A)	For materials that pose an inhalation hazard, either of the expressions "Poison - Inhalation Hazard" or "Toxic - Inhalation Hazard" and the applicable words "Zone A," "Zone B," "Zone C," or "Zone D" for gases or "Zone A" or "Zone B" for liquids, are entered on the shipping paper immediately following the shipping description.
Flammable solids, n.o.s. (sodium), 4.3, UN1325, PGII	If a hazardous material is also a hazardous substance other than a radioactive material, and the name of the hazardous material does not identify the hazardous substance by name, one of the following descriptions is entered in parentheses in the shipping description of the hazardous material: **(a)** the name of the hazardous substance; or
UN3082, Hazardous waste, liquid, n.o.s., 9, (contains trichloroethylene), PGIII (F001)[d]	**(b)** for waste streams, the hazardous waste number; or
NA3082, Hazardous waste, liquid, n.o.s., (contains toluene and methanol), 9, PGIII (EPA ignitability) or NA3082, Hazardous waste, liquid, n.o.s., (contains toluene and methanol), 9, PGIII (D001)	**(c)** for waste streams that exhibit an EPA characteristic, the letters "EPA" followed by the characteristic or the EPA hazardous waste number.[d,e]

[d]As noted in Section 1.5, the EPA hazardous waste numbers for *listed* hazardous wastes are prefixed with one of the capital letters F, K, P, or U. The nature of the listed hazardous wastes is published at 40 C.F.R. §§261.31, 261.32, 261.33(e) and 261.33(f).
[e]Four EPA *characteristics* of a hazardous waste are identified at 40 C.F.R. §§261.21 – 261.24: ignitability (Section 3.2); corrosivity (Section 8.13); reactivity (Section 5.8); and toxicity (Section 10.1). Hazardous wastes exhibiting these characteristics have been assigned the EPA hazardous waste numbers D001, D002, D003, and D004 through D043, respectively.

(*continued*)

| TABLE 6.3 | Information Required in Shipping Descriptions (*continued*) |

EXAMPLE (SELECT PHRASES)	RELEVANT REGULATION
RQ, UN1621, London purple (contains diarsenic trioxide and aniline), 6.1, PGII (Marine Pollutant) or UN1621, London purple (contains diarsenic trioxide and aniline), 6.1, PGII, (Marine Pollutant) RQ	If a hazardous material is also a hazardous substance other than a radioactive material, and the name of the hazardous material does not identify the hazardous substance by name, the letters "RQ" are entered either before or after the shipping description.
UN1193, Residue: Last contained ethyl methyl ketone, 3, PGII	The shipping description for the residue of a hazardous material contained in packaging other than a rail tankcar *may include* the words "Residue: Last contained _____ (name of hazardous material in packaging before it was emptied)."
RQ, UN1052, Hydrogen fluoride, anhydrous, 8, (6.1), PGI (Poison - Inhalation Hazard, Zone C) (DOT-SP ***) (*** is replaced with the appropriate special permit number.)	When a shipper or carrier has been assigned an approved special permit from DOT to transport a hazardous material in an unconventional fashion, the shipping description includes the notation "DOT-SP" followed by the assigned special permit number and so located that the notation is clearly associated with the description to which the exemption applies. A DOT-approved special permit allows shippers and carriers to legally transport a hazardous material in an unconventional fashion (*e.g.*, in authorized packaging) for a specified period, after which the shipper or carrier must reapply for the special permit.
UN2214, HOT Phthalic anhydride, 8, PGIII, RQ	If a liquid hazardous material other than molten sulfur and molten aluminum is to be transported at an "elevated temperature," and the fact that it is an elevated-temperature material is not disclosed in the shipping name, the word "HOT" immediately precedes the proper shipping name of the hazardous material.
6 cyl (DOT-3AA1800), RQ, Phosgene, 2.3, UN1193, (8), (Poison - Inhalation Hazard, Zone A) or 5 steel drums (UN1A1), UN1193, Ethyl methyl ketone, 3, PGII	The number and type of package with its DOT-specification number are included with the shipping description.

EXAMPLE (SELECT PHRASES)	RELEVANT REGULATION
Transportation of a Hazardous Material by Passenger-Carrying Railroad or Aircraft UN1827, Stannic chloride, anhydrous, 8, PGII (Limited Quantity) *or* UN1827, Stannic chloride, anhydrous, 8, PGII (Ltd Qty)	When DOT authorizes the transportation of a hazardous material in a limited quantity by passenger-carrying railroad or aircraft, the shipping description includes the words "Limited Quantity" or "Ltd Qty."
Transportation of a Hazardous Material by Cargo Aircraft UN2363, Ethyl mercaptan, 3, PGI (Cargo Aircraft Only) (Limited Quantity)	When DOT authorizes the transportation of a hazardous material by cargo aircraft but prohibits its transportation aboard passenger-carrying aircraft, the words "Cargo Aircraft Only" are entered in the shipping description.
UN2363, Ethyl mercaptan, 3, PGI (Cargo Aircraft Only) (Limited Quantity) *or* UN2363, Ethyl mercaptan, 3, PGI (Cargo Aircraft Only) (Ltd Qty)	When DOT authorizes the transportation of a hazardous material in a limited quantity by cargo aircraft, the shipping description includes the words "Limited Quantity" or "Ltd Qty" in addition to "Cargo Aircraft Only."
Nonbulk Transportation of a Marine Pollutant on Water and Bulk Transportation of a Marine Pollutant in All Modes UN1381, Phosphorus, white, under water, 4.2, (6.1), PGI (Marine Pollutant) (Poison)	When a shipper or carrier transports a non-bulk quantity of a marine pollutant onboard watercraft or a bulk quantity of a marine pollutant by any mode, the words "Marine Pollutant" are included with the shipping description.
UN1621, London Purple (contains diarsenic trioxide and aniline), 6.1, PGII (Marine Pollutant) (Poison)	When the name of the component causing a material to be a marine pollutant is not evident, the name of the component is identified parenthetically in the shipping description.
RQ, UN3082, Environmentally hazardous substances, liquid, n.o.s. (sodium cyanide and cupric cyanide), 9, PGIII (Marine Pollutant) (Poison)	When a material designated with an n.o.s. entry in the Hazardous Materials Table is a marine pollutant, the names of at least two of the components most predominantly contributing to the marine pollutant designation are entered in parentheses in the shipping description.

(continued)

TABLE 6.3 | Information Required in Shipping Descriptions (*continued*)

EXAMPLE (SELECT PHRASES)	RELEVANT REGULATION
Transportation of a Hazardous Material on Water	
UN1175, Ethylbenzene, 3, PGII (flashpoint = 15 °C)	When a hazardous material possesses a flashpoint equal to or less than 142 °F (61 °C) and is transported on water, the flashpoint in degrees Celsius (closed cup) is entered with the shipping description.
UN2796, Battery fluid, acid, 8, PGII (IMDG Code 1 – Acids)	When a hazardous material is designated with an n.o.s. entry in the Hazardous Materials Table and is not listed in section 3.1.4 of the IMDG Code,[f] the applicable IMDG Code segregation group is entered with the shipping description.
Transportation of a Hazardous Material by Railway	
UN1962, Ethylene, compressed, 2.1, UN1962 (DOT-113) (Do Not Hump or Cut Off Car While in Motion) Placarded: FLAMMABLE GAS (HYDX11111)	The shipping description of a flammable gas or its residue contained within a DOT-113 rail tankcar contains an appropriate notation such as "DOT-113" and the statement "Do Not Hump or Cut Off Car While in Motion."
UN2214, HOT Phthalic anhydride, 8, UN2214, III, RQ (Maximum Operating Speed 15 mph) (LCTX111111)	When DOT permits the transportation by railroad of an elevated-temperature material pursuant to certain exceptions, the shipping description contains an appropriate notation such as "Maximum Operating Speed 15 mph."
UN1193, RESIDUE: Last contained ethyl methyl ketone, 3, PGII Placarded: FLAMMABLE (ACTX11111)	The shipping description for the residue of a hazardous material contained in a rail tankcar *must include* the words "Residue: Last contained _____ (name of hazardous material in a rail tankcar before it was emptied)."
RQ, UN1076, Phosgene, 2.3, (8), (Poison - Inhalation Hazard, Zone A) Placarded: POISON GAS (DUCX11111)	When hazardous materials are transported by rail, the accompanying shipping paper bears the notation "Placarded:___," in which the name of the placard displayed on the railcar is inserted.
RQ, UN1075, Liquefied petroleum gas, 2.1, (Noncorrosive) (CSXT111111)	When hazardous materials are transported by rail, the accompanying shipping paper lists the reporting mark and number[g] that the carrier displays on the transport vehicle used for shipment.

[f]The International Maritime Dangerous Goods (IMDG) Code governs the vast majority of the shipments of hazardous materials by water.

[g]See Section 3.3-B. CSXT is the reporting mark of CSX Transportation Company.

EXAMPLE (SELECT PHRASES)	RELEVANT REGULATION
Transportation of Anhydrous Ammonia by Highway RQ, UN1005, Ammonia, anhydrous, 2.2, (0.2 percent water) (Inhalation Hazard)	When a carrier transports anhydrous ammonia having a water content equal to or greater than 0.2% by mass in a DOT Specification MC330 or MC331 tank truck, the shipping description includes "0.2 percent water" to indicate the suitability for shipping anhydrous ammonia in a cargo tank constructed from quenched and tempered steel.
RQ, UN1005 Ammonia, anhydrous, 2.2, (Not for Q and T tanks) (Inhalation Hazard)	When a carrier transports anhydrous ammonia having a water content less than 0.2% by mass in a DOT Specification MC330 or MC331 tank truck, the shipping description includes the phrase "Not for Q and T tanks."
RQ, UN1005, Ammonia, anhydrous, 2.2, (0.2 percent water) (Inhalation Hazard)	When a carrier transports anhydrous ammonia or ammonia solutions, *relative density less than 0.880 at 15 degrees C in water*, the shipping description includes the words "Inhalation Hazard" without either of the words "Poison" or "Toxic" or the designation of a hazard zone.
Transportation of Liquefied Petroleum Gas by Highway UN1075, Liquefied petroleum gas, 2.1 (Noncorrosive) *or* UN1075, Liquefied petroleum gas, 2.1 (Noncor)	When a carrier transports liquefied petroleum gas in a DOT Specification MC330 or MC331 tank truck, the shipping description includes "Noncorrosive" (or "Noncor") to indicate the suitability of transporting noncorrosive liquefied petroleum gas in a cargo tank made of quenched-and-tempered steel.
Transportation of Bulk Amounts of Oil by Highway or Railroad UN1263, Paint-related material, 3, PGIII (oil)	When petroleum oil is transported in packaging having a capacity of 3500 gal (13,250 L) or more, *or* when any oil is transported in packaging in a quantity of 42,000 gal (159,000 L) or more, the word "oil" is included on the shipping paper.

A courier transports two 15-lb cylinders of laughing gas to a dental clinic within a motor van. Laughing gas is the common name for a substance more correctly called nitrous oxide or dinitrogen monoxide. What shipping description does DOT require on the shipping paper used by the courier?

Solution: From the information in columns 2, 3, and 4 of Table 6.1, the shipping description is determined to be as follows:

Units	HM	Shipping Description (Identification Number, Proper Shipping Name, Primary Hazard Class or Division, Subsidiary Hazard Class or Division, and Packing Group)	Weight (lb)
2 cylinders	X	UN1070, Nitrous oxide, 2.2, (5.1)	30

6.2-B REPORTABLE QUANTITIES

hazardous substance
■ For purposes of DOT regulations, a substance listed in Appendix A at 40 C.F.R. §172.101 in an amount that exceeds its reportable quantity and is transported in a single package

reportable quantity
■ For purposes of DOT regulations, the amount listed at 40 C.F. R. §302.4 and 49 C.F.R. §172.101 of a "hazardous substance" within the meaning of DOT and CERCLA, the release of which triggers mandatory notification to the National Response Center (abbreviated RQ)

On occasion, the letters RQ are entered with the shipping description of a hazardous material. RQ refers to the amount of a **hazardous substance** that constitutes a **reportable quantity**. For the purposes of the Superfund law (Section 1.5-D), any substance listed at 40 C.F.R. §302.4 is a hazardous substance, regardless of the amount at issue; but for the purposes of the DOT regulations, only a substance listed at 49 C.F.R. §172.101, Appendix A, in an amount equal to or exceeding the reportable quantity is a hazardous substance. There are only five reportable quantities: 5000 lb (2270 kg); 1000 lb (454 kg); 100 lb (45.4 kg); 10 lb (4.54 kg); and 1 lb (0.45 kg). Examples of several hazardous substances and their DOT reportable quantities are provided in Table 6.4.

When a shipper intends to transport a hazardous substance in an amount equal to or greater than its reportable quantity, DOT requires RQ to be entered on the shipping paper either before or after its basic description. Suppose, for instance, that a shipper intends to transport domestically by tank truck 4500 lb of an aqueous ammonia solution containing 12% ammonia having a gross weight of 4500 lb. Although there are four shipping descriptions for ammonia solutions in Table 6.1, the description having the identification number UN2672 most adequately describes the commodity. Because the shipper intends to transport an amount of an ammonia solution exceeding the reportable quantity listed in Table 6.4 as 1000 lb (454 kg), the letters RQ must be included either before or after its proper shipping description on the accompanying waybill. If the shipper elects to enter RQ before the shipping description, the complete entry is noted as follows:

Units	HM	Shipping Description (Identification Number, Proper Shipping Name, Primary Hazard Class or Division, Subsidiary Hazard Class or Division, and Packing Group)	Weight (lb)
1 tank truck	X	RQ, UN2672, Ammonia solution (contains 12% ammonia in water), 8, PGIII	4500

The following shipping description of a hazardous material is provided on a transportation manifest:

Units	HM	Shipping Description (Identification Number, Proper Shipping Name, Primary Hazard Class or Division, Subsidiary Hazard Class or Division, and Packing Group)	Weight (gal)
1 rail tankcar	X	RQ, NA3082, Hazardous waste liquid, n.o.s. (contains toluene and xylene), 9, PGIII (EPA ignitability) (UTLX42888)	10,000

How is this shipping description useful to an emergency-response team?

Solution: There are at least four pieces of information in this shipping description that are useful to emergency-response personnel:

- The notation RQ informs first-on-the-scene responders that the commodity is a hazardous substance, the release of which could adversely affect public health and the environment. The emergency-response crew should proceed to dam or dike the area to confine the leak and prevent its widespread release into the environment.

- The reference to "EPA ignitability" indicates that the hazardous waste liquid exhibits the RCRA characteristic of ignitability (Section 3.2). This designation informs emergency-response personnel that the hazardous waste liquid possesses a flashpoint equal to or less than 140 °F (60 °C). Thus, it poses a risk of fire and explosion.

- Packing Group III indicates that the relative degree of hazard possessed by the commodity is minor. A minor degree of hazard means that although the material ignites, other groups of hazardous materials are far more hazardous.

- In Section 6.8, we shall see that NA3082 directs emergency-response personnel to certain recommended procedures for properly responding to an incident involving the release of this hazardous waste liquid.

A tear gas manufacturer wishes to ship twenty-five 1-lb aerosol spray cans of "Riot Away," a liquid tear gas product, to a police department in Cincinnati, Ohio, by motor carrier. The active component of the tear gas is the substance whose chemical name is α-chloroacetophenone. What shipping description does DOT require the carrier to enter on the accompanying shipping paper?

Solution: For "Tear gas substances, liquid, n.o.s.," the entry in column 1 of Table 6.1 is G, indicating that a technical name must be parenthetically included in this hazardous material's basic description. Because the active component of the tear gas is α-chloroacetophenone, the shipping description is entered on a shipping paper as follows:

Units	HM	Shipping Description (Identification Number, Proper Shipping Name, Primary Hazard Class or Division, Subsidiary Hazard Class or Division, and Packing Group)	Weight (lb)
25 aerosol cans	X	UN1693, Tear gas substances, liquid, n.o.s. (α-chloroaceto-phenone), 6.1, PGI	25

SOLVED EXERCISE 6.4

The deadliest industrial chemical accident on record occurred in December 1984 at a pesticide plant in Bhopal, India. During the accident, several thousand kilograms of methyl isocyanate, a poisonous and flammable liquid, was released into the environment. Since it is denser than air, the vapor of this substance flowed along the ground and infiltrated shantytowns located next to the plant, where more than 2500 people died from the exposure.

Use Tables 6.1 and 6.3 to determine the basic description that DOT requires shippers and carriers to enter on an accompanying shipping paper when a bulk volume of methyl isocyanate is transported.

Solution: The entry in column 7 of Table 6.1 reveals that methyl isocyanate is a liquid that poses an inhalation hazard. The entry 1 refers to Zone A. When shippers transport methyl isocyanate, DOT requires them to enter either of the expressions Poison - Inhalation Hazard or Toxic - Inhalation Hazard and the words Zone A on the shipping paper immediately following the shipping description. Thus, from the information listed in columns 2, 3, 4, and 7 of Table 6.1, either of the following basic descriptions for methyl isocyanate may be entered on the relevant shipping paper:

UN2480, Methyl isocyanate, 6.1, (3), PGI (Poison - Inhalation Hazard, Zone A)
UN2480, Methyl isocyanate, 6.1, (3), PGI (Toxic - Inhalation Hazard, Zone A)

6.2-C MARINE POLLUTANTS

marine pollutant
■ A substance denoted in Appendix B, 49 C.F.R. §172.101

When shippers and carriers transport a quantity of a substance in nonbulk containers onboard a ship or other form of watercraft, or in bulk by any mode, they must determine whether the substance is a **marine pollutant**. This is a material whose transportation is primarily regulated to reduce or eliminate the impact on the aquatic environment from potential releases into our nation's waterways. DOT provides a listing of the marine pollutants at 49 C.F.R. §172.101, Appendix B, an excerpt of which is reproduced in Table 6.5.

TABLE 6.4	Some Hazardous Substances and their Reportable Quantities[a]		

	Reportable Quantity	
HAZARDOUS SUBSTANCE	POUNDS	KILOGRAMS
Acetone	5000	2270
Acetonitrile	5000	2270
Ammonia solutions	1000	454
Asbestos, white	1	0.454
Cupric cyanide	10	4.54
Chlorine	10	4.54
Chlorobenzene	100	45.4
Dichloromethane	1000	454
Ethyl methyl ketone	5000	2270
Hydrazine	1	0.454
Hydrogen fluoride, anhydrous	100	45.4
Hydrogen sulfide	100	45.4
Methyl parathion	100	45.4
Nickel carbonyl	10	4.54
Nitric acid	1000	454
Phosgene	10	4.54
Phosphine	100	45.4
Phosphorus	1000	454
Phthalic anhydride	5000	2270
Sodium	10	4.54
Selenium disulfide	10	4.54
Sodium hydroxide	1000	454
1,1,2,2-Tetrachloroethane	100	45.4
Toluene	1000	454
1,1,1-Trichloroethane	1000	454
Xylene(s)	100	45.4

[a]40 C.F.R. §302.4 and 49 C.F.R. §172.101, Appendix A, Table 1.

TABLE 6.5	Examples of Some Marine Pollutants[a]
Barium cyanide	London purple
Cadmium compounds	Mercuric oxide
Carbon tetrachloride	Nickel carbonyl
Chlorine	Phosphorus, white or yellow, dry, wet, under water, or in solution
Cupric sulfate	Sodium cyanide, solid
Ferric arsenate	Triethylbenzene
Lead acetate	Zinc bromide

[a]49 C.F.R. §172.101, Appendix B.

dangerous-cargo manifest
■ For purposes of DOT regulations, a cargo manifest that lists the hazardous materials stowed aboard watercraft and the specific locations at which they are stowed

When DOT designates a material as a marine pollutant, the words Marine Pollutant are included parenthetically with the shipping description on the **dangerous-cargo manifest** or other form of a shipping paper. A dangerous-cargo manifest lists the hazardous materials stowed aboard watercraft and the specific locations at which they are stowed. For example, when a glass bottle containing 10 lb (4.54 kg) of mercury(II) oxide is overpacked in a fiberboard box and transported by watercraft, the shipper conveys that the substance is a marine pollutant on the accompanying dangerous-cargo manifest as follows:

Units	HM	Shipping Description (Identification Number, Proper Shipping Name, Primary Hazard Class or Division, Subsidiary Hazard Class or Division, and Packing Group)	Weight (lb)
1 glass bottle within a fiberboard box	X	UN1641, Mercury oxide, 6.1, PGII (Marine Pollutant) (Toxic)	10

When the name of the commodity does not include the name of the component that causes it to be a marine pollutant, DOT requires the shipper to parenthetically identify the name of the component (see Table 6.3).

PERFORMANCE GOALS FOR SECTION 6.3:

■ Identify the individual responsible for providing a shipping paper to emergency responders or the location from which emergency responders may expect to obtain the shipping paper during a transportation mishap involving hazardous materials.

6.3 Location of the Shipping Paper During Transit

During a transportation mishap, where may emergency-response personnel expect to locate the shipping paper? To avoid any confusion over its location, DOT requires the shipping paper to be carried on transport vehicles in a specific location. Because it is important to quickly locate this document when responding to incidents involving the release of a hazardous material, these requirements are briefly summarized next.

DOT requires the drivers of motor vehicles transporting hazardous materials and each carrier using the vehicles to clearly distinguish the appropriate shipping paper from other papers. When the driver is at the vehicle's controls, DOT requires the shipping paper to be within immediate reach and readily visible to any person entering the driver's compartment. For example, as shown in Figure 6.4, it may be stored in a holder mounted to the inside of the door on the driver's side of the motor vehicle. When the driver is not at the vehicle's controls, DOT requires the shipping paper to be located either in the holder or on the driver's seat inside the vehicle.

When hazardous materials are transported by railroad, DOT requires a member of the train crew to be in charge of the shipping papers that describe the consignments that are transported. A person in the train crew is also required to retain in his or her possession a document indicating the position in the train of each loaded and placarded car containing a hazardous material.

When hazardous materials are transported by aircraft, DOT requires the aircraft carrier to provide a shipping paper describing the consignment to the pilot in command. The pilot retains the shipping paper until the plane reaches its destination.

When hazardous materials are transported by watercraft, DOT requires the carrier to prepare a dangerous-cargo manifest, list, or stowage plan and submit it to the master of the vessel. This document must be contained within a designated holder on or near the vessel's bridge.

PERFORMANCE GOALS FOR SECTION 6.4:

- Describe the nature of a Hazardous Materials Safety Permit.
- Identify the location within a motor vehicle at which emergency responders may expect to locate the safety permit.

Shipping papers

FIGURE 6.4 When the driver of a motor vehicle is at the vehicle's controls, DOT requires the shipping paper to be kept within immediate reach inside a holder mounted to the inside of the door on the driver's side of the vehicle.

6.4 Hazardous Materials Safety Permit

At 49 C.F.R. §385.400, DOT requires commercial motor carriers that transport certain hazardous materials within interstate or intrastate commerce to obtain a **Hazardous Materials Safety Permit**. To obtain this permit, DOT requires the motor carrier to have a "satisfactory" safety rating issued either by the Federal Motor Carrier Safety Administration or the state in which the carrier has its principal place of business. DOT further requires the motor carrier to certify that its security program complies with requirements published at 49 C.F.R. §385.407(b). In addition, DOT requires motor carriers to keep a copy of the permit or other document showing the permit number within the vehicle and provide the permit to federal, state, or local authorities on their request.

The issuance of a Hazardous Materials Safety Permit confers authority on the motor carrier to transport in commerce certain hazardous materials in amounts equal to or exceeding published threshold values. It also requires the carrier to prepare and implement a security plan that identifies a route plan the carrier plans to use and the manner by which the safety elements are secured during transportation. The specific hazardous materials and the threshold amounts that are subject to issuance of a Hazardous Materials Safety Permit are noted in appropriate future sections of this book.

PERFORMANCE GOALS FOR SECTION 6.5:

- Describe the DOT regulations that require shippers and carriers to affix warning labels to the packaging containing hazardous materials when they are transported.
- Memorize the inscriptions on the DOT labels corresponding to all hazard classes and division numbers.
- Memorize the colors of the DOT labels in each of the nine hazard classes.

6.5 DOT Labeling Requirements

In most instances involving the transportation of a hazardous material, DOT requires shippers to affix warning **labels** corresponding to the primary and subsidiary hazard classes of the material directly to the outside surfaces of the packaging. The required labels are identified by the code entries in column 6 of the Hazardous Materials Table. Additional labeling requirements for hazard classes 1 and 7 are provided in Sections 15.4-D and 16.10-B, respectively.

The DOT labels that shippers and carriers affix to packaging containing a hazardous material are displayed in Figure 6.5. Their names are listed in Table 6.6. Each label is diamond-shaped and color-coded to signify a specific hazard class or division and may also include a pictograph to rapidly identify the potential hazards of the contents. Each label also bears the hazard class or division number in its lower corner.

The following points regarding the nature of the labels in Figure 6.5 should be noted:

- The uppercase letters that follow the division designations for the EXPLOSIVE labels are examples of the designations for the compatibility groups of explosives, whose nature is discussed later, in Section 14.4-B.
- The label at the far right of Hazard Class 1 is the EXPLOSIVE subsidiary label. It is affixed to packaging when the code listed in column 6 of the Hazardous Materials Table indicates that the hazardous material has an explosive subsidiary hazard.
- The ORGANIC PEROXIDE label is in the process of being replaced with the new label shown at the far right of Hazard Class 5. Use of the new ORGANIC PEROXIDE label becomes mandatory on January 1, 2011.

HAZARD CLASS 1

EXPLOSIVES 1.1 EXPLOSIVES 1.2 EXPLOSIVES 1.3 EXPLOSIVES 1.4 EXPLOSIVES 1.5 EXPLOSIVES 1.6

HAZARD CLASS 2   **HAZARD CLASS 3**

FLAMMABLE   NON-FLAMMABLE   POISON GAS   FLAMMABLE
GAS   GAS

HAZARD CLASS 4   **HAZARD CLASS 5**

FLAMMABLE   SPONTANEOUSLY   DANGEROUS   OXIDIZER   ORGANIC   NEW ORGANIC
SOLID   COMBUSTIBLE   WHEN WET   PEROXIDE   PEROXIDE

HAZARD CLASS 6

POISON INHALATION   POISON   INFECTIOUS
HAZARD   SUBSTANCE

HAZARD CLASS 7

RADIOACTIVE   RADIOACTIVE   RADIOACTIVE   FISSILE
WHITE-I   YELLOW-II   YELLOW-III

HAZARD CLASS 8   **HAZARD CLASS 9**

CORROSIVE   CLASS 9

FIGURE 6.5 DOT requires shippers to affix warning labels that identify the primary and subsidiary hazard classes of a hazardous material on the exterior surface of its packaging. The ORGANIC PEROXIDE label is in the process of being replaced with the new label at the far right of Hazard Class 5. DOT requires carriers to use the new ORGANIC PEROXIDE label on January 1, 2011. The INFECTIOUS SUBSTANCE label bears the following words (not shown here on label): In case of Damage or Leakage, Immediately Notify Public Health Authority. In USA, Notify Director CDC, Atlanta, GA (800) 232-0124.

TABLE 6.6 | DOT Warning Labels[a]

HAZARD CLASS OR DIVISION	LABEL NAME
1.1	EXPLOSIVE 1.1
1.2	EXPLOSIVE 1.2
1.3	EXPLOSIVE 1.3
1.4	EXPLOSIVE 1.4
1.5	EXPLOSIVE 1.5
1.6	EXPLOSIVE 1.6
2.1	FLAMMABLE GAS
2.2	NON-FLAMMABLE GAS
2.3	POISON GAS
3	FLAMMABLE LIQUID
4.1	FLAMMABLE SOLID
4.2	SPONTANEOUSLY COMBUSTIBLE
4.3	DANGEROUS WHEN WET
5.1	OXIDIZER
5.2	ORGANIC PEROXIDE
6.1 (Inhalation hazard, Zone A or B)	POISON INHALATION HAZARD
6.1 (other than inhalation hazard, Zone A or B)	POISON
6.2	INFECTIOUS SUBSTANCE
7	RADIOACTIVE WHITE-I
7	RADIOACTIVE YELLOW-II
7	RADIOACTIVE YELLOW-III
7 (fissile material)	FISSILE
7 (empty packages)[b]	EMPTY
8	CORROSIVE
9	CLASS 9

[a]49 C.F.R. §172.400.
[b]See Section 16.10-B.

DOT requires shippers and carriers to affix the specified labels on the surface of the package near the spot where they mark the proper shipping name. For instance, when shippers and carriers transport gasoline, DOT requires them to affix the FLAMMABLE LIQUID label to the outside surface of the drum adjacent to the spot where it is marked "Gasoline." When cylinders or packages have irregular surfaces on which labels cannot be satisfactorily affixed, DOT permits labels to be printed on a securely affixed tag, unless the packages contain radioactive materials.

FIGURE 6.6 DOT requires shippers to affix this CARGO AIRCRAFT ONLY label to the exterior surface of the hazardous material packaging when its transportation is approved solely onboard cargo aircraft.

DOT also requires shippers and carriers to affix *duplicate* labels corresponding to their applicable codes to certain packaging types. Duplicate labels are displayed on at least two sides or two ends other than the bottom. For multiunit car tanks, they are displayed on each end. For a freight container or aircraft unit load device, they are displayed on or near the closure. The regulations pertaining to use of duplicate labels are published at 49 C.F.R. §172.406.

When shippers and carriers transport packages containing hazardous materials authorized solely for transport on cargo aircraft, DOT requires that they display the CARGO AIRCRAFT ONLY label in Figure 6.6 in addition to any other required labels.

SOLVED EXERCISE 6.5

Identify the label(s), if any, that DOT requires a shipper to affix to each of five 55-gal steel drums containing the solvent known commercially as methyl ethyl ketone.

Solution: We first learn that DOT lists this substance in column 2 of Table 6.1 as ethyl methyl ketone. Then, the code entered in column 6 informs us that DOT requires a FLAMMABLE LIQUID label to be affixed on the exterior surface of each drum containing this solvent.

SOLVED EXERCISE 6.6

Identify the label(s), if any, that DOT requires a shipper to affix to each of five fiberboard boxes that contain eight 1-lb bottles of solid white elemental phosphorus submerged under water.

Solution: By reference to the entry in column 6 of Table 6.1, we determine that the label codes for white elemental phosphorus under water are 4.2 and 6.1, respectively. These entries mean that the shipper must affix SPONTANEOUSLY COMBUSTIBLE and POISON labels on the exterior surface of each fiberboard box. These labels must appear side by side on the surface of two of the sides or ends (other than the bottoms) of each box.

PERFORMANCE GOALS FOR SECTION 6.6:

- Identify the information that DOT requires shippers and carriers to mark on nonbulk and bulk packaging used to transport a hazardous material.
- Identify when DOT requires identification numbers to be displayed on the exterior surfaces of packaging containing hazardous materials.

6.6 DOT Marking Requirements

Two types of marking are required on the packaging used to transport a hazardous material. The first type is provided by the *manufacturers* of the packaging. DOT requires manufacturers to mark certain information on the packaging prior to its use by shippers or carriers. This information is called the *specification marking*. It serves to show that the packaging was constructed and tested in compliance with applicable DOT specifications and standards. The marking includes the name and address or symbol of the packaging manufacturer or approval agency and the letters and numerals that identify the standard or specification. We previously reviewed DOT's specification marking requirements for gases transported within cylinders in Section 3.3-B, and within rail tankcars in Section 3.3-C.

With few exceptions, DOT also requires the *shippers* of hazardous materials to legibly mark certain other information on an outer exterior surface of the packaging used for shipment. These markings are written in English and are displayed on the packaging, unobstructed by the presence of labels, advertising, and other information. Some commonly encountered marking requirements for shippers and carriers are provided in Sections 6.6-A and 6.6-B. Special marking requirements applicable to hazard classes and divisions 1, 6.2, and 7 are noted later, in Sections 15.4, 10.20-A, and 16.10-C, respectively.

6.6-A HAZARDOUS MATERIALS TRANSPORTED IN NONBULK PACKAGING

In compliance with global harmonization (Section 1.8), DOT requires *manufacturers* at 49 C.F.R. §178.503 to provide the following specification markings in the presented sequence on nonbulk packaging when it is used to transport liquid and solid hazardous materials:

- UN symbol (the lowercase letters u and n drawn within a circle)
- Packaging identification code from 49 C.F.R. §173.212 designating the type of package and its material of construction
- One of the letters X, Y, or Z, which designate the performance standard for which the packaging was successfully tested, as follows:

    X   For packages meeting PGI, PGII, and PGIII tests

    Y   For packages meeting PGII and PGIII tests

    Z   For packages meeting PGIII tests only

- A designation of the specific gravity or mass in kilograms for which the package has been tested
- For single and composite packagings intended to contain liquids, the test pressure in kilopascals rounded to the nearest 10 kPa of the hydrostatic pressure test that the packaging design type has successfully passed, and for packages intended to contain solids or inner packagings, the letter S
- The last two digits of the year during which the package was manufactured
- The letters USA indicating that the package was marked pursuant to DOT's standards
- The symbol of the manufacturer

Suppose 50 lb (22.7 kg) of cupric cyanide is transported within several glass bottles that are cushioned within a fiberboard box. Most markings on the box in Figure 6.7 are provided by its manufacturer.

However, aside from these specification markings, DOT also requires the *shipper* to mark on nonbulk packaging the information published at 49 C.F.R. §172.301. The example in Figure 6.7 shows certain information that the shipper has provided on the surface of the box. Briefly, when shippers intend to transport a hazardous material in nonbulk packaging, DOT usually requires them to mark the outer exterior surface of the packaging with the name and address of its shipper

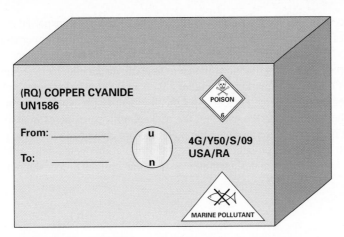

FIGURE 6.7 This fiberboard box is an approved outer PGII packaging that holds glass bottles containing copper cyanide, the proper shipping name for cupric cyanide. RQ must be included with the proper shipping name, because the reportable quantity for this commodity is 10 lb. At 49 C.F.R. §173.212, 4G is provided as the identification code for a fiberboard box. X designates the performance standard to which the box was successfully tested to retain its integrity when 50 kg of material is contained therein. S designates that cupric cyanide is a solid; 09 means the box was manufactured in 2009; and USA/RA designates that the box was marked pursuant to DOT standards by RA, the symbol of the box's manufacturer. DOT requires the shipper to affix MARINE POLLUTANT markings and POISON labels to the box.

and receiver and the proper shipping name and identification number of the commodity. When the proper shipping name in the Hazardous Materials Table is designated by an n.o.s. entry, DOT requires shippers and carriers to also mark the outer exterior surface of the packaging with the technical name of the material in parentheses immediately following or below the proper shipping name.

When shippers transport a hazardous substance in nonbulk packaging in an amount that exceeds its reportable quantity, DOT requires them to mark the outer exterior surface of the packaging with the letters RQ in association with the proper shipping name. If the proper shipping name does not specifically identify the hazardous substance, DOT requires them to indicate the name of the hazardous substance on the packaging in parentheses in association with the proper shipping name. If more than one hazardous substance is contained within the packaging, DOT requires them to indicate at least the two hazardous substances with the lowest reportable quantities.

When shippers load 8820 lb (4000 kg) or more of a *single* hazardous material at one loading facility into nonbulk packages that are carried within a transport vehicle or freight container, DOT requires them to mark the transport vehicle or freight container on each side and each end with the proper shipping name and identification number of the hazardous material.

When a poisonous material is transported within nonbulk plastic packaging, DOT requires shippers to *permanently* mark the packaging with the word POISON. Furthermore, when shippers transport a material that poses a health hazard by inhalation within nonbulk packaging, DOT requires them to mark on each side and each end of the packaging the material's proper shipping name and identification number when the material is ranked in Hazard Zone A or B *and* the transport vehicle or freight container is loaded at one facility in an aggregate gross amount equal to or exceeding 2205 lb (1000 kg).

When shippers transport *liquid* hazardous materials in nonbulk packaging, DOT requires them at 49 C.F.R. §172.312 to assure that closure is upward, and that packages other than liquefied compressed gas cylinders are marked with package orientation markings on the outer exterior surface of the packaging on two opposing sides with arrows pointing in the correct upright direction.

When an RCRA-regulated hazardous waste (Section 1.5-C) is transported in containers having a capacity of 119 gal (450 L) or less, EPA and DOT require its shipper to affix the HAZARDOUS WASTE marking in Figure 6.8 on an outer surface of each container.

When shippers transport a marine pollutant in nonbulk packaging, DOT requires them to affix the MARINE POLLUTANT marking in Figure 6.9 on an outer exterior surface of the packaging. When the proper shipping name does not specifically identify the name of the marine pollutant, DOT requires them to indicate the name on the packaging in parentheses in association with the proper shipping name. If two or more marine pollutants are contained within the packaging, DOT requires them to indicate parenthetically at least the two components most predominantly contributing to the marine pollutant designation in association with the proper shipping name.

SOLVED EXERCISE 6.7

A shipper intends to transport six 50-lb cardboard boxes of zinc bromide as cargo by watercraft.

(a) What information does DOT require the shipper to enter on the accompanying dangerous-cargo manifest?

(b) What does DOT require the shipper to affix or mark on each of the six boxes?

Solution:

(a) Table 6.5 indicates that zinc bromide is a marine pollutant. DOT requires the shipper to communicate this information on the dangerous-cargo manifest as follows:

Units	HM	Shipping Description (Identification Number, Proper Shipping Name, Primary Hazard Class or Division, Subsidiary Hazard Class or Division, and Packing Group)	Weight (lb)
Six cardboard boxes		Zinc bromide (Marine Pollutant)	300

(b) Zinc bromide is not listed in Table 6.1; nor is it listed in the complete Hazardous Materials Table at 49 C.F.R. §172.101. This means that DOT does not require the shipper to affix DOT labels on the surfaces of the six boxes. Nonetheless, DOT requires the shipper to inscribe on the surface of each box the name and address of the shipper and receiver and a description of the package contents as follows:

Zinc bromide (Marine Pollutant)

DOT also requires the shipper to affix the MARINE POLLUTANT marking on two opposing sides of each box.

6.6-B HAZARDOUS MATERIALS TRANSPORTED IN BULK PACKAGING

There are also DOT marking regulations that apply to the transportation of a hazardous material in bulk packaging. They are published at 49 C.F.R. §172.302 and are briefly summarized next.

HAZARDOUS WASTE

STATE & FEDERAL LAW PROHIBITS IMPROPER DISPOSAL.
IF FOUND, CONTACT THE NEAREST POLICE OR PUBLIC SAFETY
AUTHORITY OR THE U.S. ENVIRONMENTAL PROTECTION AGENCY
OR THE CALIFORNIA DEPARTMENT OF TOXIC SUBSTANCES CONTROL.

GENERATOR INFORMATION:

NAME _____
ADDRESS _____ PHONE _____
CITY _____ STATE _____ ZIP _____
EPA /MANIFEST
ID NO./ DOCUMENT NO. _____ /_____
EPA   CA.   ACCUMULATION
WASTE NO. _____ WASTE NO. _____ START DATE _____
CONTENTS COMPOSITION _____

PHYSICAL STATE: HAZARDOUS PROPERTIES ☐ FLAMMABLE   ☐ TOXIC
☐ SOLID ☐ LIQUID | ☐ CORROSIVE ☐ REACTIVITY ☐ OTHER _____

_____
_____
_____
_____
D.O.T. PROPER SHIPPING NAME AND UN OR NA NO. WITH PREFIX

HANDLE WITH CARE!
CONTAINS HAZARDOUS OR TOXIC WASTES

FIGURE 6.8 When a container having a capacity of 119 gal (450 L) or less is used to transport hazardous waste (Section 1.5-C), DOT and EPA require shippers at 49 C.F.R. §172.304 and 40 C.F.R. §262.32, respectively, to provide this HAZARDOUS WASTE marking to the carrier with the waste generator's name and address and the relevant manifest number.

When shippers transport a hazardous material in bulk packaging, DOT requires them to mark the packaging with the identification number of the material on each side and each end if the packaging has a capacity equal to or greater than 1000 gal (3785 L); *or* on two opposing sides if the packaging has a capacity of less than 1000 gal (3785 L); *or* for cylinders permanently installed on a tube trailer motor vehicle, on each side and each end of the motor vehicle.

When shippers transport a bulk quantity of a material that poses a health hazard by inhalation and is ranked in either Zone A or Zone B, DOT requires them to mark INHALATION HAZARD on two opposing sides of the packaging, except when these words appear on the required labels or placards.

elevated-temperature material
▪ For purposes of DOT regulations, a hazardous material that when offered for transportation or transported in a bulk packaging exists in one of the following: the liquid phase of matter at a temperature at or above 212 °F (100 °C); liquid phase of matter with a flashpoint equal to or greater than 100 °F (37.8 °C) that is intentionally heated and offered for transportation or transported at or above its flashpoint; *or* the solid phase of matter at a temperature equal to or greater than 464 °F (240 °C).

INHALATION HAZARD

When shippers transport bulk packaging containing an **elevated-temperature material** other than molten aluminum or molten sulfur, DOT requires them to display the HOT marking in Figure 6.10 on two opposing sides of the packaging. The identification number of the hazardous material may also be displayed on the HOT marking. When shippers or carriers transport molten aluminum or molten sulfur in bulk packaging, DOT requires them to mark the containers with the applicable words MOLTEN ALUMINUM or MOLTEN SULFUR.

FIGURE 6.9 When a marine pollutant is offered for transportation in nonbulk packaging by water or bulk packaging by any mode, DOT requires shippers and carriers to post this MARINE POLLUTANT marking on its packaging.

MARINE POLLUTANT

FIGURE 6.10 When an elevated-temperature material is transported in bulk packaging, DOT requires carriers at 49 C.F.R. §172.325 to mark the word HOT on two opposing sides in black and white Gothic lettering on a contrasting background. The marking may be displayed on the packaging itself or in black lettering on a white square-on-point configuration having the same outside dimensions as a DOT placard. The word HOT may also be displayed with the identification number of the elevated-temperature material as long as the word HOT appears in the upper corner of the white square-on-point configuration.

When shippers transport a bulk quantity of certain hazardous materials in a tankcar by highway or railroad, DOT requires them to mark the key word from the applicable proper shipping name on each side of the tankcar. These key words for the relevant hazardous materials are listed in Table 6.7.

When shippers transport a marine pollutant in bulk packaging, DOT requires them to affix the MARINE POLLUTANT marking on at least two opposing sides or two ends other than the bottom if the packaging has a capacity of less than 1000 gal (3785 L), *or* on each side and each end if the

TABLE 6.7	Key Words of Proper Shipping Names That Are Marked on Each Side of a Tankcar for Shipment by Highway or Rail[a]
Acrolein, stabilized	Hydrogen cyanide, stabilized (less than 3% water)
Ammonia, anhydrous, liquefied	Hydrogen fluoride
Ammonia solutions (more than 50% ammonia)	Hydrogen peroxide, aqueous solutions (greater than 10% hydrogen peroxide)
Bromine *or* Bromine solutions	Hydrogen peroxide, stabilized
Bromine chloride	Hydrogen peroxide and peroxyacetic acid mixtures
Chloroprene, stabilized	Nitric acid (other than red fuming)
Dispersant gas *or* Refrigerant gas	Phosphorus (amorphous)
Division 2.1 materials	Phosphorus, white dry *or* Phosphorus, white, under water *or* Phosphorus white, in solution, *or* Phosphorus, yellow dry *or* Phosphorus, yellow under water *or* Phosphorus, yellow in solution
Division 2.2 materials (DOT-107 tankcars only)	Phosphorus, white, molten
Division 2.3 materials	Potassium nitrate and sodium nitrate mixtures
Formic acid	Potassium permanganate
Hydrocyanic acid, aqueous solutions	Sulfur trioxide, stabilized
	Sulfur trioxide, uninhibited

[a]49 C.F.R. §172.330(a)(ii).

FIGURE 6.11 A fumigant is a pesticide that is a vapor or gas, or forms a vapor or gas on application, and whose method of pesticidal action is through inhalation of the gaseous state. DOT regards a rail tankcar, freight container, truck body, or trailer in which lading has been fumigated, or is undergoing fumigation, as a package containing a hazardous material. DOT requires carriers at 49 C.F.R. §173.9 to display either this FUMIGANT marking or the fumigant's FIFRA hazard warning label so it may be readily observed by a person attempting to enter the transport vehicle containing the lading. When the FUMIGANT marking is displayed, DOT requires carriers to identify the name of the fumigant on the marking.

packaging has a capacity equal to or exceeding 1000 gal (3785 L). When shippers or carriers transport a marine pollutant within a transport vehicle or freight container, DOT requires them to mark MARINE POLLUTANT on each side and each end of the transport vehicle or freight container.

When the lading in a railcar, freight container, truck body, or trailer has been fumigated or is undergoing fumigation, DOT requires the carrier to display the FUMIGANT marking shown in Figure 6.11 so it is observable when a person attempts to enter the interior of the transport vehicle or freight container. DOT also requires the carrier to inscribe the technical name of the fumigant and the date and time of its application on the face of the marking.

At 49 C.F.R. §§172.332 and 172.336, DOT requires carriers to mark the identification number on bulk packaging by displaying it on an orange panel, a placard, or a white square-on-point diamond on each side and each end of the bulk packaging. For example a carrier may display the identification number 1203 on an orange panel when transporting gasoline.

When the identification number is displayed on a placard or a white square-on-point diamond, it appears across their center areas as shown below:

DOT requires the carrier to post the required FLAMMABLE placard adjacent to both the orange panel and white square-on-point diamond.

When carriers transport more than one hazardous material in separate compartments within a portable tank, cargo tank, or tankcar, DOT requires them to display the identification numbers of the hazardous materials on the ends of the tank *and* on the sides in the same sequence as the compartments containing the materials they identify.

- Describe the regulations that require shippers and carriers to display DOT warning placards on bulk packaging, freight containers, unit load devices, transport vehicles, or rail tankcars used to transport hazardous materials.
- Memorize the inscriptions on the DOT placards corresponding to all hazard classes and division numbers.
- Memorize the colors of the DOT placards in each of the nine hazard classes.

6.7 DOT Placarding Requirements

placard
■ For purposes of DOT regulations, a sign displayed by the carrier on bulk packaging, freight containers, transport vehicles, unit containment devices, or railcars to rapidly communicate hazard information concerning the hazardous material being transported

DOT requires carriers at 49 C.F.R. §172.504 to display in plain view one or more warning **placards** on each side and each end of the bulk packaging, freight container, unit load device, transport vehicle, or rail tankcar used to transport hazardous materials. These DOT placards are shown in Figure 6.12. The features of the individual placards are similar to those of the corresponding labels. Both are diamond-shaped and color-coded to signify a specific hazard class or division and may also include a pictograph.

The following points regarding the nature of the placards in Figure 6.12 are relevant:

- Although there is a COMBUSTIBLE placard, there is not an analogous label. Furthermore, although there is an INFECTIOUS SUBSTANCE label, there is not an analogous placard.
- The word GASOLINE may be used in lieu of the word FLAMMABLE on a placard that is displayed on a cargo tank or a portable tank used to transport gasoline by highway.
- The words FUEL OIL may be used in lieu of the word COMBUSTIBLE on a placard that is displayed on a cargo tank used to transport fuel oil that is not classed as a flammable liquid by highway.
- When solely transporting oxygen, carriers may display OXYGEN placards (Section 7.1-F) in lieu of NON-FLAMMABLE GAS placards. There is not an analogous OXYGEN label.
- As with the EXPLOSIVE labels, the uppercase letters that follow the division designations for the EXPLOSIVES placards are examples of the designations for the compatibility groups of explosives, whose nature is discussed later, in Section 14.4-B.
- The ORGANIC PEROXIDE placard is in the process of being replaced with the new placard at the far right of hazard class 5. Use of the new ORGANIC PEROXIDE placard becomes mandatory on January 1, 2011, for transportation by rail, watercraft, and aircraft, and on January 1, 2014, for transportation by highway.

When shippers intend to ship a specific hazardous material, they select the applicable placards by referring to the code listed in column 3 of the Hazardous Materials Table. Then, shippers provide the placards to the carrier, who displays them as follows:

■ Carriers *always* display the applicable placards when transporting materials whose hazard classes are listed in Table 6.8, regardless of the amounts transported. For example, a shipper or carrier displays RADIOACTIVE placards on each side and each end of a motor van used to transport wooden boxes on which RADIOACTIVE YELLOW-III labels have been affixed, regardless of the amount of hazard class 7 material within the boxes.

■ Carriers display the applicable placards on the same transport vehicle when they transport a material whose hazard class is listed in Table 6.9 and when the aggregate gross mass of the hazardous materials in any one hazard class equals or exceeds 1001 lb (454 kg). For example, DOT requires carriers to display FLAMMABLE placards on each side and each end of a rail boxcar used to transport steel drums containing acetonitrile, a flammable liquid, only when their aggregate gross mass equals or exceeds 1001 lb (454 kg).

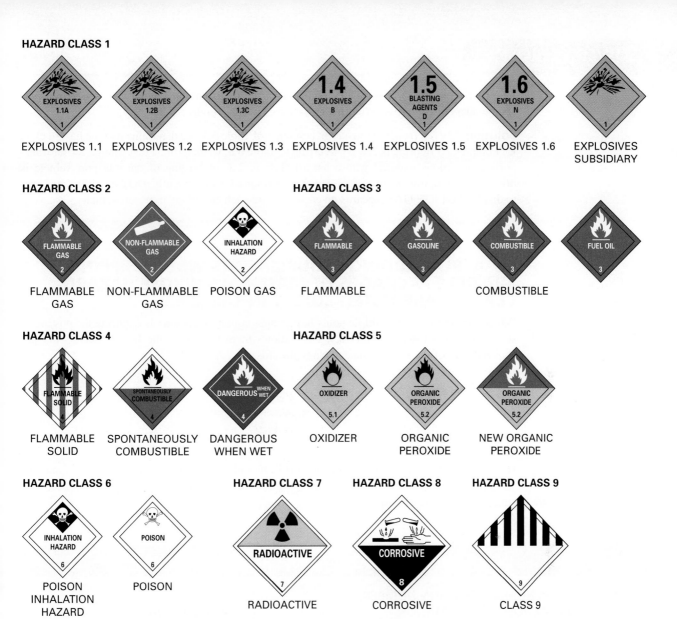

FIGURE 6.12 When carriers transport hazardous materials, DOT requires them to post warning placards on their bulk packaging. The placard at the far right in Hazard Class 1 is the EXPLOSIVES subsidiary placard. It is posted on bulk packaging when the number 1 is listed in column 6 of the Hazardous Materials Table as a subsidiary hazard. The ORGANIC PEROXIDE placard is in the process of being replaced with the new placard at the far right in Hazard Class 5. Use of the new ORGANIC PEROXIDE placard becomes mandatory on January 1, 2011, for transportation by rail, watercraft, and aircraft, and on January 1, 2014, for transportation by highway.

When carriers transport multiple hazardous materials including at least one of whose mass exceeds 1001 lb (454 kg), they display the applicable placards for *each* hazardous material on the transport vehicle. For example, if a carrier transports more than 1001 lb (454 kg) of acetonitrile with 500 lb (227 kg) of sulfuric acid, a corrosive material, DOT requires the carrier to display *both* FLAMMABLE and CORROSIVE placards on each side and each end of the transport vehicle.

Identify the placard(s), if any, DOT requires a carrier to display on a motor van used to transport six cylinders of compressed fluorine gas, each of which weighs 2.5 lb.

Solution: By referring to the entry in column 3 of Table 6.1, we determine that the hazard class for compressed fluorine is 2.3. By referring to Table 6.8, we determine that when a carrier transports a hazardous material whose hazard class is 2.3, placarding of the transport vehicle is *always* required, irrespective of the amount transported. Consequently, DOT requires the carrier to display POISON GAS placards on each side and each end of the transport vehicle.

6.7-A PLACARDING REQUIREMENTS WHEN SHIPPING MULTIPLE PACKAGES OF MATERIALS WHOSE HAZARD CLASSES ARE LISTED SOLELY IN TABLE 6.9

When carriers transport multiple nonbulk packages containing materials requiring the posting of several placards listed in Table 6.9, DOT allows them to display the DANGEROUS placard shown in lieu of the separate hazard class placards:

However, DOT prohibits the use of DANGEROUS placards when an aggregate gross mass of 2205 lb (1000 kg) of materials having a single hazard class listed in Table 6.9 is loaded at a single location. In this instance, carriers are obligated to display the placards specified in Table 6.9.

TABLE 6.8	DOT Hazard Classes That Always Require Placarding[a]
HAZARD CLASS NUMBER, DIVISION NUMBER, AND ADDITIONAL INFORMATION	**PLACARD NAME**
1.1	EXPLOSIVES 1.1
1.2	EXPLOSIVES 1.2
1.3	EXPLOSIVES 1.3
2.3	POISON GAS
4.3	DANGEROUS WHEN WET
5.2 (Organic peroxide, Type B, liquid or solid, temperature-controlled)	ORGANIC PEROXIDE
6.1 (Inhalation hazard, Zone A or B)	POISON INHALATION HAZARD
7 (RADIOACTIVE YELLOW-III label only)[b]	RADIOACTIVE

[a]49 C.F.R. §172.504, Table 1.
[b]See also Section 16.10-D.

TABLE 6.9	DOT Hazard Classes That Require Placarding Only Under Certain Conditions[a]

HAZARD CLASS NUMBER, DIVISION NUMBER, AND ADDITIONAL INFORMATION	PLACARD NAME
1.4	EXPLOSIVES 1.4
1.5	EXPLOSIVES 1.5
1.6	EXPLOSIVES 1.6
2.1	FLAMMABLE GAS
2.2	NON-FLAMMABLE GAS
3	FLAMMABLE
Combustible liquid	COMBUSTIBLE
4.1	FLAMMABLE SOLID
4.2	SPONTANEOUSLY COMBUSTIBLE
5.1	OXIDIZER
5.2 (other than organic peroxide, type B, liquid or solid, temperature-controlled)	ORGANIC PEROXIDE
6.1 (other than inhalation hazard, Zone A or Zone B)	POISON
6.2	(None)
8	CORROSIVE
9	CLASS 9
ORM-D	(None)

[a]49 C.F.R. §172.504, Table 2.

DOT never permits carriers to display DANGEROUS placards in lieu of the required placards when they transport hazardous materials whose hazardous class or division is listed in Table 6.8.

SOLVED EXERCISE 6.9

A chemical manufacturer in Charleston, West Virginia, intends to ship by motor carrier one plastic drum of selenium disulfide (SeS_2), an orange-to-red solid used in the preparation of an antidandruff shampoo, to a pharmacist in Chicago, Illinois. The contents of the drum weigh 100 lb (45.4 kg).

(a) What information does DOT require the manufacturer to provide on the accompanying shipping paper?
(b) Identify the labels DOT requires the manufacturer to affix to the plastic drum.
(c) Identify the markings, other than the drum specification markings, that DOT requires on the plastic drum.
(d) Which placards, if any, does DOT require the carrier to display on the transport vehicle?

Solution:

(a) In Table 6.4, the reportable quantity for selenium disulfide is listed as 10 lb (4.54 kg). Because 100 lb (45.4 kg) exceeds this reportable quantity, RQ must be a component of the shipping description. From the information in columns 2, 3, 4, and 5 of Table 6.1, the following information is then entered on the accompanying shipping paper:

Units	HM	Shipping Description (Identification Number, Proper Shipping Name, Primary Hazard Class or Division, Subsidiary Hazard Class or Division, and Packing Group)	Weight (lb)
1 plastic drum	X	RQ, UN2657, Selenium disulfide, 6.1, PGII (Poison)	100

(b) The entry in column 6 of the Hazardous Materials Table is 6.1, the label code for a poison. Because selenium disulfide is neither a gas nor a liquid, it does not pose an inhalation hazard. Consequently, DOT requires the POISON label to be affixed to the plastic drum.

(c) DOT requires that the plastic drum bear the markings "(RQ) Selenium disulfide" and UN2657. Because the hazardous material is transported within nonbulk plastic packaging, DOT also requires the word POISON to be permanently marked on the drum. Selenium disulfide is not listed as a marine pollutant in Table 6.5; hence, DOT does not require the MARINE POLLUTANT marking to be displayed on the drum.

(d) Because the amount of hazardous material in hazard class 6.1 does not equal or exceed 1001 lb (454 kg), DOT does not require the carrier to display POISON placards on the transport vehicle.

SOLVED EXERCISE 6.10

A carrier loads 450 lb (205 kg) of compressed nitrogen, 1250 lb (568 kg) of acid battery fluid, and 1750 lb (795 kg) of acetone into an 18-wheel motor van for delivery from a chemical distribution facility to a customer's plant site. Does DOT authorize the carrier to display DANGEROUS placards on each side and each end of the transport vehicle?

Solution: We note from Table 6.1 that the hazard classes codes for compressed nitrogen, acid battery fluid, and acetone are 2.2, 8, and 3, respectively. Because the aggregate gross mass of acid battery fluid and acetone exceeds 1001 lb (454 kg), DOT requires the carrier to display placards on the motor van for each class of hazardous material that is transported in the van; that is, the carrier would ordinarily display FLAMMABLE, CORROSIVE, and NON-FLAMMABLE GAS placards on each side and each end of the motor van.

DOT authorizes carriers to display DANGEROUS placards on a motor vehicle in lieu of the separate hazard class placards when the aggregate gross mass of the materials in any one hazard class does not exceed 2205 lb (1000 kg) and is loaded at the same facility. Hence, the carrier may display DANGEROUS placards on each side and each end of the 18-wheel motor van in lieu of the FLAMMABLE, CORROSIVE, and NON-FLAMMABLE GAS placards.

A carrier loads nonbulk containers of the following hazardous materials into a rail boxcar for delivery from a chemical warehouse to a customer's plant site: 1500 lb (682 kg) of acetone, 500 lb (227 kg) of ethyl mercaptan, 500 lb (227 kg) of barium cyanide, and 1000 lb (454 kg) of methyl isobutyl ketone. Which placards does DOT require the carrier to display on the boxcar?

Solution: We note from Table 6.1 that the hazard class codes of acetone, ethyl mercaptan, barium cyanide, and methyl isobutyl ketone are 3, 3, 6.1, and 3, respectively. The gross aggregate mass of the hazardous materials in hazard class 3 is 1500 lb + 500 lb + 1000 lb, or 3000 lb.

DOT authorizes carriers to display DANGEROUS placards on the boxcar in lieu of the separate hazard class placards *only* when the aggregate gross mass of the materials in any one hazard class does not exceed 2205 lb (1000 kg). Because the aggregate gross mass of hazardous materials in hazard class 3 is 3000 lb, DOT prohibits the carrier from displaying DANGEROUS placards and requires individual FLAMMABLE and POISON placards to be displayed on each side and each end of the boxcar.

6.7-B PLACARDING FOR SUBSIDIARY HAZARDS

In the following three circumstances, DOT requires carriers to display warning placards to account for a subsidiary hazard *in addition to* displaying other required placards:

- When carriers transport a hazardous material that poses an inhalation hazard, DOT requires them to display POISON GAS or POISON INHALATION HAZARD placards on each side and each end of the transport vehicle, freight container, portable tank, unit load device, or railcar used for transportation

- When carriers transport a hazardous material whose subsidiary hazard is dangerous-when-wet, DOT requires them to display DANGEROUS WHEN WET placards on each side and each end of the transport vehicle, freight container, portable tank, unit load device, or railcar used for transportation

- When carriers transport 1001 lb (454 kg) or more of uranium hexafluoride (UF_6), DOT requires them to display CORROSIVE and RADIOACTIVE placards on each side and each end of the transport vehicle, portable tank, or freight container used for transportation. This subject is noted again in Section 16.8-B.

When a hazardous material possesses subsidiary hazards, DOT permits carriers to display multiple placards to account for the subsidiary hazards, even when the displaying is not specifically required.

PERFORMANCE GOALS FOR SECTION 6.8:

- Identify the locations at which emergency responders may expect to locate the identification numbers of hazardous materials involved in a transportation mishap.
- Demonstrate how first-on-the-scene responders use the identification number and the *Emergency Response Guidebook* during the mitigation of a transportation mishap involving hazardous materials.

6.8 Responding to Incidents Involving the Release of Hazardous Materials

When shippers and carriers have correctly complied with the DOT regulations, the identification number of a hazardous material can be readily located during an emergency-response action by any of the following means:

- Listed on a shipping paper as a component of the shipping description of the hazardous material
- Marked on the packaging containing the hazardous material
- Displayed on bulk packaging within an orange panel, across the center area of a placard, or on a white square-on-point panel

Having located the identification number, how can first-on-the-scene responders use it during an incident involving the release of a hazardous material? The answer to this question involves consulting the *Emergency Response Guidebook* (*ERG*).* This is a DOT manual that primarily provides direction to emergency-response crews during the initial phases of a response action.

The guidebook serves as the primary reference source book for first-on-the-scene personnel. It provides them with a guide to be implemented during the initial response phase of a transportation mishap. When properly implemented, each guide listed within the guidebook provides emergency-response personnel with vital information on how to initially deal with an incident involving the release of a unique hazardous material until more specific information is obtained from the shipper, carrier, manufacturer, or CHEMTREC.

The *Emergency Response Guidebook* is organized into the following sections:

- **YELLOW SECTION**, so called because it has yellow-bordered pages. It provides a listing of the individual identification numbers and the hazardous materials having that number. A guide number that refers the user to the Orange Section is also provided.
- **BLUE SECTION** This second section has blue-bordered pages. It contains the names of the hazardous materials listed alphabetically with their corresponding identification numbers. Like the Yellow Section, it provides a guide number that refers the user to the Orange Section.
- **ORANGE SECTION** The third section has orange-bordered pages. It contains 172 individual guide numbers, each of which constitutes a two-page summary of information concerning the potential hazards of the referenced hazardous material, recommended actions that relate to public safety, and recommended emergency-response actions to be implemented during a transportation mishap involving the material in question.
- **GREEN SECTION** The fourth section has green-bordered pages. It contains information concerning initial isolation and protective-action distances for small and large spills of hazardous materials that are highlighted in the yellow and blue sections. It also provides a list of dangerous-when-wet materials.
- **WEAPONS OF MASS DESTRUCTION** The fifth section is not colored. It provides information on hazardous materials specifically used for terrorism activities.

The guidebook should always be accessible within emergency-response vehicles. Responding personnel should locate the appropriate guide number and implement the recommended actions.

DOT also requires shippers and carriers to have emergency-response information available to workers during all phases of the transportation of a hazardous material. This requirement is often met by having available a copy of the *Emergency Response Guidebook* during hazardous material loading, unloading, and transfer operations.

*In the United States, the *Emergency Response Guidebook* is provided free of charge to state and local safety authorities. It is also available from private commercial companies; the Research and Special Programs Administration, Hazardous Materials Transportation Bureau, U.S. Department of Transportation, Washington, DC 20590; the U.S. Government Printing Office, Superintendent of Documents, Mail Stop SSOP, Washington, DC 20402-9328; and the American Chemistry Council's Bookstore at (301) 617-7842. From government agencies, request ISBN 0-16-042938-2.

Suppose information is needed concerning the appropriate response action to be taken at a transportation incident involving an initially unidentified hazardous material being transported by motor carrier on a crowded highway. On arriving at the scene, personnel note that CORROSIVE placards are displayed on the exterior surface of the vehicle, adjacent to which are orange panels in which the number 1805 is inscribed. On securing the shipping paper from the driver, personnel also note that a number of items not regulated by DOT are components of the consignment as well as a hazardous material having the following shipping description:

Units	HM	Shipping Description (Identification Number, Proper Shipping Name, Primary Hazard Class or Division, Subsidiary Hazard Class or Division, and Packing Group)	Weight (lb)
10 drums	X	UN1805, Phosphoric acid solution, 8, PGIII	485

The orange panel and shipping description provide the identification number of the hazardous material as UN1805. The shipping description also provides the precise name of the hazardous material contained within the drums: phosphoric acid solution.

In the *Emergency Response Guidebook*, the number 1805 directs the reader to Guide 154, which is reproduced in Figure 6.13. This guide number provides general information concerning this substance. Most important, it notes that the hazardous material is toxic and/or corrosive, but noncombustible. Responders may then dismiss any concern as to whether the substance may ignite.

SOLVED EXERCISE 6.12

When a firefighting team arrives at the scene of a highway transportation mishap, they encounter an overturned motor van bearing POISON INHALATION HAZARD placards on its visible sides and ends. On cautiously examining the van's interior from a distance, they also observe broken boxes to which POISON INHALATION HAZARD labels are affixed. The boxes are marked "Nickel carbonyl, UN1259," "INHALATION HAZARD, ZONE A," and "MARINE POLLUTANT." Punctured cylinders of the hazardous material are scattered outside their protective boxes, and their liquid contents have spilled on the underlying floor of the van. Securing the shipping paper from the driver, the team captain notes the relevant portion of the shipping description of this consignment as follows:

Units	HM	Shipping Description (Identification Number, Proper Shipping Name, Primary Hazard Class or Division, Subsidiary Hazard Class or Division, and Packing Group)	Weight (lb)
Eight steel cylinders overpacked within wooden boxes	X	UN1259, Nickel carbonyl, 6.1, (3), PGI (Poison - Inhalation Hazard, Zone A)	8

What precautionary actions should be implemented by these first-on-the-scene responders?

Solution: From the shipping description, labeling, marking, and placarding information, the team members learn that nickel carbonyl is a toxic and flammable liquid. Because the liquid is not confined within its containers, its vapor poses a pronounced risk of inhalation toxicity *and* fire. To protect public health and the environment, they should immediately cordon off the area, direct the public from the scene, and prevent entrance of the liquid into a waterway or sewer. Even though the total quantity of this nonbulk shipment is only 8 lb, emergency-response personnel should don total-encapsulating suits with self-contained breathing apparatus. Only then should they remove the intact boxes from the van and set them aside.

For more specific guidance in responding to this incident, CHEMTREC should be contacted. The team members should also consult the *Emergency Response Guidebook* for immediate assistance. The identification number 1259 refers the responders to Guide 131. Here, personnel are provided with direction such as absorbing the spilled liquid and eliminating all potential sources of ignition. Guide 57 also provides first-aid information to help the responders who experience ill effects from inadvertent exposure to nickel carbonyl vapor.

FIGURE 6.13 *Guide 154 from DOT's Emergency Response Guidebook* (*Washington, D.C.: U.S. Department of Transportation, 2008*).

| GUIDE 154 | SUBSTANCES - TOXIC AND/OR CORROSIVE (NON-COMBUSTIBLE) | ERG2008 |

POTENTIAL HAZARDS

HEALTH
- **TOXIC**; inhalation, ingestion or skin contact with material may cause severe injury or death.
- Contact with molten substance may cause severe burns to skin and eyes.
- Avoid any skin contact.
- Effects of contact or inhalation may be delayed.
- Fire may produce irritating, corrosive and/or toxic gases.
- Runoff from fire control or dilution water may be corrosive and/or toxic and cause pollution.

FIRE OR EXPLOSION
- Non-combustible, substance itself does not burn but may decompose upon heating to produce corrosive and/or toxic fumes.
- Some are oxidizers and may ignite combustibles (wood, paper, oil, clothing, etc.).
- Contact with metals may evolve flammable hydrogen gas.
- Containers may explode when heated.

PUBLIC SAFETY
- **CALL Emergency Response Telephone Number on Shipping Paper first. If Shipping Paper not available or no answer, refer to appropriate telephone number listed on the inside back cover.**
- As an immediate precautionary meausre, isolate spill or leak area in all directions for at least 50 meters (150 feet) for liquids and at least 25 meters (75 feet) for solids.
- Keep unauthorized personnel away.
- Stay upwind.
- Keep out of low areas.
- Ventilate enclosed areas.

PROTECTIVE CLOTHING
- Wear positive pressure self-contained breathing apparatus (SCBA).
- Wear chemical protective clothing that is specifically recommended by the manufacturer. It may provide little or no thermal protection.
- Structural firefighters' protective clothing provides limited protection in fire situations ONLY; it is not effective in spill situations **where direct contact with the substance is possible.**

EVACUATION
Spill
- See Table 1-Initial Isolation and Protective Action Distances for highlighted **materials.** For non-highlighted **materials,** increase, in the downwind direction, as necessary, the isolation distance shown under "PUBLIC SAFETY".
Fire
- If tank, rail car or tank truck is involved in a fire, ISOLATE for 800 meters (1/2 mile) in all directions; also, consider initial evacuation for 800 meters (1/2 mile) in all directions.

EMERGENCY RESPONSE

FIRE

Small Fire

- Dry chemical, CO_2 or water spray.

Large Fire

- Dry chemical, CO_2, alcohol-resistant foam or water spray.
- Move containers from fire area if you can do it without risk.
- Dike fire-control water for later disposal; do not scatter the material.

Fire involving Tanks or Car/Trailer Loads

- Fight fire from maximum distance or use unmanned hose holders or monitor nozzles.
- Do not get water inside containers.
- Cool containers with flooding quantities of water until well after fire is out.
- Withdraw immediately in case of rising sound from venting safety devices or discoloration of tank.
- ALWAYS stay away from tanks engulfed in fire.

SPILL OR LEAK

- ELIMINATE all ignition sources (no smoking, flares, sparks or flames in immediate area).
- Do not touch damaged containers or spilled material unless wearing appropriate protective clothing.
- Stop leak if you can do it without risk.
- Prevent entry into waterways, sewers, basements or confined areas.
- Absorb or cover with dry earth, sand or other non-combustible material and transfer to containers.
- DO NOT GET WATER INSIDE CONTAINERS.

FIRST AID

- **Move** victim to fresh air.   • Call 911 or emergency medical service.
- **Give** artificial respiration if victim is not breathing.
- **Do not use mouth-to-mouth method if victim ingested or inhaled the substance; give artificial respiration with the aid of a pocket mask equipped with a one-way valve or other proper respiratory medical device.**
- Administer oxygen if breathing is difficult.
- Remove and isolate contaminated clothing and shoes.
- In case of contact with substance, immediately flush skin or eyes with running water for at least 20 minutes.
- For minor skin contact, avoid spreading material on unaffected skin.
- Keep victim warm and quiet.
- Effects of exposure (inhalation, ingestion or skin contact) to substance may be delayed.
- Ensure that medical personnel are aware of the material(s) involved and take precautions to protect themselves.

PERFORMANCE GOALS FOR SECTION 6.9:

■ Describe when carriers are obligated to notify the National Response Center that a hazardous substance has been released into the environment.

6.9 Reporting the Release of a Hazardous Substance

At 40 C.F.R. §302.6, EPA requires persons in charge of facilities such as transport vehicles, watercraft, and aircraft to notify the National Response Center (NRC) (Section 1.11) as soon as a hazardous substance has been released into the environment in an amount equal to or greater than its reportable quantity. Notice to the NRC is also required when oil is released into a waterway or when the release of oil may affect the quality of a waterway.

In addition, DOT requires at 49 C.F.R. §171.15 that notice be given to the NRC whenever any of the following circumstances result:

- In connection with the release of one or more hazardous materials, at least one of the following is true:
  - A person is killed.
  - A person receives injuries requiring admittance to a hospital.
  - The general public is evacuated for 1 hour or more.
  - A major transportation artery or facility is closed or shut down for 1 hour or more.
  - The operational flight pattern or routine of aircraft is altered.

- Fire, breakage, spillage, or suspected radioactive contamination occurs.
- Fire, breakage, spillage, or suspected contamination involving an infectious substance (Sections 10.20-A and -B) occurs. (Notice of the release of an infectious substance may be reported to the Centers for Disease Control and Prevention, U.S. Public Health Service, Atlanta, Georgia, at (800) 232-0124 in lieu of notice to the NRC.)
- A marine pollutant is released in a quantity exceeding 119 gal (450 L) for a liquid or 882 lb (400 kg) for a solid.
- Even when these criteria are not met, the transportation mishap is so severe that, in the judgment of the carrier, it should be reported nonetheless.

The notice provided to the NRC includes the following:

- The name of the reporter
- The name and address of the person represented by the reporter
- The telephone number where the reporter can be contacted
- The date, time, and location of the release incident
- The class or division, proper shipping name, and quantity of the hazardous material involved in the release, if such information is available
- The type of incident and nature of the hazardous material involvement and whether a continuing danger to life exists at the scene

As previously noted in Section 1.11, the NRC can initially be contacted by telephone at (800) 424-8802 or, in the District of Columbia, at (202) 426-2675. EPA and DOT require the submission of a follow-up written report to the following person:

Director, Office of Hazardous Materials Regulations
Materials Transportation Bureau
Department of Transportation
Washington, DC 20590

The informational requirements of the follow-up report are published at at 49 C.F.R. §171.16.

Shipping Papers

6.1 A liquor company intends to transport twenty 25-gal kegs of Scotch whiskey by motor van from its distillery to cellars where the kegs will be stored during aging. Use Table 6.1 to identify the shipping description that DOT requires on the accompanying shipping paper.

6.2 Potentially dangerous quantities of flammable hydrogen may be generated when lithium hydride reacts with water. Use Tables 6.1 and 6.3 to identify the basic description that DOT requires a shipper to enter on the accompanying shipping paper when transporting 25 lb (11 kg) of lithium hydride in a metal can.

6.3 Phillips Petroleum Company (PPC) intends to transport 450 m^3 of liquefied petroleum gas in five DOT-113A rail tankcars bearing the numbers PPRX111111, PPRX111112, PPRX111113, PPRX111114, and PPRX111115, respectively. PPRX is PPC's reporting mark. Use Tables 6.1 and 6.3 to identify the shipping description that DOT requires the shipper to provide on the accompanying waybill.

DOT Labeling, Marking, and Placarding Requirements

6.4 A chemical manufacturer offers a 25-L bottle of ethyl mercaptan cushioned within a wooden box to an airline carrier for domestic shipment by cargo air.

(a) Use Tables 6.1 and 6.3 to identify the shipping description that DOT requires the shipper to provide on the accompanying shipping paper.

(b) Use Table 6.1 to identify the warning labels, if any, that DOT requires the carrier to affix to the packaging.

(c) What information does DOT require the shipper to mark on the packaging?

6.5 What markings, other than specification markings, does DOT require a chemical facility to provide on the exterior surface of nonbulk plastic packaging containing 5 lb (2.3 kg) of the poison barium cyanide when it is offered to a carrier as freight by motor van?

6.6 When offering a single cylinder of compressed fluorine gas for transportation, a shipper provides the following shipping description to a private motor carrier:

Units	HM	**Shipping Description** (Identification Number, Proper Shipping Name, Primary Hazard Class or Division, Subsidiary Hazard Class or Division, and Packing Group)	**Weight (lb)**
1 cylinder	X	UN1045, Fluorine, compressed, 2.3, (5.1, 8), (Poison - Inhalation Hazard, Zone A)	2.5

(a) Using this shipping description, identify the warning labels that DOT requires to be affixed to the fluorine cylinder.

(b) What markings does DOT require on the cylinder?

6.7 Use Table 6.7 to identify the unique marking that DOT requires a rail carrier to display on the exterior surface of a DOT-105A rail tankcar used for the transportation of anhydrous ammonia.

6.8 A paint manufacturer retains CSX Transportation Company to ship 5000 gal of a solvent mixture consisting of 15% toluene, 30% methyl ethyl ketone, 25% ethylbenzene, 15% ethyl acetate, and 15% xylene by means of a rail tankcar.

(a) Use Tables 6.1 and 6.3 to determine the shipping description that DOT requires the shipper to enter on the accompanying waybill.

(b) Other than accurately describing the solvent mixture on the waybill, what additional information must the train crew document concerning this shipment?

(c) What information must be marked on the exterior surface of the tankcar?

6.9 Which placards, if any, does DOT require a carrier to display on a motor van used solely for the domestic transportation of the mixed load of hazardous materials noted on the following bill of lading:

Units	HM	Shipping Description (Identification Number, Proper Shipping Name, Primary Hazard Class or Division, Subsidiary Hazard Class or Division, and Packing Group)	Weight (lb)
2 boxes	X	UN1504, Sodium peroxide, 5.1, PG I	50
6 drums	X	UN1245, Methyl isobutyl ketone, 3, PGII	330
4 jars	X	UN1565, Barium cyanide, 6.1, PGI (toxic)	100
42 drums	X	RQ, UN1897, 1,1,2,2-Tetrachloroethane, 6.1, PGII	2250
23 sacks		UN1910, Calcium oxide, 8, PGIII	1500

Emergency-Response Actions at Transportation Mishaps

6.10 A train crew notes that a liquid is leaking from the piping connected to a rail tankcar containing a liquid hazardous material. This causes the engineer to stop the train and contact the local firefighting department. On arrival of the firefighters at the scene, a member of the train crew provides a waybill to the team captain. The relevant portion of the waybill gives the next shipping description of the leaking liquid:

Units	HM	Shipping Description (Identification Number, Proper Shipping Name, Primary Hazard Class or Division, Subsidiary Hazard Class or Division, and Packing Group)	Weight (lb)
1 tankcar	X	RQ, UN1134, Chlorobenzene, 3, PGIII (Placarded FLAMMABLE) (DUPX1111)	5,000

The captain also notes that the presence of the following orange panel on each side and end of the tankcar adjacent to a FLAMMABLE placard:

1134

Given this combination of information, what actions should these first-on-the-scene responders take to protect public health, safety, and the environment?

6.11 A firefighting team responds to a fire involving an overturned motor van on which DANGEROUS placards have been displayed. The intensity of the fire prevents the team from approaching the vehicle and retrieving the shipping paper. Despite the fact that the DANGEROUS placards are visible, why is it difficult for the team to determine the best action to take at this scene to protect public health, safety, and the environment?

6.12 Arriving at the scene of a train wreck, the members of a firefighting team note the presence of the following orange panel on at least three sides of an overturned rail tankcar that is now leaking its contents.

3028

The placards and markings on the tankcar are otherwise obscured. Using the *Emergency Response Guidebook* determine

(a) The name of the hazardous material having the identification number 3028
(b) The guide number to which the identification number 3028 is referenced
(c) Whether the relevant guide number indicates that the hazardous material is flammable
(d) Whether the relevant guide number indicates that the hazardous material is poisonous

6.13 On entering a burning garage owned by a landscaping company, the captain of a firefighting team observes the presence of 10 stainless-steel containers against an inner wall. Through the dense smoke the captain is unable to read the labels identifying their contents. Nonetheless, each container appears to be intact and placarded as shown here.

To protect public health, safety, and the environment, what information about the container contents should the captain provide to the team members?

Chemistry of Some Common Elements

In this chapter, we examine the characteristic properties of seven elements: oxygen, hydrogen, fluorine, chlorine, phosphorus, sulfur, and carbon. These particular elements have been selected here for significant reasons. The section on fluorine has been included because exposure to this element poses a degree of hazard that is higher than exposure to any other element. Fluorine is so hazardous that most chemists hesitate before using it. The reasons for including the sections on the other six elements are twofold: These elements possess certain hazardous features of which emergency responders should be aware, and they are also used more widely than other elements. This latter distinction increases the likelihood that they will be encountered during emergency-response efforts involving hazardous materials.

PERFORMANCE GOALS FOR SECTION 7.1:

- Identify the principal properties of elemental oxygen, especially noting its ability to support combustion and the life process.
- Identify the common industrial and institutional consumers of bulk quantities of oxygen.
- Discuss the unique hazards associated with the release of liquid oxygen (LOX).
- Identify the properties of ozone.
- Identify the purposes for which ozone is commercially generated.
- Describe the general manner by which ground-level ozone is generated.
- Describe how exposure to ground-level ozone negatively affects the quality of human and animal life.
- Describe how stratospheric ozone depletion negatively affects life on Earth.
- Describe the intent and purpose of the Montréal protocol.

7.1 Oxygen

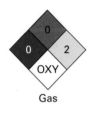

Gas

What image is conjured up in your mind when someone mentions oxygen? Most individuals first think of the air. It is true that oxygen is a constituent of air, yet oxygen makes up only about 21% of the volume of dry air. The major component—78% by volume—is nitrogen. Although oxygen is the most abundant element on Earth, most of it is not found in the free state. Instead, it is combined with other elements in chemical compounds such as water and carbon dioxide. Many oxygen compounds also exist in the chemical structures of the rocks and minerals that make up Earth's crust. A review of Table 4.1 shows that oxygen accounts for approximately 50% by mass of all substances in Earth's crust, the atmosphere, oceans, and other bodies of water.

Atmospheric oxygen is constantly consumed. It is also replenished by **photosynthesis**, a chemical process in which plants convert carbon dioxide and water into oxygen and cellulose (Section 14.5-A) in the presence of chlorophyl and sunlight. Scientists estimate that approximately 70% of the world's supply of atmospheric oxygen is generated within the huge spans of the Amazonian rainforest in South America. This natural process helps maintain the concentration of atmospheric oxygen at approximately 21% by volume.

At the ambient temperatures encountered on Earth, oxygen is a colorless, odorless, and tasteless gas that is only slightly more dense than air. At very low temperatures and under pressure, oxygen exists as a pale blue liquid or solid and has the curious feature of being attracted to a magnet. Other physical properties of oxygen are listed in Table 7.1. Chemists represent the chemical formula of elemental oxygen as O_2.

photosynthesis
■ The chemical process by which plants use carbon dioxide and water to construct cellulosic materials with the aid of sunlight and chlorophyll

TABLE 7.1	Physical Properties of Elemental Oxygen
Melting point	−360 °F (−218.4 °C)
Boiling point	−297 °F (−183 °C)
Specific gravity at 68 °F (20 °C)	1.43
Specific gravity (liquid) at boiling point	1.14
Vapor density (air = 1)	1.11
Liquid-to-gas expansion ratio	875

Oxygen is usually obtained for commercial use by liquefying air. Confined air liquefies when it is compressed at a pressure exceeding 545 psi$_a$ (3760 kPa) while being simultaneously cooled below −318 °F (−194 °C). To obtain these temperature and pressure conditions, air is passed through a series of compression, expansion, and cooling operations.

7.1-A LIQUID OXYGEN

By volume, liquid air is essentially a mixture of approximately 43% nitrogen, 54% oxygen, and less than 3% by mass of argon and other gases. When this mixture is allowed to boil at atmospheric pressure, the vapor above the liquid becomes enriched in nitrogen while, simultaneously, the remaining liquid slowly becomes enriched in oxygen. The oxygen-enriched mixture is called *liquid oxygen*, or *cryogenic oxygen*; in commerce, it is also referred to as *LOX*. When LOX vaporizes, it becomes *gaseous oxygen*, sometimes denoted as *GOX*. When 1 L of liquid oxygen vaporizes at room temperature, approximately 875 L of gaseous oxygen is produced.

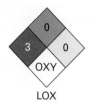

LOX

Facilities using large quantities of oxygen often store it as the cryogenic liquid within a specialized tank like that shown in Figure 7.1. Although oxygen may be discharged from the tank as either the liquid or gas, the release of gaseous oxygen is often more desirable from the standpoint of regulating its temperature and pressure.

7.1-B COMMERCIAL USES OF OXYGEN

Huge volumes of liquid oxygen are used by the aerospace industry to oxidize rocket fuels and propellants. In today's rockets, liquid oxygen and fuel are pumped into a combustion chamber, in which the mixture is ignited. The combustion reaction produces a high-pressure, high-velocity stream of hot gases that flows through the nozzle and accelerates the rocket to the tremendous velocities needed to reach outer space. Although other oxidizers could be used to oxidize rocket fuels, none has been shown to perform as effectively and economically as liquid oxygen.

Oxygen is also used in a number of less esoteric ways. For example, it is used by cold storage and food processing plants, hospitals, metal fabrication plants, electrical power plants, and the steel industry. Within hospitals, oxygen-enriched air is delivered into oxygen tents and hoods

FIGURE 7.1 Large volumes of medical oxygen are often stored at hospitals and clinics as its cryogenic liquid (LOX) in tanks. Although it is stored as the liquid, the oxygen is generally first vaporized prior to its delivery into hospital lines for subsequent use.

FIGURE 7.2 This cabinet is specifically marked to identify it as an area to be utilized solely for the storage of aluminum cylinders containing compressed oxygen. (*Courtesy of A & A Sheet Metal Products, Inc., SE-Cur-All® Products, LaPorte, Indiana.*)

for therapeutic uses. Gaseous oxygen under increased pressure, called *hyperbaric oxygen*, is also discharged within an enclosed chamber to treat patient's circulatory and respiratory problems, thermal and radiation burns, cerebral palsy, brain injuries, and carbon monoxide poisoning (Section 10.9-A). Portable aluminum cylinders of oxygen are also available for therapeutic uses, especially for individuals afflicted with emphysema. Within a hospital or clinic, emergency responders may encounter them within a specialized cabinet like the one shown in Figure 7.2, in which the cylinders have been nested to reduce or eliminate the likelihood of their movement (Section 3.3-A).

7.1-C RESPIRATION AND COMBUSTION

Two chemical properties of oxygen are especially important to emergency-response personnel:

■ In human and animal organisms, oxygen supports the life process. Oxygen is assimilated through the lungs, from which it is carried by the bloodstream to the body's cells. It is then used to oxidize food nutrients. A product of this oxidation is carbon dioxide, which is exhaled from the lungs into the atmosphere. The overall phenomenon, called **respiration**, releases energy that the body uses to maintain life. Without sufficient oxygen, survival is impossible.

■ The combustion process is *supported* by atmospheric oxygen; that is, when a substance burns in the air, it unites with atmospheric oxygen. This means that atmospheric oxygen is the oxidizing agent in ordinary combustion reactions. One means by which fires are extinguished is by smothering them, an action that prevents the burning material from contacting the oxygen required for combustion.

respiration
■ The chemical process in which the oxygen in inhaled air is conveyed from the lungs to the tissues and their cells of the body, where carbon dioxide and water are produced

Earth's atmosphere has not always been sufficiently oxygenated to support human life. Roughly 2.2 billion years ago, oxygen first began to enter the anaerobic environment by means of photosynthesis. Early microbes used oxygen as they evolved into complex life forms. As the human organism evolved, it ultimately learned to adapt to an atmosphere that contained oxygen at a concentration of 21% by volume—the concentration in today's atmosphere.

When obliged to breathe concentrations lower or higher than 21%, the human organism responds to the situation. Consider, for instance, breathing an oxygen concentration greater than 21%. Such oxygen levels are commonly administered to patients for therapeutic purposes. An oxygen concentration between 25% and 50% by volume is often delivered to patients inside oxygen tents and hoods. Breathing this elevated concentration typically helps them return to good health more rapidly.

TABLE 7.2	Adverse Health Effects Associated with Breathing Reduced Levels of Oxygen	
PERCENT OXYGEN IN THE AIR	**SIGNS AND SYMPTOMS**	
21 (normal atmospheric concentration)	No signs or symptoms	
12–15	Muscular coordination for skilled movements is lost	
10–15	Consciousness continues, but judgment is faulty, and muscular effort leads to rapid fatigue	
6–8	Collapse occurs rapidly, but quick treatment prevents fatal outcome	
<6	Death occurs within 6 to 8 minutes	

Yet, long-term breathing of an elevated concentration of oxygen does not always provide a positive outcome. Clinical observations demonstrate that persons may experience adverse health effects by inhaling elevated oxygen concentrations for extended periods. For instance, individuals who breathe an atmosphere containing 80% oxygen for more than 12 hours often suffer coughing, nasal stuffiness, sore throat, chest pain, and other respiratory problems.

It is from the mass of the overlying air that humans extract oxygen for survival. As first noted in Section 2.8, the mass of air exerts an average atmospheric pressure on planet Earth of 14.7 psi, or 101.3 kPa. At elevated heights from Earth's surface, the air still contains 21% oxygen, but the atmospheric pressure is reduced. When the atmospheric pressure is less than normal, there is less pressure pushing oxygen into the lungs. This means that it takes more effort to take in oxygen when breathing the air at elevated heights. Normal breathing may be unable to provide sufficient oxygen for survival. This occurs when mountaineers climb to great heights and experience **altitude sickness**.

Conversely, adverse health effects can also be experienced from breathing oxygen at an *increased* pressure, even for short periods of time. This is the reason clinicians routinely administer hyperbaric oxygen therapy to patients for no more than ½ to 1 hour.

Firefighters are more likely to encounter incidents in which the oxygen concentration has been *reduced* below 21% by volume. As Table 7.2 illustrates, individuals would soon be gasping for breath if the concentration of oxygen in the air fell below 17%, as occurs when fire burns within an enclosed area. Without an alternative air supply, emergency responders could collapse and die if compelled to remain within such an environment for even a few minutes.

Whereas the oxygen in the air is diluted with nitrogen and other gases, the compressed oxygen and LOX available commercially are concentrated forms of the element. When they are released from their containers at fire scenes, the oxygen enhances the rate at which nearby combustible materials burn. Put simply, matter burns more rapidly in an atmosphere of pure oxygen than it does in the air. This observation is consistent with our general understanding from Section 5.5-D, in which it was noted that the rate of a reaction increases when the concentration of a reactant is increased.

Liquid oxygen increases the rate of combustion in the following circumstances:

■ *The combustion of ordinary matter.* Substances that ordinarily burn slowly may burn spontaneously when in contact with liquid oxygen. Fuels, oils, greases, tar, asphalt, paper, and textiles are examples of matter that do not easily ignite without first being kindled. When these substances contact liquid oxygen, however, they are likely to ignite spontaneously.

altitude sickness
■ The inability of the human organism to extract sufficient oxygen for survival from the air at higher altitudes, where the atmospheric pressure is reduced

■ **The combustion of metals.**   Certain metals are likely to burn on contact with liquid oxygen. The combination of magnesium shavings and liquid oxygen is so chemically reactive that these elements explode on contact. The components of aluminum pump parts also react with liquid oxygen, especially when the reaction is initiated by friction. This explains why the use of aluminum is avoided in equipment used for storing and handling cryogenic oxygen.

■ **The combustion of porous materials.**   Wood, concrete, and asphalt are examples of porous materials. When liquid oxygen contacts them, it is readily absorbed into their pores and produces media that can be shock-sensitive. An example is an asphalt surface on which liquid oxygen has spilled. This oxygen-enriched asphalt may ignite explosively when subjected to a mechanical impact.

7.1-D CHEMICAL OXYGEN GENERATORS

Chemical oxygen generators are portable devices that can be chemically actuated to produce oxygen on demand. They are used in mines and other places with limited accessibility in which the available oxygen supply may be limited. They were once used to escape from noxious or poisonous atmospheres, but this practice is now obsolete.

One type of chemical oxygen generator consists of an apparatus with two separate compartments containing powdered iron and sodium chlorate, respectively. Oxygen is generated when these substances are mixed and heated.

$$\underset{\text{Iron}}{Fe(s)} + \underset{\text{Sodium chlorate}}{NaClO_3(s)} \longrightarrow \underset{\text{Iron(II) oxide}}{FeO(s)} + \underset{\text{Sodium chloride}}{NaCl(s)} + \underset{\text{Oxygen}}{O_2(g)}$$

A percussion cap activated by a hammer provides the heat needed to initiate the reaction. The heat that is evolved by this exothermic reaction permits it to be self-sustaining until either of the reactants is depleted.

Another type of oxygen generator utilizes the thermal decomposition of an alkali metal perchlorate. Lithium perchlorate, for example, decomposes when heated as follows:

$$\underset{\text{Lithium perchlorate}}{LiClO_4(s)} \longrightarrow \underset{\text{Lithium chloride}}{LiCl(s)} + \underset{\text{Oxygen}}{2O_2(g)}$$

The reaction is initiated through use of an igniter that is struck by a firing pin.

7.1-E WORKPLACE REGULATIONS INVOLVING BULK OXYGEN SYSTEMS

bulk oxygen system
■ For purposes of OSHA regulations, the assembly of equipment (storage tank, pressure regulators, safety devices, vaporizers, manifolds, and interconnecting piping) that has an oxygen storage capacity at normal temperature and pressure of more than 13,000 ft³ (368 m³) when connected for service, or more than 25,000 ft³ (708 m³) when available as an unconnected reserve

OSHA defines a **bulk oxygen system** as the assembly of equipment (storage tank, pressure regulators, safety devices, vaporizers, manifolds, and interconnecting piping) that possesses an oxygen storage capacity at normal temperature and pressure of more than 13,000 ft^3 (368 m^3) when connected for service, or more than 25,000 ft^3 (708 m^3) when available as an unconnected reserve.

When bulk quantities of oxygen are stored at the premises of industrial or institutional consumers, OSHA requires their owners to comply with certain regulations to prevent or minimize the risk of fire and explosion. These regulations pertain to the location of the system, its elevation, its accessibility to authorized personnel, measures associated with leakage from the system, diking, required distances between the system and nearby exposures, and other safety issues. For example, OSHA requires the location of bulk oxygen storage systems to be either outdoors aboveground *or* within a building of noncombustible construction that is adequately vented and used for that purpose exclusively. The selected location must be such that containers and associated equipment are not exposed to electric power lines, flammable or combustible liquid lines, or flammable gas lines.

OSHA requires the location of bulk oxygen storage systems to be either outdoors above-ground *or* within a building of noncombustible construction that is adequately vented and used for that purpose exclusively. The selected location must be such that containers and associated equipment are not exposed to electric power lines, flammable or combustible liquid lines, or flammable gas lines. Despite the fact that oxygen is a nonflammable gas, what is the most likely reason OSHA requires bulk oxygen storage systems to be installed in this fashion?

Solution: The most likely reason OSHA enacted this regulation is to minimize a pronounced risk of fire and explosion. Although oxygen is a nonflammable gas, it *supports* the combustion of many common materials. Because these materials burn at increased rates in an oxygen-enriched environment, it is prudent to prevent the accumulation of oxygen near flammable or combustible materials. This is accomplished by installing bulk oxygen storage systems in the indicated fashion.

OSHA advises the owners of bulk oxygen systems to locate them on ground higher than flammable or combustible liquids in nearby storage. When it is necessary to locate a bulk oxygen system on ground lower than adjacent flammable or combustible liquid storage tanks, OSHA requires the owner to provide a suitable means of diking, diversion curbing, or grading such that in the event of a release from these tanks, the liquids do not accumulate under the bulk oxygen system. Because flammable and combustible liquids burn at increased rates in an oxygen-enriched environment, it is prudent to prevent their accumulation near the components of a bulk oxygen system.

To warn workers of the presence of a bulk oxygen system within a workplace, OSHA also requires its owner at 29 C.F.R. §1910.104(b)(8)(viii) to permanently post a placard that reads as follows:

DANGER

**OXYGEN
NO SMOKING
NO OPEN FLAMES**

7.1-F TRANSPORTING OXYGEN

When shippers offer GOX or LOX for transportation, DOT requires them to identify it on the accompanying shipping paper as either of the following, as relevant:

UN1072, Oxygen, compressed, 2.2, (5.1)
or
UN1073, Oxygen, refrigerated liquid, 2.2, (5.1)

DOT also requires them to affix NON-FLAMMABLE GAS and OXIDIZER labels to the cylinders or other packaging.

When carriers transport 1001 lb (454 kg) or more of GOX or LOX, DOT requires them to display NON-FLAMMABLE GAS placards on the bulk packaging or transport vehicle used for the shipment; or, in lieu of posting NON-FLAMMABLE GAS placards, DOT allows carriers to display OXYGEN placards that look like the following:

The motor van shown in Figure 7.3 has been placarded for the transportation of cylinders and tanks containing both oxygen and nitrogen.

When carriers transport oxygen within bulk packaging, DOT also requires them to display the relevant identification number, 1072 or 1073, on orange panels, across the center area of the NON-FLAMMABLE GAS or OXYGEN placards, or on white square-on-point panels affixed on each side and each end of the packaging.

When shippers transport chemical oxygen generators or spent oxygen generators, DOT requires them to identify them in either of the following ways on the accompanying shipping paper:

UN3356, Oxygen generator, chemical, 5.1, PGII
or
NA3356, Oxygen generator, chemical spent, 9, PGIII

FIGURE 7.3 DOT authorizes carriers at 49 C.F.R. §172.504(f)(7) to display OXYGEN placards in lieu of NON-FLAMMABLE GAS placards on vehicles used for the sole shipment of oxygen in nonbulk cylinders or other containers. To emphasize that oxygen is an oxidizer, the background color of the OXYGEN placard is yellow, and the color of the symbol, text, hazard class number, and border is black. When carriers transport oxygen and other nonflammable gases in nonbulk cylinders or other containers within the same vehicle, DOT permits them to display NON-FLAMMABLE GAS placards *or* both OXYGEN and NON-FLAMMABLE GAS placards on the vehicle. (*Courtesy of Airgas Inc., Radnor, Pennsylvania.*)

When the generators are equipped with an attached means of initiation, DOT requires their carriers to obtain approval for shipment by demonstrating that they have been outfitted with two positive means of preventing their unintentional actuation.

7.1-G RESPONDING TO INCIDENTS INVOLVING A RELEASE OF OXYGEN

Although oxygen is a nonflammable gas, it does support combustion. When it is released from a storage or transportation device, the oxygen concentration is likely to become elevated within the immediate area. Then, secondary fires are likely to occur, especially when the oxygen contacts fuels. An attempt should be made to combat secondary fires only when they are located far from the area at which the oxygen is released or when the flow of oxygen from its containment vessel can be stopped without exposing personnel to risk.

In the event of a transportation mishap, DOT recommends an initial downwind evacuation of unauthorized persons to a distance of at least 0.33 mi (500 m) from the scene of the mishap. If a rail tankcar or tank truck is involved in a fire, DOT advises evacuation to a distance of 0.5 mi (800 m).

7.1-H OZONE, THE ALLOTROPE OF OXYGEN

Chemists refer to *ozone* as an **allotrope** of oxygen. An allotrope is a unique variation of an element possessing a set of physical and chemical properties that are different from the properties of any other form of the element. Four elements that have allotropes are oxygen, phosphorus, sulfur, and carbon.

allotrope
■ A form of the same element having different physical and chemical properties

Ozone is a form of elemental oxygen having three atoms of oxygen per molecule instead of the usual two; thus, its chemical formula is represented as O_3. Its Lewis structure is represented as follows:

$$:\ddot{O}:\ddot{O}::\ddot{O}: \text{ or } O-O=O$$

Ozone is produced when either an electrical discharge or ultraviolet radiation is passed through oxygen. Although it is reasonably stable at relatively low temperatures, ozone decomposes rapidly at room temperature; that is, it spontaneously reverts into "ordinary" oxygen. It is this relative instability that accounts for the fact that ozone is generally encountered at very low concentrations.

At room conditions, ozone is a pale blue gas. With a vapor density of 1.7 relative to air, ozone is more dense than normal oxygen. Whereas it possesses a sweet smell in low concentrations, ozone has an irritating, pungent, "metallic" odor at moderate to high concentrations. The sweet smell of ozone can sometimes be detected around operating electric motors or following a lightning storm.

Two characteristics make ozone one of the most hazardous materials known: a prodigious chemical reactivity and a pronounced toxicity. Ozone is a powerful oxidizing agent—considerably more reactive than oxygen itself. For example, ozone converts lead sulfide into lead sulfate, whereas oxygen reacts so slowly with lead sulfide that the rate is virtually imperceptible.

$$3PbS(s) \quad + \quad 4O_3(g) \quad \longrightarrow \quad 3PbSO_4(s)$$
$$\text{Lead sulfide} \qquad \text{Ozone} \qquad\qquad \text{Lead sulfate}$$

The ill effects experienced by humans exposed to ozone are noted in Table 7.3. Because ozone is a toxic substance, every precaution should be taken to prevent breathing it. The ozone damages the scavenger cells of the immune system. These cells, called *macrophages*, customarily destroy foreign bacteria, but when they are damaged, they are unable to effectively accomplish this task. Then, persons are more susceptible to contracting diseases.

7.1-I COMMERCIAL USES OF OZONE

microbicide
■ A chemical substance capable of destroying microorganisms

Ozone is used commercially for several purposes. It is used to bleach oils, fats, textiles, and sugar solutions. It is also used as a **microbicide** at drinking water and wastewater treatment

TABLE 7.3	Adverse Health Effects Associated with Breathing Ground-Level Ozone	
OZONE EXPOSURE	**ILL EFFECTS**	
0.18 ppm for 1 to 3 hr	*In Adults:* Reduced exercise performance	
0.12 ppm for 1 to 3 hr *or* 0.08 ppm for 6 hr	*In heavily exercising adults:* Reduced lung function, cough, shortness of breath, chest pain, and airway inflammation	
Repeated exposure to 0.12 ppm	*In laboratory animals:* Changes in lung structure, function, and biochemistry, which could signal the onset of chronic lung disease	
0.08 ppm for 3 hr	*In laboratory animals:* Increased susceptibility to bacterial respiratory infections	
0.01 to 0.15 ppm for several days	*In children and adolescents playing outdoors:* Reduced lung function, aggravation of asthma, and increased hospital visits	

plants. A microbicide is a substance that effectively kills disease-causing microorganisms. Ozone is an especially effective microbicide. It is even capable of destroying the fearsome parasite cryptosporidium that is sometimes found in chlorinated drinking water. When water contaminated with cryptosporidium is drunk, the parasite causes gastrointestinal diseases that can lead to death. Chlorination (Section 7.4-B) alone does not destroy the parasite in water.

When ozone is used for water treatment, the processes illustrated in Figure 7.4 are implemented. Air is first compressed to separate oxygen and nitrogen. Then, the oxygen is zapped with electricity to produce ozone, which is passed into less-than-pristine water under pressure. The ozone reacts with the impurities in the water and kills its constituent microorganisms. The water is then filtered and pumped into the municipal drinking water supply.

The use of ozone is also advantageous for treating wastewater. Ozone converts the constituent hydrogen sulfide (sewer gas, Section 10.12) into sulfuric acid. At the concentration produced, the sulfuric acid is benign.

$$3H_2S(s) \quad + \quad 4O_3(g) \quad \longrightarrow \quad 3H_2SO_4(aq)$$

Hydrogen sulfide 　　　Ozone 　　　　　Sulfuric acid

When ozone is used for chemical treatment, the process is called **ozonation**.

ozonation

■ A chemical reaction that involves the addition of ozone to a substance; the treatment of contaminated drinking water to kill the microorganisms (including cryptosporidium) that cause waterborne diseases

FIGURE 7.4 A simplified route that illustrates the production of ozone from air and its subsequent use for producing pure water. (*Adapted from an illustration prepared for the Southern Nevada Water Authority.*)

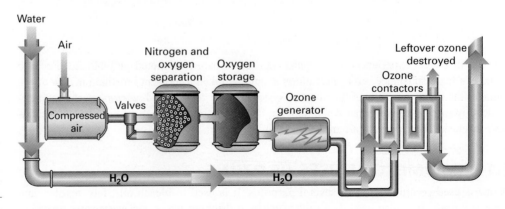

FIGURE 7.5 Following fires and floods, ozone can be used to decontaminate and restore buildings and their furnishings to their original condition. Because ozone is an unstable gas, it must be generated for use on demand. (*Courtesy of Medallion Clean Indoor Air Solutions of Las Vegas, Las Vegas, Nevada.*)

Because ozone is unstable and extremely poisonous, it is always synthesized at its intended point of use in minutely low concentrations. This synthesis is accomplished within an apparatus called an *ozone generator*, an apparatus that supplies an electrical current to oxygen or air. The ozone generator shown in Figure 7.5 is familiar to firefighters, since its use is often promoted within buildings that were damaged by fire. While the buildings are enclosed and vacated, ozone produced by the generator oxidizes the constituents of smoke. This removes the offensive odors from furniture, clothing, and other items damaged by fire. Ozone generators are also used to remove the musty odors that persist within walls and carpeting that were damaged by mold or mildew.

7.1-J GROUND-LEVEL OZONE

In certain metropolitan regions of the United States, **ground-level ozone** is often simultaneously present in the air with nitric oxide and nitrogen dioxide (Section 10.13) and certain organic compounds called **volatile organic compounds** (VOCs). The VOCs are constituents of the air emissions from petroleum refineries, fuel dispensers, chemical plants, and other industrial facilities. VOCs are typically regarded as organic compounds that boil at temperatures of less than approximately 392 °F (200 °C). The substances in gasoline vapor are examples of VOCs. Figure 7.6 shows that the majority of the atmospheric VOCs originate during nonchemical industrial processes.

The oxides of nitrogen are generated during the operation of automobiles; hence, they are constituents of vehicular exhaust. In addition, the oxides of nitrogen are constituents of the plumes emanating from the smokestacks of fossil-fuel-fired power plants. Within the lower atmosphere, these oxides react photochemically with the VOCs to produce ozone. At ground level where we live and breathe, ozone and the nitrogen oxides are often the *primary* constituents of polluted air, especially during the daylight hours of summertime, when there is ample sunlight to catalyze ozone production.

ground-level ozone
■ The ozone that forms in the lower atmosphere by photochemically catalyzed reactions between volatile organic compounds and nitrogen oxides

volatile organic compounds
■ Certain substances having a boiling point of less than approximately 392 °F (200 °C), including the constituents of gasoline vapor (VOCs); organic compounds that undergo photochemical reactions in the atmosphere to form ozone or other air pollutants

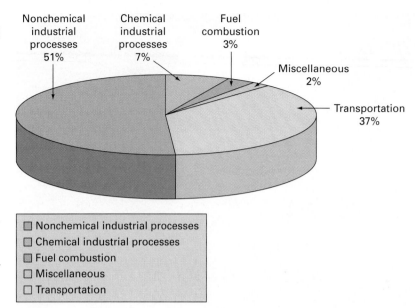

FIGURE 7.6 Ground-level ozone is produced when the volatile organic compounds (VOCs) in the air react with the nitrogen oxides discharged in vehicular exhaust. VOCs are primarily contained within the gaseous discharges of power plants, motor vehicles, and petroleum refineries. (*Courtesy of United States Environmental Protection Agency.*)

SOLVED EXERCISE 7.2

When filling the gas tank of an automobile, why do environmentalists advise us not to "top-off"?

Solution: "Topping-off" is the act of continuing to fill the gas tank beyond the point at which the gas pump has automatically stopped. During this process, fumes of volatile organic compounds (VOCs) are released into the air. When these VOCs react with the nitrogen oxides in the air produced in vehicular exhaust, ground-level ozone is photochemically produced. Because the ozone concentration in the air adversely affects health, we should take appropriate action to avoid releasing VOCs into the air. One simple way is to cease topping-off at the gas pump.

The presence of ground-level ozone is directly linked with the climatic conditions shown in Figure 7.7. When the ozone concentration exceeds approximately 100 ppb, weather forecasters declare an *ozone alert*, meaning that the ozone concentration in the air is approaching an unhealthful condition. At this concentration, affected individuals—especially the elderly—struggle for breath and feel dizzy. Their eyes, noses, throats, and lungs become irritated, and they experience fatigue, a lethargic feeling, coughing, wheezing, and hoarseness.

The inhalation of low concentrations of ozone is also likely to exacerbate the illnesses of individuals suffering from heart and lung ailments, emphysema, or chronic bronchitis. The afflicted are likely to cough more frequently and with greater intensity, experience chest pain and sinus congestion, and suffer severe headaches. Research studies have linked short-term exposure to ozone with premature death, and long-term cumulative exposure to ozone with an increased risk of dying from respiratory causes. This risk rises up to 4% for every 10 ppb increase in exposure to ozone.* This means that individuals who live in Los Angeles, Houston, and other areas

*Michael Jerrett et al., "Long-term ozone exposure and mortality," *New England Journal of Medicine* 360 (2009): 1085–95.

VERY UNHEALTHY

Ozone concentration 0.201 ppm or more

Very hazy, very hot, and humid
Stationary high atmospheric pressure
Sunny skies
Temperature, 95 °F (35 °C) or higher

UNHEALTHY

Ozone concentration 0.121 to 0.200 ppm

Hazy, hot, and humid
Stationary high atmospheric pressure
Sunny skies
Temperature, 90 °F (32 °C) or higher

APPROACHING UNHEALTHY

Ozone concentration 0.101 to 0.120 ppm

Slow-moving high atmospheric pressure
Sunny skies
Light winds
Temperature, 86 °F (30 °C) to 92 °F (33 °C)

MODERATE

Ozone concentration 0.061 to 0.100 ppm

Light to moderate wind
High pressure system
Partly cloudy to sunny skies
Temperature, 75 °F (24 °C) to 85 °F (29 °C)

GOOD

Ozone concentration 0.060 ppm or less

Passing cold front
Windy conditions
Partly sunny to cloudy skies or rain
Temperature, 75 °F (24 °C) to 80 °F (27 °C)

FIGURE 7.7 The association between ground-level ozone concentrations and the prevailing climatic conditions. (*Adapted from an illustration prepared by the Delaware Valley Regional Planning Committee for the Ozone Action Partnership.*)

that have routinely high ozone concentrations will experience a greater than 30% annual risk of dying from lung disease compared with those who live in areas having relatively low ozone concentrations.

7.1-K WORKPLACE REGULATIONS INVOLVING OZONE

The adverse health effects noted in Table 7.3 may be experienced when an individual inhales an ozone concentration exceeding 0.1 ppm for a period as short as an hour. When the use of ozone is needed in the workplace, OSHA requires employers to limit employee exposure to an ozone concentration in air of 0.1 ppm, averaged over an 8-hr workday.

7.1-L ENVIRONMENTAL REGULATIONS INVOLVING GROUND-LEVEL OZONE

EPA has evaluated the adverse health effects that result from exposure to ground-level ozone. Using the legal authority of the Clean Air Act, EPA established the primary and secondary national ambient air quality standards for this criteria air pollutant (Section 1.5-A) as 0.075 ppm (157 µg/m^3) as an 8-hr average.

SOLVED EXERCISE 7.3

On a hot, humid day, the ozone concentration in urban air can exceed 100 ppb. What actions to limit the adverse health effects associated with breathing ozone should be taken by persons who suffer from emphysema?

Solution: Emphysema is a lung ailment associated with the swelling of the alveoli and connecting tissues. Emphysema-sufferers cough frequently, experience headaches, and have trouble breathing. When they inhale air contaminated with ozone, these ill effects are exacerbated. To limit undue distress when the ozone concentration exceeds 100 ppb, emphysema-sufferers are advised to remain in an air-conditioned environment, avoid heavy work and exercise, and breathe oxygen from a portable source.

7.1-M STRATOSPHERIC OZONE

Although its presence in the lower atmosphere can be a troublesome factor for maintaining good health, ozone otherwise provides an outstanding benefit to the inhabitants of Earth. Approximately 10 to 19 mi (16 to 30 km) over Earth's surface, ozone occurs naturally in the region of the stratosphere called the **ozone layer**. As shown in Figure 7.8, oxygen is bombarded within the stratosphere by high-energy particles that originate in the sun. This bombardment causes some oxygen molecules to dissociate into their individual atoms $(\cdot\ddot{O}\cdot)$. The oxygen within the ozone layer exists as a mixture of oxygen atoms and oxygen molecules. These oxygen species collide to initiate a chemical reaction producing ozone.

ozone layer
- The region of the atmosphere rich in ozone, approximately 10 to 19 mi (16 to 30 km) from Earth's surface

$$O_2(g) \longrightarrow 2 \cdot\ddot{O}\cdot(g)$$

Oxygen molecule   Oxygen atoms

$$O_2(g) + \cdot\ddot{O}\cdot(g) \longrightarrow O_3(g)$$

Oxygen molecule Oxygen atom   Ozone molecule

When **stratospheric ozone** absorbs low-energy ultraviolet radiation, some ozone molecules revert to the ordinary form of oxygen.

stratospheric ozone
- The naturally occurring layer of ozone in the upper atmosphere that shields Earth's surface from the Sun's harmful ultraviolet rays

$$O_3(g) \longrightarrow O_2(g) + \cdot\ddot{O}\cdot(g)$$

Ozone molecule   Oxygen molecule Oxygen atom

$$O_3(g) + \cdot\ddot{O}\cdot(g) \longrightarrow 2O_2(g)$$

Ozone molecule Oxygen atom   Oxygen molecules

Under normal conditions, the continuous formation and photodecomposition of ozone compete favorably with one another. This chemical activity produces a steady-state condition in which the rate at which the stratospheric ozone forms from oxygen equals the rate at which it undergoes photodecomposition into oxygen.

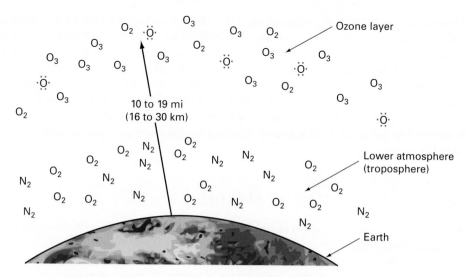

FIGURE 7.8 At the outer edge of the stratosphere is a region called the ozone layer, where oxygen is constantly converted into ozone. The ozone layer prevents much of the Sun's ultraviolet radiation from reaching Earth. Atmospheric scientists have discovered that in modern times the amount of stratospheric ozone has been depleted compared with the amount in past times. This depletion allows more harmful ultraviolet radiation to penetrate to Earth's surface, increasing the instances of skin cancer and cataracts and weakening immune systems.

In the past, the presence of the stratospheric ozone layer protected Earth and its occupants from overexposure to the harshness of ultraviolet radiation. Today, however, the former use of the halon fire extinguishers (Section 5.13), Freon refrigerants, foam-blowing agents, and coolants (Section 12.16), and similar halogen-containing compounds is continuing to adversely affect Earth and its occupants. Released into the environment, these compounds diffuse upward to the ozone layer, where their exposure to ultraviolet radiation produces halogen atoms. We note more succinctly in Section 12.16 that these atoms catalyze the decomposition of ozone and cause its overall stratospheric depletion.

In 1985, atmospheric scientists began to note a precipitous drop in stratospheric ozone, first over Antarctica and then over the Arctic. The areas of the ozone layer in which this thinning of the ozone layer has been observed are generally referred to as **ozone holes**.

Owing to the existence of ozone holes, more ultraviolet radiation from the sun now penetrates Earth's atmosphere compared with the amount that reached Earth prior to 1985. The concern among scientists is that this increased radiation exposure is causing serious health and environmental consequences. There is also evidence that the radiation has weakened plant life, which contributes to lesser agricultural outputs. It may also have caused climatic changes that could ultimately disturb the balance of aquatic and land ecosystems. The overexposure to ultraviolet radiation in humans has dramatically increased the incidence of skin cancer and cataracts and weakened immune systems. In aquatic systems, it has harmed marine life. Scientists estimate that erosion of the ozone layer will cause an additional 200,000 deaths over the next 50 years.

ozone hole
■ A thin spot in the ozone layer that has been linked to destruction of stratospheric ozone by chlorofluorocarbons and other substances

7.1-N ENVIRONMENTAL REGULATIONS INVOLVING STRATOSPHERIC OZONE

All life owes its very existence on planet Earth to the protective ozone layer. Hence, in 1987, to ensure the survival of the planet, 43 representatives of the world's industrialized nations agreed to phase out the manufacture, import, and use of ozone-depleting substances through implementation of an international agreement known as the Montréal Protocol on Substances That Deplete the Ozone Layer, or **Montréal protocol**. By 1995, 163 nations including the United States had agreed to its conditions. This was accomplished using the statutory authority of the Clean Air Act.

Montréal protocol (formally the Montréal Protocol on Substances That Deplete the Ozone Layer)
■ The international agreement that phases out the worldwide production, manufacture, and use of chlorofluorocarbons and other ozone-depleting substances

Among the substances that deplete stratospheric ozone are methyl bromide, carbon tetrachloride, 1,1,1-trichloroethane, the halon fire extinguishers, and certain chlorofluorocarbons. By January 1996, the manufacture, import, and use of carbon tetrachloride, 1,1,1-trichloroethane, and the chlorofluorocarbons were banned in the United States. The use of methyl bromide as an agricultural fumigant by U.S. farmers was provisionally allowed until 2006, only because discontinuance of its use would have seriously disrupted the American agricultural market. The use of methyl bromide today is illegal under federal law.

PERFORMANCE GOALS FOR SECTION 7.2:

- Identify the principal properties of elemental hydrogen, particularly noting its flammability, vapor density, rate of diffusion, and heat of combustion.
- Describe the ways hydrogen is produced for commercial use.
- Identify the common industrial and institutional facilities that use hydrogen.
- Identify the metals that react with acids and/or water to produce hydrogen.
- Describe the manner by which hydrogen is produced during the charging of lead–acid storage batteries.
- Describe why hydrogen is uniquely used by the aerospace industry as a rocket fuel.
- Describe the response actions to be executed when hydrogen has been released into the environment.

7.2 Hydrogen

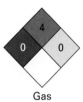

Gas

Although Table 4.1 shows that hydrogen ranks ninth in natural abundance by mass on Earth, only traces of this element exist naturally in the free state. Hydrogen is such a reactive element that it is found on Earth only within compounds. Throughout the entire universe, however, hydrogen is the most abundant element, 75.2% by mass.

Elemental hydrogen is an odorless, colorless, tasteless, and nontoxic substance. Its chemical formula is H_2. Although hydrogen exists as a gas at ordinary temperatures and pressures, the gas liquefies when compressed under a pressure greater than 294 psi (2030 kPa) and at a temperature less than $-390\ °F$ ($-234.5\ °C$). Liquid hydrogen is sometimes denoted as *LH2*.

7.2-A COMMERCIAL USES OF HYDROGEN

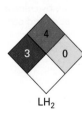

LH_2

Elemental hydrogen is commercially available as both the compressed gas and the cryogenic liquid. The chemical industry uses hydrogen as a raw material for the production and manufacture of other chemical substances. The petroleum industry uses it for the production and manufacture of petroleum fuels. The aerospace industry uses hydrogen to propel rockets into space.

Aerospace engineers chose hydrogen as a rocket fuel, since the combustion of a very small mass produces a prodigious amount of energy (see below). It is within the aerospace industry that massive volumes of hydrogen are likely to be encountered in storage. For example, at the J. F. Kennedy Space Center, Cape Canaveral, Florida, a staggering 800,000 gal (3000 m^3) of cryogenic hydrogen is held in a single storage tank.

In the not-so-distant future, hydrogen could become a popular fuel used to power automobiles and other surface vehicles. Considerable efforts are now underway to identify a means of storing compressed hydrogen onboard the cars of tomorrow in a safe, efficient, compact, and economical

FIGURE 7.9 At a hydrogen refueling station, hydrogen fuel is stored either as a liquid or gas under pressure. As hydrogen fuel is dispensed from the pump, its pressure typically ranges from 5000 psi$_g$ (34,456 kPa) to 10,000 psi$_g$ (68,911 kPa). (*Courtesy of Air Products, Allentown, Pennsylvania.*)

fashion. In a hydrogen-powered automobile, the hydrogen directly replaces the gasoline used in gasoline-powered vehicles.

Hydrogen is also used in *hydrogen fuel cells* to operate automobiles. In this case, the vehicle actually operates on electricity. The hydrogen stored in the car's fuel tank reacts with atmospheric oxygen to generate electricity and operate the motor.

The use of hydrogen for powering automobiles, whether directly or indirectly, is now limited to areas equipped with a hydrogen-refueling station like the one shown in Figure 7.9. These stations are primarily located in California, where the current use of hydrogen as an automotive fuel is greatest.

The prospective use of hydrogen as a vehicular fuel has the advantage of reducing U.S. reliance on fossil-fuel resources from foreign countries and reducing air pollution. Scientists predict that by 2040 the use of hydrogen-powered vehicles could cut greenhouse-gas emissions by more than 500 million tn/y (454 million t/y) and cut oil consumption by 11 million tn/day (10 million t/day). Although the use of hydrogen as a vehicular fuel would truly revolutionize the automotive industry and theoretically produce a cleaner environment, the hydrogen-production methods currently in use would require alteration. The methods now used to produce hydrogen (Section 7.2-B) are associated with the generation of carbon dioxide. This implies that the use of hydrogen-powered vehicles will be environmentally friendly only when the hydrogen fuel is produced without simultaneously generating carbon dioxide, or by capturing and sequestering it during production.

Physicists and engineers have also been researching ways to successfully develop a thermonuclear reactor, in which the nuclei of hydrogen atoms fuse and produce heavier nuclei by converting mass into energy. Nuclear fusion is commonplace in stars, and it was demonstrated on Earth during the detonation of the hydrogen bomb pictured in Figure 7.10. However, the energy released during the detonation of a hydrogen bomb is essentially uncontrolled. To serve as a commercial energy source, fusion reactions must be controlled on a large scale. If a thermonuclear

reactor could be successfully developed, hydrogen would most likely become *the* energy source of the future.* However, serious technical obstacles in achieving a sustainable controlled nuclear power source now exist. Commercialized nuclear fusion is not anticipated until late in this century, if at all.

When hydrogen burns in an atmosphere of pure oxygen, the accompanying heat of combustion is 61,600 Btu/lb (143 kJ/g). This is the most energy for the least mass that evolves when any substance burns. The magnitude of this energy is used for launching spacecraft and cutting and welding metals and glass. Jewelry manufacturers, for example, use the oxyhydrogen torch illustrated in Figure 7.11 to create rings and bracelets made of platinum, a metal that melts at 3191 °F (1755 °C). Compressed hydrogen and oxygen are separately stored within steel cylinders and then pressure-fed through tubing into a mixing chamber within the torch in a 2:1 ratio. This mixture is then discharged from the nozzle, where combustion of these gases occurs.

Hydrogen is also used commercially in connection with the chemical process called **hydrogenation**. At elevated temperatures and pressures, and generally in the presence of a catalyst, hydrogen combines with certain organic compounds to form commercially useful products. When vegetable oils are hydrogenated, for example, they become suitable for use as shortening and raw materials for the manufacture of soaps and lubricants.

hydrogenation
■ The addition of hydrogen to a substance

*A multinational effort, called the International Thermonuclear Experimental Reactor Consortium, aims to have a fusion-demonstration power plant in operation by 2016. The United States, European Union, Russia, India, Japan, South Korea, and China are members of the consortium that funds the project. In 2006, representatives of these seven partners agreed to build the plant at Cadarache in southern France.

The underlying technology in demonstrating fusion involves the use of a doughnut-shaped magnetic bottle that will confine hydrogen nuclei as they are heated to 180 million degrees. At this temperature, hydrogen nuclei fuse to form helium, neutrons, and sufficient energy that can be used to generate electricity. If the initial tests are successful, the consortium estimates that the construction of a commercial-scale fusion power plant could begin by midcentury.

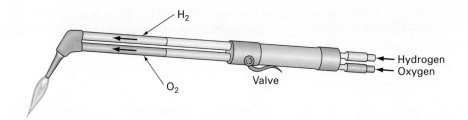

FIGURE 7.11 A simplified diagram illustrating the operation of the oxyhydrogen torch. Hydrogen and oxygen enter the mixing chamber separately from the right. A 2:1 mixture discharges from the nozzle at the left at a rate that causes the flame to burn at the tip of the torch. When this mixture is ignited, temperatures ranging between 3300 and 4400 °F (1800 and 2400 °C) are achieved.

Within the chemical industry, hydrogen is used to produce metallic and nonmetallic hydrides. Hydrogen combines with elemental sodium, for example, to produce sodium hydride (Section 9.5-A).

$$2Na(s) + H_2(g) \longrightarrow 2NaH(g)$$

Sodium   Hydrogen   Sodium hydride

It is also used to produce compounds such as hydrogen chloride and ammonia.

$$H_2(g) + Cl_2(g) \longrightarrow 2HCl(g)$$

Hydrogen   Chlorine   Hydrogen chloride

$$3H_2(g) + N_2(g) \longrightarrow 2NH_3(g)$$

Hydrogen   Nitrogen   Ammonia

7.2-B PRODUCTION OF HYDROGEN

Hydrogen is produced for commercial use by the following methods:

■ In the first method, steam is first passed over red-hot coke or coal under high temperature and pressure. This process, called **coal gasification**, forms a mixture of gases called **water gas**, **synthesis gas**, or **syngas**. This mixture consists mainly of carbon monoxide and hydrogen.

$$C(s) + H_2O(g) \longrightarrow CO(g) + H_2(g)$$

Carbon   Water   Carbon monoxide   Hydrogen

The hydrogen is isolated from the carbon monoxide, or the carbon monoxide is converted to carbon dioxide, by passing the water gas with additional steam over a catalyst such as iron(III) oxide.

$$CO(g) + H_2O(g) \longrightarrow CO_2(g) + H_2(g)$$

Carbon monoxide   Water   Carbon monoxide   Hydrogen

The carbon dioxide in the resulting mixture is removed by passing it through an alkaline solution.

■ The second industrial method of producing hydrogen involves the chemical action of steam on methane at high temperatures.

$$CH_4(g) + H_2O(g) \longrightarrow 3H_2(g) + CO(g)$$

Methane   Water   Hydrogen   Carbon monoxide

The water gas produced by the chemical reaction is treated by either of the methods previously noted for isolating carbon monoxide–free hydrogen.

■ Hydrogen is also produced by a multistep, complex reaction between propane and steam that is denoted by the following overall equation:

$$C_3H_8(g) + 6H_2O(g) \longrightarrow 3CO_2(g) + 10H_2(g)$$

Propane   Water   Carbon dioxide   Hydrogen

coal gasification
■ The chemical process that produces a mixture of carbon monoxide and hydrogen when steam is exposed to coal at high temperature and pressure

water gas (synthesis gas; syngas)
■ The mixture of carbon monoxide and hydrogen produced by blowing steam through a bed of red-hot coke

The carbon dioxide in the resulting mixture is removed by passing it through an alkaline solution.

■ Hydrogen is also produced by the reaction of steam on methanol vapor (Section 13.2-A) using a copper oxide/zinc oxide catalyst as follows:

$$CH_3OH(g) \quad + \quad H_2O(g) \quad \longrightarrow \quad CO_2(g) \quad + \quad 3H_2(g)$$

Methanol   Water   Carbon dioxide   Hydrogen

Once again, the carbon dioxide is removed by passing the mixture through an alkaline solution.

7.2-C PROPERTIES OF HYDROGEN

Elemental hydrogen possesses several interesting physical properties, some of which are listed in Table 7.4. A notable property is its vapor density—only 0.07 compared with air. Because this value is so low, it gives the confined gas a natural **lifting power**. When expressed in metric units, the lifting power of hydrogen is equal to the difference in mass between 1 L of air and 1 L of hydrogen at a given temperature and pressure. At 32 °F (0 °C) and 14.7 psi (101.3 kPa), the mass of 1 L of air is 1.2930 g, whereas the mass of 1 L of hydrogen is 0.0899 g; hence, the lifting power of hydrogen in air at this temperature and pressure is 1.2930 g/L − 0.0899 g/L, or 1.2031 g/L. In Section 7.2-G, we note that this lifting power was once used to keep dirigibles aloft.

Another important property of hydrogen is associated with the tiny size of its molecules. Hydrogen molecules are extremely small when compared with the sizes of other molecules. They are so tiny that hydrogen molecules easily leak through openings such as pipe joints and valve connections. To prevent loss during storage and transport, the steel cylinders, tubes, and tanks used to hold the compressed gas must be uniquely designed.

Another important property of hydrogen is the relatively high rate at which it diffuses into the air. When released from a tank or container, hydrogen diffuses into the air more rapidly than any other substance; that is, at the same temperature, hydrogen moves faster than other gases or vapors.

Hydrogen is a flammable gas. It burns when its concentration in air is between 4% and 75% by volume. A concentration within this wide flammable range can readily be achieved when hydrogen is released from its container or tank. Nonetheless, since it dissipates so rapidly, unconfined hydrogen does not remain in any single area for long.

Hydrogen burns in air to form water vapor.

$$2H_2(g) \quad + \quad O_2(g) \quad \longrightarrow \quad 2H_2O(g)$$

Hydrogen   Oxygen   Water

lifting power
■ The natural ability of a confined gas that is less dense than air to rise in the atmosphere, measured as the difference between the mass of a given volume of air and that of the same volume of the gas at the same temperature and pressure

TABLE 7.4	Physical Properties of Elemental Hydrogen
Melting point	−434.5 °F (−259.2 °C)
Boiling point	−423.0 °F (−252.8 °C)
Specific gravity (gas) at 68 °F (20 °C)	0.071
Vapor density (air = 1)	0.069
Autoignition point	1058 °F (570 °C)
Lower explosive limit	4%
Upper explosive limit	75%
Liquid-to-gas expansion ratio	848

Given its low vapor density and high diffusion rate, the combustion reaction occurs as the gas moves upward. Hydrogen burns with an almost nonluminous flame that is especially difficult to observe during daylight hours.

7.2-D HYDROGEN AND THE RISK OF FIRE AND EXPLOSION

The physical properties of hydrogen give rise to two potential scenarios relating to its risk of fire and explosion:

- When released indoors or into an enclosure where the gas can accumulate, the presence of the hydrogen poses a pronounced risk of fire and explosion.
- When released outdoors or in a manner that enables the hydrogen to readily dissipate, the likelihood that it will form a flammable mixture is comparatively small.

SOLVED EXERCISE 7.4

Hydrogen diffuses into the air at the rate of 0.098 in.^{2}/s (0.634 cm^2/s) at 32 °F (0 °C). What does the magnitude of this rate reveal about the likelihood that a flammable mixture of hydrogen and air will be produced under most accident scenarios that occur outside buildings?

Solution: This information indicates that hydrogen diffuses very rapidly into the surrounding air. In fact, unconfined hydrogen moves more rapidly into air than any other gas. Hydrogen is also flammable over a wide concentration range (4% to 75% by volume). Because hydrogen does not remain localized for long, a flammable mixture of hydrogen and air is not ordinarily produced under most accident scenarios that occur outside buildings.

Because hydrogen rises in the air, hydrogen-sensing devices are often installed near the ceilings of enclosures in which hydrogen could be inadvertently released, as from a leaking storage tank into a room. These devices activate exhaust fans capable of evacuating the hydrogen into the outside air at very rapid speeds. When emergency-response crews are called to such incidents, they should enter the room only after the hydrogen has been dispelled and its concentration measures less than 4% by volume.

7.2-E CHEMICAL REACTIONS THAT GENERATE HYDROGEN

Although hydrogen may be encountered in cylinders and storage tanks, it may also be inadvertently generated as the product of chemical reactions. Certain metals generate hydrogen by displacing it from water and acid solutions. These displacement reactions occur at rates that depend on the nature of the metal.

The alkali metals and alkaline earth metals react violently with both water and acids, generating hydrogen at explosive rates. Certain other metals react much more slowly with water and acids. Nonetheless, these reactions pose the risk of fire and explosion, since the hydrogen produced by them often absorbs the heat of reaction and bursts spontaneously into flame. We examine this subject in more depth in Sections 9.2 and 9.3-A.

Many metals possess the capability of releasing hydrogen from acids. Metallic tin and aluminum displace hydrogen from hydrochloric acid and sulfuric acid, respectively, as follows:

$$Sn(s) \ + \ 2HCl(aq) \ \longrightarrow \ SnCl_2(aq) \ + \ H_2(g)$$

Tin   Hydrochloric acid   Tin(II) chloride   Hydrogen

$$2Al(s) \ + \ 3H_2SO_4(aq) \ \longrightarrow \ Al_2(SO_4)_3(aq) \ + \ 3H_2(g)$$

Aluminum   Sulfuric acid   Aluminum sulfate   Hydrogen

TABLE 7.5	Activity Series of the Elements

	RATE OF REACTION INCREASES ↑
Elements that release hydrogen from water	Cesium
	Lithium
	Rubidium
	Potassium
	Barium
	Sodium
	Calcium
	Magnesium
Elements that release hydrogen from water and acids	Aluminum
	Manganese
	Zinc
	Chromium
	Iron
	Nickel
	Tin
	Lead
Elements that do not release hydrogen from water or acids	Bismuth
	Copper
	Mercury
	Silver
	Platinum
	Gold

activity series
■ An arrangement of the metals in order of their decreasing ability to generate hydrogen by chemical reaction with water and acids

The rates at which metals displace hydrogen from water and acidic solutions can be determined experimentally. When the metals are then arranged by decreasing reaction rates, the compilation in Table 7.5 is obtained. This arrangement of the metals is called an **activity series**. The significance of the compilation is summarized as follows:

■ The metals listed at the top of the table are so chemically reactive they react with *both* water and acids.
■ The metals below magnesium release hydrogen from steam and acid solutions, but not from liquid water.
■ The metals below iron in the series do not displace hydrogen from steam, even when the temperature is elevated.
■ The metals below lead possess insufficient chemical reactivity to release hydrogen from either water or acids.

Hydrogen is also displaced by certain metals from solutions of sodium hydroxide. The common metals exhibiting this chemistry are aluminum, zinc, and lead, but these reactions occur slowly.

7.2-F HYDROGEN GENERATION DURING THE CHARGING OF LEAD–ACID STORAGE BATTERIES

Most of us are familiar with the **lead–acid storage battery** as a component of virtually all vehicles on the roadways of the world. Its purpose is to produce electric current by means of either of the two reversible chemical reactions commonly referred to as charging and "discharging." A battery is said to be "charged" when an outside current has been delivered through it; then, as the battery "discharges," it produces electricity by means of chemical reactions involving the substances that were formed during the charging process. Hydrogen is one of them. When the battery is being charged, bubbles of hydrogen accumulate near the negative plate; simultaneously, bubbles of oxygen accumulate near its positive plate. As they are produced, both gases dissipate into the atmosphere.

However, when banks of lead–acid storage batteries like those shown in Figure 7.12 are simultaneously charged within an enclosure, the concentration of hydrogen is likely to exceed its lower explosive limit. Within an enclosed room, for instance, hydrogen rises and concentrates along the length of its ceiling. Because a concentration of only 4% by volume of hydrogen is sufficient to produce a flammable mixture in air, adequate ventilation should always be provided within rooms containing banks of storage batteries to prevent the accumulation of hydrogen and minimize an unreasonable risk of fire and explosion.

lead–acid storage battery
■ The collective group of individual cells, each of which is composed of lead and lead oxide plates, immersed in a solution of sulfuric acid having a specific gravity ranging from 1.25 to 1.30

7.2-G THE *HINDENBURG*

Perhaps the best-known incident involving the burning of hydrogen was the destruction of the German dirigible the *Hindenburg*. Although the combustion of diesel fuel powered its engines, hydrogen stored in large gas bags held this airship aloft. Germany had intended to use the *Hindenburg* to inaugurate a new era in fast transatlantic travel, but the airship mysteriously caught fire and exploded while attempting its landing approach at Lakehurst, New Jersey. Given the presence of hydrogen onboard, the immediate sentiment linked a hydrogen leak with the burning of the airship (see also Section 9.3-D).

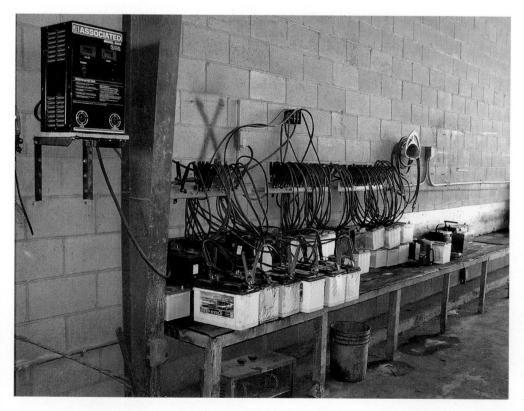

FIGURE 7.12 A battery-charging station is a location at which banks of individual batteries are simultaneously charged. Hydrogen gas is produced during the battery-charging process. When the battery-charging station is located inside a large ventilated room, the production of hydrogen is a relatively small problem, but when it is located inside a small enclosure, the hydrogen may reach a concentration within its flammable range. Then, the accumulation of hydrogen poses the risk of fire and explosion. (*Courtesy of Interstate Batteries of Las Vegas, Las Vegas, Nevada.*)

The tragedy of the *Hindenburg* caused the use of hydrogen as a buoyant gas to be discontinued in airships. Today, helium is used to provide lifting power in airships. Helium is safer to use because it is a nonflammable gas, but it possesses only about 93% of the lifting power of hydrogen.

7.2-H WORKPLACE REGULATIONS INVOLVING HYDROGEN

When it is intended for use at manufacturing and processing plants, hydrogen is generally encountered as a confined gas within cylinders or storage tanks. The capacities of the tanks used to store hydrogen range from less than 3000 ft^3 (85 m^3) to over 15,000 ft^3 (425 m^3). At 29 C.F.R. §§1910.103(b)(2)(1)(a)–(d), OSHA regulates their locations in relation to the position of buildings and other storage tanks so they are readily accessible to delivery equipment and authorized personnel. Hydrogen storage systems must be located aboveground but not beneath electric power lines and away from the piping for other flammable gases and flammable liquids. To alert individuals to the presence of this flammable gas, OSHA also requires the tanks to be permanently placarded with a sign that reads as follows:

> **DANGER**
>
> **HYDROGEN - FLAMMABLE GAS**
> **NO SMOKING**
> **NO OPEN FLAMES**

7.2-I TRANSPORTING HYDROGEN

When shippers intend to transport compressed or cryogenic hydrogen, DOT requires them to identify it on the accompanying shipping paper as either of the following, as relevant:

> **UN1049, Hydrogen, compressed, 2.1**
> *or*
> **UN1966, Hydrogen, refrigerated liquid, 2.1**

DOT also requires shippers who transport hydrogen to affix a FLAMMABLE GAS label to its cylinders or other packaging.

When carriers transport 1001 lb (454 kg) or more of hydrogen, DOT requires them to post FLAMMABLE GAS placards on the bulk packaging or transport vehicle used for shipment. DOT also requires carriers to display the relevant identification number, 1049 or 1966, on orange panels, across the center area of the FLAMMABLE GAS placards or on white square-on-point panels affixed on each side and each end of the packaging.

7.2-J RESPONDING TO INCIDENTS INVOLVING A RELEASE OF HYDROGEN

As a practical matter, combating an ongoing fire involving hydrogen is rarely successful unless the flow of hydrogen from its containment vessel can first be stopped without exposing personnel to undue risk. For this reason, it is often best to permit a hydrogen fire to burn without interference. Nonetheless, since a hydrogen fire is likely to cause secondary fires, appropriate emergency-response actions should be taken to prevent their spread.

At scenes where the release of LH$_2$ has occurred, the use of a water fog is generally warranted. A water fog reduces the hydrogen concentration in the nearby atmosphere.

PERFORMANCE GOALS FOR SECTION 7.3:

- Identify the principal properties of elemental fluorine, especially noting its pronounced toxicity and unique chemical reactivity.

7.3 Fluorine

As noted in Table 4.1, fluorine is found in nature to the extent of 0.027% by mass, but it is never encountered as the free element. Instead, it is found distributed in rocks and minerals. As a hazardous material, the element possesses a certain distinction: It is more chemically reactive than any other element. Fluorine is so reactive that it forms compounds with nearly every other element, the possible exceptions being the lighter noble gases. At room conditions, elemental fluorine is a pale yellow gas with a pungent, irritating odor. Some of its important physical properties are provided in Table 7.6.

7.3-A CHEMICAL REACTIVITY OF FLUORINE

The exceptional chemical reactivity of fluorine gives rise to considerable difficulties both in its preparation and handling. Fluorine can be prepared only by passing an electric current through a molten mixture of potassium fluoride and anhydrous (without water) hydrofluoric acid in a specially designed apparatus.

Although elemental fluorine is available commercially, its pronounced reactivity and poisonous nature generally limit its widespread use. Fluorine is used to produce uranium(VI) fluoride, which is needed in uranium-enrichment processes (Section 16.8-C).

Elemental fluorine does not burn, but like oxygen, it supports the combustion of other substances. The ability of fluorine to support combustion is apparent from its reaction with hydrogen. Fluorine and hydrogen unite explosively, forming hydrogen fluoride.

$$H_2(g) \quad + \quad F_2(g) \quad \longrightarrow \quad 2HF(g)$$
$$\text{Hydrogen} \qquad \text{Fluorine} \qquad \text{Hydrogen fluoride}$$

Ordinary materials like wood, plastics, and metals also burn spontaneously when exposed to fluorine; even "fireproof" asbestos burns in a fluorine-rich atmosphere.

The reactivity of fluorine is associated with its powerful nature as an oxidizing agent. For example, fluorine possesses the unusual ability of chemically reacting with the components of water. Although the chemistry is somewhat complicated, it may be represented as follows:

$$5F_2(g) \quad + \quad 5H_2O(l) \quad \longrightarrow \quad 8HF(aq) \quad + \quad O_2(g) \quad + \quad H_2O_2(aq) \quad + \quad OF_2(g)$$
$$\text{Fluorine} \qquad \text{Water} \qquad \text{Hydrogen fluoride} \qquad \text{Oxygen} \qquad \text{Hydrogen peroxide} \qquad \text{Oxygen difluoride}$$

The reaction is so violent that its occurrence is sufficient reason to avoid the use of water when combating fires supported by fluorine.

Not only is fluorine a powerful oxidizing agent, but it is also highly toxic when inhaled. Even small traces of fluorine can be irritating, since the element reacts with the moisture in the lungs and respiratory tract to produce other reactive and corrosive substances. On contact with the skin, fluorine causes deep, penetrating burns that can be delayed and progressively worsen with time. Prolonged exposure to elevated concentrations (>50 ppm) of fluorine often leads to the accumulation of fluid within the lungs, a condition known as **pulmonary edema**.

pulmonary edema
■ The excessive accumulation of fluid within the lungs

TABLE 7.6	Physical Properties of Elemental Fluorine
Melting point	−363.3 °F (−219.6 °C)
Boiling point	−306.6 °F (−188.1 °C)
Specific gravity (gas) at 68 °F (20 °C)	1.312
Specific gravity (liquid) at boiling point	1.505
Vapor density (air = 1)	1.31

7.3-B WORKPLACE REGULATIONS INVOLVING FLUORINE

When the use of fluorine is needed in the workplace, OSHA requires employers to regulate employee exposure by limiting the fluorine concentration to no greater than 0.1 ppm, averaged over an 8-hr workday. This low concentration reduces or eliminates the possibility that workers will experience the ill effects associated with exposure to fluorine.

7.3-C TRANSPORTING FLUORINE

When shippers intend to transport fluorine, DOT requires them to identify it on the accompanying shipping paper as follows:

UN1045, Fluorine, compressed, 2.3, (5.1, 8) (Poison - Inhalation Hazard, Zone A)

Although elemental fluorine exists as both a compressed gas and cryogenic, liquid, DOT permits only the compressed gas to be transported by any mode. The compressed gas is transported in DOT-specification cylinders made of a specially designed steel or nickel, to which fluorine is relatively unreactive. These cylinders must be seamless and equipped with a valve-protection cap, but no safety-relief devices. DOT prohibits charging of the cylinders with more than 6 lb (3 kg) of fluorine under a pressure not exceeding 400 psi_g (2760 kPa) at 70 °F (21 °C). When shippers offer fluorine gas for transportation, DOT requires them to affix POISON GAS, OXIDIZER, and CORROSIVE labels to the cylinders.

DOT requires carriers to display POISON GAS placards on the transport vehicle used to transport the cylinders. DOT also requires them to display the identification number 1045 on orange panels, across the center area of the POISON GAS placards, or on white square-on-point panels. Because the presence of fluorine poses an inhalation hazard, DOT also requires carriers to mark the packaging with the words INHALATION HAZARD.

7.3-D RESPONDING TO INCIDENTS INVOLVING A RELEASE OF FLUORINE

When fluorine is encountered during a transportation mishap as a leak from a small cylinder, DOT recommends evacuating all unauthorized persons to an initial distance of 100 ft (30 m) in all directions. DOT further requires their evacuation to a distance of 0.1 mi (0.2 km) downwind from the scene and 0.3 mi (0.5 km) during the day and nighttime hours, respectively.

Firefighters are cautioned to avoid extinguishing fires supported by elemental fluorine. These fires are generally permitted to burn until the fuel has been totally exhausted. When fluorine cylinders are exposed to flames or excessive heat, water from unmanned monitors is used to cool them only when the monitors may be positioned without exposing personnel to risk. When fluorine cylinders are exposed to prolonged, intense heat, these cylinders readily rupture, since they are not equipped with safety-relief devices.

PERFORMANCE GOALS FOR SECTION 7.4:

- Identify the principal properties of elemental chlorine, especially acknowledging its inhalation toxicity.
- Illustrate that elemental chlorine can act as an oxidizing agent.
- Identify the industries that use bulk quantities of chlorine.
- Describe the response actions to be executed when chlorine has been released into the environment.

TABLE 7.7	Physical Properties of Elemental Chlorine
Melting point	−149.8 °F (−101 °C)
Boiling point	−30.28 °F (−34.6 °C)
Specific gravity (gas) at 68 °F (20 °C)	1.56
Specific gravity (liquid) at 6.86 atm	1.41
Vapor density (air = 1)	2.49
Liquid-to-gas expansion ratio	457.6

7.4 Chlorine

Elemental chlorine is not found naturally, but chlorine is found on Earth to the extent of 0.19% by mass in a variety of compounds including sodium chloride, potassium chloride, calcium chloride, and magnesium chloride.

At room conditions, chlorine exists as a yellow-green gas with a characteristic penetrating and irritating odor. The element is about 2½ times heavier than air and is highly poisonous when inhaled. Several other physical properties of chlorine are noted in Table 7.7.

Elemental chlorine should not be confused with solid "chlorine" products used to treat the water in domestic and municipal swimming pools. Although these latter products are frequently called "chlorine," they are actually oxidizing agents that generate chlorine by chemical action within the swimming pool. The properties of these chlorine-containing products are more appropriately discussed in Chapter 11.

The principal risk associated with exposure to elemental chlorine is inhalation toxicity. The exposure initially causes coughing, dizziness, nausea, headache, and severe inflammation of the eyes, nose, and throat. Extended inhalation is likely to cause congestion of the lungs, which can result in pulmonary edema. The ill effects listed in Table 7.8 are likely to be experienced as well.

TABLE 7.8	Adverse Health Effects Associated with Breathing Chlorine

CHLORINE (ppm)	SIGNS AND SYMPTOMS
0.5	No signs or symptoms
3–8	Discomfort; stinging and burning of the eyes
50–250	Severely irritating to the eyes and mucuous membranes; coughing, choking, nausea, dizziness, headache, and difficulty in breathing
250–500 (for 3 min)	Delayed pulmonary edema
1000 (few breaths)	Death

The following information has been imprinted on the exterior surface of a rail tankcar hauling chlorine as a liquefied compressed gas:

PROX28829		
LD LMT	5000 LB	2273 KG
LT WT	2000 LB	909 KG
NEW	08 97	

Use the data in Table 7.7 to calculate the approximate volume in cubic feet of gaseous chlorine that is generated when the contents of this tank are suddenly released into the atmosphere during a transportation mishap, and ascertain whether this volume is likely to pose an immediate risk of inhalation toxicity.

Solution: First, the imprinted information indicates that the tank holds 1 tn of liquid chlorine. Using the specific gravity of liquid chlorine in Table 7.7, we calculate the density of the liquid as follows:

$$1.41 \times 62.4 \frac{\text{lb}}{\text{ft}^3} = 88.0 \text{ lb/ft}^3$$

The volume occupied by 1 tn of liquid chlorine is then

$$1 \text{ tn} \times 2000 \frac{\text{lb}}{\text{tn}} \times \frac{1 \text{ ft}^3}{88.0 \text{ lb}} = 22.7 \text{ ft}^3$$

Using the liquid-to-gas expansion ratio for chlorine listed in Table 7.7, we determine the volume of gaseous chlorine as follows:

$$22.7 \text{ ft}^3_{\text{(liquid)}} \times 457.6 = 10,388 \text{ ft}^3_{\text{(gas)}}$$

This calculation illustrates that when 1 tn of liquid chlorine is released from a rail tankcar, it expands to occupy an initial volume of 10,388 ft^3 as the gas. This volume moves swiftly from the accident scene as it mixes with the air.

One ton of chlorine per 10,388 ft^3 is equivalent to an undiluted concentration in air of 0.19 lb/ft^3. Because inhalation of one or two breaths of chlorine at this concentration is fatal, the vaporization of 1 tn of liquid chlorine poses an immediate risk of inhalation toxicity to anyone in the vicinity.

What is the most likely reason DOT prohibits carriers from accepting any quantity of elemental chlorine for domestic transport on passenger-carrying aircraft or railcar?

Solution: The primary hazard associated with exposure to elemental chlorine is inhalation toxicity. The most likely reason DOT prohibits carriers from transporting chlorine on domestic passenger-carrying aircraft or railcar is to reduce or eliminate the likelihood that passengers and transportation personnel will experience the ill effects caused by chlorine exposure.

7.4-A COMMERCIAL USES OF CHLORINE

For commercial use, elemental chlorine is prepared by passing an electric current through either molten sodium chloride or an aqueous solution of sodium chloride or magnesium chloride. When aqueous sodium chloride is used, sodium hydroxide and hydrogen are simultaneously produced as follows:

$$2NaCl(aq) \ + \ 2H_2O(l) \ \longrightarrow \ 2NaOH(aq) \ + \ H_2(g) \ + \ Cl_2(g)$$

Sodium chloride · · · Water · · · Sodium hydroxide · Hydrogen · Chlorine

The elemental chlorine is encountered commercially as a gas and a liquefied compressed gas.

Throughout the civilized world, large volumes of elemental chlorine are annually required for the following commercial uses:

- A raw material for the production and manufacture of a wide range of chlorine-containing compounds used as solvents, pesticides, dyes, bleaching agents, plastics, refrigerants, and other commercial products
- A microbicide for treatment of water and sewage
- A bactericide for sterilizing drinking water and pool water against waterborne infectious diseases
- A bleaching agent of paper pulp and certain textiles

On a somber note, chlorine was used during World War I as a chemical warfare agent. Taking advantage of its highly poisonous nature, Germany unleashed the gas as a chemical weapon of mass destruction against British and French soldiers. The Germans discharged the chlorine from pressurized cylinders into the wind, which carried it into the trenches, where unprotected soldiers inhaled the gas and died.

In 1925, the League of Nations took the first step toward eliminating the use of poisonous gases during warfare. As a consequence of its efforts, over 130 nations ratified the **Geneva protocol**, which prohibits the use of chlorine and other poisonous gases as chemical warfare agents. Known formally as the Protocol for the Prohibition of the Use in War of Asphyxiating, Poisonous, or Other Gases, and of Bacteriological Methods of Warfare, it successfully banned civilized countries from ever again using poisonous gases against their enemies.

Notwithstanding the existence of the ban, the use of chlorine during warfare continues. In 2007, insurgents used chlorine to kill and injure Iraqi civilians by exploding trucks containing chlorine tanks. These incidents prompted the U.S. Department of Homeland Security to warn American chemical plants to guard against the potential theft of chlorine tanks from their domestic facilities.

Geneva protocol (formally the Protocol for the Prohibition of the Use in War of Asphyxiating, Poisonous, or Other Gases, and of Bacteriological Methods of Warfare)
■ An international agreement established by the League of Nations in 1925 that bans the use of poisonous gases against the enemy during warfare

7.4-B CHEMICAL REACTIVITY OF CHLORINE

Chlorine is a very versatile element. Like oxygen and fluorine, it is a nonflammable gas capable of supporting combustion. The oxidizing ability of chlorine is apparent from its reaction with hydrogen. The hydrogen burns to form hydrogen chloride.

$$H_2(g) \ + \ Cl_2(g) \ \longrightarrow \ 2HCl(g)$$

Hydrogen · · · Chlorine · · · Hydrogen chloride

Elements other than hydrogen also burn when exposed to an atmosphere of chlorine. For example, finely divided copper, arsenic, antimony, phosphorus, and sulfur burn with incandescence.

$$Cu(s) \ + \ Cl_2(g) \ \longrightarrow \ CuCl_2(s)$$

Copper · · · Chlorine · · · Copper(II) chloride

$$2As(s) \ + \ 3Cl_2(g) \ \longrightarrow \ 2AsCl_3(s)$$

Arsenic · · · Chlorine · · · Arsenic trichloride

$$2Sb(s) \ + \ 3Cl_2(g) \ \longrightarrow \ 2SbCl_3(s)$$

Antimony · · · Chlorine · · · Antimony trichloride

$$P_4(s) \quad + \quad 6Cl_2(g) \quad \longrightarrow \quad 4PCl_3(l)$$

Phosphorus   Chlorine   Phosphorus trichloride

$$S_8(l) \quad + \quad 10Cl_2(g) \quad \longrightarrow \quad 2S_2Cl_2(l) \quad + \quad 4SCl_4(l)$$

Sulfur   Chlorine   Disulfur dichloride   Sulfur tetrachloride

In these instances, the chlorine acts as an oxidizing agent.

Chlorine also supports the combustion of certain organic compounds. These combustion reactions occur only in the presence of light, which acts catalytically. For instance, a mixture of elemental chlorine and gasoline vapor is essentially unreactive unless it is exposed to light.

When chlorine atoms are incorporated into a compound, the associated phenomenon is called a **chlorination** reaction. The oxidation of a substance by chlorine is an example of a chlorination reaction. The term "chlorination" is also used to describe the treatment of drinking water and wastewater with elemental chlorine and chlorine-containing oxidizing agents. When used in this fashion, chlorine acts as a microbicide by killing undesirable microorganisms. The use of chlorine for treating drinking water saves millions of lives annually from waterborne diseases like cholera and dysentery. Notwithstanding this fact, as noted in Section 7.1-I, chlorine is limited in its use as a microbicide, since it does not effectively destroy the bacterium cryptosporidium.

When it is dissolved in water, chlorine reacts to form hypochlorous acid, which is even more powerful than chlorine as an oxidizing agent.

$$Cl_2(g) \quad + \quad H_2O(l) \quad \longrightarrow \quad HClO(aq) \quad + \quad HCl(aq)$$

Chlorine   Water   Hypochlorous acid   Hydrochloric acid

During the chlorination of drinking water and wastewater, it is the presence of hypochlorous acid that aids in making chlorine an effective microbicide.

chlorination

■ A chemical reaction that involves the addition of chlorine to a substance; the treatment of contaminated drinking water to kill the microorganisms that cause diseases like dysentery, cholera, typhoid, and hepatitis

7.4-C WORKPLACE REGULATIONS INVOLVING CHLORINE

When the use of elemental chlorine is necessary in the workplace, OSHA requires employers to provide their workers with respiratory protective gear. When employees are exposed to a relatively small amount of chlorine, they should wear goggles and impermeable gloves, and whenever possible, their activities should be conducted within a fume hood. However, when working on bulk chlorine storage systems, employees should wear total-encapsulating suits and use self-contained breathing apparatus.

To avoid or minimize the ill effects to employees associated with exposure to chlorine, OSHA requires employers to limit the chlorine concentration in air to no greater than 1 ppm, averaged over an 8-hr workday.

7.4-D TRANSPORTING CHLORINE

Chlorine is transported as a liquefied compressed gas in steel cylinders, ton-containers, motorized cargo tanks, and rail tankcars at 84 psi (580 kPa) at 70 °F (21 °C). The ton-container is less frequently used for transporting chlorine; it is a welded tank having a maximum loaded mass of 3700 lb (1680 kg).

When shippers intend to transport chlorine in bulk, DOT requires them to identify it on the accompanying shipping paper as follows:

UN1017, Chlorine, 2.3, (8) (Marine Pollutant) (Poison - Inhalation Hazard, Zone B)

DOT also requires them to affix POISON GAS and CORROSIVE labels to the cylinders or other containment devices.

DOT also requires POISON GAS placards and MARINE POLLUTANT markings to be displayed on the bulk packaging or transport vehicle used for shipment. DOT requires carriers to display the identification number 1017 on orange panels, across the center area of the POISON GAS placards, or on white square-on-point panels. Because the presence of chlorine poses an inhalation hazard, DOT requires the words INHALATION HAZARD to be marked on the packaging. The implementation of these DOT requirements is shown on the rail tankcar in Figure 7.13.

The DOT-approved tanks and cylinders used for transporting chlorine are equipped with fusible plugs designed to melt in the temperature range from 158 to 165 °F (70 to 74 °C). The

FIGURE 7.13 When carriers transport any amount of chlorine in a rail tankcar, DOT requires them to post POISON GAS placards on each of its sides and to mark CHLORINE and INHALATION HAZARD on two opposing sides. DOT also requires them to display the identification number 1017 on each side. In this instance, the carrier chose to display 1017 across the center area of the POISON GAS placards. HOKX 7718 is the reporting mark and number of Occidental Chemical Corporation, Dallas, Texas. At the time this tank was photographed, DOT did not require MARINE POLLUTANT to be marked on two opposing sides. (*Courtesy of the Chlorine Institute, Arlington, Virginia and Andrew Johnson, iMed Design, Inc., Reno, Nevada.*)

plugs are located on the valve just below the valve seat. When exposed to the temperatures routinely associated with fire conditions, the melting of these plugs permits chlorine to be slowly released to the environment and prevents the vessels from rupturing.

DOT also regulates the construction and fabrication of rail tankcars and tank trucks used to transport chlorine. To maintain the confined chlorine primarily in the liquid state, DOT requires a minimum of 4 in. (10 cm) of insulation about them. As noted in Section 3.7-B, thermally insulated tankcars constructed to DOT specifications have been designed to allow external heat to be slowly transferred to the contents under the normal conditions of transport. When exposed to heat, however, these tanks lose their integrity, and rupture.

7.4-E RESPONDING TO INCIDENTS INVOLVING A RELEASE OF CHLORINE

Because elemental chlorine poses an inhalation health hazard, emergency responders cannot be too wary when they encounter chlorine during transportation incidents. In 2005, after two freight trains collided in Graniteville, South Carolina, a pressurized tankcar ruptured and spilled more than 9200 gal (35 m³) of liquid chlorine. This sole incident caused the deaths of nine people, the hospitalization of more than 250 individuals, and the evacuation of more than 5400 people.*

When an emergency-response team is called to a scene at which chlorine tanks or containers have ruptured or could potentially rupture, it is vital to acknowledge the toxicity hazard associated with the inhalation of chlorine. Because inhaling chlorine can be fatal, the use of fully encapsulated suits and self-contained breathing apparatus is absolutely essential.

---

*The adverse health effects experienced by nearby residents in connection with this incident have been examined by a team from the U.S. Centers for Disease Control and Prevention and the South Carolina Department of Health and Environmental Control. The results of the study are published in the *American Journal of Emergency Medicine* 27 (2009): 1–7.

The first action of the team responding to a chlorine release is to evacuate all unauthorized persons from the scene to a distance dependent on the amount of chlorine that has been or could be released into the environment. In connection with a transportation mishap, DOT recommends an isolation distance (Section 10.15) of 100 ft (30 m) in all directions for a "small" cylinder leak and 900 ft (275 m) for a "large" spill from a bulk container. As additional protection during the occurrence of a "small" spill, DOT also requires the evacuation of unauthorized persons to a distance of 0.2 mi (0.3 km) downwind from the scene and 0.7 mi (1.1 km) during the day and nighttime hours, respectively. For large spills, DOT advises evacuation to a distance of 1.7 mi (2.7 km) and 4.2 mi (6.8 km) during the day and nighttime hours, respectively.

An emergency-response team must also identify those who have been exposed to chlorine from leaks or ruptured containers. These individuals should be moved downwind to fresh air, where if necessary, artificial respiration can be supplied. They should also be kept warm with blankets and provided with immediate first-aid attention. To minimize skin burns, persons exposed to chlorine should remove their clothing and shower thoroughly. To prevent the impairment of vision, they should irrigate their eyes with flowing water for approximately 30 minutes. Finally, they should be transported to an emergency medical facility for follow-up examinations by physicians.

An emergency-response team is occasionally obligated to examine the physical condition of a chlorine container or rail tankcar. Unruptured vessels should be cooled with water as the team members pinpoint the specific spots from which chlorine is leaking or could potentially leak. Because containers and transport vessels contain *liquid* chlorine, it is often the liquid that drips from valves, fittings, or openings. Liquid chlorine is much more concentrated than its gas, and when unconfined, it readily evaporates. One volume of liquid chlorine evaporates into approximately 460 volumes of gas. For this reason, the area immediately surrounding a chlorine leak becomes a highly toxic environment within seconds.

Locating a chlorine leak is not always simple, especially when the gas has been escaping from its container or transport vessel for some time. One method of detecting a chlorine leak is based on the chemical reaction between chlorine and ammonia. These two substances react to form ammonium chloride and ammonium hypochlorite, each of which is a white solid.

$$2NH_3(g) \quad + \quad Cl_2(g) \quad + \quad H_2O(l) \quad \longrightarrow \quad NH_4Cl(s) \quad + \quad NH_2ClO(s)$$

Ammonia   Chlorine   Water   Ammonium chloride Ammonium hypochlorite

The method consists of tying a rag soaked with household ammonia to a broomstick and then passing the stick along the surface of the chlorine container or tank. Ammonia vaporizes from the liquid and reacts with the chlorine at the point from which it is escaping. At this location, a white cloud drifts into the air. By observing the cloud's formation, the source of the leak is easily identified. This simple procedure is ineffective when the surrounding atmosphere is heavily laden with chlorine, since under this condition, the ammonium compounds are produced virtually throughout the area.

When chlorine is leaking from a tank or container, the exit points must be closed or sealed. When sealing an opening is impossible or impractical, an attempt should be made to prevent the further escape of liquid chlorine into the environment. This can sometimes be accomplished by rolling the vessel so its opening points upward. Although chlorine gas continues to escape from the opening, the more concentrated liquid remains confined within the vessel.

When emergency responders arrive at the scene of an incident involving the release of chlorine, specialized equipment should be available for their immediate use. An essential item is a kit that contains a clamping device to seal a leak at the fusible plug and a patching device to seal a leak in the cylinder sidewall. They are components of the so-called *Chlorine Institute Emergency Kit "A,"* which is available from several commercial outlets.*

To improve the speed and effectiveness of a response action at an emergency involving the release of chlorine, the Chlorine Institute formalized a Chlorine Emergency Plan known as **CHLOREP**. Under this plan, the United States and Canada are divided into regional sectors in

CHLOREP

■ A program designed by the Chlorine Institute to aid emergency responders during incidents involving the release of chlorine into the environment during transportation mishaps or at user locations

*The Chlorine Institute is a trade association consisting primarily of company representatives interested in the safe production, distribution, and use of chlorine and other substances associated with the chlor-alkali industry. The offices of the Chlorine Institute are located at 1300 Wilson Boulevard, Arlington, Virginia 22209.

which specially trained teams are located. In the event of an emergency involving the release of chlorine, or when CHEMTREC (Section 1.11) is contacted, the caller is put into immediate contact with the closest CHLOREP team, which then oversees the handling of the incident.

PERFORMANCE GOALS FOR SECTION 7.5:

- Identify the principal properties of the red and white allotropes of elemental phosphorus, especially acknowledging the ability of white phosphorus to spontaneously ignite.
- Describe how the red and white allotropes are produced.
- Identify the industries that use bulk quantities of white phosphorus.
- Describe the response actions to be executed when red and white phosphorus have been released into the environment.

7.5 Phosphorus

Elemental phosphorus has several allotropes, two of which are especially important: *white phosphorus* and *red phosphorus*. White phosphorus is the unstable form of the element at room conditions. On standing, it acquires a yellow coloration owing to the partial conversion of the white allotrope to the more stable red allotrope. It is for this reason that white phosphorus is sometimes called *yellow phosphorus*. Red phosphorus is also known as *amorphous phosphorus*.

The physical properties of these allotropes of phosphorus are provided in Table 7.9. Their properties are so strikingly different that we discuss them separately. Both are used to manufacture important chemical products such as special alloys (e.g., phosphor bronze), rodenticides, fireworks, matches, phosphoric acid, and metallic phosphides.

In the past, elemental white and red phosphorus were used extensively as the active agent in incendiary bombs. When these bombs were dropped from military aircraft, they set fires to objects or caused burn injury to persons through the action of flame, heat, or a combination thereof. The composition of some incendiary bombs containing red phosphorus was 50% of the incendiary mixture. The use of white phosphorus in incendiary bombs also provides incidental positive results such as illumination of the battlefield during nighttime. However, the use of white phosphorus in munitions is now considered an uncivilized practice, since flaming droplets may become embedded beneath skin and seriously harm civilians.

White phosphorus has also been used in tracer, smoke, and signaling systems. The mixture of tetraphosphorus hexoxide and tetraphosphorus decoxide produced by the combustion of white phosphorus possesses the best obscuring power of any known "smoke-producing" material.

TABLE 7.9	Physical Properties of Elemental Phosphorus	
	WHITE PHOSPHORUS	**RED PHOSPHORUS (AMORPHOUS)**
Melting point	111.4 °F (44.1 °C)	1094 °F (590 °C) at 43 atm (4400 kPa)
Boiling point	535 °F (280 °C)	781 °F (416 °C) (sublimes)
Specific gravity at 68 °F (20 °C)	1.82	2.34
Vapor density (air = 1)	4.42	4.77
Autoignition point	86 °F (30 °C) (spontaneous ignition in dry air)	500 °F (260 °C)

In 1983, many of the world's civilized countries agreed to ban the use of incendiary weapons during certain warfare operations by means of Protocol III of the **Convention on Prohibitions or Restrictions of the Use of Certain Conventional Weapons Which May Be Deemed to Be Excessively Injurious or to Have Indiscriminate Effects**, also known as the Convention on Certain Conventional Weapons. Protocol III prohibits the use of incendiary weapons against civilian populations and restricts their use against military targets located within a concentration of civilians, but it does not restrict the use of incendiary bombs in illumination, tracer, smoke, or signaling systems.

Although the United States is a signatory to the Convention on Certain Conventional Weapons, it has not ratified Protocol III. In 2005, during the U.S.-led assault on Fallujah, Iraq, incendiary munitions containing white phosphorus were used to flush enemy troops from covered positions.

7.5-A WHITE PHOSPHORUS

White phosphorus is a waxy, translucent solid at ambient conditions. Because it consists of tetratomic molecules, the chemical formula of white phosphorus is P_4.

White phosphorus is industrially prepared by heating calcium phosphate rock with sand and coke in an electric furnace. The principal components of sand and coke are silicon dioxide and carbon, respectively. The production of white phosphorus is denoted by the following equation:

$$2Ca_3(PO_4)_2(s) \quad + \quad 6SiO_2(s) \quad + \quad 10C(s) \quad \longrightarrow \quad 6CaSiO_3(s) \quad + \quad 10CO(g) \quad + \quad P_4(g)$$

Calcium phosphate   Silicon dioxide   Carbon   Calcium silicate   Carbon monoxide   Phosphorus

The phosphorus vapors are vented from the furnace and condensed under water to produce a white solid. Small quantities are generally stored and transported under water in containers, but bulk quantities are often encountered as molten phosphorus under water or blanketed with nitrogen.

Because the autoignition temperature of white phosphorus is only 86 °F (30 °C), white phosphorus spontaneously ignites when exposed to air. The autoignition temperature is so low that even body heat initiates ignition. To reduce or eliminate the risk of its spontaneous combustion, white phosphorus is stored under water or a blanket of nitrogen.

White phosphorus is often said to possesses the offensive odor of a mixture of garlic and rotten fish, but this comparison is misleading. The odor is more likely associated with the presence of phosphine (Section 9.6-B), a toxic gas slowly produced by the reaction between phosphorus and cold water.

$$P_4(s) \quad + \quad 6H_2O(l) \quad \longrightarrow \quad 3H_3PO_2(aq) \quad + \quad PH_3(g)$$

Phosphorus   Water   Hypophosphorous acid   Phosphine

The combustion of white phosphorus produces two oxides, tetraphosphorus hexoxide and tetraphosphorus decoxide, as follows:

$$P_4(s) \quad + \quad 3O_2(g) \quad \longrightarrow \quad P_4O_6(s)$$

Phosphorus   Oxygen   Tetraphosphorus hexoxide

$$P_4(s) \quad + \quad 5O_2(g) \quad \longrightarrow \quad P_4O_{10}(s)$$

Phosphorus   Oxygen   Tetraphosphorus decoxide

As implied by these equations, tetraphosphorus hexoxide and tetraphosphorus decoxide are the products of incomplete and complete combustion, respectively. Both are white compounds. Consequently, when white phosphorus burns, billows of dense white, choking smoke are produced. This luminous dense smoke accounts for the use of white phosphorus as a component of military smoke and signaling systems.

Although spontaneous combustion is the principal hazard associated with white phosphorus, its vapor is highly poisonous. Prolonged exposure to the vapor of white phosphorus causes the affliction known as **phossy jaw**, or **phosphorus necrosis**, which in the worst instances, causes disintegration of the jawbone. In addition, white phosphorus burns the skin and produces wounds that are extremely painful and slow to heal. For this reason, users must always handle white phosphorus with gloves.

phossy jaw (phosphorus necrosis)
■ The disfiguring affliction caused by overexposure to phosphorus vapor

7.5-B TRANSPORTING WHITE PHOSPHORUS

When shippers offer solid or molten white phosphorus for transportation in bulk, DOT requires them to identify it on the accompanying shipping paper as either of the following, as relevant:

UN1381, Phosphorus, white, dry, 4.2, (6.1), PGI (Marine Pollutant) (Poison)
or
UN2447, Phosphorus, white, molten, 4.2, (6.1), PGI (Marine Pollutant) (Poison)

DOT also requires shippers to affix SPONTANEOUSLY COMBUSTIBLE and POISON labels to the containment devices.

When carriers transport 1001 lb (454 kg) or more of white phosphorus, DOT requires them to display SPONTANEOUSLY COMBUSTIBLE placards and MARINE POLLUTANT markings on the bulk packaging or transport vehicle used for shipment. DOT also requires carriers to display the relevant identification number, 1381 or 2447, on orange panels, across the center area of the SPONTANEOUSLY COMBUSTIBLE placards, or on white square-on-point panels.

7.5-C RESPONDING TO INCIDENTS INVOLVING A RELEASE OF WHITE PHOSPHORUS

When emergency-response crews first arrive at a scene involving a release of white phosphorus, the element is generally burning. Although water is an effective fire extinguisher on a white phosphorus fire, experts advise the use of dry sand as the acceptable extinguisher, since it blankets the element, prevents reignition, and prevents the potential exposure of personnel to phosphine.

When small quantities are burning, it is best to segregate them from nearby combustible materials. This may not be a simple matter, since phosphorus melts at a relatively low temperature and flows as it burns into nearby low areas. For this reason, appropriate action should be taken to segregate the molten material from combustible materials by constructing dams or dikes.

Firefighters responding to incidents involving elemental phosphorus should wear protective gear and use self-contained breathing apparatus. They should be particularly cautious to avoid inhaling the fumes from a phosphorus fire. Fumes contain particulates of the phosphorus oxides, which when inhaled, can seriously irritate the nose, mouth, throat, and lungs.

7.5-D RED PHOSPHORUS

The red allotrope of phosphorus is a dark red solid consisting of molecules containing numerous P_4 units. Its chemical formula is usually denoted as either P or P_x, where x is an undefined integer equal to or less than 4. Red phosphorus is produced industrially by heating white phosphorus at 482 °F (250 °C) in an iron container from which air has been excluded.

Red phosphorus is not nearly so chemically reactive as white phosphorus. Furthermore, it is neither poisonous nor *spontaneously* combustible in small quantities. Small quantities of red phosphorus may be stored in closed containers without the overlying protection of water or nitrogen to prevent its ignition.

Red phosphorus burns when it is exposed to an ignition source. The combustion produces a mixture of the two oxides, tetraphosphorus hexoxide and tetraphosphorus decoxide. Red phosphorus also combines with atmospheric oxygen in the *absence* of an ignition source, but the combustion reaction occurs very slowly and is highly exothermic. Consequently, the element is rarely stored in bulk, to prevent its self-ignition from the accumulated heat of reaction.

7.5-E TRANSPORTING RED PHOSPHORUS

When shippers intend to transport red phosphorus, DOT requires them to identify it on the accompanying shipping paper as follows:

UN1338, Phosphorus, amorphous, 4.1, PGIII

DOT also requires them to affix FLAMMABLE SOLID labels to red phosphorus containers.

When carriers transport 1001 lb (454 kg) or more of red phosphorus, DOT requires them to display FLAMMABLE SOLID placards on the bulk packaging or transport vehicle used for shipment. DOT also requires carriers to display the identification number, 1338, on orange panels, across the center area of the FLAMMABLE SOLID placards, or on white square-on-point panels.

7.5-F RESPONDING TO INCIDENTS INVOLVING A RELEASE OF RED PHOSPHORUS

Bulk quantities of red phosphorus are unlikely to be encountered, but containers holding small quantities have been involved in fires. These fires can be extinguished by the application of dry sand, foam, or dry chemicals. Because the toxic gas phosphine is produced, the use of wet sand is inadvisable for fire extinguishment.

PERFORMANCE GOALS FOR SECTION 7.6:

- Identify the principal properties of elemental sulfur, noting particularly the toxic nature of its combustion product.
- Demonstrate that elemental sulfur acts as a reducing agent in chemical reactions.
- Identify the industries that use bulk quantities of sulfur.
- Describe the response actions to be executed when sulfur has been released into the environment.

7.6 Sulfur

Elemental sulfur occurs naturally, particularly in countries bordering the Gulf of Mexico and in Japan, Mexico, and Italy. Sulfur also occurs in minerals and ores too numerous to mention, in which it is combined with metals and other nonmetals. Sulfur accounts for 0.06% by mass of all the elements found on Earth.

Most of the world's supply of elemental sulfur comes from deposits of the element called *brimstone*. These deposits are often located near hot springs and volcanoes. To isolate the sulfur, hot water under pressure is pumped into subterranean brimstone-bearing deposits, whereupon the sulfur melts and is brought to the surface by an airlift. Although brimstone has an offensive odor, pure sulfur is an odorless solid at room temperature.

Sulfur vaporizes when brimstone is heated at atmospheric pressure. When the vapor is allowed to condense on a cold surface, a finely divided powder of solid sulfur is produced, called **flowers of sulfur**. The powder is commercially used as a pesticide.

flowers of sulfur
■ The finely divided powder of elemental sulfur

The pure solid state of sulfur occurs in either of two allotropes called *orthorhombic sulfur* and *monoclinic sulfur*. Orthorhombic sulfur is the stable allotrope of solid sulfur at ambient conditions. It is a pale yellow, crystalline solid. Its physical properties are noted in Table 7.10. When orthorhombic sulfur is maintained at a temperature between 204.8 and 235.2 °F (96 and 112.8 °C), it changes into a mass of long transparent needles composed of monoclinic sulfur. But because monoclinic sulfur is not the stable allotrope, it slowly changes back into the orthorhombic form.

Both solid allotropes exist as molecules having eight sulfur atoms bonded together in a puckered-ring arrangement. Consequently, the chemical formula of solid sulfur is represented as S_8.

$$\begin{array}{ccc} & \overset{..}{S}-\overset{..}{S} & \\ :\!\overset{..}{S}: & & :\!\overset{..}{S}: \\ | & & | \\ :\!\overset{..}{S}: & & :\!\overset{..}{S}: \\ & \overset{..}{S}-\overset{..}{S} & \end{array}$$

However, molten sulfur has a highly complex molecular arrangement. Chemists denote it as S_x, where x is a small but undefined number. The chemical formula of sulfur vapor at its boiling

TABLE 7.10	Physical Properties of Elemental Sulfur (Orthorhombic)
Melting point	248 °F (120 °C)
Boiling point	832 °F (445 °C)
Specific gravity at 68 °F (20 °C)	2.07
Vapor density (air = 1)	8.9
Vapor pressure at 475 °F (246 °C)	10 mmHg
Flashpoint	405 °F (207 °C)
Autoignition point	450 °F (232 °C)
Lower explosion limit (as dust)[a]	0.002 lb/ft^3 (35 g/m^3)
Upper explosive limit (as dust)[a]	0.09 lb/ft^3 (1400 g/m^3)

[a]The lower and upper explosive limits of sulfur dust vary considerably with its particle size and dispersion.

point is also represented as S_8, but when the vapor is further heated, the cyclic arrangement breaks down, and the sulfur molecules assume the formula S_2.

Elemental sulfur burns with a blue flame. To initiate its burning, sulfur must first be heated to at least 405 °F (207 °C). The combustion produces sulfur dioxide, a poisonous gas having a suffocating, choking odor (Section 10.11).

$$S_8(s) \quad + \quad 8O_2(g) \quad \longrightarrow \quad 8SO_2(g)$$
$$\text{Sulfur} \qquad \text{Oxygen} \qquad \text{Sulfur dioxide}$$

Elemental sulfur is likely to spontaneously ignite under the following conditions:

■ When flowers of sulfur are dispersed into air, a potentially explosive mixture is produced. The spontaneous ignition of this mixture is triggered by the static electricity generated by the movement of the sulfur particles within the air. The potential for a dust explosion may be markedly reduced by electrically grounding the vessel in which the sulfur is confined.

■ Elemental sulfur reacts with many oxidizing agents. When they are activated, the resulting heat of reaction is likely to cause the ignition of the residual sulfur. Consequently, all mixtures of elemental sulfur and oxidizing agents pose the risk of fire and explosion. Every effort should be exercised to keep them segregated.

Elemental sulfur also combines with most metals. For example, when a mixture of mercury and iron is heated, the elements unite, forming mercury(II) sulfide and iron(II) sulfide, respectively.

$$8Hg(l) \quad + \quad S_8(s) \quad \longrightarrow \quad 8HgS(s)$$
$$\text{Mercury} \qquad \text{Sulfur} \qquad \text{Mercury(II) sulfide}$$

$$8Fe(s) \quad + \quad S_8(s) \quad \longrightarrow \quad 8FeS(s)$$
$$\text{Iron} \qquad \text{Sulfur} \qquad \text{Iron(II) sulfide}$$

7.6-A USES OF SULFUR

Sulfur is one of the world's most important raw materials. There is hardly a segment of the chemical industry that does not use elemental sulfur or one of its compounds during its manufacturing or production processes. Approximately 80% of the elemental sulfur produced in the United States is used as a raw material for the manufacture of sulfuric acid (Section 8.6). As Figure 7.14 illustrates,

FIGURE 7.14
Elemental sulfur is a constituent of many industrial and domestic products including gunpowder, matches, insecticides, fertilizers, and vulcanized rubber. When the sulfur burns, it is converted into the toxic gas sulfur dioxide.

sulfur is also used to produce vulcanized rubber products, fertilizers, dyes and other chemical substances, drugs and other pharmaceuticals, black gunpowder, fireworks, pesticides, and matches.

Sulfur is also used as a raw material by the chemical industry to produce of other commercial chemical products. For example, elemental sulfur is united with carbon and fluorine to produce carbon disulfide (Section 13.10) and sulfur hexafluoride, respectively:

■ Carbon disulfide is produced by passing sulfur vapor over very hot carbon in the absence of air. The presence of air is avoided, because carbon disulfide is a very flammable substance.

$$C(s) \quad + \quad S_2(g) \quad \longrightarrow \quad CS_2(g)$$
Carbon   Sulfur   Carbon disulfide

■ Sulfur hexafluoride is the predominant product formed when sulfur combines with fluorine.

$$S_8(s) \quad + \quad 24F_2(g) \quad \longrightarrow \quad 8SF_6(g)$$
Sulfur   Fluorine   Sulfur hexafluoride

It is used as an insulator in high-voltage electrical equipment.

7.6-B TRANSPORTING SULFUR

When shippers offer sulfur for domestic transportation, DOT requires them to identify it on the accompanying shipping paper as either of the following, as relevant:

> **NA1350, Sulfur, 9, PGIII**
> *or*
> **NA2448, Sulfur, molten, 9, PGIII**

DOT also requires them to affix CLASS 9 labels on containers.

However, DOT does not require carriers to display CLASS 9 placards on the bulk packaging or transport vehicle used for shipment. Instead, when they transport sulfur in bulk quantities, DOT requires them to display the relevant identification number, 1350 or 2448, on orange panels or on white square-on-point diamonds.

When molten sulfur is transported in bulk, DOT also requires the packaging to be marked MOLTEN SULFUR. The following is an acceptable example of these markings.

| MOLTEN SULFUR | | 2448 |

7.6-C RESPONDING TO INCIDENTS INVOLVING A RELEASE OF SULFUR

When emergency-response crews first arrive at a scene involving a release of sulfur, the element is generally burning. The use of water is effective as a fire extinguisher when it is applied as a fog. The fog extinguishes the fire by removing heat. Applying the water as a fog also avoids the buildup of steam under layers of the solid sulfur, which could later erupt and splatter the hot material.

Sulfur readily melts under normal fire conditions, thereby flowing into lower adjacent areas where it can initiate secondary fires. For this reason, appropriate action should be taken to segregate the molten material from combustible materials by constructing dams or dikes.

Firefighters responding to incidents involving elemental sulfur should wear protective gear and use self-contained breathing apparatus to avoid inhaling toxic sulfur dioxide.

PERFORMANCE GOALS FOR SECTION 7.7:

- Compare the principal properties of elemental carbon in its principal allotropic forms, graphite and diamond.
- Illustrate that carbon acts as a reducing agent in chemical reactions.
- Identify the industries that use bulk quantities of coal, coke, charcoal, and carbon black.
- Identify the hazardous nature of coal tar distillates and identify the industries that use them.
- Describe the response actions to be executed when coal, coke, and charcoal have been released into the environment.

7.7 Carbon

Elemental carbon occurs naturally as its two common allotropes, *graphite* and *diamond*.* It also occurs in, *coal*, *coke*, *charcoal*, and *carbon black*. In each instance, the elemental carbon is represented by its chemical symbol, C.

The natural abundance of carbon is only 0.08% by mass on Earth, yet carbon ranks much higher in terms of importance, since it is a constituent of the compounds needed by all living organisms for survival.

7.7-A COMMON ALLOTROPES OF CARBON

Although neither diamond nor graphite is a hazardous material, each is encountered so frequently that we note here some of their common characteristics.

Graphite and diamond possess substantially different physical properties, some of which are compared in Table 7.11. *Graphite* (the "lead" of pencils) is a black, greasy material, but diamond can be cut and polished to a crystalline, transparent luster. Although graphite is one of the softest known substances, diamond is the hardest substance found in nature.

The physical properties of graphite and diamond are linked to their crystalline structures. In the graphite structure shown in Figure 7.15(a), each carbon atom is bonded to other carbon atoms in planar hexagonal rings joined to one another in successive sheets. In the diamond structure

---

*In 1985, the first of several new forms of elemental carbon was discovered by vaporizing graphite with a laser. The first form possessed 60 carbon atoms per molecule, but subsequently, forms possessing 70 and more carbon atoms per molecule were identified. Collectively, they are known as *fullerenes*. Their properties are not discussed in this book.

The carbon allotrope having the formula C_{60} is called *buckminsterfullerene*, a name derived from the observation that its soccer-ball-shaped molecules resemble the celebrated geodesic domes designed by American architect and engineer R. Buckminster Fuller. C_{60} molecules are also called *buckyballs*. Each C_{60} molecule has 32 interlocking rings (20 hexagons and 12 pentagons).

TABLE 7.11 — Physical Properties of Graphite and Diamond

	GRAPHITE	DIAMOND
Melting point	6381 °F (3527 °C)	~6872 °F (~3800 °C)
Boiling point	6740 °F (3727 °C)	8726 °F (4830 °C)
Specific gravity at 68 °F (20 °C)	2.20–2.35	3.52
Vapor density (air = 1)	0.4	
Flashpoint	>200 °F (>93.3 °C)	
Autoignition point	1346 °F (730 °C)	

shown in Figure 7.15(b), each carbon atom is symmetrically bonded to four other carbon atoms in a tetrahedral arrangement.

The diamond structure is produced in nature when carbon is subjected to a pressure of approximately 0.8 million psi (5.6 million kPa) within the hot mantle of Earth, over 75 mi (120 km) below the surface. The best-grade diamonds have been found in ancient volcanic "pipes" in mines located in South Africa, Canada, Arkansas, and elsewhere.

For decades, scientists and engineers have attempted to produce diamonds in laboratories by replicating Nature's process. Although limited success was experienced, the diamonds produced at very high temperatures and pressures were small, low-grade *industrial diamonds*. In 2008, however, significant advances in producing gemstone-quality diamonds were reported by passing carbon vapor over diamond seeds inside a vacuum chamber at an approximate temperature of 2000 °F (1093 °C). The largest single-crystal diamond produced in this fashion is about 0.7 in. × 0.2 in. × 0.2 in. (1.8 cm × 0.5 cm × 0.5 cm), or 15 carats. By comparison, the largest mined diamond is the renowned *Culligan diamond*, which weighed 3000 carats before it was cut. (One carat = 200 mg.)

At moderate temperatures and pressures, graphite is the stable allotrope of carbon. On Earth's surface, diamond converts into graphite at an imperceptibly slow rate. However, when diamond is heated to approximately 1830 °F (1000 °C) in the absence of air, it rapidly converts into graphite.

$$C_{diamond} \longrightarrow C_{graphite}$$

FIGURE 7.15 The structures of two carbon allotropes, diamond and graphite. In the graphite structure shown in (a), planar hexagonal rings covalently bond to one another in successive sheets. In the diamond structure shown in (b), each carbon atom is covalently bonded to four carbon atoms in a tetrahedral arrangement.

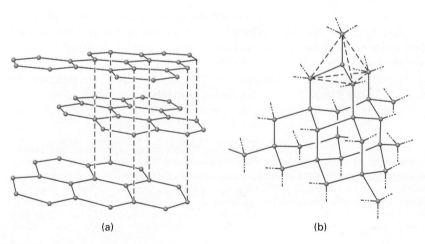

(a)   (b)

Although graphite and diamond are stable substances at the conditions experienced on Earth's surface, both burn in air. As we first noted in Section 5.6, carbon and the compounds that contain carbon are said to burn either "incompletely" or "completely." The product of the incomplete combustion is carbon monoxide, whereas the product of complete combustion is carbon dioxide.

Incomplete combustion: $2C(s) + O_2(g) \longrightarrow 2CO(g)$
                                      Carbon   Oxygen   Carbon monoxide

Complete combustion: $C(s) + O_2(g) \longrightarrow CO_2(g)$
                               Carbon   Oxygen   Carbon dioxide

7.7-B USES OF DIAMOND AND GRAPHITE

As virtually every civilized person knows, diamond companies advertise that a diamond is "a girl's best friend," the supreme token of love and affection. In many parts of the world, a woman's acceptance of a gemstone-quality diamond from her fiancé serves as a formal sign of their betrothal. Although gemstone-quality diamonds primarily serve in this exalted role, industrial diamonds now have more mundane uses in saw blades to cut marble, to prepare wire-drawing dies, and to produce drill bits, grinding wheels, and hacksaw blades.

Graphite and diamond undergo phase changes at exceptionally high temperatures. For example, Table 7.11 shows that diamond melts at 6872 °F (3800 °C). This melting point is the highest of all elements.

Graphite and diamond also possess an important anomalous property when compared with other nonmetals: They are extraordinarily good conductors of heat. If you can access a 5-carat or similar weight diamond, hold it to the tip of your tongue. The diamond feels cold, because it conducts the heat from your tongue. Diamond possesses the highest thermal conductivity of all substances.

The properties of graphite give rise to many industrial applications, of which the following are representative:

- Graphite is molded into crucibles that are used to hold molten steel and other high-melting metals.
- Graphite is used to line the walls of furnaces and other vessels where high-temperature operations are conducted. The graphite protects the underlying metal from melting or softening.
- The nose and leading edges of aircraft wings are generally coated with graphite to protect the underlying metal from the heat of friction generated when the aircraft travels at high speeds through the air.
- Graphite-based dry powder is an effective fire extinguisher for NFPA class D fires, because the carbon conducts heat away from the burning metal.

SOLVED EXERCISE 7.7

Which element of the fire tetrahedron illustrated in Figure 5.8 explains why graphite-based dry powder is an effective extinguisher of NFPA class D fires?

Solution: Graphite-based dry powder effectively extinguishes NFPA class D fires by conducting heat from the fire scene, thereby lowering the temperature of the burning material to a value less than its flashpoint. Thus, when graphite-based dry powder is used as a fire extinguisher, heat is the component of the fire tetrahedron whose removal results in extinguishment of the fire.

Carbon in the form of coke (Section 7.7-C) is also used in the chemical and metallurgical industries as a reducing agent. This chemical property is put to use during the production of iron and other metals from their naturally occurring ores. In Section 7.2-B, for example, we noted that coal or coke is used as a raw material to produce hydrogen by reacting it with steam; and in

Section 7.5-A, we observed that coke is used to produce elemental phosphorus from calcium phosphate rock.

7.7-C COAL, COKE, CHARCOAL, AND THE PRODUCTS MADE FROM THEM

Almost without exception, all forms of carbon found in nature can be traced to the giant plants that grew during the prehistoric period known as the *carboniferous age*. Millions of years ago, these plants grew more luxuriantly than they do today. As Earth evolved, the remains of these plants were ultimately buried at great depths below the planet's surface, where intense temperature and pressure converted them into coal. The process is called the **carbonization** of vegetable matter.

The forms of matter produced by the carbonization of vegetable matter—such as coal and crude petroleum (Section 12.14-A)—are fossil fuels (Section 5.6). In the United States, natural sources of coal are far more plentiful than natural sources of crude petroleum. Coal deposits occur primarily in West Virginia, Pennsylvania, Kentucky, and Wyoming. Fossil fuels are the major source of energy in the United States, but this fact may be problematic, since fossil fuels are nonrenewable natural resources, and once used, only their by-products remain.

The extent to which carbonization has occurred in nature determines the *rank* of a coal. Several major ranks are recognized, the most common of which are peat, lignite, subbituminous, bituminous, semianthracite, and anthracite. Each rank of coal in this listing is progressively older and denser than those whose names appear before it. Lignite and anthracite have specific gravities of 1.29 and 1.47, respectively. Given this wide range, these two ranks of coal are sometimes referred to as *soft coal* and *hard coal*, respectively.

Each rank also has a fixed carbon content. Lignite and anthracite have fixed carbon contents of approximately 70% and 90%, respectively. The carbon occurs as numerous compounds that become locked within the complex structure of coal. Those compounds that are volatile slowly evolve from coal as it is crushed and pulverized. They burn when their vapors are exposed to an ignition source. Before they ignite, however, a sufficient energy of activation must generally be provided to first vaporize and release them from the inner structure of coal. The evolved heat serves to vaporize more flammable vapor, which subsequently ignites. In this fashion, coal fires are self-sustaining until the flammable compounds in the coal have been entirely exhausted. After these flammable components have burned, a solid residue or ash generally remains. This ash consists of a mixture of the oxides of arsenic, barium, beryllium, boron, cadmium, chromium, thallium, selenium, molybdenum, mercury, and other elements. Although constituents of this mixture were formerly ejected as *flyash* into the atmosphere from the smokestacks of coal-fired plants, today's environmental regulations require its capture using air-pollution-control equipment. Flyash consists of the mixture of light, fine particulates captured from exhaust gases emitted by coal-fired plants.

In the United States, coal is used as the fuel at 460 power plants where electricity is generated. No federal regulations that control the fate of the ash generated by the burning of coal currently exist. Most of the combustion waste from coal-fired power plants is stored near the plants in large piles, reservoirs, or impoundments.

When coal is superheated in the absence of air within a simple closed assembly like that shown in Figure 7.16, certain volatile gases evolve. The mixture of these gases, called **coal gas**, contains ammonia, carbon monoxide, hydrogen sulfide, hydrogen cyanide, and methane, none of which condenses when exposed to the temperature of a cold water bath. Although coal gas was once used to heat and illuminate homes, its use is now obsolete.

The heating of coal also produces a viscous liquid called **coal tar**, a product used directly for waterproofing, as a roofing sealer, and as a coating on underground pipelines. It is also used as a binder during the construction of carbon electrodes for the production of elemental aluminum. In the chemical industry, coal tar is a raw material used to manufacture certain organic compounds including naphthalene (Section 12.18-A), anthracene, and cresols (Section 13.2-H).

carbonization
■ The process of converting an organic compound into carbon or a carbon-containing residue, typically conducted under intense temperature and pressure

coal gas
■ The flammable mixture of gases and vapors that form when coal is strongly heated in the absence of air

coal tar
■ The condensed black, viscous liquid produced by heating coal in the absence of air

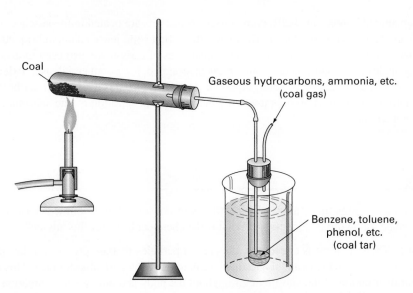

Coal

Gaseous hydrocarbons, ammonia, etc.
(coal gas)

Benzene, toluene,
phenol, etc.
(coal tar)

FIGURE 7.16 When coal is strongly heated in the absence of air, two flammable or combustible mixtures are isolated: coal gas and coal tar. When coal is exposed to an ignition source, these mixtures are generated. Their chemical components ignite and burn.

Coal tar may be subjected to a separation process called **distillation**, during which the coal tar is heated within specified temperature ranges to vaporize its components, which are then condensed to liquids. The resulting materials are called *coal tar distillates*. Although the latter term is loosely defined, the following four fractions of coal tar are commercially recognized:

- The "light" oil is so named because it floats on water. It boils near 392 °F (200 °C) and contains benzene (Section 12.12-A), toluene (Section 12.12-B), and other compounds.
- The "middle" oil boils between 392 and approximately 518 °F (200 and 270 °C). It contains naphthalene (Section 12.13-A), phenol (Section 13.2-H), and cresols (Section 13.2-I).
- The "heavy" oil boils between 518 and 662 °F (270 and 350 °C). It contains anthracene and other polynuclear aromatic hydrocarbons (Section 12.4-E). A commercially important product produced from this heavy coal tar distillate is called **creosote oil**. It is widely used to preserve railroad cross ties and utility poles against decay and to "cut" asphalt so it is suitable for application as a road and roofing tar.
- The residue remaining after coal tar is heated to 662 °F (350 °C) is called **coal tar pitch**. Products made from coal tar pitch are used chiefly as sealants and roofing and road-paving compounds.

The solid residue that remains when coal is heated in the absence of air is called *coke*. The heating process is conducted in industrial ovens. Coke is the form of carbon used by metallurgists to reduce the ores of arsenic, tin, copper, iron, zinc, phosphorus, and other elements. It is also used in the steel industry for reducing the iron oxide in iron ore in blast furnaces.

$$C(s) \ + \ FeO(s) \ \longrightarrow \ Fe(s) \ + \ CO(g)$$

Carbon   Iron(II) oxide   Iron   Carbon monoxide

This process is called **smelting**. A unique form of coke called *petroleum coke* is similarly produced when the residue from heating crude oil fractions (Section 12.14) is thermally treated.

distillation
- A physical process in which a substance or group of substances is separated from a mixture contained within a vessel by heating the mixture to a specified temperature or range of temperatures, thereby converting one or more constituents into a vapor that is subsequently condensed in another vessel

creosote oil
- An oily liquid obtained from the distillation of coal tar

coal tar pitch
- The residue remaining after coal is heated to approximately 662 °F (350 °C) in the absence of air

smelting
- A chemical process for isolating an element from its naturally occurring ore by reacting the ore with the constituent carbon in coke

adsorption

■ A physical phenomenon characterized by the adherence or occlusion of atoms, ions, or molecules of a gas or liquid to the surface of another substance

When animal bones, nut shells, corn cobs, or peach pits are heated in the absence of air, the product called *charcoal* is produced. Most individuals first experience charcoal briquettes as the fuel for backyard barbeques, but considerable quantities of carbon are also used industrially for adsorbing undesirable substances from products destined for commercial use. **Adsorption** refers to the surface retention of solid, liquid, or gaseous molecules and should not be confused with the term *absorption*, a process involving the physical penetration of one substance into the bulk of another one.

SOLVED EXERCISE 7.8

Anesthesiologists sometimes advise their patients to consume lightly burned toast during the recovery stage following major surgery. What is the chemical basis for this advice?

Solution: When the toast is consumed, the carbon particles that are present on the surface of the burned toast adsorb the residual anesthetic remaining in the patient's stomach. This prevents the further distribution of the anesthetic throughout the body and aids in its elimination through the gastrointestinal tract.

Charcoal is a very effective adsorbing agent. Because charcoal retains the skeletal cellular structure of the material from which it was made, it is highly porous. This porosity gives charcoal a very large surface area per unit weight. Charcoal can also be heated in the absence of air at 1472 to 1652 °F (800 to 900 °C) to produce a product called *activated charcoal*, or *activated carbon*. This material is often chosen as an adsorbing medium, since its average internal surface area is 284,000 ft^2/oz (929 m^2/g).

Large quantities of activated charcoal are industrially used as adsorbing agents. Activated charcoal is used to purify the atmospheric emissions from gas stacks by reducing or eliminating the concentration of undesirable toxic gases from the gas streams. It is also used in gas mask canisters and cigarette filter tips for the same purpose.

7.7-D CARBON BLACK

When petroleum products are vaporized and burned in a furnace with a limited amount of air, a finely divided form of carbon called *carbon black* is produced. Many grades of carbon black are available commercially, distinguished primarily by their particle size. They are used in a number of consumer products including inks, paints, plastics, and tires, belts, and other abrasion-resistant rubber products.

7.7-E CONSUMER PRODUCT REGULATIONS INVOLVING CHARCOAL

To inform the public that poisonous carbon monoxide is produced when burning charcoal, the U.S. Consumer Product Safety Commission requires charcoal manufacturers to affix the label shown in Figure 7.17 to charcoal packaging. In addition to the written warnings, the label contains a pictograph of a grill situated inside a tent, home, and vehicle. These drawings are enclosed in a circle with an X through it. They serve to convey the message that burning charcoal enclosed areas should be avoided, since deadly concentrations of carbon monoxide could accumulate.

7.7-F TRANSPORTING COAL TAR DISTILLATES, COKE, CHARCOAL, AND ACTIVATED CARBON

When shippers intend to transport coal tar distillates, DOT requires them to identify the distillates on the accompanying shipping paper. Depending on the flashpoint and boiling point of the distillate, the shipping description is either of the following, as relevant

⚠ WARNING	CARBON MONOXIDE HAZARD
	Burning charcoal inside can kill you. It gives off carbon monoxide, which has no odor.
	NEVER burn charcoal inside homes, vehicles or tents.

FIGURE 7.17 At 16 C.F.R. §1500.14(b)(6), the U.S. Consumer Product Safety Commission requires charcoal manufacturers to affix this label to the front and back panels of bags holding charcoal briquettes and other forms of charcoal that are intended for retail sale and use in cooking or heating.

> **UN1136, Coal tar distillates, flammable, 3, PGII**
> *or*
> **UN1136, Coal tar distillates, flammable, 3, PGIII**

DOT also requires shippers to affix FLAMMABLE LIQUID labels to their containers.

When carriers are transporting 1001 lb (454 kg) or more of these distillates, DOT requires them to display FLAMMABLE placards on the bulk packaging or transport vehicle used for shipment. DOT also requires the carriers to display the identification number, 1136, on orange panels, across the center area of the FLAMMABLE placards, or on white square-on-point panels.

When shippers offer hot coal tar pitch for transportation in a kettle or other bulk container as an elevated-temperature material, DOT requires them to display the HOT marking in Figure 6.10 on both sides and ends of the container.

DOT regulates the domestic and international transportation of charcoal and activated carbon. For domestic transportation, DOT requires shippers who offer charcoal for transportation to identify it on the accompanying shipping paper as follows:

> **NA1361, Charcoal, 4.2, PGIII**

Logically, DOT prohibits the transportation of hot coke.

DOT requires shippers who offer coke, carbon black, or activated carbon for international transportation to identify the proper shipping description on the accompanying shipping paper as one of the following, as relevant:

> **UN1361, Carbon (coke), 4.2, PGII**
> *or*
> **UN1361, Carbon (coke), 4.2, PGIII**
> **UN1361, Carbon (carbon black), 4.2, PGII**
> *or*
> **UN1361, Carbon (carbon black), 4.2, PGIII**
> **UN1362, Carbon, activated, 4.2, III**

DOT also requires them to affix SPONTANEOUSLY COMBUSTIBLE labels to the containers of these hazardous materials.

Carriers who transport 1001 lb (454 kg) or more of these forms of carbon are required to display SPONTANEOUSLY COMBUSTIBLE placards on the bulk packaging or transport vehicle used for shipment.

7.7-G RESPONDING TO INCIDENTS INVOLVING A RELEASE OF COAL, COKE, OR CHARCOAL

Most fires involving coal, coke, and charcoal can be effectively extinguished with water. When bulk quantities of these materials are burning, it is essential to use a deluging volume of water for the following reasons:

- When fires are extinguished only on the surfaces of coal, coke, and charcoal piles, combustion continues within the interior of these piles. Because the liberated heat cannot easily dissipate, the temperature of the entire bulk increases, and the coal, coke, or charcoal erupts into flame once again.

- When coal, coke, or charcoal are only lightly moistened with water, the carbon and steam chemically react to produce carbon monoxide and hydrogen.

$$C(s) \quad + \quad H_2O(g) \quad \longrightarrow \quad H_2(g) \quad + \quad CO(g)$$

Carbon   Water   Hydrogen   Carbon monoxide

This is the water gas mixture noted in Section 7.2. The ignition of the water gas serves to rekindle coal, coke, and charcoal fires.

The need to use a deluging volume of water on coal fires is underscored by the thousands of fires that are ongoing within abandoned coal mines throughout the world. For example, scientists estimate that Burning Mountain in Australia has been continuously burning for 6000 years. It constitutes the world's oldest known coal fire.

Fires burn downward in coal mines, acquiring oxygen from the air that passes through the fissures in surrounding rock. Although some attempts have been made to extinguish them, most are nearly impossible to access. Entering these old mines is highly dangerous. The fires cannot be saturated with a deluging volume of water, and it is also impossible to starve them of oxygen, since the burning coal is exposed to a new source of atmospheric oxygen almost continuously through the countless boreholes that were formerly driven into the mines to provide ventilation to the coal miners. For these reasons, these fires are only rarely extinguished and continue to quietly smolder for decades.

Oxygen

7.1 Complete and balance each of the following equations and provide an acceptable name for each product:

(a) $Mg(s) + O_2(g) \longrightarrow$
(b) $H_2(g) + O_2(g) \longrightarrow$
(c) $C(s) + O_2(g) \longrightarrow$

7.2 The OSHA regulation at 29 C.F.R. §1910.104(b)(2)(iii) requires owners of bulk oxygen systems to provide noncombustible surfacing in areas where liquid oxygen might leak during operation of the system or during the filling of a storage container. Are owners of bulk oxygen systems in compliance with this regulation if they provide an asphalt surface in areas on which oxygen could potentially leak?

Ozone

7.3. When ozone generators are used to treat smoke-damaged items following a residential fire, why is it essential for the treatment to be conducted while the home is unoccupied?

Hydrogen

7.4. Complete and balance each of the following equations, and provide an acceptable name for each product:

(a) $Li(l) + H_2(g) \longrightarrow$
(b) $C(s) + H_2(g) \longrightarrow$
(c) $S_8(s) + H_2(g) \longrightarrow$
(d) $FeO(s) + H_2(g) \longrightarrow$

7.5. The OSHA regulation at 29 C.F.R. §1910.103(b) requires employers to arrange safety-relief devices on all containers of hydrogen in the workplace, other than containers having a capacity of 2 ft^3 (0.06 m^3) or less, so they discharge upward into the open air in an unobstructed fashion and in such a manner as to prevent the impingement of escaping gas on its container, adjacent structures, or personnel.

(a) What is the most likely reason OSHA requires employers to arrange the safety-relief devices so they discharge *upward* into the open air?

(b) What is the most likely reason OSHA requires employers to arrange the safety-relief devices so the escaping hydrogen does not impinge on its container, adjacent structures, or personnel?

7.6. What is the most likely reason signs like the following are posted in areas where lead–acid storage batteries are periodically charged?

DANGER

NO SMOKING
BATTERY-CHARGING AREA

Fluorine

7.7 Complete and balance each of the following equations, and provide an acceptable name for each product:

(a) $Cu(s) + F_2(g) \longrightarrow$

(b) $H_2(g) + F_2(g) \longrightarrow$

(c) $P_4(s) + F_2(g) \longrightarrow$

Chlorine

7.8 Complete and balance each of the following equations, and provide an acceptable name for each product:

(a) $Fe(s) + Cl_2(g) \longrightarrow$

(b) $H_2(g) + Cl_2(g) \longrightarrow$

(c) $Al(s) + Cl_2(g) \longrightarrow$

7.9 The Department of Homeland Security (DHS) first established measures in 2007 to ensure the security from theft of chlorine cylinders that are stored at water treatment plants and other facilities that use chlorine. What is the most likely reason DHS took this action?

Phosphorus

7.10 Complete and balance each of the following equations, and provide an acceptable name for each product:

(a) $P_4(s) + O_2(g) \longrightarrow$

(b) $P_4(s) + Ca(s) \longrightarrow$

(c) $P_4(s) + Cl_2(g) \longrightarrow$

7.11 Why should gloves be worn while handling white phosphorus?

Sulfur

7.12 Complete and balance each of the following equations, and provide an acceptable name for each product:

(a) $C(s) + S_8(s) \longrightarrow$

(b) $Cl_2(g) + S_8(s) \longrightarrow$

(c) $Cu(s) + S_8(s) \longrightarrow$

7.13 When bulk quantities of sulfur are burning, why are nearby secondary fires likely to result?

Carbon

7.14 Complete and balance each of the following equations, and provide an acceptable name for each product:

(a) $C(s) + F_2(g) \longrightarrow$

(b) $C(s) + CuO(s) \longrightarrow$

(c) $C(s) + H_2O(g) \longrightarrow$

(d) $C(s) + SnO_2(s) \longrightarrow$

7.15 The DOT regulation at 49 C.F.R. §174.450 stipulates that when fire occurs in a domestic rail shipment of charcoal in transit, water should not be used if it is practicable to locate and remove the burning charcoal. Any charcoal that becomes wet while extinguishing a fire must be removed from the car and not reshipped, and the remainder of the charcoal must be held under observation in a dry place for at least 5 days before forwarding. What is the most likely reason DOT regulates the domestic transportation of charcoal in this fashion?

7.16 The flashpoint of creosote oil is 165 °F (74 °C). When 5000 lb (2270 kg) of creosote oil is transported at ambient temperature by means of a motorized cargo tank, which placards does DOT require the carrier to post on the vehicle?

Chemistry of Some Corrosive Materials

I n everyday practice, we say that a corrosive material is a substance that "eats into" and destroys the chemical character of metals, minerals, and body tissues. Government regulations, however, provide more specificity. For example, the DOT regulations define a **corrosive material** as a liquid or solid that causes full-thickness destruction of human skin at the site of contact within a specified period, or a liquid that chemically reacts with steel or aluminum surfaces at a rate exceeding 0.25 in./y (6.25 mm/y) at a test temperature of 130 °F (54 °C) when measured in accordance with prescribed testing procedures.

The most common examples of corrosive materials are acids. Their unique properties and the manner by which they corrode matter are the principal subjects of this chapter.

corrosive material
■ For purposes of DOT regulations, a liquid or solid that causes full-thickness destruction of human skin at the site of contact within a specified period; or a liquid that chemically reacts with steel or aluminum surfaces at a rate exceeding 0.25 in./y (6.25 mm/y) at a test temperature of 130 °F (54 °C) when measured in accordance with prescribed testing procedures

- Describe Arrhenius's theory of acids and bases.
- Distinguish between strong and weak acids, and strong and weak bases.
- Distinguish between mineral acids and organic acids, and name the common acids in each category.
- Distinguish between oxidizing and nonoxidizing acids, and name the common acids in each category.
- Distinguish between diluted and concentrated acids.
- Show that the specific strength of an acid or base solution is connected to its degree of corrosiveness.

8.1 The Nature of Acids and Bases

Several theories have been proposed to account for the properties of acids and bases, but for simplicity's sake, we use only one of them in this book. In 1887, Swedish chemist Svante Arrhenius first advocated a theory that has been modernized here to reflect current scientific knowledge. Arrhenius proposed that an **acid** is any substance that generates hydrogen ions (H^+) when dissolved in water. Hydrogen ions are nothing more than hydrogen atoms stripped of their electrons. A single hydrogen ion is the same as the nucleus of a hydrogen atom, or a single proton.

acid
- A compound that forms hydrated hydrogen ions, $H^+(aq)$, when dissolved in water

Today, chemists know that free hydrogen ions cannot exist alone in aqueous solution owing to their high charge density. Instead, they rapidly become "solvated"; that is, they bond loosely to water molecules. These solvated hydrogen ions are very complex in the manner by which they interact; we collectively represent them by the notation $H^+(aq)$.

The ionization of all acids can be represented by Arrhenius's theory. For instance, when hydrogen chloride dissolves in water, hydrogen ions and chloride ions are generated.

$$HCl(aq) \longrightarrow H^+(aq) + Cl^-(aq)$$
$$\text{Hydrogen chloride} \qquad \text{Hydrogen ion} \qquad \text{Chloride ion}$$

This solution of hydrogen chloride in water is called *hydrochloric acid*. Its chemical formula is $HCl(aq)$.

Arrhenius further proposed that a **base** is a substance that produces hydroxide ions (OH^-) when it is dissolved in water. For example, when sodium hydroxide dissolves in water, solvated sodium and hydroxide ions are generated. We represent them as $Na^+(aq)$ and $OH^-(aq)$, respectively.

base
- A compound whose units produce hydrated hydroxide ions, $OH^-(aq)$, when dissolved in water

$$NaOH(aq) \longrightarrow Na^+(aq) + OH^-(aq)$$
$$\text{Sodium hydroxide} \qquad \text{Sodium ion} \qquad \text{Hydroxide ion}$$

The solution of sodium hydroxide in water is represented as $NaOH(aq)$.

8.1-A STRONG AND WEAK ACIDS AND BASES

It is the ability of acids and bases to form ions in water that gives rise to their corrosive nature. The relative strength of an acid or base refers to the tendency of an individual substance to form hydrated hydrogen ions and hydroxide ions, respectively, when dissolved in water.

Those acids and bases that yield a relatively high concentration in water of hydrated hydrogen and hydroxide ions are called **strong acids** and **strong bases**, respectively. For instance, hydrochloric acid is an example of a strong acid, because the hydrogen chloride almost completely ionizes in water. Likewise, sodium hydroxide is an example of a strong base, since it almost completely ionizes in water.

strong acid
- Any acid that predominantly forms hydrated hydrogen ions when dissolved in water

strong base
- Any base that predominantly forms hydrated hydroxide ions when dissolved in water

In contrast, some substances essentially retain their unit formulas when they are dissolved in water. Such substances yield relatively low concentrations of hydrogen or hydroxide ions and are called **weak acids** and **weak bases**, respectively. Acetic acid is an example of a weak acid, because it primarily exists as molecules of acetic acid when it is dissolved in water, although some hydrogen and acetate ions are also produced. Ammonium hydroxide is an example of a weak base. When ammonia is dissolved in water, it continues to exist primarily as molecular ammonia and does not appreciably form ammonium and hydroxide ions.

Each acid in the group listed in Table 8.1 is ranked as a strong or a weak acid. Phosphoric acid is regarded as a moderately strong acid, whereas the acids above and below phosphoric acid in this listing are considered strong acids and weak acids, respectively.

8.1-B MINERAL ACIDS AND ORGANIC ACIDS

Acids may be classified as either mineral acids or organic acids. **Mineral acids** consist of molecules having atoms of hydrogen, an identifying nonmetal like chlorine, sulfur, or phosphorus, and sometimes oxygen. Minerals acids are most likely so named because they were initially produced from minerals existing in naturally occurring ores.

Organic acids, or **carboxylic acids**, are substances whose molecules possess carbon, hydrogen, and oxygen atoms only. All organic acids have molecular structures that contain at least one of the following group of atoms:

$$-\overset{\displaystyle O}{\overset{\displaystyle \|}{C}}\diagdown_{\displaystyle OH}$$

TABLE 8.1	Relative Strengths of Some Common Acids in Water
NAME OF ACID	**CHEMICAL FORMULA**
Perchloric acid	$HClO_4(aq)$
Sulfuric acid	$H_2SO_4(aq)$
Hydrochloric acid	$HCl(aq)$
Nitric acid	$HNO_3(aq)$
Phosphoric acid	$H_3PO_4(aq)$
Nitrous acid	$HNO_2(aq)$
Hydrofluoric acid	$HF(aq)$
Acetic acid	$CH_3COOH(aq)$
Carbonic acid[a]	$CO_2(aq)$
Hydrocyanic acid	$HCN(aq)$
Boric acid	$H_3BO_3(aq)$

[a]The chemical formula of carbonic acid sometimes appears in the literature as $H_2CO_3(aq)$. However, an acid having this chemical composition has never been isolated or identified. The chemical formula of carbonic acid is correctly denoted as $CO_2(aq)$.

This group of atoms is characteristic of organic acids and is called the **carboxyl group**. Acetic acid is the only acid noted in this chapter that is an organic acid. The others are mineral acids.

8.1-C OXIDIZING AND NONOXIDIZING ACIDS

An acid chemically reacts as either an oxidizing acid or a nonoxidizing acid. An **oxidizing acid** is any acid that participates in a chemical reaction as an oxidizing agent. In Section 5.4, we noted that an oxidizing agent is a substance that causes oxidation, sometimes by taking electrons from another substance. Nitric acid is an example of an oxidizing acid. We note in Section 8.7-A that it oxidizes metallic copper, zinc, and other substances by taking electrons from them.

A **nonoxidizing acid** is any acid that participates in a chemical reaction by some means other than oxidation. Hydrochloric acid is an example of a nonoxidizing acid, since its reactions do not involve the taking of electrons from other substances.

Hot sulfuric acid, nitric acid, and perchloric acid are oxidizing acids, but hydrochloric acid, hydrofluoric acid, phosphoric acid, and acetic acid are nonoxidizing acids. The degree to which the oxidizing acids corrode depends on how powerfully they participate as oxidizing agents. We note their characteristics when we examine their individual properties.

8.1-D CONCENTRATED AND DILUTED ACIDS

Chemists often refer to an acid as either diluted or concentrated. When they refer to a **concentrated acid**, chemists generally mean the commercially available acid that has the greatest concentration. When referring to a **diluted acid**, they mean a solution produced by adding water to the concentrated acid.

This is not meant to imply that water is absent from a concentrated acid. A concentrated acid may actually contain some amount of water. For example, concentrated hydrochloric acid consists of approximately 36% to 38% hydrogen chloride by mass in water; 62% to 64% of the solution is water. Diluted hydrochloric acid is any solution that results when additional water is added to this concentrated acid. Although concentrated hydrochloric acid is a corrosive material, diluted hydrochloric acid may or may not exhibit corrosiveness depending on its strength. When very highly diluted with water, all acids lose their corrosive character.

PERFORMANCE GOALS FOR SECTION 8.2:

- Describe the format of the pH scale.
- Identify the ranges of the pH scale at which the degree of corrosiveness of a solution is greatest.

8.2 The pH Scale

The **pH** is a number that ranges from 0 to 14 and denotes the acidity or alkalinity of an aqueous solution. Aqueous solutions having a pH in the range from 0 to 7 are said to be **acidic**. This range of values is a measure of the hydrogen ion concentration. Let's consider a simple example. In pure water, there is a very small hydrogen ion concentration derived by the dissociation of the water molecules. This is represented as follows:

$$H_2O(l) \longrightarrow H^+(aq) + OH^-(aq)$$

| Water | Hydrogen ion | Hydroxide ion |

When the hydrogen ion concentration in pure water is determined experimentally, we find that it is only 0.0000001 mol/L, or 10^{-7} mol/L. Instead of writing all these zeros, we express this

carboxyl group
- The $-\overset{\displaystyle O}{\underset{\displaystyle OH}{\overset{\|}{C}}}$ functional group, common to carboxylic acids

oxidizing acid
- An acid capable of reacting as an oxidizing agent

nonoxidizing acid
- An acid incapable of reacting as an oxidizing agent

concentrated acid
- Pertaining usually to its commercially available form that contains the greatest concentration of the substance

diluted acid
- Any form of an acid that has been produced by mixture with water

pH
- A numerical scale from 0 to 14 used to quantify the acidity of alkalinity of a solution with neutrality indicated as 7

acidic
- The property of aqueous solutions that have a pH ranging from 0 to 7; the property of any substance that corrodes steel or destroys tissue at the site of contact

concentration by indicating that the pH equals 7; that is, the pH is the degree of the negative exponent. When the hydrogen ion concentration of an aqueous solution is 0.01 mol/L, or 10^{-2} mol/L, the pH of the solution equals 2.

Aqueous solutions having a pH ranging from 7 to 14 are referred to as **basic**, **caustic**, or **alkaline**. Because water is neither acidic nor basic, the pH of pure water is 7. Table 8.2 lists the pH values of some common solutions and mixtures.

A unit change in a pH value represents a 10-fold difference in the hydrogen ion concentration of a solution, and a difference of two pH units represents a 100-fold difference. This means that the hydrogen ion concentration of an aqueous solution having a pH of 4 is 100 times greater than the hydrogen ion concentration of a solution having a pH of 6; and it is 1000 times greater than one having a pH equal of 7. In everyday practice, we say that a solution having a pH of 4 is 100 times more acidic than a solution having a pH of 6 and 1000 times more acidic than a solution having a pH of 7. Also, a solution with a pH of 8 is 10 times as alkaline as one with a pH of 7; a solution having a pH of 9 is 10 times as alkaline as one having a pH of 8 and 100 times as alkaline as one having a pH of 7, and so on.

The pH is frequently determined by means of an electrometric apparatus called a *pH meter*. The pH meter in Figure 8.1 is a voltage-measuring device that is connected to an electrode whose tip is immersed in a solution. The pH of the solution is readily determined by simply reading the display monitor.

basic (caustic)
■ The property of aqueous solutions that have a pH ranging from 8 to 14

alkaline
■ Basic, as opposed to acidic; an aqueous solution or other liquid whose pH is greater than 7

TABLE 8.2	The pH Values of Some Commonly Encountered Solutions and Mixtures

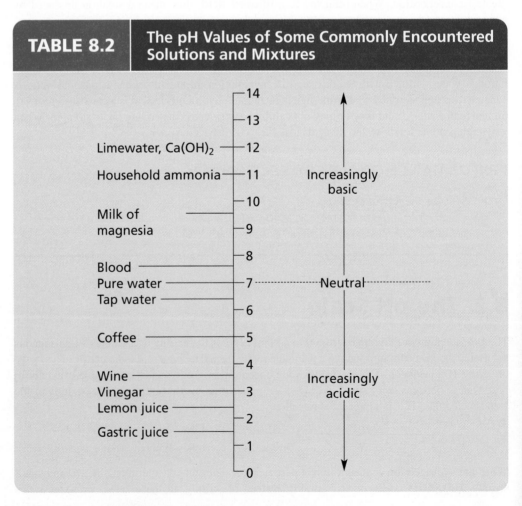

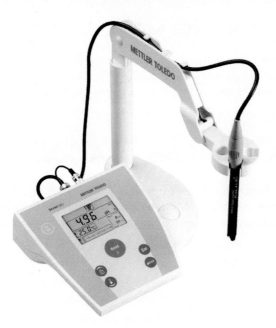

FIGURE 8.1 A pH meter is used for measuring the pH of an aqueous solution. The electrode is immersed into a sample of the solution, and its pH is read on the display monitor as 4.96. (*Courtesy of Thermo Fisher Scientific, Inc., Pittsburgh, Pennsylvania.*)

PERFORMANCE GOALS FOR SECTION 8.3:

- Identify the common properties of acids and bases.
- Describe the chemical reaction that results in the production of a salt.

8.3 Properties of Acids and Bases

All acids are associated with certain common properties. They taste sour; they cause indicator dyes to change to identifiable colors; and they react with bases to form salts and water. In contrast, bases taste bitter; they feel slippery; they also cause the colors of indicator dyes to change to identifiable colors; and they react with acids to form salts and water.

Acids and bases can be readily differentiated from one another by the colors they impart to pieces of litmus paper. Litmus is a common indicator dye derived from certain *lichens*, any of a group of mosslike plants. A solution of litmus is used to impregnate strips of paper that are subsequently dried. Individual strips are moistened with an aqueous solution of an acid or a base. Acids turn the litmus paper red, and bases turn the litmus paper blue.

When acids and bases chemically interact, they neutralize each other. An example of a neutralization reaction is represented by the following equation:

$$\underset{\text{Hydrochloric acid}}{\text{HCl}(aq)} + \underset{\text{Sodium hydroxide}}{\text{NaOH}(aq)} \longrightarrow \underset{\text{Sodium chloride}}{\text{NaCl}(aq)} + \underset{\text{Water}}{\text{H}_2\text{O}(l)}$$

The compound whose formula is listed to the immediate right of the arrow is called a **salt**. Any compound in which the hydrogen in an acid has been replaced by a metallic ion is a salt. NaCl is the chemical formula of sodium chloride, which we commonly know as ordinary table salt. It has properties that are dissimilar from those of both hydrochloric acid and sodium hydroxide.

salt
■ A compound in which the hydrogen ion from an acid has been substituted with a metallic ion

PERFORMANCE GOALS FOR SECTION 8.4:

■ Define the terms *acidic anhydride* and *basic anhydride*.
■ Identify the manner by which some acids and bases are produced by the reactions of acidic anhydrides and basic anhydrides, respectively, with water.

8.4 The Anhydrides of Acids and Bases

When a metallic or nonmetallic element combines with oxygen, the compounds produced are called **metallic oxides** and **nonmetallic oxides**, respectively. Sodium and calcium are two examples of metals that burn in oxygen to form their corresponding metallic oxides.

$$4Na(s) \ + \ O_2(g) \ \longrightarrow \ 2Na_2O(s)$$
Sodium   Oxygen   Sodium oxide

$$2Ca(s) \ + \ O_2(g) \ \longrightarrow \ 2CaO(s)$$
Calcium   Oxygen   Calcium oxide

Sulfur and phosphorus are two examples of nonmetals that burn in oxygen to form their corresponding nonmetallic oxides.

$$S_8(s) \ + \ 8O_2(g) \ \longrightarrow \ 8SO_2(g)$$
Sulfur   Oxygen   Sulfur dioxide

$$P_4(s) \ + \ 3O_2(g) \ \longrightarrow \ P_4O_6(s)$$
Phosphorus   Oxygen   Tetraphosphorus hexoxide

A metallic oxide reacts with water to produce a base.

$$Na_2O(s) \ + \ H_2O(l) \ \longrightarrow \ 2NaOH(aq)$$
Sodium oxide   Water   Sodium hydroxide

$$CaO(s) \ + \ H_2O(l) \ \longrightarrow \ Ca(OH)_2(aq)$$
Calcium oxide   Water   Calcium hydroxide

A nonmetallic oxide reacts with water to produce an acid. These combination reactions are represented by the following equations:

$$SO_2(g) \ + \ H_2O(l) \ \longrightarrow \ H_2SO_3(aq)$$
Sulfur dioxide   Water   Sulfurous acid

$$P_4O_6(s) \ + \ 6H_2O(l) \ \longrightarrow \ 4H_3PO_3(aq)$$
Tetraphosphorus hexoxide   Water   Phosphorous acid

anhydride
■ A metallic oxide or nonmetallic oxide

basic anhydride
■ A metallic oxide that chemically combines with water to produce a base

acidic anhydride
■ A nonmetallic oxide that chemically combines with water to produce an acid

The word **anhydride** refers to a substance from which the elements of water have been extracted. It is either a metallic or nonmetallic oxide. A metallic oxide that combines with water to produce a base is called a **basic anhydride**, and a nonmetallic oxide that combines with water to produce an acid is called an **acidic anhydride**. Sodium oxide and calcium oxide are examples of basic anhydrides, and sulfur dioxide and tetraphosphorus hexoxide are examples of acidic anhydrides. Some common acids and bases and their respective acidic anhydrides and basic anhydrides are listed in Table 8.3.

SOLVED EXERCISE 8.1

Write equations for the following chemical phenomena:

(a) Bubbles of hydrogen are generated when small chunks of metallic calcium are dropped into an aqueous solution of hydrochloric acid.

(b) A surface coating of zinc oxide is removed by an aqueous solution of sulfuric acid.

(c) Bubbles of carbon dioxide are generated when potassium carbonate is mixed into an aqueous solution of phosphoric acid.

Solution: The phenomena in (a), (b), and (c) are representative of the chemical reactions of acids with metals, metallic oxides, and metallic carbonates, respectively.

(a) Acids react with common metals other than copper, silver, gold, and mercury to produce hydrogen and a salt of the metal. Consequently, metallic calcium reacts with hydrochloric acid to produce hydrogen and calcium chloride according to the following equation:

$$Ca(s) \ + \ 2HCl(aq) \ \longrightarrow \ CaCl_2(aq) \ + \ 2H_2(g)$$

Calcium   Hydrochloric acid   Calcium chloride   Hydrogen

(b) Acids react with metallic oxides to produce water and a salt of the metal. Consequently, zinc oxide reacts with sulfuric acid to produce zinc sulfate and water.

$$ZnO(s) \ + \ H_2SO_4(aq) \ \longrightarrow \ ZnSO_4(aq) \ + \ H_2O(l)$$

Zinc oxide   Sulfuric acid   Zinc sulfate   Water

(c) Acids react with metallic carbonates to produce carbon dioxide, water, and a salt of the metal. Consequently, the reaction between potassium carbonate and phosphoric acid produces carbon dioxide, water, and potassium phosphate.

$$3K_2CO_3(s) \ + \ 2H_3PO_4(aq) \ \longrightarrow \ 3CO_2(g) \ + \ 3H_2O(l) \ + \ 2K_3PO_4(aq)$$

Potassium carbonate   Phosphoric acid   Carbon dioxide   Water   Potassium phosphate

TABLE 8.3	Acidic Anhydrides and Basic Anhydrides		
ANHYDRIDE	**CHEMICAL FORMULA**	**ACID/BASE**	**CHEMICAL FORMULA**
Dinitrogen trioxide	$N_2O_3(g)$	Nitrous acid	$HNO_2(aq)$
Dinitrogen pentoxide	$N_2O_5(g)$	Nitric acid	$HNO_3(aq)$
Sulfur dioxide	$SO_2(g)$	Sulfurous acid	$H_2SO_3(aq)$
Sulfur trioxide	$SO_3(g)$	Sulfuric acid	$H_2SO_4(aq)$
Tetraphosphorus hexoxide	$P_4O_6(s)$	Phosphorous acid	$H_3PO_3(aq)$
Tetraphosphorus decoxide	$P_4O_{10}(s)$	Phosphoric acid	$H_3PO_4(aq)$
Calcium oxide	$CaO(s)$	Calcium hydroxide	$Ca(OH)_2(aq)$
Magnesium oxide	$MgO(s)$	Magnesium hydroxide	$Mg(OH)_2(aq)$
Potassium oxide	$K_2O(s)$	Potassium hydroxide	$KOH(aq)$
Sodium oxide	$Na_2O(s)$	Sodium hydroxide	$NaOH(aq)$

- Describe the manner by which acids react with metals, metallic oxides, metallic carbonates, and skin tissue.
- Describe the general nature of the corrosive effects resulting from the contact of acids and bases with human skin tissue.
- Identify the manner by which bases react with metallic aluminum, lead, and zinc.

8.5 Acids and Bases as Corrosive Materials

Acids act as corrosive materials by reacting with metals, metallic oxides, metallic carbonates, and skin tissue. Bases act as corrosive materials by reacting with metals and skin tissue. We consider these phenomena separately.

8.5-A REACTIONS OF ACIDS AND METALS

Diluted acids react with all the commonly encountered metals other than copper, silver, gold, and mercury. These are simple displacement reactions that produce hydrogen and a salt of the metal. Two examples of such reactions are illustrated by the following equations:

$$Mg(s) \ + \ 2HCl(aq) \ \longrightarrow \ MgCl_2(aq) \ + \ H_2(g)$$

Magnesium   Hydrochloric acid   Magnesium chloride   Hydrogen

$$2Al(s) \ + \ 3H_2SO_4(aq) \ \longrightarrow \ Al_2(SO_4)_3(aq) \ + \ 3H_2(g)$$

Aluminum   Sulfuric acid   Aluminum sulfate   Hydrogen

Because acids and metals are chemically incompatible, it is unsafe to store acids in metal containers. This is why nonbulk quantities of acids are usually stored in glass or plastic containers.

8.5-B REACTIONS OF ACIDS AND METALLIC OXIDES

Acids react with metallic oxides to form a salt of the metal and water. Examples of this type of double-displacement reaction are illustrated by the following equations:

$$FeO(s) \ + \ 2HCl(aq) \ \longrightarrow \ FeCl_2(aq) \ + \ H_2O(l)$$

Iron(II) oxide   Hydrochloric acid   Iron(II) chloride   Water

$$Al_2O_3(s) \ + \ 6HNO_3(aq) \ \longrightarrow \ 2Al(NO_3)_3(aq) \ + \ 3H_2O(l)$$

Aluminum oxide   Nitric acid   Aluminum nitrate   Water

pickling
■ The combination of chemical reactions associated with removing surface impurities from metals by dipping them into an acid bath

An acid is sometimes beneficially used to remove metallic oxides and other impurities from the surface of metals. When used in this fashion, the acid is commonly called *pickle liquor,* and the associated phenomenon is called **pickling**. The steel industry uses large volumes of pickle liquor during the manufacture of such products as wire, rod, nuts, and bolts. The common acids used in pickle liquors are sulfuric acid, hydrochloric acid, and phosphoric acid.

SOLVED EXERCISE 8.2

Why are sulfur dioxide and sulfur trioxide the respective acidic anhydrides of sulfurous acid and sulfuric acid?

Solution: An acidic anhydride is a compound derived from an acid by elimination of its water. Sulfur dioxide and sulfur trioxide are the acidic anhydrides of sulfurous acid and sulfuric acid, respectively, since they are produced when water is lost from these acids.

$$H_2SO_3(aq) \longrightarrow SO_2(g) + H_2O(l)$$
$$\text{Sulfurous acid} \qquad \text{Sulfur dioxide} \quad \text{Water}$$

$$H_2SO_4(aq) \longrightarrow SO_3(g) + H_2O(l)$$
$$\text{Sulfuric acid} \qquad \text{Sulfur trioxide} \quad \text{Water}$$

8.5-C REACTIONS OF ACIDS AND METALLIC CARBONATES

Acids react with metallic carbonates to produce carbon dioxide, water, and a salt of the metal. Examples of these reactions are illusted by the following equations:

$$CaCO_3(s) + 2HCl(aq) \longrightarrow CaCl_2(aq) + CO_2(g) + H_2O(l)$$
$$\text{Calcium carbonate} \quad \text{Hydrochloric acid} \qquad \text{Calcium chloride} \quad \text{Carbon dioxide} \quad \text{Water}$$

$$ZnCO_3(s) + H_2SO_4(aq) \longrightarrow ZnSO_4(aq) + CO_2(g) + H_2O(l)$$
$$\text{Zinc carbonate} \qquad \text{Sulfuric acid} \qquad \text{Zinc sulfate} \qquad \text{Carbon dioxide} \quad \text{Water}$$

8.5-D REACTIONS OF ACIDS WITH SKIN TISSUE

The nature of the corrosive effect caused by prolonged exposure of skin tissue to an acid depends on the concentration of the acid. When the skin contacts a diluted acid, the site of contact may appear only reddened, whereas exposure to a concentrated acid for the same duration could cause the skin to blister. In the worst incidents, prolonged exposure of the skin to a concentrated acid causes irreversible tissue damage and permanent disfigurement at the site of contact. In either situation, the skin tissue is said to be "burned," since its appearance visually resembles a thermal burn.

8.5-E REACTIONS OF BASES AND METALS

Three common metals react with the concentrated solutions of strong bases, namely, aluminum, zinc, and lead. During the chemical reactions, hydrogen and a complex compound of the metal are produced. The following equations illustrate the chemical behavior of these three metals with a concentrated solution of sodium hydroxide:

$$2Al(s) + 6NaOH(aq) \longrightarrow 2Na_3AlO_3(aq) + 3H_2(g)$$
$$\text{Aluminum} \quad \text{Sodium hydroxide} \qquad \text{Sodium aluminate} \qquad \text{Hydrogen}$$

$$Zn(s) + 2NaOH(aq) \longrightarrow Na_2ZnO_2(aq) + H_2(g)$$
$$\text{Zinc} \quad \text{Sodium hydroxide} \qquad \text{Sodium zincate} \qquad \text{Hydrogen}$$

$$Pb(s) + 2NaOH(aq) \longrightarrow Na_2PbO_2(aq) + H_2(g)$$
$$\text{Lead} \quad \text{Sodium hydroxide} \qquad \text{Sodium plumbite} \qquad \text{Hydrogen}$$

8.5-F REACTION OF BASES WITH SKIN TISSUE

Aqueous solutions of bases corrode skin tissue in a fashion that is associated with the concentration of the base. When skin is exposed to a diluted solution, it appears reddened at the site of contact. Wounds of this type heal rapidly. However, when the skin has been exposed to a more concentrated solution of the same base, it changes the skin's texture into a thick, sticky material. Prolonged exposure causes the development of deep wounds that are very slow to heal. In this instance, exposure to the base causes irreversible tissue damage and permanent disfigurement at the site of contact.

   Exposure of the eyes to solutions of caustic substances causes injurious changes in the structure of the cornea, ultimately leading to complete opacification (clouding).

TABLE 8.4	Physical Properties of Concentrated Sulfuric Acid
Melting point	50 °F (10 °C)
Boiling point	640 °F (338 °C)
Specific gravity at 68 °F (20 °C)	1.84
Vapor density (air = 1)	2.8
Vapor pressure at 68 °F (20 °C)	<0.001 mmHg
Solubility in water	Infinitely soluble

PERFORMANCE GOALS FOR SECTION 8.6:

- Identify the properties of sulfuric acid.
- Identify the industries that use sulfuric acid.
- Describe the chemical nature of oleum.
- Identify the principal hazards associated with concentrated sulfuric acid and oleum.
- Describe the means shippers and carriers use to inform emergency responders of the hazards associated with encountering sulfuric acid or oleum during a transportation mishap.

8.6 Sulfuric Acid

Sulfuric acid is a mineral acid whose chemical formula is H_2SO_4. It is an odorless, colorless, oily liquid having a density approximately twice that of water. Impure or spent sulfuric acid is brown to black in color. An industrial grade of sulfuric acid is sometimes encountered as a clear-to-brownish-colored liquid. The brown color reflects a lower degree of purity. Some of the important physical properties of sulfuric acid are noted in Table 8.4.

If we were to list chemical substances by the amounts produced and consumed annually within the United States, we would find sulfuric acid near the top of these lists for the past five decades. In the United States alone, well over 40 million tn (36 million t) of sulfuric acid are produced annually. Sulfuric acid is so important commercially that its production and consumption rates have been used by economists to estimate the extent to which a country has industrialized.

The layperson is generally aware of sulfuric acid through its use as the electrolyte in the lead–acid storage battery shown in Figure 8.2. As noted in Section 7.2-F, this battery is the electrical

FIGURE 8.2 A lead–acid storage battery contains a group of individual cells, each of which consists of a lead plate and a lead oxide plate immersed in an aqueous solution of sulfuric acid. The specific gravity of the sulfuric acid ranges from 1.25 to 1.30. It serves as an electrolyte.

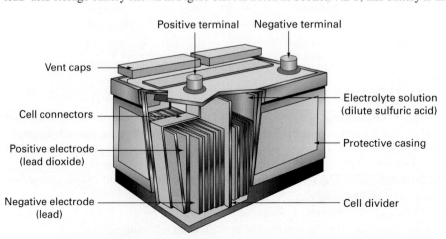

Positive terminal   Negative terminal

Vent caps

Cell connectors

Positive electrode (lead dioxide)

Negative electrode (lead)

Electrolyte solution (dilute sulfuric acid)

Protective casing

Cell divider

source in virtually all motor vehicles. However, sulfuric acid has countless other uses. In the chemical industry alone, sulfuric acid is used to manufacture explosives, fertilizers, and dozens of other compounds including other acids. Given this widespread usage, sulfuric acid has been called the workhorse of the industrial world. The popularity of its use implies that sulfuric acid is likely to be encountered more frequently than other corrosive materials when responding to emergencies involving hazardous materials.

Sulfuric acid is prepared at industrial plants by first burning sulfur to produce sulfur dioxide, and then oxidizing the sulfur dioxide further in the presence of a catalyst to produce sulfur trioxide. This is the acidic anhydride of sulfuric acid. It unites with water to form sulfuric acid.

$$S_8(s) \quad + \quad 8O_2(g) \quad \longrightarrow \quad 8SO_2(g)$$
$$\text{Sulfur} \qquad\quad \text{Oxygen} \qquad\qquad \text{Sulfur dioxide}$$

$$2SO_2(g) \quad + \quad O_2(g) \quad \longrightarrow \quad 2SO_3(g)$$
$$\text{Sulfur dioxide} \qquad \text{Oxygen} \qquad\qquad \text{Sulfur trioxide}$$

$$H_2O(l) \quad + \quad SO_3(g) \quad \longrightarrow \quad H_2SO_4(l)$$
$$\text{Water} \qquad \text{Sulfur trioxide} \qquad \text{Sulfuric acid}$$

However, sulfur trioxide does not unite with pure water readily. When sulfuric acid is industrially manufactured, the sulfur trioxide is absorbed into a solution of sulfuric acid containing 97% sulfuric acid by mass instead of pure water. Sulfur trioxide readily dissolves in this sulfuric acid solution. The final solution boils at 640 °F (338 °C) at 14.7 psi$_a$ (101.3 kPa) and contains 98.3% sulfuric acid by mass. In commerce, this solution is called *concentrated sulfuric acid*. It is encountered in numerous industries in a variety of nonbulk and bulk containers such as those illustrated in Figures 8.3(a) and 8.3(b).

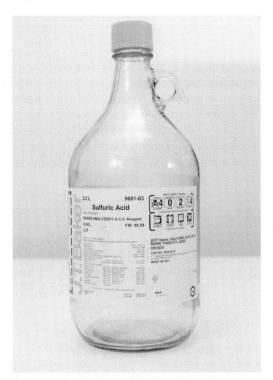

FIGURE 8.3 Sulfuric acid is often stored for future use in nonbulk containers such as the 0.6-gal (2.5-L) glass bottle shown in (a). When transported, the bottle is cushioned within an outer overpack container to which CORROSIVE labels are affixed. Sulfuric acid is also stored in bulk containers such as the plastic tote container shown in (b). In this instance, when the acid is transported, DOT requires the shipper to affix a CORROSIVE placard to the container and display the DOT identification number 1830 across its center area. (*Courtesy of Mallinckrodt Baker, Inc., Phillipsburg, New Jersey.*)

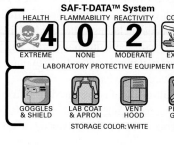

Sulfuric Acid

POISON! ☠ DANGER!

CORROSIVE. LIQUID AND MIST CAUSE SEVERE BURNS TO ALL BODY TISSUE. MAY BE FATAL IF SWALLOWED OR CONTACTED WITH SKIN. HARMFUL IF INHALED. AFFECTS TEETH. WATER REACTIVE. CANCER HAZARD. STRONG INORGANIC ACID MISTS CONTAINING SULFURIC ACID CAN CAUSE CANCER. RISK of cancer depends on duration and level of exposure. Do not get in eyes, on skin, or on clothing. Do not breathe mist. Keep container closed. Use only with adequate ventilation. Wash thoroughly after handling. Do not contact with water. FIRST AID: In all cases call a physician immediately. In case of contact, immediately flush eyes or skin with plenty of water for at least 15 minutes while removing contaminated clothing and shoes. Wash clothing before re-use. Excess acid on skin can be neutralized with a 2% bicarbonate of soda solution. If swallowed, DO NOT INDUCE VOMITING. Give large quantities of water. Never give anything by mouth to an unconscious person. If inhaled, remove to fresh air. If not breathing, give artificial respiration. If breathing is difficult, give oxygen.

SEE MATERIAL SAFETY DATA SHEET

Acide Sulfurique

TOXIQUE! DANGER!

CORROSIF. LE LIQUIDE ET LE BROUILLARD PEUVENT PROVOQUER DE GRAVES BRÛLURES A TOUT LES TISSUS DU CORPS. PEUT ETRE FATAL EN CAS D'INGESTION OU DE CONTACT AVEC LA PEAU. NUISIBLE PAR INHALATION. AFFECTE LES DENTS. REAGIT AVEC L' EAU. RISQUE DE CANCER. DES BROUILLARDS INTENSES D'ACIDE INORGANIQUE CONTENANT DE L'ACIDE SULFURIQUE PEUVENT CAUSER UN CANCER. LE risque de cancer dépend de la durée et du niveau d'exposition. Ne pas laisser venir en contact avec la peau, les yeux et les vêtements Ne pas respirer les brouillards. Maintenir le contenant fermé Utiliser avec une ventilation adéquate. Bien se laver après la manipulation Ne pas laisser au contact de l'eau. PREMIERS SECOURS: Dans tous les cas, appeler un medecin immédiatement. En cas de contact, rincer immédiatement les yeux ou la peau avec de grandes quantités d'eau pendant au moins 15 minutes tout en enlevant les vêtements et les chaussures contaminés. Laver les vêtements avant de les reutilise. Tout acide restant sur la peau peut être neutralisé avec une solution de bicarbonate de soude à 2%. En cas d'ingestion, NE PAS PROVOQUER DE VOMISSEMENTS. Faire boire de grandes quantités d'eau. Ne jamais rien faire prendre par la bouche à une personne inconsciente. En cas d'inhalation, porter la personne au grand air. En cas d'arrêt de la respiration, pratiquer la respiration artificielle. En cas de difficulté à respirer, administrer de l'oxygène.

VOIR FICHE SIGNALETIQUE

---

500 mL   **9681-00**

Sulfuric Acid

Acide Sulfurique

'BAKER ANALYZED'® A.C.S. Reagent

H₂SO₄   **FW 98.08**

LOT

Meets A.C.S. Specifications
Meets Reagent Specifications for testing USP/NF monographs

Assay (H₂SO₄)	95.0 - 98.0 %
Appearance	Passes Test
Color (APHA)	10 max.
Specific Gravity at 60°/60°F	1.84 min.
Residue after Ignition	4 ppm max.
Substances Reducing Permanganate (as SO₂)	2 ppm max.

Trace Impurities (in ppm):

Ammonium (NH₄)	1 max.
Arsenic (As)	0.004 max.
Chloride (Cl)	0.1 max.
Nitrate (NO₃)	0.5 max.

Trace Impurities (in ppb):

Aluminum (Al)	200 max.
Calcium (Ca)	200 max.
Chromium (Cr)	100 max.
Copper (Cu)	100 max.
Gold (Au)	100 max.
Heavy Metals (as Pb)	500 max.
Iron (Fe)	200 max.
Lead (Pb)	200 max.
Magnesium (Mg)	200 max.
Manganese (Mn)	100 max.
Mercury (Hg)	5 max.
Nickel (Ni)	200 max.
Potassium (K)	300 max.
Sodium (Na)	300 max.
Tin (Sm)	200 max.
Titanium (Ti)	200 max.
Zinc (Zn)	200 max.

For Laboratory, Research or Manufacturing Use
Sulfuric Acid . . . . CAS No: 7664-93-9
Water . . . . CAS No: 7732-18-5

SAF-T-DATA™ System

HEALTH	FLAMMABILITY	REACTIVITY	CONTACT
4	**0**	**2**	**4**
EXTREME	NONE	MODERATE	EXTREME

LABORATORY PROTECTIVE EQUIPMENT

GOGGLES & SHIELD	LAB COAT & APRON	VENT HOOD	PROPER GLOVES

STORAGE COLOR: WHITE

DOT Name: SULFURIC ACID (WITH MORE THAN 51% ACID) UN1830

CAS NO: 7664-93-9

J. T. Baker NEUTRASORB® or TEAM® 'Low Na⁺' acid neutralizers are recommended for spills of this product.

MADE IN USA

NFPA

R31

tyco
Specialty Products

Mallinckrodt Baker, Inc.
Phillipsburg, NJ 08865
PH (908) 859-2151
www.jtbaker.com

J.T.Baker

FIGURE 8.4 Mallinckrodt Baker, Inc., affixes this label to a bottle of concentrated sulfuric acid. Included on the label is hazard information concerning the acid. Also included is information relating to the "Saf-T-Data System," which provides the relative ratings for health, flammability, reactivity, and contact hazards associated with sulfuric acid. (*Courtesy of Mallinckrodt Baker, Inc., Phillipsburg, New Jersey.*)

As first noted in Section 1.6-A, OSHA requires every manufacturer and importer of a chemical substance to ensure that appropriate hazard warnings are provided on the containers of hazardous materials. In compliance with this requirement, the manufacturers and distributors of sulfuric acid provide hazard information on the labels affixed to containers. For example, Mallinckrodt Baker, Inc., affixes the label shown in Figure 8.4 to a glass bottle containing concentrated sulfuric acid. The information written under the heading POISON! DANGER! summarizes certain hazards, each of which is noted in more detail in the sections that follow. Several hazards associated with sulfuric acid are also illustrated in Figure 8.5.

8.6-A LIBERATION OF HEAT

Considerable heat is released when concentrated sulfuric acid is diluted with water. Because this reaction is highly exothermic, extreme caution must be exercised when diluting sulfuric acid. The recommended practice is to slowly pour the acid into the water while stirring. When concentrated sulfuric acid is diluted in the reverse manner, localized boiling and violent spattering occur.

8.6-B EXTRACTION OF WATER

Concentrated sulfuric acid can extract water from materials. Concentrated sulfuric acid so strongly extracts the elements of water from some organic compounds that carbon is often the only visibly remaining residue. It is for this reason that concentrated sulfuric acid completely destroys wood, textiles, and paper. This dehydrating phenomenon also occurs when concentrated sulfuric acid burns body tissues.

The ability of sulfuric acid to extract water is put to use during many chemical manufacturing processes. The manufacture of the explosive nitroglycerin, for example, uses sulfuric acid to extract the elements of water from glycerol and nitric acid (Section 15.6).

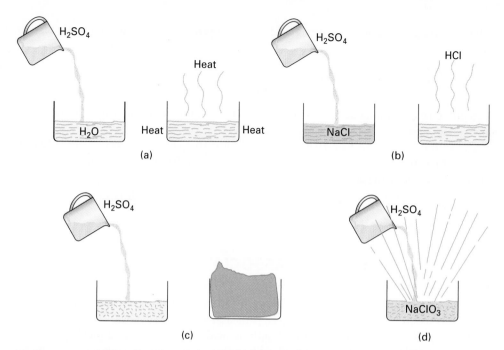

FIGURE 8.5 Some potentially hazardous features associated with concentrated sulfuric acid. In (a), the mixing of concentrated sulfuric acid with water causes heat to evolve that could trigger the self-ignition of nearby materials. In (b), the double-replacement reaction of concentrated sulfuric acid and sodium chloride produces the toxic gas hydrogen chloride. In (c), concentrated sulfuric acid removes the elements of water from sugar and leaves a residue of carbon. In (d), the reaction between concentrated sulfuric acid and sodium chlorate is explosive, clearly illustrating that these two substances are chemically incompatible.

8.6-C OXIDIZING POTENTIAL OF SULFURIC ACID

Hot concentrated sulfuric acid reacts as a strong oxidizing agent. Many acids do not generally react with copper, carbon, and lead, but hot, concentrated sulfuric acid oxidizes them.

$$Cu(s) \; + \; 2H_2SO_4(conc) \; \longrightarrow \; CuSO_4(aq) \; + \; SO_2(g) \; + \; 2H_2O(g)$$

Copper   Sulfuric acid   Copper(II) sulfate Sulfur dioxide   Water

$$C(s) \; + \; 2H_2SO_4(conc) \; \longrightarrow \; CO_2(g) \; + \; 2SO_2(g) \; + \; 2H_2O(g)$$

Carbon   Sulfuric acid   Carbon dioxide Sulfur dioxide   Water

$$Pb(s) \; + \; 3H_2SO_4(conc) \; \longrightarrow \; Pb(HSO_4)_2(s) \; + \; SO_2(g) \; + \; 2H_2O(g)$$

Lead   Sulfuric acid   Lead(II) bisulfate Sulfur dioxide   Water

However, at room temperature, concentrated sulfuric acid reacts so slowly with these elements that the rates are barely perceptible.

SOLVED EXERCISE 8.3

Sulfuric acid is an example of a "strong acid," a "mineral acid," and (when hot) an "oxidizing acid." What is meant by each of these terms?

Solution: An acid is either strong or weak depending on the extent to which it ionizes in water. Those that ionize almost completely are strong acids, whereas those that do not ionize significantly are weak acids. Sulfuric acid is a strong acid, since it ionizes almost completely when dissolved in water.

$$H_2SO_4(conc) \longrightarrow H^+(aq) + HSO_4^-(aq)$$
$$HSO_4^-(aq) \longrightarrow H^+(aq) + SO_4^{2-}(aq)$$

Sulfuric acid is a mineral acid, since a nonmetal—sulfur—is a component of its chemical composition.

Hot sulfuric acid is an oxidizing acid because it participates as a reactant during which it takes electrons from other substances. An example of this oxidizing nature of sulfuric acid is evident in its reaction with carbon.

$$C(s) + 2H_2SO_4(conc) \longrightarrow CO_2(g) + 2SO_2(g) + 2H_2O(g)$$

8.6-D HEALTH HAZARDS ASSOCIATED WITH EXPOSURE TO SULFURIC ACID

The repeated inhalation of sulfuric acid mists by workers in occupational settings has been linked with the onset of cancers of the larynx, paranasal sinuses, and lungs. These mists are generated during the manufacture and use of sulfuric acid, especially during pickling processes. Given its potential to cause cancer, all work with sulfuric acid should be conducted only within a workplace that provides adequate ventilation.

oleum
■ Concentrated sulfuric acid containing additional dissolved sulfur trioxide; fuming sulfuric acid

8.6-E OLEUM

Sulfur trioxide is soluble in concentrated sulfuric acid, producing a thick, fuming yellow liquid. When this liquid contains a higher proportion of sulfur trioxide than is found in "ordinary" sulfuric acid, the resulting material is called **oleum** or *fuming sulfuric acid* or, less commonly, *Nordhausen acid*. This acid is available commercially in concentrations ranging from 20% to 99.9% sulfur trioxide. Although oleum was formerly used in the petroleum refining industry, hydrofluoric acid (Section 8.10) now serves the same purpose. Today, oleum is primarily used within the chemical industry.

The chemical formula of oleum is often denoted as $xH_2SO_4 \cdot ySO_3$, where x and y are the number of moles of sulfuric acid and sulfur trioxide, respectively. For example, oleum that contains 1 mol of sulfur trioxide (80.1 g) for each mole of sulfuric acid (98.1 g) has the formula $H_2S_2O_7$, since in this instance, $x = y = 1$. Oleum containing 65% sulfur trioxide by mass is expressed by the formula $4H_2SO_4 \cdot 9SO_3$.

SOLVED EXERCISE 8.4

Show that oleum containing 65% sulfur trioxide by mass is correctly expressed as $4H_2SO_4 \cdot 9SO_3$.

Solution: The molecular weight of oleum as $4H_2SO_4 \cdot 9SO_3$ is 1113.3.

$$
\begin{array}{rl}
4 \times 98.1\ g = & 392.4 \\
9 \times 80.1\ g = & \underline{720.9} \\
& 1113.3
\end{array}
$$

The sulfur trioxide content in this oleum is $1113.3 \times 0.65 = 723.6$, or close to 9 mol. The remaining 35% is sulfuric acid. The sulfuric acid content is $1113.3 \times 0.35 = 389.6$, or close to 4 mol. Hence, the formula $4H_2SO_4 \cdot 9SO_3$ is the correct representation of oleum that contains 65% sulfur trioxide and 35% sulfuric acid.

Like sulfuric acid, oleum is a corrosive material. It severely burns the skin, which generally heals to produce ugly scars. In addition, oleum spontaneously releases toxic sulfur trioxide vapor, which poses the potential risk of inhalation toxicity. Consequently, oleum should be stored and used only in a well-ventilated location.

8.6-F TRANSPORTING SULFURIC ACID AND OLEUM

DOT regulates the transportation of three forms of sulfuric acid: a liquid having 51% acid or more; a liquid with less than 51% acid; and a spent (sulfuric acid) solution. When shippers offer these acids for transportation, DOT requires them to provide one of the following proper shipping descriptions on the accompanying shipping paper, as relevant:

> **UN1830, Sulfuric acid (contains 51% or more acid), 8, PGII**
> *or*
> **UN2796, Sulfuric acid (contains less than 51% acid), 8, PGII**
> *or*
> **UN1821, Sulfuric acid, spent, 8, PGII**

DOT also requires shippers to affix a CORROSIVE label to sulfuric acid containers.

When carriers transport 1001 lb (454 kg) or more of sulfuric acid, DOT requires them to display CORROSIVE placards on the bulk packaging or transport vehicle used to transport the acid. DOT also requires carriers to display the relevant identification number, 1830, 2796, or 1821, on orange panels, across the center area of the CORROSIVE placards, or on white square-on-point diamonds.

DOT also regulates the transportation of two forms of fuming sulfuric acid: a liquid having 30% or more free sulfur trioxide; and a liquid having less than 30% free sulfur trioxide. When shippers offer these acids for transportation, DOT requires them to identify the appropriate substance on the accompanying shipping paper in one of the following ways, as relevant:

> **UN1831, Sulfuric acid, fuming (contains less than 30% sulfur trioxide), 8, PGI**
> *or*
> **UN1831, Sulfuric acid, fuming (contains 30% or more sulfur trioxide), 8, (6.1), PGI (Poison - Inhalation Hazard, Zone B)**

DOT requires shippers who transport fuming sulfuric acid to affix a CORROSIVE label to its containers. DOT also requires shippers who offer fuming sulfuric acid with a concentration equal to or greater than 30% sulfur trioxide for transportation to affix POISON INHALATION HAZARD labels in addition to the CORROSIVE label to its containers.

When carriers transport 1001 lb (454 kg) or more of these forms of fuming sulfuric acid, DOT requires them to display CORROSIVE placards on the bulk packaging or transport vehicle used for shipment.

DOT also requires carriers who transport fuming sulfuric acid with a concentration equal to or greater than 30% sulfur trioxide to display POISON INHALATION HAZARD placards on the bulk packaging or transport vehicle used for shipment. When transporting 1001 lb (454 kg) or more, DOT requires carriers to display CORROSIVE placards in addition to the POISON INHALATION HAZARD placards. DOT also requires carriers to display the identification number 1831 on orange panels, across the center area of the CORROSIVE placards, or on white square-on-point diamonds.

- Identify the properties of nitric acid, including fuming nitric acid.
- Identify the industries that use nitric acid.
- Identify the principal hazards associated with nitric acid.
- Describe the means shippers and carriers use to inform emergency responders of the hazards associated with encountering nitric acid during a transportation mishap.

8.7 Nitric Acid

Nitric acid is second among the acids most commonly used throughout the United States. It is the raw material used for the manufacture of ammonium nitrate fertilizers, explosives, and nitrated organic compounds. It is also required for the production of patent leather and related fabrics (Section 14.5-A). The chemical formula of nitric acid is HNO_3. Its important physical properties are provided in Table 8.5.

Pure nitric acid is a colorless liquid. It is encountered commercially in concentrated and diluted forms. Concentrated nitric acid is an aqueous solution consisting of 68.2% nitric acid by mass. All concentrations containing less than 68.2% acid are forms of diluted nitric acid.

When encountered, nitric acid is often yellow to red-brown in color, which indicates that nitrogen dioxide is present. Nitrogen dioxide is a red-brown gas produced by the slow decomposition of nitric acid, a phenomenon catalyzed by sunlight.

$$4HNO_3(l) \longrightarrow 4NO_2(g) + 2H_2O(l) + O_2(g)$$

Nitric acid   Nitrogen dioxide   Water   Oxygen

Almost all nitric acid is industrially manufactured from ammonia by means of a series of reactions. Gaseous ammonia is first mixed with about 10 times its volume of air, and then exposed to platinum gauze. The platinum increases the rate of the reaction that converts ammonia into nitric monoxide (NO). Then, additional air is permitted to enter the reaction system so the nitrogen monoxide can be further oxidized to nitrogen dioxide.

$$4NH_3(g) + 5O_2(g) \longrightarrow 4NO(g) + 6H_2O(g)$$

Ammonia   Oxygen   Nitrogen monoxide   Water

$$2NO(g) + O_2(g) \longrightarrow 2NO_2(g)$$

Nitrogen monoxide   Oxygen   Nitrogen dioxide

TABLE 8.5	Physical Properties of Concentrated Nitric Acid
Melting point	−44 °F (−42 °C)
Boiling point	187 °F (86 °C)
Specific gravity at 68 °F (20 °C)	1.50
Vapor density (air = 1)	3.2
Vapor pressure at 68 °F (20 °C)	47.9 mmHg
Solubility in water	Infinitely soluble

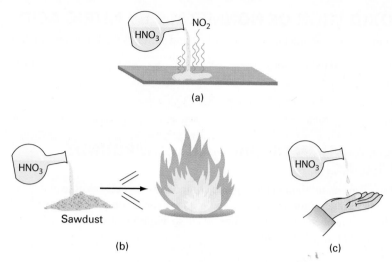

(a)

(b)   (c)

FIGURE 8.6 Some potentially hazardous features associated with concentrated nitric acid. In (a), the hot acid corrodes metal and produces a toxic oxide of nitrogen. In (b), the concentrated acid causes sawdust to ignite. The incidents illustrated in both (a) and (b) illustrate that concentrated nitric acid is a strong oxidizing agent. In (c), contact between skin tissue and the concentrated acid results in the production of ugly, yellow scars.

The nitrogen dioxide is then reacted with water to produce nitric acid.

$$3NO_2(g) \ + \ H_2O(l) \ \longrightarrow \ 2HNO_3(l) \ + \ NO(g)$$

Nitrogen dioxide   Water   Nitric acid   Nitrogen monoxide

The excess nitrogen monoxide produced in this reaction is recycled through the system.

The hazards of concentrated nitric acid are illustrated in Figure 8.6. Like hot sulfuric acid, nitric acid is an oxidizing acid. Although sulfuric acid is a powerful oxidizing agent only when it is hot, nitric acid is a powerful oxidizing agent even at room temperature.

8.7-A OXIDATION OF METALS BY NITRIC ACID

When nitric acid chemically attacks metals, the metals are oxidized to their corresponding positive ions as the nitric acid is reduced to one or more of the following: nitrogen, nitrogen monoxide, nitrogen dioxide, dinitrogen monoxide, or the ammonium ion. Nitric acid reacts with some metals to form each of these products under specific conditions. For example, it reacts with zinc to form nitrogen, nitrogen monoxide, nitrogen dioxide, dinitrogen monoxide, or ammonium nitrate in separate reactions.

$$5Zn(s) \ + \ 12HNO_3(aq) \ \longrightarrow \ 5Zn(NO_3)_2(aq) \ + \ 6H_2O(l) \ + \ N_2(g)$$

Zinc   Nitric acid   Zinc nitrate   Water   Nitrogen

$$3Zn(s) \ + \ 8HNO_3(aq) \ \longrightarrow \ 3Zn(NO_3)_2(aq) \ + \ 4H_2O(l) \ + \ 2NO(g)$$

Zinc   Nitric acid   Zinc nitrate   Water   Nitrogen monoxide

$$Zn(s) \ + \ 4HNO_3(conc) \ \longrightarrow \ Zn(NO_3)_2(aq) \ + \ 2H_2O(l) \ + \ 2NO_2(g)$$

Zinc   Nitric acid   Zinc nitrate   Water   Nitrogen dioxide

$$4Zn(s) \ + \ 10HNO_3(aq) \ \longrightarrow \ 4Zn(NO_3)_2(aq) \ + \ 5H_2O(l) \ + \ N_2O(g)$$

Zinc   Nitric acid   Zinc nitrate   Water   Dinitrogen monoxide

$$4Zn(s) \ + \ 10HNO_3(aq) \ \longrightarrow \ 4Zn(NO_3)_2(aq) \ + \ 3H_2O(l) \ + \ NH_4NO_3(aq)$$

Zinc   Nitric acid   Zinc nitrate   Water   Ammonium nitrate

When nitric acid oxidizes a metal, the most common products formed are nitrogen monoxide and nitrogen dioxide. In general, *diluted* nitric acid oxidizes metals to produce nitrogen monoxide, and *concentrated* nitric acid oxidizes metals to produce nitrogen dioxide.

8.7-B OXIDATION OF NONMETALS BY NITRIC ACID

Hot nitric acid corrodes nonmetals such as carbon and sulfur as the following equations illustrate:

$$C(s) \;+\; 4HNO_3(conc) \;\longrightarrow\; CO_2(g) \;+\; 4NO_2(g) \;+\; 2H_2O(l)$$

Carbon   Nitric acid   Carbon dioxide Nitrogen dioxide   Water

$$S_8(s) \;+\; 48HNO_3(conc) \;\longrightarrow\; 8H_2SO_4(l) \;+\; 48NO_2(g) \;+\; 16H_2O(l)$$

Sulfur   Nitric acid   Sulfuric acid   Nitrogen dioxide   Water

8.7-C OXIDATION OF ORGANIC COMPOUNDS BY NITRIC ACID

Nitric acid oxidizes many flammable organic compounds, sometimes at explosive rates. This results in their subsequent ignition. For example, the organic compounds turpentine, acetic acid, acetone, ethanol, nitrobenzene, and aniline react so vigorously when mixed with hot concentrated nitric acid that they burst into flame.

8.7-D REACTIONS OF NITRIC ACID WITH CELLULOSIC MATERIALS

Nitric acid is capable of initiating the spontaneous ignition of wood, excelsior, and other cellulosic materials, especially when these materials have been finely divided. It is for this reason that bottles of nitric acid are not cushioned with a cellulosic material when they are transported.

8.7-E REACTIONS OF NITRIC ACID WITH SKIN TISSUE

Nitric acid corrodes body tissues by reacting with the complex proteins that make up their structures. The exposure of nitric acid to the skin results in ugly, yellow burns that heal very slowly. The chemistry associated with this phenomenon involves the production of a yellow-brown-colored substance called *xanthoproteic acid*. The discoloration of the skin typically wears away in two to three weeks.

SOLVED EXERCISE 8.5

Metallic copper does not replace the hydrogen in nitric acid, but metallic copper does react with nitric acid. In these reactions the nitric acid reacts as an oxidizing agent. What are the products of the reactions in which metallic copper is oxidized by concentrated nitric acid and diluted nitric acid?

Solution: Concentrated nitric acid oxidizes metals to produce nitrogen dioxide, whereas diluted nitric acid oxidizes metals to produce nitrogen monoxide. The metal is oxidized to the copper(II) ion.

$$Cu(s) \;+\; 4HNO_3(conc) \;\longrightarrow\; Cu(NO_3)_2(aq) \;+\; 2NO_2(g) \;+\; 2H_2O(l)$$

Copper   Nitric acid   Copper(II) nitrate   Nitrogen dioxide   Water

$$3Cu(s) \;+\; 8HNO_3(dil) \;\longrightarrow\; 3Cu(NO_3)_2(aq) \;+\; 2NO(g) \;+\; 4H_2O(l)$$

Copper   Nitric acid   Copper(II) nitrate   Nitrogen monoxide   Water

8.7-F FUMING NITRIC ACID

The oxides of nitrogen are readily soluble in concentrated nitric acid. When the acid contains a higher proportion of nitrogen oxides than is contained in ordinary nitric acid, the resulting material is called *fuming nitric acid*, of which there are two forms:

■ *Red fuming nitric acid* contains more than 85% nitric acid, less than 5% water, and from 6% to 15% nitrogen oxides.

- *White fuming nitric acid* contains more than 97.5% nitric acid, less than 2% water, and less than 0.5% nitrogen oxides. This second form is not widely encountered.

Contact of the skin with fuming nitric acid is highly irritating. In addition, it spontaneously releases toxic nitrogen oxide vapors, which potentially pose a high degree of inhalation toxicity. Given these properties, fuming nitric acid should always be segregated from other chemical substances and stored and used in well-ventilated locations.

8.7-G WORKPLACE REGULATIONS INVOLVING NITRIC ACID

OSHA requires employers to limit employee exposure to a maximum nitric acid vapor concentration of 2 ppm, averaged over an 8-hr workday, and 4 ppm for a short-term exposure.

8.7-H TRANSPORTING NITRIC ACID

When shippers offer nitric acid or fuming nitric acid for transportation, DOT requires them to identify the appropriate substance on the accompanying shipping paper as one of the following, as relevant:

> **UN2031, Nitric acid, other than red fuming, with not more than 70% nitric acid, 8, PGII**
> *or*
> **UN2031, Nitric acid, other than red fuming, with more than 70% nitric acid, 8, (5.1), PGI**
> *or*
> **UN2032, Nitric acid, red fuming, 8, (5.1, 6.1), PGI (Poison - Inhalation Hazard, Zone B)**

DOT also requires shippers who transport nitric acid with a concentration of less than 70% to affix CORROSIVE labels to its containers.

When carriers transport 1001 lb (454 kg) or more of nitric acid with a concentration of less than 70%, DOT requires them to display CORROSIVE placards on the bulk packaging or transport vehicle used for shipment. DOT also requires carriers to display the identification number 2031 on orange panels, across the center area of the CORROSIVE placards, or on white square-on-point diamonds.

DOT requires shippers who offer nitric acid with a concentration greater than 70% for transportation to affix CORROSIVE *and* OXIDIZER labels to its containers. When carriers transport this acid in an amount equal to or greater than 1001 lb (454 kg), DOT requires them to display CORROSIVE placards on the bulk packaging or transport vehicle used for shipment. DOT also requires carriers to display the identification number 2031 on orange panels, across the center area of the CORROSIVE placards, or on white square-on-point diamonds.

DOT requires shippers who offer red fuming nitric acid for transportation to affix CORROSIVE, OXIDIZER, and POISON INHALATION HAZARD labels to its containers. DOT requires carriers to display POISON INHALATION HAZARD placards on the bulk packaging or transport vehicle used for shipment. When transporting 1001 lb (454 kg) or more, DOT requires carriers to display CORROSIVE placards. DOT further requires carriers to display the identification number 2032 on orange panels, across the center area of the CORROSIVE placards, or on white square-on-point diamonds, and to mark the bulk packaging or transport vehicle on two opposing sides with the words INHALATION HAZARD.

- Identify the properties of hydrochloric acid.
- Identify the industries that use hydrochloric acid.
- Identify the principal hazards associated with concentrated hydrochloric acid.
- Identify the principal hazards associated with concentrated hydrochloric acid and anhydrous hydrogen chloride.
- Describe the means shippers and carriers use to inform emergency responders of the hazards associated with encountering hydrochloric acid or anhydrous hydrogen chloride during a transportation mishap.

8.8 Hydrochloric Acid

Hydrochloric acid is another commercially important acid. It is most familiar to the general public as a constituent of certain household cleaning products like Lysol Toilet Bowl Cleaner and Lime-Away. It is also the liquid used to maintain the proper acidity of the water in residential and public swimming pools.

Outside the home, hydrochloric acid has many uses. As a component of pickle liquor, it is used for galvanizing, tinning, and enameling. In the food industry, hydrochloric acid is used as a processing agent during the production of certain food products such as corn syrup. It is used in the petroleum industry to activate petroleum wells, and it is used in the chemical industry to manufacture and produce dozens of important compounds. Hydrochloric acid is frequently encountered in educational and research facilities, where it usually stored in nonbulk plastic containers like the example shown in Figure 8.7.

Pure hydrochloric acid is a colorless, fuming, and pungent-smelling liquid composed of hydrogen chloride dissolved in water. The chemical formulas of hydrochloric acid and hydrogen chloride are $HCl(aq)$ and $HCl(g)$, respectively. Some physical properties of the concentrated acid and its vapor are noted in Table 8.6.

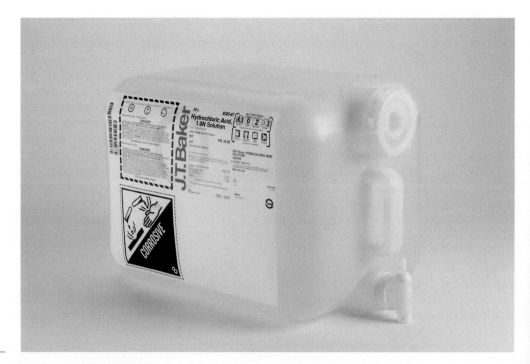

FIGURE 8.7
Hydrochloric acid may be encountered commercially as the concentrated acid or its diluted solutions. This plastic container holds 5 gal (19 L) of an aqueous solution of the acid. (*Courtesy of Mallinckrodt Baker, Inc., Phillipsburg, New Jersey.*)

TABLE 8.6	Physical Properties of Hydrochloric Acid and Its Vapor	
	HCL (*conc*)	**HCl (*g*)**
Melting point	−101 °F (−74 °C)	−174 °F (2114 °C)
Boiling point	127 °F (53 °C)	−121 °F (−85 °C)
Specific gravity at 68 °F (20 °C)	1.18	1.27
Vapor density (air = 1)	1.3	1.3
Vapor pressure at 68 °F (20 °C)	150 mmHg	30,780 mmHg
Solubility in water	85 g/100 g of H_2O	67% at 68 °F (20 °C)

Hydrochloric acid is prepared by dissolving anhydrous hydrogen chloride in water. The concentrated acid contains from 36% to 38% hydrogen chloride by mass and is colorless. When purity is not a factor, an industrial grade of hydrochloric acid, called **muriatic acid**, is often used. It is a dilute solution that is slightly yellow owing to the presence of dissolved compounds of iron.

muriatic acid
■ Technical-grade hydrochloric acid

8.8-A VAPORIZATION OF HYDROCHLORIC ACID

The principal hazardous feature of hydrochloric acid is associated with the hydrogen chloride vapor released spontaneously from the concentrated acid. Hydrogen chloride is a poisonous gas. When concentrated hydrochloric acid spills or leaks from its container, the pungent odor of its toxic vapor is immediately detectable. Because the vapor is approximately one-fifth heavier than air, it lingers in low areas, where it can pose an unreasonable risk to health and safety.

8.8-B ILL EFFECTS RESULTING FROM EXPOSURE TO HYDROGEN CHLORIDE

When hydrogen chloride vapor is encountered, individuals experience the symptoms noted in Table 8.7. In the worst situations, prolonged exposure to massive amounts of hydrogen chloride causes pulmonary edema that can completely deteriorate the tissue cells within the respiratory tract and destroy its lining.

TABLE 8.7	Adverse Health Effects Associated with Breathing Hydrogen Chloride Vapor
HYDROGEN CHLORIDE (ppm)	**SYMPTOMS**
1–5	Threshold limit for detection of smell
5–10	Mild irritation of mucous membranes, eyes, nose, and throat
35	Distinct irritation of mucous membranes, eyes, nose, and throat
5–100	Barely tolerable effects including severe coughing and panic; possible injury to the bronchial region
1000	Danger of pulmonary edema after 24-hr exposure; potentially fatal

Use the data in Tables 8.4 and 8.6 to determine whether sulfuric acid or hydrochloric acid poses the greater hazard by inhalation toxicity at 68 °F (20 °C).

Solution: To pose an inhalation-toxicity hazard, the liquid acid must produce sufficient vapor at 68 °F (20 °C) to cause illness or death when inhaled. In Table 8.4, the vapor pressure of sulfuric acid is listed as <0.001 mmHg. This low value signifies that virtually no vapor is produced at this temperature. Accordingly, exposure to sulfuric acid does not pose the risk of inhalation toxicity. However, the vapor pressure of hydrochloric acid is listed in Table 8.6 as 150 mmHg. This elevated value indicates that hydrochloric acid produces sufficient vapor at 68 °F (20 °C) to cause the ill effects noted in Table 8.7 when inhaled.

8.8-C REACTIONS OF HYDROCHLORIC ACID WITH OXIDIZING AGENTS

Hydrochloric acid is chemically incompatible with oxidizing agents such as metallic chlorates, metallic dichromates, and metallic permanganates. These reactions result in the production of toxic chlorine.

$$KClO_3(s) \ + \ 6HCl(conc) \ \longrightarrow \ KCl(aq) \ + \ 3H_2O(l) \ + \ 3Cl_2(g)$$
Potassium chlorate   Hydrochloric acid   Potassium chloride   Water   Chlorine

$$K_2Cr_2O_7(s) \ + \ 14HCl(conc) \ \longrightarrow \ 2KCl(aq) \ + \ 2CrCl_3(aq) \ + \ 7H_2O(l) \ + \ 3Cl_2(g)$$
Potassium dichromate   Hydrochloric acid   Potassium chloride   Chromium(III) chloride   Water   Chlorine

$$2KMnO_4(s) \ + \ 16HCl(conc) \ \longrightarrow \ 2MnCl_2(aq) \ + \ 2KCl(aq) \ + \ 8H_2O(l) \ + \ 5Cl_2(g)$$
Potassium permanganate   Hydrochloric acid   Manganese(II) chloride   Potassium chloride   Water   Chlorine

To reduce or eliminate the likelihood of an unwanted reaction between hydrochloric acid and chlorine-producing pool chemicals, acid manufacturers often label their containers with a message like the following:

**DO NOT STORE NEAR
CHLORINE-PRODUCING
POOL CHEMICALS**

8.8-D ANHYDROUS HYDROGEN CHLORIDE

Anhydrous hydrogen chloride is itself a commercial chemical product that is prepared by the direct combination of elemental hydrogen and chlorine.

$$H_2(g) \ + \ Cl_2(g) \ \longrightarrow \ 2HCl(g)$$
Hydrogen   Chlorine   Hydrogen chloride

Hydrogen chloride is also produced as a by-product during the chlorination of organic compounds.

8.8-E WORKPLACE REGULATIONS INVOLVING HYDROGEN CHLORIDE

OSHA requires employers to limit employee exposure to a maximum hydrogen chloride vapor concentration of 5 ppm, which should not be exceeded during any part of the working exposure.

8.8-F TRANSPORTING HYDROCHLORIC ACID AND ANHYDROUS HYDROGEN CHLORIDE

When shippers offer hydrochloric acid or anhydrous hydrogen chloride for transportation, DOT requires them to identify its proper shipping description on the accompanying shipping paper as one of the following, as relevant:

> **UN1789, Hydrochloric acid, 8, PGII**
> *or*
> **UN1050, Hydrogen chloride, anhydrous, 2.3, (8), (Poison - Inhalation Hazard, Zone C)**
> *or*
> **UN2186, Hydrogen chloride, refrigerated liquid, 2.3, (8), (Poison - Inhalation Hazard, Zone C)**

DOT also requires shippers who offer hydrochloric acid for transportation to affix a CORROSIVE label to its containers.

When carriers transport hydrochloric acid in an amount equal to or greater than 1001 lb (454 kg) or more, DOT requires them to display CORROSIVE placards on the bulk packaging or transport vehicle used for shipment. DOT also requires carriers to display the identification number 1789 on orange panels, across the center area of the CORROSIVE placards, or on white square-on-point diamonds.

DOT requires shippers who offer anhydrous hydrogen chloride for transportation to affix POISON GAS and CORROSIVE labels to its containers. Carriers must display POISON GAS placards on the bulk packaging or transport vehicle used for shipment. DOT also requires carriers to display the relevant identification number, 1050 or 2186, on orange panels, across the center area of the POISON GAS placards, or on white square-on-point diamonds and to mark the bulk packaging or transport vehicle on two opposing sides with the words INHALATION HAZARD.

PERFORMANCE GOALS FOR SECTION 8.9:

- Identify the properties of perchloric acid.
- Identify the industries that use perchloric acid.
- Identify the principal hazards associated with concentrated perchloric acid.
- Describe the means shippers and carriers use to inform emergency responders of the hazards associated with encountering perchloric acid during a transportation mishap.

8.9 Perchloric Acid

Perchloric acid is an acid used primarily by the chemical, electroplating, and incendiary (fireworks) industries. Its chemical formula is $HClO_4$.

Perchloric acid is a colorless, aqueous liquid having an approximate composition of 72% by mass. This is the composition of the concentrated perchloric acid available commercially. Although solutions of perchloric acid having an acid concentration greater than 72% are known, they are not routinely encountered, since they are explosively unstable. Some physical properties of concentrated perchloric acid are noted in Table 8.8.

TABLE 8.8	Physical Properties of Concentrated Perchloric Acid
Melting point	0 °F (−18 °C)
Boiling point	397 °F (203 °C)
Specific gravity at 68 °F (20 °C)	1.70
Vapor density (air = 1)	3.5
Vapor pressure at 68 °F (20 °C)	6.75 mmHg
Solubility in water	Very soluble

Concentrated perchloric acid is prepared by distilling a mixture of potassium perchlorate and sulfuric acid at lower than atmospheric pressure. The chemical reaction associated with its production is represented by the following equation:

$$2KClO_4(s) \quad + \quad H_2SO_4(aq) \quad \longrightarrow \quad 2HClO_4(conc) \quad + \quad K_2SO_4(aq)$$

Potassium perchlorate   Sulfuric acid   Perchloric acid   Potassium sulfate

8.9-A THERMAL DECOMPOSITION OF PERCHLORIC ACID

Concentrated perchloric acid may be safely heated to 194 °F (90 °C), but above this temperature, it is likely to explosively decompose, as the following equation illustrates:

$$4HClO_4(conc) \quad \longrightarrow \quad 2Cl_2(g) \quad + \quad 7O_2(g) \quad + \quad 2H_2O(g)$$

Perchloric acid   Chlorine   Oxygen   Water

8.9-B OXIDIZING POTENTIAL OF PERCHLORIC ACID

Concentrated perchloric acid is a powerful oxidizing acid when it is hot, but when diluted with water, perchloric acid reacts as a weak oxidizing agent even when hot. The hot concentrated perchloric acid reacts violently with organic compounds including cellulosic materials such as sawdust. In fact, a mixture of perchloric acid and sawdust ignites spontaneously. Every measure should be taken to segregate perchloric acid and organic compounds, since their combination is regarded as a fire and explosion hazard.

8.9-C TRANSPORTING PERCHLORIC ACID

When shippers offer perchloric acid for transportation, DOT requires them to identify it on the accompanying shipping paper in one of the following ways, as relevant:

UN1802, Perchloric acid (contains not more than 50% acid by mass), 8, (5.1), PGII
or
UN1873, Perchloric acid (contains more than 50% acid but not more than 72% acid by mass), 5.1, (8), PGI

DOT requires shippers who offer perchloric acid containing not more than 50% acid for transportation to affix CORROSIVE and OXIDIZER labels to its containers.

When carriers transport 1001 lb (454 kg) or more of perchloric acid containing not more than 50% acid, DOT requires them to display CORROSIVE placards on the bulk packaging or transport vehicle used for shipment. DOT also requires carriers to display the identification number 1802 on orange panels, across the center area of the CORROSIVE placards, or on white square-on-point diamonds.

DOT requires shippers who offer perchloric acid containing more than 50% acid for transportation to affix OXIDIZER and CORROSIVE labels to its containers. When carriers transport 1001 lb (454 kg) or more of this acid, DOT requires them to display OXIDIZER placards on the bulk packaging or transport vehicle used for shipment. DOT requires carriers to display the relevant identification number 1873 on orange panels, across the center area of the OXIDIZER placards, or on white square-on-point diamonds.

PERFORMANCE GOALS FOR SECTION 8.10:

- Identify the properties of hydrofluoric acid.
- Identify the industries that use hydrofluoric acid.
- Describe the properties of anhydrous hydrogen fluoride.
- Identify the principal hazards associated with concentrated hydrofluoric acid and anhydrous hydrogen fluoride.
- Describe the means shippers and carriers use to inform emergency responders of the hazards associated with encountering hydrofluoric acid or anhydrous hydrogen fluoride during a transportation mishap.

8.10 Hydrofluoric Acid

Hydrogen fluoride is a colorless, fuming liquid or vapor having the chemical formula $HF(l)$ or $HF(g)$, respectively. Hydrofluoric acid is a solution of hydrogen fluoride in water having the chemical formula $HF(aq)$.

The concentrated hydrofluoric acid of commerce is a liquid solution containing either 49% or 70% hydrogen fluoride by mass. It is a colorless, fuming liquid having the physical properties noted in Table 8.9. These forms of the acid may be encountered in bulk volumes, since they are transported by means of rail tankcars and cargo tanks.

Hydrofluoric acid is prepared by reacting sulfuric acid with calcium fluoride, a constituent of the naturally occurring ores fluorspar and fluorite. This production reaction is represented by the following equation:

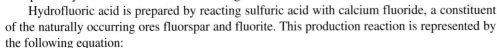

$$CaF_2(s) \quad + \quad H_2SO_4(conc) \quad \longrightarrow \quad CaSO_4(s) \quad + \quad 2HF(aq)$$

Calcium fluoride   Sulfuric acid   Calcium sulfate   Hydrofluoric acid

TABLE 8.9	Physical Properties of Hydrofluoric Acid	
	HF(conc) (38% TO 60%)	**ANHYDROUS HF(l)**
Melting point	−60 to −35 °F (−51.1 to −37.2 °C)	−118 °F (−84 °C)
Boiling point	223 to 238 °F (106 to 111 °C)	67.2 °F (19.54 °C)
Specific gravity at 68 °F (20 °C)	1.16	0.987
Vapor density (air = 1)	1.27	0.92
Vapor pressure at 68 °F (20 °C)	25 mmHg	776 mmHg
Solubility in water	Infinitely soluble	Infinitely soluble

The calcium sulfate is filtered from the resulting mixture, and the solution of hydrofluoric acid is boiled until the desired concentration is achieved.

Hydrofluoric acid solutions are useful for a variety of purposes. In the household, they are found as components of rust removers and aluminum-cleaning products. In the glass industry, they are used to polish, etch, and frost glass. In the metallurgical and steel industries, they are used to pickle brass, copper, and certain steel alloys. In the computer industry, they are used to etch silicon wafers during the manufacture of computer chips. In the chemical industry, they are used as catalysts and fluorinating agents.

8.10-A HYDROFLUORIC ACID REACTIONS WITH SKIN AND OTHER TISSUES

Although the data in Table 8.1 show that hydrofluoric acid is a very weak acid, both the liquid and its vapor severely corrode the skin and produce burns that heal very slowly. The skin damage is generally noted by external destruction to the outermost layer of the skin, but the underlying tissues may also be severely damaged. When individuals are exposed to hydrofluoric acid, it is essential that they follow the first-aid instructions provided in Figure 8.8. These instructions include the application of calcium gluconate gel, which forms calcium fluoride as it neutralizes the acid.

The pain associated with exposure to hydrofluoric acid may not be experienced immediately, and the visible signs of corrosion may not be apparent until hours following the initial exposure. At this point, the area of contact may appear blanched and bloodless. It is often necessary to seek specialized medical treatment following exposure to the acid or its fumes, since if the wound is left unattended, gangrene may develop, and the destructive action may penetrate into the bones, where decalcification can occur. This specialized medical treatment includes injection of calcium-containing substances into burned areas or into the blood.

8.10-B REACTIONS OF HYDROFLUORIC ACID WITH SILICON COMPOUNDS

The most distinguishing chemical property of concentrated hydrofluoric acid is its ability to slowly react with silicon compounds to produce gaseous silicon tetrafluoride. Silicon compounds are components of ordinary glass. Hydrofluoric acid reacts with them to produce silicon tetrafluoride. For example, hydrofluoric acid reacts with the sodium silicate and calcium silicate in glass as follows:

$$\underset{\text{Sodium silicate}}{Na_2SiO_3(s)} + \underset{\text{Hydrofluoric acid}}{6HF(conc)} \longrightarrow \underset{\text{Sodium fluoride}}{2NaF(aq)} + \underset{\text{Silicon tetrachloride}}{SiF_4(g)} + \underset{\text{Water}}{3H_2O(l)}$$

$$\underset{\text{Calcium silicate}}{CaSiO_3(s)} + \underset{\text{Hydrofluoric acid}}{6HF(conc)} \longrightarrow \underset{\text{Calcium fluoride}}{CaF_2(s)} + \underset{\text{Silicon tetrachloride}}{SiF_4(g)} + \underset{\text{Water}}{3H_2O(l)}$$

Because hydrofluoric acid reacts with the components of glass, the acid is routinely stored and transported in polyethylene or other hydrofluoric acid–resistant plastic bottles and drums.

SOLVED EXERCISE 8.7

When an aqueous solution of hydrofluoric acid is stored in a glass bottle, why does the solution appear milky after time?

Solution: When stored in a glass bottle, hydrofluoric acid reacts with the constituent sodium silicate and calcium silicate to produce silicon tetrafluoride (see text for equations). The reaction with calcium silicate forms calcium fluoride, a white compound that is insoluble in water. As the calcium fluoride forms, it disperses within the acid. This causes the acid solution to appear milky.

FIRST AID
FOR HYDROFLUORIC ACID
EXPOSURE

SEEK IMMEDIATE MEDICAL ATTENTION
CALL 911

SERIOUS TISSUE DAMAGE
WITH DELAYED ONSET

SKIN CONTACT

- **IMMEDIATELY** (within seconds) proceed to the **NEAREST SAFETY SHOWER** and wash affected area **FOR 5 MINUTES**.
- **REMOVE** all contaminated **CLOTHING** while in the shower.
- **WITH NITRILE DOUBLE-GLOVED HANDS MASSAGE CALCIUM GLUCONATE GEL** into the affected area. If calcium gluconate gel is **not available, wash area for at least 15 minutes** or until emergency medical assistance arrives.
- **REAPPLY CALCIUM GLUCONATE GEL** and massage it into affected area **EVERY 15 MINUTES** until medical assistance arrives or pain disappears.

EYE CONTACT

- **IMMEDIATELY** (within seconds) proceed **TO THE NEAREST EYEWASH STATION**.
- Thoroughly **WASH EYES WITH WATER FOR AT LEAST 15 MINUTES** while holding eyelids open.
- **DO NOT APPLY CALCIUM GLUCONATE GEL TO EYES**.

INHALATION

- **GET MEDICAL ASSISTANCE** by calling 911.

FIGURE 8.8 First-aid instructions for individuals who have been exposed to hydrofluoric acid. (*Courtesy of Environmental Health & Safety, University of Washington, Tacoma, Washington.*)

8.10-C ANHYDROUS HYDROGEN FLUORIDE

Anhydrous hydrogen fluoride, sometimes called *anhydrous hydrofluoric acid*, is itself a commercial chemical product. Like its solutions, it is prepared from calcium fluoride and sulfuric acid.

$$CaF_2(s) \quad + \quad H_2SO_4(conc) \quad \longrightarrow \quad CaSO_4(s) \quad + \quad 2HF(g)$$

Calcium fluoride   Sulfuric acid   Calcium sulfate   Hydrogen fluoride

Its important physical properties are included in Table 8.9. Note that anhydrous hydrogen fluoride readily vaporizes near room temperature.

Anhydrous hydrogen fluoride is used mainly in the chemical industry to produce chlorofluorocarbons, hydrofluorocarbons, and hydrochlorofluorocarbons (Sections 12.16). In the petroleum refining industry, it is used catalytically to produce alkylates (Section 12.14-D). It is also used during uranium-enrichment processes (Section 16.8-C) to prepare uranium hexafluoride, a gas essential for the production of uranium fuel for the nuclear power industry.

Exposure to even low concentrations (<15 ppm) of anhydrous hydrogen fluoride is irritating to the eyes, skin, and the bronchial passageways and lung surfaces. Prolonged exposure to the vapor may cause the development of fatal pulmonary edema.

As with its solutions, contact with anhydrous hydrogen fluoride severely corrodes skin tissue and produces severe burns deep beneath the skin. The tissue corrosion may become evident only hours following the initial exposure with no immediate experience of pain. Vapor burns to the eyes may result in the formation of lesions or may cause blindness.

8.10-D WORKPLACE REGULATIONS INVOLVING HYDROGEN FLUORIDE

OSHA requires employers to limit employee exposure to a maximum hydrogen fluoride vapor concentration of 3 ppm, averaged over an 8-hr workday.

8.10-E TRANSPORTING HYDROFLUORIC ACID AND ANHYDROUS HYDROGEN FLUORIDE

Hydrofluoric acid and anhydrous hydrogen fluoride are transported as liquids. When shippers intend to transport these chemical commodities, DOT requires them to provide one of the following proper shipping descriptions on the accompanying shipping paper, as relevant:

> **UN1790, Hydrofluoric acid solution (contains not more than 60% strength), 8 (6.1), PGII**
> *or*
> **UN1790, Hydrofluoric acid solution (contains more than 60% strength), 8, (6.1), PGI**
> *or*
> **UN1052, Hydrogen fluoride, anhydrous, 8, (6.1), PGI (Poison - Inhalation Hazard, Zone C)**

DOT also requires shippers who offer hydrofluoric acid solutions for transportation to affix CORROSIVE and POISON INHALATION HAZARD labels to its containers.

When transporting 1001 lb (454 kg) or more of these solutions, DOT requires carriers to display CORROSIVE placards on the bulk packaging or transport vehicle used for shipment. DOT requires the carrier to display the identification number 1790 on orange panels, across the center area of the CORROSIVE placards, or on white square-on-point diamonds.

DOT requires shippers who offer anhydrous hydrogen fluoride for transportation to affix CORROSIVE and POISON INHALATION HAZARD labels to its containers. DOT also requires carriers to display POISON INHALATION HAZARD placards on the bulk packaging or transport vehicle used for shipment, and when transporting 1001 lb (454 kg) or more, to also post CORROSIVE placards. DOT requires carriers to display the identification number 1052 on orange panels, across the center area of the CORROSIVE placards, or on white square-on-point diamonds.

- Identify the properties of phosphoric acid.
- Identify the industries that use phosphoric acid.
- Identify the principal hazards associated with concentrated phosphoric acid.
- Describe the properties of phosphoric anhydride.
- Describe the means shippers and carriers use to inform emergency responders of the hazards associated with encountering phosphoric acid or phosphoric anhydride during a transportation mishap.

8.11 Phosphoric Acid

Although there are at least eight mineral acids containing phosphorus, *phosphoric acid* is the only one that is commonly encountered. Its chemical formula is H_3PO_4.

Phosphoric acid is used as a raw material for manufacturing organophosphate pesticides (Section 10.19) and a number of commercially important metallic phosphates. For instance, phosphoric acid is used to produce superphosphate fertilizer, a synthetic fertilizer consisting of a mixture of calcium dihydrogen phosphate and calcium sulfate. Phosphoric acid is also used to produce monoammonium phosphate, the dry-chemical fire extinguisher (Section 5.13-C). It is also used to prepare the surface of steel sheets prior to painting.

Various grades of phosphoric acid are commercially available. When an aqueous solution of phosphoric acid is allowed to boil at atmospheric pressure, a syrupy solution containing about 85% phosphoric acid by mass is produced. This is the concentrated phosphoric acid of commerce, some physical properties of which are noted in Table 8.10. In addition, a commercially available food-grade phosphoric acid is used as an ingredient of certain soft drinks and other food products.

Phosphoric acid is manufactured from phosphate rock, an ore containing calcium phosphate, by either of the following methods:

- In the first method, phosphate rock is reacted with sulfuric acid.

$$Ca_3(PO_4)_2(s) \quad + \quad 3H_2SO_4(aq) \quad \longrightarrow \quad 3CaSO_4(s) \quad + \quad 2H_3PO_4(aq)$$

Calcium phosphate   Sulfuric acid   Calcium sulfate   Phosphoric acid

TABLE 8.10	Physical Properties of Concentrated Phosphoric Acid
Melting point	108 °F (42 °C)
Boiling point	500 °F (260 °C)
Specific gravity at 68 °F (20 °C)	1.69
Vapor density (air = 1)	3.4
Vapor pressure at 68 °F (20 °C)	0.0285 mmHg
Solubility in water	Very soluble

- In the second method, elemental phosphorus is produced from phosphate rock by the method noted in Section 7.5; then, the phosphorus is burned to form tetraphosphorus decoxide, which is then reacted with water.

$$P_4(s) \ + \ 5O_2(g) \longrightarrow P_4O_{10}(s)$$

Phosphorus　　　Oxygen　　Tetraphosphorus decoxide

$$P_4O_{10}(s) \ + \ 6H_2O(l) \longrightarrow 4H_3PO_4(aq)$$

Tetraphosphorus decoxide　　Water　　　Phosphoric acid

The reaction with water is violent and releases considerable heat, 41 kJ/mol.

Phosphoric acid exhibits the hazardous features of all corrosive materials: It reacts with metals, metallic oxides, and metallic carbonates, and it damages skin tissue.

8.11-A PHOSPHORIC ANHYDRIDE

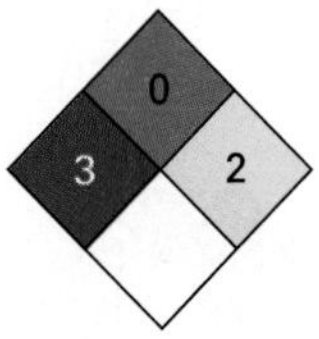

Tetraphosphorus decoxide is the acidic anhydride of phosphoric acid. It is also known as *phosphoric anhydride*. Its chemical formula is P_4O_{10}. This substance has such a strong affinity for water that it is used industrially as a drying agent.

Because its reaction with water releases 41 kJ/mol into the surroundings, phosphoric anhydride should be segregated from combustible matter to prevent its inadvertent ignition. This amount of heat also causes thermal burns when the oxide comes in contact with the skin.

SOLVED EXERCISE 8.8

The DOT regulation at 49 C. F. R. §173.188(a)(1) requires shippers to package phosphoric anhydride in hermetically sealed (airtight) bottles that are inserted into a wooden box and cushioned with noncombustible packing material. What is the most likely reason DOT regulates the shipment of phosphoric anhydride in this fashion?

Solution: Phosphoric anhydride reacts with water—including atmospheric moisture—to produce phosphoric acid, a corrosive material. The most likely reason DOT requires the use of hermetically sealed bottles for containment of phosphoric anhydride during transportation is to prevent entry of water into the containers. When phosphoric anhydride reacts with water, the heat evolved to the surroundings is 41 kJ/mol. This is sufficient heat to initiate the combustion of combustible packing material. The most likely reason DOT requires carriers to use noncombustible packing material when shipping bottles of phosphoric anhydride is to reduce or eliminate the risk of fire.

8.11-B TRANSPORTING PHOSPHORIC ACID AND PHOSPHORIC ANHYDRIDE

When shippers offer phosphoric acid or phosphoric anhydride for transportation, DOT requires them to identify it on the accompanying shipping paper as one of the following, as relevant:

UN3453, Phosphoric acid, solid, 8, PGIII
or
UN1805, Phosphoric acid solution, 8, PGIII
or
UN1807, Phosphorus pentoxide (phosphoric anhydride), 8, PGII

DOT also requires shippers who offer these substances for transportation to affix CORROSIVE labels to its containers.

When carriers transport 1001 lb (454 kg) or more of these substances, DOT requires them to display CORROSIVE placards on the bulk packaging or transport vehicle used for shipment. DOT also requires the carrier to display the relevant identification number, 3453, 1805, or 1807, on orange panels, across the center area of the CORROSIVE placards, or on white square-on-point diamonds.

PERFORMANCE GOALS FOR SECTION 8.12:

- Identify the properties of acetic acid.
- Describe how acetic acid is produced for commercial use.
- Identify the industries that use acetic acid.
- Identify the principal hazards associated with concentrated acetic acid.
- Describe the means shippers and carriers use to inform emergency responders of the hazards associated with encountering acetic acid during a transportation mishap.

8.12 Acetic Acid

Acetic acid is the most commonly encountered organic acid. Its chemical formula is CH_3COOH, or $HC_2H_3O_2$. Each molecule of acetic acid is represented by the following Lewis structure:

$$
\begin{array}{ccc}
\text{H} & & \text{O} \\
| & & \parallel \\
\text{H}-\text{C}-\text{C} & & \\
| & \backslash \\
\text{H} & \text{OH}
\end{array}
$$

Acetic acid is the substance responsible for the sour taste and sharp odor of vinegar, the common food product containing from 3% to 6% acetic acid by volume. It is used primarily by the chemical industry as a raw material for the synthesis of ethyl acetate, vinyl acetate, cellulose acetate, and other chemical and pharmaceutical products.

At room temperature, concentrated acetic acid is a colorless, pungent liquid. The commercial varieties contain from 80% to 99% acid by mass. It is a weak, nonoxidizing acid. The important physical properties of concentrated acetic acid are noted in Table 8.11.

Acetic acid is manufactured by a number of methods, the most popular of which involves the gas-phase combination of methanol and carbon monoxide.

$$
\underset{\text{Methanol}}{CH_3OH(g)} \quad + \quad \underset{\text{Carbon monoxide}}{CO(g)} \quad \longrightarrow \quad \underset{\text{Acetic acid}}{CH_3-\overset{\displaystyle O}{\overset{\parallel}{C}}(g)} \atop \backslash \text{OH}
$$

When aqueous solutions of acetic acid are cooled to temperatures near 61 °F (16 °C), a mixture of liquid and solid phases is produced. The liquid phase contains impurities and is recycled or discarded, but the solid phase typically contains greater than 99% acetic acid by mass. This component is the concentrated acetic acid of commerce, called *glacial acetic acid*. When encountered, glacial acetic acid is generally stored in glass and plastic containers like those shown in Figure 8.9.

TABLE 8.11	Physical Properties of Concentrated Acetic Acid
Melting point	61 °F (17 °C)
Boiling point	244 °F (118 °C)
Specific gravity at 68 °F (20 °C)	1.05
Vapor density (air = 1)	2.1
Vapor pressure at 68 °F (20 °C)	11 mmHg
Flashpoint	109 °F (43 °C)
Autoignition point	800 °F (426 °C)
Lower explosive limit by volume	4%
Upper explosive limit by volume	19.9%
Solubility in water	Infinitely soluble

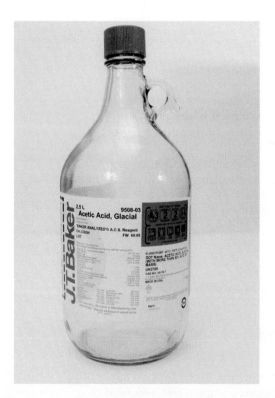

FIGURE 8.9 A small volume of acetic acid is often stored for future use in a container such as the 0.6-gal (2.5-L) glass bottle shown on the left. When transported, the bottle is cushioned within an outer overpack cardboard box to which CORROSIVE labels are affixed. Acetic acid is also stored in larger containers such as the 211-gal (200-L) plastic tote container shown on the right. (*Courtesy of Mallinckrodt Baker, Inc., Phillipsburg, New Jersey.*)

8.12-A VAPORIZATION OF ACETIC ACID

Table 8.1 shows that acetic acid is a very weak acid. Nonetheless, concentrated acetic acid releases a vapor that, when inhaled, is choking and suffocating. The vapor can readily damage the bronchial tract. Exposure to other body tissues, particularly the eyes, results in severe burns.

8.12-B COMBUSTIBLE NATURE OF ACETIC ACID

Concentrated acetic acid is a class II combustible liquid, but aqueous solutions of acetic acid containing less than 80% acid are nonflammable. The complete combustion of acetic acid vapor is represented as follows:

$$CH_3-\overset{\overset{\displaystyle O}{\|}}{\underset{\displaystyle OH}{C}}(g) \ + \ 2O_2(g) \ \longrightarrow \ 2CO_2(g) \ + \ 2H_2O(g)$$

$$\text{Acetic acid} \qquad\qquad \text{Oxygen} \qquad\qquad \text{Carbon dioxide} \qquad \text{Water}$$

Fires involving concentrated acetic acid may be extinguished by diluting the acid with water.

8.12-C WORKPLACE REGULATIONS INVOLVING ACETIC ACID

OSHA requires employers to limit employee exposure to a maximum acetic acid vapor concentration of 10 ppm, averaged over an 8-hr workday.

8.12-D TRANSPORTING ACETIC ACID

When shippers offer acetic acid or its solutions for transportation, DOT requires them to provide one of the following proper shipping descriptions on the accompanying shipping paper, as relevant:

UN2789, Acetic acid glacial (contains more than 80% acid by mass), 8, (3), PGII
or
UN2790, Acetic acid solution (contains more than 10% but not more than 50% acid by mass), 8, PGIII
or
UN2790, Acetic acid solution (contains not less than 50% but not more than 80% acid by mass), 8, PGII
or
UN2789, Acetic acid solution (contains more than 80% acid by mass), 8, (3), PGII

The actual acetic acid concentration is provided in the description in lieu of the parenthetical statement. DOT requires shippers who offer acetic acid solutions having an acid concentration of less than 80% for transportation to affix CORROSIVE labels to its containers.

When carriers transport 1001 lb (454 kg) or more of these acetic acid solutions, DOT requires them to display CORROSIVE placards on the bulk packaging or transport vehicle used for shipment. DOT requires the carrier to display the identification number 2789 on orange panels, across the center area of the CORROSIVE placards, or on white square-on-point diamonds.

When offering glacial acetic acid and its concentrated solutions (>80% by mass) for transportation, DOT requires shippers to affix CORROSIVE and FLAMMABLE LIQUID labels to their containers. When carriers transport 1001 lb (454 kg) or more of these solutions, DOT requires them to display CORROSIVE placards on the bulk packaging or transport vehicle used for shipment. DOT also requires the carrier to display the identification number 2790 on orange panels, across the center area of the CORROSIVE placards, or on white square-on-point diamonds.

SOLVED EXERCISE 8.9

When shippers transport an aqueous solution containing 65% acetic acid within a bulk cargo tank:

(a) What shipping description does DOT require them to enter on the accompanying shipping paper?

(b) How does DOT require the cargo tank to be placarded and marked?

Solution:

(a) In Table 6.3, we learned that when a concentration range of a hazardous material is a component of a proper shipping name, the actual concentration of the hazardous material, if it is within the indicated range, is used in the shipping description in lieu of the range. Consequently, the basic description of an aqueous solution containing 65% acetic acid is the following:

UN2790, Acetic acid solution (contains 65% acid by mass), 8, PGII

(b) DOT requires the carrier to display a CORROSIVE placard on each side and each end of the cargo tank. DOT also requires the identification number 2790 to be displayed within orange panels, across the center area of the CORROSIVE placards, or on white square-on-point diamonds affixed to each side and each end of the tank.

PERFORMANCE GOALS FOR SECTION 8.13:

- Identify the commercially important alkaline metallic hydroxides.
- Identify the industries that use the alkaline metal hydroxides.
- Identify the principal hazards associated with the alkaline metal hydroxides.
- Describe the means shippers and carriers use to inform emergency responders of the hazards associated with encountering an alkaline metallic hydroxide during a transportation mishap.

8.13 Alkaline Metallic Hydroxides

There are three commercially important alkaline corrosive materials: *sodium hydroxide*, *potassium hydroxide*, and *calcium hydroxide*. Their chemical formulas are $NaOH$, KOH, and $Ca(OH)_2$, respectively. We refer to them in this section as the "alkaline metallic hydroxides." Their physical properties are noted collectively in Table 8.12.

8.13-A SODIUM HYDROXIDE AND POTASSIUM HYDROXIDE

Sodium hydroxide is also known commercially as *lye* and *caustic soda*. Potassium hydroxide is called *caustic potash*, or *potash lye*.

Sodium hydroxide is used in countless applications including the purification of petroleum products, the reclaiming of rubber, and the processing of textiles and paper. The chemical industry

TABLE 8.12	Physical Properties of the Alkaline Metal Hydroxides		
	SODIUM HYDROXIDE	**POTASSIUM HYDROXIDE**	**CALCIUM HYDROXIDE**
Melting point	599 °F (315 °C)	680 °F (360 °C)	1076 °F (580 °C)
Boiling point	2534 °F (1390 °C)	2408 °F (1320 °C)	Decomposes
Specific gravity at 68 °F (20 °C)	2.13	2.04	2.50
Solubility in water	42 g/100 g of H_2O	107 g/100 g of H_2O	0.18 g/100 g of H_2O

uses large amounts of sodium hydroxide as a raw material during the manufacturing of soap, rayon, and cellophane.

To the layperson, sodium hydroxide solutions are best known as the constituents of commercial products used to unclog plumbing, such as Drāno and Liquid-Plumr. They are also used in the home in products such as oven cleaners. Sodium hydroxide solutions are also used as industrial cleaners at car- and truck-washing facilities and garages, and they are used at wastewater facilities to remove metallic compounds as water-insoluble metallic hydroxides.

Potassium hydroxide is used mainly by the chemical industry for the production of compounds like fertilizers, soft soaps, and pharmaceutical products. It is also the electrolyte in alkaline storage batteries. Potassium hydroxide was formerly used in liquid drain cleaners, but at 16 C.F.R. §1500.17, the Consumer Product Safety Commission has now banned the manufacture and sale of cleaners containing potassium hydroxide at a concentration of 10% or more by weight (Section 1.4-A).

Sodium hydroxide and potassium hydroxide are generally manufactured by passing an electric current through solutions of sodium chloride (brine) and potassium chloride, respectively.

$$2NaCl(aq) \ + \ 2H_2O(l) \ \longrightarrow \ 2NaOH(aq) \ + \ H_2(g) \ + \ Cl_2(g)$$
Sodium chloride   Water   Sodium hydroxide   Hydrogen   Chlorine

$$2KCl(aq) \ + \ 2H_2O(l) \ \longrightarrow \ 2KOH(aq) \ + \ H_2(g) \ + \ Cl_2(g)$$
Potassium chloride   Water   Potassium hydroxide   Hydrogen   Chlorine

At room temperature, they are commercially available as flakes, pellets, sticks, granules, and concentrated aqueous solutions.

8.13-B CALCIUM HYDROXIDE

Calcium hydroxide is known more commonly as *slaked lime*, or *quicklime*. It is primarily used as a raw material for the production of mortar, plaster, and cement, but it is also widely used commercially for other purposes. Calcium hydroxide is also available in a variety of grades including a food grade that is found in antacids. Because it is only sparingly soluble in water, calcium hydroxide is not available commercially as a solution.

Calcium hydroxide is industrially prepared by reacting calcium oxide and water as follows:

$$CaO(s) \ + \ H_2O(l) \ \longrightarrow \ Ca(OH)_2(s)$$
Calcium oxide   Water   Calcium hydroxide

This reaction is called **slaking**. Calcium oxide is called *unslaked lime*.

8.13-C TRANSPORTING ALKALINE SUBSTANCES

When shippers intend to transport an alkaline substance, DOT requires them to identify it on the accompanying shipping paper as one of the following, as relevant:

> **UN1910, Calcium oxide (unslaked lime), 8, PGIII**

slaking
■ The chemical process associated with reacting calcium oxide (lime) with water to form calcium hydroxide.

or

UN1813, Potassium hydroxide, solid, 8, PGII

or

UN1814, Potassium hydroxide, solution, 8, PGII

or

UN1823, Sodium hydroxide solid, 8, PGII

or

UN1824, Sodium hydroxide, solution, 8, PGII

DOT regulates the transportation of calcium oxide by air only. Hence, shippers are obligated to affix CORROSIVE and CARGO AIRCRAFT ONLY labels to its packaging. DOT requires shippers who intend to transport the compounds other than calcium oxide to affix CORROSIVE labels to their containers.

When carriers transport 1001 lb (454 kg) or more of these substances, DOT requires them to display CORROSIVE placards on the bulk packaging or transport vehicle used for shipment. DOT also requires carriers to display the relevant identification number—1813, 1814, 1923, or 1824—on orange panels, across the center area of the CORROSIVE placards as shown in Figure 8.10, or on white square-on-point diamonds.

DOT does not regulate the transportation of calcium hydroxide.

PERFORMANCE GOALS FOR SECTION 8.14:

- Describe the properties of a chemical waste that exhibits the RCRA characteristic of corrosivity.

FIGURE 8.10 When carriers transport a sodium hydroxide solution, DOT requires them to display the identification number 1824 across the center area of the CORROSIVE placards posted on the transport vehicle. DOT also permits the identification number to be displayed on orange panels or on white square-on-point diamonds. (*Courtesy of Bulk Transportation, Inc., Walnut, California.*)

8.14 RCRA Corrosivity Characteristic

As first noted in Section 1.5-C EPA uses the legal authority of RCRA to regulate the treatment, storage, and disposal of hazardous wastes that exhibit certain characteristics. One of these characteristics is corrosivity.

A waste exhibits **corrosivity** when it possesses either of the following properties:

- It is aqueous and has a pH less than or equal to 2 or greater than or equal to 12.5.
- It is a liquid and corrodes steel at a rate greater than 0.25 in./y (6.25 mm/y) at a test temperature of 130 °F (55 °C) using a specified test method.

A waste that exhibits the RCRA characteristic of corrosivity is assigned the hazardous waste number D002.

corrosivity
■ For purposes of RCRA regulations, the characteristic of a liquid waste that possesses either of the following properties (40 C.F.R. §261.22): It is an aqueous solution and has a pH less than or equal to 2 or equal to or greater than 12.5; *or* it is a liquid and corrodes steel at a rate greater than 0.250 in./y (6.35 mm/y) at a test temperature of 130 °F (54 °C) when measured in accordance with prescribed testing procedures

PERFORMANCE GOALS FOR SECTION 8.15:

- Describe the features of the accident-prevention tags, warning labels, and worded signs used by OSHA that signal the presence of corrosive materials in the workplace.

8.15 Workplace Regulations Involving Corrosive Materials

OSHA is charged with protecting workers from the ill effects caused by exposure to corrosive materials. To accomplish this aim, it requires employers to display accident-prevention tags, warning labels, and worded signs that signal the presence of corrosive materials in the workplace exposure to which could damage property or cause accidental injury to workers or the public. Some examples that relate to corrosive acids and bases are shown in Figure 8.11.

Corrosive materials are frequently used in science laboratories and in certain work environments. In all areas where individuals may be exposed to an injurious corrosive material, OSHA

FIGURE 8.11 OSHA requires employers to post worded signs that warn employees of the presence of corrosive materials in the workplace. OSHA also requires employers at 29 C.F.R. §1200(h)(3)(iv) to affix accident-prevention tags or warning labels to their in-plant containers of corrosive materials.

FIGURE 8.12 OSHA requires employers at 29 C.F.R. §1910.51(c) to provide emergency-use eyewash units and drench showers in areas where employees may be exposed to corrosive materials. Each eyewash unit and shower should deliver 4 gal/min and 20 gal/min of water, respectively, during a 15-min period.

requires the availability of suitable facilities for quick drenching and flushing of the eyes and body. An example of a suitable eyewash station and shower is shown in Figure 8.12.

SOLVED EXERCISE 8.10

A metal-plating company generates aqueous acid waste. When a pH meter is used to establish the pH of a representative sample of the waste, a pH of 1.8 is recorded.

(a) On the basis of this information, does the waste exhibit the RCRA characteristic of corrosivity?
(b) What is the proper shipping description that DOT requires a shipper to enter on a shipping manifest when offering 45 drums of the waste to a carrier for transportation to a waste-treatment facility?

Solution:

(a) An aqueous waste exhibits the RCRA characteristic of corrosivity when a sample has a pH equal to or less than 2.0, or equal to or greater than 12.5. Because the pH of the aqueous acid waste is 1.8, the waste exhibits the characteristic of corrosivity.
(b) As indicated in Table 6.3, DOT requires the shipper to enter either of the following proper shipping descriptions on the waste manifest accompanying the shipment:

> **NA3082, Hazardous waste liquid, n.o.s., 9 (EPA corrosivity)**
> *or*
> **NA3082, Hazardous waste liquid, n.o.s., 9 (D002)**

- Describe the nature of the response actions to be implemented when a corrosive material has been released into the environment.

8.16 Responding to Incidents Involving a Release of a Corrosive Material

First-on-the-scene responders may generally identify the presence of a corrosive material at the emergency scene by reference to the NFPA hazard diamond affixed or imprinted on its storage vessel. Any of the expressions ACID, ALK, or CORR may appear in the bottom quadrant of the hazard diagram. They rapidly convey the message that a corrosive material is contained within the vessel.

At a transportation mishap, emergency responders identify the presence of a corrosive material by observing the following:

- The number 8 as a component of a shipping description of a hazardous material listed on a shipping paper
- The word CORROSIVE and the number 8 printed on black-and-white labels affixed to containment devices
- The word CORROSIVE and the number 8 printed on black-and-white placards displayed on each side and each end of a transport vehicle containing 1001 lb (454 kg) or more of a corrosive material

At an emergency scene involving a corrosive material, the members of the response crew should conduct their work only while wearing fully encapsulated entry suits with clear face shields. Because these suits are fabricated from a material through which the corrosive material cannot penetrate, they protect their wearers from bodily contact with it. The firefighters shown in Figure 8.13 are able to conduct their work without the fear that exposure to a corrosive material will cause a severe burn to body tissue.

Although fully encapsulated entry suits are not themselves designed to protect their users from inhaling toxic vapors or fumes, apparel equipped with self-contained breathing apparatus (SCBA) is marketed for use by emergency responders. The suits shown in Figure 8.13 have been fabricated so SCBA may be used while they are worn. When emergency responders encounter a bulk quantity of a corrosive material that has a significant vapor pressure, they must use SCBA in addition to wearing these suits. Examples of acids that spontaneously emit harmful vapors are oleum, fuming nitric acid, concentrated hydrochloric acid, and concentrated acetic acid.

When they are called to a scene involving the release of a corrosive material, emergency responders should implement the procedures illustrated in Figure 8.14. In addition, they may take either of the following actions:

- Dilute the corrosive material with an approximate volume of water equal to at least 10 times the volume of the material that has been released into the environment. This is usually an adequate response when relatively small quantities of an acid or base have spilled or leaked from its container or storage tank. For instance, suppose that a 1-gal container of liquid sulfuric acid has inadvertently spilled on a laboratory floor and has flowed toward a drain leading to an off-site wastewater treatment plant. In this situation, dilution of the spilled acid with a copious volume of water lessens the acid's corrosive nature.

FIGURE 8.13 These firefighters are wearing fully encapsulating body suits equipped with self-contained breathing apparatus. The use of such suits is recommended when responding to certain emergency incidents, including those that involve the release of an acid or other corrosive material into the environment. The chemical nature of the fabric prevents bodily contact with the corrosive material. The use of the self-contained breathing apparatus is essential when the responding to emergencies in which the corrosive material possesses a significant vapor pressure. (*Courtesy of Lakeland Industries, Inc., Ronkonkoma, New York.*)

■ Neutralize the corrosive material. This action is recommended when a crew responds to the environmental release of a relatively large volume of a corrosive material such as a leak of 10,000 gal (38 m³) of an acid from a storage tank or during a transportation mishap. A bulk quantity of a corrosive material is often transported within a cargo tank like the one whose profile shown in Figure 8.10. It is generally impractical to dilute such a large volume of a corrosive material with water, since an even larger volume of water is needed to effectively reduce its corrosiveness.

A large volume of an acid may be effectively neutralized with solid substances such as either *lime* or *soda ash*, the common names for calcium hydroxide and anhydrous sodium carbonate, respectively. As noted in Section 8.3, an acid reacts with a base to produce a salt of the acid and water; and an acid reacts with a metallic carbonate to produce a salt of the acid, water, and carbon dioxide.

When lime or soda ash is used to neutralize a spill of hydrochloric acid, the resulting chemical reactions may be represented by the following equations:

$$\underset{\text{Calcium hydroxide}}{Ca(OH)_2(s)} \; + \; \underset{\text{Hydrochloric acid}}{2HCl(aq)} \; \longrightarrow \; \underset{\text{Calcium chloride}}{CaCl_2(aq)} \; + \; \underset{\text{Water}}{2H_2O(l)}$$

$$\underset{\text{Sodium carbonate}}{Na_2CO_3(s)} \; + \; \underset{\text{Hydrochloric acid}}{2HCl(aq)} \; \longrightarrow \; \underset{\text{Sodium chloride}}{2NaCl(aq)} \; + \; \underset{\text{Carbon dioxide}}{CO_2(g)} \; + \; \underset{\text{Water}}{H_2O(l)}$$

The use of lime and soda ash results in reducing or eliminating the corrosive nature of hydrochloric acid by chemically converting it into a group of relatively benign substances.

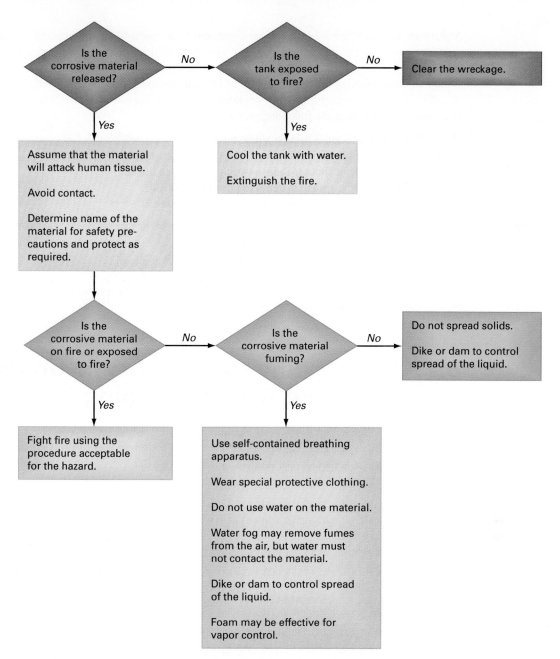

FIGURE 8.14 The recommended procedures to be implemented when responding to a transportation mishap involving the release of a corrosive material from bulk packaging. (*Adapted with permission of the American Society for Testing and Materials, from a figure in ASTM STP 825,* A Guide to the Safe Handling of Hazardous Materials Accidents, *second edition. Copyright © 1990, American Society for Testing and Materials.*)

First-on-the-scene responders arriving at a domestic transportation mishap observe that a 5000-gal (19-m³) overturned tank truck is leaking its liquid contents. The transportation manifest indicates that the consignment consists solely of concentrated hydrochloric acid. The highway on which the truck was traveling is located approximately 250 ft (76.2 m) from the edge of a lake. What procedures should these first responders implement to save lives, property, and the environment?

Solution: Hydrochloric acid is a corrosive material. Consequently, first responders must don an acid-impervious suit to avoid bodily contact with hydrochloric acid. Another hazard associated with concentrated hydrochloric acid is exposure to the vapor that spontaneously evolves from the liquid. To avoid inhaling this vapor, they must use self-contained breathing apparatus.

To save lives, property, and the environment, Figure 8.14 directs the first-on-the-scene responders to implement the following procedures:

- Use self-contained breathing apparatus.
- Wear special protective clothing.
- Use water fog or foam to reduce toxic gas fumes in the air.
- Dike or dam the spilled material to prevent its spread.
- Use lime to neutralize the acid within the diked area.

PERFORMANCE GOALS FOR SECTION 8.17:

- Describe the nature of the actions taken by paramedics when an individual has swallowed a corrosive material.

8.17 Responding to Incidents Involving Acid- and Alkali-Poisoning

Paramedics are often called to assist in situations in which individuals have inadvertently been exposed to corrosive materials. In such incidents, a member of the paramedic team should immediately contact the American Association of Poison Control Centers.* In addition, the following actions are appropriate:

- Because corrosive materials can irreversibly alter skin tissue at the site of contact, the affected area should be thoroughly flushed with water.
- When a corrosive material has been inadvertently splashed into an individual's eyes, pain, swelling, corneal erosion, and blindness can rapidly ensue. Consequently, it is vital to immediately flush the eyes with a gentle stream of running water for at least 30 min (lifting the upper and lower lids occasionally). If the individual is wearing contact lenses, the eyes should first be irrigated for several minutes, and then, the lenses should be removed and the eyes again irrigated. The individual should be advised to promptly contact an ophthalmologist for professional eye treatment.
- When a corrosive material has been swallowed, the individual may experience difficulty in swallowing, nausea, intense thirst, shock, difficulty in breathing, and death. Paramedics should *not* induce vomiting unless given direction to do so by a physician. The stomach wall is relatively tough and normally capable of withstanding the presence of gastric juices having a pH of less than 2. Vomiting should not be induced, because the individual's stomach contents could inadvertently be channeled into the bronchial tract, where they could cause serious damage.
- When a corrosive material has been consumed, paramedics may attempt to neutralize the individual's stomach contents. *Milk of magnesia*, a white suspension of magnesium hydroxide in water, neutralizes acids. Acidic foods such as vinegar and citrus fruit juices neutralize bases. To ensure that the corrosive material has been completely neutralized, the individual should consume a volume of the neutralizing agent at least equal to the amount swallowed.

---

*The American Association of Poison Control Centers can be accessed by telephone at (800) 222-1222. It is essentially a library staffed by nonmedical personnel. Be prepared to give the name of the poisonous product involved in the emergency and any information provided on the label.

Chemical Nature of Acids and Bases

8.1 Complete and balance each of the following equations, and provide an acceptable name for each reactant and product:

(a) $Cr_2O_3(s) + H_2SO_4(aq) \longrightarrow$
(b) $PbCO_3(s) + HCl(aq) \longrightarrow$
(c) $Zn(s) + HClO_4(aq) \longrightarrow$
(d) $NaOH(aq) + H_3PO_4(aq) \longrightarrow$

pH Scale

8.2 A solution having a pH of 4 is how many times more acidic than a solution having a pH of 5?

Corroding Nature of Substances

8.3 Write an equation that illustrates the corrosive effect, if any, that occurs when each substance listed under the A heading reacts with the substance listed under the B heading:

	A	B
(a)	sulfuric acid	iron
(b)	hydrochloric acid	cadmium oxide
(c)	sulfuric acid	chromium(III) carbonate

8.4 How can the storage of a diluted acid within a steel tank connected to metal fittings and pipes constitute a fire and explosion hazard?

Sulfuric Acid

8.5 Complete and balance each of the following equations, and provide an acceptable name for each reactant and product:

(a) $NaCN(s) + H_2SO_4(aq) \longrightarrow$
(b) $K(s) + H_2SO_4(aq) \longrightarrow$
(c) $Cu(s) + H_2SO_4(conc, hot) \longrightarrow$
(d) $NaCl(s) + H_2SO_4(conc) \longrightarrow$

8.6 Why does an aqueous solution of sulfuric acid conduct an electric current, but concentrated sulfuric acid does not?

Nitric Acid

8.7 Complete and balance each of the following equations, and provide an acceptable name for each reactant and product:

(a) $BaCO_3(s) + HNO_3(aq) \longrightarrow$
(b) $Fe_2O_3(s) + HNO_3(aq) \longrightarrow$

(c) $PbS(s) + HNO_3(aq) \longrightarrow$

(d) $Ag(s) + HNO_3(conc) \longrightarrow$

8.8 Following work at a transportation scene involving the release of nitric acid, the members of an emergency-response crew note a yellowing of the skin on their feet. What is the most likely cause of the discoloration?

Hydrochloric Acid

8.9 Complete and balance each of the following equations, and provide an acceptable name for each reactant and product:

(a) $BaO(s) + HCl(aq) \longrightarrow$

(b) $CaSO_3(s) + HCl(aq) \longrightarrow$

(c) $SnO_2(s) + HCl(aq) \longrightarrow$

Perchloric Acid

8.10 Complete and balance each of the following equations, and provide an acceptable name for each reactant and product:

(a) $ZnO(s) + HClO_4(aq) \longrightarrow$

(b) $Al_2O_3(s) + HClO_4(aq) \longrightarrow$

(c) $CdCO_3(s) + HClO_4(aq) \longrightarrow$

(d) $Zn(s) + HClO_4(aq) \longrightarrow$

Hydrofluoric Acid

8.11 Complete and balance each of the following equations, and provide an acceptable name for each reactant and product:

(a) $NiO(s) + HF(aq) \longrightarrow$

(b) $Mg(s) + HF(aq) \longrightarrow$

(c) $Al_2(CO_3)_3(s) + HF(aq) \longrightarrow$

8.12 Why does a sealed glass bottle containing hydrofluoric acid generally burst?

Phosphoric Acid

8.13 Complete and balance each of the following equations, and provide an acceptable name for each reactant and product:

(a) $Fe_2O_3(s) + H_3PO_4(aq) \longrightarrow$

(b) $Sn(s) + H_3PO_4(aq) \longrightarrow$

(c) $CuCO_3(s) + H_3PO_4(aq) \longrightarrow$

Acetic Acid

8.14 Complete and balance each of the following equations, and provide an acceptable name for each reactant and product:

(a) $BaO(s) + CH_3COOH(aq) \longrightarrow$

(b) $Na(s) + CH_3COOH(aq) \longrightarrow$

(c) $MgCO_3(s) + CH_3COOH(aq) \longrightarrow$

Alkaline Metal Hydroxides

8.15 When responding to a domestic transportation mishap involving a spill of caustic soda near a lake, why should emergency responders take precaution to prevent the spilled material from entering the lake?

DOT Regulations Relating to the Transportation of Corrosive Materials

8.16 Nalco Chemical Company (NCC) offers to transport 5000 lb of caustic soda solution by rail tankcar from Chicago, Illinois, to a customer in Baltimore, Maryland. NALX is NCC's reporting mark. What shipping description does DOT require NCC to enter on the accompanying waybill?

Responding to Incidents Involving a Release of Corrosive Materials

8.17 When first-on-the-scene responders arrive at a transportation mishap, they discover that a motor van has overturned, and liquids are leaking from a ruptured stainless-steel drum and broken glass bottles. The accompanying bill of lading reads as follows:

Units	HM	Shipping Description (Identification Number, Proper Shipping Name, Primary Hazard Class or Division, Subsidiary Hazard Class or Division, and Packing Group)	Volume (gal)
1 drum	X	UN2031, Nitric acid, other than red fuming, with more than 70% nitric acid, 8, (5.1), PGI	55
4 glass bottles	X	UN1805, Phosphoric acid solution, 8, PGIII	4

What procedures should be implemented by these responders to save lives, property, and the environment?

Chemistry of Some Water-Reactive Substances

The use of water as a fire extinguishing agent has always been a common practice, but on occasion, the application of water at an emergency-response scene may be inappropriate and potentially dangerous. Water reacts with certain hazardous materials to produce spontaneously flammable products whose ignition can cause secondary fires and make it more difficult to extinguish the primary fire. Furthermore, water reacts with some hazardous materials to produce toxic or corrosive products that can endanger emergency responders and the general public by exposure. We collectively refer to these types of hazardous materials as **water-reactive substances**.

When is it inappropriate for emergency-response crews to use water when responding to an incident involving a release of a hazardous material? To assist in answering this question, this chapter is devoted to the study of substances that react with water at such rapid rates they can pose an unreasonable risk to the health and safety of emergency-response personnel and the general public.

water-reactive substance
■ A substance that by its chemical reaction with water is likely to become spontaneously flammable emit flammable or toxic gases, or generate sufficient heat to self-ignite or cause the ignition of nearby combustible materials

- Describe how consignments of water-reactive substances are identified at transportation mishaps.
- Describe the procedures NFPA uses to warn firefighters of the presence of water-reactive substances at locations where they are stored.
- Describe the general nature of the DOT regulations as they apply to the transportation of water-reactive substances.

9.1 When Is a Substance Water-Reactive?

DOT regulates the transportation of many water-reactive substances as **dangerous-when-wet materials**. These are substances that react with water to produce a spontaneously flammable and/or toxic product. DOT also regulates the transportation of certain other water-reactive substances as spontaneously combustible materials, flammable solids, and corrosive materials. Individual water-reactive substances may be members of one or more of these hazard classes.

When water-reactive substances are transported, DOT requires their shippers and carriers to comply with relevant labeling, marking, and placarding requirements. In particular, DOT requires shippers to affix the relevant labels to packages containing water-reactive substances. When any amount of a dangerous-when-wet material is transported, DOT requires carriers to display DANGEROUS WHEN WET placards on the bulk packaging or transport vehicle used for shipment. When 1001 lb (454 kg) or more of a spontaneously combustible material, flammable solid, or corrosive material is transported, DOT requires their carriers to display the appropriate SPONTANEOUSLY COMBUSTIBLE, FLAMMABLE SOLID, or CORROSIVE placards. When the subsidiary hazard class of a hazardous material is "dangerous-when-wet," DOT also requires carriers to display DANGEROUS WHEN WET placards on the bulk packaging or transport vehicle used for shipment.

For apparent reasons, fires involving water-reactive materials are fought using special fire extinguishers, not water. It is often appropriate to segregate these materials from the remaining cargo so they can burn under controlled conditions.

When water reacts with another substance, the chemical phenomenon is called **hydrolysis**. Chemists represent this process by using the following general equation:

$$A + H_2O(l) \longrightarrow C + D$$

Here, A represents a water-reactive substance and C and D are the substances produced when A reacts with water. The application of water should always be avoided during emergency-response actions at which A is present when either C or D is a flammable or toxic substance.

Consider the metals that react chemically with water to produce flammable hydrogen. These metals constitute the fuels of NFPA class D fires (Section 1.3-D). Several of them are so chemically reactive that they spontaneously ignite in air without exposure to an ignition source. These metals form hydrogen when they react with the moisture in the air. The hydrogen absorbs the heat of reaction and self-ignites. It is this combustion of hydrogen that activates the combustion of the metal.

A substance that self-ignites in air without exposure to an ignition source is said to be **pyrophoric**. OSHA defines this term more specifically as follows: A substance that ignites spontaneously in air at a temperature of 130 °F (54.4 °C) or below.

Many pyrophoric substances react vigorously with water, including atmospheric humidity. To prevent their spontaneous ignition, pyrophoric substances are generally stored and handled under suitable nonaqueous liquids that prevent their contact with air. When pyrophoric substances are also water-reactive, they are processed in enclosed, oxygen-free, dry atmospheres. To avoid their ignition during storage and transportation, their manufacturers seal them in airtight containers.

dangerous-when-wet material
- For purposes of DOT regulations, a material that by interaction with water is likely to become spontaneously flammable or to release a flammable or toxic gas or vapor at a rate greater than 28 in.^{3}/lb (1 L/kg) per hour when subjected to prescribed test procedures

hydrolysis
- The chemical reaction between a substance and water

pyrophoric
- Capable of igniting instantaneously on exposure to air

TABLE 9.1 — Some Classes of Water-Reactive Substances

CLASS OF SUBSTANCE	EXAMPLE	CHEMICAL FORMULA	HYDROLYSIS PRODUCT
Acetyl halides	Acetyl bromide	$CH_3-\overset{\overset{O}{\parallel}}{\underset{Br}{C}}$	Hydrogen bromide
	Acetyl chloride	$CH_3-\overset{\overset{O}{\parallel}}{\underset{Cl}{C}}$	Hydrogen chloride
Acids	Fluorosulfonic acid	$F-SO_2-OH$	Hydrogen fluoride
	Nitrosylsulfuric acid	$O=N-O-SO_2-O-H$	Nitrogen dioxide
Chlorosilanes	Methyldichlorosilane	$CH_3-Si-Cl_2H$	Hydrogen chloride
	Methyltrichlorosilane	$CH_3-Si-Cl_3$	Hydrogen chloride
	Trichlorosilane	Cl_3SiH	Hydrogen chloride
Metallic amides	Lithium amide	$LiNH_2$	Ammonia
	Magnesium diamide	$Mg(NH_2)_2$	Ammonia
Metallic cyanides	Potassium cyanide	KCN	Hydrogen cyanide
	Sodium cyanide	$NaCN$	Hydrogen cyanide
Metallic halides	Aluminum bromide, anhydrous	$AlBr_3$	Hydrogen bromide
	Aluminum chloride, anhydrous	$AlCl_3$	Hydrogen chloride
	Antimony pentafluoride, anhydrous	SbF_5	Hydrogen fluoride
Metallic hypochlorites	Calcium hypochlorite	$Ca(ClO)_2$	Chlorine, hydrogen chloride
	Lithium hypochlorite	$LiClO$	Chlorine, hydrogen chloride
Metallic oxychlorides	Chromium oxychloride	$Cr(OCl)_3$	Hydrogen chloride
	Aluminum phosphide	AlP	Phosphine
Metallic phosphides	Calcium phosphide	Ca_3P_2	Phosphine
	Magnesium aluminum phosphide	$Mg_3P_2 \cdot AlP$	Phosphine
	Magnesium phosphide	Mg_3P_2	Phosphine
	Potassium phosphide	K_3P	Phosphine
	Sodium phosphide	Na_3P	Phosphine
	Stannic phosphide	Sn_3P_4	Phosphine
	Strontium phosphide	Sr_3P_2	Phosphine
	Zinc phosphide	Zn_3P_2	Phosphine

CLASS OF SUBSTANCE	EXAMPLE	CHEMICAL FORMULA	HYDROLYSIS PRODUCT
Nonmetallic halides	Iodine pentafluoride	IF_5	Hydrogen fluoride
	Phosphorus pentachloride	PCl_5	Hydrogen chloride
	Silicon tetrachloride	$SiCl_4$	Hydrogen chloride
	Thionyl chloride	$SOCl_2$	Hydrogen chloride, sulfur dioxide
Sulfides	Ammonium hydrosulfide	NH_4HS	Hydrogen sulfide, ammonia
	Ammonium sulfide	$(NH_4)_2S$	Hydrogen sulfide, ammonia
Others	Chloride dioxide (hydrate)[a]	ClO_2	Chlorine
	Uranium hexafluoride	UF_6	Hydrogen fluoride

[a]See Section 11.8.

The chemical reaction between a substance and water may also form a product that is toxic or corrosive. Some dangerous-when-wet substances produce enough toxic vapor when in contact with water that they pose a health risk to individuals who are located within 0.3–6.0 mi (0.5–10 km) downwind. A list of these substances is provided in DOT's *Emergency Response Guidebook*, which was used to produce Table 9.1. When these substances are involved in emergencies, first-on-the-scene responders should give special attention to them and select appropriate actions to prevent illness or death.

SOLVED EXERCISE 9.1

What information does an NFPA hazard diamond displayed on the exterior wall of a burning shed immediately convey to responding firefighters regarding the water reactivity of the chemical products stored therein?

Solution: Information regarding the water reactivity of a substance is conveyed on an NFPA hazard diamond in two ways. As first noted in Section 1.10, the relative degree of the health, fire, and reactivity hazards of a substance is conveyed through the number that appears in the three topmost quadrants of a diamond. The relative degree of water reactivity is conveyed by the number that appears in the rightmost yellow quadrant as follows:

- A 3 means the substance reacts violently with water.
- A 2 means the substance may react violently with water or may form potentially explosive mixtures with water.
- A 1 means the substance may react with water with some release of energy, but not violently.
- A 0 means the substance does not react with water.

In addition, a W with a line drawn through its center in the bottommost quadrant of the diamond cautions firefighters against applying water when fighting a fire in the shed.

Responding firefighters use this combination of information to select an action appropriate to the incident at hand.

PERFORMANCE GOALS FOR SECTION 9.2:

- Identify and compare the physical and chemical properties of the alkali metals, especially noting their water reactivity.
- Identify the industries that use bulk quantities of the alkali metals.

9.2 Alkali Metals

We consider the properties of three alkali metals in this section: lithium, sodium, and potassium. The properties of these three metals illustrate the uniqueness of their reactions. Some of their physical properties are noted in Table 9.2.

The alkali metals spontaneously ignite. Furthermore, they displace hydrogen from water as the following equations illustrate:

$$2Li(s) \ + \ 2H_2O(l) \ \longrightarrow \ 2LiOH(aq) \ + \ H_2(g)$$
Lithium · Water · Lithium hydroxide · Hydrogen

$$2Na(s) \ + \ 2H_2O(l) \ \longrightarrow \ 2NaOH(aq) \ + \ H_2(g)$$
Sodium · Water · Sodium hydroxide · Hydrogen

$$2K(s) \ + \ 2H_2O(l) \ \longrightarrow \ 2KOH(aq) \ + \ H_2(g)$$
Potassium · Water · Potassium hydroxide · Hydrogen

When metallic lithium reacts with water, the hydrogen produced does not immediately burn (Section 9.2-A), but when metallic sodium and potassium react with water, the hydrogen produced spontaneously bursts into flame.

9.2-A METALLIC LITHIUM

Lithium is a soft, silvery metal and is the least dense solid element at normal conditions. Metallic lithium is so light that its pieces float on the surface of liquid petroleum products like kerosene and gasoline.

Metallic lithium and lithium compounds are valuable raw materials used to manufacture porcelain, ceramics, castings, fungicides, bleaching agents, pharmaceuticals, and greases. In contemporary times, they have become increasingly popular within the chemical industry as raw materials that are used to synthesize organic compounds. Metallic lithium is itself a component of a lightweight magnesium alloy. It is also used in a type of lithium battery that powers laptop computers, cell phones, and digital cameras. Lithium batteries store more energy than traditional

TABLE 9.2	Physical Properties of the Alkali Metals		
	LITHIUM	**SODIUM**	**POTASSIUM**
Melting point	354 °F (179 °C)	208 °F (97.8 °C)	147 °F (64 °C)
Boiling point	2437 °F (1337 °C)	1618 °F (881.4 °C)	1425 °F (774 °C)
Specific gravity at 68 °F (20 °C)	0.53	0.97	0.86
Flashpoint	354 °F (178.9 °C)		
Autoignition point		250 °F (121 °C)	

batteries. When they discharge suddenly during a short circuit, intense heat is generated that can cause fire.

When lithium metal reacts with water, the hydrogen is slowly displaced from the water. It dissipates into the surrounding environment without ever achieving a concentration equal to or greater than 4% by volume, its lower explosive limit.

During its reaction with water, metallic lithium remains in the solid state of matter. Although the heat of the reaction is initially absorbed by the metal, it is transmitted to the surrounding water. The temperature of the metal remains below the boiling point of water.

When metallic lithium is left exposed to the air at room conditions, it does not spontaneously ignite. Although the metal oxidizes in the air, it does so very slowly. Even molten lithium oxidizes so slowly it can be poured from a container in the open air without losing its bright luster.

In an atmosphere of absolutely dry air, lithium metal does not spontaneously burn. When exposed to an ignition source, however, the metal burns in the air with a characteristic crimson color, forming a mixture of lithium oxide and lithium nitride.

$$4Li(s) \quad + \quad O_2(g) \quad \longrightarrow \quad 2Li_2O(s)$$
Lithium   Oxygen   Lithium oxide

$$6Li(s) \quad + \quad N_2(g) \quad \longrightarrow \quad 2Li_3N(s)$$
Lithium   Nitrogen   Lithium nitride

To reduce its potential for ignition, lithium is generally stored under kerosene.

9.2-B METALLIC SODIUM

Like metallic lithium, sodium is also a soft, silvery bright metal. It is the most commonly encountered alkali metal and the only one produced in bulk. Sodium metal is generally available for commercial use in the form of solid bricks.

The majority of the metallic sodium produced in the United States was formerly used as a raw material for the manufacture of the vehicular fuel additives tetraethyllead and tetramethyllead. This use of metallic sodium was sharply curtailed in 1975, when EPA banned the use of leaded gasoline in vehicular fuels. Today, metallic sodium is primarily used as a raw material for the production of highly reactive sodium compounds such as sodium peroxide and sodium hydride. In addition, metallic sodium is used as a catalyst during the production of certain types of synthetic rubber.

Metallic sodium is sometimes encountered commercially in alloys such as the sodium/potassium alloy, sodium/lead alloy, and sodium amalgam. An **amalgam** is a special alloy of one or more metals that have been mixed with elemental mercury. When a sodium alloy is used instead of sodium, the reactions of metallic sodium proceed less vigorously. Consequently, chemical manufacturers have used sodium alloys when the rate of a reaction required careful control; however, amalgams have lost their popularity owing to the toxicity of mercury.

amalgam
■ An alloy of mercury with one or more metals

Metallic sodium reacts very rapidly with water. In fact, the reaction occurs so rapidly that the hydrogen produced is unable to dissipate before it ignites. Instead, it concentrates in the immediate vicinity of the metal, where, induced by the heat of reaction, it self-ignites and spontaneously burns. The metallic sodium absorbs the heat of reaction and melts, thereby exposing an underlying surface of the solid metal for further reaction.

Unlike lithium, metallic sodium does not react with atmospheric nitrogen. Metallic sodium burns in an atmosphere of oxygen, producing a mixture of sodium oxide and sodium peroxide.

$$4Na(s) \quad + \quad O_2(g) \quad \longrightarrow \quad 2Na_2O(s)$$
Sodium   Oxygen   Sodium oxide

$$2Na(s) \quad + \quad O_2(g) \quad \longrightarrow \quad Na_2O_2(s)$$
Sodium   Oxygen   Sodium peroxide

Because sodium peroxide is a powerful oxidizer (Section 11.16-A), it is itself a hazardous material.

In air, metallic sodium ignites spontaneously at room temperature with a characteristic yellow flame. In absolutely dry air, however, this oxidation does not occur at an appreciable rate, which suggests that the oxidation of metallic sodium in air is triggered by the reaction of metallic sodium and atmospheric water vapor. To reduce its potential for ignition, sodium is generally stored under kerosene.

9.2-C METALLIC POTASSIUM

Potassium is a soft, silvery metal. Although it was formerly used with sodium as a heat-exchanger fluid in nuclear reactors, metallic potassium has such relatively few commercial uses now that it is rarely encountered.

Metallic potassium reacts with water even more rapidly than does sodium. The vigorous nature of this reaction is most likely due to the presence of minute amounts of potassium superoxide (see further in the section) that are produced when potassium burns. The hydrogen produced by the reaction of potassium and water initially concentrates around the metal, where it self-ignites. Then, the heat of reaction triggers the burning of the potassium metal.

The combustion of potassium metal in air is associated with the production of a characteristic purple flame. The combustion product is primarily potassium oxide.

$$4K(s) \quad + \quad O_2(g) \quad \longrightarrow \quad 2K_2O(s)$$
$$\text{Potassium} \qquad \text{Oxygen} \qquad \text{Potassium oxide}$$

However, when potassium burns in an atmosphere of pure oxygen, a mixture of potassium oxide, potassium peroxide, and potassium superoxide is produced. A *superoxide*, more properly called a *hyperoxide*, is a compound containing the superoxide ion, whose chemical formula is O_2^-. The following equation illustrates the production of potassium superoxide:

$$K(s) \quad + \quad O_2(g) \quad \longrightarrow \quad KO_2(s)$$
$$\text{Potassium} \qquad \text{Oxygen} \qquad \text{Potassium superoxide}$$

metallic superoxide (metallic hyperoxide)
■ An inorganic compound composed of metallic and superoxide ions

Metallic superoxides (metallic hyperoxides) are extraordinarily reactive oxidizing agents. It is their exceptional chemical reactivity that accounts for the pronounced rate of reaction between metallic potassium and water. Potassium superoxide reacts with atmospheric moisture to produce oxygen and hydrogen peroxide.

$$2KO_2(s) \quad + \quad 2H_2O(l) \quad \longrightarrow \quad 2KOH(aq) \quad + \quad H_2O_2(aq) \quad + \quad O_2(g)$$
$$\text{Potassium superoxide} \qquad \text{Water} \qquad \text{Potassium hydroxide} \quad \text{Hydrogen peroxide} \qquad \text{Oxygen}$$

As potassium burns in moist air, this added presence of oxygen in the burning zone enhances the rate at which the metal burns.

The reaction between potassium superoxide and atmospheric moisture occurs with such ease it has been utilized commercially as a means of supplying oxygen in self-contained breathing apparatus.

9.2-D TRANSPORTING ALKALI METALS

When shippers offer an alkali metal or its alloys, amalgams, or dispersions for transporttion, DOT requires them to identify the appropriate material on the accompanying shipping paper by using the relevant basic description listed in Table 9.3. If shippers offer for transportation a liquid alkali metal alloy that is not listed in Table 9.3, DOT requires them to identify its basic description with the name of the specific alloy entered parenthetically as follows:

UN1421, Alkali metal alloys, liquid, n.o.s. (lithium/potassium), 4.3, PGI (Dangerous When Wet)

DOT also requires shippers to affix a DANGEROUS WHEN WET label on containers of alkali metals and their alloys, amalgams, and dispersions. Furthermore, regardless of the amount transported, DOT requires carriers to display DANGEROUS WHEN WET placards on the bulk packaging or transport vehicle used for their shipment.

TABLE 9.3	Basic Descriptions of the Alkali Metals, Their Amalgams, Dispersions, and Alloys

ALKALI METAL OR ITS AMALGAMS, DISPERSIONS, OR ALLOYS	BASIC DESCRIPTION
Alkali metal amalgams (liquid)	UN1389, Alkali metal amalgams, liquid, 4.3, PGI (Dangerous When Wet)
Alkali metal amalgams (solid)	UN3401, Alkali metal amalgams, solid, 4.3, PGI (Dangerous When Wet)
Alkali metal dispersions	UN1391, Alkali metal dispersions, 4.3, PGI (Dangerous When Wet)
Cesium	UN1407, Cesium, 4.3, PGI (Dangerous When Wet)
Lithium	UN1415, Lithium, 4.3, PGI (Dangerous When Wet)
Potassium	UN2257, Potassium, 4.3, PGI (Dangerous When Wet)
Potassium metal alloys (liquid)	UN1420, Potassium metal alloys, liquid, 4.3, PGI (Dangerous When Wet)
Potassium metal alloys (solid)	UN3403, Potassium metal alloys, solid, 4.3, PGI (Dangerous When Wet)
Potassium sodium alloys (liquid)	UN1422, Potassium sodium alloys, liquid, 4.3, PGI (Dangerous When Wet)
Potassium sodium alloys (solid)	UN3404, Potassium sodium alloys, solid, 4.3, PGI (Dangerous When Wet)
Rubidium	UN1423, Rubidium, 4.3, PGI (Dangerous When Wet)
Sodium	UN1428, Sodium, 4.3, PGI (Dangerous When Wet)

PERFORMANCE GOALS FOR SECTION 9.3:

- Compare the physical and chemical properties of magnesium, zirconium, titanium, aluminum, and zinc.
- Discuss the fire and explosion hazards associated with these metals when they are produced as dusts, powders, chips, turnings, flakes, punchings, borings, ribbons, and shavings during metal-forging and metal-machining operations, particularly when they are coated with combustible cutting oils.
- Identify the industries that use bulk quantities of these combustible metals.

9.3 Combustible Metals

Magnesium, titanium, zirconium, aluminum, and zinc possess a common hazardous feature. Although bulk pieces of these metals are typically difficult to ignite, their finely divided forms readily burn. They are called **combustible metals**. When they burn, these metals are the fuels of NFPA class D fires.

combustible metals
■ Metals including magnesium, aluminum, titanium, and zinc that can readily ignite under certain conditions to produce NFPA class D fires

TABLE 9.4	Physical Properties of Several Combustible Metals		
	MAGNESIUM	**ZIRCONIUM**	**TITANIUM**
Melting point	1200 °F (649 °C)	3326 °F (1830 °C)	3034 °F (1668 °C)
Boiling point	2012 °F (1100 °C)	7910.6 °F (4377 °C)	5948 °F (3260 °C)
Specific gravity at 68 °F (20 °C)	1.74	6.49	4.51
Autoignition point	883 °F (472.78 °C) (powder) 950 °F (510 °C) (ribbons and shavings) 1202 °F (650 °C) (massive chunks)	662 °F (350 °C) (powder)	482 °F (250 °C) (3.175-mm-thick plate); >2192 °F (>1200 °C) (6.35-mm-diameter rod)
	ALUMINUM	**ZINC**	
Melting point	1220 °F (660 °C)	786 °F (419 °C)	
Boiling point	4221 °F (2327 °C)	1665 °F (907 °C)	
Specific gravity at 68 °F (20 °C)	2.70	7.14	
Autoignition point	1400 °F (760 °C) (powder)	860 °F (460 °C) (powder)	

The finely divided forms of the combustible metals noted in this section are regarded as pyrophoric and water-reactive substances to varying degrees. They include dusts, powders, chips, turnings, flakes, punchings, borings, ribbons, and shavings. These forms are commonly produced during metal-forging and metal-machining operations. Often, the heat retained from these processes is sufficient to cause the metals to spontaneously ignite. The finely divided forms of metals generated during their machining, grinding, boring, and other fabrication processes are also likely to be coated with the cutting oils used as lubricants, which can ignite as the primary fuel.

The finely divided forms of combustible metals react with water to produce hydrogen. The spontaneous ignition of the hydrogen triggers the burning of the underlying metal. The rate of hydrogen production is affected by a number of factors including the particle size, distribution and dispersion, purity, and ignition temperature of the metal, as well as the moisture content of the surrounding atmosphere. Some relevant physical properties of these metals are provided in Table 9.4.

9.3-A METALLIC MAGNESIUM

As previously noted in Table 4.1, magnesium occurs to the extent of 1.9% by mass on Earth's surface, where it is typically found in such ores as *magnesite*, *dolomite*, *soapstone*, and *brucite*. Magnesium is also found extensively in underground brines, mainly as magnesium chloride. It is also present in seawater as magnesium chloride and magnesium sulfate.

Magnesium metal is mainly produced for commercial use by the electrolysis of a molten mixture of anhydrous magnesium chloride and potassium chloride.

$$MgCl_2(l) \longrightarrow Mg(s) + Cl_2(g)$$

Magnesium chloride   Magnesium   Chlorine

The potassium chloride increases the conductivity of the salt mixture and reduces its melting point. At the temperature of the electrolytic cell, molten magnesium floats on the salt mixture and is periodically removed through a trough and poured into molds.

Magnesium is an exceptionally lightweight metal. It is therefore often utilized in the construction of aircraft, racing cars, transportable machinery, engine parts, automobile frames and bumpers, and other items for which the mass of the object is pertinent. Because of its popularity, magnesium is commercially available in a variety of sizes ranging from a dust or powder to massive ingots.

As illustrated by the following examples, magnesium is a very reactive metal:

- Although it reacts slowly with cold water, magnesium reacts rapidly with warm and hot water, producing hydrogen.

$$Mg(s) \ + \ 2H_2O(l) \ \longrightarrow \ Mg(OH)_2(s) \ + \ H_2(g)$$

   Magnesium   Water   Magnesium hydroxide   Hydrogen

- Metallic magnesium is also a strong reducing agent. This property is put to use in the metal manufacturing industry, where molten metallic magnesium is used to reduce the metals in certain ores such as those containing titanium and zirconium compounds (Sections 9.3-B and 9.3-C). Magnesium powder is also used as a reducing agent in many fireworks (Section 11.9), in which its reactions contribute to the production of brilliant displays of light.

The most well-known chemical property of magnesium metal is its combustibility. As elemental magnesium burns, approximately 75% combines with atmospheric oxygen to form magnesium oxide.

$$2Mg(s) \ + \ O_2(g) \ \longrightarrow \ 2MgO(s)$$

     Magnesium   Oxygen   Magnesium oxide

The remaining 25% combines with atmospheric nitrogen to form magnesium nitride.

$$3Mg(s) \ + \ N_2(g) \ \longrightarrow \ Mg_3N_2(s)$$

     Magnesium   Nitrogen   Magnesium nitride

It was the combustion of magnesium powder that produced the brilliant burst of light when camera photoflash bulbs were formerly energized.

The burning of bulk pieces of magnesium is very hazardous. When raised to a temperature of 1200 °F (649 °C), massive ingots, castings, and other bulk forms of metallic magnesium melt and burn vigorously with the production of brilliant, blinding white flames. Burning as it flows, the liquid metal drops from its initial location to lower levels, where it ignites most combustible materials encountered in its pathway.

9.3-B METALLIC TITANIUM

As noted in Table 4.1, titanium occurs on Earth's surface to the extent of 0.58% by mass. The element occurs primarily as titanium(IV) oxide. One such ore, *rutile*, is abundant in beach sands in Australia, South Africa, and Sri Lanka.

The titanium manufacturing process consists of the following two steps:

- First, rutile or a similar ore is reacted with chlorine and carbon at approximately 1112 °F (600 °C) to produce titanium(IV) chloride (Section 9.8-H).

$$TiO_2(s) \ + \ 2C(s) \ + \ 2Cl_2(g) \ \longrightarrow \ TiCl_4(g) \ + \ 2CO(g)$$

 Titanium(IV) oxide   Carbon   Chlorine   Titanium(IV) chloride Carbon monoxide

- Then, titanium(IV) chloride is reacted with molten magnesium within a steel vessel under an atmosphere of an inert gas like helium or argon at approximately 1472 °F (800 °C).

$$TiCl_4(g) \ + \ 2Mg(l) \ \longrightarrow \ Ti(s) \ + \ 2MgCl_2(s)$$

  Titanium(IV) chloride Magnesium   Titanium   Magnesium chloride

The magnesium chloride formed as a by-product is leached from the reaction mixture, leaving the basic form of the metal known as *titanium sponge*, so called because its physical appearance resembles the shape of a sponge.

These production steps are very costly, which currently hinders the widespread use of titanium.

Although it is 45% lighter in mass than steel, metallic titanium is just as strong as steel. Because titanium possesses this combination of lightness and strength, titanium is often alloyed with aluminum and vanadium, which is then used to manufacture aircraft and automotive parts, jet engines, and missiles. In modern-day commercial and military jet aircraft, titanium is replacing aluminum in blades, discs, rings, and engine cases, as well as bulkheads, tail sections, landing gear, wing supports, and fasteners.

In the automotive industry, titanium metal is now used in several consumer car and motorcycle applications. It is primarily used for exhaust systems, suspension springs, engine valves, connecting rods, and turbocharger compressor wheels.

Metallic titanium is also found in everyday items such as jewelry, skis, and golf equipment. Titanium prostheses for hip and knee replacements are often selected by orthopedic surgeons.

Titanium metal has a great affinity for oxygen. Once exposed to air, the surface of the metal is quickly passivated with a thin layer of titanium(IV) oxide, which serves to protect the underlying metal from further chemical attack. This passivated metal is highly resistant to corrosion by most acids, chlorine, oxidizing agents, and seawater.

The resistance of passivated titanium to chemical attack is put to good use in the following ways:

■ Because it is resistant to corrosion by seawater, metallic titanium is used in the structure of submarine hulls and underwater machinery. Most likely for the same reason, Russia chose in 2007 to mount a titanium flag on the ocean floor at the North Pole from a deep submergence vehicle.

■ Because it is resistant to corrosion by most acids and chlorine, metallic titanium is also used in the construction of vessels in which these raw materials will be stored, transported, or reacted. One notable exception to this general observation, however, is that hydrofluoric acid cannot be stored in titanium vessels. Hydrofluoric acid chemically attacks titanium.

Although the bulk forms of titanium are not considered hazardous, finely divided titanium is a pyrophoric metal that poses a dangerous risk of fire and explosion. This form of titanium is generated when the metal is fabricated and its parts are cut, formed, and welded.

When metallic titanium burns in air, a mixture of titanium(IV) oxide and titanium(III) nitride is produced.

$$\underset{\text{Titanium}}{\text{Ti}(s)} + \underset{\text{Oxygen}}{\text{O}_2(g)} \longrightarrow \underset{\text{Titanium(IV) oxide}}{\text{TiO}_2(s)}$$

$$\underset{\text{Titanium}}{2\text{Ti}(s)} + \underset{\text{Nitrogen}}{\text{N}_2(g)} \longrightarrow \underset{\text{Titanium(III) nitride}}{2\text{TiN}(s)}$$

The reactions may be initiated by the combustion of the hydrogen produced when the finely divided metal reacts with atmospheric moisture.

$$\underset{\text{Titanium}}{\text{Ti}(s)} + \underset{\text{Water}}{2\text{H}_2\text{O}(l)} \longrightarrow \underset{\text{Titanium(IV) oxide}}{\text{TiO}_2(s)} + \underset{\text{Hydrogen}}{2\text{H}_2(g)}$$

SOLVED EXERCISE 9.2

The OSHA regulation at 29 C.F.R. §1910.109(g)(2)(iv)(e) stipulates that metal powders such as titanium must be kept dry and stored in containers or bins that are moisture-resistant or weathertight. What is the most likely reason OSHA regulates the storage of metal powders in this fashion?

Solution: Various-sized pieces of hot metal are produced when titanium parts are mechanically fabricated. These pieces include titanium dust, powder, chips, turnings, flakes, ribbons, and shavings, each of which is prone to self-ignition. The risk of their ignition must be acknowledged when these pieces are first generated, since at that time, they retain the heat of friction. Once the metal ignites, it is essential to extinguish the fire as quickly as possible to prevent its subsequent spread to adjoining areas. This can effectively be accomplished through use of an automatic sprinkling system that applies water to the fire in an amount sufficient to rapidly cool the burning metal.

9.3-C METALLIC ZIRCONIUM

Metallic zirconium is primarily produced from *zirconia*, a naturally occurring ore containing zirconium(IV) oxide (ZrO_2), by the method previously noted for production of titanium. Today, the metal is used almost exclusively in the nuclear and steel industries. In the nuclear industry, zirconium is used in reactor cores and to clad uranium fuel rods, and in the steel industry, zirconium is used to remove oxygen from molten steel. Zirconium dust was formerly used as the active component of specialized camera photoflash bulbs. It was also used militarily as an incendiary agent.

Zirconium dust constitutes a risk of fire and explosion. When it is transferred from one container to another, the dust particles absorb the heat generated by friction as the particles move against one another. When these particles are hot, as when they are first produced, the temperature of the dust can exceed its autoignition point, whereupon it spontaneously ignites.

The ease of ignition of zirconium dust is associated with its former use as an incendiary agent in a shotgun round known as *Dragon's Breath*. The round was inserted into the magazine of a shotgun that when fired, burst into flame and shot from the gun's barrel like a flamethrower. However, zirconium dust is extremely expensive. Despite its fearsomeness in application, the routine military use of Dragon's Breath is cost-prohibitive.

When zirconium dust burns in air, the resulting fire provides an exceedingly brilliant, white flame. The combustion results in the production of a mixture of zirconium oxide and zirconium nitride.

$$Zr(s) \quad + \quad O_2(g) \quad \longrightarrow \quad ZrO_2(s)$$
Zirconium   Oxygen   Zirconium oxide

$$2Zr(s) \quad + \quad N_2(g) \quad \longrightarrow \quad 2ZrN(s)$$
Zirconium   Nitrogen   Zirconium nitride

The reactions may be initiated by the combustion of hydrogen, produced when the dust reacts with atmospheric moisture.

$$Zr(s) \quad + \quad 2H_2O(g) \quad \longrightarrow \quad ZrO_2(s) \quad + \quad 2H_2(g)$$
Zirconium   Water   Zirconium oxide   Hydrogen

9.3-D METALLIC ALUMINUM

Table 4.1 lists aluminum as the most abundant metal on Earth's surface, 7.5% by mass. As the element, aluminum is too reactive to be found in an uncombined form in nature. Instead, it is found in minerals like cryolite and in such materials as clay and feldspar, in which it is combined with silicon and oxygen.

Aluminum metal is produced by the electrolysis of aluminum oxide dissolved in fused cryolite, which serves as the electrolyte.

$$2Al_2O_3(s) \quad \longrightarrow \quad 4Al(l) \quad + \quad 3O_2(g)$$
Aluminum oxide   Aluminum   Oxygen

The aluminum metal is denser than molten cryolite; therefore, it collects at the bottom of the electrolytic cell, from which it is tapped and cast into ingots.

Aluminum is an example of a **malleable** metal; that is, it can be rolled into a relatively thin sheet or foil. Aluminum foil is a popular kitchen item. Firefighters encounter aluminum foil bonded to fabric in the aluminized protective coats commonly called *silvers*. Figure 9.1 shows their use when firefighters must approach fires that release high levels of radiant heat.

malleable
■ A physical property of metals characterized by a capability of being hammered into sheets

Because it is lightweight and durable, aluminum sheeting is used to produce a wide variety of commercial products including the common soda can. In building construction, aluminum sheeting is used as siding, eaves, screens, and window and door frames. During building fires, these aluminum components can melt and collapse, since the temperature attained often exceeds the melting point of aluminum, 1220 °F (660 °C).

The principal material employed as the metal skin of most standard aircraft is an aluminum or aluminum alloy sheeting. The metal cannot be used as the outer skin of supersonic aircraft,

FIGURE 9.1 When responding to fires involving aluminum alkyl compounds and their halide and hydride derivatives, firefighters should wear special protective clothing. This aluminum suit reflects heat from the body of its wearer and provides protection against bodily contact with the reactive substances as they burn. (*Courtesy of Lakeland Industries, Inc., Ronkonkoma, New York.*)

however, since it becomes too hot and softens from the friction generated by the fast movement through the air.

Aluminum is also a **ductile** substance; that is, it can be drawn into wires. Although aluminum wire is twice as effective as copper wire for conducting electricity, the use of aluminum electrical wiring is undesirable. This unfavorable feature is linked to the heavy deposit of aluminum oxide produced on the surface of aluminum wiring, which restricts the flow of electrical current and causes the metal to become overheated. This situation constitutes a fire hazard.

The deposition of aluminum oxide on the surface of aluminum wiring is associated with the extraordinary affinity that aluminum and oxygen have for each other. Aluminum exposed to air is covered with a thin, tenacious coating of aluminum oxide that gives the metal a dull, white luster. Although this oxide coating protects the underlying metal from further oxidation, the coating does not protect the aluminum from other forms of chemical attack. Seawater, for instance, corrodes metallic aluminum.

This chemical affinity of metallic aluminum for oxygen is evident from the chemical reaction noted in Figure 9.2 involving powdered aluminum and iron(III) oxide. The mixture of 27% powdered aluminum and 73% iron(III) oxide is commonly called **thermite**. When the mixture is activated by a magnesium fuse, a reaction producing molten iron and aluminum oxide occurs.

$$2Al(s) \quad + \quad Fe_2O_3(s) \quad \longrightarrow \quad 2Fe(l) \quad + \quad Al_2O_3(s)$$

Aluminum · · · · · · · Ferric oxide · · · · · · · Iron · · · · · · · Aluminum oxide

This phenomenon is called the **thermite reaction**. It releases such considerable heat that temperatures of approximately 3990 °F (2199 °C) result. Because this temperature is above the melting point of iron [2800 °F (1538 °C)], it is produced by this chemical reaction as a molten, white-hot liquid. The thermite reaction cannot be stopped with water.

In the days of the old West, the thermite reaction was used to weld rails together during the construction of railroads. During World War II, thermite was used extensively as the incendiary agent in bombs, especially against the British during the London Blitz. However, in contemporary warfare, the use of thermite in incendiary weapons is essentially banned by Protocol III of the Convention on Certain Conventional Weapons (Section 7.5). Protocol III limits the use of all incendiary weapons against civilian targets.

ductile
■ The physical property of metals associated with their ability to be stretched into wires

thermite
■ The mixture of 27% powdered aluminum and 73% iron(III) oxide

thermite reaction
■ The chemical reaction used in some welding operations and incendiary weapons during which elemental iron is produced when elemental aluminum reduces iron(III) oxide

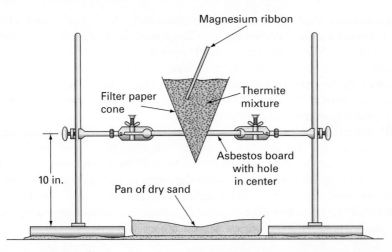

FIGURE 9.2 The thermite reaction may be illustrated in a chemical laboratory as follows: A mixture of powdered aluminum and iron(III) oxide is inserted into a cone over a pan of dry sand, which protects the tabletop from possible damage. A magnesium ribbon is inserted into the thermite mixture and ignited. The heat of combustion initiates the reaction between aluminum and iron(III) oxide. Molten iron spits from the reaction mixture and drips into the sand.

Although elemental aluminum is very stable in the form of foil and sheets, aluminum dust and powder are pyrophoric materials that pose the risk of fire and explosion. The aluminum burns violently in air with an intensely bright, white and orange flame producing a mixture of aluminum oxide and aluminum nitride.

$$4Al(s) \quad + \quad 3O_2(g) \quad \longrightarrow \quad 2Al_2O_3(s)$$

Aluminum   Oxygen   Aluminum oxide

$$2Al(s) \quad + \quad N_2(g) \quad \longrightarrow \quad 2AlN(s)$$

Aluminum   Nitrogen   Aluminum nitride

These reactions may be initiated by the combustion of hydrogen, produced when the dust and powder react with atmospheric moisture.

$$2Al(s) \quad + \quad 3H_2O(l) \quad \longrightarrow \quad Al_2O_3(s) \quad + \quad 3H_2(g)$$

Aluminum   Water   Aluminum oxide   Hydrogen

Powdered aluminum burns spontaneously on contact with liquid oxygen. Aluminum oxide is the sole product of combustion.

The reactivity of aluminum powder is put to use in many fireworks (Section 11.9), in which the metal reacts to produce a brilliant display of orange light. It is also incorporated into some paints and varnishes for decorative purposes as well as for its heat-reflective property; but consideration must be given to their use, since these coatings may behave as flammable solids once the paint solvent has evaporated. Aluminum powder is also a component of solid rocket fuels, in which it is mixed with ammonium nitrate and ammonium perchlorate. The mixture of powdered aluminum and ammonium nitrate is an explosive called *ammonal*.

The catastrophe of the German dirigible *Hindenburg* may have been linked with the combustion of aluminum powder. The exterior surface of the dirigible consisted of a cloth cover impregnated with a doping mixture of aluminum powder and ferric oxide. The presence of aluminum powder provided a surface having high reflectivity. The cover was intended to serve an important purpose: The aluminum particles reflected heat off the vessel and prevented the hydrogen from expanding. The prevailing theory is that the aluminum powder first caught fire at an isolated location, perhaps triggered by static electricity or lightning. Once initiated, the fire then rapidly spread across the entire covering, ultimately igniting the reserves of hydrogen. The resulting inferno consumed the vessel.

In circumstances where the temperature is substantially elevated compared with the norm, even bulk aluminum acts as a fast-burning fuel. The skin of shuttle aircraft, for example, must be armored with heat shielding to protect the shuttle when it reenters Earth's atmosphere from outer space, where it can experience temperatures in excess of 3000 °F (1650 °C). If this shielding is pierced in any way, the underlying aluminum becomes superheated. Aluminum melts at 1220 °F (660 °C) and vaporizes at 4221 °F (2327 °C). At these temperatures, aluminum fires occur when oxygen is available to support the combustion.

In 2003, the space shuttle *Columbia* exploded on reentry into Earth's atmosphere, killing the seven astronauts onboard. The shuttle was covered with more than 20,000 interlocking ceramic tiles designed to protect the aluminum alloy shell from the heat of reentry. Experts who examined debris from the accident wreckage observed droplets of aluminum and stainless steel. This observation suggested that the cause of the accident was linked with the loss of tiles on the left wing, especially along its leading edge. Without its protective covering, the underlying aluminum alloy most likely burned, ultimately destroying the entire shuttle.

9.3-E METALLIC ZINC

Zinc is produced primarily by means of the following two-step thermal process:

- First, the zinc sulfide ore *sphalerite*, or *zinc blende*, is roasted in air to produce zinc oxide.

$$2ZnS(s) \ + \ 3O_2(g) \ \longrightarrow \ 2ZnO(s) \ + \ 2SO_2(g)$$
Zinc sulfide   Oxygen   Zinc oxide   Sulfur dioxide

- Then, the oxide is reduced with carbon monoxide.

$$ZnO(s) \ + \ CO(g) \ \longrightarrow \ Zn(g) \ + \ CO_2(g)$$
Zinc oxide   Carbon monoxide   Zinc   Carbon dioxide

The zinc vapor produced by the reaction is then distilled, condensed, and cast into ingots. The zinc deposits on the walls of the distillation apparatus as a gray, finely divided powder known as *zinc dust*.

The zinc manufacturing process is complicated by the presence of the metal impurities silver, lead, copper, arsenic, antimony, and manganese, all of which occur naturally in sphalerite. These metals are removed by a combination of chemical processes. The manufacturing process is also complicated by the simultaneous production of the pollutant sulfur dioxide (Section 10.11), which must be scrubbed from the off-gas plume generated during the roasting process.

Metallic zinc is used for several purposes. The metal is coated on iron products to protect them from corrosion by the air. This zinc-coated iron is said to be **galvanized**. Zinc is also used as a component of several alloys; for example, zinc and copper are combined in the molten state to produce *brass*. Metallic zinc is also used in the manufacture of dry-cell batteries and a variety of structural materials. Zinc dust is a component of certain primers and rust-resistant paints.

galvanized
■ The process of coating a metal with a protective layer of elemental zinc

Zinc is hazardous only as its dust. Especially when it is hot, zinc dust is a pyrophoric material that poses a fire and explosion hazard. It ignites spontaneously in air with a green flame, producing zinc oxide as the sole combustion product.

$$2Zn(s) \ + \ O_2(g) \ \longrightarrow \ 2ZnO(s)$$
Zinc   Oxygen   Zinc oxide

The reaction may be initiated by the combustion of hydrogen, produced when the dust reacts with atmospheric moisture.

$$Zn(s) \ + \ H_2O(l) \ \longrightarrow \ ZnO(s) \ + \ H_2(g)$$
Zinc   Water   Zinc oxide   Hydrogen

9.3-F TRANSPORTING COMBUSTIBLE METALS

When shippers offer a combustible metal for transportation, DOT requires them to identify the appropriate material on the accompanying shipping paper by selecting the relevant entry from Table 9.5. DOT also requires shippers who offer a combustible metal for transportation to affix the

TABLE 9.5	Basic Descriptions of Several Combustible Metals
COMBUSTIBLE METALS	**BASIC DESCRIPTION**
Aluminum, molten	NA9260, Aluminum, molten, 9, PGI
Aluminum powder	UN1309, Aluminum powder, coated, 4.1, PGII *or* UN1396, Aluminum powder, uncoated, 4.3, PGII (Dangerous When Wet)
Aluminum silicon powder	UN1398, Aluminum silicon powder, 4.3, PGIII (Dangerous When Wet)
Magnesium (with more than 50% magnesium in pellets, turnings, or ribbons)	UN1869, Magnesium, 4.1, PGIII
Magnesium alloys (with more than 50% magnesium in pellets, turnings, or ribbons)	UN1869, Magnesium alloys, 4.1, PGIII
Magnesium alloys, powder	UN1418, Magnesium alloys, powder, 4.3, (4.2), PGI (Dangerous When Wet) *or* UN1418, Magnesium alloys, powder, 4.3, (4.2), PGII (Dangerous When Wet) *or* UN1418, Magnesium alloys, powder, 4.3, (4.2), PGIII (Dangerous When Wet)
Magnesium granules (particle size not less than 149 microns)	UN2950, Magnesium granules, coated, 4.3, PGIII (Dangerous When Wet)
Magnesium powder	UN1418, Magnesium powder, 4.3, (4.2), PGI (Dangerous When Wet) *or* UN1418, Magnesium powder, 4.3, (4.2), PGII (Dangerous When Wet) *or* UN1418, Magnesium powder, 4.3, (4.2), PGIII (Dangerous When Wet)
Magnesium scrap	UN1869, Magnesium scrap, 4.1, PGIII
Titanium powder	UN2546, Titanium powder, dry, 4.2, PGI
Titanium (powder), wetted with not less than 25% water (a visible excess of water must be present) **(a)** mechanically produced, particle size less than 53 microns; **(b)** chemically produced, particle size less than 840 microns	UN1352, Titanium powder, 4.1, PGII
Titanium sponge granules	UN2878, Titanium sponge granules, 4.1, PGIII
Titanium sponge powders	UN2878, Titanium sponge powders, 4.1, PGIII
Zinc ashes	UN1435, Zinc ashes, 4.3, PGIII (Dangerous When Wet)

(continued)

TABLE 9.5 | Basic Descriptions of Several Combustible Metals (*continued*)

COMBUSTIBLE METALS	BASIC DESCRIPTION
Zinc dust	UN1436, Zinc dust, 4.3, (4.2), PGI (Dangerous When Wet)
Zinc powder	UN1436, Zinc powder 4.3, (4.2), PGI (Dangerous When Wet)
Zirconium, dry [coiled wire, finished metal sheets, strip (thinner than 254 microns but not thinner than 18 microns)]	UN2858, Zirconium, dry, 4.1, PGIII
Zirconium, dry (finished sheets, strip, or coil wire)	UN2009, Zirconium, dry, 4.1, PGIII
Zirconium powder	UN2008, Zirconium powder, dry, 4.2, PGI *or* UN2008, Zirconium powder, dry, 4.2, PGII *or* UN2008, Zirconium powder, dry, 4.2, PGIII
Zirconium powder, wetted with not more than 25% water [(a visible excess of water must be present) **(a)** mechanically produced, particle size less than 53 microns; **(b)** chemically produced, particle size less than 840 microns)]	UN1358, Zirconium powder, wetted, 4.1, PGII
Zirconium scrap	UN1932, Zirconium scrap, 4.2, PGIII
Zirconium suspended in a liquid	UN1308, Zirconium suspended in a liquid, 3, PGI *or* UN1308, Zirconium suspended in a liquid, 3, PGII *or* UN1308, Zirconium suspended in a liquid, 3, PGIII

appropriate FLAMMABLE SOLID, SPONTANEOUSLY COMBUSTIBLE, or DANGEROUS WHEN WET label on its container. When they transport zinc powder or zinc dust, shippers affix SPONTANEOUSLY COMBUSTIBLE *and* DANGEROUS WHEN WET labels on its container.

When 1001 lb (454 kg) or more of a flammable solid or a spontaneously combustible material is transported, DOT requires carriers to display the appropriate FLAMMABLE SOLID or SPONTANEOUSLY COMBUSTIBLE placards on the bulk packaging or transport vehicle used for shipment. When any amount of a dangerous-when-wet material is transported, DOT requires carriers to display DANGEROUS WHEN WET placards.

When shippers or carriers transport a bulk quantity of molten aluminum, DOT requires them to mark the packaging with the words MOLTEN ALUMINUM. DOT also requires the identification number 9260 to be displayed on an orange panel or on a white square-on-point diamond on each side and each end of the packaging.

MOLTEN ALUMINUM		9260

DOT does not require carriers to display CLASS 9 placards on the packaging.

9.4 Aluminum Alkyl Compounds and Their Derivatives

The **aluminum alkyls** are a group of compounds whose molecules consist of an aluminum atom covalently bonded to carbon atoms that are components of three alkyl groups (Section 12.2). An example of an aluminum alkyl compound is triethylaluminum, whose chemical formula is $Al(CH_2CH_3)_3$, or $Al(C_2H_5)_3$. In this instance, the alkyl group is named *ethyl*, which is designated by the formula $-CH_2CH_3$. In the chemical industry, triethylaluminum is known as TEA. In this section, we consider it as representative of all aluminum alkyl compounds.

Two special groups of aluminum alkyl compounds are the *aluminum alkyl halides* and *aluminum alkyl hydrides*. These compounds are the halide and hydride derivatives of aluminum alkyl compounds, respectively, in which one or two halide or hydrogen atoms substitute for an alkyl group. Examples of these derivatives are diethylaluminum chloride and diisobutylaluminum hydride, whose formulas are $(C_2H_5)_2AlCl$ and $[(CH_3)_2CHCH_2]_2AlH$, respectively. The alkyl group having the formula $(CH_3)_2CHCH_2-$ is named *isobutyl*. In the chemical industry, these compounds are sometimes known as DEAC and DIBAH, respectively. In this section, we consider them as representative of the halide and hydride derivatives of all aluminum alkyl compounds.

Table 9.6 provides some physical properties of triethylaluminum, diethylaluminum chloride, and diisobutylaluminum hydride.

aluminum alkyl
- A compound whose molecules are composed of an aluminum atom covalently bonded to three carbon atoms, each of which is a component of an alkyl group

9.4-A USES OF THE ALUMINUM ALKYL COMPOUNDS AND THEIR DERIVATIVES

The aluminum alkyls are primarily used by the chemical industry as polymerization catalysts, one of which is a mixture of titanium(IV) chloride and an aluminum alkyl. It is called a

TABLE 9.6	Physical Properties of an Aluminum Alkyl Compound and Two Derivatives		
	TRIETHYLALUMINUM	**DIETHYLALUMINUM CHLORIDE**	**DIISOBUTYLALUMINUM HYDRIDE**
Melting point	−62 °F (−52 °C)	−121 °F (−85 °C)	−112 °F (−80 °C)
Boiling point	367 °F (186 °C)	260.6 °F (127 °C)[a]	237 °F (114 °C)[b]
Specific gravity	0.837[c]	0.961[d]	0.798[d]
Vapor pressure	0.0147 mmHg[c]	0.17 mmHg[d]	
Flashpoint	−63 °F (−53 °C)	−9.4 °F (−23 °C)	
Autoignition temperature	Spontaneously flammable in air	Spontaneously flammable in air	Spontaneously flammable in air

[a]At 50 mmHg (6.67 kPa).
[b]At 3 mmHg (0.3 kPa).
[c]At 68 °F (20 °C).
[d]At 77 °F (25 °C).

Ziegler–Natta catalyst, after Karl Ziegler and Giulio Natta, the chemists who first discovered its catalytic capability. Aluminum alkyl halides and aluminum alkyl hydrides are also primarily used as catalysts in the chemical industry.

Aluminum alkyl compounds have also been used by the military, albeit rarely, as incendiary agents. For example, triethylaluminum has been used in flamethrowers as their active component. Trimethylaluminum has also been used to produce a luminous trail in the upper atmosphere when tracking the location of rockets.

9.4-B PROPERTIES OF THE ALUMINUM ALKYL COMPOUNDS AND THEIR DERIVATIVES

The aluminum alkyl compounds and their derivatives are spontaneously combustible, pyrophoric, violently water-reactive, and highly toxic liquids. They are commercially available as individual compounds and as solutions in which they are dissolved in organic solvents. When triethylaluminum, diethylaluminum chloride, and diisobutylaluminum hydride spontaneously ignite, their combustion reactions are represented as follows:

$$2(C_2H_5)_3Al(l) \ + \ 21O_2(g) \ \longrightarrow \ Al_2O_3(s) \ + \ 12CO_2(g) \ + \ 15H_2O(g)$$

Triethylaluminum (TEA) Oxygen Aluminum oxide Carbon dioxide Water

$$2(C_2H_5)_2AlCl(l) \ + \ 14O_2(g) \ \longrightarrow \ Al_2O_3(s) \ + \ 8CO_2(g) \ + \ 9H_2O(g) \ + \ 2HCl(g)$$

Diethylaluminum chloride (DEAC) Oxygen Aluminum oxide Carbon dioxide Water Hydrogen chloride

$$2[(CH_3)_2CHCH_2]_2AlH(s) \ + \ 27O_2(g) \ \longrightarrow \ Al_2O_3(s) \ + \ 16CO_2(g) \ + \ 19H_2O(g)$$

Diisobutylaluminum hydride (DIBAH) Oxygen Aluminum oxide Carbon dioxide Water

When triethylaluminum and diethylaluminum chloride react with water, the flammable gas ethane (Section 12.2) is produced as a hydrolysis product.

$$Al(C_2H_5)_3(l) \ + \ 3H_2O(l) \ \longrightarrow \ Al(OH)_3(s) \ + \ 3C_2H_6(g)$$

Triethylaluminum (TEA) Water Aluminum hydroxide Ethane

$$(C_2H_5)_2AlCl(l) \ + \ 3H_2O(l) \ \longrightarrow \ Al(OH)_3(s) \ + \ 2C_2H_6(g) \ + \ HCl(g)$$

Diethylaluminum chloride (DEAC) Water Aluminum hydroxide Ethane Hydrogen chloride

However, diisobutylaluminum hydride is a reducing agent. When it reacts with water, the flammable gases isobutene and hydrogen are produced.

$$[(CH_3)_2CHCH_2]_2AlH(s) \ + \ 3H_2O(l) \ \longrightarrow \ Al(OH)_3(s) \ + \ 2(CH_3)_2CH{=}CH_2(g) \ + \ 2H_2(g)$$

Diisobutylaluminum hydride (DIBAH) Oxygen Aluminum hydroxide Isobutene Hydrogen

When water is applied to these reactive substances, the gaseous hydrolysis products immediately burst into flame as they are generated. The considerable heat evolved to the environment often triggers secondary fires. Bulk quantities of aluminum alkyl compounds burn so vigorously and persistently that they pose a dangerous risk of fire and explosion. The heat of combustion that evolves necessitates that firefighters wear special protective gear like the silvers shown in Figure 9.1 when combating these fires.

To warn users of the inordinately high degree of reactivity associated with the aluminum alkyl compounds, chemical manufacturers often label their containers as follows:

> SPONTANEOUSLY FLAMMABLE
> IN AIR
> HIGHLY FLAMMABLE LIQUID
> REACTS VIOLENTLY WITH WATER
> MOISTURE SENSITIVE
> STORE UNDER INERT GAS
> INCOMPATIBLE WITH AIR, MOISTURE, WATER,
> OXIDIZING AGENTS, AND ALCOHOLS

To prevent their accidental ignition, the aluminum alkyl compounds and their halide and hydride derivatives are often stored within electrically grounded containers under an atmosphere of nitrogen in a cool, well-ventilated area.

SOLVED EXERCISE 9.3

During the Vietnam conflict, the U.S. military used triethylaluminum as the active component of a portable encapsulated-flame system for controlling the extent of unwanted vegetation and to rout the enemy from jungles and other densely vegetated areas. What is the most logical reason the Department of Defense selected triethylaluminum as the incendiary agent for this use?

Solution: When discharged into the open air, triethylaluminum spontaneously bursts into flame. This property gives rise to its potential use in a system intended to rapidly discharge a stream of fire on demand, like a portable encapsulated-flame system. The Department of Defense thus most likely selected triethylaluminum as the active component of a portable flamethrower because this substance met its needs.

9.4-C TRANSPORTING ALUMINUM ALKYL COMPOUNDS AND THEIR DERIVATIVES

When shippers offer an aluminum alkyl compound or any halide or hydride derivative for transportation, DOT requires them to identify it on the accompanying shipping paper as one of the following, as relevant:

> UN3051, Aluminum alkyls, 4.2, (4.3), PGI (Dangerous When Wet)
> UN3052, Aluminum alkyl halides, liquid, 4.2, (4.3), PGI (Dangerous When Wet)
> UN3461, Aluminum alkyl halides, solid, 4.2, (4.3), PGI (Dangerous When Wet)
> UN3076, Aluminum alkyl hydrides, solid, 4.2, (4.3), PGI (Dangerous When Wet)

Because the proper shipping name is listed generically in the Hazardous Materials Table, shippers must enter the name of the specific compound parenthetically. DOT also requires shippers to affix SPONTANEOUSLY COMBUSTIBLE and DANGEROUS WHEN WET labels on its containers.

DOT requires carriers of an aluminum alkyl compound or its halide or hydride derivatives to display DANGEROUS WHEN WET placards on the bulk packaging or transport vehicle used for shipment. When 1001 lb (454 kg) or more of these compounds are transported, DOT requires the displaying of SPONTANEOUSLY COMBUSTIBLE placards *in addition to* the DANGEROUS WHEN WET placards. The implementation of these DOT placarding requirements is illustrated in Figure 9.3.

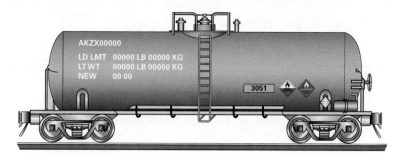

FIGURE 9.3 When a carrier transports an aluminum alkyl compound or its halide or hydride derivative in an amount exceeding 1001 lb (454 kg), DOT requires SPONTANEOUSLY COMBUSTIBLE and DANGEROUS WHEN WET placards to be displayed on the transport vehicle. AKZX is the reporting mark of AKZO Nobel Chemicals, Inc., Chicago, Illinois, a distributor of triethylaluminum and other class 4 compounds.

SOLVED EXERCISE 9.4

When 387 gal (829 L) of liquid triisobutylaluminum is transported in a 400-gal (857-L) portable tank by highway:

(a) What shipping description does DOT require the shipper to enter on the accompanying shipping paper?

(b) How does DOT require the carrier to placard and mark the tank?

Solution:

(a) There are two regulations in Table 6.3 that are pertinent to preparing the basic description. First, when a hazardous material is described with a generic description in the Hazardous Materials Table, shippers must include the name of the substance in parentheses in the basic description. Second, when a hazardous material, by chemical interaction with water, is liable to become spontaneously flammable or give off flammable gases in dangerous quantities, the words "Dangerous When Wet" must be included with the basic description. Consequently, the shipping description of triisobutylaluminum is the following:

Units	HM	Shipping Description (Identification Number, Proper Shipping Name, Primary Hazard Class or Division, Subsidiary Hazard Class or Division, and Packing Group)	Volume (gal)
1 portable tank	X	UN3051, Aluminum alkyls (triisobutylaluminum), 4.2, (4.3), PGI (Dangerous When Wet)	387

(b) DOT requires carriers to display side by side a SPONTANEOUSLY COMBUSTIBLE *and* DANGEROUS WHEN WET placard on each side and each end of the cargo tank. Because the tank has a capacity of less than 1000 gal (3785 L), DOT requires them to mark the tank with the identification number 3051 on two opposing sides on orange panels, across the center area of the SPONTANEOUSLY COMBUSTIBLE placards, or on white square-on-point diamonds.

- Note the water reactivity of the three types of ionic hydrides.
- Identify the major industry in which ionic hydrides are used.

9.5 Ionic Hydrides

Approximately ten **ionic hydrides** are encountered commercially. They are compounds consisting of metallic ions bonded to simple or complex hydride ions. Some metallic hydrides are not ionic hydrides. For example, although tin(IV) hydride is a metal hydride, it is composed of molecules. Each molecule consists of a tin atom *covalently* bonded to four hydrogen atoms. Its chemical formula is SnH_4.

ionic hydride
■ A compound composed of a metallic ion and a simple or complex hydride ion

Ionic hydrides are used as powerful reducing agents by the chemical industry. They can be classified according to their general chemical composition as simple ionic hydrides, ionic borohydrides, and ionic aluminum hydrides.

9.5-A SIMPLE IONIC HYDRIDES

Simple ionic hydrides are compounds consisting of metallic ions bonded to hydride ions (H^-). They are lithium hydride, sodium hydride, calcium hydride, magnesium hydride, and aluminum hydride, whose chemical formulas are LiH, NaH, CaH_2, MgH_2, and AlH_3, respectively. They are produced by reactions between the corresponding metal and hydrogen. For example, sodium hydride is a simple ionic hydride produced by the union of sodium metal and hydrogen.

$$2Na(s) \ + \ H_2(g) \ \longrightarrow \ 2NaH(s)$$
Sodium   Hydrogen   Sodium hydride

9.5-B IONIC BOROHYDRIDES

Ionic borohydrides are ionic hydrides in which metallic ions are bonded to borohydride ions (BH_4^-). The commercially important ionic borohydrides are lithium borohydride, sodium borohydride, and aluminum borohydride, whose chemical formulas are $LiBH_4$, $NaBH_4$, and $Al(BH_4)_3$, respectively.

The ionic borohydrides are produced by relatively complex chemical reactions. Sodium borohydride, for example, is prepared from sodium hydride and trimethyl borate.

$$4NaH(s) \ + \ (CH_3)_3BO_3(l) \ \longrightarrow \ NaBH_4(s) \ + \ 3NaOCH_3(s)$$
Sodium hydride   Trimethyl borate   Sodium borohydride   Sodium methoxide

9.5-C IONIC ALUMINUM HYDRIDES

Ionic aluminum hydrides are ionic hydrides in which metallic ions are bonded to aluminum hydride ions (AlH_4^-). Two commercially important ionic aluminum hydrides are lithium aluminum hydride and sodium aluminum hydride, whose chemical formulas are $LiAlH_4$ and $NaAlH_4$, respectively. These substances are produced by reactions between the relevant metallic hydride and anhydrous aluminum chloride (Section 9.8-A). Lithium aluminum hydride, for example, is prepared from lithium hydride and anhydrous aluminum chloride.

$$4LiH(s) \ + \ AlCl_3(s) \ \longrightarrow \ LiAlH_4(s) \ + \ 3LiCl(s)$$
Lithium hydride   Aluminum chloride   Lithium aluminum hydride   Lithium chloride

The reaction is conducted in diethyl ether (Section 13.3-B), which acts as a solvent.

9.5-D WATER REACTIVITY OF THE IONIC HYDRIDES

Although the ionic hydrides are relatively stable compounds, they possess several common hazardous features. Of special interest here is the fact that they react with water to produce flammable hydrogen.

To prevent their contact with atmospheric moisture, all metallic hydrides are stored in tightly sealed containers. When the metallic hydrides are encountered commercially, they are often covered with petroleum oil. The presence of the oil lends an element of safety when handling and storing them. However, these compounds are also encountered as ethereal solutions; that is, they are dissolved in diethyl ether, a highly flammable liquid. The combination of diethyl ether and a metallic hydride poses the risk of fire and explosion.

The following equations illustrate the water reactivity of several representative metallic hydrides:

$$LiH(s) \ + \ H_2O(l) \ \longrightarrow \ LiOH(aq) \ + \ H_2(g)$$

Lithium hydride   Water   Lithium hydroxide   Hydrogen

$$3NaBH_4(s) \ + \ 6H_2O(l) \ \longrightarrow \ 3NaBO_2(aq) \ + \ 12H_2(g)$$

Sodium borohydride   Water   Sodium borate   Hydrogen

$$LiAlH_4(s) \ + \ 4H_2O(l) \ \longrightarrow \ Al(OH)_3(s) \ + \ LiOH(aq) \ + \ 4H_2(g)$$

Lithium aluminum hydride   Water   Aluminum hydroxide   Lithium hydroxide   Hydrogen

$$Al(BH_4)_3(s) \ + \ 12H_2O(l) \ \longrightarrow \ Al(OH)_3(s) \ + \ 3H_3BO_3(aq) \ + \ 12H_2(g)$$

Aluminum borohydride   Water   Aluminum hydroxide   Boric acid   Hydrogen

As these ionic hydrides react with water, the evolved hydrogen absorbs the heat of reaction and spontaneously bursts into flame.

Because the ionic hydrides are water-reactive substances, precautions should be exercised to avoid exposing them to humid air or other potential sources of water. Experts recommend that firefighters use water as a fire extinguisher only when they encounter small spills of these substances. This latter statement is supported by the section titled "Firefighting Measures" of the material safety data sheet for sodium aluminum hydride shown in Figure 9.4.

SODIUM ALUMINUM HYDRIDE

Firefighting Measures

No standard methods have been developed for extinguishing large-scale sodium aluminum hydride fires. Small fires may be extinguished using dry powder, provided that a large excess of dry powder is used. Water should be used only on small amounts of sodium aluminum hydride. Water-based foams, chemical foams, and halogenated extinguishers should NOT be used.

The following general procedures are recommended for spills involving portable containers of sodium aluminum hydride:

(a) Water may be used (if one is thoroughly familiar with the violent reaction and if personnel are properly protected from possible explosion) to decompose sodium aluminum hydride from small spills and to keep adjacent tanks or equipment cool.

(b) If spillage from a container is burning, move flammables away or move the container to a safe place, if possible. Avoid increasing the spill. Keep water, materials wet with water, and liquid fire extinguishers away. Small fires can be controlled with dry materials or with water, if one is thoroughly familiar with the violent reaction and if personnel are properly protected from possible explosion. If possible use sandbags to contain and isolate the burning material.

(c) If material is spilled from a container but is not burning, ventilate the area or move the container out in the open; otherwise, proceed as in (b).

The following procedure is recommended for handling sodium aluminum hydride when there are fires from an external source in the vicinity:

If a container is threatened by fire from an external source (not from the chemical substance itself), extinguish the external fire by conventional means, or move the container, whichever can be done quicker and with less danger. If this cannot be done, keep the container cool by spraying water on it. Water reacts violently with sodium aluminum hydride but has been found to be an effective fire control treatment if one is thoroughly familiar with the violent reaction and personnel and properly protected from explosion.

FIGURE 9.4 This excerpt from the material safety data sheet for sodium aluminum hydride cautions emergency responders to use water *only* when encountering a small spill of the substance from a portable container. (*Courtesy of Albemarle Corporation, Baton Rouge, Louisiana.*)

9.5-E TRANSPORTING IONIC HYDRIDES

When shippers offer an ionic hydride for transportation, DOT requires them to provide the appropriate basic description listed in Table 9.7 on the accompanying shipping paper. When the basic description is not listed in Table 9.7, DOT requires them to identify the commodity generically on a shipping paper as one of the following:

UN3182, Metal hydrides, flammable, n.o.s., 4.1, PGI
or
UN3182, Metal hydrides, flammable, n.o.s., 4.1, PGII
UN1409, Metal hydrides, water-reactive, n.o.s., 4.3, PGI (Dangerous When Wet)
or
UN1409, Metal hydrides, water-reactive, n.o.s., 4.3, PGII (Dangerous When Wet)

The basic description includes the name of the specific compound entered parenthetically. When shippers offer a water-reactive ionic hydride for transportation, DOT requires them to affix a DANGEROUS WHEN WET label on its containers.

DOT also requires carriers to display DANGEROUS WHEN WET placards on the bulk packaging or transport vehicle used for shipment. When carriers transport 1001 lb (454 kg) or more of aluminum borohydride, including aluminum borohydride in devices, they must display SPONTANEOUSLY COMBUSTIBLE *and* DANGEROUS WHEN WET placards.

TABLE 9.7	Basic Descriptions of Some Ionic Hydrides
IONIC HYDRIDE	**BASIC DESCRIPTION**
Aluminum borohydride	UN2870, Aluminum borohydride, 4.2, (4.3), PGI (Dangerous When Wet) *or* UN2870, Aluminum borohydride in devices, 4.2, (4.3), PGI (Dangerous When Wet)
Aluminum hydride	UN2463, Aluminum hydride, 4.3, PGI (Dangerous When Wet)
Calcium hydride	UN1404, Calcium hydride, 4.3, PGI (Dangerous When Wet)
Lithium aluminum hydride	UN1410, Lithium aluminum hydride, 4.3, PGI (Dangerous When Wet)
Lithium aluminum hydride dissolved in ether	UN1411, Lithium aluminum hydride, ethereal, 4.3, (3), PGI (Dangerous When Wet)
Lithium borohydride	UN1413, Lithium borohydride, 4.3, PGI (Dangerous When Wet)
Lithium hydride	UN1414, Lithium hydride, 4.3, PGI (Dangerous When Wet)
Lithium hydride, fused solid	UN2805, Lithium hydride, fused solid, 4.3, PGII (Dangerous When Wet)
Potassium borohydride	UN1870, Potassium borohydride, 4.3, PGI (Dangerous When Wet)
Sodium aluminum hydride	UN2835, Sodium aluminum hydride, 4.3, PGII (Dangerous When Wet)
Sodium borohydride	UN1426, Sodium borohydride, 4.3, PGI (Dangerous When Wet)
Sodium hydride	UN1427, Sodium hydride, 4.3, PGI (Dangerous When Wet)
Titanium hydride	UN1871, Titanium hydride, 4.1, PGII
Zirconium hydride	UN1437, Zirconium hydride, 4.1, PGII

When shippers offer either titanium hydride or zirconium hydride for transportation, DOT requires them to affix a FLAMMABLE SOLID label on its containers. When carriers transport 1001 lb (454 kg) or more of either substance, DOT also requires them to display FLAMMABLE SOLID placards on the bulk packaging or transport vehicle used for shipment.

PERFORMANCE GOALS FOR SECTION 9.6:

- Note the water reactivity of the metallic phosphides.
- Note the flammable and toxic features of phosphine.

9.6 Metallic Phosphides

metallic phosphide
- An inorganic compound composed of metallic and phosphide ions

Metallic phosphides are produced by combination reactions in which a given metal unites with elemental phosphorus. Calcium phosphide, for example, is formed by heating calcium and phosphorus.

$$6Ca(s) \ + \ P_4(s) \ \longrightarrow \ 2Ca_3P_2(s)$$
Calcium   Phosphorus   Calcium phosphide

These compounds were once popular fumigants used on grain and other postharvest crops, but in the United States, their use is not nearly as popular now as it was in the past.

The metallic phosphides function as fumigants by reacting with atmospheric moisture to produce the toxic gas *phosphine*.

$$Ca_3P_2(s) \ + \ 6H_2O(l) \ \longrightarrow \ 3Ca(OH)_2(s) \ + \ 2PH_3(g)$$
Calcium phosphide   Water   Calcium hydroxide   Phosphine

When calcium phosphide is applied within an enclosure used for the storage of crops, it is the phosphine produced by hydrolysis that actually kills mice and other unwanted pests.

9.6-A TRANSPORTING METALLIC PHOSPHIDES

When shippers offer a metallic phosphide for transportation, DOT requires them to identify the appropriate material on the accompanying shipping paper by selecting the relevant entry listed in Table 9.8. DOT also requires shippers to affix DANGEROUS WHEN WET and POISON labels on its containers.

DOT requires carriers to display DANGEROUS WHEN WET placards on the bulk packaging or transport vehicle used for shipment. When the metallic phosphides are transported in an amount exceeding 1001 lb (454 kg), DOT requires carriers to display POISON placards *in addition to* DANGEROUS WHEN WET placards.

SOLVED EXERCISE 9.5

What is the most likely reason DOT requires shippers to affix DANGEROUS WHEN WET and POISON labels to packages of stannic phosphide?

Solution: DOT assigns two hazard codes, 4.3 and 6.1, to stannic phosphide because the properties of this substance comply with the defining criteria for both a dangerous-when-wet substance and a poison. These criteria are noted in the definitions of a dangerous-when-wet substance and poison listed in the glossary. Stannic phosphide is a solid compound that reacts with water to produce phosphine, a flammable and toxic gas.

$$Sn_3P_4(s) \quad + \quad 12H_2O(l) \quad \longrightarrow \quad 3Sn(OH)_4(s) \quad + \quad 4PH_3(g)$$

Stannic phosphide   Water   Stannic hydroxide   Phosphine

When shippers affix DANGEROUS WHEN WET and POISON labels to packages of stannic phosphide, these hazard warning labels quickly warn emergency-response personnel that the substance is simultaneously water-reactive and toxic.

9.6-B PHOSPHINE

As noted in Section 9.6-A, phosphine is generated when the metallic phosphides react with water. This substance possesses the following hazardous features:

- Phosphine is a poisonous gas. The gas may be detected by its exceptionally offensive odor, which has been described as a mixture of garlic and rotten fish. The odor threshold for phosphine is only 0.15 ppm. When inhaled, it primarily attacks the cardiovascular and respiratory systems, causing pulmonary edema and massive destruction of the lung tissues. Long-term exposure to lesser concentrations causes the bones to soften. Exposure to a concentration of 50 ppm is immediately dangerous to an individual's life and health.

TABLE 9.8	Basic Descriptions of Some Metallic Phosphides
METALLIC PHOSPHIDE	**BASIC DESCRIPTION**
Aluminum phosphide	UN1397, Aluminum phosphide, 4.3, (6.1), PGI (Dangerous When Wet) (Poison)
Aluminum phosphide (pesticides)	UN3048, Aluminum phosphide pesticides, 6.1, PGI (Poison)
Calcium phosphide	UN1360, Calcium phosphide, 4.3, (6.1), PGI (Dangerous When Wet) (Poison)
Magnesium aluminum phosphide	UN1419, Magnesium aluminum phosphide, 4.3, (6.1), PGI (Dangerous When Wet) (Poison)
Magnesium phosphide	UN2011, Magnesium phosphide, 4.3, (6.1), PGI (Dangerous When Wet) (Poison)
Potassium phosphide	UN2012, Potassium phosphide, 4.3, (6.1), PGI (Dangerous When Wet) (Poison)
Sodium phosphide	UN1432, Sodium phosphide, 4.3, (6.1), PGI (Dangerous When Wet) (Poison)
Stannic phosphide	UN1433, Stannic phosphide, 4.3, (6.1), PGI (Dangerous When Wet) (Poison)
Strontium phosphide	UN2013, Strontium phosphide, 4.3, (6.1), PGI (Dangerous When Wet) (Poison)
Zinc phosphide	UN1714, Zinc phosphide, 4.3, (6.1), PGI (Dangerous When Wet) (Poison)

- Phosphine is also a spontaneously flammable gas. Its autoignition temperature is only 100 °F (37.8 °C), a value readily attained in most environments. Phosphine burns in air to produce a dense white cloud of tetraphosphorus decoxide and water vapor.

$$4PH_3(g) + 8O_2(g) \longrightarrow P_4O_{10}(s) + 6H_2O(g)$$

Phosphine   Oxygen   Tetraphosphorus decoxide   Water

Phosphine is also available as a commercial chemical product. It is used in the semiconductor and chemical industries.

We examine the methods recommended for extinguishing fires involving gases that are simultaneously flammable and poisonous in Section 10.14.

9.6-C WORKPLACE REGULATIONS INVOLVING PHOSPHINE

When phosphine is present in the workplace, OSHA requires employers to limit employee exposure to a concentration of 0.3 ppm, averaged over the 8-hr workday.

9.6-D TRANSPORTING PHOSPHINE

When shippers offer phosphine for transportation, DOT requires them to identify the gas on the accompanying shipping paper as follows:

UN2199, Phosphine, 2.3, (2.1) (Poison - Inhalation Hazard, Zone A)

DOT also requires them to affix a POISON GAS and FLAMMABLE GAS label on its cylinders or other containment devices. Furthermore, since phosphine potentially poses a health hazard by inhalation, DOT requires them to mark the bulk packaging or transport vehicle with the words INHALATION HAZARD.

DOT requires carriers of phosphine to display POISON GAS placards on the bulk packaging or transport vehicle used for shipment.

PERFORMANCE GOALS FOR SECTION 9.7:

- Note the water reactivity of the metallic carbides.
- Identify the industries in which metallic carbides are used.

9.7 Metallic Carbides

Metals bond with carbon to form compounds having either ionic or covalent units. Our concern here is solely with the compounds composed of metallic ions and carbon ions that exist as C_2^{2-} or C^{4-}. They are called **metallic carbides**. There are only two commercially important metallic carbides: aluminum carbide and calcium carbide. Both are water-reactive substances.

metallic carbide
■ An inorganic compound composed of metallic and carbide ions

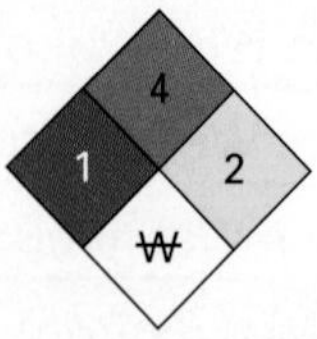

9.7-A ALUMINUM CARBIDE

Aluminum carbide is prepared by heating aluminum oxide with coke in an electric furnace. It is used as a catalyst by the chemical industry. When aluminum carbide reacts with water, flammable methane is produced as a hydrolysis product.

$$Al_4C_3(s) + 12H_2O(l) \longrightarrow 4Al(OH)_3(s) + 3CH_4(g)$$

Aluminum carbide   Water   Aluminum hydroxide   Methane

Aluminum carbide that has been exposed to humid air poses a flammable and explosive hazard.

9.7-B CALCIUM CARBIDE

Calcium carbide is manufactured by heating a mixture of coke and lime at an elevated temperature in an electric furnace.

$$CaO(s) \quad + \quad 3C(s) \quad \longrightarrow \quad CaC_2(s) \quad + \quad CO(g) \quad [t > 3600\ °F\ (1982\ °C)]$$

Calcium oxide   Carbon   Calcium carbide   Carbon monoxide

The production process is very energy-intensive.

Calcium carbide was formerly used as a raw material for the production of industrial-grade acetylene.

$$CaC_2(s) \quad + \quad 2H_2O(l) \quad \longrightarrow \quad Ca(OH)_2(s) \quad + \quad C_2H_2(g)$$

Calcium carbide   Water   Calcium hydroxide   Acetylene

Today, this method of producing and manufacturing acetylene is used to a very limited extent.

Because of its water-reactive nature, calcium carbide that has been exposed to humid air poses a flammable and explosive hazard. For this reason, it must be stored and handled in a dry environment that is free of ignition sources.

SOLVED EXERCISE 9.6

Identify the function of the water-reactive metallic carbides that the military formerly used in incendiary munitions.

Solution: The water-reactive metallic carbides produce flammable gases like methane and acetylene, each of which burns readily, sets fires to objects, and causes burn injury to enemies through the action of flame and heat. This is why metallic carbides were formerly used as the active components of incendiary munitions.

9.7-C TRANSPORTING METALLIC CARBIDES

When shippers offer aluminum carbide or calcium carbide for transportation, DOT requires them to identify it on the accompanying shipping paper as one of the following, as relevant:

UN1394, Aluminum carbide, 4.3, PGII (Dangerous When Wet)
UN1402, Calcium carbide, 4.3, PGI (Dangerous When Wet)

DOT also requires shippers to affix a DANGEROUS WHEN WET label on containers.

DOT requires carriers to display DANGEROUS WHEN WET placards on the bulk packaging or transport vehicle used for shipment.

SOLVED EXERCISE 9.7

Why do chemical manufacturers recommend protection against the buildup of electrostatic charges for steel drums containing calcium carbide?

Solution: Calcium carbide is a water-reactive material. Atmospheric moisture sealed inside the steel drums could react with the product to produce a concentration of acetylene sufficient to pose the risk of fire and explosion. The presence of an electrostatic charge could ignite the acetylene when the drums are unsealed. To prevent such an incident, chemical manufacturers urge users of calcium carbide to ground the drums and prevent the buildup of electrostatic charges.

- Note that certain substances react with water to produce hydrogen chloride vapor.
- Identify the industries in which these water-reactive substances are used.
- Use Table 8.7 to recall the adverse health effects caused by the inhalation of various concentrations of hydrogen chloride.

9.8 Water-Reactive Substances That Produce Hydrogen Chloride

Certain substances react with water to produce hydrogen chloride vapor or hydrochloric acid as products of their hydrolysis. When they are encountered, these substances routinely have the suffocating, pungent odor of hydrogen chloride, which fumes in air and limits visibility. As first noted in Section 8.8-B, hydrogen chloride is a toxic, irritating gas, and hydrochloric acid is a corrosive liquid. In this section, the hydrolysis product is denoted solely as hydrogen chloride vapor.

9.8-A ANHYDROUS ALUMINUM CHLORIDE

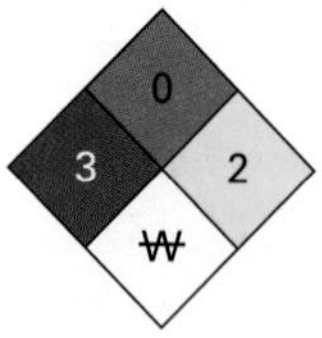

Anhydrous aluminum chloride is a white-to-yellow solid whose chemical formula is $AlCl_3$. In the chemical industry substantial quantities are used as catalysts and as a raw material for the production of aluminum alkyl compounds and lithium aluminum hydride. It is also used to produce antiperspirants.

Anhydrous aluminum chloride reacts violently with water to produce hydrogen chloride.

$$2AlCl_3(s) \; + \; 3H_2O(l) \; \longrightarrow \; Al_2O_3(s) \; + \; 6HCl(g)$$

Aluminum chloride   Water   Aluminum oxide   Hydrogen chloride

For this reason, the manufacturers and distributors of this substance caution potential users that it can irritate the skin, eyes, and respiratory tract. This warning is provided on the manufacturer's material safety data sheet for this product as the following excerpt illustrates:

ANHYDROUS ALUMINUM CHLORIDE

HAZARDS IDENTIFICATION

CAUTION!

MAY CAUSE IRRITATION TO SKIN, EYES, AND RESPIRATORY TRACT

STABILITY AND REACTIVITY

Stability: Stable under ordinary conditions of use and storage.

Hazardous Decomposition Products: May emit toxic fumes when heated to decomposition.

Hazardous Polymerization: Will not occur.

Incompatibilities: Water and moist air.

Conditions to Avoid: Water

9.8-B PHOSPHORUS OXYCHLORIDE

Phosphorus oxychloride, also called *phosphoryl chloride*, is a colorless, fuming liquid whose chemical formula is $POCl_3$. It is used primarily by the chemical industry as a chlorinating agent.

Phosphorus oxychloride reacts violently with water to produce hydrogen chloride.

$$POCl_3(l) \ + \ 3H_2O(l) \ \longrightarrow \ H_3PO_4(aq) \ + \ 3HCl(g)$$

Phosphorus oxychloride   Water   Phosphoric acid   Hydrogen chloride

9.8-C PHOSPHORUS TRICHLORIDE

This substance is a colorless, fuming liquid whose chemical formula is PCl_3. It is used in the chemical industry as a chlorinating agent and catalyst. Phosphorus trichloride is used, for example, as a raw material for producing acetyl chloride (Section 9.9-C).

Phosphorus trichloride reacts with water to form phosphorous acid and hydrogen chloride.

$$PCl_3(l) \ + \ 3H_2O(l) \ \longrightarrow \ H_3PO_3(aq) \ + \ HCl(g)$$

Phosphorus trichloride   Water   Phosphorous acid   Hydrogen chloride

9.8-D PHOSPHORUS PENTACHLORIDE

This substance is a yellow-to-green solid whose chemical formula is PCl_5. It is primarily used as a chlorinating and dehydrating agent by the chemical industry.

Phosphorus pentachloride is decomposed by water in a multistep process, summarized in the following equation:

$$PCl_5(s) \ + \ 4H_2O(l) \ \longrightarrow \ H_3PO_4(aq) \ + \ 5HCl(g)$$

Phosphorus pentachloride   Water   Phosphoric acid   Hydrogen chloride

9.8-E SILICON TETRACHLORIDE

Silicon tetrachloride is a colorless, fuming liquid whose chemical formula is $SiCl_4$. It is used as a raw material to manufacture liquid or semisolid silicon-containing polymers known as *silicones*, substances widely used in electrical insulation. Silicon tetrachloride is also used in the semiconductor manufacturing industry.

Silicon tetrachloride reacts vigorously with water to form silicic acid and hydrogen chloride.

$$SiCl_4(l) \ + \ 4H_2O(l) \ \longrightarrow \ H_4SiO_4(aq) \ + \ 4HCl(g)$$

Silicon tetrachloride   Water   Silicic acid   Hydrogen chloride

9.8-F SULFURYL CHLORIDE

Sulfuryl chloride, also called *sulfonyl chloride*, is a colorless liquid whose chemical formula is SO_2Cl_2. Sulfuryl chloride is mainly used in the chemical industry as a chlorinating and dehydrating agent.

Sulfuryl chloride reacts very slowly with water to form sulfuric acid and hydrogen chloride.

$$SO_2Cl_2(l) \ + \ 2H_2O(l) \ \longrightarrow \ H_2SO_4(aq) \ + \ 2HCl(g)$$

Sulfuryl chloride   Water   Sulfuric acid   Hydrogen chloride

9.8-G THIONYL CHLORIDE

Thionyl chloride is a red-to-yellow liquid whose chemical formula is $SOCl_2$. Thionyl chloride is used mainly within the chemical industry.

Thionyl chloride reacts vigorously with water to form sulfurous acid and hydrogen chloride.

$$SOCl_2(l) \ + \ 2H_2O(l) \ \longrightarrow \ H_2SO_3(aq) \ + \ 2HCl(g)$$

Thionyl chloride   Water   Sulfurous acid   Hydrogen chloride

9.8-H ANHYDROUS TIN(IV) CHLORIDE

Tin(IV) chloride also known as *stannic tetrachloride*, is a colorless, fuming liquid whose chemical formula is $SnCl_4$. It is used to manufacture blueprint and similarly sensitized types of paper.

Anhydrous tin(IV) chloride reacts slowly with water to form tin(IV) oxide and hydrogen chloride.

$$SnCl_4(l) \quad + \quad 2H_2O(l) \quad \longrightarrow \quad SnO_2(s) \quad + \quad 4HCl(g)$$

Tin(IV) chloride · · · · · · Water · · · · · · Tin(IV) oxide · · · Hydrogen chloride

9.8-I ANHYDROUS TITANIUM(IV) CHLORIDE

Titanium(IV) chloride, also called *titanium tetrachloride*, is a colorless, volatile liquid whose chemical formula is $TiCl_4$. Titanium(IV) chloride is the intermediate compound produced during the production of metallic titanium (Section 9.3-B) and the white paint pigment titanium dioxide. The mixture of titanium(IV) chloride and an aluminum alkyl compound is an important polymerization catalyst (Section 9.4).

Titanium(IV) chloride reacts slowly with water to produce titanium(IV) dioxide and hydrogen chloride.

$$TiCl_4(l) \quad + \quad 2H_2O(l) \quad \longrightarrow \quad TiO_2(s) \quad + \quad 4HCl(g)$$

Titanium(IV) chloride · · · · · Water · · · · · Titanium(IV) oxide · Hydrogen chloride

silane
■ An organic compound whose molecules are composed of silicon and hydrogen atoms

chlorosilane
■ A chlorinated derivative of silane (SiH_4)

9.8-J TRICHLOROSILANE

The hydrides of silicon are called **silanes**. The simplest silane, itself called *silane*, is a substance having the formula SiH_4. When one or more of the hydrogen atoms in silane are replaced with chlorine atoms, the resulting substances are called **chlorosilanes**. Some chlorosilanes are water-reactive substances.

Trichlorosilane is widely used to manufacture the polysilicon employed for the production of solar cells and solar wafers. Some of its physical properties are noted in Table 9.9. It is a water-reactive, flammable, and corrosive liquid having the chemical formula $SiHCl_3$. Trichlorosilane reacts violently with water, producing choking vapors of hydrogen chloride and trihydroxysilane.

$$H-SiCl_3(l) \quad + \quad 3H_2O(l) \quad \longrightarrow \quad H-Si(OH)_3(s) \quad + \quad 3HCl(g)$$

Trichlorosilane · · · · · · Water · · · · · · Trihydroxysilane · · · Hydrogen chloride

Trichlorosilane suppliers often mark their containers to provide immediate information regarding the hazards of this substance to its user. The following label is illustrative:

DANGER!

FLAMMABLE, CORROSIVE LIQUID AND VAPOR

HARMFUL IF INHALED

CAN CAUSE EYE, SKIN, AND RESPIRATORY TRACT BURNS

CAN FORM EXPLOSIVE MIXTURES WITH AIR

WATER CAN CAUSE VIOLENT REACTION

CONTACT WITH WATER OR MOIST AIR LIBERATES

IRRITATING GAS

SELF-CONTAINED BREATHING APPARATUS AND PROTECTIVE

CLOTHING MUST BE WORN BY RESCUE WORKERS

UNDER AMBIENT CONDITIONS, THIS IS A COLORLESS LIQUID

WITH AN IRRITATING, CHOKING ODOR

TABLE 9.9	Physical Properties of Some Chlorosilanes		
	METHYLDICHLOROSILANE	**METHYLTRICHLOROSILANE**	**TRICHLOROSILANE**
Melting point	−135.4 °F (−93 °C)	−130 °F (−90 °C)	−195.9 °F (−126.6 °C)
Boiling point	105.8 °F (41 °C)	149 °F (66.4 °C)	89.4 °F (31.9 °C)
Specific gravity at 68 °F (20 °C)	1.1	1.27	1.33
Vapor density (air = 1)	3.97	5.2	4.7
Vapor pressure at 68 °F (20 °C)	321 mmHg	134 mmHg	500 mmHg
Flashpoint	−18.4 °F (−28 °C)	14 °F (−10 °C)	6.8 °F (−14 °C)
Autoignition point	471.2 °F (244 °C)	759 °F (404 °C)	219 °F (104 °C)
Lower explosive limit	3.4%	3.4%	1.2%
Upper explosive limit	55%	>55%	90.5%

Other chlorosilanes, including methyldichlorosilane and methyltrichlorosilane, possess similar hazardous properties. The important physical properties of these two chlorosilanes are included in Table 9.9.

9.8-K TRANSPORTING SUBSTANCES THAT REACT WITH WATER TO PRODUCE HYDROGEN CHLORIDE VAPOR

When shippers transport any substance noted in this section, DOT requires them to identify the compound on the accompanying shipping paper by selecting the appropriate basic description listed in Table 9.10. DOT also requires shippers to affix the appropriate CORROSIVE, FLAMMABLE LIQUID, POISON, and/or DANGEROUS WHEN WET label on its containers. When the substance poses a health hazard by inhalation, DOT requires shippers to mark the words INHALATION HAZARD on its containers.

DOT requires carriers who transport methyl dichlorosilane or trichlorosilane to display DANGEROUS WHEN WET placards on the bulk packaging or transport vehicle used for shipment. When carriers transport any of the other substances listed in Table 9.10 in an amount exceeding 1001 lb (454 kg), DOT requires them to display the appropriate CORROSIVE or FLAMMABLE placards on the bulk packaging or transport vehicle.

Trichlorosilane is unique among the substances listed in Table 9.10. When shippers offer trichlorosilane for transportation, DOT requires them to containerize it within packaging that is *not* equipped with pressure-relief devices. DOT also requires them to affix three labels to packaging containing trichlorosilane: DANGEROUS WHEN WET, FLAMMABLE LIQUID, and CORROSIVE.

TABLE 9.10	Basic Descriptions of Substances That Generate Hydrogen Chloride When They React with Water

WATER-REACTIVE SUBSTANCE	BASIC DESCRIPTION
Aluminum chloride, anhydrous	UN1726, Aluminum chloride, anhydrous, 8, PGII
Methyl dichlorosilane	UN1242, Methyl dichlorosilane, 4.3, (8, 3), PGI (Dangerous When Wet)
Methyltrichlorosilane	UN1250, Methyltrichlorosilane, 3, (8), PGI
Phosphorus oxychloride	UN1810, Phosphorus oxychloride, 8, (6.1), PGII (Poison - Inhalation Hazard, Zone B)
Phosphorus pentachloride	UN1806, Phosphorus pentachloride, 8, PGII
Phosphorus trichloride	UN1809, Phosphorus trichloride, 6.1, (8), PGI (Poison - Inhalation Hazard, Zone B)
Silicon tetrachloride	UN1818, Silicon tetrachloride, 8, PGII
Stannic chloride	UN1827, Stannic chloride, anhydrous, 8, PGII
Sulfuryl chloride	UN1834, Sulfuryl chloride, 8, (6.1), PGI (Poison - Inhalation Hazard, Zone A)
Thionyl chloride	UN1836, Thionyl chloride, 8, PGI
Titanium tetrachloride	UN1838, Titanium tetrachloride, 8, (6.1), PGII (Poison - Inhalation Hazard, Zone B)
Trichlorosilane	UN1295, Trichlorosilane, 4.3, (3, 8), PGI (Dangerous When Wet)

SOLVED EXERCISE 9.8

Why should emergency-response personnel use protective gear including self-contained breathing apparatus when investigating a transportation mishap involving a massive spill of liquid trichlorosilane?

Solution: As indicated by the data in Table 9.9 as well as its basic description in Table 9.10, trichlorosilane is a water-reactive, flammable, and corrosive liquid. On its reaction with water, trichlorosilane forms hydrogen chloride vapor. Table 8.7 shows that the inhalation of this vapor causes exposed individuals to experience a variety of adverse health effects. Because hydrogen chloride poses a health hazard by inhalation, the use of self-contained breathing apparatus is essential when emergency responders are investigating a transportation mishap involving trichlorosilane.

DOT requires carriers to display DANGEROUS WHEN WET placards on the bulk packaging or transport vehicle used to transport trichlorosilane, and when transporting an amount exceeding 1001 lb (454 kg), to display FLAMMABLE and CORROSIVE placards as well.

Which substances listed in Table 9.10 are regarded by DOT as hazardous materials that pose a health hazard by inhalation?

Solution: As Table 6.3 shows, when a hazardous material poses a health hazard by inhalation, DOT requires shippers to include the words "Poison-Inhalation Hazard" and to designate the applicable hazard zone in its basic description. The basic descriptions of phosphorus oxychloride, phosphorus trichloride, and titanium tetrachloride are listed in Table 9.10. They indicate that these substances can pose a health hazard by inhalation.

PERFORMANCE GOALS FOR SECTION 9.9:

- Discuss the water reactivities of acetic anhydride and acetyl chloride.

9.9 Water-Reactive Compounds That Produce Acetic Acid Vapor

When acetic acid is formed as a product of a chemical reaction, its highly irritating and pungent odor is immediately evident. This can pose serious consequences, as the inhalation of acetic acid vapor is suffocating, and exposure to the eyes and nose is severely irritating.

We briefly note here two organic compounds that react with water to produce acetic acid vapor: acetic anhydride and acetyl chloride. Some physical properties of these compounds are provided in Table 9.11.

9.9-A ACETIC ANHYDRIDE

Acetic anhydride is a colorless, fuming liquid whose condensed chemical formula is $(CH_3CO)_2O$. Acetic anhydride is principally used by the chemical, pharmaceutical, and polymer industries for the manufacture of aspirin, cellulose acetate (Section 14.5-A), and related products.

When acetic anhydride combines with water, the sole product produced is acetic acid.

$$CH_3-C \underset{CH_3-C}{\overset{O}{\diagup}} O(l) + H_2O(l) \longrightarrow 2CH_3-\underset{OH}{\overset{O}{C}}(g)$$

Acetic anhydride   Water   Acetic acid

TABLE 9.11 | Physical Properties of Two Water-Reactive Organic Compounds

	ACETIC ANHYDRIDE	ACETYL CHLORIDE
Melting point	−99 °F (−73 °C)	−170 °F (−112 °C)
Boiling point	284 °F (140 °C)	124 °F (51 °C)
Specific gravity at 68 °F (20 °C)	1.08	1.10
Vapor density (air = 1)	3.52	2.7
Vapor pressure at 68 °F (20 °C)	4 mmHg	249 mmHg
Flashpoint	130 °F (54 °C)	40 °F (4.44 °C)
Autoignition point	734 °F (390 °C)	734 °F (390 °C)
Lower explosive limit	2.7%	7.3%
Upper explosive limit	10.3%	19%

On contact with the skin or eyes acetic anhydride causes severe burns. Inhalation of its vapor is suffocating and causes irritation of the respiratory tract. To warn users of these hazards, chemical manufacturers often mark acetic anhydride containers as follows:

DANGER: CORROSIVE

CAUSES BURNS TO ANY AREA OF CONTACT
FLAMMABLE LIQUID AND VAPOR
WATER REACTIVE
HARMFUL IF SWALLOWED OR INHALED
VAPOR CAUSES RESPIRATORY TRACT
IRRITATION AND SEVERE EYE IRRITATION

9.9-B WORKPLACE REGULATIONS INVOLVING ACETIC ANHYDRIDE

When acetic anhydride is used in the workplace, OSHA requires employers to limit employee exposure to a concentration of 5 ppm, averaged over the 8-hr workday.

9.9-C ACETYL CHLORIDE

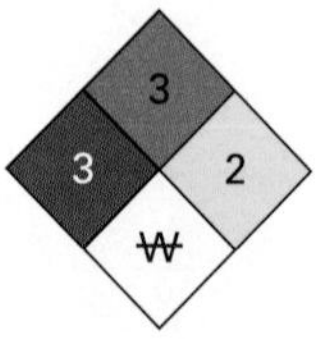

Acetyl chloride is a colorless, fuming liquid with the chemical formula $CH_3-C{\overset{O}{\underset{Cl}{}}}$. Acetyl chloride is used principally in the chemical industry. It is produced by various means, one of which involves the reaction between phosphorus trichloride and acetic acid.

$$PCl_3(l) \;+\; 3CH_3-\overset{O}{\underset{OH}{C}}(l) \longrightarrow 3CH_3-\overset{O}{\underset{Cl}{C}}(l) \;+\; H_3PO_3(l)$$

Phosphorus trichloride   Acetic acid   Acetyl chloride   Phosphorous acid

On contact with water, acetyl chloride reacts violently, producing acetic acid and hydrochloric acid vapor. This vapor poses the risk of inhalation toxicity.

$$CH_3-\overset{\displaystyle O}{\underset{\displaystyle Cl}{C}}(l) \quad + \quad H_2O(l) \quad \longrightarrow \quad CH_3-\overset{\displaystyle O}{\underset{\displaystyle OH}{C}}(g) \quad + \quad HCl(g)$$

Acetyl chloride   Water   Acetic acid   Hydrogen chloride

Table 9.11 reveals that acetyl chloride has a relatively low flashpoint. Consequently, acetyl chloride also poses a dangerous risk of fire and explosion.

9.9-D TRANSPORTING ACETIC ANHYDRIDE AND ACETYL CHLORIDE

When shippers offer either acetic anhydride or acetyl chloride for transportation, DOT requires them to identify it on the accompanying shipping paper as one of the following, as relevant:

> **UN1715, Acetic anhydride, 8, (3), PGII**
> **UN1717, Acetyl chloride, 3, (8), PGII**

DOT also requires them to affix CORROSIVE and FLAMMABLE LIQUID labels on containers of these substances.

DOT requires carriers to display CORROSIVE placards on the bulk packaging or transport vehicle used for shipment of acetic anhydride in an amount exceeding 1001 lb (454 kg). DOT requires carriers to display FLAMMABLE placards when transporting acetyl chloride in an amount exceeding 1001 lb (454 kg).

PERFORMANCE GOAL FOR SECTION 9.10:

- Describe the means shippers and carriers use to inform emergency responders of the hazards associated with encountering a flammable solid, spontaneously combustible material, or water-reactive substance during a transportation mishap.
- Identify the response actions to be implemented when a flammable solid, spontaneously combustible material, or water-reactive substance has been released into the environment.
- Identify the recommended procedures for extinguishing fires that involve the reactive metals.

9.10 Responding to Incidents Involving the Release of a Material in Hazard Classes 4.1, 4.2, and 4.3

When a flammable solid, spontaneously combustible material, or water-reactive substance is involved in a transportation mishap, first-on-the-scene responders identify it by noting any of the following:

(a) A flammable solid

- The number 4.1 as a component of a shipping description of a hazardous material listed on a shipping paper
- The words FLAMMABLE SOLID and the number 4 printed on white-and-red-striped labels affixed to packaging

- The words FLAMMABLE SOLID and the number 4 printed on white-and-red-striped placards displayed on each side and each end of a transport vehicle containing 1001 lb (454 kg) or more of a flammable solid

(b) A spontaneously combustible material

- The number 4.2 as a component of a shipping description of a hazardous material listed on a shipping paper
- The words SPONTANEOUSLY COMBUSTIBLE and the number 4 printed on white-and-red labels affixed to packaging
- The words SPONTANEOUSLY COMBUSTIBLE and the number 4 printed on white-and-red placards displayed on each side and each end of a transport vehicle containing 1001 lb (454 kg) or more of a spontaneously combustible material

(c) A water-reactive material

- The number 4.3 as a component of a shipping description of a hazardous material listed on a shipping paper
- The words DANGEROUS WHEN WET and the number 4 printed on blue labels affixed to their packages
- The words DANGEROUS WHEN WET and the number 4 printed on blue placards displayed on each side and each end of the transport vehicle

Although it is always desirable to know the chemical identity of a hazardous material involved in any transportation mishap, this statement has special meaning to the first responders arriving at a scene involving a substance that could spontaneously burst into flame or readily react with water. These responders are confronted with a unique challenge, since the use of most common extinguishers would exacerbate the ongoing emergency.

Experts recommend that when emergency responders are called to a scene involving the release of a hazardous material in hazard classes 4.1 and 4.2, they should implement the procedures provided in Figure 9.5. When called to a scene involving the release of a hazardous material in hazard class 4.3, emergency responders should implement the procedures provided in Figure 9.6.

Because special procedures are required when responding to fires involving the reactive substances noted in this chapter, we briefly note the recommended practices for the following groups of substances.

The Alkali Metals

When first-on-the-scene responders consider a response action involving an alkali metal, it is appropriate to recall that it reacts with two common fire extinguishers, water and carbon dioxide. Alkali metal fires cannot be extinguished with water, since the alkali metals displace flammable hydrogen from water. Furthermore, alkali metal fires cannot be extinguished with carbon dioxide, since the alkali metal reacts with it to produce carbon particulates.

$$4Na(s) \quad + \quad CO_2(g) \quad \longrightarrow \quad 2Na_2O(s) \quad + \quad C(s)$$

Sodium   Carbon dioxide   Sodium oxide   Carbon

Because this reaction is exothermic, the underlying metal usually erupts into flame as the carbon dioxide dissipates.

The use of a dry-chemical or dry-powder fire extinguisher is frequently recommended for extinguishing or controlling the spread of alkali metal fires. Nonetheless, caution needs to be exercised when using graphite-based dry-powder to extinguish an alkali metal fire. Graphite effectively extinguishes the fire by a smothering action, thereby limiting the amount of atmospheric oxygen and moisture available to the metal. However, at the high temperatures accompanying alkali metal fires, the graphite may react with the metal to produce metallic carbides. As previously noted in Section 9.7, these compounds are water-reactive substances. Even the moisture in the air could cause the alkali metal fire to reignite.

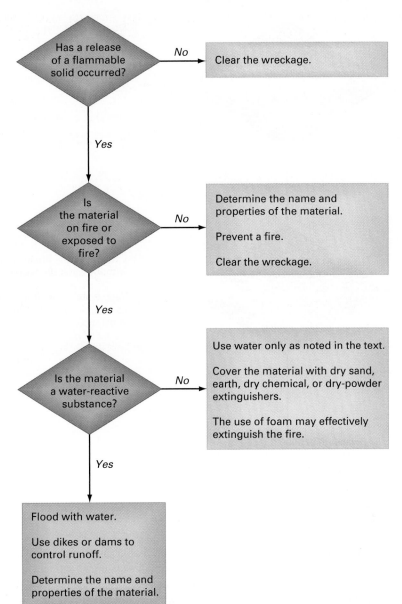

FIGURE 9.5 The recommended procedures when responding to a transportation mishap involving the release of a flammable solid or spontaneously combustible material from its packaging. (*Adapted with permission of the American Society for Testing and Materials, from a figure in ASTM STP 825*, A Guide to the Safe Handling of Hazardous Materials Accidents, *second edition. Copyright© 1990, American Society for Testing and Materials.*)

When combating fires involving metallic lithium, two fire extinguishing agents are uniquely recommended for use: lithium chloride, a dry-chemical fire extinguisher; and *LITH-X*, the graphite-based dry-powder extinguisher illustrated in Figure 9.7. Both effectively function by smothering the lithium fires. The use of LITH-X is also recommended for extinguishing magnesium, sodium, potassium, and zirconium fires.

Combustible Metals

Firefighters are frequently warned against using water on combustible metal fires. Notwithstanding this generally sound advice, water effectively extinguishes combustible metal fires when the following two conditions are met:

- The water is discharged in a volume that totally deluges the fire scene and cools the metal.
- The water is discharged rapidly soon after the metal first ignites.

Firefighters must consider not only whether enough water is available at the fire scene but also whether an appropriate means is available to rapidly apply it to the fire.

FIGURE 9.6 The recommended procedures when responding to a transportation mishap involving the release of a dangerous-when-wet material from its packaging. (*Adapted with permission of the American Society for Testing and Materials, from a figure in ASTM STP 825,* A Guide to the Safe Handling of Hazardous Materials Accidents, *second edition. Copyright© 1990, American Society for Testing and Materials.*)

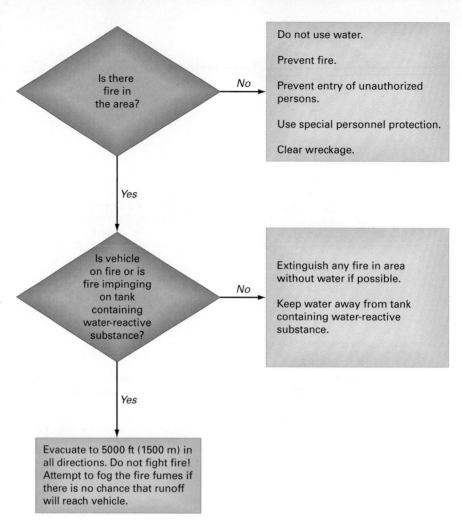

When a deluging volume of water is unavailable for extinguishing a combustible metal fire, experts recommend the use of dry sand, earth, dry-chemical extinguishers, or dry powders. The application of a special extinguishing agent is recommended for use on specific combustible metal fires. For example, the use of *MET-L-X* dry powder is often recommended to extinguish a magnesium fire. The dry powder used in the MET-L-X extinguisher, shown in Figure 9.8, is primarily sodium chloride. The use of MET-L-X is also recommended for extinguishing titanium and zirconium fires.

The application of carbon dioxide is not recommended for extinguishing combustible metal fires, since these hot metals react with carbon dioxide. Magnesium, for example, reacts with carbon dioxide to produce a sooty plume of carbon.

$$2Mg(s) \quad + \quad CO_2(g) \quad \longrightarrow \quad 2MgO(s) \quad + \quad C(s)$$

Magnesium   Carbon dioxide   Magnesium oxide   Carbon

Because this reaction is exothermic, the use of carbon dioxide does not cool the burning metal, and the magnesium fire is not extinguished.

The environment of an NFPA class D fire can be extremely caustic because of the formation of the corresponding metallic oxides, hydroxides, and carbonates. The particulates of such compounds are constituents of the smoke accompanying alkali metal and combustible metal fires. Their inhalation can cause adverse health effects ranging from minor irritation and congestion of

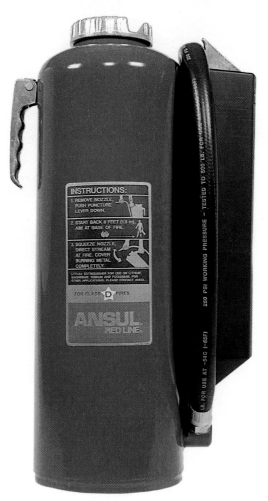

FIGURE 9.7 LITH-X dry powder, a graphite-based fire extinguisher. Although it was initially developed to extinguish lithium fires, LITH-X also extinguishes magnesium, sodium, potassium, and zirconium fires. (*Courtesy of Ansul/ Tyco Fire Suppression and Building Products, Marinette, Wisconsin.*)

the nose, throat, and bronchi to severe lung injury. Because combustible metal fires frequently burn with exceptionally brilliant flames, firefighters should be aware that the evolved radiant energy could damage the retinas of their eyes. They should also avoid breathing the smoke evolved during these fires, as it contains tiny particulates of caustic metallic oxides. When inhaled, exposure to these particulates causes considerable discomfort and localized injury to the respiratory tract and inflammation of the eyes.

Aluminum Alkyl Compounds, Metallic Hydrides, Metallic Phosphides, and Metallic Carbides

When first-on-the-scene responders encounter the release of aluminum alkyl compounds, metallic hydrides, metallic phosphides, or metallic carbides from its packaging, experts recommend the use of vermiculite, dry sand, or dry powder pressurized with nitrogen to extinguish fires.

Water-Reactive Substances That Generate Hydrogen Chloride

When responding to a transportation mishap involving the release of any substance listed in Table 9.10, emergency responders should implement the procedures previously cited in Figure 8.14, but use water only sparingly and cautiously. Due to the corrosiveness of these products and their potential to form hydrogen chloride, responders must wear fully encapsulated protective clothing and use self-contained breathing apparatus.

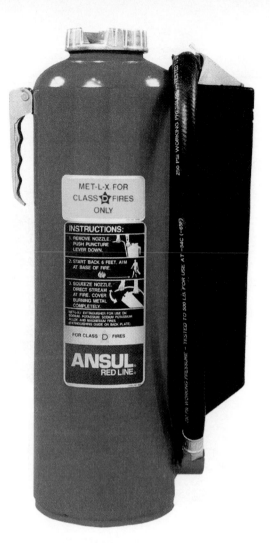

FIGURE 9.8 MET-L-X, a sodium chloride–based fire extinguisher for use on NFPA class D fires, especially magnesium fires. (*Courtesy of Ansul/Tyco Fire Suppression and Building Products, Marinette, Wisconsin.*)

Acetic Anhydride and Acetyl Chloride

To extinguish fires involving acetic anhydride or acetyl chloride, experts recommend the use of carbon dioxide or dry chemical. When emergency responders are called to a scene involving a release of acetic anhydride or acetyl chloride, they should wear fully encapsulated protective clothing, use self-contained breathing apparatus, and totally avoid the use of water as a fire extinguisher.

Metallic Lithium

9.1. At 49 C.F.R. §173.185, DOT prohibits the transportation of bulk shipments of lithium metal (nonrechargeable) batteries onboard passenger-carrying aircraft. What is the most likely reason DOT promulgated this regulation?

Metallic Sodium

9.2. When metallic sodium is briefly exposed to moist air, three white compounds form on its surface. Identify these compounds, and write balanced equations that illustrate their formation.

Metallic Potassium

9.3. Why does metallic potassium react at an explosive rate with kerosene, whereas metallic lithium and sodium can be covered with kerosene during storage in bottles and cans without incident?

Metallic Magnesium

9.4. Individuals who inadvertently inhale the fumes arising from metallic magnesium fires generally exhibit influenza-type symptoms. What substances most likely cause these individuals to experience these symptoms?

Metallic Titanium

9.5. Most combustible materials can be stored in an enclosed area containing an atmosphere of nitrogen to prevent their ignition. Nonetheless, this practice is inadvisable for storing freshly generated titanium chips, turnings, and other fines as they are generated. Why are these forms of titanium metal still likely to ignite when stored in a fully enclosed area containing nitrogen?

Metallic Zirconium

9.6. Why are zirconium strips frequently fabricated with the metal submerged under water?

Metallic Aluminum

9.7. The paint formerly used on some storage tanks contained aluminum powder, which reflected heat from the surface of the tanks. Why were these tanks grounded and used for the sole storage of noncombustible liquids?

Metallic Zinc

9.8. Which is more hazardous as a flammable solid: zinc dust at room temperature or zinc dust as it is removed after its production within a clay retort?

Aluminum Alkyl Compounds and Their Halide and Hydride Derivatives

9.9. The DOT regulation at 49 C.F.R. §173.181 requires shippers and carriers to transport pyrophoric liquids in specific types of packaging such as metal cans and steel drums that remain sealed by means of friction-free closures until use. What is the most likely reason DOT requires friction-free closures on the packaging used for the domestic shipment of triethylaluminum?

Metallic Hydrides

9.10. What is the most likely reason DOT limits the quantity of sodium aluminum hydride transported on domestic cargo aircraft to 25 lb (11 kg)?

Metallic Phosphides

9.11. What is the most likely reason DOT prohibits railroad carriers from accepting for domestic transport any quantity of aluminum phosphide within a passenger-carrying railcar?

Metallic Carbides

9.12. What is the most likely reason DOT prohibits the domestic transportation of any quantity of calcium carbide within a passenger-carrying railcar?

Substances That React with Water to Produce Hydrogen Chloride

9.13. Anhydrous titanium tetrachloride is used in offensive and defensive operations as a means of concealing the movement of troops and installations in a combat zone.

   (a) Describe how the use of anhydrous titanium tetrachloride effectively deceives the enemy and prevents them from locating the military's presence. (*Hint:* Titanium dioxide is a white solid.)

   (b) When using anhydrous titanium tetrachloride, why are military personnel concerned about the humidity of the air in the combat zone?

   (c) Which compound is responsible for the acrid smell in the air when anhydrous titanium tetrachloride is used?

9.14. What is the most logical reason DOT prohibits airline carriers from accepting for transport any quantity of phosphorus oxychloride by passenger-carrying aircraft?

Substances That React with Water to Produce Acetic Acid

9.15. Why do chemical manufacturers recommend grounding a rail tankcar containing acetyl chloride before transferring the product into a storage tank?

10

Chemistry of Some Toxic Substances

**toxic substance
(toxicant; poison)**
■ A substance that can
negatively affect the
life processes and can
cause death, temporary
incapacitation, or per-
manent harm to
humans or animals

Certain substances cause death, illness, injury, or incapacitation when the body is exposed to them in relatively small quantities. They are called **toxic substances**, **toxicants**, or **poisons**. Throughout the remainder of this book, these terms are used interchangeably.

Toxic substances are routinely encountered by firefighters and other emergency responders. For example, the toxic gas carbon monoxide is encountered whenever firefighters enter a burning building. Other toxic substances such as asbestos and lead are encountered in the dust that is generated as old buildings burn or are otherwise wracked by the forces associated with fires. It is for this reason that emergency responders need to study the commonly encountered toxic substances in some detail.

In this chapter, we focus primarily on the following four topics relating to the study of toxic substances:

■ The general features of all toxic substances, especially the factors that cause them to be poisonous
■ The properties of the specific toxic gases most likely to be encountered by emergency responders—carbon monoxide, hydrogen cyanide, sulfur dioxide, hydrogen sulfide, the nitrogen oxides, and ammonia
■ The properties of asbestos, lead, and pesticides
■ The practices recommended for effectively responding to accidents during which toxic substances are released

PERFORMANCE GOALS FOR SECTION 10.1:

■ Describe the means used by OSHA and DOT to identify toxic substances when they are encountered in the workplace and during transit, respectively.
■ Describe the concept of *toxicity* as EPA uses this term in the RCRA regulations.
■ Describe the general nature of the DOT regulations as they apply to the transportation of toxic substances.

10.1 Toxic Substances and Government Regulations

Although the layperson has a general perception of the meaning of the term *poison*, it is important to acknowledge the specific way the term is defined and used in government regulations. We briefly review how toxicity affects transportation, workplace, and environmental regulations.

10.1-A WORKPLACE REGULATIONS INVOLVING TOXIC SUBSTANCES

When toxic substances are stored or used in the workplace, OSHA requires employers to ensure their employees that their exposure to toxic substances will be limited to no more than certain maximum allowable concentrations averaged over the 8-hr workday. OSHA also requires employers to tag, label, or otherwise identify the presence of toxic substances in the workplace by the use of accident-prevention tags, warning labels, and worded signs. These tags, labels, and signs warn observers to exercise adequate precaution to reduce or eliminate their exposure to toxic substances. Most bear the word POISON in addition to an imprint of a skull and crossbones, which has long served as the internationally recognized symbol for a poison.

10.1-B THE RCRA TOXICITY CHARACTERISTIC

As first noted in Section 1.5-C, EPA regulates the treatment, storage, and disposal of a special type of waste known as a *hazardous waste*. A given waste is a hazardous waste if a representative sample exhibits any of the following four characteristics: ignitability, corrosivity, reactivity, or toxicity. In this section, we are concerned with the toxicity characteristic.

A chemical analyst determines in a laboratory whether a representative sample of a waste exhibits the toxicity characteristic. In practice, the analyst conducts certain specified test procedures to determine whether the sample contains one or more of the constituents listed in Table 10.1 at

TABLE 10.1	Waste Contaminants Subject to the RCRA Toxicity Characteristic[a]				
EPA HAZARDOUS WASTE NUMBER	CONTAMINANT	THRESHOLD LEVEL (mg/L)	EPA HAZARDOUS WASTE NUMBER	CONTAMINANT	THRESHOLD LEVEL (mg/L)
D004	Arsenic	5.0	D032	Hexachlorobenzene	0.13
D005	Barium	100.0	D033	Hexachlorobutadiene	0.5
D018	Benzene	0.5	D034	Hexachloroethane	3.0
D006	Cadmium	1.0	D008	Lead	5.0
D019	Carbon tetrachloride	0.5	D013	Lindane	0.4
D020	Chlordane	0.03	D009	Mercury	0.2
D021	Chlorobenzene	100.0	D014	Methoxychlor	10.0
D022	Chloroform	6.0	D035	Methyl ethyl ketone	200.0
D007	Chromium	5.0	D036	Nitrobenzene	2.0
D023	o-Cresol	200.0	D037	Pentachlorophenol	100.0
D024	m-Cresol	200.0	D038	Pyridine	5.0
D025	p-Cresol	200.0	D010	Selenium	1.0
D026	Cresol (total)	200.0	D011	Silver	5.0
D016	2,4-Dichlorophenoxy-acetic acid (2,4-D)	10.0	D039	Tetrachloroethylene	0.7
D027	2,4-Dichlorobenzene	7.5	D015	Toxaphene	0.5
D028	1,2-Dichloroethane	0.5	D040	Trichloroethylene	0.5
D029	1,1-Dichloroethylene	0.7	D041	2,4,5-Trichlorophenol	400.0
D030	2,4-Dinitrotoluene	0.13	D042	2,4,6-Trichlorophenol	2.0
D012	Endrin	0.02	D017	2,4,5-Trichlorophen-oxyacetic acid (2,4,5-TP, Silvex)	1.0
D031	Heptachlor (and its epoxide)	0.008	D043	Vinyl chloride	0.2

[a]40 C.F.R. §261.24.

concentrations equal to or greater than the relevant threshold values. Each listed substance adversely affects public health and the environment. When the analyst determines that the waste sample contains at least one constituent at a concentration equal to or above the value listed in Table 10.1, the waste is said to exhibit the characteristic of **toxicity**.

10.1-C TRANSPORTATION OF TOXIC SUBSTANCES

In Chapter 6, we first noted that DOT regulates the transportation of several types of toxic substances: **gases poisonous-by-inhalation (poison gases), poisonous materials**, and infectious substances. These toxic substances are designated as hazardous materials in hazard classes 2.3, 6.1, and 6.2, respectively. All hazardous materials in hazard class 2.3 pose a health threat by inhalation, and the hazardous materials in hazard class 6.1 that are identified by a 1, 2, 3, or 4 in column 7 of the Hazardous Materials Table also pose a health threat by inhalation.

When toxic substances are transported, DOT requires shippers and carriers to comply with applicable labeling, marking, and placarding requirements. When preparing the basic description of a toxic substance, shippers must include the expression "Poison - Inhalation Hazard" and the relevant hazard zone in the description. Shippers must also affix the applicable POISON GAS or POISON INHALATION HAZARD labels to the relevant packaging.

DOT requires carriers to display the applicable POISON GAS or POISON INHALATION HAZARD placards on the bulk packaging or transport vehicle used for shipment. Carriers are required to display the POISON INHALATION HAZARD placards when transporting a toxic substance that is classed in hazard Zone A or Zone B. When a material that poses a health hazard by inhalation is shipped in bulk packaging, DOT requires the packaging to be marked on two opposing sides with the words INHALATION HAZARD.

**INHALATION
HAZARD**

When carriers transport toxic substances whose hazard classes are 2.3 *and* 6.1 in nonbulk packages in the same transport vehicle, DOT allows them to post POISON GAS placards only.

Even when a hazardous material in division 6.1 does not pose a health hazard by inhalation, DOT may obligate shippers and carriers to disclose that it is poisonous. They do so by entering "Poison" or "Toxic" in the basic description on a shipping paper, affixing POISON labels to the packaging, marking POISON or TOXIC on the packaging, and displaying POISON placards on the transport vehicle when the aggregate gross mass equals or exceeds 1001 lb (454 kg).

When carriers transport more than 1.06 qt (1 L) per package of a material that poses a health hazard by inhalation and meets the criteria for Zone A, DOT also requires them to prepare and implement a *security plan* whose components comply with the requirements of 49 C.F.R. §172.802. Motor carriers are also required to obtain a Hazardous Materials Safety Permit prior to transporting any of the following:

- A material that poses a health hazard by inhalation and meets the criteria for Zone A in an amount more than 1.08 lb (1 L) per package
- A material that poses a health hazard by inhalation and meets the criteria for Zone B in bulk packaging [capacity greater than 119 gal (460 L)]
- A material that poses a health hazard by inhalation and meets the criteria for Zone C or Zone D in packaging having a capacity equal to or greater than 3500 gal (13,248 L)

Table 6.2 notes that DOT requires shippers to identify any hazardous waste that exhibits the toxicity characteristic in its basic description by using the term "EPA toxicity" or the relevant EPA hazardous waste number listed in Table 10.1. For example, when shippers offer for domestic transportation a liquid waste that exhibits the toxicity characteristic owing to its arsenic

---

toxicity

■ For purposes of RCRA regulations, the characteristic of a waste, as determined by implementing prescribed test procedures, in which a representative sample of the waste is found to contain one or more of certain constituents at concentrations equal to or greater than the levels designated at 40 C.F.R. §261.24

gas poisonous-by-inhalation (poison gas)

■ For purposes of DOT regulations, a gas at 68 °F (20 °C) or less and a pressure of 14.7 psi_a (101.3 kPa) that is known to be so toxic to humans as to pose a health hazard during transportation; *or* in the absence of adequate data on human toxicity, is presumed to be toxic to humans because when tested on laboratory animals, it has an LC_{50} of less than 5000 mL/m³

poisonous material (poison)

■ For purposes of DOT regulations, a material other than a gas that is known to be so toxic to humans as to afford a hazard to health during transportation, *or that* in the absence of adequate data on human toxicity, is presumed to be toxic to humans because of data obtained from prescribed oral, dermal, and inhalation toxicity tests performed on animals; *or* is an irritating material with properties similar to those of tear gas that causes extreme irritation, especially within confined spaces concentration, DOT requires the waste characteristic to be properly noted on the accompanying waste manifest by use of either "EPA toxicity" or D004. The basic description for this waste is then either of the following:

> **NA3082, Hazardous waste, liquid, n.o.s. (arsenic), 9, PGIII (EPA toxicity)**
> *or*
> **NA3082, Hazardous waste, liquid, n.o.s. (arsenic), 9, PGIII (D004)**

As a second example, when shippers offer for domestic transportation a solid hazardous waste that exhibits the toxicity characteristic owing to its mercury concentration, they note either of the following basic descriptions on the waste manifest:

> **NA3077, Hazardous waste, solid, n.o.s. (mercury), 9, PGIII (EPA toxicity)**
> *or*
> **NA3077, Hazardous waste, solid, n.o.s. (mercury), 9, NA3077, PGIII (D009)**

SOLVED EXERCISE 10.1

Sulfuryl fluoride is often used as a fumigant to kill insects and parasites on fruits and vegetables. It is also used to exterminate the resident rodents and cockroaches within buildings. When purchased from a company in Syracuse, New York, sulfuryl fluoride is shipped to a chemical treatment company in Baton Rouge, Louisiana, in five 100-lb DOT-3E1800 cylinders by railcar.

(a) Identify the shipping description that DOT requires on the accompanying shipping paper.
(b) Which labels does DOT require to be affixed to the exterior surface of each cylinder?
(c) What markings does DOT require on the exterior surface of each cylinder?
(d) Which placards does DOT require to be displayed on the railcar?

Solution:

(a) Because sulfuryl fluoride is not listed at 49 C.F.R. §172.101, Appendix A, we do not enter RQ with the shipping description. From the information in columns 2, 3, 4, and 7 of Table 6.1, we write the shipping description as follows:

Units	HM	Shipping Description (Identification Number, Proper Shipping Name, Primary Hazard Class or Division, Subsidiary Hazard Class or Division, and Packing Group)	Weight (lb)
5 DOT-3E1800 gas cylinders loaded within rail boxcar	X	UN2191, Sulfuryl fluoride, 2.3 (Poison - Inhalation Hazard, Zone D) (Placarded POISON GAS)	500

DOT also requires the reporting mark and number that the carrier displays on the railcar to be entered in the shipping description.

(b) From the information in column 6 of the Hazardous Materials Table, the label code is determined to be 2.3. This means that DOT requires the shipper to affix a POISON GAS label to the outer exterior surface of each cylinder.

(c) DOT requires the manufacturer to mark DOT-3E1800 (the specification and service pressure) on the outer exterior surface of each cylinder. In addition, DOT requires the shipper to mark the outer exterior surface of each cylinder with the proper shipping name and identification number of the hazardous material: SULFURYL FLUORIDE and UN2191. The surface of each cylinder must also be marked with the words INHALATION HAZARD.

(d) DOT requires the carrier to affix POISON GAS placards on the railcar used to transport the cylinders.

PERFORMANCE GOALS FOR SECTION 10.2

- Describe the common means by which toxic substances enter the body.

10.2 How Toxic Substances Enter the Body

A toxic substance may enter the body by various routes, but we are concerned here with only three: **ingestion**, **skin absorption**, and **inhalation**.

10.2-A INGESTION

ingestion
- The taking in of a substance through the mouth and into the digestive system

Ingestion refers to the swallowing of a substance through the mouth into the stomach and its subsequent movement through the gastrointestinal tract as shown in Figure 10.1. Once it is ingested, the substance may pass through the intestinal walls and into the circulatory system, where its molecules are further disseminated to the organs and tissues of the body. The substance may undergo chemical changes in the cells of these organs and tissues. The combination of these chemical reactions that occur in the cells is the phenomenon called **metabolism**.

skin absorption
- The passage through the epidermis into the dermis or subcutaneous tissue

10.2-B SKIN ABSORPTION

inhalation
- The taking in of a substance in the form of a gas, vapor, fume, mist, or dust into the respiratory system

The skin, a cross section of which is illustrated in Figure 10.2, constitutes the largest single organ of the human body. An average adult's skin spans an area of 21 ft^2 (2 m^2), weighs 9 lb (4.1 kg), and contains more than 11 mi (18 km) of blood vessels. The skin helps the body maintain a normal temperature. It also protects the internal organs and prevents direct contact between them and foreign substances. Although the skin acts as an organ of defense in this manner, it can also act as a permeable membrane. Some foreign substances penetrate the epidermis, the outermost layer, and enter the underlying dermis or subcutaneous tissue, from which they may be further absorbed into the circulatory system and spread throughout the body.

metabolism
- The processes that occur in the body's cells to break down absorbed foods or other ingested substances

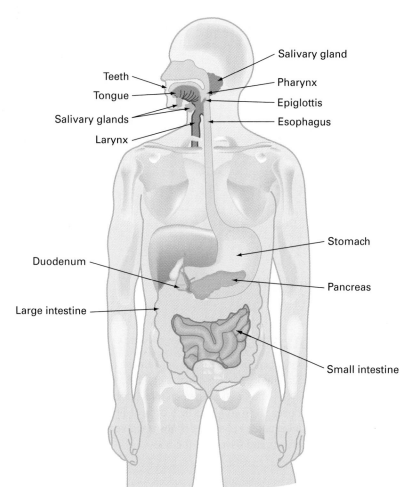

FIGURE 10.1 The major components of the human digestive system. A substance taken orally passes through the mouth, into the esophagus, and then into the stomach. Partial degradation or complete chemical alteration of the substance occurs in the mouth and stomach. Molecules of the substance may then pass through the stomach wall directly into the bloodstream. More generally, however, absorption into the bloodstream occurs after the molecules of the substance or its degradation products pass into the small intestine.

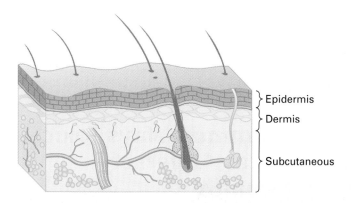

FIGURE 10.2 The cross section of human skin showing some of its principal components. The outer layer of the skin is the epidermis, a thin surface membrane of dead cells. Before a substance may be absorbed into the bloodstream, it must first penetrate past the epidermis and enter the dermis, a collection of cells that collectively act as a porous medium. From the dermis, the substance is absorbed into the bloodstream.

FIGURE 10.3 The major components of the human respiratory system. Inhaled gases, fumes, and vapors—including air—first pass through either the mouth or nose, through the pharynx, and into the trachea (commonly called the windpipe) at the larynx. These inhaled substances then enter either of two bronchi, each of which leads to a lung. The individual divisions of each bronchus are called *bronchioles*.

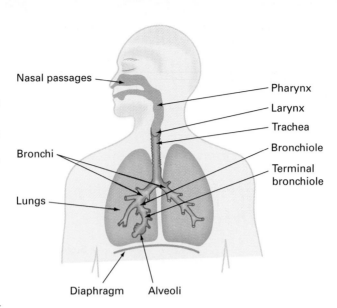

Nasal passages

Pharynx

Larynx

Trachea

Bronchiole

Terminal bronchiole

Bronchi

Terminal bronchiole

Lungs

Diaphragm   Alveoli

10.2-C INHALATION

Inhalation is the route responsible for the movement of gases, vapors, and fumes through the components of the respiratory system, illustrated in Figure 10.3: the nasal passageways, pharynx, larynx, trachea, bronchi, and lungs. When an inhaled substance enters the lungs, it is exposed to blood vessels that cover an average surface area of approximately 90 yd^2 (75 m^2). Given this massive area of exposure, the substance is absorbed very rapidly into the bloodstream and then circulated to other organs and tissues of the body.

PERFORMANCE GOALS FOR SECTION 10.3:

- Describe generally the ways a toxic substance may adversely affect one's health.
- Identify the nature of the following toxicants: hemotoxicants, hepatotoxicants, nephrotoxicants, neurotoxicants, and reproductive toxicants.
- Discuss the mechanisms that cause asphyxia in humans who inhale elevated concentrations of certain gases.
- Describe how exposure to certain substances may irritate the tissues of the respiratory system, eyes, or skin.

10.3 Some Common Ways Toxic Substances Adversely Affect Health

Once a toxic substance has been absorbed into the body, it acts in the following general ways:

systemic
■ Referring to the body as a whole

- It may cause immediate impairment or death.
- It may cause a localized effect at the site of contact.
- It may target the body as a whole. This phenomenon is said to be a **systemic effect**.

- It may target only a specific organ that undergoes localized dysfunction or impairment. When a toxic substance targets a specific organ, the substance is denoted accordingly as one or more of the following:
  - **Hemotoxicant**, a substance that decreases the function of the blood's hemoglobin and deprives the tissues of oxygen
  - **Hepatotoxicant**, a substance that causes liver damage
  - **Nephrotoxicant**, a substance that causes kidney damage
  - **Neurotoxicant**, a substance that adversely affects the central nervous system
  - **Respiratory toxicant**, a substance that causes adverse effects on the nasal passages, pharynx, trachea, bronchi, and lungs
  - **Reproductive toxicant**, a substance that adversely affects an individual's reproductive capabilities
- Ingested substances that are not foods metabolize in the cells of the body, and the metabolic by-products may cause the host victim to experience ill effects. For instance, when methanol and ethylene glycol are ingested, they metabolize to form toxic by-products that are responsible for the onset of ill effects including death (Sections 13.2-A and 13.2-G).

10.3-A ASPHYXIANTS

One way that a gas or vapor adversely affects health is by inducing unconsciousness when inhaled. The gas or vapor is called an **asphyxiant**. Common asphyxiants are nitrogen and carbon dioxide.

There are three common means by which an asphyxiant causes unconsciousness in humans, as follows:

- ***Denial of sufficient oxygen.*** As first noted in Section 2.5-B, a gas having a vapor density greater than 1 displaces the air within the lower regions of an enclosure. An example of such a gas is carbon dioxide. A person who inhales carbon dioxide loses consciousness, since the body is denied sufficient oxygen. When provided fresh air or oxygen, the individual can regain consciousness. If unattended, however, the individual can die from suffocation or experience damage to the central nervous system.

hemotoxicant
- A substance that decreases the function of the blood's hemoglobin and deprives the tissues of oxygen

hepatotoxicant
- A substance that causes liver damage

nephrotoxicant
- A substance that causes kidney damage

neurotoxicant
- A substance that adversely affects the central nervous system

respiratory toxicant
- A substance capable of inflaming the airways or otherwise decreasing lung function

reproductive toxicant
- A substance capable of producing a negative impact on a fetus on exposure of the mother to the substance

asphyxiant
- A gas or vapor that dilutes or displaces air and, when inhaled, causes unconsciousness or death

SOLVED EXERCISE 10.2

Before an underground gasoline storage tank is removed from its location, the American Petroleum Institute recommends purging it with nitrogen or carbon dioxide. Immediately following this purging:

(a) Is it generally safe to implement cutting or welding operations on the tank in connection with its removal?

(b) Is it generally safe for an individual to enter the tank?

Solution:

(a) When gasoline vapor has been purged from a storage tank with nitrogen or carbon dioxide, its concentration is no longer within the flammable range. For this reason, cutting or welding operations on the tank immediately after conducting the purging can generally be conducted without fear of fire or explosion. Nonetheless, gasoline vapor may enter a supposedly empty tank from connecting pipelines. For this reason, immediately prior to conducting any work on the tank, it is always prudent to check the vapor concentration therein using a portable combustible gas monitor.

(b) A tank that has been purged with nitrogen or carbon dioxide is deficient in air. An individual who inhales the atmosphere of nitrogen or carbon dioxide experiences asphyxiation. It is rarely safe for an individual to enter the storage tank immediately following the purging operation, since the individual could die from suffocation or experience damage to the central nervous system.

■ *Action as a hemotoxicant.* An asphyxiant like carbon monoxide causes unconsciousness primarily by preventing the normal transmission of oxygen in the bloodstream. Although the exposed individual can regain consciousness when provided fresh air or oxygen, we learn in Section 10.9 that a short-term exposure to certain concentrations of carbon monoxide is fatal.

■ *Inability to utilize cellular oxygen.* An asphyxiant such as hydrogen cyanide causes unconsciousness primarily by hindering the cells' ability to utilize oxygen. Subsequent exposure to fresh air or oxygen may not revive an individual who has inhaled hydrogen cyanide. An individual who has inhaled hydrogen cyanide generally survives only when promptly administered an appropriate antidote such as amyl nitrite.

10.3-B IRRITANTS

irritant
■ A substance capable of injuring the body's tissues by causing inflammation at the site of contact

The inhalation of a gas or vapor may also affect the body as an **irritant** by injuring the tissues that it contacts. When a low concentration of an irritant is inhaled, the result is often minor inflammation of the tissues that line the respiratory passageways. This happens when a relatively small amount of ammonia vapor is inhaled. However, inhalation of an elevated concentration of the same substance can cause respiratory failure. Whether a substance causes respiratory irritation or failure is often associated with the degree to which it corrodes the nose, pharynx, larynx, trachea, bronchi, and lungs.

PERFORMANCE GOALS FOR SECTION 10.4:

■ Describe the general nature of the acute, chronic, short-term, and latent effects that can result from exposure of the human organism to toxic substances.

10.4 Types of Toxicological Effects

The harmful health effects associated with exposure to substances are often classified as acute, chronic, short-term, or latent. When these substances are constituents of commercial chemical products, these effects are noted by the manufacturer on labels and MSDSs (Section 1.6-B).

10.4-A ACUTE HEALTH EFFECT

acute health effect
■ Disease or impairment that manifests itself rapidly on exposure to a hazardous substance but usually subsides when the exposure ceases

An **acute health effect** is manifested when an injury, disease, or death that is caused by exposure to a substance develops rapidly and quickly comes to a crisis. An example of an acute effect is the tearing of the eyes that immediately occurs from a short-term exposure to ammonia vapor.

10.4-B CHRONIC HEALTH EFFECT

A **chronic health effect** is manifested when an injury, disease, or death that is caused by exposure to a substance develops slowly over a few days, weeks, or longer periods. An example of a chronic effect is coronary heart disease.

10.4-C SHORT-TERM HEALTH EFFECT

A **short-term health effect** is manifested when an injury or disease of relatively short duration is caused by exposure to a substance *and* from which recovery occurs rapidly. For example, when an individual becomes asphyxiated from exposure to carbon dioxide but recovers when provided with oxygen or air, the asphyxiation is called a short-term health effect.

10.4-D LATENT HEALTH EFFECT

A **latent health effect** is manifested when an injury or disease caused by exposure to a substance occurs *only* after an extensive time period has passed. For example, the development of liver angiosarcoma by an individual exposed to vinyl chloride vapor is a latent health effect, since the appearance of malignant tumors on the liver occurs decades after inhalation of the vapor. A number of cancerous and noncancerous afflictions are also examples of latent effects associated with exposure to certain types of asbestos fibers. These diseases usually emerge 30 or more years after initial exposure to the fibers.

chronic health effect
■ Disease or impairment resulting from exposure to a hazardous substance that manifests itself over an extended period (months to years)

short-term health effect
■ An injury or disease of relatively short duration caused by exposure to a substance and from which recovery occurs rapidly

latent health effect
■ An injury or disease that manifests itself only after considerable time has passed following one or more initial exposures to a substance

---

PERFORMANCE GOALS FOR SECTION 10.5:

■ Describe the following factors that affect the degree of toxicity: the quantity of a substance; the duration of exposure; the rate at which a substance is absorbed into the bloodstream; the age, sex, ethnicity, and health of the afflicted individual; individual sensitivities; and, for women, exposure to certain toxic substances during pregnancy.

10.5 Factors Affecting the Degree of Toxicity

The human body is a very complex and delicately balanced system. Its cells assimilate nutrients from foods, resist biological attack, reproduce, and generate the substances that the body needs for its survival. Chemical reactions in the cells are responsible for life itself; but when toxic substances are absorbed across cellular membranes, they can upset this delicate chemistry. When the cells malfunction, impaired health, disease, or death is experienced.

Fortunately, the body has natural mechanisms for protecting itself against foreign substances. One such mechanism involves the action of specific organs. For instance, the liver is particularly effective at converting many harmful substances into harmless ones or into substances that can be rapidly excreted. As a result of these metabolic changes, foreign substances are modified in chemical structure, temporarily stored in specific organs, and/or directly eliminated from the body.

Notwithstanding these natural protective mechanisms, the body is sometimes incapable of protecting itself from invasion by toxic substances. Whether exposure to a toxic substance actually causes death, disease, or injury to the human organism depends on several factors, the most important of which are the following: the quantity of substance, the duration of exposure; the rate at which a substance is absorbed into the bloodstream; the age, sex, ethnicity, and health of the afflicted person; and individual sensitivities. Exposure to a toxic substance by women during pregnancy can also affect the health of their unborn children. We examine each factor independently.

10.5-A QUANTITY OF SUBSTANCE

dose
■ The quantity of a substance per body weight administered directly to and absorbed by an organism

Although all substances have the potential to be toxic, we are concerned with those substances that cause an adverse impact on an organism in relatively small amounts. The amount of the substance administered or absorbed per body weight is called the **dose**. Some substances are so toxic that exposure to the very tiniest dose is lethal. For instance, the bacterium *Clostridium botulinum*, the causative agent of botulism, is a single cell that releases a bacterium so potent that just 1 g can cause the death of more than 100,000 individuals. Fortunately, there are relatively few substances that exhibit this pronounced degree of toxicity.

The impact that a specific substance has on an organism is a function of the dose. For example, the effects that alcoholic beverages exert on the body are a function of dose. When an individual drinks enough alcohol to achieve a blood alcohol concentration of 50 mg/dL, he or she may feel slightly subdued, relaxed, perhaps even elevated in spirit; but when the blood alcohol concentration reaches 150 mg/dL, the person is patently intoxicated. [One deciliter (dL) is 100 milliliters, or one-tenth of a liter.]

10.5-B DURATION OF EXPOSURE

As a general policy, the longer or more frequently an individual is exposed to low doses of a toxic substance, the more likely that person will experience ill effects. Consider the mixture of the approximately 4000 substances known to be contained in tobacco smoke. Although the inhalation of tobacco smoke has irrefutably been shown to increase the risk of developing cancers of the lungs, mouth, larynx, kidney, and bladder, most of us know people who smoke regularly but have never contracted cancer in any organ. If exposure to the substances in tobacco smoke causes these cancers, why do some smokers appear to be shielded from contracting them?

Although the answer to this question is not entirely straightforward, it is apparent that many individuals who become afflicted with cancer have generally smoked tobacco products for many years. The manifestation of cancer occurs only after many repeated exposures to the toxic substances in tobacco smoke. The assimilation of each dose lengthens the total period that the body is exposed to them.

Also, individuals do not respond in the same fashion when exposed to a given concentration of a toxic substance over the same length of time. Some individuals are capable of tolerating toxic substances in their environment more than others. Some even appear to adapt to their presence. Individuals respond to toxic substances over a given length of time in the following ways:

■ Although some individuals experience ill effects from exposure to a toxic substance, others experience the same effects only after they have been exposed to the same concentration of the same substance for a longer period of time.

■ Although some individuals experience adverse health effects from exposure to a given concentration of a substance, they may not experience the same effects on subsequent exposure to the same concentration of the same substance.

■ Some individuals who do not experience ill effects from a short-term exposure to one concentration of a substance may be unable to tolerate a long-term exposure to a smaller concentration of the same substance. These individuals require this longer period of exposure to reach their *body burden*, that is, the concentration that causes them to experience ill effects.

10.5-C RATE AT WHICH A SUBSTANCE IS ABSORBED INTO THE BLOODSTREAM

When a toxic substance is inhaled, the adverse effect caused by its presence may not be evident immediately. Inhaling a lethal concentration of chlorine or carbon monoxide, for example, causes death within seconds or minutes, whereas inhaling a lethal concentration of hydrogen chloride or nitrogen dioxide can result in death hours or days later. These observations demonstrate that the manifestation of an effect caused by exposure to a toxic substance depends not only on the concentration to which an individual is exposed but on the rate of the mechanism that causes the impairment or death.

When a toxic substance is ingested or absorbed through the skin, sufficient time must then elapse for the substance to be absorbed into the bloodstream. When ingested, the substance must usually pass from the stomach into the small intestine, where it is then absorbed into the bloodstream. When exposed to skin, the substance must first pass into the dermis, where it is then absorbed into the bloodstream.

It is fortuitous that time is required for the absorption of a substance into the bloodstream. The time needed for this process to occur provides a window of opportunity during which emergency-medical personnel can administer an antidote, induce vomiting, or implement other appropriate first-aid measures.

10.5-D AGE, SEX, ETHNICITY, AND HEALTH OF INDIVIDUALS

The degree to which toxic substances affect individuals often depends on their age, sex, ethnicity, or general state of health. Young children and the elderly are typically more susceptible than middle-aged individuals to the effects of exposure to toxic substances. Children are vulnerable, since their average body weight is relatively low and their immune systems are still developing. Newborn infants are the most vulnerable, since they must rely primarily on the immune factors acquired from their mothers prior to birth.

The sex and ethnicity of a person also affects the susceptibility to contracting certain diseases. For example, even though the liver and heart are anatomically the same in both men and women, men are more likely to experience primary liver cancer (cancer that arises in the liver) than women, and women are more prone than men to contract types of heart disease that are fundamentally different from the types contracted by men. Furthermore, even though the composition of the blood is the same in all races, black people of African descent are more prone to developing sickle-cell anemia. Why is one sex or one ethnic group more prone to be stricken with certain diseases? Although the answer to this question is unknown, genetics most likely plays a significant role.

Likewise, those persons already weakened from disease are more likely than healthy people to be susceptible to the effects of exposure to toxic substances. This sensitivity is particularly evident in individuals suffering from heart and respiratory ailments. Consider individuals suffering from heart ailments. Although these persons are usually incapable of tolerating low concentrations of ground-level ozone, healthy people could be oblivious to the presence of this air pollutant.

A well-acknowledged group of unhealthy individuals are tobacco smokers. As a general rule, nonsmokers are able to tolerate the presence of toxic substances better than smokers, whose immune systems have not been compromised. Nonsmokers are usually less likely to develop certain illnesses compared with smokers. For example, nonsmokers do not fall victim as easily as do smokers to asbestosis (Section 10.18-A), a progressive, irreversible lung disease associated with exposure to asbestos fibers. It is safe to conclude that emergency responders who are nonsmokers are less likely than emergency responders who smoke to experience the ill effects caused by exposure to carbon monoxide and other toxic gases.

10.5-E INDIVIDUAL SENSITIVITIES

There are vast differences in the ways that individuals respond to exposure to toxic substances. Often, these differences are not even associated with age or health. For example, although some people with asthma are seemingly able to tolerate exposure to ground-level ozone, others experience breathing problems from exposure to lesser doses for a shorter period of time. How is this variability explained?

The answer to this question probably has something to do with inherited traits. Each year, 7.9 million children are born with serious birth defects caused at least partly by genetic flaws. Researchers have shown that the nature of the DNA in the genes predisposes some individuals to contracting diseases such as coronary heart disease and some types of cancer. As noted in Section 10.5-D, genetics may explain why one sex or ethnic group is more likely to contract certain diseases. Genes may also play a role in whether exposure to a given concentration of a toxic substance is tolerable or causes ill effects.

The mechanism by which genes influence vulnerability to contracting disease from exposure to toxic substances is an active area of ongoing medical research. One result of such studies is the discovery that exposure to the constituents of secondhand tobacco smoke damages DNA. The result is that individuals who routinely breathe secondhand smoke put not only themselves but also *their offspring* at increased risk of contracting certain diseases and other ailments. This vulnerability can be traced directly to the genetic information programmed into their DNA.

10.5-F EXPOSURE TO TOXIC SUBSTANCES BY WOMEN DURING PREGNANCY

Pregnant women and their unborn children may experience adverse health effects caused by the women's exposure to toxic substances like mercury (Section 1.4-B), ethanol (Section 13.2-C), Bisphenol-A (Section 13.2-L), alkyl glycol ethers (Section 13.3-D), and diethylhexyl phthalate (Section 13.7-B). As a prudent practice, pregnant women should probably take special measures to limit or eliminate their exposure to hazardous substances altogether, especially when the exposure is elective, as in the workplace.

As first noted in Section 10.5-F, a mother's exposure to pollutants during pregnancy can put her unborn child at a greater risk of contracting cancer and other diseases later in life. We also note in Section 13.4-A that even a father's exposure to dioxin can cause the onset of birth defects in his children who are born after the exposure.

PERFORMANCE GOALS FOR SECTION 10.6:

- Describe the methods by which toxicity is quantified from studies conducted on animals: lethal dose, 50% kill (LD_{50}); lethal concentration, 50% kill (LC_{50}); up-and-down dose; and threshold limit value (TLV).
- Demonstrate how the lethal dose and lethal concentration of a substance can be approximated for humans from the LD_{50} and LC_{50} values obtained from animal studies.
- Define the short-term exposure limit (STEL), permissible exposure limit (PEL), and immediately-dangerous-to-life-and-health level (IDLH).
- Describe how emergency responders use the information conveyed by these limits when they encounter a toxic substance during the line of duty.

10.6 Measuring Toxicity

Toxicologists have devised procedures for measuring the concentration of a substance that causes an organism to experience injury, disease, or death. These procedures involve exposing groups of laboratory animals to concentrations of a substance, observing the effects caused by the exposure, and extrapolating the results to humans. When exposed to a unique concentration of a substance, animals and humans do not always respond in the same way. Consequently, these parameters may have only limited relevance.

In the workplace, OSHA and NIOSH require employers to reduce or eliminate the likelihood that employees will experience adverse health effects from exposure to toxic substances. To do so, employers consider three relevant exposure limits: the short-term exposure limit, the permissible exposure limit, and the immediately-dangerous-to-life-and-health level. Firefighters also use these limits at emergency-response scenes to evaluate the impact of an inhalation hazard posed by a known concentration of a toxic substance.

10.6-A LETHAL DOSE, 50% KILL

The **lethal dose, 50% kill (LD$_{50}$)** is the amount of a substance that kills half of a group of laboratory animals to which the substance is administered during a preestablished time. The LD$_{50}$ is expressed in milligrams of administered substance per body mass of the animal in kilograms (mg/kg). When a substance affects a specific organ, the LD$_{50}$ may be measured by considering the mass of that organ.

lethal dose, 50% kill (LD$_{50}$)
■ The dose of a substance that is lethal to 50 percent of the organisms tested during a specified time

The lethal dose of a substance to humans is calculated using the LD$_{50}$ measurement obtained from animal studies. For an average person having a mass of w kilograms, the lethal dose is the product of the LD$_{50}$ and the mass:

$$\text{Lethal dose} = \text{LD}_{50} \times w$$

10.6-B LETHAL CONCENTRATION, 50% KILL

The **lethal concentration, 50% kill (LC$_{50}$)** is the concentration of a substance that kills half of a group of laboratory animals to which the substance is administered during a preestablished time. A LC$_{50}$ is typically expressed in parts per million (ppm) by volume. These measurements are used in different disciplines for a variety of purposes. DOT uses LC$_{50}$ values to characterize materials in hazard classes 2.3 and 6.1 to establish the hazard zone for a substance that poses a health hazard by inhalation. We review this process in Section 10.7.

lethal concentration, 50% kill (LC$_{50}$)
■ The concentration of a substance that when administered to laboratory animals kills half of them

The lethal concentration of a substance to humans is calculated using the LC$_{50}$ measurement obtained from animal studies. For an average person having a mass of w kilograms, the lethal concentration is the product of the LC$_{50}$ and the mass:

$$\text{Lethal concentration} = \text{LC}_{50} \times w$$

10.6-C UP-AND-DOWN DOSE

The **up-and-down dose** is the concentration in milligrams per kilogram (mg/kg) of a substance that kills a laboratory animal within 48 hours. First, a single dose of the substance is administered to the animal. If it remains alive after 48 hours, a single, slightly higher dose is administered to a second animal. If the first animal dies within 48 hours, a slightly lower dose is administered to the second animal. This process is then repeated by administering higher or lower doses to additional animals until the specific concentration that kills a laboratory animal is identified.

up-and-down dose
■ The concentration in milligrams per kilogram of a substance that kills a laboratory animal within 48 hours when administered according to a prescribed procedure

10.6-D THRESHOLD LIMIT VALUE

The **threshold limit value** (TLV) is the upper limit of a concentration to which an average healthy person can be repeatedly exposed on an all-day, everyday basis without suffering adverse health effects. The TLV for gaseous substances in the air is usually expressed in ppm. The TLV for fumes or mists in air is expressed in milligrams per cubic meter (mg/m^3). The TLV is the level of exposure at which the probability of the occurrence of adverse health effects is deemed negligible. These values are established by the American Conference of Governmental Industrial Hygienists (ACGIH).*

threshold limit value
■ A guideline standard established by the American Conference of Governmental Industrial Hygienists (ACGIH) for airborne concentrations to which an average worker may be exposed day after day without experiencing adverse health effects (abbreviated TLV)

*TLVs are published by the American Conference of Governmental Industrial Hygienists in *Documentation of the Threshold Limit Values and Biological Exposure Indices*, Cincinnati, Ohio (2008).

SOLVED EXERCISE 10.3

When researchers expose rabbits to phenol by skin contact, they determine an LD_{50} of 630 mg/kg. Assuming that human skin similarly responds to an exposure to phenol, determine the dose of phenol that is lethal to a 200-lb individual by skin contact.

Solution: The lethal dose of phenol for a 200-lb person by skin contact is calculated as follows:

$$200 \text{ lb} \times \frac{1 \text{ kg}}{2.2 \text{ lb}} \times 630 \frac{\text{mg}}{\text{kg}} \times \frac{1 \text{ g}}{1000 \text{ mg}} = 57 \text{ g}$$

Thus, skin exposure to 57 g or more of phenol by of a 200-lb individual is lethal.

SOLVED EXERCISE 10.4

Experiments conducted at the National Institute of Standards and Technology, Gaithersburg, Maryland, established the following inhalation LC_{50} values for hydrogen cyanide by exposing laboratory animals to this toxic gas for the indicated time periods:

Exposure Time (min)	LC_{50} (ppm)
1	3000
2	1600
5	570
10	290
20	170
30	110
60	90

Animal deaths were determined during both the exposure and postexposure periods. What do these data demonstrate about the inhalation toxicity of hydrogen cyanide?

Solution: These data are interpreted as follows: When exposed to hydrogen cyanide, half of the animals died following a 1-min exposure to a concentration of 3000 ppm; half died following a 2-min exposure to a concentration of 1600 ppm; half died following a 5-min exposure to a concentration of 570 ppm; and so forth. Overall, these data demonstrate that death can result not only from a short-term exposure to a high concentration of a toxicant but also from a long-term exposure to a lesser concentration of the same substance.

10.6-E PERMISSIBLE EXPOSURE LIMIT (PEL)

permissible exposure limit
■ For purposes of OSHA regulations, the time-weighted average threshold limit value of substances listed at 29 C.F.R. §1910.1000 to which workers can be exposed continuously during an 8-hr work shift without suffering ill effects (abbreviated PEL)

The **permissible exposure limit** (PEL) is the time-weighted average threshold limit value of a toxic substance to which workers can be exposed continuously for 8 hours in conformance with relevant OSHA regulations. Employers calculate the cumulative employee exposure to a listed toxic substance over an 8-hr work shift and compare the exposure with values established by NIOSH.

Ceiling values are also denoted for some substances by NIOSH. These values represent the maximum concentrations to which workers can safely be exposed.

10.6-F SHORT-TERM EXPOSURE LIMIT (STEL)

The **short-term exposure limit** (STEL) is the concentration of a substance to which workers can be exposed continuously for a period of approximately 15 minutes without suffering irritation, chronic or irreversible tissue damage, or narcosis sufficient to increase the risk of accidental injury, impair self-rescue, or materially reduce work-related efficiency. This assumes that the daily time-weighted average threshold limit value for the substance has not been exceeded. STELs for specific substances are published when toxicological effects from relatively elevated short-term exposures to either humans or animals have been reported. They are established by NIOSH.*

10.6-G IMMEDIATELY-DANGEROUS-TO-LIFE-AND-HEALTH LEVEL (IDLH)

The **immediately-dangerous-to-life-and-health level** (IDLH) is the atmospheric concentration of any substance that poses an immediate threat to life, causes irreversible or delayed adverse health effects, or interferes with an individual's ability to escape during a 30-min period from a dangerous atmosphere. These values are established by NIOSH.

The relative differences among several of the toxicity measurements noted in this section are listed in Table 10.2 for some common gases and vapors.

PERFORMANCE GOALS FOR SECTION 10.7:

- Describe how DOT determines the hazard zones of gases and liquids that pose an inhalation hazard when released to the environment.

10.7 The Hazard Zone

In Chapter 6, it was noted that DOT assigns one of four hazard zones to gases and one of two hazard zones to liquids that pose a health hazard by inhalation. DOT determines the hazard zones for all toxic gases and liquids by reference to their LC_{50}s as follows:

- When the LC_{50} is equal to or less than 200 ppm, DOT assigns Zone A to the gas or liquid.
- When the LC_{50} is greater than 200 ppm but equal to or less than 1000 ppm, DOT assigns Zone B to the gas or liquid.
- When the LC_{50} is greater than 1000 ppm but equal to or less than 3000 ppm, DOT assigns Zone C to the gas.
- When the LC_{50} is greater than 3000 ppm but equal to or less than 5000 ppm, DOT assigns Zone D to the gas.

Emergency-response personnel use the hazard zone designation to determine the degree of toxicity of a given gas or liquid. Clearly, although exposure to a gas with a Zone A designation is likely to be immediately dangerous to life, exposure to a gas with a Zone D designation could be momentarily tolerable for some individuals. This distinction notwithstanding, the warning to emergency-response personnel at transportation mishaps involving the spill or leak of a gas or liquid to which any of the four hazard zones has been assigned is to don protective clothing and use self-contained breathing apparatus prior to executing a response action.

PERFORMANCE GOALS FOR SECTION 10.8:

- Describe the general atmosphere of the typical fire scene, especially the routine presence of smoke and carbon monoxide.
- Describe the physical composition of smoke and note the specific ways it hampers performance during emergency-response activities.
- Describe how the inhalation of smoke negatively affects the human organism.

short-term exposure limit
- The concentration to which workers can be exposed continuously for a short period of time without suffering irritation, chronic, or irreversible tissue damage, or narcosis of sufficient degree to increase the likelihood of accidental injury, impair self-rescue, or materially reduce work efficiency, and provided that the daily threshold limit value, time-weighted average is not exceeded (abbreviated STEL)

immediately-dangerous-to-life-and-health level
- The atmospheric concentration of any substance that poses an immediate threat to life, causes an irreversible or delayed adverse health effect, or interferes with an individual's ability to escape during a 30-min period (abbreviated IDHL)

---

*STELs, PELs, and IDLHs are published by the National Institute of Occupational Safety and Health in *NIOSH Pocket Guide to Chemical Hazards*, U.S. Department of Health and Human Services, Public Health Service, Centers for Disease Control and Prevention, Washington, DC (2007). PELs are also published at 29 C.F.R. §1910.1000.

TABLE 10.2		Toxicity Measurements of Some Common Gases and Vapors		
TOXIC SUBSTANCE	**LC_{50} (ppm)**	**PERMISSIBLE EXPOSURE LIMIT**	**SHORT-TERM EXPOSURE LIMIT**	**IMMEDIATELY-DANGEROUS-TO-LIFE-AND-HEALTH LEVEL (ppm)**
Ammonia	4000	25 ppm (18 mg/m^3) (NIOSH)	50 ppm (35 mg/m^3) (OSHA) 35 ppm (27 mg/m^3) (NIOSH)	300
Carbon monoxide	3760	50 ppm (55 mg/m^3) (OSHA) 35 ppm (40 mg/m^3) Ceiling 200 ppm (229 mg/m^3)(NIOSH)		1200
Chlorine	293	Ceiling 1 ppm (3 mg/m^3) (OSHA) Ceiling 0.5 ppm (1.45 mg/m^3), 15 min (NIOSH)		10
Fluorine	185	0.1 ppm (0.2 mg/m^3) (OSHA/NIOSH)		25
Hydrogen chloride	2810	Ceiling 5 ppm (7 mg/m^3) (OSHA/NIOSH)		50
Hydrogen cyanide		10 ppm (11 mg/m^3), skin absorption (OSHA)	4.7 ppm (5 mg/m^3), skin absorption (NIOSH)	50
Hydrogen sulfide	712	Ceiling 20 ppm (28.4 mg/m^3) 50 ppm/ 10-min maximum peak (OSHA) 10 ppm (15 mg/m^3), 10-min maximum peak (NIOSH)		100
Nitric oxide	115	25 ppm (30 mg/m^3) (OSHA/NIOSH)		100
Nitrogen dioxide	115	Ceiling 5 ppm (9 mg/m^3) (OSHA)	1 ppm (1.8 mg/m^3) (NIOSH)	20
Phosgene	5	0.1 ppm (0.2 mg/m^3) (OSHA) 0.1 ppm (0.4 mg/m^3) Ceiling 0.2 ppm (0.8 mg/m^3), 15-min exposure (NIOSH)		2
Phosphine	20	0.3 ppm (0.4 mg/m^3) (OSHA/NIOSH)	1 ppm (1 mg/m^3) (NIOSH)	50
Sulfur dioxide	2520	5 ppm (13 mg/m^3) (OSHA) 2 ppm (5 mg/m^3)(NIOSH)	5 ppm (13 mg/m^3) (NIOSH)	100

10.8 Toxicity of the Fire Scene

The fire scene is usually an extremely dangerous environment. Even if all its physical hazards could be eliminated, the atmosphere would still be filled with smoke, dust, toxic gases, and other toxic substances. The potential for illness and death at a fire scene most often results when individuals are exposed to the smoke and toxic gases produced during the fire.

What is smoke? Put simply, smoke is the gray-to-black plume of matter that emanates from NFPA class A and class B fires. It consists of an airborne suspension of water droplets and finely divided particulates of carbon and ash.

During most fires, carbon is initially produced as microscopic particles, which rapidly agglomerate into black particulates collectively referred to as **soot**. Because these particulates can have diameters of less than 10 μm, they are small enough to be inhaled into the lungs. The larger particulates in smoke are usually filtered within the nasal passageways, but these smaller ones can be drawn into the bronchi and lungs.

soot
■ The agglomeration of carbon particulates generated during the incomplete combustion of carbonaceous materials

The ultrafine soot particles—those having diameters of less than 2.5 μm—penetrate even deeper. They may pose the greater risk to health, since they lodge deeply and are not cleared by coughing. Trapped within the surrounding tissue of the lungs, they obstruct the proper functioning of the lungs and contribute to the onset of pulmonary illnesses and premature death.

The long-term inhalation of particulate matter has been linked with an increase in cardiovascular and pulmonary diseases in susceptible individuals. In particular, the inhalation of particulates has been linked with hastening the deaths of sick, elderly individuals by contributing to the premature onset of heart attacks, strokes, and emphysema. It is safe to infer that soot particulates can also contribute to the inception of adverse health effects in those firefighters who regularly inhale soot while combating fires.

Figure 10.4 illustrates that the production of smoke is directly linked to the incomplete combustion of matter. When produced at a fire scene, the smoke and combustion products like carbon monoxide often represent a greater hazard to life and a more serious hindrance to firefighting efforts than the fire itself. Individuals who are unable to escape from a fire scene often die from exposure to smoke and combustion products, even before the fire reaches them.

Fires are not the sole source of soot particles, although they are generally the indirect source. Other sources include the flyash in stack emissions from fossil-fuel-fired power plants and industrial manufacturing plants and the tailpipe emissions from heavy-duty vehicles and machinery.

10.8-A THE IMPACT OF SMOKE ON VISION

When the eyes are exposed to the irritant gases in smoke, they sting and involuntarily tear and close. Zero visibility results particularly in burning buildings from which smoke cannot readily escape. Certain irritants in smoke are called **lacrimators** (Section 13.12). Their vapors act on the sensitive nerve endings of the mucous membrane of the eye and cause an involuntary unleashing of a flood of tears. Examples of lacrimators often found in smoke are formaldehyde and acrolein.

lacrimator
■ A substance that causes the eyes to involuntarily tear and close

10.8-B THE IMPACT OF SMOKE ON BREATHING

The trachea and bronchi are lined with tiny hairlike projections called **cilia** that act in a wavelike manner to force deposited particles upward towards the esophagus, where they are generally swallowed. Inhaled smoke, however, greatly impairs the ability of the cilia to effectively remove carbon particulates from the respiratory passages, which causes the onset of choking, gagging, and respiratory injuries and contributes to the onset of other adverse health effects.

cilia
■ Fine hairlike projections such as those along the exterior of the respiratory tract that move in unison to help prevent the passage of fluids and fine particles into the lungs

10.8-C LATENT HEALTH EFFECTS CAUSED BY INHALING SMOKE

Aside from linking the inhalation of particulate matter to an increase in cardiovascular and pulmonary diseases, researchers have also demonstrated that particulate matter increases the

FIGURE 10.4 Smoke, carbon monoxide, and water vapor are typically produced during the incomplete combustion of carbon-bearing fuels. Depending on the nature of the fuel, the atmosphere surrounding a fire may also contain vapors of the unburned fuel, its decomposition products, and other oxidation products. The inhalation of this combination of substances is generally more injurious to the health of firefighters and fire victims than the accompanying flames or heat.

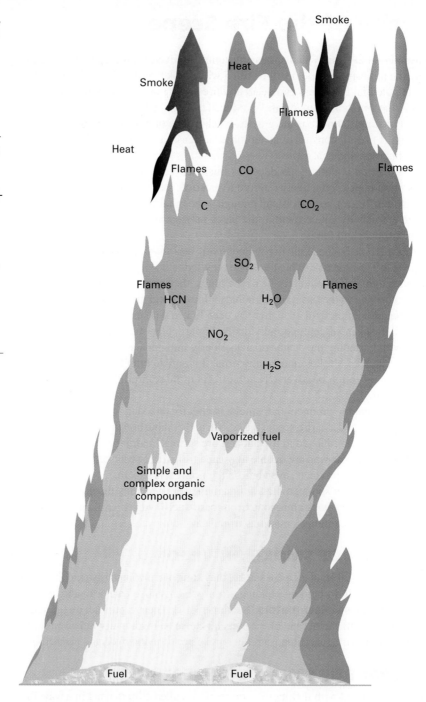

rate at which individuals are likely to die from lung cancer. The research information indicates that those individuals who regularly inhale particulate matter are 8% more likely to develop lung cancer compared with unexposed individuals. The onset of latent health effects like cancer is linked to the exposure to certain compounds that are adsorbed on smoke particulates. These compounds include the polynuclear aromatic hydrocarbons, which we visit in Section 12.13.

10.8-D THE ADSORPTION OF TOXIC GASES ON THE SURFACES OF CARBON PARTICULATES

The carbon particulates in smoke act as adsorbents for toxic gases. This means that each time smoke is inhaled, the carbon particulates aid in drawing carbon monoxide and other undesirable gases into the bronchi and lungs. As these particulates deposit on their surfaces, the expulsion of the gases from the body is simultaneously hindered.

NFPA class A and class B fires often involve the combustion of wood, plastics, heating fuels, and other carbon-rich materials. When such materials burn, the constituent carbon unites with oxygen to form carbon monoxide or carbon dioxide, both of which are colorless, odorless gases. The presence of carbon monoxide at a fire scene is a matter of special concern, since carbon monoxide is highly poisonous.

To understand how these gases are produced at fire scenes, it is necessary to revisit two processes introduced earlier: incomplete and complete combustion. The nature of these processes is schematically described in Figure 10.5.

Incomplete combustion is the burning phenomenon that occurs when the supply of air or oxygen, or access to either air or oxygen, is limited; that is, incomplete combustion occurs in fuel-rich fires. When carbon-bearing fuels burn, incomplete combustion processes produce carbon monoxide and soot. The following equations illustrate the production of these substances during the incomplete combustion of methane:

$$2CH_4(g) \ + \ 3O_2(g) \ \longrightarrow \ 2CO(g) \ + \ 4H_2O(g)$$

Methane   Oxygen   Carbon monoxide   Water

$$CH_4(g) \ + \ O_2(g) \ \longrightarrow \ C(s) \ + \ 2H_2O(g)$$

Methane   Oxygen   Carbon   Water

The carbon monoxide produced during incomplete combustion enters the atmosphere, where it slowly oxidizes to carbon dioxide.

By contrast, complete combustion is the burning process that occurs when plenty of air or oxygen is available. When carbon-bearing fuels burn, complete combustion is associated with the production of carbon dioxide. The complete combustion of methane is illustrated by the following equation:

$$CH_4(g) \ + \ 2O_2(g) \ \longrightarrow \ CO_2(g) \ + \ 2H_2O(g)$$

Methane   Oxygen   Carbon dioxide   Water

Complete combustion occurs when the flow of fuel and air can be regulated, as in a normal heating system.

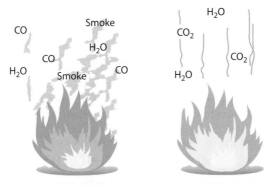

FIGURE 10.5 The incomplete combustion of carbon-bearing fuels, shown on the left, produces carbon monoxide, carbon particulates (smoke), and water vapor. This situation is typical of a fuel-rich or oxygen-starved fire. The complete combustion of carbon-bearing fuels, shown on the right, produces carbon dioxide and water vapor. This situation is typical of a fuel that was mixed with air or oxygen prior to its ignition.

Although carbon monoxide and carbon dioxide are present at virtually all fire scenes, several other gaseous combustion products can also be present, including hydrogen cyanide, ammonia, sulfur dioxide, nitrogen dioxide, hydrogen chloride, and acrolein. Whether one or more of these substances are actually present at a given fire scene depends on the chemical nature of the material that burns, smolders, or undergoes thermal decomposition.

Studies with laboratory animals demonstrate that exposure to a mixture of toxic gases may adversely affect health differently than exposure to a single gas. Our understanding with respect to how the combination toxicologically interacts when inhaled is the subject of ongoing research.

10.8-E ENVIRONMENTAL REGULATIONS INVOLVING PARTICULATE MATTER

Using the authority of the Clean Air Act, EPA regulates the concentration of fine particulate matter (other than windblown dust or soils) in the ambient air as a criteria air pollutant (Section 1.5-A). For particulates having a diameter ranging between 2.5 and 10 μm, EPA set the primary national air quality 24-hr standard at 150 $\mu g/m^3$. For particulates having a diameter of 2.5 μm or less, EPA set the primary national ambient air quality 24-hr standard at 35 $\mu g/m^3$ and the annual average at 15 $\mu g/m^3$ of air.

PERFORMANCE GOALS FOR SECTION 10.9:

- Discuss the technical basis for associating the presence of carbon monoxide with virtually all NFPA class A and class B fires.
- Identify the adverse effects experienced when humans are exposed to different concentrations of carbon monoxide.
- Describe how exposure to carbon monoxide causes the blood's hemoglobin to impede the proper transfer of oxygen.
- Identify the industries that use bulk volumes of carbon monoxide.
- Describe the means shippers and carriers use to inform emergency responders of the hazards associated with encountering carbon monoxide during transportation mishaps.
- Describe the response actions to be executed when carbon monoxide has been released into the environment.

10.9 Carbon Monoxide

Carbon monoxide is an odorless, colorless, tasteless, and nonirritating gas at ordinary room conditions. Some other important physical properties are provided in Table 10.3.

10.9-A HOW DOES CARBON MONOXIDE KILL?

Inhalation toxicity is the primary risk associated with exposure to carbon monoxide. To understand why it is poisonous, we must first examine the chemistry that occurs during respiration. When air is inhaled into the lungs, a supply of atmospheric oxygen is assimilated into the bloodstream. The oxygen is carried throughout the body by a complex component of the blood called **hemoglobin**. Each red blood cell contains about 300 million hemoglobin molecules. The molecular structure of hemoglobin is very complex. We represent it by the symbol Hb.

hemoglobin
■ The component of red blood cells that transports oxygen to the tissues of the body

TABLE 10.3	Physical Properties of Carbon Monoxide	
Melting point		$-341\,°F\,(-207\,°C)$
Boiling point		$-314\,°F\,(-192\,°C)$
Specific gravity at 68 °F (20 °C)		0.81
Vapor density (air = 1)		0.969
Autoignition point		1128 °F (609 °C)
Lower explosive limit		12.5%
Upper explosive limit		74.2%

When hemoglobin unites with oxygen, the compound called **oxyhemoglobin** is produced. Its molecules are also complex, so we represent them as O_2Hb. Respiration is then represented in the following simplified fashion:

$$Hb(aq) \;+\; O_2(g) \longrightarrow O_2Hb(aq)$$
$$\text{Hemoglobin} \qquad \text{Oxygen} \qquad \text{Oxyhemoglobin}$$

oxyhemoglobin
■ The compound that forms when oxygen combines with the blood's hemoglobin

The oxyhemoglobin molecules move within the circulatory system. As they arrive at the various tissues and organs of the body, they release their oxygen at the cellular level for biochemical use and again become hemoglobin molecules. The hemoglobin molecules then return through the circulatory system to the lungs, where they secure a new supply of oxygen. This process occurs over and over with every breath.

Carbon monoxide interrupts this normal act of respiration in part by interfering with the transport of oxygen. Carbon monoxide can also bind to hemoglobin. The reaction produces the substance called **carboxyhemoglobin**, represented as COHb.

$$Hb(aq) \;+\; CO(g) \longrightarrow COHb(aq)$$
$$\text{Hemoglobin} \quad \text{Carbon monoxide} \quad \text{Carboxyhemoglobin}$$

carboxyhemoglobin
■ The compound that forms when carbon monoxide reacts with hemoglobin

When the blood's hemoglobin is bound within the structure of carboxyhemoglobin, it cannot perform its normal bodily function. This is particularly serious, since the chemical affinity of hemoglobin for carbon monoxide is about 200 times greater than its affinity for oxygen. Because carboxyhemoglobin forms so readily, an encounter with carbon monoxide may be deadly.

The ill effects caused by inhalation of carbon monoxide are not solely associated with the production of carboxyhemoglobin. Studies show that the carbon monoxide inhaled into the lungs does not always bond to the blood's hemoglobin; instead, it is assimilated into the cells of the body's tissues, where it may interfere with enzymatic processes. For this reason, long-term exposure to a low concentration of carbon monoxide may cause ill effects that are not otherwise experienced by a short-term exposure to the same concentration.

Individuals who inhale carbon monoxide experience the combination of the signs and symptoms that are listed in Table 10.4 and illustrated in Figure 10.6. These signs and symptoms are linked with the conversion of hemoglobin into the biologically useless carboxyhemoglobin. Inhalation of air containing 0.1% carbon monoxide converts more than half the hemoglobin in the blood to carboxyhemoglobin. Under such conditions, death generally occurs within 1 hour.

The presence of carboxyhemoglobin in the bloodstream kills the host victim by hindering the transport of oxygen. The host victim suffers from **anoxia**, a lack of oxygen, which causes the victim's skin to assume a bright pink color. As the tissues are deprived of oxygen, the afflicted person experiences difficulty in breathing, drowsiness, headache, dizziness, and chest pains. When deprived of oxygen for an extended period, the victim may be unable to survive.

anoxia
■ The absence or near absence of a supply of oxygen to the body's organs or tissues

TABLE 10.4	Adverse Health Effects Resulting from the Production of Various Concentrations of Carboxyhemoglobin Within the Bloodstream
PERCENT CARBOXYHEMOGLOBIN	**SIGNS AND SYMPTOMS**
0–10	No signs or symptoms
10–20	Tightness across forehead, possible slight headache, dilation of the cutaneous blood vessels
20–30	Headache and throbbing in the temples
30–40	Severe headache, weakness, dizziness, dimness of vision, nausea, vomiting, and collapse
40–50	Same as immediately above, greater possibility of collapse, cerebral anemia, and increased pulse and respiratory rates
50–60	Cerebral anemia, increased pulse and respiratory rates, coma, and intermittent convulsions
60–70	Coma, intermittent convulsions, depressed heart action and respiratory rate, and possible death
70–80	Weak pulse and slow respiration leading to death within hours
80–90	Death in less than 1 hour
90–100	Death within a few minutes

10.9-B CARBON MONOXIDE AT FIRE AND OTHER SCENES

Carbon monoxide becomes a component of the atmosphere at virtually every fire scene. Exposure to this gas is linked with the majority of the illnesses and deaths that arise in connection with fires. However, Figure 10.7 shows that carbon monoxide is not associated solely with the burning of

100ppm   1000ppm   1300ppm   >2000ppm

FIGURE 10.6 Carbon monoxide is a toxic combustion product to which all firefighters are exposed during their routine work-related activities. Most adults may tolerate carbon monoxide at a concentration of 100 ppm without noticeably suffering adverse health effects. However, when adults are exposed for 1 hour to carbon monoxide at a concentration of 1000 ppm, they generally experience a mild headache and the simultaneous development of a reddish coloration of the skin. When adults are exposed for 1 hour to carbon monoxide at a concentration of 1300 ppm, they usually experience a throbbing headache. Exposure for 1 hour to carbon monoxide at a concentration of 2000 ppm is likely to either cause death or severe damage to the respiratory and nervous systems.

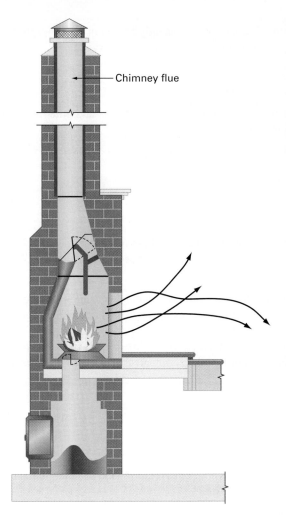

FIGURE 10.7 The combustion products produced within an operating fireplace may enter the home when a chimney flue becomes blocked. Under this circumstance, the fireplace serves as a potential source of carbon monoxide poisoning. To prevent a poisonous atmosphere from developing in the home, occupants should always be assured that chimney flues are unblocked, so the noxious combustion products properly vent up the flue and into the outside atmosphere.

buildings and other structures. In the United States, the inhalation of carbon monoxide generated by the operation of faulty furnaces, stoves, space heaters, and blocked chimneys and flues is considered the leading cause of accidental poisoning within the home.

Although the primary risk associated with exposure to carbon monoxide is inhalation toxicity, bulk quantities of the gas also pose the risk of fire and explosion. When ignited, carbon monoxide oxidizes to carbon dioxide.

$$2CO(g) \;+\; O_2(g) \;\longrightarrow\; 2CO_2(g)$$

Carbon monoxide   Oxygen   Carbon dioxide

SOLVED EXERCISE 10.5

The carbon monoxide concentration in the lower atmosphere normally ranges from 0.2 to 0.5 ppm. Because human activities (e.g., the burning of fuels) regularly produce carbon monoxide, why doesn't the carbon monoxide concentration become so elevated that the lower atmosphere becomes poisonous?

Solution: Carbon monoxide does not permanently remain in the atmosphere. It slowly converts into carbon dioxide by chemical union with atmospheric oxygen.

$$2CO(g) \quad + \quad O_2(g) \quad \longrightarrow \quad 2CO_2(g)$$

Carbon monoxide   Oxygen   Carbon dioxide

Although some carbon dioxide is absorbed into the oceans and other bodies of water, most remains unaltered in the atmosphere.

SOLVED EXERCISE 10.6

The carbon monoxide concentration in a car's exhaust is approximately 8% by volume when the car is idling, but is reduced to approximately 4% when the car is moving. Why are these concentrations so dissimilar?

Solution: When the supply of air is restricted, incomplete combustion occurs. Because this situation exists within the car's combustion chambers, carbon monoxide is produced when vehicular fuels burn. A comparatively lesser volume of air is drawn into the car's combustion chambers during idling than when the car is moving. This causes a correspondingly larger amount of carbon monoxide to be produced during idling compared with when the car is moving.

SOLVED EXERCISE 10.7

Why are a growing number of cities and states requiring the installation in homes of devices that detect the presence of carbon monoxide?

Solution: A carbon monoxide detector like the simple device shown can alert individuals to the presence of this odorless, tasteless, colorless, and poisonous gas.

City and state legislators are hopeful that the widespread use of carbon monoxide detectors will result in saving the lives of family members, which occurred when they required smoke detectors to be installed in homes. The presence of carbon monoxide detectors does not replace the necessity for the installation of smoke detectors; in fact, carbon monoxide detectors are not designed to detect smoke or fire. Both should be installed in homes.

10.9-C USES OF CARBON MONOXIDE

Carbon monoxide is encountered as a commercial chemical product. Because it is a component of water gas (Section 7.2), large volumes of carbon monoxide are produced by the chemical industry during the manufacture of hydrogen. Its isolation from water gas serves as the primary means by which carbon monoxide is produced for commercial use. Although the chemical industry is the primary user of carbon monoxide for the manufacture of methanol (Section 13.2-A), metallurgists also use it to reduce metallic oxides to their corresponding metals.

10.9-D WORKPLACE REGULATIONS INVOLVING CARBON MONOXIDE

When the use of carbon monoxide is necessary in the workplace, OSHA requires employers to limit employee exposure to a maximum concentration of 50 ppm ($55 \ mg/m^3$), averaged over an 8-hr workday.

10.9-E ENVIRONMENTAL REGULATIONS INVOLVING CARBON MONOXIDE

Relatively low concentrations of carbon monoxide are natural components of polluted air. Using the authority of the Clean Air Act, EPA regulates the carbon monoxide concentration in the ambient air as a criteria air pollutant. EPA has also set the primary national ambient air quality standard for carbon monoxide at 9 ppm ($10 \ mg/m^3$).

10.9-F CONSUMER PRODUCT REGULATIONS INVOLVING CARBON MONOXIDE

The death toll from inhaling carbon monoxide in the home is considerable. This fact has caused the CPSC to take certain steps to inform the public that exposure to the gas is potentially lethal. These steps center on warning people about two relatively common ways of generating carbon monoxide: the burning of charcoal indoors in grills, hibachis, and similar items; and the operation indoors of portable generators. The production of carbon monoxide in both instances can kill a home's occupants within minutes.

The CPSC uses appropriate labeling to warn the public about the lethal nature of carbon monoxide. As noted earlier in Section 7.7-E, the CPSC requires charcoal manufacturers to affix the label shown in Figure 7.17 on charcoal packaging. The CPSC and FEMA also require portable generator manufacturers to affix the label shown in Figure 10.8 on portable generators to warn the unsuspecting public that their indoor use could be deadly.

10.9-G TRANSPORTING CARBON MONOXIDE

Carbon monoxide is commercially available as a nonliquefied compressed gas and cryogenic liquid. When shippers offer either commodity for transportation, they identify it on the accompanying shipping paper in the following manner, as relevant:

> **UN1016, Carbon monoxide, compressed, 2.3, (2.1), (Poison - Inhalation Hazard, Zone D)**

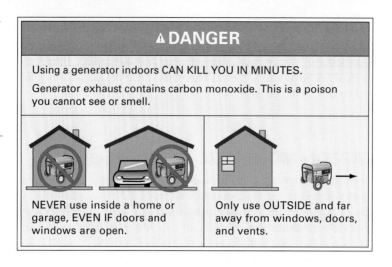

▲DANGER

Using a generator indoors CAN KILL YOU IN MINUTES.

Generator exhaust contains carbon monoxide. This is a poison you cannot see or smell.

NEVER use inside a home or garage, EVEN IF doors and windows are open.

Only use OUTSIDE and far away from windows, doors, and vents.

NA9202, Carbon monoxide, refrigerated liquid, 2.3, (2.1), (Poison - Inhalation Hazard, Zone D)

DOT also requires shippers to affix POISON GAS and FLAMMABLE GAS labels on its packaging.

DOT requires carriers to display POISON GAS placards and INHALATION HAZARD markings on the bulk packaging or transport vehicle used for shipment.

10.9-H RESPONDING TO INCIDENTS ASSOCIATED WITH EXPOSURE TO CARBON MONOXIDE

When individuals have lost consciousness from inhaling carbon monoxide, it is essential to move them swiftly to an open space where a means of artificial respiration and fresh oxygen can be administered. Breathing oxygen for 1 hour reduces the concentration of carbon monoxide in the victim's blood to approximately half its initial concentration.

hyperbaric oxygen therapy
■ The treatment of various afflictions including carbon monoxide poisoning, during which the patient breathes oxygen under increased pressure within a sealed steel chamber

Individuals who are seriously afflicted by exposure to carbon monoxide, especially when they are comatose, should be provided oxygen and rapidly transported to a hospital or clinic capable of administering **hyperbaric oxygen therapy**. This treatment process involves subjecting the person to a high-oxygen atmosphere inside a pressurized chamber of the type shown in Figure 10.9. The inhalation of 100% oxygen at a pressure of 2 atm (202.6 kPa) for ½ to 1 hour results in the faster conversion of carboxyhemoglobin to oxyhemoglobin compared with the use of 100% oxygen at 1 atm (101.3 kPa) for the same period. For this reason, hyperbaric oxygen treatment may successfully reverse the adverse impact caused by inhaling carbon monoxide.

Breathing oxygen under increased pressure allows the blood to carry oxygen deeper into the body's tissues. Under pressure, the capillaries diffuse the blood farther, allowing oxygen-rich blood to be carried into tissues that may otherwise be inaccessible. Even with effective treatment, however, individuals who survive exposure to carbon monoxide may still suffer long-term damage to the heart and brain.

The use of hyperbaric oxygen therapy has also been used to treat the victims of hydrogen cyanide (Section 10.10) and hydrogen sulfide (Section 10.12) poisoning, but in these instances, only limited success has been achieved.

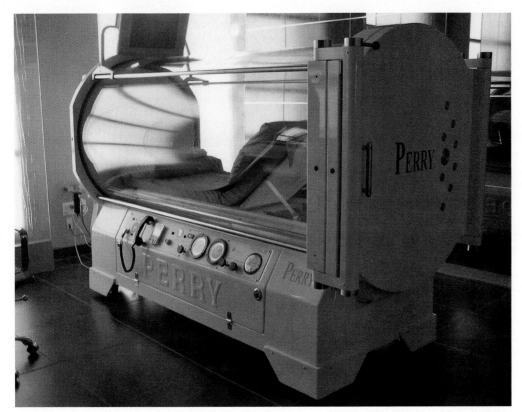

FIGURE 10.9
Hyperbaric oxygen therapy involves administering 100% oxygen at a pressure up to three times the normal atmospheric pressure for 30 to 60 minutes to a patient while he or she lies inside an enclosed chamber. The utilization of hyperbaric oxygen therapy at 2 atm (202.6 kPa) accelerates the removal of carbon monoxide from the bloodstream. Thus, hyperbaric oxygen therapy can save the lives of emergency responders who have been overexposed to carbon monoxide. (*Courtesy of Perry Baromedical Corporation, Riviera Beach, Florida.*)

PERFORMANCE GOALS FOR SECTION 10.10:

- Identify the consumer products that decompose to produce hydrogen cyanide when they smolder or are exposed to intense heat.
- Identify the adverse effects experienced on exposure to hydrogen cyanide.
- Identify the industries that use bulk quantities of hydrogen cyanide, hydrocyanic acid, and metallic cyanides.
- Describe the means shippers and carriers use to inform emergency responders of the hazards associated with encountering hydrogen cyanide, hydrocyanic acid, or metallic cyanides during transportation mishaps.

10.10 Hydrogen Cyanide

Hydrogen cyanide is a colorless gas that has the odor of bitter almonds. It is a highly toxic gas, with a TLV of only 10 ppm. Some other physical properties are provided in Table 10.5. Hydrogen cyanide may also be liquefied, but the liquid is so highly unstable that it is not useful as a commercial chemical product.

Most people—including terrorists—are well aware of the poisonous nature of cyanides. Given the unorthodox ways by which terrorists may use hazardous materials to kill massive numbers of people, law enforcement agencies should be provided with an accounting of the cyanide products potentially available to terrorists within their individual jurisdictions.

TABLE 10.5	Physical Properties of Hydrogen Cyanide
Melting point	6.8 °F (−14 °C)
Boiling point	79 °F (26 °C)
Specific gravity at 68 °F (20 °C)	0.69
Vapor density (air = 1)	0.938
Flashpoint	0 °F (−18 °C)
Autoignition point	~1000 °F (~538 °C)
Lower explosive limit	6%
Upper explosive limit	41%

10.10-A HOW DOES HYDROGEN CYANIDE KILL?

Hydrogen cyanide enters the body primarily by inhalation, whereupon individuals experience the ill effects listed in Table 10.6. The initial symptoms include dizziness, headache, and diarrhea, which individuals experience on exposure to many poisons.

cyanosis

■ The adverse affliction resulting from exposure to cyanides and certain other substances, exhibited physically by the presence of a bluish tinge in the fingernail beds, lips, ear lobes, conjunctiva, mucous membranes, and tongue, exhibited by dizziness, headache, diarrhea, and anemia

The combination of effects associated with exposure to hydrogen cyanide is called **cyanosis**. The mechanism of cyanosis is associated with the ability of hydrogen cyanide to inhibit the normal biological activity of *cytochrome oxidase*, an enzyme essential for cellular respiration and energy production. Although the supply of oxygen in the cells of the body's tissues may be plentiful, the hydrogen cyanide inhibits the ability of the cells to utilize the oxygen effectively.

10.10-B HYDROGEN CYANIDE AT FIRE SCENES

Hydrogen cyanide is generated at fire scenes by the thermal decomposition of consumer products composed of substances whose molecules have carbon atoms bonded to nitrogen atoms ($C \equiv N$). These products include wool, silk, polyacrylonitrile, polyurethane, and nylon.

TABLE 10.6	Adverse Health Effects Associated with Breathing Hydrogen Cyanide
HYDROGEN CYANIDE (PPM)	**SIGNS AND SYMPTOMS**
0.2–5.0	Odor threshold
10	Threshold limit value
18–36	Dizziness, headache, diarrhea, and anemia after several hours
45–54	Tolerated for 0.5 to 1 hour without difficulty
100	Death within 1 hr
110–135	Fatal in 0.5 to 1 hr
181	Fatal after 10 min
280	Immediately fatal

When carpeting and other textiles composed of these materials smolder or are exposed to intense heat, hydrogen cyanide may be released into the surrounding environment along with carbon monoxide.

Although hydrogen cyanide may form during fires, scientists believe that it generally does not survive because hydrogen cyanide is easily consumed by combustion. The incomplete and complete combustion of hydrogen cyanide is respectively represented as follows:

$$4HCN(g) \ + \ 5O_2(g) \ \longrightarrow \ 4CO(g) \ + \ 4NO(g) \ + \ 2H_2O(g)$$

Hydrogen cyanide Oxygen Carbon monoxide Nitric oxide Water

$$4HCN(g) \ + \ 9O_2(g) \ \longrightarrow \ 4CO_2(g) \ + \ 4NO_2(g) \ + \ 2H_2O(g)$$

Hydrogen cyanide Oxygen Carbon dioxide Nitrogen dioxide Water

The ease of combustion implies that hydrogen cyanide is likely to be present at its highest concentration during the early stages of a fire, when sufficient heat is present for thermal decomposition but before the gas oxidizes.

10.10-C USES OF HYDROGEN CYANIDE

Hydrogen cyanide is a commercial chemical product. It is produced by the catalytic reaction between ammonia and air with methane or natural gas.

$$2NH_3(g) \ + \ 3O_2(g) \ + \ 2CH_4(g) \ \longrightarrow \ 2HCN(g) \ + \ 6H_2O(g)$$

Ammonia Oxygen Methane Hydrogen cyanide Water

The hydrogen cyanide molecules produced by the reaction tend to react at an explosive rate with other hydrogen cyanide molecules. To prevent this reaction, the commercial grades of hydrogen cyanide are stabilized with water and 0.05% phosphoric acid. Today, they are used primarily in the chemical industry to produce methacrylates (Section 14.6-E).

Because hydrogen cyanide and other cyanides are well-known poisons, their nonchemical uses have been associated with morbid examples of human cruelty. Perhaps the most infamous use of hydrogen cyanide for the taking of human life was authorized by Adolf Hitler throughout most of World War II. The Nazis used the cyanide-containing insecticide Zyklon B to exterminate European Jews, gypsies, homosexuals, and other people deemed to be "undesirable." At Nazi concentration camps, pellets of Zyklon B were dropped into the vents of locked rooms that served as gas chambers. These pellets evolved hydrogen cyanide on exposure to the air and killed the camp prisoners within 20 minutes.

Perhaps the most infamous concentration camp was the one operated in Auschwitz, Poland. Historians estimate that inhalation of hydrogen cyanide caused the deaths of 1 million to 2 million people at this single camp alone.

Ultimately, after having been confronted with defeat at the end of World War II, Hitler chose cyanide ingestion himself to commit suicide. To test its efficiency, he first dosed his pet dog.

Hydrogen cyanide inhalation has been authorized for implementing death-sentence penalties by four states—Arizona, California, Missouri, and Wyoming. The lethal gas is generated by dropping pellets of a metallic cyanide into acid within an enclosed gas chamber.

$$NaCN(s) \ + \ HCl(aq) \ \longrightarrow \ NaCl(aq) \ + \ HCN(g)$$

Sodium cyanide Hydrochloric acid Sodium chloride Hydrogen cyanide

In the past, hydrogen cyanide was also used as a fumigant to kill rodents and insects, especially onboard ships, but given the accompanying human danger, this practice is no longer popular.

Terrorists are also aware that the use of hydrogen cyanide instills fear. Information suggests that terrorists planned to use hydrogen cyanide in April 2003 within the confines of the New York City subway system. However, they canceled the attack for unknown reasons.*

*Ron Suskind, *The One Percent Doctrine* (New York, NY: Simon & Schuster, 2007).

10.10-D WORKPLACE REGULATIONS INVOLVING HYDROGEN CYANIDE

When the use of hydrogen cyanide is needed in the workplace, OSHA requires employers to limit employee exposure to a dermal concentration of 10 ppm (11 mg/m^3) averaged over an 8-hr workday.

10.10-E TRANSPORTING HYDROGEN CYANIDE

When shippers offer hydrogen cyanide for transportation, DOT requires them to identify it on an accompanying shipping paper as one of the following, as relevant:

> **UN1051, Hydrogen cyanide, stabilized (contains less than 3% water), 6.1, (3), PGI (Marine Pollutant) (Poison Inhalation Hazard, Zone A)**

> **UN1614, Hydrogen cyanide, stabilized (contains less than 3% water; absorbed into a porous inert material), 6.1, PGI (Marine Pollutant)**

With regard to the first form, DOT also requires them to affix POISON INHALATION HAZARD and FLAMMABLE LIQUID labels to its packaging. DOT requires carriers to display POISON INHALATION HAZARD placards and MARINE POLLUTANT and INHALATION HAZARD markings on the bulk packaging or transport vehicle used for shipment.

When shippers offer the second form of hydrogen cyanide for transportation, DOT requires them to affix POISON labels to the packaging. Furthermore, DOT requires carriers to display POISON placards and MARINE POLLUTANT and INHALATION HAZARD markings on the bulk packaging or transport vehicle used for shipment.

10.10-F HYDROCYANIC ACID

Hydrogen cyanide dissolves in water to form a solution known as *hydrocyanic acid*, or *prussic acid*. The acid solution is so weak it is incapable of turning litmus paper red. It is colorless, but very volatile. It is used to fumigate ships, warehouses, and greenhouses.

10.10-G TRANSPORTING HYDROCYANIC ACID

Three hydrocyanic acid solutions are available in commerce. When shippers intend to transport a hydrocyanic acid solution in bulk, DOT requires them to identify the appropriate solution on an accompanying shipping paper as one of the following:

> **UN1613, Hydrocyanic acid aqueous solutions (contains not more than 20% hydrogen cyanide), 6.1, PGI (Marine Pollutant) (Poison - Inhalation Hazard, Zone B)**

> **NA1613, Hydrocyanic acid aqueous solutions (contains less than 5% hydrogen cyanide), 6.1, PGII**

> **UN3294, Hydrogen cyanide, solution in alcohol (contains not more than 45% hydrogen cyanide), 6.1, (3), PGI (Marine Pollutant) (Poison Inhalation Hazard, Zone B)**

With regard to the two solutions that pose an inhalation hazard, DOT requires shippers to affix POISON INHALATION HAZARD labels to their packaging. DOT requires carriers to display POISON INHALATION HAZARD placards and MARINE POLLUTANT and INHALATION HAZARD markings on the bulk packaging or transport vehicle used to ship them.

When they intend to transport a solution containing less than 5% hydrogen cyanide, DOT requires shippers to affix POISON labels to its packaging. When carriers transport 1001 lb

(454 kg) or more of this solution, DOT requires them to display POISON placards on the bulk packaging or transport vehicle used for shipment.

10.10-H METALLIC CYANIDES

Metallic cyanides are substances like sodium cyanide, potassium cyanide, copper(II) cyanide, and zinc cyanide. They are ionic compounds composed of metallic and cyanide ions. Sodium cyanide and potassium cyanide are available commercially as solids and aqueous solutions, whereas copper(II) cyanide and zinc cyanide are available solely as solids. All are used to prepare aqueous solutions from which copper and zinc are respectively electroplated.

As noted earlier in Section 1.4-A, the Consumer Product Safety Commission has banned products containing soluble metallic cyanides as constituents of consumer products intended for use in the United States. This regulatory ban is published at 16 C.F.R. §1500.17.

10.10-I TRANSPORTING METALLIC CYANIDES

When shippers transport a metallic cyanide in bulk, DOT requires them to identify the appropriate substance on an accompanying shipping paper as one of the following, as relevant:

> **UN1565, Barium cyanide, 6.1, PGI (Marine Pollutant) (Poison)**
> **UN1587, Copper cyanide, 6.1, PGII (Marine Pollutant) (Poison)**
> **UN1620, Lead cyanide, 6.1, PGII (Marine Pollutant) (Poison)**
> **UN1680, Potassium cyanide, solid, 6.1, PGI (Marine Pollutant) (Poison)**
> **UN3413, Potassium cyanide solution, 6.1, PGI (Marine Pollutant) (Poison)**
> *or*
> **UN3413, Potassium cyanide solution, 6.1, PGII (Marine Pollutant) (Poison)**
> *or*
> **UN3413, Potassium cyanide solution, 6.1, PGIII (Marine Pollutant) (Poison)**
> **UN1689, Sodium cyanide, solid, 6.1, PGI (Marine Pollutant) (Poison)**
> **UN3414, Sodium cyanide solution, 6.1, PGI (Marine Pollutant) (Poison)**
> *or*
> **UN3414, Sodium cyanide solution, 6.1, PGII (Marine Pollutant) (Poison)**
> *or*
> **UN3414, Sodium cyanide solution, 6.1, PGIII (Marine Pollutant) (Poison)**
> **UN1713, Zinc cyanide, 6.1, PGI (Marine Pollutant) (Poison)**

When the name of a metallic cyanide or its solution is not listed in the Hazardous Materials Table, its basic description is identified generically as follows:

> **UN1935, Cyanide solutions, n.o.s., 6.1, PGI (Marine Pollutant) (Poison)**
> *or*
> **UN1935, Cyanide solutions, n.o.s., 6.1, PGII (Marine Pollutant) (Poison)**
> *or*
> **UN1935, Cyanide solutions, n.o.s., 6.1, PGIII (Marine Pollutant) (Poison)**
> **UN1588, Cyanide, inorganic, solid, n.o.s., 6.1, PGI (Marine Pollutant) (Poison)**
> *or*
> **UN1588, Cyanides, inorganic, solid, n.o.s., 6.1, PGII (Marine Pollutant) (Poison)**

Additionally, shippers must provide the name of the specific compound parenthetically. DOT also requires shippers to affix a POISON label and MARINE POLLUTANT marking on the packaging containing a metallic cyanide.

When carriers transport 1001 lb (454 kg) or more of a metallic cyanide, DOT requires them to display POISON placards on the bulk packaging or transport vehicle used for shipment.

PERFORMANCE GOALS FOR SECTION 10.11:

- Identify the products that produce sulfur dioxide when they burn.
- Identify the potential hazard experienced when humans are exposed to sulfur dioxide. Identify the industries that use bulk quantities of sulfur dioxide.
- Describe the means shippers and carriers use to inform emergency responders of the hazards associated with encountering sulfur dioxide during transportation mishaps.

10.11 Sulfur Dioxide

Sulfur dioxide is a colorless, toxic gas having the sharp, pungent odor associated with burning tires. Some other physical properties of this gas are noted in Table 10.7.

Sulfur dioxide is produced when coal and other fossil fuels are burned. Fossil-fuel-fired power plants in the United States emit almost two-thirds of the sulfur dioxide found in the air.

Sulfur dioxide is also produced naturally. The majority of the sulfur dioxide that occurs naturally is linked with the eruption of volcanoes. In Hawaii, for example, the Kilauea volcano shown in Figure 10.10 has continuously spewed massive amounts of sulfur dioxide into the atmosphere since at least January 3, 1983, when scientists first began studying its eruption. According to the American Lung Association, the volcano has emitted an average of 1100 tn/day (1000 t/day) of sulfur dioxide into the atmosphere.

10.11-A ENVIRONMENTAL ISSUES ASSOCIATED WITH SULFUR DIOXIDE

Sulfur dioxide released into the air slowly oxidizes to sulfur trioxide, which in turn reacts with atmospheric moisture to form sulfuric acid.

$$2SO_2(g) \ + \ O_2(g) \ \longrightarrow \ 2SO_3(g)$$

Sulfur dioxide   Oxygen   Sulfur trioxide

$$SO_3(g) \ + \ H_2O(g) \ \longrightarrow \ H_2SO_4(aq)$$

Sulfur trioxide   Water   Sulfuric acid

acid rain
■ Precipitation having an approximate pH of 5.6 or less and caused by dissolved nitrogen oxides and sulfur dioxide

The presence of sulfuric acid in the atmosphere gives rise to a type of **acid rain**, that is, precipitation having a pH of 5.6 or less. Acid rain has severely damaged building materials made of concrete, marble, mortar, and limestone. A startling example of the negative impact that acid rain has had on stone structures is evident from observing Cleopatra's Needle, an obelisk that was moved from Egypt to New York City in the late 19th century. It has deteriorated more in 100 years from exposure to acid rain than it did during the 3000 years it stood in Egypt.

TABLE 10.7	Physical Properties of Sulfur Dioxide
Melting point	−105 °F (−76 °C)
Boiling point	14 °F (−10 °C)
Specific gravity at 68 °F (20 °C)	1.4
Vapor density (air = 1)	2.22

FIGURE 10.10 The eruption of Kilauea, the active volcano on the island of Hawaii, serves as the major natural source of sulfur dioxide emitted into the atmosphere within the United States. *(Courtesy of the Hawaiian Volcano Observatory, United States Geological Survey, Hawaii National Park, Hawaii; photograph by J. D. Griggs.)*

The release of sulfur dioxide to the atmosphere is also associated with worldwide cooling. When volcanoes eject substantial quantities of sulfur dioxide into the atmosphere, the gas agglomerates as aerosol particles containing sulfuric acid. These particles reflect the Sun's rays away from our planet and reduce the amount of radiation that reaches Earth's surface. The result is a severe drop in temperature. Scientists propose that the sulfur dioxide produced during the 1815 eruption of the Tambora volcano in Indonesia caused the Little Ice Age, the approximate period from 1550 to 1850 during which bitterly cold winters were experienced in many parts of the world.

10.11-B HOW DOES SULFUR DIOXIDE KILL?

Inhalation toxicity is the primary hazard associated with exposure to sulfur dioxide. As shown in Table 10.8, coughing, chest pains, shortness of breath, and constriction of the airways are immediate symptoms associated with exposure to a concentration of 10 ppm of this gas. Many individuals are able to tolerate very low concentrations of sulfur dioxide—such as those routinely found in polluted air—for short periods of time. An exposure to a sulfur dioxide concentration of 100 ppm in air is considered immediately dangerous to life and health; death can result immediately. Individuals who have been exposed to elevated concentrations of sulfur dioxide should seek immediate medical attention.

10.11-C SULFUR DIOXIDE AT FIRE SCENES

Sulfur-containing compounds are constituents of many products including the following:

- Coal, natural gas, and crude oil
- Complex proteins present in wool, hair, and animal hides (Section 14.5-B)
- Several natural and synthetic polymers, including vulcanized rubber

TABLE 10.8 | Adverse Health Effects Associated with Breathing Sulfur Dioxide

SULFUR DIOXIDE (ppm)	SIGNS AND SYMPTOMS
<0.1	No observable effect
≥0.1	Bronchoconstriction in sensitive, exercising asthmatics
0.3–1	Generally detectable by taste and smell
1–2	Lung function changes in healthy, nonasthmatic individuals
6–12	Possible nasal and throat irritation
10	Upper respiratory irritation with coughing, chest pains, shortness of breath, and some nosebleeds
20	Irritation of the eyes; development of chronic respiratory symptoms
50–100	Maximum tolerable exposure limit for 30–60 min
≥100	Immediately dangerous to life

When these products burn, the sulfur is converted into sulfur dioxide. Firefighters encounter the sulfur dioxide when they respond to fires involving their combustion or the burning of other products to which sulfur was added.

10.11-D USES OF SULFUR DIOXIDE

Sulfur dioxide is also a commercial chemical product. It is primarily produced by burning sulfur.

$$S_8(g) \;+\; 8O_2(g) \;\longrightarrow\; 8SO_2(g)$$

Sulfur   Oxygen   Sulfur dioxide

Although sulfur dioxide was once a popular refrigerant, its inherent toxicity has resulted in its replacement with nontoxic alternatives.

Today, sulfur dioxide is used mainly as a bleaching agent by the pulp and paper industry. It is also used by the agricultural industry to whiten refined sugar and to lengthen the shelf life of dried fruits. When grapes are dried and processed in an atmosphere of sulfur dioxide, raisins with a golden color are produced. When they are not exposed to sulfur dioxide, the raisins are dark brown.

Sulfur dioxide is also used in the chemical industry to produce metallic sulfites, which are popular reducing agents at water treatment plants.

10.11-E WORKPLACE REGULATIONS INVOLVING SULFUR DIOXIDE

When the use of sulfur dioxide is needed in the workplace, OSHA requires employers to limit employee exposure to a maximum concentration of 5 ppm ($13 \, mg/m^3$), averaged over an 8-hr workday.

10.11-F ENVIRONMENTAL REGULATIONS INVOLVING SULFUR DIOXIDE

Using the authority of the Clean Air Act, EPA regulates the sulfur dioxide concentration in the ambient air as a criteria air pollutant. EPA has set the primary national ambient air quality standard for sulfur dioxide as an annual arithmetic mean equal to 0.030 ppm ($80 \, \mu g/m^3$) and a 24-hr average equal to 0.14 ppm ($365 \, \mu g/m^3$). The secondary standard has been set at 0.50 ppm ($1300 \, \mu g/m^3$) as a 3-hr average.

10.11-G TRANSPORTING SULFUR DIOXIDE

Sulfur dioxide is available commercially as a liquefied compressed gas. When shippers transport sulfur dioxide, DOT requires them to identify it on an accompanying shipping paper as follows:

UN1079, Sulfur dioxide, 2.3, (8), (Poison - Inhalation Hazard, Zone C)

DOT also requires shippers to affix POISON GAS and CORROSIVE labels to the packaging. DOT requires carriers to display POISON GAS placards and INHALATION HAZARD markings on the bulk packaging or transport vehicle used for shipment.

PERFORMANCE GOALS FOR SECTION 10.12:

- Identify the products that produce hydrogen sulfide when they smolder or are exposed to intense heat.
- Discuss the reason hydrogen sulfide is encountered only at trace concentrations at fire scenes.
- Identify the potential hazards associated with exposure to hydrogen sulfide.
- Identify the facilities in which hydrogen sulfide is generated and the industries in which bulk quantities of hydrogen sulfide are stored.
- Describe the means shippers and carriers use to inform emergency responders of the hazards associated with encountering hydrogen sulfide during transportation mishaps.
- Describe the nature of the response actions to be executed when hydrogen sulfide has been released into the environment.

10.12 Hydrogen Sulfide

Hydrogen sulfide is a colorless, toxic gas having the physical properties noted in Table 10.9. Perhaps its most apparent feature is a disagreeable odor, that of rotten eggs. Humans detect this smell at just 2 ppb. The stench is truly offensive at concentrations as low as 3 to 5 ppm. Sometimes individuals describe the odor of a material by saying that it smells like sulfur. Because elemental sulfur is an odorless solid (Section 7.6), what they actually mean to say is that the material smells like hydrogen sulfide.

TABLE 10.9	Physical Properties of Hydrogen Sulfide
Melting point	$-117\ °F\ (-85.49\ °C)$
Boiling point	$-76\ °F\ (-60.33\ °C)$
Specific gravity at 68 °F (20 °C)	1.54
Vapor density (air = 1)	1.18
Flashpoint	$-116\ °F\ (-82.4\ °C)$
Autoignition point	500 °F (260 °C)
Lower explosive limit	4.3%
Upper explosive limit	46%

Hydrogen sulfide occurs naturally. It is produced by the decay of organisms in swamps, sewers, and other anaerobic (nonoxygen) environments from which the gas seeps into the atmosphere. The decomposition of sewage gives rise to the term *sewer gas*, a term used by sanitary engineers who work in sewers and at sewage treatment facilities. The odor is also encountered at petroleum refineries, where the gas is recovered by the desulfurization of crude oil. Hydrogen sulfide is also a constituent of mammalian flatulence. Trace amounts are used by the body to regulate metabolism rates.

Because hydrogen sulfide has an objectionable odor, most people immediately become aware of its presence. Nonetheless, long-term exposure to low concentrations of hydrogen sulfide temporarily deadens an individual's sense of smell. This phenomenon is called **olfactory fatigue**. Continued exposure causes individuals to become oblivious to the presence of the gas. When this occurs, they can unknowingly inhale a lethal concentration.

olfactory fatigue
▪ The temporary inability to identify the odor of an airborne substance after prolonged exposure

10.12-A HOW DOES HYDROGEN SULFIDE KILL?

Table 10.10 indicates that inhalation toxicity is the primary hazard associated with exposure to hydrogen sulfide. Exposure initially gives rise to dizziness and the onset of a headache, but unconsciousness and respiratory paralysis can follow immediately. Inhalation of air having a hydrogen sulfide concentration of 1000 ppm is regarded as fatal. Death results from asphyxiation unless breathing is restored before heart action ceases.

10.12-B HYDROGEN SULFIDE AT FIRE SCENES

Hydrogen sulfide is a flammable gas. When ignited in air, it readily burns as follows:

$$2H_2S(g) \ + \ 3O_2(g) \ \longrightarrow \ 2H_2O(g) \ + \ 2SO_2(g)$$

Hydrogen sulfide $\qquad$ Oxygen $\qquad\qquad$ Water $\qquad$ Sulfur dioxide

At fire scenes, the presence of hydrogen sulfide—although possible—is generally improbable. Although small concentrations may evolve during the thermal decomposition of certain materials manufactured from animal products (including leather items and wool carpeting), the gas is easily consumed by combustion. For this reason, hydrogen sulfide is not generally present at the typical fire scene.

TABLE 10.10	Adverse Health Effects Associated with Breathing Hydrogen Sulfide
HYDROGEN SULFIDE (PPM)	**SIGNS AND SYMPTOMS**
0.0005–0.3	Odor threshold
1	Headache, nausea, vomiting, reduction in ability to reason effectively
10	Same as above; prolonged exposure may possibly damage the cornea of the eyes
20–50	Onset of paralysis of the sense of smell
50–250	Exposure for several hours may cause pulmonary edema
100–1000	Possible respiratory paralysis, sudden loss of consciousness, or death after a single breath or two; in survivors, brief exposure may cause permanent or long-term effects such as headache, poor attention span, poor memory, and poor motor function

10.12-C USES OF HYDROGEN SULFIDE

Although hydrogen sulfide is available as a liquefied compressed gas, it is not a popular commercial product. For its limited use, it is primarily derived from petroleum refineries and natural gas wells. In commerce, hydrogen sulfide is used to produce elemental sulfur and sulfuric acid, to process mineral ores, to produce metallic sulfides such as nickel sulfide and molybdenum sulfide, and to manufacture phosphors used in television tubes.

10.12-D WORKPLACE REGULATIONS INVOLVING HYDROGEN SULFIDE

When the use of hydrogen sulfide is needed in the workplace, OSHA requires employers to limit employee exposure to a maximum concentration of 50 ppm in a 10-min maximum period.

10.12-E TRANSPORTING HYDROGEN SULFIDE

Hydrogen sulfide is transported in steel cylinders and bulk transport vehicles. When shippers offer hydrogen sulfide for transportation, DOT requires them to identify it on an accompanying shipping paper as follows:

UN1053, Hydrogen sulfide, 2.3, (2.1), (Poison - Inhalation Hazard, Zone A)

DOT also requires shippers to affix POISON GAS and FLAMMABLE GAS labels to the packaging.

DOT requires carriers to display POISON GAS placards and INHALATION HAZARD markings on the bulk packaging or transport vehicle used for shipment.

10.12-F RESPONDING TO INCIDENTS INVOLVING A RELEASE OF HYDROGEN SULFIDE

Most fatalities associated with exposure to hydrogen sulfide have occurred when safety practices were ignored at petroleum refineries and chemical manufacturing facilities, within sewer systems, and other locations at which hydrogen sulfide was stored or generated. Petroleum refiners must always be wary of the presence of hydrogen sulfide, since the gas is a by-product of refining crude oil.

Special attention must also be given to enclosures in which hydrogen sulfide may be unknowingly generated. For example, onboard modern ships, chambers are used for the storage of sewage until the contents can be biologically and chemically treated on shore. Hydrogen sulfide can accumulate within these chambers at lethal concentrations. When it is necessary to repair fractures in the walls or connecting pipes of these chambers, workers must avoid exposure to this deadly gas by wearing fully encapsulated suits and breathing air from self-contained sources. No work should ever be conducted in any area where hydrogen sulfide can accumulate without first monitoring its concentration.

PERFORMANCE GOALS FOR SECTION 10.13:

- Identify the products that produce nitrogen oxides when they burn.
- Identify the adverse effects experienced when humans are exposed to nitrogen dioxide.
- Identify the industries that use bulk quantities of the nitrogen oxides.
- Describe the means shippers and carriers use to inform emergency responders of the hazards associated with encountering nitric oxide or nitrogen dioxide during transportation mishaps.

10.13 Nitrogen Oxides

Nitric oxide

Nitrogen dioxide

Eight oxides of nitrogen are known, but we are concerned here only with two, namely, *nitric oxide*, also known as nitrogen monoxide, and *nitrogen dioxide*. Their chemical formulas are NO and NO_2, respectively. The molecules of nitric oxide and nitrogen dioxide are stable free radicals.

$$:N::\ddot{O}: \qquad \cdot\ddot{O}:N::\ddot{O}:$$

Because they are often produced in combination, nitric oxide and nitrogen dioxide are represented jointly as NO_x. Nitric oxide is a colorless gas with an irritating odor, but nitrogen dioxide is a dark red-brown gas having a pungent, acrid odor. Both are poisonous, nonflammable, and corrosive substances. Their physical properties are listed in Table 10.11.

SOLVED EXERCISE 10.8

Why do two nitrogen dioxide molecules combine to produce a single molecule of dinitrogen tetroxide?

Solution: The Lewis symbols for the nitrogen and oxygen atoms are $\cdot\dot{N}\cdot$ and $\cdot\ddot{O}\cdot$, respectively. Using them, the following Lewis structure (Section 4.10) for nitrogen dioxide is written:

$$\cdot\ddot{O}:N::\ddot{O}: \quad \text{or} \quad O{-}N{=}O$$

Because this structure has an unpaired electron, it actually represents a free-radical molecule (Section 5.10). To complete the unfilled octet and achieve electronic stability, two nitrogen dioxide molecules combine to produce a single dinitrogen tetroxide molecule.

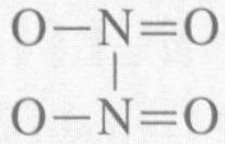

$$\begin{array}{c} O{-}N{=}O \\ | \\ O{-}N{=}O \end{array}$$

10.13-A ENVIRONMENTAL ISSUES ASSOCIATED WITH NO$_X$

In the atmosphere, nitrogen and oxygen combine very slowly to produce nitric oxide and nitrogen dioxide at ambient temperatures. However, the combination occurs rapidly at elevated temperatures.

$$\underset{\text{Nitrogen}}{N_2(g)} \; + \; \underset{\text{Oxygen}}{O_2(g)} \; \longrightarrow \; \underset{\text{Nitric oxide}}{2NO(g)}$$

$$\underset{\text{Nitric oxide}}{2NO(g)} \; + \; \underset{\text{Oxygen}}{O_2(g)} \; \longrightarrow \; \underset{\text{Nitrogen dioxide}}{2NO_2(g)}$$

TABLE 10.11	Physical Properties of Nitric Oxide and Nitrogen Dioxide	
	NITRIC OXIDE	**NITROGEN DIOXIDE**
Melting point	−263 °F (−164 °C)	12 °F (−11 °C)
Boiling point	−243 °F (−153 °C)	68 °F (20 °C)
Specific gravity at 68 °F (20 °C)	1.34	1.49
Vapor density (air = 1)	1.04	1.59

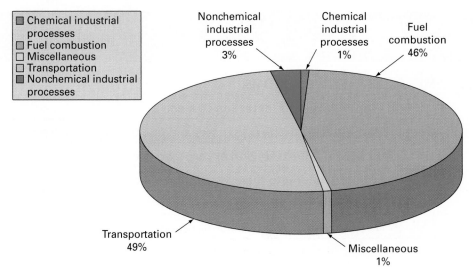

Chemical industrial processes
Fuel combustion
□ **Miscellaneous**
□ **Transportation**
Nonchemical industrial processes

Nonchemical industrial processes 3%

Chemical industrial processes 1%

Fuel combustion 46%

Transportation 49%

Miscellaneous 1%

FIGURE 10.11 Nearly all the NO$_x$ in the ambient air is generated during transportation and fuel-combustion activities. (*Courtesy of United States Environmental Protection Agency, Washington, DC.*)

The oxidation of nitric oxide to nitrogen dioxide occurs rapidly. At fossil-fuel-fired power plants and in the combustion chambers of motor vehicles, nitrogen dioxide is produced at the concentrations shown in Figure 10.11. For this reason, the gas is regarded as a major air pollutant. It is the principal culprit responsible for the brown atmospheric haze that hangs over many cities.

The presence of NO$_x$ in the lower atmosphere is directly linked with the production of ground-level ozone (Section 7.1-H) and smog. Concerned about the adverse impact on public health posed by NO$_x$, especially in urban areas, EPA required automobile manufacturers to build cars that reduced the amount of NO$_x$ emitted in vehicular exhaust. In response, the manufacturers installed catalytic converters in new automobiles. One chemical reaction that occurs on the surface of the converter is the reduction of nitric oxide into environmentally friendly nitrogen and oxygen. Use of these converters during the operation of modern-day motor vehicles reduces the amount of NO$_x$ that would otherwise be ejected into the atmosphere.

10.13-B HOW DOES NO$_x$ KILL?

Inhalation toxicity is the primary hazard associated with exposure to both nitric oxide and nitrogen dioxide. Initial exposure produces irritation of the mucous membranes of the eyes, throat, nose, and lungs. As Table 10.12 shows, when an individual is exposed to nitrogen dioxide, these initial symptoms are followed by coughing, choking, headache, nausea, and fatigue.

Both nitric oxide and nitrogen dioxide are readily absorbed into the bloodstream, where they combine with hemoglobin to form **methemoglobin**. For example, when nitrogen dioxide is inhaled, methemoglobin is produced as follows:

$$Hb(aq) + NO_2(g) \longrightarrow (NO_2)Hb(aq)$$

Hemoglobin   Nitrogen dioxide   Methemoglobin

Like carboxyhemoglobin, methemoglobin cannot effectively transport oxygen to the body's tissues, so the host suffers from the ailment called **methemoglobinemia**. Individuals who inhale nitric oxide or nitrogen dioxide exhibit the initial symptoms of cyanosis (dizziness, headache, diarrhea, and anemia), but their blood also becomes chocolate brown in color. It is this color that initially signals the severity of NO$_x$ poisoning.

Hyperbaric oxygen therapy (Section 10.9-D) is not generally useful in helping individuals overcome the impact of methemoglobinemia. Instead, exchange transfusions of the blood are usually needed to decrease the methemoglobin concentration in the bloodstream. Latent effects may also be experienced, and death can occur days after the initial exposure.

methemoglobin
▪ The substance produced when nitrogen dioxide reacts with the blood's hemoglobin

methemoglobinemia
▪ The disorder resulting from exposure to nitrogen dioxide, metallic nirtrites, metallic nitrates, and similar substances, caused by the presence of methemoglobin in the blood, and exhibited by the symptoms of cyanosis (dizziness, headache, diarrhea, and anemia)

TABLE 10.12	Adverse Health Effects Associated with Breathing Nitrogen Dioxide

NITROGEN DIOXIDE (PPM)	SIGNS AND SYMPTOMS
5	Threshold limit for detection by smell
10–20	Mild irritation to the eyes, nose, and upper respiratory tract
25–40	No adverse effects to workers exposed over a period of years
50	Distinct irritation to the eyes, nose, and upper respiratory tract
80	Tightness in the chest after 3- to 5-min exposure
100–200	Threat to life after 20- to 60-min exposure
250	Danger of death following short-term exposure

The toxicity of nitrogen dioxide is exacerbated by the ability of the gas to react with water. A mixture of nitrous acid and nitric acid is produced as follows:

$$2NO_2(g) \quad + \quad H_2O(g) \quad \longrightarrow \quad HNO_2(aq) \quad + \quad HNO_3(aq)$$

Nitrogen dioxide   Water   Nitrous acid   Nitric acid

When it is inhaled, nitrogen dioxide reacts with atmospheric moisture to produce a mixture of these acids directly within the respiratory system. Because nitric acid is a strong acid, it damages the lining of the lungs and connective tissues, and even at low concentrations, its presence can cause pulmonary edema.

10.13-C NITROGEN OXIDES AT FIRE SCENES

When nitrogen-containing materials burn, nitric oxide and nitrogen dioxide are produced as the products of their incomplete and complete combustion, respectively. Because nitric oxide rapidly oxidizes in air, nitrogen dioxide is the predominant nitrogen oxide identified at fire scenes.

Examples of nitrogenous materials that burn to produce nitrogen dioxide include the polyurethane products (Section 14.8-A) used in certain types of building insulation, bedding, and furniture cushioning. Although the presence of this gas is signaled by its red-brown color, it is generally impossible to detect this color within a predominantly black, smoky plume.

Certain commercial explosives also produce nitrogen dioxide when they detonate. Examples of these explosives include trinitrotoluene (Section 15.9), cyclonite (Section 15.10), tetryl (Section 15.11), and HMX (Section 15.13).

10.13-D USES OF NITRIC OXIDE AND NITROGEN DIOXIDE

Nitric oxide and nitrogen dioxide are available commercially as chemical products. These commercial gases are regarded as nonflammable and corrosive oxidizers.

Nitric oxide and nitrogen dioxide are produced for commercial use by the catalytic oxidation of ammonia as follows:

$$4NH_3(g) \quad + \quad 5O_2(g) \quad \longrightarrow \quad 4NO(g) \quad + \quad 6H_2O(g)$$

Ammonia   Oxygen   Nitric oxide   Water

$$2NO(g) \quad + \quad O_2(g) \quad \longrightarrow \quad 2NO_2(g)$$

Nitric oxide   Oxygen   Nitrogen dioxide

When confined within a cylinder at room temperature, the nitrogen dioxide combines with itself, mole for mole, to become dinitrogen tetroxide.

$$2NO_2(g) \longrightarrow N_2O_4(g)$$

Nitrogen dioxide   Dinitrogen tetroxide

Although dinitrogen tetroxide exists at room conditions, it is commonly still referred to as nitrogen dioxide.

Perhaps surprisingly, minute concentrations of nitric oxide are produced by the body and are involved in sexual arousal and certain other bodily functions. Despite its toxic nature, supplies of nitric oxide are found in modern hospitals and clinics, where they are used for the treatment of emphysema and other pulmonary diseases. When very low doses of nitric oxide are mixed with oxygen and administered to respiratory patients, the nitric oxide reduces inflammation and regulates the dilation of blood vessels. When it is administered to premature babies, nitric oxide reduces the risk of fatalities from such pulmonary diseases as bronchopulmonary dysplasia.

Nitrogen dioxide has been used by the aerospace industry to oxidize rocket fuels. For example, the Apollo astronauts used it with unsymmetrical dimethylhydrazine to leave the Moon's surface.

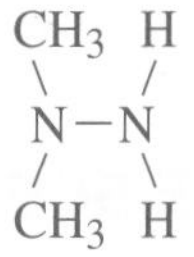

$$
\begin{array}{c}
\mathrm{CH_3 \quad H} \\
\backslash \quad / \\
\mathrm{N - N} \\
/ \quad \backslash \\
\mathrm{CH_3 \quad H}
\end{array}
$$

unsym-Dimethylhydrazine

A mixture of nitrogen dioxide and *unsym*-dimethylhydrazine is self-reactive, or **hypergolic**; that is, the substances react immediately on contact. The nitrogen dioxide and *unsym*-dimethylhydrazine act as an oxidizing agent and a reducing agent, respectively.

hypergolic
■ Pertaining to the ability of a substance to ignite spontaneously on contact with another substance

10.13-E WORKPLACE REGULATIONS INVOLVING THE NITROGEN OXIDES

When the use of nitric oxide and nitrogen dioxide is needed in the workplace, OSHA requires employers to limit employee exposure to maximum concentrations of 25 ppm (30 mg/m^3) and 5 ppm (9 mg/m^3), respectively.

10.13-F ENVIRONMENTAL REGULATIONS INVOLVING THE NITROGEN OXIDES

EPA regulates the concentration of NO_x in the ambient air as a criteria air pollutant. EPA has set both the primary and secondary national ambient air quality standards for NO_x at 0.053 ppm (100 mg/m^3) as annual arithmetic means.

10.13-G TRANSPORTING THE NITROGEN OXIDES

When shippers transport nitrogen dioxide (dinitrogen tetroxide) or nitric oxide, DOT requires them to identify the gas by means of either of the following basic descriptions, as relevant:

UN1067, Dinitrogen tetroxide, 2.3, (5.1, 8), (Poison - Inhalation Hazard, Zone A)
UN1660, Nitric oxide, compressed, 2.3, (5.1, 8), (Poison - Inhalation Hazard, Zone A)

DOT also requires them to affix POISON GAS, OXIDIZER, *and* CORROSIVE labels on its packaging.

DOT requires carriers to display POISON GAS placards and INHALATION HAZARD markings on the bulk packaging or transport vehicle used for shipment.

- Identify the common consumer products that contain ammonia.
- Identify the adverse effects experienced when humans are exposed to ammonia.
- Identify the industries that use bulk volumes of ammonia.
- Describe the means shippers and carriers use to inform emergency responders of the hazards associated with encountering anhydrous ammonia or ammonia solutions during transportation mishaps.
- Describe the actions that emergency responders execute when anhydrous ammonia has been released into the environment.

10.14 Ammonia

Ammonia is a colorless, pungent gas that is easily detected and recognized by its familiar pungent odor. It has been known since the days of alchemy, when it was produced by heating either coal or animal hoofs and horns. Some of its important physical properties are noted in Table 10.13.

10.14-A AMMONIA AT FIRE SCENES

Ammonia is a flammable gas within the range from 16% to 25% by volume. When it is released outdoors, ammonia dissipates with the wind, and a flammable mixture in air is not ordinarily attained. Under these circumstances, an ammonia fire cannot occur.

Nonetheless, ammonia concentrations within the flammable range *are* attained during some accident scenarios like the indoor release of ammonia into a confined area from which the ammonia cannot escape. In this circumstance, the occurrence of an ammonia fire is highly probable. When ammonia burns in air, the combustion evolves 9650 Btu/lb (22.5 kJ/g) and produces nitrogen and water vapor.

$$4NH_3(g) \quad + \quad 3O_2(g) \quad \longrightarrow \quad 2N_2(g) \quad + \quad 6H_2O(g)$$

Ammonia   Oxygen   Nitrogen   Water

TABLE 10.13	Physical Properties of Ammonia
Melting point	−108 °F (−78 °C)
Boiling point	−28 °F (−33 °C)
Specific gravity (liquid) at 68 °F (20 °C)	0.77
Specific gravity (gas) at 68 °F (20 °C)	0.68
Vapor density (air = 1)	0.589
Flashpoint	52 °F (11 °C)
Autoignition point	1204 °F (651 °C)
Lower explosive limit	16%
Upper explosive limit	25%

This heat of combustion is low compared with the values listed in Table 2.8 for other substances. It is for this reason that ammonia fires are not sustainable unless they are exposed to an additional heat source.

Ammonia may be generated at a fire scene in at least two ways:

■ Materials manufactured from animal products—like leather items and wool carpeting—thermally decompose when they are exposed to intense heat. When these materials are present at fire scenes, the odor of ammonia is often perceptible, but the concentration is generally too low to cause harm.

■ Fertilizers containing ammonium compounds decompose on exposure to heat. For example, when ammonium sulfate fertilizer is heated to high temperatures, it thermally decomposes to ammonia, sulfur dioxide, nitrogen, and water as follows:

$$3(NH_4)_2SO_4(s) \longrightarrow 4NH_3(g) + N_2(g) + 3SO_2(g) + 6H_2O(g)$$

Ammonium sulfate   Ammonia   Nitrogen   Sulfur dioxide   Water

10.14-B PRODUCTION AND USES OF AMMONIA

For commercial use, ammonia is produced from atmospheric nitrogen and hydrogen at elevated temperature and pressure. The production reaction is represented as follows:

$$N_2(g) + 3H_2(g) \longrightarrow 2NH_3(g)$$

Nitrogen   Hydrogen   Ammonia

The ammonia produced in the United States is directly used as a fertilizer or coolant in large refrigeration systems. Over 80% is used for agricultural purposes. Although less than 2% is used in refrigeration, ammonia was once the principal refrigerant used for very large refrigeration systems and still remains the most economical coolant choice for meatpacking plants, dairies, and similar facilities.

Because the nitrogen content of ammonia is 82% by mass, it is an ideal agricultural fertilizer. When used by farmers, the ammonia is dispensed directly to the soil or into irrigation waters. Farmers discharge the ammonia through a distribution pod from a tractor saddle tank or **nurse tank** like that shown in Figure 10.12. A nurse tank is a specific type of portable tank with a capacity of 3000 gal (11 m^3) or less, is painted white or aluminum, and is never filled completely. It is mounted behind the tillage tool.

When ammonia is used as a refrigerant or fertilizer, it is encountered as the liquefied compressed gas called *anhydrous ammonia*. This product is transported by public highway, railway, and watercraft and transferred by pipeline. In Alaska, anhydrous ammonia is the most prevalent hazardous substance throughout the state. It is used as a refrigerant at seafood-processing plants and in radiators associated with the Trans Alaska Pipeline System. The ammonia disperses the heat that evolves from the hot crude oil as it travels through elevated parts of the 799-mi (1242-km) pipeline. It is necessary to disperse the heat to prevent the melting of the underlying permafrost in which the supports for the elevated portions of the pipeline are embedded.

nurse tank
■ For purposes of DOT regulations, a cargo tank having a capacity of 3000 gal (11 m^3) or less, painted white or aluminum, never filled to capacity, and used solely for the transportation of anhydrous ammonia

10.14-C WORKPLACE REGULATIONS INVOLVING ANHYDROUS AMMONIA

When the use of anhydrous ammonia is needed in a workplace, OSHA requires employers to limit employee exposure to a concentration of 50 ppm (35 mg/m^3), averaged over an 8-hr workday. OSHA also regulates how anhydrous ammonia is stored and handled in all workplaces other than ammonia manufacturing facilities and refrigeration plants using anhydrous ammonia solely as a refrigerant.

FIGURE 10.12 DOT requires a nurse tank used for the application of anhydrous ammonia to agricultural soils to be marked ANHYDROUS AMMONIA on all four sides and INHALATION HAZARD on two opposing sides. The DOT identification number 1005 must also be displayed across the center of green NON-FLAMMABLE GAS placards posted on all four sides of the tanks. (*Courtesy of The Fertilizer Institute, Washington, DC.*)

10.14-D TRANSPORTING ANHYDROUS AMMONIA

When shippers offer anhydrous ammonia for transportation, DOT requires them to identify the gas on the accompanying shipping paper for domestic and international shipments as follows:

For domestic shipments:

UN1005, Ammonia anhydrous, 2.2 (Inhalation Hazard)

For international shipments:

UN1005, Ammonia, anhydrous, 2.3, (8) (Poison - Inhalation Hazard, Zone D)

For domestic shipments of anhydrous ammonia, DOT requires shippers to affix a NON-FLAMMABLE GAS label to its packaging.

Carriers often transport anhydrous ammonia in a rail tankcar or portable tank from its production and manufacturing facility to its point of distribution. When carriers transport 1001 lb (454 kg) or more of anhydrous ammonia, DOT requires them to post NON-FLAMMABLE GAS placards and to mark INHALATION HAZARD on the bulk packaging or transport vehicle used for shipment. Motor carriers who ship anhydrous ammonia in an amount equal to or exceeding 3500 gal (13,248 L) must possess a Hazardous Materials Safety Permit (Section 6.4).

For international shipments of anhydrous ammonia, DOT requires shippers to affix POISON GAS and CORROSIVE labels to its packaging. DOT also requires carriers to display POISON GAS placards and INHALATION HAZARD markings on the bulk packaging used for international shipment.

Certain other requirements apply when anhydrous ammonia is transported by highway. As first noted in Table 6.3, when anhydrous ammonia is transported domestically by means of a DOT MC330 or MC 331 tank truck, DOT requires shippers to provide information regarding the constituent water in the basic description. For example, when 2500 lb of anhydrous ammonia containing water at a concentration of 0.2% or more by mass is transported within an MC331 tank truck, the shipping description is listed on the accompanying shipping paper as follows:

Units	HM	Shipping Description (Identification Number, Proper Shipping Name, Primary Hazard Class or Division, Subsidiary Hazard Class or Division, and Packing Group)	Weight (lb)
1 MC331 tank truck	X	RQ, UN1005, Anhydrous ammonia, 2.2 (0.2 percent water) (Inhalation Hazard)	2500

When carriers transport anhydrous ammonia having a water content of less than 0.2% by mass in an MC330 or MC331 tank truck, DOT also requires shippers and carriers to evaluate the nature of the MC330 or MC331 tank truck used for transporting anhydrous ammonia and to provide this information in the basic description. In particular, carriers may transport anhydrous ammonia having a water content of less than 0.2% by mass in an MC330 or MC331 tank truck only when it has been constructed from a specially "quenched and tempered steel." Consequently, when 2500 lb of anhydrous ammonia containing water at a concentration of less than 0.2% is transported domestically in an MC330 tank truck, the proper shipping description is the following:

Units	HM	Shipping Description (Identification Number, Proper Shipping Name, Primary Hazard Class or Division, Subsidiary Hazard Class or Division, and Packing Group)	Weight (lb)
1 MC330 tank truck	X	RQ, UN1005, Anhydrous ammonia, 2.2 (Not for Q & T tanks) (Inhalation Hazard)	2500

In each of the latter examples, DOT also requires "Inhalation Hazard" as a component of the basic description (without the word "Poison" and without a hazard zone designation).

10.14-E RESPONDING TO INCIDENTS INVOLVING A RELEASE OF ANHYDROUS AMMONIA

Ammonia injures the human organism through two different mechanisms:

- As a very cold liquid at -28 °F (-33 °C), anhydrous ammonia freezes skin and other tissues at the points of contact.
- When individuals breathe ammonia vapor, they experience the ill effects listed in Table 10.14. The eyes, skin, and mucous membranes are particularly susceptible to severe irritation from exposure to ammonia. Elevated ammonia concentrations in the blood cause death by suffocation. Anhydrous liquid ammonia is also corrosive and severely irritates the skin and eyes.

Given the nature of the ill effects resulting from exposure to ammonia, the use of self-contained breathing apparatus is essential when emergency responders arrive at a scene involving a release of ammonia. When released directly from a liquid storage tank into the atmosphere, ammonia is generally visible as a white fog consisting of droplets of condensed atmospheric moisture. Although its presence severely limits visibility within the immediate environment, the initial formation of the fog helps workers locate the spot at which the ammonia is leaking from its storage tank or container.

TABLE 10.14	Adverse Health Effects Associated with Breathing Ammonia

AMMONIA (PPM)	SIGNS AND SYMPTOMS
5–10	Detectable limit by odor
50	No chronic effects
150–200	General discomfort; eye tearing; irritation and discomfort of exposed skin; irritation of mucous membranes
400–700	Pronounced irritation and discomfort to the eyes, ears, nose, and throat
2000	Barely tolerable for more than a few moments; serious blistering of the skin; danger of pulmonary edema, asphyxia, and death following prolonged exposure
5000	Death by suffocation within minutes

The vapor density listed in Table 10.13 is an indication that ammonia is lighter than air. When it is released outdoors into dry air, ammonia quickly disperses into the atmosphere, especially under windy conditions. When anhydrous liquid ammonia is released into moist air, however, small liquid droplets of ammonia are produced along with the gas. Under such conditions, these droplets behave as a dense gas, traveling along the surface of the ground instead of rising into the air and dispersing.

Water absorbs a very large volume of ammonia: At room conditions, one volume of water absorbs 1176 volumes of ammonia gas. This physical property is used to advantage when responding to incidents involving a release of ammonia. The recommended practice is to establish a water curtain downwind from the point at which the ammonia is being released. The water dissolves the ammonia and reduces the concentration that can travel farther.

SOLVED EXERCISE 10.9

To protect public health, safety, and the environment, what immediate action should first-on-the-scene responders take when they encounter a strong odor of ammonia within an enclosed room at an ice cream plant?

Solution: Ammonia cannot ignite in air unless a concentration within its flammable range is exposed to an ignition source. The flammable range of ammonia is 16% to 25% by volume. A concentration within this range is readily achieved when ammonia escapes from large refrigeration units into an enclosed room. When the members of an emergency-response team first detect a strong odor of ammonia, their primary concern should be aimed at eliminating the risk of its ignition. They can do this by opening doors and windows to vent the ammonia from the enclosure while simultaneously taking precautions to avoid generating a spark.

Because an exposure to ammonia burns the eyes and nasal passageways, and the inhalation of elevated concentrations causes immediate asphyxiation, the implementation of any emergency-response action involving ammonia should be undertaken only while wearing a full-body suit and using self-contained breathing apparatus. No activity should ever be conducted in an enclosure in which the ammonia concentration is within its flammable range, as survival is unlikely if the ammonia subsequently ignites.

When emergency responders are called to the scene of a transportation mishap involving bulk shipments of anhydrous ammonia, they must acknowledge that their own safety as well as the safety of transportation personnel and the general public is at risk. The largest spill of ammonia in the United States occurred in 2002, when 31 rail tankcars derailed near Minot, North Dakota, during the wreck of a Canadian Pacific Railway train. Five of the 31 tank cars catastrophically ruptured and instantly released 146,700 gal (555 m³) of ammonia into the atmosphere. Exposure to the ammonia resulted in the death of a local resident. More than 1400 others experienced some degree of respiratory impairment.*

10.14-F AMMONIA SOLUTIONS

Most individuals have encountered ammonia as the gas that escapes from the commercial cleaning product called *household ammonia*. This is an aqueous solution of ammonia primarily used for cleaning glass, marble, and porcelain surfaces. Chemists refer to any aqueous solution of ammonia as *ammonium hydroxide* or *aqueous ammonia*, but in common practice, it is called *ammonia solution*. Its chemical formula is $NH_3(aq)$, or NH_4OH. The concentration of ammonia dissolved in household ammonia ranges from 2% to 5% by volume. The vapor is perceptible at concentrations as low as approximately 5 ppm, but most individuals do not find this concentration irritating.

Commercial ammonia solutions contain dissolved ammonia that ranges in concentration from 10% to more than 50%. These aqueous solutions are hazardous primarily owing to the ammonia that rapidly escapes from them.

10.14-G TRANSPORTING AMMONIA SOLUTIONS

DOT regulates the transportation of ammonia solutions according to their ammonia concentrations. When shippers offer an aqueous ammonia solution containing more than 50% ammonia for transportation, the basic descriptions of the commodity are differentiated for domestic and international shipments as follows:

For domestic transportation:

> **UN3318, Ammonia solution (contains more than 50% ammonia in water; relative density less than 0.880 at 15 °C), 2.2, (Inhalation Hazard)**

For international transportation:

> **UN3318, Ammonia solution (contains more than 50% ammonia in water; relative density less than 0.880 at 15 °C), 2.3, (8), (Poison - Inhalation Hazard, Zone D)**

When shippers offer this solution for transportation in domestic commerce, DOT requires them to affix a NON-FLAMMABLE GAS label to the packaging. Carriers who transport this ammonia solution in an amount exceeding 1001 lb (454 kg) are required to display NON-FLAMMABLE GAS placards and INHALATION HAZARD markings on the bulk packaging or transport vehicle used for shipment.

When shippers offer the same solution for transportation in international commerce, DOT requires them to affix POISON GAS and CORROSIVE labels to the packaging. Carriers must display POISON GAS placards on the bulk packaging or transport vehicle used for shipment.

*Railroad Accident Report, "Derailment of Canadian Pacific Railway Freight Train 292-16 and Subsequent Release of Anhydrous Ammonia Near Minot, North Dakota, January 18, 2002," NTSB/RAR-04/01, PB2004-916301 (National Transportation Safety Board, Washington, DC, 2004).

When shippers offer an aqueous ammonia solution containing ammonia in either the range 10% to 35% or 35% to 50% for transportation, DOT requires them to enter one of the following basic descriptions on the accompanying shipping paper, as relevant:

UN2672, Ammonia solution (contains more than 10% ammonia but not more than 35% ammonia in water; relative density between 0.880 and 0.957 at 15 °C), 8, PGIII

UN2073, Ammonia solution (contains more than 35% ammonia but not more than 50% ammonia in water; relative density less than 0.880 at 15 °C), 2.2

When shippers offer these solutions for transportation, DOT requires them to affix the appropriate CORROSIVE or NON-FLAMMABLE GAS label to the packaging. When carriers transport either solution in an amount greater than 1001 lb (454 kg), they must display the appropriate CORROSIVE or NON-FLAMMABLE GAS placards on the bulk packaging or transport vehicle used for shipment.

PERFORMANCE GOALS FOR SECTION 10.15:

- Describe the response actions to be executed when encountering the release of a toxic substance into the environment.
- Describe how emergency responders use the *Emergency Response Guidebook* to establish the initial isolation and protective-action zones associated with large and small spills of a toxic substance.

10.15 Response Actions at Scenes Involving a Release of Toxic Substances

At the scene of transportation mishaps, an emergency-response crew may establish that a toxic substance has been released to the environment by observing any of the following:

- The hazard class numbers 2.3 or 6.1 are included in the shipping description of a toxic substance on a shipping paper
- Either POISON or INHALATION HAZARD is marked on its packaging
- White POISON GAS, POISON INHALATION HAZARD, or POISON labels are affixed to the packaging
- A white POISON GAS placard is displayed on each side and each end of the bulk packaging or transport vehicle containing any amount of a toxic substance
- A white POISON INHALATION HAZARD placard is displayed on each side and each end of the bulk packaging or transport vehicle containing any amount of a toxic substance that poses a health hazard by inhalation and is classed in hazard Zone A or Zone B.
- A white POISON placard is displayed on each side and each end of the bulk packaging or transport vehicle used to transport 1001 lb (454 kg) or more of a toxic substance in hazard class 6.1 other than those that pose an health hazard by inhalation in hazard Zone A or Zone B.

When a toxic substance has spilled or leaked during a transportation mishap, DOT provides emergency-response personnel with a table of initial isolation and protective-action distances in the *Emergency Response Guidebook* (Section 6.8). Excerpts of this table are provided in Table 10.15 for several toxic substances noted in this book. To use this table effectively, first-on-the-scene

TABLE 10.15 | Initial Isolation and Protective-Action Distances[a]

ID NO.	NAME OF MATERIAL	Small Spills (From a Small Package or Small Leak from a Large Package)						Large Spills (From a Large Package or from Many Small Packages)					
		First ISOLATE in All Directions		Then PROTECT Persons Downwind During				First ISOLATE in All Directions		Then PROTECT Persons Downwind During			
				DAY		NIGHT				DAY		NIGHT	
		m	ft	km	mi	km	mi	m	ft	km	mi	km	mi
1005 1005	Ammonia, anhydrous Anhydrous ammonia	30	100	0.1	0.1	0.2	0.1	150	500	0.8	0.5	2.3	1.4
1016 1016	Carbon monoxide Carbon monoxide, compressed	30	100	0.1	0.1	0.1	0.1	150	500	0.7	0.5	2.7	1.7
1017	Chlorine	60	200	0.4	0.3	1.6	1.0	600	2000	3.5	2.2	8.0	5.0
1023 1023	Coal gas Coal gas, compressed	30	100	0.1	0.1	0.1	0.1	60	200	0.3	0.2	0.4	0.3
1045 1045	Fluorine Fluorine, compressed	30	100	0.1	0.1	0.3	0.2	150	500	0.8	0.5	3.1	1.9
1050	Hydrogen chloride, anhydrous	30	100	0.1	0.1	0.4	0.2	60	200	0.3	0.2	1.4	0.9
1051 1051 1051	Hydrocyanic acid, aqueous solutions, with more than 20% hydrogen cyanide Hydrogen cyanide, anhydrous, stabilized Hydrogen cyanide, stabilized	60	200	0.2	0.1	0.6	0.4	400	1250	1.6	1.0	4.1	2.5
1052	Hydrogen fluoride, anhydrous	30	100	0.1	0.1	0.5	0.3	300	1000	1.7	1.1	3.6	2.2
1053	Hydrogen sulfide	30	100	0.1	0.1	0.4	0.3	300	1000	2.0	1.3	6.2	3.9
1062	Methyl bromide	30	100	0.1	0.1	0.2	0.1	150	500	0.7	0.4	2.2	1.4
1067 1067	Dinitrogen tetroxide Nitrogen dioxide	30	100	0.1	0.1	0.4	0.2	400	1250	1.1	0.7	3.0	1.9

[a]Adapted from *Emergency Response Guidebook* (Washington, DC: U.S. Department of Transportation), 2008.

(continued)

TABLE 10.15   Initial Isolation and Protective-Action Distances (*continued*)

		Small Spills (From a Small Package or Small Leak from a Large Package)						Large Spills (From a Large Package or from Many Small Packages)						
		First ISOLATE in All Directions		Then PROTECT Persons Downwind During				First ISOLATE in All Directions		Then PROTECT Persons Downwind During			
				DAY		NIGHT				DAY		NIGHT	
ID NO.	NAME OF MATERIAL	m	ft	km	mi	km	mi	m	ft	km	mi	km	mi
1076	Phosgene	100	300	0.7	0.4	2.6	1.6	500	1500	3.3	2.0	9.7	6.1
1079	Sulfur dioxide	60	200	0.3	0.2	1.2	0.7	400	1250	2.1	1.3	5.7	3.6
1242	Methyldichlorosilane (when spilled in water)	30	100	0.1	0.1	0.3	0.2	60	200	0.8	0.5	2.5	1.6
1250	Methyltrichlorosilane (when spilled in water)	30	100	0.1	0.1	0.2	0.2	60	200	0.6	0.4	2.0	1.3
1295	Trichlorosilane (when spilled in water)	30	100	0.1	0.1	0.3	0.2	60	200	0.7	0.5	2.3	1.4
1613	Hydrocyanic acid, aqueous solution, with not more than 20% hydrogen cyanide	30	100	0.1	0.1	0.1	0.1	100	300	0.5	0.3	1.1	0.7
1613	Hydrogen cyanide, aqueous solution, with not more than 20% hydrogen cyanide												
1614	Hydrogen cyanide, stabilized (absorbed)	60	200	0.2	0.1	0.6	0.4	150	500	0.6	0.4	1.7	1.1
1660 1600	Nitric oxide Nitric oxide, compressed	30	100	0.1	0.1	0.6	0.4	100	300	0.6	0.4	2.2	1.4
1680	Potassium cyanide (when spilled in water)	30	100	0.1	0.1	0.2	0.1	100	300	0.3	0.2	1.2	0.8
1680	Potassium cyanide, solid (when spilled in water)												

ID NO.	NAME OF MATERIAL	Small Spills (From a Small Package or Small Leak from a Large Package)						Large Spills (From a Large Package or from Many Small Packages)						
		First ISOLATE in All Directions		Then PROTECT Persons Downwind During				First ISOLATE in All Directions		Then PROTECT Persons Downwind During			
				DAY		NIGHT				DAY		NIGHT	
		m	ft	km	mi	km	mi	m	ft	km	mi	km	mi
1689 1680	Sodium cyanide (when spilled in water) Sodium cyanide, solid (when spilled in water)	30	100	0.1	0.1	0.2	0.1	100	300	0.4	0.3	1.4	0.9
1717	Acetyl chloride	30	100	0.1	0.1	0.3	0.2	100	300	0.9	0.6	2.8	1.8
1829	Sulfur trioxide, inhibited	60	200	0.4	0.2	1.0	0.6	300	1000	2.9	1.8	5.7	3.6
1829 1829	Sulfur trioxide, stabilized Sulfur trioxide uninhibited												
1831 1831	Sulfuric acid, fuming Sulfuric acid, fuming, with not less than 30% free sulfur trioxide	60	200	0.4	0.2	1.0	0.6	300	1000	2.9	1.8	5.7	3.6
1953	Compressed gas, flammable, poisonous, n.o.s. (Inhalation Hazard Zone A)	100	300	0.6	0.4	2.5	1.5	800	2500	4.4	2.7	8.9	5.6
1953	Compressed gas, flammable, poisonous, n.o.s. (Inhalation Hazard Zone B)	30	100	0.2	0.1	0.8	0.5	400	1250	1.9	1.2	4.8	3.0
1953	Compressed gas, flammable, poisonous, n.o.s. (Inhalation Hazard Zone C)	30	100	0.1	0.1	0.3	0.2	300	1000	1.3	0.8	4.1	2.6

(continued)

TABLE 10.15 — Initial Isolation and Protective-Action Distances (continued)

ID NO.	NAME OF MATERIAL	Small Spills (From a Small Package or Small Leak from a Large Package)						Large Spills (From a Large Package or from Many Small Packages)						
		First ISOLATE in All Directions		Then PROTECT Persons Downwind During				First ISOLATE in All Directions		Then PROTECT Persons Downwind During			
				DAY		NIGHT				DAY		NIGHT	
		m	ft	km	mi	km	mi	m	ft	km	mi	km	mi
1953	Compressed gas, flammable, poisonous, n.o.s. (Inhalation Hazard Zone D)	30	100	0.1	0.1	0.2	0.1	150	500	0.7	0.5	2.7	1.7
1953	Compressed gas, flammable, toxic, n.o.s. (Inhalation Hazard Zone A)	100	300	0.6	0.4	2.5	1.5	800	2500	4.4	2.7	8.9	5.6
1953	Compressed gas, flammable, toxic, n.o.s. (Inhalation Hazard Zone B)	30	100	0.2	0.1	0.8	0.5	400	1250	1.9	1.2	4.8	3.0
1953	Compressed gas, flammable, toxic, n.o.s. (Inhalation Hazard Zone C)	30	100	0.1	0.1	0.3	0.2	300	1000	1.3	0.8	4.1	2.6
1953	Compressed gas, flammable, toxic, n.o.s. (Inhalation Hazard Zone D)	30	100	0.1	0.1	0.2	0.1	150	500	0.7	0.5	2.7	1.7
2199	Phosphine	100	300	0.6	0.4	2.5	1.5	800	2500	4.4	2.7	8.9	5.6

responders must first determine whether a "small" or a "large" spill of the toxic substance has occurred. This determination can ordinarily be made by speaking with the individual most immediately responsible for transportation of the hazardous material. A "small" spill is one that involves a single package such as a 55-gal (208-L) drum, a small cylinder, or a small leak from a large package. A "large" spill involves a release from bulk packaging or multiple releases from several small packages.

Once a determination regarding the size of the release has been established, the guidebook is used to determine each of the following:

- The **initial isolation distance** is the distance from an emergency scene to which everyone should be quickly moved in a *crosswind* direction to establish the **initial isolation zone** shown in Figure 10.13(a). The initial isolation zone is the area surrounding an emergency scene in which persons may be exposed to dangerous, life-threatening concentrations of a toxic gas or vapor that poses an inhalation hazard. Emergency-response personnel should first direct persons to areas outside the initial isolation zone.

- The **protective-action distance** is the distance downwind from the scene that establishes the **protective-action zone** shown in Figure 10.13(b). The protective-action zone is the area in which persons may become incapacitated and unable to take protective action and/or incur serious, irreversible health effects. After they direct persons outside the initial isolation zone, emergency-response personnel should then direct others outside the protective-action zone, beginning with those individuals nearest the right-hand semicircular component of the initial isolation zone.

initial isolation distance
- For purposes of DOT regulations, the distance from a transportation mishap involving the release of a hazardous material to which everyone should be quickly moved in a crosswind direction

initial isolation zone
- For purposes of DOT regulations, the area surrounding a transportation mishap involving the release of a hazardous material in which persons may be exposed to dangerous, life-threatening concentrations of a gas or vapor that poses an inhalation hazard

SOLVED EXERCISE 10.10

Using Table 10.15, identify the initial isolation and protective-action distances recommended by DOT for protection of public health when responding to an 11:00 P.M. domestic transportation mishap involving a leaking 75-lb (34-kg) cylinder of hydrogen sulfide.

Solution: It is first necessary to identify whether this incident involves a "small" or a "large" spill of hydrogen sulfide. It is prudent to acknowledge that 75 lb (34 kg) constitutes a "large" spill of a toxic substance. Because it is dark at 11:00 P.M. in most parts of the world, Table 10.15 is examined under the heading "night" for the initial isolation and protective-action distances for potential exposure to hydrogen sulfide. These distances are listed in Table 10.15 as 1000 ft (300 m) and 3.9 mi (1.9 km), respectively.

When emergency-response personnel are required to encounter a toxic substance, they must first don appropriate protective clothing to prevent the possibility of skin contact. This protective clothing includes impermeable coveralls or similar fully encapsulating body suit, gloves, head coverings, and self-contained breathing apparatus. Because toxic substances may enter the body not only by inhalation but also by skin absorption, the use of a fully encapsulating body suit minimizes or eliminates the risk of exposure to the substance by both routes. Once a toxic substance has come in contact with clothing, it is essential to thoroughly wash the clothing with water prior to its removal.

It is usually critical to quickly provide professional medical attention to individuals who have been exposed to toxic substances, even if they do not exhibit apparent signs or symptoms of an injury. Whenever a team responds to any accident involving the release of a toxic substance, information concerning the unique properties of the poison may be essential for the successful protection of lives, property, and the environment.

protective-action distance
- For purposes of DOT regulations, the minimum distance downwind from a transportation mishap involving the release of a hazardous material at which emergency responders and the public can reasonably anticipate that their health, safety, and welfare will be preserved

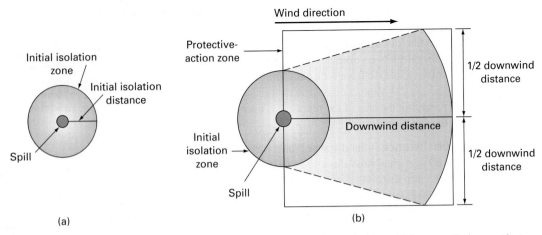

FIGURE 10.13 The initial isolation zone and the protective-action zone shown in (a) and (b), respectively, constitute regions in which individuals risk exposure to a hazardous material that poses a health hazard by inhalation. From a practical viewpoint, the initial isolation zone is the area of a circle whose radius equals the initial isolation distance, and the protective-action zone is the area of a square each of whose sides equals the protective-action (downwind) distance. The initial isolation distance and protective-action distance are published in the green section of the *Emergency Response Guidebook* (Washington, DC: U.S. Department of Transportation, 2008) (Section 6.8) for small and large spills of materials that pose a health hazard by inhalation.

protective-action zone
■ For purposes of DOT regulations, the area in which persons may become incapacitated, unable to take protective action, and/or incur serious, irreversible health effects

When an accident has occurred within an enclosed area, the use of a confined-space entry monitor (Figure 3.1) can quickly provide an estimate of the vapor concentration to which the workers or other individuals were exposed. CHEMTREC and the American Association of Poison Control Centers may also be utilized as informational sources,* and when an injury has occurred in connection with the use of a commercial product, the U.S. Consumer Product Safety Commission should be contacted.[†]

When an emergency-response team is called to a transportation mishap involving the release of a toxic substance, the procedures in Figures 10.14, 10.15, and 10.16 should be implemented. As a basis for making technical decisions, firefighters may compare the reading from a confined-space entry monitor with the STEL, PEL, and IDLH for the toxic substance. If the decision to fight a fire involving a toxic substance is made, firefighters should also observe the following basic principles:

■ Use impermeable coveralls or similar fully encapsulated body suit, gloves, head covering, and self-contained breathing apparatus.
■ Avoid direct exposure to the smoke or fumes that evolve from these fires by attacking them from an upwind location.
■ Use fog instead of direct streams of water to limit the generation of poisonous dust.
■ Keep runoff water to a minimum, and channel runoff water into a temporary reservoir to prevent its entrance into local sewers or waterways.
■ Notify the operators of the relevant storm and sanitary sewer system and water treatment plant of the ongoing fire.

PERFORMANCE GOALS FOR SECTION 10.16:

■ Describe the nature by which carcinogens cause cancer.
■ Describe the manner by which chemical carcinogens are generally classified.
■ Describe the means OSHA uses to signal the presence of carcinogens in the workplace.

---

*See footnote on p. 340.
[†]The U.S. Consumer Product Commission may be contacted at (800) 638-CPSC.

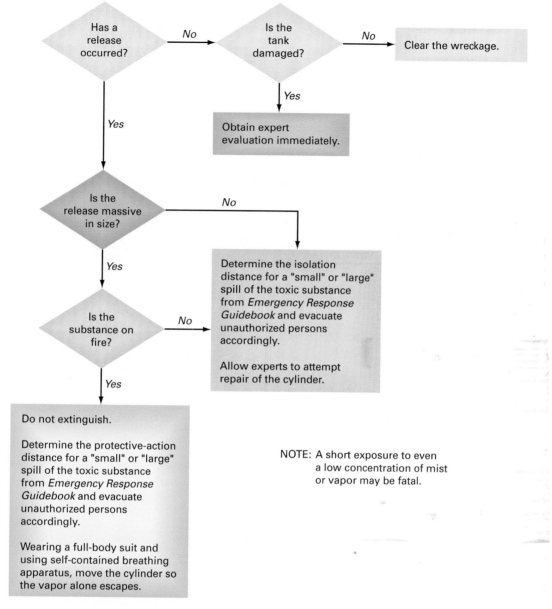

FIGURE 10.14 The recommended procedures when responding to a transportation mishap involving the release of a poisonous gas from bulk packaging. (*Adapted with permission of the American Society for Testing and Materials, from a figure in ASTM STP 825, A Guide to the Safe Handling of Hazardous Materials Accidents, second edition. Copyright © 1990, American Society for Testing and Materials.*)

10.16 Carcinogenesis

Millions of individuals are annually diagnosed with cancer. In many instances, the cause can be traced to exposure to certain agents called **carcinogens**. Chemical carcinogens, physical carcinogens (like ultraviolet radiation), and viruses cause cancer in living tissues, but this section is concerned primarily with chemical carcinogens.

Cancer starts as a disease of malfunctioning cells. It is caused by exposure to agents external to the body that enter or penetrate the body (as radiation) and begin to corrupt its cells. Chemical

carcinogen
■ A substance or agent that contributes to the causation of a tumor in a living organism

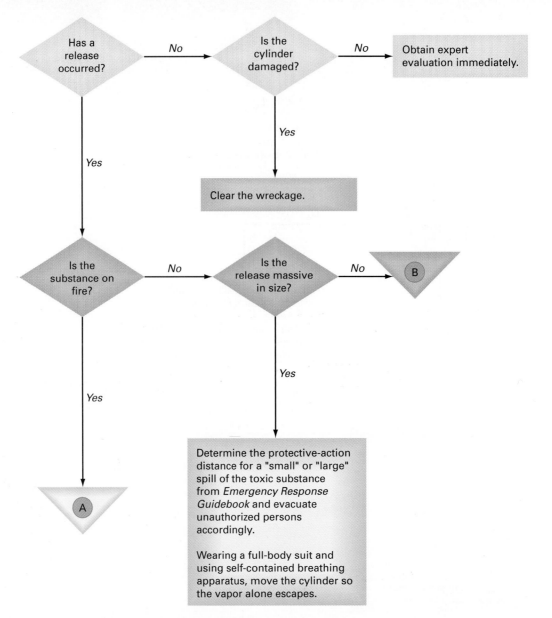

FIGURE 10.15 The recommended procedures when responding to a transportation mishap involving the release of a poisonous gas from nonbulk packaging. (*Adapted with permission of the American Society for Testing and Materials, from a figure in ASTM STP 825,* A Guide to the Safe Handling of Hazardous Materials Accidents, *second edition. Copyright © 1990, American Society for Testing and Materials.*)

carcinogens act with vastly different potencies, differing in their abilities to cause cancer by as much as a-million-fold.

Although the mechanism of carcinogenesis—the production of cancer—is not completely understood, scientists now know that it consists of a multistepped cellular phenomenon that can be viewed in three stages: initiation, promotion, and progression.

initiation
■ The stage of cancer during which one or more healthy cells are triggered to subsequently respond to the actions of a promoter

■ During the **initiation** stage, the genetic material within a healthy cell or group of healthy cells becomes altered. Sometimes, this cellular alteration results from a single exposure to a carcinogen, whereas on other occasions it results from repeated exposures. Carcinogens that are initiators transform or mutate cellular DNA. Thus, the molecular pathway by

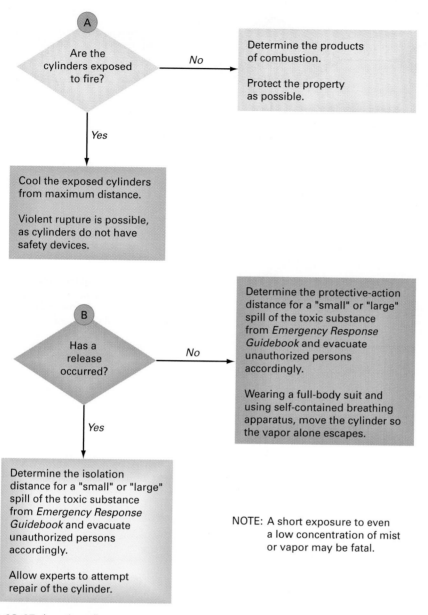

A

Are the cylinders exposed to fire?

No → Determine the products of combustion.

Protect the property as possible.

Yes ↓

Cool the exposed cylinders from maximum distance.

Violent rupture is possible, as cylinders do not have safety devices.

B

Has a release occurred?

No → Determine the protective-action distance for a "small" or "large" spill of the toxic substance from *Emergency Response Guidebook* and evacuate unauthorized persons accordingly.

Wearing a full-body suit and using self-contained breathing apparatus, move the cylinder so the vapor alone escapes.

Yes ↓

Determine the isolation distance for a "small" or "large" spill of the toxic substance from *Emergency Response Guidebook* and evacuate unauthorized persons accordingly.

Allow experts to attempt repair of the cylinder.

NOTE: A short exposure to even a low concentration of mist or vapor may be fatal.

FIGURE 10.15 *(continued)*

which cancer can occur is established during the initiation stage. Overall, however, this initial damage rarely results in cancer, because there are many cellular mechanisms for repairing the damaged DNA. It is when the cell is unable to repair the damage that it is more prone to becoming cancerous.

■ During the subsequent **promotion** stage, the damaged cell divides into new but similarly damaged cells, each of which subsequently divides repeatedly into new, similarly damaged cells. In most instances, cell division is accelerated when the damaged cells are exposed to carcinogens; that is, these carcinogens act as promoters rather than initiators. When cell division occurs over an extended period, the aggregates of damaged cells often create a mass of tissue called a **tumor**. (Leukemia, is cancer of the blood, in which cancerous white blood cells form a diffuse liquid tumor throughout the circulatory system.) It is mind-boggling to realize that each tumor descends from a single ancestral cell or group of cells that go awry. Some carcinogens are so

promotion
■ The second stage of cancer, during which the formation of malfunctioning cells is accelerated to produce tumors

tumor
■ A mass of tissue that persists and grows independently of its surrounding structures and has no known physiological function

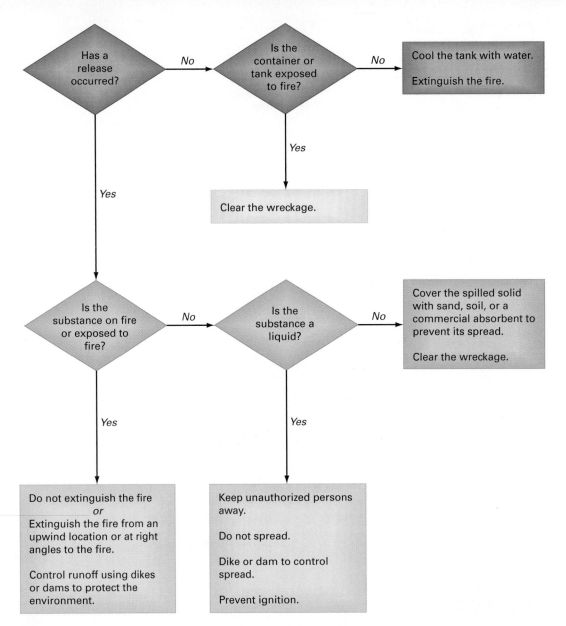

FIGURE 10.16 The recommended procedures when responding to a transportation mishap involving the release of a liquid or solid poison that does not pose a toxic inhalation hazard. (*Adapted with permission of the American Society for Testing and Materials, from a figure in ASTM STP 825,* A Guide to the Safe Handling of Hazardous Materials Accidents, *second edition. Copyright © 1990, American Society for Testing and Materials.*)

progression
■ The third stage of cancer, during which tumors evolve into malignant growths in the body

powerful that they cause cancer without the need for promotion. Exposure to ionizing radiation (Section 16.5), for example, directly initiates a variety of cancers.

■ During the final **progression** stage, the damaged cells invade nearby tissues and migrate to other tissues and organs in the body, where they may provoke the emergence of tumors. Although tumors may be either *benign* or *malignant*, it is generally during the progression stage that these growths may become malignant. Cancer is viewed, in part, as the transformation of healthy tissue cells into such malignant growths. The phenomenon associated with the formation of a malignant growth at one site in the body from cells derived from a malignancy located elsewhere in the body is called **metastasis**. A tumor that is localized and noninvasive is called benign, whereas a tumor that is invasive and metastatic is said to be malignant.

Individuals who are heavy tobacco smokers assume a lifetime risk of lung cancer that is more than 20-fold higher than that of nonsmokers. It seems likely that the cells of the mouth, throat, lungs, and other organs in smokers undergo the first stage of cancer when they are exposed to the carcinogenic substances in tobacco smoke. The subsequent exposure of these damaged cells to certain other carcinogens provokes the emergence of tumors by accelerating the rates at which cancer develops in these organs. Asbestos, ethanol, and radon are characteristic of carcinogens that cause cancer more rapidly in smokers than in nonsmokers. They act as promoters, and the related phenomena are examples of **synergism**; that is, the effects of both carcinogens are greater than the effects of each acting separately.

metastasis
■ The formation of a malignant growth at one site in the body from cells derived from a malignancy located elsewhere in the body

synergism
■ The aspect of two substances that interact to produce an effect that is greater than the sum of the substances' individual effects of the same type

10.16-A TYPES OF CHEMICAL CARCINOGENS

Epidemiologists recognize the following three groups of chemical carcinogens:

- *Known human carcinogens.* These are substances for which there is sufficient evidence from epidemiological studies to unquestionably conclude that exposure causes cancer in humans.
- *Probable human carcinogens.* These are substances for which there is sufficient evidence to conclude that exposure causes cancer in experimental animals but no direct evidence that exposure causes cancer in humans. It is safe to say that exposure to a probable human carcinogen can reasonably be anticipated to cause cancer in humans.
- *Possible human carcinogens.* These are substances for which there is reason to suspect that exposure causes cancer in animals or humans, but no studies have yet affirmed this suspicion.

The classification of a substance as a known, probable, or possible human carcinogen is a necessary one, since the existing data do not always enable epidemiologists to clearly ascertain whether the substance causes cancer in the human organism. Research studies in carcinogenesis are complicated by the observation that certain substances appear incapable of directly causing cancer, but they can affect the rate of cancerous tumor formation when exposure to them occurs in combination with exposure to other substances.

Individuals can be exposed to cancer-causing substances in a variety of ways. Carcinogens are contained in many commercial products used routinely, including crude petroleum and petroleum products. When the presence of a carcinogen in a commercial product is known, certain state laws require that the consumer be advised. Conveying this information is the responsibility of the seller, who displays the information at the locations where the product is sold.

10.16-B WORKPLACE REGULATIONS INVOLVING CARCINOGENS

When the use of carcinogens is needed in the workplace, OSHA requires employers to provide employees with protective clothing, self-contained breathing apparatus, and a segregated area in which the work with carcinogens can be safely conducted. OSHA also requires employers to post this area with a sign that reads as follows:

> **WARNING**
>
> **CANCER-SUSPECT AGENT EXPOSED IN THIS AREA IMPERVIOUS SUIT INCLUDING GLOVES, BOOTS, AND AIR-SUPPLIED HOOD REQUIRED AT ALL TIMES AUTHORIZED PERSONNEL ONLY**

To reduce or eliminate their exposure to carcinogens, emergency-response teams should heed this warning when they enter buildings in which carcinogens are stored or used.

We identify the carcinogens of immediate concern to emergency-response personnel as they are encountered in this and future chapters.*

PERFORMANCE GOALS FOR SECTION 10.17:

- Identify the metals and metalloids required in the diet for good health.
- Identify the negative effects experienced by the human organism from exposure to lead compounds.
- Identify the ways in which emergency-response personnel are likely to be exposed to lead compounds.
- Describe the means OSHA uses to signal the presence of lead in the workplace.
- Describe the means shippers and carriers use to inform emergency responders of the hazards associated with encountering lead compounds during transportation mishaps.

10.17 Compounds of Toxic Metals

From a nutritional viewpoint, metallic compounds may be classified as follows:

- ***Essential elements.*** Trace amounts of 14 metals and metalloids are essential in the diet for the proper functioning and survival of the human organism. For this reason, nutritionists refer to them as the **essential elements**. They are calcium, potassium, sodium, magnesium, iron, zinc, copper, tin, vanadium, chromium, manganese, molybdenum, cobalt, and nickel. Their proper balance and availability are essential for proper biological function. This is not meant to imply that the essential elements are beneficial in all doses. The ingestion of excessive concentrations of the essential elements may cause illness or death.

essential elements
■ The elements essential for the proper functioning and survival of the human organism

- ***Toxic metals.*** Certain other metals and metalloids are not known to offer any dietary benefit to the human organism. These **toxic metals** include arsenic, barium, beryllium, lead, mercury, and thallium. The consumption of even minute amounts of their compounds can cause illness or death.

toxic metal
■ Any of certain metals or metalloids whose consumption, even in minute amounts, can cause illness or death

Certain metals and metalloids may be either toxic or beneficial, depending on their concentration and other factors. They include the compounds of selenium and tungsten. They can either exert a toxic effect or perform an essential biological function in the human body.

10.17-A LEAD AND ITS COMPOUNDS

Of all the toxic metals, the *compounds of lead* are most likely to be regularly encountered by emergency responders. For example, firefighters may be exposed to them by inhaling the lead-laden dust or smoke that is generated when old structures burn.

Lead compounds were formerly used as pigments in surface coatings to provide color and prevent corrosion. Examples of these pigments are lead acetate, lead chromate, and lead oxide. The surface coatings are called *lead-based paints.*

Although EPA banned the use of lead-based paint for most purposes in 1978, millions of apartments and other dwellings with lead-based paint still exist today. When these structures collapse during building fires, lead-laden dust is dispersed into the atmosphere. Firefighters are then

---

*Various scientific groups conduct research relating to the ability or tendency of substances to produce cancer in normal tissues. Unless otherwise indicated, the designations in this book are those denoted by the U.S. Department of Health and Human Services and the World Health Organization's International Agency for Research on Cancer (IARC).

put at risk of inhaling lead particulates. The use of respiratory protection equipment significantly eliminates or reduces exposure to this contaminated dust.

However, children who live in the dwellings formerly coated with lead-based paint are not as fortunate. Lead-based paint is considered the greatest remaining source of childhood lead exposure. Hungry youngsters, especially those forced to live in substandard housing, sometimes have no other alternative but to eat flaking paint and plaster.

When ingested or inhaled, lead primarily affects the human blood-forming, nervous, and kidney systems, but it also harms the reproductive, endocrine, hepatic, cardiovascular, immunologic, and gastrointestinal processes. Lead interferes with the body's mechanism for absorbing the essential elements calcium and iron. Elevated lead concentrations can cause severe health problems, including brain disease, colic, palsy, and anemia. Damage to the central nervous system in general, and to the brain in particular, is one of the most severe consequences of chronic lead poisoning. Exposure to certain lead compounds may also cause latent health effects. In particular, lead acetate and lead phosphate are regarded as probable human carcinogens.

Children are especially susceptible to the adverse effects caused by exposure to this neurotoxic metal. In particular, lead impairs their ability to learn and execute mental processes effectively. Research studies have equated a 7.4-point drop in the intelligence quotient (IQ) of children with an increase from 1 to 10 µg of lead per deciliter of blood (1 to 10 µg/dL).

10.17-B CONSUMER PRODUCT REGULATIONS INVOLVING LEAD

The Consumer Product Safety Commission has taken action to reduce or eliminate the risk of consumer exposure to lead. At 16 C.F.R. §1303.4, the CPSC bans lead in paint and other similar surface-coating materials; toys and other articles intended for use by children that bear lead-based paint; and furniture articles that bear lead-based paint. As first noted in Section 1.4-A, CPSC bans at 16 C.F.R. §1500.17 the sale of paint having a lead content above 0.06% by dry weight (other than artist's paint) in the United States.

Notwithstanding the CPSC's efforts, lead-based paint has been identified on the surfaces of toys manufactured outside the United States. Children who play with these toys are at risk of lead exposure.

10.17-C WORKPLACE REGULATIONS INVOLVING LEAD

When the use of lead compounds is needed in the workplace, OSHA requires employers to limit employee exposure to a concentration of 0.050 mg/m^3, averaged over an 8-hr workday. OSHA also requires the posting of the following warning sign in work areas where airborne lead is generated.

WARNING
LEAD WORK AREA
POISON
NO SMOKING OR EATING

10.17-D ENVIRONMENTAL REGULATIONS INVOLVING LEAD

EPA regulates the concentration of lead in the ambient air as a criteria air pollutant. EPA has set the primary and secondary national ambient air quality standards for lead at 0.15 µg/m^3 as a quarterly average.

Using the legal authority of TSCA, EPA also regulates the manner by which certified contractors may remove or treat lead-based paint from dwellings. These regulations are published at 40 C.F.R. §745.227.

When a representative sample of the lead-laden waste exhibits the RCRA characteristic of toxicity, EPA regulates the treatment, storage, and disposal of the waste.

10.17-E TRANSPORTING LEAD

When shippers offer a nonoxidizing lead compound for transportation, DOT requires them to identify it in the basic description on an accompanying shipping paper. Two examples of such basic descriptions are the following:

> **UN1616, Lead acetate, 6.1, PGIII (Poison)**
> **UN1620, Lead cyanide, 6.1, PGII (Poison)**

DOT also requires shippers to affix a POISON label on its packaging.

When carriers transport nonoxidizing lead compounds in an amount exceeding 1001 lb (454 kg), DOT requires them to display POISON placards on the bulk packaging or transport vehicle used for shipment.

We note the DOT requirements for transporting oxidizing lead compounds in Chapter 11.

PERFORMANCE GOALS FOR SECTION 10.18:

- Identify the chemical nature of asbestos.
- Identify the general type of asbestos products likely to contain friable asbestos.
- Describe the latent effects associated with exposure to friable asbestos fibers.
- Describe the means OSHA uses to signal the presence of asbestos in the workplace.
- Identify the asbestos abatement procedures required by EPA when buildings containing friable asbestos are demolished or renovated.
- Describe the means shippers and carriers use to inform emergency responders of the hazards associated with encountering friable asbestos during transportation mishaps.
- Describe the nature of the response actions to be executed when friable asbestos has been released into the environment.

10.18 Asbestos

asbestos
■ The generic name of a class of naturally occurring fibrous silicates

Asbestos is the generic name of a complex mineral composed of several naturally occurring magnesium silicates that contain SiO_4^{4-} groups linked into chains. Several types of asbestos are recognized, all of which are nonflammable and excellent insulators. They can be woven into strong, flexible fibers that can be spun into threads and woven into fireproof fabric.

When mixed with magnesium oxide, asbestos was formerly used for a variety of fireproofing, insulating, soundproofing, and decorative purposes. It was a component of products like asbestos pipe coverings, flooring- and ceiling-tile products, paper products, thermal system insulation, antifriction materials (e.g., brake linings and clutch facings), roofing materials, fire walls or doors, and surface-coating and patching compounds. Although these asbestos products were once prized for such qualities, their availability in today's market has been sharply curtailed or entirely eliminated.

10.18-A ILL EFFECTS RESULTING FROM EXPOSURE TO ASBESTOS

The virtual absence of asbestos products in today's marketplace is linked with the recognition that asbestos exposure causes serious health problems including cancer. Unless the asbestos has been completely sealed into a product, as in asbestos floor tiles, it often breaks apart into tiny

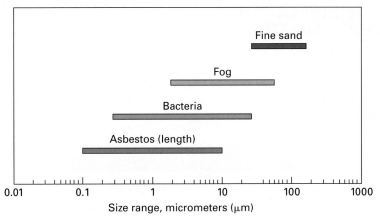

FIGURE 10.17 The typical size of asbestos fibers ranges from 0.1 to 10μm, shown here relative to the size of some other materials. This size is too small to be visible to the human eye. (*Courtesy of the U.S. Environmental Protection Agency, "Asbestos-Containing Materials in School Buildings: A Guidance Document," Office of Toxic Substances, Washington, DC, EPA 450/2-78-014, March 1978.*)

fibers, much smaller and more buoyant than ordinary dust. This occurs when the asbestos-containing material is damaged or disintegrates with age. It is said to be **friable**. It is this friable asbestos that poses severe health hazards to those who inhale or swallow its fibers.

Figure 10.17 illustrates that asbestos fibers may range from 0.1 to 10 μm in length and have a diameter of only 30 nm. They float almost indefinitely in the air and may be readily inhaled or swallowed. When they are inhaled, asbestos fibers are sometimes removed by the cilia lining the respiratory tract. Then, they move into the throat, where they can be swallowed. Once swallowed, they pose the threat of causing cancer of the esophagus, stomach, intestines, and rectum.

Exposure to asbestos fibers potentially cause the following illnesses:

- *Lung cancer*, the growth of malignant tumors in the lungs.
- **Asbestosis**, a noncancerous scarring of lung tissue that occurs when sharp asbestos fibers deposit deep within the respiratory tract.
- **Mesothelioma**, a unique and characteristic cancer of the membranes lining the chest and abdomen. Mesothelioma is characteristic of asbestos exposure, since it rarely occurs in individuals who have not been exposed to asbestos. It is virtually always fatal.

Because epidemologists have established that exposure to asbestos causes lung cancer and mesothelioma in humans, asbestos is denoted as a known human carcinogen.

When asbestos fibers are inhaled by cigarette smokers, they act synergistically with the carcinogenic substances in tobacco smoke. The result is that a smoker exposed to asbestos fibers has up to 50 times the chance of developing lung cancer compared with a nonexposed nonsmoker.

friable
- Brittle or capable of readily being crumbled, as in friable asbestos

asbestosis
- A scarring of lung tissue by asbestos fibers

mesothelioma
- A cancer of the membranes lining the chest and abdominal cavities resulting from exposure to asbestos fibers

10.18-B CONSUMER PRODUCT REGULATIONS INVOLVING ASBESTOS

The CPSC has promulgated regulations to eliminate or reduce asbestos in consumer products sold for use in the United States. For example, at 16 C.F.R. §1304.4, the use of consumer patching compounds containing friable asbestos is banned; and at 16 C.F.R. §§1305.4 and 1500.17, the use of general-use garments, tape joint compounds, and artificial fireplace ash and embers containing friable asbestos is banned. The manufacturers of handheld hair dryers voluntarily ceased using asbestos in hair dryers.

10.18-C WORKPLACE REGULATIONS INVOLVING ASBESTOS

When the use of asbestos is needed in the workplace, OSHA requires employers at 29 C.F.R. §1910.1001 to limit employee exposure to an airborne concentration of asbestos fibers of 0.1 fiber per cubic centimeter of air as an 8-hr, time-weighted average. When asbestos is present in the workplace, OSHA also requires the employer to post the area so as to identify the potential harm as follows:

> **CAUTION**
>
> **AVOID ASBESTOS FIBERS**
> **AVOID CREATING DUST**
> **BREATHING ASBESTOS**
> **DUST MAY CAUSE SERIOUS**
> **BODILY HARM**

10.18-D ENVIRONMENTAL REGULATIONS INVOLVING ASBESTOS

Using the authority of the Clean Air Act, EPA banned the use of asbestos in pipe and boiler coverings as well as all uses of asbestos materials that are applied by spraying. Using the authority of TSCA, EPA also banned the use of asbestos in corrugated paper, rollboard, commercial paper, specialty paper, and flooring felt.

To safeguard the health of school children and others who work in schools, EPA launched a program at 40 C.F.R. §763.91 that obligates each local educational agency to accomplish the following within their jurisdictions:

- Inspect for the presence of friable asbestos in its school buildings
- Determine whether friable asbestos fibers are being released into the air
- Remove and repair all damaged asbestos-containing material

EPA also regulates at 40 C.F.R. §61.145 how the asbestos is to be stripped from walls, ceilings, and other locations. The proper abatement procedures include the following:

- Sealing a work site before the asbestos-containing material is removed
- Preventing the asbestos fibers from becoming airborne by keeping them wet
- Removing, encapsulating, repairing, enclosing, or encasing the asbestos-containing material
- Placing the asbestos debris in leak-tight containers that are subsequently sealed

10.18-E TRANSPORTING ASBESTOS

When shippers offer asbestos (including asbestos waste) for transportation, DOT requires them to comply with certain packaging requirements published at 49 C.F.R. §173.216, to transport the asbestos by means of closed freight containers, motor vehicles, or railcars, and to affix a CLASS 9 label to the packaging. DOT also requires them to identify the material on an accompanying shipping paper as follows:

NA2212, Asbestos, 9, PGIII

When bulk quantities of asbestos are transported, DOT requires carriers to display the identification number 2212 on each side and each end of an orange panel or on a white square-on-point diamond displayed on the packaging. DOT does not require the posting of CLASS 9 placards on the packaging.

10.18-F RESPONDING TO INCIDENTS INVOLVING A RELEASE OF ASBESTOS

Despite efforts to eliminate asbestos in buildings, emergency responders are among the individuals who are most likely to be exposed to asbestos fibers. As firefighters attack fires in old buildings, they may be exposed to airborne asbestos, especially as dust is generated when ceilings and roofs collapse. To reduce or eliminate this potential exposure, firefighters should combat these fires from upwind locations and use self-contained breathing apparatus.

PERFORMANCE GOALS FOR SECTION 10.19:

- Identify the four groups of organic pesticides using their chemical structures as the basis for classification.
- Describe the advisory information pesticide manufacturers provide on their product labels.
- Describe the means shippers and carriers use to inform emergency responders of the hazards associated with encountering pesticides during transportation mishaps.

10.19 Pesticides

As first noted in Section 1.4-C, EPA defines a pesticide in part at 40 C.F.R. §152.3 as any substance or mixture of substances intended for preventing, destroying, repelling, or mitigating any pest, or intended for use as a plant regulator, defoliant, or desiccant. In more general terms, a pesticide is generally regarded as a commercial product that has been intentionally designed to destroy or control the spread of insects, fungi, rodents, weeds, mold, mildew, algae, and other pests.

There are two chemical types of pesticides: inorganic pesticides and organic pesticides. Inorganic pesticides are generally identified either by the name of their active component or the use of a generic term like "arsenical pesticide." Because their application leaves a hazardous substance in the environment to pose long-term damage, few inorganic pesticides are encountered in today's commercial markets.

Although organic pesticides may be classified in several ways, we divide them into the following four groups based on their chemical structures:

- Carbamate pesticides, derivatives of carbamic acid, $H_2N-\overset{\displaystyle O}{\underset{\displaystyle \backslash OH}{\overset{\|}{C}}}$

- Organochlorine pesticides, chlorinated derivatives of certain complex hydrocarbons

- Organophosphorus pesticides, derivatives of phosphoric acid, $HO-\overset{\displaystyle O}{\underset{\displaystyle OH}{\overset{\|}{P}}}-OH$

- Thiocarbamate pesticides, derivatives of thiocarbamic acid, $H_2N-\overset{\displaystyle S}{\underset{\displaystyle \backslash OH}{\overset{\|}{C}}}$

This chemical classification scheme is employed by DOT for generically describing organic pesticides (Section 10.19-B).

Which organophosphorus insecticide is more toxic: malathion or parathion? Their LD_{50} values are 2800 and 2 mg/kg, respectively.

Malathion   Parathion

Solution: In a comparison of the toxicities of malathion and parathion, their LD_{50} values signify that half of a group of laboratory animals succumbed during a preestablished time from a dose containing a lesser amount of parathion. Consequently, parathion is more toxic than malathion.

10.19-A ENVIRONMENTAL REGULATIONS PERTAINING TO PESTICIDES

To advise consumers of the hazards posed by exposure to specific pesticides, EPA uses the authority of FIFRA to compel manufacturers to provide advisory information on pesticide product labels. The nature of this information was previously noted in Section 1.4-C.

10.19-B TRANSPORTING PESTICIDES

DOT requires shippers and carriers to identify the shipping description of a pesticide on the shipping paper accompanying the shipment. The Hazardous Materials Table at 40 C.F.R. §172.101 lists specific chemical names as well as generic descriptions of pesticides.

When the shipping description is listed generically, DOT requires its technical name or the names of the principal constituents that cause the pesticide to meet the definition of a Division 6.1, Packing Group I or II, to be entered in parentheses in the shipping description. Some examples of these generic shipping descriptions follow:

> UN2757, **Carbamate pesticides, solid, toxic (ethyl carbamate), 6.1, PGI (Poison)**
> UN2761, **Organochlorine pesticides, solid, toxic (Endosulfan), 6.1, PGI (Poison)**
> UN2783, **Organophosphorus pesticides, solid, toxic (Chlorpyrifos-methyl), 6.1, PGI (Poison)**
> UN2771, **Thiocarbamate pesticides, solid, toxic (ethylenebisdithiocarbamic acid), 6.1, PGI (Poison)**
> UN2588, **Pesticides, solid, toxic, n.o.s. (ammonium sulfamate), 6.1, UN2588, PGI (Poison)**

DOT also require shippers to affix a POISON label to the pesticide packaging. When carriers transport these pesticides in an amount exceeding 1001 lb (454 kg), DOT requires them to display POISON placards on the bulk packaging or transport vehicle used for shipment. Emergency responders should heed these shipping descriptions, labels, and placards, since pesticide exposure may adversely affect the human organism—despite the fact that pesticide use is focused on application to nonhuman pests.

- Describe how terrorists may use biological warfare agents to injure, incapacitate, or cause mass casualties.
- Identify some common biological warfare agents and the adverse health effects they cause.

10.20 Biological Warfare Agents

Certain microorganisms and toxicants are so highly poisonous to humans that during wartime, their distribution could serve to sicken, injure, incapacitate, or otherwise cause mass casualties. These are called **biological warfare agents**, some forms of which are listed in Table 10.16. Although their use during wartime could cause impairment, disease, death, or biological malfunctions among the enemy on a large scale, this practice has been condemned by civilized countries as unethical and inhumane. In 1925, the United Nations introduced the Protocol for the Prohibition of the Use in War of Asphyxiating, Poisonous, or Other Gases, and of Bacteriological

biological warfare agent
- An infectious substance that inflicts disease or causes mass casualties when disseminated throughout a population

TABLE 10.16	Some Potential Biological Warfare Agents
TYPE OF BIOLOGICAL WARFARE AGENT	**ADVERSE HEALTH EFFECTS CAUSED BY INFECTION**
Bacillus anthracis, a spore-forming bacterium	Anthrax, usually fatal in the inhaled form (for symptoms, see text)
Bacillus cereus	Food-poisoning-like symptoms
Bacillus licheniformis	Produces a protease enzyme that interferes with protein metabolism
Bacillus megaterium, a spore-forming bacterium	Unknown
Bacillus subtilis	Unknown
Brucella abortus (multiple biotypes)	Brucellosis, an infectious disease that can cause an infected pregnant woman to abort her fetus
Brucella melitensis	Fatigue, loss of appetite, and other symptoms
Burkholderia mallei	Glanders, a disease acquired by humans from infected animals; characterized by fever, muscle aches, chest pain, muscle tightness, weakness, light sensitivity, pus-forming skin infections, diarrhea, and headache
Burkholderia pseudomallei	Meliodosis, a disease similar to glanders and characterized by the same symptoms
Clostridium botulinum (seven biotypes)	Botulism, characterized by vomiting, paralysis of the muscles, constipation, thirst, headache, fever, dizziness, double or blurred vision, deadening of the nerves, and dilation of the pupils
Clostridium perfringens	Gas gangrene, respiratory distress, and failure of the body's internal organs

(continued)

| TABLE 10.16 | Some Potential Biological Warfare Agents (*continued*) |

TYPE OF BIOLOGICAL WARFARE AGENT	ADVERSE HEALTH EFFECTS CAUSED BY INFECTION
Clostridium tetani	Tetanus (lockjaw)
Corynebacterium diphtheriae	Diphtheria, an often fatal respiratory disease
Coxiella burnetii	Q fever, characterized by chills, fever, coughing, headache, fatigue, hallucination, and sometimes chest pain
Francisella tularensis	Tularemia, a disease often contracted by rodents and transmitted to humans; characterized by sudden chills, fever, coughing, severe weakness, and the development of large skin lesions; can lead to fatal pneumonia
Listeria monocytogenes	Fever; muscle ache, nausea, and diarrhea; can lead to meningitis, encephalitis, and intrauterine and cervical infections in infected women
Plasmodium malaria	Malaria
Pneumocystis carinii	Pneumonia
Poxvirus variola	Smallpox, a highly infectious, often fatal disease characterized by high fever, headache, body ache, vomiting, and the appearance of a rash on the tongue and in the mouth that turns to blisters
Rickettsia prowazekii	Typhus, a disease caused by louse-borne bacteria; characterized by severe headache; sustained high fever; severe muscle pain; rash; sensitivity to light; and delirium
Salmonella enteritidis	Salmonellosis (see *Salmonella typhimurium*), but not typhoid fever
Salmonella typhimurium[a]	Salmonellosis, or food poisoning, a disease associated with nausea, vomiting, chills, abdominal cramps, and diarrhea; in severe cases, accompanied by typhoid fever, a disease associated with intestinal inflammation and ulceration
Staphylococcus aureus (multiple biotypes)	Staphylococcus food poisoning, characterized by sudden severe nausea, vomiting, abdominal cramps, muscle pain, severe diarrhea, headache, fever, and chills
Streptococcus pneumoniae	Pneumonia

[a]The first act of bioterrorism on American soils was carried out in 1984 by members of the Rajneeshees, an Indian religious cult. The bioterrorism act involved contaminating the foods displayed on salad bars in several Oregonian restaurants with *Salmonella typhimurium*. More than 750 people who consumed food from the salad bars were infected with salmonellosis. [Judith Miller, Stephen Engelberg, and William Broad, *Germs: Biological Weapons and America's Secret War* (New York: Simon & Schuster, 2001), pp. 15–33.]

TYPE OF BIOLOGICAL WARFARE AGENT	ADVERSE HEALTH EFFECTS CAUSED BY INFECTION
Vibrio cholerae	Cholera, an acute intestinal disease characterized by a profuse watery diarrhea
Yersinia pestis[b,c]	Bubonic plague, a disease associated with the production of ugly black welts and bulges called *buboes* in the groin or near the armpits

[b]Bubonic plague has often been associated with the so-called Black Death that historians credited with killing 20 million Europeans during the 14th century. The conventional belief is that this loss of life was caused by deadly bacilli that were transmitted from fleas carried on rodents to humans. In contemporary times, however, scientists are seriously questioning whether the Black Death was actually caused by the human consumption of beef tainted with an anthraxlike bacillus. [See e.g., Norman F. Cantor, *In the Wake of the Plague* (New York: Free Press, 2001).]

[c]During World War II, thousands of people died from consuming rice and wheat that the Japanese dropped from aircraft on Chinese and Manchurian towns and cities. These foods were contaminated with fleas that had been infected with *Yersinia pestis*.

Methods of Warfare (Section 7.4-A), to which most member countries have consented. Nonetheless, mistrust and fear among nations have prompted the production and stockpiling of biological warfare agents for defensive purposes. Today, the threat centers more on the likelihood that bioterrorists will deploy them to sicken or kill Americans or its allies.

10.20-A WHAT IS A BIOLOGICAL WARFARE AGENT?

Table 10.16 shows that a biological warfare agent is either an infectious substance or a toxin.

- An **infectious substance** usually consists of living bacteria, viruses, or fungi exposure to which causes such diseases as anthrax, botulism, bubonic plague, cholera, and smallpox.
- A **toxin** is a poisonous substance uniquely produced by certain plants and animals. Examples of organisms that produce toxins include certain mushrooms, spiders, frogs, toads, jellyfish, and snakes. On entry into the body, toxins typically act on either the central nervous system or the circulatory system; that is, they act as neurotoxins or cytotoxins, respectively.

10.20-B DISSEMINATION OF A BIOLOGICAL WARFARE AGENT

A biological warfare agent may be covertly disseminated by several means that include the following:

- Containerization in canisters or cluster bombs that are then inserted into modified missile heads or artillery shells
- Aerosolized dispersal from aircraft (e.g., crop dusters, helicopters, or low-flying robotic craft) in an enclosed space or over a city
- Aerosolized dispersal using generators carried on trucks or boats in an enclosed space or over a city
- Direct introduction into food or drinking water supplies
- Inoculation of fabric, paper, and similar items

To maximize the adverse impact from exposure to a pathogen, a period of time must pass following its initial dissemination. The period may be hours, days, weeks, or even years. During this time the agent enters the bodies of its victims and multiplies exponentially until the symptoms and ailments of a disease are clearly evident. During this gap in time, or *incubation period*, the disease-causing microorganism may be passed to other people, who may also transmit it to others. Thus, an incubation period allows an epidemic to become established before medical authorities are fully aware the disease has spread.

infectious substance
■ For purposes of DOT regulations, a material known to contain or suspected of containing a pathogen

toxin
■ A poisonous substance produced by a living organism or bacterium

Medical authorities categorically recommend the isolation of infected individuals from the general population until the bacterium or toxin is identified as contagious or noncontagious. Contagious infections may be transmitted from person to person by inhaling bacteria spewed into the air by coughing or sneezing, contacting body fluids, and simply touching something the infected person previously touched.

The isolation of persons infected with a contagious disease serves to prevent the spread of the disease to others. During a bioterrorist attack involving numerous individuals, state and local governments would most likely use their legal authorities to quarantine infected persons. Then, the infected persons could be medically treated to prevent further spread of the disease.

10.20-C RESPONDING TO INCIDENTS INVOLVING A RELEASE OF A BIOLOGICAL WARFARE AGENT

The typical emergency responder is untrained to identify biological warfare agents. This is the job of medical personnel. In fact, doctors and nurses may be the first individuals who clearly identify a bioterrorism act by associating identical symptoms and ailments of multiple sick people with the outbreak of a pathogen.

The identification of a disease-causing bacterium or its antibody can be confirmed only by conducting clinical tests on samples of blood, skin lesions, or respiratory tissue sections taken from exposed individuals. Similarly, the presence of a toxin can be detected by conducting appropriate clinical tests on samples of material removed from infected individuals. Once the bacterium or toxin has been specifically identified, an appropriate antibiotic or vaccine can be administered to curb the impact of the pathogen and protect the infected victim against death.

When a bioterrorism act has been committed, emergency-response personnel need to cordon off the areas potentially contaminated with a biological warfare agent from the general public. While working in these areas personnel need to wear fully encapsulating body suits and use self-contained breathing apparatus. They may also make individuals in the communities surrounding the contaminated areas aware not only of the importance of seeking medical treatment but that failure to do so could well be fatal and adversely affect those with whom they live and work.

Areas contaminated with a pathogen may be decontaminated by knowledgeable professionals, typically through use of a microbiocide such as chlorine dioxide (Section 11.8). For individual protection, these workers also need to wear fully encapsulating body suits and use self-contained breathing apparatus.

10.20-D TRANSPORTING INFECTIOUS SUBSTANCES

Although the general public fears exposure to biological warfare agents, microbiologists often study microorganisms as a routine component of their jobs. Consequently, it is necessary to transport bacteria and viruses for legitimate purposes from their point of production to clinical laboratories and elsewhere. DOT regulates the transportation of infectious substances as division 6.2 hazardous materials. Certain special rules also apply that we note here.

When shippers and carriers intend to transport an infectious substance, they must first become aware of the applicable risk group. A **risk group** is a ranking of a microorganism's ability to cause injury through disease as defined by criteria developed by the World Health Organization (WHO). These criteria are based on the severity of the disease caused by the microorganism, the mode and relative ease of transmission, the degree of risk to both an individual and a community, and the reversibility of the disease through the availability of known and effective preventative agents and treatment. The risk groups are identified by the numbers 1 through 4, and the criteria for each rank are provided in Table 10.17.

DOT regulates the transportation of hazardous materials in division 6.2 having the proper shipping names, "infectious substances, affecting humans," "infectious substances, affecting animals," and "diagnostic specimen." A **diagnostic specimen** is any human or animal material,

risk group
■ For purposes of DOT regulations, a ranking of a microorganism's ability to cause injury through development of a disease as defined by criteria developed by the World Health Organization (WHO) based on the severity of the disease caused by the organism, the mode and relative ease of transmission, the degree of risk to both an individual and a community, and the reversibility of the disease through the availability of known and effective preventive agents and treatment

diagnostic specimen
■ For purposes of DOT regulations, a human or animal material, including excreta, blood and its components, tissue, and tissue fluids being transported for diagnostic or investigational purpose (but excluding live infected humans or animals), when the source human or animal has or may have a serious human or animal disease from a Risk Group 4 pathogen

TABLE 10.17 | Risk Groups for Pathogens

RISK GROUP	PATHOGEN	RISK TO INDIVIDUALS	RISK TO COMMUNITY
4	A pathogen that usually causes serious human or animal disease and that can be readily transmitted from one individual to another, directly or indirectly, and for which effective treatments and preventive measures are not usually available	High	High
3	A pathogen that usually causes serious human or animal disease but does not ordinarily spread from one infected individual to another, and for which effective treatments and preventive measures are available	High	Low
2	A pathogen that can cause human or animal disease but is unlikely to be a serious hazard, and although capable of causing serious infection on exposure, for which there are effective treatments and preventive measures available and the risk of spread of infection is limited	Moderate	Low
1	A microorganism that is unlikely to cause human or animal disease[a]	None or very low	Low

[a]Not subject to DOT regulations.

including excreta, blood and its components, tissue, and tissue fluids being transported for diagnostic or investigational purposes (but excluding live infected humans or animals), when the source human or animal has or may have a serious human or animal disease from a Risk Group 4 pathogen.

When shippers prepare the applicable shipping paper, they are obligated to include the technical name of the infectious substance in its shipping description. For example, when shippers transport 1 g of *Bacillus anthracis*, they enter the following shipping description on the accompanying shipping paper:

Units	HM	Shipping Description (Identification Number, Proper Shipping Name, Primary Hazard Class or Division, Subsidiary Hazard Class or Division, and Packing Group)	Weight (g)
1 glass tube	X	UN2814, Infectious substances, affecting humans (*Bacillus anthracis*), 6.2	1

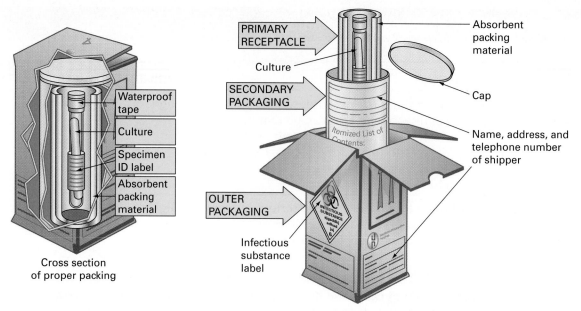

FIGURE 10.18 When shippers offer an infectious substance for transportation, DOT requires them to use triple packaging comprising a primary receptacle, water-tight secondary packaging, and rigid outer packaging.

Although an infectious substance is not assigned a packing group, it is packaged to meet rigorous DOT performance tests. Figure 10.18 illustrates the triple packaging that is generally encountered when an infectious substance is transported. It consists of a primary receptacle, water-tight secondary packaging, and rigid outer packaging.

DOT requires shippers and carriers to mark the outer packaging with the following information:

- Shipper or carrier's full name, address, and telephone number
- Proper shipping name, including the name of the infectious substance
- UN number
- Precautionary statement
- 24-hr emergency-response telephone number
- The net quantity of infectious substance

In addition, DOT requires shippers to display the BIOHAZARD marking shown in Figure 10.19 on two opposing sides of the package.

DOT also requires INFECTIOUS SUBSTANCE labels to be affixed to packaging containing an infectious substance. The INFECTIOUS SUBSTANCE label was previously illustrated in Figure 6.5. DOT also requires shippers to affix the applicable NON-FLAMMABLE GAS or CLASS 9 labels to the packaging *in addition to* the INFECTIOUS SUBSTANCE labels when the infectious substance is preserved in dry ice or nitrogen, respectively. When 50 mL or 50 g of the infectious substance is to be transported by cargo aircraft, DOT requires them also to affix CARGO AIRCRAFT ONLY labels to the packaging.

Carriers are obligated to notify the consignee and either the National Response Center (Section 6.10) or the Centers for Disease Control and Prevention at (800) 232-0124 of an incident involving the release of an infectious substance. This telephone number is printed on the INFECTIOUS SUBSTANCE label.

DOT does not require carriers to placard the transport vehicle used to ship an infectious substance.

FIGURE 10.19 At 49 C.F.R. §172.323, DOT requires shippers to display this BIOHAZARD marking on two opposing sides or two ends other than the bottom of packaging if it has a capacity of less than 1000 gal (3785 L) *or* on each side and each end if the packaging has a capacity equal to or greater than 1000 gal (3785 L).

10.20-E ANTHRAX

Special concern about potential exposure to *anthrax* was first raised in the United States in 2001, following bioterrorist acts during which anthrax-laden mailings were discovered in postal areas and throughout the Hart Senate Office Building in Washington, DC. Seven anthrax-laden letters were mailed to U.S. senators and media outlets. The incidents constituted the first major acts of bioterrorism in our nation. They caused the deaths of five people and sickened 17 others. In 2008, an Army scientist was identified as the individual most likely responsible for these mailings. He committed suicide as the FBI was about to arrest him.

The anthrax pathogen is not new to bioterrorism. During World War II, both German and Allied scientists developed varieties of the anthrax bacillus, although neither side ever actually employed them as biological weapons.

The bacterium *Bacillus anthracis* causes anthrax. The spores of the bacterium can be absorbed by the skin ingested, and inhaled, giving rise to three forms of anthrax infection: skin anthrax, intestinal anthrax, and inhalation anthrax. The most potentially lethal form of the disease is *inhalation anthrax*.

The initial symptoms of exposure to the anthrax bacillus are similar to those experienced by an individual having a common cold: high fevers, fatigue, coughing, hard breathing, and chest discomfort. As time progresses, the victims of anthrax poisoning develop severe respiratory distress, hemorrhage of the lungs, heavy convulsions, and perspiration, followed by the onset of shock. Anthrax victims may be treated by antimicrobial therapy. Untreated victims usually die within 24 to 36 hours after the first onset of symptoms.

10.20-F RICIN

Ricin is a toxin produced by the common castor plant, *Ricinus communis*, the beans of which also serve as a source of castor oil, the well-known laxative. Ricin is produced from the waste generated when castor beans are processed.

The chemical analysis of ricin reveals that it is a solid polymer (Chapter 13). In warfare jargon, it is known as *Agent W*. Its LD_{50} is estimated to be approximately 3 mg/kg. Ricin can be inhaled as a mist or powder, swallowed as a pellet or dissolved in a liquid, or injected into muscle. Its use is banned by chemical and biological weapons treaties.

The following incidents involving the use of ricin toxin as a biological warfare agent have been documented:

■ In 1978, Georgi Markov, a Bulgarian exiled journalist living in London, was jabbed in the thigh with an umbrella by an individual suspected of being a KGB agent. Two days later, he died from cardiac arrest. An autopsy revealed the presence of a small metal ball with four holes in his thigh.

- In Paris, a similar platinum–iridium ball was removed from the back of a friend of the journalist. The wax used to seal the holes in the ball had not yet melted. Chemical analysis of the residue in the ball revealed the presence of ricin. The friend recalled being jabbed with an umbrella on the Paris Metro.
- In 1980, a Soviet citizen visiting Tyson Corners, Virginia, was also jabbed. He died en route to the local hospital. A similar platinum–iridium ball was removed from the deceased.
- In 2003, London antiterrorism police and security service agents arrested seven terrorists who were plotting to use ricin to kill and terrify the local population. The details of their methods were never made public.

When ricin is assimilated into the body, its victims experience fever, cough, shortness of breath, chest tightness, and low blood pressure within 8 hours. Death can occur within 36 to 72 hours after the initial exposure. Ricin acts as a hemotoxin. There is no specific treatment for ricin exposure.

Basic Elements of Toxicology

10.1 When emergency-response personnel are obliged to work in an atmosphere known to contain an elevated concentration of a toxic gas, why is it essential that they don protective clothing that has been designed to prevent contact of the gas with their skin?

Measuring Toxicity

10.2 One constituent of hardwood smoke is *acrolein*, exposure to which can cause the eyes to tear. Acrolein is also a commercial chemical product. It is a toxic gas whose inhalation LD_{50} is 0.4 mg/kg. Using this measurement, determine whether the inhalation of 0.05 mg of acrolein by a 180-lb unprotected firefighter is likely to be fatal.

10.3 Governmental hygienists report the IDLH value for sulfur dioxide as 100 ppm.

  (a) Is it safe for an emergency-response team without respiratory protection to enter a room in which the concentration of sulfur dioxide is 750 ppm?

  (b) For what approximate period of time can an individual using respiratory protection remain in an environment containing sulfur dioxide at a concentration of 100 ppm without experiencing any escape-impairing symptoms or irreversible health effects?

Toxicity of the Fire Scene

10.4 Why are active on-duty firefighters likely to experience the adverse health effects from inhaling 300 ppm of carbon monoxide faster than nonactive off-duty firefighters who inhale the same concentration of the same substance?

Carbon Monoxide

10.5 In the United States, the number of carbon monoxide–poisoning incidents increases during cold weather. What factor is most likely associated with this rise in carbon monoxide–poisoning incidents?

10.6 Using the following partial sketch of a single-floored home, identify the locations at which carbon monoxide detectors should be installed to provide maximum protection for its occupants.

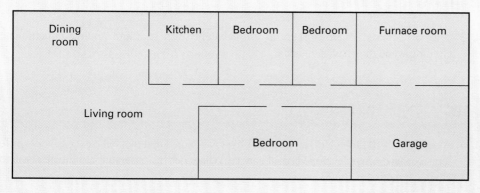

10.7 Why should carbon monoxide detectors be positioned at least 15 ft (4.6 m) from fuel-burning appliances?

Hydrogen Cyanide

10.8 Although hydrogen cyanide is detected at many residential fire scenes, it generally does not survive long within the immediate environment. Why is its survival appear to be limited?

Sulfur Dioxide

10.9 Why are vulcanologists required to use self-contained breathing apparatus when studying the characteristics of an active volcano?

Hydrogen Sulfide

10.10 Is a danger of fire and explosion present in a segment of a sanitary sewer in which the hydrogen sulfide concentration measures 2.5% by volume in the surrounding air?

Nitrogen Oxides

10.11 Why do safety experts recommend separating nitric oxide from flammable gases during storage by a minimum distance of 20 ft (6.1 m) or by a barrier of noncombustible material at least 5 ft (1.5 m) high, having a fire-resistance rating of at least ½ hour.

Ammonia

10.12 Although ammonia has a vapor density of 0.6 (air = 1), why does released ammonia sometimes linger near the surface of the ground rather than rising into the air and dispersing?

Metallic Cyanides

10.13 Why do experts advise workers to immediately wash skin and remove contaminated clothing when they have contacted metallic cyanides?

Carcinogenesis

10.14 As early as 1761, John Hill, an English physician, noted that men who excessively sniffed tobacco snuff often developed nasal cancer. What does his observation indicate about the chemical nature of snuff?

Emergency-Response Actions Involving Toxic Substances

10.15 On responding to a 2:00 A.M. domestic transportation mishap, a firefighting team determines that a gas is leaking from a 5-lb (2.3-kg) steel cylinder. The chemical commodity is described on the accompanying shipping paper as follows:

Units	HM	**Shipping Description** (Identification Number, Proper Shipping Name, Primary Hazard Class or Division, Subsidiary Hazard Class or Division, and Packing Group)	Weight (lb)
1 gas cylinder	X	UN1953, Compressed gas, toxic, flammable, n.o.s. (carbon monoxide/butane mixture), 2.3, (2.1) (Poison - Inhalation Hazard, Zone D)	5

Use Table 10.15 to identify the initial isolation and protective-action distances recommended by DOT when responding to this mishap.

10.16 Use Table 10.15 to identify the initial isolation and protective-action distances recommended by DOT when responding to each of the following domestic transportation mishaps:

(a) A rail tankcar leaking anhydrous ammonia at 9:30 A.M.

(b) A truck that overturned at 7:45 P.M. and from which a drum containing aqueous (25%) hydrocyanic acid has burst

Toxic Metallic Compounds

10.17 In 1992, Congress passed the Residential Lead-Based Paint Hazard Reduction Act, which requires housing owners to notify their tenants of the presence or suspicion of the presence of lead-based paint. What is the most likely reason Congress took this action?

Asbestos

10.18 The DOT regulation at 49 C.F.R. §173.216 requires that asbestos be transported in "rigid, leak-tight packaging" such as dust- and sift-proof bags. What is the most logical reason DOT requires the use of this special packaging?

Pesticides

10.19 While combating a fire involving the pesticide Aldicarb, a 150-lb firefighter inadvertently swallowed 0.5 g of pesticide dust, despite the use of protective gear. The oral LD_{50} (rat) for Aldicarb is 2.5 mg/kg. The firefighter did not survive.

(a) What is the most likely reason this fatality occurred?

(b) Describe the first-aid action that if rapidly implemented, could have saved the firefighter's life?

Biological Warfare Agents

10.20 When emergency responders or civilians suspect they may have been exposed to a biological warfare agent, why is it essential they seek medical attention within hours of the suspected exposure?

Transporting Toxic Substances

10.21 Chlorine trifluoride (ClF_3) is contained within two small DOT-3E1800 steel cylinders as a greenish yellow, liquefied compressed gas. It is a highly reactive chemical substance, even attacking water. Although it is nonflammable, chlorine trifluoride has the properties of a corrosive and toxic oxidizer. Even on short exposure, inhalation of the gas causes serious temporary or residual injury to humans. The OSHA permissible exposure limit for chlorine trifluoride as a time-weighted average is only 0.1 ppm.

(a) Which integer or symbol should be inserted in each of the four quadrants of the NFPA hazard diamond displayed outside buildings where cylinders of chlorine trifluoride are stored?

(b) When chlorine trifluoride is transported within steel cylinders by in a rail boxcar, which DOT warning labels should be affixed to the exterior surface of the cylinders?

(c) Which DOT placards should be affixed to a railcar in which the steel cylinders of chlorine trifluoride are transported to their use destination?

(d) Identify the markings that DOT requires on the exterior surface of cylinders used to hold chlorine trifluoride.

(e) When chlorine trifluoride is offered for transportation by rail, what shipping description does DOT require the shipper to enter on the accompanying waybill?

(f) Determine the initial isolation distance and protective-action distance recommended by DOT for use by emergency-response crews when they arrive at the scene of transportation mishaps involving a slow leak of chlorine trifluoride from a 5-lb (2.3-kg) steel cylinder during daytime hours.

Certain chemical reactions are of great benefit to the world's civilized societies. They include the combustion of fuels, the chlorination of water, the explosion of dynamite, the bleaching of fabrics, and the flaring of recreational fireworks. These specific phenomena are associated with oxidation–reduction reactions, also called *redox reactions*. These reactions and the hazardous materials associated with them constitute the subject matter of this chapter.

When oxidation–reduction reactions are conducted in a controlled fashion, the energy they release can be harnessed to advantage. For example, the reactions that occur in devices such as batteries, dry cells, and fuel cells provide a supply of portable electrical energy. However, when redox reactions occur in an uncontrolled fashion, the generated energy is released into the immediate environment, where it can initiate fire and explosion and cause the loss of life and property.

This potential for destructiveness necessitates the examination of redox reactions as a component of the study of hazardous materials.

PERFORMANCE GOALS FOR SECTION 11.1:

- Describe the chemical nature of an oxidizer, especially noting that the presence of oxygen in the chemical composition is unnecessary.
- Describe how DOT defines an oxidizer.

11.1 What Is an Oxidizer?

It was noted in Chapter 7 that hydrogen burns in oxygen, fluorine, and chlorine to produce water, hydrogen fluoride, and hydrogen chloride, respectively. These combustion reactions are represented by the following equations:

$$2H_2(g) + O_2(g) \longrightarrow 2H_2O(g)$$

Hydrogen   Oxygen   Water

$$H_2(g) + F_2(g) \longrightarrow 2HF(g)$$

Hydrogen   Fluorine   Hydrogen fluoride

$$H_2(g) + Cl_2(g) \longrightarrow 2HCl(g)$$

Hydrogen   Chlorine   Hydrogen chloride

oxidizer
- For purposes of DOT regulations, a substance that may enhance or support the combustion of other materials, generally by yielding its oxygen

The substances supporting these independent combustion processes are examples of **oxidizers**, **oxidants**, or **oxidizing agents**, whereas hydrogen is an example of a reducing agent. Oxidizing agents always react with reducing agents in concert.

DOT defines an oxidizer as a material that may enhance or support the combustion of other materials, generally by yielding oxygen. Elemental oxygen is an example of an oxidizer, but oxygen is not an essential component of an oxidizer. Furthermore, the occurrence of an oxidation–reduction reaction does not necessarily require the participation of elemental oxygen or compounds containing oxygen. The fact that hydrogen burns in fluorine and chlorine atmospheres demonstrates that oxygen is not the sole substance that enhances or supports the combustion of other materials. Like oxygen, fluorine and chlorine are oxidizers.

oxidizing agent (oxidant)
- An oxygen-rich substance capable of readily yielding some of its oxygen; the substance reduced during an oxidation–reduction reaction

PERFORMANCE GOALS FOR SECTION 11.2:

- Describe the concept of the oxidation number.
- For a given substance, determine the oxidation numbers of its atoms and ions.

oxidation number
- A number assigned to an atom or ion by following certain established rules to reflect its capacity for combining with other atoms or ions

11.2 Oxidation Numbers

Chemists use the concept of an **oxidation number** as a means of describing the combining capability of one ion for another ion or of one atom for another atom. In practice, an oxidation number is formally assigned to the atoms that make up a substance by using the following rules:

- The oxidation number of each atom in an element is zero; thus, the oxidation number of each atom in the elements H_2, O_2, Na, and Mg is 0.
- The algebraic sum of the oxidation numbers of the atoms in any substance is zero.
- The hydrogen atom in hydrogen-containing compounds other than metallic hydrides has an oxidation number of $+1$. The hydrogen in a hydride ion (H^-) has an oxidation number of -1.
- The oxygen atom in oxygen-containing compounds other than peroxides and superoxides has an oxidation number of -2. Each oxygen atom in a peroxide ion (O_2^{2-}) and a superoxide ion (O_2^-) has an oxidation number of -1 and -0.5, respectively.
- The oxidation number of a monatomic ion (i.e., having one atom) is the same as its net ionic charge; thus, the oxidation number of sodium is $+1$; of magnesium, $+2$; of the chloride ion, -1; and of the sulfide ion, -2.
- The algebraic sum of the oxidation numbers of the atoms in a polyatomic ion is equal to its ionic charge.

The chemical formula of a substance must be known to use these rules and determine the oxidation numbers of its component atoms. As an example, let's use them to determine the oxidation number of each atom in sodium chlorate, an oxidizing agent used in fireworks (Section 11.9) and signaling flares (Section 11.10).

The chemical formula of sodium chlorate is $NaClO_3$. This formula indicates that the compound is composed of two ions, the sodium ion (Na^+) and the chlorate ion (ClO_3^-). Using the rules, we establish the oxidation number of each element in sodium chlorate as follows:

CONSTITUENT	OXIDATION NUMBER	RELEVANT RULE
Sodium ion, Na^+	$+1$	The oxidation number of a monatomic ion is the same as its ionic charge.
Oxygen atom	-2	Other than in peroxides, the oxidation number of oxygen is -2.
Chlorate ion, ClO_3^-	-1	The algebraic sum of the oxidation numbers of the atoms in a polyatomic ion is equal to its ionic charge; that is, $-1 = +5 + [3 \times (-2)] = +5 + (-6)$.
Chlorine atom	$+5$	The algebraic sum of the oxidation numbers in any substance is zero; that is, $0 = +1 + 5 + [3 \times (-2)] = +6 + (-6)$.

Additional examples of determining the oxidation numbers of the atoms in some compounds are noted in Table 11.1.

TABLE 11.1 | Examples of Oxidation Numbers

SYMBOL OF ELEMENT	OXIDATION NUMBER	CHEMICAL FORMULA OF A REPRESENTATIVE COMPOUND
F	-1	HF
O	-2	SO_2
	-1	H_2O_2
N	-3	NH_3
	-2	N_2H_4
	-1	NH_2OH
	$+1$	N_2O
	$+2$	NO
	$+3$	HNO_2
	$+4$	NO_2
	$+5$	HNO_3
Cl	-1	HCl
	$+1$	HClO
	$+3$	$KClO_2$
	$+5$	$KClO_3$
	$+7$	$HClO_4$
S	-2	H_2S
	$+4$	H_2SO_3
	$+6$	H_2SO_4
P	-3	AlP
	$+3$	H_3PO_3
	$+5$	H_3PO_4

PERFORMANCE GOALS FOR SECTION 11.3:

■ For a given equation, use the change in oxidation number for an atom or ion to identify the oxidizing agent, reducing agent, the substance oxidized, and the substance reduced.

11.3 Oxidation–Reduction Reactions

Because a redox reaction occurs between an oxidizing agent and a reducing agent, the equation illustrating a redox reaction is always written in the following generalized form:

$$\text{Oxidizing agent} + \text{Reducing agent} \longrightarrow \text{Products}$$

A specific example of this is the reaction between iron(III) chloride and tin(II) chloride. We write the equation for this reaction as follows:

$$2FeCl_3(aq) \quad + \quad SnCl_2(aq) \quad \longrightarrow \quad 2FeCl_2(aq) \quad + \quad SnCl_4(aq)$$

Iron(III) chloride   Tin(II) chloride   Iron(II) chloride   Tin(IV) chloride

Because the chloride ions do not participate in this redox reaction, they are often called **spectator ions**. We can eliminate them and write an **ionic equation** for the reaction as follows:

$$2Fe^{3+}(aq) + Sn^{2+}(aq) \longrightarrow 2Fe^{2+}(aq) + Sn^{4+}(aq)$$

spectator ion
■ One or more of the ions that do not participate in an oxidation–reduction reaction

ionic equation
■ An equation that depicts only the ions that participate in an oxidation–reduction reaction

In an ionic equation, only the symbols of the ions contributing to the reaction are written. In this representation, electrons have simply been transferred between the iron ions and tin ions. The information conveyed by this equation is summarized as follows:

- The iron(III) ions become iron(II) ions. The oxidation number of iron decreases from +3 to +2; thus, the iron(III) ions are *reduced*. Iron(III) chloride is called the *oxidizing agent*, because it oxidizes tin(II) chloride.
- The tin(II) ions become tin(IV) ions. The oxidation number of tin increases from +2 to +4; thus, the tin(II) ions are *oxidized*. Tin(II) chloride is called the *reducing agent*, since it reduces iron(III) chloride.

Because a simultaneous increase and decrease in oxidation numbers always accompanies oxidation and reduction, respectively, we can summarize the nature of this process in ionic systems as follows:

- *Oxidation* is the phenomenon associated with an *increase* in oxidation number and *loss* of electrons from an ion, atom, or group of atoms.
- *Reduction* is the phenomenon associated with a *decrease* in oxidation number and *gain* of electrons from an ion, atom, or group of atoms.
- The substance that *accepts* electrons during a redox reaction is called the *oxidizing agent*.
- The substance that *loses* electrons during a redox reaction is called the *reducing agent*.

Oxidation and reduction phenomena can also be represented separately by equations like the following:

$$Sn^{2+}(aq) \longrightarrow Sn^{4+}(aq) + 2e^-$$
$$Fe^{3+}(aq) + e^- \longrightarrow Fe^{2+}(aq)$$

The phenomenon illustrated by each equation is called a **half-reaction**. One half-reaction represents oxidation, whereas the other represents reduction. The processes represented by half-reactions always occur simultaneously.

half-reaction
■ An equation that separately depicts either oxidation or reduction

Several other examples of oxidation–reduction phenomena are provided in Table 11.2.

TABLE 11.2	Some Common Oxidizing Agents			
OXIDIZING AGENT	ELEMENT THAT CHANGES OXIDATION NUMBER	OXIDATION NUMBER		EQUATION ILLUSTRATING HALF-REACTION
		IN REACTANT	IN PRODUCT	
Sodium peroxide	Oxygen	−1	−2	$Na_2O_2(aq) + 2H_2O(l) + 2e^-$ $\longrightarrow 2Na^+(aq) + 4OH^-(aq)$
Metallic hypochlorites	Chlorine	+1	−1	$ClO^-(aq) + 2H^+(aq) + 2e^-$ $\longrightarrow Cl^-(aq) + H_2O(l)$
Metallic chlorates	Chlorine	+5	−1	$ClO_3^-(aq) + 6H^+(aq) + 6e^-$ $\longrightarrow Cl^-(aq) + 3H_2O(l)$
Nitric acid (concentrated)	Nitrogen	+5	+4	$NO_3^-(aq) + 2H^+(aq) + e^-$ $\longrightarrow NO_2(g) + H_2O(l)$
Nitric acid (dilute)	Nitrogen	+5	+2	$NO_3^-(aq) + 4H^+(aq) + 3e^-$ $\longrightarrow NO(g) + 2H_2O(l)$
Metallic peroxydisulfates	Sulfur	+7	+6	$S_2O_8^{2-}(aq) + 2e^-$ $\longrightarrow 2SO_4^{2-}(aq)$

Identify the element oxidized, the element reduced, the oxidizing agent, and the reducing agent in the redox reaction denoted by the following equation:

$$6FeSO_4(aq) \quad + \quad Na_2Cr_2O_7(aq) \quad + \quad 7H_2SO_4(aq) \quad \longrightarrow$$

Iron(II) sulfate   Sodium dichromate   Sulfuric acid

$$3Fe_2(SO_4)_3(aq) \quad + \quad Na_2SO_4(aq) \quad + \quad Cr_2(SO_4)_3(aq) \quad + \quad 7H_2O(l)$$

Iron(III) sulfate   Sodium sulfate   Chromium(III) sulfate   Water

Sultion: In this equation, the sodium and sulfate ions are identified as spectator ions, because the oxidation numbers of sodium, sulfur, and oxygen are the same on each side of the arrow. When the spectator ions are eliminated, the following ionic equation is written:

$$6Fe^{2+}(aq) + Cr_2O_7^{2-}(aq) + 14H^+(aq) \longrightarrow 6Fe^{3+}(aq) + 2Cr^{3+}(aq) + 7H_2O(l)$$

To determine the oxidation number of chromium in the dichromate ion, recall that the algebraic sum of the oxidation numbers of the atoms in a polyatomic ion is equal to its ionic charge. We thus determine that the oxidation number of chromium is +6.

$$-2 = [2 \times (+6)] + [7 \times (-2)]$$
$$-2 = -2$$

The oxidation number of the monatomic ions is the same as their ionic charge.

In the ionic equation, we readily observe that the iron(II) ions become iron(III) ions. The oxidation number of iron increases from +2 to +3, as each iron(II) ion loses an electron. Because an *increase* in oxidation number is associated with oxidation, the iron(II) ions are *oxidized*. Iron(II) sulfate is the *reducing agent*.

The ionic equation also notes that dichromate ions become chromium(III) ions. The oxidation number of chromium decreases from +6 in $Cr_2O_7^{2-}$ to +3 in Cr^{3+}, as each chromium ion gains three electrons. Because a *decrease* in oxidation number is associated with reduction, the dichromate ions are *reduced*. Thus, sodium dichromate is the *oxidizing agent*.

PERFORMANCE GOALS FOR SECTION 11.4:

- Describe the hazards associated with oxidizers listed in each NFPA class in Table 11.4.
- Describe the ability of oxidizers containing oxygen to produce oxygen when they thermally decompose.

11.4 Common Features of Oxidizers

Oxidizers are generally perceived as relatively powerful chemical substances, since they can react rapidly, even at explosive rates, with other substances. The latter substances include fuels, lubricants, greases, oils, cotton, animal and vegetable fats, paper, coal, coke, straw, sawdust, and wood shavings. Because they are powerfully reactive substances, oxidizers have been chosen by terrorists to cause destruction. In 1995, the ignition of an ammonium nitrate/fuel oil mixture destroyed the Murrah Federal Building in Oklahoma City, shown in Figure 11.1. The incident killed 168 people and injured 850.

Different oxidizers have different strengths. This variation in the strength of individual oxidizers can be approximated by examining the listing in Table 11.3. In this series, oxidizing

FIGURE 11.1 The criminal act of detonating a weapon of mass destruction at the Oklahoma City federal building caused the deaths of 168 workers and other individuals and injured 850 people. The weapon was a 4000-lb (1800-kg) mixture of ammonium nitrate fertilizer and fuel oil (ANFO). The detonation occurred at the rate of approximately 13,000 ft/s (4000 m/s). (*Courtesy of the Oklahoma City National Memorial & Museum, Oklahoma City, Oklahoma.*)

TABLE 11.3	Relative Strength of Oxidizing Agents[a]
	Fluorine
	Ozone
	Hydrogen peroxide
	Hypochlorous acid
	Metallic chlorates[b]
	Lead(II) dioxide
	Metallic permanganates[b]
	Metallic dichromates[b]
	Nitric acid (concentrated)
	Chlorine
	Sulfuric acid (concentrated)
	Oxygen
	Metallic iodates
	Bromine
	Iron(III) (Fe^{3+}) compounds
	Iodine
	Sulfur
	Tin(IV) (Sn^{4+}) compounds

[a]Listed in descending order of oxidizing power.
[b]In an acidic environment.

agents are arranged according to their decreasing oxidizing power. Any substance whose name appears on this list is a stronger oxidizing agent than the substances whose names are listed below it. It is noteworthy that many substances are stronger oxidizing agents than oxygen itself.

NFPA distinguishes the degree of hazard posed by specific oxidizers by assigning them to **classes of oxidizers** as follows:

NFPA classes of oxidizers
■ Any of four groups (class 1, class 2, class 3, and class 4) into which individual oxidizers are assigned based on their ability to affect the burning rate of combustible materials or to undergo self-sustained decomposition

NFPA class 1 oxidizers	These are substances that *slightly* increase the burning rate of combustible materials with which they come in contact. They do not cause the materials to undergo spontaneous combustion.
NFPA class 2 oxidizers	These are substances that *moderately* increase the burning rate of combustible materials. They may cause the spontaneous ignition of the combustible materials with which they come in contact.
NFPA class 3 oxidizers	These are substances that *severely* increase the burning rate of combustible materials with which they come in contact; or they may undergo violent, self-sustained decomposition owing to contamination or exposure to heat.
NFPA class 4 oxidizers	These are substances that enhance the burning rate of combustible materials with which they come in contact *at an explosive rate* and may cause them to undergo spontaneous combustion. They may also decompose explosively on exposure to thermal or physical shock.

This NFPA classification is noted in Table 11.4 for some specific oxidizers.

SOLVED EXERCISE 11.2

Magnesium perchlorate and sodium chlorate are inserted into separate containers and mixed with an arbitrary but equal volume of turpentine. Which oxidizer poses the greater risk of increasing the rate at which the turpentine will spontaneously ignite?

Solution: In Table 11.4, magnesium perchlorate and sodium chlorate are identified as NFPA class 1 and class 3 oxidizers, respectively. Based on this information, sodium chlorate poses the greater risk of increasing the rate at which turpentine will spontaneously ignite.

Oxidizers with oxygen atoms in their ionic structures do not burn in the traditional sense, but they may support combustion by providing a supply of oxygen to the materials burning at a fire scene. For example, potassium chlorate decomposes to produce oxygen when it is exposed to intense heat.

$$2KClO_3(s) \longrightarrow 2KCl(s) + 3O_2(g)$$

Potassium chlorate   Potassium chloride   Oxygen

The oxygen produced by this reaction supports the combustion of nearby materials at least as well as atmospheric oxygen. When the oxidizer is confined while it undergoes decomposition, the oxygen may generate excessive internal pressure within its container and cause the container to rupture.

TABLE 11.4	Some Typical Oxidizers by NFPA Classification

CLASS 1

Ammonium nitrate	Potassium nitrate
Ammonium persulfate	Potassium persulfate
Barium chlorate	Silver nitrate
Barium nitrate	Sodium carbonate peroxide
Calcium chlorate	Sodium dichloro-*sym*-triazine-2,4,6-trione
Calcium nitrate	dihydrate
Calcium peroxide	Sodium dichromate
Cupric nitrate	Sodium nitrate
Hydrogen peroxide solutions, over 8%	Sodium nitrite
but not exceeding 27.5% by mass	Sodium perborate
Lead nitrate	Sodium perborate tetrahydrate
Lithium hypochlorite	Sodium perchlorate monohydrate
Lithium peroxide	Sodium persulfate
Magnesium nitrate	Strontium chlorate
Magnesium perchlorate	Strontium nitrate
Magnesium peroxide	Strontium peroxide
Nickel nitrate	Thorium nitrate
Nitric acid, 70% or less	Uranium nitrate
Perchloric acid, less than 60% by mass	Zinc chlorate
Potassium dichromate	Zine peroxide

CLASS 2

Calcium hypochlorite, 50% or less by mass	Sodium chlorite, 40% or less
Chromium trioxide (chromic acid)	Sodium peroxide
Halane (1,3-dichloro-5,5-dimethyl hydantoin)	Sodium permanganate
Hydrogen peroxide, 27.5% to 52% by mass	Trichloro-*sym*-triazine-2,4,6-trione
Nitric acid, more than 70%	(trichloroisocyanuric acid)
Potassium permanganate	

CLASS 3

Ammonium dichromate	Potassium bromate
Calcium hypochlorite, over 50% by mass	Potassium chlorate
Hydrogen peroxide, 52% to not more than	Potassium dichloro-*sym*-triazine-2,4,6-trione
91% by mass	(potassium dichloroisocyanurate)
Mono-(trichloro)tetra-(monopotassium	Sodium chlorate
dichloro)-penta-*sym*-triazine-2,4,6-trione	Sodium chlorite, over 40% by mass
Perchloric acid solutions, 60% to 72.5%	Sodium dichloro-*sym*-triazine-2,4,6-trione
by mass	(sodium dichloroisocyanurate)

CLASS 4

Ammonium perchlorate	Perchloric acid solutions, more than 72.5%
Ammonium permanganate	by mass
Guanidine nitrate	Potassium superoxide
Hydrogen peroxide solutions, more than 91%	
by mass	

- Describe the hazardous properties of the aqueous solutions of hydrogen peroxide, noting that their degree of hazard increases with the H_2O_2 concentration.
- Identify the industries that use bulk volumes of hydrogen peroxide.
- Describe the means shippers and carriers use to inform emergency responders of the hazards associated with encountering hydrogen peroxide during a transportation mishap.
- Describe the response actions to be executed when hydrogen peroxide has been released from its packaging into the environment.

11.5 Hydrogen Peroxide

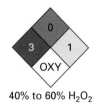

40% to 60% H_2O_2

Hydrogen peroxide is an important substance having the chemical formula and molecular structure H_2O_2 and $H-O-O-H$, respectively. The substance is encountered commercially in aqueous solutions having the following approximate compositions and uses:

1% to 3% H_2O_2	This solution is a topical antiseptic used on minor cuts and wounds. It is commonly available in drug stores.
6% H_2O_2	This solution is used for bleaching hair. In the process, hydrogen peroxide oxidizes the dark-colored pigment called melanin to colorless products.
30% to 50% H_2O_2	This solution is used in the chemical industry for the synthesis of peroxo-organic compounds, which are discussed in Section 13.9. The 30% solution is also used by manufacturing and process industries for bleaching cotton, flour, wool, straw, leather, gelatin, and paper, and as a component of a cleaning solution (Section 11.16). To a limited extent, the solution appears to be replacing the use of chlorine for treating drinking water. Contemporaneously, the solution has also achieved a certain notoriety in connection with its misuse by terrorists for the production of the explosive triacetone triperoxide (Section 13.9-C).
70% H_2O_2	This is the most popular shipping grade. In the chemical industry, it can be used directly for oxidation–reduction reactions, but it is often diluted prior to use and stored as 35% to 50% solutions.
90% H_2O_2	In the aerospace industry, this solution has been used to oxidize fuels such as hydrazine that helped launch *Apollo 17* and other payloads into space. Russia uses the solution to launch Soyuz rockets into space. The 90% solution has also been used for U.S. military applications.

All these solutions visibly resemble water in physical appearance, although they may have slightly pungent, irritating odors. The 90% solution is a thick, oily liquid with a density about 1½ times that of water.

Although anhydrous hydrogen peroxide has been produced, it is unavailable commercially, since it decomposes violently near the boiling point of water.

A hydrogen peroxide solution at a specified concentration is a member of each NFPA class of oxidizer. As the hydrogen peroxide concentration increases in the solution, so also does its degree of reactivity.

The world's supply of hydrogen peroxide is manufactured by a number of methods. One involves the oxidation of isopropyl alcohol, during which acetone (Section 13.5-C) is coproduced.

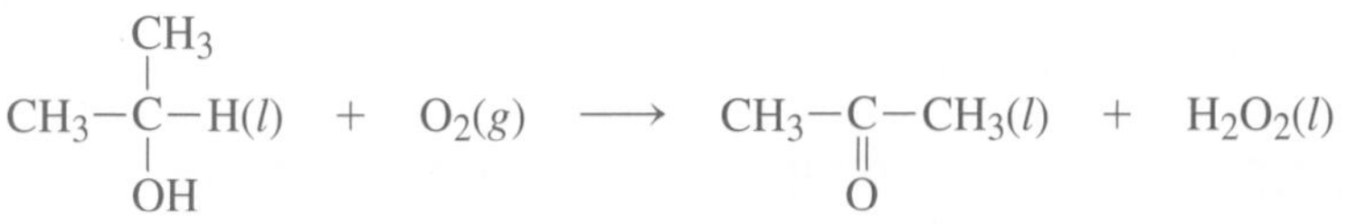

$$CH_3\!-\!\overset{\overset{\displaystyle CH_3}{|}}{\underset{\underset{\displaystyle OH}{|}}{C}}\!-\!H(l) \;+\; O_2(g) \;\longrightarrow\; CH_3\!-\!\overset{\overset{\displaystyle}{\|}}{\underset{\underset{\displaystyle O}{}}{C}}\!-\!CH_3(l) \;+\; H_2O_2(l)$$

$$\;\;\text{Isopropyl alcohol}\qquad\quad\text{Oxygen}\qquad\qquad\qquad\text{Acetone}\qquad\qquad\text{Hydrogen peroxide}$$

The hydrogen peroxide and acetone are separated by distillation. The hydrogen peroxide solution may then be diluted to produce the desired concentration for commercial sale.

Hydrogen peroxide is an inherently unstable substance. It slowly decomposes as follows:

$$2H_2O_2(aq) \;\longrightarrow\; 2H_2O(l) \;+\; O_2(g)$$
$$\text{Hydrogen peroxide}\qquad\quad\text{Water}\qquad\;\text{Oxygen}$$

The rate of the decomposition reaction is catalyzed by sunlight as well as certain metals, most notably, iron, copper, chromium, and silver. An aqueous solution of 8% hydrogen peroxide is completely decomposed following a 10-month exposure to light, whereas a similar solution kept in darkness for the same length of time remains virtually unaltered. In solutions of concentrations of less than 30%, the decomposition of hydrogen peroxide occurs so slowly when stored in dark glass bottles that it is virtually imperceptible.

Concentrated solutions of hydrogen peroxide ($>6\%$ H_2O_2) become intensely heated when they decompose. These hot solutions then vaporize. To prevent the decomposition from posing a hazard before the intended use of the oxidant, all commercial forms of hydrogen peroxide are stabilized by addition of a substance that retards its decomposition.

Concentrated hydrogen peroxide solutions have the following properties:

- Concentrated hydrogen peroxide solutions cause combustible materials to burn spontaneously at explosive rates.
- Hydrogen peroxide solutions having a strength in excess of 20% are highly corrosive. Exposure to the skin causes severe irritation. Exposure to the eyes can cause blindness.

These hazardous properties are often identified on container labels as follows (see also Figure 1.7):

DANGER

STRONG OXIDIZER
CONTACT WITH OTHER MATERIAL MAY CAUSE FIRE

CORROSIVE

CAUSES BURNS TO SKIN, EYES, AND RESPIRATORY TRACT
HARMFUL WHEN SWALLOWED OR INHALED

11.5-A THE *KURSK*

Hydrogen peroxide has been linked with a naval disaster on the Russian submarine *Kursk*. In 2000, the *Kursk* exploded and sank in the Barents Sea. Aboard the *Kursk* were a number of torpedoes, each of whose fuel systems consisted in part of highly concentrated hydrogen peroxide. Investigators proposed that the cause of the disaster was linked to a leak of hydrogen peroxide from a single torpedo. The oxidizer interacted with the torpedo's stainless steel casing, which catalyzed the decomposition of the hydrogen peroxide. The subsequent buildup of oxygen resulted in overpressurization of the torpedo and its subsequent explosion. This first explosion then initiated the detonation of other torpedoes within the storage compartment. The hull of the submarine burst, and the *Kursk* foundered and sank. There were no survivors.

11.5-B WORKPLACE REGULATIONS INVOLVING HYDROGEN PEROXIDE

When used in the workplace, OSHA requires employers to limit employee exposure to an inhalation concentration of 1 ppm, averaged over an 8-hr workday.

11.5-C TRANSPORTING HYDROGEN PEROXIDE

Nonbulk quantities of hydrogen peroxide solutions are often transported by means of motor vans on public highways or by rail in boxcars. They are usually contained in darkened glass or plastic bottles. Bulk quantities of hydrogen peroxide are often contained in plastic containers like that shown in Figure 11.2. These containers are also transported in motor vans and rail boxcars. In

FIGURE 11.2 In compliance with DOT regulations, the shipper has affixed OXIDIZER and CORROSIVE labels to a bulk plastic container holding a hydrogen peroxide solution consisting of 30% to 32% hydrogen peroxide. The shipper has also posted OXIDIZER placards that display the DOT identification number 2014 across their center areas. A CORROSIVE placard is also shown to account for the subsidiary hazard of this hydrogen peroxide solution, although DOT does not specifically require CORROSIVE placards to be displayed. (*Courtesy of Mallinckrodt Baker, Inc., Phillipsburg, New Jersey.*)

FIGURE 11.3 In the foreground, workers at FMC Corporation's hydrogen peroxide manufacturing facility prepare a cargo tank for transporting an aqueous solution containing more than 60% hydrogen peroxide to a customer. As required by DOT regulations, the identification number 2015 is displayed across the center area of OXIDIZER placards posted on each side and each end of the cargo lank. The warning label and material safety data sheet for this chemical product were previously provided in Figures 1.7 and 1.8, respectively. (*Courtesy of FMC Corporation, Hydrogen Peroxide Division, Philadelphia, Pennsylvania.*)

addition, however, hydrogen peroxide may be transported in bulk in a tank truck like that shown in Figure 11.3.

When shippers offer a hydrogen peroxide solution for transportation, DOT requires them to describe it on an accompanying shipping paper by one of the entries listed in Table 11.5. DOT also requires them to affix either an OXIDIZER label, or OXIDIZER *and* CORROSIVE labels, as relevant, on its packaging.

When carriers transport a hydrogen peroxide solution in an amount exceeding 1001 lb (454 kg), DOT requires them to post OXIDIZER placards on the bulk packaging or transport vehicle used for shipment.

TABLE 11.5	Shipping Descriptions of Aqueous Hydrogen Peroxide Solutions

CONCENTRATION OF HYDROGEN PEROXIDE SOLUTION BY VOLUME	SHIPPING DESCRIPTION[a]
8% to 20%	UN2984, Hydrogen peroxide, aqueous solution (contains _____% hydrogen peroxide), 5.1, PGIII
20% to 40%	UN2014, Hydrogen peroxide, aqueous solution (stabilized) (contains _____% hydrogen peroxide), 5.1, (8), PGII
40% to 60%	UN2014, Hydrogen peroxide, aqueous solution (stabilized) (contains _____% hydrogen peroxide), 5.1, (8), PGII
>60%	UN2015, Hydrogen peroxide, aqueous solution (stabilized) (contains _____% hydrogen peroxide), 5.1, (8), PGI or UN2015, Hydrogen peroxide, stabilized (contains _____% hydrogen peroxide), 5.1, (8), PGI

[a]When a concentration range is a component of a shipping name, the actual concentration of the hazardous material, if it is within the indicated range, is used in lieu of the range.

SOLVED EXERCISE 11.3

When a shipper offers a carrier 2500 lb (1135 kg) of an aqueous solution containing 30% to 32% hydrogen peroxide for transportation in a cargo tank by highway:

(a) What shipping description does DOT require the shipper to provide for this chemical commodity?
(b) Which DOT placards and markings does DOT require to be displayed on the cargo tank?

Solution:

(a) DOT requires the shipper to provide the following shipping description of an aqueous solution containing 30% to 32% hydrogen peroxide solution on the accompanying shipping paper:

Units	HM	Shipping Description (Identification Number, Proper Shipping Name, Primary Hazard Class or Division, Subsidiary Hazard Class or Division, and Packing Group)	Weight (lb)
1 cargo tank	X	UN2014, Hydrogen peroxide, aqueous solution (stabilized) (contains 30% to 32% hydrogen peroxide), 5.1, (8), PGII	2500

(b) Because more than 1001 lb (454 kg) of the hydrogen peroxide solution is offered for transportation, DOT requires the carrier to display OXIDIZER placards on each side and each end of the cargo tank. The carrier may also display CORROSIVE placards on each side and each end of the cargo tank to account for hydrogen peroxide's subsidiary hazard, but displaying the CORROSIVE placards is not specifically required. DOT also requires the carrier to display the identification number 2014 on an orange panel, across the center area of the OXIDIZER placards, or on white square-on-point panels on each side and each end of the tank. DOT prohibits the display of the identification number on the subsidiary CORROSIVE placards. Finally, because the solution contains more than 10% hydrogen peroxide, DOT requires the carrier to mark each side and each end of the tank HYDROGEN PEROXIDE, AQUEOUS SOLUTION *or* HYDROGEN PEROXIDE, STABILIZED (see Table 6.7).

11.5-D RESPONDING TO INCIDENTS INVOLVING A RELEASE OF HYDROGEN PEROXIDE

When emergency-response teams encounter hydrogen peroxide solutions at accident scenes, an attempt should be made not only to extinguish ongoing fires but to dam or dike the water runoff generated while combating them. When a fire has not occurred, the solutions should be segregated from combustible materials.

PERFORMANCE GOALS FOR SECTION 11.6:

- Describe the meaning of the term *available chlorine*.
- Describe the hazardous properties of sodium hypochlorite and calcium hypochlorite.
- Identify the industries that use bulk quantities of sodium hypochlorite and calcium hypochlorite.
- Describe the means shippers and carriers use to inform emergency responders of the hazards associated with encountering an inorganic hypochlorite during a transportation mishap

11.6 Metallic Hypochlorites

Metallic hypochlorites are compounds composed of a metallic ion and a hypochlorite ion (ClO^-). The most common commercial examples are sodium hypochlorite and calcium hypochlorite, each of which is encountered as the active component of several household and commercial bleaching and sanitation products. Their chemical formulas are $NaClO$ and $Ca(ClO)_2$, respectively. During laundering, bleaching agents remove undesirable stains from fabrics and other textiles. The hypochlorite-containing bleaching agents can be safely used only on cotton and linen fabrics, but not on wool, silk, and nylon. Nonchemists generally contend that their bleaching action is accomplished by chlorine, but it is actually the oxidizing power of the hypochlorite ion that accomplishes the bleaching action.

metallic hypochlorite
■ An inorganic compound composed of metallic and hypochlorite ions

The effectiveness of hypochlorite bleaching agents is commercially described by their **available chlorine**. This term is used to compare the effectiveness of bleaching agents with elemental chlorine. A bleaching agent having 99.2% available chlorine has the same bleaching power as a solution containing 99.2% chlorine by mass. Substances having 99.2% available chlorine are useful for sanitizing water, whereas substances having an available chlorine of less than approximately 35% are useful as laundry bleaches.

available chlorine
■ A measure of the effectiveness of commercial bleaching agents compared with elemental chlorine

The available chlorine in a bleaching agent is determined by utilizing the reaction between its active ingredient and an acid; for instance, the available chlorine in a bleaching agent containing calcium hypochlorite is determined by reacting the substance with hydrochloric acid as follows:

$$Ca(ClO)_2(s) \ + \ 4HCl(aq) \ \longrightarrow \ CaCl_2(aq) \ + \ 2H_2O(l) \ + \ 2Cl_2(g)$$

Calcium hypochlorite Hydrochloric acid   Calcium chloride   Water   Chlorine

The percentage of chlorine produced from the given amount of calcium hypochlorite is then determined.

Some hypochlorite bleaching agents are solids at room temperature, but they are also available commercially as aqueous solutions. When they are heated, inorganic hypochlorites decompose to produce oxygen. For instance, the thermal decomposition of calcium hypochlorite is represented as follows:

$$Ca(ClO)_2(s) \ \longrightarrow \ CaCl_2(s) \ + \ O_2(g)$$

Calcium hypochlorite   Calcium chloride   Oxygen

When the decomposition reaction occurs at a fire scene, oxygen is supplied to support combustion reactions.

11.6-A SODIUM HYPOCHLORITE

Solid *sodium hypochlorite* is an unstable compound, but its aqueous solutions are sufficiently stable when their contact with light is prevented. Even the solutions of the inorganic hypochlorites decompose when exposed to the ultraviolet radiation in sunlight.

We are most familiar with sodium hypochlorite solutions that are used as bleaches and as components of other household laundry products. Perhaps the most obvious example is the bleaching agent known as *Clorox*, an aqueous solution containing approximately 6% sodium hypochlorite.

Solutions that contain from 10% to 12.5% sodium hypochlorite are also available commercially. The solution known commercially as *Multi-chlor* is a sodium hypochlorite solution used as a disinfectant and sanitizer for the treatment of water supplies, sewage effluents, and the water in swimming pools, spas, and hot tubs.

Sodium hypochlorite is manufactured for commercial use by passing gaseous chlorine into an aqueous solution of sodium hydroxide. A mixture of sodium hypochlorite and sodium chloride is produced.

$$2NaOH(aq) \ + \ Cl_2(g) \ \longrightarrow \ NaClO(aq) \ + \ NaCl(aq) \ + \ H_2O(l)$$

Sodium hydroxide   Chlorine   Sodium hypochlorite Sodium chloride   Water

The sodium chloride formed as a coproduct does not interfere with the sodium hypochlorite when the latter is intended for use as a bleaching or sanitizing agent.

Sodium hypochlorite may also be generated for use on demand. In the system shown in Figure 11.4, for example, it is prepared by electrolyzing a portable brine solution.

$$NaCl(aq) \ + \ H_2O(l) \ \longrightarrow \ NaClO(aq) \ + \ H_2(g)$$

Sodium chloride   Water   Sodium hypochlorite Hydrogen

11.6-B CALCIUM HYPOCHLORITE

high-test hypochlorite
■ A commercial oxidizer capable of producing chlorine in acidic water (abbreviated HTH)

Calcium hypochlorite is commonly used for large-scale bleaching operations, sanitizing municipal drinking water, disinfecting domestic and municipal swimming pools, and sewage treatment. It is encountered in products known commercially by the tradename **high-test hypochlorite** (HTH).

Calcium hypochlorite is manufactured by reacting chlorine with a lime slurry. A mixture of calcium hypochlorite and calcium chloride is produced.

$$2Ca(OH)_2(aq) \ + \ 2Cl_2(g) \ \longrightarrow \ Ca(ClO)_2(aq) \ + \ CaCl_2(aq) \ + \ 2H_2O(l)$$

Calcium hydroxide   Chlorine   Calcium hypochlorite Calcium chloride   Water

Sodium chloride is then added to the solution. This causes the calcium hypochlorite to precipitate as the solid.

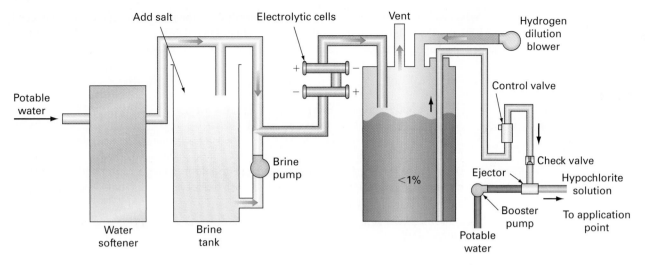

FIGURE 11.4 This schematic illustrates the manner by which sodium hypochlorite may be electrolytically produced on-site from a brine (sodium chloride) solution. The oxidizer is used by hospital personnel to disinfect surfaces and equipment contaminated with microbial pathogens. It may also be used to control algae growth in large bodies of standing water.

11.6-C TRANSPORTING METALLIC HYPOCHLORITES

When shippers offer a solid metallic hypochlorite for transportation, DOT requires them to identify the appropriate substance on an accompanying shipping paper. Barium hypochlorite, lithium hypochlorite, and calcium hypochlorite are the only common commercial chemical products that are stable solids at room temperature. They are identified by the following shipping descriptions:

> **UN2741, Barium hypochlorite, 5.1, (6.1), PGII (Poison)**
> **UN1748, Calcium hypochlorite, dry (contains more than 39% available chlorine), 5.1, PGII**
> **UN2208, Calcium hypochlorite (contains more than 10% but not more than 39% available chlorine), 5.1, PGIII**
> **UN2880, Calcium hypochlorite, hydrated (contains not less than 5.5% but not more than 16% water), 5.1, PGII**
> **UN1471, Lithium hypochlorite, dry (contains more than 39% available chlorine), 5.1, PGII**

The shipping descriptions of all other metallic hypochlorites and their solutions are identified generically in the Hazardous Materials Table as one of the following:

> **UN1791, Hypochlorites, inorganic, n.o.s., 8, PGII**
> **UN1791, Hypochlorite solutions (contains sodium hypochlorite), 8, PGII**
> *or*
> **UN1791, Hypochlorite solutions (contains sodium hypochlorite), 8, PGIII**

Shippers must include the name of the specific compounds parenthetically in the proper shipping name. For example, when shippers offer zinc hypochlorite for transportation, they enter the following shipping description on a shipping paper:

> **UN1791, Hypochlorites, inorganic, n.o.s. (contains zinc hypochlorite), 8, PGII**

Although DOT requires shippers to affix an OXIDIZER label on packaging containing calcium hypochlorite and lithium hypochlorite, it requires OXIDIZER and POISON labels on packaging containing barium hypochlorite. When shippers offer a solution of a metallic hypochlorite for transportation, DOT requires them to affix a CORROSIVE label on the packaging.

When carriers transport the metallic hypochlorites and their solutions in an amount exceeding 1001 lb (454 kg), DOT requires them to post either OXIDIZER or CORROSIVE placards on the bulk packaging or transport vehicle used for shipment, as relevant.

11.7 Di- and Trichloroisocyanuric Acids and Their Salts

Although calcium hypochlorite continues to be a popular component of industrial sanitation products, it has been largely replaced in modern products with certain chlorinated derivatives of isocyanuric acid. These compounds include *dichloroisocyanuric acid* and *trichloroisocyanuric acid*. The molecular structures of these compounds are relatively complex.

Dichloroisocyanuric acid, or
Dichloro-*sym*-triazine-2,4,6-trione

Trichloroisocyanuric acid, or
Trichloro-*sym*-triazine-2,4,6-trione

We study the nature of the bonding in these compounds more fully in Chapter 12.

Di- and trichloroisocyanuric acids are also available commercially as their sodium and potassium salts. They are encountered as white tablets or powders having an available chlorine that ranges from 90% to 99%.

11.7-A USES OF DI- AND TRICHLOROISOCYANURIC ACIDS AND THEIR SALTS

The major use of the di- and trichloroisocyanuric acid products is to chlorinate the water in swimming pools. For this purpose, these compounds are usually referred to as *dichlor* and *trichlor*, respectively. The chlorinating agent shown in Figure 11.5 contains trichloro. Di- and trichloroisocyanuric acids and their salts release their chlorine slowly into swimming pool water. They have much longer lifespans when dissolved in water than either free chlorine or the inorganic hypochlorites, all of which rapidly decompose when exposed to the ultraviolet radiation in sunlight. Thus, di- and trichloroisocyanuric acids and their salts are considered more desirable

FIGURE 11.5 A commercially available product consisting of 99.4% trichloro-*sym*-triazine-2,4,6-trione, or trichloroisocyanuric acid. It is used primarily for sanitizing the water used in spas and swimming pools. (*Courtesy of Leslie's Pools, Las Vegas, Nevada.*)

for the purpose of chlorination, especially in areas where swimming pools are exposed for extended periods to intense sunlight.

Di- and trichloroisocyanuric acids are also components of dry laundry bleaches, detergents and other dishwashing compounds, scouring powders, and bactericides. They have also been used for the purification of drinking water and the nonshrinking treatment of wool.

11.7-B TRANSPORTING DI- AND TRICHLOROISOCYANURIC ACIDS AND THEIR SALTS

When shippers offer dichloroisocyanuric acid, trichloroisocyanuric acid, or the salts of these compounds for transportation, DOT requires them to enter the appropriate description on an accompanying shipping paper as one of the following, as relevant:

> **UN2465, Dichloroisocyanuric acid, dry, 5.1, PGII**
> *or*
> **UN2465, Dichloroisocyanuric acid salts, dry, 5.1, PGIII**
> **UN2468, Trichloroisocyanuric acid, 5.1, PGII**
> *or*
> **UN2468, Trichloroisocyanuric acid salts, 5.1, PGIII**

DOT also requires shippers to affix an OXIDIZER label on its packaging.

When carriers transport any of these hazardous materials in an amount exceeding 1001 lb (454 kg), DOT requires them to post OXIDIZER placards on the bulk packaging or transport vehicle used for shipment.

- Describe the hazardous properties of chlorine dioxide.
- Describe the primary use of chlorine dioxide.
- Describe the means shippers and carriers use to inform emergency responders of the hazards associated with encountering chlorine dioxide hydrate during a transportation mishap.

11.8 Chlorine Dioxide

Although chemists recognize three oxides of chlorine, only one is commercially important: *chlorine dioxide*. At room temperature, this substance is a red-yellow gas having the chemical formula ClO_2. It is a highly unstable substance, decomposing into its elements at an explosive rate.

$$2ClO_2(g) \longrightarrow Cl_2(g) + 2O_2(g)$$
Chlorine dioxide $\qquad$ Chlorine $\qquad$ Oxygen

Chlorine dioxide is routinely produced near the point of its intended use. In the simplified generator in Figure 11.6, it is produced in a two-step process, each of which is conducted under vacuum conditions. In the first step, sodium hypochlorite is reacted with hydrochloric acid to generate chlorine.

$$NaClO(aq) + 2HCl(aq) \longrightarrow NaCl(aq) + H_2O(l) + Cl_2(g)$$
Sodium hypochlorite $\quad$ Hydrochloric acid $\qquad$ Sodium chloride $\qquad$ Water $\qquad$ Chlorine

In the second step, the chlorine is reacted with a sodium chlorite solution, which results in the generation of the chlorine dioxide.

$$2NaClO_2(aq) + Cl_2(g) \longrightarrow 2ClO_2(g) + 2NaCl(aq)$$
Sodium chlorite $\qquad$ Chlorine $\qquad$ Chlorine dioxide $\qquad$ Sodium chloride

When chlorine dioxide is dissolved in water, an aqueous solution is produced known as *chlorine dioxide hydrate*. In commerce, this solution is frozen prior to its being shipped to a point of intended use. When frozen, the aqueous solution survives for a long time, but gentle heating causes it to decompose.

As with chlorine itself, the principal risk associated with exposure to chlorine dioxide is inhalation toxicity. The exposure causes severe respiratory and eye damage.

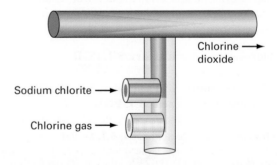

FIGURE 11.6 Chlorine dioxide is usually generated at the location where it is to be used. In this simplified schematic, two independent chemical reactions work in unison to generate the gas. Although its primary use is to bleach paper pulp, in modern times, chlorine dioxide has been frequently selected over other microbicides to disinfect areas contaminated with mold, bacteria, and viruses.

11.8-A USES OF CHLORINE DIOXIDE

Chlorine dioxide and elemental chlorine are generally used for the same purposes. By comparison, however, chlorine dioxide is approximately 2½ times more powerful than chlorine as an oxidizing agent. For this reason, chlorine dioxide is sometimes selected over chlorine for bleaching paper pulp and producing white wheat flour, even though chlorine dioxide is more costly.

Chlorine dioxide is sometimes also used as a microbicide. For example, it is used to disinfect water in the dairy, beverage, and other food industries. At water-treatment plants, its use is generally supplementary to the use of chlorine for killing bacteria. Because it is capable of killing cryptosporidium, the use of chlorine dioxide to disinfect drinking water ranks favorably with ozone use.

In 2002, when anthrax-laden mail was discovered in postal areas and in the Hart Senate Office Building, EPA contractors chose chlorine dioxide to destroy the anthrax. The mail packages and Senate offices were decontaminated by discharging the gas into the atmosphere of sealed-off areas, where it remained for several hours before neutralization. The hard surfaces within the Senate building were disinfected using liquid chlorine dioxide.

In 2005, following the Hurricane Katrina disaster along the Gulf coast, chlorine dioxide was also used to eradicate the mold that grew in homes that had been inundated with water. It has also been used to control the infestation of the pipe-clogging, thumbnail-sized quagga mussel in waterways.

11.8-B TRANSPORTING CHLORINE DIOXIDE

DOT allows chlorine dioxide to be transported only when it is frozen as its hydrate. When shippers offer the frozen hydrate for transportation, DOT requires them to maintain the gas in the solid state using dry ice or other means and to identify the gas on an accompanying shipping paper as follows:

NA9191, Chlorine dioxide, hydrate, frozen, 5.1, (6.1), PGII (Poison)

DOT also requires shippers to affix an OXIDIZER and POISON label on the chlorine dioxide packaging.

When carriers transport chlorine dioxide in an amount exceeding 1001 lb (454 kg), DOT requires them to post OXIDIZER placards on the bulk packaging or transport vehicle used for shipment.

PERFORMANCE GOALS FOR SECTION 11.9:

- Describe the general chemical composition of all fireworks.
- Identify the common oxidizers used in fireworks.
- Describe the chemical reactions that occur when fireworks are actuated.
- Describe the means shippers and carriers use to inform emergency responders of the hazards associated with encountering fireworks during a transportation mishap.
- Describe the means shippers and carriers use to inform emergency responders of the hazards associated with encountering an inorganic chlorite, chlorate, or perchlorate during a transportation mishap.

11.9 Oxidizers in Fireworks

Fireworks are items designed primarily for the purpose of producing audible and visible pyrotechnic effects by combustion, explosion, deflagration, or detonation reactions that occur without the intervention of atmospheric oxygen. Some familiar examples include firecrackers, Roman candles, pinwheels, flares, serpents, sparklers, skyrockets, and toy torpedoes. Special-effects fireworks include comets, fountains, strobes, and kaleidoscope shells. To choreograph a recreational fireworks program, professionals use computers, flame generators, and firing systems that carefully actuate a set of redox reactions, each of which is precisely timed to provide lavish aerial scenes such as chrysanthemum shells. In a modern program, lighting consoles synchronized with the pyrotechnics are integrated into the performance.

fireworks
■ Pyrotechnic devices intentionally designed for the purpose of producing a visible or audible effect by means of combustion, explosion, deflagration, or detonation reactions

common fireworks
■ Relatively small firework devices designed primarily for use by the general public

special fireworks
■ Relatively large firework devices designed primarily for use by experts

Fireworks are divided into two types: common fireworks and special fireworks. **Common fireworks** are relatively small devices containing fireworks designed primarily for use by the general public; **special fireworks** are relatively large firework devices designed primarily for use by fireworks experts. At 16 C.F.R. §1500.17, the Consumer Product Safety Commission has banned the use of certain fireworks devices intended to produce audible effects.

Although state and local laws attempt to prevent the widespread sale of fireworks to unauthorized individuals, incidents involving the misuse of fireworks annually cause numerous secondary fires, blindness, deafness, the mutilation of fingers, and death. In most states, to reduce or eliminate the likelihood of these incidents, only a licensed retail outlet that posts a permit is allowed to sell fireworks to the public. To further reduce or eliminate the risk of incidents caused by their misuse, state and local laws require that common fireworks be actuated in accordance with the manufacturer's warnings and instructions and that special fireworks be actuated only by trained experts. Even under these latter conditions, the actuation of fireworks can constitute a risk of fire and explosion.

One component of all fireworks is a pyrotechnic mixture of an oxidizing agent and a reducing agent. The most commonly encountered oxidizers are sodium chlorite, sodium chlorate, and sodium perchlorate, all of which are also components of vehicular air bags, solid rocket fuels, and certain fertilizers. Sodium perchlorate is the oxidizing agent of choice for most pyrotechnic displays. The reducing agents include charcoal, sulfur, pulverized magnesium, and aluminum flakes. To provide color when fireworks are actuated, one or more metallic compounds are added to fireworks formulations.

Sodium chlorite, sodium chlorate, and sodium perchlorate decompose to produce oxygen when they are heated.

$$\underset{\text{Sodium chlorite}}{NaClO_2(s)} \longrightarrow \underset{\text{Sodium chloride}}{NaCl(s)} + \underset{\text{Oxygen}}{O_2(g)}$$

$$\underset{\text{Sodium chlorate}}{2NaClO_3(s)} \longrightarrow \underset{\text{Sodium chloride}}{2NaCl(s)} + \underset{\text{Oxygen}}{3O_2(g)}$$

$$\underset{\text{Sodium perchlorate}}{NaClO_4(s)} \longrightarrow \underset{\text{Sodium chloride}}{NaCl(s)} + \underset{\text{Oxygen}}{2O_2(g)}$$

Although the oxygen produced by these reactions aids in actuating the display, more complex redox reactions are responsible for the brilliant lighting and sound effects associated with fireworks displays. When the mixture of an oxidizer, charcoal, sulfur, and finely divided magnesium or aluminum is activated, the resulting redox reactions occur at explosive rates. The presence of magnesium powder enhances the brilliance of a fireworks display, whereas the addition of coarse aluminum flakes produces luminous tails. The sets of equations in Table 11.6 illustrate some of the redox reactions associated with the actuation of the reactive mixture.

The unique colors observed during the display of fireworks result as certain compounds vaporize or decompose. These compounds are almost exclusively metallic chlorides, which fluoresce strongly in the visible wavelengths. Some examples and the colors they exhibit on vaporization are noted here:

Barium chloride	yellow-green	Cupric chloride	blue-green
Lithium chloride	crimson red	Calcium chloride	orange red
Potassium chloride	lavender	Sodium chloride	golden yellow
Strontium chloride	carmine red		

11.9-A TRANSPORTING FIREWORKS

When shippers offer fireworks for transportation, DOT requires them to describe the fireworks generically as explosives in hazard classes 1.1G, 1.2G, 1.3G, 1.4G, or1.4S on a shipping paper as one of the following, as relevant:

UN0333, Fireworks, 1.1G, PGII (EX-xxxxx)
UN0334, Fireworks, 1.2G, PGII (EX-xxxxx)

UN0335, Fireworks, 1.3G, PGII (EX-xxxxx)
UN0336, Fireworks, 1.4G, PGII (EX-xxxxx)
UN0337, Fireworks, 1.4S, PGII (EX-xxxxx)

DOT requires shippers to include the applicable EX-number (designated here as EX-xxxxx), Department of Defense (DOD) national stock number, product code, or other identifying information in the shipping description. The nature of these requirements is discussed in Section 15.4-C. Shippers are also required to affix the applicable EXPLOSIVE 1.1, 1.2, 1.3, or 1.4 label to containers.

When carriers transport explosives in hazard classes 1.1, 1.2, and 1.3, and when shipping explosives in hazard class 1.4 in an amount exceeding 1001 lb (454 kg), DOT requires them to post the relevant EXPLOSIVES 1.1, 1.2, 1.3, or 1.4 placard on the bulk packaging or transport vehicle used for shipment. DOT also requires shippers and carriers to display the appropriate compatibility group (Section 15.4-B) on the applicable labels and placards. DOT also requires shippers to affix an EXPLOSIVE 1.4 label with a G displayed on containers of consumer fireworks intended for use by the general public. DOT requires shippers to affix an EXPLOSIVE 1.1, EXPLOSIVE 1.2, or EXPLOSIVE 1.3 label with a G displayed, or an EXPLOSIVE 1.4 label with an S displayed, on containers of the fireworks intended for professional use.

TABLE 11.6	Examples of the Chemical Phenomena Associated with the Actuation of Fireworks	

OXIDIZING AGENT	REDUCING AGENT	EQUATION REPRESENTING THE REDOX REACTION
Sodium chlorite	Charcoal	$C(s) + NaClO_2(s) \longrightarrow NaCl(s) + CO_2(g)$ Carbon Sodium chlorite Sodium chloride Carbon dioxide
	Sulfur	$S_8(s) + 8NaClO_2(s) \longrightarrow 8NaCl(s) + 8SO_2(g)$ Sulfur Sodium chlorite Sodium chloride Sulfur dioxide
	Magnesium	$2Mg(s) + NaClO_2(s) \longrightarrow 2MgO(s) + NaCl(s)$ Magnesium Sodium chlorite Magnesium oxide Sodium chloride
	Aluminum	$4Al(s) + 3NaClO_2(s) \longrightarrow 2Al_2O_3(s) + 3NaCl(s)$ Aluminum Sodium chlorite Aluminum oxide Sodium chloride
Sodium chlorate	Charcoal	$3C(s) + 2NaClO_3(s) \longrightarrow 2NaCl(s) + 3CO_2(g)$ Carbon Sodium chlorate Sodium chloride Carbon dioxide
	Sulfur	$3S_8(s) + 16NaClO_3(s) \longrightarrow 16NaCl(s) + 24SO_2(g)$ Sulfur Sodium chlorate Sodium chloride Sulfur dioxide
	Magnesium	$3Mg(s) + NaClO_3(s) \longrightarrow 3MgO(s) + NaCl(s)$ Magnesium Sodium chlorate Magnesium oxide Sodium chloride
	Aluminum	$2Al(s) + NaClO_3(s) \longrightarrow Al_2O_3(s) + NaCl(s)$ Aluminum Sodium chlorate Aluminum oxide Sodium chloride
Sodium perchlorate	Charcoal	$2C(s) + NaClO_4(s) \longrightarrow NaCl(s) + 2CO_2(g)$ Carbon Sodium perchlorate Sodium chloride Carbon dioxide
	Sulfur	$S_8(s) + 4NaClO_4(s) \longrightarrow 4NaCl(s) + 8SO_2(g)$ Sulfur Sodium perchlorate Sodium chloride Sulfur dioxide
	Magnesium	$4Mg(s) + NaClO_4(s) \longrightarrow 4MgO(s) + NaCl(s)$ Magnesium Sodium perchlorate Magnesium oxide Sodium chloride
	Aluminum	$8Al(s) + 3NaClO_4(s) \longrightarrow 4Al_2O_3(s) + 3NaCl(s)$ Aluminum Sodium perchlorate Aluminum oxide Sodium chloride

11.9-B DISPLAYING FIREWORKS

Local ordinances regulate how fireworks are displayed, whether individually or as public presentations. Regardless of their nature, fireworks should never be actuated when climatic conditions are dangerously dry or windy, that is, when the wind speed exceeds 30 mi/hr (48 km/hr).

NFPA notes the following two recommendations for professionals who display special fireworks publicly:

- Professionals who publicly display fireworks must be experienced, responsible, and bonded.
- Professionals may actuate fireworks *no nearer than* 150 ft (45.7 m) from any part of an audience, 200 ft (61 m) from the nearest building, highway, or railroad, and 50 ft (15 m) from any electric wire line, tree, or overhead obstruction.

11.9-C TRANSPORTING METALLIC CHLORITES, CHLORATES, AND PERCHLORATES

The oxidizers used in fireworks are available as commercial chemical products. When shippers offer these individual compounds for transportation, DOT requires them to provide the relevant shipping description as shown in Table 11.7 on an accompanying shipping paper.

When shippers offer barium chlorate, barium perchlorate, and lead perchlorate for transportation, DOT requires them to affix OXIDIZER and POISON labels to the packaging. When shippers offer the other oxidizers listed in Table 11.7 for transportation, DOT requires them to affix an OXIDIZER label on their packaging.

When carriers transport any compound in Table 11.7 in an amount exceeding 1001 lb (454 kg), DOT requires them to post OXIDIZER placards on the bulk packaging or transport vehicle used for shipment.

Metallic chlorites, chlorates, and perchlorates that are not listed in Table 11.7 are described generically as shown:

> **UN1462, Chlorites, inorganic, n.o.s., 5.1, PGII**
> **UN1908, Chlorite solution, 8, PGII**
> **UN1908, Chlorite solution, 8, PGIII**
> **UN1461, Chlorates, inorganic, n.o.s., 5.1, PGII**
> **UN3210, Chlorates, inorganic, aqueous solution, n.o.s., 5.1, PGII**
> *or*
> **UN3210, Chlorates, inorganic, aqueous solution, n.o.s., 5.1, PGIII**
> **UN1481, Perchlorates, inorganic, n.o.s., 5.1, PGII**
> **UN3211, Perchlorates, inorganic, aqueous solution, n.o.s., 5.1, PGII**
> *or*
> **UN3211, Perchlorates, inorganic, aqueous solution, n.o.s., 5.1, PGIII**

When shippers prepare the shipping descriptions, DOT requires them to name the specific compound parenthetically. When shippers offer these compounds for transportation, DOT requires that they affix the relevant OXIDIZER or CORROSIVE label to the packaging.

When carriers transport these compounds in an amount exceeding 1001 lb (454 kg), DOT requires them to post OXIDIZER or CORROSIVE placards, as relevant, on the bulk packaging or transport vehicle used for shipment.

11.9-D ILL EFFECTS RESULTING FROM EXPOSURE TO METALLIC PERCHLORATES

Considerable concern about the health effects likely to result from exposure to perchlorate ions has been publicly voiced, particularly because perchlorate has been identified in many of our nation's groundwater sources. Widespread perchlorate contamination in groundwater is associated with the former military use of inorganic perchlorates as rocket fuels. Scientists have detected perchlorate in the water supplies of 35 states.

TABLE 11.7	**Shipping Descriptions of Metallic Chlorites, Chlorates, and Perchlorates**

METALLIC CHLORITE, CHLORATE, OR PERCHLORATE	SHIPPING DESCRIPTION
Barium chlorate	UN1445, Barium chlorate, solid, 5.1, (6.1), PGII (Poison)
Barium chlorate solution	UN3405, Barium chlorate, solution, 5.1, (6.1), PGII (Poison) *or* UN3405, Barium chlorate, solution, 5.1, (6.1), PGIII (Poison)
Barium perchlorate (solid)	UN1447, Barium perchlorate, solid, 5.1, (6.1), PGII (Poison)
Barium perchlorate solution	UN3406, Barium perchlorate, solution, 5.1, (6.1), PG II (Poison)
Calcium chlorate	UN1452, Calcium chlorate, 5.1, UN1452, PGII
Calcium chlorate solution	UN2429, Calcium chlorate, aqueous solution, 5.1, PGII *or* UN2429, Calcium chlorate, aqueous solution, 5.1, PGIII
Calcium chlorite	UN1453, Calcium chlorite, 5.1, PGII
Calcium perchlorate	UN1455, Calcium perchlorate, 5.1, PGII
Lead perchlorate (solid)	UN1470, Lead perchlorate, solid, 5.1, (6.1), PGII (Poison)
Lead perchlorate solution	UN3408, Lead perchlorate, solution, 5.1, (6.1), PGII (Poison) *or* UN3408, Lead perchlorate, solution, 5.1, (6.1), PGIII (Poison)
Magnesium chlorate	UN2723, Magnesium chlorate, 5.1, PGII
Magnesium perchlorate	UN1475, Magnesium perchlorate, 5.1, PGII
Potassium chlorate	UN1485, Potassium chlorate, 5.1, PGII
Potassium chlorate solution	UN2427, Potassium chlorate, aqueous solution, 5.1, PGII *or* UN2427, Potassium chlorate, aqueous solution, 5.1, PGIII
Potassium perchlorate	UN1489, Potassium perchlorate, 5.1, PGII
Sodium chlorate	UN1495, Sodium chlorate, 5.1, PGII
Sodium chlorate solution	UN2428, Sodium chlorate, aqueous solution, 5.1, PGII *or* UN2428, Sodium chlorate, aqueous solution, 5.1, PGIII
Sodium chlorite	UN1496, Sodium chlorite, 5.1, PGII
Sodium perchlorate	UN1502, Sodium perchlorate, 5.1, PGII
Strontium chlorate	UN1506, Strontium chlorate, 5.1, PGII
Strontium perchlorate	UN1508, Strontium perchlorate, 5.1, PGII
Zinc chlorate	UN1513, Zinc chlorate, 5.1, PGII

Does human exposure to perchlorate pose a health risk? Animal studies reveal a link between the consumption of perchlorate-contaminated water and damage to the thyroid gland, neurological problems, and the inability to properly produce growth- and fetal-developmental hormones. Perchlorate pollution potentially affects the health of tens of millions of American women and the children they conceive, especially since perchlorate prevents the proper production of the hormones needed for developing fetuses. In 2005, EPA set a daily dose of perchlorate that people can safely ingest at 0.7 µg/kg of body weight.

Perchlorate contamination has become a hotly contested political issue in California and Nevada, two states in which perchlorate pollution is especially pronounced. Owing to public pressure, it is likely that additional information on the health effects relating to perchlorate exposure will be forthcoming.

PERFORMANCE GOALS FOR SECTION 11.10:

- Identify the oxidizing agent and the reducing agent used in most flares and signaling smokes.
- Describe the means shippers and carriers use to inform emergency responders of the hazards associated with encountering packages of flares and smoke bombs during a transportation mishap.

11.10 Oxidizers in Flares, Signaling Smokes, and Smoke Bombs

flare
■ An article containing a pyrotechnic substance designed to illuminate, identify, signal, or warn

signaling smoke
■ A type of smoke bomb whose actuation communicates prearranged information to troops or others who subsequently observe the signal from a distance

smoke bomb
■ A device containing substances that produce smoke when actuated, generally for use in concealing military operations

Flares, **signaling smokes**, and **smoke bombs** are pyrotechnic mixtures of an oxidizing agent and reducing agent used by civilian police and the military to identify and obscure scenes of interest, to cordon off accident sites, and to coordinate activities during local assault operations. When they are actuated by the military, these pyrotechnics are helpful in locating friendly units and targets and controlling the laying and lifting of artillery.

11.10-A CHEMICAL ACTUATION OF FLARES, SIGNALING SMOKES, AND SMOKE BOMBS

The active chemical component of flares, signaling smokes, and smoke bombs is a pyrotechnic mixture of an oxidizing agent and reducing agent that has been combined with sodium bicarbonate and an organic dye. Most commonly, the oxidizing and reducing agents are potassium chlorate and elemental sulfur, respectively. The mixture of components is compressed into cartridges, hand grenades, and canisters.

When the chemical mixture in flares, signaling smokes, and smoke bombs is activated by ignition, at least two chemical reactions occur:

- The oxidation–reduction reaction between potassium chlorate and sulfur

$$16KClO_3(s) \quad + \quad 3S_8(s) \quad \longrightarrow \quad 16KCl(s) \quad + \quad 24SO_2(g)$$

Potassium chlorate $\qquad$ Sulfur $\qquad$ Potassium chloride $\qquad$ Sulfur dioxide

- The thermal decomposition of sodium bicarbonate

$$2NaHCO_3(s) \quad \longrightarrow \quad Na_2CO_3(s) \quad + \quad CO_2(g) \quad + \quad H_2O(g)$$

Sodium bicarbonate $\qquad$ Sodium carbonate $\qquad$ Carbon dioxide $\qquad$ Water

Carbon dioxide and sulfur dioxide evolve into the air and disperse the dye in the immediate area. Signaling flares also contain barium nitrate, strontium nitrate, or other metallic nitrates that display color when they thermally decompose. For example, the carmine red color of a signaling flare is produced when strontium nitrate decomposes.

$$2Sr(NO_3)_2(s) \quad \longrightarrow \quad 2SrO(s) \quad + \quad 4NO_2(g) \quad + \quad O_2(g)$$

Strontium nitrate $\qquad$ Strontium oxide $\qquad$ Nitrogen dioxide $\qquad$ Oxygen

Smoke bombs are used by the military to visually mask the movement of troops and vehicles during combat activities. A popular smoke bomb used by the U.S. Army during the Korean conflict was a pyrotechnic mixture containing 6.68% granulated aluminum, 46.66% zinc oxide, and 46.66% (C_2H_6, or CCl_3-CCl_3) hexachloroethane by mass. When the components of this mixture were actuated with a fuse, the products that appeared in the smoke were aluminum oxide, zinc chloride, and carbon (soot).

$$2Al(s) \quad + \quad CCl_3-CCl_3(s) \quad + \quad 3ZnO(s) \quad \longrightarrow \quad 3ZnCl_2(s) \quad + \quad Al_2O_3(s) \quad + \quad 2C(s)$$

| Aluminum | Hexachloroethane | Zinc oxide | Zinc chloride | Aluminum oxide | Carbon |

The zinc chloride attracted atmospheric moisture to form a fog, the aluminum oxide deflected light rays, and the carbon colored the smoke cloud gray.

11.10-B TRANSPORTING FLARES, SIGNALING SMOKES, AND SMOKE BOMBS

When shippers offer flares, signaling smokes, and smoke bombs for transportation, DOT requires them to identify the appropriate item generically as one of the following, as relevant:

> **UN0093, Flares, aerial, 1.3G, PGII (EX-xxxxx)**
> **UN0403, Flares, aerial, 1.4G, PGII (EX-xxxxx)**
> **UN0404, Flares, aerial, 1.4S, PGII (EX-xxxxx)**
> **UN0421, Flares, aerial, 1.2G, PGII (EX-xxxxx)**
> **UN0092, Flares, surface, 1.3G, PGII (EX-xxxxx)**
> **UN0418, Flares, surface, 1.1G, PGII (EX-xxxxx)**
> **UN0419, Flares, surface, 1.2G, PGII (EX-xxxxx)**
> **UN0196, Signals, smoke, 1.1G, PGII (EX-xxxxx)**
> **UN0197, Signals, smoke, 1.4G, PGII (EX-xxxxx)**
> **UN0313, Signals, smoke, 1.2G, PGII (EX-xxxxx)**
> **UN0487, Signals, smoke, 1.3G, PGII (EX-xxxxx)**
> **UN2028, Bombs, smoke, nonexplosive, 8, PGII**

When shippers offer flares and signaling smokes in hazard classes 1.1, 1.2, or 1.3, DOT requires them to affix the applicable EXPLOSIVE 1.1, 1.2, or 1.3 label to their containers. Carriers must post the relevant EXPLOSIVES 1.1, 1.2, or 1.3 placard on the bulk packaging or transport vehicle used for shipment. DOT requires the compatibility group G to be displayed on these labels and placards.

When shippers offer flares and smoke signals in hazard class 1.4, DOT requires them to affix EXPLOSIVE 1.4 labels to their packaging. When carriers transport flares and smoke signals in hazard class 1.4 in an amount exceeding 1001 lb (454 kg), they must post EXPLOSIVES 1.4 placards on the bulk packaging or transport vehicle used for shipment. DOT also requires the relevant compatibility group, G or S, to be displayed on these labels and placards.

When shippers offer to transport smoke bombs that contain nonexplosive components but a corrosive liquid, DOT requires them to affix CORROSIVE labels to their containers. DOT also requires carriers to post CORROSIVE placards on the bulk packaging or transport vehicle used for their shipment when they transport an amount exceeding 1001 lb (454 kg).

PERFORMANCE GOALS FOR SECTION 11.11:

- Identify the ways by which ammonium compounds decompose when they are heated.
- Describe the means shippers and carriers use to inform emergency responders of the hazards associated with encountering oxidizing ammonium compounds during a transportation mishap.

11.11 The Thermal Stability of Ammonium Compounds

Ammonium compounds are substances whose units contains an ammonium ion (NH_4^+) and some negative ion. All ammonium compounds are thermally unstable. When they are heated, ammonium compounds decompose in either of the following ways:

■ Ammonium compounds may decompose into the compounds from which they were initially made. For instance, when ammonium chloride is heated, it thermally decomposes to form ammonia and hydrogen chloride.

$$NH_4Cl(s) \longrightarrow NH_3(g) + HCl(g)$$
Ammonium chloride   Ammonia   Hydrogen chloride

The ammonium compounds that decompose in this simple manner are not generally regarded as hazardous materials.

■ Ammonium compounds may also decompose by oxidation–reduction. Examples of ammonium compounds that decompose in this manner are illustrated by the equations in Table 11.8. The majority of them decompose at explosive rates when heated. For this reason,

TABLE 11.8	Examples of Ammonium Compounds That Are Hazardous Materials
AMMONIUM COMPOUND	**EQUATION ILLUSTRATING THERMAL DECOMPOSITION**
Ammonium azide[a]	$NH_4N_3(s) \longrightarrow 2N_2(g) + 2H_2(g)$ Ammonium azide Nitrogen Hydrogen
Ammonium bromate[a]	$2NH_4BrO_3(s) \longrightarrow 2NH_4Br(s) + 3O_2(g)$ Ammonium bromate Ammonium bromide Oxygen
Ammonium chlorate[a]	$2NH_4ClO_3(s) \longrightarrow 2NH_4Cl + 3O_2(g)$ Ammonium chlorate Ammonium chloride Oxygen
Ammonium dichromate	$(NH_4)_2Cr_2O_7(s) \longrightarrow Cr_2O_3(s) + N_2(g) + 4H_2O(g)$ Ammonium dichromate Chromium(III) oxide Nitrogen Water
Ammonium fulminate[a]	$2NH_4NCO(s) \longrightarrow 2N_2(g) + 4H_2(g) + 2CO(g)$ Ammonium fulminate Nitrogen Hydrogen Carbon monoxide
Ammonium nitrite[a]	$NH_4NO_2(s) \longrightarrow N_2(g) + 2H_2O(g)$ Ammonium nitrite Nitrogen Water
Ammonium perchlorate	$2NH_4ClO_4(s) \longrightarrow N_2(g) + Cl_2(g) + 2O_2(g) + 4H_2O(g)$ Ammonium perchlorate Nitrogen Chlorine Oxygen Water
Ammonium permanganate[a]	$2NH_4MnO_4(s) \longrightarrow 2MnO(s) + N_2(g) + O_2(g) + 4H_2O(g)$ Ammonium permanganate Manganese(II) oxide Nitrogen Oxygen Water
Ammonium peroxydisulfate[b]	$3(NH_4)_2S_2O_8(s) \longrightarrow 4NH_3(s) + N_2(g) + 3O_2(g) + 6SO_2(g) + 6H_2O(g)$ Ammonium peroxydisulfate Ammonia Nitrogen Oxygen Sulfur dioxide Water

[a]DOT prohibits the transportation of these materials.
[b]Also named ammonium persulfate.

DOT prohibits carriers to transport them by any mode. Examples of these ammonium compounds are ammonium azide, ammonium bromate, ammonium chlorate, ammonium fulminate, ammonium nitrite, and ammonium permanganate.

11.11-A USES OF AMMONIUM COMPOUNDS

Because few compounds containing the ammonium ion are thermally stable, the industrial uses of these compounds are fairly limited. Ammonium nitrate is by far the most important compound containing the ammonium ion. In Section 11.12, we observe that it is a major fertilizer and explosive.

Ammonium perchlorate is the only other ammonium compound used in the United States in significant amounts. The U.S. aerospace industry uses it as a solid propellant. Ammonium perchlorate accounts for 70% of the solid propellant in the U.S. space shuttle's booster rockets. As it thermally decomposes, ammonium perchlorate generates enough gas to propel the shuttle.

11.11-B TRANSPORTING AMMONIUM COMPOUNDS

When ammonium compounds are transported, DOT requires shippers to identify the appropriate substance on an accompanying shipping paper by selecting the relevant entry from Table 11.9.

TABLE 11.9	Shipping Descriptions of Ammonium Compounds
AMMONIUM COMPOUND	**SHIPPING DESCRIPTION**
Ammonium dichromate	UN1439, Ammonium dichromate, 5.1, UN1439, II
Ammonium nitrate fertilizer	NA2071, Ammonium nitrate based fertilizer, 9, NA2071, III
Ammonium nitrate emulsion	UN2607, Ammonium nitrate emulsion, 5.1, PGIII
Ammonium nitrate suspension	UN2067, Ammonium nitrate suspension, 5.1, PGIII
Ammonium nitrate gel (intermediate for blasting explosives)	UN3375, Ammonium nitrate gel, 5.1, PGII
Ammonium nitrate/fuel oil mixtures (contains prilled ammonium nitrate and fuel oil)	NA0331, Ammonium nitrate - fuel oil mixtures, 1.5D, PGII (EX-xxxxx) *or* UN0331, Agent, blasting type B, 1.5D, PGII (EX-xxxx) *or* UN0331, Explosive, blasting, type B, 1.5D, PGII (EX-xxxxx)
Ammonium nitrate, liquid (hot concentrated solution)	UN2426, Ammonium nitrate, liquid, 5.1
Ammonium nitrate (contains more than 0.2% combustible substances)	UN0222, Ammonium nitrate, 1.1D, PGII (EX-xxxxx) *or* UN1942, Ammonium nitrate, 5.1, PGIII *or* UN0081, Explosive, blasting, type A, 1.1D, PGII (EX-xxxxx)
Ammonium perchlorate	UN0402, Ammonium perchlorate, 1.1D, PGII (EX-xxxxx) *or* UN1442, Ammonium perchlorate, 5.1, PGII
Ammonium persulfate	UN1444, Ammonium persulfate, 5.1, UN1444, PGIII

When shippers offer an ammonium compound that is classified as an oxidizer, DOT requires them to affix an OXIDIZER label to its packaging.

When carriers transport 1001 lb (454 kg) or more of these ammonium compounds, DOT requires them to display OXIDIZER placards on the bulk packaging or transport vehicle used for shipment.

DOT regulates ammonium nitrate and ammonium perchlorate as either an oxidizer or an explosive. DOT requires manufacturers to test samples of their product to establish the relevant hazard class. The determination is related in part to the product's particle size. When shippers transport ammonium perchlorate that is classified as an explosive, DOT requires them to affix an EXPLOSIVE 1.1 label to its packaging. When carriers transport this explosive, DOT requires them to post EXPLOSIVES 1.1 placards on the relevant bulk packaging or transport vehicle used for shipment. The shippers and carriers also display the compatibility group D on these labels and placards.

When shippers transport ammonium nitrate that is classified in hazard class 1.1, DOT requires them to affix an EXPLOSIVE 1.1 label to the packaging. When carriers transport this explosive, DOT requires them to post EXPLOSIVES 1.1 placards on the bulk packaging or transport vehicle used for shipment. When shipping ammonium nitrate that is classified in hazard class 1.5, DOT requires shippers to affix an EXPLOSIVE 1.5 label to its packaging. When carriers transport this explosive in an amount exceeding 1001 lb (454 kg), DOT requires them to post EXPLOSIVES 1.5 placards. Shippers and carriers must also display the compatibility group D on all EXPLOSIVE 1.1 or 1.5 labels and placards used when shipping ammonium nitrate classified as an explosive.

When shippers transport ammonium nitrate fertilizer onboard aircraft or watercraft, DOT requires them to affix a CLASS 9 label to the packaging.

PERFORMANCE GOALS FOR SECTION 11.12:

- Describe how ammonium nitrate/fuel mixtures are used.
- Identify the major industries in which bulk quantities of ammonium nitrate are used.
- Describe the OSHA, DHS, and DOT regulations that pertain to the handling, storage, stowing, loading, unloading, discharge, or transporting bulk quantities of ammonium nitrate.
- Describe the means shippers and carriers use to inform emergency responders of the hazards associated with encountering ammonium nitrate during a transportation mishap.

11.12 Ammonium Nitrate

Ammonium nitrate is by far the most important commercial chemical product containing the ammonium ion. Its chemical formula is NH_4NO_3. Several grades are available commercially including a fertilizer, dynamite, nitrous oxide, and technical grade. All are available as crystals, flakes, grains, and prills (small round aggregates). The characteristics of each grade are briefly noted next.

- Fertilizer-grade ammonium nitrate is commonly encountered at farms and farm distribution centers. Fertilizer-grade ammonium nitrate is a formulation of ammonium nitrate with ammonium sulfate or calcium carbonate, each of which reduces the potential risk of its self-decomposition. As its name implies, it is intended for use as an agricultural fertilizer. It is commonly sold in 50-lb (23-kg) bags.

- Dynamite-grade ammonium nitrate is widely used during road construction and blasting operations in mines and quarries, where it is the active component of blasting agents. It is also a component of unique explosives like ammonal, a mixture of ammonium nitrate and powdered aluminum. Ammonium nitrate has also been combined with TNT in a bursting charge used in demolition bombs. It is also available commercially as a mixture with diesel fuel called *ammonium nitrate/fuel oil mixture*, or ANFO. Although it can be safely handled and transported, ANFO detonates with a powerful force when it is ignited.

- Nitrous oxide–grade ammonium nitrate is used to produce dinitrogen monoxide, or nitrous oxide, by gentle heating. It is known commonly as *laughing gas*, an anesthetic mainly used by dentists.

$$\text{NH}_4\text{NO}_3(s) \longrightarrow \text{N}_2\text{O}(g) + 2\text{H}_2\text{O}(g)$$
Ammonium nitrate   Dinitrogen monoxide   Water

- Technical-grade ammonium nitrate is produced for a variety of uses. For example, it is a component of cold packs used for treating minor athletic injuries. Because the dissolving of ammonium nitrate in water is an endothermic process, the solution is cold when first produced and provides relief from pain when applied to bruises.

11.12-A AMMONIUM NITRATE AT FIRE SCENES

Ammonium nitrate melts at 337 °F (170 °C) and decomposes between 350 and 410 °F (177 and 210 °C). When a bulk quantity of ammonium nitrate is present at a fire scene, it may undergo thermal decomposition at an explosive rate. The decomposition produces an array of products including those illustrated by the following two equations:

$$4\text{NH}_4\text{NO}_3(s) \longrightarrow 3\text{N}_2(g) + 2\text{NO}_2(g) + 8\text{H}_2\text{O}(g)$$
Ammonium nitrate   Nitrogen   Nitrogen dioxide   Water

$$2\text{NH}_4\text{NO}_3(s) \longrightarrow 2\text{N}_2(g) + \text{O}_2(g) + 4\text{H}_2\text{O}(g)$$
Ammonium nitrate   Nitrogen   Oxygen   Water

These decomposition products are gaseous.

One hazard associated with encountering ammonium nitrate at a fire scene is the risk that it will decompose explosively. Another hazard is the potential for exposure to a dangerous, life-threatening concentration of nitrogen dioxide (Section 10.13-C).

When the gases produced by the decomposition of ammonium nitrate are vented from the fire scene, the threat of explosive decomposition is reduced or eliminated. However, when the decomposition products are confined or cannot be vented from the storage area, a major explosion is likely to occur. This was the fate of the French cargo ship SS *Grandcamp*, a portion of whose features are illustrated in Figure 11.7.

In 1947, the *Grandcamp* was docked at the harbor in Texas City, Texas, when it caught fire with nearly 2280 tn (2070 t) of fertilizer-grade ammonium nitrate onboard stored within two holds. The heat generated by the fire caused the ammonium nitrate to melt and decompose.

Although fires involving oxidizing agents can effectively be extinguished through application of water, this attempt was never made onboard the *Grandcamp*. Instead, orders were issued to seal the hatches to the holds to eliminate the air supply to the fire and prevent damage to the cargo by water. Supported by oxidizing agents, however, the fire continued to burn in the absence of air. Internal pressure within the holds increased until a major explosion released the confined gases.

The explosion of the *Grandcamp* is now regarded as one of the worst maritime incidents involving a chemical substance. It serves to illustrate dramatically the potentially hazardous nature of oxidizers in general and ammonium nitrate in particular.

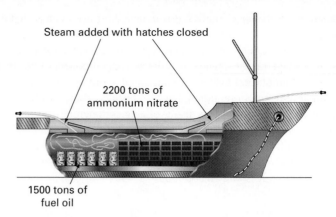

Steam added with hatches closed

2200 tons of
ammonium nitrate

1500 tons of
fuel oil

FIGURE 11.7 On April 15, 1947, fire was detected in two holds containing 2200 tn (1998 t) of ammonium nitrate onboard the SS *Grandcamp,* a French cargo ship docked in Texas City, Texas. An order was given to close the hatches to the holds and apply steam. As the fire continued to burn, the ammonium nitrate decomposed into a mixture of nitrogen, nitrogen dioxide, oxygen, and water vapor. This production of gases caused the buildup of internal pressure within the holds that was relieved only when the cargo ship subsequently exploded. Six hundred people were killed and another 3500 were injured. The property damage was comparable to that experienced during a major wartime bombing incident. The total property loss was estimated as $33 million based on 1947 costs.

11.12-B WORKPLACE REGULATIONS INVOLVING THE BULK STORAGE OF AMMONIUM NITRATE

OSHA has promulgated regulations relating to the storage of bulk quantities of ammonium nitrate at 29 C.F.R. §1910.109. These regulations apply to farmers and other employers who store and use ammonium nitrate in workplaces, as well as the owner and lessee of any building, premise, or structure in which the ammonium nitrate is stored in quantities of 1000 lb (454 kg) or more.

The OSHA regulations pertain to three subject areas: the nature of the building in which storage of ammonium nitrate is intended; the storage of ammonium nitrate in containers or piles within an approved building; and the storage of ammonium nitrate with other substances. We note each subject area separately to appreciate the extent of the precautions that must be undertaken to prevent mishaps.

The Nature of the Building in Which the Storage of Ammonium Nitrate Is Intended

- A building used to store ammonium nitrate may have a basement only when the basement has been constructed so it is open on at least one side.
- The building used to store ammonium nitrate is limited to one story in height.
- The building used for the storage of ammonium nitrate must be adequately ventilated or be of a construction that allows self-ventilating in the event of a fire.
- The wall on the exposed side of an ammonium nitrate storage building that is within 50 ft (15 m) of a combustible building, forest, piles of combustible materials and similar exposure hazards must be of fire-resistive construction. In lieu of a fire-resistive wall, other suitable means of exposure protection, such as a freestanding wall, may be used.
- All flooring in storage and handling areas within an ammonium nitrate storage building must be of noncombustible material or protected against impregnation of ammonium nitrate, and must be without open drains, traps, tunnels, pits, or pockets into which any molten ammonium nitrate could flow and be confined in the event of fire.

- The ammonium nitrate storage building and its structures must be dry and free from water seepage through the roof, walls, and floors.
- Unless constructed of noncombustible material, or unless adequate facilities for fighting a roof fire are available, bulk storage structures within an ammonium nitrate storage building cannot exceed a height of 40 ft (12 m).
- Suitable fire-control devices, such as small-hose or portable fire extinguishers, must be provided throughout an ammonium nitrate storage building and in the loading and unloading areas.
- Water supplies and fire hydrants must be located nearby an ammonium nitrate storage building in accordance with recognized good practices.

The Storage of Ammonium Nitrate in Containers

- Containers of ammonium nitrate cannot be accepted for storage when the temperature of the product exceeds 130 °F (54 °C).
- Bags of ammonium nitrate cannot be stored within 30 in. (76 cm) of the walls and partitions of the ammonium nitrate storage building.
- No more than 2500 tn (2270 t) of bagged ammonium nitrate may be stored in an ammonium nitrate storage building that is not equipped with an automatic sprinkler system.

The Storage of Ammonium Nitrate in Piles

- Ammonium nitrate cannot be accepted for storage in piles when the temperature of the product exceeds 130 °F (54 °C).
- Piles of ammonium nitrate cannot ordinarily exceed 20 ft (6 m) in height, 20 ft (6 m) in width, and 50 ft (15 m) in length. When the piles are stored within a building of noncombustible construction or protected with automatic sprinklers, the length of the piles is unlimited. Ammonium nitrate piles cannot be stacked closer than 36 in. (91 cm) below the roof or supporting and spreader beams.
- Aisles must be provided to separate ammonium nitrate piles by a clear space of not less than 3 ft (0.9 m) in width. At least one service or main aisle in the storage area that is not less than 4 ft (1.2 m) in width must also be provided.
- Bins must be kept clean and free of materials that can contaminate ammonium nitrate.
- Owing to the corrosive and reactive properties of ammonium nitrate and to avoid contamination, galvanized iron, copper, lead, and zinc cannot be used in the construction of a bin unless they are suitably protected. Although aluminum and wooden bins protected against impregnation by ammonium nitrate are permissible, the partitions dividing the ammonium nitrate storage area from the storage of other products that could contaminate the ammonium nitrate must be of "tight" construction, as illustrated in Figure 11.8.
- Ammonium nitrate storage bins and piles must be clearly marked with signs reading as follows:

AMMONIUM NITRATE

- The piles of ammonium nitrate must be sized and arranged so that all material in the pile is moved periodically to minimize caking of the ammonium nitrate during its storage.
- The height or depth of the ammonium nitrate piles is limited by the pressure-settling tendency of the product; however, in no case can the ammonium nitrate be piled higher at any point than 36 in. (91 cm) below the roof or the supporting and spreader beams overhead.
- Dynamite, other explosives, and blasting agents are prohibited for loosening ammonium nitrate that has caked.

The Storage of Ammonium Nitrate with Other Substances

- Ammonium nitrate must be stored in a separate building, separated by approved-type firewalls of not less than 1-hr fire-resistance rating, or separated by a space of at least

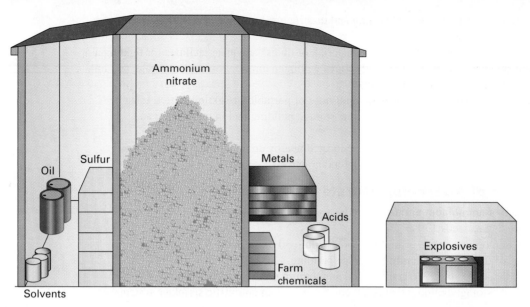

FIGURE 11.8 At 29 C.F.R. §1910.109, OSHA requires employers to store bulk quantities [>1000 lb (454 kg)] of ammonium nitrate within a bin that is clean and free of contaminants. When the ammonium nitrate is stored as a pile, it must be sized in a manner that all material within the pile may be periodically moved. The bin must be separated by fire walls from flammable and combustible materials, corrosive materials, other oxidizers, and substances with which the ammonium could react. Employers may not store explosives or blasting agents within the same building. (*Courtesy of The Fertilizer Institute, Washington, DC*)

30 ft (9 m) from the storage of organic chemicals, acids, or other corrosive materials; materials that may require blasting during processing or handling; compressed flammable gases; and flammable and combustible materials or other contaminating substances, including but not limited to animal fats, baled cotton, baled rags, baled scrap paper, bleaching powder, burlap or cotton bags, caustic soda, coal, coke, charcoal, cork, camphor, excelsior, fibers of any kind, fish oils, fish meal, foam rubber, hay, lubricating oil, linseed oil, or other oxidizable or drying oils, naphthalene, oakum, oiled clothing, oiled paper, oiled textiles, paint, straw, sawdust, wood shavings, or vegetable oils.

■ Flammable liquids (such as gasoline, kerosene, solvents, and light fuels), liquefied petroleum gas, sulfur, and finely divided metals cannot be stored on the same premises where bulk ammonium nitrate is stored.

11.12-C HOMELAND SECURITY MEASURES INVOLVING AMMONIUM NITRATE

Since 9-11 law enforcement agencies have taken measures to reduce the potential for the criminal misuse of ammonium nitrate that is stored, handled, or transported within their jurisdictions. Given the unorthodox ways by which terrorists may use ANFO, law enforcement agencies should be provided with sufficient information to account for the ammonium nitrate potentially available within their jurisdictions to terrorists. This information becomes available to law enforcement agencies by means of regulations promulgated by the Department of Homeland Security (DHS) pursuant to the Secure Handling of Ammonium Nitrate Act of 2007. These regulations require all ammonium nitrate producers, sellers, and purchasers who take custody of the ammonium nitrate to register with DHS.

DHS has been especially sensitive to the entry of bulk quantities of ammonium nitrate at the nation's ports. At 49 C.F.R. §176.415, DHS requires the owner or operator of a watercraft loaded with bulk quantities of ammonium nitrate, or a watercraft on which bulk quantities are intended to be loaded, to receive written permission from the Captain of the Port before the loading or unloading operations occur. The permit stipulates that the owner or operator must load or unload the ammonium nitrate at a facility removed from congested areas or high-value or high-hazard industrial facilities and at which an abundance of water for firefighting purposes is available. DHS also requires the watercraft to be moored bow to seaward and maintained in a mobile status by the presence of tugs and the readiness of its engines so it may pass unrestricted to open water in the event of an emergency.

SOLVED EXERCISE 11.4

The OSHA regulation at 29 C.F.R. §1910.109 stipulates that ammonium nitrate cannot ordinarily be stored in the same building with either sulfur or finely divided metals. What is the most likely reason OSHA requires ammonium nitrate to be segregated from these substances?

Solution: Ammonium nitrate is chemically incompatible with both elemental sulfur and finely divided metals. When exposed to an ignition source, the resulting combustion reactions are violent and occur at explosive rates. These reactions are denoted as follows, where metallic magnesium has been selected as the combustible metal:

$$S_8(s) \ + \ 32NH_4NO_3(s) \ \longrightarrow \ 8SO_2(g) \ + \ 8NO_2(g) \ + \ 64H_2O(g) \ + \ 28N_2(g)$$

Sulfur   Ammonium nitrate   Sulfur dioxide   Nitrogen dioxide   Water   Nitrogen

$$12Mg(s) \ + \ 16NH_4NO_3(s) \ \longrightarrow \ 12MgO(s) \ + \ 2NO_2(g) \ + \ 32H_2O(g) \ + \ 15N_2(g)$$

Magnesium   Ammonium nitrate   Magnesium oxide   Nitrogen dioxide   Water   Nitrogen

OSHA most likely requires the segregation of ammonium nitrate from sulfur and finely divided metals to reduce or eliminate the risk of fire and explosion.

At 33 C.F.R. §126.28, DHS has also promulgated regulations relating to the handling, storage, stowing, loading, unloading, discharging, and transportation of bulk quantities of ammonium nitrate at a waterfront facility. These regulations are paraphrased as follows:

- The proper shipping name of the ammonium nitrate must be marked on the outside surface of all containers.
- The buildings on a waterfront facility used for the storage of ammonium nitrate must be constructed so as to afford good ventilation.
- The storage of ammonium nitrate must occur at a safe distance from electric wiring, steam pipes, radiators, or any other heating mechanism.
- The ammonium nitrate must be separated by a fire-resistive wall or by a distance of at least 30 ft (9 m) from organic materials or other substances that could cause contamination, such as flammable liquids, combustible liquids, corrosive liquids, inorganic chlorates, inorganic permanganates, finely divided metals, caustic soda, charcoal, sulfur, cotton, coal, fats, fish oils, or vegetable oils.
- Ammonium nitrate must be stored on a clean wooden or concrete floor or on pellets over a clean floor. When ammonium nitrate is stored on a concrete floor, the floor must first be covered with a moisture barrier such as a polyethylene sheet or asphaltic-laminated paper.
- Any spilled ammonium nitrate must be promptly and thoroughly cleaned up and removed from the waterfront facility. If ammonium nitrate has remained in contact with a wooden floor for any length of time, the floor must be scrubbed with water, and all spilled material must be thoroughly dissolved and flushed away.
- An abundance of water for firefighting must be readily available.
- Open drains, traps, pits, or pockets that could be filled with molten ammonium nitrate in the event of a fire must be eliminated or plugged.

PERFORMANCE GOALS FOR SECTION 11.13:

- Identify the oxidizing chromium compounds and describe their hazardous properties.
- Identify the industries that use bulk quantities of oxidizing chromium compounds.
- Describe the means shippers and carriers use to inform emergency responders of the hazards associated with encountering an oxidizing chromium compound during a transportation mishap.

11.13 Oxidizing Chromium Compounds

hexavalent-chromium compound

■ A compound containing chromium in the +6 oxidation state, the absorption of which has been shown to cause certain adverse health ailments including cancer

metallic chromate

■ An inorganic compound composed of metallic and chromate ions

metallic dichromate

■ An inorganic compound composed of metallic and dichromate ions

Chromium-containing compounds in which the chromium assumes the +6 oxidation state are called *oxidizing chromium compounds*. The +6 oxidation state of chromium is called the *hexavalent-chromium ion*. The oxidizing chromium compounds are also called **hexavalent-chromium compounds**. They are **metallic chromates**, **metallic dichromates**, *chromium trioxide*, and *chromyl chloride*.

The metallic chromates and metallic dichromates are compounds composed of a positive ion and the chromate ion (CrO_4^{2-}) and dichromate ion ($Cr_2O_7^{2-}$), respectively. Potassium chromate and potassium dichromate are specific examples. Their chemical formulas are K_2CrO_4 and $K_2Cr_2O_7$, respectively.

The oxidizing chromium compounds are manufactured for commercial use from chromite ore, which we represent here as $FeCr_2O_4$. For example, to produce potassium dichromate the ore is first heated at kiln temperatures with potassium carbonate to produce potassium chromate as follows:

$$4FeCr_2O_4(s) \ + \ 8K_2CO_3(s) \ + \ 7O_2(g) \ \longrightarrow \ 8K_2CrO_4(s) \ + \ 2Fe_2O_3(s) \ + \ 8CO_2(g)$$

Chromite   Potassium carbonate   Oxygen   Potassium chromate   Iron(III) oxide   Carbon dioxide

Potassium dichromate is then produced by the reaction of sulfuric acid with potassium chromate.

$$2K_2CrO_4(aq) \ + \ H_2SO_4(aq) \ \longrightarrow \ K_2Cr_2O_7(aq) \ + \ H_2O(l) \ + \ K_2SO_4(aq)$$

Potassium chromate   Sulfuric acid   Potassium dichromate   Water   Potassium sulfate

Potassium chromate may be produced from potassium dichromate, and vice versa, by altering the pH conditions as follows:

$$K_2Cr_2O_7(aq) \ + \ 2KOH(aq) \ \longrightarrow \ 2K_2CrO_4(aq) \ + \ H_2O(l)$$

Potassium dichromate   Potassium hydroxide   Potassium chromate   Water

$$2K_2CrO_4(aq) \ + \ 2HCl(aq) \ \longrightarrow \ K_2Cr_2O_7(aq) \ + \ H_2O(l) \ + \ 2KCl(aq)$$

Potassium chromate   Hydrochloric acid   Potassium dichromate   Water   Potassium chloride

The metallic chromates and metallic dichromates are solid yellow and orange compounds, respectively. Consequently, compounds such as lead chromate, strontium chromate, and zinc chromate have been useful as pigments in industrial paints. The oxidizing chromium compounds have also been used to manufacture textile dyes and preserve wood and leather. They are also the major components of chrome electroplating bath solutions.

Prior to the 1990s, hexavalent-chromium compounds were commonly used as water-treatment agents to prevent the formation of a mineral scale that would otherwise form in the circulating waters used in air conditioners and industrial cooling towers. The production of this scale reduced the efficiency by which heat was transferred to the water. Cooling towers are often huge structures located at petroleum refineries, power plants, chemical plants, and elsewhere, where they are used to cool water by evaporation. The most readily recognizable type of cooling tower is probably the rounded hourglass structure used at nuclear power plants, but cooling towers are also located wherever large heating and ventilation, air-conditioning, and refrigeration systems are needed. The cooling tower shown in Figure 11.9 is a component of a air-conditioning system that cools a major hotel.

FIGURE 11.9 Cooling towers are often massive water-recirculation structures used to remove heat from the water by contacting the fluid with air. An agent containing a hexavalent-chromium compound was formerly added to the water in industrial cooling towers to prevent the formation of a mineral scale on grates and other tower components. Because hexavalent chromium is a known carcinogen, EPA promulgated regulations at 40 C.F.R. §749.68 that prohibited the continued use of a hexavalent-chromium compound for this purpose.

In the past, the waste generated in connection with using hexavalent-chromium-based water-treatment agents at petroleum refineries, power plants, chemical plants was often discharged into on-site unlined surface impoundments. Rainwater caused the hexavalent-chromium compound to seep into the subsurface and contaminate the underlying groundwater aquifer.

Scientific research has revealed that the absorption of a hexavalent-chromium compound into the body can cause a number of health ailments. In particular, the inhalation of the dust or fines of hexavalent-chromium compounds has been linked with the onset of lung cancer in humans. Thus, epidemiologists rank these compounds as known human carcinogens by the inhalation route of exposure.

Research studies concerned with establishing a link between various diseases and the consumption of hexavalent-chromium compounds in water are in progress. Erin Brockovich, the heroine in the blockbuster movie of the same name, alleged in the film that the long-term consumption of water contaminated with elevated levels of hexavalent chromium causes cancer and other health ailments. At the time of the movie's release in 2000, there was scant evidence to definitively link these health problems with the consumption of hexavalent-chromium-contaminated water. Today, however, scientists have determined that drinking water that has been contaminated with hexavalent-chromium compounds not only causes the onset of stomach ulcers and stomach and intestinal cancers but may also damage the kidneys and liver.

EPA has used the authorities of the Toxic Substances Control Act and the Clean Air Act to promulgate regulations prohibiting the use of water-treatment agents containing hexavalent-chromium in heating and ventilation and air-conditioning and refrigeration systems, to reduce or eliminate the potential of exposing individuals to this cancer-causing substance.

11.13-A POTASSIUM DICHROMATE

Potassium dichromate is frequently used commercially as a strong oxidizer; for example, it is a component of the bath solutions used for chromium plating. When heated, potassium dichromate decomposes to generate oxygen as follows:

$$2K_2Cr_2O_7(s) \longrightarrow 2Cr_2O_3(s) + 2K_2O(s) + 3O_2(g)$$

Potassium dichromate   Chromic oxide   Potassium oxide   Oxygen

Because it is a strong oxidizer, potassium dichromate is chemically incompatible with many other substances. For example, potassium dichromate and hydrochloric acid react to produce chlorine.

$$K_2Cr_2O_7(aq) + 14HCl(aq) \longrightarrow 2CrCl_3(aq) + 2KCl(aq) + 7H_2O(l) + 3Cl_2(g)$$

Potassium dichromate Hydrochloric acid   Chromic chloride Potassium chloride   Water   Chlorine

The emergency incidents involving this hazardous material are generally associated with its tendency to oxidize the substances with which it makes contact.

11.13-B CHROMIUM TRIOXIDE

The acids associated with the inorganic chromates and inorganic dichromates exist only in aqueous solutions, but when the water is evaporated from them, a compound formally called *chromium(VI) oxide* remains. This anhydrous compound is also known as *chromium trioxide, chromium anhydride*, and *chromic acid*. Its chemical formula is CrO_3.

Chromium trioxide is a red solid prepared industrially by adding concentrated sulfuric acid to a saturated solution of potassium dichromate.

$$K_2Cr_2O_7(aq) + 2H_2SO_4(conc) \longrightarrow 2KHSO_4(aq) + H_2O(l) + 2CrO_3(s)$$

Potassium dichromate   Sulfuric acid   Potassium bisulfate   Water   Chromium trioxide

Chromium trioxide is used for metal-treatment processes such as chrome plating, copper stripping, aluminum anodizing, and corrosion prevention.

11.13-C CHROMIUM OXYCHLORIDE

Chromium oxychloride is a dark red liquid that forms when a mixture of concentrated hydrochloric acid and sulfuric acid is added to a saturated solution of potassium dichromate. It is also called *chromyl chloride*, and its chemical formula is CrO_2Cl_2.

Chromium oxychloride is not only an oxidizer, but it also reacts very violently with water forming chromium(III) chloride, chromic acid, and chlorine.

$$6CrO_2Cl_2(l) + 4H_2O(l) \longrightarrow 2CrCl_3(aq) + 4H_2CrO_4(aq) + 3Cl_2(g)$$

Chromium oxychloride   Water   Chromium(III) chloride   Chromic acid   Chlorine

In this instance, the chromic acid is the "true" chromic acid (not chromium trioxide), which exists only in solution.

11.13-D AMMONIUM DICHROMATE

Ammonium dichromate is an orange solid used as a component of certain pyrotechnics. It is also used to dye fabrics and to alter the texture of animal hides during the production of leather. Its chemical formula is $(NH_4)_2Cr_2O_7$. When heated, ammonium dichromate decomposes to form nitrogen.

$$(NH_4)_2Cr_2O_7(s) \longrightarrow Cr_2O_3(s) + 4H_2O(g) + N_2(g)$$

Ammonium dichromate   Chromic oxide   Water   Nitrogen

11.13-E WORKPLACE REGULATIONS INVOLVING THE OXIDIZING CHROMIUM COMPOUNDS

When used in the workplace, OSHA requires employers to limit employee exposure to a hexavalent-chromium inhalation concentration of 5 μg/m^3, averaged over an 8-hr workday.

11.13-F TRANSPORTING THE OXIDIZING CHROMIUM COMPOUNDS

When shippers offer an oxidizing chromium compound for transportation, DOT requires them to describe the substance on an accompanying shipping paper as one of the following, as relevant:

> **UN1755, Chromic acid solution, 8, PGII**
> *or*
> **UN1755, Chromic acid solution, 8, PGIII**
> **UN2720, Chromium nitrate, 5.1, PGIII**
> **UN1758, Chromium oxychloride, 8, PGI**
> **UN1463, Chromium trioxide, anhydrous, 5.1, (8), PGII**
> **UN2240, Chromosulfuric acid, 8, PGI**

DOT requires shippers to affix either the appropriate OXIDIZER or CORROSIVE label, or both, on the packaging used to transport it.

When carriers transport any of these chromium compounds in an amount exceeding 1001 lb (454 kg), DOT requires them to post the applicable OXIDIZER or CORROSIVE placards on the bulk packaging or transport vehicle used for shipment.

PERFORMANCE GOALS FOR SECTION 11.14:

- Describe the hazardous properties of the metallic permanganates.
- Identify the industries that use bulk quantities of the metallic permanganates.
- Describe the means shippers and carriers use to inform emergency responders of the hazards associated with encountering a metallic permanganate during a transportation mishap.

11.14 Sodium Permanganate and Potassium Permanganate

Metallic permanganates are compounds in which manganese is in the +7 oxidation state. They are generally encountered as dark violet, iridescent crystals. There are two commercially important members of this class of compounds: *sodium permanganate* and *potassium permanganate*. Their chemical formulas are $NaMnO_4$ and $KMnO_4$, respectively.

metallic permanganate
- An inorganic compound composed of metallic and permanganate ions

11.14-A PRODUCTION AND USES

The metallic permanganates are manufactured from manganese ore containing about 70% manganese dioxide. As the ore is heated with either sodium hydroxide or potassium hydroxide the manganese undergoes oxidation to produce the corresponding alkali metal manganate. For example, potassium manganate is formed from the ore and potassium hydroxide.

$$2MnO_2(s) \;+\; 4KOH(l) \;+\; O_2(g) \longrightarrow 2K_2MnO_4(s) \;+\; 2H_2O(g)$$

Manganese dioxide   Potassium hydroxide   Oxygen   Potassium manganate   Water

A solution of the potassium manganate is then oxidized electrolytically to produce potassium permanganate.

$$2K_2MnO_4(aq) \;+\; 2H_2O(l) \longrightarrow 2KMnO_4(aq) \;+\; 2KOH(aq) \;+\; H_2(g)$$

Potassium manganate   Water   Potassium permanganate Potassium hydroxide   Hydrogen

The potassium permanganate is then separated from the potassium hydroxide by crystallization.

You may be familiar with dilute solutions of sodium permanganate and potassium permanganate that are used pharmaceutically to cure dermatitis having a bacterial or fungal origin ("athlete's foot"). These same solutions are used to treat fish diseases. The concentrated solutions are used to remove objectionable matter from chemical and biological process wastes by oxidation. The effluent gases from many industrial sources are also often eliminated or reduced in concentration through redox reactions involving the use of sodium permanganate or potassium permanganate.

Potassium permanganate has also been used to control the infestation of the quagga mussel (Section 11.8-A) in waterways. Potassium permanganate solutions are injected into pipes where the mussels congregate and thereby disturb the flow of water in the intake and discharge systems. The oxidizing agent kills the pesky mussels.

11.14-B TRANSPORTING METALLIC PERMANGANATES

When shippers offer a metallic permanganate for transportation, DOT requires them to identify the relevant substance on an accompanying shipping paper as one of the following, as relevant:

> **UN1448, Barium permanganate, 5.1, (6.1), PGII (Poison)**
> **UN1457, Calcium permanganate, 5.1, PGII**
> **UN1490, Potassium permanganate, 5.1, PGII**
> **UN1503, Sodium permanganate, 5.1, PGII**
> **UN1515, Zinc permanganate, 5.1, PGII**

When shippers transport barium permanganate, DOT requires them to affix OXIDIZER and POISON labels to its packaging. When they transport the permanganates of calcium, potassium, sodium, or zinc, DOT requires them to affix an OXIDIZER label to their packaging.

When shippers offer any other metallic permanganate for transportation, DOT requires them to identify the compound generically as follows:

> **UN1482, Permanganates, inorganic, n.o.s., 5.1, PGII**
> *or*
> **UN1482, Permanganates, inorganic, n.o.s., 5.1, PGIII**
> **UN3214, Permanganates, inorganic, aqueous solution, n.o.s., 5.1, PGII**

DOT also requires the name of the specific compound to be entered parenthetically in the shipping description. DOT also requires shippers to affix an OXIDIZER label to its packaging.

When carriers transport any metallic permanganates in an amount exceeding 1001 lb (454 kg), DOT requires them to post OXIDIZER placards on the bulk packaging or transport vehicle used for shipment.

PERFORMANCE GOALS FOR SECTION 11.15:

- Describe the hazardous properties of the metallic nitrites and nitrates.
- Describe the means shippers and carriers use to inform emergency responders of the hazards associated with encountering a metallic nitrite or nitrate during a transportation mishap.

metallic nitrite
■ An inorganic compound composed of metallic and nitrite ions

metallic nitrate
■ An inorganic compound composed of metallic and nitrate ions

11.15 Metallic Nitrites and Metallic Nitrates

Metallic nitrites and **metallic nitrates** are compounds composed of metallic ions and the nitrite ion (NO_2^-) and nitrate ion (NO_3^-), respectively. In the metallic nitrites, the nitrogen atom assumes an oxidation number of $+3$, whereas in the metallic nitrates, the oxidation number is $+5$.

Metallic nitrates are especially common oxidizers; for example, sodium nitrate and potassium nitrate are common components of blasting agents and other explosives (Section 15.2). Several metallic nitrates are also used to produce color in activated signaling flares (Section 11.10).

Sodium nitrite and sodium nitrate have been added to raw meat (bacon, hot dogs, and luncheon meats), poultry, and fish for decades to fix their color and inhibit the growth of *Clostridium botulinum*, the bacterial spores that cause botulism. This use was limited by the U.S. Food and Drug Administration when scientists demonstrated that nitrites can produce carcinogenic N-*nitrosamines* in the stomach. The nitrites react with stomach acid to form nitrous acid, which in turn reacts with the protein in meat to form N-nitrosamines, which are organic compounds containing the group of atoms $-N-N=O$. Small quantities have been identified in smoked fish, cured meat, beer, and cheese. Because test animals exposed to N-nitrosamines develop cancer, epidemiologists classify N-nitrosamines as probable human carcinogens.

When individuals ingest or inhale sodium nitrite or sodium nitrate in excess, they are likely to suffer from methemoglobinemia (Section 10.13-B). In rare instances, this ailment can also be contracted by absorption through the skin, e.g., from exposure to molten sodium nitrite or sodium nitrate.

11.15-A SOME PROPERTIES OF METALLIC NITRITES

Metallic nitrites can act as either oxidizing agents or reducing agents. They are oxidized to metallic nitrates and reduced to nitric oxide (NO) as follows:

$$NaNO_2(aq) \ + \ Na_2O_2(aq) \ + \ H_2O(l) \longrightarrow \ NaNO_3(aq) \ + \ 2NaOH(aq)$$

Sodium nitrite   Sodium peroxide   Water   Sodium nitrate   Sodium hydroxide

$$2NaNO_2(aq) \ + \ Na_2SO_3(aq) \ + \ 2HCl(aq) \longrightarrow$$

Sodium nitrite   Sodium sulfite   Hydrochloric acid

$$Na_2SO_4(aq) \ + \ 2NaCl(aq) \ + \ 2NO(g) \ + \ H_2O(l)$$

Sodium sulfate   Sodium chloride   Nitric oxide   Water

Although metallic nitrites are generally stable to heat, they decompose at elevated temperatures. For example, sodium nitrite decomposes as follows:

$$2NaNO_2(s) \longrightarrow Na_2O(s) \ + \ NO_2(g) \ + \ NO(g)$$

Sodium nitrite   Sodium oxide   Nitrogen dioxide   Nitric oxide

11.15-B SOME PROPERTIES OF METALLIC NITRATES

Metallic nitrates react only as oxidizing agents. In the presence of acids, they are converted to nitric oxide, nitrogen dioxide, or ammonia. This chemical behavior is exemplified by the chemical reactions of zinc with concentrated nitric acid noted in Section 8.7.

When heated, metallic nitrates decompose in the following ways:

■ The alkali metal nitrates decompose to form the respective alkali metal nitrite and oxygen. For example, sodium nitrate thermally decomposes to form oxygen.

$$2NaNO_3(s) \longrightarrow 2NaNO_2(s) \ + \ O_2(g)$$

Sodium nitrate   Sodium nitrite   Oxygen

■ Nitrates of the noble metals decompose to produce the metal, nitrogen dioxide, and oxygen. For example, silver nitrate thermally decomposes to produce silver, nitrogen dioxide, and oxygen.

$$2AgNO_3(s) \longrightarrow 2Ag(s) \ + \ 2NO_2(s) \ + \ O_2(g)$$

Silver nitrate   Silver   Nitrogen dioxide   Oxygen

■ Metal nitrates other than the alkali metal nitrates and the noble metal nitrates decompose to form the respective metallic oxide, nitrogen dioxide, and oxygen. For example, lead(II) nitrate thermally decomposes to produce lead oxide, nitrogen dioxide, and oxygen.

$$2Pb(NO_3)_2(s) \longrightarrow 2PbO(s) \ + \ 4NO_2(g) \ + \ O_2(g)$$

Lead nitrate   Lead oxide   Nitrogen dioxide   Oxygen

Write equations that illustrate the thermal decomposition of the oxidizers copper(II) nitrate and chromium(III) nitrate.

Solution: Metallic nitrates other than the alkali metal nitrates and noble metal nitrates decompose when they are heated. These decompositions produce the respective metallic oxide, nitrogen dioxide, and oxygen. Thus, the thermal decompositions of copper(II) nitrate and chromium(III) nitrate are denoted as follows:

$$2Cu(NO_3)_2(s) \longrightarrow 2CuO(s) + 4NO_2(g) + O_2(g)$$

Copper(II) nitrate · · · Copper(II) oxide · Nitrogen dioxide · · Oxygen

$$4Cr(NO_3)_3(s) \longrightarrow 2Cr_2O_3(s) + 12NO_2(g) + 3O_2(g)$$

Chromium(III) nitrate · · Chromium(III) oxide · Nitrogen dioxide · · Oxygen

11.15-C FDA REGULATIONS INVOLVING SODIUM NITRITE AND SODIUM NITRATE

At 21 C.F.R. §§172.170 and 172.175, FDA publishes maximum concentrations of sodium nitrite and/or sodium nitrate that are permissible in certain foods as follows:

- When used or intended for use as a preservative and color fixative, 200 ppm of sodium nitrite and 500 ppm of sodium nitrate in smoked, cured sablefish, salmon, and shad
- When used or intended for use as a preservative and color fixative, 200 ppm of sodium nitrite and 500 ppm of sodium nitrate in meat-curing preparations for the home curing of meat and meat products including poultry and wild game
- When used or intended for use as a color fixative, 10 ppm of sodium nitrite in smoked, cured tunafish products

11.15-D TRANSPORTING METALLIC NITRITES AND NITRATES

When shippers offer a metallic nitrite or metallic nitrate for transportation, DOT requires them to describe the appropriate substance on an accompanying shipping paper by selecting the relevant entry from Table 11.10. When they offer to transport a metallic nitrate or metallic nitrite that is not listed in Table 11.10, DOT requires them to identify the compound generically as one of the following:

> **UN3219, Nitrites, inorganic, aqueous solution, n.o.s., 5.1, PGII**
> *or*
> **UN3219, Nitrites, inorganic, aqueous solution, n.o.s., 5.1, PGIII**
> **UN2627, Nitrites, inorganic, n.o.s., 5.1, PGII**
> *or*
> **UN2627, Nitrites, inorganic, n.o.s., 5.1, PGIII**
> **UN3218, Nitrates, inorganic, aqueous solution, n.o.s., 5.1, PGII**
> **UN3218, Nitrates, inorganic, aqueous solution, n.o.s., 5.1 PGIII**
> **UN1477, Nitrates, inorganic, n.o.s., 5.1, PGII**
> *or*
> **UN1477, Nitrates, inorganic, n.o.s., 5.1, PGIII**

DOT also requires the name of the specific compound to be entered parenthetically in the shipping description. DOT requires shippers who offer barium nitrate, beryllium nitrate, or lead nitrate for transportation to affix OXIDIZER and POISON labels to their packaging. For shipping a metallic nitrate other than barium nitrate, beryllium nitrate, and lead nitrate, DOT requires shippers to affix an OXIDIZER label to the packaging.

TABLE 11.10	Shipping Descriptions of Metallic Nitrites and Nitrates
METALLIC NITRITE OR NITRATE	**SHIPPING DESCRIPTION**
Aluminum nitrate	UN1438, Aluminum nitrate, 5.1, UN1438, PGIII
Barium nitrate	UN1446, Barium nitrate, 5.1, (6.1), PGII (Poison)
Beryllium nitrate	UN2464, Beryllium nitrate, 5.1, (6.1), PGII (Poison)
Calcium nitrate	UN1454, Calcium nitrate, 5.1, PGIII
Ferric nitrate	UN1466, Ferric nitrate, 5.1, PGIII
Lead nitrate	UN1469, Lead nitrate, 5.1, (6.1), PGII (Poison)
Lithium nitrate	UN2722, Lithium nitrate, 5.1, PGIII
Nickel nitrate	UN2725, Nickel nitrate, 5.1, PGIII
Nickel nitrite	UN2726, Nickel nitrite, 5.1, PGIII
Potassium nitrate	UN1498, Potassium nitrate, 5.1, PGII
Potassium nitrite	UN1488, Potassium nitrite, 5.1, PGII
Silver nitrate	UN1493, Silver nitrate, 5.1, PGII
Sodium nitrate	UN1498, Sodium nitrate, 5.1, PGII
Sodium nitrite	UN1500, Sodium nitrite, 5.1, PGII
Strontium nitrate	UN1507, Strontium nitrate, 5.1, PGII
Zinc nitrate	UN1514, Zinc nitrate, 5.1, PGII
Zirconium nitrate	UN1515, Zirconium nitrate, 5.1, PGII

When carriers transport any metallic nitrite or metallic nitrate in an amount exceeding 1001 lb (454 kg), DOT requires them to post OXIDIZER placards on the bulk packaging or transport vehicle used for shipment.

PERFORMANCE GOALS FOR SECTION 11.16

- Describe the hazardous properties of the metallic peroxides and superoxides.
- Describe the means shippers and carriers use to inform emergency responders of the hazards associated with encountering a metallic peroxide or superoxide during a transportation mishap.

11.16 Metallic Peroxides and Superoxides

The compounds composed of metallic ions and peroxide ions [O_2^{2-} or $(-O-O-)^{2-}$] are called **metallic peroxides**, and the compounds composed of metallic ions and superoxide ions [O_2^- or $(-O-O-)^-$] are called *metallic superoxides*. The commercially important metallic peroxides

metallic peroxide
■ An inorganic compound composed of metallic and peroxide ions

are compounds that contain an alkali metal ion or an alkaline earth metal ion. Examples are sodium peroxide and barium peroxide, whose chemical formulas are Na_2O_2 and BaO_2, respectively. There is only one commercially important metallic superoxide: potassium superoxide. Its chemical formula is KO_2.

11.16-A PROPERTIES OF METALLIC PEROXIDES

Sodium peroxide

The metallic peroxides are thermally unstable, highly reactive compounds. When heated, they decompose to the corresponding metallic oxide and oxygen.

$$2Na_2O_2(s) \longrightarrow 2Na_2O(s) + O_2(g)$$
Sodium peroxide   Sodium oxide   Oxygen

$$2BaO_2(s) \longrightarrow 2BaO(s) + O_2(g)$$
Barium peroxide   Barium oxide   Oxygen

They also oxidize finely divided combustible metals like powdered aluminum.

$$2Al(s) + 3Na_2O_2(s) + 3H_2O(l) \longrightarrow Al_2O_3(s) + 6NaOH(s)$$
Aluminum   Sodium peroxide   Water   Aluminum oxide   Sodium hydroxide

Barium peroxide

The metallic peroxides are also water-reactive substances (Section 9.1).

$$2Na_2O_2(s) + 2H_2O(l) \longrightarrow 4NaOH(aq) + O_2(g)$$
Sodium peroxide   Water   Sodium hydroxide   Oxygen

To prevent metallic peroxides from absorbing atmospheric moisture, they should be stored in tightly closed, moisture-proof containers.

11.16-B PROPERTIES OF METALLIC SUPEROXIDES

The metallic superoxides are thermally unstable, water-reactive compounds. Their water reactivity is put to use in a type of chemical oxygen generator that provides breathable oxygen to its user for a period ranging from 5 minutes to several hours. The moisture and carbon dioxide in the user's breath are absorbed by potassium superoxide contained in a canister. The chemical action of the moisture and the superoxide generates oxygen. The oxygen enters a breathing bag for subsequent inhalation, and the exhaled breath repeats the cycle.

$$12KO_2(s) + 6H_2O(g) \longrightarrow 12KOH(s) + 9O_2(g)$$
Potassium superoxide   Water   Potassium hydroxide   Oxygen

$$K_2O(s) + CO_2(g) \longrightarrow K_2CO_3(s)$$
Potassium oxide   Carbon dioxide   Potassium carbonate

11.16-C TRANSPORTING METALLIC PEROXIDES AND SUPEROXIDES

DOT requires shippers who offer a metallic peroxide or superoxide for transportation to identify the relevant substance on an accompanying shipping paper as shown in Table 11.11. DOT generally requires them to affix OXIDIZER labels to its packaging, but DOT requires OXIDIZER *and* POISON labels on packaging containing barium peroxide.

When carriers transport a metallic peroxide or superoxide in an amount exceeding 1001 lb (454 kg), DOT requires them to display OXIDIZER placards on the bulk packaging or transport vehicle used for shipment.

TABLE 11.11 Shipping Descriptions of Metallic Peroxides and Superoxides

METALLIC PEROXIDE OR METALLIC SUPEROXIDE	SHIPPING DESCRIPTION
Barium peroxide	UN1449, Barium peroxide, 5.1, (6.1), PGII (Poison)
Calcium peroxide	UN1457, Calcium peroxide, 5.1, PGII
Lead dioxide[a]	UN1872, Lead dioxide, 5.1, PGIII (Poison)
Lithium peroxide	UN1472, Lithium peroxide, 5.1, PGII
Magnesium peroxide	UN1476, Magnesium peroxide, 5.1, PGII
Potassium peroxide	UN1491, Potassium peroxide, 5.1, PGI
Potassium superoxide	UN1492, Potassium superoxide, 5.1, PGIII
Sodium peroxide	UN1504, Sodium peroxide, 5.1, PGI
Sodium superoxide	UN1505, Sodium superoxide, 5.1, PGIII
Strontium peroxide	UN1509, Strontium peroxide, 5.1, PGII
Zinc peroxide	UN1516, Zinc peroxide, 5.1, PGII

[a]Lead dioxide is the compound having the chemical formula PbO_2. It is correctly named lead(IV) oxide. It is not named lead(II) peroxide, because peroxide ions are not components of this substance.

When shippers offer a metallic peroxide or its solutions that are not listed in Table 11.11 for transportation, DOT requires them to identify the compounds or solutions generically as follows:

UN1483, Peroxides, inorganic, n.o.s., 5.1, PGII
or
UN1483, Peroxides, inorganic, n.o.s., 5.1, PGIII
UN3214, Peroxides, inorganic, aqueous solution, n.o.s., 5.1, PGII
or
UN3214, Peroxides, inorganic, aqueous solution, n.o.s., 5.1, PGIII

DOT also requires the names of the specific compounds to be entered parenthetically in the shipping description. DOT also requires shippers to affix an OXIDIZER label to their packaging.

When carriers transport metallic peroxides in an amount exceeding 1001 lb (454 kg), DOT requires them to post OXIDIZER placards on the bulk packaging or transport vehicle used for shipment.

SOLVED EXERCISE 11.6

What basic description does DOT require shippers to enter on a shipping paper when transporting beryllium peroxide in nonbulk containers?

Solution: Beryllium peroxide is neither listed in Table 11.11 nor the complete Hazardous Materials Table at 49 C.F.R. §172.101. Consequently, DOT requires shippers to identify it generically on a shipping manifest. In Section 10.17, we learned that beryllium compounds are poisonous. Depending on the nature of the packaging used to containerize beryllium peroxide, shippers identify the basic description as either of the following, as relevant:

UN1483, Peroxides, inorganic, n.o.s. (beryllium peroxide), 5.1, (6.1), PGII (Poison)
or
UN1483, Peroxides, inorganic, n.o.s. (beryllium peroxide), 5.1, (6.1), PGIII (Poison)

PERFORMANCE GOALS FOR SECTION 11.17:

- Describe the hazardous properties of the metallic persulfates.
- Describe the means shippers and carriers use to inform emergency responders of the hazards associated with encountering an inorganic persulfate during a transportation mishap.

11.17 Potassium Persulfate and Sodium Persulfate

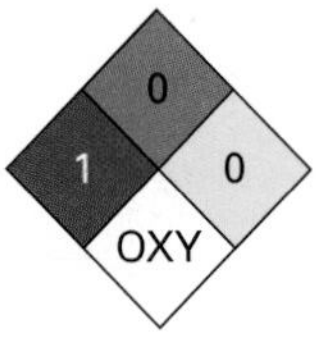

Metallic persulfates are compounds composed of a metallic ion and the *persulfate ion*, also called the *peroxydisulfate ion* ($S_2O_8^{2-}$). Three examples of metallic persulfates are ammonium persulfate, potassium persulfate, and sodium persulfate, whose chemical formulas are $(NH_4)_2S_2O_8$, $K_2S_2O_8$, and $Na_2S_2O_8$, respectively. They are white solids used mainly in the computer industry.

11.17-A PROPERTIES AND USES

The metallic persulfates are the salts of peroxydisulfuric acid, whose chemical formula is $H_2S_2O_8$. The latter substance is a liquid prepared by mixing 30% hydrogen peroxide and 98% sulfuric acid in a 1:3 ratio by volume.

$$2H_2SO_4(l) \;+\; H_2O_2(l) \;\longrightarrow\; H_2S_2O_8(l) \;+\; 2H_2O(l)$$

$\quad$ Sulfuric acid $\qquad$ Hydrogen peroxide $\quad$ Peroxydisulfuric acid $\qquad$ Water

This highly corrosive solution is popular for several technical applications. Most notably, in the computer-chip industry, where it is known as "piranha," peroxydisulfuric acid is used to clean the wafers that are processed into chips.

The metallic persulfates are thermally unstable. For example, oxygen and sulfur dioxide are generated when sodium persulfate is heated.

$$2Na_2S_2O_8(s) \;\longrightarrow\; 2Na_2O(s) \;+\; 4SO_2(g) \;+\; 3O_2(g)$$

$\quad$ Sodium persulfate $\qquad$ Sodium oxide $\qquad$ Sulfur dioxide $\qquad$ Oxygen

11.17-B TRANSPORTING METALLIC PERSULFATES

When shippers offer potassium persulfate or sodium persulfate for transportation, DOT requires them to identify the substance on an accompanying shipping paper as either of the following, as relevant:

UN1492, Potassium persulfate, 5.1, PGII
UN1505, Sodium persulfate, 5.1, PGIII

DOT also requires them to affix an OXIDIZER label to its packaging.

When carriers transport potassium persulfate or sodium persulfate in an amount exceeding 1001 lb (454 kg), DOT requires them to post OXIDIZER placards on the bulk packaging or transport vehicle used for shipment.

When shippers offer metallic persulfates other than potassium persulfate or sodium persulfate and their solutions for transportation, DOT requires them to identify the substances generically as follows:

UN3215, Persulfates, inorganic, n.o.s., 5.1, PGIII
UN3216, Persulfates, inorganic, aqueous solution, n.o.s., 5.1, PGIII

DOT also requires the names of the specific compounds to be entered parenthetically in the shipping description. DOT also requires shippers to affix an OXIDIZER label to their packaging.

When carriers transport potassium persulfate or sodium persulfate in an amount exceeding 1001 lb (454 kg), DOT requires them to post OXIDIZER placards on the bulk packaging or transport vehicle used for shipment.

PERFORMANCE GOALS FOR SECTION 11.18:

- Identify the components of strike-anywhere matches and safety matches.
- Describe the chemical reactions that occur when matches burn.
- Describe the means shippers and carriers use to inform emergency responders of the hazards associated with encountering packages of matches during a transportation mishap.

11.18 Matches

Matches have served as one of the earliest commercial products in which redox reactions were used to provide fire on demand. Even today, they are the most common items used to intentionally initiate fires.

The two types of matches shown in Figure 11.10 are commercially popular. They are called **strike-anywhere matches** and **safety matches**. It is appropriate to examine the chemistry associated with their burning, especially insofar as these processes involve the use of oxidizers.

11.18-A STRIKE-ANYWHERE MATCHES

The head of a strike-anywhere match consists of a mixture of tetraphosphorus trisulfide (also called phosphorus sesquisulfide), sulfur, lead(IV) oxide, powdered glass, and glue. This mixture is mounted on a small stick of wood and covered with paraffin wax. Tetraphosphorus trisulfide ignites as the match head is struck against a hard surface. The evolved heat of combustion initiates the combustion of the sulfur and wood.

$$P_4S_3(s) \quad + \quad 8O_2(g) \quad \longrightarrow \quad P_4O_{10}(s) \quad + \quad 3SO_2(g)$$

Tetraphosphorus trisulfide Oxygen Tetraphosphorus decoxide Sulfur dioxide

11.18-B SAFETY MATCHES

The active components of a safety match consists of a mixture of antimony trisulfide, sulfur, and potassium chlorate, glued to a piece of cardboard. A safety match ignites by means of friction against a prepared surface consisting of red phosphorus and powdered glass. This surface is

match
- An item used to intentionally initiate fire when the mixture of substances contained in its heads is rubbed against a hard surface

strike-anywhere match
- A match having a "bulls-eye" head that consists of a white tip containing tetraphosphorus trisulfide that is actuated by rubbing it against an abrasive strip

safety match
- A match designed to ignite only when its head is rubbed on a prepared surface consisting of red phosphorus and powdered glass

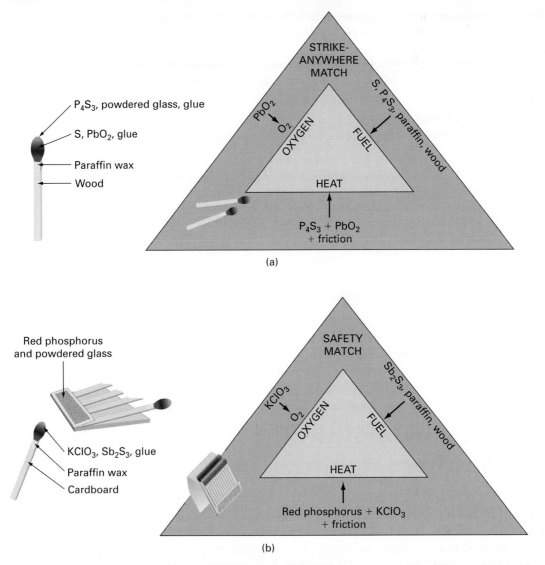

FIGURE 11.10 The strike-anywhere match (a) ignites when the surface of the match head is rubbed against a hard surface. The safety match (b) ignites when the surface of a match head is struck against a surface containing red phosphorus and powdered glass. The fire triangles to the right of each match illustrate the composition of the fuels, the substances that react to provide the initial heat, and the relevant oxidizer.

usually on the box, book, or card. When the match head is rubbed on this surface, the heat initiates a redox reaction as follows:

$$16KClO_3(s) \quad + \quad 3S_8(s) \quad \longrightarrow \quad 16KCl(s) \quad + \quad 24SO_2(g)$$

Potassium chlorate   Sulfur   Potassium chloride   Sulfur dioxide

Then, the energy evolved from the oxidation of the sulfur ignites the antimony trisulfide.

$$2Sb_2S_3(s) \quad + \quad 9O_2(g) \quad \longrightarrow \quad 2Sb_2O_3(g) \quad + \quad 6SO_2(g)$$

Antimony trisulfide   Oxygen   Antimony trioxide   Sulfur dioxide

11.18-C TRANSPORTING MATCHES

When shippers offer matches for transportation, DOT requires them to describe the appropriate type on an accompanying shipping paper as one of the following, as relevant:

UN1944, Matches, safety book, card, 4.1, PGIII
UN1331, Matches, strike anywhere, 4.1, PGIII

DOT also requires shippers to affix a FLAMMABLE SOLID label to their packaging.

When carriers transport matches in an amount exceeding 1001 lb (454 kg), DOT requires them to post FLAMMABLE SOLID placards on the bulk packaging or transport vehicle used for shipment.

PERFORMANCE GOALS FOR SECTION 11.19:

- Describe the response actions to be executed when an oxidizer has been released into the environment.

11.19 Responding to Incidents Involving a Release of Oxidizers

First-on-the-scene responders can generally identify the presence of a metallic oxidizer at a transportation mishap by observing at least one of the following:

- The number 5.1 as a component of a shipping description of a hazardous material listed on a shipping paper
- The word OXIDIZER and the number 5.1 printed on yellow labels affixed to packaging
- The word OXIDIZER and the number 5.1 printed on yellow placards posted on each side and each end of a transport vehicle containing 1001 lb (454 kg) or more of an oxidizer

Figure 11.11 shows that the recommended method of extinguishing fires supported by most liquid or solid oxidizers is to deluge them with water. Metallic oxidizers are generally soluble in water. On dilution, their chemical reactivity is sharply reduced or eliminated.

Emergency responders need to exercise caution when using water on metallic oxidizers that are present during fires. Many solid oxidizers melt before they decompose. These hot molten materials flow to adjoining areas, where they may mix with combustible material and support its ignition. Although water may be applied to fires supported by oxidizers, it should never be applied to bulk quantities of molten oxidizers. Steam is rapidly generated and may cause the hot molten material to splatter. When emergency responders encounter bulk quantities of molten oxidizers, experts recommend the use of sand to prevent their spread.

SOLVED EXERCISE 11.7

Why is it essential to dam or dike the water runoff generated during a response action involving a solid oxidizer?

Solution: When water is used during a response action involving a solid oxidizer, its primary function is to dilute the oxidizer appreciably so that the rate of an oxidation–reduction reaction is slowed. When the runoff is left unattended, however, the water evaporates and leaves the oxidizer in a dry state, whereupon it is again susceptible to supporting the combustion of matter. For this reason, the water runoff generated during a response action involving a solid oxidizer should always be dammed or diked so the oxidizer may be neutralized prior to its ultimate discharge. Sodium sulfite is a typical neutralizing agent.

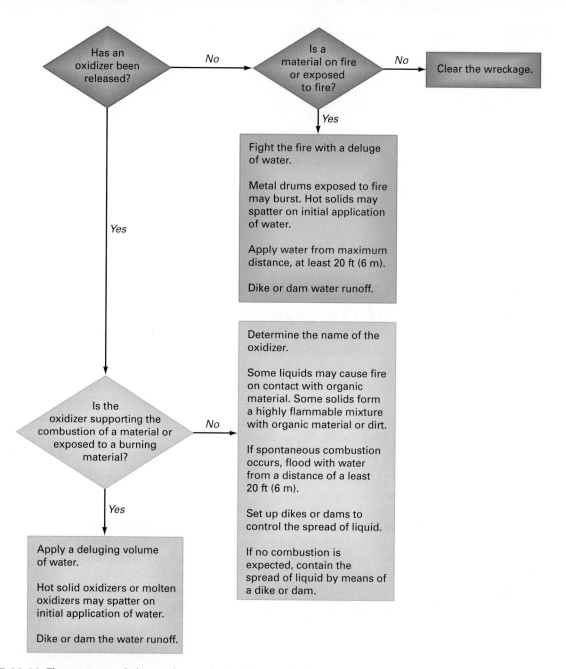

FIGURE 11.11 The recommended procedures to be implemented when responding to a transportation mishap involving the release of an oxidizer other than an organic peroxide from its packaging. (*Adapted with permission of the American Society for Testing and Materials, from a figure in ASTM STP 825,* A Guide to the Safe Handling of Hazardous Materials Accidents, *second edition. Copyright © 1990, American Society for Testing and Materials.*)

Basic Concepts Involving Oxidation–Reduction

11.1 What is the oxidation number of each underlined atom or ion in the following chemical formulas?

(a) Ba<u>O</u> and Ba<u>O</u>$_2$
(b) K<u>Mn</u>O$_4$ and <u>Mn</u>O$_2$
(c) H$_2$<u>S</u>O$_3$ and H$_2$<u>S</u>O$_4$
(d) <u>Pb</u>O and <u>Pb</u>O$_2$
(e) <u>F</u>$_2$ and H<u>F</u>
(f) <u>O</u>$_2$ and H$_2$<u>O</u>
(g) Na<u>H</u> and <u>H</u>$_2$O

11.2 Identify the oxidizing agent, the reducing agent, the substance oxidized, and the substance reduced in the redox reaction noted by the following equation:

$$8Al(s) + 3NaClO_4(s) \longrightarrow 4Al_2O_3(s) + 3NaCl(s)$$

Hydrogen Peroxide

11.3 What is the most likely reason DOT prohibits the transportation on passenger-carrying aircraft and railcars of any amount of a hydrogen peroxide solution that contains more than 40% hydrogen peroxide by mass?

11.4 Why is the use of water generally the best means of extinguishing fires in which hydrogen peroxide is involved?

Hypochlorite Oxidizers

11.5 Before the advent of plastic bottles, sodium hypochlorite solutions were stored in brown glass bottles. Why was a container of this type chosen?

11.6 Why do acid manufacturers typically affix the following label to containers of muriatic acid intended for use by homeowners in their swimming pools?

> **CAUTION**
>
> **DO NOT STORE NEAR
> CHLORINE-PRODUCING
> POOL CHEMICALS**

Chlorination Agents Other Than Inorganic Hypochlorites

11.7 Why must carriers take precaution to segregate trichloroisocyanuric acid from combustible matter when loading it onboard railcars for shipment?

Oxidizers in Fireworks, Flares, and Signaling Smokes

11.8   Strontium perchlorate is the component of railroad flares responsible for the carmine red light emitted when they are ignited. Write the equation denoting this thermal decomposition.

Oxidizing Ammonium Compounds

11.9   The OSHA regulation at 29 C.F.R. §1910.109(c)(4)(iii)(*a*) permits employers to store ammonium nitrate in a building having a basement but *only* when the basement is open on at least one side. What is the most likely reason OSHA requires one side of the basement to be open?

11.10 The OSHA regulation at 29 C.F.R. §1910.109(c)(4)(iii)(*d*) requires employers to protect the flooring and handling areas in buildings against impregnation by ammonium nitrate and requires that the floors be constructed without open drains, traps, tunnels, pits, or pockets into which any molten ammonium nitrate could flow and be confined in the event of fire. What is the most likely reason OSHA mandates these structural features?

Other Oxidizers

11.11 A chemical manufacturer in Denver, Colorado, intends to ship sixteen 0.5-lb bottles of sodium peroxide to a customer in Dallas, Texas, by cargo aircraft. Use Table 6.1 to determine what shipping description DOT requires the manufacturer to enter on the accompanying shipping paper.

11.12 Write the equation that illustrates the thermal decomposition of each of the following inorganic nitrates:

   (a) lithium nitrate
   (b) zinc nitrate
   (c) nickel nitrate

Fighting Fires Involving Oxidizers

11.13 When properly posted on the exterior wall of a building, what entry on an NFPA hazard diamond conveys that an oxidizer is stored therein to on-duty emergency responders?

11.14 Why is the use of carbon dioxide generally ineffective for extinguishing fires supported by oxidizers?

Transporting Oxidizers Other Than Organic Peroxides

11.15 What markings, other than specification markings, does DOT require a chemical facility to provide on the exterior surface of packaging containing a plastic bottle of each of the following substances when it is offered to a carrier as freight by rail in an amount less than its reportable quantity:

   (a) barium hypochlorite
   (b) sodium hypochlorite solutions

Organic compounds are constituents of many commercial products including heating and motor fuels, solvents, plastics, resins, fibers, surface coatings, refrigerants, textiles, and explosives. Their relative commonplace underscores the need to study them in some detail. Many organic compounds of commercial interest are flammable gases or flammable liquids. It is hardly surprising to learn that many organic compounds are burning when they are encountered at transportation mishaps and other emergency scenes. Although the potential for fire and explosion is generally their primary hazard, exposure to many organic compounds also poses a health risk. Prolonged exposure to certain organic compounds causes a variety of acute and chronic health effects that includes damage to the liver, kidneys, and heart, depression of the central nervous system, and the onset of cancer.

This chapter is an introduction to the study of organic compounds and the properties of some hydrocarbons and halogenated hydrocarbons that are commonly encountered commercially. Chapters 13, 14, and 15 discuss the properties of more complex organic compounds.

PERFORMANCE GOALS FOR SECTION 12.1:

- Discuss how a carbon atom covalently bonds to nonmetallic atoms including other carbon atoms.
- Describe the nature of carbon–carbon single bonds, carbon–carbon double bonds, and carbon–carbon triple bonds.

12.1 What Are Organic Compounds?

The molecules of all organic compounds have one common feature: the presence of one or more carbon atoms. In most instances, the carbon atoms share electrons with other nonmetallic atoms. As shown by their molecular structures in Figure 12.1, methane, carbon tetrachloride, carbon monoxide, and carbon disulfide are examples of organic compounds having molecules in which the carbon atoms are bonded to other nonmetallic atoms.

H–C–H   Cl–C–Cl   C≡O   S=C=S

Methane   Carbon tetrachloride   Carbon monoxide   Carbon disulfide

FIGURE 12.1 A carbon atom may share its electrons with the electrons of hydrogen, chlorine, oxygen, and sulfur. The compounds that result from this electron sharing are methane, carbon tetrachloride, carbon monoxide, and carbon disulfide, respectively.

Carbon atoms can also mutually share electrons with other carbon atoms. When the molecular structures of such compounds are examined, we find that two carbon atoms can share electrons to form any of the following: **carbon–carbon single bonds** (C—C); **carbon–carbon double bonds** (C=C); and **carbon–carbon triple bonds** (C≡C). Carbon–carbon single, double, and triple bonds consist of one, two, and three pairs of shared electrons, respectively, between two carbon atoms.

Figure 12.2 illustrates the bonding in molecules of ethane, ethylene, and acetylene, each of which has two carbon atoms. The molecules of ethane have carbon–carbon single bonds, molecules of ethylene have carbon–carbon double bonds, and molecules of acetylene have carbon–carbon triple bonds.

The covalent bonds between carbon atoms in the molecules of more complex organic compounds may be linked into chains, including branched chains, or into rings. Consider just the carbon skeleton of the following two molecules noted as (a) and (b):

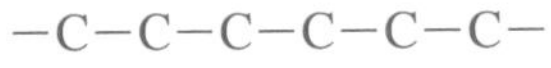

$$-C-C-C-C-C-C-$$

(a)

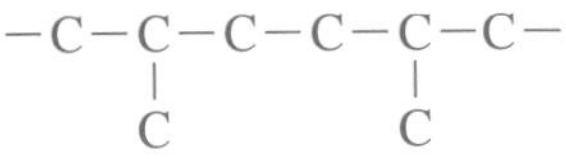

$$-C-C-C-C-C-C-$$
$$\qquad\quad | \qquad\qquad\quad |$$
$$\qquad\quad C \qquad\qquad\quad C$$

(b)

carbon–carbon single bond
■ A shared pair of electrons between two carbon atoms

carbon–carbon double bond
■ Two shared pairs of electrons between two carbon atoms

carbon–carbon triple bond
■ Three shared pairs of electrons between two carbon atoms

In (a), the carbon atoms are bonded to one another in a continuous chain, whereas in (b), the carbon atoms are bonded to one another in a branched pattern. This means that carbon atoms bond not only to other carbon atoms in long chains but also to groups of other carbon atoms as side chains attached to the main chain.

Aside from being bonded to other carbon atoms, each carbon atom in either a straight-chain or branched-chain arrangement is typically bonded to one or more atoms of hydrogen, oxygen, nitrogen, sulfur, or a halogen. For instance, when the carbon atoms in the compound having the carbon skeleton (a) are bonded to hydrogen atoms, the molecular structure of the compound is written as follows:

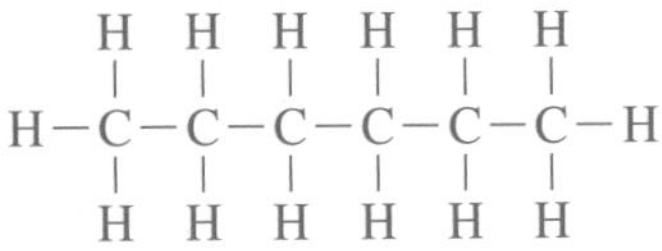

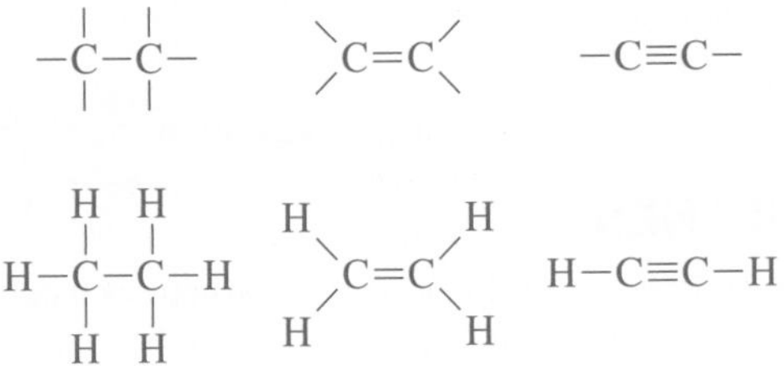

$$\text{H–C–C–C–C–C–C–H}$$

-C-C-   C=C   -C≡C-

H H   H   H
H–C–C–H   C=C   H–C≡C–H
H H   H   H

FIGURE 12.2 Two carbon atoms may share their electrons in either of three ways, resulting in the formation of carbon–carbon single bonds, carbon–carbon double bonds, and carbon–carbon triple bonds. When carbon atoms unite with hydrogen atoms, the compounds that result are ethane, ethene, and acetylene, respectively.

When the carbon atoms in (b) are bonded to hydrogen atoms, the molecular structure of the compound is written as follows:

$$
\begin{array}{cccccc}
& H & H & H & H & H & H \\
& | & | & | & | & | & | \\
H- & C- & C- & C- & C- & C- & C-H \\
& | & | & | & | & | & | \\
& H & | & H & H & | & H \\
& & H-C-H & & H-C-H & \\
& & | & & | & \\
& & H & & H &
\end{array}
$$

hydrocarbon
■ A compound whose molecules are composed solely of carbon and hydrogen atoms

aliphatic hydrocarbon
■ A compound containing only carbon and hydrogen atoms and whose molecules are not composed of benzene or benzene-like structures

aromatic hydrocarbon
■ A compound whose molecules are composed solely of carbon and hydrogen atoms, at least some of which are structurally similar to benzene

These structures illustrate that many organic compounds consist of continuous or branched chains of carbon atoms. The actual number of such compounds appears to be virtually unlimited. More than 2 million organic compounds are already known.

We begin the study of organic compounds by examining the simplest group, the **hydrocarbons**. These are compounds whose molecules are composed of only carbon and hydrogen atoms. All hydrocarbons are broadly divided into two groups: aliphatic and aromatic hydrocarbons. **Aliphatic hydrocarbons** are nonaromatic hydrocarbons. The **aromatic hydrocarbons** are best characterized by a common feature of their molecular structures: They contain benzene rings. We examine this feature in more detail in Section 12.12.

PERFORMANCE GOALS FOR SECTION 12.2:

■ Describe the chemical bonds that exist in molecules of the alkanes and cycloalkanes.
■ Illustrate that alkanes may have structural isomers.
■ Memorize the rules for naming simple alkanes.
■ Describe the means shippers and carriers use to inform emergency responders of the hazards associated with encountering alkanes and cycloalkanes during transportation mishaps.

12.2 Alkanes and Cycloalkanes

alkane
■ A hydrocarbon whose molecules are composed of only carbon–carbon single bonds (C—C)

saturated hydrocarbon
■ A hydrocarbon having molecules that are composed solely of C—C and C—H bonds

An **alkane** is a hydrocarbon whose molecules have only carbon–carbon single bonds. The general chemical formula of an alkane is C_nH_{2n+2}, where n is a nonzero integer. When the number of carbon atoms is only 1, the number of hydrogen atoms is 4. The corresponding compound is named methane. Its chemical formula is CH_4. When the number of carbon atoms is 2, the number of hydrogen atoms is 6. The corresponding compound is named ethane, and its chemical formula is C_2H_6. The alkanes are called **saturated hydrocarbons**, since the four bonding electrons of each carbon atom are shared with the bonding electrons of four other atoms. Several examples of alkanes having carbon atoms from one to eight are noted in Table 12.1. Their names and formulas should be memorized.

12.2-A FORMULAS OF THE ALKANES

Although the straight-chain arrangements have been provided for the molecular structures listed in Table 12.1, several molecular structures can correctly be written for the alkanes beginning with butane. Butane molecules have four carbon atoms per molecule, which can be represented by their Lewis structures as follows:

```
    H  H  H  H              H  H  H
    |  |  |  |              |  |  |
 H—C—C—C—C—H           H—C—C—C—H
    |  |  |  |              |  |  |
    H  H  H  H              H  |  H
                               H—C—H
                                  |
                                  H
```

TABLE 12.1	Simple Alkanes Having a Continuous Chain of Carbon Atoms		
NAME	**MOLECULAR FORMULA**	**LEWIS STRUCTURE**	**CONDENSED FORMULA**
Methane	CH_4	$H-C(-H)(-H)-H$	CH_4
Ethane	C_2H_6	$H-C-C-H$	CH_3CH_3
Propane	C_3H_8	$H-C-C-C-H$	$CH_3CH_2CH_3$
Butane	C_4H_{10}	$H-C-C-C-C-H$	$CH_3CH_2CH_2CH_3$
Pentane	C_5H_{12}	$H-C-C-C-C-C-H$	$CH_3CH_2CH_2CH_2CH_3$
Hexane	C_6H_{14}	$H-C-C-C-C-C-C-H$	$CH_3CH_2CH_2CH_2CH_2CH_3$
Heptane	C_7H_{16}	$H-C-C-C-C-C-C-C-H$	$CH_3CH_2CH_2CH_2CH_2CH_2CH_3$
Octane	C_8H_{18}	$H-C-C-C-C-C-C-C-C-H$	$CH_3CH_2CH_2CH_2CH_2CH_2CH_2CH_3$

In the first structure, the carbon atoms are bonded to one another in a continuous chain. In the second structure, only three of the carbon atoms are bonded in a continuous chain. The fourth carbon atom is bonded to the carbon atom in the middle of the chain.

Because there are two correct ways to write molecular structures for the formula C_4H_{10}, there are two distinct compounds having this formula. To distinguish them by name, the compound having the carbon atoms bonded in a continuous chain is called n-*butane*, whereas the compound with the branched structure is named *isobutane*.

The phenomenon associated with two or more compounds that have the same molecular formula but different structural arrangements of their atom is called **structural isomerism**. One way by which structural isomers are identified is with an *n-* (for "normal") in front of the name of the compound that contains a continuous chain of carbon atoms, and *iso-* (meaning "the same") in front of the name of the compound that has a methyl group (CH_3-) bonded to a carbon atom next to the terminal (end) carbon atom. This system is generally referred to as the **common system of nomenclature**, and the compound names are called **common names**.

Although a molecular structure may be written for any alkane, it is generally convenient to condense it by simply writing the symbols of the atoms next to the symbol of the carbon atom to which they are bonded. When writing the formula for an alkane, the symbols of the hydrogen atoms are written next to the symbols of the carbon atoms to which they are bonded. The formulas derived in this manner are called **condensed formulas**. For example, the condensed formulas for *n*-butane and isobutane are noted here:

$CH_3CH_2CH_2CH_3$ $CH_3-CH-CH_3$

                                     CH_3

(*n*-Butane)   Isobutane

Condensed formulas convey the same bonding information as the more complete Lewis structures, but the dashes are either entirely omitted or used only in a limited fashion.

12.2-B FORMULAS OF THE CYCLOALKANES

It is possible to write formulas for the alkanes in which the first and last carbon atoms in a continuous chain are chemically linked to each other in a cyclic arrangement. These compounds are called **cycloalkanes**. Their general chemical formula is C_nH_{2n}. They are named by placing the prefix *cyclo-* in front of the name of the parent hydrocarbon. The first four cycloalkanes are cyclopropane, cyclobutane, cyclopentane, and cyclohexane.

Sometimes the structural formulas for cyclopropane, cyclobutane, and cyclopentane are represented by writing a triangle, square, and pentagon for C_3H_6, C_4H_8, and C_5H_{10}, respectively. The structural formula of cyclohexane (C_6H_{12}) is represented by two unique conformations called the *boat and chair conformations*.

Cyclopropane   Cyclobutane   Cyclopentane   (boat)   Cyclohexane   (chair)

In this book, the chair conformation solely is used to represent cyclohexane.

structural isomerism
■ The phenomenon associated with compounds whose molecules have the same atoms but differ in the manner in which the atoms are structurally arranged

common system of nomenclature
■ The system used historically for the naming of organic compounds

common name
■ The historical name for an organic compound

condensed formula
■ The formula of a compound in which symbols of the atoms are written next to the symbol of the atoms to which they are bonded and which dashes are omitted or used only in a limited fashion

cycloalkane
■ A hydrocarbon whose molecules are composed of a cyclic ring of carbon–carbon single bonds

TABLE 12.2 | Some Common Alkyl Substituents

NAME[a]	CHEMICAL FORMULA
Methyl	$-CH_3$
Ethyl	$-CH_2CH_3$, or C_2H_5-
n-Propyl	$-CH_2CH_2CH_3$, or C_3H_7-
Isopropyl	CH_3-CH-, or $(CH_3)_2CH-$ $\qquad \overset{\displaystyle |}{CH_3}$
n-Butyl	$-CH_2CH_2CH_2CH_3$, or C_4H_9-
Isobutyl	$\qquad \overset{\displaystyle CH_3}{\overset{\displaystyle |}{}}$ $-CH_2-CH$, or $(CH_3)_2CHCH_2-$ $\qquad \overset{\displaystyle |}{CH_3}$
sec-Butyl[b]	$-CH-CH_2CH_3$, or CH_3CH_2-CH- $\;\; \overset{\displaystyle |}{CH_3} \qquad\qquad\qquad\quad \overset{\displaystyle |}{CH_3}$
tert-Butyl[b]	$\qquad \overset{\displaystyle CH_3}{\overset{\displaystyle |}{}}$ CH_3-C-, or $(CH_3)_3-C-$ $\qquad \overset{\displaystyle |}{CH_3}$
n-Pentyl[c]	$-CH_2CH_2CH_2CH_2CH_3$, or $C_5H_{11}-$

[a]The alkyl groups are named by replacing the -ane suffix of the parent hydrocarbon with -yl.
[b]The carbon atom to which a substituent is bonded may be identified as primary, secondary, or tertiary. If it is bonded to one other carbon atom, it is primary; if bonded to two carbon atoms, it is secondary (sec-); and if bonded to three carbon atoms, it is tertiary (tert-).
[c]Also called n-amyl.

12.2-C THE IUPAC SYSTEM OF NOMENCLATURE

When a hydrogen atom is removed from an alkane, the resulting group is called an **alkyl group** or **alkyl substituent**. Their general chemical formula is C_nH_{2n+1}. The most familiar examples are the **methyl group**, **ethyl group**, and **n-propyl group**.

$$-CH_3 \qquad -CH_2CH_3 \qquad -CH_2CH_2CH_3$$
$$\text{Methane} \qquad \text{Ethyl group} \qquad \text{n-Propyl group}$$

The names of the common alkyl substituents are provided in Table 12.2. These names and their molecular formulas should be memorized.

Although the simple hydrocarbons are generally referred to by their common names, the complex hydrocarbons are difficult to name using the common system of nomenclature. To overcome this hurdle, a second system of nomenclature was devised that is both simple and systematic. It was adopted by the International Union of Pure and Applied Chemistry (IUPAC) and is called the **IUPAC system of nomenclature**. The names of organic compounds derived from use of the IUPAC system are called **IUPAC names**.

---

alkyl substituent (alkyl group)
■ Any group of atoms having the general chemical formula C_nH_{2n+1}, obtained by removing a hydrogen atom from the formula for an alkane

methyl group
■ The designation for the group of atoms CH_3-

ethyl group
■ The designation for the group of atoms CH_3CH_2-

n-propyl group
■ The designation for the group of atoms $CH_3CH_2CH_2-$

IUPAC system of nomenclature
■ The internationally recognized system adopted by the International Union of Pure and Applied Chemistry for the naming of organic compounds

IUPAC names
■ The names of organic compounds derived from use of the system adopted by the International Union of Pure and Applied Chemistry

In the IUPAC system, the following rules apply to the naming of alkanes from their molecular structures:

- Locate the longest chain of carbon atoms in the structure. This is called the "main chain." It is not necessary that the main chain be written horizontally to be continuous.
- Assign numbers consecutively to each carbon atom in the main chain starting from the end that gives the alkyl substituents attached to the chain the smaller numbers.
- Designate the position of each substituent by the number of the carbon atom along the main chain to which it is attached.
- Name the substituents alphabetically (ethyl before methyl, and so on) and place the names of the substituents as prefixes on the name of the main chain.
- If several substituents occur in the same compound, indicate the number of identical groups by the use of the following prefixes before the name of the substituent: *di-* for two identical groups, *tri-* for three identical groups, *tetra-* for four identical groups, and so on. The prefixes *di*, *tri*, *tetra*, *sec*, and *tert* are ignored when alphabetizing the substituents, but the prefixes *iso* and *cyclo* are not ignored.

The use of these IUPAC rules of nomenclature is illustrated in Table 12.3.

TABLE 12.3	Examples of Using the IUPAC Rules for Naming Alkanes When Their Chemical Formulas Are Known
ALKANE	**CHEMICAL FORMULA**
3-Methylheptane	$CH_3CH_2-CH-CH_3$ $\mid$ CH_2 $\mid$ CH_2 $\mid$ CH_2 $\mid$ CH_3
2,2-Dimethylbutane	CH_3 $\mid$ $CH_3-C-CH_2CH_3$ $\mid$ CH_3
2,4-Dimethylhexane	$CH_3-CH-CH_2-CH-CH_3$ $\mid \qquad\quad \mid$ $CH_3 \qquad CH_2$ $\mid$ CH_3
3-Ethyl-4-isopropylheptane	CH_3 $\mid$ $CH_3-CH-CH-CH_2CH_2CH_3$ $\mid$ $CH_3CH_2-CH-CH_2CH_3$
5-Ethyl-2,5-dimethylheptane	CH_2CH_3 $\mid$ $CH_3CH_2-C-CH_2CH_2-CH-CH_3$ $\mid \qquad\qquad\qquad \mid$ $CH_3 \qquad\qquad\quad CH_3$

Using the IUPAC system, name the alkane having the following condensed formula:

$$CH_3-CH-CH_2-CH-CH_2-CH_3$$
$$\quad\quad\;\; |\quad\quad\quad\quad\; |$$
$$\quad\quad\; CH_3\quad\quad\quad CH_2$$
$$\quad\quad\quad\quad\quad\quad\quad\quad |$$
$$\quad\quad\quad\quad\quad\quad\quad\quad CH_3$$

Solution: First, identify the longest continuous chain of carbon atoms. By examination, we see that the longest chain contains six carbon atoms, which signifies that the compound is a derivative of hexane. Next, assign a number to each carbon atom of this longest chain.

$$\overset{1}{CH_3}-\overset{2}{CH}-\overset{3}{CH_2}-\overset{4}{CH}-\overset{5}{CH_2}-\overset{6}{CH_3}$$
$$\quad\quad\;\; |\quad\quad\quad\quad\; |$$
$$\quad\quad\; CH_3\quad\quad\quad CH_2$$
$$\quad\quad\quad\quad\quad\quad\quad\quad |$$
$$\quad\quad\quad\quad\quad\quad\quad\quad CH_3$$

We see that the methyl group is bonded to the carbon atom numbered 2, and the ethyl group is bonded to the carbon atom numbered 4. Because the alkyl substituents are named in alphabetical order, the compound is correctly named 4-ethyl-2-methylhexane.

It is incorrect to number the carbon atoms from right to left along the horizontal chain of carbon atoms, since this numbering scheme gives larger numbers to the alkyl groups bonded to the main chain (5 for the methyl group and 3 for the ethyl group.) However, it is correct to number the longest chain of carbon atoms in either of the following ways:

$$CH_3-\overset{2}{CH}-\overset{3}{CH_2}-\overset{4}{CH}-\overset{5}{CH_2}-\overset{6}{CH_3}$$
$$\quad\quad\;\; |\quad\quad\quad\quad\; |$$
$$\quad\quad\; {}_1CH_3\quad\quad\quad CH_2$$
$$\quad\quad\quad\quad\quad\quad\quad\quad |$$
$$\quad\quad\quad\quad\quad\quad\quad\quad CH_3$$

$$CH_3-\overset{2}{CH}-\overset{3}{CH_2}-\overset{4}{CH}-CH_2-CH_3$$
$$\quad\quad\;\; |\quad\quad\quad\quad\; |$$
$$\quad\quad\; {}_1CH_3\quad\quad\quad {}_5CH_2$$
$$\quad\quad\quad\quad\quad\quad\quad\quad |$$
$$\quad\quad\quad\quad\quad\quad\quad\quad {}_6CH_3$$

The use of either of these numbering schemes also gives 4-ethyl-2-methylhexane as the correct name of this compound.

The detailed analysis of a commercial rubber solvent provides the following approximate chemical composition:

(a) 50% by volume is a mixture of 2,3-dimethylbutane, 2,3-dimethylpentane, and 3,3-dimethylpentane;

(b) 30% is a mixture of 2,2-dimethylbutane, 3-methylpentane, 2,2-dimethylpentane, methyl-cyclopentane, methylcyclohexane, 1,2-dimethylcyclopentane, *n*-hexane, and *n*-heptane; and

(c) 20% is a mixture of *n*-butane, *n*-pentane, and 2,4-dimethylhexane.

Write the chemical formula for each substance.

Solution: The solvent's constituents have the following condensed formulas:

(a)

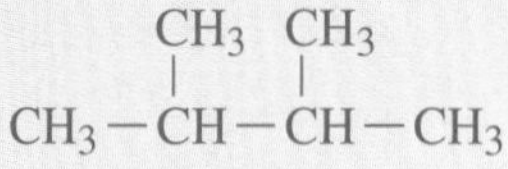

$$CH_3-\overset{\overset{\displaystyle CH_3}{|}}{CH}-\overset{\overset{\displaystyle CH_3}{|}}{CH}-CH_3$$

2,3-Dimethylbutane

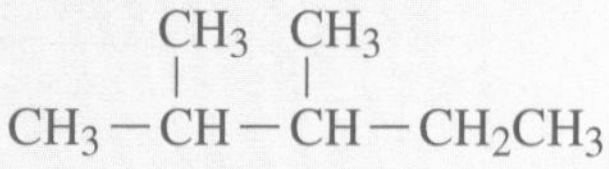

$$CH_3-\overset{\overset{\displaystyle CH_3}{|}}{CH}-\overset{\overset{\displaystyle CH_3}{|}}{CH}-CH_2CH_3$$

2,3-Dimethylpentane

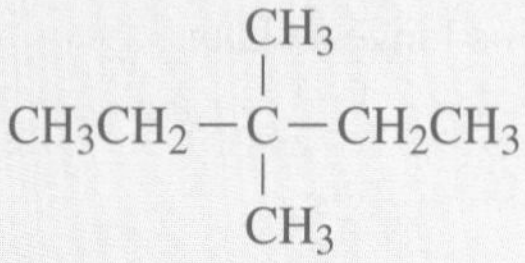

$$CH_3CH_2-\overset{\overset{\displaystyle CH_3}{|}}{\underset{\underset{\displaystyle CH_3}{|}}{C}}-CH_2CH_3$$

3,3-Dimethylpentane

(b)

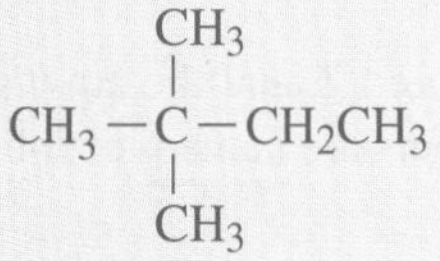

$$CH_3-\overset{\overset{\displaystyle CH_3}{|}}{\underset{\underset{\displaystyle CH_3}{|}}{C}}-CH_2CH_3$$

2,2-Dimethylbutane

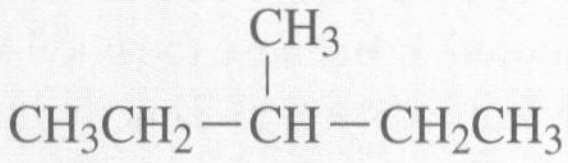

$$CH_3CH_2-\overset{\overset{\displaystyle CH_3}{|}}{CH}-CH_2CH_3$$

3-Methylpentane

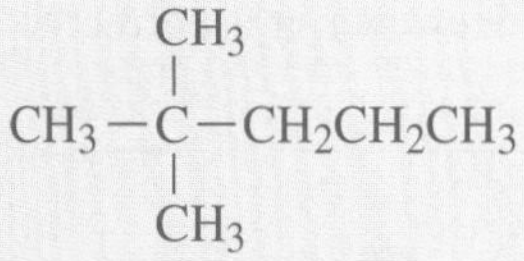

$$CH_3-\overset{\overset{\displaystyle CH_3}{|}}{\underset{\underset{\displaystyle CH_3}{|}}{C}}-CH_2CH_2CH_3$$

2,2-Dimethylpentane

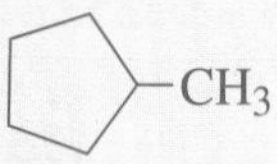

Methylcyclopentane

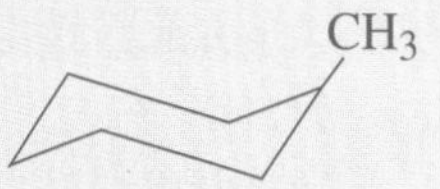

Methylcyclohexane

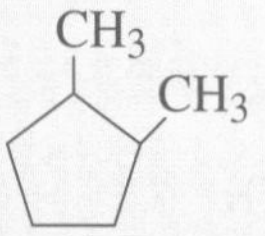

1,2-Dimethylcyclopentane

(c) $CH_3CH_2CH_2CH_3$
n-Butane

$CH_3CH_2CH_2CH_2CH_3$
n-Pentane

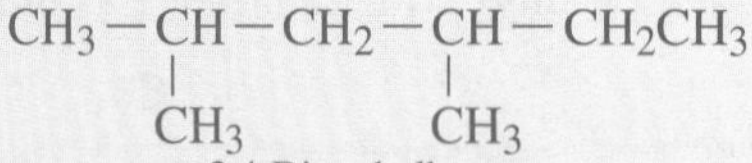

$$CH_3-\overset{\overset{\displaystyle CH_3}{|}}{CH}-CH_2-\overset{\overset{\displaystyle CH_3}{|}}{CH}-CH_2CH_3$$

2,4-Dimethylhexane

12.2-D TRANSPORTING ALKANES AND CYCLOALKANES

When shippers offer an alkane or cycloalkane for transportation, DOT requires them to enter the relevant basic description from Table 12.4 on an accompanying shipping paper. DOT also requires them to affix either a FLAMMABLE GAS or FLAMMABLE LIQUID label on its packaging.

When carriers transport an alkane or cycloalkane in an amount exceeding 1001 lb (454 kg), DOT requires them to post the applicable FLAMMABLE GAS or FLAMMABLE placards on the bulk packaging or transport vehicle used for shipment.

TABLE 12.4	Basic Descriptions of Some Alkanes and Cycloalkanes

ALKANE, CYCLOALKANE, OR GROUPS THEREOF	BASIC DESCRIPTION
Butane	UN1011, Butane, 2.1
Cyclobutane	UN2601, Cyclobutane, 2.1
Cycloheptane	UN2241, Cycloheptane, 3, PGII
Cyclohexane	UN1145, Cyclohexane, 3, PGII
Cyclopentane	UN1146, Cyclopentane, 3, PGII
Cyclopropane	UN1027, Cyclopropane, 2.1
n-Decane	UN2247, n-Decane, 3, PGII
2,3-Dimethylbutane	UN2457, 2,3-Dimethylbutane, 3, PGII
Dimethylcyclohexanes	UN2263, Dimethylcyclohexanes, 3, PGII
2,2-Dimethylpropane	UN2044, 2,2-Dimethylpropane, 2.1
Ethane	UN1035, Ethane, 2.1
Ethane, cryogenic	UN1961, Ethane, refrigerated liquid, 2.1
Heptanes	UN1206, Heptanes, 3, PGII
Hexanes	UN1208, Hexanes, 3, PGII
Isobutane	UN1969, Isobutane, 2.1
Methane (compressed)	UN1971, Methane, compressed, 2.1
Methane, cryogenic	UN1972, Methane, refrigerated liquid, 2.1
Methylcyclohexanes	UN2296, Methylcyclohexane, 3, PGII
Methylcyclopentane	UN2298, Methylcyclopentane, 3, PGII
Natural gas (compressed)	UN1971, Natural gas, compressed, 2.1
Natural gas, cryogenic	UN1972, Natural gas, refrigerated liquid, 2.1
Nonanes	UN1920, Nonanes, 3, PGII
Octanes	UN1262, Octanes, 3, PGII
Pentamethylheptane	UN2286, Pentamethylheptane, 3, PGII
Pentanes	UN1265, Pentanes, 3, PGI *or* UN1265, Pentanes, 3, PGII
Propane	UN2330, Propane, 2.1
Undecane[a]	UN2330, Undecane, 3, PGIII

[a]Undecane is the alkane having the chemical formula $C_{11}H_{24}$.

PERFORMANCE GOALS FOR SECTION 12.3:

- Describe the chemical bonds that exist in alkene, diene, triene, cycloalkene, cyclodiene, and cyclotriene molecules.
- Illustrate that alkenes may have structural isomers and geometrical isomers.
- Memorize the rules for naming alkenes, dienes, trienes, cycloalkenes, cyclodienes, and cyclotrienes.
- Describe the means shippers and carriers use to inform emergency responders of the hazards associated with encountering alkenes, dienes, trienes, cycloalkenes, cyclodienes, and cyclotrienes during transportation mishaps.

12.3 Alkenes, Dienes, Trienes, Cycloalkenes, Cyclodienes, and Cyclotrienes

alkene (olefin)
■ A hydrocarbon whose molecules are composed of at least one carbon–carbon double bond (C=C)

unsaturated hydrocarbon
■ A hydrocarbon whose molecules contain C—H bonds and at least one carbon–carbon double bond or one carbon–carbon triple bond

cycloalkene
■ Any alkene whose carbon atoms are bonded to each other in a cyclic fashion

diene
■ An organic compound each of whose molecules has two carbon–carbon double bonds

triene
■ An organic compound, each of whose molecules has three carbon–carbon double bonds

Hydrocarbons whose molecules contain one or more carbon–carbon double bonds are called **alkenes**, or **olefins**. Because the molecular structure of each alkene is deficient in hydrogen atoms relative to the molecular structure of the corresponding alkane, the alkenes are said to be **unsaturated hydrocarbons**. Like the cycloalkanes, the alkenes have the general chemical formula C_nH_{2n}.

The simplest alkene is named ethene, or ethylene. Its molecular formula is C_2H_4, and its Lewis structure is the following:

$$
\begin{array}{c}
\text{H} \quad\quad \text{H} \\
\diagdown \quad\quad \diagup \\
\text{C}=\text{C} \\
\diagup \quad\quad \diagdown \\
\text{H} \quad\quad \text{H}
\end{array}
$$

The presence of the carbon–carbon double bond in the structure restricts the movement of the atoms about the bond. This means that the six atoms in the ethylene molecule spend their average time in one plane.

There are also alkenes in which the first and last carbon atoms in a continuous chain are joined to each other in a cyclic arrangement. They are called **cycloalkenes**. The general chemical formula of the cycloalkenes is C_nH_{2n-2}. They are named by placing the prefix *cyclo-* in front of the name of the parent alkene. The three simplest cycloalkenes are cyclobutene, cyclopentene, and cyclohexene. Their chemical formulas are often represented by using appropriate geometrical designs as shown here:

Cyclobutene   Cyclopentene   Cyclohexene

The molecules of alkenes can also have multiple carbon–carbon double bonds. When they have two and three double bonds, the compounds are called **dienes** and **trienes**, respectively.

12.3-A STRUCTURAL ISOMERISM IN ALKENES

Ethene and propene do not have structural isomers. For example, the condensed formula for propene can be correctly represented in either of the following ways:

$$CH_2=CH-CH_3 \quad\quad CH_3-CH=CH_2$$

Because either formula is the other one turned around end for end, it is a representation of the same substance.

However, we can write the following three condensed formulas for the structural isomers of the alkene having the formula C_4H_8.

$$CH_2{=}CHCH_2CH_3 \qquad CH_3CH{=}CHCH_3 \qquad CH_2{=}\overset{\displaystyle CH_3}{\underset{\displaystyle |}{C}}{-}CH_3$$

In the first two formulas, the carbon atoms are displayed in a continuous chain. The molecular structures differ only by the position of the carbon–carbon double bond. The third formula differs in that the carbon atoms are bonded to one another in a branched pattern.

12.3-B NAMING ALKENES, DIENES, TRIENES, CYCLOALKENES, CYCLODIENES, AND CYCLOTRIENES

Chemists use the -ene suffix to name alkenes. Consequently, in the IUPAC system, the first four members of the alkene series are named ethene, propene, butene, and pentene. Their common names are ethylene, propylene, butylene, and amylene, respectively.

In the IUPAC system, not only is the -ene suffix used to identify an alkene, but the position of the double bond in the main chain of its molecules is also indicated. The following rules are used to name alkenes:

- Consecutively number the carbon atoms in the main chain of continuous carbon atoms by beginning at the end that is *nearer* the double bond. The carbon–carbon double bond must always be included within the main chain of continuous carbon atoms.
- Indicate the position of the carbon–carbon double bond by the appropriate numerical prefix.
- Designate the position of each substituent by the number of the carbon atom along the main chain to which it is bonded.

For example, the compounds having the formulas $CH_2{=}CHCH_2CH_3$ and $CH_3CH{=}CHCH_3$ are named 1-butene and 2-butene, respectively. The compound having the formula $CH_2{=}\overset{\displaystyle }{\underset{\displaystyle |}{C}}{-}CH_3$
$$\underset{\displaystyle CH_3}{}$$
is named isobutene, isobutylene, or 2-methylpropene. The name used exclusively in commerce is isobutylene.

When dienes and trienes are named, the positions of the carbon–carbon double bonds are denoted with appropriate numbers, and *diene* or *triene* is used as the relevant suffix. The following examples illustrate the naming of a diene and a triene:

$$CH_2{=}CH{-}CH{=}CH_2 \qquad CH_3{-}CH{=}CH{-}CH{=}CH{-}CH{=}CH_2$$

1,3- Butadiene $\qquad\qquad\qquad$ 1,3,5- Heptatriene

An alkyl derivative of a cycloalkene, cyclodiene, and cyclotriene is named by using one or more numbers to identify the location of the alkyl group relative to location of the double bond. For example, the compounds designated by the following molecular structures are named 1-methylcyclopentene and 3-methylcyclopentene:

1-Methylcyclopentene $\qquad\qquad$ 3-Methylcyclopentene

The atoms in the carbon–carbon double bond of a cycloalkene are always numbered 1 and 2. Consequently, methylcyclopentene may be preceded only by a 1-, 3-, or 4-.

alkenyl group

■ Any noncyclic group of atoms having the general chemical formula C_nH_{2n-1} other than the methylene group, whose formula is $-CH_2-$

methylene group

■ The designation for the group of atoms $-CH_2-$

vinyl group

■ The designation for the group of atoms $CH_2=CH-$

allyl group

■ The designation for the group of atoms $CH_2=CH-CH_2-$

geometrical isomerism

■ The phenomenon associated with compounds whose molecules have the same atoms but differ in the manner in which the atoms are spatially arranged

Finally, just as there are alkyl groups, there are also **alkenyl groups**. The three most commonly encountered alkenyl groups are the **methylene group**, **vinyl group**, and **allyl group**.

$$-CH_2- \qquad CH_2=CH- \qquad CH_2=CH-CH_2-$$

Methylene group   Vinyl group   Allyl group

The methylene group is the only alkenyl group that does not contain a carbon–carbon double bond. The alkenyl groups are used in the common system to name simple organic compounds such as the chlorinated derivatives of methane, ethene, and propene.

$$CH_2Cl_2 \qquad CH_2=CH-Cl \qquad CH_2=CH-CH_2-Cl$$

Methylene chloride   Vinyl chloride   Allyl chloride
(Dichloromethane)   (Chloroethene)   (3-Chloropropene)

12.3-C GEOMETRICAL ISOMERISM IN ALKENES

The relative rigidity of the carbon–carbon double bond leads to a new kind of isomerism called **geometrical isomerism**. This phenomenon arises when the two substituents or atoms are different on each carbon atom that constitute the carbon–carbon double bond. For example, two Lewis structures can be written for 2-butene as follows:

$$\begin{array}{cc} CH_3 & H \\ \diagdown & \diagup \\ C & = & C \\ \diagup & \diagdown \\ H & CH_3 \end{array} \qquad\qquad \begin{array}{cc} CH_3 & CH_3 \\ \diagdown & \diagup \\ C & = & C \\ \diagup & \diagdown \\ H & H \end{array}$$

trans-2-Butene   *cis*-2-Butene

These structures illustrate that a hydrogen atom and a methyl group are bonded to each carbon atom in the carbon–carbon double bond. Because they are bonded to the same carbon atoms in each Lewis structure, they do not represent structural isomers, yet the arrangement of their atoms differs in space. When two compounds have the same structural formula but differ by the spatial arrangement of their atoms they are called *geometrical isomers*.

Geometrical isomers are named by using either of the prefixes *trans*- (meaning on the opposite side of the double bond) or *cis*- (meaning on the same side of the bond). As previously shown, the two geometrical isomers of 2-butene are named *trans*-2-butene and *cis*-2-butene, respectively.

SOLVED EXERCISE 12.3

Using the IUPAC system, name the alkene having the following condensed formula:

$$\begin{array}{c} CH_3CH_2-C=CH_2 \\ | \\ CH_3CH_2 \end{array}$$

Solution: First, we identify the longest continuous chain of carbon atoms that contains the carbon–carbon double bond. The longest chain contains four carbon atoms, which signifies that the compound is a derivative of butene. Assigning a number to each carbon atom from the right to the left of this longest chain, we see that the two carbon atoms in the carbon–carbon double bond are numbered 1 and 2, respectively.

$$\begin{array}{cccc} 4 & 3 & 2 & 1 \\ CH_3CH_2 & - & C & = & CH_2 \\ & & | \\ & & CH_3CH_2 \end{array}$$

This means that the compound is a derivative of 1-butene. The ethyl group is bonded to the carbon atom numbered 2. The compound is then correctly named 2-ethyl-1-butene. (We do not assign a number to each carbon atom from the left to the right of the longest chain, since then the carbon atoms in the carbon–carbon double bond would be numbered 3 and 4. We always assign numbers to the carbon atoms so that the *smaller* numbers are assigned to the carbon atoms in the carbon–carbon double bond.)

12.3-D TRANSPORTING ALKENES, DIENES, TRIENES, CYCLOALKENES, CYCLODIENES, AND CYCLOTRIENES

When shippers offer an alkene, diene, triene, cycloalkene, cyclodiene, or cyclotriene for transportation, DOT requires them to enter the relevant basic description from Table 12.5 on an accompanying shipping paper. DOT also requires shippers to affix either a FLAMMABLE GAS or FLAMMABLE LIQUID label on its packaging.

When carriers transport any of these compounds in an amount exceeding 1001 lb (454 kg) DOT requires them to post the applicable FLAMMABLE GAS or FLAMMABLE placards on the bulk packaging or transport vehicle used for shipment.

PERFORMANCE GOALS FOR SECTION 12.4:

- Describe the chemical bonds that exist in alkyne molecules.
- Illustrate that alkynes may have structural isomers.
- Memorize the rules for naming alkynes.
- Describe the means shippers and carriers use to inform emergency responders of the hazards associated with encountering alkynes during transportation mishaps.

12.4 Alkynes

Hydrocarbons having one or more carbon–carbon triple bonds are called **alkynes**. Like the alkenes, they are unsaturated hydrocarbons. The general chemical formula of an alkyne is C_nH_{2n-2}.

The simplest member of the alkyne series has the formula C_2H_2. It is named ethyne in the IUPAC system, but it is known more generally as acetylene, its common name. The Lewis structure of acetylene is denoted as follows:

$$H-C\equiv C-H$$

In the common system, alkynes are named as derivatives of acetylene. If an alkyne is represented by either of the general formulas $R-C\equiv C-H$ or $R-C\equiv C-R'$, the corresponding compound is named by identifying the alkyl groups, R and R' in their formulas. Thus, the derivatives of acetylene having one and two methyl groups are named methylacetylene and dimethylacetylene, respectively.

$$CH_3-C\equiv C-H \qquad CH_3-C\equiv C-CH_3$$
Methylacetylene   Dimethylacetylene

In the IUPAC system, the alkynes are named by replacing the *-ane* or *-ene* suffix on the associated alkane or alkene, respectively, with the suffix *-yne*. A numerical prefix is used to indicate the position of the carbon–carbon triple bond in the main chain of continuous carbon atoms.

alkyne
- A hydrocarbon whose molecules are composed of at least one carbon–carbon triple bond ($C\equiv C$)

TABLE 12.5 — Basic Descriptions of Some Alkenes, Dienes, Trienes, and Cycloalkenes

ALKENE, DIENE, TRIENE, OR CYCLOALKENE	BASIC DESCRIPTION
Bicyclo[2,2,1]hepta-2,5-diene, stabilized[a]	UN2251, Bicyclo[2,2,1]hepta-2,5-diene, stabilized, 3, PGII
Butadienes, inhibited[b]	UN1010, Butadienes, inhibited, 2.1
1-Butene[c]	UN1012, Butylene, 2.1
cis-2-Butene[c]	UN1012, Butylene, 2.1
trans-2-Butene[c]	UN1012, Butylene, 2.1
1,5,9-Cyclododecatriene	UN2518, 1,5,9-Cyclododecatriene, 6.1, PGIII
Cycloheptatriene	UN2603, Cycloheptatriene, 3, PGII
Cycloheptene	UN2242, Cycloheptene, 3, PGII
Cyclohexene	UN2256, Cyclohexene, 3, PGII
Cyclooctadienes	UN2520, Cyclooctadienes, 3, PGIII
Cyclopentene	UN2246, Cyclopentene, 3, PGII
Dicyclopentadiene	UN2048, Dicyclopentadiene, 3, PGIII
Diisobutylene, isomeric compounds[d]	UN2050, Diisobutylene, isomeric compounds, 3, PGII
Dipentene	UN2052, Dipentene, 3, PGIII
Ethylene	UN1962, Ethylene, 2.1
n-Heptene	UN2278, n-Heptene, 3, PGII
Hexadienes	UN2458, Hexadienes, 3, PGII
1-Hexene	UN2370, 1-Hexene, 3, PGII
Isobutene	UN1055, Isobutylene, 2.1
Isoheptenes	UN2287, Isoheptenes, 3, PGII
Isohexenes	UN2288, Isohexenes, 3, PGII
Isooctenes	UN1216, Isooctenes, 3, PGII
Isopentenes	UN2371, Isopentenes, 3, PGI
2-Methyl-1-butene	UN2459, 2-Methyl-1-butene, 3, PGII
2-Methyl-2-butene	UN2460, 2-Methyl-2-butene, 3, PGII
3-Methyl-1-butene	UN2561, 3-Methyl-1-butene, 3, PGI
Methylpentadienes	UN2461, Methylpentadienes, 3, PGII
Octadiene	UN2309, Octadiene, 3, PGII
1-Pentene	UN1108, 1-Pentene, 3, PGI
Propylene	UN1077, Propylene, 2.1

[a]Bicyclo[2,2,1]hepta-2,5-diene is also known as 2,5-norbornadiene.

[b]Prior to their shipment, DOT requires the addition of a substance to 1,2- and 1,3-butadiene to inhibit their autopolymerization (Section 14.3).

[c]For purposes of DOT regulations, 1-butene, cis-2-butene, and trans-2-butene are designated "butylene."

[d]For purposes of DOT regulations, diisobutylene refers to a mixture of the compounds 2,4,4-trimethyl-1-pentene and 2,4,4-trimethyl-2-pentene.

Consider the two isomers of the alkyne having the molecular formula C_4H_6. The compound named 1-butyne is the isomer in which the carbon–carbon triple bond is located between a terminal carbon atom and the one immediately adjacent to it. The compound named 2-butyne is the C_4H_6 isomer in which the carbon–carbon triple bond is located between the two nonterminal carbon atoms.

$$H-C\equiv C-CH_2CH_3 \qquad CH_3-C\equiv C-CH_3$$

1-Butyne   2-Butyne

Although an alkyne may have structural isomers, it does not have geometrical isomers. In the IUPAC system, the alkynes are named by using the following rules:

- Consecutively number the carbon atoms in the main chain of continuous carbon atoms by beginning at the end of the chain that is *nearer* the triple bond. The carbon–carbon triple bond must always be included within the main chain of continuous carbon atoms.
- Indicate the position of the triple bond by the appropriate numerical prefix.
- Designate the position of each substituent by the number of the carbon atom along the main chain to which it is bonded.

SOLVED EXERCISE 12.4

Using the IUPAC system, name the alkyne having the following condensed formula:

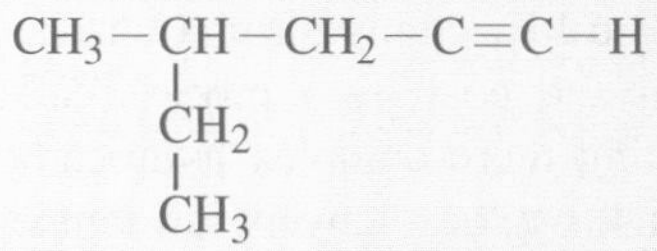

$$CH_3-CH-CH_2-C\equiv C-H$$
$$|$$
$$CH_2$$
$$|$$
$$CH_3$$

Solution: First, we determine the longest continuous chain of carbon atoms that contains the carbon–carbon triple bond. The longest chain contains six carbon atoms. This signifies that the compound is a derivative of 1-hexyne. Next, we assign numbers to the carbon atoms from the right to the left and then downward. The two carbon atoms in the carbon–carbon triple bond are then numbered 1 and 2, respectively.

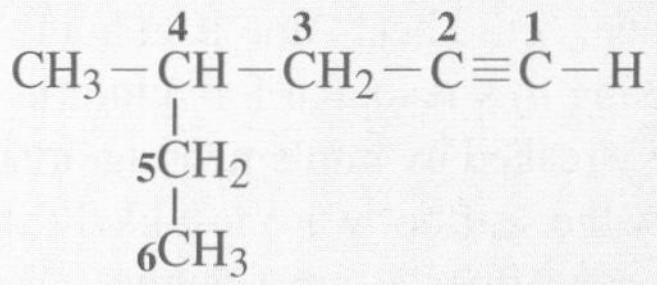

$$CH_3-\overset{4}{C}H-\overset{3}{C}H_2-\overset{2}{C}\equiv\overset{1}{C}-H$$
$$|$$
$$_5CH_2$$
$$|$$
$$_6CH_3$$

Because the methyl group is bonded to the carbon atom numbered 4, the compound is correctly named 4-methyl-1-hexyne.

PERFORMANCE GOALS FOR SECTION 12.5:

- Identify the hazardous properties of natural gas (methane).
- Discuss the most probable means by which methane became available in nature.
- Identify the main uses of methane.

12.5 Natural Gas (Methane)

natural gas
■ The flammable gas—primarily consisting of methane—formed in nature by the decomposition of animal and plant life

cracking
■ The industrial process typically conducted at a petroleum refinery during which the covalent bonds in hydrocarbon molecules are cleaved into smaller ones

thermal cracking
■ Cracking induced by heat

catalytic cracking
■ Cracking accomplished in the presence of a catalyst

The simple aliphatic hydrocarbons are most commonly encountered as gaseous domestic and industrial fuels. The simplest is *methane*, whose chemical formula is CH_4. Methane is the primary component of **natural gas**, roughly 70% by volume. Although pure methane is odorless, colorless, and tasteless, natural gas is commercially encountered with a slightly offensive smell caused by the presence of *methyl mercaptan* (CH_3-SH). Suppliers intentionally add methyl mercaptan to natural gas to assist personnel in detecting gas leaks. Some physical properties of methane are provided in Table 12.6.

In nature, methane occurs as a result of the decay and alteration of animal and plant remains deep below Earth's surface. Consequently, it often accompanies nearby deposits of crude petroleum (Section 12.14-A), which is believed to be produced by the same mechanism. Methane is also abundant in the atmosphere in coal mines, where it is called *firedamp*. When the mud at the bottom of stagnant pools, swamps, and rice paddies is disturbed, methane bubbles to the surface. Here, it is called *marsh gas*. Methane is also the primary constituent of mammalian flatulence.

As they digest their food, cattle and other bovines also produce methane. There are 1.8 billion bovines worldwide. On average, a single cow exhales 634 qt (600 L) of methane into the air each day. EPA has estimated that approximately 25% of the methane emitted into the air is linked to the belching of livestock. As amusing as this may appear, the production of methane by livestock has become a serious matter, because methane is a greenhouse gas (Section 5.8). Methane constitutes 10% to 15% of the gases linked to global warming. The ability of methane to warm the atmosphere is exceeded only by that of carbon dioxide. To reduce this impact on the atmosphere, agricultural scientists are testing ways of altering the volume of methane that livestock produce.

Approximately 85% of the natural gas used in the United States is now obtained from domestic sources, with another 13% to 14% coming from Canada and Mexico by pipeline. It is also produced from crude petroleum and coal by a process called **cracking**. During cracking, methane is produced by subjecting more complex hydrocarbons to high temperatures under moderate pressure (**thermal cracking**) or relatively low temperatures in the presence of a catalyst (**catalytic cracking**).

The domestic and international sources of natural gas are generally retrieved from underground, porous reservoirs where it has accumulated for centuries. These underground reserves are seriously dwindling in the United States, but there are substantial reserves of natural gas worldwide, even at locations off U.S. shores beneath the seafloor. For example, the U.S. Department of Interior has estimated that the outer continental shelf holds 420 trillion ft^3 (12 trillion m^3) of unrecovered natural gas, which if tapped, could heat American homes for the next 80 years. Although scientists see a benefit in harnessing this resource for future use by humankind, environmental activists fear that havoc would be wreaked by implementing offshore drilling operations.*

The methane that exists below the seafloor was most likely produced by microbial action. The microbes are located below the ocean floor at great depths, where the prevailing pressure is 30 times the atmospheric norm. Under this condition, methane and water molecules coexist in a unique manner: The methane molecules are trapped within "cages" made up of six water molecules. Many cages link together to form white crystals of a substance that physically resembles ice. The substance is called *methane hydrate* and is extremely unstable under normal pressure conditions. Chemists represent each of its units by the formula $CH_4 \cdot 6H_2O(s)$. When pieces of methane hydrate are brought to the ocean's surface, these units collapse, and methane wafts into the air.

Large volumes of natural gas are directly transferred from petroleum fields or refineries in the gas-producing southern states by a network of pipelines to individual homes, apartment complexes, and business buildings throughout most of the United States. Approximately 60% of U.S.

*To protect the coastal waters of the United States, a presidential order signed in 1981 forbade offshore drilling until 2012 on 85% of the outer continental shelf. The ban covered from 3 mi (4.8 km) offshore, where federal waters begin, to 200 mi (322 km) out, where U.S. jurisdiction ends. In 2008, this order was rescinded, but it is unclear when leases for offshore natural gas drilling will be granted by the U.S. Department of the Interior.

TABLE 12.6	Physical Properties of Methane
Melting point	−297 °F (−183 °C)
Boiling point	−258 °F (−161 °C)
Specific gravity at 68 °F (20 °C)	0.42
Vapor density (air = 1)	0.553
Flashpoint	−366 °F (−221 °C)
Autoignition point	999 °F (537 °C)
Lower explosive limit	5%
Upper explosive limit	15%
Heat of combustion	23,800 Btu/lb (55.5 kJ/g)

households use natural gas to provide the heat on the kitchen stovetop and to fire the home furnace, water heater, and clothes dryer. In addition, natural gas is used by the chemical industry both as a feedstock and energy source.

The pipelines that deliver natural gas appear to be buried almost everywhere. Firefighters are often called to scenes at which these pipelines have been inadvertently damaged during a routine landscaping operation. To help avoid such incidents, local gas companies recommend that they be notified of an intent to dig so the locations at which the pipelines are buried can be flagged. Gas companies also recommend that everyone know the location of the incoming natural gas shutoff valves into their homes and other buildings so this information can be rapidly conveyed during fires to first-on-the-scene responders.

Aside from being transferred by pipeline, natural gas is also transported as a compressed gas and cryogenic fluid called **liquefied natural gas** (LNG). The latter means of transportation is becoming increasingly popular as the nation's sources of natural gas decrease. Table 2.10 illustrated that when methane is cooled to −258 °F (−161 °C), the liquid volume is reduced to 1/650th of its gaseous volume. This implies that transporting of LNG is more cost-effective than transporting the compressed gas. LNG is received in the United States from foreign sources such as Tobago, Nigeria, and Trinidad. It is transported by barge from their overseas ports to one of five domestic ports. It is then stored and transported on land by means of refrigerated tank trucks, or it is vaporized and transferred as the gas by pipeline to homes and businesses for use.

liquefied natural gas
■ Methane as a cryogenic liquid (abbreviated LNG)

When mixed with air, methane burns at concentrations between approximately 5% and 15% by volume. When methane burns, a slightly luminous flame is produced that accompanies the release of 23,800 Btu/lb (55.5 kJ/g) to the surroundings. This high heat value is the basis for regarding fire and explosion as the primary hazards associated with methane. When compared with gasoline, methane is a desirable vehicular fuel, not only because its combustion provides this high heat value but because fewer volatile organic compounds (Figure 7.5) are components of the exhaust emissions. For this reason major municipalities have elected to use compressed natural gas as the fuel in city-operated vehicles.

PERFORMANCE GOALS FOR SECTION 12.6:

- Identify the hazardous properties of liquefied petroleum gas (LPG).
- Identify the types of LPG that are available for commercial use as bottled gas.
- Identify how these types are used, as well as the limitations in their use.

12.6 Liquefied Petroleum Gas

Propane

Butane

liquefied petroleum gas
■ The compressed and liquefied mixture obtained as a by-product during the refining of petroleum and consisting primarily of a mixture of propane and *n*-butane

Most people are first introduced to *propane* or *n*-butane as the substances used to fuel a backyard barbeque grill. The mixture of these substances as well as each of its two primary components is called **liquefied petroleum gas**, or LPG. As noted earlier, propane and *n*-butane are alkanes having the chemical formulas C_3H_8 and C_4H_{10}, respectively.

LPG is isolated as a by-product from rectifiers used for treating natural gas. Although the main constituents of LPG are propane and *n*-butane, small amounts of ethane, ethene, propene, butene, isobutane, isobutene, and isopentene may also be present.

There are three types of LPG: a *propane/butane mixture*, *commercial propane*, and *commercial butane*. The propane/butane mixture consists of approximately 60% propane and 40% butane by volume; commercial propane is approximately 92% propane; and commercial butane is approximately 85% *n*-butane.

At most ambient conditions, propane and *n*-butane are colorless, odorless, and tasteless gases, but an odorant is usually added to each commercial type so workers may easily detect when the gases leak from their containers. When they burn, their heats of combustion range from 21,200 Btu/lb (49.5 kJ/g) to 21,600 Btu/lb (50.3 kJ/g). Consequently, all LPG types are highly desirable as domestic and industrial fuels.

Some important physical properties of propane and *n*-butane are noted in Table 12.7. These data show that they readily liquefy under moderate pressure. Hence, the three types of LPG exist as liquefied compressed gases when pressurized in steel cylinders. *n*-Butane has a particularly interesting property. Because this substance liquefies near the freezing point of water, its use as a fuel is limited when the temperature of the surroundings is colder than 31 °F (−0.5 °C).

SOLVED EXERCISE 12.5

Why is propane generally more suitable than butane for fueling mobile appliances and heaters during very cold weather?

Solution: It is the gaseous state of a fuel that burns. The information in Table 12.7 tells us that propane liquefies when the surrounding temperature is −49 °F (−45 °C), whereas butane liquefies at 31 °F (−0.5 °C). This fact severely limits the use of butane as a fuel in cold climates. Because there are very few areas of the world where temperatures colder than −49 °F (−45 °C) are experienced, propane is generally more suitable than butane for fueling mobile appliances and heaters.

bottled gas
■ Broadly, any gas stored under pressure in a portable gas cylinder; specifically, propane, butane, or a propane/butane mixture stored under pressure in a portable gas cylinder

The commercial types of LPG are usually shipped under pressure in a motor or rail tankcar from petroleum gas wells and refineries to filling stations, where they are transferred into a bulk storage tank. Distributors then transfer them into horizontal or vertical dispensers from which they are further dispensed into far smaller portable, thick-walled cylinders and tanks. The contents are then said to be "bottled" and the gas is commonly referred to as **bottled gas**. These portable bottled gas cylinders and tanks are delivered to rural areas and other locations where natural gas cannot be supplied economically by pipeline. They are always stored outdoors, since local codes and ordinances prohibit indoor storage.

Care must be exercised when storing a portable LPG tank. When it is rapidly subjected to intense heat, an LPG tank can rupture. The tank may also be struck by lightning or be exposed to

TABLE 12.7 | Physical Properties of Propane and *n*-Butane

	PROPANE	*n*-BUTANE
Melting point	−305 °F (−187 °C)	−216 °F (−138 °C)
Boiling point	−49 °F (−45 °C)	31 °F (−0.5 °C)
Specific gravity at 68 °F (20 °C)	0.58	0.60
Vapor density (air = 1)	1.52	2.04
Flashpoint	−156 °F (−104 °C)	−76 °F (−60 °C)
Autoignition point	874 °F (468 °C)	761 °F (405 °C)
Lower explosive limit	2.2%	1.9%
Upper explosive limit	9.5%	8.5%
Heat of combustion	21,600 Btu/lb (50.3 kJ/g)	21,200 Btu/lb (49.5 kJ/g)

another ignition source. In these instances, the tank contents immediately ignite as a fireball as they expand into the atmosphere.

LPG is also transported by truck to fill permanently installed tanks like the one shown in Figure 12.3. These tanks are always situated away from homes and other buildings at prescribed distances from other major buildings and connected by means of copper tubing to appliances and furnaces located indoors.

FIGURE 12.3 In areas where natural gas lines have not been installed, buildings are often heated using liquefied petroleum gas (LPG) as the fuel. Prior to its use for this purpose, the LPG is stored within a permanently installed tank that is situated away from major buildings at distances prescribed by local codes and ordinances. This LPG tank has a volume of 2000 gal (7.5 m^3.)

TABLE 12.8 | Physical Properties of Ethylene and Propylene

	ETHYLENE	PROPYLENE
Melting point	−272 °F (−169 °C)	−301 °F (−185 °C)
Boiling point	−155 °F (−104 °C)	−54 °F (−48 °C)
Specific gravity at 68 °F (20 °C)	0.001	0.51
Vapor density (air = 1)	1.0	1.5
Flashpoint	−213 °F (−135 °C)	−162 °F (−108 °C)
Autoignition point	842 °F (450 °C)	860 °F (460 °C)
Lower explosive limit	3.1%	2%
Upper explosive limit	32%	11.1%
Heat of combustion	21,600 Btu/lb (50.3 kJ/g)	21,000 Btu/lb (48.9 kJ/g)

The largest volumes of LPG are found at LPG-filling stations, at least one of which is located in virtually every major city. At the filling station, the LPG is generally contained within a large bulk storage tank having a capacity of less than 90,000 gal (340 m^3).

PERFORMANCE GOALS FOR SECTION 12.7:

- Identify the primary risks associated with exposure to ethylene and propylene.
- Identify the principal uses of these gases.

12.7 Ethylene and Propylene

Ethylene

Propylene

Ethylene and *propylene*, or *ethene* and *propene*, respectively, are the common names of the simplest alkenes. As noted earlier, their chemical formulas are C_2H_4, or $CH_2{=}CH_2$, and C_3H_6, or $CH_3{-}CH{=}CH_2$, respectively.

The physical properties of ethylene and propylene are provided in Table 12.8. The tabulated information indicates that ethylene and propylene pose the risk of fire and explosion when either gas is discharged within an enclosed area.

Ethylene and propylene are produced in the United States in larger amounts than all other substances. In the chemical industry, ethylene is used primarily for the manufacture of polyethylene (Section 14.6-A), but it is also used as a raw material for the manufacture of a variety of chemical products including vinyl chloride, vinylidine chloride, styrene, and ethylene glycol. Ethylene is also used outside the chemical industry; for example, it is used as a welding fuel and to hasten the ripening of produce including apples, bananas, berries, tomatoes, and citrus fruits.

Propylene is used primarily for the manufacture of polypropylene (Section 14.6-B). It is also used to produce isopropanol (Section 13.2-F), propylene glycol (Section 13.2-G), and other substances.

PERFORMANCE GOALS FOR SECTION 12.8:

- Identify the primary risks associated with exposure to butadiene.
- Identify the main uses of butadiene.
- Identify the potential hazards associated with inhaling butadiene.

TABLE 12.9	Physical Properties of 1,3-Butadiene
Melting point	−164 °F (−109 °C)
Boiling point	23 °F (−4.7 °C)
Specific gravity at 68 °F (20 °C)	1.88
Vapor density (air = 1)	1.97
Flashpoint	−105 °F (−76 °C)
Autoignition point	804 °F (429 °C)
Lower explosive limit	2%
Upper explosive limit	11.5%

12.8 Butadiene

1,3-Butadiene is a colorless, highly flammable gas with a mild sweet odor. Its physical properties are provided in Table 12.9. The gas is often manufactured at petroleum refineries by the dehydrogenation of 1-butene and 2-butene.

$$CH_3CH_2CH{=}CH_2(g) \longrightarrow CH_2{=}CH{-}CH{=}CH_2(g) \ + \ H_2(g)$$

  1-Butyne   1,3-Butadiene   Hydrogen

$$CH_3CH_2{=}CHCH_3(g) \longrightarrow CH_2{=}CH{-}CH{=}CH_2(g) \ + \ H_2(g)$$

  2-Butyne   1,3-Butadiene   Hydrogen

1,3-Butadiene is a very reactive substance. It even reacts spontaneously with itself, molecule for molecule, to form polybutadiene (Section 14.10-B). To prevent the reaction, the gas must be mixed with an inhibitor immediately following its production.

The commercial product is called *butadiene*, without the numerical prefixes. It is often transferred by means of an airtight pipeline to nearby petrochemical plants. Approximately 80% of the butadiene produced in the United States is used as a raw material to manufacture various types of synthetic rubbers (Section 14.10).

PERFORMANCE GOALS FOR SECTION 12.9:

- Identify the primary risks associated with exposure to acetylene.
- Discuss how acetylene is manufactured.
- Identify the main uses of acetylene.

12.9 Acetylene

Acetylene is the simplest alkyne and the sole member of this class of compounds having commercial importance. As noted earlier, its chemical formula is C_2H_2, or $H{-}C{\equiv}C{-}H$. The physical properties of acetylene are provided in Table 12.10.

TABLE 12.10	Physical Properties of Acetylene
Melting point	Sublimes
Boiling point	−118 °F (−83 °C)
Specific gravity at 68 °F (20 °C)	0.91
Vapor density (air = 1)	0.899
Flashpoint	0 °F (−17.7 °C)
Autoignition point	635 °F (335 °C)
Lower explosive limit	3%
Upper explosive limit	82%
Heat of combustion	21,400 Btu/lb (49.9 kJ/g)

In the United States, acetylene is produced mainly by cracking natural gas.

$$2CH_4(g) \longrightarrow C_2H_2(g) + 3H_2(g)$$

$$\text{Methane} \qquad \text{Acetylene} \qquad \text{Hydrogen}$$

dehydrogenation
■ A chemical process typically conducted at a high temperature and pressure and during which molecular hydrogen is removed from a compound

In this instance, the cracking involves **dehydrogenation**. Although pure acetylene is a colorless gas with an ethereal odor, industrial-grade acetylene can be foul-smelling. As noted in Section 9.7-B, industrial-grade acetylene is sometimes produced by hydrolyzing calcium carbide. Because calcium carbide is generally contaminated with calcium phosphide (Section 9.6-B), the foul-smelling phosphine is simultaneously produced.

Acetylene is an innately unstable substance. Even without exposure to air or oxygen, acetylene decomposes at an explosive rate into its elements when it has been compressed in excess of approximately 2 atm (202.6 kPa).

$$C_2H_2(g) \longrightarrow 2C(s) + H_2(g)$$

$$\text{Acetylene} \qquad \text{Carbon} \qquad \text{Hydrogen}$$

The decomposition is especially prone to occur when the compressed acetylene has been subjected to thermal or mechanical shock.

Not only is acetylene flammable and innately unstable, but it reacts with compounds containing the copper(I) ion, especially in moist air. The product of this chemical reaction is copper(I) acetylide, which when dry, can decompose explosively.

$$H{-}C{\equiv}C{-}H(g) + Cu^+(s) \longrightarrow H{-}C{\equiv}C{-}Cu(s) + H^+(aq)$$

$$\text{Acetylene} \qquad \text{Copper(I) ion} \qquad \text{Copper(I) acetylide} \qquad \text{Hydrogen ion}$$

To reduce or eliminate the potential for this explosive reaction, welders and other individuals who work with acetylene must avoid the use of copper fittings or tubing on acetylene cylinders.

When sparked, a mixture of acetylene and oxygen burns with an intensely hot flame, reaching temperatures as high as 5400 °F (2982 °C). This combustion process yields 21,400 Btu/lb (49.9 kJ/g), which is put to use when welding and cutting steel and cladding metals. This use of acetylene is shown in Figure 12.4.

Careless welding practices have caused many major fires. Heat sufficient to weld steel can also trigger the ignition of many other commonly encountered flammable and combustible materials. When acetylene is used to generate heat for welding and similar purposes, the risk that its heat of combustion will be transmitted to nearby combustible materials should always be acknowledged.

FIGURE 12.4 An intensely hot flame is produced when a mixture of acetylene and oxygen burns. This combustion reaction is put to good use by welders who use the heat to weld and cut metals.

To counteract its potential decomposition, a special method is utilized to safely compress and store acetylene within steel cylinders. This procedure takes advantage of the solubility of acetylene in acetone (Section 13.5-C). One volume of acetone dissolves approximately 25 volumes of acetylene at 1 atm (101.3 kPa) and 300 volumes at 12 atm (1216 kPa).

When gas manufacturers dissolve acetylene in acetone, they increase the amount of the gas that may safely be compressed into a cylinder by means of the two-step process shown in Figure 12.5. First, they pack a porous medium consisting of a mixture of monolithic filler and balsa wood into the cylinder. Then, they saturate this medium with liquid acetone. When acetylene at a prescribed pressure is charged into the cylinder, it dissolves in the acetone.

SOLVED EXERCISE 12.6

The DOT regulation at 40 C.F.R. §173.303 stipulates that when acetylene is offered for domestic shipment, the pressure in its cylinders cannot exceed 250 psi (1720 kPa) at 70°F (21°C). What is the most likely reason DOT limits the pressure to which acetylene can be compressed in cylinders during its domestic transportation?

Solution: Acetylene is an innately unstable substance. Its tendency to undergo self-decomposition increases as acetylene is compressed, such as when it is stored under excessive pressure within cylinders. DOT most likely limits the pressure to which acetylene can be compressed within cylinders to reduce or eliminate the risk of its self-decomposition.

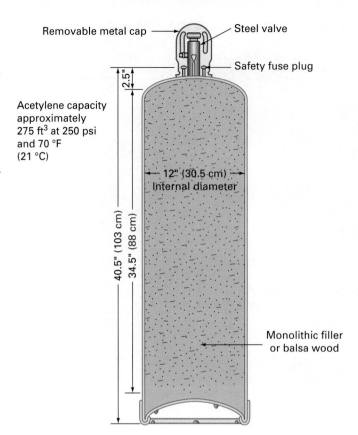

FIGURE 12.5 The cutaway of this acetylene cylinder illustrates its unique features. The monolithic filler or balsa wood is a porous material that is charged with acetone, into which the acetylene is dissolved.

Removable metal cap — Steel valve

— Safety fuse plug

2.5"

Acetylene capacity approximately 275 ft³ at 250 psi and 70 °F (21 °C)

← 12" (30.5 cm) →
Internal diameter

40.5" (103 cm)
34.5" (88 cm)

Monolithic filler or balsa wood

When acetylene is released from its container, as during equipment failure, the principal hazard is its potential for fire and explosion. This is evident from the magnitude of its lower and upper explosive limits: When sparked, acetylene ignites in air when its concentration ranges from 3% to 82% by volume.

PERFORMANCE GOALS FOR SECTION 12.10:

- Identify the primary health hazards associated with breathing an atmosphere containing methane, LPG, ethylene, propylene, butadiene, or acetylene.

12.10 Ill Effects Resulting from Breathing the Simple Gaseous Hydrocarbons

The human body responds in the following ways when acetylene, butadiene, ethylene, LPG (propane/butane), methane, or propylene is inhaled:

- Acetylene, ethylene, LPG, methane, and propylene pose health hazards as asphyxiants. Individuals who inhale these gases lose consciousness, because they are denied sufficient oxygen. Prolonged exposure to them can constitute a direct threat to life by suffocation.
- The inhalation of low concentrations of butadiene causes severe irritation of the eyes, nose, and throat.

- Long-term repeated exposure to butadiene is linked with the incidence of cancer. Epidemiologists classify it as a known human carcinogen. OSHA requires employers to limit employee exposure to a maximum butadiene concentration of 1 ppm, averaged over an 8-hr workday.

The OSHA Butadiene Standard at 29 C.F.R. §1910.1051(i) requires employers to provide and ensure the use of protective clothing, eye protection, and respiratory-protective gear to prevent employees from inhaling butadiene and to limit their dermal exposure in the workplace. What is the most likely reason OSHA promulgated this regulation?

Solution: Because butadiene irritates the eyes, nose, and throat and is classified as a known human carcinogen, OSHA requires employers to ensure that workers avoid inhaling or otherwise contacting this substance. This is accomplished by wearing appropriate protective clothing and eyeglasses and utilizing self-contained respiratory devices to eliminate or minimize contact with butadiene.

PERFORMANCE GOALS FOR SECTION 12.11:

- Describe the means shippers and carriers use to inform emergency responders of the hazards associated with encountering a gaseous hydrocarbon during transportation mishaps.

12.11 Transporting the Simple Gaseous Hydrocarbons

When shippers offer acetylene, butadiene, ethylene, methane, or propylene for transportation, DOT requires them to identify the chemical commodity on an accompanying shipping paper as one of the following, as relevant:

UN1001, Acetylene, dissolved, 2.1
UN1010, Butadiene, stabilized, 2.1
UN1962, Ethylene, 2.1
UN1971, Methane, compressed
UN1971, Natural gas, compressed, 2.1
UN1972, Methane, refrigerated liquid, 2.1
UN1972, Natural gas, refrigerated liquid, 2.1
UN1077, Propylene, 2.1

Motor carriers are also required to obtain a Hazardous Materials Safety Permit prior to transporting methane either as a compressed gas or cryogenic liquid with a methane content of at least 85% in bulk packaging having a capacity equal to or greater than 3500 gal (13,248 L). Regardless of the manner by which the methane is transported, DOT also requires carriers to prepare and implement *a security plan* whose components comply with the requirements published at 49 C.F.R. §172.802.

When LPG is transported in a DOT Specification M 330 or M 331 tank truck, DOT requires carriers to determine whether the commodity is corrosive and to indicate the appropriate information in the basic description as either of the following:

RQ, UN1075, Liquefied petroleum gas, 2.1, (Non-corrosive)
or
RQ, UN1075, Petroleum gases, liquefied, 2.1 (Noncor)
RQ, UN1075, Liquefied petroleum gas, 2.1 (Not for Q and T tanks)

Noncorrosive LPG may be transported in tank trucks that have been constructed from a special steel that is "quenched and tempered," or "Q and T."

When shippers offer LPG for transportation in bulk, DOT requires them to add to it a small amount of an odoriferous substance such as ethyl mercaptan, thiophane, or amyl mercaptan. When shippers offer nonbulk quantities of unodorized LPG for transportation, DOT requires the packaging to be marked NON-ODORIZED or NOT ODORIZED near the spot where the proper shipping name is marked. DOT also requires shippers to affix a FLAMMABLE GAS label to its cylinders or other containers.

When carriers transport acetylene, butadiene, ethylene, methane, or propylene, DOT requires them to post FLAMMABLE GAS placards on the bulk packaging or transport vehicle used for shipment.

PERFORMANCE GOALS FOR SECTION 12.12:

- Describe the molecular structures of benzene, toluene, and xylene(s).
- Name the derivatives of the simple aromatic hydrocarbons when provided with their molecular structures.
- Identify the primary and secondary risks associated with exposure to benzene, toluene, and xylene(s).
- Describe the means shippers and carriers use to inform emergency responders of the hazards associated with encountering the simple aromatic hydrocarbons during transportation mishaps.

12.12 Aromatic Hydrocarbons

The word *aromatic* suggests that aromatic hydrocarbons are compounds possessing fragrant odors. The odors of some simple aromatic hydrocarbons are actually fragrant, but otherwise, this perception is misleading. Aromatic hydrocarbons are regarded as compounds whose molecules are composed of one or more special rings of carbon atoms. These compounds are typified by the substance called *benzene*, the simplest aromatic hydrocarbon. Its chemical formula is C_6H_6. Although the molecular structure of benzene is represented by either of the hexagons shown on the left, it is more commonly represented by a hexagon with a circle inside, as shown on the right:

Aromatic and aliphatic hydrocarbons are thus differentiated by the fact that their molecular structures either resemble or do not resemble the molecular structure of benzene. Hereafter, the molecular structure of benzene will be represented by the hexagonal structural formula with the inscribed circle. It is called the *benzene ring*. Although the symbols of the carbon atoms are not written as part of the hexagon, it is always understood that a carbon atom occupies each corner of the hexagon, where it is bonded to two other carbon atoms and a hydrogen atom.

When any one of the six hydrogen atoms in benzene is substituted with a methyl group, the compound that results is called *methylbenzene*, or more commonly, *toluene*. The chemical formula of toluene is $C_6H_5-CH_3$, and its molecular structure is represented as follows:

When two hydrogen atoms in the benzene molecule are substituted with methyl groups, the resulting compound has the chemical formula $C_6H_4-(CH_3)_2$. Because it is possible to position the methyl groups in three different ways on the benzene ring, three structural isomers for the formula $C_6H_4-(CH_3)_2$ exist. These three compounds are the structural isomers of *dimethylbenzene*, or *xylene*. Their molecular structures are written as follows:

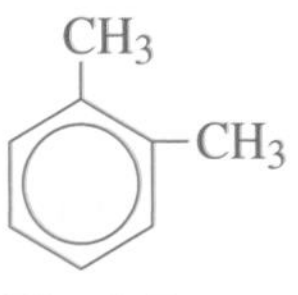

| 1,2-Dimethylbenzene | 1,3-Dimethylbenzene | 1,4-Dimethylbenzene |
| (*o*-Xylene) | (*m*-Xylene) | (*p*-Xylene) |

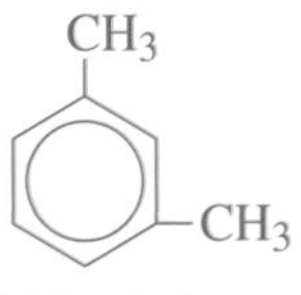

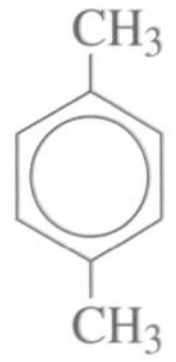

In the common system of nomenclature, the prefixes **ortho- (o-)**, **meta- (m-)**, and **para- (p-)** are employed to differentiate these structural isomers from one another. *Ortho-* means "straight ahead," *meta-* means "beyond," and *para-* means "opposite." Isomeric mixtures of xylene are sometimes referred to as either "xylene" or "xylene(s)" with total disregard for their isomeric distinctions.

The simplest derivatives of an aromatic hydrocarbon are compounds in which a single alkyl substituent has replaced a hydrogen atom on the benzene ring. These compounds are named by identifying the substituent followed by the word "benzene." For example, the compounds having the following molecular structures are named ethylbenzene and *n*-propylbenzene.

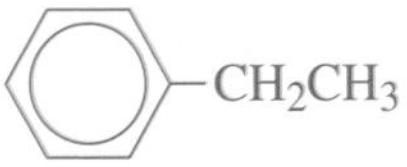

| Ethylbenzene | *n*-Propylbenzene |

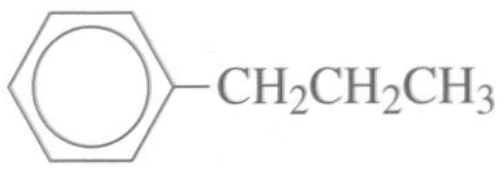

When two substituents have replaced two hydrogen atoms on the benzene ring, the locations at which they are bonded must be identified. The molecular structures of the three compounds in Figure 12.6 illustrate two arbitrary substituents, A and B, positioned on the benzene ring. B is located in the ortho-, meta-, and para- positions relative to the position of A. When naming substituents, the italicized letters *o-*, *m-*, and *p-* are used as prefixes to identify the location of one substituent relative to the other one. When one of the two substituents is a methyl group, the compound is often named as a derivative of toluene as shown by the following examples:

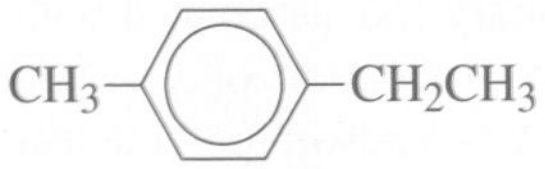

| *p*-Ethyltoluene | *o*-Isobutyltoluene |

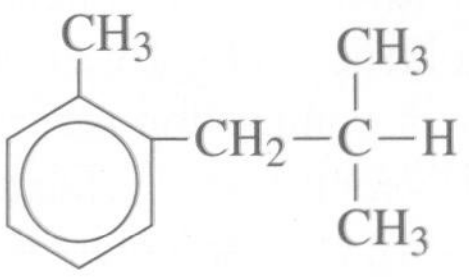

ortho-(o-)
- Characterized by the substitution of the hydrogen atoms on adjacent (the first and second) carbon atoms on the benzene ring

meta- (m-)
- Characterized by the substitution of the hydrogen atoms on the first and third carbon atoms on the benzene ring

para- (p-)
- Characterized by the substitution of the hydrogen atoms on the first and fourth carbon atoms on the benzene ring

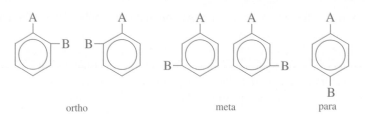

ortho   meta   para

FIGURE 12.6 The ortho, meta, and para positions of B, an arbitrary substituent, relative to a fixed position of another substituent on the benzene ring. The chemical identity of A and B may be identical or different. The two configurations noted for the ortho and meta positions of B relative to A are equivalent ways of denoting the same compound.

phenyl group
■ The designation for the benzene ring when it is named as a substituent on another molecule (C_6H_5—)

benzyl group
■ The designation for the seven-carbon atom unit consisting of a benzene ring and a methylene group ($C_6H_5CH_2$—)

aryl substituent (aryl group)
■ Typically the phenyl or benzyl group of atoms, obtained by removing a hydrogen atom from the chemical formulas of benzene and toluene, i.e., C_6H_5— and $C_6H_5CH_2$—, respectively

BTX
■ The acronym for benzene, toluene, and xylene

In the IUPAC system, when two or more alkyl groups are bonded to the benzene ring, the resulting hydrocarbon is named by listing the alkyl groups alphabetically and numbering the first group with a 1. Each of the six carbon atoms in the benzene ring is then numbered from 1 to 6. Two examples that illustrate the use of this rule for naming the derivatives of aromatic hydrocarbons are noted next:

CH₂CH₃ —CH₂CH₂CH₃
1-Ethyl-2-*n*-propylbenzene

CH₃CH₂CH₂——CH₂—CH—CH₃ / CH₃
1-Isobutyl-3-*n*-propylbenzene

Aromatic hydrocarbons are sometimes named by identifying the C_6H_5— and $C_6H_5CH_2$— groups of atoms. C_6H_5— and $C_6H_5CH_2$— are called the **phenyl group** and **benzyl group**, respectively. They are examples of **aryl substituents** or **aryl groups**.

Phenyl group   —CH₂— Benzyl group

Thus, the compounds having the following molecular structures are named phenylethylene and dibenzylacetylene, respectively.

Phenylethylene   Dibenzylacetylene

12.12-A BENZENE, TOLUENE, AND XYLENE(s)

Benzene, toluene, and xylene are the three most commonly encountered aromatic hydrocarbons. They are sometimes referred to collectively by the acronym **BTX**. Benzene, toluene, and xylene(s) are colorless, water-insoluble, highly volatile liquids. Some important physical properties of benzene, toluene, and xylene(s) are provided in Table 12.11. Their flashpoints, flammable ranges, and heats of combustion indicate that the primary risk associated with them is fire and explosion. Their combustion is characterized by the production of very smoky flames. Researchers have shown the particulate matter in these flames adsorb polynuclear aromatic hydrocarbons (Section 12.13).

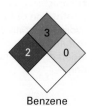

Benzene

Toluene

TABLE 12.11	Physical Properties of Benzene, Toluene, and the Isomeric Xylenes				
	BENZENE	**TOLUENE**	**o-XYLENE**	**m-XYLENE**	**p-XYLENE**
Melting point	41 °F (5.4 °C)	−139 °F (−95 °C)	−15 °F (−26 °C)	−54 °F (−48 °C)	55 °F (13 °C)
Boiling point	176 °F (80 °C)	231 °F (111 °C)	291 °F (144 °C)	282 °F (139 °C)	280 °F (138 °C)
Specific gravity at 68 °F (20 °C)	0.88	0.87	0.90	0.87	0.86
Vapor density (air = 1)	2.8	3.1	3.7	3.7	3.7
Vapor pressure at 68 °F (20 °C)	75 mmHg	22 mmHg	8 mmHg	8 mmHg	9 mmHg
Flashpoint	12 °F (−11 °C)	40 °F (4.4 °C)	90 °F (32 °C)	84 °F (29 °C)	81 °F (27 °C)
Autoignition point	1044 °F (562 °C)	997 °F (536 °C)	867 °F (464 °C)	982 °F (528 °C)	984 °F (529 °C)
Lower explosive limit	1.4%	1.4%	1.0%	1.1%	1.1%
Upper explosive limit	8%	6.7%	6.0%	7.0%	7.0%
Heat of combustion	18,184 Btu/lb (42.3 kJ/g)	18,200 Btu/lb (42.3 kJ/g)	18,400 Btu/lb (42.8 kJ/g)	18,400 Btu/lb (42.8 kJ/g)	18,400 Btu/lb (42.8 kJ/g)

Benzene, toluene, and xylene are manufactured by several processes including the catalytic reforming of light naphtha (Section 12.14-C), a fraction of crude petroleum containing hydrocarbons with 5 to 11 carbon atoms per molecule. When naphtha is subjected to catalytic reformation, some of its constituent hydrocarbons are converted into a compound within the BTX group. Examples of these reforming reactions are illustrated by the following equations:

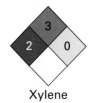

Xylene

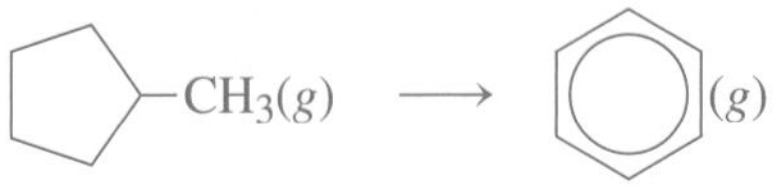

Methylcyclopentane → Benzene

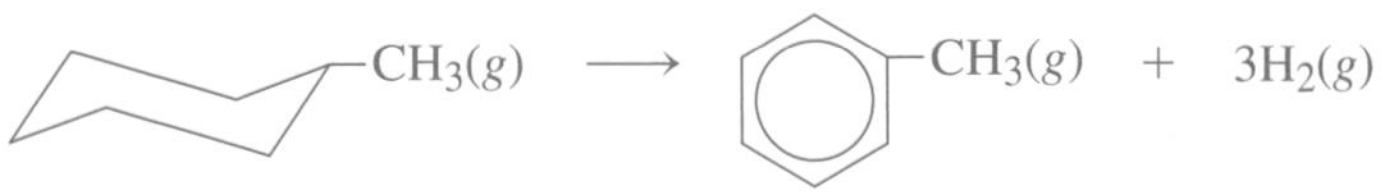

Methylcyclohexane → Toluene + 3H₂(g) Hydrogen

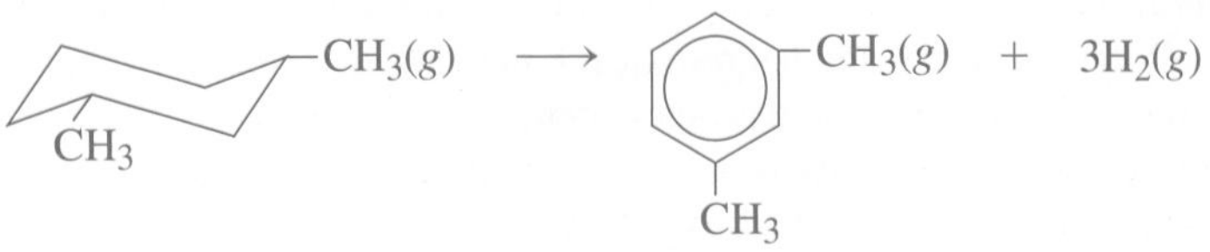

1,3-Dimethylcyclohexane → m-Xylene + 3H₂(g) Hydrogen

The production of benzene, toluene, or xylene(s) may be maximized by regulating the temperature and pressure conditions.

In the chemical industry, *p*-xylene is the most commercially important xylene isomer. It competes with naphthalene (Section 12.13-A) as the raw material needed to manufacture phthalic anhydride, a substance widely employed for the production of certain resins and polyesters such as poly(ethylene terephthalate) (Section 14.2-B). The boiling points of the *m*- and *p*-xylene are very similar. This similarity permits *o*-xylene to be separated from the other two by fractional distillation. When the distillate cools, the *p*- isomer crystallizes, which allows its isolation from *m*-xylene.

Although benzene was once a very popular industrial solvent, it is no longer widely used for this purpose. Researchers discovered that worker exposure to benzene vapor was linked with the onset of several human blood disorders including acute myelogenous leukemia and aplastic anemia (Section 12.12-A). Following this discovery, industrial workers turned to toluene and xylene as replacement solvents.

Notwithstanding that worker exposure poses the risk of contracting cancer, the BTX group of compounds is still used by the chemical industry in very substantial quantities as a raw material for the manufacture of other organic compounds. For example, benzene is used to produce ethylbenzene and isopropylbenzene.

| Benzene | | Ethylene | | Ethylbenzene |

$$\text{Benzene}(g) \; + \; CH_2{=}CH_2(g) \; \longrightarrow \; \text{Ethylbenzene} \; {-}CH_2CH_3(g)$$

$$\text{Benzene}(g) \; + \; CH_3CH{=}CH_2(g) \; \longrightarrow \; \text{Isopropylbenzene} \; {-}CH(CH_3)_2(g)$$

Benzene   Propene   Isopropylbenzene (Cumene)

Each BTX compound is transported commercially in nonbulk and bulk containers. A common nonbulk container is the 5-gal (19-L) metal can shown in Figure 12.7. Considerable quantities of toluene and xylene were formerly used as components of solvent-based paints and other coatings, adhesives, inks, and cleaning agents, but in contemporary times, EPA regulations have caused a reduction in use of these volatile organic compounds for these purposes.

Toluene is also the raw material needed for the production of the explosive trinitrotoluene (Section 14.9) and toluene diisocyanate (Section 14.8), which is used to produce polyurethane. The xylene isomers are valuable constituents of motor and aviation gasolines because of their high octane numbers (Section 12.14). They are also used in the chemical industry as feedstocks for the production of compounds needed in the polymer industry.

12.12-B ILL EFFECTS RESULTING FROM INHALING BTX VAPORS

Of the simplest hydrocarbons, benzene poses the most serious health risk when its vapor is inhaled. Short-term effects include dizziness, confusion, and asphyxiation. Moreover, in humans, benzene is a well-known hemo- and neurotoxicant; that is, when benzene vapor is inhaled, it negatively affects the blood and central nervous systems.

acute myelogenous leukemia
■ A cancer of the blood that develops in white blood cells, often associated with chronic exposure to benzene

Exposure to benzene is also associated with the potential onset in humans of four types of leukemia, one of which is **acute myelogenous leukemia**, a generally fatal cancer during which the immature white blood cells in the bone marrow multiply uncontrollably. Most blood

FIGURE 12.7 Toluene vapor is highly flammable. Although liquid toluene is commercially available in bulk, emergency responders are more likely to encounter it in 55-gal (208-L) steel drums. When toluene is transported, DOT requires the shipper to affix a FLAMMABLE LIQUID warning label to the container. (*Courtesy of Spectrum Chemical Manufacturing Corporation, Gardena, California.*)

cells are manufactured in the bone marrow. Myelogenous leukemia usually develops after a latency period of approximately 15 years. Its onset is usually accompanied by a second illness known as **aplastic anemia**, during which an individual's bone marrow is irreversibly injured. The presence of white blood cells in the blood is important, since these cells fight infection in the body. Consequently, even with treatment, individuals who have contracted benzene-induced acute myelogenous leukemia are often more susceptible to infection and have a poor prognosis for full recovery.

aplastic anemia
■ Bone marrow injury, often associated with exposure to benzene

Because research studies have clearly linked benzene exposure with the incidence of **leukemia** in humans, epidemiologists classify it as a known human carcinogen. EPA estimates that a lifetime exposure to 4 ppb benzene in air results in one additional case of leukemia in a population of 10,000 exposed individuals.

leukemia
■ Cancer of the blood-forming tissues and organs

Exposure to toluene and xylene is not linked with the onset of cancer. However, the inhalation of their vapors irritates the respiratory system and depresses the central nervous system. Initially, the inhalation of toluene and xylene vapors causes dizziness and nausea, and long-term inhalation can have a narcotic impact on the human body. The long-term inhalation of toluene can also be addictive. The dangerous practice colloquially known as *glue sniffing* involves voluntarily inhaling the toluene vapor emitted by certain glue products like hobby glue. People who regularly sniff glue risk permanent damage to the liver, heart, and lungs.

12.12-C WORKPLACE REGULATIONS INVOLVING THE BTX HYDROCARBONS

OSHA publishes regulations at 29 C.F.R. §§1910.1028 and 1910.1000 that aim to protect worker exposure to benzene and xylene, respectively.

OSHA requires employers to limit employee exposure to airborne vapors of benzene, toluene, and xylene(s) as follows:

- The maximum benzene vapor concentration has been established at 1 ppm, averaged over an 8-hr workday. This value reflects the cancer-causing potential of benzene.
- The maximum toluene vapor concentration has been established at 200 ppm, averaged over an 8-hr workday.
- The maximum xylene vapor concentration has been established at 100 ppm, averaged over an 8-hr workday.

When benzene is present in the workplace, OSHA requires employers to establish regulated areas where the concentration of benzene vapor exceeds the permissible exposure limit of 1 ppm as a time-weighted average limit, or 5 ppm as an average concentration over a 15-min period. OSHA also requires employers to post the following warning sign at the entrances to these regulated areas:

DANGER

BENZENE
CANCER HAZARD
FLAMMABLE - NO SMOKING
AUTHORIZED PERSONNEL ONLY
RESPIRATOR REQUIRED

TABLE 12.12	Basic Descriptions of Some Simple Aromatic Hydrocarbons
AROMATIC HYDROCARBON OR GROUPS THEREOF	**BASIC DESCRIPTION**
Benzene	UN1114, Benzene, 3, PGII
Butylbenzenes	UN2709, Butylbenzenes, 3, PGIII
Butyltoluenes	UN2267, Butyltoluenes, 6.1, PGIII (Poison)
Cymenes[a]	UN2046, Cymenes, 3, UN2046, PGIII
Diethylbenzene	UN2049, Diethylbenzene, 3, PGIII
Dimethylcyclohexanes	UN2263, Dimethylcyclohexanes, 3, PGII
Ethylbenzene	UN1175, Ethylbenzene, 3, PGII
Isopropylbenzene[a]	UN1918, Isopropylbenzene, 3, PGIII
n-Propylbenzene	UN2364, *n*-Propylbenzene, 3, PGIII
Toluene	UN1294, Toluene, 3, PGII
1,3,5-Trimethylbenzene	UN2325, 1,3,5-Trimethylbenzene, 3, PGIII
Xylenes	UN1307, Xylenes, 3, PGII *or* UN1307, Xylenes, 3, PGIII

[a]Cumene and cymene are the common names for isopropylbenzene and isopropyltoluene, respectively.

12.12-D TRANSPORTING AROMATIC HYDROCARBONS

When shippers offer a simple aromatic hydrocarbon or any its derivatives for transportation, DOT requires them to enter the relevant basic description from Table 12.12 on an accompanying shipping paper. When shippers offer a simple aromatic hydrocarbon or any its derivatives for transportation, DOT usually requires them to affix a FLAMMABLE LIQUID label on its packaging.

When carriers transport a simple aromatic hydrocarbon or any its derivatives in an amount exceeding 1001 lb (454 kg), DOT usually requires them to post FLAMMABLE placards on the bulk packaging or transport vehicle used for shipment.

However, the primary hazard associated with the butyltoluenes is their toxicity. When shippers offer a butyltoluene for transportation, DOT requires them to identify the substance on an accompanying shipping paper as a poison in hazard class 6.1. DOT also requires them to affix POISON labels on their packaging.

When carriers transport a butyltoluene in an amount exceeding 1001 lb (454 kg), DOT requires them to post POISON placards on the bulk packaging or transport vehicle used for shipment.

PERFORMANCE GOALS FOR SECTION 12.13:

- Describe the bonding in molecules of the polynuclear aromatic hydrocarbons.
- Identify the means by which emergency responders are likely to encounter PAHs.
- Identify the primary risks associated with exposure to the PAHs.
- Identify the sources from which PAHs may be released.

12.13 Polynuclear Aromatic Hydrocarbons

There are more than 100 substances commonly called **polynuclear aromatic hydrocarbons** (PAHs). The simplest members of this class of compounds are a group of structurally similar hydrocarbons having two or more mutually fused benzene rings per molecule. Benzene rings are said to be "mutually fused" when they share a pair of carbon atoms and the bond between them. The PAH having two mutually fused benzene rings is called *naphthalene*. The PAHs having three mutually fused benzene rings are called *anthracene* and *phenanthrene*.

polynuclear aromatic hydrocarbon
■ An aromatic hydrocarbon whose molecules have two or more mutually fused benzene or other rings (abbreviated PAH)

Naphthalene   Anthracene   Phenanthrene

The molecules of more complex PAHs may consist of benzene rings fused to cyclopentenyl (C_5H_7-) and other rings.

12.13-A NAPHTHALENE

Naphthalene is the only PAH of commercial importance. It is a white to colorless solid generally described as having the odor of mothballs. Although naphthalene was once the main constituent of a type of mothball, this specific use is now discouraged. As noted in Section 7.7-C, naphthalene may be isolated from coal tar, and as noted in Section 12.12-A, it is primarily used in the chemical industry as a raw material for the manufacture of phthalic anhydride.

12.13-B ILL EFFECTS RESULTING FROM BREATHING PAHs

Although naphthalene is the only PAH likely to be encountered commercially, it is still possible to unknowingly experience the adverse effects that result from the simultaneous exposure to several PAHs. Many processes that involve the incomplete combustion of organic materials generate mixtures of PAHs. Manufacturing and process industries, for example, often spew PAHs from their smokestacks into the air. PAHs are also dissolved constituents of coal tar distillates (Section 7.7-C), crude petroleum, and certain petroleum fractions including heavy diesel fuel, heating oils, motor oil, heavy naphtha, and asphalt. They are also components of the emissions and residues associated with the burning of coal, petroleum oil, wood, tobacco, and peat. They are even found in minute amounts on bread crusts, burnt toast, and the burnt surfaces of meats cooked on barbecue grills at high temperatures.

The presence of PAH mixtures is especially prominent in diesel-powered vehicular exhaust, where they adsorb to the constituent carbon particulates of soot. Because many heavy-duty trucks

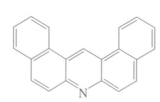

Dibenz[*a,j*]acridine

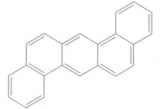

Dibenz[*a,h*]anthracene

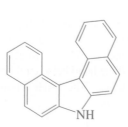

7*H*-Dibenzo[*c,g*]carbazole

Dibenzo[*a,e*]pyrene

Dibenzo[*a,h*]pyrene

Dibenzo[*a,i*]pyrene

FIGURE 12.8 The molecular structures of the 15 polynuclear aromatic hydrocarbons now classified as probable human carcinogens. (Report on Carcinogens, *11th edition*, U.S. Department of Health and Human Services, Public Health Service, National Toxicology Program: Washington, D.C., 2005.)

and buses are diesel-powered, the PAHs produced during their use become components of the environments in which they are driven.

During incomplete combustion processes, PAHs are produced simultaneously with soot by a multistep mechanism similar to that depicted in Section 5.10. When simple hydrocarbons are heated to high temperatures, molecular fragments—such as free radicals—are produced. These fragments undergo reactions that result in the production of new molecules. The original molecules and their fragments dehydrogenate while the new molecules similarly fragment, amalgamate, and continue to dehydrogenate. When dehydrogenation has occurred to its maximum extent, all that remains is soot or carbon black.

All PAHs are solid compounds at room temperature. The most common form of exposure to them is inhalation of the particulate matter contained in polluted air, wood smoke, tobacco smoke, and vehicular exhaust.

When they are inhaled, the PAHs collectively pose a chronic respiratory hazard to humans by inducing asthma and activating allergies. Moreover, the following information has persuasively linked exposure to the 15 PAHs whose molecular structures are represented in Figure 12.8 with the emergence of cancerous growths:

- Massive numbers of mutated cells have been observed in the bronchial passageways of individuals repeatedly exposed in their occupational settings to diesel-engine exhaust.
- Epidemiological tests conducted on animals demonstrate that malignant tumors are produced systemically at the physiological sites where the PAHs were directly applied.

The combination of these observations caused scientists to classify soot as a known human carcinogen.

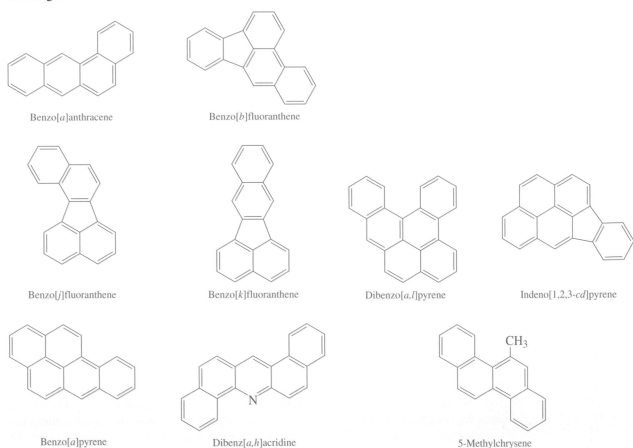

Benzo[*a*]anthracene   Benzo[*b*]fluoranthene

Benzo[*j*]fluoranthene   Benzo[*k*]fluoranthene   Dibenzo[*a,l*]pyrene   Indeno[1,2,3-*cd*]pyrene

Benzo[*a*]pyrene   Dibenz[*a,h*]acridine   5-Methylchrysene

FIGURE 12.8 (*continued*)

Epidemiological test results also served as the basis for classifying the 15 PAHs in Figure 12.8 as probable human carcinogens. Each is a constituent of tobacco smoke or tobacco tar, which implies that the lung, mouth, and throat cancers experienced by tobacco smokers are linked with exposure to them.

The presence of these compounds is relatively widespread. As expected, they are particularly prominent in areas located near heavily trafficked public highways. Benzo[*a*]pyrene, one of the more potent of the PAHs, has even been identified on the surfaces of charcoal-broiled meats.

The molecules of 7*H*-dibenzo[*c,g*]carbazole and the two isomers of dibenzacridine have a ring nitrogen atom (that is, a nitrogen atom substitutes for a ring carbon atom in these molecules). Those substances whose molecules have one or more ring atoms other than carbon are called **heterocyclic compounds**.

heterocyclic compound
■ A compound whose molecules contain one or more ring atoms other than carbon

Scientists have concluded that a lifetime exposure to diesel-engine exhaust could result in the onset of 450 cases of lung cancer in an exposed population of 1 million people. This conclusion generated substantial concern regarding the potential harmful effects caused by inhaling the carcinogenic PAHs and other hazardous components of diesel-engine exhaust. EPA used the information as the technical basis for setting the primary national ambient air quality standard for particulate matter, noted in Section 10.8-E.

In occupational settings, career-oriented firefighters are one of the major groups of individuals most vulnerable to *routinely* inhaling PAHs. It is safe to conclude that they are regularly exposed to airborne PAHs while combating fires. These repeated exposures could constitute the basis for long-term health concerns.

SOLVED EXERCISE 12.8

Since 2007, the state of California has required domestic and foreign cargo and cruise ships to switch from using heavy diesel to light diesel whenever they are within 24 mi (39 km) of the California coast. What is the most likely reason the state adopted this regulation?

Solution: Diesel-powered ships and engines emit soot to which PAHs and sulfur dioxide adhere. Soot and the 15 PAHs whose molecular structures are represented in Figure 12.8 are probable human carcinogens, and sulfur dioxide is a criteria air pollutant. Because states are obligated to protect the health of their citizens, the most likely reason California adopted this regulation was to lessen the health threat caused by inhaling diesel fuel emissions to which the people living in and near its seaport communities are exposed.

SOLVED EXERCISE 12.9

A building having an asphalt covering on a metal deck roof is located within an industrialized area of a major city. Identify the conditions that are likely to cause the ignition of the asphalt during a three-alarm fire involving this building.

Solution: When a fire occurs within a building having an asphalt covering on a metal deck roof, the evolved heat rises upward to the roof and ultimately increases the temperature of the asphalt covering to above 400 °F (204 °C). At this elevated temperature the asphalt softens and ignites when exposed to flames. The softened, burning asphalt also moves downward by gravity and contributes to the spread of fire from the roof to the areas below it.

12.13-C WORKPLACE REGULATIONS INVOLVING PAHs

When the use of naphthalene is required in the workplace, employers must limit employee exposure to a maximum naphthalene vapor concentration of 10 ppm (50 mg/m^3), averaged over an 8-hr workday.

OSHA has also established permissible exposure limits for anthracene, benzo[a]pyrene, chrysene, phenanthrene, and pyrene vapor concentrations at 0.2 mg/m^3.

PERFORMANCE GOALS FOR SECTION 12.14:

- Name the common products produced by the petroleum industry.
- Describe the procedures that experts recommend for extinguishing fires involving bulk volumes of crude petroleum or petroleum products.
- Identify the petroleum products that are produced by the fractionation of crude petroleum.
- Identify the procedures used to chemically treat certain petroleum fractions.
- Describe the nature of the octane number and cetane number assigned to certain petroleum products.
- Describe the means shippers and carriers use to inform emergency responders of the hazards associated with encountering petroleum and petroleum products during transportation mishaps.
- Describe the chemical nature of petrochemicals.

12.14 Petroleum and Petroleum Products

The word *petroleum* is derived from the Greek *petra*, meaning "rock," and Latin *oleum*, meaning "oil." The name "rock oil" is reminiscent of humans' earliest contacts with this material: a liquid that oozed from fissures in rocks.

12.14-A THE NATURE AND NATURAL ORIGIN OF CRUDE OIL

Petroleum is a highly complex mixture consisting of many thousands of organic compounds, approximately 75% of which are hydrocarbons, each of whose molecules have from 3 to 60 carbon atoms. The actual number of individual hydrocarbons in petroleum has been estimated to be between 50,000 and 2 million. Petroleum occurs naturally deep below Earth's surface in certain areas around the world. Wells are drilled through the rock to the oil-bearing stratum, through which the petroleum is then pumped to the surface. In this form, it is called **crude oil, crude petroleum**, or **crude**. The crude petroleum is often transported in oil tankers to plant sites known as **petroleum refineries**, where it is treated to produce *petroleum products*. In the United States, it is also transferred by pipeline to refineries from oil fields located in Texas, Oklahoma, and other states.

Although the mechanism by which crude oil formed in the earth is open to some debate, most scientists believe that it originated from the partial decomposition of animals and plants that lived millions of years ago. Geological forces caused the position of Earth's crust to change over the millennia, causing these materials to be buried at great depths. The temperature and pressure at these depths then caused them to decompose into fossil fuels like crude oil and natural gas. Crude oil formed when the temperature was approximately 150 °F (76 °C), whereas natural gas formed when the temperature rose to approximately 200 °F (93 °C). Scientists estimate that approximately 13.0 lb (5.9 kg) of crude oil resulted from the decomposition of 98 tn (89 t) of prehistoric matter.

crude petroleum (crude oil; crude)
- The liquid mixture consisting primarily of hydrocarbons formed in nature by the decomposition of plant life, and which has not been processed in a refinery

petroleum refinery
- A facility primarily engaged in producing on a commercial scale gasoline, kerosene, fuel oils, lubricants, and other products through fractionation, alkylation, redistillation of unfinished derivatives, cracking, blending, and other processes

U.S. dependence on foreign supplies of crude oil as a source of energy became increasingly apparent during the past decade. A group of Middle Eastern and South American countries controls a large portion of the world's supply of crude oil. They formed a cartel called the *Organization of Petroleum Exporting Countries* (OPEC) that exerts control over both the supply and price of crude oil. Considerable efforts are now ongoing to replace petroleum fuels with alternatives and shed U.S. dependence on these sources of crude oil.

12.14-B FIGHTING FIRES INVOLVING CRUDE PETROLEUM

Crude petroleum is a highly flammable liquid, since its flashpoint ranges from 20 to 90 °F (−7 to 32 °C). Fires involving this material have occurred at oil fields, in transit, and during storage. Capping burning oil wells is a job for specially trained experts, but firefighters have been called on to combat fires involving crude oil at storage facilities and during its transit.

Although small crude oil fires can be extinguished using a deluging volume of water, most experts recommend the use of aqueous film-forming foam (AFFF) (Section 5.11-B) as the most practical means of extinguishing a crude oil fire inside a bulk storage tank. Here, the use of water is recommended solely for cooling purposes. Water and petroleum are immiscible liquids, and petroleum is less dense than water. When discharged on a petroleum fire, the water sinks to the bottom of the storage tank, where it settles as a separate layer.

The presence of a water layer at the bottom of a bulk storage tank represents a special concern when combating a petroleum fire. Although the fire occurs at the tank's surface, where flammable vapor is emitted, the accompanying heat of combustion can be slowly transmitted by convection and radiation through the underlying petroleum to the water layer. Absorption of the intense heat causes the temperature of the petroleum and the underlying water to rise until ultimately, the water boils. The water vapor then forces its way upward, pushing the burning petroleum up and over the walls of the storage tank. This phenomenon is appropriately called a **boilover**.

When a crude petroleum fire occurs inside a storage tank, every attempt should be made to extinguish the fire before the boilover occurs. The burning, frothing petroleum that spills outside the storage tank has been known to flow substantial distances from the tank, triggering numerous secondary fires. The firefighters who combat these fires should be particularly wary, as the flowing, burning petroleum could overtake unsuspecting firefighters in its pathway.

12.14-C FRACTIONATION OF CRUDE PETROLEUM

One of the major operations occurring at petroleum refineries is the fractional isolation of the substances in crude petroleum. The process is called *fractionation*, or **fractional distillation** and is accomplished by heating the crude within preestablished temperature ranges. The vapors of the constituent compounds thus formed are then condensed and collected in separate receivers. They are referred to as **petroleum distillates**.

A single petroleum fraction may be used directly as a commercial petroleum product, or it may be stored until it can be further processed. Bulk volumes of petroleum products are transported to major distribution centers, where they are often stored in multiple tanks in a tank farm.

The fractions most commonly isolated during the fractionation of crude petroleum are represented in Figure 12.9. They consist of the following:

- *A gaseous mixture of methane (65% to 90%), ethane, propane, and butane.* This fraction separates from crude petroleum at a temperature below 70 °F (21 °C). It can be used directly as a *feedstock* for the production of other substances, or its components can be isolated to produce petroleum products such as bottled gas (Section 12.6).

- *A liquid mixture consisting mainly of straight-chain and cyclic aliphatic hydrocarbons having from 5 to 11 carbon atoms per molecule.* This fraction, called *light naphtha* or **ligroin**, is generally isolated within the boiling point range 158 to 284 °F (70 to 140 °C). At one time, light naphtha was used directly as the fuel called *straight-run gasoline*. However, the use of light naphtha is now considered environmentally undesirable for direct use, since it can

boilover
- The phenomenon associated with the expulsion of crude oil and the production of steam during a fire within a crude oil storage tank

fractional distillation (fractionation)
- A method of separating multiple components of a mixture based on the different boiling points of its constituents, during which the vapors are collected at specified temperatures or within specified temperature ranges and subsequently condensed to liquids

petroleum distillate
- A fraction of crude petroleum obtained by the vaporization of its components within a specific temperature range followed by its condensation

ligroin
- A fraction of crude petroleum; light naphtha

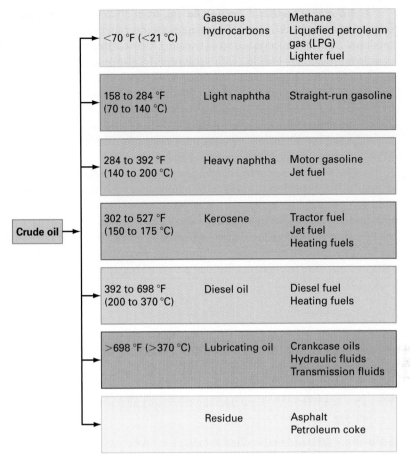

<70 °F (<21 °C)	Gaseous hydrocarbons	Methane Liquefied petroleum gas (LPG) Lighter fuel
158 to 284 °F (70 to 140 °C)	Light naphtha	Straight-run gasoline
284 to 392 °F (140 to 200 °C)	Heavy naphtha	Motor gasoline Jet fuel
302 to 527 °F (150 to 175 °C)	Kerosene	Tractor fuel Jet fuel Heating fuels
392 to 698 °F (200 to 370 °C)	Diesel oil	Diesel fuel Heating fuels
>698 °F (>370 °C)	Lubricating oil	Crankcase oils Hydraulic fluids Transmission fluids
	Residue	Asphalt Petroleum coke

Crude oil

FIGURE 12.9 Some products derived by the fractionation of crude petroleum. Liquefied petroleum gas (LPG) is marketed in the form that comes directly from the distillation tower, but the other fractions require further refinement to remove undesirable components, most notably, benzene and sulfurous and nitrogenous compounds. The refinement processes include vacuum distillation, catalytic reforming, hydrocracking, and catalytic cracking. Additives are also blended with gasoline and other petroleum products to improve their performance.

contain as much as 3% benzene by volume. (In 1995, EPA used its authority under the Clean Air Act to require refineries to produce petroleum products containing no more than 1% benzene.) Today, light naphtha is used as the feedstock either for the production of benzene, toluene, and xylene by catalytic reformation or for the production of ethylene and propylene by catalytic cracking.

   ■ *A liquid mixture consisting mainly of straight-chain and cyclic aliphatic hydrocarbons having from 7 to 11 carbon atoms per molecule.* This fraction, called *heavy naphtha*, is generally isolated within the boiling point range 284 to 392 °F (140 to 200 °C). It is typically used as the feedstock for the production of gasoline. The production of gasoline involves catalytic cracking, during which the hydrocarbons having from 7 to 11 carbon atoms per molecule are reduced to hydrocarbons having a smaller number of carbon atoms per molecule. This fraction is also the feedstock for the production of toluene.

   ■ *A liquid petroleum product consisting of a mixture of hydrocarbons having from 9 to 16 carbon atoms per molecule.* This fraction, called **kerosene**, is often isolated within the boiling point range 302 to 527 °F (150 to 275 °C). Today, kerosene is used primarily for the production of jet fuels having names like JP-4, JP-7, and JP-8, where "JP" is the acronym for jet propulsion.

   ■ *A liquid mixture of compounds having from 10 to 15 carbon atoms per molecule.* This fraction, called *diesel fuel* or **diesel oil**, is usually isolated within the boiling point range 392 to 698 °F (200 to 370 °C). It is blended with additives and used to heat buildings and to power passenger cars, pickup and heavy-duty trucks, buses, ships, and certain heavy equipment. More than 50% of the passenger cars now used by Europeans are diesel-powered; more than 90% of the cars in Italy alone are diesel-powered. In the United States, the use of diesel-powered passenger vehicles has fluctuated over past decades.

kerosene
■ An oily liquid obtained by the fractional distillation of crude petroleum and often used as a tractor fuel, jet fuel, and heating fuel

diesel oil
■ A petroleum product commonly used as the fuel for operating diesel engines

lubricating oil (lubricant)
■ Any petroleum oil capable of producing a film that coats the surfaces of moving metal parts

asphalt (asphaltic bitumen)
■ The residue remaining from the fractional distillation of crude petroleum, usually combined with mineral matter

asphalt kettle
■ A vessel or container used to process, treat, hold for heating, or dispense flammable or combustible roofing materials in the form of viscous liquids

■ *A liquid to semisolid fraction having from 30 to 45 carbon atoms per molecule.* This mixture of compounds is typically isolated at temperatures above 698 °F (370 °C). The products produced from this fraction are broadly called **lubricating oils** or **lubricants**. For example, the petroleum fraction is blended with various performance additives, corrosion inhibitors, and detergents and then used as motor oils like crankcase oil or hydraulic fluid. It is also the feedstock from which other motor and non-vehicle lubricating oils are produced. Typically, this fraction is hydrotreated and subjected to other chemical processes to remove its undesirable constituents containing metals, nitrogen, and sulfur and then blended to achieve a specified viscosity, stability, and lubricity. When they are used, lubricating oils produce an oily film that coats the surfaces of moving metal parts, thereby reducing friction and wear.

■ *The solid residue of the distillation process.* The fractionation residue is called **asphalt** or **asphaltic bitumen**, a mixture of compounds with molecules having 40 or more carbon atoms. Asphalt ignites at approximately 400 °F (204 °C). It is used commercially as a component of protective coating, waterproofing, and adhesive products. Prior to its use for paving, road, and roofing construction, hot asphalt is often held in crucibles, **asphalt kettles**, and portable tank trucks. DOT regulates the transportation of hot asphalt as an elevated-temperature material (see the glossary). The fractionation residue is also treated to produce petroleum coke.

SOLVED EXERCISE 12.10

Why should firefighters avoid inhaling the smoke that evolves when tar paper and roofing shingles smolder and burn?

Solution: Tar paper and roofing shingles are often manufactured from asphalt, the residue that remains after crude petroleum has been fractionated. The constituents of asphalt include the group of high-molecular-weight compounds called polynuclear aromatic hydrocarbons (PAHs) (Section 12.13). When asphalt-containing products burn, these compounds either oxidize or become components of the particulate matter evolved in the associated smoke plume. Firefighters should avoid breathing this smoke to reduce the risk of stimulating respiratory ailments or of contracting throat and lung cancer.

(Scientists have verified the presence of dibenzo[*a,h*]pyrene, dibenzo[*a,i*]pyrene, and indeno[1,2,3-*cd*]pyrene in the matter that volatilizes from molten asphalt. These three PAHs are among the 15 PAHs represented in Figure 12.8 that cause cancer.)

12.14-D CHEMICAL TREATMENT OF PETROLEUM FRACTIONS

alkylation
■ The combination of an alkane and an alkene, resulting in a branched-chain alkane

alkylate
■ Any branched-chain alkane resulting from the combination of an alkane and an alkene

A major chemical treatment process that is conducted at petroleum refineries is **alkylation**, that is, the chemical reaction during which a branched-chain hydrocarbon is produced by the chemical union of an alkane and an alkene. Alkylation is generally conducted in the presence of hydrofluoric acid, which acts catalytically. The product of an alkylation reaction is called an **alkylate**. Bulk quantities of alkylates are often present at petroleum refineries in bulk storage tanks. They are highly flammable substances and pose a fire and explosion hazard.

The alkylate commonly called *isooctane* is produced from the chemical union of isobutane and isobutene.

$$CH_3-CH-CH_3(g) \quad + \quad CH_2=C-CH_3(g) \quad \longrightarrow \quad CH_3-C-CH_2-C-CH_3(g)$$

Isobutane   Isobutene   "Isooctane"

The name isooctane is actually a misnomer; the correct name of this compound is 2,2,4-trimethylpentane.

Alkylation reactions also produce trimethylpentanes, dimethylhexanes, and other branched-chain hydrocarbons. These compounds are desirable additives in gasoline, since they improve the completeness of its combustion. In their absence, the phenomenon called **knocking** occurs. This term refers to the audible pings and rattling noises that are generated within the engine as the fuel burns. Fuel additives that serve to reduce the degree of knocking are called **antiknock agents**.

knocking
- The noises associated with the incomplete combustion of a petroleum fuel, especially in internal combustion engines

antiknock agent
- A petroleum fuel additive that reduces or eliminates the noises generated within a combustion chamber when the fuel burns

SOLVED EXERCISE 12.11

Several chemical processes conducted at petroleum refineries involve the production of antiknock agents, which increase the octane rating of the fuels with which they are blended. For example, propene that has been isolated from crude petroleum by fractionation is induced to react with itself to produce the antiknock agents 2-methyl-1-pentene, 4-methyl-1-pentene, and 4-methyl-2-pentene. Write the equations that illustrate the production of these branched alkenes.

Solution: Propene reacts *with itself* to produce 2-methyl-1-pentene, 4-methyl-1-pentene, and 4-methyl-2-pentene. Consequently, propene is the reactant, and 2-methyl-1-pentene, 4-methyl-1-pentene, and 4-methyl-2-pentene are the reaction products. The equations illustrating these production reactions are the following:

$$2CH_3CH=CH_2 \longrightarrow CH_3CH_2CH_2-\overset{\overset{\displaystyle CH_3}{|}}{C}=CH_2$$

Propene   2-Methyl-1-pentene

$$2CH_3CH=CH_2 \longrightarrow CH_2=CH-CH_2-\overset{\overset{\displaystyle CH_3}{|}}{CH}-CH_3$$

Propene   4-Methyl-1-pentene

$$2CH_3CH=CH_2 \longrightarrow CH_3-CH=CH-\overset{\overset{\displaystyle CH_3}{|}}{CH}-CH_3$$

Propene   4-Methyl-2-pentene

12.14-E GASOLINE

Gasoline is a petroleum product used mainly as a motor fuel. Its flashpoint is approximately $-40\ °F$ ($\sim-40\ °C$). It is produced by blending heavy naphtha or other petroleum fractions with oxygenates (Section 13.2-D) and other performance additives.

The different commercial types of gasoline are distinguished by their inherent octane numbers. The **octane number**, or *octane rating*, is a representation of the antiknock properties of a fuel under laboratory or test conditions. Octane numbers of zero and 100 are arbitrarily assigned to *n*-heptane (a high "knocker") and 2,2,4-trimethylpentane (a low "knocker"), respectively.

gasoline
- The complex mixture of volatile hydrocarbons commonly used as a vehicular fuel

Heptane is an undesirable fuel, since it causes engine knocking when it burns in a combustion chamber, but 2,2,4-trimethylpentane is a desirable fuel, since its combustion does not cause motor engines to knock.

$$CH_3CH_2CH_2CH_2CH_2CH_2CH_3$$

n-Heptane
Octane number = 0

$$CH_3-\overset{\overset{\displaystyle CH_3}{|}}{\underset{\underset{\displaystyle CH_3}{|}}{C}}-CH_2-\overset{\overset{\displaystyle CH_3}{|}}{CH}-CH_3$$

2,2,4-Trimethylpentane
Octane number = 100

The octane number of a fuel is determined by comparing its knocking with the knocking of various mixtures of *n*-heptane and 2,2,4-trimethylpentane. When a sample of a fuel is found by testing to knock like a mixture of 85 parts 2,2,4-trimethylpentane and 15 parts *n*-heptane, the fuel is assigned an octane number of 85.

Two types of gasoline are marketed: **aviation gasoline** and *motor gasoline*. Aviation gasoline has an octane rating over 100 and is intended for fueling aircraft. Motor gasoline has a typical octane rating in the range of 87 to 91 and is most commonly used to fuel motor vehicles like passenger cars and small trucks. There are three grades of motor gasoline:

- **Regular gasoline**, also called *regular unleaded*, which has an octane rating of 87
- **Mid-grade gasoline**, which has an octane rating of 89
- **Premium gasoline**, also called *supreme* and *super unleaded*, which has an octane rating of 91

At service stations, customers may obtain each grade from fuel pumps like those shown in Figure 12.10.

FIGURE 12.10 Gasoline fuels are differentiated by their octane numbers, which are established by subjecting a gasoline sample to certain test conditions to ascertain how well it knocks when ignited in a combustion chamber. In the test, octane numbers of 0 and 100 are assigned to *n*-heptane and 2,2,4-trimethylpentane, respectively.

All gasoline sold in the United States since 1975 is **unleaded gasoline**, meaning that it does not contain tetramethyllead or tetraethyllead. These compounds were formerly added to gasoline to function as antiknock agents in operating vehicles. Using the authority of the Clean Air Act, EPA banned the manufacture and use of lead compounds in gasoline owing to their inherent toxicity (Section 10.17-A).

unleaded gasoline
■ Gasoline that does not contain a lead compound such as tetraethyl lead, formerly an additive that functioned as an antiknock agent

SOLVED EXERCISE 12.12

What is meant by the term "high octane" in reference to vehicular and aviation fuels?

Solution: "High octane" refers to a type of gasoline that was manufactured by blending a base gasoline stock with certain substances that increase the octane rating of the fuel. These additives include 2,2,4-trimethylpentane, other trimethylpentane isomers, dimethylhexanes, and other branched-chain hydrocarbons, each of which serves as an antiknock agent. 2,2,4-Trimethylpentane is commonly called "isooctane" in the petroleum industry. Because the trimethylpentanes and dimethylhexanes are isomers of octane, the blended gasoline contains elevated concentration of these compounds. This gave rise the expression "high octane" fuel, since each of these compounds has eight carbon atoms per molecule.

The chemical composition of petroleum fuels varies not only in different parts of the world but also at different times of the year. The chemical composition of a given fuel produced at a refinery can even change daily. Despite this variability, motor gasoline typically consists of a mixture of hydrocarbons having the following approximate composition by volume: 55% normal and branched alkanes; 34% aromatic hydrocarbons; 5% alkenes; and 5% cycloalkanes and cycloalkenes.

12.14-F DIESEL OIL

Diesel oil is the fuel used to power diesel engines in certain vehicles and machinery. Its flashpoint ranges from 110 to 190 °F (43 to 88 °C). Diesel oil is produced from the untreated fraction of crude oil that refiners call **sour crude**. Because this fraction contains sulfurous and nitrogenous compounds, it must be treated to reduce or eliminate the sulfur and nitrogen content to comply with today's Clean Air Act standards. This is accomplished by hydrogenation. The process, called **hydrotreatment**, produces a diesel oil consisting of alkanes, cycloalkanes, and aromatic compounds. During hydrotreating, the sulfur and nitrogen atoms are converted into hydrogen sulfide and ammonia, respectively, both of which are isolated for use during other refinery operations.

sour crude
■ Untreated crude oil or the fraction thereof that contains sulfurous and nitrogenous compounds

hydrotreatment
■ The processing of gasoline and other petroleum products with hydrogen to reduce or eliminate their nitrogenous and sulfurous content

Untreated and hydrotreated diesel oil are sometimes referred to as *heavy diesel oil* and *light diesel oil*, respectively. The combustion of untreated diesel oil yields a sooty smoke consisting of carbon particulates to which polynuclear aromatic hydrocarbons adhere. As noted in Section 12.13-B, the production of this plume poses an unreasonable risk to public health and the environment, since PAH exposure is linked with the development of malignant tumors. The combustion of light diesel oil yields an emission that contains fewer pollutants. To achieve the Clean Air Act standard for fine particulates (Section 10.8-E) in the United States, the diesel industry was obligated to change from heavy to light diesel oil in the mid-2000s. Today, the diesel-powered vehicles that drive on U.S. roadways use mainly light diesel oil as their fuel.

The different commercial types of light diesel oil are rated by a system similar but not equivalent to the octane-number system used for rating different types of gasoline, namely, the **cetane number**. Whereas the octane number measures the ability of a gasoline fuel to reduce engine knocking, the cetane number gauges the ease with which a diesel fuel autoignites when compressed in a diesel engine.

cetane number
■ The percentage of cetane ($C_{16}H_{34}$) that must be mixed with 2,2,4,4,6,8,8-heptamethylnonane to yield the same ignition performance as a sample of a diesel oil

Cetane is the common name for *n*-hexadecane, an alkane having the formula $C_{16}H_{34}$. For test purposes, the cetane number of 2,2,4,4,6,8,8-heptamethylnonane and cetane are arbitrarily set at 15 and 100, respectively.

$$CH_3-\underset{\underset{CH_3}{|}}{\overset{\overset{CH_3}{|}}{C}}-CH_2-\underset{\underset{CH_3}{|}}{\overset{\overset{CH_3}{|}}{C}}-CH_2-\overset{\overset{CH_3}{|}}{CH}-CH_2-\underset{\underset{CH_3}{|}}{\overset{\overset{CH_3}{|}}{C}}-CH_3$$

2,2,4,4,6,8,8-Heptamethylnonane
Cetane number = 15

$$CH_3CH_2CH_2CH_2CH_2CH_2CH_2CH_2CH_2CH_2CH_2CH_2CH_2CH_2CH_2CH_3$$

n-Hexadecane (cetane)
Cetane number = 100

The cetane number of a diesel fuel is determined in special variable-compression-ratio test engines.

Generally, diesel engines operate well when the diesel fuel has a cetane number between 40 and 55. The two diesel fuel types that are most commonly available are referred to as **regular diesel oil** and **premium diesel oil**, having cetane numbers ranging from 40 to 60 and 45 to 50, respectively.

In today's marketplace, use is also made of the **alternative fuel** called **biodiesel oil**. The most basic biodiesel fuel is produced directly from soybean oil, cottonseed oil, or recycled restaurant grease. The use of biodiesel as a fuel is becoming increasingly popular. For example, San Francisco uses a biodiesel blend called B20 in its entire array of diesel-operated vehicles from ambulances to street sweepers. B20 is a mix of 20% soybean-oil-based biofuel and 80% petroleum-based diesel fuel. The combination of Diesel #1, Diesel #2, or JP-8 with 7.7% to 15% ethanol (Section 13.2-D) and other additives is another popular biodiesel blend. The use of biodiesel as a fuel is encouraged as a means of reducing the nation's dependence on foreign oil supplies. The Energy Independence and Security Act of 2007 calls for the production of 36 billion gal/y (36×10^9 gal/y, or 14×10^7 m^3/y) of ethanol from corn and cellulose by 2022, to be blended with petroleum-based fuels.

12.14-G HEATING FUELS

Some petroleum products are useful as heating fuels. They are manufactured by fractionation, followed by blending with specified additives. For example, diesel fuel and heavy naphtha are often processed to produce heating oils, either by generating fractions within a narrower distillation range or by adding a specified amount of other petroleum fractions until the final blend possesses desirable ignition temperatures and heat (Btu) values. These products are utilized not only as home heating oils but also as the fuels in boilers for factories, apartment and office buildings, and schools.

Six grades of heating oils are commercially recognized in the United States, each designated by the words "Fuel oil" followed by a number from 1 to 6. The grades differ by the temperature at which each fuel ignites, progressively increasing from 444 °F (229 °C) to 765 °F (407 °C), the ignition temperatures of Fuel Oil No. 1 and Fuel Oil No. 6, respectively.

12.14-H TRANSPORTING CRUDE PETROLEUM AND PETROLEUM PRODUCTS

More petroleum products are transported by public highway than any other hazardous materials. When shippers offer crude petroleum or a petroleum product for transportation, DOT requires them to select the relevant basic description from Table 12.13 for entry on an accompanying

regular diesel oil
■Diesel oil having a cetane number ranging from 40 to 60

premium diesel oil
■Diesel oil having a cetane number ranging from 45 to 50

alternative fuel
■ A motor fuel that replaces ordinary gasoline in whole or part

biodiesel oil
■Vehicular fuel produced from vegetable products and animal fats, sometimes mixed with petroleum diesel oil

TABLE 12.13	Basic Descriptions of Some Petroleum Products
PETROLEUM PRODUCT	**BASIC DESCRIPTION**
Asphalt (at or near its flashpoint)	NA1999, Asphalt, 3, PGIII
Crude oil	UN1267, Petroleum crude oil, 3, PGI *or* UN1267, Petroleum crude oil, 3, PGII
Diesel fuel, regular or premium (domestic transportation)	NA1993, Diesel fuel, 3, PGIII
Diesel fuel, regular or premium (international transportation)	UN1202, Diesel fuel, 3, PGIII
Fuel, aviation	UN1863, Fuel, aviation, turbine engine, PGI *or* UN1863, Fuel, aviation, turbine engine, PGII *or* UN1863, Fuel, aviation, turbine engine, PGIII
Fuel oil (No. 1, 2, 4, 5, or 6)	NA1993, Fuel oil, 3, PGIII
Gas oil	UN1202, Gas oil, 3, PGIII
Gasoline	UN1203, Gasoline, 3, PGII
Kerosene	UN1223, Kerosene, 3, PGIII
Liquefied petroleum gas	Petroleum gases, liquefied, 2.1, UN1075
Petroleum oil	NA1270, Petroleum oil, 3, PGI *or* NA1270, Petroleum oil, 3, PGII *or* NA1270, Petroleum oil, 3, PGIII
Petroleum products	NA1270, Petroleum products, 3, PGI *or* NA1270, Petroleum products, 3, PGII *or* NA1270, Petroleum products, 3, PGIII
Tars, liquid (road asphalt and oils, bitumen, and cutbacks)	UN1999, Tars, liquid, 3, PGII

shipping paper. DOT also requires them to affix the applicable FLAMMABLE GAS or FLAM-MABLE LIQUID label on packaging containing crude oil and petroleum products.

When carriers transport crude petroleum or a petroleum product in an amount exceeding 1001 lb (454 kg), DOT requires them to display the applicable FLAMMABLE GAS, FLAMMABLE, or COMBUSTIBLE placards on the bulk packaging or transport vehicle used for shipment.

When shippers offer hot asphalt for transportation in a crucible, kettle, or other appropriate bulk container as an elevated-temperature material, DOT requires them to display the HOT marking in Figure 6.10 on both sides and ends of the container.

12.14-I PETROCHEMICALS

petrochemical
■ A substance obtained directly or indirectly from natural gas, crude petroleum, or a petroleum fraction

During refinery operations, a petroleum fraction can be subjected to further distillation to isolate its individual components, or it can be treated to yield substances called **petrochemicals**. Broadly speaking, petrochemicals are substances produced in bulk directly from natural gas, crude petroleum, or a petroleum fraction. The most familiar petrochemicals are ethylene, propylene, 1,3-butadiene, benzene, toluene, xylene(s), and methanol. They serve as indispensable raw materials for the plastics, synthetic rubber, and synthetic fiber industries.

In the United States, by far the most important petrochemicals are ethylene and propylene. They are routinely manufactured by cracking the ethane and propane extracted from natural gas.

$$C_2H_6(g) \longrightarrow CH_2{=}CH_2(g) + H_2(g)$$

Ethane　　　　　　　Ethylene　　　　　　Hydrogen

$$2C_3H_8(g) \longrightarrow CH_4(g) + CH_2{=}CH_2(g) + CH_2{=}CHCH_3(g) + H_2(g)$$

Propane　　　　　Methane　　　　　Ethylene　　　　　　Propylene　　　　　　Hydrogen

PERFORMANCE GOALS FOR SECTION 12.15:

- Describe the molecular nature of the halogenated hydrocarbons having one and two carbon atoms per molecule and provide acceptable names for them.
- Identify the primary risks associated with exposure to the simplest halogenated hydrocarbons.
- Describe the means shippers and carriers use to inform emergency responders of the hazards associated with encountering the simple halogenated hydrocarbons during transportation mishaps.

12.15 Simple Halogenated Hydrocarbons

halogenated hydrocarbon (alkyl/aryl halide)
■ A hydrocarbon derivative in whose molecules one or more halogen atoms are substituted for hydrogen atoms

When one or more hydrogen atoms in the molecules of hydrocarbons are substituted with halogen atoms, the resulting compounds are called **halogenated hydrocarbons**, **alkyl halides**, or **aryl halides**. When one hydrogen atom is substituted with a halogen atom, the general chemical formula of the resulting halogenated hydrocarbon is $R{-}X$, where R is the formula of an arbitrary alkyl or aryl group, and X is the chemical symbol of a specific halogen. When the halogen is chlorine, the compounds are called **chlorinated hydrocarbons**, **alkyl chlorides**, or **aryl chlorides**.

chlorinated hydrocarbon (alkyl/aryl chloride)
■ A hydrocarbon derivative in whose molecules one or more chlorine atoms are substituted for hydrogen atoms

Consider the following Lewis structures of the chlorinated derivatives of methane:

| Methyl chloride (Chloromethane) | Methylene chloride (Dichloromethane) | Chloroform (Trichloromethane) | Carbon tetrachloride (Tetrachloromethane) |

From left to right, each structure identifies a substance in which one, two, three, or four hydrogen atoms, respectively, have been substituted with chlorine atoms. These compounds are known mainly by their common names: methyl chloride, methylene chloride, chloroform, and carbon tetrachloride, respectively.

In the IUPAC system, the halogen atoms are named as substituents of the compound having the longest continuous chain of carbon atoms. As first noted in Section 5.13, the halogen atoms are named as substituents by replacing the *-ine* suffix on the name of the halogen with *-o*. The number of halogen atoms is indicated by the use of *mono-*, *di-*, *tri-*, *tetra-*, and so forth, for one, two, three, four, or more atoms of the same halogen, respectively. Hence, the IUPAC names of the chlorinated derivatives of methane are chloromethane (the mono- prefix is dropped), dichloromethane, trichloromethane, and tetrachloromethane.

As noted in Section 12.16-C, the major component in the fire suppressant known as Pyro-Chem FM-200 is 1,1,1,2,3,3,3-heptafluoropropane. Show how this name is used to determine the structural formula for this substance.

Solution: The name of this fire suppressant indicates that it is a fluorinated derivative of propane, whose formula is C_3H_8. "Hepta" in the name of the fire suppressant means that *seven* of the eight hydrogen atoms in propane have been replaced with fluorine atoms. The notation "1,1,1,2,3,3,3" identifies precisely which hydrogen atoms along the carbon–carbon chain in the structural formula of propane have been replaced with fluorine atoms. Three fluorine atoms replace three hydrogen atoms that are bonded to each of the first and third carbon atom, and one fluorine atom replaces the hydrogen atom bonded to the second carbon atom. Consequently, the structural formula of Pyro-Chem FM-200 is the following:

$$
\begin{array}{ccccc}
 & F & F & F & \\
 & | & | & | & \\
F - & C - & C - & C - & F \\
 & | & | & | & \\
 & F & H & F &
\end{array}
$$

This formula may be condensed to $CF_3-CHF-CF_3$.

The molecular structures and names of several chlorinated derivatives of ethane are noted here:

CH_3-CH_2Cl
Chloroethane

$Cl-CH-CH_3$
$\quad\quad |$
$\quad\quad Cl$
1,1-Dichloroethane

CH_2Cl-CH_2Cl
1,2-Dichloroethane

Cl
$|$
$Cl-C-CH_3$
$|$
Cl
1,1,1-Trichloroethane

$Cl-CH-CH_2Cl$
$\quad\quad |$
$\quad\quad Cl$
1,1,2-Trichloroethane

$Cl-CH-CH-Cl$
$\quad\quad |\quad\quad |$
$\quad\quad Cl\quad Cl$
1,1,2,2-Tetrachloroethane

Cl
$|$
$Cl-C-CH_2Cl$
$|$
Cl
1,1,1,2-Tetrachloroethane

The molecular structures and names of the chlorinated derivatives of ethene are noted next:

$$\begin{array}{cc} \text{Cl} & \text{H} \\ \diagdown & \diagup \\ \text{C} & = & \text{C} \\ \diagup & \diagdown \\ \text{H} & \text{H} \end{array}$$

Chloroethene

$$\begin{array}{cc} \text{Cl} & \text{Cl} \\ \diagdown & \diagup \\ \text{C} & = & \text{C} \\ \diagup & \diagdown \\ \text{H} & \text{H} \end{array}$$

cis-1,2-Dichloroethene

$$\begin{array}{cc} \text{Cl} & \text{H} \\ \diagdown & \diagup \\ \text{C} & = & \text{C} \\ \diagup & \diagdown \\ \text{H} & \text{Cl} \end{array}$$

trans-1,2-Dichloroethene

$$\begin{array}{cc} \text{Cl} & \text{H} \\ \diagdown & \diagup \\ \text{C} & = & \text{C} \\ \diagup & \diagdown \\ \text{Cl} & \text{H} \end{array}$$

1,1-Dichloroethene

$$\begin{array}{cc} \text{Cl} & \text{Cl} \\ \diagdown & \diagup \\ \text{C} & = & \text{C} \\ \diagup & \diagdown \\ \text{Cl} & \text{H} \end{array}$$

Trichloroethene

$$\begin{array}{cc} \text{Cl} & \text{Cl} \\ \diagdown & \diagup \\ \text{C} & = & \text{C} \\ \diagup & \diagdown \\ \text{Cl} & \text{Cl} \end{array}$$

Tetrachloroethene

Each of these simple chlorinated hydrocarbons is a colorless, water-insoluble, highly volatile liquid; most are nonflammable. For decades, this combination of physical properties was the technical basis for their popular use by many manufacturing and process industries. For example, trichloroethene, 1,1,1-trichloroethane, and 1,1,2,2-tetrachloroethane were once popular industrial **degreasers**. These are liquids that effectively clean oil, grease, wax, and other undesirable material from metallic, textile, and glass surfaces. Carbon tetrachloride was once used as a fire extinguisher. Methylene chloride was once widely used as an aerosol, and is still used as a solvent in the pharmaceutical industry. Popular methylene chloride containers are illustrated in Figure 12.11. Tetrachloroethene, also called perchloroethylene and "perc," is still a popular dry-cleaning agent.

However, when inhaled, the vapors of these simple chlorinated hydrocarbons can cause cancer. They also deplete stratospheric ozone. One by one, each has been replaced with another

degreaser
■ Any organic compound used as a solvent to remove grease and oil from surfaces

FIGURE 12.11 Because it is a very volatile liquid, methylene chloride is generally stored and used in relatively small volumes like this 1-gal (4-L) brown-glass bottle and 4.7-gal (18-L) recyclable tote container. The glass bottle is transported within a recyclable stainless-steel overpack container, which provides a degree of safety when handled and transported. (*Courtesy of Mallinckrodt Baker, Inc., Phillipsburg, New Jersey.*)

substance that performs as well for a specific purpose but does not cause cancer or deplete ozone. Chemical companies are actively searching for a substance to replace tetrachloroethene in the dry-cleaning industry. Hence, the days of using tetrachloroethane as a dry-cleaning solvent are probably numbered.

12.15-A ILL EFFECTS RESULTING FROM INHALING THE VAPORS OF THE SIMPLE HALOGENATED HYDROCARBONS

When the vapors of the halogenated methanes and ethanes are inhaled at relatively low concentrations, they usually cause lightheadedness, dizziness, and fatigue. However, when these vapors are inhaled at concentrations above their permissible exposure limits, they cause a lowering of consciousness that can be life-threatening.

Epidemiologists have determined that exposure to the following halogenated hydrocarbons causes cancer in laboratory animals: bromodichloromethane, carbon tetrachloride, chloroform, 1,2-dichloroethane, hexachloroethane, methylene chloride, tetrachloroethene, tetrafluoroethene, and trichloroethene. They are classified as probable human carcinogens.

12.15-B TRANSPORTING THE HALOGENATED HYDROCARBONS

When shippers offer a halogenated hydrocarbon for transportation, DOT requires them to enter the relevant basic description from Table 12.14 on an accompanying shipping paper. DOT also requires them to affix on its packaging a label that corresponds to the hazard class or division code of the halogenated hydrocarbon.

When carriers transport a halogenated hydrocarbon in an amount exceeding 1001 lb (454 kg), DOT requires them to post the applicable placards corresponding to the hazard class or division code on the bulk packaging or transport vehicle used for shipment.

PERFORMANCE GOALS FOR SECTION 12.16:

- Describe the molecular nature of the common chlorofluorocarbons.
- Identify the principal ways by which chlorofluorocarbons were formerly used.
- Describe the chemistry associated with depletion of stratospheric ozone by chlorofluorocarbons.
- Describe the nature of the marking that EPA requires to be posted when all ozone-depleting substances are stored in bulk.

12.16 Chlorofluorocarbons and Their Related Compounds

A special group of halogenated hydrocarbons are the **chlorofluorocarbons** (CFCs). These substances were formerly important commercial compounds, but for reasons to be discussed, their use has been sharply curtailed worldwide.

The commercially important chlorofluorocarbons used in the past were primarily members of the following classes of substances:

chlorofluorocarbon
■ A compound whose molecules are composed solely of carbon, chlorine, and fluorine atoms (abbreviated CFC)

- **Chlorofluoromethanes.** These compounds have the general chemical formula CF_nCl_{x-n}, where x and n are whole numbers less than 4.
- **Chlorofluoroethanes.** These compounds have the general chemical formula $C_2F_nCl_{x-n}$, where x and n are whole numbers less than 6.

TABLE 12.14	Basic Descriptions of Some Simple Halogenated Hydrocarbons
HALOGENATED HYDROCARBON OR GROUPS THEREOF	**BASIC DESCRIPTION**
Allyl chloride	UN1100, Allyl chloride, 3, (6.1), PGI (Poison)
Amyl chlorides	UN1107, Amyl chlorides, 3, PGII
Benzyl chloride	UN1738, Benzyl chloride, 6.1, (8), PGII (Poison)
Bromobenzene	UN2514, Bromobenzene, 3, PGIII
1-Bromobutane	UN1126, 1-Bromobutane, 3, PGII
Bromotrifluoroethylene	UN2419, Bromotrifluoroethylene, 2.1
Carbon tetrachloride	UN1846, Carbon tetrachloride, 6.1, PGII (Marine Pollutant) (Poison)
Chlorobenzene	UN1134, Chlorobenzene, 3, PGIII
Chlorobenzyl chlorides, liquid	UN2235, Chlorobenzyl chlorides, liquid, 6.1, PGIII
Chlorobenzyl chlorides, solid	UN3427, Chlorobenzyl chlorides, solid, PGIII
Chlorobutanes	UN1127, Chlorobutanes, 3, PGII
Chlorodifluoromethane	UN1018, Chlorodifluoromethane, 2.2
Chloroform	UN1888, Chloroform, 6.1, PGIII (Poison)
2-Chloropropane	UN2356, 2-Chloropropane, 3, PGI
1-Chloro-1,2,2,2-tetrafluoroethane	UN1021, 1-Chloro-1,2,2,2-tetrafluoroethane, 2.2
Chlorotrifluoromethane	UN1022, Chlorotrifluoromethane, 2.2
Dibromomethane	UN2264, Dibromomethane, 6.1, PGIII (Poison)
o-Dichlorobenzene	UN1591, o-Dichlorobenzene, 6.1, PGIII (Poison)
Dichlorodifluoromethane/ difluoroethane mixture	UN2602, Dichlorodifluoromethane, 2.2
1,1-Dichloroethane	UN2362, 1,1-Dichloroethane, 3, PGII
Dichloropentanes	UN1152, Dichloropentanes, 3, PGIII
1,2-Dichloropropane	UN1279, 1,2-Dichloropropane, 3, PGII
Dichloropropenes	UN2047, Dichloropropenes, 3, PGII
Ethylene dichloride[a]	UN1184, Ethylene dichloride, 3, PGII
Hexachlorobenzene	UN2729, Hexachlorobenzene, 6.1, PGIII (Poison)
Hexachlorobutadiene	UN2279, Hexachlorobutadiene, 6.1, PGIII (Poison)
Methyl bromide	UN1062, Methyl bromide, 2.3 (Poison - Inhalation Hazard, Zone C)
Methyl chloride	UN1063, Methyl chloride, 2.1

[a]Ethylene dichloride and 1,2-dichloroethane are synonyms.

HALOGENATED HYDROCARBON OR GROUPS THEREOF	BASIC DESCRIPTION
Methylene chloride	UN1593, Dichloromethane, 6.1, PGIII (Poison)
Polychlorinated biphenyls, liquid	UN2315, Polychlorinated biphenyls, liquid, 9, PGII (Marine Pollutant)
Polychlorinated biphenyls, solid	UN3432, Polychlorinated biphenyls, solid, 9, PGII (Marine Pollutant)
1,1,2,2-Tetrachloroethane	UN1702, 1,1,2,2-Tetrachloroethane, 6.1, PGII (Poison)
Tetrachloroethylene	UN1897, Tetrachloroethylene, 6.1, PGIII (Poison)
1,1,1-Trichloroethane	UN2831, 1,1,1-Trichloroethane, 6.1, PGIII (Poison)
Trichloroethylene	UN1710, Trichloroethylene, 6.1, PGIII (Poison)
Trifluoromethane	UN3136, Trifluoromethane, 2.2
Vinyl chloride	UN1086, Vinyl chloride, stabilized, 2.1, UN1086

As a class of organic compounds, the chlorofluorocarbons are relatively inert substances. They readily vaporize at room temperature. Although many are nonflammable gases, others are flammable.

The chlorofluorocarbons were typically prepared by reacting hydrofluoric acid with a commercially available chlorinated hydrocarbon. For example, a mixture of chlorofluoromethanes was prepared from carbon tetrachloride as follows:

$$2CCl_4(g) + 3HF(l) \longrightarrow CCl_2F_2(g) + CCl_3F(g) + 3HCl(g)$$

Carbon tetrachloride   Hydrofluoric acid   Dichlorodifluoromethane   Trichlorofluoromethane   Hydrogen chloride

The two chlorofluorocarbons were then separated by distillation.

Some examples of the CFCs and their related compounds (hydrofluorocarbons and hydrochlorofluorocarbons, see further) are noted in Table 12.15. Their commercial names are denoted as *CFC-wxyz*, or *R-wxyz*, where *w*, *x*, *y*, and *z* are digits derived from their chemical formulas as follows:

w = the number of carbon–carbon double bonds per molecule
x = the number of carbon atoms per molecule minus one
y = the number of hydrogen atoms per molecule plus one
z = the number of fluorine atoms per molecule

When *w*, *x*, *y*, or *z* is zero, the digit is omitted. For example, *w*, *x*, *y*, and *z* for the chlorofluorocarbon having the formula $CFCl_3$ are 0, 0, 1, and 1, respectively. It is denoted commercially by the product names CFC-11 or R-11. Chemists give it the IUPAC name trichlorofluoromethane.

Sometimes, it is necessary to distinguish between two or more structural isomers of the chlorofluoroethanes in their commercial names. The compound within a group of these isomers having the smallest atomic mass difference on each of the two carbon atoms is denoted without a letter. A lowercase *a*, *b*, *c*, or *d* is appended to *wxyz* to differentiate the isomers as their masses diverge from this difference.

As individual compounds and blends thereof, the chlorofluorocarbons were once widely used for the following purposes:

- **Refrigerant gases** and coolants in residential and commercial refrigeration and air-conditioning equipment, including refrigerators, freezers, dehumidifiers, water coolers, ice machines, and air-conditioning units (including automotive air-conditioning units). The CFCs

refrigerant gas
■ For purposes of DOT regulations, an acceptable name for any chlorofluorocarbon or other halogenated hydrocarbon whose intended commercial use is as a rerigerant

CFC OR HCFC DESIGNATION	CHEMICAL NAME	MOLECULAR FORMULA
CFC-11	Trichlorofluoromethane	$\begin{array}{c} \text{Cl} \\ | \\ \text{Cl}-\text{C}-\text{Cl} \\ | \\ \text{F} \end{array}$
CFC-12	Dichlorodifluoromethane	$\begin{array}{c} \text{F} \\ | \\ \text{Cl}-\text{C}-\text{Cl} \\ | \\ \text{F} \end{array}$
CFC-13	Chlorotrifluoromethane	$\begin{array}{c} \text{F} \\ | \\ \text{Cl}-\text{C}-\text{F} \\ | \\ \text{F} \end{array}$
CFC-21	Dichlorofluoromethane	$\begin{array}{c} \text{H} \\ | \\ \text{Cl}-\text{C}-\text{Cl} \\ | \\ \text{F} \end{array}$
HCFC-22	Chlorodifluoromethane	$\begin{array}{c} \text{F} \\ | \\ \text{Cl}-\text{C}-\text{H} \\ | \\ \text{F} \end{array}$
CFC-113	1,1,2-Trichloro-1,2,2-trifluoroethane	$\begin{array}{c} \text{F} \quad \text{F} \\ | \quad | \\ \text{Cl}-\text{C}-\text{C}-\text{F} \\ | \quad | \\ \text{Cl} \quad \text{Cl} \end{array}$
CFC-114	1,2-Dichloro-1,1,2,2-tetrafluoroethane	$\begin{array}{c} \text{F} \quad \text{F} \\ | \quad | \\ \text{Cl}-\text{C}-\text{C}-\text{Cl} \\ | \quad | \\ \text{F} \quad \text{F} \end{array}$
HCFC-123	2,2-Dichloro-1,1,1-trifluoroethane	$\begin{array}{c} \text{Cl} \quad \text{F} \\ | \quad | \\ \text{Cl}-\text{C}-\text{C}-\text{F} \\ | \quad | \\ \text{H} \quad \text{F} \end{array}$
HCFC-123a	1,2-Dichloro-1,1,2-trifluoroethane	$\begin{array}{c} \text{F} \quad \text{F} \\ | \quad | \\ \text{Cl}-\text{C}-\text{C}-\text{Cl} \\ | \quad | \\ \text{F} \quad \text{H} \end{array}$

were first commercially introduced into the U.S. market in 1931 as safe alternatives to the toxic substances commonly used then for cooling, especially sulfur dioxide (Section 10.11) and ammonia (Section 10.14). In commerce, these refrigerant gases and coolants are known as **Freons**, or *Freon agents*.

- Cleaning fluids for electric, precision electronic, and photographic equipment and for maintaining aircraft
- **Foam-blowing agents** by extruded-polystyrene (Section 14.2-A) and rigid-polyurethane-foam manufacturers (Section 14.8). Because the CFCs have low boiling points, they readily vaporized when mixed with the hot plastics. The bubbles of their vapors caused the plastics to expand until they resembled frothy mixtures that solidified as foams. These foams were then marketed commercially as insulation.
- **Aerosols** for dispensing consumer products containerized in metal cans
- When mixed with ethylene oxide, sterilants of heat-sensitive reusable hospital apparatus
- A general inhalation anesthetic in hospitals and veterinary clinics, especially the compound known commercially as *halothane*, or *fluothane*

$$Br-\underset{\underset{H}{|}}{\overset{\overset{Cl}{|}}{C}}-\underset{\underset{F}{|}}{\overset{\overset{F}{|}}{C}}-F$$

2-Bromo-2-chloro-1,1,1-trifluoroethane
(Halothane)

Although the CFCs served well for this combination of purposes, they also have a less desirable feature: They destroy stratospheric ozone.

12.16-A HOW THE CFCs DESTROY STRATOSPHERIC OZONE

The CFCs are extremely stable compounds. When their vapors enter the atmosphere, they decompose into simpler substances only over very long periods. Their atmospheric lifetimes are hundreds of years—long enough to be transported to the stratospheric ozone layer (Section 7.1-M). Here, the CFC molecules absorb certain wavelengths of the ultraviolet radiation from the Sun, which causes their carbon–chlorine bonds to rupture, resulting in the production of chlorine atoms ($:\ddot{C}l\cdot$). When a chlorofluoromethane molecule absorbs ultraviolet radiation, for example, chlorine atoms are produced.

$$CF_nCl_{x-n}(g) \longrightarrow CF_nCl_{x-n-1}(g) + :\ddot{C}l\cdot(g)$$

A chlorofluorocarbon   A chlorofluorocarbon radical   Chlorine atom

When these chlorine atoms are exposed to ozone molecules in the ozone layer, they react by the following three-step process:

$$O_3(g) + :\ddot{C}l\cdot(g) \longrightarrow Cl\ddot{O}\cdot(g) + O_2(g)$$

Ozone   Chlorine atom   Chlorine monoxide radical   Oxygen

$$Cl\ddot{O}\cdot(g) + O_3(g) \longrightarrow ClO_2(g) + O_2(g)$$

Chlorine monoxide radical   Ozone   Chlorine dioxide   Oxygen

$$ClO_2(g) \longrightarrow :\ddot{C}l\cdot(g) + O_2(g)$$

Chlorine dioxide   Chlorine atom   Oxygen

The first two steps of this process destroy ozone. The actual ozone-destroying agents are the chlorine atom and chlorine monoxide radical.

The continuous repetition of this process has contributed to the formation of ozone holes through which ultraviolet radiation penetrates. The radiation then travels into the troposphere where humans live. To protect the integrity of the ozone layer, EPA used the authority of the Clean Air Act in 1966 to ban the use of chlorofluorocarbons as refrigerants, foam-blowing agents, and aerosol propellants. Furthermore, the CPSC used the authority of the Federal Hazardous Substances Act to ban the use of consumer products containing a CFC propellant. The impact of these two actions has led to the disappearance of virtually all CFCs from the commercial marketplace in the United States.

Freon
- The trademark of the chlorofluorocarbons, hydrochlorofluorocarbons, and hydrofluorocarbons used as foam-blowing agents and refrigerants

foam-blowing agent
- A low-boiling liquid mixed with polymers such as polystyrene and rigid polyurethane to produce a hard foam for use as insulation

aerosol
- A fine mist or spray consisting of minutely sized particles, generally expelled from a nonrefillable receptacle containing contents under pressure

In Section 5.13, it was noted that the ban on the manufacture and use of ozone-depleting substances is positively affecting recovery of the ozone layer. With regard to the ban on CFCs, this ban is environmentally beneficial in a second way, since their nonuse contributes to a reduction in global warming. The CFCs are potent greenhouse gases. In fact, the global warming potential of the chlorofluorocarbons is far greater than that of an equivalent amount of carbon dioxide (Section 5.8). Carbon dioxide is the greenhouse gas of primary concern only because it is discharged into the atmosphere in such huge amounts from burning fossil fuels. Because the presence of CFCs in our atmosphere seriously exacerbates global warming, the restriction in their worldwide manufacture and use is viewed as a positive move toward reducing global warming.

12.16-B HYDROFLUOROCARBONS AND HYDROCHLOROFLUOROCARBONS

hydrofluorocarbon
■ A compound whose molecules are composed solely of carbon, hydrogen, and fluorine atoms (abbreviated HFC)

hydrochlorofluorocarbons
■ A compound whose molecules are composed solely of carbon, hydrogen, chlorine, and fluorine atoms (abbreviated HCFC)

When the use of chlorofluorocarbons was banned, chemical manufacturers sought to find environmentally friendly replacements for use as refrigerants, coolants, and foam-blowing agents. Two groups of such substances were identified: **hydrofluorocarbons** (HFCs) and **hydrochlorofluorocarbons** (HCFCs). HFC molecules have only $C-H$ and $C-F$ bonds, but HCFC molecules have $C-H$, $C-F$, and $C-Cl$ bonds. Examples of substances in these two groups are difluoromethane and chlorodifluoromethane, respectively.

$$
\begin{array}{cc}
\overset{\displaystyle F}{\underset{\displaystyle F}{H-C-H}} & \overset{\displaystyle F}{\underset{\displaystyle Cl}{F-C-H}} \\[2mm]
\text{Difluoromethane} & \text{Chlorodifluoromethane} \\
\text{(HFC-32)} & \text{(HCFC-22)}
\end{array}
$$

Initial scientific research suggested that most HFCs and HCFCs were destroyed in the atmosphere at low altitudes and did not significantly diffuse upward to the ozone layer. When this initial information could not be corroborated, however, EPA proceeded also to phase out the manufacture and importation of HFCs and HCFCs for use as refrigerant gases, coolants, and foam-blowing agents. A ban on the manufacture and importation of HCFC-141b, or 1,1-dichloro-1-fluoroethane, is already in place. In 2007, to mark the 10th anniversary of the Montréal protocol, 190 countries agreed to phase out the use of all HCFCs by January 2020.

Several replacements for the HCFCs that do not deplete stratospheric ozone have been approved by EPA. 1,1,1,3,3-Pentafluoropropane, or HFC-245a, has been approved for use as a refrigerant, coolant, and foam-blowing agent.

$$
\overset{\displaystyle F\quad H\quad F}{\underset{\displaystyle F\quad H\quad F}{F-C-C-C-H}}
$$

1,1,1,3,3-Pentafluoropropane
(HFC-245a)

In 2007, EPA approved the use of RS-45, RS-428a, and KDD5 as replacements for HCHC-22 in new and existing refrigeration and air-conditioning equipment. RS-45 is a blend of 1,1,1-trifluoroethane, pentafluoroethane, 1,1,1,2-tetrafluoroethane, and isobutane. R-428a is a blend of pentafluoroethane, 1,1,1,2-tetrafluoroethane, propane, and isobutane. KDD5 is a proprietary blend.

bromofluorocarbon
■ Any halogenated hydrocarbon whose molecules are composed solely of carbon, bromine, and fluorine atoms (abbreviated BFC)

12.16-C ALTERNATIVES TO HALON FIRE EXTINGUISHERS

As first noted in Section 5.13, certain substances including Halon 1301 and Halon 1211 were once popular fire extinguishers. They included bromotrifluoromethane and bromochlorodifluoromethane, respectively. These compounds are examples of **bromofluorocarbons** (BFCs) and

bromochlorofluorocarbons (BCFCs), respectively. Their production and use in the United States ceased in January 1994.

Chemical manufacturers have identified substances that serve as alternatives to these halon products. At least two alternative fire extinguishers are now marketed: Pyro-Chem FM (Fire Master)-200 and 3M Novec 1230.

1,1,1,2,3,3,3-Heptafluoropropane
(Pyro-Chem FM-200)

1,1,1,2,2,4,5,5,5-Nonafluoro-4-(trifluoromethyl)-3-pentanone
(3M Novec 1230)

Both are nonflammable, nontoxic substances, and neither depletes stratospheric ozone nor contributes to global warming.

12.16-D STORING OZONE-DEPLETING SUBSTANCES

Ozone-depleting substances, including the CFCs, HFCs, and HCFCs, are still likely to be encountered today within storage containers and tanks, especially at facilities that maintain, repair, or dispose of refrigeration and air-conditioning equipment. When an ozone-depleting substance is stored, EPA regulations require its owners to mark their containers and tanks with a warning statement that reads as follows: "WARNING: Contains [or manufactured with, if applicable] [insert name of substance], a substance which harms public health and environment by destroying ozone in the upper atmosphere." For example, at 40 C.F.R. §82.106, EPA requires owners of HCFC-22 to mark their containers and tanks of chlorodifluoromethane as follows:

> **WARNING: CONTAINS CHLORODIFLUOROMETHANE (HCFC-22),
> A SUBSTANCE WHICH HARMS PUBLIC HEALTH AND THE
> ENVIRONMENT BY DESTROYING OZONE IN THE
> UPPER ATMOSPHERE**

This statement serves to assist first-on-the-scene responders who encounter ozone-depleting substances in storage. To protect the environment, the responders should take precautionary action to prevent the rupture of these containers and tanks, for example, during an ongoing fire by cooling them with a prodigious volume of water. When rupture has occurred, emergency responders should implement appropriate measures to confine the contaminated water for subsequent collection.

bromochlorofluoro-
carbons
■ Any halogenated hydrocarbon whose molecules are composed solely of carbon, bromine, chlorine, and fluorine atoms (abbreviated BCFC)

SOLVED EXERCISE 12.14

While combating a fire at an auto service shop, firefighters encounter a large storage tank labeled as follows:

> **WARNING: CONTAINS DICHLORODIFLUOROMETHANE
> (CFC-12), A SUBSTANCE WHICH HARMS PUBLIC
> HEALTH AND THE ENVIRONMENT BY DESTROYING
> OZONE IN THE UPPER ATMOSPHERE**

What special action should the firefighters take to protect public health, safety, and the environment?

Solution: Dichlorodifluoromethane is an example of a *chlorofluorocarbon*, a substance that when released into the environment is known to contribute to the destruction of the stratospheric ozone layer. The EPA regulation at 40 C.F.R. §82.106 requires the owner of this product to post the referenced label on the storage tank in such a fashion as to be clearly legible and conspicuous to observers. To prevent the potential impact on the ozone layer, firefighters should take precautionary action to prevent the rupture of the tank and the subsequent unintended release of the product into the environment. They can do this by cooling the tank with a prodigious volume of water until the fire in the auto service shop has been successfully extinguished.

12.16-E TRANSPORTING CFCs, HFCs, AND HCFCs

When shippers offer a CFC, HFC, or HCFC for transportation, DOT requires them to identify the appropriate commodity on an accompanying shipping paper either by using its IUPAC name or "Refrigerant gas R-*wxyz*," where the nature of the suffix *wxyz* was noted earlier. Some examples of these basic descriptions are provided in Table 12.16.

TABLE 12.16	Basic Descriptions of Some CFCs, HFCs, and HCFCs
CFC, HFC, OR HCFC	**BASIC DESCRIPTION**
Bromotrifluoromethane	UN1009, Bromotrifluoromethane, 2.2 *or* UN1009, Refrigerant gas R-13B1
1-Chloro-1,1-difluoroethane	UN2517, 1-Chloro-1,1-difluoroethane, 2.1 *or* UN2517, Refrigerant gas R-500, 2.1
1-Chloro-1,2,2,2-tetrafluoroethane	UN1021, 1-Chloro-1,2,2,2-tetrafluoroethane, 2.2 *or* UN1021, Refrigerant gas R-124, 2.2
Chlorotrifluoroethane	UN1022, Chlorotrifluoromethane, 2.2 *or* UN1022, Refrigerant gas R-13, 2.2
1,1,1,2-Tetrafluoroethane	UN3159, 1,1,1,2-Tetrafluoroethane, 2.2 *or* UN3159, Refrigerant gas R-134a, 2.2
1,1,1-Trifluoroethane	UN2035, 1,1,1-Trifluoroethane, 2.1 *or* UN2035, Refrigerant gas R-143a, 2.1

DOT also requires shippers to affix the relevant FLAMMABLE or NON-FLAMMABLE GAS label on the packaging.

When carriers transport a CFC, HFC, or HCFC in an amount exceeding 1001 lb (454 kg), DOT requires them to post the relevant FLAMMABLE or NON-FLAMMABLE GAS placards on the bulk packaging or transport vehicle used for shipment.

PERFORMANCE GOALS FOR SECTION 12.17:

- Describe the molecular nature of the polychlorinated biphenyls (PCBs).
- Identify the primary risks associated with exposure to the PCBs, especially noting that they are major environmental pollutants.
- Identify how the PCBs were formerly used.
- Identify the ways in which PCBs are still likely to be encountered by emergency responders.
- Identify the locations in a typical community where emergency responders are likely to encounter PCB transformers.
- Describe the nature of the markings that EPA requires to be posted on PCB-containing electrical equipment.

12.17 Polychlorinated Biphenyls

Another group of chlorinated hydrocarbons are the **polychlorinated biphenyls** (PCBs). They are the chlorinated derivatives of the complex hydrocarbon *biphenyl*, which has the following molecular structure:

There are 209 structural isomers of the PCBs, each of which has the chemical formula $C_{12}H_xCl_y$, where x ranges from 0 to 9, and y equals $10 - x$. Approximately 130 isomers were components of the PCB products formerly manufactured and marketed throughout the world.

At ordinary ambient conditions, the products containing PCBs are usually viscous, sticky liquids. Even at elevated temperatures, their individual components are nonflammable, incredibly stable compounds that can be heated to their respective boiling points without undergoing decomposition or bursting into flame. This combination of properties was formerly considered ideal for implementing certain industrial practices that required high temperatures and fire resistance.

12.17-A PCBs IN ELECTRICAL EQUIPMENT

From approximately 1930 to 1977, products containing PCBs were marketed for commercial utilization within certain types of electrical equipment such as transformers, voltage regulators, circuit breakers, capacitors, motors, electromagnets, reclosers, cables, and fluorescent light ballasts. They were also manufactured as heat-transfer, hydraulic, lubricating, and cutting fluids. PCBs were also used to manufacture carbonless copy paper, and they were incorporated as constituents in certain plastic, paint, wax, and rubber products. In all these instances, the primary purpose of the PCBs was to provide these commercial products with a greater degree of fire resistance.

In the 1960s, however, the PCBs became the focus of several scientific studies conducted on animals. From these studies, scientists concluded that exposure to PCBs could cause cancer, birth defects, liver damage, chloracne, impotence, and death. **Chloracne** is a disfiguring skin

polychlorinated biphenyl
■ Any chlorinated derivative of biphenyl, the manufacture, sale, and distribution of which have been banned in the United States; the major constituent of certain dielectric fluids formerly contained within capacitors, transformers, and other electrical equipment (abbreviated PCB)

chloracne
■ An acnelike skin affliction caused by overexposure to certain halogenated hydrocarbons, including the polychlorinated biphenyls

hormone
■ A chemical substance produced by an endocrine gland and carried through the bloodstream to target organs or a group of cells of the body, where it serves to regulate certain body functions

endocrine glands
■ Ductless glands, such as the adrenal, thyroid, and pituitary, that secrete fluids directly into the general circulatory system

endocrine disrupters
■ Compounds that mimic hormones or interfere with hormonal action when ingested into the body

dielectric fluid
■ The fluid used in certain types of electrical equipment to increase their resistance to arcing and serve as a heat-transfer medium

PCB transformer
■ For purposes of TSCA regulations, a transformer whose dielectric fluid contains polychlorinated biphenyls at a concentration equal to or exceeding 500 ppm

mineral oil transformer
■ An electrical transformer containing mineral oil as its dielectric fluid

PCB-contaminated transformer
■ For purposes of TSCA regulations, a transformer whose dielectric fluid contains polychlorinated biphenyls at a concentration greater than 50 ppm but less than 500 ppm

condition that resembles severe acne. The scientists also discovered that once ingested, PCBs migrate into various receptor tissues of the thymus, lungs, spleen, kidneys, liver, brain, muscle, and testes. The PCBs primarily bioaccumulate in the body's fat, but their periodic release has been linked with a variety of neurobehavioral disorders and birth abnormalities. In particular, children exposed to high levels of PCBs during their mother's pregnancy often have reduced motor skills and short-term memories. From studies conducted on animals, scientists have also noted that PCB exposure over long periods induces the formation of malignant tumors in the liver, mouth, adrenal glands, and lungs.

Epidemiological studies have shown that the PCBs cause physiological disorders by mimicking the action of the body's own **hormones** or by interfering with the hormones's normal functions. Hormones are substances secreted by the **endocrine glands**, the organs that control such activities as growth, development, metabolism, and reproduction. When absorbed into the body, the PCBs react like these hormones; hence, they are referred to as **endocrine disrupters**. Given this combination of adverse properties, PCBs are ranked by epidemiologists as probable human carcinogens.

Because PCBs pose an unreasonable risk to public health and the environment, EPA banned their manufacture, sale, import, and distribution in 1978 by using the legal authority of the Toxic Substances Control Act (TSCA) (Section 1.4-D). The relevant regulations are published at 40 C.F.R. §761 et seq. Although their manufacture is banned in the United States, EPA permits the indefinite use of PCBs in systems that were operating prior to 1978, albeit under specific regulatory conditions. Even today, PCBs can be legally used in a "totally enclosed manner," that is, within equipment that has been designed and constructed so as to ensure that no human or environmental exposure to PCBs occurs. When this equipment is withdrawn from service or repaired, EPA requires replacement of the PCB fluid with a non-PCB-containing alternative product.

An electrical transformer is an example of a totally enclosed system. One type can often be observed near the tops of electric poles. It is a static device whose function is to transfer electrical energy between circuits of different voltage. Each transformer has a magnetic core around which there are several windings or coils of a metal such as copper. This assembly of core and current-carrying coils is then immersed in a fluid within a tank. The purpose of the fluid, called a **dielectric fluid**, is to increase the resistance of the unit to sustained arcing and serve as a heat-transfer medium. Large numbers of transformers are located at electrical substations operated by municipal utility companies.

Within the context of the TSCA regulations, EPA defines a **PCB transformer** at 40 C.F.R. §761.3 as a transformer that contains PCBs at a concentration of 500 ppm or greater. A typical PCB transformer contains approximately 200 gal (1500 L) of fluid with a PCB concentration of approximately 50% to 60% by volume, but a large PCB transformer may hold as much as 1000 gal (7500 L) of PCB fluid. Forty percent to 50% of the fluid volume consists of diluents, typically a mixture of trichlorobenzene and tetrachlorobenzene isomers.

The vast majority of transformers in service today were manufactured with the intent of using mineral oil as their dielectric fluid. Mineral oil is a petroleum oil. The units are called **mineral oil transformers**. Although PCBs were never intentionally added to the mineral oil used in these transformers, the oil often became contaminated with PCBs when workers conducted routine equipment maintenance on both transformer types using the same hoses and pumping equipment. EPA defines a unit as a **PCB-contaminated transformer** when sampling and analysis indicates that its fluid contains PCBs at a concentration greater than 50 ppm but less than 500 ppm. A **non-PCB transformer** is a unit containing PCBs at a concentration equal to or less than 50 ppm.

12.17-B PCBs AND FIREFIGHTING

Although PCB fluids do not readily burn, they may ignite within an energized transformer when high-energy arcing, electrical overloading, or short-circuiting occurs. These fires are characterized by large amounts of oily, black soot in the accompanying smoke plume. When the energized

FIGURE 12.12 There are two EPA marking formats, called the M_L and M_s markings. They differ only by their wording and size. The letters and striping on both markings are black on a yellow background and include the warning CAUTION CONTAINS PCBs and instructions to be implemented in the event of a PCB release. The M_L marking also includes the telephone number of the National Response Center. The TSCA regulation at 40 C.F.R. §761.45 requires a marking to be affixed to PCB transformers and PCB capacitors in service and PCB items containing or contaminated with PCBs in storage for disposal when the PCB concentration is equal to or greater than 50 ppm. The marking must also be affixed to all doors or accesses leading from the outermost door to the door immediately accessing the PCB unit.

transformer is located within a building, the soot and PCB vapors travel with the smoke plume through ventilation shafts, building ductwork, and other open-construction areas.

Today's firefighters respond rarely to PCB transformer fires, but since the service lifetime of a transformer is 40 to 85 years, such an action is not outside the realm of possibility. Firefighters are faced with an unwarranted health risk when they are exposed to the smoke from these NFPA class C fires. To reduce the magnitude of the risk, EPA requires the owners of the PCB transformers to prominently affix the M_L marking shown in Figure 12.12 so it is easily read by firefighting personnel who respond to an electrical-equipment fire. EPA has also imposed mandatory reporting requirements at 40 C.F.R. §761.30 to ensure that the National Response Center (Section 1.12) is contacted immediately on the occurrence of an accidental fire involving PCBs. EPA requires the M_s mark, also shown in Figure 12.12, to be affixed to PCB equipment that cannot accommodate the larger M_L mark.

non-PCB transformer
■ For purposes of TSCA regulations, a transformer whose dielectric fluid contains polychlorinated biphenyls at a concentration equal to or less than 50 ppm

SOLVED EXERCISE 12.15

Once fire personnel have been informed of an ongoing fire within a PCB transformer, why does EPA strongly urge notification of the operators of the relevant storm and sanitary sewer system and treatment plants of the incident?

Solution: Combating a PCB transformer fire can generate a relatively large volume of contaminated water, the discharge of which can subsequently cause widespread pollution of sewer systems and associated treatment plants. The subsequent cleanup of this contamination is very costly. For these reasons, the operators of sewer systems and treatment plants should quickly be notified of an ongoing PCB transformer fire and of the likelihood that the water used to combat the fire is contaminated with PCBs and their incomplete combustion products. The operators of these plants can then implement appropriate measures to confine the contaminated water and minimize the adverse environmental impact.

polychlorinated dibenzofurans
■ Any chlorinated derivative of furan, generated as an unwanted by-product during certain uncontrolled incineration, paper pulp bleaching, and chemical manufacturing operations (abbreviated PCDFs)

polychlorinated dibenzo-*p*-dioxin
■ Any chlorinated derivative of dibenzo-*p*-dioxin, generated as an unwanted by-product during certain uncontrolled incineration, paper pulp bleaching, and chemical manufacturing operations (abbreviated PCDDs)

EPA has also imposed reporting requirements on the owners of facilities that still use electrical equipment containing PCBs. For example, EPA requires the registration of PCB transformers, including out-of-service transformers held in storage for reuse. Fire department personnel should identify the location of these transformers within their area of jurisdiction and plan a response action in case of an accidental PCB release.

When PCBs undergo incomplete combustion during transformer and other electrical-equipment fires, they produce the highly toxic compounds called **polychlorinated dibenzofurans** and **polychlorinated dibenzo-*p*-dioxins** (Section 13.4-A). During electrical-equipment fires involving PCBs, these substances may also be released into the environment. Firefighters should be especially wary when combating electrical-equipment fires, since exposure to these toxic substances could cause long-term ill effects. Regardless of the magnitude of the fire, firefighters should always utilize self-contained breathing apparatus to eliminate or reduce the potential for exposure to these substances. Public health can also be effectively protected by constructing suitable diking, diversion curbing, or grading to confine the spread of these contaminants and prevent their drainage into storm sewers.

A fire involving electrical equipment is not the only type during which PCBs, polychlorinated dibenzofurans, and polychlorinated dibenzo-*p*-dioxins may be released. Prior to 1977, PCBs were components of some protective and decorative paints, varnishes, lacquers, sealants, adhesives, caulking, and similar construction products. Some of them contained 5% to 30% PCBs. Although these products were used several decades ago, the PCBs still remain in surface coatings, joint sealants around windows, caulking between masonry panels, and elsewhere. As fires occur in the buildings in which these PCB-containing products were once applied or used, the PCBs and their incomplete combustion products are released into the environment.

12.17-C THE ADVERSE IMPACT OF PCBs ON THE ENVIRONMENT

Why are the PCBs regarded as ultrahazardous substances? The answer to this question is linked with the horrific impact they could have not only on public health but also on a wide segment of the environment. When PCBs are consumed by an animal, they adversely affect the health of the animal throughout its lifetime. When the animal is eaten by another animal, the PCBs pass progressively through the food chain from animal to animal: worms to fish and birds, fish to birds, and birds and fish to mammals. The PCBs can thereby affect the well-being of many individual species including humans who consume fish or meat contaminated with PCBs. In the United States alone, EPA identified 703 bodies of water where health advisories have been posted to warn people against consuming PCB-ridden fish.

Extensive studies have been conducted on fish and birds to identify how PCBs affect the health of these organisms. They have revealed that the immune systems of fish who have consumed PCBs become dysfunctional. The number of mature ovaries are reduced, and even when the ovaries are mature, the fish experience a reduced ability to spawn. Both fish and birds who have consumed PCBs experience growth and reproductive impairment. In birds alone, these problems include eggshell thinning, lower egg production, decreased egg hatching, deformities in the juvenile birds that do emerge from eggs, and reduction in their growth and survival rates.

The environmental impact from exposure to PCBs is especially evident in aquatic organisms. Within the ocean hierarchy, for example, killer whales are contaminated with PCBs at concentrations higher than those observed in other marine animals, since they rank unrivaled at the top of the food chain. Although herring may carry PCBs at an average concentration of only 1 ppm, the seals that eat them may contain 20 ppm, and the killer whales that eat these contaminated seals may have PCB levels as high as 250 ppm. The male killer whales carry the PCBs throughout their lives, but the females periodically rid these pollutants from their bodies by passing them to their calves. This could explain why the males live only half as long as the females, and why many calves die within several months of their birth.

SOLVED EXERCISE 12.16

PCB pollution generated in Europe, North America, and even Asia is carried to Norway's remote Svalbard Islands by the normal movement of air and ocean currents. Researchers have shown that the tissues of virtually all animals who live in this environment are contaminated with PCBs and other pollutants. The concentration of PCBs is especially elevated in tissue samples collected from polar bears, *Ursus maritimus*, that live on the Svalbard Islands, compared with the norm.

(a) How did the polar bears acquire the PCBs at concentrations that are elevated compared with the levels observed in other animals?

(b) What adverse health effects are likely to be experienced by them from the PCB exposure?

Solution:

(a) Polar bears rank near the top of the food chain. They eat fish and mammals whose fat stores PCBs acquired by eating Arctic fish. Once the polar bears have consumed these animals, the PCBs dissolve in their fatty tissues, where they accumulate. Because few polar bears are eaten by other animals, their tissues tend to become contaminated with PCBs at elevated concentrations compared with the levels identified in the tissues of other animals who live in the same environment.

(b) Researchers have shown that PCBs are endocrine disrupters. This means that polar bears who have consumed animals that carry PCBs are likely to be negatively affected in functions that are routinely influenced by the normal secretion of hormones. These adverse health effects include growth and reproductive impairment, which are controlled by thyroid and sex hormones.

12.17-D FDA REGULATIONS INVOLVING PCBs

Because PCBs have become an unavoidable contaminant in the environment, FDA publishes the following tolerance limits for PCB residues in foods, at 21 C.F.R. §109.30:

- 1.5 ppm in milk (fat basis)
- 1.5 ppm in manufactured dairy products (fat basis)
- 3 ppm in poultry (fat basis)
- 0.3 ppm in eggs
- 0.2 ppm in finished animal feed for food-producing animals (other than feed concentrates, feed supplements, and feed premixes)

- 2 ppm in animal feed components of animal origin, including fishmeal and other by-products of marine origin and in animal feed concentrates, feed supplements, and feed premixes intended for food-producing animals
- 2 ppm in the edible portions of fish and shellfish
- 0.2 ppm in infant and junior foods

12.17-E TRANSPORTING PCBs

When shippers offer polychlorinated biphenyls for transportation, DOT requires them to identify the PCBs on an accompanying shipping paper as one of the following:

UN2315, Polychlorinated biphenyls, liquid, 9, PGII
UN2315, Polychlorinated biphenyls, solid, 9, PGII

DOT also requires them to affix CLASS 9 labels on its packaging.

When bulk quantities of PCBs are transported, DOT requires the identification number 2315 to be on displayed on orange panels or on white square-on-point diamonds displayed on the packaging. DOT does not require CLASS 9 placards to be displayed.

When shippers offer a PCB-containing electrical unit including a PCB-contaminated transformer for transportation, EPA requires them to affix the appropriate marking in Figure 12.12 on the unit when it is transported. EPA requires this marking to be affixed on an electrical unit even when the fluid has been drained from it, as the unit typically contains residual PCB fluid.

PERFORMANCE GOALS FOR SECTION 12.18:

- Describe the molecular nature of the organochlorine pesticides.
- Identify the primary risks associated with exposure to the organochlorine pesticides, especially noting that they are major environmental pollutants.

12.18 Organochlorine Pesticides

organochlorine pesticides
■ Any chlorinated hydrocarbon used exclusively for controlling, destroying, preventing, repelling, or mitigating any pest

Certain chlorinated hydrocarbons used as pesticides are called **organochlorine pesticides** (Section 10.19). The most recognizable of them is the insecticide known as DDT.

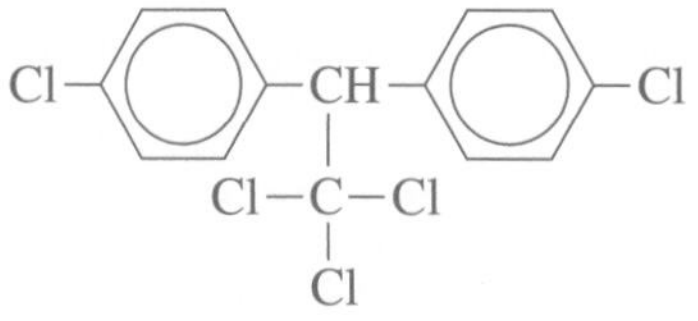

Dichlorodiphenyltrichloroethane
(DDT)

DDT was first developed for use as an insecticide in 1939. It was used during World War II to eradicate legions of insects that carried typhoid fever, malaria, and typhus. Its use has been credited for saving the lives of many American and Allied soldiers who fought in the jungles of the Far East. The insecticide was subsequently used throughout Asia, Africa, Latin America, southern Europe, and the southern United States, and is still used today in some Third World countries in which insect-borne diseases continue to threaten the quality of human life. In sub-Saharan Africa, for example, where millions of Africans have perished from the ravages of malaria, the United Nations recommends the use of DDT for controlling mosquitoes that transmit malaria by spraying the insecticide on interior house walls. Countries that use DDT for malaria control follow guidelines established in 1995 through an international pact known as the **Stockholm Convention on Persistent Organic Pollutants,** or *POPs treaty.*

Notwithstanding its exceptional insecticidal value, DDT accumulates in the environment, where it has been linked with eggshell thinning in bald eagles, ospreys, and other birds and the decline of many other animal species. It was this observation that caused EPA in 1972 to ban the use of DDT in the United States by using the legal authority of FIFRA (Section 1.4-C).

In 2006, epidemiologists reported the results of research studies indicating that DDT poses a chronic health hazard to humans. Infants exposed to DDT in the womb as well as babies and toddlers of female farmworkers formerly exposed to DDT exhibit neurological problems sufficiently severe to slow their mental and physical development. These farmworkers were exposed to DDT in Mexico, where the use of DDT was permitted on farms until 1995 and for mosquito control until 2000. Researchers determined that for each 10-fold increase in DDT levels in the mothers, the mental development scores of their babies decreased by 2% to 3%. This result is being considered by those who are attempting to eradicate malaria-bearing mosquitoes in African countries.

There are approximately a dozen organochlorine pesticides other than DDT that were formerly available in the commercial market. They include aldrin, chlordane, dieldrin, endrin, heptachlor, mirex, and toxaphene. Over time, their use has also been discontinued by countries that have ratified the Stockholm Convention.

Stockholm Convention on Persistent Organic Pollutants
■ The United Nations treaty that bans or limits the use of DDT and other organochlorine pesticides

REVIEW **EXERCISES**

Aliphatic Hydrocarbons

12.1 Determine whether each of the following chemical formulas represents an alkane, cycloalkane, alkene, diene, triene, cycloalkene, or alkyne:

(a) C_4H_8
(b) $CH_3-(CH_2)_4-CH_3$
(c) $CH_3-CH=CH-CH_3$
(d) $CH_3-(CH_2)_4-C\equiv C-CH_2CH_3$
(e) $CH_3-CH=CH-CH=CH-CH=CH_2$
(f)

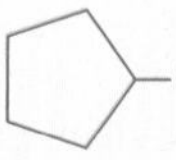

12.2 Write the condensed chemical formula of each aliphatic hydrocarbon represented by the following carbon skeletons:

(a) C−C−C−C−C−C−C−C

(b)
```
C−C−C−C−C−C−C
    |       |
    C       C
            |
            C
```

(c) C=C−C−C−C−C−C

(d)
```
C=C−C−C−C−C−C
            |
            C
            |
            C
```

12.3 Using the IUPAC system, name the alkanes having the following condensed formulas:

(a)
```
CH3CH2−CH−CH2CH3
        |
        CH3
```

(b)
```
CH3−CH−CH2−CH−CH3
     |        |
     CH2       CH3
     |
     CH3
```

(c)
```
            CH2CH2CH3
            |
CH3CH2−CH−CH2CH3
```

(d)
```
                        CH3
                        |
CH3CH2CH2CH2−C−CH2CH2CH3
                 |
            CH3−C−CH3
                 |
                CH3
```

12.4 Using the IUPAC system, name the alkenes having the following condensed formulas:

(a)
```
CH3   H
  \   /
   C=C
  /   \
 H     H
```

(b)
```
CH3CH2    H
     \   /
      C=C
     /   \
 CH3CH2    H
```

(c)
```
CH3CH2CH2    CH2CH2CH2CH3
        \   /
         C=C
        /   \
     CH3     H
```

(d)
```
CH3    CH2CH3
  \   /
   C=C
  /   \
 H     H
```

12.5   Using the IUPAC system, name the alkynes having the following condensed formulas:

(a) $H-C\equiv C-CH_3$ (b) $CH_3CH_2-C\equiv C-CH_2CH_3$

12.6   Write the condensed formulas for each of the following aliphatic hydrocarbons:

(a) acetylene   (b) isobutane
(c) *n*-hexane   (d) 2-pentene
(e) 3-methylpentane   (f) 2-ethyl-1-butene
(g) *trans*-2-butene   (h) cyclohexane
(i) cyclobutane   (j) methylcyclopentane
(k) 3-methyloctane   (l) isopentane
(m) 3-ethyl-3-methylheptane   (n) 1-ethylcyclopentene
(o) 2,3-dimethyl-2-butene   (p) 3-hexene
(q) 3-ethyl-2-methylhexane   (r) diisopropylacetylene
(s) 4-methyl-1,3-pentadiene   (t) 3,3-dimethyl-1-butene

Natural Gas (Methane)

12.7   The Mine Safety and Health Administration (MSHA) of the U.S. Department of Labor mandates the following at 30 C.F.R. §75.323(b)(1)(i): "When 1.0% or more methane is present in a working place or an intake air course . . . electrically powered equipment in the affected area must be deenergized, and other mechanized equipment shall be shut off." What is the most likely reason MSHA adopted this regulation?

12.8   Using the relevant data in Table 2.11, calculate the approximate volume in cubic feet of gaseous methane that is generated when each ton of LNG is released from its transport vessel.

Liquefied Petroleum Gas

12.9   The DOT regulation at 49 C.F.R. §173.315(b)(1) stipulates that liquefied petroleum gas must be odorized by the addition of a small amount of a substance such as ethyl mercaptan, thiophane, or amyl mercaptan. What is the most likely reason DOT requires liquefied petroleum gas to be odorized in this fashion?

Ethylene and Propylene

12.10 Why is the U.S. production of ethylene and propylene far greater than the production of all other chemical substances?

Butadiene

12.11 When shippers offer butadiene for transportation, DOT requires them to enter the word "stabilized" with the basic description. What is the most likely reason DOT requires the inclusion of this word in the basic description?

Acetylene

12.12 Methyl acetylene and propadiene are minor constituents produced when acetylene is manufactured from natural gas.

    (a) Write the chemical formulas of these two substances.

    (b) Write the equations that illustrate their complete combustion.

Petroleum Products

12.13 Using standard methods of testing, *n*-butane is found to have an octane number of 99. What does this number mean?

12.14 When it burns, which produces the greater concentration of sulfur dioxide: light diesel fuel or heavy diesel fuel?

Aromatic Hydrocarbons

12.15 Name the benzene derivatives having the following chemical formulas:

    (a) CH_2CH_3 (benzene ring)

    (b) CH_3, CH_3 (benzene ring)

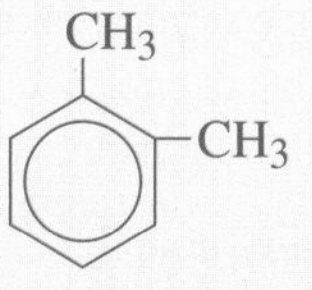

    (c) $Cl-$(benzene ring)$-Cl$

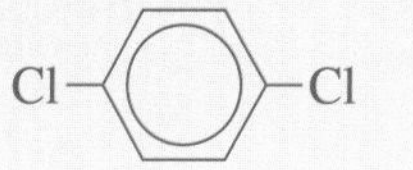

    (d) CH_3, CH_2CH_3 (benzene ring)

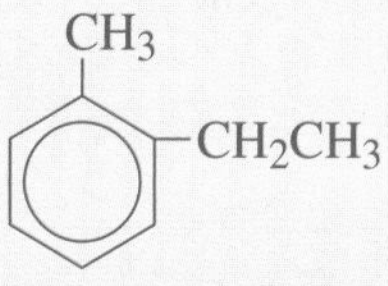

    (e) CH_2Cl (benzene ring)

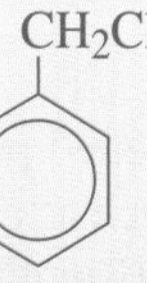

    (f) $CH_3-CH-CH_3$ (benzene ring)

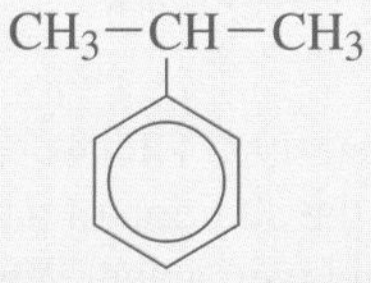

12.16 Write the molecular formula for each of the following aromatic compounds:

    (a) *p*-xylene

    (b) *p*-diethylbenzene

    (c) 1,3,5-trimethylbenzene

    (d) *sec*-butylbenzene

    (e) diphenylacetylene

    (f) 1,2,4,5-tetramethylbenzene

Simple Halogenated Hydrocarbons

12.17 Using the IUPAC system, name the halogenated hydrocarbons having the following condensed formulas:

    (a) $Br-CH_2CH_2-Br$

    (b) $CH_3-CH-Br$ with Br

(c) $CHCl_3$

(d) 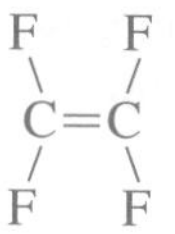
$$F \quad F$$
$$\backslash \quad /$$
$$C = C$$
$$/ \quad \backslash$$
$$F \quad F$$

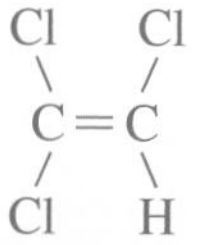

(e)
$$Cl \quad Cl$$
$$\backslash \quad /$$
$$C = C$$
$$/ \quad \backslash$$
$$Cl \quad H$$

(f)
$$CH_2CH_2Cl$$
$$|$$
$$CH_3 - CH - CH_3$$

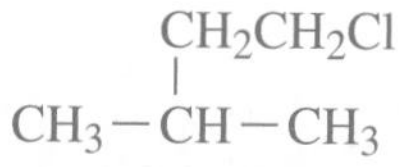

12.18 Write the condensed formulas for each of the following halogenated hydrocarbons:

(a) carbon tetrachloride
(b) *cis*-2,3-dichloro-2-butene
(c) 2-bromopropane
(d) 2,2-dichlorobutane
(e) dichlorodifluoromethane
(f) trichlorofluoromethane

Chlorofluorocarbons, Hydrofluorocarbons, and Hydrochlorofluorocarbons

12.19 Chlorofluorocarbons were formerly used as refrigerants in soft-drink vending machines and ice-cream display cases. Why has their use been curtailed and, in most cases, eliminated?

12.20 Show that CFC-114 is the commercial name for the chlorofluorocarbon having the chemical name dichlorotetrafluoroethane.

12.21 What are the molecular structures and chemical names of the two structural isomers of dichlorotrifluoroethane?

Polychlorinated Biphenyls

12.22 The EPA regulation published at 40 C.F.R. §761.40(j)(ii) stipulates that owners of PCB electrical transformers must comply with the following: Mark the vault door, machinery room door, fence, hallway, or other means of access to these transformers with a marking shown in Figure 12.12; coordinate with the primary fire department; and document that the fire department knows, accepts, and recognizes the meaning of this marking. What is the most likely reason EPA requires these steps to be taken?

12.23 What special action should firefighters take when responding to a smoldering fire at a metal salvage yard where employees dismantled electrical transformers to retrieve their copper content?

Organochlorine Pesticides

12.24 A shipping company intends to transport by rail boxcar the organochlorine pesticide methoxychlor within PGI packaging. The total amount of methoxychlor in the shipment is 30 boxes, or 1500 lb (682 kg).

(a) What shipping description does DOT require a shipping company representative to enter on the accompanying waybill?

(b) Does DOT obligate the company to affix POISON placards to the boxcar?

(c) Does DOT obligate the company to mark the boxcar in any way?

Responding to Incidents Involving a Release of Organic Compounds

12.25 Methyl bromide is an odorless, colorless, and poisonous gas. It was formerly used with chloropicrin (Section 13.11-D) as an agricultural fumigant by forest tree nurseries and tomato and strawberry growers to exterminate plant parasites and wipe out disease in soils. However, methyl bromide is an ozone-depleting substance. Its production and use have been largely phased out in the United States in compliance with the Montréal protocol (Section 7.1-N), but methyl bromide may still be used in agriculture if a critical-use exemption is received from EPA when no feasible alternative exists.

When first-on-the-scene responders arrive at the scene of a railway accident at 11:30 P.M. in San Diego, California, they note a ruptured, overturned tankcar and the following shipping description on the accompanying railroad waybill:

Units	HM	Shipping Description (Identification Number, Proper Shipping Name, Primary Hazard Class or Division, Subsidiary Hazard Class or Division, and Packing Group)	Weight (lb)
1 tankcar	X	UN1062, Methyl bromide, 2.3 (Inhalation Hazard, Zone C) (Placarded POISON GAS)	2500

Use Table 10.15 to identify the isolation and protective-action distances recommended by DOT to protect public health, safety, and the environment.

12.26 When first-on-the-scene responders arrive at the scene of a highway accident, they note the HOT marking displayed on the exterior surface of a kettle that has been ruptured by impact with a motor van. The hot contents have spilled on the underlying highway. The responders note the following shipping description of the kettle contents:

Units	HM	Shipping Description (Identification Number, Proper Shipping Name, Primary Hazard Class or Division, Subsidiary Hazard Class or Division, and Packing Group)	Weight (lb)
1750-gal kettle	X	UN3257, Elevated-temperature liquid, n.o.s., 9, PGIII (asphalt)	1200

What immediate actions should the responders take to protect public health, safety, and the environment?

KEY TERMS

Certain organic compounds are composed of molecules in which one or more of their carbon atoms are bonded directly to an oxygen atom. These compounds are called alcohols, ethers, aldehydes, ketones, organic acids, esters, and peroxo-organic compounds. Other organic compounds, called amines, are composed of molecules in which a carbon atom is bonded to a nitrogen atom. We study the properties of these compounds in this chapter.

CLASS OF ORGANIC COMPOUND	GENERAL FORMULA	FUNCTIONAL GROUP
Halogenated hydrocarbon	$R-CH_2-X$ $R-\underset{\underset{R'}{\vert}}{CH}-X$ $R-\underset{\underset{R''}{\vert}}{\overset{\overset{R'}{\vert}}{C}}-X$	$-X$ (X=F, Cl, Br, or I, referred to as fluoro, chloro, bromo, and iodo, respectively)
Alcohol	$R-CH_2-OH$ $R-\underset{\underset{R'}{\vert}}{CH}-OH$ $R-\underset{\underset{R''}{\vert}}{\overset{\overset{R'}{\vert}}{C}}-OH$	$-OH$, hydroxyl
Ether	$R-O-R'$	$-O-$, oxy
Aldehyde	$R-\overset{\overset{O}{\Vert}}{C}-H$	$-\overset{\overset{O}{\Vert}}{C}-H$, carbonyl
Ketone	$R-\underset{\underset{O}{\Vert}}{C}-R'$	$-\underset{\underset{O}{\Vert}}{C}-$, carbonyl
Carboxylic acid	$R-\overset{\overset{O}{\Vert}}{C}-OH$	$-\overset{\overset{O}{\Vert}}{C}-O-$, carboxyl
Ester	$R-\overset{\overset{O}{\Vert}}{C}-O-R'$	$-\overset{\overset{O}{\Vert}}{C}-O-$, carboxyl
Amine	$R-NH_2$ $R-NH-R'$ $R-\underset{\underset{R''}{\vert}}{N}-R'$	$-NH_2$, amino
Hydroperoxide	$H-O-O-R$	$H-O-O-$, hydroperoxy
Peroxide	$R-O-O-R'$	$-O-O-$, peroxy

[a]R, R′, and R″ are arbitrary alkyl or aryl substituents.

13.1 Functional Groups

One or more hydrogen atoms in the molecular structure of a hydrocarbon may be substituted with another atom or group of atoms. When this substitution occurs, a new organic compound is produced. The atom or group of atoms that substitutes for the hydrogen atom is called the **functional group**, since it represents the location on the molecule at which chemical reactions often occur. Multiple functional groups are components of complex molecules.

The functional groups that are components of the molecular structures of many important organic compounds are listed in Table 13.1. The groups of atoms associated with each class of organic compound should be memorized. The properties of hydrocarbons and halogenated hydrocarbons were examined in Chapter 12; we study the other classes of organic compounds in this chapter.

functional group
■ An atom or group of atoms that determines many of an organic compound's characteristic chemical properties and identifies it as an alcohol, aldehyde, ester, ether, ketone, carboxylic acid, and so forth

SOLVED EXERCISE 13.1

Aspirin, or acetylsalicylic acid, is a well-known drug used to relieve minor aches and pains. Its molecular formula is noted below:

Identify the functional groups in the aspirin molecule.

Solution: Table 13.1 indicates that $-\overset{\displaystyle O}{\underset{\displaystyle O-}{C}}$ is the carboxyl group. The aspirin molecule contains two carboxyl groups.

■ Identify the functional group that characterizes alcohols.
■ Write the molecular formulas for the alcohols having from one to four carbon atoms per molecule.
■ Recognize how the alcohols having from one to four carbon atoms per molecule are named.
■ Identify the primary risks associated with exposure to methanol, ethanol, isopropanol, phenol, and the cresols.

Chapter 13 Chemistry of Some Hazardous Organic Compounds: Part II   591

13.2 Alcohols

alcohol
■ An organic compound each of whose molecules has a hydroxyl group ($-OH$)

hydroxyl group
■ The $-OH$ functional group, as in an alcohol

Alcohols are organic compounds derived by substituting one or more hydrogen atoms in the hydrocarbon molecule with the **hydroxyl group** ($-OH$). Thus, the general chemical formula of the simple alcohols is $R-OH$, where R is the formula of an arbitrary alkyl or aryl group.

When one hydrogen atom is substituted with the hydroxyl group in molecules of methane and ethane, the resulting compounds are methyl alcohol and ethyl alcohol, respectively.

$$
\begin{array}{cc}
\text{H} & \text{H}\ \ \text{H} \\
| & |\ \ \ \ | \\
\text{H}-\text{C}-\text{OH} & \text{H}-\text{C}-\text{C}-\text{OH} \\
| & |\ \ \ \ | \\
\text{H} & \text{H}\ \ \text{H} \\
\text{Methyl alcohol} & \text{Ethyl alcohol}
\end{array}
$$

These are the common names of the two simplest alcohols.

In the IUPAC system, the simple alcohols are named by replacing the *-e* in the name of the corresponding alkane with *-ol*; hence, the compounds having the formulas CH_3OH and CH_3CH_2OH are also named methanol and ethanol, respectively.

To name more complex alcohols, it is necessary to indicate the position of the hydroxyl group by a number immediately preceding the name of the alcohol, using the following rules:

- Locate the lowest possible number of a carbon atom in the *longest chain of carbon atoms that contains the hydroxyl group*.
- Assign the lowest possible number to the carbon atom bonded to the hydroxyl group.

These rules are applied in the following examples:

$$
\overset{1}{CH_3}-\overset{2}{CH}-\overset{3}{CH_3} \qquad \overset{1}{CH_3}-\overset{2}{CH}-\overset{3}{CH_2}\overset{4}{CH_3} \qquad \overset{1}{CH_3}-\overset{2}{CH}-\overset{3}{CH_2}-\overset{4}{CH}-\overset{5}{CH_2}
$$
$$
\ \ \ \ \ \ \ | \ | \ | \ \ \ \ \ \ \ \ \ |
$$
$$
\ \ \ \ \ \ OH \ \ \ \ \ \ \ \ \ \ \ \ \ \ \ \ \ \ OH \ \ \ \ \ \ \ \ \ \ \ \ \ \ \ \ \ OH \ \ \ \ \ \ CH_3
$$
$$
\text{2-Propanol} \qquad\qquad \text{2-Butanol} \qquad\qquad \text{4-Methyl-2-pentanol}
$$

Table 13.2 lists some physical properties of three simple aliphatic alcohols: methanol, ethanol, and isopropanol. These data indicate that when exposed to an ignition source in air, their vapors easily burn. Fire and explosion are considered their primary risks. Because oxygen is a component of their molecular structures, these alcohols predominantly produce the products of complete combustion when they burn.

$$
\begin{array}{ccccccc}
CH_3OH(g) & + & 2O_2(g) & \longrightarrow & CO_2(g) & + & 2H_2O(g) \\
\text{Methanol} & & \text{Oxygen} & & \text{Carbon dioxide} & & \text{Water}
\end{array}
$$

$$
\begin{array}{ccccccc}
CH_3CH_2OH(g) & + & 3O_2(g) & \longrightarrow & 2CO_2(g) & + & 3H_2O(g) \\
\text{Ethanol} & & \text{Oxygen} & & \text{Carbon dioxide} & & \text{Water}
\end{array}
$$

$$
\begin{array}{ccccccc}
2(CH_3)_2-CHOH(g) & + & 9O_2(g) & \longrightarrow & 6CO_2(g) & + & 8H_2O(g) \\
\text{Isopropanol} & & \text{Oxygen} & & \text{Carbon dioxide} & & \text{Water}
\end{array}
$$

TABLE 13.2 | Physical Properties of Simple Alcohols

	METHANOL	ETHANOL	ISOPROPANOL
Melting point	−144 °F (−98 °C)	−173 °F (−114 °C)	−128 °F (−89 °C)
Boiling point	149 °F (65 °C)	174 °F (79 °C)	180 °F (82 °C)
Specific gravity at 68 °F (20 °C)	0.79	0.79	0.79
Vapor density (air = 1)	1.11	1.59	2.07
Vapor pressure at 68 °F (20 °C)	98 mmHg	50 mmHg	33 mmHg
Flashpoint (without water) (with 15% water by volume) (with 24% water by volume)	54 °F (12 °C)	54 °F (12 °C) 68 °F (20 °C) 97 °F (36 °C)	53 °F (12 °C)
Autoignition point	867 °F (464 °C)	793 °F (423 °C)	750 °F (399 °C)
Lower explosive limit	6%	3.3%	2.3%
Upper explosive limit	36.5%	19%	12.7%
Heat of combustion	9740 Btu/lb (22.7 kJ/g)	12,900 Btu/lb (30.0 kJ/g)	14,200 Btu/lb (33.0 kJ/g)

When aliphatic alcohols burn, they also produce flames that are nearly imperceptible to the naked eye. When methanol, ethanol, and isopropanol burn, for example, pale blue, virtually invisible flames are produced without the accompaniment of soot. This absence of visible flames and particulate matter is also characteristic of the burning of other aliphatic compounds that contain oxygen atoms in their molecular structures. Because alcohol flames cannot be detected with the naked eye, firefighters often encounter unique difficulties when combating fires involving them.

SOLVED EXERCISE 13.2

Why is very little soot produced when methanol and ethanol burn in air? How does the absence of soot affect observation of the accompanying flames?

Solution: Methanol and ethanol are examples of organic compounds whose molecules contain oxygen atoms in addition to carbon and hydrogen atoms. This oxygen is utilized together with the available atmospheric oxygen when methanol and ethanol burn. Provided with this "supplemental" oxygen, methanol and ethanol burn predominantly by complete combustion, producing carbon dioxide instead of carbon monoxide and soot. Because flames are manifested only when minute particulates of matter can be heated to incandescence (Section 5.6), the absence of soot causes the flames accompanying the combustion of these simple alcohols to be nearly imperceptible.

Fires involving bulk quantities of methanol are among the most difficult to successfully combat.

(a) What unique characteristic of these fires contributes to this problem?
(b) What special procedures can be implemented to assist firefighters in effectively extinguishing them?

Solution:

(a) Bulk methanol fires are often difficult to successfully combat, since the evolving flames are virtually invisible to the naked eye. This situation hinders the firefighters' ability to observe the magnitude of these fires and rapidly assess their severity.

(b) Methanol vapor generally ignites near the point where the liquid initially leaks from its bulk container or tank. Firefighters must locate and seal this leak before attempting to extinguish the fire. Approaching a bulk storage or transportation vessel, however, could constitute an unwarranted risk owing to the likelihood that the vessel will rupture. Unable to directly view the flame evolving from a methanol fire, firefighters could search from a suitable distance for telltale signs of combustion, such as blistering paint or weakened metal. However, specialists recommend the use of a thermal-imaging camera (Section 5.6) to locate the leak more rapidly.

Because the flames associated with methanol fires are nearly invisible to the naked eye, it is difficult to determine not only where liquid methanol is leaking but whether efforts to extinguish the fire were successful. Specialists again recommend using a thermal-imaging camera to establish that the fire has been successfully extinguished.

13.2-A METHANOL

wood alcohol
■ The common name for methanol produced by heating wood in the absence of air

At room conditions, *methanol* is a colorless, water-soluble, and highly volatile liquid. It was once called **wood alcohol**, a term that acknowledges methanol as a constituent of the mixture that results when wood is heated in the absence of air.

In the United States, methanol is manufactured mainly by the high-temperature, high-pressure hydrogenation of carbon monoxide in the presence of an appropriate catalyst.

$$CO(g) \quad + \quad 2H_2(g) \quad \longrightarrow \quad CH_3OH(g)$$

Carbon monoxide   Hydrogen   Methanol

Industrially, methanol is used as a solvent for shellac and gums; as a raw material for the manufacture of acetic acid (Section 8.12) and formaldehyde (Section 13.5-A); and directly as an antifreeze, aircraft fuel-injection fluid, windshield washer fluid, automobile racing fuel, and heat source for chafing dishes.

Fire and explosion are considered the primary risks associated with methanol. Its complete combustion is represented as follows:

$$2CH_3OH(g) \quad + \quad 3O_2(g) \quad \longrightarrow \quad 2CO_2(g) \quad + \quad 4H_2O(g)$$

Methanol   Oxygen   Carbon dioxide   Water

A secondary risk is also associated with methanol. The ingestion of as little as 0.06 pt (30 mL) of methanol can cause death. Lesser amounts have been known to cause the onset of optic diseases including retinal edema that further lead to either partial or irreversible blindness. The toxicity of methanol is directly linked with its stepwise transformation into the toxic metabolic by-products formaldehyde and formic acid.

$$CH_3OH(aq) \longrightarrow \underset{\text{Formaldehyde}}{H-\overset{\overset{\displaystyle O}{\|}}{\underset{\displaystyle OH}{C}}(aq)} \longrightarrow \underset{\text{Formic acid}}{H-\overset{\overset{\displaystyle O}{\|}}{\underset{\displaystyle OH}{C}}(aq)}$$

Methanol Formaldehyde Formic acid

The formaldehyde production occurs in the retina of the eyes, where the ultimate accumulation of formic acid causes the onset of optic diseases.

The production of formic acid during the metabolism of methanol also causes the pH of the blood, normally about 7.4, to decline. It is this condition, called *acidosis*, that causes victims of methanol poisoning to initially experience hyperventilation and disorders of the central nervous system. These symptoms are similar to those associated with ethanol intoxication. Thereafter, the victims of methanol poisoning experience blurred vision and decreased visual acuity. When left untreated, these individuals may lose their sense of sight.

13.2-B CONSUMER PRODUCT REGULATIONS INVOLVING METHANOL

Because death and blindness may result from the ingestion of methanol, the U.S. Consumer Product Safety Commission requires that the label affixed to methanol containers include the following statements:

DANGER

**POISON
VAPOR HARMFUL
MAY BE FATAL OR CAUSE BLINDNESS IF SWALLOWED
CANNOT BE MADE NONPOISONOUS**

13.2-C ETHANOL IN ALCOHOLIC BEVERAGES

Ethanol is also a colorless, water-soluble, highly volatile liquid. It is most commonly encountered as the active constituent of beer, wine, and the so-called hard liquors. The ethanol in wine, champagne, and various brandies is produced by fermenting the sugars naturally present in ripe fruits. For example, the ethanol in wine is produced by fermenting the sugar in grapes. When the wine is charged with carbon dioxide, champagne is produced. When the ethanol is distilled from fermented peaches, apricots, apples, and other fruits, brandies are produced.

Ethanol intended for consumption is also produced by fermenting the sugars generated by the enzymatic conversion of the starches in corn, potatoes, barley, rye, and wheat. The ethanol in beer is produced by fermenting malted barley. The ethanol in whiskey is produced by fermenting corn, barley, or wheat; the ethanol in vodka is produced by fermenting potato mash; and the ethanol in gin is produced by fermenting rye. Since it is produced by fermenting cereal grains, the ethanol present in beers, whiskeys, gin, and vodka is sometimes called **grain alcohol**.

The concentration of ethanol in alcoholic beverages is frequently expressed by use of the term *proof*. In the 17th century, people believed that beverages contained spirits who caused the physiological effects in individuals who consumed them. A procedure devised in England to test for the presence of these spirits consisted of preparing a mixture of gunpowder and a sample of the liquor being tested. If the gunpowder ignited after the alcohol had burned away, the event was taken as "proof" or confirmation that spirits were actually present. If the gunpowder did not ignite, the event was taken to mean that spirits were absent. It actually meant that the liquor had been diluted with so much water it could not burn.

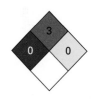

grain alcohol
■ The common name of ethanol produced by the fermentation of a grain

FIGURE 13.1 The
Alcotest 7110 machine
used by law enforce-
ment personnel to
determine the concen-
tration of ethanol in
the blood. In 2007, the
Supreme Court of the
State of New Jersey
ruled that this machine
provides sufficiently
reliable readings that
can be admitted into
evidence without the
need for an expert
witness. (*Courtesy of
Draeger Safety Diagnos-
tics, Inc., Irving, Texas.*)

The percentage by volume of ethanol is half the proof of an alcoholic beverage. Thus, al-
cholic beverages containing 95% and 90% ethanol are 190 proof and 180 proof solutions, respectively. A "proof-gallon" of an alcoholic beverage is defined as a standard gallon containing 50%
ethanol by volume.

When an individual consumes an alcoholic beverage, the ethanol is absorbed into the bloodstream. The rate of absorption is faster than the rates of metabolism and subsequent elimination
from the body. An estimate of the amount of alcohol that has been absorbed into the bloodstream
can be approximated with an instrument like the one shown in Figure 13.1.

The metabolism of ethanol involves the stepwise oxidation to acetaldehyde and acetic acid.

$$CH_3CH_2OH(aq) \longrightarrow \underset{\text{Acetaldehyde}}{CH_3-\overset{\displaystyle O}{\underset{\displaystyle H}{\overset{\|}{C}}}(aq)} \longrightarrow \underset{\text{Acetic acid}}{CH_3-\overset{\displaystyle O}{\underset{\displaystyle OH}{\overset{\|}{C}}}(aq)}$$

Ethanol

**blood alcohol
concentration**
■ The amount of
alcohol in a specified
volume of an individ-
ual's blood, typically
measured in milligrams
per deciliter (mg/dL)
(abbreviated BAC)

The process occurs primarily in the liver. The average rate of oxidation is equivalent to approximately two-thirds to three-fourths of an ounce of whiskey, or about a glass of beer, per hour.
When more alcohol is ingested than the liver can oxidize in a timely fashion, the **blood alcohol
concentration** (BAC) increases.

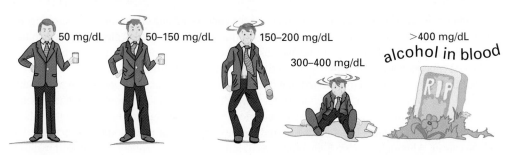

50 mg/dL   50–150 mg/dL   150–200 mg/dL   >400 mg/dL
                                             alcohol in blood
                            300–400 mg/dL

FIGURE 13.2 The adverse effects experienced by a person having different blood alcohol concentrations: sedation or tranquility, 50 mg/dL; lack of coordination, 50 to 150 mg/dL; intoxication (delirium), 150 to 200 mg/dL; unconsciousness, 300 to 400 mg/dL; and death, >400 mg/dL. (*Adapted from an illustration by Joe Orlando from* Chemistry, *by Harvey A. Yablonsky, copyright 1973 by Thomas T. Crowell; used by permission of the publisher.*)

When ethanol is consumed, it acts simultaneously as a brain stimulant and a central nervous system depressant. The stimulation is responsible for the pleasurable, sociable, and deinhibiting effects commonly associated with drinking alcohol, whereas the depression accounts for the anxiety, tension, and dulling of normal cognitive and motor processes. At elevated altitudes, a person has to breathe harder to get ample oxygen into the blood (Section 7.1-C); and when alcoholic beverages are drunk at high altitudes, the alcohol is absorbed more quickly into the bloodstream. This speeds up the effects that consumption of a couple of drinks would normally have.

When alcoholic beverages are consumed, individuals experience physical, behavioral, and speech changes. The metabolic by-product acetaldehyde is responsible for these unpleasant physiological effects (including the dreaded "hangover"), but it is easier to measure the ethanol rather than the acetaldehyde in the consumer's blood. The combination of the experiences illustrated in Figure 13.2 can be correlated with an individual's blood alcohol concentration. Individuals who have blood alcohol concentrations between 50 and 150 mg/dL, or between 0.05% and 0.15%, typically experience a complete lack of coordination. Because this condition prevents them from safely operating a motor vehicle, 100 mg/dL, or 0.1%, has been selected as the legal limit above which a person is prohibited from driving on roadways in all states. [One deciliter (dL) is 100 milliliters, or 1/10 of a liter.] In *most* states, the legal limit is 50 mg/dL, or 0.05%.

Alcoholic beverages should be consumed only in moderation—between one and three drinks per day. "Binge drinking," the indiscriminate consumption of multiple drinks, one immediately after the other, is known to have caused death within a few short hours of consumption.

Because alcohol consumption dulls the senses and causes reflexes to be sluggish, emergency responders who consume alcohol in amounts beyond moderation cannot expect to perform to the best of their ability. Emergency responders serve their communities best by avoiding alcohol or drinking only in moderation. As a general practice, everyone should choose to drink responsibly at all times.

Individuals who consume alcohol in relatively large amounts over long periods of time often become afflicted with chronic health problems, the most serious of which is **alcoholism**. This disease results from an addiction to alcohol. Such chronic alcohol misuse forces the liver to use alcohol as an energy source. Following aggravated misuse, the liver is unable to function properly. These heavy alcohol drinkers develop either or both of two diseases: *alcoholic hepatitis*, an inflammation of the liver; or **cirrhosis**, a disease in which normal liver tissue is replaced by scar tissue. Because cirrhosis causes serious impairment of liver function, it is often fatal.

Studies have also linked alcohol abuse with mouth, larynx, esophagus, liver, and breast cancers. They show that women who consume two or more drinks a day are 41% more likely to experience breast cancer compared with women who are nondrinkers. One study concluded that as little as one drink per day can increase the risk of bowel cancer by 70%. Although it is not clear how ethanol causes cancer, the prevailing theory is that DNA is damaged by acetaldehyde in such a manner that it cannot be repaired.

alcoholism
■ A disease resulting from the misuse of ethanol

cirrhosis
■ A chronic disease of the liver characterized by inflammation, pain, and jaundice (a yellowishness of the skin)

Studies have also shown that alcohol enhances the cancer-causing effects of other carcinogens, particularly those found in tobacco smoke. The frequent consumption of alcohol by a habitual smoker leads to as much as a 100-fold increase in the risk for contracting mouth, tracheal, or esophageal cancers compared with people who neither smoke nor drink.

Researchers have also demonstrated that drinking alcohol stimulates the growth of an existing cancer in mice. When provided with either water or water and alcohol for a month and then exposed to melanoma cells, mice that drank the water and alcohol developed tumors twice as large as those mice that drank only water. The alcohol spurred the development of new blood vessels needed to feed the cancer. Although the metabolic processes of humans and mice are not entirely comparative, individuals with cancer are probably better off to avoid the consumption of alcohol altogether.

In combination with medications containing barbiturates, tranquilizers, or narcotics, ethanol produces cumulative effects that can slow the cardiac and respiratory systems to the point of causing coma and death. For this reason, individuals on medications should avoid drinking alcoholic beverages.

Pregnant women should be especially wary of consuming alcoholic beverages, since alcohol consumption during pregnancy may cause their offspring to subsequently experience mental disabilities, physical deficiencies, and other birth disorders. In particular, prenatal exposure to alcohol injures the neurological function of a developing fetus, ultimately causing a reduction in the child's intellectual potential for its entire life. Studies link the mother's consumption of as little as one drink per week with mental impairment in her offspring. This affliction is known as **fetal alcohol syndrome**. The acetaldehyde formed during metabolism is responsible for the onset of fetal alcohol syndrome, because acetaldehyde crosses the placental barrier and accumulates in the liver of the fetus.

fetal alcohol syndrome
■ The disorder manifested by a reduction of mental capabilities in children of women who consumed alcoholic beverages during pregnancy

One-third of the population of the mentally handicapped children in the Western world most likely acquired their handicap in conjunction with their mother's consumption of alcohol and medications during pregnancy. Even moderate or social drinking may be harmful to the developing fetus. The U.S. Surgeon General and Britain's Royal College of Psychiatrists uncompromisingly advise total abstinence from alcohol during pregnancy.

SOLVED EXERCISE 13.4

The governments of several European countries have requested the manufacturers of alcoholic beverages to include the following expression on their product labels:

AVOID ALCOHOL IF PREGNANT OR TRYING TO CONCEIVE

What factor most likely contributed to this request?

Solution: The consumption of alcohol during pregnancy gives rise in the mother's offspring to the disorder known as fetal alcohol syndrome. Children born with this affliction experience a range of neurological and other problems throughout their lives. Examples of these neurological problems include poor memory, the inability to understand certain concepts such as time and money, poor language skills, and poor problem-solving skills. Children born with this affliction are also likely to experience growth differences compared with the norm such as facial and limb abnormalities. European governments most likely requested alcohol manufacturers to include the referenced statement on their product labels to protect the health of the unborn and inform pregnant women that alcoholic consumption could cause their children to experience this disorder.

To apprise the U.S. public of the potential dangers associated with alcohol consumption, the Alcohol and Tobacco Tax and Trade Bureau requires at 27 C.F.R. §16.21 labeling of all alcohol containers with the health-warning message shown here:

Posting of this label has been required on all alcohol containers sold in the United States since November 18, 1989.

Notwithstanding the combination of adverse effects associated with the consumption of alcohol, research studies reveal that *moderate* drinking by mature adults contributes to the prevention of heart attacks, regardless of the nature of the alcoholic beverage consumed. Preliminary studies show that individuals who drink alcoholic beverages three times a week experience approximately one-third fewer heart attacks than nondrinkers. This may mean that drinking a small amount of alcohol each day is beneficial to one's health, but drinking in moderation is always the recommended guide.

13.2-D INDUSTRIAL-GRADE ETHANOL

Aside from its use in alcoholic beverages, ethanol is also widely used for various industrial purposes. Manufacturing and process industries use large volumes of ethanol, most typically as a solvent in toiletries, cosmetics, pharmaceuticals, and surface coatings; a raw material in the manufacture of other substances; and an oxygenate in vehicular fuels (see below).

Ethanol is sometimes produced for industrial use by the acid-catalyzed vapor-phase reaction between ethylene and water.

$$CH_2{=}CH_2(g) \quad + \quad H_2O(g) \quad \longrightarrow \quad CH_3CH_2OH(g)$$

Ethylene   Water   Ethanol

In the United States, substantial volumes are also produced by fermenting the sugar generated by the enzymatic conversion of the starch in corn. First, corn kernels are ground into flour and slurried with water; then, enzymes are added to the mixture to convert the starch into sugar. Yeast then converts the sugar into ethanol, which is distilled from the mixture. This process was once used by moonshiners when Prohibition was in effect from 1920 to 1933. The essential difference is that today, alcohol manufacturers must add a **denaturant** (see below), usually gasoline, to ethanol intended for purposes other than consumption.

Industrial-grade ethanol may be produced not only from corn but also from various other starch and cellulosic materials including switchgrass (Section 14.5-A) and wood residues from the forest products industry. When it is used as a component of vehicular fuels, the ethanol is an example of a **biofuel**. In Brazil, pure ethanol—without gasoline—made from sugarcane is used directly as a biofuel.

Two forms of industrial-grade alcohol are available commercially:

- **Denatured alcohol** is an aqueous solution containing approximately 95% ethanol by volume. Denatured alcohol is intended for use as a solvent. To impede its consumption, a small amount of methanol, pyridine, or aviation gasoline is added to denatured alcohol. The presence of these substances causes the alcohol to be not only unpalatable but poisonous. Each of these additives is called a **denaturant**.
- **Absolute alcohol** is ethanol without water (anhydrous) or a denaturant.

Whereas the sale of absolute alcohol is closely regulated by governments throughout the world, denatured alcohol is sold without the imposition of an excise tax.

denaturant
- A substance added to ethanol to make it unpalatable

biofuel
- A vehicular fuel made from a domestic farm crop such as corn kernels or soybeans

denatured alcohol
- Ethanol that has been made unfit for consumption as a beverage by the addition of a toxic or noxious substance

absolute alcohol
- Pure ethanol

In 1990, following the mandate of the Clean Air Act, EPA required petroleum refineries in certain areas of the United States to blend gasoline with a substance that improved the completeness of the ensuing combustion. This substance is a type of antiknock agent called an **oxygenate**. When ethanol is added to gasoline, it specifically functions as an oxygenate. During the 1970s, it was called **gasohol**. When gasoline containing an oxygenate is used as a vehicular fuel, the concentrations of carbon monoxide and particulates are reduced in the tailpipe emissions. For this reason EPA requires its use as a vehicular fuel in certain urban areas where compliance with the mandates of the Clean Air Act is otherwise difficult.

The use of ethanol in gasoline became even more commonplace during 2005 and thereafter, when the cost of crude oil from the Middle East rose excessively. To lower energy prices for U.S. consumers, Congress overhauled the nation's energy policies through the introduction of a biofuels program that requires the blending of a biofuel like ethanol into gasoline. Under the mandate of the Energy Independence and Security Act of 2007, refineries, blenders, and importers are collectively responsible for blending 15 billion gal/y (15×10^9 gal/y, or 3.4×10^9 m^3/y) of ethanol and other biofuels into gasoline by 2015.

In today's marketplace, two types of ethanol-containing alternative motor fuels comply with the intent of this act:

- *E10*, a blend of 10% ethanol with 90% gasoline by volume. It may be used as the fuel in any vehicle without engine modification. Because most petroleum fuels used in the United States now contain ethanol as an additive, they are correctly denoted as E10.

- *E85*, a blend of 85% ethanol and 15% gasoline by volume. It may be used only in specially built vehicles called *flex-fuel vehicles*. For example, people who drive GM FlexFuel vehicles can use E85 as the motor fuel. These cars and trucks have sensors to adjust the timing of spark plugs and fuel injectors so the engine runs smoothly regardless of the fuel's ethanol concentration.

Given the mandate of the Energy Independence and Security Act, it is apparent that the demand for biofuel use in U.S. vehicular fuels—including ethanol use—is likely to substantially increase.

The sole use of ethanol as a biofuel is decidedly unpromising in the United States and is not without its problems. First, ethanol has only about 66% of the energy content of gasoline. A second problem is ethanol's ability to easily attract moisture from the air. The presence of water in a fuel causes corrosion within fuel pipelines, requiring that ethanol be transported from rail or barge terminals by a separate distributor to regional fuel terminals or gasoline-blending racks. Third, the corn that is used for ethanol production affects its availability throughout the food industry and causes the price of certain other agricultural products to rise.

The growing market for ethanol in the United States has also resulted in more transportation incidents involving ethanol. For example, in 2006, twenty-three railcars containing ethanol in an 86-tank-long train derailed while crossing a bridge, resulting in an explosion and fire near New Brighton in rural Pennsylvania.*

13.2-E EXTINGUISHING ETHANOL FIRES

Fire and explosion are considered the primary risks associated with ethanol. The combustion process, which at times can be nearly imperceptible, is represented as follows:

$$CH_3CH_2OH(g) \quad + \quad 3O_2(g) \quad \longrightarrow \quad 2CO_2(g) \quad + \quad 3H_2O(g)$$

Ethanol · · · · · · Oxygen · · · · · · Carbon dioxide · · · Water

Because ethanol is a water-soluble substance, the use of water as a fire extinguisher on burning ethanol should bring its fires under control. However, the flashpoint data in Table 13.2 reveal that even when ethanol is diluted with water to produce a solution containing 24% water, the solution is still flammable. From a practical viewpoint, the sole use of water can extinguish only a

---

*Railroad Accident Report, "Derailment of Norfolk Southern Railway Company Train 68QB119 with Release of Hazardous Material and Fire, October 20, 2006," NTSB/RAR-08/02, PB2008-916302 (National Transportation Safety Board, Washington, DC, 2008).

oxygenate
- An additive that promotes the complete combustion of vehicular fuels, thereby generating lesser amounts of carbon monoxide and other pollutants

gasohol
- A blend of gasoline and ethanol used as a vehicular fuel

nonbulk ethanol fire. Testing reveals that when water alone is used on a bulk ethanol fire, a great volume (>5:1) is needed to extinguish the fire. In most instances, it is simply impractical to have this volume available at a fire scene.

Experts concur that among the firefighting foams that are now commercially available, the use of an alcohol-resistant aqueous-film-forming foam (AR-AFFF) (Section 5.11-C) serves best to extinguish a bulk ethanol fire. However, the experts are also quick to point out that the profession needs a foam performing even more effectively than AR-AFFF.

13.2-F ISOPROPANOL

There are two propyl alcohols, *n-propyl alcohol* and *isopropanol*. Both are commercially important, but isopropanol is used in larger volume. Isopropanol is also known as isopropyl alcohol, 2-propanol, and rubbing alcohol. Most people recognize it as the most commonly encountered alcohol after ethanol. The formula of isopropanol is $(CH_3)_2$—$CHOH$.

In the chemical industry, isopropanol is manufactured by the acid-catalyzed vapor-phase reaction between propene and water.

$$CH_3CH{=}CH_2(g) \quad + \quad H_2O(g) \quad \longrightarrow \quad (CH_3)_2{-}CHOH(g)$$

Propene   Water   Isopropanol

Isopropanol is used in many ways. In the chemical industry, large volumes are needed for the production of hydrogen peroxide (Section 11.5), acetone (Section 13.5-C), and other substances. Most people recognize it as *rubbing alcohol*, a 60% to 70% solution (with methanol and ethanol) that is applied externally to relieve muscle and joint pains. As it evaporates, it cools and soothes the skin at the point of contact. Isopropanol is also used as a gasoline additive, where it serves as an oxygenate and a "deicer." In the latter case, it dissolves water in the fuel line. Isopropanol is also used as a solvent for many lotions, oils, and other commercial products.

Although the major hazard associated with isopropanol is the risk of fire and explosion, the liquid is also a poison. The consumption of isopropanol causes permanent disabling illnesses, and when consumed in excess, it causes death. The ingested isopropanol metabolizes primarily to acetone.

13.2-G GLYCOLS

Glycols are alcohols having two hydroxyl groups per molecule. They are also called *diols*. The two most commonly encountered glycols are *ethylene glycol* and *propylene glycol*. They are produced by various methods including the gas-phase hydration of ethylene oxide and propylene oxide, respectively.

glycol
■ An alcohol each of whose molecules has two hydroxyl groups; a diol

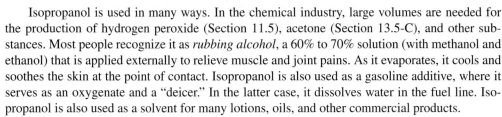

Both are used as liquid antifreeze agents and coolants in motor vehicles and deicing fluids for aircraft. Propylene glycol is also used as a raw material for the manufacture of certain polyester resins.

Although the glycols are considered nontoxic substances, they can be the source of *glycol poisoning*, the medical condition associated with the ingestion of radiator fluids or other products containing glycols. The by-products of the metabolism of glycols are toxic substances that destroy tissues and upset the normal pH of the blood. The metabolism of ethylene glycol, for example, occurs within the cells of the liver and involves the stepwise oxidation to four toxic metabolites: glycoaldehyde, glycolic acid, glyoxalic acid, and oxalic acid.

$$HO-CH_2CH_2-OH(aq) \longrightarrow \underset{\underset{H}{|}}{\overset{\overset{O}{\|}}{C}}-CH_2-OH(aq) \longrightarrow HO-CH_2-\underset{\underset{H}{|}}{\overset{\overset{O}{\|}}{C}}(aq)$$

Ethylene glycol   Glycoaldehyde   Glycolic acid

$$\longrightarrow \underset{\underset{H}{|}}{\overset{\overset{O}{\|}}{C}}-\underset{\underset{OH}{|}}{\overset{\overset{O}{\|}}{C}}(aq) \longrightarrow \underset{\underset{HO}{|}}{\overset{\overset{O}{\|}}{C}}-\underset{\underset{OH}{|}}{\overset{\overset{O}{\|}}{C}}(aq)$$

Glyoxalic acid   Oxalic acid

Following the initial ingestion of ethylene glycol, the victims of glycol poisoning experience inebriation, but as the metabolism progresses, these acids cause extensive cellular damage, especially renal (kidney) damage. Furthermore, the victims of glycol poisoning experience the symptoms of acidosis including hyperventilation and disorders of the central nervous system. In severe instances, death is likely to follow. Individuals who demonstrate the symptoms of glycol poisoning should be quickly rushed to emergency facilities for antidotal treatment.

13.2-H PHENOL

The *phenols*, or *phenolic compounds*, are derivatives of benzene in which one or more hydrogen atoms have been substituted with the hydroxyl group of atoms ($-OH$). The simplest phenolic compound is itself called **phenol**, or *carbolic acid*. Its chemical formula is C_6H_5OH.

phenol
■ An organic compound whose molecules have a hydroxyl group ($-OH$) bonded directly to the benzene ring

C_6H_5-O- is named the *phenoxy group*.

At room conditions, phenol is a colorless to white-pink crystalline solid that often darkens to red on exposure to light. Because phenol readily absorbs atmospheric moisture, it is often encountered as a liquid. Phenol has the sweet acrid odor characteristic of disinfectants.

Although phenol can be distilled from the middle coal tar distillate (Section 7.7-C), it is generally produced in the chemical industry using other raw materials. The dominant method involves the production and decomposition of cumene hydroperoxide. Cumene is oxidized using air to cumene hydroperoxide, which decomposes in situ to phenol and acetone.

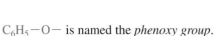

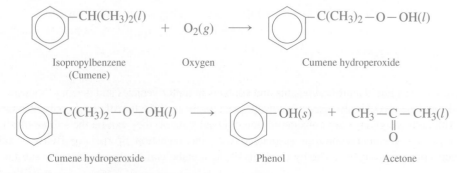

TABLE 13.3 | Physical Properties of Phenol and the Isomeric Cresols

	PHENOL	o-CRESOL	m-CRESOL	p-CRESOL
Melting point	104 °F (40 °C)	88 °F (31 °C)	54 °F (12 °C)	95 °F (35 °C)
Boiling point	358 °F (181 °C)	376 °F (191 °C)	397 °F (203 °C)	396 °F (202 °C)
Specific gravity at 68 °F (20 °C)	1.07	1.05	1.03	1.04
Vapor density (air = 1)	3.24	3.7	3.7	3.7
Vapor pressure at 68 °F (20 °C)	0.357 mmHg	0.3 mmHg	<1 mmHg	<1 mmHg
Flashpoint	175 °F (79 °C)	178 °F (81 °C)	202 °F (94 °C)	202 °F (94 °C)
Autoignition point	1319 °F (715 °C)	1110 °F (599 °C)	1038 °F (559 °C)	1038 °F (559 °C)
Lower explosive limit	1.5%	1.35%	1.06%	1.06%
Upper explosive limit			1.35% @302 °F (150 °C)	1.4% @302 °F (150 °C)

This production method is economically desirable, since the coproduct acetone (Section 13.5-C) is also a commercially important chemical product.

Phenol is an important industrial substance, since it is the raw material from which a number of derivatives and phenolic resins are manufactured. Phenolic derivatives are also used in surgical antiseptics and other germicidal solutions. Phenol itself was the *first* surgical antiseptic. A solution of one part phenol in 850 parts of water by mass prevents the multiplication of certain bacteria; for this reason, phenol derivatives are common constituents of mouthwashes, gargles, and sprays.

Some important physical properties of phenol are provided in Table 13.3. These data show that phenol burns, but its flashpoint is fairly elevated: 175 °F (79 °C). For this reason, phenol does not generally pose the risk of fire and explosion.

When dissolved in water, phenol dissociates into hydrogen ions and phenoxide ions, causing the phenol solution to be very acidic.

Thus, phenol is highly irritating to exposed skin, mucous membranes, and the eyes. Because it is a corrosive substance, plastic surgeons use dilute solutions of phenol with soap and vegetable oil for chemical facial peels to produce a partial-thickness controlled burn of predictable depth when removing wrinkles and irregular facial pigmentation.

Phenol is also a poisonous substance. Although it is absorbed into the body by all routes of exposure, the ingestion and absorption of phenol through the skin can be especially damaging, since phenol impairs liver and kidney function and profoundly disturbs the central nervous system. Individuals exposed to elevated concentrations of phenol often experience coma and may die from respiratory failure. Because its dermal LD_{50} is 630 mg/kg (rabbits), phenol is also considered a severe skin irritant.

13.2-I CRESOLS

A commercially important group of phenols are the hydroxy derivatives of toluene, called **cresols**. There are three isomeric cresols, whose formulas and names follow:

o-Cresol   *m*-Cresol   *p*-Cresol

Their mixture is called *cresylic acid*. The physical properties of the cresol isomers are included in Table 13.3.

In the chemical industry, the cresol isomers are generally isolated as a mixture from the middle coal tar distillate (Section 7.7-C), after which the mixture is subjected to fractional distillation to separate the individual isomers. *o*-Cresol boils at 191 °F (376 °C); thus, it easily separates from a mixture of the other two isomers, which boils at approximately 394 °F (201 °C). This mixture of *m*- and *p*-cresol is chemically treated with an acid to produce compounds that can be isolated more easily. The individual isomers are also produced by unique synthetic procedures.

Of the three isomers, the *p*- isomer is most important commercially. For instance, *p*-cresol is used as a raw material for the manufacture of the fire retardants tricresyl phosphate and cresyl diphenyl phosphate.

Tri-*p*-cresyl phosphate   *p*-Cresyldiphenyl phosphate

In these structures, the group of atoms $CH_3-C_6H_4-O-$ is designated as the *p*-cresyl group when the methyl group is bonded in the para position to the $-O-$ group.

When individuals are exposed to the cresol isomers, they are likely to experience the same adverse health effects as those noted for exposure to phenol. For example, both phenol and the cresols impair liver and kidney function and disturb the central nervous system when ingested or absorbed through the skin. The cresols are more corrosive to skin than phenol. The dermal LD_{50} (*o*-, *m*-, and *p*-) is 301 mg/kg (rabbits).

13.2-J WORKPLACE REGULATIONS INVOLVING ALCOHOLS

When methanol is used in the workplace, OSHA requires employers to limit employee exposure to a maximum vapor concentration of 200 ppm, averaged over an 8-hr workday.

When ethanol is used in the workplace, OSHA requires employers to limit employee exposure to a maximum vapor concentration of 1000 ppm, averaged over an 8-hr workday.

When phenol and the cresols are used in the workplace, OSHA requires employers to limit employee exposure by dermal contact to a maximum concentration of 5 ppm, averaged over an 8-hr workday.

13.2-K TRANSPORTING ALCOHOLS

Although alcohols may be transported in bulk, they are usually encountered in nonbulk containers like those shown in Figure 13.3. When shippers offer an alcohol for transportation, DOT requires them to enter the relevant basic description from Table 13.4 on an accompanying shipping paper. The following DOT regulations also apply:

- When shippers offer an alcohol in hazard class 3 for transportation, DOT also requires them to affix a FLAMMABLE LIQUID label on its packaging.
- When carriers transport an alcohol in hazard class 3 in an amount exceeding 1001 lb (454 kg), DOT requires them to post FLAMMABLE placards on the bulk packaging or transport vehicle used for shipment.
- When shippers offer allyl alcohol for transportation, DOT requires them to affix POISON INHALATION HAZARD labels on its packaging.
- When carriers transport allyl alcohol in any amount, DOT requires them to display POISON INHALATION HAZARD placards on the bulk packaging or transport vehicle used for shipment.
- When shippers offer an alcohol in hazard class 6.1 other than allyl alcohol for transportation, DOT requires them to affix a POISON label on its packaging.
- When carriers transport an alcohol in hazard class 6.1 other than allyl alcohol in an amount exceeding 1001 lb (454 kg), DOT requires them to post POISON placards on the bulk packaging or transport vehicle used for shipment.
- When shippers offer an alcohol in hazard class 8 for transportation, DOT requires them to affix a CORROSIVE label on the packaging.
- When carriers transport an alcohol in hazard class 8 in an amount exceeding 1001 lb (454 kg), DOT requires them to post CORROSIVE placards on the bulk packaging or transport vehicle used for shipment.

FIGURE 13.3
Emergency responders are likely to encounter a nonbulk quantity of an alcohol in either a 55-gal (208-L) steel drum or 5-gal (19-L) metal can. (*Courtesy of Spectrum Chemical Manufacturing Corporation, Gardena, California.*)

TABLE 13.4	Basic Descriptions of Some Alcohols

ALCOHOL OR GROUP THEREOF	BASIC DESCRIPTION
Alcoholic beverages	UN3065, Alcoholic beverages, 3, UN3065, PGII *or* UN3065, Alcoholic beverages, 3, PGIII
Allyl alcohol	UN1098, Allyl alcohol, 6.1, (3), PGI (Poison - Inhalation Hazard, Zone B)
Butanols	UN1120, Butanols, 3, PGII *or* UN1120, Butanols, 3, PGIII
Cresols, liquid	UN2076, Cresols, liquid, 6.1, (8), PGII (Poison)
Cresols, solid	UN3455, Cresols, solid, 6.1, (8), PGII (Poison)
Cresylic acid	UN2022, Cresylic acid, 6.1, (8), PGII (Poison)
Cyclopentanol	UN2244, Cyclopentanol, 3, PGIII
Denatured alcohol	NA1987, Denatured alcohol, 3, PGII
Ethanol	UN1170, Ethanol, 3, PGII
2-Ethylbutanol	UN2275, 2-Ethylbutanol, 3, PGIII
Gasohol (<20% ethanol)	NA1203, Gasohol (contains less than 20% ethanol), 3, PGII
Isopropanol	UN1219, Isopropanol, 3, PGII
Methanol (international transportation)	UN1230, Methanol, 3, (6.1), PGII (Poison)
Methanol (domestic transportation)	UN1230, Methanol, 3, PGII
Methylcyclohexanols	UN2617, Methylcyclohexanols, flammable, 3, PGIII
Methyl isobutyl carbinol[a]	UN2053, Methyl isobutyl carbinol, 3, PGIII
Phenol, molten	UN2312, Phenol, molten, 6.1, PGII (Poison)
Phenol, solid	UN1671, Phenol, solid, 6.1, PGII (Poison)
Phenol solutions	UN2821, Phenol solutions, 6.1, PGII (Poison)
n-Propanol	UN1274, *n*-Propanol, 3, PGII
Xylenols, liquid	UN3430, Xylenols, liquid, 6.1, PGII (Poison)
Xylenols, solid	UN2261, Xylenols, solid, 6.1, PGII (Poison)

[a]Methyl isobutyl carbinol is an alternative common name for methyl amyl alcohol.

- When shippers offer methanol for transportation, DOT requires them to affix FLAMMABLE LIQUID and POISON labels on the packaging. For a domestic shipment, DOT requires them to affix only a FLAMMABLE LIQUID label to the packaging.
- When carriers transport methanol in an amount exceeding 1001 lb (454 kg), DOT requires them to post FLAMMABLE placards on the bulk packaging or transport vehicle used for shipment.

- When shippers offer to transport a flammable liquid alcohol or a flammable and toxic liquid alcohol whose name is not listed in Table 13.4, its shipping description is identified generically as follows:

> **UN1987, Alcohols, n.o.s., 3, PGI**
> *or*
> **UN1987, Alcohols, n.o.s., 3, PGII**
> *or*
> **UN1987, Alcohols, n.o.s., 3, PGIII**
> **UN1986, Alcohols, flammable, toxic, n.o.s., 3, PGI**
> *or*
> **UN1986, Alcohols, flammable, toxic, n.o.s., 3, PGII**
> *or*
> **UN1986, Alcohols, flammable, toxic, n.o.s., 3, PGIII**

The basic description also includes the name of the specific compound entered parenthetically. For example, shippers describe a shipment of 2-octanol in PG III packaging as follows:

> **UN1987, Alcohols, n.o.s. (contains 2-octanol), 3, PGIII**

SOLVED EXERCISE 13.5

When shippers offers gasoline and ethanol for transportation within separate compartmented cargo tanks by highway:

(a) What proper basic descriptions does DOT require them to enter on the accompanying shipping paper?

(b) How does DOT require the tanks to be placarded and marked?

Solution:

(a) DOT requires shippers to enter the following proper basic descriptions of gasoline and ethanol on the accompanying shipping paper:

> **UN1203, Gasoline, 3, PGII**
> **UN1170, Ethanol, 3, PGII**

(b) DOT requires the carrier to display a FLAMMABLE placard on each side and each end of the cargo tank. Because more than one hazardous material is transported in separate compartments within a portable tank, DOT requires the carrier to display the identification numbers 1203 and 1170 on the ends of the tank *and* on the sides in the same sequence as the compartments containing the materials they identify. They may be displayed either within orange panels or on white square-on-point panels on each side and each end of the tank near the FLAMMABLE placard, but two numbers are never displayed across the center area of a single placard.

13.2-L BISPHENOL-A

When phenol reacts with acetone, the product predominantly produced is *p,p'*-isopropylidene diphenol, more commonly known by the product name *Bisphenol-A* (BPA). The production reaction is catalyzed by anhydrous hydrogen chloride.

$$2\ \text{C}_6\text{H}_5\text{OH}(g) + \text{CH}_3-\overset{\text{O}}{\underset{\|}{\text{C}}}-\text{CH}_3(g) \longrightarrow \text{HO}-\text{C}_6\text{H}_4-\overset{\text{CH}_3}{\underset{\text{CH}_3}{\text{C}}}-\text{C}_6\text{H}_4-\text{OH}(s) + \text{H}_2\text{O}(g)$$

| Phenol | Acetone | Bisphenol-A | Water |

Bisphenol-A is used as an intermediate in the polymer industry to manufacture epoxy resins (Section 14.7) and polycarbonate plastics. These resins and plastics are used to manufacture electrical, electronic, and sports-safety equipment. The epoxy resins are also used as protective coatings inside food, milk, and beverage cans and baby bottles. It is this use that primarily gives rise to potential health concerns, since BPA residues leach from these epoxy-resin coatings and become constituents of the container contents. When the foods and beverages are consumed, BPA is unknowingly ingested.

The question that naturally arises is whether BPA ingestion adversely affects human health. The weight of scientific evidence indicates that the answer is most likely yes. Studies conducted on rodents show that exposure in the womb to low concentrations of BPA causes an increase in the rates of prostate and breast cancers, reproductive abnormalities, lowered sperm counts, early onset of puberty in females, obesity, and insulin-dependent diabetes. Researchers have also established that BPA exposure causes chromosomal abnormalities in mice.

When ingested, Bisphenol-A acts as an endocrine disrupter (Section 12.17) by blocking the normal biochemical action of estrogen in women. Estrogen is a hormone that promotes and regulates the development of female characteristics. BPA exposure could potentially put these women and their unborn and newborn babies at risk. In particular, scientists have linked in utero BPA exposure with the development of breast cancer in the adult daughters of an exposed mother. They concluded that a woman's exposure to BPA during pregnancy could predispose her developing daughters to an increased risk of breast cancer during their adulthood.

In recognition of this combination of information, Canada banned the import, sale, and advertising of polycarbonate baby bottles that contain BPA.

PERFORMANCE GOALS FOR SECTION 13.3:

- Identify the functional group that characterizes ethers.
- Identify the primary risks associated with exposure to diethyl ether and methyl *tert*-butyl ether.
- Memorize the rules for naming ethers.
- Note the risks associated with encountering dangerous concentrations of peroxo-organic compounds within ether containers.
- Describe the means shippers and carriers use to inform emergency responders of the hazards associated with encountering the commercially available ethers during transportation mishaps.

13.3 Ethers

ether
■ An organic compound, each of whose molecules consists of an oxygen atom bonded between two alkyl and/or aryl groups

An **ether** is an organic compound whose molecules have one or more oxygen atoms bridged between two alkyl or aryl groups. The simple ethers have the general chemical formula $\text{R}-\text{O}-\text{R}'$, where R and R' are the formulas of arbitrary alkyl or aryl groups.

The common names of the ethers are determined by alphabetically noting the names of the alkyl or aryl groups bonded to the oxygen atom, followed by the word "ether." For example, the compound having the chemical formula $\text{CH}_3-\text{O}-\text{CH}_2\text{CH}_3$ is named *ethyl methyl ether*. Here,

R is the methyl group and R′ is the ethyl group. The simplest ether having a single aryl group is the substance whose formula is $C_6H_5-O-CH_3$. This compound is most frequently encountered by the name *anisole*.

In the IUPAC system, ethers are named by replacing the *-yl* suffix of the relevant alkyl group with *-oxy*, and naming the ether as a substituted alkane. Thus, ethyl methyl ether and anisole are named *methoxyethane* and *methoxybenzene*, respectively.

$$CH_3-O-CH_2CH_3$$

Ethyl methyl ether
(Methoxyethane)

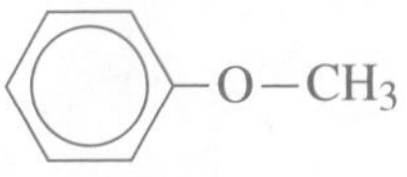

Anisole
(Methoxybenzene)

13.3-A REACTIONS OF ETHERS WITH ATMOSPHERIC OXYGEN

The primary hazard associated with most ethers is that they are highly volatile, flammable liquids; hence, they constitute dangerous fire and explosion hazards. Ethers are also hazardous because they produce potentially unstable peroxo-organic compounds (Section 13.9) by slowly reacting with atmospheric oxygen. For example, the following equations illustrate successive reactions that occur between diethyl ether and oxygen:

$$CH_3CH_2-O-CH_2CH_3(l) \longrightarrow CH_3CH_2-O-CH_2CH_2-O-OH(s)$$

Diethyl ether   2-Ethoxyethyl hydroperoxide

$$\longrightarrow CH_3CH_2-O-O-CH_2CH_3(s)$$

Diethyl peroxide

2-Ethoxyethyl hydroperoxide and diethyl peroxide are examples of peroxo-organic compounds that decompose at explosive rates.

The eight ethers having the following molecular formulas are the most vulnerable to forming peroxo-organic compounds:

$$CH_3-CH-O-CH-CH_3$$
$$\quad\quad | \quad\quad\quad |$$
$$\quad\quad CH_3 \quad\quad CH_3$$

Diisopropyl ether

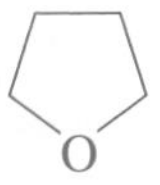

Tetrahydrofuran

1,4-Dioxane

$$CH_3CH_2-O-CH_2CH_3$$

Diethyl ether

$$CH_3-O-CH_2CH_2-O-CH_3$$

Ethylene glycol dimethyl ether
(Glyme)

$$CH_3-O-CH_2CH_2-O-CH_2CH_2-O-CH_3$$

Diethylene glycol dimethyl ether
(Diglyme)

$$C_4H_9-O-CH=CH_2$$

Butyl vinyl ether

$$CH_2=CH-O-CH=CH_2$$

Divinyl ether

These ethers are popular solvents. To warn users of their potential reactivity with oxygen, ether manufacturers and distributors typically affix warning labels like the following on ether containers:

**PROPERLY DISCARD 30 DAYS
AFTER OPENING OR
AFTER ONE YEAR IF UNOPENED**

The reactions between ethers and oxygen are catalyzed by light. Consequently, the rates at which peroxo-organic compounds are formed may be effectively altered by storing the ethers in metal cans or brown glass bottles, each of which prevents the penetration of light (Figure 13.4).

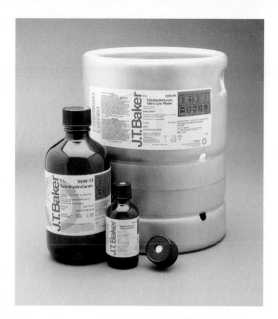

FIGURE 13.4 Many liquid hazardous materials are stored within brown glass bottles, but this practice serves a unique purpose when storing tetrahydrofuran and other ethers. Light cannot penetrate through their glass walls to catalyze the reactions with atmospheric oxygen that produce explosively unstable products; hence, the practice of storing ethers within brown glass bottles serves to retard the rate at which these undesirable products are produced. (*Courtesy of Mallinckrodt Baker, Inc., Phillipsburg, New Jersey.*)

Because the peroxo-organic compounds are produced slowly, ethers are likely to be more susceptible to decomposition if they were purchased years ago. Because ethers are also very volatile, the organic peroxides concentrate within the final residual volume of the liquid. Hence, the risk of an explosive reaction may be greatest in ether containers in which small liquid residues remain.

Safety experts recommend storing ethers in a cooled room and marking the date received and the date first opened on a label affixed to their containers. An example of this label is shown:

Peroxo-Organic Compound–Forming Ether
Date Received: _____
Date Opened: _____

13.3-B DIETHYL ETHER

Diethyl ether is a colorless, water-soluble, flammable, and highly volatile liquid. In the chemical industry, it is produced by the dehydration of ethanol using concentrated sulfuric acid.

$$2CH_3CH_2OH(l) \longrightarrow CH_3CH_2-O-CH_2CH_3(l) + H_2O(l)$$
  Ethanol   Diethyl ether   Water

On inhalation, the vapor of diethyl ether acts on the body as a short-lived muscle relaxant. Owing to this feature, it was once a well-known inhalation anesthetic. A generation ago, the odor of diethyl ether was commonplace in the surgical rooms of medical clinics and hospitals, where it was known simply as *ether*. The use of ether allowed surgeons to perform surgical operations while the patient was unconscious. However, the patient's recovery from exposure to ether is slow and unpleasant. Anesthesiologists now generally select other alternatives including halothane (Section 12.16), isoflurane, and enflurane (Section 13.4). Today, you are more likely to detect the odor of ether when using an automotive starting-fluid during cold weather.

TABLE 13.5	Physical Properties of Diethyl Ether
Melting point	−189 °F (−123 °C)
Boiling point	94 °F (34 °C)
Specific gravity at 68 °F (20 °C)	0.71
Vapor density (air = 1)	2.55
Vapor pressure at 68 °F (20 °C)	442 mmHg
Flashpoint	−49 °F (−45 °C)
Autoignition point	320 °F (160 °C)
Lower explosive limit	1.85%
Upper explosive limit	48%

The physical properties of diethyl ether are provided in Table 13.5. The very low flashpoint, wide flammable range, and relatively high vapor density attest to the flammable nature of diethyl ether. As demonstrated in Figure 13.5, its vapor is heavy and moves downward.

Diethyl ether burns with the production of a virtually invisible, pale blue flame and no accompanying soot.

$$CH_3CH_2{-}O{-}CH_2CH_3(g) \quad + \quad 6O_2(g) \quad \longrightarrow \quad 4CO_2(g) \quad + \quad 5H_2O(g)$$

Diethyl ether   Oxygen   Carbon dioxide   Water

The presence of oxygen in the molecular structure of this ether accounts for the generation of a virtually imperceptible flame.

13.3-C METHYL *tert*-BUTYL ETHER

Methyl tert-*butyl ether* (MTBE) was first introduced in the U.S. fuel market in 1995 to boost the oxygen content of gasoline, since it has an antiknock rating of 116. MTBE was formerly a major organic compound manufactured in the United States.

FIGURE 13.5 In this illustration, a cloth is saturated with diethyl ether and placed at the top end of a trough that has been arranged at a 45° angle. Because the ether has a vapor density of 2.55 (air = 1), it moves *down* the trough, replacing the air, until it reaches the lighted candle. Then, the vapor ignites and flashes up the trough to the fuel source.

$$CH_3-O-\underset{\underset{\displaystyle CH_3}{|}}{\overset{\overset{\displaystyle CH_3}{|}}{C}}-CH_3$$

Methyl *tert*-butyl ether
(*tert*-Butoxymethane)

The use of MTBE as an oxygenate caused gasoline to burn more cleanly and produce less carbon monoxide than the petroleum fuels containing nonoxygenated antiknock agents.

During the late 1990s, the presence of MTBE was identified as a low-level constituent of a drinking water source in California. Its origin was linked with leaking gasoline storage tanks. Even at very low concentrations (40 µg/L), the presence of MTBE causes the water to smell and taste foul. To ensure that drinking water supplies are simultaneously palatable and unlikely to have harmful constituents, EPA phased out its use as a gasoline additive. Today, MTBE is no longer a component of the existing gasoline pool.

Based on a scientific review of available data, EPA concluded that when MTBE is consumed in high doses, it causes cancer. Hence, it is classified as a probable human carcinogen.

13.3-D ALKYL ETHERS OF ETHYLENE GLYCOL

The *alkyl ethers of ethylene glycol* have the following general chemical formula, in which R is an alkyl group and R′ is a hydrogen atom or alkyl group:

$$R-O-CH_2CH_2-O-R'$$

These compounds are commercially known by the tradename *Cellosolve*. Examples of several alkyl ethers of ethylene glycol are noted in Table 13.6, among which are three examples of glycol ether acetates. In the formulas of these latter compounds, $-O-R'$ is the acetate group.

The alkyl monoethylene glycol ethers are widely used as industrial solvents. For example, ethylene glycol monoethyl ether is a common solvent used in the printing industry. Several alkyl glycol ethers are also ingredients in commercial products like surface coatings, inks, adhesives, and cleaning fluids. All are flammable liquids.

Human exposure to the alkyl glycol ethers occurs via of either skin absorption or vapor inhalation. Pregnant women are primarily at risk: Those who inhale their vapors experience an increase in miscarriages. Human exposure to ethylene glycol monomethyl ether has been specifically linked with adverse developmental, neurological, and reproductive abnormalities. When inhaled, its vapor also damages red blood cells and bone marrow, which can potentially cause **anemia**. The combination of these adverse health risks prompted Canada to ban the manufacture, import, sale, and use of this chemical product.

anemia
■ The condition represented by a below-normal reduction in the body's quantity of hemoglobin

SOLVED EXERCISE 13.6

Why is it prudent for pregnant women to avoid exposure to lacquers, varnishes, and latex paints containing an alkyl monoethylene glycol ether?

Solution: Certain alkyl monoethylene glycol ethers are ingredients of surface coatings. Pregnant women are likely to be exposed to their vapors as these substances evaporate during application. Because the inhalation of the vapors of these substances during pregnancy has been linked with the occurrence of reproductive problems in the developing fetus, it is prudent that pregnant women avoid the use of surface coatings and other products in which these substances are contained as ingredients.

TABLE 13.6	Some Monoethylene Glycol Ethers

COMMERCIAL NAME	CHEMICAL NAME	CHE MICAL FORMULA
Butyl Cellosolve	Ethylene glycol monobutyl ether, or 2-butoxyethanol	$C_4H_9-O-CH_2CH_2-OH$
Butyl Cellosolve acetate	Ethylene glycol monobutyl ether acetate	$CH_3-\overset{\displaystyle O}{\overset{\displaystyle \|}{C}}$ $O-CH_2CH_2-O-C_4H_9$
Cellosolve acetate	Ethylene glycol monoethyl ether acetate	$CH_3-\overset{\displaystyle O}{\overset{\displaystyle \|}{C}}$ $O-CH_2CH_2-O-C_2H_5$
Cellosolve solvent	Ethylene glycol monoethyl ether, or 2-ethoxyethanol	$C_2H_5-O-CH_2CH_2-OH$
Dibutyl Cellosolve	Ethylene glycol dibutyl ether, or 1,2-dibutoxyethane	$C_4H_9-O-CH_2CH_2-O-C_4H_9$
Dimethyl Cellosolve	Ethylene glycol dimethyl ether, or 1,2-dimethoxyethane	$CH_3-O-CH_2CH_2-O-CH_3$
Methyl Cellosolve	Ethylene glycol monomethyl ether, or 2-methoxyethanol	$CH_3-O-CH_2CH_2-OH$
Methyl Cellosolve acetate	Ethylene glycol monomethyl ether acetate	$CH_3-\overset{\displaystyle O}{\overset{\displaystyle \|}{C}}$ $O-CH_2CH_2-O-CH_3$

13.3-E TRANSPORTING ETHERS

When shippers offer an ether for transportation, DOT requires them to enter the relevant basic description from Table 13.7 on an accompanying shipping paper. DOT also requires shippers to affix the appropriate label on its packaging. When shippers offer dimethyl ether or ethyl methyl ether for transportation, DOT requires them to post FLAMMABLE GAS labels on the packaging.

When carriers transport any amount of dimethyl ether or ethyl methyl ether, DOT requires them to post FLAMMABLE GAS placards on the bulk packaging or transport vehicle used for shipment. When they transport the ethers other than diallyl ether, dimethyl ether, and ethyl methyl ether in an amount exceeding 1001 lb (454 kg), DOT requires the carriers to display FLAMMABLE placards.

When shippers offer diallyl ether for transportation, DOT requires them to affix FLAMMABLE LIQUID and POISON labels on its packaging. When carriers transport diallyl ether in an amount exceeding 1001 lb (454 kg), DOT requires them to post FLAMMABLE placards on the bulk packaging or transport vehicle used for shipment.

TABLE 13.7 | Basic Descriptions of Some Ethers

ETHER OR GROUP THEREOF	BASIC DESCRIPTION
Anisole	UN2222, Anisole, 3, PGIII
Butyl methyl ether	UN2350, Butyl methyl ether, 3, PGII
Butyl vinyl ether	UN2352, Butyl vinyl ether, stabilized, 3, PGII
Diallyl ether	UN2360, Diallyl ether, 3, (6.1), PGI (Poison)
Dibutyl ethers	UN1149, Dibutyl ethers, 3, PGII
Diethoxymethane	UN2350, Diethoxymethane, 3, PGII
Diethyl ether	UN2373, Diethyl ether, 3, PGII
Diisopropyl ether	UN1155, Diisopropyl ether, 3, PGI
1,1-Dimethoxyethane	UN2377, 1,1-Dimethoxyethane, 3, PGII
1,2-Dimethoxyethane	UN2252, 1,2-Dimethoxyethane, 3, PGII
Dimethyl ether	UN1033, Dimethyl ether, 2.1
Dioxane	UN1165, Dioxane, 3, PGII
Di-*n*-propyl ether	UN2384, Di-*n*-propyl ether, 3, PGII
Divinyl ether	UN1167, Divinyl ether, stabilized, 3, PGI
Ethyl butyl ether	UN1179, Ethyl butyl ether, 3, PGII
Ethyl methyl ether	UN1039, Ethyl methyl ether, 2.1
Ethyl propyl ether	UN2615, Ethyl propyl ether, 3, PGII
Methyl *tert*-butyl ether	UN2398, Methyl *tert*-butyl ether, 3, PGII
Tetrahydrofuran	UN2056, Tetrahydrofuran, 3, PGII

When shippers offer a flammable liquid ether other than those listed in Table 13.7 for transportation, DOT requires them to identify the commodity generically on a shipping paper as either of the following, as relevant:

UN3271, Ethers, n.o.s., 3, PGII
or
UN3271, Ethers, n.o.s, 3, PGIII

In both instances, the basic description includes the name of the specific ether entered parenthetically.

When shippers offer the alkyl ethers of ethylene glycol for transportation, DOT requires them to identify the commodity on a shipping paper as one of the following:

UN1153, Ethylene glycol diethyl ether, 3, PGII
or
UN1153, Ethylene glycol diethyl ether, 3, PGIII
UN1171, Ethylene glycol monoethyl ether, 3, PGIII
UN1172, Ethylene glycol monoethyl ether acetate, 3, PGIII
UN1181, Ethylene glycol monomethyl ether, 3, PGIII
UN1171, Ethylene glycol monomethyl ether acetate, 3, PGIII

DOT also requires them to affix a FLAMMABLE LIQUID label on its packaging.

When carriers transport an alkyl ether of ethylene glycol in an amount exceeding 1001 lb (454 kg), DOT requires them to post FLAMMABLE placards on the bulk packaging or transport vehicle used for shipment.

PERFORMANCE GOALS FOR SECTION 13.4:

- Identify the functional groups that characterize the halogenated ethers.
- Identify the hazardous properties of the halogenated ethers, including the PCDFs, PCDDs, PBDFs, PBDDs, and PBDEs, and identify the ways by which emergency-response personnel are likely to be exposed to them.

13.4 Halogenated Ethers

A **halogenated ether** is a substance whose molecules possess halogen, hydrogen, and oxygen atoms. Two examples of hazardous materials that are chlorinated ethers are *methyl chloromethyl ether* and *bis(chloromethyl) ether*.

halogenated ether
- Any halogenated derivative of an ether

CH_3-O-CH_2Cl
Methyl chloromethyl ether

$ClCH_2-O-CH_2Cl$
Bis(chloromethyl) ether

The prefix *bis* is used when naming the second substance to signify the presence of the chloromethyl group of atoms *twice* in its chemical formula.

Both methyl chloromethyl ether and bis(chloromethyl) ether are colorless, volatile liquids. They have a common flashpoint of $-2.2\,°F$ ($-19\,°C$). This property alone implies that they are highly flammable liquids. The vapors of methyl chloromethyl ether are also poisonous, and both substances are classified as known human carcinogens. Fortunately, neither is produced in substantial quantities. Chemists use them while implementing certain unique synthetic procedures.

The hazardous properties of methyl chloromethyl ether and bis(chloromethyl) ether are evident from their DOT basic descriptions:

UN1239, Methyl chloromethyl ether, 6.1, (3), PGI (Poison - Inhalation Hazard, Zone A)
UN2249, Dichlorodimethyl ether, symmetrical, 6.1, (3), PGI

Certain halogenated ethers are useful as anesthetics. The halogenated ethers known commercially as isoflurane and enflurane, when mixed with air or oxygen, are widely used as general inhalation anesthetics.

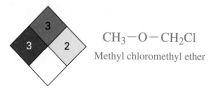

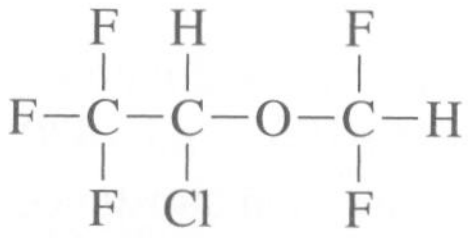

Isoflurane
(1-Chloro-2,2,2-trifluoroethyl difluoromethyl ether)

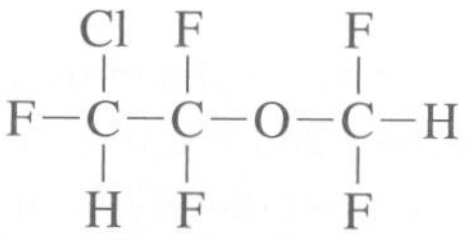

Enflurane
(2-Chloro-1,1,2-trifluoroethyl difluoromethyl ether)

An advantage to using them in a hospital setting is that neither substance is flammable.

13.4-A POLYCHLORINATED DIBENZOFURANS AND POLYCHLORINATED DIBENZO-*p*-DIOXINS

The **polychlorinated dibenzofurans** and **polychlorinated dibenzo-*p*-dioxins** are, respectively, the chlorinated derivatives of the heterocyclic ethers dibenzofuran and dibenzo-*p*-dioxin.

Dibenzofuran   Dibenzo-*p*-dioxin

Their general molecular structures are noted below:

Polychlorinated dibenzofurans   Polychlorinated dibenzo-*p*-dioxins

In these general molecular structures, x and y are integers, and the lines drawn from Cl_x and Cl_y to the benzene rings are bonded at any available position. Using the indicated numbering system, of greatest interest here are the 17 compounds chlorinated at least at the 2, 3, 7, and 8 positions, since epidemiologists associate them with a relatively high degree of toxicity.

There are 135 polychlorinated dibenzofuran isomers and 75 polychlorinated dibenzo-*p*-dioxin isomers. For convenience, we designate them as PCDFs and PCDDs, respectively. Among them, the PCDF and PCDD having the following molecular structures have the highest degree of toxicity:

2,3,7,8-Tetrachlorodibenzofuran   2,3,7,8-Tetrachlorodibenzo-*p*-dioxin

2,3,7,8-Tetrachlorodibenzo-*p*-dioxin is commonly denoted as **dioxin**, or *2,3,7,8-TCDD*, but the name "dioxin" is chemically imprecise.

The PCDFs and PCDDs were never intentionally manufactured as commercial products, but they were generated as unwanted by-products during certain uncontrolled incineration, paper pulp bleaching, and chemical manufacturing operations. The latter included the production and manufacture of trichlorophenol, tetrachlorophenol, pentachlorophenol, and 2,4,5-trichlorophenoxyacetic acid.

Interest in dioxin was first stimulated by public health officials when the substance was identified as a trace contaminant in herbicides formerly used as weed and brush killers. These herbicides had primarily been used by the U.S. military during the Vietnam conflict to defoliate jungle terrain and remove cover for the enemy. Between 1962 and 1970, the U.S. Air Force sprayed approximately 18 million gal ($68,100$ m^3) of one such substance, called *Agent Orange*, on southern Vietnam. This herbicide was a 50:50 mixture of 2,4-dichlorophenoxyacetic acid and dioxin-contaminated 2,4,5-trichlorophenoxyacetic acid. 2,4-Dichlorophenoxyacetic acid and 2,4,5-trichlorophenoxyacetic acid are commonly called 2,4-D and 2,4,5-TP, respectively.

2,4-Dichlorophenoxyacetic acid
(2,4-D)

2,4,5-Trichlorophenoxyacetic acid
(2,4,5-TP)

Military personnel and local civilians who became exposed to Agent Orange subsequently contracted horrific diseases. In addition, their children were at times born with serious birth defects. Researchers linked Agent Orange exposure in Vietnam veterans with the inception of 10 diseases including cancers of the lung, bronchi, larynx, and trachea. Agent Orange exposure has also been linked with spina bifida, a congenital birth defect, in veterans' offspring.

The LD_{50} value for 2,3,7,8-TCDD is 0.0006 mg/kg for guinea pigs, which makes dioxin an extremely toxic substance. In comparison, the LD_{50} value for sodium cyanide (Section 10.10-H), an acknowledged poison, is "only" 15 mg/kg. Research studies conducted by epidemiologists have characterized the high level of toxicity associated with 2,3,7,8-TCDD. These studies show that 2,3,7,8-TCDD functions as an endocrine disrupter by upsetting the natural order of the neurological, immunological, and reproductive systems. The studies also note that humans who have been exposed to dioxin often suffer from skin diseases, muscle dysfunctions, and nervous system disorders, and their offspring can be born with birth defects. Because dioxin causes immune-system disorders, the exposure causes the natural defenses toward contracting other diseases to be weakened.

Even today, exposure to PCDFs and PCDDs is possible. In Section 12.17-A it was noted that these substances are generated during certain electrical-equipment fires; and in Section 14.6 it is noted that they are generated as products of the incomplete combustion of chlorinated plastics such as poly(vinyl chloride) (PVC). Under these circumstances PCDFs and PCDDs can pose a pronounced health risk to emergency responders.

SOLVED EXERCISE 13.7

The chemical analysis of soot samples collected from various locations following a 1981 PCB transformer fire within a Binghamton, New York, office building revealed the presence of 2,3,7,8-TCDD at concentrations ranging from 0.6 to 2.8 µg/g. How is the detection of this substance most logically explained, and why is its presence at a fire scene an issue of concern to the on-duty firefighters?

Solution: During PCB transformer fires, PCBs undergo incomplete combustion reactions in which PCDFs and PCDDs are formed. This production explains the presence of 2,3,7,8-TCDD in soot samples collected from an office building in which a PCB transformer fire occurred. The formation of 2,3,7,8-TCDD is worrisome for emergency responders who fight such fires, since exposure to this substance has been linked with the onset of cancer.

The U.S. Department of Health and Human Services has ranked dioxin as a known human carcinogen and the remaining PCDDs and PCDFs whose molecules have chlorine atoms in the 2, 3, 7, and 8 positions as probable human carcinogens. To avoid adverse health effects from dioxin exposure, EPA designates 6.4 femtograms per kilogram of body weight as the acceptable daily dosage for humans. (One femtogram is one quadrillionth (10^{-15}) of a gram.)

13.4-B POLYBROMINATED DIBENZOFURANS AND POLYBROMINATED DIBENZO-*p*-DIOXINS

The *polybrominated dibenzofurans* and *polybromated dibenzo-p-dioxins* (PBDFs) and (PBDDs) are the polybrominated derivatives of dibenzofuran and dibenzo-*p*-dioxin. They were once incorporated into certain plastic products to serve as **fire retardants**.

Polybrominated dibenzofurans

Polybrominated dibenzo-*p*-dioxins

Here again, x and y are integers, and the lines drawn from Br_x and Br_y to the benzene rings are bonded at any available position. Like the PCDFs and PCDDs, there are 135 PBDFs and 75 PBDDs.

When they are exposed to heat, the PBDFs and PBDDs undergo thermal decomposition and produce bromine atoms. These bromine atoms consume the free radicals produced when polymers decompose and thereby retard the development of fire (Section 5.13).

Research studies using animals have revealed that the PBDFs and PBDDs are toxic substances, since they disrupt the normal biological function of the thyroid and sex hormones. They also damage developing brains, impair motor skills and mental abilities, weaken the immune system, and alter bone structure. Although studies have not been performed on humans, scientists fear that exposure to the PBDFs and PBDDs negatively affects the human organism in a similar fashion.

Like the PCDFs and PCDDs, the compounds associated with elevated toxicities are those whose molecules are brominated in at least the 2, 3, 7, and 8 positions. These PBDFs and PBDDs are regarded as probable human carcinogens.

It is assumed that emergency responders are at risk from exposure to polybrominated dibenzofurans and polybrominated dibenzo-*p*-dioxins. During fires, plastics containing brominated fire retardants undergo incomplete combustion and release low concentrations of PBDFs and PBDDs to the surroundings. The inhalation of these substances can subject firefighters and others to a health risk.

13.4-C POLYBROMINATED DIPHENYL ETHERS

The *polybrominated diphenyl ethers* (PBDEs) are the polybrominated derivatives of diphenyl ether, whose chemical formula is $C_6H_5-O-C_6H_5$.

Polybrominated diphenyl ethers
(PBDEs)

In this general molecular structure, x and y are integers less than 5. There are 209 possible isomers of the PBDEs.

Like the polybrominated dibenzofurans and dibenzo-*p*-dioxins, the PBDEs were formerly manufactured for incorporation as fire retardants into the polymers used to make carpets and carpet padding and automobile and furniture cushioning. When used commercially, they were often manufactured as mixtures of the penta-, octa-, and decabromodiphenyl ether isomers. The pentabromodiphenyl ether isomers, collectively called the *penta- group*, were once formulated into the polyurethane foam used in upholstery (Section 14.8-A) to make it fire resistant. Also popular were the octa- and decabromodiphenyl ether isomers called the *octa-* and *deca- groups*, respectively, which were once used in the housings for business machines and electrical appliances.

U.S. production and manufacture of the PBDEs ceased in 2004 when the presence of these compounds in human serum and breast milk was revealed. Nonetheless, despite their nonproduction, stockpiles of these compounds were formulated into U.S. products after 2004. Furthermore, carpets, blankets, upholstery, fabrics, and other items manufactured prior to 2004 remain in U.S. homes. It is safe to assume that firefighters are exposed to the PBDEs when they battle ordinary house fires.

Animal exposure to PBDEs has been shown to cause the same adverse health effects previously noted for exposure to PBDFs and PBDDs. The available information suggests that PBDEs may cause neurodevelopmental problems, endocrine disruption, and cancer. Epidemiologists rank the PBDEs as probable human carcinogens. The research studies linking these adverse health effects with PBDE exposure in humans are ongoing.

EPA has shown that exposure to PBDE-laden dust was the key means by which household cats acquired high levels of certain PBDEs. The cats became stricken with overactive thyroids. Although hyperthyroidism is treatable in both cats and humans, the health concern is that youngsters are just as likely as cats to be exposed to the PBDEs in normal household dust. Youngsters are more likely to be exposed to PBDEs than adults, since they crawl on carpeted floors and chew blankets. Prudent parents can reduce the risk of child exposure to PBDEs by frequently vacuuming carpets, laundering blankets, and cleaning other items containing these flame retardants.

The PBDEs are now considered major environmental contaminants not only in the general U.S. population but also in animals. PBDEs enter the environment when the products in which they were incorporated are either incinerated as components of municipal waste or buried in landfills. The impact of their environmental presence is especially worrisome for the survival of the animals at the top of the food chain, like Arctic polar bears. PBDE concentrations are more elevated in polar bears than in ringed seals or Arctic foxes, for example, since the compounds accumulate as they pass from prey to predator. Whether these animals survive in the wild may well be connected with international efforts taken to continue the ban against the use of PBDEs and other brominated fire retardants.

PERFORMANCE GOALS FOR SECTION 13.5:

- Identify the functional group that characterizes aldehydes and ketones.
- Identify the primary risks associated with exposure to formaldehyde and acetone.
- Identify the industries that use aldehydes and ketones.
- Describe the means shippers and carriers use to inform emergency responders of the hazards associated with encountering aldehydes or ketones during transportation mishaps.

13.5 Aldehydes and Ketones

Aldehydes and **ketones** are organic compounds whose molecules contain the **carbonyl group** of atoms ($\overset{\backslash}{\underset{/}{C}}{=}O$). In aldehydes, the carbonyl group is located at the end of a chain of carbon atoms, whereas in ketones it is located at a nonterminal position within the chain. Thus, aldehydes and ketones have the following general chemical formulas, where R and R′ are arbitrary alkyl or aryl groups.

$$R-\overset{\overset{O}{\parallel}}{\underset{\underset{H}{\backslash}}{C}} \qquad R-\overset{}{\underset{\underset{O}{\parallel}}{C}}-R'$$

An aldehyde   A ketone

aldehyde
- An organic compound each of whose molecules is composed of an alkyl or aryl group, a carbonyl group, and a hydrogen atom

ketone
- An organic compound each of whose molecules is composed of a carbonyl group bonded to alkyl and/or aryl groups

carbonyl group
- The ($\overset{\backslash}{\underset{/}{C}}{=}O$) functional group, common to both aldehydes and ketones

Naming Aldehydes

In the common system, an aldehyde is named by combining the word "aldehyde" with the prefix of the name of the acid to which it is oxidized. The aldehydes having one, two, three, and four carbon atoms per molecule are named formaldehyde, acetaldehyde, propionaldehyde, and *n*-butyraldehyde, respectively, since they oxidize to formic acid, acetic acid, propionic acid, and *n*-butyric acid, respectively.

In the IUPAC system, an aldehyde is named by replacing the *-e* in the name of the corresponding hydrocarbon with *-al*. The position of the carbonyl carbon atom does not have to be designated, since it is always located on a terminal carbon atom. The aldehydes having one, two, three, and four carbon atoms per molecule are then named methanal, ethanal, propanal, and butanal, respectively, by replacing the *-e* in methane, ethane, propane, and butane with *-al*.

| Formaldehyde | Acetaldehyde | Propionaldehyde | *n*-Butyraldehyde |
| (Methanal) | (Ethanal) | (Propanal) | (Butanal) |

The simplest aromatic aldehyde is called *benzaldehyde*. Its chemical formula is C_6H_5-CHO.

Benzaldehyde

Naming Ketones

In the common system, a ketone is named by alphabetically identifying the alkyl or aryl groups of which the substance is composed. Thus, the compound having the formula $CH_3-C-CH_2CH_2CH_3$ is properly named methyl *n*-propyl ketone.

In the IUPAC system, a ketone is named by replacing the *-e* in the name of the corresponding hydrocarbon with *-one*; then, the chain of continuous carbon atoms is numbered so as to assign the *smallest possible number* to the carbonyl group. Hence, the IUPAC name for methyl *n*-propyl ketone is 2-pentanone.

$$\overset{1}{C}H_3-\overset{2}{C}-\overset{3}{C}H_2\overset{4}{C}H_2\overset{5}{C}H_3$$

Methyl *n*-propyl ketone
(2-Pentanone)

The ketone whose formula is $C_6H_5-C-CH_3$ is named methyl phenyl ketone, but it is more commonly known as *acetophenone*.

Methyl phenyl ketone
(Acetophenone)

Identify the functional group in the compound having the following condensed formula, and name the compound having this formula.

$$CH_3CH_2-\underset{\underset{O}{\|}}{C}-\underset{\underset{CH_3}{|}}{CH}CH_2CH_3$$

Solution: By examination of Table 13.1, we determine that the functional group represented as $C=O$ is the carbonyl group. When it bonded to a carbon chain at a nonterminal position, the carbonyl group is characteristic of the organic compounds called ketones. We number each carbon atom in the longest continuous chain of carbon atoms that contains the carbonyl group from left to right. Then, we see that this particular ketone is a "hexanone." This name is derived by replacing the -*e* with -*one* in hexane, the alkane having six carbon atoms per molecule.

$$\overset{1\quad 2\quad\ \ 3\ \ \,4\ \ 5\ \ \,6}{CH_3CH_2-\underset{\underset{O}{\|}}{C}-\underset{\underset{CH_3}{|}}{CH}CH_2CH_3}$$

The correct IUPAC name of this compound is 4-methyl-3-hexanone. In the common system, the compound is named ethyl *sec*-butyl ketone.

It is incorrect to number the longest chain of continuous carbon atoms from right to left as follows, since a higher number (4) would then be assigned to the carbonyl group.

$$\overset{6\quad 5\quad\ \ 4\ \ \,3\ \ 2\ \ \,1}{CH_3CH_2-\underset{\underset{O}{\|}}{C}-\underset{\underset{CH_3}{|}}{CH}CH_2CH_3}$$

13.5-A FORMALDEHYDE

The simplest aldehyde is *formaldehyde*, or *methanal*. Its chemical formula is HCHO.

$$H-\overset{\overset{O}{\|}}{\underset{\underset{H}{\backslash}}{C}}$$

Formaldehyde is a colorless, flammable gas at room conditions that has a pungent, highly irritating odor that is detectable at concentrations as low as 1 ppm. Some of its important physical properties are provided in Table 13.8.

In the presence of water vapor, formaldehyde produces a solid substance called *paraformaldehyde*. This substance has an indefinite composition. Its chemical formula is $HO(CH_2O)_nH$, where n ranges from approximately 8 to 100.

$$n\,H-\overset{\overset{O}{\|}}{\underset{\underset{H}{\backslash}}{C}}(g)\ +\ H_2O(g)\ \longrightarrow\ HO-[-(CH_2)_n-O-]-H(s)$$

| Formaldehyde | Water | Paraformaldehyde |

self-reactive material
■ For purposes of DOT regulations, a material that is thermally unstable and can undergo a strongly exothermic decomposition even without the participation of atmospheric oxygen

Formaldehyde is a **self-reactive material**. For this reason, formaldehyde gas is unavailable commercially. However, formaldehyde is very soluble in water and alcohol, producing colorless, corrosive solutions whose properties vary with the formaldehyde concentration. These solutions are widely available commercially.

TABLE 13.8	Physical Properties of Formaldehyde
Melting point	−134 °F (−92 °C)
Boiling point	−3 °F (−19 °C)
Specific gravity at 68 °F (20 °C)	0.82
Vapor density (air = 1)	1.08
Vapor pressure at 68 °F (20 °C)	1.3 mmHg
Flashpoint	
Autoignition point	806 °F (430 °C)
Lower explosive limit	7.0%
Upper explosive limit	73%

formalin

■ A solution of formaldehyde in water and/or methanol

The solution called **formalin** is an aqueous solution contain approximately 40% water and 5% to 12% methanol. It is widely used as a disinfecting, sterilizing, and embalming agent. A formalin solution containing 37% water and 15% methanol flashes at 122 °F (50 °C), whereas an aqueous solution of formaldehyde without methanol flashes at 185 °F (85 °C) and is very corrosive.

For industrial use, formaldehyde may be generated as the gas from *paraformaldehyde* and sym-*trioxane*.

Paraformaldehyde consists of several formaldehyde units bonded one to another in a short chain. Its formula is $HO-[-(CH_2)_n-O-]-H$, where n equals 8 or more. When it is mildly heated, paraformaldehyde decomposes into formaldehyde and water.

$$HO-[-(CH_2)_n-O-]-H(s) \longrightarrow nH-\overset{\overset{O}{\|}}{\underset{H}{C}}(g) + H_2O(g)$$

Paraformaldehyde   Formaldehyde Water

sym-Trioxane is a combustible solid whose molecular structure resembles three joined formaldehyde molecules. When mixed with a strong acid, sym-trioxane decomposes into formaldehyde.

$$(s) \longrightarrow 3H-\overset{\overset{O}{\|}}{\underset{H}{C}}(g)$$

sym-Trioxane   Formaldehyde

Formaldehyde is also unintentionally released into the air during certain combustion processes. Aldehydes are produced as incomplete combustion products when organic substances burn within confined areas. Materials that produce formaldehyde on burning include tobacco, wood, diesel fuel, kerosene, natural gas, and ethanol. Researchers have identified aldehydes as the major class of organic compounds contained in diesel-engine exhaust, with formaldehyde being the most abundant among them.

Most persons first encounter formaldehyde while studying the anatomy of frogs and fetal pigs during a high school biology class. Formalin solutions were formerly used to preserve biological specimens. They are also used by morticians as embalming agents. The bulk of the formaldehyde produced in the United States is used as a raw material for production of phenol–formaldehyde and urea–formaldehyde plastics (Section 14.2), from which building-construction materials are manufactured.

Prolonged exposure to formaldehyde occurs inside residential and commercial buildings constructed, insulated, and furnished with these materials, especially when these buildings are newly constructed and infrequently vented. Certain components of mobile homes are constructed from materials made of phenol–formaldehyde and/or urea–formaldehyde plastics. Consequently, the presence of formaldehyde may be especially prominent inside them. Formaldehyde leaches from these plastics, some specific sources of which are the following:

- Kitchen and bathroom cabinetry and furniture
- Glues and adhesives that were used to bind wood pieces and fragments into particleboard, hardwood plywood, and fiberboard composites
- Foam insulation
- Synthetic padding and carpets

Figure 13.6 shows that low concentrations of formaldehyde often leach from these sources into the surrounding environment. These concentrations become elevated when the doors and windows of a mobile home are shut for extensive periods. The rate of formaldehyde release decreases as the age of these products increases and is dependent on the prevailing temperature and humidity. Formaldehyde is released into the air at its maximum rate when the temperature and humidity are elevated.

The concern about formaldehyde exposure is linked with the ill effects that it causes. Formaldehyde is toxic by both skin contact and inhalation. Individuals who inhale formaldehyde

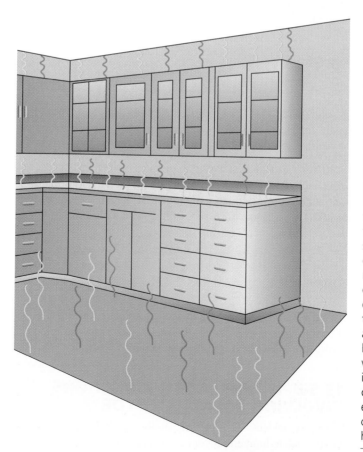

FIGURE 13.6
Formaldehyde gas escapes very slowly from the following construction sources: kitchen and bathroom cabinetry and furniture; glue and adhesives used to bind wood particles in particleboard, hardwood plywood, and fiberboard; foam insulation; and synthetic padding and carpeting. When a mobile home is newly constructed, unvented, or heavily insulated, the interior air contains an elevated formaldehyde concentration, which when inhaled by its occupants, may cause them to experience respiratory discomfort and other health problems.

TABLE 13.9	Inhalation Effects of Formaldehyde in Humans

FORMALDEHYDE CONCENTRATION IN AIR (ppm)	SYMPTOMS
1	Odor threshold
2–3	Mild irritation of the eyes, nose, throat, and windpipe
4–5	Mild lacrimation (tearing)
>10	Difficult breathing; severe burning sensations in the nose, throat, and windpipe; wheezing; coughing; nausea; skin rashes; severe lacrimation
>50	Possible pulmonary edema, pneumonitis, and death

experience the combination of signs and symptoms listed in Table 13.9. In air at a concentration of only 0.1 to 5 ppm, formaldehyde causes irritation of the eyes and bronchial passageways. When concentrations in excess of 10 ppm are inhaled, formaldehyde causes problems that include mild irritation of the eyes, nose, throat, and bronchial passageways. This exposure also produces localized blisters within the nose, throat, and lungs and causes allergic dermatitis in susceptible individuals.

SOLVED EXERCISE 13.9

A solution of 2% to 5% phenol in ethanol with antimicrobial agents has replaced the use of formalin for preserving biological specimens. What advantage does this replacement provide the students who study these specimens?

Solution: Students studying biological specimens preserved in formalin were exposed to formaldehyde as it vaporized from the preservative. Formaldehyde vapor is an eye and skin irritant and a known human carcinogen. Although phenol is corrosive to skin, it does not cause cancer. By replacing formalin with a phenol solution in ethanol, students are prevented from exposure to a cancer-causing substance.

In research studies conducted on animals, the inhalation of formaldehyde gas has been shown to cause nasal cancer. Based on these results, epidemiologists rank formaldehyde as a probable human carcinogen by inhalation.

13.5-B WORKPLACE REGULATIONS INVOLVING FORMALDEHYDE

When formaldehyde is present in the workplace, OSHA requires employers at 29 C.F.R. §1910.1048 to limit employee exposure to a maximum concentration of 0.75 ppm in air, averaged over an 8-hr workday, with a ceiling limit of 2 ppm. The NIOSH recommended exposure limit is 0.016 ppm with a ceiling limit of 0.1 ppm, and the short-term exposure limit is 2 ppm over a 15-min period. These exposure limits are so low that those who routinely use formalin have been encouraged to select an alternative substance that accomplishes the same task. Morticians, for example, still use formalin as an embalming agent, but many have switched to using glutaraldehyde as a preferred substitute.

$$\underset{H}{\overset{O}{\underset{\diagup}{\overset{\diagdown}{C}}}} - CH_2CH_2CH_2 - \underset{H}{\overset{O}{\underset{\diagdown}{\overset{\diagup}{C}}}}$$

Glutaraldehyde
(Pentanedial)

As noted, the glutaraldehyde molecule has two carbonyl groups that are bonded to the two terminal carbon atoms in a chain of five carbon atoms. Glutaraldehyde is a liquid with a lower acute toxicity than formaldehyde. Although exposure to glutaraldehyde causes asthma and skin irritation, research studies do not link it with the onset of cancer.

At 29 C.F.R. §1910.1048(e)(1), OSHA requires employers to establish regulated areas where the concentration of airborne formaldehyde exceeds the time-weighted average value or the short-term exposure limit and to post all entrance and accessways with warning signs that bear the following information:

> **DANGER**
>
> **FORMALDEHYDE**
> **IRRITANT AND POTENTIAL CANCER**
> **HAZARD**
> **AUTHORIZED PERSONNEL ONLY**

13.5-C ACETONE

The simplest ketone is called *acetone*, *propanone*, or *dimethyl ketone*. Its chemical formula is $(CH_3)_2-C=O$.

$$CH_3 - \underset{\overset{\|}{O}}{C} - CH_3$$

Acetone

It is a colorless, water-soluble, and highly volatile liquid with a sweet odor.

The data in Table 13.10 indicate that acetone and other simple ketones pose the risk of fire and explosion. When they burn, the products of combustion are carbon dioxide and water vapor.

$$(CH_3)_2-C=O(g) \quad + \quad 4O_2(g) \quad \longrightarrow \quad 3CO_2(g) \quad + \quad 3H_2O(g)$$

Acetone   Oxygen   Carbon dioxide   Water

In the chemical industry, acetone is manufactured by several methods. As noted in Section 13.2-H, acetone is produced together with phenol by the decomposition of cumene hydroperoxide. It is also produced together with hydrogen peroxide by the oxidation of isopropanol. The most common method of production, however, involves the catalytic dehydrogenation of isopropanol.

$$(CH_3)_2-CH-OH(g) \quad \longrightarrow \quad CH_3 - \underset{\overset{\|}{O}}{C} - CH_3(g) \quad + \quad H_2(g)$$

Isopropanol   Acetone   Hydrogen

	ACETONE	METHYL ETHYL KETONE (MEK)	METHYL ISOBUTYL KETONE (MIBK)
TABLE 13.10 Physical Properties of Acetone, Methy Ethyl Ketone, and Methyl Isobutyl Ketone			
Melting point	−137 °F (−94 °C)	−124 °F (−87 °C)	−121 °F (−85 °C)
Boiling point	133 °F (56 °C)	175 °F (80 °C)	243 °F (117 °C)
Specific gravity at 68 °F (20 °C)	0.79	0.81	0.80
Vapor density (air = 1)	2.0	2.5	3.5
Vapor pressure at 68 °F (20 °C)	181.7 mmHg	71 mmHg	15.7 mmHg
Flashpoint	0 °F (−18 °C)	20 °F (−7 °C)	73 °F (23 °C)
Autoignition point	1000 °F (538 °C)	960 °F (515 °C)	860 °F (460 °C)
Lower explosive limit	3%	2%	1.4%
Upper explosive limit	13%	10%	7.5%

Acetone is largely utilized commercially as a solvent in varnishes, lacquers, paints, fingernail polish remover, and acetylene (Section 12.9). It is also used as a raw material for the manufacture of methyl methacrylate (Section 14.6-E), methyl isobutyl ketone (Section 13.5-D), and other substances. A nonbulk quantity of acetone is frequently stored within laboratory, research, and industrial settings in a metal container like that shown in Figure 13.7.

FIGURE 13.7 A 5-gal (19-L) metal container is often used to store acetone for future use in laboratory, research, and industrial settings. When it is transported, DOT requires the shipper to affix a FLAMMABLE LIQUID warning label to the container. (*Courtesy of Mallinckrodt Baker, Inc., Phillipsburg, New Jersey.*)

13.5-D OTHER KETONES

Other commercially important ketones have physical properties similar to those of acetone. Two are ethyl methyl ketone and methyl isobutyl ketone.

$$CH_3-\overset{\underset{\displaystyle \|}{O}}{C}-CH_2CH_3$$

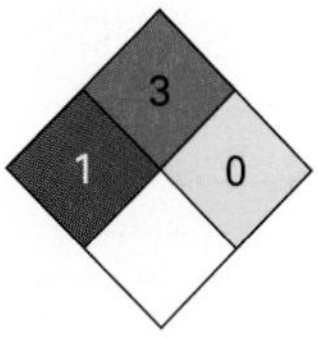

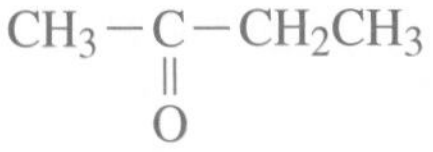

Ethyl methyl ketone
Methyl ethyl ketone
(2-Butanone)

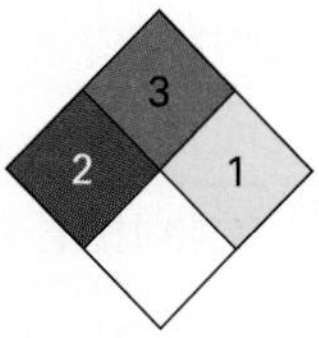

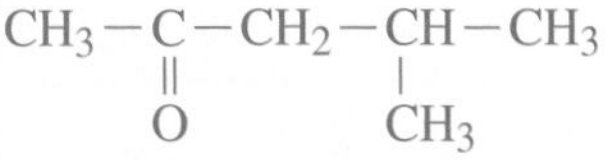

$$CH_3-\overset{\underset{\displaystyle \|}{O}}{C}-CH_2-\overset{\underset{\displaystyle CH_3}{|}}{CH}-CH_3$$

Isobutyl methyl ketone
Methyl isobutyl ketone
(4-Methyl-2-pentanone)

In commerce, they are known more widely by their acronyms, MEK and MIBK, respectively. Both are flammable, water-soluble, highly volatile liquids at room conditions. They were formerly encountered as constituents of solvent mixtures, especially solvent-based paints.

Several commercially important ketones containing multiple carbonyl groups are also known. The compound called *diacetyl* is a liquid whose molecules have two carbonyl groups. They are bonded to the two nonterminal carbon atoms making up a chain of four carbon atoms. Its IUPAC name is *butanedione*. (Note that the *-e* in the name of the alkane is not dropped when naming a dione.)

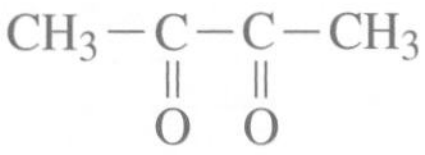

$$CH_3-\overset{\underset{\displaystyle O}{\|}}{C}-\overset{\underset{\displaystyle O}{\|}}{C}-CH_3$$

Diacetyl
(Butanedione)

Diacetyl was once used as a butter-flavoring agent and an aroma carrier in food products like microwave popcorn until it became apparent that its inhalation could cause the lung disease *bronchiolitis obliterans*. Popcorn manufacturers voluntarily eliminated it from microwave popcorn in 2008, but its use as a butter-flavoring agent in other foods continues.

13.5-E TRANSPORTING ALDEHYDES AND KETONES

When shippers offer an aldehyde or ketone for transportation, DOT requires them to enter the relevant basic description from Table 13.11 on an accompanying shipping paper. When the name of the aldehyde or ketone is not listed in the Hazardous Materials Table, its basic description is identified generically as follows:

UN1989, Aldehydes, n.o.s., 3, PGI
or
UN1989, Aldehydes, n.o.s., 3, PGII
or
UN1989, Aldehydes, n.o.s., 3, PGIII
UN1988, Aldehydes, flammable, toxic, n.o.s., 3, (6.1), PGI
or
UN1988, Aldehydes, flammable, toxic, n.o.s., 3, (6.1), PGII
UN1224, Ketones, liquid, n.o.s., 3, PGI
or
UN1224, Ketones, liquid, n.o.s., 3, PGII
or
UN1224, Ketones, liquid, n.o.s., 3, PGIII

DOT requires shippers to include the name of the aldehyde or ketone parenthetically in the basic description. Except when they offer benzaldehyde or paraformaldehyde for transportation, DOT also requires shippers to affix a FLAMMABLE LIQUID label on the packaging. When they

TABLE 13.11 — Basic Descriptions of Some Aldehydes and Ketones

ALDEHYDE, KETONE, OR GROUP THEREOF	BASIC DESCRIPTION
Acetaldehyde	UN1089, Acetaldehyde, 3, PGI
Acetone	UN1090, Acetone, 3, PGII
n-Amyl methyl ketone	UN1110, n-Amyl methyl ketone, 3, PGIII
Benzaldehyde	UN1990, Benzaldehyde, 9, PGIII
Butyraldehyde	UN1129, Butyraldehyde, 3, PGII
Cyclohexanone	UN1915, Cyclohexanone, 3, PGIII
Diacetyl	UN2346, Butanedione, 3, PGII (Marine Pollutant)
Diethyl ketone	UN1156, Diethyl ketone, 3, PGII
Diisobutylketone	UN1157, Diisobutylketone, 3, PGIII
Ethyl amyl ketone	UN2271, Ethyl amyl ketone, 3, PGIII
Ethyl methyl ketone	UN1193, Ethyl methyl ketone, 3, PGII
2-Ethylbutyraldehyde	UN1178, 2-Ethylbutyraldehyde, 3, UN1178, PGII
Formaldehyde solutions (flammable)	UN1198, Formaldehyde solutions, flammable, 3, PGIII
Formaldehyde solutions (>25% formaldehyde)	UN2209, Formaldehyde solutions, 8, PGIII
n-Heptaldehyde	UN3056, n-Heptaldehyde, 3, PGIII
Hexaldehyde	UN1207, Hexaldehyde, 3, PGIII
Isobutyraldehyde	UN2045, Isobutyraldehyde, 3, PGII
2-Methylbutanal	UN3371, 2-Methylbutanal, 3, PGII
Methylcyclohexanone	UN2997, Methylcyclohexanone, 3, PGII
Methyl isobutyl ketone	UN1245, Methyl isobutyl ketone, 3, PGII
Methyl isopropyl ketone	UN2397, 3-Methylbutan-2-one, 3, PGIII
Octyl aldehydes	UN1191, Octyl aldehydes, 3, PGIII
Paraformaldehyde	UN2213, Paraformaldehyde, 4.1, PGIII
Pentanal	UN2058, Valeraldehyde, 3, UN2058, PGII
Propionaldehyde	UN1275, Propionaldehyde, 3, PGII

offer benzaldehyde or paraformaldehyde for transportation, DOT requires shippers to affix CLASS 9 and FLAMMABLE SOLID labels to their respective packaging.

When carriers transport most aldehydes or ketones in an amount exceeding 1001 lb (454 kg), DOT requires them to post FLAMMABLE placards on the bulk packaging or transport vehicle used for shipment. However, when carriers transport a bulk quantity of benzaldehyde, DOT requires the identification number 1990 to be displayed on orange panels or on white square-on-point diamonds. DOT does not require CLASS 9 placards to be posted.

When carriers transport paraformaldehyde in an amount exceeding 1001 lb (454 kg), DOT requires them to pose FLAMMABLE SOLID placards on the bulk packaging or transport vehicle used for shipment.

PERFORMANCE GOALS FOR SECTION 13.6:

- Identify the functional group that characterizes organic acids.
- Identify the primary risks associated with exposure to the simple organic acids.
- Memorize the names of the organic acids having from one to three carbon atoms per molecule.
- Identify the industries that use organic acids.
- Describe the means shippers and carriers use to inform emergency responders of the hazards associated with encountering organic acids during transportation mishaps.

13.6 Organic Acids

Organic acids are compounds containing the carboxyl group, $-\overset{\overset{\displaystyle O}{\|}}{\underset{\displaystyle OH}{C}}$ or $-COOH$.

They are also called **carboxylic acids**. The general chemical formula of a carboxylic acid is $R-COOH$, where R is an arbitrary alkyl or aryl group.

carboxylic acid
■ An organic compound containing the carboxyl group (formic acid, acetic acid, etc.)

The simple organic acids are identified by their common historical names: formic acid, acetic acid, propionic acid, and so forth. In the IUPAC system, they are named by replacing the final *-e* in the name of the corresponding alkane with *-oic acid*. When necessary, the position of a substituent along a chain of carbon atoms is designated by a number. The carbon atom in the carboxyl group is assigned the number 1.

The organic acids having from one to four carbon atoms per molecule are most commonly encountered. These acids are named as follows:

Formic acid (Methanoic acid)   Acetic acid (Ethanoic acid)   Propionic acid (Propanoic acid)

n-Butyric acid (Butanoic acid)   Isobutyric acid (2-Methylpropanoic acid)

They are colorless, water-soluble liquids with characteristic odors. Formic acid, acetic acid, and propionic acid have pungent but not disagreeable odors, whereas the odors of butyric acid and isobutyric acid are highly disagreeable.

As reflected by the data in Table 13.12, organic acids are inherently corrosive liquids. They are weak acids (Section 8.1-A), meaning that they only partially ionize when dissolved in water. Although corrosiveness is considered their primary hazard, concentrated isobutyric acid also poses the risk of fire and explosion. Furthermore, concentrated formic acid and acetic acid easily burn at elevated temperatures. When they burn, these acids produce the products of complete

TABLE 13.12 | Physical Properties of the Simple Organic Acids

	FORMIC ACID	ACETIC ACID	PROPIONIC ACID
Melting point	47 °F (8.2 °C)	61 °F (17 °C)	−8 °F (−22 °C)
Boiling point	101 °F (213 °C)	244 °F (118 °C)	286 °F (141 °C)
Specific gravity at 68 °F (20 °C)	1.22	1.05	0.99
Vapor density (air = 1)	1.59	2.1	2.56
Vapor pressure at 68 °F (20 °C)	44.8 mmHg	11 mmHg	3 mmHg
Flashpoint	156 °F (69 °C)	109 °F (43 °C)	130 °F (54 °C)
Autoignition point	1114 °F (601 °C)	800 °F (426 °C)	955 °F (513 °C)
Lower explosive limit	18%	4%	2.9%
Upper explosive limit	57%	16%	
	n-BUTYRIC ACID	ISOBUTYRIC ACID	
Melting point	17.5 °F (−7.9 °C)	−51 °F (−47 °C)	
Boiling point	327 °F (164 °C)	309 °F (153 °C)	
Specific gravity at 68 °F (20 °C)	0.96	0.949	
Vapor density (air = 1)	3.0	3.04	
Vapor pressure at 68 °F (20 °C)	0.84 mmHg	1.5 mmHg	
Flashpoint	151 °F (66 °C)	132 °F (55 °C)	
Autoignition point	846 °F (452 °C)	824 °F (439 °C)	
Lower explosive limit	2%	2%	
Upper explosive limit	10%	10%	

combustion. For instance, it was noted in Section 8.12-B that carbon dioxide and water vapor are produced when acetic acid burns. When the organic acids are diluted with water, they do not vaporize. Then, the acids are nonflammable.

The simplest aromatic acid is named *benzoic acid*. It is a white solid having the formula C_6H_5-COOH. Benzoic acid and its sodium salt have the structures shown:

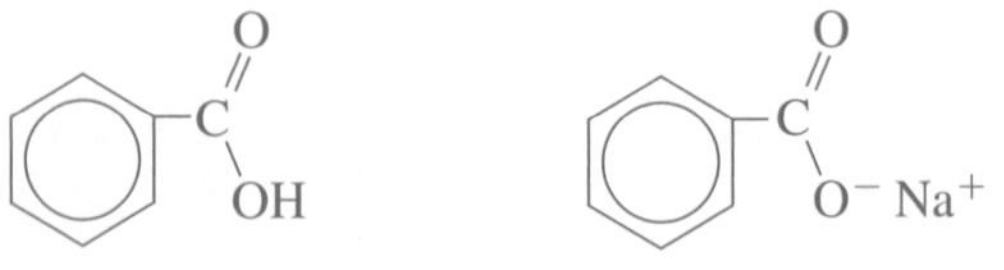

Benzoic acid   Sodium benzoate

Sodium benzoate is identified on the labels of many food containers to indicate its presence as a preservative and antimicrobial agent.

13.6-A THE FORMYL, ACETYL, AND BENZOYL GROUPS

The *formyl*, *acetyl*, and *benzoyl* groups are derived from formic acid, acetic acid, and benzoic acid, respectively. Their molecular structures are represented as follows:

Formyl group   Acetyl group   Benzoyl group

These groups are encountered in many compounds. Acetyl chloride (Section 9.9-C) and benzoyl chloride, for instance, are the compounds having the formulas CH_3-COCl and C_6H_5-COCl, respectively.

Acetyl chloride   Benzoyl chloride

Formyl chloride is unknown. When chemists attempt to prepare it, carbon monoxide and hydrogen chloride are produced instead. The acetyl and benzoyl groups are common components of peroxo-organic compounds (Section 13.9).

13.6-B PERFLUOROOCTANOIC ACID

Perfluorooctanoic acid is best known by its acronym PFOA. It is the totally fluorinated derivative of octanoic acid.

Perfluorooctanoic acid

Two other compounds are often associated with the acronym PFOA: ammonium perfluorooctanoate, which is also denoted as APFO; and perfluorooctane sulfonate, which is also denoted as PFOS.

Ammonium perfluorooctanoate   Perfluorooctane sulfonate

Although the use of the term PFOA can refer to any of these three compounds, it is more generally used to denote perfluorooctanoic acid.

PFOA was formerly used as an essential processing aid during the manufacture of fluoropolymers like *Teflon* (Section 14.9-A) and as a surfactant in aqueous-film-forming foam fire extinguishers (Section 5.11-B). The PFOAs were also widely used to impart fire-resistance and oil-, stain-, grease-, and water-repellence to carpets, textiles, and paper. They were also used to make a low-friction coating for bearings, gears, and weapons components. PFOA was used to provide nonstick surfaces on cookware and waterproof, breathable membranes for clothing. Consequently, the PFOAs have been widely used for decades in dozens of commonly used commercial products.

There is no evidence that normal use of nonstick Teflon cookware is harmful. Nonetheless, scientists discovered that when Teflon is heated to temperatures greater than approximately 1000 °F (600 °C), its coating decomposes and releases perfluorooctanoic acid. Exposure to PFOA may be harmful to individuals who heat Teflon cookware to these very high temperatures.

In the late 1990s, EPA first raised concern over the fact that PFOAs had been identified on a widespread basis in samples of blood withdrawn from virtually all "average" Americans. Approximately 95% of the American population is believed to have at least a trace of the PFOAs in their blood. EPA also noted that these compounds persist in the environment and bioaccumulate within the fatty tissues of animals.

Initial studies indicate that animal exposure to PFOA causes developmental problems, immune-system damage, and other adverse health effects. This finding was so startling that many major American PFOA manufacturers voluntarily ceased their production of this substance. In 2005, EPA reported that additional studies performed on laboratory animals linked exposure to PFOA with breast, testes, pancreas, and liver cancer. Based on the results from these studies, PFOA is now regarded as a probable human carcinogen. EPA has asked PFOA manufacturers to phase out the production of this substance.

13.6-C TRANSPORTING ORGANIC ACIDS

When shippers offer an organic acid for transportation, DOT requires them to enter the relevant description from Table 13.13 on a shipping paper. When shippers offer a liquid organic acid whose name is not listed in Table 13.13 for transportation, DOT requires them to identify the commodity generically on a shipping paper as one of the following:

> **UN3265, Corrosive liquid, acidic, organic, n.o.s., 8, PGI**
> *or*
> **UN3265, Corrosive liquid, acidic, organic, n.o.s., 8, PGII**
> *or*
> **UN3265, Corrosive liquid, acidic, organic, n.o.s., 8, PGIII**

The basic description includes the name of the specific organic acid entered parenthetically. DOT also requires shippers to affix the appropriate CORROSIVE and/or FLAMMABLE LIQUID labels on its packaging.

When carriers transport an organic acid in an amount exceeding 1001 lb (454 kg), DOT requires them to post the appropriate CORROSIVE or FLAMMABLE placards on the bulk packaging or transport vehicle used for shipment.

SOLVED EXERCISE 13.10

Why does DOT regulate the transportation of isobutyric acid as a flammable liquid?

Solution: In the glossary we see that DOT defines a liquid as a material, other than an elevated-temperature material, with a melting point or initial melting point of 68 °F (20 °C) or lower at a standard pressure of 14.7 psi$_a$ (101.3 kPa). We also see that DOT defines a flammable liquid as either of the following:

- Any liquid having a flashpoint of not greater than 141 °F (60.5 °C)
- Any material in a liquid phase with a flashpoint at or above 100 °F (37.8 °C) that is intentionally heated and offered for transportation or transported at or above its flashpoint within bulk packaging

The data in Table 13.12 indicate that isobutyric acid has a melting point of −51 °F (−47 °C). Because the melting point is less than 68 °F (20 °C), DOT classifies isobutyric acid as a liquid. These data also indicate that isobutyric acid has a flashpoint of 132 °F (55 °C). Consequently, DOT further classifies isobutyric acid as a flammable liquid and regulates its transportation as a flammable liquid (hazard class = 3).

TABLE 13.13 — Basic Descriptions of Some Organic Acids

ORGANIC ACID OR SOLUTIONS THEREOF	CONCENTRATION	BASIC DESCRIPTION[a]
Acetic acid, glacial		UN2789, Acetic acid, glacial, 8, (3), PGII
Acetic acid solutions	More than 80% acetic acid by mass	UN2789, Acetic acid solution (contains more than 80% acid by mass), 8, (3), PGII
Acetic acid solutions	Not less than 50% but not more than 80% acetic acid by mass	UN2790, Acetic acid solution (contains not less than 50% but not more than 80% acid by mass), 8, PGII
Acetic acid solutions	More than 10% but less than 50% acetic acid by mass	UN2790, Acetic acid solution (contains more than 10% but less than 50% acid by mass), 8, PGIII
Butyric acid		UN2820, Butyric acid, 8, PGII
Caproic acid		UN2829, Caproic acid, 8, PGIII
Formic acid solutions	Not less than 10% but not more than 85% formic acid by mass	UN3412, Formic acid (contains not less than 10% but not more than 85% formic acid by mass), 8, PGII
Formic acid solutions	Not less than 5% but less than 10% formic acid by mass	UN3412, Formic acid (contains not less than 5% but less than 10% acid by mass), 8, PGIII
Formic acid	More than 85% formic acid by mass	UN1779, Formic acid (contains more than 85% acid by mass), 8, (3), PGII
Isobutyric acid		UN2529, Isobutyric acid, 3, PGIII
Propionic acid	Not less than 90% propionic acid by mass	UN3463, Propionic acid (contains not less than 90% acid by mass), 8, (3), PGII
Propionic acid solutions	Not less than 10% and less than 90% propionic acid by mass	UN1848, Propionic acid (contains not less than 10% and less than 90% acid by mass), 8, PGIII

[a]DOT requires shippers to provide the actual acetic acid concentration, when known, in lieu of the parenthetical statement.

PERFORMANCE GOALS FOR SECTION 13.7:

- Identify the functional group that characterizes the esters, including the phthalate esters.
- Identify the primary risks associated with exposure to ethyl acetate and di(2-ethylhexyl) phthalate.
- Identify the industries that use these esters.
- Describe the means shippers and carriers use to inform emergency responders of the hazards associated with encountering esters during transportation mishaps.

ester

■ An organic compound having molecules in which an alkyl or aryl group has replaced the hydrogen atom in the carboxylate group

esterification

■ The process of producing an ester by the reaction of an organic acid with an alcohol

13.7 Esters

An **ester** is an organic compound having the general chemical formula $R-\overset{\displaystyle O}{\overset{\displaystyle \|}{C}}\overset{}{\underset{O-R'}{}}$. It is the

compound produced when an organic acid reacts with an alcohol. This type of chemical reaction is called **esterification**.

Esters are most commonly denoted by their IUPAC names. In the IUPAC system, an ester is named by changing the *-ic* suffix of the organic acid to *-ate*, preceded by the name of the alkyl or aryl group of the alcohol used to produce it. In other words, the R′ group is named first followed by the name of the acid with *-ic acid* replaced by *-ate*.

The simple esters are colorless, highly volatile, and flammable liquids that are only partially soluble in water. Many possess very pleasant, fruity odors. For example, the compounds responsible for the odors of pineapples, bananas, and apples are ethyl butyrate, isoamyl acetate, and ethyl-2-methylbutyrate, respectively. Several synthetic esters are added to foods as flavoring agents.

Ethyl butyrate   Isoamyl acetate   Ethyl-3-methylbutyrate

13.7-A ETHYL ACETATE

Ethyl acetate is a commonly encountered constituent of nitrocellulose-based lacquers and other protective coatings. It is produced by reacting ethanol and acetic acid.

$$CH_3CH_2OH(aq) \ + \ CH_3-\overset{\displaystyle O}{\overset{\displaystyle \|}{C}}\underset{OH}{}(aq) \ \longrightarrow \ CH_3-\overset{\displaystyle O}{\overset{\displaystyle \|}{C}}\underset{O-CH_2CH_3}{}(aq) \ + \ H_2O(l)$$

Ethanol   Acetic acid   Ethyl acetate   Water

Ethyl acetate is a colorless liquid with a pleasing, fragrant odor. The data in Table 13.14 indicate that it poses the risk of fire and explosion. When ethyl acetate burns, carbon dioxide and water vapor are the products of combustion.

$$CH_3-\overset{\displaystyle O}{\overset{\displaystyle \|}{C}}\underset{O-CH_2CH_3}{}(g) \ + \ 5O_2(g) \ \longrightarrow \ 4CO_2(g) \ + \ 4H_2O(g)$$

Ethyl acetate   Oxygen   Carbon dioxide   Water

13.7-B DI(2-ETHYLHEXYL) PHTHALATE

phthalate

■ An ester of *o*-, *m*-, or *p*-phthalic acid

The **phthalates** are esters of 1,2-, 1,3-, and 1,4-dicarboxylic acid, commonly called *o*-, *m*-, and *p*-phthalic acid. The *o*- isomers are commercially important. Their chemical formula is generally represented as follows, where R and R′ are arbitrary alkyl or aryl groups:

TABLE 13.14	Physical Properties of Ethyl Acetate
Melting point	−119 °F (−84 °C)
Boiling point	171 °F (77 °C)
Specific gravity at 68 °F (20 °C)	0.90
Vapor density (air = 1)	3.04
Vapor pressure at 68 °F (20 °C)	76 mmHg
Flashpoint	24 °F (−4.4 °C)
Autoignition point	800 °F (427 °C)
Lower explosive limit	2.18%
Upper explosive limit	9%

The individual phthalates are identified by naming R and R', followed by the word phthalate. The commercially important phthalates include di-n-butyl phthalate (DBP), benzyl butyl phthalate (BBzP), di(2-ethylhexyl) phthalate (DEHP), diisononyl phthalate (DINP), diisodecyl phthalate (DIDP), and diisooctyl phthalate (DIOP). They are produced by reacting o-phthalic acid with appropriate alcohols.

DEHP is representative of the commercial phthalates. It is produced by reacting o-phthalic acid with 2-ethylhexanol.

o-Phthalic acid   2-Ethylhexanol

o-Di-(2-ethylhexyl) phthalate (DEHP)   Water

DEHP is a component of various rubber and other polymeric formulations, wherein it increases the flexibility and toughness of the related products. When used in this fashion, the DEHP is called a **plasticizer**. DEHP is the most widely used plasticizer for the manufacture of products made from the polymer poly(vinyl chloride) (Section 14.6-C). Persons most likely encounter DEHP while searching for a new car. It is the substance that provides the "new-car smell" when the windows and doors of the car are shut. DEHP slowly vaporizes from the vinyl upholstery in new cars. DEHP is also used as a plasticizer with certain high explosives (Section 15.1), where its presence produces more malleable forms and increases their stability.

plasticizer
■ A substance, such as a phthalate ester, that when added in a prescribed amount to a polymeric formulation, facilitates processing and provides flexibility and toughness to the end product

When DEHP is used during the manufacture of plasticized products, it is mixed with the polymers or explosives but does not chemically bond with them. Hence, DEHP is capable of leaching from plasticized products. It is for this reason that a potential health problem may arise.

In human studies, health concerns have been raised over the impact of leaching DEHP from plasticized medical products such as intravenous bags, tubing, and syringes. When blood is stored in DEHP-plasticized bags, the phthalate ester leaches from the plastic and dissolves in the blood. The prevailing fear is that phthalates are endocrine disrupters that may interfere with development.

In research studies, elevated levels of DEHP have been shown to interfere with the normal kidney and liver function of laboratory animals. They have also caused physiological abnormalities in the reproductive systems of animals, especially their male offspring.

Whether phthalates can interfere with the hormones generated by humans is a controversial topic. FDA initially concluded that patient exposures to DEHP are generally well below the levels expected to cause adverse effects, although it did concede that children, especially infants, may represent a special population at increased risk from exposure. In recognition of this guarded view, the FDA proposed voluntary restrictions on the use of DEHP-plasticized devices by pregnant women and newborn infants. In 2009, the FDA banned the manufacture, sale, and distribution of any toy or other object that contains DEHP, DBP, or BBP and is intended for use by children under 12 years of age.

The European Parliament also banned the use of certain phthalate esters, including DEHP, in children's toys and child-care items including pacifiers and teething rings.

SOLVED EXERCISE 13.11

During medical treatment, why is it prudent for pregnant women and nursing mothers to request the use of devices that do not contain DEHP?

Solution: Although scientists have determined that exposure to DEHP is linked with health problems in animals, they do not presently concur as to whether DEHP causes health problems in humans. Until this medical debate has been resolved, the prudent practice for pregnant women and nursing mothers receiving medical treatment is to request the use of devices that were manufactured without DEHP. The primary aim of this practice is to protect the health of the unborn and infants, who represent the group at highest risk from exposure to DEHP.

13.7-C TRANSPORTING ESTERS

When shippers offer an ester for transportation, DOT requires them to identify the relevant basic description listed in Table 13.15 and enter it on an accompanying shipping paper. When they transport a flammable ester whose name is not provided in the Hazardous Materials Table, DOT requires them to identify the commodity generically on a shipping paper as either one of the following:

> **UN3272, Esters, n.o.s., 3, PGII**
> *or*
> **UN3272, Esters, n.o.s., 3, PGIII**

The basic description includes the name of the specific ester entered parenthetically.

When carriers transport an ester in an amount exceeding 1001 lb (454 kg), DOT requires them to post FLAMMABLE plcards on the bulk packaging or transport vehicle used for shipment.

TABLE 13.15 | Basic Descriptions of Some Esters

ESTER OR GROUPS THEREOF	BASIC DESCRIPTION
Amyl acetates	UN1104, Amyl acetates, 3, PGIII
Amyl butyrates	UN2620, Amyl butyrates, 3, PGIII
Amyl formates	UN1109, Amyl formates, 3, PGIII
Butyl acetates	UN1123, Butyl acetates, 3, PGII *or* UN1123, Butyl acetates, 3, PGIII
n-Butyl formate	UN1128, *n*-Butyl formate, 3, PGII
Cyclohexyl acetate	UN2243, Cyclohexyl acetate, 3, PGIII
Ethyl acetate	UN1173, Ethyl acetate, 3, PGII
Ethyl butyrate	UN1180, Ethyl butyrate, 3, PGIII
Ethyl formate	UN1190, Ethyl formate, 3, PGII
Ethyl propionate	UN1195, Ethyl propionate, 3, PGII
Isobutyl acetate	UN1213, Isobutyl acetate, 3, PGII
Isobutyl isobutyrate	UN2406, Isobutyl isobutyrate, 3, PGII
Isobutyl propionate	UN2394, Isobutyl propionate, 3, PGII
Isopropyl butyrate	UN2405, Isopropyl butyrate, 3, PGIII
Isopropyl isobutyrate	UN2406, Isopropyl isobutyrate, 3, PGII
Isopropyl propionate	UN2409, Isopropyl propionate, 3, PGII
Methyl butyrate	UN1237, Methyl butyrate, 3, PGII
Methyl formate	UN1243, Methyl formate, 3, PGI
Methyl propionate	UN1248, Methyl propionate, 3, PGII
Methylamyl acetate	UN1233, Methylamyl acetate, 3, UN1233, PGIII
n-Propyl acetate	UN1276, *n*-Propyl acetate, 3, UN1276, PGII
Propyl formates	UN1281, Propyl formates, 3, PGII

PERFORMANCE GOALS FOR SECTION 13.8:

- Identify the functional group that characterizes the amines.
- Identify the industries that use amines.
- Identify the primary risks associated with exposure to the common amines.
- Describe the means shippers and carriers use to inform emergency responders of the hazards associated with encountering amines during transportation mishaps.

13.8 Amines

amine
■ An alkylated analog of ammonia; an organic compound derived by the substitution of one or more of the hydrogen atoms in a hydrocarbon molecule with the $-NH_2$ group of atoms

The **amines** are organic derivatives of ammonia that is, compounds that contain one or more alkyl or aryl groups bonded to a nitrogen atom. Their general formulas are $R-NH_2$, R_2NH, and R_3N, where R is an arbitrary alkyl or aryl group.

$$R-N\begin{array}{c}H\\\\H\end{array} \qquad R-N\begin{array}{c}H\\\\R\end{array} \qquad R-N\begin{array}{c}R\\\\R\end{array}$$

For example, the amines whose molecules have one, two, and three methyl groups bonded to the nitrogen atom are the compounds whose structures follow:

$$CH_3-N\begin{array}{c}H\\\\H\end{array} \qquad CH_3-N\begin{array}{c}CH_3\\\\H\end{array} \qquad CH_3-N\begin{array}{c}CH_3\\\\CH_3\end{array}$$

Methylamine   Dimethylamine   Trimethylamine

Table 13.16 provides their physical properties, which reflect the fact that the primary risk associated with these three amines is fire and explosion. Their incomplete combustion produces nitric oxide, carbon monoxide, and water vapor. Their complete combustion produces nitrogen dioxide, carbon dioxide, and water vapor.

The more complex amines are usually flammable and corrosive liquids. Their corrosivity is linked with the similarity in chemical behavior that they share with ammonia. Although the gaseous amines have the odor of ammonia, the liquid amines have the characteristic odor of rotten fish.

The simple amines are generally used by manufacturing and process industries as coagulants, flocculating agents, corrosion inhibitors, bactericides, and fungicides. In the pharmaceutical industry, they are used to manufacture drugs, and in the chemical industry, they are used to manufactures dyes and other substances.

There are several commercially important chemical commodities that contain multiple amino groups per molecule. Two examples are hexamethylenediamine and hexamethylenetetramine.

TABLE 13.16	Physical Properties of the Simple Amines		
	METHYLAMINE	**DIMETHYLAMINE**	**TRIMETHYLAMINE**
Melting point	−137 °F (−94 °C)	−134 °F (−92 °C)	−179 °F (−117 °C)
Boiling point	21 °F (−6 °C)	45 °F (7.4 °C)	39 °F (4 °C)
Specific gravity at 68 °F (20 °C)	0.69	0.68	0.66
Vapor density (air = 1)	1.08	1.65	2.0
Flashpoint	34 °F (1 °C)	−58 °F (−50 °C)	8 to 18 °F (−13 to −8 °C)
Autoignition point	806 °F (430 °C)	755 °F (402 °C)	374 °F (190 °C)
Lower explosive limit	4.9%	2.8%	2%
Upper explosive limit	20.7%	14.4%	11.6%

$$H_2N-CH_2CH_2CH_2CH_2CH_2CH_2-NH_2$$

Hexamethylenediamine
(1,6-Diaminohexane)

Hexamethylenetetramine

Hexamethylenediamine is a solid used as a feedstock in the polymer industry for the production of nylon 66 (Table 14.2) and other nylons. Hexamethylenetetramine is a flammable solid used to manufacture the explosives cyclonite and HMX (Sections 15.10 and 15.13). It is also the primary component of the urinary antiseptic urotropin. The molecular structure of hexamethylenetetramine consists of four six-membered rings of alternating carbon and nitrogen atoms.

The simple amines are named by identifying the names of the alkyl or aryl group bonded to the nitrogen atom, followed by the suffix *-amine*. The names of the simple amines—methylamine, dimethylamine, and trimethylamine—are examples. Although *di-* and *tri-* are respectively used to name two and three identical groups bonded to a nitrogen atom, the prefixes *bis* and *tris*, instead of *di* and *tri*, are also used when complex groups are components of a formula.

The simplest amines are also named as aminoalkanes, alkylaminoalkanes, and dialkylaminoalkanes. A number is used to show the position of the amino group along the carbon–carbon chain. For example, the compounds whose formulas are $CH_3CH_2CH_2NH_2$, $(CH_3)CH-NH_2$, and $(CH_3)_2CH-NHCH_3$ are then named as follows:

$$CH_3CH_2CH_2-NH_2$$

1-Aminopropane

$$CH_3-\underset{\underset{NH_2}{|}}{C}-CH_3$$

2-Aminopropane

$$CH_2-\underset{\underset{CH_3}{|}}{CH}-NHCH_3$$

2-Methylaminopropane

When chemists use the IUPAC system to name amines, they replace the *-e* in the name of the alkane with *-amine*. A number is again assigned to show the position of the amino group along the chain. The prefix *N-*(in italics) is used to designate the bonding of an alkyl or aryl substituent on the nitrogen atom. Then, the substituents are listed alphabetically regardless of whether they are bonded to the nitrogen atom or along the chain of carbon atoms.

$$CH_3CH_2-\underset{\underset{NHCH_2CH_3}{|}}{CH}-CH_2CH_2CH_3$$

N-Ethyl-3-hexanamine

$$CH_3-\underset{\underset{CH_3}{|}}{N}-CH_2CH_2CH_3$$

N,N-Dimethylpropanamine

The common name of the simplest aromatic amine is *aniline*; its chemical formula is $C_6H_5-NH_2$. The substances having the chemical formulas $C_6H_5-NHCH_3$ and $C_6H_5-N(CH_3)_2$ are named *N*-methylaniline and *N,N*-dimethylaniline, respectively.

Aniline

N-Methylaniline

N,N-Dimethylaniline

13.8-A WORKPLACE REGULATIONS INVOLVING THE SIMPLE AMINES

Inhalation of the simple amines has been linked with eye, nose, and throat irritation. When these compounds are present in the workplace, OSHA requires employers to limit employee exposure to them at the following maximum concentrations in air, averaged over an 8-hr workday:

methylamine, 10 ppm; dimethylamine, 10 ppm; trimethylamine, 10 ppm; ethylamine, 10 ppm; and aniline, 2 ppm (skin).

13.8-B TRANSPORTING AMINES

When shippers offer an amine for transportation, DOT requires them to identify the relevant substance from Table 13.17 on a shipping paper. When shippers offer an amine whose name is not listed in Table 13.17 for transportation, DOT requires them to determine whether the commodity is solely corrosive, or corrosive and flammable. Then, they identify the commodities generically on a shipping paper as one of the following:

> **UN2733, Amines, flammable, corrosive, n.o.s., 3, (8), PGI**
> *or*
> **UN2733, Amines, flammable, corrosive, n.o.s., 3, (8), PGII**
> *or*
> **UN2733, Amines, flammable, corrosive, n.o.s., 3, (8), PGIII**
> **UN2734, Amines, liquid, corrosive, flammable, n.o.s., 8, (3), PGI**
> *or*
> **UN2734, Amines, liquid, corrosive, flammable, n.o.s., 8, (3), PGII**
> **UN2735, Amines, liquid, corrosive, n.o.s., 8, PGI**
> *or*
> **UN2735, Amines, liquid, corrosive, n.o.s., 8, PGII**
> *or*
> **UN2735, Amines, liquid, corrosive, n.o.s., 8, PGIII**
> **UN3259, Amines, solid, corrosive, n.o.s., 8, PGI**
> *or*
> **UN3259, Amines, solid, corrosive, n.o.s., 8, PGII**
> *or*
> **UN3259, Amines, solid, corrosive, n.o.s., 8, PGIII**

This basic description includes the name of the specific amine entered parenthetically.

DOT also requires shippers to affix the applicable label corresponding to the hazard class and division code of the hazardous material on its packaging.

When carriers transport dimethylamine, ethylamine, methylamine, or trimethylamine, DOT requires them to post FLAMMABLE GAS placards on the bulk packaging or transport vehicle used for shipment. When carriers transport an amine other than dimethylamine, ethylamine, methylamine, and trimethylamine in an amount exceeding 1001 lb (454 kg), DOT requires them to post the relevant FLAMMABLE, CORROSIVE, or POISON placards.

SOLVED EXERCISE 13.12

Identify the combustion products that are produced when methylamine burns in air.

Solution: The chemical formula of methylamine is CH_3NH_2, which indicates that each methylamine molecule is composed of carbon, hydrogen, and nitrogen atoms. When methylamine burns in air, these carbon, hydrogen, and nitrogen atoms unite with atmospheric oxygen. On incomplete combustion, carbon monoxide, nitrogen monoxide, and water vapor are produced. On complete combustion, carbon dioxide, nitrogen dioxide, and water vapor are produced. These respective processes are denoted as follows:

$$4CH_3NH_2(g) \ + \ 9O_2(g) \ \longrightarrow \ 4CO(g) \ + \ 4NO(g) \ + \ 10H_2O(g)$$

Methylamine · · · · · · · Oxygen · · · · · · Carbon monoxide · Nitric oxide · · · · · Water

$$4CH_3NH_2(g) \ + \ 13O_2(g) \ \longrightarrow \ 4CO_2(g) \ + \ 4NO_2(g) \ + \ 10H_2O(g)$$

Methylamine · · · · · · · Oxygen · · · · · · Carbon dioxide · · Nitrogen dioxide · · · Water

TABLE 13.17 | Basic Descriptions of Some Amines

AMINE OR GROUPS THEREOF	BASIC DESCRIPTION
Amylamines	UN1106, Amylamines, 8, (3), PGII *or* UN1106, Amylamines, 8, (3), PGIII
Aniline	UN1547, Aniline, 6.1, PGII
n-Butylamine	UN1125, *n*-Butylamine, 3, (8), PGII
N-Butylaniline	UN2738, *N*-Butylaniline, 6.1, PGII
Cyclohexylamine	UN2357, Cyclohexylamine, 8, (3), PGII
Diallylamine[a]	UN2359, Diallylamine, 3, (6.1, 8), PGII
Di-*n*-amylamine	UN2841, Di-*n*-amylamine, 3, (6.1), PGIII
Di-*n*-butylamine	UN2248, Di-*n*-butylamine, 8, (3), PGII
Diethylamine	UN1154, Diethylamine, 3, (8), PGII
N,N-Diethylaniline	UN2432, *N,N*-Diethylaniline, 6.1, PGIII
Diisobutylamine	UN2361, Diisobutylamine, 3, (8), PGIII
Dimethylamine (anhydrous)	UN1032, Dimethylamine, anhydrous, 2.1
Dimethylamine solutions	UN1160, Dimethylamine solutions, 3, (8), PGII
Dipropylamine	UN2383, Dipropylamine, 3, (8), PGII
Ethylamine	UN1036, Ethylamine, 2.1
2-Ethylaniline	UN2273, 2-Ethylaniline, 6.1, PGIII
Hexamethylenediamine	UN2280, Hexamethylenediamine, solid, 8, PGIII
Hexamethylenediamine solutions	UN1783, Hexamethylenediamine, solution, 8, PGII *or* UN1783, Hexamethylenediamine, solution, 8, PGIII
Hexamethylenetetramine	UN1328, Hexamethylenetetramine, 4.1, PGIII
Methylamine, anhydrous	UN1061, Methylamine, 2.1
Methylamine solutions	UN1235, Methylamine, aqueous solution, 3, (8), PGII
N-Methylaniline	UN2994, *N*-Methylaniline, 6.1, PGIII
N-Methylbutylamine	UN2945, *N*-Methylbutylamine, 3, (8), PGII
Tributylamine	UN2542, Tributylamine, 6.1, PGII
Triethylamine	UN1296, Triethylamine, 3, (8), PGII
Trimethylamine, anhydrous	UN1083, Trimethylamine, anhydrous, 2.1
Trimethylamine solutions (contains not more than 50% trimethylamine by mass)	UN1297, Trimethylamine solutions, 3, PGI
Tripropylamine	UN2260, Tripropylamine, 3, (8), PGIII

[a]Diallylamine is the substance having the formula $H_2C{=}CHCH_2-NH-CH_2CH{=}CH_2$.

- Identify the functional groups that characterize the peroxo-organic compounds.
- Identify the primary industry that uses bulk quantities of the peroxo-organic compounds.
- Identify the primary risks associated with exposure to peroxo-organic compounds.
- Identify the unique DOT requirements associated with preparing the basic description of peroxo-organic compounds.
- Describe the means shippers and carriers use to inform emergency responders of the hazards associated with encountering peroxo-organic compounds during transportation mishaps.
- Describe how terrorists have misused triacetone triperoxide.
- Describe the response actions that emergency responders execute when peroxo-organic compounds have been released into the environment.

13.9 Peroxo-Organic Compounds

organic hydroperoxide
- A derivative of hydrogen peroxide in which one hydrogen atom in the H_2O_2 molecule has been substituted with an alkyl or aryl group; an organic compound having the general chemical formula $H-O-O-R$, where R is an alkyl or aryl group

Peroxo-organic compounds may be classified as **organic hydroperoxides** and **organic peroxides**. The compounds in both classes are derivatives of hydrogen peroxide (Section 11.5). They respectively differ by the substitution of one or both of the hydrogen atoms in the H_2O_2 molecule with alkyl or aryl groups (R and R'). Consequently, their chemical formulas are $R-O-O-H$ and $R-O-O-R'$, respectively.

The simplest organic hydroperoxides are identified by the name of the alkyl or aryl group bonded to the $H-O-O-$ group followed by the word "hydroperoxide." Examples are *tert*-butyl hydroperoxide and isopropylbenzene hydroperoxide, whose chemical formulas are $(CH_3)_3-C-O-OH$ and $C_6H_5-C(CH_3)_2-O-OH$, respectively.

tert-Butyl hydroperoxide

Isopropylbenzene hydroperoxide
(Cumene hydroperoxide)

organic peroxide
- A derivative of hydrogen peroxide in which both hydrogen atoms in the H_2O_2 molecule have been substituted with alkyl or aryl groups; an organic compound having the general chemical formula $R-O-O-R'$, where R and R' are alkyl or aryl groups; for purposes of DOT regulations, an organic compound containing oxygen in the bivalent $-O-O-$ structure and that may be considered a derivative of hydrogen peroxide in which one or more of the hydrogen atoms have been replaced by organic radicals

The simplest organic peroxides are identified by the names of the alkyl or aryl groups bonded to the $-O-O-$ group followed by the word "peroxide." When the alkyl or aryl group is the same $(R=R')$, the name of the group is preceded by the prefix *di*. Examples are diacetyl peroxide and di-*tert*-butyl peroxide, whose chemical formulas are $(CH_3CO_2)_2$ and $[(CH_3)_3-C-]_2O_2$, respectively.

Diacetyl peroxide

Di-*tert*-butyl peroxide

Di-*tert*-butyl peroxide is a substance sometimes added in low concentration to diesel oil to improve its cetane number (Section 12.14-F).

Some commercially important peroxo-organic compounds are perketones, peracids, and peresters, respectively. These substances are oxidized acids, ketones, and esters. The peracids are easily named by inserting the prefix *per* before the name of the acid that has been oxidized. The common "peracids" have from 1 to 4 carbon atoms per molecule. For example, perpropionic acid is the compound with the following molecular structure:

$$CH_3CH_2-\overset{\displaystyle O}{\underset{\displaystyle O-OH}{C}}$$

Perpropionic acid
(Perpropanoic acid)

The perketones are commonly named by inserting the word "peroxide" after the name of the ketone from which it was produced. These substances generally contain multiple functional groups, one of which is the peroxide group, as illustrated by the following molecular structures for the two perketones ethyl methyl ketone peroxide and cyclohexanone peroxide.

$$HO-O-\underset{\underset{\displaystyle CH_3}{|}}{\overset{\overset{\displaystyle CH_2CH_3}{|}}{C}}-O-O-\underset{\underset{\displaystyle CH_3}{|}}{\overset{\overset{\displaystyle CH_2CH_3}{|}}{C}}-O-OH$$

Bis(2-hydroperoxy-*sec*-butyl)peroxide
(Ethyl methyl ketone peroxide)
(Methyl ethyl ketone peroxide)

1-Hydroxy-1′-hydroperoxydicyclohexyl peroxide
(Cyclohexanone peroxide)

A perester is often identified by inserting the term "peroxy" before the name of alkyl or aryl group that is bonded to the carboxyl group, as in *tert*-butylperoxybenzoate.

$$CH_3-\underset{\underset{\displaystyle CH_3}{|}}{\overset{\overset{\displaystyle CH_3}{|}}{C}}-O-O-\overset{\displaystyle O}{C}-\text{(phenyl)}$$

tert-Butylperoxybenzoate

Sometimes, however, the perester is named as a dialkylperoxydicarbonate, whose general chemical formula is the following:

$$R-O-\overset{\displaystyle O}{\underset{\displaystyle O-O}{C}}\qquad\overset{\displaystyle O}{\underset{\displaystyle }{C}}-O-R$$

For example, when R is the isopropyl group, $(CH_3)_2CH-$, the peroxyester is named diisopropyl-peroxydicarbonate.

$$CH_3-\underset{\underset{\displaystyle CH_3}{|}}{CH}-O-\overset{\displaystyle O}{\underset{\displaystyle O-O}{C}}\qquad\overset{\displaystyle O}{C}-O-\underset{\underset{\displaystyle CH_3}{|}}{CH}-CH_3$$

Diisopropylperoxydicarbonate

The peroxo-organic compounds are commonly utilized in the chemical industry as raw materials in synthetic processes. Peroxo-organic compounds are also used widely to induce polymerization (Section 14.2), a process essential to the production and manufacture of plastics. Although they are either liquids or solids at room conditions, peroxo-organic compounds are often encountered dissolved in water or an appropriate organic solvent These solvents serve to stabilize peroxo-organic compounds against premature thermal decomposition.

Although the peroxo-organic compounds are flammable substances, they are also oxidizers containing active oxygen within their molecular structures. When they burn, peroxo-organic compounds support their own combustion. This combination of properties poses a particularly pronounced risk of fire and explosion. When ignited, the peroxo-organic compounds often burn furiously and more intensely than other combustible substances. This burning phenomenon is called **deflagration**. Sufficient energy is generated during deflagration to enable the burning to

deflagration
▪ The chemical process during which a substance intensely burns instead of detonating

FIGURE 13.8 The Firefighting Measures and Handling and Storage excerpts from the material safety data sheet for dibenzoyl peroxide (*Adapted from the* Material Safety Data Sheet for Benzoyl Peroxide, *Mallinckrodt Baker, Inc., Phillipsburg, New Jersey.*)

DIBENZOYL PEROXIDE

Firefighting Measures:

Flashpoint: 104 °F (40 °C)
Autoignition temperature: 176 °F (80 °C)
Extremely flammable!
Substance is a strong oxidizer and a strong supporter of combustion. Its heat of reaction with combustibles and reducing agents may cause ignition.

Explosion:

Explosive! Extremely explosion-sensitive to shock, heat, and friction. May explode spontaneously when dry. Sensitive to mechanical impact. Sensitive to static discharge.

Handling and Storage:

Protect containers against physical damage. Isolate in a well-detached, fire-resistant, cool and well-ventilated building with no other materials stored therein. Shield containers from direct sunlight and maintain their temperature at less than 100 °F (38 °C). Employ grounding, bonding, venting, and explosion-relief provisions in accord with accepted engineering practices in any process capable of generating an explosion due to static discharge, shock, impact, heat, friction, or blows. Provide explosion-venting in a safe direction and prohibit any electrical installation or heating facilities. Dibenzoyl peroxide should be stored in and used from original containers. Do not return product that has been taken out of original container. Never mix unless at least 33% water is present. KEEP PURE! Impurities are hazardous in organic peroxides. Do not add accelerators. Containers of this material may be hazardous when empty since they retain product residues (dust, solids); observe all warnings and precautions listed for the product. DO NOT attempt to clean empty containers since residue is difficult to remove. Do not pressurize, cut, weld, braze, solder, grill, grind, or expose such containers to heat, sparks, flames, static electricity, or other sources of ignition: They may explode and cause injury or death.

proceed and accelerate without the input of additional energy from another source. An example is illustrated by the following equation:

$$\underset{\text{Di-}\textit{tert}\text{-butyl peroxide}}{[(CH_3)_3{-}C{-}]_2O_2(s)} + \underset{\text{Oxygen}}{9O_2(g)} \longrightarrow \underset{\text{Carbon dioxide}}{3CO_2(g)} + \underset{\text{Carbon monoxide}}{5CO(g)} + \underset{\text{Water}}{9H_2O(g)}$$

Most peroxo-organic compounds are unstable. When exposed to an ignition source, they are just as likely to undergo rapid, autoaccelerated decomposition as they are to burn. Some are so sensitive to friction, heat, or shock that they cannot be safely handled unless maintained at low temperatures and diluted within an inert solid material or solvent. These hazardous features are illustrated in the excerpts from the MSDS for dibenzoyl peroxide that are provided in Figure 13.8.

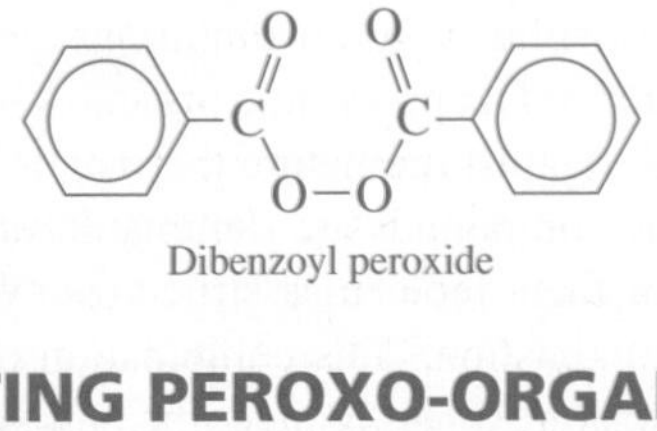

Dibenzoyl peroxide

13.9-A TRANSPORTING PEROXO-ORGANIC COMPOUNDS

Before a peroxo-organic compound may be offered for transportation, DOT requires its manufacturer or other party to test a sample of the substance using prescribed procedures to determine

TABLE 13.18 | Generic Types of Organic Peroxides[a]

TYPE	IDENTIFYING FEATURES
Type A	Can detonate or rapidly deflagrate as packaged for transport. The transportation of type A organic peroxides is forbidden by DOT.
Type B	Neither detonates nor deflagrates rapidly when correctly packaged for transportation but can undergo a thermal explosion
Type C	Neither detonates nor deflagrates rapidly when correctly packaged for transportation and cannot undergo a thermal explosion
Type D	Detonates only partially, but does not deflagrate rapidly, and is not affected by heat when confined; *or* does not detonate, deflagrates slowly, and manifests no violent effect if heated when confined; *or* does not detonate or deflagrate and manifests a medium effect when heated under confinement
Type E	Neither detonates nor deflagrates and manifests either a low or no effect when heated under confinement
Type F	Will not detonate in a cavitated state, does not deflagrate, manifests only a low or no effect if heated when confined, and possesses either a low or no explosive power
Type G[b]	Will not detonate in a cavitated state, will not deflagrate at all, shows no effect when heated under confinement, and manifests no explosive power

[a]49 C.F.R. §173.128.
[b]A type G organic peroxide is not subject to DOT's transportation requirements for organic peroxides if it is "thermally stable," i.e., its self-accelerating decomposition temperature is 122 °F (50 °C) or higher for a 110-lb (50-kg) package. An organic peroxide that possesses all characteristics of type G other than thermal stability and requires temperature control is classed as a type F, temperature-controlled organic peroxide.

whether it will undergo decomposition under the temperature conditions likely to be encountered during transportation. DOT then evaluates these test data and either authorizes or prohibits its transportation. When the testing data show that the substance is likely to decompose with a self-accelerated decomposition temperature of 122 °F (54 °C) or less, DOT forbids its transportation.

When shippers prepare the basic description of a peroxo-organic compound, they consult the Hazardous Materials Table and the Organic Peroxides Table published at 49 C.F.R. §§172.101 and 173.225, respectively. DOT does not distinguish between organic hydroperoxides and organic peroxides in the hazardous materials regulations. For the purpose of their transportation, each is referred to as an "organic peroxide."

The shipping descriptions of organic peroxides are listed generically in the Hazardous Materials Table. Several examples are provided in Table 6.1. Each listing denotes a type of organic peroxide and its state of matter (liquid or solid). There are seven organic peroxide types, each of which is designated by a capital letter A through G, which collectively form a continuum of decreasing hazard. Their features are noted in Table 13.18.

The proper shipping names of the generic peroxo-organic compounds noted in the Hazardous Materials Table refer only to types A through F, but DOT forbids the transportation of type A organic peroxides. Some also include the words "temperature-controlled."

The letter G appears in column 1 of the Hazardous Materials Table for each proper shipping name of an organic peroxide. As noted in Section 6.2-A, this signifies that the technical name of the organic peroxide must be entered in parentheses in its shipping description. This name is included in the shipping information provided in the Organic Peroxides Table, an excerpt of which is provided in Table 13.19.

TABLE 13.19   Organic Peroxides Table[a,b]

TECHNICAL NAME (1)	IDENTIFICATION NUMBER (2)	CONCENTRATION (MASS %) (3)	DILUENT (MASS %)			CONCENTRATION OF WATER (MASS %) (5)	TEMPERATURE (° C)	
			(4a)	(4b)	(4c)		CONTROL (7a)	EMERGENCY (7b)
Acetyl acetone peroxide	UN3105	≤42	≥48			≥8		
tert-Butyl hydroperoxide	UN3103	>79 to 90				≥10		
Cumyl hydroperoxide	UN3107	>90 to 98	≤10					
Cyclohexanone peroxide(s)	Exempt	≤32			≥68			
Diacetyl peroxide	UN3115	≤27		≥73			20	25
Dibenzoyl peroxide	UN3102	>77 to 94				≥6		
Dicumyl peroxide	UN3110	>52 to 100	≤48					
Di-(2-methylbenzoyl) peroxide	UN3112	≤87				≥13	30	35
Methylcyclohexanone peroxide	UN3115	≤67		≥33			35	40

Peroxyacetic acid, Type D		UN3105		≤43	
1,1,3,3-Tetramethyl-butyl hydroperoxide		UN3105		≤100	

[a]Excerpted from the DOT Organic Peroxides Table, 49 C.F.R. §173.225. The information tabulated in columns 6 and 8 of the original reference has been redacted.
[b]The columnar entries noted in this table signify the following:

Column 1 The technical name of the compound

Column 2 The DOT identification number. When the word "Exempt" appears in column 2, the transportation of the applicable substance is not regulated as an organic peroxide.

Column 3 The concentration limits, if any

Column 4 The type and concentration of a diluent or inert solid, when required, that must be mixed with the peroxo-organic compound for effective desensitization.

- A "Diluent type A" is an organic liquid that does not detrimentally affect the thermal stability or increase the hazard of the peroxo-organic compound and possesses a boiling point of not less than 302 °F (150 °C).
- A "Diluent type B" is an organic liquid that is compatible with the peroxo-organic compound and possesses a boiling point of less than 302 °F (150 °C), but at least 140 °F (60 °C), and a flashpoint greater than 41 °F (5 °C).
- An "inert solid" is a solid that does not detrimentally affect the thermal stability or increase the hazard of the peroxo-organic compound.

Column 5 The minimum concentration of water in mass percent that must be present in the formulation of the peroxo-organic compound.

Column 7 The control temperature and emergency temperature, when temperature controls are required

- The control temperature is the temperature above which DOT prohibits carriers from transporting the peroxo-organic compound.
- The emergency temperature is the temperature at which DOT prohibits carriers who are transporting a peroxo-organic compound to implement emergency measures owing to the imminent danger that the compound may undergo thermal decomposition.

When shippers prepare the shipping description of a peroxo-organic compound, they select the applicable proper shipping name from the generic group in the Hazardous Materials Table, parenthetically identify the name of the specific compound offered for transportation, and include the concentration or the concentration range of the substance. For example, shippers use the following shipping description when they transport 50 lb (110 kg) of 80% dibenzoyl peroxide within a plastic-lined cardboard box:

Units	HM	Shipping Description (Identification Number, Proper Shipping Name, Primary Hazard Class or Division, Subsidiary Hazard Class or Division, and Packing Group)	Weight (lb)
1 box	X	UN3012, Organic peroxide type B, solid, 5.2, (1), PGII (dibenzoyl peroxide, paste, 80%)	50

When shippers offer for transportation a peroxo-organic compound whose control temperature and emergency temperature are listed in column 7 of the Organic Peroxides table, DOT requires them to include these temperatures on the shipping paper as shown in the following example:

Units	HM	Shipping Description (Identification Number, Proper Shipping Name, Primary Hazard Class or Division, Subsidiary Hazard Class or Division, and Packing Group)	Weight (lb)
2 boxes	X	UN3115, Organic peroxide type D, liquid, temperature-controlled, 5.2, PGII (diacetyl peroxide, 27%, and dimethyl phthalate, 73%) [Control temperature 68 °F (20 °C)] [Emergency temperature, 77 °F (25 °C)]	60

When shippers offer a peroxo-organic compound for transportation, DOT requires them to affix an ORGANIC PEROXIDE label to its packaging. DOT also requires shippers to mark each package containing a peroxo-organic compound with the following information:

- Name and address of the shipper and its receiver
- The proper shipping name
- Identification number of the commodity
- Applicable specifications, instructions, and precautions

In addition, when shippers offer the packaging for transportation by aircraft, DOT requires them to affix the KEEP AWAY FROM HEAT handling mark shown on the packaging. This handling mark is shown in Figure 13.9.

Keep away from heat

FIGURE 13.9 When packages containing self-reactive substances of Division 4.1 or organic peroxides of Division 5.2 are transported by aircraft, they must be marked with this KEEP AWAY FROM HEAT handling mark. The color of the starburst is red, but the other symbols, letters, and border are black on a white background.

When warranted, DOT requires carriers to post ORGANIC PEROXIDE placards on the bulk packaging or transport vehicle used for shipment. Carriers display ORGANIC PEROXIDE placards when they ship by highway or rail any amount of a liquid or solid type B, temperature-controlled organic peroxide or 1001 lb (454 kg) or more of the organic peroxides other than type B.

13.9-B RESPONDING TO INCIDENTS INVOLVING A RELEASE OF PEROXO-ORGANIC COMPOUNDS

Emergency responders identify the presence of peroxo-organic compounds at transportation mishaps by observing any of the following:

- The number 5.2 as a component of a proper shipping description of a hazardous material on a shipping paper
- The words ORGANIC PEROXIDE and the number 5.2 on yellow labels affixed to their packaging
- The words ORGANIC PEROXIDE and the number 5.2 on yellow placards posted on the vehicle used to transport the packages

Because many organic compounds are flammable gases or flammable liquids, their fires are fought using the general techniques noted in Sections 3.5 and 3.8. However, this statement does not ordinarily apply to fires involving the peroxo-organic compounds, since they are likely to undergo thermal decomposition. The generally recommended practice for fighting fires involving peroxo-organic compounds is to implement either of the following procedures:

- Use unmanned monitors to cool the area where these reactive materials are stored and assure that fire does not reach them.
- When a fire has engulfed the immediate area where peroxo-organic compounds are located, evacuate all personnel and do not combat the fire.

These recommended procedures are illustrated in Figure 13.10.

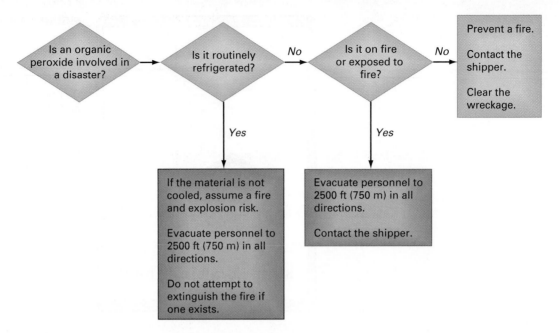

FIGURE 13.10 The recommended procedures when responding to a transportation mishap involving the release of an organic peroxide from its packaging. (*Adapted with permission of the American Society for Testing and Materials, from a figure in ASTMSTP 825, A Guide to the Safe Handling of Hazardous Materials Accidents, second edition. Copyright © 1990, American Society for Testing and Materials.*)

13.9-C TERRORISTS' MISUSE OF TRIACETONE TRIPEROXIDE

Triacetone triperoxide (TATP) is a peroxo-organic compound produced by the acid-catalyzed union of acetone and hydrogen peroxide. This substance is named 3,3,6,6,9,9-hexamethyl-1,2,4,5,7,8-hexaoxocyclononane using the IUPAC system. For obvious reasons, the substance is more widely known as triacetone triperoxide.

$$3(CH_3)_2-C{=}O(l) \quad + \quad 3H_2O_2(l) \quad \longrightarrow \quad [(CH_3)_2-C-O-O]_3(s) \quad + \quad 3H_2O(l)$$

Acetone   Hydrogen peroxide   Triacetone triperoxide   Water

The molecular formula of TATP consists of a nine-membered ring containing three peroxo groups, as shown:

Triacetone triperoxide (TATP)
3,3,6,6,9,9-Hexamethyl-1,2,4,5,7,8-hexaoxocyclononane

Given its molecular complexity, this formula is usually condensed to $[(CH_3)_2-C-O-O]_3$.

Although TATP is produced as crystals, it is generally encountered as a white powder that readily deflagrates and explodes when exposed to an ignition source. The deflagration and explosion are respectively expressed as follows:

$$2[(CH_3)_2-C-O_2]_3(s) \; + \; 21O_2(g) \; \longrightarrow \; 18CO_2(g) \; + \; 18H_2O(g)$$

Triacetone triperoxide   Oxygen   Carbon dioxide   Water

$$8[(CH_3)_2-C-O_2]_3(s) \; \longrightarrow \; 21(CH_3)_2-C=O(g) \; + \; 9CO_2(g) \; + \; 9H_2O(g)$$

Triacetone triperoxide   Acetone   Carbon dioxide   Water

Terrorists have used or attempted to use TATP clandestinely. The following terrorist acts are well acknowledged:

▪ In December 2001, while traveling to the United States onboard an international aircraft, the so-called shoe-bomber attempted to use triacetone triperoxide to accelerate the detonation of the explosive pentaerythritol tetranitrate (PETN) (Section 15.12). The terrorist had packed a blend of TATP and PETN within the soles of his shoes but was overtaken by passengers before he could successfully activate the mixture.

▪ Triacetone triperoxide was also used by four terrorists responsible for the London transit bombings on July 7, 2005 (p. 2). Packing the unstable substance and chemical explosives within backpacks, the terrorists detonated the mixture inside London's bus and underground subway system. Quantities of triacetone triperoxide were subsequently identified in the apartment used by one of the terrorists.

▪ Triacetone triperoxide was probably connected with an attempted terrorist plot in August 2006, during which terrorists sought to bomb multiple trans-Atlantic passenger jets bound for the United States from England. The terrorists intended to bring liquids onboard the planes in water bottles stored within their hand luggage. The nature of these liquids has never clearly been identified. British authorities were forewarned of the terrorists's plans and captured the conspirators before they were able to execute their plans.

▪ In September 2006, triacetone triperoxide was identified during the arrest of seven suspected terrorists in Vollsmose, Denmark.

▪ During September 2007, triacetone triperoxide was also identified during the arrest of eight suspected terrorists in Copenhagen, Denmark.

▪ Also during September 2007, German antiterrorist forces foiled plans initiated by three German terrorists who aimed to bomb the U.S. airbase at Ramstein as well as the U.S. and Uzbek consulates in Germany. The German forces uncovered more than 1600 lb (727 kg) of a 35% hydrogen peroxide solution, which the terrorists presumably intended to use for production of triacetone triperoxide.

PERFORMANCE GOALS FOR SECTION 13.10:

▪ Identify the primary risks associated with exposure to carbon disulfide.
▪ Identify the industries that use carbon disulfide.
▪ Describe the means shippers and carriers use to inform emergency responders of the hazards associated with encountering carbon disulfide during transportation mishaps.

13.10 Carbon Disulfide

Carbon disulfide is a colorless, water-insoluble, and highly volatile liquid whose chemical formula is CS_2. Whereas pure carbon disulfide has a pleasant odor, the industrial grades of carbon disulfide are generally yellow and exhibit cabbagelike odors. Some other physical properties of carbon disulfide are listed in Table 13.20.

TABLE 13.20	Physical Properties of Carbon Disulfide
Melting point	−169 °F (−112 °C)
Boiling point	115 °F (46 °C)
Specific gravity at 68 °F (20 °C)	1.26
Vapor density (air = 1)	2.6
Vapor pressure at 68 °F (20 °C)	300 mmHg
Flashpoint	−22 °F (−30 °C)
Autoignition point	212 °F (100 °C)
Lower explosive limit	1%
Upper explosive limit	44%

Carbon disulfide is usually prepared by reacting methane with vaporized sulfur at 1200 °F (650 °C).

$$2CH_4(g) \ + \ S_8(g) \ \longrightarrow \ 2CS_2(g) \ + \ 4H_2S(g)$$

Methane   Sulfur   Carbon disulfide   Hydrogen sulfide

The compound is used commercially as a solvent and as a raw material for the manufacture of viscose rayon, cellophane, and other textiles.

Because its flashpoint is −22 °F (−30 °C), carbon disulfide is a highly flammable liquid. Its flammable range extends from 1% to 44% by volume, and its vapor is 2.6 times heavier than air. The autoignition point of carbon disulfide is extremely low, 212 °F (100 °C), and the ignition of its vapor may be initiated by exposure to a hot steam pipe or an electric lightbulb located a considerable distance away. This combination of properties constitutes cause for grave concern to firefighters who identify carbon disulfide at an emergency-response scene.

Sulfur dioxide and carbon monoxide form as products of incomplete combustion when carbon disulfide burns.

$$2CS_2(g) \ + \ 5O_2(g) \ \longrightarrow \ 4SO_2(g) \ + \ 2CO(g)$$

Carbon disulfide   Oxygen   Sulfur dioxide   Carbon monoxide

To protect against inhaling these toxic substances, the use of self-contained breathing apparatus is essential when firefighters respond to major fires involving carbon disulfide.

The inhalation of carbon disulfide vapor is also harmful. The repeated inhalation of the vapor damages the liver and kidneys and permanently affects the central nervous system. The prolonged contact of carbon disulfide with the skin is also harmful, as the liquid is absorbed through the skin. In the worst case, the absorption is fatal.

13.10-A WORKPLACE REGULATIONS INVOLVING CARBON DISULFIDE

When carbon disulfide is present in the workplace, OSHA requires workers to limit employee exposure to a maximum vapor concentration of 20 ppm, averaged over an 8-hr workday.

13.10-B TRANSPORTING CARBON DISULFIDE

When shippers offer carbon disulfide for transportation, DOT requires them to describe the substance on a shipping paper as follows:

UN1131, Carbon disulfide, 3, (6.1), PGI

DOT also requires shippers to affix FLAMMABLE LIQUID and POISON labels on its packaging.

When carriers transport carbon disulfide in an amount exceeding 1001 lb (454 kg), DOT requires them to post FLAMMABLE placards on the bulk packaging or transport vehicle used for shipment.

PERFORMANCE GOALS FOR SECTION 13.11:

- Describe the chemical nature of the chemical warfare agents.
- Identify the general types of chemical warfare agents.
- Describe the response actions executed by emergency-response teams when a chemical warfare agent has been dispersed during a terrorist act.

13.11 Chemical Warfare Agents

Some substances are so highly toxic to humans that during wartime, their use could serve to injure, incapacitate, or cause mass casualties among the enemy. They are called **chemical warfare agents**. Their use during wartime has been disdained by civilized countries as an inhumane military act, but in the past, mutual mistrust and fear prompted their production and stockpiling. Signatories to the 1997 Chemical Weapons Convention Treaty are now prohibited from producing or stockpiling chemical warfare agents.

An example of a chemical warfare agent is chlorine (Section 7.4). Because it is a poisonous gas with a vapor density of 2.49 (air = 1), the discharge of chlorine by Germany during World War I caused mass casualties by maintaining a ground-level vapor concentration capable of causing death when inhaled.

Most substances potentially useful as chemical warfare agents are liquids that can be sealed into canisters or charged into missile warheads and other explosive devices. Because they vaporize when released into the atmosphere, these substances readily provide a lethal concentration on inhalation.

chemical warfare agent
■ A nerve gas, vesicant, or other substance that inflicts harm or causes mass casualties when disseminated

13.11-A NERVE AGENTS

Nerve agents are volatile liquids at room conditions. Most are complex organofluorophosphorus compounds of which the following molecular structures are representative:

Sarin, GB, or O-Isopropyl methylphosphonofluoridate

Cyclosarin, or O-Cyclohexylmethylfluorophosphonate

Tabun, or Ethyl N,N-dimethylphosphoroamidocyanidate

VX agent, or O-Ethyl-S-(2-diisopropylaminoethyl)-methyl phosphonothiolate

These formulas are similar to those of organophosphorus pesticides (Section 10.19), although the latter substances are not fluorinated.

Nerve agents are fairly simple to synthesize from readily available raw materials. For this reason, law enforcement authorities are sensitive to the fact that they could easily be obtained and used by terrorists as weapons of mass destruction.

SOLVED EXERCISE 13.13

Why is the use of ordinary self-contained breathing apparatus generally an ineffective defense when responding to a release of a nerve agent?

Solution: A nerve agent causes death when it is inhaled or absorbed through the skin. Although an ordinary self-contained breathing apparatus may protect an individual to some degree from inhaling a nerve agent, it is totally ineffective at preventing absorption of its vapor through the skin. Respiratory-protective gear rated for use in a nerve-agent environment with full protective clothing is the only suitable means by which emergency responders can be protected from exposure to a release of a nerve agent.

The physical properties of sarin, cyclosarin, tabun, and VX are noted in Table 13.21. These data illustrate that the vapors of these nerve agents are much heavier than air. Consequently, when the liquid agents are initially released from their containers, their vapors seek out low-lying areas. Sarin is the most volatile of them, and VX is the most lethal.

Nerve agents generally cause their ill effects by two mechanisms: inhalation and absorption through the skin. The initial exposure often occurs by penetration through the eyes. Once in the body, they interact with substances that cause the nerves to transmit impulses to nearby muscles. This action paralyzes the muscles, which in turn, causes convulsions, respiratory failure, and immediate death. The nerve agents thus act as neurotoxicants.

Although the U.S. military produced and stockpiled nerve agents during the Cold War, their use against an enemy has never been authorized. In contrast, Iraq used the nerve agent sarin in its war against Iran during 1984–1988, as well as against its own citizens in 1988. Sarin was also released in the Tokyo subway system in 1955 by members of the Aum Shinrikyo, a Japanese religious movement. The release caused the deaths of 12 persons and injured 5500 other individuals.

vesicant
■ A substance used in chemical warfare to blister the skin and body tissues of the enemy

13.11-B VESICANTS

Vesicants are substances that blister skin and damage the eyes, mucous membranes, and respiratory tract on exposure. The best-known vesicant is *mustard gas*, a yellowish brown,

TABLE 13.21	Physical Properties of Some Nerve Agents			
	SARIN	**CYCLOSARIN**	**TABUN**	**VX**
Melting point	−68.8 °F (−57 °C)	−22 °F (−30 °C)	−58 °F (−50 °C)	−58 °F (−50 °C)
Boiling point	316 °F (158 °C)	462 °F (239 °C)	464 °F (240 °C)	568 °F (298 °C)
Specific gravity at 68 °F (20 °C)	1.11	1.12	1.07	1.01
Vapor density (air = 1)	4.86	6.2	5.63	9.2
Vapor pressure at 68 °F (20 °C)	1.48 mmHg	0.044 mmHg	0.037 mmHg	0.00044 mmHg

oily liquid having a faint odor of garlic or mustard. Its chemical name is 2,2'-dichloroethyl sulfide.

$$ClCH_2CH_2 - S - CH_2CH_2Cl$$

2,2'-Dichloroethyl sulfide
(Mustard gas)

Mustard gas may be poured on the ground, sprayed into the air, or loaded into artillery shells and dropped from planes to serve as a chemical weapon of mass destruction.

The grisly success of mustard gas as a vesicant is due in part to its low vapor density of 5.4 (air = 1). The vapor hovers for a relatively long period at ground level, where it harms the enemy by causing painful, long-lasting blisters. Prolonged exposure to the eyes is highly irritating and causes temporary blindness. When it is inhaled, mustard gas may cause cancer of the respiratory tract.

The use of mustard gas by Germany against enemy troops during World War I is well documented. Discharged as a liquid, its vapor moved in the direction of the wind along the surface of the ground and into the trenches where soldiers were seeking a level of protection against live artillery. It was within the trenches that the deadly vapor inflicted the maximum harm on unsuspecting troops.

More than 6000 tn (5455 t) of mustard gas that was produced for use by the U.S. military still remains in storage, mainly in Utah. Destruction of this vesicant began in 2006.

Vesicants other than mustard gas include the compounds known in the military as the *nitrogen mustards*. These compounds are similar in molecular structure to 2,2'-dichloroethyl sulfide in that an amino nitrogen atom replaces the sulfur atom. There are three well-known nitrogen mustards: methyl bis(2-chloroethyl)amine, ethyl bis(2-chloroethyl)amine, and tris(2-chloroethyl)amine.

$$
\underset{\text{Methyl bis(2-chloroethyl)amine}}{ClCH_2CH_2 - \overset{\overset{\displaystyle CH_3}{|}}{N} - CH_2CH_2Cl}
\qquad
\underset{\text{Ethyl bis(2-chloroethyl)amine}}{ClCH_2CH_2 - \overset{\overset{\displaystyle CH_2CH_3}{|}}{N} - CH_2CH_2Cl}
\qquad
\underset{\text{Tris(2-chloroethyl)amine}}{ClCH_2CH_2 - \overset{\overset{\displaystyle CH_2CH_2Cl}{|}}{N} - CH_2CH_2Cl}
$$

These compounds are liquids whose vapors are especially heavy [vapor densities = 5.9, 5.4, and 7.1 (air = 1), respectively.]

The Geneva protocol (Section 7.4-A) now prohibits the use of mustard gas, the nitrogen mustards, and other poisonous gases during warfare.

13.11-C BLOOD AGENTS

Blood agents are highly volatile liquids that cause seizures, respiratory failure, and cardiac arrest by interfering with the blood's ability to absorb atmospheric oxygen. An example of a blood agent is hydrocyanic acid (Section 10.10-F), whose presence in the bloodstream inhibits the effective utilization of oxygen at the cellular level.

13.11-D PULMONARY AGENTS

Pulmonary agents are liquid compounds that cause pulmonary edema (Section 7.4) and damage the respiratory tract when their vapors are inhaled. An example of a pulmonary agent is chloropicrin, whose chemical formula is CCl_3NO_2.

$$
Cl - \overset{\overset{\displaystyle Cl}{|}}{\underset{\underset{\displaystyle Cl}{|}}{C}} - NO_2
$$

Nitrotrichloromethane
(Chloropicrin)

blood agent
■ A chemical warfare agents that cause seizures, respiratory failure, and cardiac arrest by interfering with the blood's ability to absorb atmospheric oxygen

pulmonary agent
■ A chemical warfare agent that damages the respiratory tract and cause pulmonary edema

Although chloropicrin is a chemical warfare agent, it has been used more frequently in low concentration as a soil fumigant to kill insects and other agricultural pests, especially by forest tree nurseries and tomato and strawberry growers.

13.11-E RESPONDING TO INCIDENTS INVOLVING A RELEASE OF A CHEMICAL WARFARE AGENT

One can only imagine how gruesome it must be to encounter a scene at which a chemical warfare agent has been dispersed. In particular, death comes quickly to those who are exposed to a nerve agent, blood agent, or pulmonary agent. Hence, when emergency-response personnel arrive at a scene where a chemical warfare agent has been dispersed, their major job may be limited to bringing a sense of calm and order to the prevailing pandemonium and providing immediate help to those people who were fortunate to survive the ordeal.

Most major cities are now prepared to respond to a terrorist act involving the release of a chemical warfare agent. The recommended response action involves the participation of both emergency-response and medical personnel. The major job of these first-on-the-scene responders is quickly to move the exposed individuals to an agent-free environment while wearing fully encapsulated suits and breathing air from self-contained sources. Victims must be encouraged to quickly remove their contaminated clothing and physically wash any exposed areas with soap and water. To ensure complete removal of the agent, experts recommend washing three times, giving special attention to shampooing the hair, to which the agent may cling. Emergency responders must also collect the contaminated clothing and seal it within bags for ultimate disposition.

It is only after the victims have been completely decontaminated that appropriate medical attention is given by health-care providers. The victims should have access to respirators, given an antidote designed to counteract the impact of the specific agent to which they were exposed, and monitored at a health-care facility for at least 24 hours.

Emergency responders must also tend to the deceased. The local coroner can provide appropriate directions for removing clothing and decontaminating the bodies prior to their transferral to the local morgue.

PERFORMANCE GOALS FOR SECTION 13.12:

- Describe the chemical nature of the commercially available lacrimators.

13.12 Lacrimators

lacrimator
- A substance that causes the eyes to involuntarily tear and close

riot-control agent
- A substance that rapidly produces sensory irritation or a disabling physical effect in humans and animals and disappears within a short time following termination of exposure

Lacrimators (Section 10.8-A) are substances that cause the eyes to involuntarily tear and close. Although lacrimators have been used against the enemy during warfare, they are also dispersed by law enforcement personnel to control unruly crowds and discourage unlawful acts. During their use in law enforcement and military actions, they render the recipient temporarily incapable of resistance or flight. They are commonly known as **riot-control agents**, or *tear gases*.

Riot-control agents are used defensively to control the action of activists and troops by temporarily impairing their vision and causing them to choke and breathe painfully. It is the *temporary* aspect that causes the discriminate use of lacrimators by authorized individuals to be considered humane.

Some representative molecular formulas of lacrimators are shown below:

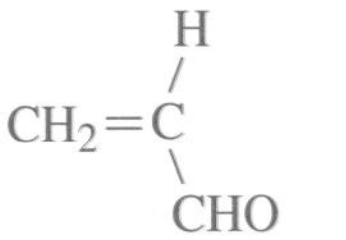

Acrolein, or 2-Propenal

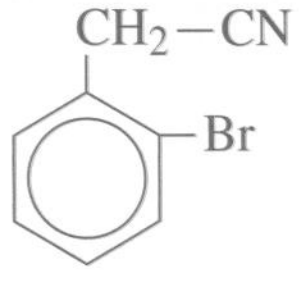

o-Bromobenzyl cyanide

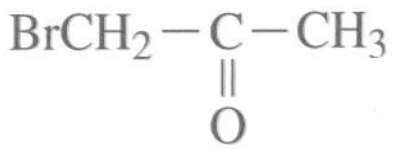

Bromoacetone

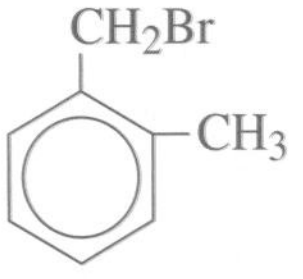

α-Bromoxylene

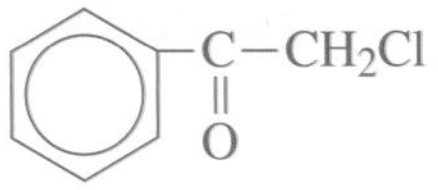

α-Chloroacetophenone
(Mace)

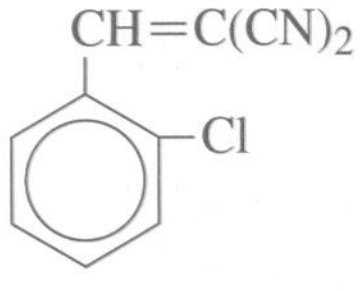

o-Chlorobenzylidene malononitrile
(CS gas)

During World War I, o-bromobenzyl cyanide was extensively used as a lacrimator, most likely because uncontrollable tearing resulted from an exposure to a concentration in air of only 0.0000003 g/L (3.0×10^{-7} g/L).

PERFORMANCE GOALS FOR SECTION 13.13:

- Describe the chemical nature of an incendiary agent such as napalm.

13.13 Napalm

Certain substances are regularly used during warfare to intentionally initiate fires. They are called **incendiary agents**. The use of white phosphorus in incendiary bombs, and triethylaluminum in flamethrowers, was previously noted in Sections 7.5 and 9.4, respectively.

During warfare, several materials have been used to thicken petroleum products for use as incendiary agents. The first incendiary agent of this type was produced in 1942. It was called **napalm** and consisted of a mixture of aluminum compounds made from several triglycerides including those found naturally in coconut oil. The following substances are representative:

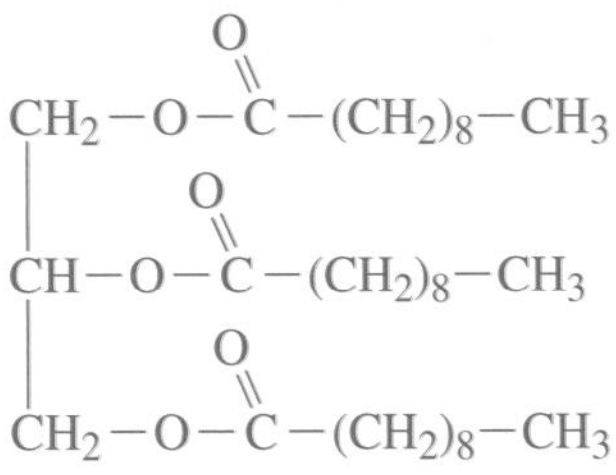

Glyceryl tricaprate

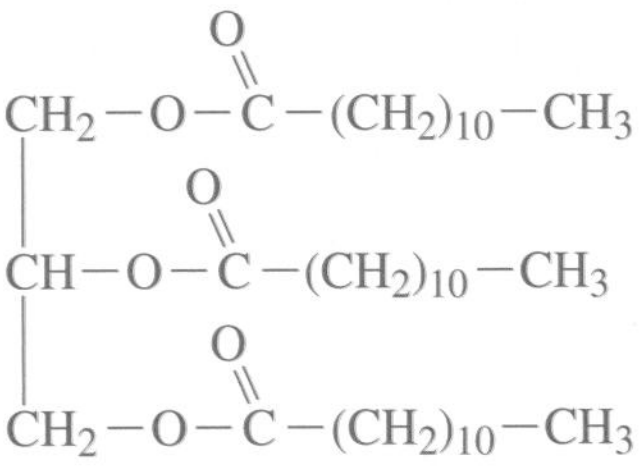

Glyceryl trilaurate

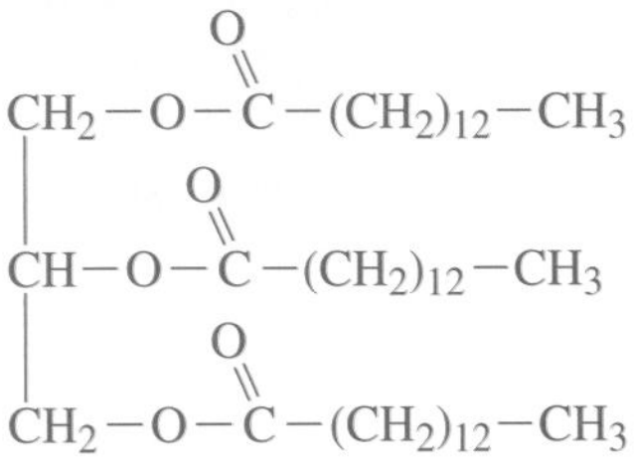

Glyceryl trimyristate

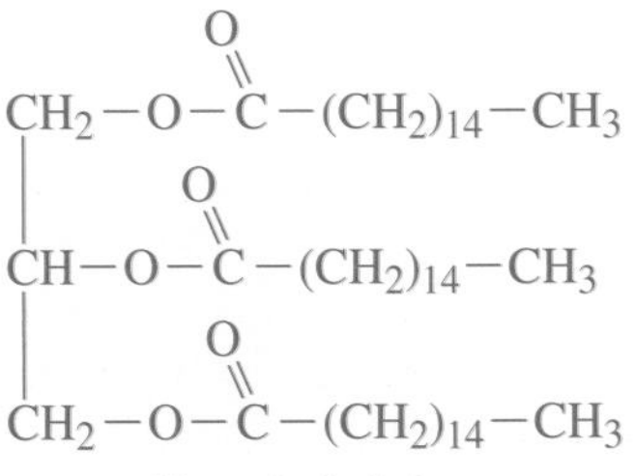

Glyceryl tripalmitate

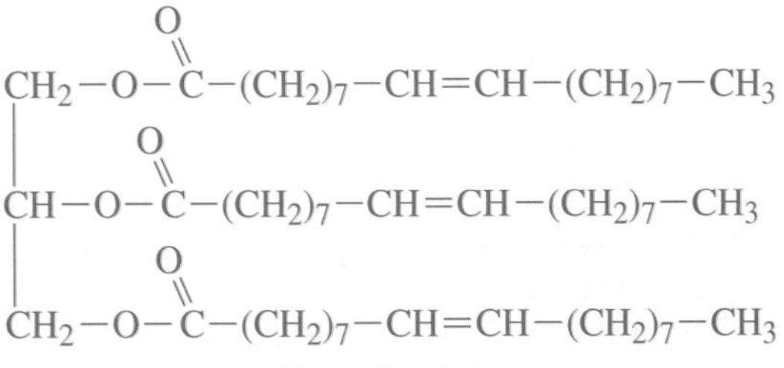

Glyceryl trioleate

incendiary agent
- A substance such as elemental white phosphorus and napalm, primarily used during warfare to intentionally set fire to objects or to cause burn injury to persons through the action of flames, heat, or a combination thereof

napalm
- The mixture of aluminum compounds produced by reacting aluminum hydroxide with certain substances such as those found in coconut oil

The conversion of these triglycerides into napalm is accomplished by saponification (Section 5.16) with aluminum hydroxide and excess sodium hydroxide. The reaction of glyceryl tripalmitate with aluminum hydroxide is typical.

$$3 \begin{array}{c} CH_2-O-\overset{\overset{\displaystyle O}{\|}}{C}-(CH_2)_{14}-CH_3 \\[1em] CH-O-\overset{\overset{\displaystyle O}{\|}}{C}-(CH_2)_{14}-CH_3 \\[1em] CH_2-O-\overset{\overset{\displaystyle O}{\|}}{C}-(CH_2)_{14}-CH_3 \end{array}(l) \; + \; Al(OH)_3(l) \; \longrightarrow \; Al[CH_3-(CH_2)_{14}-COO]_3(s) \; + \; 3\begin{array}{c} CH_2OH \\ CHOH \\ CH_2OH \end{array}(l)$$

Glyceryl tripalmitate   Aluminum hydroxide   Aluminum palmitate   Glycerol

Napalm thickens gasoline until a mixture containing approximately 4% napalm by volume is produced. The resulting material is jellylike in consistency. When napalm or a napalmlike substance is used as the active agent in firebombs, an explosive substance triggers the burning of the gasoline, jet fuel, or kerosene. When these firebombs are activated, massive fireballs are often produced.

For conducting military activities, the use of fuel thickened with napalm is associated with results that are provided by few other materials. Napalm increases the range of flamethrowers, imparts slower-burning properties compared with gasoline alone, adds a "clinging" feature, and causes the burning flames to move around corners and rebound off walls and other surfaces. Its use during warfare can psychologically affect the enemy.

Upgraded versions of the napalm formulation have now been produced. Modern versions consist of jet fuel or kerosene thickened with a polystyrene-based gel. During World War II and the Korean and Vietnam conflicts, napalm formulations were used as incendiary agents in firebombs and portable and mechanical flamethrowers. The burning of napalm controlled the extent of unwanted vegetation and routed the enemy from hiding places within jungles and other densely vegetated terrains.

Today, however, most countries have agreed to limit the use of all incendiary weapons when they are likely to affect civilian populations. Protocol III of the Convention on Certain Conventional Weapons (Section 7.5) effectively limits the future use of napalm and napalmlike incendiary agents. Because the United States did not ratify this protocol, the U.S. military continues to use incendiary agents in warfare. For example, during the initial advance into Baghdad in 2003, U.S. pilots dropped napalmlike incendiary bombs called *Mark 77 firebombs* on Iraqi troops; later, these firebombs were again used during an attack on an observation post at Safwan Hill, a location near the Iraq–Kuwait border.

Classes of Organic Compounds

13.1 Use Table 13.1 to identify the class of organic compound represented by each of the following molecular formulas:

(a) $CH_3CH_2-O-CH_2CH_3$

(b) 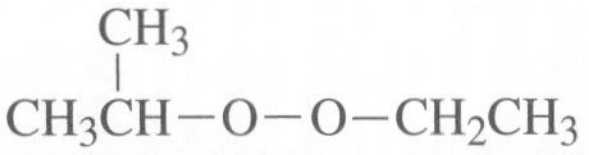 $CH_3\overset{\overset{\displaystyle CH_3}{|}}{CH}-O-O-CH_2CH_3$

(c) 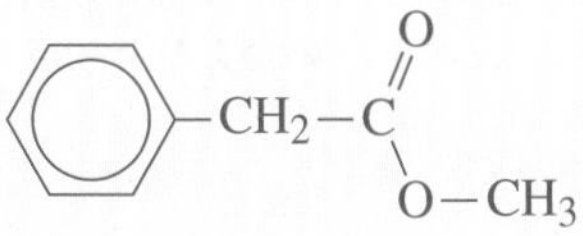 $-CH_2-\overset{\overset{\displaystyle O}{||}}{C}\underset{O-CH_3}{}$

(d) 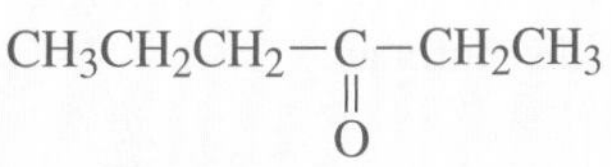 $CH_3CH_2CH_2-\underset{\underset{\displaystyle O}{||}}{C}-CH_2CH_3$

Alcohols

13.2 Provide the IUPAC name for the alcohol having each of the following condensed formulas:

(a) $CH_3CH_2CH_2CH_2OH$

(b) 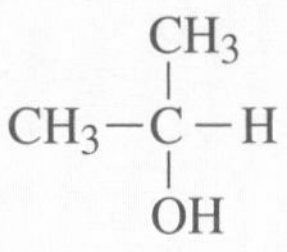 $CH_3-\overset{\overset{\displaystyle CH_3}{|}}{\underset{\underset{\displaystyle OH}{|}}{C}}-H$

(c) $H-\overset{\overset{\displaystyle CH_3}{|}}{\underset{\underset{\displaystyle CH_3}{|}}{C}}-CH_2OH$

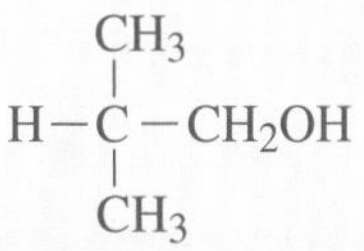

(d) 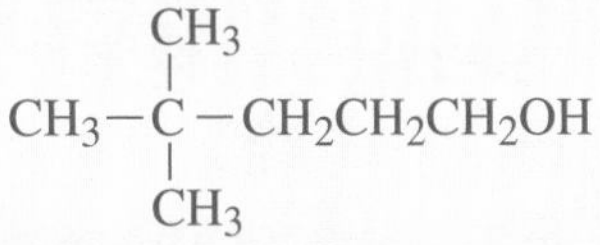 $CH_3-\overset{\overset{\displaystyle CH_3}{|}}{\underset{\underset{\displaystyle CH_3}{|}}{C}}-CH_2CH_2CH_2OH$

13.3 Write the chemical formula for the alcohol having each of the following names:

(a) isopropyl alcohol
(b) *p*-cresol
(c) *tert*-butyl alcohol
(d) 3-ethylphenol
(e) 2,2-dimethyl-1-propanol
(f) 1-methylcyclopentanol

13.4 Why is the flame associated with the burning of methanol almost invisible to the naked eye?

Ethers

13.5 Provide an acceptable name for the ether having each of the following condensed formulas:

(a) $CH_3CH_2-O-CH_2CH_2CH_3$

(b) 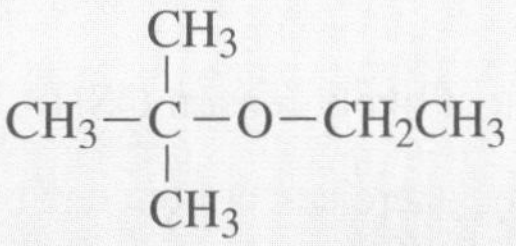 $CH_3-\overset{\overset{\displaystyle CH_3}{|}}{\underset{\underset{\displaystyle CH_3}{|}}{C}}-O-CH_2CH_3$

(c) CH_3CH_2-O-

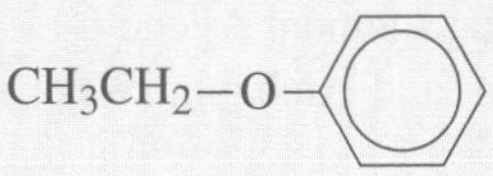

(d) 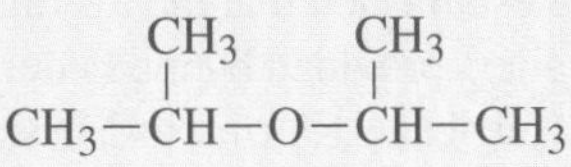 $CH_3-\overset{\overset{\displaystyle CH_3}{|}}{CH}-O-\overset{\overset{\displaystyle CH_3}{|}}{CH}-CH_3$

13.6   Provide molecular structures for each of the following ethers:

(a) methoxybenzene
(b) methyl *tert*-butyl ether
(c) ethyl phenyl ether
(d) 2-methoxybutane

13.7   Why should ethers be purchased in small quantities, kept in tightly sealed containers, and used promptly?

Halogenated Ethers

13.8   In the past, wooden railroad ties were soaked in a solution of pentachlorophenol, which served to kill the microorganisms responsible for decay of the wood. Why has this procedure been discontinued?

Aldehydes and Ketones

13.9   Provide an acceptable name for the aldehyde or ketone having each of the following condensed formulas:

(a) 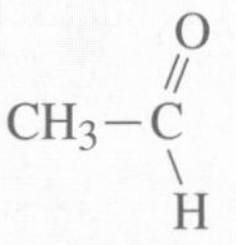 $CH_3-\overset{\overset{O}{\parallel}}{C}\underset{H}{\diagdown}$

(b) $CH_3CH_2-\overset{\overset{O}{\parallel}}{C}\underset{H}{\diagdown}$

(c) $CH_3-\underset{\underset{CH_3}{|}}{CH}-CH_2-\overset{\overset{O}{\parallel}}{C}\underset{H}{\diagdown}$

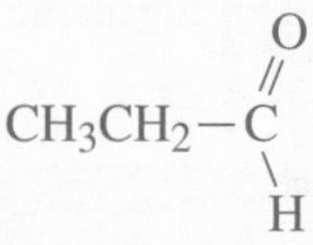

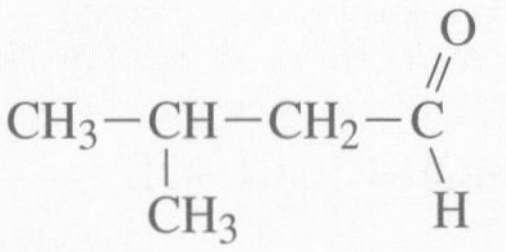

(d) 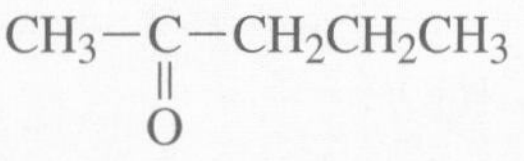 $CH_3-\underset{\underset{O}{\parallel}}{C}-CH_2CH_2CH_3$

13.10   Write the chemical formula for each aldehyde or ketone named as follows:

(a) 2-methylpropanal
(b) pentanal
(c) methyl *n*-propyl ketone
(d) 3-hexanone

13.11   The OSHA Formaldehyde Standard at 29 C. F. R. §1910.1048(h)(2)(i) requires employers to post the following sign in areas where formaldehyde-contaminated clothing or equipment are ventilated:

> **DANGER**
>
> **FORMALDEHYDE-CONTAMINATED [CLOTHING EQUIPMENT]**
>
> **AVOID INHALATION AND SKIN CONTACT**

What is the most likely reason OSHA promulgated this regulation?

13.12   Air-monitoring exercises conducted by the Centers for Disease Control and Prevention during 2008 established that the average concentration of formaldehyde was 77 ppb in the air inside trailers provided by the Federal Emergency Management Agency (FEMA) to the victims of Hurricane Katrina in Mississippi, Louisiana, and Alabama. What is the most logical technical basis for explaining how formaldehyde originated within these trailers?

Organic Acids

13.13 Provide an acceptable name for the organic acid having each of the following condensed formulas:

(a) $CH_3-\overset{\displaystyle O}{\underset{\displaystyle OH}{C}}$

(b) $CH_3-\overset{\displaystyle CH_3}{\underset{\displaystyle CH_3}{C}}-\overset{\displaystyle O}{\underset{\displaystyle OH}{C}}$

(c) $CH_3CH_2-\overset{\displaystyle CH_3}{CH}-CH_2-\overset{\displaystyle O}{\underset{\displaystyle OH}{C}}$

(d) $CH_3-\bigcirc-\overset{\displaystyle O}{\underset{\displaystyle OH}{C}}$

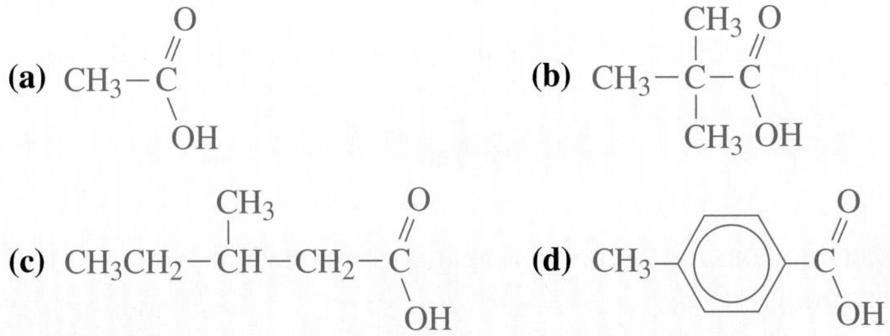

13.14 Write the chemical formula for each organic acid named as follows:

(a) acetic acid
(b) propionic acid
(c) methanoic acid
(d) phenoxyacetic acid

Esters

13.15 Provide an acceptable name for the ester having each of the following condensed formulas:

(a) $H-\overset{\displaystyle O}{\underset{\displaystyle O-CH_2CH_3}{C}}$

(b) $CH_3-\overset{\displaystyle O}{\underset{\displaystyle O-CH_2-\bigcirc}{C}}$

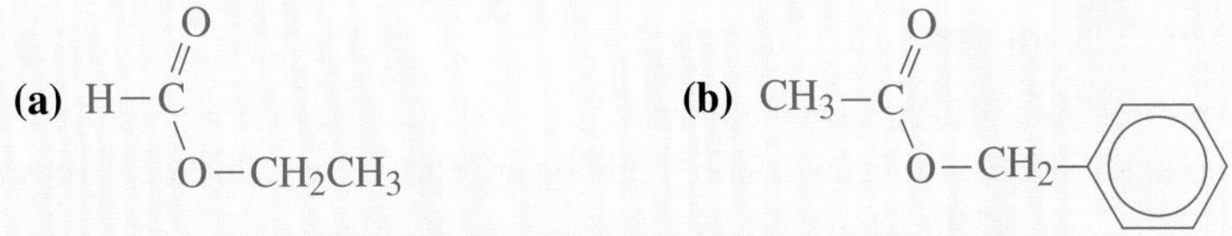

13.16 Write the chemical formulas for each of the following esters:

(a) *n*-propyl formate
(b) isopropyl benzoate
(c) ethyl propanoate
(d) cyclohexyl propanoate

13.17 On entering an enclosed warehouse where ethyl butyrate is known to be stored, firefighters note an intensely strong odor of pineapples. Why does the perception of pineapples warn firefighters of the risk of fire and explosion within the warehouse?

Amines

13.18 Provide an acceptable name for the amine having each of the following condensed formulas:

(a) $CH_3CH_2CH_2NH_2$

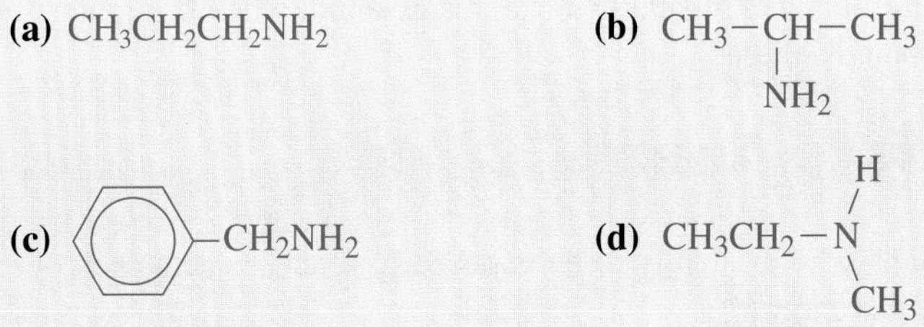

(b) $CH_3-\overset{\displaystyle CH}{\underset{\displaystyle NH_2}{CH}}-CH_3$

(c) $\bigcirc-CH_2NH_2$

(d) $CH_3CH_2-\overset{\displaystyle H}{\underset{\displaystyle CH_3}{N}}$

13.19 Write the chemical formula for each of the following amines:

(a) *N*-ethylaniline
(b) 2-aminobutane
(c) *tert*-butylamine
(d) diisopropylamine
(e) 2-butanamine
(f) dimethylethylamine

Peroxo-Organic Compounds

13.20 Provide an acceptable name for the peroxo-organic compounds having each of the following condensed formulas:

(a) 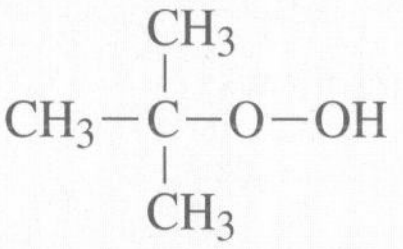
$$CH_3-\underset{\underset{CH_3}{|}}{\overset{\overset{CH_3}{|}}{C}}-O-OH$$

(b)

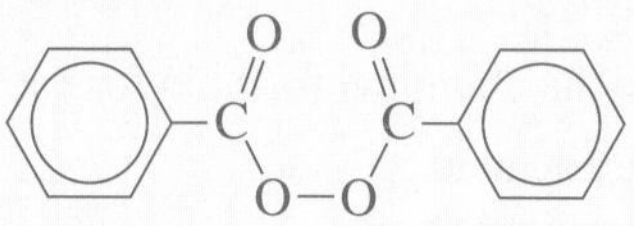

(c) 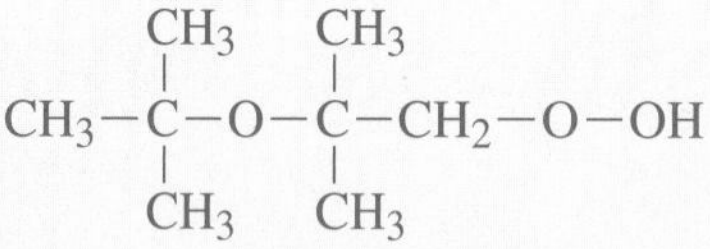
$$CH_3-\underset{\underset{CH_3}{|}}{\overset{\overset{CH_3}{|}}{C}}-O-\underset{\underset{CH_3}{|}}{\overset{\overset{CH_3}{|}}{C}}-CH_2-O-OH$$

(d) 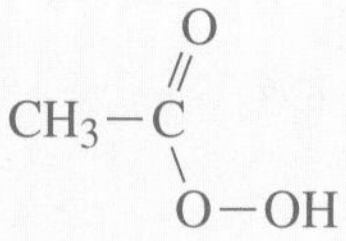
$$CH_3-\overset{\overset{O}{\|}}{C}\overset{\diagdown}{_{O-OH}}$$

13.21 Write the condensed formula for each of the following peroxo-organic compounds:

(a) di-*n*-hexyl peroxide
(b) bis-*p*-chlorobenzoyl peroxide
(c) diisopropylbenzene hydroperoxide
(d) acetyl benzoyl peroxide

13.22 What information on a NFPA hazard diamond rapidly informs firefighters that peroxo-organic compounds stored within an isolated building are particularly hazardous substances?

13.23 A chemical manufacturer desires to transport 400 kg of liquid 1,1,3,3-tetramethylbutyl hydroperoxide by motor vehicle within 18 plastic-lined fiberboard boxes. Testing of this compound establishes that it neither detonates nor deflagrates rapidly, and when confined and heated, it manifests no violent effect.

(a) What shipping description does DOT require the manufacturer to enter on the accompanying shipping paper?
(b) Which labels does DOT require the manufacturer to affix on the outside surface of each package containing 1,1,3,3-tetramethylbutyl hydroperoxide?
(c) Which markings does DOT require the manufacturer to print on the outside surface of each package?
(d) Which placards, if any, does DOT require the carrier to post on the motor vehicle used to transport the packages?

Carbon Disulfide

13.24 Why is the use of self-contained breathing apparatus essential when firefighters respond to fires involving carbon disulfide?

Chemical Warfare Agents

13.25 Why should individuals who are likely to be exposed to a nerve agent seek out refuge above the point of its discharge?

Responding to Incidents Involving a Release of Organic Compounds

13.26 Why should first-on-the-scene responders to a highway transportation mishap exercise special precaution on reading the following entry on the bill of lading accompanying the shipment?

Units	HM	Shipping Description (Identification Number, Proper Shipping Name, Primary Hazard Class or Division, Subsidiary Hazard Class or Division, and Packing Group)	Weight (lb)
25 plastic bottles	X	UN3114, Organic peroxide type C, solid, temperature-controlled, 5.2, PGII	150

13.27 When first-on-the-scene responders arrive at the scene of a rail freight accident, they note that at least one side of an overturned boxcar is posted with a yellow ORGANIC PEROXIDE placard. There is not an ongoing fire. A member of the train crew provides the responders with the waybill, the relevant portion of which reads as follows:

Units	HM	Shipping Description (Identification Number, Proper Shipping Name, Primary Hazard Class or Division, Subsidiary Hazard Class or Division, and Packing Group)	Weight (lb)
75 plastic bottles	X	UN3115, Organic peroxide type D, liquid, temperature-controlled, 5.2, PGII (methylcyclohexanone peroxide, 67%) (Placarded ORGANIC PEROXIDE)	1050
		(Control temperature, 35 °C; Emergency temperature, 40 °C)	

What immediate actions should the emergency responders execute to protect public health, safety, and the environment?

Chemistry of Some Polymeric Materials

Over the past 75 years, our way of life has been dramatically changed by the polymer industry. It is unlikely that a modern civilized society could long survive without the wares this industry provides. In today's world, we regularly use products manufactured from both natural and synthetic polymers. Our clothing has been made from polymeric fibers, including cotton, polyesters, nylon, and polyacrylics. Our homes are constructed from wood, insulated with polystyrene, carpeted with polypropylene, coated with polyacrylic paints, and decorated with polyacrylonitrile and other polymeric fabrics.

An appreciable portion of our automobiles have also been manufactured from polymers. Often, the bumpers are made of an acrylonitrile–butadiene–styrene copolymer, the roofs are made of poly(vinyl chloride), the upholstery is cushioned with polyurethane, and the rubber tires are manufactured from a styrene–butadiene copolymer.

Because they are stable at ambient conditions and do not routinely pose a health risk, polymers are not ordinarily considered hazardous materials. However, most polymeric products burn and generate toxic gases on combustion. Not only do most polymeric products burn, but their combustion is involved in virtually all common fires. For these reasons, the burning of polymers is a topic of great concern to firefighters.

In this chapter, to understand why they pose special hazards during fires, we examine the features and structural characteristics of several commonly encountered polymers.

PERFORMANCE GOALS FOR SECTION 14.1:

- Note the commonplace of polymeric materials in certain commercial products and relate the importance of this fact to emergency responders.
- Distinguish between thermoplastic and thermosetting plastics.
- Describe the method used to name simple polymers.

14.1 What Are Polymers?

Polymers are substances that are best characterized by the relatively sizable nature of their molecules. Because these molecules are substantially larger than those otherwise encountered, chemists call them **macromolecules**. Each polymer macromolecule comprises a number of repeating smaller units; a polymer is a compound typically composed of hundreds or thousands of repeating units.

Polymers are sometimes described by their response to heat. Some polymers soften when exposed to heat but then return to their original shape when cooled. They are said to be **thermoplastic polymers**. The following are examples of thermoplastic polymers: polyethylene, polypropylene, polystyrene, poly(ethylene terephthalate), and poly(vinyl chloride).

Other polymers solidify or set irreversibly when they are heated. They are called **thermosetting polymers**. Polyurethane is an example of a thermosetting polymer.

polymer
- A high-molecular-weight substance produced by the linkage and cross-linkage of its multiple subunits (monomers)

macromolecule
- The giant molecule of which polymers are composed, comprising an aggregation of hundreds or thousands of atoms and typically consisting of repeating chemical units linked together into chains and cross-linked into complex three-dimensional networks

thermoplastic polymer
- Any polymer that softens when heated but returns to its original condition on cooling to ambient temperature

thermosetting polymer
- Any polymer that cannot be remolded once it has solidified

SOLVED EXERCISE 14.1

Both the white and yolk of an uncooked egg consist of a mixture of various proteins, all of which are polymers. Why is a cooked egg an example of a natural thermosetting polymer?

Solution: The uncooked egg white and yolk consist of protein strands in a semiliquid state. Although the protein strands are tightly curled and entwined when the egg is raw, the strands relax, unfold, and then interlink with one another until they are firmly set when the egg is cooked. This process is called *coagulation*. [Typically, the protein strands in the uncooked egg begin to coagulate in the temperature range from 145 to 150 °F (63 to 66 °C)]. This is an irreversible process. No physical activity returns the linked strands to their initial semiliquid condition. For this reason, a cooked egg is an example of a thermosetting polymer.

Polymers are often classified according to the ways they are used. For instance, some polymers are used to produce plastic products. **Plastics** are polymers that can be shaped by means of molding, casting, extrusion, calendering, laminating, foaming, and blowing. Examples of some common plastics are poly(vinyl chloride), polystyrene, and poly(methyl methacrylate). The polymers in plastic products are usually combined with other materials including fillers and reinforcing agents.

plastics
- Any of a variety of synthetic polymeric substances that can be molded and shaped

fiber
■ A polymeric substance that can be separated into threads or thread-like structures

textile
■ A fabric that has been produced by weaving fibers

elastomer
■ A synthetic polymer that elongates or stretches under strain but is incapable of retaining the deformation when the strain is released

Some natural and synthetic polymers are used as threads or yarns collectively called **fibers**. These polymers are characterized by a high tenacity and a high ratio of length to diameter (several hundred to one). They differ widely in form, flexibility, durability, and porosity.

Natural fibers are a class of polymers derived from vegetable and animal sources. They include cotton and wool, respectively. *Synthetic fibers* are a class of polymers made by some non-natural means. Two examples of synthetic fibers are nylon and poly(acrylonitrile). Synthetic fibers also include polymers that have been chemically modified in some way. They are used to manufacture numerous goods including rope, woven cloth, matted fabrics, brushes, shingles and other building and insulating materials, as well as stuffing for pillows and upholstery.

Natural and synthetic fibers are woven or knitted to produce **textiles**. These are items such as garments, carpets, carpet padding, towels, curtains, blankets, mattresses, and upholstery fabrics.

A polymer may also be an **elastomer**. This is a type of synthetic polymer that is characterized by the ability of its molecules to elongate when strained and to reversibly assume their original shape when the tension has been released. Examples of elastomers are neoprene and *cis*-1,4-polybutadiene. Products made from them include rubber bands, belts, footwear, vehicular tires, and the inner tubes of tires.

Various means have been used to name polymers. Although several common names are used, chemists often place the prefix *poly-* before the name of the substances used to produce specific polymers. Polyethylene, poly(vinyl chloride), and polystyrene are so named because they are produced from ethylene, vinyl chloride, and styrene, respectively. When a polymer is manufactured from a number of substances, all their names may be included in the name of the polymer. The acrylonitrile–butadiene–styrene copolymer commonly used in plastic sewer pipes is so named because it is made from acrylonitrile, butadiene, and styrene. In this instance, the name has thankfully been abbreviated for simplicity to *ABS* copolymer, where each letter represents a monomer used in its production.

PERFORMANCE GOALS FOR SECTION 14.2:

■ Distinguish between the chemical nature of a polymer and a copolymer.
■ Distinguish between addition and condensation polymerization reactions.
■ Explain how cross-linking and the use of plasticizers affect the properties of polymers.

14.2 Polymerization

Polymerization is a unique type of chemical reaction involving the union of certain substances called **monomers**. The equations for such reactions are generally denoted as follows, where A and B are arbitrary monomers:

$$A + B \longrightarrow Polymer$$

Only a select number of compounds undergo polymerization.

A polymerization reaction is characterized by the nature of the macromolecules it produces. These macromolecules can be visualized as follows:

■ Figure 14.1 illustrates one type of polymeric macromolecule. Its structure resembles the midsection of a freight train having identical boxcars. Like a boxcar, the repeating unit in this portion of the macromolecule has couplings at its front and rear ends, here symbolized by chemical bonds.
■ Figure 14.2 illustrates a second type of polymeric macromolecule. Its structure resembles the midsection of a freight train that has alternating dissimilar railcars. The substance having such macromolecules is called a **copolymer**. Although the portion of the macromolecule shown in Figure 14.2 has regularly repeating units, a copolymer may also be composed of units that alternate irregularly.

When chemists examine the three-dimensional structures of polymers, they find that these chains of repeating units are invariably cross-linked. Figure 14.3 illustrates a macromolecule in

polymerization
■ The chemical reaction in which monomer molecules are linked and cross-linked into macromolecules

monomer
■ One or more of the single substances that combine to produce a polymer

copolymer
■ A polymer produced from two or more different monomers

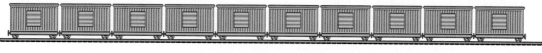

$$-CH_2-CH_2-CH_2-CH_2-CH_2-CH_2-$$

FIGURE 14.1 This midsection of a freight train consists of identical interconnected railcars. It resembles the partial structure of polyethylene, whose macromolecules have the repeating unit $\sim\sim CH_2-CH_2\sim\sim$.

which one chain has attached itself to another chain by means of a chemical bond. Plastics manufacturers often attempt to intentionally increase the degree of **cross-linking** in thermosetting plastics. The resulting polymers are more dense, and thus stronger and more durable, than those whose macromolecules have unlinked chains of atoms. Cross-linking the chains within macromolecules also potentially makes the products manufactured from them more elastic.

Macromolecular chains within polymers can also be folded, coiled, stacked, looped, or intertwined into definite three-dimensional shapes. Although these complex configurations give polymers unique properties, we require only the information conveyed by their one-dimensional patterns.

Polymer manufacturers sometimes discover that their products are too stiff and brittle for their intended use. These undesirable features can often be overcome by adding a *plasticizer* to the polymer. This is usually a liquid that manufacturers dissolve within the polymer. It causes the polymer to become flexible by lowering the attraction between the polymer chains. The most common plasticizers are phthalates (Section 13.7-B).

The production of polymers is a major activity in the chemical industry. These processes are accomplished primarily by unique chemical reactions called *addition* and *condensation*. The polymers resulting from these reactions are called **addition polymers** and **condensation polymers**, respectively. We review them independently.

cross-linking
■ The production of chemical bonds in multiple dimensions that is typically associated with the strength, durability, and elasticity within the macromolecules of certain polymers

addition polymer
■ A polymer resulting from the addition of a molecule, one by one, to a growing polymer chain

condensation polymer
■ A polymer produced by a chemical reaction along with a small molecule like water or ammonia

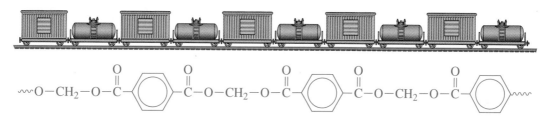

FIGURE 14.2 This midsection of a freight train consists of alternating interconnected railcars and tankcars. It resembles the partial structure of the styrene–butadiene copolymer, whose macromolecules have the following repeating unit:

$$\sim\sim (CH_2-CH=CH-CH_2)_x-CH-CH_2\sim\sim$$

FIGURE 14.3 A cross-linked polymer. The colored circles designate an arbitrary monomer, not its individual atoms. Cross-linking within and between macromolecules gives the polymer extra strength and durability.

14.2-A ADDITION POLYMERIZATION

Polystyrene is an example of an addition polymer. It is produced by the polymerization of the monomer styrene, or phenylethene, as follows:

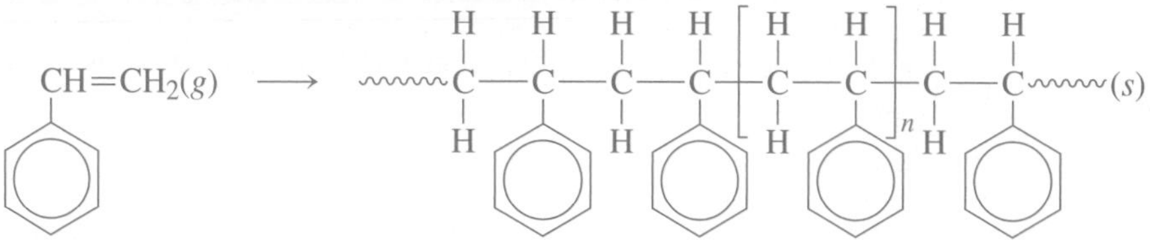

Styrene   Polystyrene

The chemical formula on the left of the arrow represents styrene, the monomer. The formula on the right of the arrow represents a section of the polystyrene macromolecule. The portion of this formula represented in parenthesis is repeated over and over again (n times, where n is a very large integer). The symbol $\sim\!\sim\!\sim$ denotes that the unit is repeated.

In the chemical industry, polymerization is initiated in a controlled fashion. One way to initiate polymerization is to use substances capable of forming free radicals (Section 5.10) when they are exposed to heat or light. Examples of such chemical initiators are peroxo-organic compounds (Section 13.9) and the Ziegler–Natta catalysts (Section 9.4). Once free radicals are produced by the thermal decomposition of a peroxo-organic compound, they combine with neutral monomer molecules to form more complex free radicals that react with other monomer molecules until the supply of the monomer has been exhausted.

For illustrative purposes, consider the polymerization of styrene induced by free radicals resulting from the dissociation of dibenzoyl peroxide. This polymerization occurs by means of a number of independent steps, some of which are represented as follows:

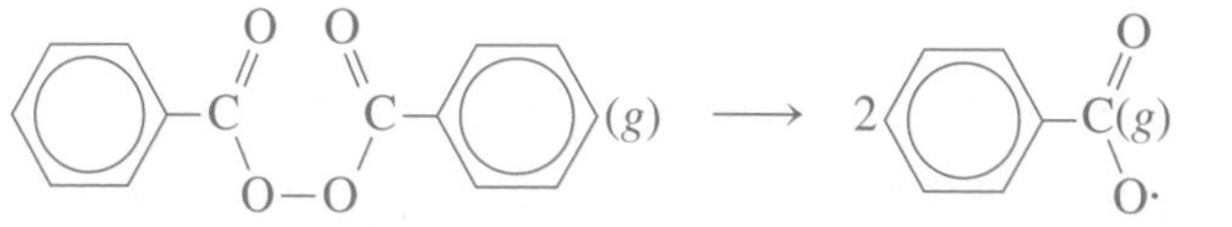

Dibenzoyl peroxide   Benzoylperoxy radical

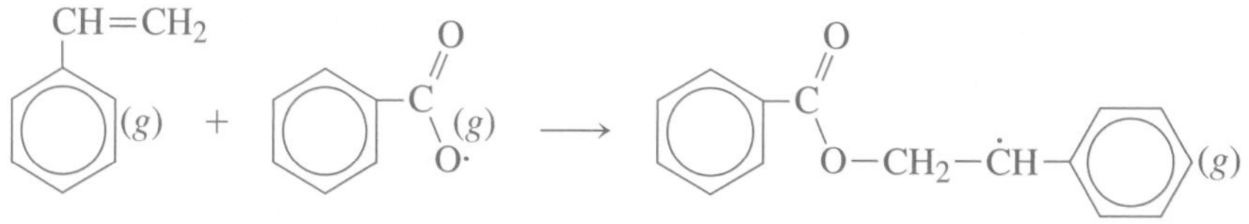

Styrene   Benzoylperoxy radical   Polymer fragment

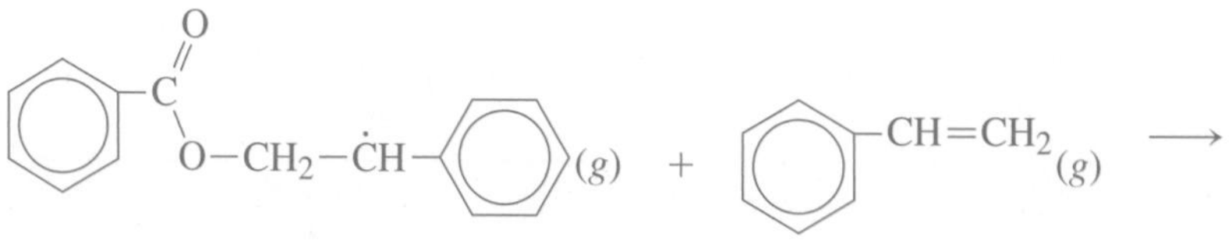

Polymer fragment   Styrene

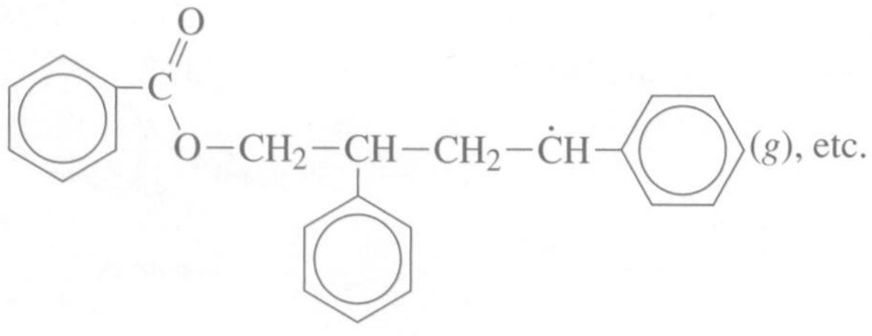

Polymer fragment

In the first equation, the oxygen–oxygen bond in dibenzoyl peroxide is ruptured, resulting in the production of benzoylperoxy free radicals. In the second equation, a benzoylperoxy free radical reacts with a styrene molecule forming a more complex free radical. In the third equation, this free radical reacts with another molecule of styrene to form a still more complex free radical. Additional steps beyond those illustrated add successively more units to the chain in a self-propagating fashion until a long chain of the polymer has been produced and the monomer source has been consumed.

The substance produced by the polymerization of styrene is called *polystyrene*. The repeating unit in this polymer is the following:

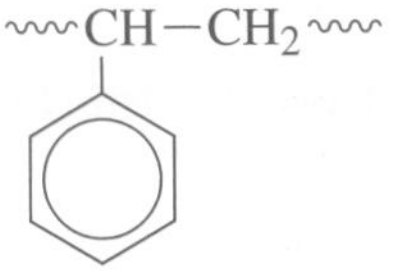

Polystyrene is commercially used to manufacture products such as brushes, combs, disposable coffee cups, thermally insulated equipment, building and electrical insulation, coaxial television cable, compact disk cases, yogurt cups, refrigerator interiors, and the "peanuts" used in packaging. For several of these purposes, polystyrene is mixed with a foam-blowing compound (Section 12.16) during processing and manufacturing.

Polystyrene is also recyclable. Its recycling symbol is the familiar arrowed triangle enclosing the number 6, beneath which appear the letters PS.

$$\triangle$$
6
PS

Several other examples of addition polymers are provided in Table 14.1.

SOLVED EXERCISE 14.2

Poly(methyl methylacrylate) is produced by the addition polymerization of methyl methacrylate, a substance having the following molecular formula:

$$
\begin{array}{c}
CH_3 \\
| \\
CH_2{=}C \\
| \\
C{=}O \\
| \\
O \\
| \\
CH_3
\end{array}
$$

Dibenzoyl peroxide is used to initiate polymerization. Write the stepwise equations for this polymerization reaction.

Solution: Addition polymerization reactions occur by a mechanism that includes the production of free radicals. Free radicals are first produced when a peroxo-organic compound such as dibenzoyl peroxide is heated. The heat causes the dibenzoyl peroxide to melt and decompose.

$$
\underset{\text{Dibenzoyl peroxide}}{C_6H_5{-}\overset{O}{\overset{\|}{C}}\underset{O{-}O}{\diagdown\diagup}\overset{O}{\overset{\|}{C}}{-}C_6H_5(l)} \longrightarrow \underset{\text{Benzoylperoxy free radical}}{2C_6H_5{-}\overset{O}{\overset{\|}{C}}{\cdot}(g)} + \underset{\text{Oxygen}}{O_2(g)}
$$

Then, by adding to a molecule of methyl methacrylate, a benzoylperoxy radical initiates polymerization.

$$C_6H_5-\overset{\overset{\displaystyle O}{\|}}{C}\cdot(g) \quad + \quad \underset{\underset{\displaystyle O}{\underset{\displaystyle \diagdown}{CH_3}}}{\overset{\overset{\displaystyle CH_3}{\diagup}}{CH_2=C}}(g) \quad \longrightarrow \quad C_6H_5-\overset{\overset{\displaystyle O}{\|}}{C}-CH_2-\underset{\underset{\displaystyle O}{\underset{\displaystyle \diagdown}{CH_3}}}{\overset{\overset{\displaystyle CH_3}{\diagup}}{C}}\cdot(g)$$

The latter radical reacts with another molecule of methyl methacrylate to produce an even larger free radical.

$$C_6H_5-\overset{O}{\overset{\|}{C}}-CH_2-\overset{CH_3}{\underset{C=O}{C}}\cdot(g) \quad + \quad \overset{CH_3}{\underset{C=O}{CH_2=C}}(g) \quad \longrightarrow \quad C_6H_5-\overset{O}{\overset{\|}{C}}-CH_2-\overset{CH_3}{\underset{C=O}{C}}-CH_2-\overset{CH_3}{\underset{C=O}{C}}\cdot(g)$$

This process repeats until hundreds or thousands of the repeating unit are produced within the macromolecule.

14.2-B CONDENSATION POLYMERIZATION

Condensation polymers are produced by a chemical reaction called a *condensation polymerization reaction*. A common example of their production involves the reaction between certain alcohols and organic acids. Specifically, a glycol (Section 13.2-G) is reacted with an acid whose molecules have two carboxyl groups. When heated with a catalyst, they combine with the simultaneous elimination of water. The polymer that results is called a **polyester**.

Consider the chemical reaction between the monomers ethylene glycol and succinic acid. The first step in this reaction is illustrated by the following equation:

$$\underset{\text{Ethylene glycol}}{\overset{\displaystyle OH}{\underset{\displaystyle OH}{\overset{\displaystyle |}{\underset{\displaystyle |}{\overset{\displaystyle CH_2}{\underset{\displaystyle CH_2}{|}}}}}}(l) \quad + \quad \underset{\text{Succinic acid}}{\overset{\displaystyle COOH}{\underset{\displaystyle COOH}{\overset{\displaystyle |}{\underset{\displaystyle |}{\overset{\displaystyle CH_2}{\underset{\displaystyle CH_2}{|}}}}}}(s) \quad \longrightarrow \quad \underset{\substack{\text{An intermediate, not yet}\\\text{a polymer}}}{\overset{\displaystyle COOCH_2CH_2OH}{\underset{\displaystyle COOH}{\overset{\displaystyle |}{\underset{\displaystyle |}{\overset{\displaystyle CH_2}{\underset{\displaystyle CH_2}{|}}}}}}(s) \quad + \quad \underset{\text{Water}}{H_2O}(l)$$

The intermediate compound resulting from this reaction has potentially reactive groups at both ends of the carbon–carbon chain. It can react with two molecules of ethylene glycol, eliminating two molecules of water and producing an even more complex intermediate as follows:

$$\overset{\displaystyle COOCH_2CH_2OH}{\underset{\displaystyle COOH}{\overset{\displaystyle |}{\underset{\displaystyle |}{\overset{\displaystyle CH_2}{\underset{\displaystyle CH_2}{|}}}}}}(s) \quad + \quad \overset{\displaystyle OH}{\underset{\displaystyle OH}{\overset{\displaystyle |}{\underset{\displaystyle |}{\overset{\displaystyle CH_2}{\underset{\displaystyle CH_2}{|}}}}}}(l) \quad + \quad \overset{\displaystyle COOH}{\underset{\displaystyle COOH}{\overset{\displaystyle |}{\underset{\displaystyle |}{\overset{\displaystyle CH_2}{\underset{\displaystyle CH_2}{|}}}}}}(s) \quad \longrightarrow \quad \overset{\displaystyle COOCH_2CH_2OH}{\underset{\displaystyle COOCH_2CH_2COOCH_2CH_2COOH}{\overset{\displaystyle |}{\underset{\displaystyle |}{\overset{\displaystyle CH_2}{\underset{\displaystyle CH_2}{|}}}}}}(s) \quad + \quad 2H_2O(l)$$

TABLE 14.1	Examples of Addition Polymers

MONOMER	REPEATING UNIT
Ethylene	Polyethylene
Vinyl chloride	Poly(vinyl chloride)
Acrylonitrile	Polyacrylonitrile
Tetrafluoroethylene	Poly(tetrafluoroethylene)
Styrene	Polystyrene
Methyl methacrylate	Poly(methyl methacrylate)

Both ends of this intermediate molecule have reactive groups. This intermediate can also combine with more ethylene glycol and succinic acid, thus increasing the length of the macromolecular chain. Finally, when the amount of either monomer is exhausted, the polyester that remains can be described by the following formula:

$$\text{H} \sim\!\!\sim \Big[\sim\!\!\sim \text{O}-\text{CH}_2\text{CH}_2-\overset{\displaystyle O \atop \|}{\text{O}} \quad\quad \overset{\displaystyle O \atop \|}{\underset{\displaystyle |}{\text{C}}}-\text{CH}_2\text{CH}_2-\overset{\displaystyle O \atop \|}{\text{C}} \quad \text{O} \sim\!\!\sim \Big]_n \sim\!\!\sim \text{H}$$

Additional examples of some condensation polymers are provided in Table 14.2.

SOLVED EXERCISE 14.3

Poly(ethylene terephthalate) is a polyester produced by the condensation polymerization of ethylene glycol and terephthalic acid. The chemical formulas of these substances are provided in Table 14.2. Write the stepwise equations that illustrate how this condensation polymerization occurs.

Solution: The chemical reaction between an alcohol and an acid produces an ester. The production of a polyester occurs by a stepwise process, each of which involves the elimination of water as follows:

Poly(ethylene terephthalate)

TABLE 14.2 | Examples of Condensation Polymers

REACTANTS	REPEATING UNIT
Ethylene glycol Terephthalic acid 	Poly(ethylene terephthalate)
Phenol Formaldehyde 	Phenol–formaldehyde
Hexamethylenediamine Adipic acid 	Nylon 66

A commonly encountered polyester is the tongue twister *poly(ethylene terephthalate)*, more commonly known as PETE, PET, or the DuPont trademark Mylar. The mechanism for the production of PETE is noted in Solved Exercise 14.3. PETE is used to manufacture a wide range of industrial and consumer products. It is predominantly used during the manufacturing of plastic bottles and fibers, but it is also used to manufacture audio, video, and computer tapes; sailboat sails; telephone and electric cable wires; and dozens of types of food packaging like ketchup and mustard bottles.

Poly(ethylene terephthalate) is also recyclable. Its recycling symbol is an arrowed triangle enclosing the number 1, beneath which appear the letters PETE.

PETE

Coca-Cola now recycles the PETE used to manufacture the plastic bottles that contain its soft drinks.

PERFORMANCE GOALS FOR SECTION 14.3:

- Describe the nature of an autopolymerization reaction.
- Describe the need for the use of inhibitors to reduce or eliminate the degree of autopolymerization during the storage and transportation of monomers.

14.3 Autopolymerization

autopolymerization
■ Spontaneous polymerization

Some monomers are capable of undergoing spontaneous polymerization. This phenomenon is called **autopolymerization**. The autopolymerization of a substance within a tank or container can result in a rise in temperature or pressure, which increases the risk that it may rupture.

TABLE 14.3	Examples of Monomers That Autopolymerize Unless Effectively Stabilized
MONOMER	**SHIPPING DESCRIPTION**
Acrylonitrile	UN1903, Acrylonitrile, stabilized, 3, (6.1), PGI (Poison - Inhalation Hazard)
Chloroprene	UN1991, Chloroprene, stabilized, 3, (6.1), PGI (Poison - Inhalation Hazard)
Ethyl acrylate	UN1917, Ethyl acrylate, stabilized, 3, PGII
Isoprene	UN1218, Isoprene, stabilized, 3, PGI
Methyl acrylate	UN1919, Methyl acrylate, stabilized, 3, PGII
Methyl methacrylate	UN1247, Methyl methacrylate monomer, stabilized, 3, PGII
Styrene	UN1247, Styrene, stabilized, 3, PGII
Vinyl acetate	UN1301, Vinyl acetate, stabilized, 3, PGII
Vinyl bromide	UN1085, Vinyl bromide, stabilized, 2.1
Vinyl butyrate	UN2838, Vinyl butyrate, stabilized, 3, PGII
Vinyl chloride	UN1086, Vinyl chloride, stabilized, 2.1
Vinylidene chloride	UN1303, Vinylidene chloride, stabilized, 3, PGI

The autopolymerization of a substance can be prevented so the substance can ultimately be utilized for its intended purpose. When the substance is gaseous, autopolymerization is hindered by the addition of nitrogen, carbon dioxide, or other inert diluents. Although inert compounds may also be used to dilute liquids that are prone to autopolymerize, the common practice used by chemical manufacturers is to add an **inhibitor** to them. As the name implies, an inhibitor is a substance that chemically reacts in such a fashion as to inhibit, retard, or stabilize the liquid against autopolymerization.

inhibitor
■ A substance or mixture of substances used to arrest or retard the speed of polymerization

Although there is not a consistent mechanism by which substances autopolymerize, some are initiated by the reaction of a peroxo-organic compound that results when the substance reacts with dissolved atmospheric oxygen. In this instance, the inhibitor selected for retarding autopolymerization is an antioxidant, which immediately reacts with the peroxo-organic compound as it forms.

DOT regulates the transportation of substances that are prone to autopolymerization, some examples of which are provided in Table 14.3. DOT also requires shippers to indicate in the basic shipping description that their products have been stabilized against autopolymerization. Examples of the basic shipping names for such substances are "ethyl acrylate, stabilized" and "vinylidene chloride, stabilized."

To warn firefighters of the presence of a substance that is susceptible to autopolymerization, chemical manufacturers and users sometimes insert a P in the bottom quadrant of the NFPA hazard diamond posted in the relevant storage area.

PERFORMANCE GOALS FOR SECTION 14.4:

- Discuss the general phenomena that occur when polymers burn on exposure to heat.
- Identify the toxic gases produced when polymers thermally decompose or burn.
- Describe the physical nature of a flashover.

14.4 Polymer Decomposition and Combustion

Most products produced from natural and synthetic polymers are combustible when exposed to an ignition source. Their combustion is associated with the following general features:

- Polymeric products often melt and thermally decompose into the monomers from which they were made or a mixture of simpler substances.
- The surfaces of some polymeric products tend to char as they burn.
- Burning polymeric products release considerable heat.
- Burning polymeric products can evolve voluminous amounts of smoke, carbon monoxide, and other hazardous gases, vapors, and fumes.

The melting of polymeric products at a typical fire scene is associated with both beneficial and detrimental effects. The melting often causes the polymer to drip from its source, as from ceiling tile to an underlying floor. Dripping molten polymer closely resembles dripping hot candle wax. The dripping serves as a cooling mechanism, removing heat from the immediate site of combustion and hindering the continued combustion of the polymer at that site. However, when they remain in the fire zone, polymers begin to thermally decompose in their molten state. Then, their decomposition products ignite and the fire spreads.

Prior to their ignition, polymers frequently undergo thermal degradation into simpler chemical species; that is, when they are exposed to heat, polymers decompose into relatively simpler substances. The decomposition of unique organic polymers occurs by different mechanisms, several of which involve scission of the macromolecular chains. The following two types of thermal decomposition are characteristic of how polymers decompose:

■ When heated, some polymers predominantly produce the monomers from which they were initially produced. They are said to *depolymerize*. The process is called **depolymerization**. For example, when poly(methyl methacrylate) thermally decomposes, its monomer, methyl methacrylate, accounts for 91% to 98% by mass of the substances produced. The same is true of poly(ethylene terephthalate). When heated, it reverts into the substances from which it was produced: terephthalic acid and ethylene glycol. Polymers that depolymerize are desirable candidates for recycling.

■ When certain other polymers are heated, they produce an array of gaseous decomposition products. For example, when polypropylene is heated, it decomposes to form methane, ethane, propane, butane, pentane, propene, 2-methylpropene, 1- and 2-pentene, 2-methyl-1-butene, and other hydrocarbons.

The vapors produced by thermal decomposition initially diffuse to the surface of the polymer, where they mix with atmospheric oxygen and ignite. As the heat increases in intensity, these vapors can also migrate from the immediate burning area and accumulate elsewhere, such as near the ceiling of a room, where they mix with air and burn.

Heat may also be conducted or radiated through a polymeric material, thereby causing the polymer to decompose at a location isolated from the heat source. Consider the section of a wall shown in Figure 14.4. It is constructed of wooden support beams to which polymeric paneling has been affixed. Although the heat from a fire impinges on only one side of the wall, it can conduct or radiate *through* the wall, causing the polymer in the paneling to decompose. The mixture of combustible gases subsequently produced by thermal decomposition readily ignites.

This generation of flammable vapors at a location isolated from the source of heat is associated with a **flashover**, the phenomenon largely responsible for the spread of fire from one room into an adjacent room. At flashover, the entire contents of a room are ignited simultaneously by radiant heat. This situation makes living conditions within the room untenable. Safe exit for residents is impossible. At flashover, room temperatures typically range from 1100 to 1470 °F (~600 to 800 °C).

Fires pose special problems in large public buildings that have been constructed in part from plastic products, since the polymers in these products usually burn differently than burning wood or other natural materials. The thermal characteristics of some common polymers in air are provided in Table 14.4. By comparison, hardwoods self-ignite in air at an average temperature of 781 °F (416 °C) and have an average heat of combustion of 8500 Btu/lb. Although hardwoods

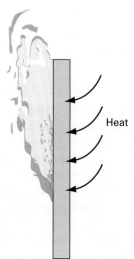

Heat

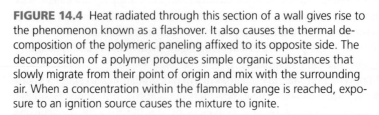

FIGURE 14.4 Heat radiated through this section of a wall gives rise to the phenomenon known as a flashover. It also causes the thermal decomposition of the polymeric paneling affixed to its opposite side. The decomposition of a polymer produces simple organic substances that slowly migrate from their point of origin and mix with the surrounding air. When a concentration within the flammable range is reached, exposure to an ignition source causes the mixture to ignite.

TABLE 14.4 | Thermal Characteristics of Some Common Polymers

POLYMER	SPECIFIC GRAVITY	DECOMPOSITION RANGE	SELF-IGNITION TEMPERATURE	HEAT OF COMBUSTION
Polyethylene (high-density)	0.965	644–824 °F (340–440 °C)	662 °F (350 °C)	20,050 Btu/lb (46,500 kJ/kg)
Polypropylene	0.91	626–770 °F (330–410 °C)	734–770 °F (390–410 °C)	19,800 Btu/lb (46,000 kJ/kg)
Polystyrene	1.05	572–752 °F (300–400 °C)	914 °F (490 °C)	18,100 Btu/lb (42,000 kJ/kg)
Poly(methyl methacrylate)	1.18	338–572 °F (170–300 °C)	842 °F (450 °C)	11,210 Btu/lb (26,000 kJ/kg)
Poly(vinyl chloride)	1.40	392–572 °F (200–300 °C)	851 °F (455 °C)	8620 Btu/lb (20,000 kJ/kg)

and synthetic polymers begin to burn at approximately the same temperature, polymers like polyethylene, polypropylene, and polystyrene emit more than twice as much heat as the same amount of burning wood.

14.4-A THE CHEMICAL NATURE OF THE GASES AND VAPORS PRODUCED DURING POLYMERIC FIRES

The fatalities that occur at fire scenes often result when individuals are exposed to the gases and vapors produced during the fires. Because each polymeric macromolecule is generally composed of hundreds of carbon atoms, burning polymers produce massive concentrations of carbon monoxide. Put simply, the larger the fire, the more carbon monoxide produced. Within an enclosure, carbon monoxide and other gases soar within a matter of seconds to life-threatening concentrations.

Fires involving burning polymers can affect areas far removed from where the fire originated when hot gases and thermal degradation products travel by convection through ventilation systems, trash chutes, and similar openings. Although this movement spreads the fire, it also causes people to be unsuspectingly exposed to toxic fumes generated elsewhere within a building.

The burning of products made from synthetic polymers often produces a different mixture of gases and fumes from that generated from the burning of nonplastic products. Carbon monoxide is still the most prevalent gas at a fire scene, but other gases associated with burning plastics are also produced. They include hydrogen chloride, ammonia, hydrogen cyanide, sulfur dioxide, and nitrogen dioxide.

At a fire scene, the origin of the nitrogenous and sulfurous gases can be traced to the chemical nature of the polymeric materials that have burned or undergone thermal decomposition. Thermal decomposition occurs primarily when the polymers are exposed to the heat emitted during slow-burning processes. The polymers that produce hydrogen cyanide, ammonia, nitric oxide, and nitrogen dioxide are nitrogenous organic compounds. They are natural and synthetic polymers that have one or more of the functional groups listed in Table 14.5, including **amines**, **amides**, **nitriles**, and **nitro compounds**. Their thermal decomposition generates hydrogen cyanide and ammonia, and their combustion produces nitric oxide and nitrogen dioxide.

amine
■ An alkylated analog of ammonia; an organic compound derived by the substitution of one or more of the hydrogen atoms in a hydrocarbon molecule with the —NH_2 group of atoms

amide
■ An organic compound whose molecules contain the following group of atoms:

nitrile
■ An organic compound whose molecules contain the —$C\equiv N$ group of atoms

nitro compound
■ An organic compound whose molecules contain the —NO_2 group of atoms

TABLE 14.5 | Some Nitrogenous Organic Compounds

CLASS OF ORGANIC COMPOUND	FUNCTIONAL GROUP	EXAMPLE
Amine	$-NH_2$	$CH_3CH_2CH_2-NH_2$ *n*-Propylamine, or 1-Aminopropane
Amide	$-\overset{\displaystyle O}{\overset{\|\|}{C}}\diagdown NH_2$	$CH_3-\overset{\displaystyle O}{\overset{\|\|}{C}}\diagdown NH_2$ Acetamide
Amino acid	$-\underset{NH_2}{CH}-\overset{\displaystyle O}{\overset{\|\|}{C}}\diagdown OH$	$CH_3-\underset{NH_2}{CH}-\overset{\displaystyle O}{\overset{\|\|}{C}}\diagdown OH$ 2-Aminopropanoic acid
Isocyanate	$-N=C=O$	$CH_3CH_2-N=C=O$ Ethyl isocyanate
Nitrate	$-O-NO_2$	$CH_3CH_2CH_2CH_2-O-NO_2$ *n*-Butyl nitrate
Nitrile, or cyanide	$-C\equiv N$	$CH_2=\overset{H}{\underset{C\equiv N}{C}}$ Acrylonitrile, or Propenenitrile, or Vinyl cyanide
Nitro compounds	$-NO_2$	⬡$-NO_2$ Nitrobenzene

amino acid

■ A carboxylic acid whose molecules contain the $-NH_2$ group of atoms

Polymers containing sulfur atoms are usually natural polymers having an animal origin. Their nonwater matter is composed of proteins, which in turn, are composed of **amino acids**. Although there are 21 amino acids that make up the structures of nearly all proteins, of concern here are only the two whose molecules contain sulfur atoms in their composition: methionine and cysteine.

$$CH_3-S-CH_2CH_2-\underset{NH_2}{CH}-\overset{\displaystyle O}{\overset{\|\|}{C}}\diagdown OH \qquad HS-CH_2-\underset{NH_2}{CH}-\overset{\displaystyle O}{\overset{\|\|}{C}}\diagdown OH$$

Methionine   Cysteine

Because sulfur atoms are components of these amino acids, they are also constituents of the proteins biologically produced from them in leather, wool, and animal hair. Their thermal decomposition produces hydrogen sulfide and ammonia, and their combustion produces carbon monoxide, carbon dioxide, sulfur dioxide, nitric oxide, nitrogen dioxide, and water.

Another toxic substance whose presence has been detected in smoke is the unsaturated aldehyde known as *acrolein*, or *acrylic aldehyde*. Its chemical formula is $CH_2{=}CHCHO$.

$$CH_2{=}CH{-}\overset{\displaystyle O}{\overset{\displaystyle \|}{C}}{\diagdown}H$$

Acrolein
(Acrylic aldehyde)

This is a pungent-smelling, intensely irritating lacrimator. A 1-min exposure to air containing as little as 1 ppm of acrolein causes nasal and eye irritation. A concentration of acrolein in air ranging from 2500 to 5000 ppm has been established as the lethal dose to laboratory animals. During World War I, acrolein was offensively used as a lacrimator (Section 10.8-A).

The impact of human exposure to acrolein has been investigated. Although researchers have not found acrolein to be carcinogenic in laboratory animals, they have discovered that it causes DNA mutations similar to those identified in the biopsies of lung samples collected from individuals with lung cancer. This suggests that acrolein may play an unknown role in the development of lung cancer despite the fact that direct exposure to acrolein does not cause cancer.

14.4-B SMOKE PRODUCED DURING POLYMERIC FIRES

The character of the smoke produced in fires involving polymeric materials varies with the chemical nature of the polymer. In particular, the amount of smoke produced in fires involving polymers produced from aromatic monomers is typically far greater than the amount produced during the burning of polymers produced from aliphatic monomers. A fire involving materials produced from polystyrene, for instance, produces considerably more soot than one involving materials produced from polyethylene.

As was first noted in Section 10.8-D, the surfaces of carbon particulates in smoke adsorb toxic gases. When smoke is inhaled, the particulates serve as the vehicles that draw toxic gases into the bronchi and lungs. Because considerable smoke is produced during the burning of materials produced from aromatic monomers, these materials could be more hazardous to health compared with materials produced from aliphatic monomers.

14.4-C CPSC REGULATIONS PERTAINING TO TEXTILES INTENDED FOR CONSUMER USE

Polymers have been used to manufacture hundreds of different consumer products. Because most polymers burn, the consumer products made from them are now evaluated to determine whether the likelihood of their ignition is delayed or eliminated when their fibers have been treated with fire retardants or flameproofing agents.

Congress has directed the Consumer Product Safety Commission to reduce injuries and deaths caused by consumer products in a variety of settings. In the course of one of its studies, the CPSC used thermal-test results to promulgate regulations that when implemented, potentially minimize or delay flashover during typical mattress/bedding fires within the home. The result of one such study is shown in Figure 14.5, which compares the open-flame burning of a mattress made from untreated fibers with one made from fibers treated with a flameproofing agent. It is apparent that a mattress made from treated fibers is more desirable when it is exposed to an open flame, since it provides the occupants with time to discover the fire and escape from the home.

Figure 14.5 also illustrates that although treatment procedures may improve the safety of products for consumer use, they do not totally eliminate their flammability. After polymeric fibers have been treated, the products made from them still burn, especially when they are exposed to the intense heat experienced during major fires. However, the treatment process

FIGURE 14.5 The open-flame burning of two mattresses in a mock bedroom fire ignited at the same time. The mattress on the top was made from untreated fibers, and the mattress on the bottom was made from fibers treated with a flameproofing agent. Although ignited simultaneously, the mattress on the top is entirely engulfed in flames, whereas the mattress on the bottom burns at a substantially slower growth rate. These thermal-test results were used by the CPSC to develop a standard that addresses the open-flame ignition of mattresses. The standard is published at 16 C.F.R. §§1633.1–1633.13. (*Courtesy of the California Bureau of Home Furnishings and Thermal Insulation, North Highlands, California.*)

produces consumer products that are more resistant to ignition and more prone to self-extinguish once fire has been initiated.

State and federal laws require manufacturers to provide consumers with materials that are unlikely to easily ignite. Their combined use has helped reduce injuries when plastics and textiles are involved in fires. The CPSC has used the legal authority of the **Flammable Fabrics Act** and other statutes to respond to public concern over accidents involving the use of products like brushed rayon in high-pile sweaters, and children's cowboy chaps that flash-burned when ignited. The CPSC now requires manufacturers to submit samples of apparel fabrics like children's sleepwear to an ignition test and a rate-of-burn test. The use of these tests also applies to the manufacture of carpets, rugs, and other home furnishings.

Flammable Fabrics Act
■ The federal statute that empowers the Consumer Products Safety Commission to establish flammability standards for clothing textiles as well as interior furnishings, including paper, plastic, foam and other materials used in wearing apparel and interior furnishings

PERFORMANCE GOALS FOR SECTION 14.5:

■ Describe and compare the macromolecular structures of cellulosic and other vegetable fibers and the common animal fibers.

14.5 Vegetable and Animal Fibers

Many common textiles have been produced from naturally occurring vegetable and animal fibers. **Cotton** and **linen** are examples of vegetable fibers, whereas wool and silk are examples of animal fibers. These naturally occurring fibers can directly be used to produce textiles, or they can be chemically altered to produce synthetic fibers.

As they occur naturally vegetable and animal fibers are often mixed with combustible oils; for example, cotton contains cottonseed oil, and wool contains lanolin. DOT regulates the transportation of the vegetable and animal fibers in Table 14.6 as hazardous materials. We note the properties of some common fibers and pay special attention to their combustible nature.

cotton
■ A naturally occurring fiber of vegetable origin

linen
■ The naturally occurring fiber of the flax plant

14.5-A CELLULOSE AND CELLULOSIC DERIVATIVES

Cellulose is the polymer that remains when the natural binding agent *lignin* is removed from wood. This polymer serves as the primary structural component of the cell walls in plants and thus is generally regarded as nature's most important polymer. Wood is approximately 50% cellulose by mass, whereas cotton and linen are nearly 100% cellulose.

Cellulose is the raw material of the paper, wood, cotton textiles, and cellulose plastics industries. It is also the primary structural component of *switchgrass*, which is among the plants used for the production of cellulosic ethanol (Section 13.2-D). Switchgrass, or *Panicum virgatum*, is a perennial summer grass native to North America.

Cellulose fiber is encountered naturally in several forms, two of which are cotton and linen. Cotton comes from any of four species of *Gossypium* that grow in warm climates throughout the world. To produce it as a yarn, raw cotton is first boiled in a dilute solution of sodium hydroxide to remove any wax that naturally adheres to the fibers. Then, it is bleached with chlorine, sodium hypochlorite, or a similar substance. Next, it is passed through a vat of diluted sulfuric acid to neutralize any remaining alkaline materials. Finally, the cotton is washed with water. At this point, it is ready to be spun into a yarn and woven into cloth.

Linen is nearly pure cellulose. Its fibers are derived from the stalk of the flax plant *Linum usitatissimum*. Linen fibers are among the strongest naturally occurring fibers. Linen fabric absorbs moisture faster than any other fabric; however, linen lacks resiliency, the ability to spring back when stretched. This lack of resiliency causes it to wrinkle easily.

The difference between the physical properties of cotton and linen is associated with the nature of their fibers. When examined under a microscope, individual cotton fibers resemble short,

cellulose
■ The substance that forms the cell walls of all plants; the major constituent of wood and cotton; a natural polymer of β-glucose

TABLE 14.6 | Some Animal and Vegetable Products

PRODUCT	PROPER SHIPPING NAME
Cotton	NA1365, Cotton, 9[a]
Cotton, waste, oily	UN1364, Cotton waste, oily, 4.2, PGIII[b]
Cotton, wet	UN1365, Cotton, wet, 4.2, PGIII[c]
Fibers *or* fabrics, animal *or* vegetable *with animal or vegetable oil*	UN1353, Fibers, animal, n.o.s., 4.2, PGIII UN1353, Fabrics, animal, n.o.s., 4.2, PGIII UN1353, Fibers, vegetable, n.o.s., 4.2, PGIII UN1353, Fabrics, vegetable, n.o.s., 4.2, PGIII
Fibers *or* fabrics impregnated with weakly nitrated nitrocellulose	UN1353, Fibers impregnated with weakly nitrated nitrocellulose, n.o.s., 4.1, PGIII UN1353, Fabrics impregnated with weakly nitrated nitrocellulose, n.o.s., 4.1, PGIII
Paper, unsaturated oil-treated *incompletely dried (including carbon paper)*	UN1379, Paper, unsaturated oil-treated, 4.2, PGIII
Wool waste	UN1387, Wool waste, wet, 4.2, PGIII[c]

[a]For domestic transportation only on water
[b]For transportation by air or on water
[c]For international transportation only by air or on water

twisted flattened tubes; linen fibers resemble long transparent tubes that have periodic junctions for cross-linking.

The chemical formula of cellulose is often abbreviated $(C_6H_{10}O_5)_n$ or $[(C_6H_7O_2(OH)_3]_n$, where n is around 5000. The more complete chemical formula is denoted as follows:

The repeating unit shown in this structure is a substance called *β-glucose*.

Cotton is commercially available not only in its natural form but also as *mercerized cotton*. Cotton fibers swell when they are immersed in a concentrated solution of sodium hydroxide, a process called *mercerizing*, after the discoverer John Mercer. When each cotton flattened fiber is mercerized, it becomes rounded and more lustrous and acquires additional strength. Mercerizing also allows dyes to penetrate the fabric more easily.

Cotton textiles can be chemically treated to produce fabrics that are wrinkleproof, as well as fabrics that can be washed and drip-dried with few wrinkles. Cotton is often combined with synthetic fibers to produce fabrics that are commercially available as blends with names such as "50% cotton, 50% acrylics."

When triggered by an ignition source, all cellulosic materials burn without melting. Because the fires of cellulosic materials are NFPA class A fires, they are routinely extinguished with water.

Synthetic fibers have also been derived from cellulose. For example, cellulose acetate is the fiber produced when wood cellulose is reacted with acetic acid or acetic anhydride. In commerce, the fiber is called *acetate rayon*. Although it has a pronounced strength, acetate rayon can be ignited with a hot iron and destroyed by some dry-cleaning solvents. Given these adverse features, acetate rayon did not retain popularity as a fabric of choice, but it is still used today in motion-picture film, airplane wings, and safety glasses.

Several synthetic products are derived from cellulose. An example is *cellulose xanthate*, commonly called *rayon*. The production of rayon involves reacting highly purified wood pulp with sodium hydroxide, followed by chemical treatment with carbon disulfide. The resulting solution is extruded through a *spinneret*, a metal disk having numerous tiny holes, and then into an acid solution that regenerates the cellulose as tiny transparent fibers.

Another example of a synthetic product derived from cellulose is *nitrocellulose*. This substance is manufactured by reacting cellulose with nitric acid. Although highly nitrated cellulose is a chemical explosive (Section 15.8), the lesser grades of nitrated cellulose can be dissolved in a solvent and applied to cloth to produce the fabric known as *patent leather*. The products produced from this artificial leather are strong and flexible. They include vehicle seat covers and convertible tops. Nitrocellulose is also used as a film-producing agent in lacquers, printing inks, and enamel nail polishes.

14.5-B WOOL AND SILK

The most common animal fabrics employed in textiles are **wool** and **silk**. Wool is the curly hair of sheep, goats, and llamas. Under a microscope, wool fibers resemble tiny, overlapping scales, much like those of fish. These fibers bend and conform to a variety of physical shapes. They also possess resiliency and tend to hold their shape.

wool
- The naturally occurring fiber obtained from the coats of sheep, llamas, and several other animals

Silk is the soft, shiny fiber produced by silkworms to form their cocoons. Silk fibers are very strong, elastic, and smooth. Under a microscope, silk fibers appear semitransparent, which accounts for their lustrous sheen. Silk is resilient; its fibers readily spring back to their original position when stretched or folded. These qualities have made silk one of the most useful fibers for the textile market. Unwinding the long, delicate silk threads of a cocoon is a tedious process; hence, the fabric woven from silk fibers is relatively expensive.

silk
- A naturally occurring protein produced by the action of silkworms

Like all other forms of animal hair, wool and silk are composed of proteins, which are biological substances whose molecules have recurring amino groups. The condensed formulas of wool and silk are $C_{42}H_{157}N_5SO_{15}$ and $C_{15}H_{23}N_5O_6$, respectively. A portion of the macromolecular structure of wool is shown in Figure 14.6.

Silk consists of a mixture of two relatively simple proteins called *silk fibroin* and *sericin*. The macromolecular structure of silk fibroin resembles the structure of wool in which X, Y, and Z are invariably $-CH_3$, $HO-C_6H_4-CH_2-$, or $HO-CH_2-$. Sericin has a similar macromolecular structure, but the primary recurring group in the protein is the following:

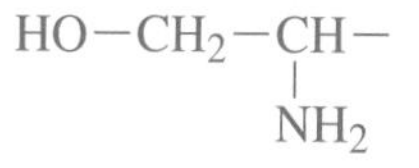

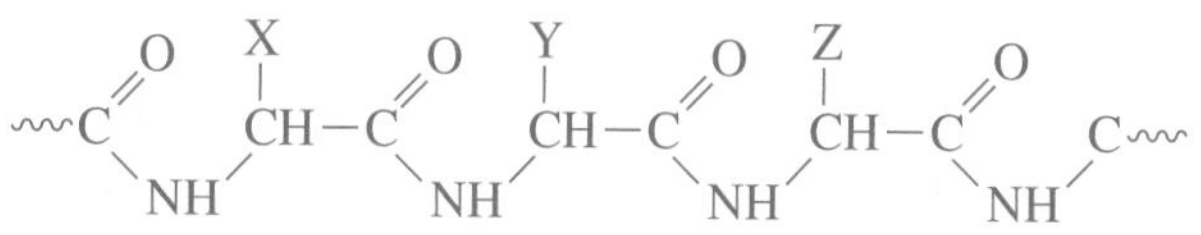

$$HO-CH_2-\underset{\underset{NH_2}{|}}{CH}-$$

FIGURE 14.6 A portion of the macromolecular structure of wool. In this protein, X, Y, and Z represent specific substances called amino acids, compounds that have the following chemical formula:

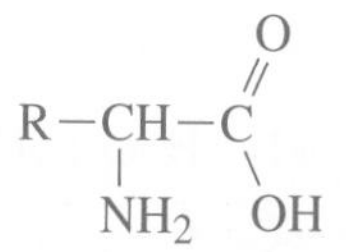

$$R-\underset{\underset{NH_2}{|}}{CH}-\overset{\overset{O}{\|}}{C}\diagdown OH$$

Only 21 amino acids are commonly found in naturally occurring proteins. The interested reader may consult more advanced chemistry textbooks for additional information concerning proteins.

Because wool and silk are proteins, both have similar properties. Each has an ignition temperature in excess of approximately 1058 °F (570 °C); hence, wool and silk are difficult to ignite and when ignited, burn very slowly. When tightly woven as in rugs, woolen textiles tend to smolder and char when burning. They absorb great quantities of water, thus permitting their fires to be easily extinguished.

Because the macromolecules of animal fibers contain bonding groups such as $-\overset{\displaystyle O}{\underset{\displaystyle NH_2}{\overset{\|}{C}}}$,

$-NH-$, and $-S-S-$, the presence of ammonia, hydrogen cyanide, and sulfur dioxide at fire scenes involving textiles is possible. Although they may be produced, ammonia and hydrogen cyanide do not appear to survive long at fire scenes, most likely because they burn.

PERFORMANCE GOALS FOR SECTION 14.6:

- Identify the commercially important vinyl polymers.
- Discuss the general properties of polyethylene, polypropylene, poly(vinyl chloride), polyacrylonitrile, and poly(methyl methacrylate), and identify the common products made from each of them.
- Identify the toxic gases produced when products made from the vinyl polymers thermally decompose or burn.

14.6 Vinyl Polymers

As noted in Section 12.3-B, the vinyl group of atoms is $CH_2=CH-$. For our purposes, a substance containing the vinyl group is a *vinyl compound*. Examples of vinyl compounds are listed in Table 14.1 under the heading "Monomer." A polymer produced from one or more vinyl compounds is called a **vinyl polymer**.

vinyl polymer
- A polymer produced from one or more vinyl compounds

The following equations illustrate the production of several commercially important vinyl polymers from their respective vinyl compounds:

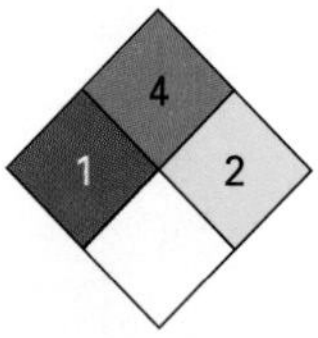

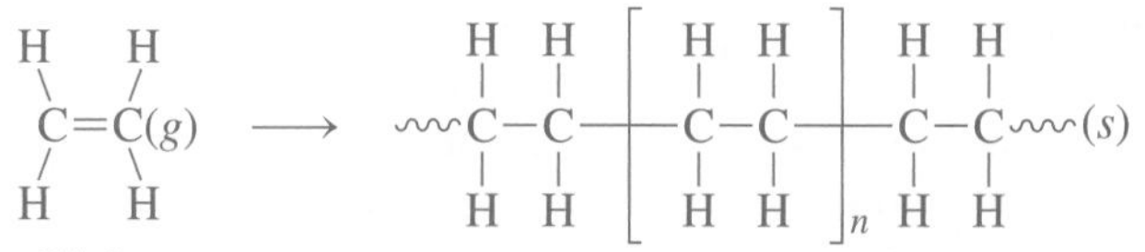

Ethylene   Polyethylene

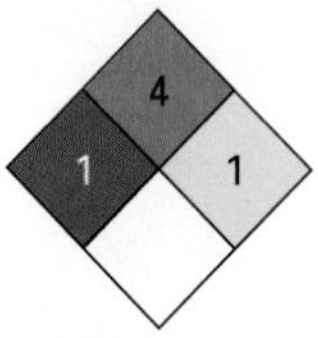

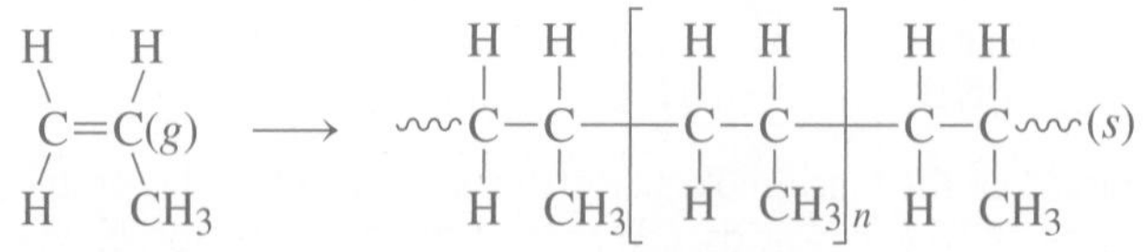

Propylene   Polypropylene

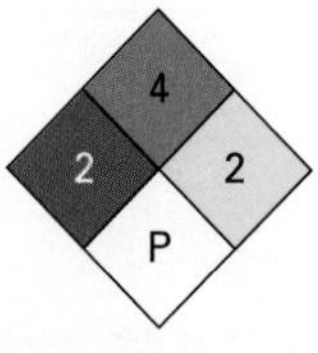

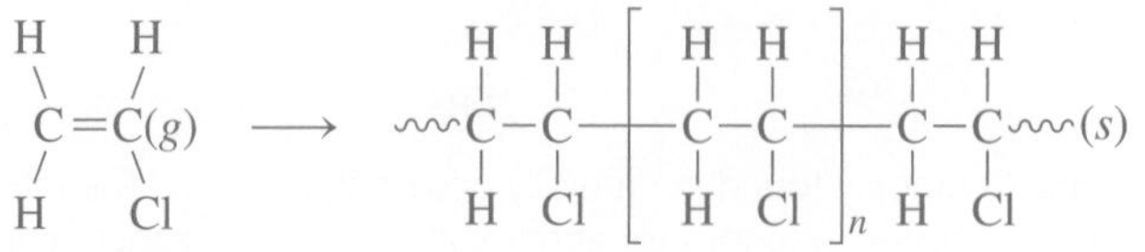

Vinyl chloride   Poly(vinyl chloride)

Vinyl acetate   Poly(vinyl acetate)

The referenced NFPA hazard diamonds designate the relevant properties of the monomers from which the vinyl polymers are produced.

A vinyl polymer may also be produced from multiple vinyl compounds. For example, the vinyl chloride–vinylidene chloride copolymer is produced from vinyl chloride and vinylidene chloride. Vinyl chloride and vinylidene chloride are synonyms for chloroethene and 1,1-dichloroethene, respectively.

Vinyl chloride   Vinylidene chloride   Vinyl chloride–vinylidene
                                                chloride copolymer

Commercially, the vinyl chloride–vinylidene chloride copolymer is known as *Saran*, or poly(vinylidene dichloride). Manufacturers form it into sheets, tubes, rods, fibers, and other molded items. Its fibers are used to produce a number of textiles, including carpets, curtains, and upholstery fabrics.

14.6-A POLYETHYLENE

As noted in Section 12.7, ethylene is prepared in the petrochemical industry, primarily by cracking ethane and propane. When ethylene is polymerized, the resulting product is called *polyethylene*. It is the most widely manufactured polymer in the United States. Polyethylene macromolecules have the repeating unit $\sim\sim CH_2 - CH_2 \sim\sim$.

Polyethylene is primarily encountered in two forms: *low-density (cross-linked)* (LDPE) and *high-density (linear)* (HDPE). A third form, low-molecular-weight polyethylene, is uniquely used in coatings and polishes. Both LDPE and HDPE are white solids. The low-density form is a thermosetting polymer, whereas the high-density form is a thermoplastic polymer. The density of LDPE may be as low as 0.915 g/mL, and of HDPE as high as 0.965 g/mL.

Low-density polyethylene is used largely for making baby diapers and molded products such as toys. For example, the "noodles" that children use in swimming pools are manufactured from low-density polyethylene. High-density polyethylene is used mainly to manufacture filmed products and hardy storage containers. The polyethylene films are used in the building industry as vapor and moisture barriers and in the agricultural industry for mulching, silage covers, greenhouse glazings, pond liners, and animal shelters. Included among the containers are plastic milk jugs, detergent and bleach containers, sandwich bags, and liners for trash cans, drums, and other containers.

Low-density and high-density polyethylene can be recycled. Their recycling symbols are arrowed triangles enclosing the numbers 4 and 2, respectively, and beneath which appear the letters LDPE and HDPE, as follows:

LDPE   HDPE

Products made from both LDPE and HDPE burn when they are exposed to fire. The data in Table 14.4 show that polyethylene is distinctive among burning polymers insofar as it releases more heat per unit mass than any other commercially popular polymer. During combustion, polyethylene products tend to disintegrate into numerous burning, molten globules or liquid pools. Because the macromolecules of polyethylene are composed of only carbon and hydrogen atoms, carbon monoxide, carbon dioxide, and water vapor are produced as combustion products.

14.6-B POLYPROPYLENE

It was noted in Section 12.7 that the cracking of propane produces methane, ethylene, and propylene. In the petrochemical industry, the propylene is separated from this mixture of gases and polymerized to manufacture the polymer called *polypropylene*. Its macromolecules have the repeating unit $\sim\!\!\sim CH_2-CH\!\!\sim\!\!\sim$ with CH_3.

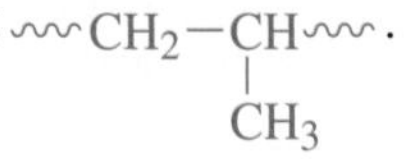

Polypropylene is used to manufacture commercial products ranging from silky fibers to rigid containers. Examples are outdoor tables and chairs, shatterproof glasses, artificial grass and turf, pipes, ropes, nets, twines, carpets, and carpet padding. The plastic items in motor vehicles are largely made from polypropylene. They include the dashboards, bumpers, carpeting, and occasionally, the upholstery.

By using a Ziegler–Natta catalyst to initiate polymerization, manufacturers have been able to produce polypropylene so the branching methyl group is arrayed regimentally along the carbon–carbon backbone, giving rise to the following three types of polypropylene macromolecules:

isotactic polymer
■ A polymer whose macromolecules have side chains positioned on the same side of a chain of carbon atoms

- **Isotactic polymer**, in which the methyl groups are all pointing in the same direction

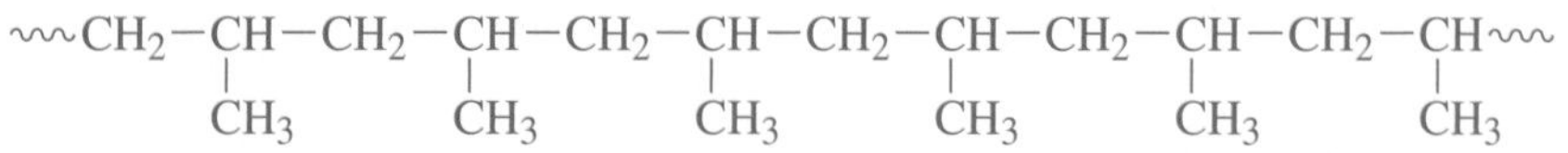

$$\sim\!\!\sim CH_2-CH-CH_2-CH-CH_2-CH-CH_2-CH-CH_2-CH-CH_2-CH\!\!\sim\!\!\sim$$
$$\quad\quad CH_3 \quad\quad CH_3 \quad\quad CH_3 \quad\quad CH_3 \quad\quad CH_3 \quad\quad CH_3$$

syndiotactic polymer
■ A type of polymer whose macromolecules have alternately positioned side chains along a chain of carbon atoms

- **Syndiotactic polymer**, in which alternate methyl groups point in opposite directions

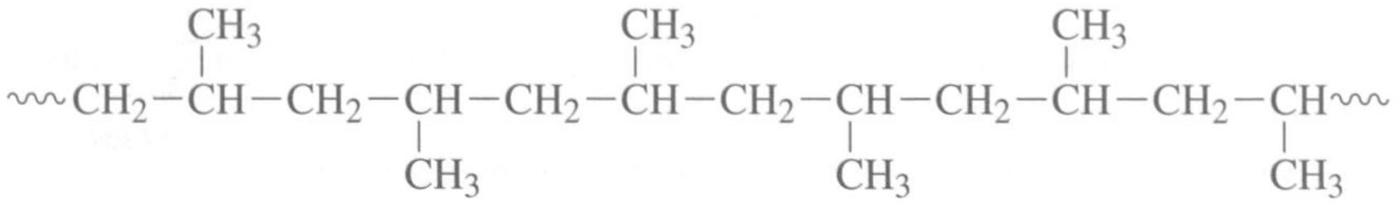

$$\quad\quad CH_3 \quad\quad\quad\quad\quad CH_3 \quad\quad\quad\quad\quad CH_3$$
$$\sim\!\!\sim CH_2-CH-CH_2-CH-CH_2-CH-CH_2-CH-CH_2-CH-CH_2-CH\!\!\sim\!\!\sim$$
$$\quad\quad\quad\quad\quad CH_3 \quad\quad\quad\quad\quad CH_3 \quad\quad\quad\quad\quad CH_3$$

atactic polymer
■ A type of polymer whose macromolecules have randomly positioned side chains along a chain of carbon atoms

- **Atactic polymer**, in which the methyl groups are randomly oriented

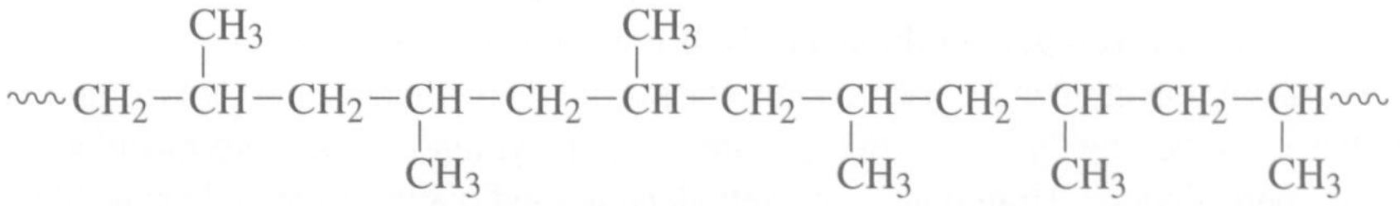

$$\quad\quad CH_3 \quad\quad\quad\quad\quad CH_3$$
$$\sim\!\!\sim CH_2-CH-CH_2-CH-CH_2-CH-CH_2-CH-CH_2-CH-CH_2-CH\!\!\sim\!\!\sim$$
$$\quad\quad\quad\quad\quad CH_3 \quad\quad\quad\quad\quad CH_3 \quad\quad CH_3 \quad\quad CH_3$$

These individual polypropylene types may be cast into shapes, drawn into sheets, or extruded into fibers, thus producing a range of diversified products.

Polypropylene is a recyclable polymer. The recycling symbol for polypropylene is an arrowed triangle enclosing the number 5, beneath which appear the letters PP.

PP

Polypropylene does not ignite easily. However, on exposure to intense heat, polypropylene thermally decomposes into a mixture of hydrocarbon vapors. This mixture catches fire and burns.

14.6-C POLY(VINYL CHLORIDE)

Vinyl chloride, or chloroethene, is a gas primarily manufactured by the catalytic dehydrochlorination of 1,2-dichloroethane.

$$ClH_2C-CH_2Cl(g) \longrightarrow CH_2=\underset{\underset{Cl}{|}}{\overset{\overset{H}{|}}{C}}(g) \quad + \quad HCl(g)$$

1,2-Dichloroethane   Vinyl chloride   Hydrogen chloride
                            (Chloroethene)

It is a flammable gas that is subject to autopolymerization.

Vinyl chloride is regarded as a known human carcinogen, since human exposure by inhalation has been affirmatively linked with the onset of liver angiosarcoma. The latency period for the onset of this cancer ranges from 15 to 40 y. Acknowledging its potential to cause cancer, the CPSC prohibits the use of vinyl chloride in self-pressurized consumer products (Section 1.4-A). OSHA limits the vinyl chloride concentration in the workplace at 1 ppm, averaged over an 8-hr workday.

Poly(vinyl chloride) is produced when vinyl chloride polymerizes. It is commonly recognized by its acronym, *PVC*. The unit that recurs in its macromolecules is the following:

The presence of the chlorine atoms gives PVC some unique properties. For example, the data in Table 14.4 show that when PVC burns, less heat is emitted per mass compared with other common polymers that burn.

PVC is the polymer used in many home-construction products like floor tiles, lighting fixtures, vinyl panels and siding, conduits, and wall coverings. Figure 14.7 shows that PVC is also the polymer used to manufacture pipe and pipe fittings. PVC is the plastic insulation sheath used in virtually all modern electrical wiring. PVC is also encountered in imitation leather, shower curtains, upholstery material, vinyl raincoats, plastic packaging materials, garden hoses, and medical products. Approximately 70% of the poly(vinyl chloride) produced in the United States is used in building construction materials. PVC copolymers are also commercially available in products such as films, fibers, sheeting, and moldings. Their mechanical properties vary from rigid to elastomeric.

PVC is recyclable. Its recycling symbol is an arrowed triangle enclosing the number 3, beneath which appears the letter V.

$$\underset{V}{\triangle_3}$$

PVC is highly susceptible to degradation, which results in unsightly discoloring and loss of mechanical properties. For example, the vinyl upholstery in an automobile is initially soft and supple, but it becomes brittle and cracks as the automobile ages. As time passes, the plasticizer vaporizes, and the upholstery takes on an entirely new character.

Although PVC does not easily ignite, the same is untrue of the plasticizers added to PVC products. The plasticizer most commonly added to PVC products is DEHP (Section 13.7-B). The ease with which PVC products ignite increases as more and more plasticizer is incorporated into them.

At elevated temperatures, PVC thermally decomposes and burns. Given the abundance of PVC products that firefighters are likely to encounter during a normal fire, the decomposition of poly(vinyl chloride) may constitute a potential health concern for the following reasons:

- Hydrogen chloride is produced when PVC burns. As noted in Section 8.8-B, hydrogen chloride poses the hazard of inhalation toxicity.

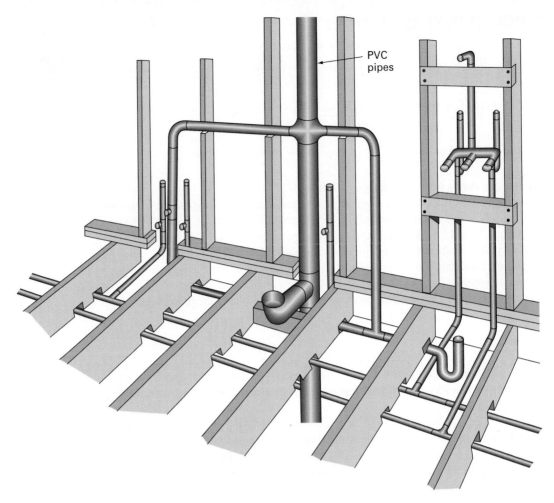

FIGURE 14.7 PVC plumbing pipes are often used within bathrooms and kitchens for hot- and cold-water delivery to fixtures, to carry drainage and waste, and to vent odors. The PVC pipes are usually joined to copper or galvanized steel pipes with transition fittings.

- Polychlorinated dibenzofurans and polychlorinated dibenzo-*p*-dioxins are produced during the incomplete combustion of poly(vinyl chloride). As noted in Section 13.4-A, exposure to minutely low dioxin concentrations causes a variety of illnesses in individuals and their offspring.
- All PVC products contain some concentration of the vinyl chloride monomer, usually less than 10 ppm, that did not become a component of the polymer. It is prudent to assume that some vinyl chloride monomer is released to the environment whenever PVC products are exposed to elevated temperatures. This may be an epidemiological concern, since the monomer is a human carcinogen.

14.6-D POLYACRYLONITRILE

Acrylonitrile is a colorless liquid produced from propylene, ammonia, and oxygen. It is a very volatile, toxic, and flammable liquid.

$$2CH_3CHC{=}CH_2(g) \quad + \quad 2NH_3(g) \quad + \quad 3O_2(g) \quad \longrightarrow \quad CH_2{=}\overset{\displaystyle H}{\underset{\displaystyle CN}{C}}(g) \quad + \quad 6H_2O(g)$$

       Propylene   Ammonia   Oxygen   Acrylonitrile   Water

Polyacrylonitrile is the polymer that results when acrylonitrile is polymerized. The unit that recurs in its macromolecules is the following:

$$\sim\!\sim\!CH_2CH\!\sim\!\sim$$
$$|$$
$$CN$$

This synthetic polymer was the first to become commercially popular in the form of *acrylic fibers*, a term referring to those fibers composed of at least 85% by mass of acrylonitrile units. Today, acrylic fibers are primarily used in carpets and other textiles, since they possess an outstanding resistance to sunlight, are quick-drying, and are easy to launder. They are popularly known by the tradenames *Orlon* and *Acrilan*. Polyacrylonitrile is also used to manufacture plastics and nitrile rubber.

As first noted in Section 10.10-B, hydrogen cyanide is produced when materials made of polyacrylonitrile undergo thermal decomposition. Although life-threatening concentrations of hydrogen cyanide could be generated at fire scenes, the available evidence indicates that the presence of hydrogen cyanide does not significantly contribute to many fire-related fatalities. It is possible that the hydrogen cyanide does not itself survive within the fire environment.

14.6-E POLY(METHYL METHACRYLATE)

Methyl methacrylate is a colorless liquid that is manufactured by the three-step process illustrated by the following equations:

$$CH_3-C-CH_3(l) \ + \ HCN(g) \ \longrightarrow \ (CH_3)_2-C-CN(l)$$
$$\quad\quad \overset{\|}{O} \quad\quad\quad\quad\quad\quad\quad\quad\quad\quad \underset{OH}{|}$$

| Acetone | Hydrogen cyanide | Acetone cyanohydrin |

$$(CH_3)_2-C-CN(l) \ + \ H_2SO_4(l) \ \longrightarrow \ \begin{matrix} CH_3 \\ | \\ CH_2\!=\!C \\ \backslash \\ C\!=\!O \\ / \\ NH_3\!\cdot\!HSO_4(l) \end{matrix}$$
$$\quad\quad\quad \underset{OH}{|}$$

| Acetone cyanohydrin | Sulfuric acid | Methylacrylamide bisulfate |

$$\begin{matrix} CH_3 \\ / \\ CH_2\!=\!C \quad (l) \\ \backslash \\ C\!=\!O \\ / \\ NH_3\!\cdot\!HSO_4 \end{matrix} \ + \ CH_3OH(l) \ \longrightarrow \ \begin{matrix} CH_3 \\ / \\ CH_2\!=\!C \quad (l) \\ \backslash \\ C\!=\!O \\ / \\ O \\ \backslash \\ CH_3 \end{matrix} \ + \ NH_4HSO_4(aq)$$

| Methylacrylamide bisulfate | Methanol | Methyl methylacrylate | Ammonium bisulfate |

Methyl methacrylate is very volatile and highly flammable.

Several polyvinyl polymers are produced from the esters of acrylic acid and methacrylic acid. An example is the commercially popular *poly(methyl methacrylate)*, which is frequently designated as *PMMA*. This polymer is produced by polymerizing methyl methacrylate. The repeating unit in PMMA is the following:

$$\sim\sim\sim CH_2-\underset{\underset{\underset{\underset{CH_3}{O}}{\overset{|}{O}}}{\overset{|}{\underset{\overset{|}{C=O}}{C}}}}{\overset{\overset{CH_3}{|}}{C}}\sim\sim\sim$$

SOLVED EXERCISE 14.4

Which combustion products are produced when materials made from polyacrylonitrile smolder and burn?

Solution: The macromolecules of polyacrylonitrile have hundreds or thousands of the following unit:

The constituent atoms of these units serve as potential source of combustion products. When the polymer smolders, it is exposed to the intense heat evolved as the polymer slowly burns. Hydrogen cyanide is produced during the smoldering process. Carbon, hydrogen, and nitrogen atoms are the only constituents that oxidize when polyacrylonitrile burns. When the carbon atoms oxidize, carbon monoxide and carbon dioxide are produced; when the nitrogen atoms oxidize, nitric oxide and nitrogen dioxide are produced; and when the hydrogen atoms are oxidized, water is produced. Consequently, the combustion products generated during the combustion of acrylonitrile are hydrogen cyanide, carbon monoxide, carbon dioxide, nitric oxide, nitrogen dioxide, and water.

PMMA is a component of the plastic products known commercially as Plexiglas and Lucite. It is as clear as glass but can be manufactured as a transparent, translucent, or opaque material. Because PMMA is virtually unbreakable under normal conditions of stress and tension, it is useful in windshields, windows, and other products simultaneously requiring weather resistance, strength, and transparency. The largest use of PMMA is associated with the manufacture of displays and advertising signs, but the polymer is also used to make lighting fixtures, building panels, and plumbing and bathroom fixtures. The military uses PMMA in cockpit canopies, windows, gun turrets, and bombardier enclosures. It is also used as a component of latex and enamel paints, a drying oil for varnishes, and a finishing compound on leather.

As noted earlier, when PMMA is exposed to heat, it depolymerizes and forms its monomer, methyl methacrylate. Within a fire environment, the monomer readily burns.

PERFORMANCE GOALS FOR SECTION 14.7:

- Describe the general properties of epoxy resins, and provide examples of their commercial use.

14.7 Epoxy Resins

epoxy resin
■ The polymer resulting from the condensation of Bisphenol-A and epichlorohydrin

In Section 13.2-L it was noted that Bisphenol-A is used in the polymer industry to manufacture **epoxy resins**. These are compounds resulting from the condensation of Bisphenol-A and epichlorohydrin.

Bisphenol-A   Epichlorohydrin

Epoxy resin

When the resin is cross-linked, the resulting polymers are especially hard, chemically resistant, and noncorrosive.

Epoxy resins are used in many commercial applications, including protective coatings for sports equipment, the hulls of ships, metal containers, and kitchen appliances.

PERFORMANCE GOALS FOR SECTION 14.8:

- Describe the general properties of polyurethane, and identify the common products produced from it.
- Identify the toxic gases produced when polyurethane thermally decomposes or burns.
- Identify the potential hazard associated with the use of untreated polyurethane as insulation material.

14.8 Polyurethane

A *polyurethane* is a substance produced by the condensation of a glycol and an organic diisocyanate. The chemical formula of a simple **isocyanate** is $R-N=C=O$. The organic diisocyanates are compounds whose molecules have two isocyanate groups ($-N=C=O$). The compounds produced by the condensation of a glycol and an organic diisocyanate contain the following functional group of atoms called the **carboxamide group**:

$$-\overset{\displaystyle O}{\overset{\displaystyle \|}{C}}-NH_2$$

isocyanate
■ An organic compound whose molecules contain the $-N=C=O$ group of atoms

carboxamide group
■ The $-\overset{O}{\overset{\|}{C}}-NH_2$ functional group, common to amides

Carboxamide groups are components of all polyurethanes.

A popular commercial polyurethane product is a solid foam produced by adding a foam-blowing agent to the reaction mixture of toluene 2,4-diisocyanate, and ethylene glycol.

O=C=N—[benzene ring]—N=C=O(*l*) + HO—CH$_2$CH$_2$—OH(*l*) ⟶

Toluene 2,4-diisocyanate   Ethylene glycol

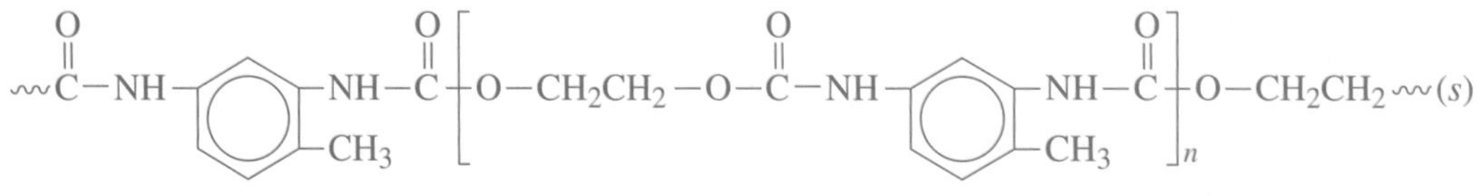

An example of a polyurethane

The heat released during the polymerization reaction causes the blowing agent to vaporize. Bubbles of vapor are subsequently captured within the viscous liquid as it polymerizes and expands. Polyurethane foam results when the frothy mixture solidifies.

14.8-A USES OF POLYURETHANE

Polyurethane foam is available commercially as products that are either rigid or flexible plastics. These foam products are manufactured with a wide range of additives like stabilizers, dyes, fire retardants, and fungicides. These two types of polyurethane products include the following uses:

- The rigid foam is primarily used as insulation, sound-deadening boards, and wall panels. The insulation is conveniently sprayed on the surface of concrete, between wall and roof joists, and around window and door frames. Rigid polyurethane is also used to insulate external fuel tanks on space shuttles.
- The flexible polyurethane foams are used in carpet padding and bedding, pillows, furniture, shoe soles, medical splints, and automobile seat cushioning. In these foam products, the softness and resiliency of polyurethane is the desirable feature.

SOLVED EXERCISE 14.5

Which toxic gases are produced when piles of polyurethane jackets smolder?

Solution: Polyurethane macromolecules have hundreds or thousands of a unit such as the following:

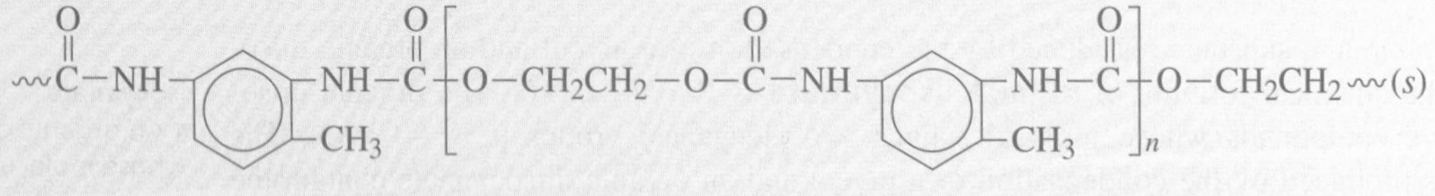

When the polyurethane jackets smolder, the toxic gases produced are hydrogen cyanide, carbon monoxide, nitric oxide, and nitrogen dioxide.

14.8-B FIGHTING POLYURETHANE FIRES

Polyurethane fires can effectively be extinguished with the application of water, but since they retain considerable heat, it is essential to check for total fire extinguishment within the foam products to prevent their reignition.

In the past, it was well acknowledged that caution had to be exercised when fighting fires involving polyurethane products. Their combustion was associated with the production of prodigious quantities of nitrogen dioxide, the presence of which posed an increased risk of inhalation toxicity to building occupants and firefighters. This risk was considered so high that safety engineers often recommended the use of a warning label such as the following on polyurethane building products:

> ## FIRE HAZARD WHEN USED INSIDE BUILDINGS
>
> **USE ONLY ON EXTERIOR APPLICATIONS OR, WHEN CONFINED, BETWEEN WALLS OF MATERIALS CONSIDERED ACCEPTABLE FOR CONSTRUCTION**

The polyurethane products encountered in the contemporary market have been formulated with fire retardants like the polybrominated diphenyl ethers (Section 13.4-C) or treated in an analogous fashion to reduce their ease of combustion. Although the burning pattern of polyurethane products may be altered by special designing, does this mean that the concern over the use of polyurethane products is primarily a matter of the past? The answer is no, since many polyurethane products still exist in today's consumer products.

In particular, the polyurethane foam in upholstered furniture constitutes a serious fire hazard. For example, in just 4 minutes, a sofa fire may engulf a living room in flames and produce smoke and toxic gases. In this short time, the temperature of the surroundings can reach a staggering 1400 °F (760 °C). Only California requires furniture using polyurethane foam to be treated with flame retardants prior to sale in the state; no other state has followed suit. Yet, the smoldering and open-flame ignition of consumer products containing polyurethane foam is a primary cause of death by fire within the home.

PERFORMANCE GOALS FOR SECTION 14.9:

- Identify the common heat- and fire-resistant polymers and the common products produced from them.

14.9 Heat- and Fire-Resistant Polymers

Research chemists have devoted considerable time learning how to make polymers that are heat and fire resistant. Their research efforts have resulted in the discovery of several synthetic polymers with these sought-after qualities. Heat- and fire-resistant polymers are very desirable for potential use in certain types of commercial products, especially building construction materials and products for home use.

14.9-A POLY(TETRAFLUOROETHYLENE)

Tetrafluoroethylene is a gas that is produced by a two-step process. First, chloroform is reacted with hydrofluoric acid to form chlorodifluoromethane.

$$CHCl_3(g) \;+\; 2HF(l) \;\longrightarrow\; CHClF_2(g) \;+\; 2HCl(g)$$

Chloroform   Hydrofluoric acid   Chlorodifluoromethane   Hydrogen chloride

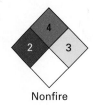

Nonfire

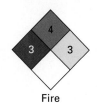

Fire

Then, this chlorofluorocarbon is passed through a tube heated to 1300 °F (700 °C), whereupon a mixture of tetrafluoroethylene and hexafluoropropylene is produced.

$$2CHClF_2(g) \longrightarrow CF_2{=}CF_2(g) \ + \ 2HCl(g)$$
Chlorodifluoromethane   Tetrafluoroethylene   Hydrogen chloride

$$2CHClF_2(g) \longrightarrow F_3C{-}CF{=}CF_2(g) \ + \ 3HCl(g)$$
Chlorodifluoromethane   Hexafluoropropylene   Hydrogen chloride

Tetrafluoroethylene is isolated from this mixture using perfluorooctanoic acid (Section 13.6-C).

Tetrafluoroethylene is used to manufacture *poly(tetrafluoroethylene)* (PTFE), whose commercial name is *Teflon*. The recurring unit in its macromolecules is $\sim\!\!\sim\!\!CF_2{-}CF_2\!\!\sim\!\!\sim$.

Teflon is well known as the polymer used to produce nonstick cookware. It is also used to produce virtually indestructible tubing, gaskets, valves, and cable insulation. Chemists often use Teflon tubing in laboratories, and cardiologists use it as artificial veins and arteries.

Teflon is extraordinarily heat resistant. Teflon products perform well when exposed to temperatures ranging from −400 to 482 °F (−240 to 250 °C). Nonstick cookware coated with Teflon can be heated to temperatures in excess of 932 °F (500 °C) without burning.

Teflon is also extraordinarily unreactive when exposed to hot corrosive acids. For this reason, Teflon is coated on the interior walls of tanks intended for acid storage. It is also the polymer used in the waterproof fabric called *GORE-TEX*, which golfers may recognize as the material commonly used in sportswear.

14.9-B NOMEX

Nomex is the tradename for a condensation polymer made from *m*-diaminobenzene and isophthaloyl chloride as follows:

$$H_2N\text{---}\bigcirc\text{---}NH_2(s) \ + \ \underset{\text{Isophthaloyl chloride}}{Cl{-}\overset{O}{\overset{\|}{C}}\text{---}\bigcirc\text{---}\overset{O}{\overset{\|}{C}}{-}Cl(l)} \longrightarrow \left[NH\text{---}\bigcirc\text{---}NH{-}\overset{O}{\overset{\|}{C}}\text{---}\bigcirc\text{---}\overset{O}{\overset{\|}{C}} \right]_n (s)$$

m-Diaminobenzene   Isophthaloyl chloride   Nomex

From this polymer, a heat-resistant fabric is produced. Emergency responders are familiar with this polymer, since it is often used in the uniforms of firefighters and race car drivers. The fabric provides an extra element of thermal protection to firefighters, since it carbonizes and thickens when exposed to intense heat. This increases the protective barrier between the heat source and the skin, thus minimizing burn injury.

14.9-C KEVLAR

Kevlar is the commercial name for a condensation polymer made from *p*-diaminobenzene and terephthaloyl chloride.

$$\underset{\text{Terephthaloyl chloride}}{\overset{O}{\overset{\|}{C}}{-}\bigcirc\text{---}\overset{O}{\overset{\|}{C}}(l)} \ + \ H_2N\text{---}\bigcirc\text{---}NH_2(s) \longrightarrow \left[\overset{O}{\overset{\|}{C}}\text{---}\bigcirc\text{---}\overset{O}{\overset{\|}{C}}{-}NH\text{---}\bigcirc\text{---}NH \right]_n (s)$$

Terephthaloyl chloride   *p*-Diaminobenzene   Kevlar

Kevlar is flameproof and may be drawn into fibers having a strength five times stronger than steel. Given this astounding resiliency, fabrics made from Kevlar fibers are used as the reinforcement in bullet-resistant vests and helmets. A 16.4-lb (7.5-kg) vest that has been lined with Kevlar and ceramic plates can stop armor-piercing bullets shot from high-powered rifles. A 4-lb (1.8-kg) helmet lined with up to 24 layers of Kevlar is approximately 40% more resistant to shrapnel than the steel helmets formerly used by the military.

No single product compares with Kevlar in terms of the number of lives saved through its use. Thousands of active police officers have saved their lives by wearing Kevlar vests and helmets during the line of duty.

Aside from its use in military and police body armor, Kevlar is also used in manufacturing rip-resistant jeans, work gloves, skis, hockey sticks, golf balls, ropes, cables, boat hulls, and aircraft structural parts. Some tire manufacturers are using Kevlar to reinforce the tread of radial automobile tires. Because Kevlar is stable at very high temperatures, it is also used in some brands of the protective clothing worn by firefighters.

PERFORMANCE GOALS FOR SECTION 14.10:

- Describe the general nature of the macromolecules in natural rubber.
- Describe how natural rubber is vulcanized.
- Relate the general nature of the macromolecules in natural and synthetic rubbers with the toxic gases that are produced when rubber products burn.
- Identify the principal substances that are components of rubber formulations.
- Describe the response actions to be executed when emergency responders encounter burning rubber products.

14.10 Rubber and Rubber Products

The term **rubber** refers to any of the natural or synthetic polymers having two main properties: deformation under strain and elastic recovery after vulcanization (defined in Section 14.10-A). As noted in Section 14.1, the rubberlike polymers are called elastomers.

rubber
■ A polymer that is simultaneously elastic, airtight, water-resistant, and long-wearing

14.10-A NATURAL RUBBER

Natural rubber is produced from *natural latex*, a white fluid that exudes from cuts in the bark of the South American rubber tree *Hevea brasiliensis*. The natural latex consists of approximately 30% to 35% by mass of *cis*-1,4-polyisoprene. Its general chemical formula is $[\sim\sim CH_2-C=CH-CH_2\sim\sim]_n$ where n is a large integer. When geometrical isomerism is
$$\underset{}{\overset{|}{\underset{CH_3}{}}}$$
considered, the macromolecular structure is represented as follows:

natural rubber
■ The polymer produced from the latex that exudes from cuts in the bark of the South American rubber tree

$$
\begin{array}{ccccccc}
H & CH_3 & H & CH_3 & H & CH_3 \\
\diagdown & \diagup & \diagdown & \diagup & \diagdown & \diagup \\
C=C & & C=C & & C=C \\
\diagup & \diagdown & \diagup & \diagdown & \diagup & \diagdown \\
\sim\sim CH_2 & CH_2 & CH_2 & CH_2 & CH_2 & CH_2\sim\sim
\end{array}
$$

This structure illustrates that the methyl groups are oriented in one direction about the carbon–carbon double bonds. Nature selectively produces only the cis isomer of 1,4-polyisoprene.

By itself, natural rubber is not entirely desirable for production of commercial products. It is soft and sticky, especially in warm climates. In 1839, however, Charles Goodyear accidentally discovered that these undesirable features could be eliminated by heating the natural rubber with elemental sulfur. This discovery led to the ultimate development of synthetic rubbers and the industrial process called **vulcanization**.

At the macromolecular level, vulcanization is a chemical process wherein the *cis*-1, 4-polyisoprene chains are cross-linked with disulfide ($-S-S-$) bonds. The structure of the vulcanized macromolecule may be illustrated two-dimensionally as follows:

$$\text{\tiny $\wedge\wedge\wedge$}CH-\underset{\underset{\underset{S}{|}}{\overset{\overset{CH_3}{|}}{C}}}{=}CH-CH_2-CH-\underset{\underset{\underset{S}{|}}{\overset{\overset{CH_3}{|}}{C}}}{=}CH-CH_2\text{\tiny $\wedge\wedge\wedge$}$$

SOLVED EXERCISE 14.6

Why do foam rubber products catch fire easily within a hot clothes dryer or under a heating blanket, whereas vehicular rubber tires are comparatively difficult to ignite even on exposure to intense heat?

Solution: Foam rubber products are produced by blowing and entrapping an inert gas within the complex structure of rubber until it hardens. This process causes an expansion of the rubber and an increase in its surface area. By contrast, an inert gas is not entrapped within the rubber used for the production of rubber tires. According to the principles outlined in Section 5.5, an increase in the surface area of the reactants increases the reaction rate. Because the rubber in foam rubber products have a greater surface area than the rubber in rubber tires, foam rubber products ignite and burn at a comparatively faster rate than rubber tires.

Virtually all rubber formulations include elemental sulfur or a sulfur-bearing compound. The rate at which the vulcanization process occurs is usually controlled through the use of peroxo-organic compounds or zinc dithiocarbamate. The resulting products are called *vulcanized rubbers*. They are tougher, less plastic and sticky, and more elastic than natural rubber. Vulcanized rubbers are capable of retaining firm shapes over a relatively wide temperature range. Vulcanization also permits the casting of natural rubber into shapes such as automobile tires. A mixture of rubber and sulfur is inserted into a mold that when sealed and heated, produces a tire carcass.

14.10-B SYNTHETIC RUBBERS

Natural rubber still accounts for approximately 35% of the demand for rubber in the United States. Today, however, the world no longer relies solely on the latex from rubber trees for its source of rubber. Chemists have synthesized elastomers called **synthetic rubbers**. They have physical properties that are similar to or better than those of natural rubber. We briefly note the properties of five synthetic rubbers: *cis*-1,4-polybutadiene rubber, styrene–butadiene rubber, acrylonitrile–butadiene rubber, neoprene, and polysulfide rubber.

■ *cis-1,4-Polybutadiene* is the simplest synthetic rubber. It is produced by the polymerization of 1,4-butadiene using a Ziegler–Natta catalyst. In the two-dimensional segment of the macromolecule shown below, the hydrogen atoms are located on the same side of the carbon–carbon double bonds.

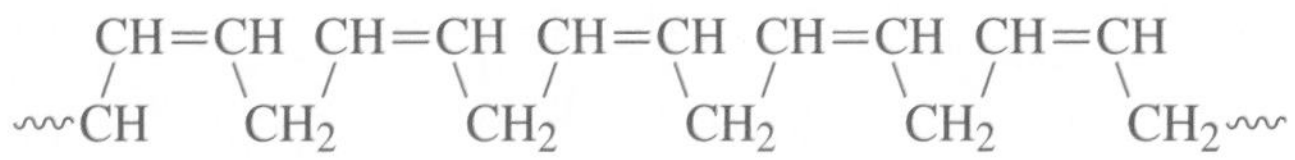

$$\begin{array}{ccccccccccc}
& \text{CH}{=}\text{CH} & & \text{CH}{=}\text{CH} & & \text{CH}{=}\text{CH} & & \text{CH}{=}\text{CH} & & \text{CH}{=}\text{CH} \\
\diagup & & \diagup\;\diagdown & & \diagup\;\diagdown & & \diagup\;\diagdown & & \diagup\;\diagdown & & \diagdown \\
\text{\textasciitilde CH} & & \text{CH}_2 & & \text{CH}_2 & & \text{CH}_2 & & \text{CH}_2 & & \text{CH}_2\text{\textasciitilde}
\end{array}$$

cis-1,4-Polybutadiene has many of the properties of natural rubber and may be vulcanized with elemental sulfur.

■ **Styrene–butadiene rubber** (SBR) was first known as *GRS* (government rubber styrene), since it was produced on behalf of the U.S. military during World War II. Because it is a copolymer of 1,3-butadiene and styrene, it became known outside government circles as *Buna S* ("Bu" for butadiene, "na" for sodium, and "S" for styrene). (The production reaction is catalyzed by sodium.) The repeating unit of SBR is represented as follows:

$$[\text{\textasciitilde}(\text{CH}_2{-}\text{CH}{=}\text{CH}{-}\text{CH}_2)_x{-}\underset{|}{\text{CH}}{-}\text{CH}_2\text{\textasciitilde}]_n$$

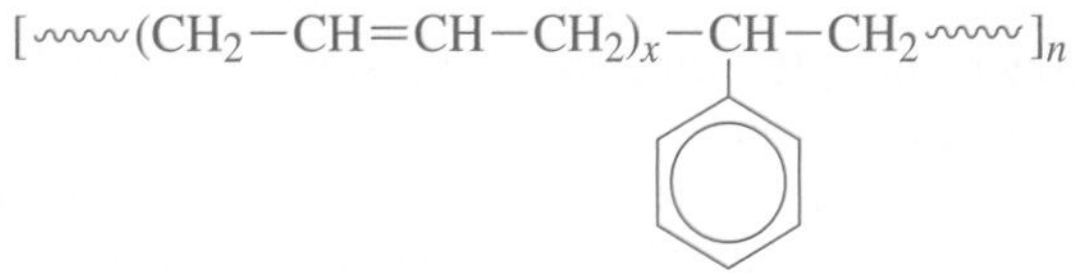

SBR is vulcanized using elemental sulfur. Because it resists wear more than any other synthetic rubber, SBR is used to manufacture most tire treads. It is also the major component of many adhesives.

■ *Buna N rubber* is a copolymer of butadiene and acrylonitrile (*N* stands for *nitrile*). It is also known as *acrylonitrile–butadiene rubber*. The repeating unit in Buna N rubber is the following:

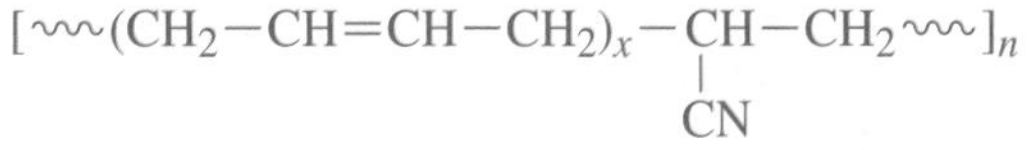

$$[\text{\textasciitilde}(\text{CH}_2{-}\text{CH}{=}\text{CH}{-}\text{CH}_2)_x{-}\underset{\underset{\text{CN}}{|}}{\text{CH}}{-}\text{CH}_2\text{\textasciitilde}]_n$$

Buna N rubber has the unique feature of withstanding heat up to 350 °F (177 °C); other synthetic rubbers soften, melt, or burn at lower temperatures.

■ *Polychloroprene* is more commonly known as **neoprene**. It is produced by heating chloroprene, or 2-chloro-1,3-butadiene, a known human carcinogen. The simplest structure of the repeating unit in neoprene is the following:

$$[\text{\textasciitilde}\text{CH}_2{-}\underset{\underset{\text{Cl}}{|}}{\text{C}}{=}\text{CH}{-}\text{CH}_2\text{\textasciitilde}]_n$$

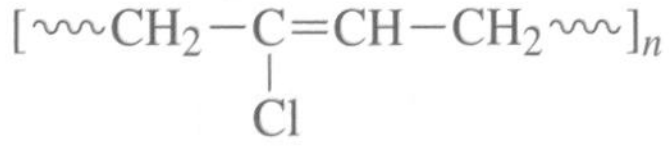

Neoprene is vulcanized with zinc oxide, during which single neoprene strands cross-link with oxygen atoms and replace some of the chlorine atoms. Macromolecular segments of the vulcanized neoprene have two-dimensional structures such as the following:

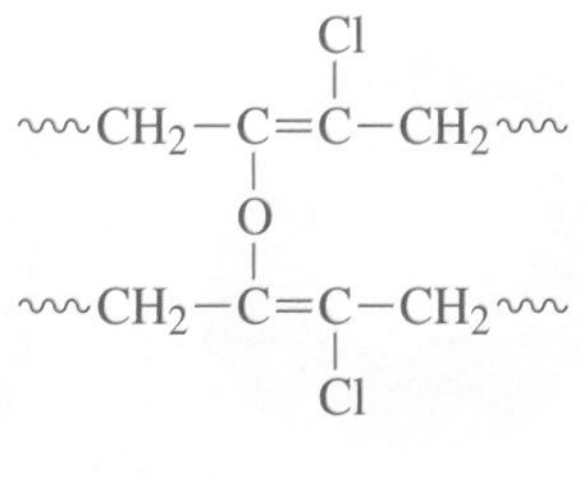

$$\begin{array}{c}
\overset{\text{Cl}}{\underset{|}{}} \\
\text{\textasciitilde}\text{CH}_2{-}\text{C}{=}\text{C}{-}\text{CH}_2\text{\textasciitilde} \\
| \\
\text{O} \\
| \\
\text{\textasciitilde}\text{CH}_2{-}\text{C}{=}\text{C}{-}\text{CH}_2\text{\textasciitilde} \\
\underset{\text{Cl}}{|}
\end{array}$$

Neoprene rubber does not possess the resilience necessary for use in tires, but it is chemically resistant to petroleum products and ozone. This feature makes neoprene suitable for specialized uses such as roofing tile and flexible hose.

■ *Thiokol* is also known as *polysulfide rubber*. Although there are several types, the most common polysulfide rubber has repeating units like the following:

$$[\sim\!\sim\!\text{CH}_2\!-\!\text{CH}_2\!-\!\underset{\underset{S}{\|}}{S}\!-\!\underset{\underset{S}{\|}}{S}\!\sim\!\sim]_n$$

It is primarily used to manufacture hoses for loading gasoline and oil.

14.10-C PRODUCTS PRODUCED FROM SYNTHETIC RUBBERS

Synthetic rubbers vulcanized with sulfur or sulfur compounds are used to manufacture a variety of products, some of which are illustrated in Figure 14.8. Their manufacture requires controlling the amount of sulfur used for vulcanization, as demonstrated by the following:

■ Natural rubber mixed with only 3% sulfur by mass is soft and elastic. It is used for manufacturing inner tubes and rubber bands.

■ With additional sulfur, vulcanized rubber becomes hard and loses some of its elasticity. Natural rubber having up to 10% sulfur by mass is used to manufacture automobile tires.

■ When rubber has been vulcanized to the extent of 68% by mass, it becomes a black solid called **ebonite**, or *hard rubber*. It is used to make black piano keys and the casings for lead–acid storage batteries.

ebonite
■ A form of hard rubber used primarily to make the casings for automobile storage batteries

Substances other then sulfur are also components of rubber formulations. Carbon black (Section 7.7-D) is often added as a reinforcing filler to make rubber stronger, abrasive, and easier to elongate, and to prevent it from breaking and tearing. A specified amount of previously used rubber (reclaimable rubber) is usually recycled by adding it to a new batch of rubber. Depending on the projected use of the rubber product, air or ammonium carbonate may also be mixed into the rubber latex. When this mixture is subsequently heated, the ammonium carbonate decomposes to ammonia and carbon dioxide. These gases become entrapped within the complex rubber structure as it hardens. The product, called *foam rubber*, is used to manufacture cushions and mattresses.

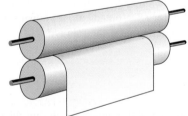

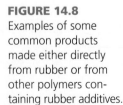

FIGURE 14.8
Examples of some common products made either directly from rubber or from other polymers containing rubber additives.

FIGURE 14.9 A fire within a massive pile of vehicle tires may be initiated by a lightning strike. As the tires burn, the smoke that evolves is intensely dense and black. To bring this type of fire under control, the use of special fire extinguishers is generally required. (*Courtesy of Pyrocool Technologies, Inc., Monroe, Virginia.*)

14.10-D RESPONDING TO INCIDENTS INVOLVING THE BURNING OF RUBBER

To understand the nature of the substances produced when rubber products burn, it is necessary to recollect the general features of their chemical formulations. During the combustion process, the constituents of these formulations combine with atmospheric oxygen in the following ways:

- As they burn, rubber products vulcanized with sulfur or sulfur-bearing compounds produce carbon monoxide, sulfur dioxide, and water vapor.
- The smoke associated with rubber fires is extraordinarily dense and black, as in the scene depicted in Figure 14.9. Both features of these fires are linked with the presence of carbon black in the rubber formulation.
- As neoprene burns, carbon monoxide, water vapor, and hydrogen chloride are produced.
- As polysulfide rubber burns, carbon monoxide, sulfur dioxide and water vapor are produced.

Because the atmosphere near burning rubber consists of a mixture of these gases, the toxic environment of the scene must be acknowledged when determining a proper firefighting strategy. To prevent fatalities, the use of self-contained breathing apparatus is always warranted.

Nature of Polymerization

14.1 Write the chemical formula for macromolecular segments of each of the following polymers:

(a) polyethylene having eight recurring units
(b) polypropylene having five recurring units
(c) poly(methyl methacrylate) having three recurring units

14.2 Poly(vinylidene chloride) is produced by the addition polymerization of vinylidene chloride, or 1,1-dichloroethene. Write the stepwise equations that illustrate how this addition polymerization occurs.

Autopolymerization

14.3 DOT prohibits shippers from offering acrolein for transportation unless it has been "stabilized." DOT then denotes its proper shipping name as "acrolein, stabilized."

(a) What specific property of acrolein must be stabilized?
(b) Use Tables 6.1 and 6.3 to determine the basic description DOT requires a shipper to enter on the accompanying waybill when offering acrolein for transportation by rail tankcar.
(c) Use Tables 6.1 and 6.7 to determine the information that DOT requires the carrier to mark on both sides and ends of the tankcar.

Combustion of Polymers

14.4 Identify the combustion products that result when polymers having each of the following recurring units burn in air:

(a) $\sim\sim\sim CH_2-\underset{\underset{\displaystyle CH_3}{\overset{\displaystyle |}{\underset{|}{O}}}{\overset{\displaystyle |}{\underset{|}{C=O}}}\underset{|}{\overset{\displaystyle H}{\underset{|}{C}}}\sim\sim\sim$

(b) $\sim\sim\sim NH-(CH_2)_5-\overset{\displaystyle O}{\overset{\displaystyle \|}{C}}\sim\sim\sim$

(c) $\sim\sim\sim CH_2CH_2CH_2CH_2-O-\overset{\displaystyle O}{\overset{\displaystyle \|}{C}}-\bigcirc-\overset{\displaystyle O}{\overset{\displaystyle \|}{C}}-O\sim\sim\sim$

(d) $\sim\sim\sim CH_2-\underset{\underset{\displaystyle Cl}{|}}{\overset{\overset{\displaystyle Cl}{|}}{C}}\sim\sim\sim$

(e) $\sim\sim\sim CH_2CH_2CH_2-S-S-CH_2CH_2\sim\sim\sim$

Vegetable and Animal Fibers

14.5 What thermal decomposition products are produced when a woolen sweater is exposed to intense heat?

Vinyl Polymers

14.6 Which requires the application of a larger volume of water for effective total extinguishment: 1000 lb (454 kg) of burning polyethylene or the same mass of burning polystyrene?

14.7 *Dacron* fabrics are produced from poly(ethylene terephthalate) fibers. The recurring unit in the macromolecules of this polyester is the following:

$$\sim\!\!\sim\!\!\sim O-CH_2CH_2-O-\overset{\displaystyle O}{\overset{\displaystyle \|}{C}}\!\!-\!\!\langle\!\bigcirc\!\rangle\!\!-\!\!\overset{\displaystyle O}{\overset{\displaystyle \|}{C}}\!\sim\!\!\sim\!\!\sim$$

Identify the combustion products that are produced when Dacron fabrics burn.

14.8 Poly(vinylidene dichloride) decomposes when heated to a temperature ranging from 437 to 527 °F (225 to 275 °C). The recurring unit in poly(vinylidene dichloride) macromolecules is $\sim\!\!\sim CH_2-\underset{\displaystyle |}{\underset{\displaystyle Cl}{CH}}-CH_2-CCl_2\sim\!\!\sim$.

(a) Identify the toxic gases that are most likely to be produced when the polymer decomposes.

(b) In what ways could the presence of these gases at a fire scene adversely affect the health of emergency responders?

Heat- and Fire-Resistant Polymers

14.9 Why are Kevlar fibers incorporated into the body armor used by emergency-response personnel?

Rubber

14.10 Air monitoring provided the following contaminant concentrations at the approximate center of a fire scene at which acres of automobile tires are burning:

	Concentration (ppm)
Carbon monoxide	1700
Carbon dioxide	7000
Sulfur dioxide	600

Ignoring synergistic effects between these gases, which individual concentrations of these gases are considered life threatening to the firefighters at the scene?

CHAPTER

Chemistry of Some Explosives

A n **explosive** is a chemical or nuclear material that may be initiated to undergo a very rapid, self-propagating decomposition that results in the formation of more stable substances, the liberation of heat, and the sudden development of pressure. We examine chemical explosives in this chapter and nuclear explosives in Chapter 16.

Chemical explosives are generally encountered as components of each of the following:

- The ammunition used for hunting and other sporting activities
- The charges implanted during the mining of ores, drilling of oil wells, tunneling through mountains, clearing land, loosening of underground rock formations, creating firebreaks, and imploding unwanted structures

explosive
■ For purposes of DOT regulations, a substance or article, including a device, that is designed to function by explosion, or that, by chemical reaction with itself, is able to function in a similar manner even if not designed to function by explosion

- The artillery and munitions used during wartime to destroy cities, sink ships, and disable the enemy

Unwanted incidents involving explosives were once primarily associated with events in which the procedures designed to safely transport, store, and handle these materials had not been rigorously followed. These incidents were once rare compared with those linked with other hazardous materials.

Today, however, reports are all too frequent about homicide terrorists who are willing to kill themselves by detonating an improvised explosive device in the presence of innocent bystanders. An **improvised explosive device** (IED) is any explosive article that has been fabricated and designed to destroy or incapacitate personnel or vehicles during unconventional warfare or terrorist actions. An example of an IED is a roadside bomb. IEDs are frequently used by terrorists during the wars in Iraq and Afghanistan.

Explosives are the active components of the weapons of mass destruction most commonly chosen by terrorists. We previously noted in Figure 11.1, for example, that an explosive mixture of ammonium nitrate and fuel oil was used to destroy the Oklahoma City Federal Building in 1995. In 1998, terrorists downed Pan Am Flight 103 over Lockerbie, Scotland, by detonating an explosive embedded within a suitcase. In this incident, 259 passengers and crew were killed along with another 11 people on the ground. In the same year, nearly simultaneous car bombings destroyed the U.S. embassies in Kenya and Tanzania, killing 231 people including 12 Americans. Also in 1998, suicide attackers rammed an explosives-laden boat into the destroyer USS *Cole* in Yemen, killing 17 American sailors. International terrorists have also used explosives to carry out acts of violence in Afghanistan, Britain, Colombia, Djerba, India, Indonesia, Iraq, Israel, Kenya, Lebanon, Morocco, Pakistan, Philippines, Russia, Saudi Arabia, Spain, Turkey, and elsewhere.

Individuals who use explosives for peaceful or wartime purposes must receive specialized training in their handling. Although such training requires more rigorous instruction than may be provided here, we learn in this chapter about the chemical composition of some explosives, their different types, and the procedures that are recommended for responding to mishaps involving them.

improvised explosive device
- Any explosive device that has been fabricated and designed to destroy or incapacitate personnel or vehicles (abbreviated IED)

PERFORMANCE GOALS FOR SECTION 15.1:

- Distinguish between the combustion and the detonation of an explosive.
- Distinguish between an explosive substance and an explosive article.
- Describe how the brisance of an explosive is determined.

15.1 General Characteristics of Explosives

There are two general types of explosives:

- **Explosive substances** are unique compounds that may be made to explode.
- **Explosive articles** are manufactured items that contain explosive substances as components, like cartridges and military bombs. Some specific examples are provided in Table 15.1. Under ordinary circumstances, the chemical substance in an explosive article is intentionally initiated to detonate by mechanical, electrical, chemical, or hydrostatic means as needed.

An explosive substance may be either an inorganic or organic compound. In both instances, it simultaneously represents the oxidizing agent *and* the reducing agent. Parts of its molecules are oxidized while other parts are reduced. An explosive substance undergoes a chemical transformation

explosive substance
- For purposes of DOT regulations, a substance like black powder or smokeless powder contained in packaging other than an article

explosive article
- For purposes of DOT regulations, a commercially available item like a detonator or flare containing an explosive substance as a component

TABLE 15.1 | Some Explosive Articles[a]

Ammunition, illuminating	Cord, detonating
Ammunition, incendiary	Cord igniter
Ammunition, practice	Fireworks
Ammunition, proof	Flares
Ammunition, smoke	Flash powder
Ammunition, tear-producing	Fuse[b]
Ammunition, toxic	Fuze[b]
Black powder (gunpowder)	Igniters
Bombs, military	Lighters, fuse
Boosters	Mines
Cartridges, blank	Powder cake (powder paste)
Cartridges, flash	Powder, smokeless
Cartridges for weapons	Primers, cap-type
Cartridges, oil well	Primers, tubular
Cartridges, power device	Projectiles
Cartridges, signal	Propellants, liquid
Cartridges, small-arms	Propellants, solid
Charges, bursting	Rocket motors
Charges, demolition	Signals
Charges, depth	Sounding devices
Charges, expelling	Torpedoes
	Warheads

[a]49 C.F.R. §173.59.
[b]"Fuse" refers to a cordlike igniting device, whereas "fuze" refers to a device used in ammunition that incorporates mechanical, electrical, chemical, or hydrostatic components to initiate an explosive train by deflagration or detonation.

detonation
■ The chemical transformation of an explosive, characterized by the essentially instantaneous passage of an energy wave at a supersonic rate, the production of large volumes of gases and vapors, the evolution of energy, and a rise in pressure

very suddenly when compared with the rates at which other chemical reactions occur. Large volumes of gases and vapors are produced as well as an extraordinary amount of energy.

The decomposition of an explosive is called a **detonation**, which may be expressed by an equation similar to the following:

$$A(s,l) \longrightarrow B(g) + C(g)$$

Here, A is the chemical formula of an explosive, and B and C are the chemical formulas of its decomposition products.

A detonation is associated with the passing of an energy wave through the body of the explosive at a supersonic rate (i.e., exceeding the speed of sound). The movement of this wave is called the **detonation velocity**. Most explosive substances are powerful not solely because of the great amount of energy released when they detonate but because of the very high rate at which this energy is propagated. On average, the detonation velocity of an explosive is a staggering 4 mi/s (7 km/s).

detonation velocity
■ The speed at which an energy wave moves through an explosive

An explosive also detonates at a unique temperature. This property is appropriately called the **detonation temperature**. The magnitude of a detonation temperature is difficult to accurately measure, since the value is heavily dependent on the explosive's density, geometrical shape, and other factors. It generally approximates an explosive's melting point.

detonation temperature
■ The temperature at which an explosive detonates

When an explosive detonates, its decomposition products are routinely carbon monoxide, carbon dioxide, nitrogen, oxygen, and/or water vapor. Carbon particulates (soot) and nitrogen oxides may also be produced. These gases and vapors absorb some of the energy that is generated. They are then heated to extremely high temperatures, typically around 5400 °F (3000 °C). This increase in temperature causes the substances simultaneously to expand so rapidly that the passage of the accompanying shock wave exerts an exceedingly high pressure on the nearby surroundings.

Consider the detonation of the explosive substance nitroglycerin. When it detonates, nitroglycerin transforms in less than one-millionth of a second into carbon dioxide, nitrogen, oxygen, and water vapor as follows:

$$\begin{array}{c} CH_2-O-NO_2 \\ | \\ 4CH-O-NO_2\ (l) \\ | \\ CH_2-O-NO_2 \end{array} \longrightarrow 12CO_2(g) + 6N_2(g) + O_2(g) + 10H_2O(g)$$

Nitroglycerin   Carbon dioxide   Nitrogen   Oxygen   Water

The volume increase is comparatively massive: Twenty-nine gaseous molecules are produced for every four nitroglycerin molecules that detonate. The energy of detonation travels at the rate of 4.8 mi/s (7.8 km/s).

Unlike combustion, a detonation does not require the interplay of atmospheric oxygen or any other oxidizer with the explosive. Instead, a detonation is an oxidation–reduction reaction involving the rapid decomposition of a single substance that simultaneously acts as both an oxidizing agent and a reducing agent. At the elevated temperatures accompanying the detonation of any explosive, the heated gases expand so rapidly they can occupy approximately 10,000 times their initial volume. This prodigious expansion is accompanied not only by the simultaneous production of a shock wave but also by a luminescence and an extraordinarily loud sound.

The shock waves produced during the detonation of an explosive substance are linked with their potential shattering power. This property is called the **brisance** of an explosive. The brisance is an important factor when an explosive substance is selected for a specific purpose. For example, when constructing a roadway, an engineer considers whether the brisance generated by a detonation is likely to clear a given mass of rock. Scientists measure the brisance of an explosive by determining the weight of graded sand that is crushed when a given amount of the explosive is detonated within the device shown in Figure 15.1.

brisance
■ The shattering power of an explosive substance

Emergency responders should always recall that explosives possess a high degree of hazard. Exposure to the shock waves that develop during the detonation of an explosive may cause eardrums to rupture and lungs to collapse. Death may result when the intensity of the shock waves is sufficiently high.

Although the detonation of an explosive substance does not require the presence of an oxidizer, most commercial explosives are actually mixed with oxidizers that provide additional oxygen and chemical energy. Some so-called explosives are actually pyrotechnic mixtures of multiple substances, at least one of which is an oxidizer like ammonium nitrate (Section 11.12). These mixtures undergo oxidation–reduction reactions so vigorously that they *appear* to detonate like true explosives. They include the formulations used in blasting agents and gunpowder (Sections 15.2 and 15.5, respectively). Mixtures of pyrotechnic substances are also used as the active components of certain illuminating, incendiary, tear-producing, and smoke-producing devices.

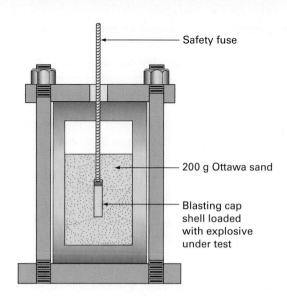

Safety fuse

200 g Ottawa sand

Blasting cap
shell loaded
with explosive
under test

FIGURE 15.1 The type of equipment for conducting a sand test, which is used to determine the brisance on an explosive. The sand test measures the quantity of Ottawa sand a chemical explosive crushes inside a heavy, thick-walled confining bomb. Ottawa sand is characterized by its inability to pass through a 30-mesh screen. After the chemical explosive is detonated, the sand is rescreened and the amount that then passes through a 30-mesh screen is weighed. The result is measured in grams of sand. Of the explosives noted in this book, PETN (Section 15.12) possesses the highest brisance, 62.7 g of sand.

PERFORMANCE GOALS FOR SECTION 15.2:

- Identify the common terms used to describe explosives.
- Describe the chemical nature of a blasting agent.
- Describe the composition of the five DOT types of blasting agents.
- Identify the function of each component of a typical round of artillery ammunition.

high explosive (detonating explosive)
■ An explosive that is capable of sustaining a detonation shock wave to produce a powerful blast

low explosive (deflagrating explosive)
■ A substance that usually deflagrates rather than detonates, such as black powder and igniter cord

blasting cap
■ A small charge of a detonator designed to be embedded in dynamite and ignited by means of a spark or burning fuse

15.2 Classification of Explosives

Individual explosives are usually classified into the following two groups, depending on the rapidity and sensitivity with which they detonate:

- **High explosives**, or **detonating explosives**
- **Low explosives**, or **deflagrating explosives**

High and low explosives are distinguishable in a broad sense by the differences in their detonation velocities. High explosives undergo chemical transformations into hot, rapidly expanding gases that move at supersonic speeds (>3300 ft/s, or 1006 m/s). Examples are triacetone triperoxide (Section 13.9-C), nitroglycerin (Section 15.6), and dynamite (Section 15.7). Their detonations are accompanied by the production of powerful blasts.

In contrast, low explosives deflagrate when unconfined; that is, they burn furiously and persistently. Although hot, rapidly expanding gases are also produced, they move at subsonic speeds. Examples are black powder and safety fuses.

A low explosive is also a component of a **blasting cap**, a small charge of a detonator designed to be embedded in a high explosive and ignited by a spark or burning fuse. One type is

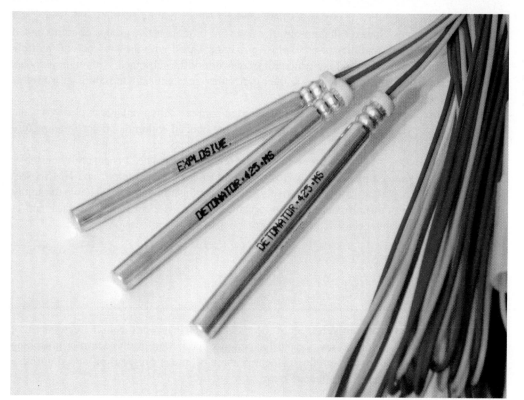

FIGURE 15.2 A common type of blasting cap that when activated, initiates the explosion of another explosive.

shown in Figure 15.2. The activation of a few milligrams of the low explosive within the blasting cap is generally sufficient to initiate the detonation of a high explosive.

The classification of explosives as high or low is not an entirely satisfactory one. Some explosives may be categorized into either class, depending on their state of subdivision, density, and degree of confinement. Smokeless powder (Section 15.8), for instance, is typically considered a low explosive, but under appropriate conditions, its detonation produces the physical effects of a detonating high explosive.

Although blasting agents have the features of low explosives, they are sometimes regarded as a separate class of explosives. For example, regulations promulgated by the Bureau of Alcohol, Tobacco, Firearms and Explosives at 27 C.F.R. §55.202 cite three classes of explosives: high explosives, low explosives, and blasting agents.

Explosives are also classified into one of the following two groups based on their susceptibility to detonate: **primary explosives** and **secondary explosives**. Both are dangerous hazardous materials, since a considerable brisance is associated with their detonations. Primary explosives are substances that are extremely sensitive to heat, mechanical shock, and friction; hence, they develop shock waves within an extremely short time. Typical primary explosives are mercury fulminate, lead azide, and lead styphnate (Sections 15.14-A, -B, and -C, respectively).

By comparison, secondary explosives are substances that are relatively insensitive to heat, mechanical shock, and friction and typically require the use of a booster to bring about their detonation. Typical secondary explosives are cyclonite, tetryl, PETN, and HMX (Sections 15.10, 15.11, 15.12, and 15.13, respectively).

Low explosives or secondary explosives should never be regarded as substances or articles that are lesser hazards when they are respectively compared with high explosives or primary explosives. In particular, a low explosive usually deflagrates, but it *may* also detonate when confined. Although it first burns, the heat generated during its combustion serves to initiate the subsequent detonation.

primary explosive
■ An explosive that is highly susceptible to detonation by exposure to heat, mechanical shock, and friction

secondary explosive
■ An explosive whose susceptibility to detonation occurs only under certain circumstances such as ignition by a primary explosive

The chemical instability of explosives may often be understood by examining the composition of their molecules. Some of these molecules have one or more active groups of atoms called **explosophores**. The instability of an explosive may be linked with the presence of multiple numbers of explosophores within the structural framework of the substance. Although many organic compounds having explosophores in their molecules are chemically unstable, only a select few are suitable as commercial explosives.

The explosophore most commonly encountered in commercially available explosives is the *nitro group* ($-NO_2$). When one or more nitro groups are bonded within the molecular structure of an organic compound, chemists refer to the resulting substance as a nitro derivative of an organic compound. When multiple nitro groups are present in a substance, it is generally unstable. Examples include the commercial explosives trinitrotoluene, cyclonite, tetryl, and HMX (Sections 15.9, 15.10, 15.11, and 15.13, respectively). Analytical instruments are available to rapidly identify the presence of the nitro derivatives of organic compounds. For example, the scanners at airports use these instruments to identify these compounds on the surfaces of luggage.

explosophore
■ The group of atoms bonded within the molecular or unit structure of a substance that provides its explosive nature (e.g., $-N_3$ or $-NO_2$)

SOLVED EXERCISE 15.1

Individuals who survive the initial impact of the shock waves that result from the detonation of an explosive within a confined area may still experience the risk of asphyxiation or poisoning from exposure to the gaseous by-products. Identify the most prevalent gases produced during a typical explosion, and name those that are toxic.

Solution: Most organic explosives contain carbon, hydrogen, oxygen, and nitrogen atoms. When they detonate, the products of decomposition include carbon monoxide, carbon dioxide, nitrogen, oxygen, water, carbon particulates, nitric monoxide, and nitrogen dioxide. There are three toxic gases in this group of substances: carbon monoxide, nitric oxide, and nitrogen dioxide.

15.2-A BLASTING AGENTS

blasting agent
■ For purposes of DOT regulations, a material designated for blasting that has been subjected to prescribed tests and has been shown to have little probability of initiating an explosion or of burning and then exploding

A **blasting agent** is any single material or mixture that has been designated for blasting, subjected to prescribed tests, and shown to have little probability of initiating an explosion or of burning and then exploding. The term generally refers to ammonium nitrate mixtures that have been sensitized with other substances. Figure 15.3 illustrates a commercially available blasting agent.

DOT regulates the transportation of the following five types of blasting agents:

■ *Type A blasting agents* are liquid nitrate esters, such as nitroglycerin, or a mixture of ingredients with one or more of the following: nitrocellulose, ammonium nitrate, other metallic nitrates, nitro derivatives of organic compounds, or combustible materials such as wood meal and aluminum powder. Type A blasting agents exist in powdery, gelatinous, plastic, and elastic forms. They include dynamite, blasting gelatin, and gelatin dynamite (Section 15.7).

■ *Type B blasting agents* are mixtures of ammonium nitrate or other metallic nitrate with an explosive such as trinitrotoluene, with or without other substances such as wood meal or aluminum powder, or a mixture of ammonium nitrate or other metallic nitrates with other nonexplosive, combustible substances. Type B blasting agents do not contain nitroglycerin, similar liquid nitrate esters, or metallic chlorates.

■ *Type C blasting agents* are mixtures of either potassium chlorate or sodium chlorate, and potassium perchlorate, sodium perchlorate, or ammonium perchlorate, with a nitro derivative of an organic compound or combustible material such as wood meal, aluminum powder, or a hydrocarbon. Such explosives do not contain nitroglycerin or any similar liquid organic nitrate.

FIGURE 15.3 A blasting agent is a division 1.5 explosive that is typically stored and transported in 50-lb (23-kg) bags. A DOT Type B blasting agent is a mixture of ammonium nitrate and a nonexplosive, combustible substance that is intended exclusively for the purpose of blasting. ANFO is the acronym for an ammonium nitrate/fuel oil mixture. (*Courtesy of Dyno Nobel Inc., Salt Lake City, Utah.*)

■ *Type D blasting agents* are mixtures of an organic nitrate and combustible materials such as hydrocarbons and aluminum powder. They do not contain nitroglycerin, any similar liquid organic nitrate, metallic chlorate, or ammonium nitrate. This type generally includes plastic explosives (Section 15.10).

■ *Type E blasting agents* are mixtures of nitro derivatives of organic compounds, hydrocarbons, or aluminum powder with a high proportion of ammonium nitrate or other oxidizer, some or all of which is dissolved in an aqueous solution. The term includes explosives that have been formulated as emulsions, slurries, and water gels.

15.2-B ARTILLERY AMMUNITION

The sensitivity of individual explosives may be appreciated by examining the role performed by each component of a round of artillery ammunition. The structure of a typical round is shown in Figure 15.4. Each of its six components performs the following functions:

■ *Primer.* In a round of artillery ammunition, the primer is generally a mixture of an oxidizer and some other arbitrary substance that readily burns. This mixture is usually confined within a cap or tube at one end of the round and is activated by a percussion action, such as a firing pin.

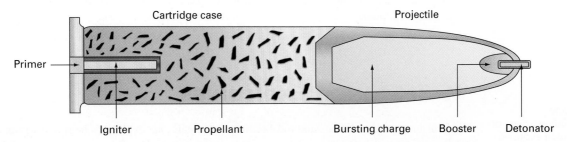

FIGURE 15.4 The component parts of a typical round of artillery ammunition. Each part serves a distinct function in the detonation of the round. An explosive is sometimes identified by its function (e.g., primer or booster).

igniter
■ Any of several nitro-cellulose explosives made by reacting cotton with nitric and sulfuric acids

propellant
■ The component of an explosive train whose purpose is to propel or provide thrust to a round of ammunition

detonator
■ The component of an explosive train that detonates from exposure to heat or mechanical impact and causes the actuation of the booster

booster
■ The component of an explosive train that is actuated by the detonator and whose purpose is to trigger the detonation of the bursting charge

bursting charge
■ The main component of an explosive train whose purpose is to detonate with maximum force at one time

explosive train
■ The sequence of charges beginning with actuation of the igniter and progressing to actuation of the bursting charge

■ *Igniter.*   In a round of ammunition, the **igniter** is almost always black powder (Section 15.5). The oxidation–reduction reaction between the components of black powder is initiated by the heat released from burning the primer.

■ *Propellant.*   The **propellant** is a material such as smokeless powder (Section 15.8) that deflagrates and produces a relatively large volume of gases and vapors. In a round of artillery ammunition, the quantity is intentionally limited to prevent a major explosion but is sufficiently sizable to generate an adequate volume of gases and vapors to propel the projectile forward. The explosive that performs the role of the propellant is physically separated from the other explosives within the round by means of a thin wall.

■ *Detonator.*   The **detonator** is the component of the round explodes from the mechanical shock it experiences when it hits a target. In a round of artillery, the detonator is present in a limited quantity.

■ *Booster.*   The **booster** is the component of the round is induced to explode by the heat generated from the explosion of the detonator. As with the detonator itself, the quantity of the booster is limited so as to generate a prescribed brisance. The function of the booster is to intensify the brisance resulting from the explosion of the detonator.

■ *Bursting charge, or main charge.*   The **bursting charge** is typically a high explosive whose detonation is initiated by the shock resulting from the booster explosion. It is the main charge in a round of artillery ammunition.

The chain of events from initiation of the primer to detonation of the bursting charge is called an **explosive train**. Each component of the train must be activated in the proper order for a round of ammunition to activate effectively.

The components of all types of ammunition available today may be regarded as modifications of those found in a typical round of artillery ammunition. A rifle cartridge, for instance, contains only a primer and propellant, or a primer, igniter, and propellant; the projectile is generally a mass of metal such as lead or steel pellets. A bomb that is dropped from a plane is also a modification of a round of artillery ammunition; generally, however, such a bomb lacks a propellant, since the detonator explodes from the force achieved by dropping the bomb from a high altitude.

When explosives are used for nonmilitary purposes such as blasting, the presence of a propellant is ordinarily unnecessary. In these instances, a blasting cap holds a prescribed amount of a high explosive that is generally detonated by means of a booster. The assembly containing the bursting charge and booster is connected to several hundred yards of electric wire and is activated from a safe distance by using an electrical detonator or other detonating device.

PERFORMANCE GOALS FOR SECTION 15.3:

■ Identify the features of the five types of magazines that are used to store commercial explosives.

■ Identify the requirements that the Bureau of Alcohol, Tobacco, Firearms and Explosives (ATF) imposes on parties who store commercial explosives.

15.3 Storing Explosives

magazine
■ A building, room, or vessel used exclusively for receiving, storing, and dispensing of explosives

Federal and state regulations and local ordinances require that explosives be stored within a specially constructed building or device called a **magazine**. The Bureau of Alcohol, Tobacco, Firearms and Explosives (ATF) regulates how magazines are constructed and maintained at 27 C.F.R. §555.213, as well as how explosives must be stored in them.

FIGURE 15.5 A day box acceptable for use as a type 2, 3, 4, or 5 magazine for storing certain explosives. (*Courtesy of U.S. Explosive Storage, Boone, North Carolina.*)

There are five types of magazines:

- ***Type 1 magazines.*** These are permanent structures intended for the storage of high explosives, subject to certain prescribed limitations, although low explosives and blasting caps may also be stored within them. A type 1 magazine is a building, igloo or "Army-type structure," tunnel, or dugout that is bullet, fire, weather, and theft resistant and ventilated.
- ***Type 2 magazines.*** These are mobile and portable indoor and outdoor structures intended for the storage of high explosives, subject to certain prescribed limitations, although low explosives and blasting agents may also be stored within them. A common type 2 magazine is a day box like that shown in Figure 15.5, but it may also be a trailer, semitrailer, or other mobile facility.
- ***Type 3 magazines.*** These are portable outdoor structures for the temporary storage of high explosives while attended, subject to certain prescribed limitations, although low explosives and blasting agents may also be stored within them. A type 3 magazine is a day box or a portable building.
- ***Type 4 magazines.*** These are structures intended for the storage of low explosives, subject to certain prescribed limitations. They are also used to store detonators that do not mass detonate. A type 4 magazine is a building, igloo or "Army-type structure," tunnel, or dugout, day box, trailer, or a semitrailer or other mobile building.
- ***Type 5 magazines.*** These are structures solely intended for the storage of blasting agents, subject to certain prescribed limitations. A type 5 magazine is a building, igloo or "Army-type structure," tunnel, dugout, bin, day box, trailer, or a semitrailer or other mobile building.

The following list of some specific regulations promulgated by ATF concerns how commercial explosives must be stored. Their implementation constitutes a combination of measures that aim to prevent explosives from accidentally detonating or being directly involved in a fire.

- Explosives may be stored only within the appropriate type of magazine that has been constructed and maintained according to specifications published by the Bureau of Alcohol, Tobacco, Firearms and Explosives.
- To maintain the integrity of the explosives stored within them, magazines must be kept clean, dry, and free of grit, paper, empty packages and containers, and rubbish at all times.
- The land surrounding a magazine must be kept clear of brush, dried grass, leaves, and dead trees at a distance of at least 25 ft (7.6 m) in all directions.

- The relevant placard shown in Figure 6.12 that DOT requires carriers to post on transport vehicles and freight containers must be displayed on an outer surface of type 5 magazines. Similar placarding is not required on type 1, type 2, type 3, and type 4 magazines.
- The lighting within magazines must be either battery-activated safety lights or lanterns or electric lighting meeting the standards prescribed by the National Electrical Code. All electrical switches must be located outside the magazines.
- Smoking or the use of matches, open flames, and spark-producing devices is prohibited in any magazine, within 50 ft (15 m) of any outdoor magazine, or within any room containing an indoor magazine.
- An explosive may be stored in a magazine only in a quantity that conforms with the requirements published by the Bureau of Alcohol, Tobacco, Firearms and Explosives at 27 C.F.R. §555.213.
- The doors of the magazine must be constructed of solid wood or metal. Hinges and hasps must be attached to them by welding, riveting, or bolting and must be installed so they cannot be removed when the doors are closed or locked. Doors must also be equipped with locks or padlocks of a type specified in regulations published by the Bureau of Alcohol, Tobacco, Firearms and Explosives at 27 C.F.R. §555.207–§555.211.
- The contents of a magazine must be inventoried daily. This inventory must include the manufacturer's name or brand name, the total quantity received into and removed from each magazine during each day, and the total remaining on hand at the end of each day.
- Magazines must be located away from inhabited buildings, passenger railways, public highways, and other magazines in conformance with prescribed distances established by the Institute of Makers of Explosives* and published at 27 C.F.R. §555.218.

PERFORMANCE GOALS FOR SECTION 15.4:

- Identify the unique DOT requirements that pertain to preparing the shipping description of an explosive for transportation.
- Identify the DOT labeling, marking, and placarding requirements that pertain to the transportation of explosives.

15.4 Transporting Explosives

forbidden explosive
■ For purposes of DOT regulations, an explosive for which shippers and carriers have not received DOT authorization to transport; an explosive listed in the Hazardous Materials Table at 49 C.F.R. §172.101 with the word "forbidden" in column 3, signifying that the material may not be transported by any mode

DOT regulates the transportation of the specific types of explosive articles *and* explosive substances listed in the Hazardous Materials Table. For purposes of DOT regulations, these explosives are divided into six divisions that form a continuum of decreasing hazard from 1.1 to 1.6.

The word "forbidden" sometimes appears in column 3 of the Hazardous Materials Table. This designation signifies a **forbidden explosive**, one that DOT prohibits from being transported by any mode. Nonetheless, DOT may allow the same explosive substance to be transported when it has been suitably "desensitized" or reduced in its ability to detonate. Manufacturers desensitize an explosive by mixing it with an inert material called a *stabilizer*.

*The Institute of Makers of Explosives is the safety association of the commercial explosives industry in the United States and Canada. It promotes the safety and protection of employees, users, the public, and the environment throughout all aspects of the manufacture and use of explosives in industrial blasting and other essential operations.

15.4-A EX-NUMBERS

Before an explosive may be offered for transportation, DOT requires its manufacturer or other party to test and evaluate the explosive following prescribed criteria and procedures to determine the division in which it should be classified. Then, the manufacturer or other party must formally apply for an **EX-number** by submitting information regarding the physical and chemical composition of the explosive, thermal-stability-test data, relevant background data, and other technical information for evaluation. When DOT approves this application, it assigns an EX-number corresponding to the specific explosive at issue. This assignment signifies that DOT approves the explosive for transportation when the shippers and carrier follow all applicable regulations. DOT also uses the information to assign a shipping description, division number, and compatibility group for the explosive.

> **EX-number**
> ▪ A number assigned by DOT to identify an explosive that has been evaluated by prescribed methods and approved for transportation

15.4-B COMPATIBILITY GROUPS

The **compatibility group** of an explosive is one of the 12 capital letters A, B, C, D, E, F, G, H, J, K, L, or S. The assignment of the compatibility group is based on the relevant description of the explosive provided in Table 15.2. DOT enters the compatibility group after the hazard class and division number for a specific explosive in column 3 of the Hazardous Materials Table. Each division number followed by a compatibility group is called the **classification code** of the explosive.

DOT requires shippers to display the compatibility group on the applicable EXPLOSIVE 1.1, 1.2, 1.3, 1.4, 1.5, or 1.6 label that they affix to packaging containing the explosive. For explosives in hazard classes 1.1, 1.2, and 1.3, the compatibility group is designated as a component of the classification code in the lower quadrant of the label and above hazard class 1. For explosives in hazard classes 1.4, 1.5, and 1.6, the compatibility group is written separately above hazard class 1 in the lower quadrant of the label, and the hazard class and division are written in the upper quadrant. DOT also requires the carrier to display the compatibility group on the EXPLOSIVES 1.1, 1.2, 1.3, 1.4, 1.5, or 1.6 placards posted on a transport vessel or freight container. These placards and labels display the compatibility group in an identical fashion.

> **compatibility group**
> ▪ For purposes of DOT regulations, a designated alphabetical letter used to classify an explosive for control during its storage and transportation

> **classification code**
> ▪ The division number of an explosive followed by its compatibility group

SOLVED EXERCISE 15.2

What is the most likely incendiary agent contained within firebombs ("ammunition, incendiary") whose DOT hazard class is each of the following?

  (a) 1.3J
  (b) 1.2H

Solution:

  (a) By referring to Table 15.2, we see that a flammable liquid or gel is contained within explosive articles that are assigned the compatibility group J. Consequently, gasoline, jet fuel, or kerosene thickened with either napalm or a napalmlike gel (Section 13.13) is the most likely incendiary agent contained within a firebomb that has been assigned the DOT hazard class 1.3J.

  (b) By again referring to Table 15.2, we see that white phosphorus (Section 7.5) is contained within explosive articles that are assigned the compatibility group H. Consequently, white phosphorus is the most likely incendiary agent that is contained within a firebomb that has been assigned the DOT hazard class 1.2H.

TABLE 15.2	Compatibility Group Assignments[a]		
EXPLOSIVE		**COMPATIBILITY GROUP**	**CLASSIFICATION CODE**
Primary explosive substance.		A	1.1A
Article containing a primary explosive substance and not containing two or more effective protective features. Some articles, such as detonators for blasting/detonator assemblies for blasting and cap-type primers, are included, even though they do not contain primary explosive substances.		B	1.1B, 1.2B, *or* 1.4B
Propellant explosive substance or other deflagrating explosive substance or article containing such explosive substance.		C	1.1C, 1.2C, 1.3C, *or* 1.4C
Secondary detonating explosive substance or black powder or article containing a secondary detonating explosive substance, in each case without means of initiation and without a propelling charge; or article containing a primary explosive substance and containing two or more effective protective features.		D	1.1D, 1.2D, 1.4D, *or* 1.5D
Article containing a secondary detonating explosive substance, without means of initiation, with a propelling charge (other than one containing flammable liquid gel or hypergolic liquid).		E	1.1E, 1.2E, *or* 1.4E
Article containing a secondary detonating explosive substance with its means of initiation, with a propelling charge (other than one containing flammable liquid gel or hypergolic liquid) or without a propelling charge.		F	1.1F, 1.2F, 1.3F, *or* 1.4F
Pyrotechnic substance or article containing a pyrotechnic substance, or article containing both an explosive substance and an illuminating, incendiary, tear-producing, or smoke-producing substance (other than a water-activated article or one containing white phosphorus, phosphide, or flammable liquid or gel or hypergolic liquid).		G	1.1G, 1.2G, 1.3G, *or* 1.4G
Article containing both an explosive substance and white phosphorus.		H	1.2H *or* 1.3H
Article containing both an explosive substance and flammable liquid or gel.		J	1.1J, 1.2J, *or* 1.3J
Article containing both an explosive substance and a toxic chemical agent.		K	1.2K *or* 1.3K
Explosive substance or article-containing substance and presenting a special risk (e.g., owing to water activation or presence of hypergolic liquids, phosphides, or pyrophoric substances) needing isolation of each type.		L	1.1L, 1.2L, *or* 1.3L
Articles containing only extremely insensitive detonating substances.		N	1.6N
Substance or article so packed or designed that any hazardous effects arising from accidental functioning are limited to the extent that they do not significantly hinder or prohibit firefighting or other emergency-response efforts in the immediate vicinity of the package.		S	1.4S

[a]49 C.F.R. §173.52.

15.4-C SHIPPING DESCRIPTIONS

When shippers offer an explosive for transportation, DOT requires them to identify the explosive in its shipping description and to mark the identification on the packaging used for shipment. The explosive may be identified through use of its assigned EX-number or, in lieu of it, the national stock number issued by the Department of Defense, product code, or other identifying information.

The *national stock number* (NSN) is a 13-digit number that identifies a variety of commercial products including the types of explosive. NSNs are primarily used by the U.S. Department of Defense (DOD) for purposes of procurement. The NSN corresponding to individual explosives may be obtained from DOD publications and other sources.

When shippers offer an explosive for transportation, they also include the net explosive mass in the shipping description. When the explosive is an article, the net explosive mass is expressed in terms of the net mass of either the article or the explosive contained within the article.

15.4-D LABELING, MARKING, AND PLACARDING REQUIREMENTS

When an explosive is transported, DOT requires shippers and carriers to comply with its labeling, marking, and placarding requirements. In particular, DOT requires shippers to affix the relevant label shown in Figure 6.5 to its packaging. DOT also requires shippers to mark the packaging with the information specified in Section 6.6-A, including the following: proper shipping name, identification number, instructions, precautions, and the applicable EX-number, national stock number, product code, or other identifying information.

When carriers transport by highway or rail any quantity of an explosive in hazard class 1.1, 1.2, or 1.3, DOT requires them to post the applicable EXPLOSIVES 1.1, EXPLOSIVES 1.2, or EXPLOSIVES 1.3 placards shown in Figure 6.12 on the transport vehicle or freight container and to display on the placard the applicable compatibility group. When carriers transport 1001 lb (454 kg) or more of an explosive in hazard class 1.4, 1.5, or 1.6, DOT requires them to post EXPLOSIVES 1.4, EXPLOSIVES 1.5, and EXPLOSIVES 1.6 placards with the applicable compatibility group on the transport vehicle or freight container.

When carriers transport more than 55 lb (25 kg) of a division 1.1, 1.2, or 1.3 explosive in a motor vehicle, railcar, or freight container, DOT requires them to prepare and implement a *security plan* whose components comply with the requirements of 49 C.F.R. §172.802. Motor carriers are also required to obtain a Hazardous Materials Safety Permit prior to transporting more than 55 lb (25 kg) of a division 1.1, 1.2, or 1.3 explosive or more than 1001 lb (454 kg) of a division 1.5 explosive. Issuance of the safety permit requires motor carriers to prepare a written route plan that complies with the requirements of 49 C.F.R. §397.67 when transporting a division 1.1, 1.2, or 1.3 explosive. Other special controls may also apply.

When more than one explosive article or explosive substance having *different* compatibility groups is transported, DOT requires carriers to display only one placard based on the following scheme:

- Explosive articles of compatibility groups C, D, or E are placarded displaying compatibility group E.
- Explosive articles of compatibility groups C, D, or E, when transported with those in compatibility group N, are placarded displaying compatibility group D.
- Explosive substances of compatibility groups C and D are placarded displaying compatibility group D.
- Explosive articles of compatibility groups C, D, E, or G, except for fireworks, are placarded displaying compatibility group E.

A fireworks manufacturer in Nitro, West Virginia, offers for transportation by motor vehicle six identical boxes of consumer fireworks having a net explosive mass of 165 lb (75 kg). The fireworks are to be transported to a distributor in Memphis, Tennessee. These fireworks were tested by DOT-prescribed methods and assigned the combined hazard code and compatibility group 1.4G. DOT also assigned EX-99887 to the manufacturer for this type of fireworks.

(a) What is the shipping description that DOT requires the fireworks manufacturer to enter on the accompanying shipping paper?
(b) Which labels does DOT require the fireworks manufacturer to affix to the outside surface of each box?
(c) Which markings does DOT require on the outside surface of each box?
(d) Which placards, if any, does DOT require the carrier to post on the motor vehicle used to transport these boxes?

Solution:

(a) By referring to Table 6.1, we see that DOT requires the shipping description to be prepared as follows:

Units	HM	**Shipping Description** (Identification Number, Proper Shipping Name, Primary Hazard Class or Division, Subsidiary Hazard Class or Division, and Packing Group)	Weight (lb)
165 lb, total explosive mass (6 boxes)	X	UN0335, Fireworks, 1.4G, PGII (EX-99887)	25 (per box)

(b) By referring to column 6 of Table 6.1, we see that DOT requires the fireworks manufacturer to affix EXPLOSIVE 1.4 labels bearing the compatibility group G to opposite sides of each box.
(c) DOT requires the box manufacturer to mark each box with the following information (Section 6.6-A): UN symbol; packaging identification code from 49 C.F.R. §173.212; one of the letters X, Y, or Z that designates the performance standard for which the packaging was successfully tested; a designation of the specific gravity or mass in kilograms for which the package has been tested; the last two digits of the year during which the package was manufactured; the letters USA indicating that the package was marked pursuant to DOT standards; and the symbol of the manufacturer.

DOT requires the fireworks manufacturer to mark each box with the name and address of the manufacturer and distributor that serve as the shipper and receiver, respectively; the designation "UN0335, Fireworks, PGII"; the EX-number; and appropriate instructions and precautionary statements like the following

CAUTION: KEEP DRY
DO NOT STORE NEAR FLAMMABLE MATERIALS

Because each box weighs less than 110 lb (50 kg), DOT does not require the shipper to mark the weight on the boxes.

(d) Because the total amount of explosives in the six boxes is less than 1001 lb (454 kg), DOT does not require the carrier to post placards on the motor vehicle used to transport them.

SOLVED EXERCISE 15.4

A consignment consists solely of explosive articles having the following shipping descriptions, where EX-xxxxx are four different EX-numbers:

> **UN0028, Black powder, compressed, 1.1D, PGII, EX-xxxxx**
> **UN0006, Cartridges for weapons, with bursting charge, 1.1E, PGII, EX-xxxxx**
> **UN0271, Charges, propelling, 1.1C, PGII, EX-xxxxx**
> **UN0408, Fuzes, detonating, with protective features, 1.1D, PGII, EX-xxxxx**

Which placards, if any, does DOT require the carrier to display when these items are transported by motor carrier from Lexington, Kentucky, to Baltimore, Maryland?

Solution: When carriers simultaneously transport explosive articles having compatibility groups C, D, and E, DOT requires them to display the compatibility group E on the appropriate placards that are posted on the motor vehicle. Because the division number for each of the referenced explosive articles is 1.1, DOT requires carriers to display EXPLOSIVES 1.1 placards bearing the compatibility group E on each side and each end of the motor vehicle used to transport them.

SOLVED EXERCISE 15.5

The DOT regulation at 49 C. F. R. §174.101(h) requires carriers of packages containing division 1.1 and division 1.2 explosives for detonators and detonating primers to securely block and brace them within their transport vehicle so they are prevented from changing position, falling to the floor, or sliding into each other under normal transportation conditions. What is the most likely reason DOT regulates the loading of these explosives in this specific fashion?

Solution: Because the detonation of explosives may be initiated by friction, shock, and mechanical impact, DOT most likely requires packaged explosives to be loaded into a transport vehicle in the indicated fashion to reduce or eliminate the likelihood that any of these means will initiate their unintended detonation. Additional measures of safety are afforded by preventing the packages from changing position, falling to the floor, or sliding into each other.

15.4-E EXPLOSIVE WASTES

It was noted in Section 5.9 that an explosive waste may exhibit the RCRA characteristic of reactivity, and Table 6.3 noted that DOT requires shippers to identify the nature of this waste when it is offered for domestic transportation. For example, shippers who offer for transportation waste dynamite that exhibits the reactivity characteristic identify its basic description on an accompanying waste manifest by using either of the following, where EX-xxxxx is the appropriate EX-number:

NA3077, Hazardous waste, solid, n.o.s. (dynamite), 9, PGIII (EPA reactivity), EX-xxxxx
or
NA3077, Hazardous waste, solid, n.o.s. (dynamite), 9, PGIII (D003), EX-xxxxx

15.4-F OTHER DOT REQUIREMENTS

When an explosive is moved by public highway, DOT may designate the routing or impose curfews, time-of-travel restrictions, lane restrictions, or route/weight restrictions on the shipper or carrier. The identity of such routing designations must be made available to the public in the form of a map, list, and/or a road sign. The route or highway DOT designates for the domestic transportation is called a **preferred route**, or **preferred highway**. Its selection complies with the regulations published at 49 C.F.R. §397.71.

DOT also requires special precautions to be implemented when explosives are loaded, unloaded, or handled. Onboard watercraft in particular, a fire hose of sufficient length to reach every part of the loading area with an effective stream of water must be laid and connected to the water main, ready for use. In the event of fire, firefighting personnel proceed immediately to prevent a premature cargo explosion.

preferred route (preferred highway)
■ For purposes of DOT regulations, the designated route or highway on which a shipper or carrier is required to transport an explosive or radioactive material

---

PERFORMANCE GOALS FOR SECTION 15.5:

- Identify the chemical composition of black powder and black gunpowder.
- Discuss the nature of the oxidation–reduction that occurs when they are activated.

15.5 Black Powder

black powder
■ A mixture of charcoal and sulfur with potassium nitrate or sodium nitrate in prescribed amounts

Black powder is the oldest known explosive.* It was discovered by the Chinese before 1000 A.D. Although its chemical composition varies, black powder generally consists of an intimate mixture of charcoal, sulfur, and either potassium nitrate or sodium nitrate. It is regarded as a low explosive. Black powder was formerly employed as a blasting agent, but today, it is more often encountered as a component of fireworks, signal flares, and certain forms of ammunition.

A physical variation of black powder is called *black gunpowder*. It is a mixture of 15 parts charcoal, 10 parts sulfur, and 75 parts potassium nitrate by mass.

When activiated, the components of black powder and black gunpowder undergo an oxidation–reduction reaction. Although the nature of the reaction is open to some debate, the following equation most likely expresses the phenomenon:

$$3S_8(s) + 16C(s) + 32KNO_3(s) \longrightarrow 16K_2O(s) + 16CO_2(g) + 16N_2(g) + 24SO_2(g)$$

Sulfur   Carbon   Potassium nitrate   Potassium oxide   Carbon dioxide   Nitrogen   Sulfur dioxide

Because they are composed of such a reactive mixture of substances, black powder and black gunpowder may be activated by a spark or static electricity. They autoignite at 867 °F (464 °C).

---

*Black powder formulations were known to the ancient Chinese, who used them in warfare and pyrotechnic displays. Although the color of black powder is indeed black, the material is not named for its color but for the direct translation of the German word *Schwarzpulver*, named for Berthold Schwarz, who experimented with black powder formulations in the 14th century.

15.5-A RESPONDING TO INCIDENTS INVOLVING THE RELEASE OF BLACK POWDER OR GUNPOWDER

Under routine conditions, responding to an emergency incident involving explosives should be undertaken only by competent, experienced individuals who have received special training in the handling of explosives. Perhaps the single exception to this general rule involves encountering black powder or gunpowder. Because black powder contains an oxidizer, the mixture is easily desensitized by diluting it with water. Neither black powder nor gunpowder detonates when soaked with water.

15.5-B TRANSPORTING BLACK POWDER/GUNPOWDER

When shippers offer black powder or gunpowder for transportation, DOT requires them to identify it on an accompanying shipping paper as one of the following, as relevant:

> **UN0028, Black powder, 1.1D, PGII, EX-xxxxx**
> *or*
> **UN0028, Gunpowder, compressed, 1.1D, PGII, EX-xxxxx**
> *or*
> **UN0028, Black powder, in pellets, 1.1D, PGII, EX-xxxxx**
> *or*
> **UN0028, Gunpowder, in pellets, 1.1D, PGII, EX-xxxxx**
> **UN0027, Black powder, 1.1D, PGII, EX-xxxxx**
> *or*
> **UN0027, Gunpowder, 1.1D, PGII, EX-xxxxx**
> **NA0027, Black powder for small arms, 1.1D, PGII, EX-xxxxx**

DOT also requires shippers to affix EXPLOSIVE 1.1 labels bearing the compatibility group D on packages of black powder or gunpowder. When carriers transport black powder or gunpowder in any amount, DOT requires them to post EXPLOSIVES 1.1 placards bearing the compatibility group D on the bulk packaging or transport vehicle used for shipment.

PERFORMANCE GOALS FOR SECTION 15.6:

- Describe the chemical reaction that occurs when nitroglycerin detonates.
- Identify the hazardous properties of nitroglycerin other than its explosive nature.

15.6 Nitroglycerin

When alcohols react with nitric acid, the compounds called **nitrate esters** are produced. The relevant reactions may be denoted by the following general equation, where R is an arbitrary alkyl or aryl group:

$$R-OH(l) + HNO_3(l) \longrightarrow R-O-NO_2(l) + H_2O(l)$$

Nitroglycerin is the best known nitrate ester. Its proper chemical name is glyceryl trinitrate.

$$
\begin{array}{l}
CH_2-O-NO_2 \\
\ \ | \\
CH-O-NO_2 \\
\ \ | \\
CH_2-O-NO_2
\end{array}
$$

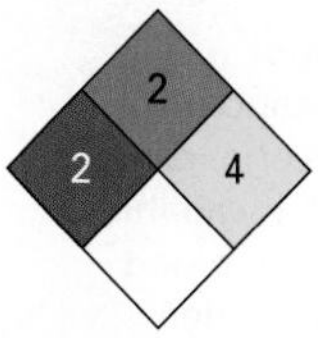

nitrate ester
■ An organic compound whose molecules contain one or more groups of atoms designated as $-O-NO_2$

TABLE 15.3	Physical Properties of Nitroglycerin
Melting point	55 °F (13.1 °C)
Detonation temperature	424 °F (218 °C)
Specific gravity at 68 °F (20 °C)	1.60
Brisance (g of sand)	51.5
Detonation velocity	4.8 mi/s (7.8 km/s)
Sensitivity	Very high; almost a primary explosive

It is painstakingly prepared by drizzling glycerol, a trihydroxy alcohol, into a cooled mixture of nitric acid and sulfuric acid. The following equation summarizes the nature of this reaction:

$$\begin{array}{c} CH_2-OH \\ | \\ CH-OH(l) \\ | \\ CH_2-OH \end{array} + 3HNO_3(l) \longrightarrow \begin{array}{c} CH_2-O-NO_2 \\ | \\ CH-O-NO_2(l) \\ | \\ CH-O-NO_2 \end{array} + 3H_2O(l)$$

Glycerol   Nitric acid   Nitroglycerin   Water

The sulfuric acid functions by extracting the elements of water from glycerol and nitric acid.

Pure nitroglycerin is a thick, oily liquid whose physical appearance otherwise resembles water. Other physical properties are noted in Table 15.3. When it is generally encountered, however, nitroglycerin is a viscous, pale yellow liquid that is extremely sensitive to spontaneous decomposition. Even a slight jarring or dropping it on a hard surface may trigger its premature detonation. Given its sensitivity to decomposition, pure nitroglycerin cannot be safely transported and is impractical for general use as an explosive. Notwithstanding this fact, nitroglycerin is often mixed with other explosives as a component of ammunition (see e.g., Section 15.8-A).

It was noted in Section 15.1 that when nitroglycerin detonates, it decomposes into carbon dioxide, water vapor, nitrogen, and oxygen. The resulting brisance of this high explosive is approximately three times that of an equivalent amount of gunpowder and occurs roughly 25 times faster.

Nitroglycerin hydrolyzes when it is exposed to atmospheric moisture. The hydrolysis produces a mixture of glycerol and nitric acid. When it has absorbed moisture, nitroglycerin becomes extremely sensitive to spontaneous decomposition. For this reason, the repeated exposure of nitroglycerin to a humid atmosphere is likely to produce a highly dangerous mixture that could detonate with the slightest provocation.

Aside from its potentially explosive nature, nitroglycerin is also a toxic substance by ingestion, inhalation, and skin absorption. When absorbed into the blood system, nitroglycerin causes the small veins, capillaries, and coronary vessels to dilate. These effects are actually caused by nitric oxide (Section 10.13-A), which is slowly released from the nitroglycerin. The nitric oxide causes the development of severe headaches, flushing of the face, and a drop in blood pressure. In severe instances, exposure to nitroglycerin may also cause death.

15.6-A USE OF NITROGLYCERIN OTHER THAN AS AN EXPLOSIVE

vasodilator
■ A substance, like nitroglycerin, exposure to which is capable of causing the blood vessels to widen

Notwithstanding its toxic nature, nontoxic doses of nitroglycerin are used medicinally for the treatment of heart and blood-circulation diseases. A compound like nitroglycerin that causes dilation of the blood vessels is called a **vasodilator**. Afflicted individuals put a tiny nitroglycerin tablet under the tongue when they experience chest pain, or they consume a small amount of the medication called *spirits of nitroglycerin*, which consists of nitroglycerin dissolved in a solvent. The nitroglycerin slowly releases nitric oxide within the bloodstream as it decomposes. The nitric oxide relaxes nearby muscle cells and temporarily lowers blood pressure.

15.6-B WORKPLACE REGULATIONS INVOLVING NITROGLYCERIN

When nitroglycerin is present in the workplace, OSHA requires employers to limit employee exposure by skin contact to a concentration of $0.2 \ mg/m^3$, averaged over an 8-hr workday.

SOLVED EXERCISE 15.6

DOT prohibits the transportation of several explosives, including nitroglycerin, unless they have been wetted with a specified amount of water or alcohol or mixed with a specified amount of a plasticizer. What is the most likely reason DOT requires these explosives to be wetted or mixed with a plasticizer prior to acceptance for transportation?

Solution: As first noted in Section 5.5, the rate of a chemical reaction is dependent on the concentration of its reactants. The mixture of an explosive with water, alcohol, or a plasticizer effectively dilutes the material, thereby reducing or eliminating the likelihood that its component molecules may interact. Consequently, the explosives are sufficiently stabilized when either wetted with water or alcohol or mixed with a plasticizer so they may be transported without prematurely exploding.

15.6-C TRANSPORTING NITROGLYCERIN

DOT prohibits the domestic transportation of nitroglycerin unless the shipper successfully demonstrates that it has been desensitized against unwanted decomposition. Nitroglycerin is usually desensitized by dissolving it in a simple alcohol such as methanol, ethanol, or isopropanol. Because these alcohols are flammable liquids, DOT may require shippers and carriers to assign these commercially available forms of nitroglycerin to hazard class 3.

When nitroglycerin is transported, DOT requires its shipper or carrier to prepare the proper shipping description and to implement the marking, labeling, and placarding requirements noted in Chapter 6. Desensitized nitroglycerin intended for use as an explosive is commercially available in several forms. When shippers offer nitroglycerin for transportation, DOT requires them to identify it on an accompanying shipping paper as one of the following, as relevant:

UN1043, Nitroglycerin, desensitized, 1.1D, (6.1), PGII, EX-xxxxx
UN1043, Nitroglycerin, solution in alcohol, 1.1D, PGII, EX-xxxxx

DOT also requires shippers to affix EXPLOSIVE 1.1 labels bearing the compatibility group D on packages containing desensitized nitroglycerin. When carriers transport the nitroglycerin in any amount, DOT requires them to post EXPLOSIVES 1.1 placards bearing the compatibility group D on the bulk packaging or transport vehicle used for shipment.

15.6-D TERRORISTS' MISUSE OF NITROGLYCERIN

It appears that most terrorists have avoided the use of nitroglycerin owing to its sensitivity to detonation. Nonetheless, during the early 2000s, Palestinian forces routinely used nitroglycerin in suicide bombs directed against the Israelis. Furthermore, a Muslim extremist planned to use nitroglycerin to blow up several U.S. planes, but the plan was foiled when police located the explosive in the terrorist's apartment.

PERFORMANCE GOALS FOR SECTION 15.7:

■ Compare the safety of nitroglycerin and dynamite.
■ Identify the three commercial types of dynamite and describe their composition.

15.7 Dynamite

dynamite
■ A detonating explosive containing nitroglycerin, similar organic nitrate esters, and one or more oxidizers, all of which are mixed into a stabilizing absorbent

In 1867, Swedish engineer Alfred Nobel discovered that nitroglycerin could be absorbed into a porous material such as siliceous earth. The resulting mixture became known as **dynamite**. Nobel acquired a substantial fortune from his discovery and the subsequent manufacture of dynamite. He used the fortune to establish a monetary fund for the world-famous Nobel prizes.

Dynamite may be handled more safely than nitroglycerin, a property that allows it to be transported and used with less risk of spontaneous decomposition. In practice, dynamite requires a detonating cap to activate its detonation. Notwithstanding this fact, dynamite is a high explosive that is sensitive to heat, shock, and friction. The equation denoting the detonation of nitroglycerin was previously noted in Section 15.1.

Today, dynamite is produced by absorbing a mixture of nitroglycerin and diethyleneglycol dinitrate into wood pulp, sawdust, flour, starch, or similar carbonaceous materials. Diethyleneglycol dinitrate is also an explosive, but its function in the production of dynamite is to depress the solution's freezing point. Calcium carbonate is often added to dynamite to neutralize the nitric acid produced by hydrolysis. Oxidizers are routinely added as a source of additional internal energy. This mixture of substances is then packed into cylindrical cartridges made of waxed paper that resemble those shown in Figure 15.6. They vary in size from 7/8 to 8 in. (2 to 20 cm) in diameter and from 4 to 30 in. (10 to 76 cm) in length. They routinely weigh approximately 0.5 lb (230 g).

FIGURE 15.6 The cylindrical cartridges into which dynamite has been packed are called *sticks* of dynamite. They are available commercially in a number of sizes.

TABLE 15.4	Physical Properties of Some Commercial Forms of Dynamite		
	STRAIGHT DYNAMITE	**AMMONIA DYNAMITE**	**GELATIN DYNAMITE**
Specific gravity at 68 °F (20 °C)	1.3	0.8–1.2	1.3–1.6
Chemical composition	20%–60% nitroglycerin depending on grade, sodium nitrate, carbonaceous materials, antacid, and moisture	20%–60% nitroglycerin depending on grade, ammonium nitrate, carbonaceous materials, sulfur, antacid, and moisture	20%–90% nitroglycerin depending on grade, sodium nitrate, and moisture, gelatinized in nitrocellulose
Detonation velocity	2.5–4 mi/s (4–6 km/s)	0.5–0.8 mi/s (0.8–1.2 km/s)	0.47–0.68 mi/s (0.75–1.1 km/s)
Sensitivity	High	High	High

When they are involved in fires, small quantities of dynamite usually burn with a bluish flame without detonating. Nonetheless, the heat generated during these fires may activate the detonation of the remaining nitroglycerin. Acknowledging this possibility, experts recommend that firefighters avoid fighting fires involving dynamite.

From the 1920s through the 1930s, dynamite was the most popular explosive used for peacetime purposes, but accidental detonations involving dynamite occurred frequently. Since the 1930s, new ways to containerize dynamite that are far safer to transport, store, and use have been developed. Even though these products are commercially available, some explosives experts still prefer to use dynamite sticks for unique demolition assignments.

15.7-A FORMS OF DYNAMITE

Three commercial forms of dynamite are commonly encountered: *ammonia dynamite*, *straight dynamite*, and *gelatin dynamite*. Information about them is provided in Table 15.4. Although nitroglycerin and diethyleneglycol dinitrate are contained in each form, ammonia dynamite and straight dynamite also contain ammonium nitrate and sodium nitrate, respectively. Gelatin dynamite contains about 1% nitrocellulose, which serves to thicken the nitroglycerin.

When their constituent nitroglycerin detonates, the three forms of dynamite produce more brisance than an equivalent amount of nitroglycerin, because additional energy is provided through the chemical action of ammonium nitrate, sodium nitrate, and nitrocellulose.

15.7-B TRANSPORTING DYNAMITE

When shippers offer dynamite for transportation, DOT requires them to prepare the proper shipping description. The shipping description of dynamite is a generic one based on the type of blasting agent, as noted in Section 15.2-A. The type is a component of the shipping description. Shippers select the relevant basic description for entry on the accompanying shipping paper from the following list:

UN0081, Explosive, blasting, type A, 1.1D, PGII (dynamite), EX-xxxxx
UN0082, Explosive, blasting, type B, 1.1D, PGII (dynamite), EX-xxxxx
UN0331, Explosive, blasting, type B, 1.5D, PGII (dynamite), EX-xxxxx
UN0083, Explosive, blasting, type C, 1.1D, PGII (dynamite), EX-xxxxx
UN0084, Explosive, blasting, type D, 1.1D, PGII (dynamite), EX-xxxxx

UN0241, Explosive, blasting, type E, 1.1D, PGII (dynamite), EX-xxxxx
UN0332, Explosive, blasting, type E, 1.5D, PGII (dynamite), EX-xxxxx

DOT also requires shippers to affix EXPLOSIVE labels bearing the compatibility group D on packages of type A, type C, and type D blasting agents. DOT requires the shippers to affix either an EXPLOSIVE 1.1 or EXPLOSIVE 1.5 label bearing the compatibility group D as relevant, on packages of type B or type E blasting agents.

When carriers transport any quantity of blasting agents in hazard classes 1.1 or 1.3, DOT requires them to post EXPLOSIVES 1.1 or EXPLOSIVES 1.3 placards bearing the compatibility group D on the bulk packaging or transport vehicle used for shipment, as relevant. When carriers transport more than 1001 lb (454 kg) of blasting agents in hazard class 1.5, DOT requires them to post EXPLOSIVES 1.5 placards bearing the compatibility group D on the bulk packaging or transport vehicle used for shipment.

PERFORMANCE GOALS FOR SECTION 15.8:

- Describe the chemical nature of nitrocellulose.
- Compare the detonation and deflagration of nitrocellulose.

15.8 Nitrocellulose

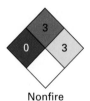

Nonfire

Fire

As first noted in Section 14.5-A, nitrocellulose is a polymer produced by reacting the cellulose in cotton with nitric acid. Because cellulose may be nitrated to different degrees, varying forms of nitrocellulose may be produced. One molecular structure of nitrocellulose is noted here:

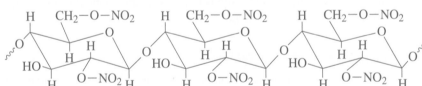

In this structure, the cellulose is almost completely nitrated. Although highly nitrated cellulose has been used as a rocket propellant, the lesser grades of nitrocellulose are utilized for other purposes. In the United States, the explosive is now produced almost exclusively by the federal government.

Condensing the chemical formulas of cellulose and nitrocellulose to $[C_6H_7O_2(OH)_3]_n$ and $[C_5H_5(CH_2ONO_2)O_2(ONO_2)_2]_n$, respectively, results in the following representation of the nitration of cellulose, where n is approximately 5,000:

$$[C_6H_7O_2(OH)_3]_n(s) + 3nHNO_3(l) \longrightarrow [C_5H_5(CH_2ONO_2)O_2(ONO_2)_2]_n(s) + 3nH_2O(l)$$

Cellulose · · · · · · Nitric acid · · · · · · Nitrocellulose · · · · · · Water

When the nitration produces a substance having a nitrogen content exceeding 13.2% by mass, the result is called **guncotton**.

guncotton
■ Any of several nitrocellulose explosives made by reacting cotton with nitric and sulfuric acids

Nitrocellulose is a white solid that resembles cotton in physical appearance. When intended for use as an explosive, it may be blocked, gelled, flaked, granulated, or powdered. To desensitize it for domestic transportation and storage, nitrocellulose is usually wetted with either water or an aqueous solution of ethanol. Notwithstanding the presence of a desensitizing agent, every nitrocellulose-based explosive is inherently unstable. The nitrocellulose slowly decomposes, producing traces of nitrogen oxides.

When nitrocellulose is activated, it usually deflagrates. Because its flashpoint is only 55 °F (13 °C), nitrocellulose burns readily and produces carbon dioxide, water vapor, and nitrogen, but not nitrogen dioxide. The deflagration of nitrocellulose is a furiously burning phenomenon represented as follows, where n is a nonzero integer:

$$2[C_5H_5(CH_2ONO_2)O_2(ONO_2)_2]_n(s) + 10nO_2(g) \longrightarrow 12nCO_2(g) + 7nH_2O(g) + 3nN_2(g)$$

Nitrocellulose   Oxygen   Carbon dioxide Water   Nitrogen

Nitrocellulose burns at a faster rate than virtually any other flammable solid. Tons of nitrocellulose may be consumed by fire within minutes.

Because the products of its combustion are colorless, nitrocellulose is often called **smokeless powder**. The absence of a visible smoke provides a tactical advantage when it is used during warfare. It prevents the enemy from easily identifying the location from which artillery was fired.

smokeless powder
■ A form of nitrocellulose

Nitrocellulose detonates when it is confined or accumulated in large quantities. The detonation reaction may be expressed as follows:

$$2[C_5H_5(CH_2ONO_2)O_2(ONO_2)_2]_n(s) \longrightarrow nCO_2(g) + 11nCO(g) + 7nH_2O(g) + 3nN_2(g)$$

Nitrocellulose   Carbon dioxide Carbon monoxide Water   Nitrogen

Nitrocellulose is probably most commonly encountered as a component of **small-arms ammunition**, although it is also used as the propellant in artillery ammunition. It is also sometimes selected for demolition work, wherein it is usually combined with a second explosive. This combination of explosives is called a *double-base formulation* (Section 15.8-A).

small-arms ammunition
■ Shotgun, rifle, pistol, or revolver cartridges

15.8-A FORMULATIONS OF SMOKELESS POWDER

Three formulations of nitrocellulose are called the *single-*, *double-*, and *triple-base smokeless powders*. Together with nonexplosive ingredients, these formulations contain one, two, or three explosives, respectively. The single-base formulation contains only nitrocellulose. The most popular variation is the double-base formulation, which consists of a mixture of nitrocellulose and nitroglycerin. It is used as a component of handgun and rifle ammunition. This formulation has been encountered in illegally improvised devices such as pipe bombs.

The triple-base smokeless powder usually contains the three explosives nitrocellulose, nitroglycerin, and nitroguanidine.

Nitroguanidine

It is most commonly encountered as a component of the ammunition used in artillery guns.

A stabilizer is added to all nitrocellulose-based formulations. It is usually diphenylamine, which functions by reacting with the nitrogen oxides to form innocuous compounds.

Diphenylamine

15.8-B TRANSPORTING NITROCELLULOSE

When shippers offer desensitized nitrocellulose for transportation, DOT requires them to identify it on an accompanying shipping paper as one of the following, as relevant:

  UN0340, Nitrocellulose, 1.1D, PGII, EX-xxxxx
  UN0343, Nitrocellulose, 1.3C, PGII, EX-xxxxx
  UN0341, Nitrocellulose, 1.1D, PGII, EX-xxxxx
  UN0342, Nitrocellulose, 1.3C, PGII, EX-xxxxx
  NA3178, Smokeless powder for small arms, 4.1, PGI

When shippers offer nitrocellulose in hazard classes 1.1 or 1.3 for transportation, DOT requires them to affix the relevant EXPLOSIVE 1.1 or EXPLOSIVE 1.3 labels bearing the compatibility group D and C, respectively, on the packages. When shippers offer smokeless powder for small arms for transportation, DOT requires them to affix a FLAMMABLE SOLID label on the packages.

When carriers transport nitrocellulose in hazard classes 1.1 or 1.3 in any amount, DOT requires them to post the relevant EXPLOSIVES 1.1 or EXPLOSIVES 1.3 placards bearing the compatibility group D and C, respectively, on the bulk packaging or transport vehicle used for shipment. When carriers transport smokeless powder for small arms in an amount exceeding 1001 lb (454 kg), DOT requires them to post FLAMMABLE SOLID placards on the bulk packaging or transport vehicle used for shipment.

PERFORMANCE GOALS FOR SECTION 15.9:

- Describe the chemical nature of trinitrotoluene.
- Describe the detonation of trinitrotoluene.
- Identify why the military uses TNT as a component of many explosive articles.
- Describe the nature of a TNT equivalent.

15.9 Trinitrotoluene

2,4,6-Trinitrotoluene is a pale yellow solid, although its commercial grade is generally yellow to dark brown. This explosive substance is known more commonly by its acronym, *TNT*.

TNT was used on a very large scale as a military explosive during World War II. Movie buffs are especially familiar with its use in depth charges catapulted from destroyers to disable enemy submarines. During peacetime, it is still used for demolition during construction and mining projects.

The equation denoting the detonation of TNT is provided in Solved Exercise 15.7. Some physical properties of TNT are provided in Table 15.5.

TABLE 15.5	Physical Properties of 2,4,6-Trinitrotoluene
Melting point	178 °F (81 °C)
Detonation temperature	450 °F (232 °C)
Specific gravity at 68 °F (20 °C)	1.65
Brisance (g of sand)	48
Detonation velocity	3.2–4.3 mi/s (5.1–6.9 km/s)
Sensitivity	Low

Write the balanced equation for the detonation of TNT by using its condensed formula, $C_7H_5(NO_2)_3$.

Solution: Before we can write the equation representing the detonation of TNT, we need first to identify the decomposition products produced. During any detonation, the hydrogen and oxygen atoms combine to form water, and the carbon and oxygen atoms combine to form either carbon monoxide or carbon dioxide, or both. If carbon and oxygen remain after production of carbon monoxide and/or carbon dioxide, they are released during the detonation as their respective elements. Because there are seven carbon atoms in each TNT molecule, some elemental carbon is inevitably produced during the detonation of TNT. The nitrogen atoms unite to form elemental nitrogen, or they react with oxygen atoms to produce either nitrogen monoxide or nitrogen dioxide, or both.

After some consideration, it becomes apparent that the TNT detonation products are nitrogen, water vapor, carbon monoxide, and elemental carbon. The unbalanced equation is then written as follows:

$$C_7H_5(NO_2)_3(s) \longrightarrow CO(g) + H_2O(g) + N_2(g) + C(s)$$

| Trinitrotoluene | Carbon monoxide | Water | Nitrogen | Carbon |

Finally, this equation is balanced.

$$2C_7H_5(NO_2)_3(s) \longrightarrow 7CO(g) + 5H_2O(g) + 3N_2(g) + 7C(s)$$

| Trinitrotoluene | Carbon monoxide | Water | Nitrogen | Carbon |

TNT is prepared by nitrating toluene using a mixture of nitric acid and sulfuric acid.

| Toluene | Nitric acid | 2,4,6-Trinitrotoluene | Water |

In the reaction, sulfuric acid acts as a catalyst. TNT is produced only at military arsenals.

TNT has primarily been used as a military explosive, since it has the following features:

- TNT is safe to handle. Although it is a high explosive, TNT is unusually insensitive to heat, shock, and friction. When it is engulfed in fire, relatively small quantities often burn but do not detonate. TNT ordinarily detonates only when confined or when relatively large amounts are intentionally activated.
- TNT does not react with atmospheric moisture.
- TNT is not susceptible to spontaneous decomposition, even after it has been kept in storage for years.
- TNT may be melted using steam with little fear of explosive decomposition. This feature is utilized when commercial explosive mixtures containing TNT are prepared. The molten TNT may be safely mixed with other explosives or oxidizers, cast into the shape of blocks, or poured into ammunition shells.

Commercial explosives are often prepared by mixing TNT with other substances. A popular explosive containing TNT is *amatol*, a mixture consisting of 80% ammonium nitrate and 20% TNT by mass. It has been widely used as a blasting agent and military and industrial explosive. Two other commercial explosives containing TNT are cyclonite (Composition B) and tetrytol (Sections 15.10 and 15.11, respectively).

Caution must be exercised when handling and using TNT, since it is toxic by ingestion, inhalation, and skin absorption. Because animals exposed to TNT have contracted cancer, the explosive is ranked as a probable human carcinogen.

15.9-A TNT EQUIVALENTS

It is customary to compare the mass of an explosive substance to the mass of TNT that produces the same explosive power. This comparison is called a **TNT equivalent**. For example, when 1 lb (454 g) of amatol detonates, the same explosive power is produced as when 0.59 lb (268 g) of TNT detonates. When 1 lb (454 g) of nitroglycerin detonates, it produces the same explosive power as when 1.49 lb (676 g) of TNT detonates.

TNT equivalents are used to compare the relative effectiveness of different explosives to accomplish a specified task.

15.9-B WORKPLACE REGULATIONS INVOLVING TNT

When TNT is present in the workplace, OSHA requires employers to limit employee exposure by skin contact to a concentration of 1.5 mg/m^3, averaged over an 8-hr workday. NIOSH recommends limiting employee exposure to a concentration of 0.5 mg/m^3 over a 40-hr workweek.

15.9-C TRANSPORTING TNT

When shippers offer desensitized TNT for transportation, DOT requires them to identify it on an accompanying shipping paper as one of the following, as relevant:

> **UN0209, Trinitrotoluene, 1.1D, PGII (EX-xxxxx)**
> **UN0388, Trinitrotoluene and trinitrobenzene mixtures, 1.1D, PGII, EX-xxxxx**
> **UN0389, Trinitrotoluene mixtures containing trinitrobenzene and hexanitrostilbene, 1.1D, PGII, EX-xxxxx**

DOT also requires shippers to affix EXPLOSIVE 1.1 labels bearing the compatibility group D on the TNT packaging. When carriers transport TNT in any amount, DOT requires them to post EXPLOSIVES 1.1 placards bearing the compatibility group D on the bulk packaging or transport vehicle used for shipment.

PERFORMANCE GOALS FOR SECTION 15.10:

- ■ Describe the chemical nature of cyclonite.
- ■ Describe the detonation of cyclonite.
- ■ Identify the commercial compositions of cyclonite.

15.10 Cyclonite

Cyclonite is the common name for the explosive whose chemical name is cyclotrimethylene-trinitramine, or hexahydro-1,3,5-trinitro-1,3,5-triazine.

Another name for cyclonite is *RDX*, the acronym for Royal Deutsche Explosive or Royal Demolition Explosive.

Cyclonite is an important member of a class of explosives called **nitramines**, organic compounds having the following group of atoms:

$$-\overset{|}{N}-NO_2$$

nitramine
- An organic compound whose molecules contain one or more groups of atoms designated as $-N-NO_2$

Other nitramines noted in this chapter include tetryl and cyclotetramethylenetetranitramine (Sections 15.11 and 15.13, respectively).

Cyclonite is prepared from hexamethylenetetramine, nitric acid, ammonium nitrate, and acetic anhydride.

Hexamethylenetetramine   Nitric acid   Ammonium nitrate   Acetic anhydride

Cyclonite   Acetic acid

It is a white crystalline solid. Other physical properties of cyclonite are listed in Table 15.6.

When the chemical formula for cyclonite is condensed to $(CH_2)_3N_3(NO_2)_3$, its detonation is represented as follows:

$$(CH_2)_3N_3(NO_2)_3(s) \longrightarrow 3CO(g) + 3N_2(g) + 3H_2O(g)$$

       Cyclonite   Carbon monoxide   Nitrogen   Water

One pound of cyclonite produces the same explosive power as 1.2 lb (545 g) of TNT. Although it is used by itself as an explosive, it is also mixed with other explosives for use. For instance, the

TABLE 15.6	**Physical Properties of Cyclotrimethylenetrinitramine**
Melting point	396 °F (204 °C)
Detonation temperature	Detonates as it melts
Specific gravity at 68 °F (20 °C)	1.82
Brisance (g of sand)	60.2
Detonation velocity	4.2–5.0 mi/s (6.8–8.0 km/s)
Sensitivity	High

mixture of cyclonite, TNT, and aluminum fines called *torpex* has been used in warfare as the explosive component of mines, depth charges, and torpedo warheads.

Although cyclonite is extremely sensitive to explosive decomposition, it may be desensitized by mixing it with beeswax, which stabilizes it even when exposed to high temperatures. The fact that cyclonite is thermally stable when desensitized makes it potentially useful when the need arises to use explosives during firefighting. Mixed with beeswax in varying proportions, cyclonite was also used widely through World War II as the bursting charge in aerial bombs, mines, and torpedoes.

The following formulations of cyclonite are encountered:

- *Composition A* is a mixture of 91% cyclonite and 9% beeswax by mass. This explosive substance is often selected by explosive experts because it detonates so rapidly. A charge of Composition A in a 1-tn bomb detonates in approximately 0.25 ms. This relatively high rate of detonation yields an extraordinary brisance.
- *Composition B* is a mixture of 60% cyclonite, 40% trinitrotoluene, and 1% beeswax by mass. This mixture has largely replaced Composition A in artillery shells.
- *Composition C-4* is a military explosive consisting of a mixture of 91% cyclonite, 5.3% di(2-ethylhexyl) sebacate, 2.1% polyisobutylene, and 1.6% 20-weight motor oil.

Sebacic acid is the common name of the substance whose proper name is 1,8-octane dicarboxylic acid. Its chemical formula is $HOOC-(CH_2)_8-COOH$. Di(2-ethylhexyl) sebacate is an ester having the following chemical formula, where C_8H_{17} is the 2-ethylhexyl group:

$$C_8H_{17}-O \underset{/}{\overset{O \atop \|}{C}} -(CH_2)_8-\underset{O-C_8H_{17}}{\overset{O \atop \|}{C}}$$

Composition C-4 is sometimes mixed with a gummy binder and hand-molded into a putty-like shape for easy use. This flexible or malleable mixture is an example of a **plastic explosive**. When mixed with the binder, the explosive substance retains its brisance when detonated but is more suitable for use when a specific shape is desired. The main disadvantage in using a plastic explosive is that it may become brittle in cold weather.

Caution must be exercised when using cyclonite, since it is toxic when ingested or inhaled. Exposure to high concentrations of the explosive has been linked with the onset of seizures. Animal exposure to cyclonite has been shown to cause cancer; hence, the explosive is ranked as a probable human carcinogen.

plastic explosive
■ A high explosive that has been mixed with a gummy binder and manufactured in a flexible, hand-malleable form for an intended use at 77 °F (25 °C) such as demolition

15.10-A WORKPLACE REGULATIONS INVOLVING CYCLONITE

When cyclonite is present in the workplace, OSHA requires workers to limit employee exposure to a concentration of 1.5 mg/m^3, averaged over an 8-hr workday. NIOSH recommends a short-term exposure limit of 1.5 mg/m^3 over a 40-hr workweek.

15.10-B TRANSPORTING CYCLONITE

When shippers offer desensitized cyclonite for transportation, DOT requires them to identify it on an accompanying shipping paper as one of the following, as relevant:

UN0483, Cyclotrimethylenetrinitramine, desensitized, 1.1D, PGII, EX-xxxxx
UN0072, Cyclotrimethylenetrinitramine, wetted, 1.1D, PGII, EX-xxxxx

DOT also requires shippers to affix EXPLOSIVE 1.1 labels bearing the compatibility group D on packages containing cyclonite. When carriers transport cyclonite in any amount, DOT requires them to post EXPLOSIVES 1.1 placards bearing the compatibility group D on the bulk packaging or transport vehicle used for shipment.

15.10-C TERRORISTS' MISUSE OF CYCLONITE

Terrorists have used cyclonite as a weapon of mass destruction. Cyclonite and PETN (Section 15.12) were identified as the active ingredients in the plastic explosive used to destroy Pan Am Flight 103 in 1998. Residues of these explosives were also identified following a terrorist action in India in 2003.

PERFORMANCE GOALS FOR SECTION 15.11:

- Describe the chemical nature of tetryl.
- Describe the detonation of tetryl.

15.11 Tetryl

Tetryl is the commercial explosive whose proper chemical name is 2,4,6-trinitrophenyl-*N*-methylnitramine.

This substance is produced in several ways, one of which involves the nitration of 2,4-dinitrochlorobenzene in a stepwise process as follows:

2,4-Dinitrochlorobenzene + Methylamine $CH_3NH_2(g)$ ⟶ *N*-Methyl-2,4-dinitrobenzene + Hydrogen chloride $HCl(g)$

N-Methyl-2,4,-dinitrobenzene + Nitric acid $HNO_3(l)$ $\xrightarrow{H_2SO_4}$ Tetryl + Water $H_2O(l)$

Tetryl is a yellow solid; other physical properties are provided in Table 15.7.

When the chemical formula for tetryl is condensed to $(NO_2)_3C_6H_2N(NO_2)(CH_3)$, its detonation is represented as follows:

$$2(NO_2)_3C_6H_2N(NO_2)(CH_3)(s) \longrightarrow 3C(s) + 11CO(g) + 5N_2(g) + 5H_2O(g)$$

Tetryl — Carbon — Carbon monoxide — Nitrogen — Water

The detonation of tetryl produces approximately the same explosive power as the same mass of TNT.

TABLE 15.7	Physical Properties of 2,4,6-Trinitrophenyl-*N*-Methylnitramine	
Melting point		266 °F (130 °C)
Detonation temperature		369 °F (187 °C)
Specific gravity (pressed) at 68 °F (20 °C)		1.60
Brisance (g of sand)		54.2
Detonation velocity		4 mi/s (7 km/s)
Sensitivity		High

Tetryl is well known for its exceptionally high brisance. Since World War II it has been the standard explosive used by the military as the booster in artillery ammunition.

When tetryl is mixed with molten trinitrotoluene and a small amount of graphite, the popular explosive substance called *tetrytol* is produced. Tetrytol is sometimes used by the military as the bursting charge in artillery ammunition.

Caution must be exercised when handling and using explosives containing tetryl, since tetryl is toxic by inhalation, ingestion, and skin absorption. Exposed individuals experience a host of symptoms, including coughing, fatigue, headache, nosebleed, nausea, vomiting, and skin rashes. Research studies suggest that exposure to tetryl may also affect kidney, liver, and spleen function.

15.11-A WORKPLACE REGULATIONS INVOLVING TETRYL

When tetryl is present in the workplace, OSHA requires employers to limit employee exposure to a concentration of 1.5 mg/m^3, averaged over an 8-hr workday. NIOSH recommends a maximum exposure limit of 1.5 mg/m^3 over a 40-hr workweek.

15.11-B TRANSPORTING TETRYL

When shippers offer tetryl for transportation, DOT requires them to identify it on an accompanying shipping paper as follows:

UN0208, Trinitrophenylmethylnitramine, 1.1D, PGII, EX-xxxxx

DOT also requires shippers to affix EXPLOSIVE 1.1 labels bearing the compatibility group D on packages containing tetryl. When carriers transport tetryl in any amount, DOT requires them to post EXPLOSIVES 1.1 placards bearing the compatibility group D on the bulk packaging or transport vehicle used for shipment.

PERFORMANCE GOALS FOR SECTION 15.12:

- Describe the chemical nature of PETN.
- Describe the detonation of PETN.

15.12 PETN

PETN (pronounced "pettin") is the active component of the commercial explosive substance whose proper chemical name is pentaerythritol tetranitrate.

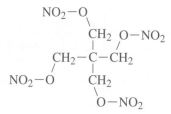

The condensed formula for this substance is $C(CH_2-O-NO_2)_4$. Like nitroglycerin, PETN is an example of a nitrate ester.

PETN is produced by the following two-step process:

- Preparation of pentaerythritol:

$$4H-\overset{H}{\underset{O}{C}}\ (aq)\ +\ CH_3-\overset{H}{\underset{O}{C}}\ (aq)\ +\ NaOH(aq)\ \longrightarrow$$

Formaldehyde   Acetaldehyde   Sodium hydroxide

$$HO-CH_2-\overset{CH_2-OH}{\underset{CH_2-OH}{C}}-CH_2-OH\,(s)\ +\ HCOONa(aq)$$

Pentaerythritol   Sodium formate

- Nitration of pentaerythritol:

$$HO-CH_2-\overset{CH_2-OH}{\underset{CH_2-OH}{C}}-CH_2-OH\,(s)\ +\ 4HNO_3(l)\ \longrightarrow$$

Pentaerythritol   Nitric acid

$$O_2N-O\diagdown$$

$$NO_2-O\ \ \ \ \ CH_2\ O-NO_2$$
$$CH_2-C-CH_2\ \ \ (s)\ +\ 4H_2O(l)$$
$$NO_2-O\ \ \ \ \ CH_2$$
$$O-NO_2$$

Pentaerythritol tetranitrate (PETN)   Water

PETN is a white solid; some of its other physical properties are provided in Table 15.8.

Although PETN is sometimes used as the booster in artillery ammunition, it is probably most commonly encountered as a form of *primacord*, a detonation fuse consisting of a core of PETN wrapped in a fabric sheath. PETN is also a key ingredient in the explosive called *Semtex*. When the chemical formula for PETN is condensed to $C(CH_2-O-NO_2)_4$, its detonation is represented as follows:

$$C(CH_2-O-NO_2)_4(s)\ \longrightarrow\ 3CO_2(g)\ +\ 2CO(g)\ +\ 2N_2(g)\ +\ 4H_2O(g)$$

PETN   Carbon dioxide   Carbon monoxide   Nitrogen   Water

TABLE 15.8	Physical Properties of Pentaerythritol Tetranitrate	
Melting point	282 °F (139 °C)	
Boiling point	437 °F (225 °C)	
Detonation temperature	374–410 °F (190–210 °C)	
Specific gravity at 68 °F (20 °C)	1.75	
Brisance (g of sand)	62.7	
Detonation velocity	4.92 mi/s (7.92 km/s)	
Sensitivity	High	

When 1 lb (454 g) of PETN detonates, the same explosive power is produced as when 1.29 lb (586 g) of TNT detonates.

15.12-A USE OF PETN OTHER THAN AS AN EXPLOSIVE

PETN acts as a vasodilator when consumed internally. For this reason, it is used medicinally for the treatment of heart and circulatory diseases.

15.12-B TRANSPORTING PETN

When shippers offer PETN for transportation, DOT requires them to identify it on an accompanying shipping paper as one of the following, as relevant:

> **UN3344, Pentaerythritol tetranitrate mixture, desensitized, solid, n.o.s, 4.1, PGII, EX-xxxxx**
> **UN0411, Pentaerythritol tetranitrate, 1.1D, PGII, EX-xxxxx**
> **UN0150, Pentaerythritol tetranitrate, wetted, 1.1D, PGII, EX-xxxxx**

Dry PETN is a forbidden explosive.

DOT also requires shippers to affix FLAMMABLE SOLID or EXPLOSIVE 1.1 labels bearing the compatibility group D, as relevant, on packages containing PETN. When carriers transport PETN in hazard class 4.1 in an amount exceeding 1001 lb (454 kg), DOT requires them to post FLAMMABLE SOLID placards on the bulk packaging or transport vehicle used for shipment. When carriers transport PETN in hazard class 1.1 in any amount, DOT requires them to post EXPLOSIVES 1.1 placards bearing the compatibility group D on the bulk packaging or transport vehicle used for shipment.

15.12-C TERRORISTS' MISUSE OF PETN

As noted in Section 13.9-C, PETN has been identified as an active ingredient in explosive mixtures used by terrorists.

PERFORMANCE GOALS FOR SECTION 15.13:

- Describe the chemical nature of HMX.
- Describe the detonation of HMX.

15.13 HMX

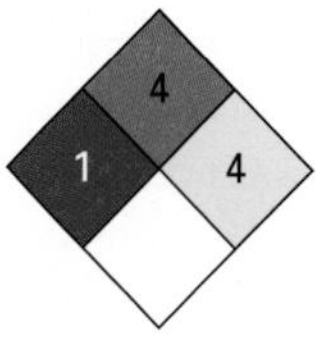

HMX is the secondary high explosive whose proper chemical name is cyclotetramethylenetetra-nitramine.

$$
\begin{array}{ccc}
& NO_2 & NO_2 \\
& | & | \\
& N & N \\
CH_2 & CH_2 & CH_2 \\
& N - CH_2 - N & \\
& | & | \\
& NO_2 & NO_2
\end{array}
$$

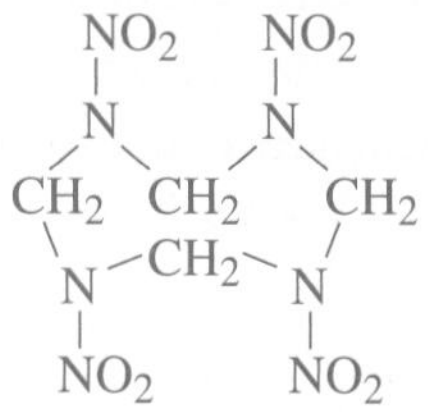

The origin of the acronym HMX is most likely linked with either of the names "high-molecular-weight explosive" or "high-melting explosive." Other commercial names for this explosive are *Octogen* and *Rowanex 2000*.

The molecular structure of HMX may be condensed to $(CH_2)_4(N-NO_2)_4$. Some of its important physical properties are provided in Table 15.9.

The detonation of HMX can be represented as follows:

$$
(CH_2)_4(N-NO_2)_4(s) \longrightarrow 4CO(g) + 4N_2(g) + 4H_2O(g)
$$

HMX   Carbon monoxide   Nitrogen   Water

When 1 lb (454 g) of HMX detonates, the same explosive power is produced as when 1.26 lb (572 g) of TNT detonates. HMX is regularly used in the military warfare actions of today, generally as shaped-charge warhead explosives and rocket propellants.

Like cyclonite, HMX is produced from hexamethylenetetramine, nitric acid, ammonium nitrate, and acetic anhydride. HMX and cyclonite are chemically similar. Their molecules are eight- and six-membered-ring-shaped nitramines, respectively.

Caution must be exercised when using HMX, since studies conducted on rats, mice, and rabbits reveal that HMX may be harmful to the hepatic (liver) and central nervous systems when ingested or absorbed through the skin.

15.13-A USE OF HMX WITH OTHER EXPLOSIVES

An explosive called *Octol 70/30* is a mixture of 70% HMX and 30% TNT by mass. The military has used it as a rocket propellant. The detonation of HMX produces approximately the same explosive power as the same mass of TNT.

TABLE 15.9	Physical Properties of Cyclotetramethylene-tetranitramine
Melting point	527 °F (275 °C)
Detonation temperature	319 °F (159 °C)
Specific gravity at 68 °F (20 °C)	1.89
Brisance (g of sand)	60.4
Detonation velocity	5.66 mi/s (9.11 km/s)
Sensitivity	High

15.13-B TRANSPORTING HMX

When shippers offer HMX for transportation, DOT requires them to identify it on an accompanying shipping paper as one of the following, as relevant:

> **UN0484, Cyclotetramethylenetetranitramine, desensitized, 1.1D, PGII, EX-xxxxx**
> **UN0226, Cyclotetramethylenetetranitramine, wetted, 1.1D, PGII, EX-xxxxx**

DOT also requires shippers to affix EXPLOSIVE 1.1 labels bearing the compatibility group D on packages containing HMX. When carriers transport HMX in any amount, DOT requires them to post EXPLOSIVES 1.1 placards bearing the compatibility group D on the bulk packaging or transport vehicle used for shipment.

PERFORMANCE GOALS FOR SECTION 15.14:

- Describe the chemical nature of the common primary explosives.
- Describe the chemical nature of their individual detonations.

15.14 Primary Explosives

As first noted in Section 15.14, primary explosives are substances whose detonation is very sensitive to a stimulus like heat, shock, or friction. For this reason, they are extremely dangerous to handle. In practice, primary explosives are used in very small quantities to initiate the detonation of a larger charge of a main explosive. When used for this purpose, they are called **initiators, initiating explosives**, or **primers**.

initiator (initiating explosive; primer)
- An explosive material used to detonate the main charge

Three primary explosives are frequently used as initiators in percussion caps, shells cartridges, detonators, and fuses. When activated, they produce the detonation wave that initiates the booster, or bursting charge. These primary explosives are mercuric fulminate, lead azide, and lead styphnate. As compounds of mercury and lead they are highly toxic. For this reason, their use by explosives experts is declining. Their physical properties are provided in Table 15.10.

15.14-A MERCURY FULMINATE

The name *mercury fulminate* is synonymous with mercury(II) cyanate, a substance whose chemical formula is $Hg(CNO)_2$.

$$O-N\equiv C-Hg-C\equiv N-O$$

TABLE 15.10	Physical Properties of the Primary Explosives		
	MERCURY FULMINATE	**LEAD AZIDE**	**LEAD STYPHNATE**
Melting point	Detonates	Detonates	Detonates
Detonation temperature		660 °F (350 °C)	500 °F (260 °C)
Specific gravity at 68 °F (20 °C)	4.42		2.9 (anhydrous)
Brisance (g of sand)	23.4	19	24
Detonation velocity	2.9 mi/s (4.7 km/s)	3.2 mi/s (5.1 km/s)	3.0 mi/s (4.8 km/s)
Sensitivity	High	High	High

It is a white-to-gray solid produced by pouring a nitric acid solution of mercury(II) nitrate into ethanol, but the mechanism of the chemical reaction is not entirely understood.

When it detonates, mercury fulminate decomposes into mercury, carbon monoxide, and nitrogen.

$$Hg(CNO)_2(s) \longrightarrow Hg(g) + 2CO(g) + N_2(g)$$

Mercury fulminate   Mercury   Carbon monoxide   Nitrogen

It is used to manufacture caps and detonators for military, industrial, and sporting purposes.

15.14-B LEAD AZIDE

Lead azide is unique among the commercial explosives in that it is the only one whose chemical composition does not include oxygen. Its chemical formula is $Pb(N_3)_2$.

It is a colorless solid prepared by reacting aqueous solutions of sodium azide and lead acetate.

$$2NaN_3(aq) + Pb(C_2H_3O_2)_2(aq) \longrightarrow Pb(N_3)_2(s) + 2NaC_2H_3O_2(aq)$$

Sodium azide   Lead acetate   Lead azide   Sodium acetate

Lead azide is often packed into aluminum detonator capsules. When lead azide detonates, it decomposes into lead and nitrogen.

$$Pb(N_3)_2(s) \longrightarrow Pb(s) + 3N_2(g)$$

Lead azide   Lead   Nitrogen

15.14-C LEAD STYPHNATE

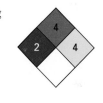

Lead styphnate and lead trinitroresorcinate are synonyms for the substance having the following formula:

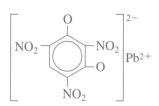

It is a yellow-orange solid made from styphnic acid, or trinitroresorcinol. Lead styphnate is produced when lead acetate reacts with styphnic acid.

Styphnic acid   Lead acetate   Lead styphnate   Acetic acid

When the formula for lead styphnate is condensed to $Pb[C_6HO_2(NO_2)_3]$, its detonation is represented as follows. The products are a mixture of gases and vapors.

$$2Pb[C_6HO_2(NO_2)_3](s) \longrightarrow Pb(s) + 3N_2(g) + H_2O(g) + 5CO(g) + 5CO_2(g) + 2C(s)$$

Lead styphnate   Lead   Nitrogen   Water   Carbon monoxide   Carbon dioxide   Carbon

15.14-D TRANSPORTING PRIMARY EXPLOSIVES

When shippers offer a primary explosive for transportation, DOT requires them to identify it on an accompanying shipping paper as one of the following, as relevant:

UN0135, Mercury fulminate, wetted, 1.1A, PGII, EX-xxxxx
UN0129, Lead azide, wetted, 1.1A, PGII, EX-xxxxx
UN0130, Lead styphnate, wetted, 1.1A, PGII, EX-xxxxx

Dry lead azide and lead styphnate are forbidden explosives.

DOT also requires shippers to affix EXPLOSIVE 1.1 labels bearing the compatibility group A on packages containing a primary explosive. When carriers transport a primary explosive in any amount, DOT requires them to post EXPLOSIVES 1.1 placards bearing the compatibility group A on the bulk packaging or transport vehicle used for shipment.

PERFORMANCE GOALS FOR SECTION 15.15:

- Describe the response actions to be executed when explosives have been released into the environment.

15.15 Responding to Incidents Involving a Release of Explosives

There are five potential types of emergency incidents affecting a release of explosives: a transportation mishap involving explosives; a fire within a magazine; disowned or abandoned ordnance; the detection of an improvised explosive device; and the treatment or disposal of a waste

TABLE 15.11	Surface Explosion Information[a]		
TNT Equivalent		**Recommended Evacuation Distances**	
lb	**kg**	**ft**	**m**
5	2.3	11–17	4–6
50	23	24–37	8–12
500	227	51–79	16–24
1000	454	64–99	20–31
4000	1814	101–157	31–48
10,000	4536	137–213	42–65
30,000	13,608	198–307	61–94
60,000	27,216	249–387	76–118

[a]The evacuation distances from the scene of a potential explosion were generated through the use of Sandia National Laboratory's BLAST computer model. Excerpted with permission from Michael B. Dillon, Ronald L. Baskett, Kevin T. Foster, and Connee S. Foster, *The NARC Emergency Response Guide to Initial Airborne Hazard Estimates*, UCRL-TM-202990 (Livermore, CA: U.S. Department of Energy, 2004), p. 8.

explosive. *Responding to any one of these emergency incidents should generally be undertaken only by competent, experienced individuals who have received special training in the handling of explosives.* To save lives at a scene where an explosive may be detonated with felonious intent, untrained responders must often limit their activities to evacuating unnecessary individuals to the distances specified in Table 15.11.

The presence of an explosive at a transportation mishap may usually be verified by observing the following:

- The number 1 as a component of a shipping description of a hazardous material on a shipping paper
- Either of the expressions EXPLOSIVE or BLASTING AGENT and the number 1 on orange labels affixed to packaging
- The expression EXPLOSIVES and the number 1 on orange placards posted on each side and each end of a transport vehicle containing an explosive in hazard classes 1.1, 1.2, or 1.3; the expression 1.4 EXPLOSIVES or 1.6 EXPLOSIVES and the number 1 on orange placards posted on each side and each end of a transport vehicle containing 1001 lb (454 kg) or more of an explosive in hazard classes 1.4 or 1.6; or the expression 1.5 BLASTING AGENTS and the number 1 on orange placards posted on each side and each end of a transport vehicle containing 1001 lb (454 kg) or more of a blasting agent in hazard class 1.5.

When explosives are involved in a transportation mishap, DOT advises first-on-the-scene responders to implement the procedures in Figures 15.7, 15.8, and 15.9 when the emergency incident involves a division 1.1 explosive, a division 1.2 or 1.3 explosive, and a division 1.4, 1.5, or 1.6 explosive, respectively. When the identity of the DOT division is unknown, or when an

SOLVED EXERCISE 15.8

The first-on-the-scene responders to a transportation mishap note the presence of an overturned motor van along a route specifically designated for the transportation of explosives. From a distance of approximately 300 ft (91 m), they also observe a small fire that appears to be spreading upward from a tire on the vehicle toward the overlying cargo van. Orange placards have been posted on the visible sides of the vehicle, although the DOT division number is not readily discernible. What immediate actions should these responders execute to protect public health, safety, and the environment?

Solution: DOT requires the posting of orange placards on a transport vehicle to warn that packages of explosives are onboard for transit. Given that the transport vehicle is on fire, the crew should promptly recognize that the highest degree of hazard is associated with potential detonation of the explosives. Because the DOT division of the explosive cannot be discerned, it is prudent to assume that it is 1.1.

Using the information summarized in Figure 15.10, the responding crew should do the following:

- Evacuate all individuals to 5000 ft (1500 m) in all directions from the transportation incident. This includes stopping traffic in all directions.
- Do not attempt to extinguish or control the spread of the fire *until* the driver, shipper, or carrier accurately confirms the identity of the division number of the explosive onboard.
- Without knowledge of the division number of the explosive, acknowledge that the risk to the lives of the responders is so great that combating the fire is absolutely unwarranted.

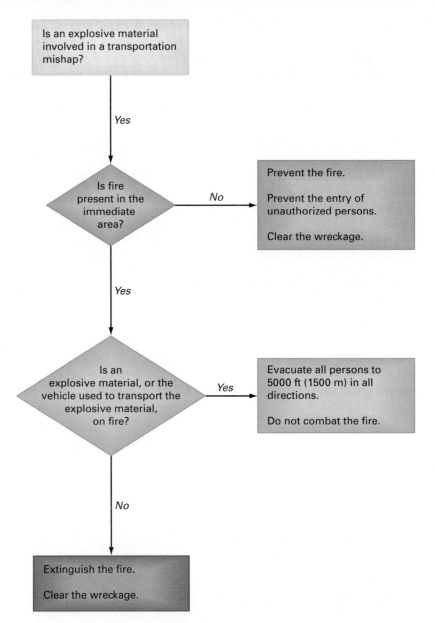

FIGURE 15.7 The recommended procedures when responding to transportation mishap involving the release of a division 1.1 explosive from its packaging. (*Adapted with permission of the American Society for Testing and Materials, from a figure in ASTM STP 825,* A Guide to the Safe Handling of Hazardous Materials Accidents, *second edition. Copyright © 1990, American Society for Testing and Materials.*)

explosive has been disowned or abandoned, all unnecessary individuals should be evacuated to a distance of 0.5 mi (800 m).

When called to an emergency incident involving an explosive magazine, the Institute of Makers of Explosives advises first-on-the-scene responders to implement the procedures in Figure 15.10. Fires outside a magazine should be controlled to ensure that they do not reach the exterior perimeter of the magazine, but fires inside the magazine should not be fought.

When emergency responders are called to incidents involving material that has been disowned or abandoned, perhaps by activists or terrorists, it is prudent to suspect that this material

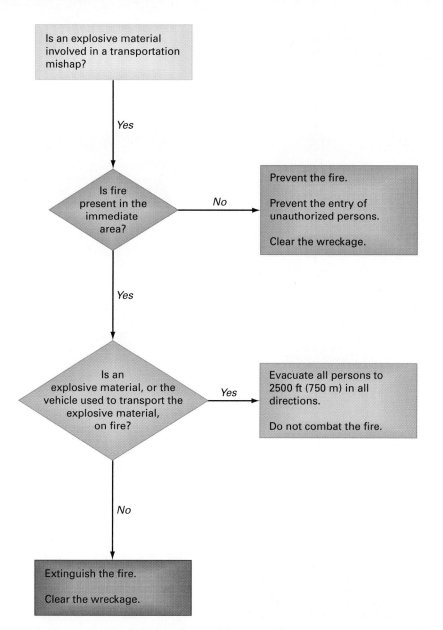

FIGURE 15.8 The recommended procedures when responding to a transportation mishap involving the release of a division 1.2 or 1.3 explosive from its packaging. (*Adapted with permission of the American Society for Testing and Materials, from a figure in ASTM STP 825,* A Guide to the Safe Handling of Hazardous Materials Accidents, *second edition. Copyright © 1990, American Society for Testing and Materials.*)

consists of unexploded ordnance. In this type of incident, the suspect material is generally removed remotely into a total-explosive-containment vessel such as that shown in Figure 15.11. Once the suspect material has been transferred, the vessel is driven to a remote location at which the chemical nature of the material may be determined by experts.

Disowned or abandoned waste explosives are RCRA-regulated hazardous wastes that exhibit the characteristic of reactivity. When a decision is made to treat or destroy them, EPA should be contacted for guidance relating to the manner of treatment or disposal that best protects public health and the environment.

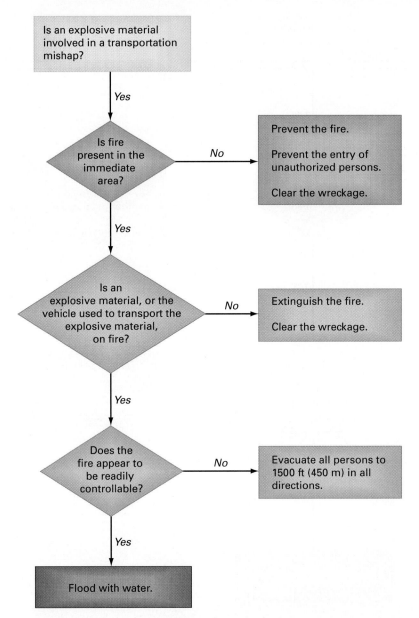

FIGURE 15.9 The recommended procedures when responding to a transportation mishap involving the release of division 1.4, 1.5, or 1.6 explosive from its packaging. (*Adapted with permission of the American Society for Testing and Materials, from a figure in ASTM STP 825,* A Guide to the Safe Handling of Hazardous Materials Accidents, *second edition. Copyright © 1990, American Society for Testing and Materials.*)

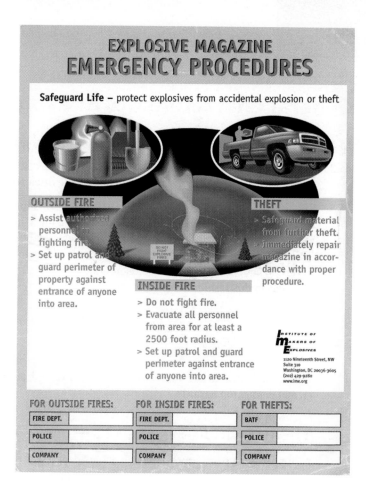

FIGURE 15.10 The "Emergency Procedures" poster published by the Institute of Makers of Explosives. (*Courtesy of Institute of Makers of Explosives, Washington, DC.*)

FIGURE 15.11 This total-explosive-containment vessel is a thermal-treatment unit that provides bomb squads with a means for destroying live ammunition, pyrotechnics, and other potentially hazardous chemical agents. This vessel is marketed to withstand the detonation of 10 lb (4.5 kg) of Composition C-4. (*Courtesy of Nabco, Inc., Cononsburg, Pennsylvania.*)

General Characteristics of Explosives

15.1   Why is a shock wave generated when an explosive detonates?

15.2   Following the detonation of an explosive, a plume often appears, much like a smoke plume during a fire. Why is this detonation plume often red?

15.3   The OSHA regulation at 29 C.F.R. §1910.109(e)(2) stipulates that empty boxes, paper, and fiber packing materials that previously contained high explosives cannot be used again for any purpose; instead, they must be destroyed by burning them at an approved, isolated location outdoors. No person may be nearer than 100 ft (31 m) after the burning begins. What is the most likely reason OSHA requires the destruction of the packaging materials in this fashion?

Transporting Explosives

15.4   Which explosive article in each of the pairs labeled as follows potentially poses the greater detonation hazard?

(a) EXPLOSIVE 1.4S or EXPLOSIVE 1.2G
(b) EXPLOSIVE 1.1D or EXPLOSIVE 1.4S

15.5   A shipper intends to offer fourteen 1-lb (0.5-kg) boxes of trinitrotoluene for transportation by a private carrier in a sole-use truck.

(a) Use Table 6.1 to determine the shipping description the carrier must use on the accompanying shipping paper if the material is wetted with less than 30% water by mass.
(b) Use Table 6.1 to determine the shipping description the carrier must use on the accompanying shipping paper if the material is wetted with more than 30% water by mass.
(c) For each situation noted in parts a and b, identify the labels, if any, that DOT requires the shipper to affix to the boxes.
(d) For each situation noted in parts a and b, identify the placards, if any, that DOT requires the carrier to post on the truck.

15.6   An ammunition manufacturer in Massachusetts desires to offer incendiary ammunition for transportation by motor vehicle to a military facility in Virginia. The ammunition consists of an explosive encased within a flammable gel. The manufacturer packs the ammunition into 10 cylindrical cartridges made of waxed cardboard, which are then packed into two wooden boxes. The ammunition explosive has a net explosive mass of 24 lb (11 kg). DOT has assigned EX-1325867 to the manufacturer.

(a) What shipping description of the ammunition does DOT require the manufacturer to enter on the accompanying shipping paper?
(b) Which labels does DOT require the manufacturer to affix to the outside surface of each wooden box?
(c) Which markings does DOT require the manufacturer to display on the outside surface of each wooden box?
(d) Which placards, if any, does DOT require the carrier to post on the motor vehicle used to transport the wooden boxes to the military facility?

Black Powder and Black Gunpowder

15.7 Black powder consists of a mixture of substances whose components, when ignited, undergo an oxidation–reduction reaction. One of the components is an oxidizer. What is the most likely reason DOT regulates the transportation of black powder as an explosive rather than as an oxidizer?

15.8 When emergency responders encounter black gunpowder while combating a fire at a sporting goods store, how may they substantially reduce its explosive potential?

Nitroglycerin and Dynamite

15.9 When dynamite sticks are discolored, excessively soft, or crumbly, or when there are visible signs of exterior crystallization, why should they be handled remotely?

15.10 Although nitroglycerin is an explosive substance, cardiologists often prescribe low doses of this substance to their patients. What is the general purpose of these prescriptions?

Nitrocellulose

15.11 Why do explosive experts strongly advise against storing nitrocellulose in bulk quantities?

Trinitrotoluene

15.12 During World War II, TNT was often used as a military explosive. What combination of properties caused the military to select TNT over other explosives?

Cyclonite, Tetryl, PETN, and HMX

15.13 Which explosive is a nitrate ester: cyclonite, tetryl, PETN, or HMX?

15.14 Military explosives containing cyclonite, tetryl, PETN or HMX are sometimes "aluminized"; that is, they are mixed with aluminum dust and an oxidizing agent. What is the role of these latter substances in the explosive articles?

Primary Explosives

15.15 Why is lead azide selected for use in virtually *all* blasting caps and other hot-wired initiated detonators?

Responding to Emergencies Involving a Release of Explosives

15.16 When first-on-the-scene responders arrive at the scene of an accident on a heavily trafficked highway, they note that orange EXPLOSIVES 1.4 placards bearing the compatibility group

D are posted on each side and each end of a vehicle that has been struck by another vehicle. Fuel escaping from the second vehicle has ignited. A representative of the carrier provides the accompanying shipping paper, which reads as follows:

Units	HM	Shipping Description (Identification Number, Proper Shipping Name, Primary Hazard Class or Division, Subsidiary Hazard Class or Division, and Packing Group)	Weight (lb)
1500 lb, net explosive mass contained within 30 boxes	X	UN0104, Cord, detonating, mild effect, 1.4D, PGII (EX-9806054) (DOT-SP 4850)	50 (per box)

What immediate actions should the responders execute to protect public health, safety, and the environment?

When we hear the term *radioactive,* two fearsome incidents generally come to mind: the accidental release of material from a nuclear power plant and the deployment of nuclear weapons.* We discover in this chapter that these events are associated with the occurrence of one or more nuclear processes. The forms of matter that display these processes are called **radioactive materials**.

We also learn in this chapter that certain health risks are linked with exposure to radioactive materials. To eliminate or minimize the impact of these risks, the U.S. Congress delegated the following responsibilities to the regulatory bodies listed:

- The U.S. Nuclear Regulatory Commission (NRC) regulates the civilian nuclear energy industry by licensing the construction and operation of the commercial nuclear power plants that generate electric power.
- The Department of Energy (DOE) oversees the research and development of new and creative means for reducing the burgeoning supply of nuclear waste. In addition, it oversees the construction and operation of nuclear waste disposal sites and responds to a release of radioactive material. DOE also certifies the integrity of the unique packages in which radioactive material is transported, provided they are for the purpose of national security.
- EPA is responsible for establishing radiation-exposure limits to protect public health. These limits apply to radiation arising in the environment, including natural radiation and the radiation from spent radioactive materials in storage.
- OSHA is responsible for establishing radiation-exposure limits that protect employees who use radioactive materials within their workplaces.
- DOT is responsible for ensuring that radioactive materials are transported safely.

In combination, these agencies serve to provide a degree of protection against the hazards associated with the inadvertent exposure to radioactive materials. Nonetheless, when compared with the other classes of hazardous materials, radioactive materials represent a very high degree of hazard. What are the properties of radioactive materials that give rise to their particularly hazardous nature? What may be done to minimize the adverse effects caused by exposure to them when they are encountered while implementing emergency-response actions? The answers to these questions are provided in this final chapter.

PERFORMANCE GOALS FOR SECTION 16.1:

- Identify the major constituents of atomic nuclei.
- Distinguish among the hydrogen isotopes.
- Show that atomic nuclei of atoms may have different numbers of protons and neutrons.
- Describe the phenomenon of radioactivity.
- Describe the concept of half-life.

16.1 Features of Atomic Nuclei

In Section 4.4 it was noted that there are two primary constituents of an atomic nucleus: protons and neutrons. Although the nuclei of all atoms of the same element have the same number of protons, they may have different numbers of neutrons. These different nuclei are called *isotopes.*

The number of protons found in the nucleus of an atom is called the *atomic number*. The number of protons equals the number of electrons in a neutral atom. Because the atomic numbers of the elements are compiled in the periodic table, it may be used to readily identify the number of protons in a given nucleus. The total number of protons and neutrons in a particular isotope is called its **mass number**. It corresponds to the atomic mass of the isotope, rounded off to the nearest whole number.

radioactive material
- A material containing an isotope that spontaneously emits ionizing radiation; for purposes of DOT regulations, a material containing one or more radioisotopes for which both the activity concentration and the total activity in the consignment exceed the applicable A_1 and A_2 values (see Section 16.10E)

mass number
- The total number of protons and neutrons in a given nucleus; the atomic mass of a specific nucleus rounded off to the nearest whole number

*Nuclear weapons, or nuclear bombs, were initially called *atomic bombs*. The term *nuclear* is preferable to *atomic,* since the phenomenon responsible for their detonation is a nuclear one.

Hydrogen has an atomic number of 1; this means that every hydrogen atom has one proton. Each hydrogen atom has one electron, but when a hydrogen atom is ionized, it is stripped of its electron, and only its nucleus remains.

Hydrogen atoms exist in any of three isotopic forms having the following unique names and compositions:

- **Protium** is the simplest of the hydrogen isotopes and, in fact, the simplest of all atoms. When a protium atom is ionized, only a proton remains.
- The nucleus of the second hydrogen isotope, **deuterium**, is composed of one proton and one neutron. When a deuterium atom is ionized, the remaining nucleus is composed of one proton and one neutron. This nucleus is called a **deuteron**.
- The nucleus of the third hydrogen isotope, **tritium**, is composed of one proton and two neutrons. When a tritium atom is ionized, the remaining nucleus is composed of one proton and two neutrons. It is called a **triton**.

Each isotope is designated by the symbol $^A_Z X$, where Z and A are the atomic number and mass number, respectively; X is the element's chemical symbol. Using this format, the three hydrogen isotopes are designated by the symbols $^1_1 H$, $^2_1 H$, and $^3_1 H$, respectively. For each, the symbol Z equals 1, the number of protons. The number of neutrons is the difference between the superscript and the subscript: for protium, the number is 0; for deuterium, the number is 1; and for tritium, the number is 2. Deuterium and tritium are also represented as D and T, respectively.

Only the isotopes of hydrogen have unique names. The isotopes of other elements are named by identifying the name or symbol of the element and the mass number of the isotope at issue. Thus, nuclei designated as $^{12}_6 C$, $^{235}_{92} U$, and $^{40}_{19} K$ are named carbon-12, uranium-235, and potassium-40, or C-12, U-235, and K-40, respectively.

All nuclei have form 3 to approximately 25 isotopes. Many isotopes are stable; that is, they retain their structure and do not undergo spontaneous changes. However, many other nuclei *are* subject to spontaneous transformations, or **transmutations**. The nuclei are said to decay, or disintegrate. The phenomenon is called **radioactivity**, and these unstable nuclei are said to be *radioactive*. They are called **radioisotopes**, or **radionuclides**. Protium and deuterium are stable, nonradioactive isotopes of hydrogen, but tritium is a radioisotope.

Radioactivity is not generally affected by any physical or chemical change in a substance. Hence, when radioactive materials are subjected to changes in pressure, temperature, or chemical nature, the spontaneous disintegration of the relevant radioisotope is not usually altered.

When a radioisotope undergoes a change, it usually emits a particle; less commonly, it absorbs an electron. Both processes are frequently accompanied by the simultaneous emission of energy. When the transformation occurs, the radioisotope is converted into a new nucleus, which is either stable or radioactive itself. Radioisotopes often undergo several transformations before they are converted into stable nuclei.

Each radioactive transformation is associated with a specific period of time. The time during which an arbitrary number of nuclei are reduced to half the number is called the **half-life** of that radioisotope. For instance, suppose a volume of tritium gas containing 500,000 molecules is set aside for 12.5 years. Because the tritium molecule is diatomic, there are 1 million atoms of tritium in this volume. After 12.5 years lapse, only 500,000 tritium atoms remain; and after another 12.5 years lapse, only 250,000 tritium atoms remain. Hence, 12.5 y is the half-life of tritium.

Half-lives vary appreciably. The half-life of uranium-238, a naturally occurring radioisotope, is 4.5 billion years, or 4.5×10^9 y, but the half-life of astatine-216, a radioisotope of an artificially produced halogen, is only three ten-thousandths of a second, or 0.0003 s.

Each element has at least one radioisotope. The hundred or so elements collectively have nearly 1200 known radioisotopes. Table 16.1 shows that few radioisotopes are present in nature. If radioisotopes were present when Earth was initially formed, most of them disappeared long ago, since the age of Earth is 3.6 billion years.

Some commercially available radioisotopes are listed in Table 16.2. Although they have many potential purposes, most commercial radioisotopes are used during the detection and treatment

protium
- The isotope of hydrogen consisting of one proton

deuterium
- The isotope of hydrogen consisting of one proton and one neutron

deuteron
- The nucleus of the deuterium atom

tritium
- The radioisotope of hydrogen consisting of one proton and two neutrons

triton
- The nucleus of the tritium atom

transmutation
- The transformation of a radioisotope to another isotope as the result of a nuclear reaction

radioactivity
- The nuclear property associated either with the spontaneous emission of alpha, beta, and/or gamma radiation, or the capture of an extranuclear electron

radioisotope (radionuclide)
- An atomic nucleus that undergoes a spontaneous change by emitting a particle or by absorbing an extranuclear electron

half-life
- The time period during which an arbitrary amount of a radioisotope is transformed into half that amount

Cesium-137 is a radioisotope often used as a source of gamma radiation for the treatment of cancer. If a clinic purchases a source containing 7.60×10^{15} atoms of cesium-137 today, how many atoms will remain in 132 years?

Solution: The half-life of cesium-137 is listed in Table 16.2 as 33 y. The number of half-lives in 132 years is then determined by division:

$$\text{Number of half-lives} = 132 \text{ y}/33 \text{ y} = 4.00$$

This means that after the passage of 132 years, cesium-137 will have disintegrated through four half-lives, and one-sixteenth of the atoms will remain.

$$\tfrac{1}{2} \times \tfrac{1}{2} \times \tfrac{1}{2} \times \tfrac{1}{2} = (\tfrac{1}{2})^4 = 1/16$$

Multiplying 7.60×10^{15} by 1/16 gives 4.75×10^{14} atoms.

$$7.60 \times 10^{15} \text{ atoms} \times 1/16 = 4.75 \times 10^{14} \text{ atoms}$$

Thus, after the passage of 132 years, 4.75×10^{14} atoms of cesium-137 will remain in the gamma radiation source.

TABLE 16.1	Naturally Occurring Radioisotopes		
RADIOISOTOPE	**TYPE OF DISINTEGRATION**	**HALF-LIFE (y)**	**RELATIVE ISOTOPIC ABUNDANCE**
Tritium	β^-	12.26	0.00013
Carbon-14	β^-	5570	
Potassium-40	β^-, EC[a]	1.2×10^9	0.012
Rubidium-87	β^-	6.2×10^{10}	27.8
Indium-115	β^-	6×10^{14}	95.8
Lanthanum-138	β^-, EC	$\sim 2 \times 10^{11}$	0.089
Neodymium-144	α	$\sim 5 \times 10^{15}$	23.9
Samarium-147	α	1.3×10^{11}	15.1
Lutetium-176	β^-	4.6×10^{10}	2.60
Rhenium-187	β^-	$\sim 5 \times 10^{10}$	62.9
Platinum-190	α	$\sim 1 \times 10^{12}$	0.012
Radium-226	α	1622	
Thorium-232	α	1.4×10^{10}	100
Uranium-235	α	7.13×10^8	0.72
Uranium-238	α	4.5×10^9	99.28

[a]EC = electron capture.

TABLE 16.2	Some Commercially Available Radioisotopes	
RADIOISOTOPE	**HALF-LIFE**	**APPLICATION**
Cesium-137	33 y	Radiation source for treatment of cancer; sealed radiation source for irradiation of foods
Chromium-51	27.8 d	Determination of red blood cell volume and total blood volume
Cobalt-57	271.8 d	Instrument calibration; determination of the effectiveness of the body's uptake of vitamin B_{12}
Cobalt-60	5.26 y	Inducement of cross-linking within polyethylene and rubber macromolecules; determination of the effectiveness of the body's uptake of vitamins; sterilization of medical devices; component of the "seeds" implanted for the treatment of prostate cancer; sealed radiation source for treatment of cancer and irradiation of foods
Fluorine-18	1.8 hr	Brain- and bone-imaging drug; positron emission tomography for tumor imaging
Gallium-67	3.3 d	Diagnostic drug for tumor detection
Indium-111	2.8 d	Diagnostic drug for tumor detection; imaging of the gastric and cardiac systems
Iodine-123	13 hr	Imaging of the brain, thyroid, and renal systems
Iodine-125	59.4 d	Cancer therapeutic drug; brain, blood, and metabolic-function diagnostic drug
Iodine-131	8 d	Brain, pulmonary, and thyroid diagnostic drug
Iridium-192	73.8 d	Cancer therapeutic drug
Iron-59	44.5 d	Measurement of the rate of formation and lifetime of red blood cells
Palladium-103	17 d	Cancer therapeutic drug
Phosphorus-32	14.3 d	Detection of skin cancer
Plutonium-238	88 y	Power source for thermoelectric generators in spacecraft
Radium-223	11.4 d	Radioimmunotherapeutic drug
Radium-226	1590 y	Radiation source for treatment of cancer
Selenium-75	128 d	Pancreatic cancer diagnostic drug
Sodium-24	15.0 hr	Detection of obstructions within the circulatory system
Technetium-99m	6.0 hr	Imaging of the brain, thyroid, liver, kidney, lung, and cardiovascular systems
Thallium-201	3 d	Cardiac diagnostic drug
Tritium, or H-3	12.3 y	Determination of total body water
Yttrium-90	2.7 d	Radioimmunotherapeutic drug

of cancer. During the treatment of prostate cancer, for example, a surgeon may implant cobalt-60 encased in tiny pellets or "seeds" within the prostate. The cobalt-60 kills cells in the nearby region, including those that are malignant. Once treatment is complete, the seeds may be surgically removed.

PERFORMANCE GOALS FOR SECTION 16.2:

- Describe each mode by which radioisotopes decay.
- Describe the differentiating features of alpha, beta, and gamma radiation.
- Write equations denoting the changes that occur when radioisotopes decay.

alpha (α) radiation
■ The radiation emitted from certain radioisotopes and composed of alpha particles

beta (β) radiation
■ Negatrons or positrons

gamma (γ) radiation
■ Electromagnetic radiation of very short wavelength emitted from some radioisotopes

ionizing radiation
■ The high-energy electromagnetic radiation emitted by radioisotopes (alpha radiation, beta radiation, and gamma radiation)

alpha particle
■ A particle (α or $_2^4$He) emitted from certain radioisotopes and having the properties of a doubly ionized helium atom

16.2 Types of Radiation and Modes of Nuclear Decay

Radioisotopes undergo nuclear transformations by one or more of the six mechanisms illustrated in Figure 16.1: alpha-particle emission, negatron emission, positron emission, electron capture, gamma-ray emission, and spontaneous fission. Most decay exclusively by one means, and only the very heavy radioisotopes ($Z > 92$) undergo spontaneous fission as a mode of decay.

The transformations of radioisotopes are associated with the emission from the nucleus of one or more of three types of radiation: **alpha (α) radiation**, **beta (β) radiation**, and **gamma (γ) radiation**. Because their passage through matter results in its ionization, they are examples of **ionizing radiation**.

The energy associated with the emission of alpha, beta, and gamma radiation is typically cited in multiples of a unit called an *electron-volt* (eV). One electron-volt is the amount of energy acquired by an electron when it is accelerated by an electric potential of one volt. It is equivalent to 1.602×10^{-19} J. Nuclear radiation is typically associated with energies of millions of electron-volts. One million electron-volts (1 MeV) is equivalent to 1.602×10^{-13} J.

16.2-A ALPHA RADIATION

Many isotopes having atomic numbers greater than 83 disintegrate by emitting particles consisting of two protons and two neutrons. These particles are called **alpha particles**; they are the

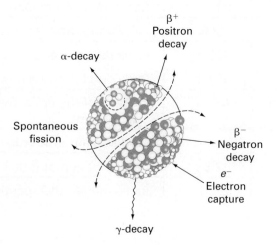

FIGURE 16.1 The modes by which radioisotopes decay, wherein the open circles represent protons, and the solid red circles represent neutrons. The most common modes of decay are those associated with the production of beta and gamma radiation. The rarest mode of decay is spontaneous fission.

nuclei of doubly ionized helium atoms. Alpha particles are symbolized as ^4_2He, but generally, they are denoted by the Greek letter α. When the nuclei of many such atoms decay, numerous alpha particles are correspondingly emitted. They are collectively called *alpha radiation*.

Alpha radiation is associated with a relatively large amount of energy that ranges from 4 to 8 MeV; but since alpha particles are doubly ionized, this energy is readily dissipated by its passage through a few centimeters of air or by absorption in a thin piece of matter. For instance, alpha radiation is absorbed by the thickness of this page.

When a radioisotope emits an alpha particle, its atomic number decreases by 2 and its mass number decreases by 4. An example of a radioisotope that disintegrates by alpha-particle emission is uranium-238. This phenomenon is represented by either of the following equations:

$$^{238}_{92}\text{U} \longrightarrow {}^{234}_{90}\text{Th} + {}^4_2\text{He}$$
$$^{238}_{92}\text{U} \longrightarrow {}^{234}_{90}\text{Th} + \alpha$$

Both equations show that the uranium-238 nucleus changes into the thorium-234 nucleus by emitting an alpha particle. The particle is written to the right of the arrow, designating that it has been emitted from the uranium-238 nucleus.

Equations denoting nuclear phenomena are not balanced in the chemical sense. Instead, a nuclear equation is balanced when each of the following is fulfilled: The sums of the charges and the sums of the mass numbers are the same on each side of the arrow.

16.2-B BETA RADIATION

The second mode of radioactive disintegration is associated with three different processes. The first is equivalent in result to the emission of an electron from the nucleus. When electrons are encountered in nuclear phenomena, they are called **negatrons** and designated as β^-, or $^0_{-1}e$. When the nuclei of many such atoms decay, a substantial number of negatrons are emitted. This is one form of beta radiation.

When negatron emission occurs, the mass numbers of the associated nuclei remain unchanged, but the atomic number increases by 1.* An example of a radioisotope that disintegrates by emitting a negatron is thorium-234. This is the nucleus produced when uranium-238 disintegrates. On emitting a negatron, thorium-234 becomes protoactinium-234. This phenomenon is represented by either of the following equations:

$$^{234}_{90}\text{Th} \longrightarrow {}^{234}_{91}\text{Pa} + {}^0_{-1}e$$
$$^{234}_{90}\text{Th} \longrightarrow {}^{234}_{91}\text{Pa} + \beta^-$$

The emission of a negatron from a nucleus raises a basic question: How can it be emitted *from* the nucleus when the electron is not a component *of* the nucleus? The process is apparently more involved than a single equation represents. Nuclear scientists have determined that during negatron emission, each neutron within the unstable nucleus transforms into a proton and an electron. The proton then becomes part of the new nucleus, and the electron is simultaneously emitted. This conversion of the neutron (^1_0n) into a proton and an electron is designated as follows:

$$^1_0n \longrightarrow {}^1_1\text{H} + {}^0_{-1}e$$

Negatrons possess a range of energies, but these energies are generally no greater than 4 MeV. They are usually absorbed by a 1-cm-thick sheet of aluminum.

A second process associated with the production of beta radiation involves the emission from the nucleus of a particle called the **positron**. It is like an electron in most features, but it is positively charged. A positron is symbolized as $^0_{+1}e$, or β^+.

negatron
■ A particle (β^-) emitted from certain radioisotopes that has all the properties of an electron

positron
■ A positively charged particle (β^+) emitted by some radioisotopes but otherwise having the characteristics of an electron

---

*Beta-decay processes are also associated with the production of neutral subatomic particles called *neutrinos* and *antineutrinos*. These particles are of no interest here.

Radioisotopes emitting positrons retain their mass numbers but decrease in atomic number by 1. This nuclear event may be envisioned as the conversion of a proton into a neutron and positron as follows:

$$_1^1H \longrightarrow {_0^1n} + {_{+1}^0e}$$

An example of a radioisotope that disintegrates by positron emission is sodium-22. This nucleus is spontaneously converted into neon-22. The event is denoted by either of the following equations:

$$_{11}^{22}Na \longrightarrow {_{10}^{22}Ne} + {_{+1}^0e}$$
$$_{11}^{22}Na \longrightarrow {_{10}^{22}Ne} + \beta^+$$

Like negatrons, positrons possess a range of energies, but they are generally no greater than 3 MeV. Positrons are usually absorbed by a 1-cm-thick sheet of aluminum. Positron emission represents a second form of beta radiation.

A third process associated with β radiation involves the union of an unstable nucleus with an extranuclear electron. This phenomenon is called **electron capture**. A radioisotope that transforms by electron capture decreases in atomic number by 1 but retains its mass number. Each electron captured by the nucleus reacts with a proton, thereby forming a neutron, which then becomes part of the structure of the new nucleus. This nuclear event is represented by the following equation:

electron capture
■ A spontaneous mode of radioisotopic decay in which the nucleus captures an extranuclear electron

$$_1^1H + {_{-1}^0e} \longrightarrow {_0^1n}$$

An example of a radioisotope that disintegrates by electron capture is oxygen-15. The phenomenon is represented as follows:

$$_8^{15}O + {_{-1}^0e} \longrightarrow {_7^{15}N}$$

SOLVED EXERCISE 16.2

Indium-111 is a radioisotope used in medicine as a tumoral diagnostic drug and for gastric and cardiac imaging. It decays by electron capture and has a half-life of 2.8 d.

(a) How many protons and neutrons are present in the indium-111 nucleus?
(b) Identify the product of its radioactive decay.
(c) What percentage of indium-111 remains in the bloodstream 8.4 days after administration of the drug?

Solution: Referring to either Figure 4.3 or the appendix at the back of this book, we see that the chemical symbol and atomic number of indium are determined to be In and 49, respectively. Thus, the symbol for the indium-111 nucleus is $_{49}^{111}In$.

(a) Because the atomic number is the number of protons in a nucleus, there are 49 protons in the indium-111 nucleus. Because the atomic mass of $_{49}^{111}In$ is 111, the total number of protons and neutrons in this nucleus is 111. The number of neutrons in $_{49}^{111}In$ is $111 - 49$, or 62.

(b) Indium-111 decays by capturing an electron, a process represented as follows:

$$_{49}^{111}In + {_{-1}^0e} \longrightarrow {_{48}^{111}Cd}$$

The product of the decay is cadmium-111.

(c) For indium-111, a period of 8.4 days represents three half-lives. After that time, 12.5% of the administered drug remains in the bloodstream.

$$\text{Final percentage} = 100\% \times \tfrac{1}{2} \times \tfrac{1}{2} \times \tfrac{1}{2} = 12.5\%$$

16.2-C GAMMA RADIATION

Nuclear transformations are frequently accompanied by the simultaneous emission of the third form of radiation: gamma (γ) radiation. This is a form of **electromagnetic radiation**. Like X radiation, infrared radiation, and ultraviolet radiation, gamma radiation has no mass or charge. In the portion of the electromagnetic spectrum shown in Figure 16.2, the various forms of radiant energy are characterized by their wavelengths. Ultraviolet, infrared, and radio waves are said to have "long" wavelengths, whereas gamma radiation and X radiation have "short" wavelengths. The components of the electromagnetic spectrum with short wavelengths are very energetic, so much so that they ionize the matter through which they pass. The components with long wavelengths are relatively nonenergetic and are unable to ionize matter.

electromagnetic radiation
▪ The entire range of energy that travels as waves through space

The individual components of gamma radiation are called *gamma rays*, or **photons** and are represented by the symbol γ. Because they do not possess a charge, they are extremely penetrating and may be absorbed only by dense forms of matter such as thick blocks of lead.

photon
▪ A massless packet of electromagnetic energy

When gamma rays are emitted from a radioisotope, no change occurs in either the atomic number or mass number. Instead, some fraction of the energy of excitation that causes the nucleus to be unstable is removed as the radioisotope undergoes internal conversion. Imagine a radioisotope that exists in only two energy states. The more energetic form, called the *excited state*, may emit one or more gamma rays from the nucleus. The phenomenon may be illustrated by the following equation, where the excited state is represented by an asterisk:

$$(_{Z}^{A}X)^* \longrightarrow {_{Z}^{A}X} + \gamma$$

In this process, the radioisotope gives up a fraction of its excitation energy to become a more stable form of the *same* radioisotope.

When the excited state of a nucleus cannot be readily measured, it is referred to as a *metastable state* of the radioisotope. The metastable state is designated by adding an *m* following the isotope's mass number. For example, technetium-99*m* is an excited state of technetium-99 that has a half-life of 6.0 hr. It decays by gamma-ray emission as follows:

$$\text{Tc}^{99m} \longrightarrow \text{Tc}^{99} + \gamma$$

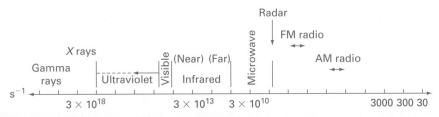

FIGURE 16.2 The components of the electromagnetic spectrum as a function of radiation frequency. Gamma rays, shown to the far left of the spectrum, are forms of energy associated with short wavelengths and high frequencies. Because gamma rays have sufficient energies to ionize matter, gamma radiation is a form of ionizing radiation.

As noted in Table 16.2, the gamma rays emitted from technetium-99*m* are specifically used to image the organs of the body.

The phenomenon of gamma-ray emission is not always represented by an equation. It is also represented by the following general diagram, where each horizontal line designates a discrete energy state of the atomic nucleus:

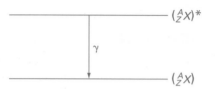

16.3 Sealed Radiation Sources

When a radioisotope listed in Table 16.2 is used commercially, it is likely to be encountered as a *sealed radiation source*. Typically, a manufacturer seals high levels of the radioisotope within double-skinned steel tubes, which are then housed within a medical or other device. The radioisotope remains sealed within these tubes throughout the period of its use.

Cobalt-60 is an example of a radioisotope that is used as a sealed source. It decays by negatron and gamma-ray emission. First, the nucleus emits a negatron and transforms into an excited state of nickel-60, which then emits two gamma rays having energies of 1.173 MeV and 1.332 MeV. We represent this phenomenon as follows:

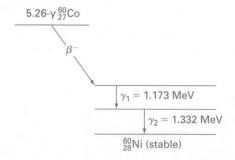

irradiation
■ Intentional exposure to ionizing radiation, usually X rays or gamma rays, for medical treatment, sterilization or preservation of foods, and other purposes

When cobalt-60 is used for a specific purpose, the sealed source is positioned so that the gamma rays emitted by the radioisotope pass through its steel container and penetrate a material. This process is referred to as **irradiation**. Gamma rays are used to irradiate polyethylene or rubber to induce cross-linking within their macromolecules. They are also used to irradiate spices and foods and sterilize medical devices.

When gamma rays are used to irradiate foods, they disrupt the fast-growing cells of insects, molds, and microbes on perishable meat, poultry, and produce. Irradiation also destroys the

FIGURE 16.3 At 21 C.F.R. §179.26(c), FDA requires this international Radura symbol to be posted on irradiated packaged foods, bulk containers of unpackaged foods, on placards at the point of purchase for fresh produce, and on invoices for irradiated ingredients and products sold to food processors. The logo is dark green and displayed on a white background.

microorganisms that cause food spoilage, but it does not render the food radioactive or cause it to lose its nutritive value, flavor, or texture.

A primary advantage associated with irradiating foods is its success at killing the bacteria within raw meat and produce. It is the only known method of eliminating the potentially deadly bacteria *Escherichia coli*, *Salmonella*, and *Campylobacter* from these foods. Fruits and vegetables may tolerate an irradiation of 1.0 kGy (Section 16.5-G), which typically inactivates 99.999% of the bacteria.

Although gamma-ray irradiation eliminates pests in fresh foods and extends their shelf life, irradiation also creates free radicals, the presence of which could negatively affect the inherent quality of foods by producing small amounts of undesirable substances. Although there is no technical basis for concluding that these irradiated foods are unsafe to consume, the widespread use of gamma-ray irradiation of foods remains a controversial subject.

FDA's approval is required to sell irradiated foods in American stores. As a component of its approval, FDA requires manufacturers and distributors to affix the *Radura symbol* in Figure 16.3 on packages of irradiated foods. FDA also requires food distributors to mark packages of irradiated foods with either of the following statements:

TREATED WITH RADIATION
TREATED BY IRRADIATION

Sealed radiation sources are highly dangerous if they become unsealed. Acknowledging the need to warn people of the presence of radioisotopes in sealed sources, the International Atomic Energy Agency (IAEA) and the International Organization for Standardization (ISO) introduced the ionizing radiation symbol shown in Figure 16.4. The symbol shows waves radiating from a three-bladed propeller called a *trefoil*, a skull-and-crossbones symbol, and a running person.

IAEA and ISO recommend affixing this symbol to devices that house high-level sealed radiation sources that can cause death or serious injury on exposure. These sources are normally visible only when attempts are made to disassemble them. The intent of the symbol is to warn

FIGURE 16.4 The IAEA/ISO ionizing radiation symbol used to warn individuals that a dangerous level of ionizing radiation in a sealed radioactive source is nearby. The triangular symbol is red with a black border and has black and white waves radiating from a trefoil, a skull-and-crossbones symbol, and a running person.

people to distance themselves from a radiation source. Because it is not affixed to building access doors, containers, or transport vehicles, this symbol supplements rather than replaces the trefoil that OSHA and DOT require on signs, labels, and placards.

PERFORMANCE GOALS FOR SECTION 16.4:

- Identify the common means for detecting radioactivity.

16.4 Detection and Measurement of Radioactivity

Several radiation-detection instruments are commercially available. The type shown in Figure 16.5 is often used by emergency responders to detect the presence of a radiation source. The operator may specifically determine whether a radiation source is nearby, how close it is, its identity, and its intensity.

The total amount of radiation to which an individual has been exposed is often determined through the use of personal-monitoring equipment like the film badges shown in Figure 16.6(a). Film badges consist of radiation-sensitive photographic emulsions that are worn for a specified period, usually one each day, and then professionally developed. The radiation exposure is ascertained by comparing the developed film against other developed films that were previously exposed to known doses of radiation.

The pocket or pen dosimeters illustrated in Figure 16.6(b) are also used to determine total radiation exposure. They are small ionization chambers that are read with an auxiliary instrument after they have been exposed to radiation for a specified time.

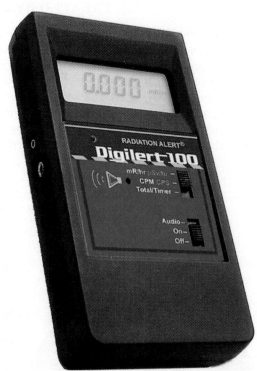

FIGURE 16.5 A portable handheld radiation detector commonly used by first-on-the-scene responders to measure the intensity of alpha, beta, gamma, and X radiation. The indicating meter may be calibrated to record a radiation intensity in milliröentgens per hour (mR/hr), microsieverts per hour (μSv/hr), or counts per minute (c/min). (*Courtesy of S.E. International, Inc., Summertown, Tennessee.*)

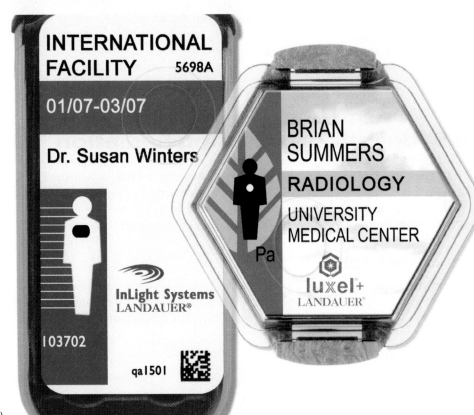

(a)

FIGURE 16.6 (a) Two types of personal radiation-monitoring badges and (b) pocket or pen dosimeters. The use of these badges and dosimeters provides a measurement of the total amount of radiation to which an individual has been exposed. (*Courtesy of Landauer, Inc., Glenwood, Illinois, and S.E. International, Inc., Summertown, Tennessee.*)

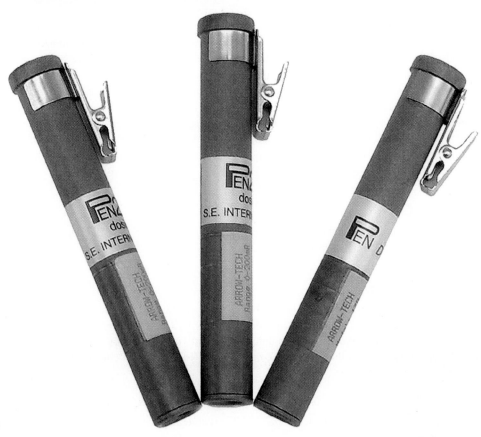

(b)

16.5 Units for Measuring Radiation

The total amount of ionizing radiation that a radioisotope emits is called its *activity*. The activity per unit mass is its *specific activity*. As different scientists developed radiation-detection instruments, they simultaneously defined units of radiation measurement to serve as a means of accounting for the activity of a specific radioisotope. We cite them in the following sections.

16.5-A RÖENTGEN

röentgen
- The amount of radiation that produces 2.1×10^9 units of charge in 1 mL of dry air at standard atmospheric pressure

The **röentgen** (R) is the international unit of radiation quantity for gamma radiation, but the unit is also used to measure alpha and beta radiation. One röentgen is the amount of radiation that produces 2.1×10^9 units of charge in 1 mL of dry air at standard atmospheric pressure.

16.5-B RAD

rad
- The amount of radiation absorbed per gram of body tissue

The term **rad** (D) is the acronym for *radiation absorbed dose*. It is the amount of radiation absorbed per gram of body tissue. The röentgen and the rad are so close in magnitude that they are considered identical. For practical purposes, 1 R = 1 D.

16.5-C REM

rem
- The amount of ionizing radiation that produces the same biological effect as one **röentgen** of X rays or gamma rays

Rem is the acronym for *röentgen equivalent man*. It is a measure of the dose of any ionizing radiation to body tissue in terms of its estimated biological effect relative to a dose of 1 R of X rays or gamma rays. The relation of the rem to other dose units depends on the biological effect under consideration. One rem is the amount of radiation that produces the same damage as 1 R of X rays or gamma rays.

The *millirem (mrem)*, or one-thousandth of a rem, is a commonly encountered fraction of the rem. Approximately 10 mrem of radiation is needed to produce a single chest X ray, whereas approximately 30 mrem is needed for a single mammogram. The average American receives around 360 mrem of radiation per year, roughly the equivalent of 36 X rays, from natural and human-made radiation sources.

16.5-D GRAY

gray
- The unit of an absorbed dose of radiation equivalent to 1 joule per kilogram of living tissue

One **gray** (Gy) is the amount of radiation equivalent to the transfer of 1 J of energy to 1 kg of living tissue. One gray is equal to 100 D.

16.5-E CURIE

curie
- The amount of a radioisotope that decays at the rate of 3.7×10^{10} disintegrations per second

The **curie** (Ci) is a measurement of the number of radioactive disintegrations occurring each second in a sample. One curie is the amount of radiation equal to 3.7×10^{10} disintegrations per second. Because the curie is an extremely large unit, the intensity of a sample of a radioactive material is usually encountered in millicuries (mCi), microcuries (μCi), and picocuries (pCi).

$$1 \text{ mCi} = 3.7 \times 10^7 \text{ disintegrations/s}$$
$$1 \text{ μCi} = 3.7 \times 10^4 \text{ disintegrations/s}$$
$$1 \text{ pCi} = 3.7 \times 10^{-2} \text{ disintegrations/s}$$

16.5-F BECQUEREL

The **becquerel** (Bq) is the SI unit used to measure radioactive disintegrations per second. One Bq is equivalent to one disintegration per second, or

$$1 \text{ Ci} = 3.7 \times 10^{10} \text{ Bq}$$

A convenient unit for measuring radioactivity is the terabecquerel (TBq), which equals a trillion becquerels.

$$1 \text{ TBq} = 10^{12} \text{ Bq}$$

becquerel
■ The SI unit of radioactivity equivalent to one disintegration per second

PERFORMANCE GOALS FOR SECTION 16.6:

- ■ Identify the adverse health effects experienced by individuals who have been exposed to radiation.
- ■ Identify the adverse health effects caused uniquely by exposure to the radioisotopes of iodine, strontium, and radium.

16.6 Ill Effects Resulting from Exposure to Radiation

All persons are constantly exposed to cosmic radiation and the inescapable low-level ionizing radiation emitted from the naturally occurring radioisotopes listed in Table 16.1. Everyone is also exposed to artificially produced radioisotopes during the use of radioactive pharmaceuticals to detect the presence of tumors or to therapeutically kill the cells within diseased tissue.

The combination of ionizing radiation from these natural and artificial sources is called **background radiation**. A person living in the United States receives an average exposure to 300 mrem of background radiation each year (300 mrem/y). Figure 16.7 illustrates the sources of this radiation. The largest single contribution, 200 mrem/y, results from exposure to the naturally occurring radon radioisotopes radon-220 and radon-222 which are discussed in Section 16.12.

When human tissue is exposed to ionizing radiation, of concern is the *dose*, or quantity of radiation absorbed by an individual per unit of mass. The adverse health effects that result from exposure of the human organism to different short-term doses of ionizing radiation are provided in Table 16.3. The manifestation of these adverse effects is called **radiation sickness**. The consequences to specific body organs, especially the reproductive and active blood-forming organs, are likely to be chronic effects.

In addition to these specific effects, the onset of cancer is a likely consequence of exposure to *any* dose of radiation. Among them are cancers of the skin, blood (leukemia), lung, stomach, esophagus, bone, thyroid, and brain. There is no threshold of exposure below which a dose of ionizing radiation is harmless—although the likelihood of acquiring cancer from radiation exposure diminishes as the dosage declines. Cancerous cells develop because the radiation-damaged cells incorrectly repair themselves.

Ionizing-radiation exposure is measured in coulombs per kilogram (C/kg), which equals approximately 3876 R. The SI unit of dose-equivalent is the **sievert** (Sv), which equals 100 rem. The average background radiation is 3 mSv/y. The taking of a single chest X ray consists of an exposure of 0.1 mSv. Scientists estimate that one out of 100 individuals is likely to develop

background radiation
■ Cosmic radiation and inescapable low-level ionizing radiation that is emitted from naturally occurring and artificially produced radioisotopes; 300 mrem/y, or 3 mSv/y

radiation sickness
■ The following adverse health effects resulting from short- or long-term exposure to radiation: nausea, vomiting, diarrhea, weakness, shock, skin sores, hair loss, and longer-term decrease in white blood cell count

sievert
■ The SI unit of dose-equivalent, equal to 100 rem

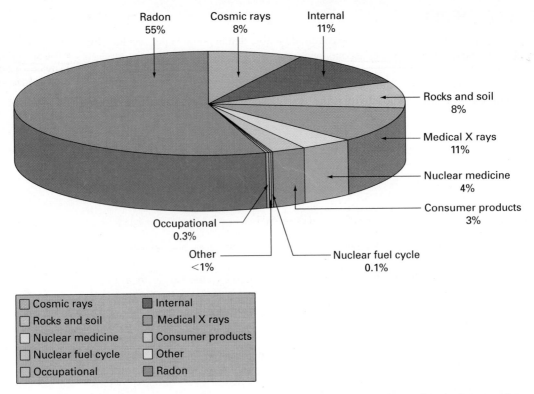

FIGURE 16.7 Sources of radiation to which the U.S. population is exposed. There is concern that the sources of exposure from medical sources, nuclear power plants, and radon are increasing. (*Reprinted with permission of the National Council on Radiation Protection and Measurements, NCRP Report No. 93.*)

cancer from a lifetime accumulated exposure of 100 mSv of radiation. The radiation dose-equivalent rate is called the *radiation level*, which is measured in millisieverts per hour (mSv/hr).

The biological impact of radiation exposure depends on the type of radiation absorbed, its energy, the nature of specific radioisotopes, and the age of the exposed individual. We note the impact of these factors separately in the sections that immediately follow.

TABLE 16.3	Adverse Health Effects Associated with Exposure to Short-Term Doses of Radiation
DOSE (rem)	**SIGNS AND SYMPTOMS**
0–25	No detectable effect
25–100	Temporary decrease in white blood cell count but no serious injury
100–200	Acute radiation sickness (nausea, vomiting, diarrhea, weakness, shock, skin sores, hair loss, and longer-term decrease in white blood cell count); possible disability
200–400	Acute radiation sickness; disability certain; possible death without treatment
400–500	50% death rate without treatment
>600	100% death rate

EPA proposes maximum exposure limits from radiation sources in the environment to protect public health. Presently, EPA proposes 15 mrem/y as the radiation limit for a whole-body dose from all exposure pathways. If an individual has been exposed to a whole-body dose of 200 μSv/y, has the person been exposed to radiation in excess of EPA's recommended limit?

Solution: Because one sievert equals 100 rem, an annual dose of 200 μSv is equivalent to 20 mrem.

$$200 \ \mu\text{Sv/y} \times \text{Sv}/10^6 \ \mu\text{Sv} \times 100 \ \text{rem/Sv} \times 10^3 \ \text{mrem/rem} = 20 \ \text{mrem/y}$$

This means that the individual has been exposed to a radiation dose in excess of EPA's recommended limit of 15 mrem/y.

16.6-A EXPOSURE TO INDIVIDUAL TYPES OF RADIATION

In an evaluation of the potential health risks from exposure to ionizing radiation, the specific nature of the radiation is relevant. As demonstrated in Figure 16.8, alpha, beta, and gamma radiation penetrate matter differently.

■ Alpha radiation is very energetic, but it is easily absorbed externally by the epidermis, the outer layer of the skin. Because the epidermis consists of dead skin cells, external exposure to alpha radiation does not represent a hazard. If it is ingested, however, the energetic alpha radiation may localize in a minute area of bone matter, where the subsequent biological damage may be severe.

■ Beta radiation penetrates deeper into tissue than alpha radiation. A 3-MeV negatron, for instance, travels through 0.6 in. (15 mm) of tissue before being absorbed. On external exposure, beta radiation passes through the epidermis into the dermis, a layer of skin tissue containing living cells, where it may cause biological damage.

■ Gamma radiation readily passes through tissue and ionizes the various substances it encounters. Internal and external exposure to gamma radiation causes more severe biological damage than exposure to either alpha or beta radiation. The adverse health effects noted in Table 16.4 illustrate that exposure to gamma radiation affects the entire organism.

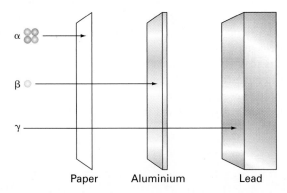

FIGURE 16.8 The relative penetrating power of alpha, beta, and gamma radiation. Alpha radiation is absorbed by a sheet of paper, beta radiation by aluminum foil, and gamma radiation by a thick block of lead.

TABLE 16.4 — Adverse Health Effects Associated with Exposure to Short-Term Doses of Gamma Rays

DOSE (rad)	SIGNS AND SYMPTOMS
0.3 weekly	No observable effect
60 (whole body)	Reduction of lymphocytes (white blood cells formed in lymphoid tissues such as the lymph nodes, spleen, thymus, and tonsils)
100 (whole body)	Nausea, vomiting, and fatigue
200 (whole body)	Reduction of all blood elements
400 (whole body)	50% of an exposed group will probably die
500 (testicles)	Sterilization
1000 (skin)	Erythema (reddening of the skin)

SOLVED EXERCISE 16.4

When a woman who may be pregnant must be subjected to an X-ray diagnostic examination, what special precautions should she exercise?

Solution: Fetal tissues are very sensitive to damage from radiation exposure. Consequently, the woman should inform her physician and X-ray technician of the likelihood of her pregnancy so her unborn child may be shielded to prevent radiation exposure.

16.6-B EXPOSURE TO SPECIFIC RADIOISOTOPES

Aside from the general effects caused by radiation exposure, specific radioisotopes often cause damage in the human organism in relatively localized ways. Of special interest are the radioisotopes of strontium and radium, each of which may replace calcium in the structure of the bones.

As first noted in Section 12.12-B, inside some bones are spaces occupied by *marrow*, a tissue containing the blood-forming cells. When radioisotopes of strontium or radium deposit within these spaces, they can irradiate these cells and cause the host victim to develop leukemia. This is especially problematic, since several radioisotopes of strontium and radium have relatively long half-lives; for example, the half-lives of strontium-90 and radium-226 are 28.1 y and 1590 y, respectively. Leukemia, or cancer of the blood, is very difficult to treat therapeutically.

Iodine radioisotopes may also cause localized damage within the body, specifically to the thyroid. The thyroid is a butterfly-shaped gland located near the base of the neck. It manufactures and secretes the hormones responsible for regulating the rates of body growth and metabolism. Two such hormones are thyroxine and triiodothyronine, whose molecular formulas are noted here:

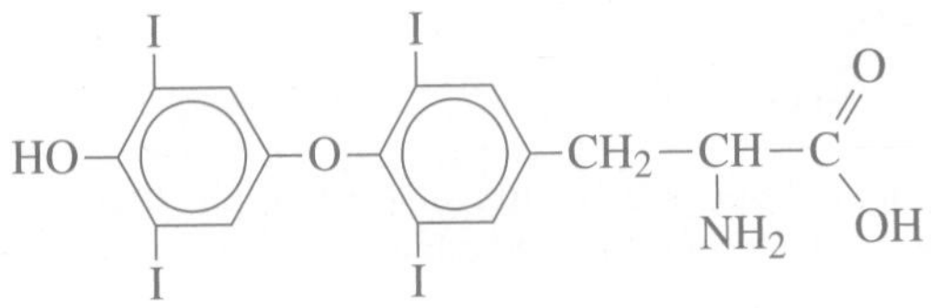

Thyroxine

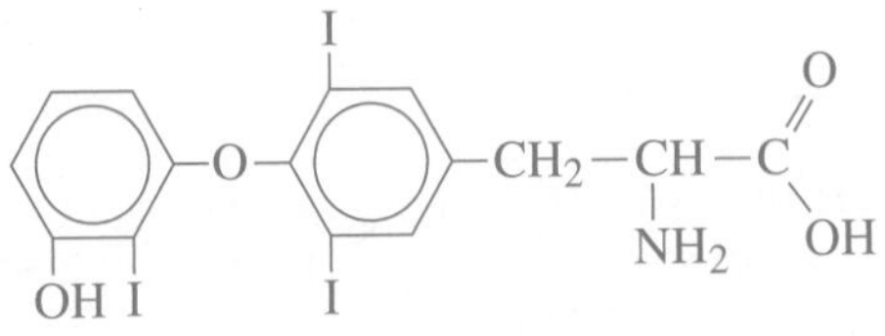

Triiodothyronine

The human body produces these hormones by reactions between tyrosine and iodine. Tyrosine is a naturally occurring amino acid found in meats, whereas iodine is acquired mainly from the iodized salt used in the diet.

Tyrosine

A supply of thyroxine and triiodothyronine remains stored in the thyroid for use as the body demands. When the hormones are flooded with iodine radioisotopes, they serve as an irradiation source.

SOLVED EXERCISE 16.5

In 2003, the American Academy of Pediatrics recommended that households, schools, and childcare centers located near nuclear power plants keep potassium iodide tablets on hand. What is the most logical reason the academy recommended this action?

Solution: Table 16.6 indicates that iodine-129 and iodine-131 are among the radioisotopes produced during fission. They are among the fission products released into the environment during a nuclear power plant accident. When exposed to the radioactivity emitted during the accident, the thyroid gland preferentially absorbs these iodine radioisotopes, since the gland biochemically manufactures iodine-bearing hormones (Section 16.6-B). During the subsequent decay of these radioisotopes, the energy that evolves disrupts normal cellular processes and triggers the onset of thyroid cancer.

The American Academy of Pediatrics most likely recommended the availability of potassium iodide tablets for immediate use in the event of a nuclear power plant accident.* Consuming an acceptable dose of these tablets saturates the thyroid hormones with nonradioactive iodine, thereby blocking absorption of the iodine radioisotopes. This practice saves children from the major impact of a radiation release and lessens the risk that they will contract thyroid cancer from the exposure.

Individuals exposed to the radioactivity emitted during a nuclear power accident should also *immediately* evacuate the area on learning of the incident. Increasing an individual's distance from the plant may be even more important than taking a dose of potassium iodide, since the latter reduces only the chance of developing thyroid cancer. Exposure to radiation causes not only thyroid cancer but leukemia and other forms of cancer on which potassium iodide has no impact.

*Although a prescription is unnecessary for the purchase of potassium iodide tablets, individuals should consult their doctor or pharmacist for the proper dosage for each family member.

16.6-C IMPACT OF AN INDIVIDUAL'S AGE

The chronic effect on the body's tissues depends on the number and nature of the cells whose biological functions have been altered. Because cell division occurs more frequently in children than in adults, exposure to radiation has an especially serious effect on youngsters.

PERFORMANCE GOALS FOR SECTION 16.7:

■ Identify the nature of the OSHA regulations that require employers to limit radiation exposure in the workplace.

16.7 Workplace Regulations Involving Radiation Exposure

OSHA regulates employee exposure to radiation in the workplace at 29 C.F.R. §1910.96 by requiring the following:

■ No employer shall possess, use, or transfer sources of ionizing radiation in such a manner as to cause any individual to receive during one calendar quarter a dose in excess of the limits quoted in Table 16.5 from sources in the employer's possession or control. These are called *permissible exposure limits* (Section 10.6-F).

■ An employer may permit an employee to receive a dose greater than those provided in Table 16.5 as long as:

a. The dose to the whole body during any calendar quarter does not exceed 3 rem, *and*

b. The dose to the whole body, when added to the accumulated occupational dose to the whole body, does not exceed $5 \times (N - 18)$ rem, where N is the individual's current age in years.

The expression *dose to the whole body* includes any dose to the entire body or a specific organ (gonads, active blood-forming organs, head and trunk, or lens of the eye).

■ No employer may permit any employee who is under 18 years of age to receive in any one calendar quarter a dose in excess of 10% of the limits in Table 16.5.

In the workplace, OSHA also requires employers to conspicuously post the applicable sign illustrated in Figure 16.9(a), (b), and (c), each of which warns individuals that a given area is any of the following:

■ A **radiation area** is an area that is accessible to personnel in which the ionizing radiation exists at a level in which the body could receive in any one hour a dose in excess of 0.05 mSv; or in any 5 consecutive days, a dose in excess of 1 mSv.

■ A **high radiation area** is an area that is accessible to personnel in which the radiation exists at such levels that a major portion of the body could receive in any one hour a dose in excess of 1 mSv.

■ An **airborne radioactivity area** is any room, enclosure, or operating area in which radioactive material exists at concentrations in excess of prescribed levels published in column 1 of Table 1 of Appendix B to 10 C.F.R. Part 20.

These signs bear a trefoil along with the word CAUTION and the applicable expression RADIATION AREA, HIGH RADIATION AREA, or AIRBORNE RADIOACTIVITY AREA.

OSHA also requires the conspicuous posting of a sign like that in Figure 16.9(d) within any area or room in which certain designated quantities of radioactive material are stored or used,

radiation area
■ For purposes of OSHA regulations, an area accessible to personnel in which the ionizing radiation exists at a level in which the body could receive a dose in excess of 0.05 mSv in any one hour, or a dose in excess of 1 mSv in any 5 consecutive days

high radiation area
■ For purposes of OSHA regulations, any area accessible to personnel in which the radiation exists at such levels that a major portion of the body could receive in any one hour a dose in excess of 1 mSv

airborne radioactivity area
■ For purposes of OSHA regulations, any room, enclosure, or operating area in which radioactive material exists at concentrations in excess of prescribed levels published in column 1 of Table 1 of Appendix B to 10 C.F.R. Part 20

TABLE 16.5	Permissible Exposure Limits in the Workplace[a]
rem/y	**PERMISSIBLE DOSE**
1.25	Whole body; head and trunk; active blood-forming organs; lens of the eyes; or gonads
18.75	Hands and forearms; feet and ankles
7.5	Skin of the whole body

[a]29 C.F.R. §1910.96(b).

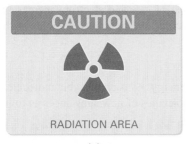

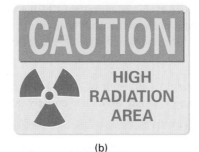

(a)

(b)

CAUTION

AIRBORNE RADIOACTIVITY
AREA

(c)

CAUTION

RADIOACTIVE MATERIALS

(d)

FIGURE 16.9 At 29 C.F.R. §§1910.1096(e)(2), 1910.1096(e)(3)(i), 1910.1096(e)(4)(ii), and 1910.1096(e)(5)(i), respectively, OSHA requires employers to conspicuously post one or more of these caution signs in areas where radioactive materials are stored or otherwise located. The conventional colors of the trefoil are magenta or purple on a yellow background. At 29 C.F.R. §1910.1096(e)(6), OSHA also requires employers to affix labels resembling the caution sign in (d) to storage containers holding certain quantities of radioactive material.

as well as on labels affixed to the containers used to store them. Although these signs and labels were intended to warn workers that radioactive materials are present at the locations where they are posted, they also serve to inform emergency responders of the presence of radioactive materials.

PERFORMANCE GOALS FOR SECTION 16.8:

■ Describe what occurs when ionizing radiation passes into or through matter.

16.8 Effects of Ionizing Radiation on Matter

When ionizing radiation passes through matter, its energy is dissipated by the ionization or excitation of the atoms or molecules of which the matter is composed. The initial action of the radiation is to ionize the substances through which it passes. The ionization is represented for a molecule of an arbitrary substance A by the following equation:

$$A \rightsquigarrow A^+ + e^-$$

The wiggly arrow indicates that the relevant reaction was induced by ionizing radiation, e^- symbolizes an electron, and A^+ is the symbol for a molecule–ion.

When such events occur repeatedly, the number of ions and electrons increases. Then, several secondary phenomena usually occur. In particular, the ions may combine with any of the electrons to form an excited state of A as follows:

$$A^+ + e^- \longrightarrow A^*$$

Here, A^* refers to a molecule of A that has excess energy. A^* is an unstable molecule. Possessing excessive energy, it may dissociate entirely into molecules of new substances as follows:

$$A^* \longrightarrow B + C$$

The ions that form when radiation initially strikes a substance may also react with neutral molecules. Such reactions produce new ions and free radicals. The direct exposure of a substance to radiation may also result in the production of free radicals. As noted before, free radicals are highly reactive chemical species that may trigger a variety of chemical reactions.

Let's consider specifically the phenomena associated with exposing water to ionizing radiation. Water constitutes three-fourths of all the body's tissues; hence, considerable research efforts have been devoted to examining the nature of the chemical species produced when water is exposed to radiation. When foreign substances are formed in this water, the cellular biochemistry can be significantly altered, thereby causing damage or death to the affected cells.

When water is exposed to ionizing radiation, primary ionization occurs first as follows:

$$H_2O(l) \rightsquigarrow (H_2O)^+ (l) + e^-$$

Then, the water molecule–ion reacts with a neutral water molecule to form a hydroxyl free radical.

$$(H_2O)^+ (l) + H_2O(l) \longrightarrow (H_3O)^+ (aq) + \cdot OH(aq)$$

The exposure of water to radiation can also produce hydrogen atoms and hydroxyl free radicals.

$$H_2O(l) \longrightarrow H \cdot (aq) + \cdot OH(aq)$$

When this latter event occurs repeatedly, the concentration of hydroxyl radicals increases. Ultimately two hydroxyl radicals unite to form hydrogen peroxide.

$$2 \cdot OH(aq) \longrightarrow H_2O_2(l)$$

The production of these highly reactive chemical species within living cells causes serious disruptions in the mechanistic steps associated with biochemical processes. When radiation damages the cellular genetic material, the consequences are even more far-reaching, since the damage leads to mutations, cancer, and cell death. Even subtle changes profoundly affect the human organism by modifying an individual's ability to adapt and survive within its environment.

PERFORMANCE GOALS FOR SECTION 16.9:

- Describe the phenomenon of nuclear fission, including spontaneous fission.
- Describe the mechanisms by which a nuclear reactor operates and a nuclear bomb detonates.
- Write equations denoting individual fission processes.
- Describe each step associated with the uranium-enrichment process.
- Identify the major reasons critics fear the use of radioisotopes for nuclear arms or nuclear power.

16.9 Nuclear Fission

Scientists have been able to synthetically produce numerous radioisotopes that are not found naturally. One method involves irradiating stable or long-lived isotopes in specially designed machines such as cyclotrons and bevatrons. Samples of such isotopes are generally irradiated with alpha particles, protons, deuterons, and other particles that have been accelerated to relatively high velocities. These high-velocity particles possess substantial energies. When they are used to bombard nuclei, nuclear reactions often occur. A *nuclear reaction* is a phenomenon in which one nucleus is artificially transformed into another. Many of the artificially produced radioisotopes used in medicine are prepared in this fashion.

SOLVED EXERCISE 16.6

Why do the workers at nuclear weapons facilities represent a group at a higher risk of contracting cancer compared with workers at nonnuclear facilities?

Solution: Workers at nuclear weapons facilities are exposed to radioactive materials more frequently than workers at nonnuclear facilities. These radioactive materials include the dozens of fission products produced when the nuclear fuel fissions, some examples of which are given in Table 16.6. Low concentrations of these fission products, especially the gaseous ones, leak through tiny openings in the reactor's structure into the surroundings. Because exposure to radioactive materials may induce various types of cancer, workers at nuclear weapons facilities are at a higher risk of contracting cancer compared with workers who are not occupationally exposed to radioactive materials. Among workers at nuclear weapons facilities there are documented instances of above-average levels of cancers in the lung, brain, bladder, stomach, respiratory tract, larynx, trachea, as well as the body's blood-forming tissues (bone marrow and the lymph system).

A second method of production involves utilizing the nuclear reactions that occur within a nuclear reactor. A nuclear reactor is the heart of a nuclear power plant. It functions because of the phenomenon called **nuclear fission**. As illustrated in Figure 16.10, fission is a process whereby a nucleus splits into two new nuclei, accompanied by the release of a relatively large amount of energy and several neutrons.

In the nuclear power industry, engineers harness this energy and convert it into electrical energy in the manner illustrated in Figure 16.11. The heat generated during fission is absorbed by

nuclear fission
■ The nuclear process during which certain nuclei are split into two lighter-mass nuclei with the simultaneous emission of one or more neutrons and a large amount of energy

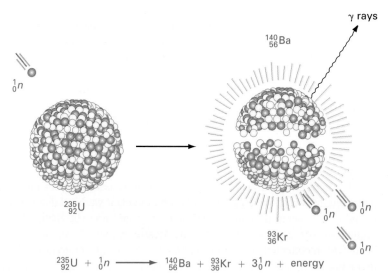

$$^{235}_{92}U + {}^{1}_{0}n \longrightarrow {}^{140}_{56}Ba + {}^{93}_{36}Kr + 3{}^{1}_{0}n + \text{energy}$$

FIGURE 16.10 The fission of a uranium-235 nucleus induced by a low-velocity neutron shown at the upper left. In this instance, barium-140, krypton-93, and three neutrons are produced.

FIGURE 16.11 In a nuclear power plant, the energy generated by fission is used to produce electricity. A nuclear power plant does not generate air pollutants like a fossil-fuel-fired power plant, but it does generate spent fuel and high-level radioactive waste, both of which require disposal.

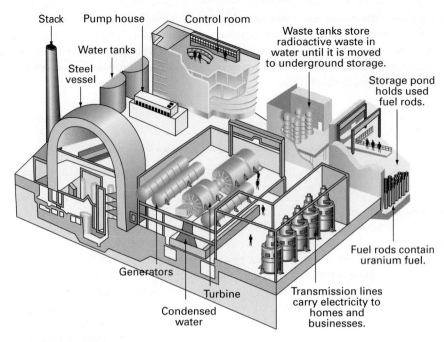

Stack Pump house Control room

Water tanks

Steel vessel

Waste tanks store radioactive waste in water until it is moved to underground storage.

Storage pond holds used fuel rods.

Fuel rods contain uranium fuel.

Generators

Turbine

Condensed water

Transmission lines carry electricity to homes and businesses.

Inside the steel vessel

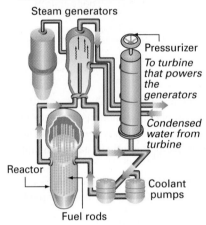

Steam generators

Pressurizer

To turbine that powers the generators

Condensed water from turbine

Reactor

Coolant pumps

Fuel rods

the water that is piped through the reactor. Next, the heated water is piped from the reactor to a steam generator, where the water vaporizes and is sent to a turbine. The steam spins the turbine blades, which generate electricity much like they do during the operation of a fossil-fuel-fired power plant.

The neutrons generated by fission may also be used to bombard other nuclei. This serves as a means of artificially producing radioisotopes that do not exist naturally. An example is cobalt-60, which may artificially be produced by radiating cobalt-59 with the neutrons produced during fission.

$$^{59}_{27}\text{Co} + ^{1}_{0}n \longrightarrow ^{60}_{27}\text{Co} + \gamma$$

Cobalt-59 is the only naturally occurring stable isotope of cobalt.

The neutrons generated by fission may also be used to *induce* nuclear fission. This is the phenomenon that occurs when a nuclear bomb or nuclear warhead detonates. Only certain nuclei may be induced by neutrons to undergo fission; they are said to be **fissile**. Uranium-233, uranium-235, plutonium-239, and plutonium-241 are **fissile nuclei**. Uranium-235 and plutonium-239 have been used in both nuclear reactors and nuclear warheads. Plutonium-239 is more desirable for use in nuclear bombs, since it is 1.7 times as likely as uranium-235 to undergo fission.

fissile

■ The ability of certain nuclei to undergo fission, during which neutrons and a relatively large amount of energy are released

fissile nuclei

■ Uranium-233, uranium-235, plutonium-239, and plutonium-241

Fissile nuclei are induced to undergo fission by exposing them to low-energy neutrons. For example, when uranium-235 captures a low-energy neutron, unstable uranium-236 forms. To achieve stability, each uranium-236 nucleus divides into two other nuclei, simultaneously releasing energy and several neutrons. One such event is represented as follows, where an asterisk is used to denote the unstable nucleus:

$$^{235}_{92}U + ^1_0n \longrightarrow [^{236}_{92}U]^* \longrightarrow ^{89}_{35}Br + ^{145}_{57}La + 2^1_0n + 192 \text{ MeV}$$

Dozens of individual fission events occur as the uranium-235 fuel undergoes fission within a reactor. The combination of these events produces numerous radioactive by-products, some of which are listed in Table 16.6. These radioisotopes are distributed among the elements from zinc to terbium; that is, their Z values range from 30 to 65, and their A values range from 72 to 161. Within a reactor, this mixture of fission products constitutes its waste; in a nuclear bomb, they form the components of atmospheric fallout. In both instances, they are highly radiotoxic.

TABLE 16.6	Some Radioactive Fission Products	
FISSION PRODUCT	**HALF-LIFE**	**MECHANISM OF DECAY**
Barium-140	12.8 d	β^-, γ
Cesium-137	33 y	β^-, γ
Cerium-141	32.5 d	β^-, γ
Cerium-144	590 d	β^-, γ
Iodine-129	1.7×10^7 y	β^-, γ
Iodine-131	8 d	β^-, γ
Krypton-85	4.4 hr	β^-, γ
Lanthanum-140	40 hr	β^-, γ
Niobium-95	90 hr	β^-, γ
Praseodymium-143	13.8 d	β^-, γ
Praseodymium-144	17 min	β^-
Promethium-147	2.26 y	β^-
Rhodium-106	30 s	β^-, γ
Ruthenium-103	39.8 d	β^-, γ
Ruthenium-106	1 y	β^-
Strontium-89	54 d	β^-
Strontium-90	27 y	β^-
Technetium-99*m*	5.9 hr	β^-
Tellurium-129	34 d	β^-, γ
Xenon-133	2.3 d	β^-, γ
Xenon-135	9.2 hr	β^-, γ
Zirconium-95	65 d	β^-, γ

Identify the isotope produced when the fission phenomenon represented by the following equation occurs:

$$^{235}_{92}U + ^1_0n \longrightarrow {}^{82}_{35}Br + \underline{\hspace{1.5cm}} + 2^1_0n$$

Solution: A nuclear equation is balanced when two conditions are fulfilled: the sums of the charges are the same on each side of the arrow, and the sums of the mass numbers are the same on each side of the arrow. On the left, the sum of the charges is $92 + 0 = 92$; on the right, the sum is $35 + x + 0 = 35 + x$, where x is the charge of the second isotope.

$$35 + x = 92$$
$$x = 57$$

Referring to either Figure 4.3 or the appendix at the back of this book, we see that the element having an atomic number of 57 is lanthanum.

On the left, the sum of the mass numbers is $235 + 1 = 236$; on the right, the sum is $82 + 2 + y = 84 + y$, where y is the mass number of the second isotope.

$$236 = 84 + y$$
$$y = 152$$

Hence, the symbol of the isotope produced by this fission event is $^{152}_{57}La$.

Only 0.71% by mass of naturally occurring uranium is uranium-235. Uranium-238 is the more abundant isotope, but it is not fissile. The neutron-induced fission of uranium-235 is a particularly unusual phenomenon, since for each single neutron that is consumed by reaction, additional neutrons are produced. These newly formed neutrons may be utilized to induce the fission of other uranium-235 nuclei. The process thereby initiates a self-sustaining **chain reaction** such as the one illustrated in Figure 16.12. In a nuclear reactor, the chain reaction is controlled by workers, and the energy is harnessed for the peaceful production of energy. By contrast, in the nuclear bomb or nuclear-armed missile, the chain reaction builds at an explosive rate, and the energy is released into the environment with devastating results.

16.9-A CRITICALITY

Can the nuclear reactor at a power plant detonate like a nuclear bomb or nuclear-armed missile? Answering this question requires understanding certain details about the fission phenomenon.

A certain minimum mass of fissile material must always be present for fission to occur; this is called the **critical mass**. When the mass of the fissile material is less than the critical mass, most of the neutrons formed during fission events escape from the mass and do not induce the fission of other nuclei. The perpetuation of a chain reaction requires that at least as many neutrons be retained by the system as were consumed in the fission reaction from which they were produced. In this way, the neutrons may react in further fission events.

chain reaction
■ A multistep reaction in which a reactive intermediate formed in one step reacts during a subsequent step to generate the species needed in the first step; a self-sustaining nuclear reaction that once started, steadily provides the energy and matter necessary to continue the reaction

critical mass
■ The minimum mass of fissionable material that must be present before low-energy, neutron-induced fission becomes self-sustaining

---

In 2006, researchers at the University of Utah showed that young children living in Utah and Nevada during the 1950s and 1960s downwind from aboveground weapons-testing sites were 7.5 times more likely to develop thyroid neoplasms, the precursors to thyroid cancer, compared with individuals in the general population. What is the most likely cause of this childhood cancer?

Solution: Table 16.6 shows that iodine-129 and iodine-131 are produced during fission. When humans are exposed to airborne releases of these radioisotopes, localized damage to the thyroid occurs. Thyroid cancer is a commonly observed abnormality associated with human exposure to airborne releases of iodine radioisotopes produced during the testing of nuclear weapons. The iodine radioisotopes are discharged into the air, from which they are subsequently removed by rainwater. The rain falls on pastures where grazing dairy cattle consume the grass and produce iodine-contaminated milk. Children who drink this milk are more likely to develop thyroid cancer than individuals who do not drink iodine-contaminated milk. Thus, those children who live in areas where iodine radioisotopes tend to accumulate are more likely to develop thyroid cancer than children who live in the general population.

To end World War II, the nuclear bombs shown in Figure 16.13 were detonated over the Japanese cities of Hiroshima and Nagasaki. They contained an amount of fissionable material less than their critical mass. This material was first compressed into a relatively small volume by the detonation of a mantle of explosive charges that caused the nuclear fuel to increase in density and achieve a supercritical mass. Then, chain reactions occurred that released a cataclysmic amount of energy into the environment.

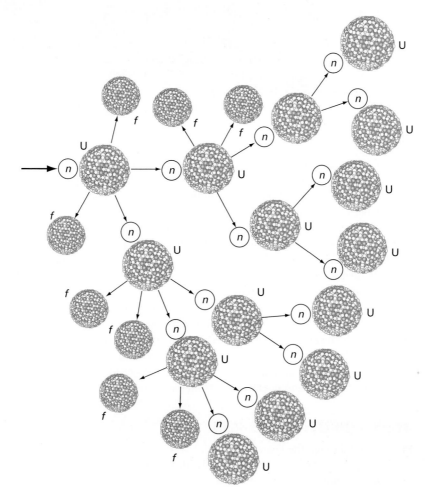

FIGURE 16.12 A nuclear chain reaction. The neutron (*n*) adjacent to the arrow is absorbed by a uranium-235 nucleus, which subsequently undergoes fission. The fission byproducts (*f*) and additional neutrons are thus produced. These neutrons induce the fission of other uranium-235 nuclei. The chain reaction is self-perpetuating until either the fuel is consumed or the neutrons are absorbed by matter other than the fuel.

FIGURE 16.13 Top: "Little Boy," the nuclear bomb detonated over Hiroshima, Japan, on August 6, 1945. This bomb contained fissile uranium-235, was 28 in. (71 cm) in diameter, 120 in. (3 m) long, weighed about 9000 lb (4100 kg), and killed more than 200,000 people. Bottom: "Fat Man," the nuclear bomb dropped on Nagasaki, Japan, on August 9, 1945. The bomb contained fissile plutonium-239, was 60 in. (150 cm) in diameter, 128 in. (3.2 m) long, weighed about 10,000 lb (4500 kg), and killed more than 74,000 people. (*Courtesy of the Los Alamos National Laboratory, Los Alamos, New Mexico.*)

By contrast, an amount of fissionable material just slightly larger than a critical mass is used as the fuel within a nuclear reactor. Far less fissionable material per unit volume is used in a reactor compared with the amount used in the nuclear bombs detonated over Japan. Furthermore, the number of neutrons available for future fission events are restricted in a nuclear reactor by means of *control rods*. These are long rods made from cadmium that are inserted into the fuel, or withdrawn therefrom, to effectively absorb extra neutrons and control their propagation. With the use of these control rods, there is little danger that a reactor will explode. Even if the chain reaction in a nuclear reactor were to become uncontrollable, the energy generated by the reaction would cause the uranium fuel to melt.

Notwithstanding its improbability, an emergency situation is likely to occur whenever individuals are exposed to fissionable material in an amount that is equal to or greater than its critical mass. Such incidents occur when safety practices are ignored. In 1999, for example, a critical mass of uranium-235 was unintentionally accumulated at a uranium-reprocessing facility in Tokaimura, Japan. The formation of this critical mass resulted in exposing three workers to neutrons, two of whom subsequently died from the exposure.

16.9-B PLUTONIUM-239

Prior to and through the years of the Cold War, vast amounts of the fissile plutonium-239 were produced as a deterrent to stabilize worldwide political power. The plutonium-239 was produced in nuclear reactors by using ordinary uranium (99.3% uranium-238 and 0.71% uranium-235) as the fuel. The initial nuclear reaction produces uranium-239, which decays in a stepwise process to neptunium-239 and plutonium-239.

$$^{238}_{92}U + ^{1}_{0}n \longrightarrow ^{239}_{92}U + \gamma$$
$$^{239}_{92}U \longrightarrow ^{239}_{93}Np + \beta^{-}$$
$$^{239}_{93}Np \longrightarrow ^{239}_{94}Pu + \beta^{-}$$

The plutonium-239 always contains some plutonium-240, which is produced when the plutonium-239 absorbs a neutron.

$$^{239}_{94}Pu + ^{1}_{0}n \longrightarrow ^{240}_{94}Pu + \gamma$$

Although the political scene is now markedly different since the days of the Cold War, an arsenal of 260 tn (236 t) of weapons-grade plutonium-239 remains scattered at spots throughout the world. This amount is sufficient to build 85,000 warheads. The U.S. arsenal of at least 90 tn (82 t) is stored within sealed bunkers at the Pantex DOE facility near Armadillo, Texas.

There are at least three reasons why storage of the world's supply of plutonium-239 may be troublesome:

■ Despite all precautions, there is always the fear that plutonium may be acquired by terrorists.
■ The half-life of plutonium-239 is 24,110 y. It is evident that, once released from its confinement, plutonium-239 will be present for a very long time.
■ Plutonium-239 is one of the world's most toxic substances. If released to the environment in virtually any amount, it would cause untold health damage to countless individuals.

16.9-C URANIUM-ENRICHMENT PROCESSES

Fissile uranium-235 is the fuel that serves as the source of the power generated at most nuclear power plants, but to effectively serve in this capacity, uranium must be enriched in uranium-235 to approximately 5% by mass. For use in a nuclear bomb, uranium-235 must be enriched to more than 80% by mass. Although the basic nature of the **uranium-enrichment process** is not secret, the engineering details are highly classified.

uranium-enrichment processes
■ Isolation techniques for concentrating fissile U-235

The uranium-enrichment process involves the production of uranium(VI) fluoride, frequently called *uranium hexafluoride*. Although it does not react appreciably with oxygen, nitrogen, or carbon dioxide, uranium hexafluoride reacts vigorously with water to produce uranyl fluoride and hydrogen fluoride.

$$UF_6(g) \quad + \quad 2H_2O(g) \quad \longrightarrow \quad UO_2F_2(s) \quad + \quad 4HF(g)$$

Uranium(VI) fluoride   Water   Uranyl fluoride   Hydrogen fluoride

Because the inhalation of hydrogen fluoride is highly dangerous to health (Section 8.9-C), uranium hexafluoride is considered hazardous primarily because of its corrosivity, not its radiological nature. As first noted in Section 6.7-B, DOT acknowledges this property of uranium hexafluoride by requiring shippers and carriers to post RADIOACTIVE *and* CORROSIVE placards on the transport vehicle or freight container used to transport 1001 lb (454 kg) or more of this substance.

The production of uranium hexafluoride is a multistep process that begins with uranium ore containing triuranium octoxide. Each step in this production process is briefly summarized as follows:

■ The triuranium octoxide is mixed with nitric acid, resulting in the production of *uranyl nitrate*.

$$U_3O_8(s) \quad + \quad 8HNO_3(aq) \quad \longrightarrow \quad 3UO_2(NO_3)_2(aq) \quad + \quad 2NO_2(g) \quad + \quad 4H_2O(l)$$

Triuranium octoxide   Nitric acid   Uranyl nitrate   Nitrogen dioxide   Water

■ Uranyl nitrate is thermally decomposed to produce uranium(VI) oxide, commonly called **orange oxide**.

$$2UO_2(NO_3)_2(aq) \quad \longrightarrow \quad 2UO_3(s) \quad + \quad 4NO_2(g) \quad + \quad O_2(g)$$

Uranyl nitrate   Uranium(VI) oxide   Nitrogen dioxide   Oxygen

orange oxide
■ Uranium(VI) oxide (UO_3), an intermediate in the production of uranium hexafluoride

■ Uranium(VI) oxide is reduced with hydrogen, which produces uranium(IV) oxide, called **yellowcake**.

yellowcake
■ Uranium(IV) oxide (UO_2), an intermediate in the production of uranium hexafluoride

$$UO_3(s) \quad + \quad H_2(g) \quad \longrightarrow \quad UO_2(s) \quad + \quad H_2O(g)$$
Uranium(VI) oxide $\qquad$ Hydrogen $\qquad\qquad$ Uranium(IV) oxide $\qquad$ Water

■ Uranium(IV) oxide is reacted with anhydrous hydrogen fluoride at 932 °F (500 °C) forming uranium(IV) fluoride, called **green salt**.

green salt
■ Uranium(IV) fluoride (UF_4), an intermediate in the preparation of uranium hexafluoride

$$UO_2(s) \quad + \quad 4HF(g) \quad \longrightarrow \quad UF_4(s) \quad + \quad 2H_2O(g)$$
Uranium(IV) oxide $\qquad$ Hydrogen fluoride $\qquad\qquad$ Uranium(IV) fluoride $\qquad$ Water

■ Uranium(IV) fluoride is further fluorinated at 482 °F (250 °C) to produce uranium hexafluoride, which is a gas above its sublimation temperature of 147 °F (64 °C).

$$UF_4(s) \quad + \quad F_2(g) \quad \longrightarrow \quad UF_6(g)$$
Uranium(IV) fluoride $\qquad$ Fluorine $\qquad$ Uranium(VI) fluoride

As it is produced, the uranium hexafluoride is an isotopic mixture of $^{235}UF_6$ and $^{238}UF_6$. Although these isotopic forms may be separated by several means, the least expensive procedure is a gas-centrifugation process that takes into account the mass difference of the two components. In essence, the mixture of $^{235}UF_6$ and $^{238}UF_6$ is spun at very high speeds, thereby creating a centrifugal force. As the centrifuge operates, the heavier $^{238}UF_6$ concentrates outward. The periodic removal of $^{238}UF_6$ leaves behind the fissile $^{235}UF_6$. It is called **enriched uranium**.

enriched uranium
■ Uranium and its compounds in which the concentration of uranium-238 has been reduced or eliminated by processing

When a process is conducted to reduce or eliminate the concentration of uranium-235 in uranium metal or a uranium compound, the substance that remains is called **depleted uranium**. Depleted uranium hexafluoride, or DUF_6, is a waste product generated during the uranium-enrichment process. In the United States, approximately 825,000 tn (750,000 t) of DUF_6 is now stored in steel cylinders at three locations: Paducah, Ohio; Portsmouth, Ohio; and East Tennessee Technology Park, Tennessee. The majority of this material was generated during past decades in connection with the U.S. military defense program. Its management is a responsibility of DOE.

depleted uranium
■ Uranium and its compounds in which the concentration of uranium-235 has been reduced or eliminated by processing

Peaceful nations are severely limited in their use of fissile uranium for military purposes. Nonetheless, in contemporary times, they are obliged to implement appropriate diplomatic steps aimed at preventing rogue nations from acquiring nuclear fuel for military purposes. The fear is that rogue nations operate nuclear power plants as a guise for energy production when their actual intention is to maximize the production of fissile material for use in the construction of nuclear-armed ballistic missiles. The diplomatic efforts of peaceful nations are directed at preventing them from conducting uranium-enrichment processes or acquiring fissile material from other nations for military purposes.

Clandestine operations associated with the production of uranium hexafluoride may often be identified by inspecting the sites where the production is suspected. Such inspections are conducted by representatives of the IAEA, Vienna, Austria, an intergovernmental organization that promotes the development of peaceful uses of nuclear technology and discourages military applications.

SOLVED EXERCISE 16.9

Use Table 16.10 to determine whether DOT regulates the domestic transportation of a cylindrical glass capsule having a length of 3.0 in. (7.6 cm) and a radius of 0.2 in. (5 mm) into which 20 Ci of potassium-40 has been sealed.

Solution: DOT regulates the transportation of any radioisotopic material where both the activity concentration and the total activity in the consignment exceed the radioisotope's A_1 and A_2 values. By referring to Table 16.10, we see that 20 Ci is less than the A_1 and A_2 values published for potassium-40. Consequently, this shipment of 20 Ci of potassium-40 is exempt from the DOT transportation regulations.

16.9-D SPONTANEOUS FISSION

As first noted in Section 16.2, spontaneous fission occurs only when certain very heavy radioisotopes decay. These radioisotopes have an atomic number greater than 92. As their nuclei undergo spontaneous fission, they split into two nuclei and simultaneously generate neutrons.

An example of a radioisotope that decays by spontaneous fission is curium-250. When it decays, curium-250 spontaneously fissions into two other nuclei. One such event is denoted as follows:

$$^{250}_{96}\text{Cm} \longrightarrow {}^{137}_{52}\text{Te} + {}^{112}_{44}\text{Ru} + {}^{1}_{0}n$$

16.9-E THE GLOBAL FEAR OF NUCLEAR BOMBS

Throughout the history of civilization, the use of nuclear weapons was necessitated only twice—when the United States detonated them over Japan and ended World War II. Thereafter, especially through the years of the Cold War, it was the *threat* of nuclear weapon deployment that facilitated the balance of worldwide political power. There was not a target on Earth that the United States and the former Soviet Union could not hit with long-range nuclear-armed ballistic missiles.

Even today, strategically stockpiled throughout the world are massive numbers of nuclear warheads. In January 2007, the United States, Russia, France, the United Kingdom, and China held more than 26,000 nuclear warheads, of which 11,530 were prepared for activation by missile or aircraft. Despite a 2002 accord with Russia that commits the United States to reducing the number of such warheads to between 1700 and 2200 by 2012, new missiles and warheads with smaller yields are continually being developed. The omnipresent fear is that their flagrant use would decimate individual civilizations and alter life on the entire planet.

Nine nations now possess nuclear weapons: the United States, Russia, the United Kingdom, France, China, Israel, India, Pakistan, and North Korea. Iran is probably close to possessing a nuclear weapon. To encourage international nuclear disarmament and reduce the further proliferation of nuclear weapons, 71 nations agreed in 1996 to the conditions of the Comprehensive Nuclear-Test-Ban Treaty (CTBT). One condition of this treaty prohibits the testing of a nuclear device of any size within any environment. As of November 2008, the CTBT had been signed by 180 states and ratified by 148.

Inspectors for the IAEA periodically monitor nuclear facilities worldwide to ensure that nuclear material has not been diverted to military uses. Notwithstanding the intent of the agency, the testing of nuclear weapons continues. In 1995 and 1996, for instance, France conducted six nuclear tests in the Pacific Ocean, and in 1998, India and Pakistan separately conducted five tests involving nuclear weapons. In 2006, 2007, and 2009, North Korea demonstrated that it could detonate a nuclear bomb. Even today, the world fears that Iran has the capability of developing nuclear weapons. The combination of these incidents signals all too clearly that the threat of mass killing and destruction from the deployment of nuclear weapons may again be realized.

The energy released by the detonation of a nuclear bomb or the engagement of a nuclear warhead far surpasses the energy associated with the detonation of ordinary chemical explosives. The explosion of a nuclear bomb, for instance, releases more energy than that involved in the detonation of approximately 1 million tn ($\sim$909,000 t) of TNT. The release of this amount of energy is globally fearsome for the following combination of reasons:

- This amount of energy could instantly kill tens of thousands of people.
- The survivors of the initial blast would remain at risk for acquiring long-term health-related problems like cancer from exposure to the bomb's residues.
- Climatic changes would occur. Computer modeling illustrates that the detonation of 100 Hiroshima-sized bombs, for example, would dim the sunlight reaching Earth for 10 or more years, reduce rainfall, and cool the globe to temperatures resembling those experienced

during the Little Ice Age (Section 10.11-A). The reduced temperatures would also cause crop failures, malnutrition, and famine throughout most of the world and cause the onset of diseases.

16.9-F NUCLEAR POWER PLANTS

Radioactive materials are not exclusively used in nuclear weapons. They are also used for many peaceful purposes. Most notably, certain radioactive materials are used by the nuclear power industry to generate electricity. Four hundred thirty-five nuclear power plants produce approximately 17% of the electrical energy used worldwide. One hundred four, roughly one-fourth of the world's plants, operate within the United States. Most are approaching their projected operating lives of 40 years. Thirty new plants are contemplated for construction during the next decade, with the first two scheduled to come online in 2014 and 2015.

The operation of a nuclear power plant provides certain environmental benefits compared with the operation of a fossil-fuel-fired power plant. In particular, a nuclear power plant does not emit the pollutants blamed for climate change. It also generates far less waste than a fossil-fuel-fired power plant. The 104 reactors in the United States generate roughly 2000 tn (1818 t) of spent nuclear fuel annually, whereas the combustion of coal generates approximately 100 million tn (90.9×10^6 t) of waste.

DOE subjects a nuclear power plant to especially rigid regulatory controls and licensing requirements. Nonetheless, a nuclear power plant could release radioactive materials from its confinement vessels into the environment. A component of the fear associated with the operation of a nuclear power plant is that its fuel or radioactive by-products could be inadvertently released into the environment through mismanagement or the failure to follow established safety practices. There is also the potential risk that nuclear power plants could be sabotaged by terrorists.

The first of these latter scenarios is not merely a hypothetical possibility. There have been well-documented instances of airborne releases of radioactive materials from nuclear power plants in the United States, the countries of the former Soviet Union, and elsewhere.

The worst disaster of this type occurred in April 1986 at the Chernobyl nuclear power plant in Russia. A series of operator errors unleashed a power surge that triggered an explosion and partial meltdown of the fuel. Radioactive fuel and by-products were discharged into the environment, the majority of which spread over Russia, Ukraine, Belarus, and parts of western Europe. This single incident turned bustling villages and towns into unpopulated areas and directly threatened the health of more than 5 million people. The deaths of 56 people, mostly rescue workers, were linked with exposure to the radiation released during the accident.

In addition, soon after the explosion, 4000 people developed thyroid cancer from exposure to airborne radiation containing iodine radioisotopes. Many survived owing to prompt medical treatment. Even years after the occurrence, thyroid cancer was identified as the most commonly identified ill effect associated with exposure to the airborne radiation from the Chernobyl reactor. In mid-2005, between 4000 and 5000 people in Ukraine and Belarus, all of whom were children at the time of the explosion, were diagnosed with the disease.

A nuclear-power-related accident was also experienced in the United States at the Three Mile Island plant in southeastern Pennsylvania in March 1979. The coolant used to modulate the generated heat was inadvertently discharged into the environment. Operators quickly shut down the plant, and little radioactivity was released. Although no one died as an immediate result of the accident, the cancer rates in the region surrounding the plant remain points of fierce debate even today.

In the United States, the growth of the defense and nuclear power industries has spawned the fear of contracting cancer from exposure to their by-product wastes. How can they safely be disposed of? In the United States, the majority of these wastes continue to be stored at the plants where they were initially generated, especially Rocky Flats, Colorado, and Hanford, Washington, where the nation's nuclear weapons were formerly produced. At the Hanford plant alone, 440 billion gal (1.7×10^9 m^3) of liquid waste was generated and intentionally dumped into the soil. An additional 53 million gal (2.0×10^5 m^3) of waste is still stored within 177 underground tanks. DOE plans to ultimately seal the radioactive components of these wastes within glasslike logs and dispose of them in an underground nuclear waste repository. This processing is scheduled to begin in 2019 and will take 20 to 25 years to complete.

Certain nuclear wastes are now stored at temporary storage sites in 35 states. These nuclear wastes were generated not only within the United States but outside the country as well. The latter wastes are stored to keep nuclear fuel from reaching international terrorist organizations or rogue nations.

Acknowledging the magnitude of the nuclear waste disposal problem, Congress directed the DOE to locate and construct a national site in which high-level nuclear wastes can permanently be stored. An essential element of the site is an acceptable means for protecting the next 25,000 generations of Americans from exposure to the nuclear waste generated contemporaneously.

The site selected by DOE and approved by Congress as a geologic repository for high-level nuclear wastes is a volcanic-rock ridge located within Yucca Mountain, approximately 100 mi (161 km) northwest of Las Vegas, Nevada. A tunnel within the mountain in which nuclear wastes will be entombed is shown in Figure 16.14. The wastes from nuclear power plants and defense-related activities were initially scheduled to arrive at the repository for permanent entombment by 2010. However, the plans for operating the repository have long been plagued by technical and legal problems that have resulted in pushing the date to 2017. Furthermore, DOE's Office of Civilian Radioactive Waste now projects that the Yucca Mountain repository will hold at least 120,130 tn (109,200 t) of nuclear waste and will remain accessible until 2133.

FIGURE 16.14
Although the U.S. Department of Energy selected Yucca Mountain, Nevada, as a permanent repository for high-level nuclear waste, the political debate surrounding the construction and use of the repository has been so heated that alternative methods of treating and disposing of the waste are now under consideration. (*Courtesy of U.S. Department of Energy, Yucca Mountain Project, Las Vegas, Nevada.*)

PERFORMANCE GOALS FOR SECTION 16.10:

- Identify the unique DOT requirements that pertain to the preparation of the shipping description of a radioactive material.
- Identify the unique labeling, marking, and placarding requirements that DOT imposes on shippers and carriers who transport radioactive materials.

16.10 Transporting Radioactive Materials

A stringent set of regulations apply to the transportation of a radioactive material. The portion of these regulations most immediately applicable to emergency responders is summarized in the subsections that follow.

Some radioisotopes are transported as "hazardous substances" within the meaning of CERLCA (Section 6.2-B), examples of which are provided in Table 16.7 along with their **reportable quantities**. There are six reportable quantities for radioisotopes: 0.00037 TBq; 0.0037 TBq; 0.037 TBq; 0.37 TBq; 3.7 TBq; and 37 TBq.

reportable quantity
- The amount of a radioisotope listed in Appendix B at 40 C.F.R. §302.4 and 49 C.F.R. §172.101, the release of which triggers mandatory notification to the National Response Center (abbreviated RQ)

16.10-A SHIPPING DESCRIPTIONS

When shippers offer a radioactive material for transportation, DOT requires them to include the following information in its shipping description:

- Proper shipping name, hazard class 7, and identification number. The proper shipping names of the hazardous materials in class 7 always begin with the words "Radioactive material." Some examples are listed in Table 6.1.
- The letters RQ when an activity equal to or greater than the reportable quantity is offered for transportation.

TABLE 16.7	Reportable Quantities of Some Selected Radioisotopic Hazardous Substances[a]		
HAZARDOUS SUBSTANCE	**RQ [TBq]**	**HAZARDOUS SUBSTANCE**	**RQ [TBq]**
Arsenic-77	37	Plutonium-239	0.0037
Copper-64	37	Potassium-40	0.037
Iodine-131	0.00037	Promethium-147	0.37
Iridium-137	3.7	Radium-228	0.0037
Lead-212	0.37	Radon-222	0.0037
Mercury-195	3.7	Strontium-90	0.0037
Molybdenum-99	3.7	Uranium-233	0.0037
Phosphorus-32	0.0037	Uranium-235	0.0037
Platinum-189	3.7	Zinc-71	3.7

[a]The reportable quantities of radioisotopes that are hazardous substances are provided in Table 2 following 49 C.F.R. §172.101.

- The name of the radioisotope.
- A description of the physical and chemical form of the material.
- The activity of the radioisotope expressed in becquerels or its multiples or fractions. The activity is determined by using appropriate radiation-detection equipment. The activity may also be expressed in curies or other units and entered parenthetically immediately next to the measurement in becquerels.
- Identification of the labels that DOT requires to be affixed to the relevant packaging.
- The transport index, when DOT requires either RADIOACTIVE YELLOW-II or RADIOACTIVE YELLOW-III labels to be affixed to the packaging. The **transport index** is a dimensionless number used to designate the degree of control to be exercised during transportation of a nonfissile radioactive material. It is determined by multiplying the maximum radiation level in mSv/hr at 3.3 ft (1 m) from the external surface of the package by 100.
- The criticality safety index (Section 16.10-F), when DOT requires FISSILE labels to be affixed to the packaging.
- The expression "highway route–controlled quantity," or HRCQ, when shippers offer for transportation a "highway route–controlled quantity" (Section 16.10-G) of a class 7 hazardous material [49 C.F.R. §172.203(d)(10)].

transport index
■ For purposes of DOT regulations, a dimensionless number that shippers and carriers assign to certain radioactive materials to designate the degree of control to be exercised during their transit

16.10-B LABELING REQUIREMENTS

When shippers offer packages of radioactive materials for transportation, DOT requires them to affix on their opposite sides other than the bottom the applicable RADIOACTIVE WHITE-I, RADIOACTIVE YELLOW-II, RADIOACTIVE YELLOW-III, or FISSILE labels. RADIOACTIVE YELLOW-III labels are always affixed on the opposite sides of packages containing a "highway route–controlled quantity" of a class 7 hazardous material, and FISSILE labels are affixed adjacent to the applicable RADIOACTIVE YELLOW-II or RADIOACTIVE YELLOW-III labels. The features of these labels were previously illustrated in Figure 6.5.

DOT requires shippers to choose the applicable RADIOACTIVE label by measuring the radiation level at any point on the external surface of the packaging and determining the transport index. The applicable label is then chosen from Table 16.8. DOT also requires shippers to inscribe the following information within the contents-entry space:

- The name of the radioisotope to be transported and its activity on the applicable RADIOACTIVE WHITE-I, RADIOACTIVE YELLOW-II, or RADIOACTIVE YELLOW-III label
- LSA-I material, or LSA-I (Section 16.10-C), on the applicable RADIOACTIVE WHITE-I, RADIOACTIVE YELLOW-II, or RADIOACTIVE YELLOW-III label, when shippers offer LSA-I material for transportation
- The transport index on RADIOACTIVE-II and RADIOACTIVE-III labels
- The criticality safety index on the FISSILE label when shippers offer subcritical masses of uranium-233, uranium-235, plutonium-239 or plutonium-241 for transportation

When shippers intend to offer packages that have been emptied of their class 7 contents for transportation, DOT may require them to affix EMPTY labels on opposite sides of the packages. The EMPTY label is a white square with black lettering.

EMPTY

DOT requires EMPTY labels to be affixed to the packages when the radiation levels measured outside them reveal the presence of internal contamination following the removal of their radioactive contents.

TABLE 16.8 | Categories of the Class 7 Labels[a]

TRANSPORT INDEX[b]	MAXIMUM RADIATION LEVEL AT ANY POINT ON A PACKAGE'S EXTERNAL SURFACE	LABEL CATEGORY
0	Less than or equal to 0.005 mSv/hr (0.5 mrem/hr)	RADIOACTIVE WHITE-I
More than 0 but not more than 1	Greater than 0.005 mSv/hr (0.5 mrem/hr) but less than or equal to 0.5 mSv/hr (50 mrem/hr)	RADIOACTIVE YELLOW-II
More than 1 but not more than 10	Greater than 0.5 mSv/hr (50 mrem/hr) but less than or equal to 2 mSv/hr (200 mrem/hr)	RADIOACTIVE YELLOW-III
More than 10	Greater than 2 mSv/hr (200 mrem/hr) but less than or equal to 10 mSv/hr (1000 mrem/hr)[c]	RADIOACTIVE YELLOW-III

[a]49 C.F.R. §172.403.

[b]When the transport index is calculated as a number less than 0.05, the value is considered zero.

[c]DOT subjects shippers to the exclusive-use provisions set forth at 49 C.F.R. §173.441 when transporting radioactive material having this maximum radiation level.

16.10-C MARKING REQUIREMENTS

At 49 C.F.R. §§173.471 - 173.473, DOT requires shippers to mark each package containing a class 7 hazardous material with the following information:

- The name and address of the shipper and receiver of the package
- The proper shipping name and UN identification number of the radioactive material
- RQ, if the package contains an amount equal to or greater than the reportable quantity of the radioactive material
- The gross mass when the package weighs over 110 lb (50 kg)
- Orientation arrows, if the contents are liquid
- "Radioactive—LSA" or "Radioactive—SCO," if applicable. **Low-specific-activity material** (LSA) material, refers to radioactive material that complies with the DOT descriptions and activity limitations provided in Table 16.9. **Surface-contaminated object** (SCO) refers to a solid object that is not itself radioactive but has radioactive material distributed on its surface.
- The package types TYPE A or TYPE B, as appropriate (Section 16.10-E)
- The trefoil when the class 7 hazardous material is contained within a type B package
- USA, if destined for export
- For industrial packages (Section 16.10-E), TYPE IP-1, TYPE IP-2, or TYPE IP-3, as applicable, with the international vehicle registration code of the country of origin of the design
- For type B packages (Section 16.10-E), TYPE B(U) or TYPE B(M), when applicable, and if destined for export, USA in conjunction with the specification marking or other package-certificate identification

low-specific-activity material
■ For purposes of DOT regulations, certain radioactive material, including uranium and thorium ores and tritium-labeled water, that complies with the descriptions and activity limitations published at 49 C.F.R. §173.403 (abbreviated LSA)

surface-contaminated object
■ For purposes of DOT regulations, a solid object that is not itself radioactive but has radioactive material distributed on its surface (abbreviated SCO)

16.10-D PLACARDING REQUIREMENTS

DOT requires carriers to post the RADIOACTIVE placard shown in Figure 6.12 on each side and each end of a transport vehicle or freight container used to ship the following by highway or rail:

TABLE 16.9	Low-Specific-Activity Materials[a]
GROUP	**FEATURES**
LSA-I	Uranium and thorium ores, concentrates of uranium and thorium ores, and other ores containing naturally occurring radionuclides that are intended to be processed for the use of the radionuclides, *or* Solid unirradiated natural uranium or thorium or depleted uranium or natural thorium or their solid or liquid compounds or mixtures, *or* Radioactive material other than fissile material for which the A_2 value is unlimited, *or* Other radioactive material, excluding fissile material in quantities not excepted under §173.453, in which the radioactive material is essentially uniformly distributed and the average specific activity does not exceed 30 times the values for activity concentration specified in §173.436, or 30 times the default values listed in Table 8 of §173.433.
LSA-II	Water with a tritium concentration of up to 0.8 TBq/L (20.0 Ci/L), *or* Other radioactive material in which the activity is distributed throughout and the average specific activity does not exceed $10^{-4}A_2$/g for solids and gases and $10^{-5}A_2$/g for liquids.
LSA-III	Solids (e.g., consolidated wastes, activated materials), excluding certain powders in which: The radioactive material is distributed throughout a solid or a collection of solid objects, or is essentially uniformly distributed in a solid compact binding agent such as concrete, bitumen, or ceramic; The radioactive material is relatively insoluble, or it is intrinsically contained in a relatively insoluble material, so that, even under loss of packaging, the loss of radioactive material per package by leaching when placed in water for seven days would not exceed 0.1 A_2; and The estimated average specific activity of the solid, excluding any shielding material does not exceed $2 \times 10^{-3}A_2$/g.

[a]49 C.F.R. §173.403.

- A package to which DOT requires RADIOACTIVE YELLOW-III labels to be affixed
- "Exclusive-use" LSA materials or SCOs within "excepted packages" (Section 16.10-E)
- A "highway route–controlled quantity" of a class 7 hazardous material

16.10-E PACKAGING TYPES

When shippers offer a radioactive material for transportation, DOT requires them to containerize it within one of the following types of packaging: excepted package, industrial package (IP), type A package, or type B package. These packages have been designed to protect workers, the public, and the environment from the inherent risks associated with transporting radioactive materials. Each is described next:

- An **excepted package** refers to a type of packaging that DOT authorizes when LSA materials and SCOs are transported by exclusive-use means. *Exclusive-use* refers to the sole use by a single consignor of a conveyance for which all initial, intermediate, and final loading and unloading are carried out in accordance with the consignor's or consignee's written instructions. An example of an excepted package is a cardboard box that complies with certain specified test

excepted package
- For purposes of DOT regulations, a package type in which LSA materials and SCOs are transported by exclusive-use means

conditions. It is "excepted" from the routine labeling and marking requirements but is subject to other requirements published at 49 C.F.R. §§173.421–173.426. Shippers must also comply with the conveyance activity limits published at 49 C.F.R. §173.427(e).

■ An *industrial package* is another type of authorized packaging in which DOT permits the shipment of certain quantities of radioactive material, instruments, or articles manufactured from natural or depleted uranium or natural thorium. The DOT regulations require shippers to use one of three types designated as IP-1, IP-2, and IP-3. Each type must be designed and fabricated to retain the integrity of containment and shielding when it is subjected to normal conditions of transport. In addition, IP-2 and IP-3 satisfy certain prescribed DOT tests published at 49 C.F.R. §§178.603 and 178.606 relating to their ability to remain intact when dropped and stacked, respectively.

■ A **type A package** has been designed to retain the integrity of containment and shielding when subjected to normal conditions of transport. Four examples of a type A package are shown in Figure 16.15. They are a fiberboard box, wooden box, steel drum, and lead canister.

■ A **type B package** has been designed to retain the integrity of containment and shielding when subjected to normal conditions of transport and to retain the integrity of containment and shielding when subjected to prescribed hypothetical accident–test conditions. Three examples are shown in Figure 16.16: a steel outer drum with an inner container, a concrete and steel cask, and a massive transport container shielded with lead.

When shippers offer a radioactive material for transportation, DOT requires them to select a type A or type B package based in part on the magnitude of two values, called A_1 and A_2.

■ A_1 is the maximum activity of a special-form class 7 material permitted in a type A package. **Special-form class 7 material** refers to an indispersible solid radioactive material that is either a single solid piece or a sealed capsule, at least one dimension of which is not less than 0.2 in. (5 mm), and in the case of the capsule, contains radioactive material that may be opened only by destroying it. A radioactive material that is not a special-form class 7 material is a **normal-form class 7 material**, or **non-special-form class 7 material**.

Package must withstand normal conditions of transport only without loss or dispersal of the radioactive control contents.

DOT Specification
fiberboard box

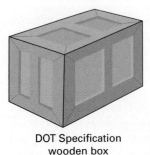

DOT Specification
wooden box

DOT Specification
steel drum

DOT Specification
type A package

FIGURE 16.15 These typical forms of type A packaging have been shown to withstand certain DOT test procedures that were designed to maintain the integrity of containment and shielding under the normal conditions of transport.

Package must stand both normal and accident test conditions without loss of contents.

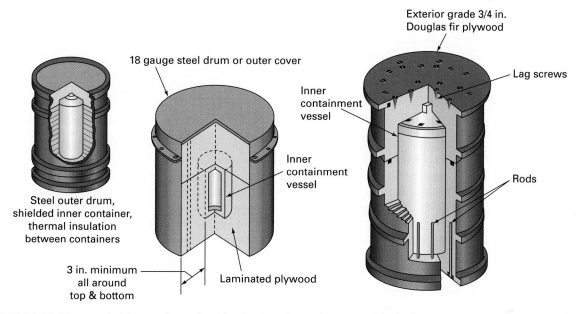

FIGURE 16.16 These typical forms of type B packaging have been shown to withstand certain DOT test procedures that were designed to maintain the integrity of containment and shielding under normal conditions of transport and the damaging conditions of a transportation accident.

■ A_2 is the maximum activity of a class 7 material, *other than* a special-form class 7 material, LSA, or SCO, that is permitted for transportation within a type A package.

DOT regulates the transportation of any radioactive material when both the activity concentration and the total activity in the consignment exceed the applicable A_1 and A_2 values. DOT publishes the A_1 and A_2 values at 49 C.F.R. §173.435, some excerpts of which are provided for several selected radioisotopes in Table 16.10.

normal-form class 7 material (non-special-form class 7 material)
■ For purposes of DOT regulations, radioactive material that does not satisfy the definition of a special-form class 7 material

| TABLE 16.10 | \multicolumn{6}{l}{A_1 and A_2 Values for Some Selected Radioisotopes[a]} |

Radioisotope	A_1		A_2		Specific Activity	
	TBq	Ci	TBq	Ci	TBq/g	Ci/g
Bismuth-207	0.7	19	0.7	19	1.9	52
Cobalt-60	0.4	11	0.4	11	42	1100
Cesium-137	2	54	0.6	16	3.2	87
Molybdenum-99	1.0	2.7	0.74[b]	20[b]	1.8×10^4	4.8×10^5
Potassium-40	0.9	24	0.9	24	2.4×10^{-7}	6.4×10^{-6}

[a]Excerpted from 49 C.F.R. §173.435 and 10 C.F.R. Part 71, Appendix A, Table A-1.
[b]Domestic use only.

A shipper offers for domestic transportation a type A package weighing 65 lb (29.5 kg) and containing $^{99}_{42}$Mo-tagged molybdenum dioxide having an activity of 2.9 TBq. The use of radiation-detection equipment reveals a maximum radiation level of 0.005 mSv/hr on the surface of the package and a radiation level of 0.0002 mSv/hr at 3.3 ft (1 m) from the surface. What is the basic description of this radioactive material, and which DOT labels, markings, and placards, if any, are required when the package is transported by means of a motor vehicle on public highways?

Solution: It is first appropriate to determine the transport index. We calculate it as 0.0002 mSv/hr × 100, or 0.02 (dimensionless). Footnote b to Table 16.8 indicates that this calculated value may be taken as 0. By referring to Table 16.8, we determine that a maximum surface radiation level of 0.005 mSv/hr and a transport index of 0 indicate that the shipper must affix RADIOACTIVE WHITE-I labels on two opposite sides other than the bottom of the package. Then, by reference to Tables 6.1 and 16.7 and Section 16.10-A, the shipper prepares the following shipping description to enter on the accompanying shipping paper:

Units	HM	Shipping Description (Identification Number, Proper Shipping Name, Primary Hazard Class or Division, Subsidiary Hazard Class or Division, and Packing Group)	Activity (TBq)
1 steel drum weighing 65 lb	X	UN2915, Radioactive material Type A package, 7 (Molybdenum-99, tagged in solid molybdenum dioxide) (RADIOACTIVE WHITE-I label)	2.9

Table 16.7 reveals that the shipper is not required to enter the letters RQ in the shipping description. Then, following information in Section 16.10-D, the shipper marks the package with the name and address of the shipper and receiver, the designations "Radioactive material, n.o.s.," Type A package, and the identification number UN2915. Because the package weighs less than 110 lb (50 kg), DOT does not require the shipper to mark the weight on the package. When DOT requires RADIOACTIVE WHITE-I labels to be affixed to the packaging, the carrier is not obligated to placard the motor vehicle used to transport the package.

DOT uses A_1 and A_2 values to define type A and type B packages. These definitions are paraphrased as follows:

- A *type A package* refers to a package designed to transport an A_1 or A_2 quantity of radioactive material, as appropriate. It is designed to retain the integrity of containment and shielding when subjected to normal conditions of transport.
- A *type B package* refers to a package designed to transport greater than an A_1 or A_2 quantity of radioactive material, as appropriate. It is designed to retain the integrity of containment and shielding when subjected to normal conditions of transport *and* DOE-prescribed hypothetical accident–test conditions.
- A *type B(U) package* refers to a type B package that together with its radioactive contents for international shipment requires unilateral approval only of the package design and of any stowage provisions that may be necessary for heat dissipation.

- A *type B(M) package* refers to a type B package that together with its radioactive material contents for international shipment requires unilateral approval of the package design and may require approval of the conditions of shipment. A type B(M) package has a maximum normal operating pressure of more than 100 psi$_g$ (700 kPa/cm^2) or a relief device that allows the release of class 7 radioactive material into the environment under accident conditions.

When exporting fissile material, shippers are required to use the package types designated as AF, B(U)F, and B(M)F. These package types have been designed to meet stringent DOE criteria for shipping fissile material.

16.10-F CRITICALITY SAFETY INDEX

Although DOT and NRC require the use of the criticality safety index for several purposes, of primary concern here is its use in connection with the transportation of fissile material. Although NRC bears the responsibility of issuing general licenses to those who transport fissile material, DOT promulgates the regulations to which the licensees must adhere when transporting it. As noted earlier, one DOT regulation is the requirement to inscribe the criticality safety index on the FISSILE labels affixed to opposite sides of packages containing fissile material.

The **criticality safety index**, or CSI, is a dimensionless number used to designate the degree of control to be exercised during transportation of a package or group of packages containing fissile material. Shippers calculate the CSI using formulas and tables published at 10 C.F.R. §71.22.

DOT and NRC allow licensed carriers to transport only a subcritical mass of fissile material. A subcritical mass has a CSI value of 10. When multiple packages containing fissile material are transported, carriers must assure that the sum of the CSIs is less than or equal to 50 for shipment on a nonexclusive-use conveyance and less than or equal to 100 for shipment on an exclusive-use conveyance. When an array of packages containing fissile material is transported, the carrier must segregate the packages on different conveyances or stack them in the same conveyance in such an arrangement that the array remains subcritical. When an array of packages containing fissile material is stored in transit in any one storage area, DOT and NRC require the carrier to limit their number so the total sum of their CSIs does not exceed 50.

criticality safety index
■ For purposes of DOT and NRC regulations, a dimensionless number that designates the degree of control to be exercised when shippers and carriers accumulate multiple packages containing fissile material (abbreviated CSI)

16.10-G HIGHWAY ROUTE–CONTROLLED QUANTITY

Under certain unique circumstances, DOT allows carriers to transport a specified amount of a radioactive material by using a special route. It is called a **highway route–controlled quantity** (HRCQ). DOT defines HRCQ as 3000 times the A$_1$ value for special-form class 7 material, 3000 times the A$_2$ value for normal-form class 7 radioactive material, or 1000 TBq (27,000 Ci), whichever is less.

When carriers intend to transport a highway route–controlled quantity of a radioactive material, DOT requires them to ensure that the public's risk of radiation exposure is eliminated or minimized, by considering accident rates, the transit time, population density, anticipated activities, and the time of day and week during which the transportation occurs. DOT then directs the carrier to transport the radioactive material on the *preferred route*, or *preferred highway* (Section 15.4-F). DOT permits the carrier to deviate from using the preferred route or preferred highway only under emergency conditions that would make continued use of the preferred route or preferred highway unsafe or, when necessary, to stop for rest, fuel, or vehicle repairs.

When carriers are directed to use a preferred route or preferred highway, DOT requires them to post on each side and each end of the motor vehicle RADIOACTIVE placards within a square having a white background and surrounded by a black border. The presence of the square signifies that the shipment contains a highway route–controlled quantity of a radioactive material.

When carriers transport a highway route–controlled quantity of a radioactive material in a motor vehicle, freight container, or railcar, DOT also requires them to prepare and implement a security plan whose components comply with the requirements of 49 C.F.R. §172.802. Motor carriers are also required to obtain a Hazardous Materials Safety Permit prior to transporting a highway route–controlled quantity of a radioactive material. Issuance of the safety permit requires

highway route–controlled quantity
■ For purposes of DOT regulations, the least of the following amounts: 3000 times the A$_1$ value for a special-form class 7 (radioactive) material; 3000 times the A$_2$ value for a normal-form class 7 material; or 1000 TBq (27,000 Ci) (abbreviated HRCQ)

motor carriers to prepare a written route plan that complies with the requirements of 49 C.F.R. §397.101. Motor carriers may also be subject to other special controls.

16.11 Responding to Incidents Involving a Release of Radioactive Material

When radiation sources are located nearby, government regulations require the authorized regulatory body to warn individuals of their presence. For example, OSHA requires the posting of the signs previously noted in Figure 16.9 in the workplace to warn employees that exposure to radiation is likely. Emergency responders are also warned of the presence of radiation sources when they encounter these signs.

A radioactive material may also be encountered at transportation-mishap scenes. Their presence is rapidly verified by observing the following:

- The number 7 as a component of a shipping description of a hazardous material on a shipping paper
- The word RADIOACTIVE and the number 7 on yellow-and-white labels or the word FISSILE and the number 7 on white-and-black labels affixed to packages
- The word RADIOACTIVE and the number 7 on yellow-and-white placards posted on the relevant transport vehicle

Once the presence of a radioactive material has been verified, how may responders protect themselves from undue exposure to ionizing radiation? The answer to this question requires the implementation of three basic principles: shielding, time, and distance.

Because moving behind a barrier or other shield may be impractical during an emergency-response action, the best means of personal protection for responders is to limit the time of exposure and to maintain a position as far removed from the source as practical so as to receive the minimum radiation dose.

Emergency responders may appreciate the importance of separation from a radioactive source by noting the **inverse square law of radiation**. This law is arithmetically expressed as follows:

inverse square law of radiation
■ The observation that the intensity of radiation decreases as the inverse square of the distance from its source increases

$$I = \frac{I_o}{r^2}$$

In this equation, I_o is the original intensity of a radiation source, and I is the intensity at a distance r. It summarizes the observation that the intensity of ionizing radiation decreases as the inverse square of the distance from the source increases. For example, if a radioactive material registers 1000 R on a Geiger counter held 1 ft (0.3 m) from the source, the counter will register only 250 R when it is held 2 ft (0.6 m) from the source. Hence, doubling the distance between an individual and the source of radiation reduces the radiation exposure to one-fourth the original value.

When a radioactive material has spilled, or when the package used for containment of a radioactive material is leaking or damaged, emergency responders should implement the procedures shown in Figure 16.17 to protect public health and the environment. Furthermore, the Green Section of DOT's *Emergency Response Guidebook* advises the following:

- For small and large spills, initial isolation distances of 100 ft (30 m) and 300 ft (95 m) in all directions, respectively

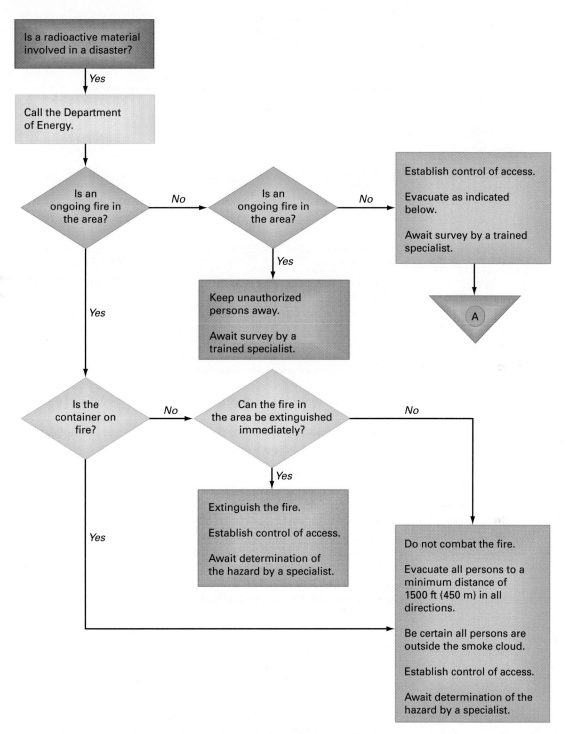

FIGURE 16.17 The recommended procedures to be implemented when responding to an emergency involving the release of a radioactive material. (*Adapted with permission of the American Society for Testing and Materials, from a figure in ASTM STP 825,* A Guide to the Safe Handling of Hazardous Materials Accidents, *second edition. Copyright © 1990, American Society for Testing and Materials.*)

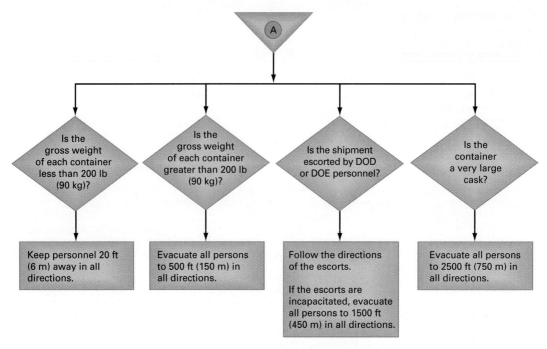

FIGURE 16.17 *(continued)*

- For small spills, protective-action distances of 0.1 mi (0.2 km) and 0.3 mi (0.5 km) during day and nighttime hours, respectively.
- For large spills, protective-action distances of 0.6 mi (1.0 km) and 1.9 mi (and 3.1 km) during day and nighttime hours, respectively

Because special training in the handling of radioactive materials is essential, the regional offices of DOE's National Nuclear Security Administration (NNSA) or the appropriate state or local radiological authorities should be notified when there is a need for an emergency-response effort involving a radioactive material. NNSA is prepared to respond to any type of radiological accident or terrorist incident occurring anywhere in the world.

When firefighters and other emergency-response personnel are called to such incidents, care should be exercised to avoid possible inhalation, ingestion, or contact with radioactive materials. Loose radioactive material and its associated package should be segregated in an area pending disposal instructions from EPA or responsible radiological authorities. If a fire is ongoing, it should be extinguished only from a distance that is considered suitable for protection of personnel. All appropriate measures should be exercised to prevent the spread of radioactivity by minimizing the quantity of runoff water.

PERFORMANCE GOALS FOR SECTION 16.12:

- Identify the health concern that radon poses when it accumulates within buildings.

16.12 Residential Radon

Radioisotopes of radium continuously form from the decay of naturally occurring uranium and thorium radioisotopes. They are produced in areas where uranium and thorium are components of metal-bearing ores, granite, or black shale. For example, radium-228 and radium-224 are produced when the naturally occurring thorium-232 in a metal-bearing ore decays by the following series of steps:

$$^{232}_{90}\text{Th} \longrightarrow {}^{228}_{88}\text{Ra} \longrightarrow {}^{228}_{89}\text{Ac} \longrightarrow {}^{228}_{90}\text{Th} \longrightarrow {}^{224}_{88}\text{Ra}$$

Although geologic formations containing uranium and thorium are relatively uncommon, they do exist below Earth's upper stratum in specific areas of the United States and elsewhere.

Buried deep beneath Earth's surface, the presence of radium radioisotopes presents little problem. However, the decay of some radium radioisotopes produces the colorless, odorless noble gas called *radon*, as illustrated by the decay of radium-224.

$$^{224}_{88}\text{Ra} \longrightarrow {}^{220}_{86}\text{Rn} + \alpha$$

Thus formed, the radon slowly percolates upward through the rock and soil and enters the atmosphere, where it dissipates. The average concentration of the radon radioisotopes in the atmosphere is 0.4 pCi/L.

In the basement of homes, radon may find its way through mortar joints, the cracks in foundation floors, gaps between wall-to-wall joints, and pores in concrete blocks and other building materials. Radon may also be dissolved within groundwater aquifers from which it may enter a home via the water supply and open sump pumps.

The radon that enters a basement and other components of a dwelling is called **residential radon**. It consists primarily of radon-220 and radon-222, which have half-lives of 54.5 s and 3.82 d, respectively. Each decays by alpha emission. The transmutation products resulting from their decay are polonium radioisotopes, which also decay by alpha emission. The average year-round residential-radon concentration is approximately 1.3 pCi/L.

Within the building shown in Figure 16.18, there is little exchange between the inside and outside air. Under this condition, the residential-radon concentration can become so high that its

residential radon
■ The radon that accumulates within a dwelling from the decay of naturally occurring uranium and thorium radioisotopes

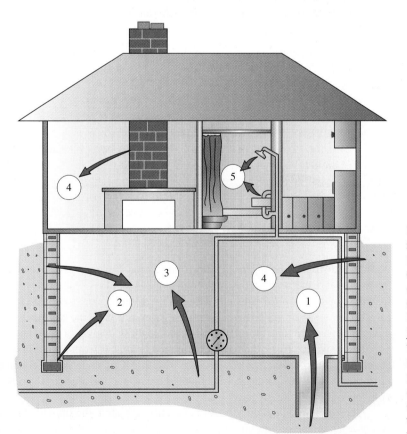

FIGURE 16.18 The six most common routes by which radon enters a home: (1) an open sump pump; (2) the gaps between basement wall-to-wall joints; (3) the foundation cracks in the basement floor; (4) the pores in concrete blocks and mortar joints; and (5) the residential water supply.

presence poses an inhalation health hazard. This is especially serious, since exposure to radon has been shown to cause lung cancer by mutating the genes of normal cells.

The risk of contracting cancer from exposure to radon is proportional to the amount that enters a home and the length of time it remains within living areas. The majority of radon that is inhaled is immediately exhaled back to the atmosphere, but since radon-220 and radon-222 have very short half-lives, some radon decays *before* it is exhaled. The polonium by-products then lodge within lung tissue and irradiate the nearby cells with alpha particles. The energy deposited in these cells is believed to initiate carcinogenesis.

Because the inhalation of radon has been shown to cause lung cancer in humans, radon is denoted as a known human carcinogen. The specific radon radioisotopes radon-220 and radon-222 are also denoted as known human carcinogens. The risk of developing lung cancer from exposure to a radon-enriched atmosphere is exacerbated for those individuals who are smokers; that is, in combination, radon and the carcinogenic components of tobacco smoke act synergistically. Researchers have determined that smoking increases an individual's risk of acquiring cancer from radon exposure by as much as 15 times. Exposure to radon is the leading cause of lung cancer in nonsmokers and the second leading cause overall. Smoking tobacco products is the leading cause of lung cancer.

It is apparent that to reduce the number of radon-induced deaths due to lung cancer, the levels of residential radon must be minimized or eliminated within dwellings. As an action guideline, EPA recommends implementing steps to achieve a residential-radon concentration of no more than 4 pCi/L. Appropriate acts include increasing the air flowing throughout residences by opening windows and using fans, and when the dwellings have a crawl space, keeping the vents in this area open throughout the year. The implementation of these measures assures home occupants that the radon concentration inside the home is no higher than that found in the ambient air outside the home.

In the late 1990s, scientific researchers estimated that residential-radon exposure causes 15,400 to 21,800 cases of lung cancer each year. They also estimated that 260 individuals die each year from the exposure to household water containing dissolved radon. Approximately one-third of these incidents could have been prevented if the residential-radon concentration had been reduced to a level below EPA's action guideline.

PERFORMANCE GOALS FOR SECTION 16.13:

- Describe the nature of a radiological dispersal device.
- Describe the response action to be executed when a radiological dirty bomb has been activated.

16.13 Radiological Dispersal Device

dirty bomb
■ A clandestine explosive device charged with a radioactive or other hazardous material that is disseminated into the environment as the bomb detonates

A **dirty bomb** may be regarded as any conventional explosive device charged with a hazardous material that disperses into the environment as the explosive detonates. In theory, the hazardous material could be a poisonous gas, flammable material, corrosive material, biological or chemical warfare agent, or radioactive material. However, it is by the widespread dissemination of a radioactive material that terrorists instill the highest degree of fear and panic in the affected population by increasing their risk of contracting cancer from radiation exposure. This situation constitutes the *radiological dirty bomb*. The threat of undergoing contact with a radiological dirty bomb is highest in densely populated areas, such as during a sports event at which large numbers of people congregate.

When a radiological dirty bomb is initially activated, the immediate area of the incident becomes contaminated with one or more radioisotopes. This area itself could constitute a sizable swath of land that would remain uninhabitable for years. Following the dispersal of radioactive material by explosive action, the radioisotopes could then diffuse throughout the air, where currents could carry them to far-removed locations including street canyons, subway tunnels, and inside buildings. In time, the radioactive material could even disperse worldwide. It is for this reason that a radiological dirty bomb is also called a *radiological dispersal device*.

The detonation of a radiological dispersal device would not ordinarily cause the immediate mass casualties or devastation that are associated with the detonation of nuclear weapons. Nonetheless, the level of radiation could induce regulatory agencies to cordon off sections of entire cities for long periods. The detonation of a radiological dirty bomb would also have the ultimate affect of increasing the background radiation to which the entire human population would be exposed. Depending on the nature of the radioactive material, the background radiation could remain elevated for hundreds or even thousands of years. During this time, the exposure to this increased radiation level would dramatically increase the incidence of cancer.

Experts fear that terrorists are most likely to load a radioisotope listed in Table 16.11 into a radiological dispersal device. This choice is based on the most common radioisotopes used in industrial and medical settings. For illustrative purposes, suppose terrorists load cesium chloride powder tagged with the radioisotope cesium-137 into a device. Cesium-137 has a half-life of 33 y. Because the powder is used as a sealed source to irradiate food, it is reasonably accessible to terrorists intent on killing massive numbers of people. If they load cesium chloride into a bomb that is subsequently detonated, the cesium-137 readily disperses into the immediate environment as a powder.

What may an emergency-response team do to best serve the public when terrorists have detonated a radiological dispersal device? The three basic tenets associated with radiation protection—shielding, time, and distance—come into play when answering this question. Translated into action, this means that the affected population should be removed as quickly as feasible from the areas where the concentrations of radioactive material are highest. These individuals may then seek out official protective information from local radio and television broadcasts.

TABLE 16.11	Radioisotopes That Terrorists Are Most Likely to Load into a Radiological Dispersal Device[a]	
RADIOISOTOPE	**HALF-LIFE (y)**	**RADIATION EMITTED**
Americium-241	4300	α, β^-, γ
Californium-252	2.645	α, n
Cesium-137	33	β^-, γ
Cobalt-60	5.26	β^-, γ
Iodine-131	0.022	β^-, γ
Iridium-192	0.21	β^-, γ
Plutonium-238	88	α, γ
Radium-226	1590	α, β^-, γ
Strontium-90	28.1	β^-

[a]Charles D. Ferguson, William C. Potter, Any Sands, Leonard S. Spector, and Fred L. Wehling, *The Four Faces of Nuclear Terrorism* (Monterey, California: Center for Nonproliferation Studies, 2005), p. 8.

EPA has issued protective-action guidelines for the personnel who respond to emergencies involving the activation of a radiological dispersal device.* Not only must responders decide how to accomplish an action, they must also acknowledge that implementing a response action involving radiation exposure could jeopardize their health. The adverse health effects that potentially result from radiation exposure were previously cited in Tables 16.3 and 16.4.

The protective-action guidelines are based on the nature of activities to be conducted at a site that is contaminated with various levels of radiation. They may be paraphrased as follows:

- When the prevailing radiation level is equal to or less than 5 rem, emergency responders may initiate operations in a normal fashion.
- When the radiation level ranges from 5 to 10 rem, careful consideration must be given to the relative importance of implementing response actions at once or waiting until the radiation level has subsided. Will multiple lives be saved, or will large populations be protected only if the actions are conducted at once?
- When the radiation level ranges from 10 to 25 rem, emergency-response workers are at risk of severe personal damage from radiation exposure, especially when the execution of a response action requires exposure over a significant time. Once again, the unique conditions of the incident at hand must be evaluated to decide whether the responders are likely to immediately save multiple lives or protect large populations.
- When the level exceeds 25 rem, EPA recommends the use of robotic equipment to the maximum extent feasible. The use of robotic equipment minimizes the radiation-exposure period that would ordinarily be experienced by the emergency responders. This could prevent the onset of radiation sickness.

*Manual of Protective-Action Guides and Protective Actions for Nuclear Incidents (Washington, DC: U.S. Environmental Protection Agency, May 1992), EPA-400-R-92-001; *Federal Register* 71 (January 3, 2006): 173–96.

Features of Atomic Nuclei

16.1 Identify the number of protons and neutrons contained in each of the following radioisotopes:

(a) tritium
(b) fluorine-18

16.2 What fraction of a radioisotope remains after the passage of five half-lives?

16.3 A radiologist determines that exposure to a particular radioisotope, X, is considered safe at concentrations of less than 1 mg/mi^2. The concentration of X at a nuclear explosion testing site is measured to be 10 mg/mi^2 immediately following the explosion of a nuclear bomb. If the half-life of X is 24 hr, will the site be safe to enter in 4 days? If not, how many days must lapse before the concentration of X has been reduced to a level that entry to the site is considered safe?

Modes of Nuclear Decay

16.4 Identify the nucleus produced when each of the following nuclear transmutations occurs:

(a) $^{33}_{15}P \longrightarrow \underline{\hspace{1cm}} + \beta^-$
(b) $^{212}_{86}Rn \longrightarrow \underline{\hspace{1cm}} + \alpha$
(c) $^{18}_{9}F \longrightarrow \underline{\hspace{1cm}} + \beta^+$
(d) $^{71}_{32}Ge + ^{0}_{-1}e \longrightarrow \underline{\hspace{1cm}}$
(e) $^{197m}_{79}Au \longrightarrow \underline{\hspace{1cm}} + \gamma$

Measurement of Radioactivity

16.5 The average annual dose from medical and dental X rays is approximately 1 rad. When expressed in millisieverts (mSv), what is the average annual X-ray dose that a person receives?

16.6 What is the specific activity in becquerels per gram of 10 g of $^{14}_{6}C$-tagged sodium bicarbonate whose activity is measured to be 1000 pCi?

16.7 To protect public health, EPA proposes 4 mrem/y as the maximum radiation limit an individual should receive from consuming groundwater contaminated with radiation. If an individual receives a dose of 5 μSv/mo through the consumption of drinking water, has the person been exposed to radiation in excess of EPA's recommended limit?

16.8 One type of smoke detector is an ionization device containing americium-241. This radioisotope ionizes the air molecules between a pair of electrodes, thereby permitting the passage of a small electrical current. When smoke particles enter this space, they adhere to the ionized molecules, thereby reducing the flow of current. This reduction activates an alarm. Ion detectors are especially useful for recognizing fast-moving fires, such as wastepaper basket and grease fires. If the specific activity of a source of americium-241 for use in manufacturing smoke detectors is measured as 3.4 Ci/g, what is its specific activity in terabecquerels per gram?

III Effects Resulting from Exposure to Radiation

16.9　The OSHA regulation at 29 C.F.R. §1910.96(b)(3) prohibits an employer from exposing individuals under 18 years of age to a dose of ionizing radiation greater than 750 mrem per calendar quarter. What is the most likely reason OSHA selected 18 as the maximum age to which radiation exposure should be strictly limited?

16.10　What biological effects are likely to be experienced by emergency responders who have received a short-term radiation dose of 400 mSv from exposure to cesium-137?

Nuclear Fission

16.11　Identify the second isotope produced during the fission phenomenon represented by the following equation:

$$^{239}_{94}\text{Pu} + ^{1}_{0}n \longrightarrow ^{91}_{36}\text{Kr} + \underline{\hspace{1cm}} + ^{1}_{0}n$$

16.12　What is the most likely reason the United States contributes low-level enriched uranium to a multinational fuel bank, which provides the fuel to countries that refrain from enriching uranium?

Transporting Radioactive Materials

16.13　The DOT regulation at 49 C.F.R. §173.417 requires carriers to limit packages of fissile radioactive material to certain prescribed amounts per transport vehicle intended for domestic shipment. What is the most likely reason DOT limits the amount of fissile material contained within a transport vehicle?

16.14　The DOT regulation at 49 C.F.R. §177.842 stipulates that shippers and carriers cannot place packages of radioactive materials bearing the RADIOACTIVE YELLOW-II and RADIOACTIVE YELLOW-III labels within a transport vehicle, storage location, or any other place closer than certain prescribed distances to any area that may be occupied by a passenger, employee, or animal. What is the most likely reason DOT requires the placement of these packages in this fashion?

16.15　What shipping description does DOT require a shipper to enter on the accompanying shipping paper when offering each of the following radioisotopes for domestic transportation by cargo air within the same aircraft:

(a) Arsenic trichloride oil containing 30 TBq of arsenic-77 within a package having a transport index of 0 and a maximum surface radiation level of 0.004 mSv/hr.

(b) Solid cesium chloride containing 1.2 Ci of cesium-137 within a package having a transport index of 0.7 and having a maximum surface radiation level of 0.4 mrem/h

Responding to Emergencies Involving a Release of Explosives

16.16　When first-on-the-scene responders arrive at the scene of a highway traffic accident, they note that yellow-and-white RADIOACTIVE placards are posted on each side and each end of an overturned motor van. A representative of the carrier provides the accompanying shipping paper, which reads as follows:

Units	HM	Shipping Description (Identification Number, Proper Shipping Name, Primary Hazard Class or Division, Subsidiary Hazard Class or Division, and Packing Group)	Activity (TBq)
1 reinforced steel drum with inner containment vessel	X	RQ, UN2917, Radioactive material Type B(M) package, 7 (Cobalt-60 as metallic cobalt) (RADIOACTIVE YELLOW-III label) (Transport Index = 7.8)	37.5

What immediate actions should the responders execute to protect public health, safety, and the environment?

Residential Radon

16.17 Why is it mandatory in most states that homeowners retain a professional to determine the residential-radon concentration and disclose this level to potential home buyers?

Table of Elements and Atomic Weights

	SYMBOL	ATOMIC NUMBER	ATOMIC WEIGHT		SYMBOL	ATOMIC NUMBER	ATOMIC WEIGHT
Actinium	Ac	89	227*	Einsteinium	Es	99	247*
Aluminum	Al	13	26.9815	Erbium	Er	68	167.26
Americium	Am	95	243*	Europium	Eu	63	151.96
Antimony	Sb	51	121.75	Fermium	Fm	100	254*
Argon	Ar	18	39.948	Fluorine	F	9	18.9984
Arsenic	As	33	74.9216	Francium	Fr	87	223*
Astatine	At	85	210*	Gadolinium	Gd	64	157.25
Barium	Ba	56	137.34	Gallium	Ga	31	69.72
Berkelium	Bk	97	245*	Germanium	Ge	32	72.59
Beryllium	Be	4	9.0122	Gold	Au	79	196.967
Bismuth	Bi	83	208.980	Hafnium	Hf	72	178.49
Bohrium	Bh	107	262*	Hassium	Hs	108	
Boron	B	5	10.811	Helium	He	2	4.0026
Bromine	Br	35	79.909	Holmium	Ho	67	164.930
Cadmium	Cd	48	112.40	Hydrogen	H	1	1.00797
Calcium	Ca	20	40.48	Indium	In	49	114.82
Californium	Cf	98	248*	Iodine	I	53	126.9044
Carbon	C	6	12.01115	Iridium	Ir	77	192.2
Cerium	Ce	58	140.12	Iron	Fe	26	55.847
Cesium	Cs	55	132.905	Krypton	Kr	36	83.80
Chlorine	Cl	17	35.453	Lanthanum	La	57	138.91
Chromium	Cr	24	51.996	Lawrencium	Lw	103	259*
Cobalt	Co	27	58.9332	Lead	Pb	82	207.19
Copper	Cu	29	63.54	Lithium	Li	3	6.939
Curium	Cm	96	245*	Lutetium	Lu	71	174.97
Darmatadtium	Ds	110		Magnesium	Mg	12	24.312
Dubnium	Db	105	262.11	Manganese	Mn	25	54.9381
Dysprosium	Dy	66	162.50	Meitnerium	Mt	109	

	SYMBOL	ATOMIC NUMBER	ATOMIC WEIGHT		SYMBOL	ATOMIC NUMBER	ATOMIC WEIGHT
Mendelevium	Md	101	256*	Rutherfordium	Rf	104	261*
Mercury	Hg	80	200.59	Samarium	Sm	62	150.35
Molybdenum	Mo	42	95.94	Scandium	Sc	21	44.956
Neodymium	Nd	60	144.24	Seaborgium	Sg	106	263*
Neon	Ne	10	20.183	Selenium	Se	34	78.96
Neptunium	Np	93	237*	Silicon	Si	14	28.086
Nickel	Ni	28	58.71	Silver	Ag	47	107.870
Niobium	Nb	41	92.906	Sodium	Na	11	22.9898
Nitrogen	N	7	14.0067	Strontium	Sr	38	87.62
Nobelium	No	102	253*	Sulfur	S	16	32.064
Osmium	Os	76	190.2	Tantalum	Ta	73	180.948
Oxygen	O	8	15.9999	Technetium	Tc	43	99*
Palladium	Pd	46	105.4	Tellurium	Te	52	127.60
Phosphorus	P	15	30.9738	Terbium	Tb	65	158.924
Platinum	Pt	78	195.09	Thallium	Tl	81	204.37
Plutonium	Pu	94	242*	Thorium	Th	90	232.038
Polonium	Po	84	210*	Thulium	Tm	69	168.934
Potassium	K	19	39.102	Tin	Sn	50	118.69
Praseodymium	Pr	59	140.907	Titanium	Ti	22	47.90
Promethium	Pm	61	145*	Tungsten	W	74	183.85
Protactinium	Pa	91	231*	Uranium	U	92	238.03
Radium	Ra	88	226*	Vanadium	V	23	50.942
Radon	Rn	86	222*	Xenon	Xe	54	131.30
Rhenium	Re	75	186.2	Ytterbium	Yb	70	173.04
Rhodium	Rh	45	102.905	Yttrium	Y	39	88.905
Rubidium	Rb	37	85.47	Zinc	Zn	30	65.37
Ruthenium	Ru	44	101.07	Zirconium	Zr	40	91.22

*Mass number of isotope of longest known half-life.

GLOSSARY

A

A₁ For purposes of DOT regulations, the maximum activity of special-form class 7 (radioactive) material permitted in a type A package

A₂ For purposes of DOT regulations, the maximum activity of class 7 (radioactive) material, other than special-form class 7 material, LSA, and SCO, that is permitted in a type A package

ABC fire extinguisher A multipurpose extinguisher containing monoammonium phosphate, which functions primarily by removing heat and free radicals from the scene of an NFPA class A, B, or C fire

absolute alcohol (*see* **alcohol, absolute**)

absolute pressure The pressure exerted on a material; for confined gases, the sum of the gauge pressure and the atmospheric pressure

absolute zero The lowest temperature attainable by a substance, equal to $-459.67\ ^\circ\text{F}\ (-273.15\ ^\circ\text{C})$

accelerant A flammable or combustible material used to initiate or promote fire

acetylene The common name of the simplest alkyne; the hydrocarbon having the chemical formula C_2H_2

acid A compound that forms hydrated hydrogen ions, $H^+(aq)$, when dissolved in water

acidic The property of aqueous solutions that have a pH ranging from 0 to 7; the property of any substance that corrodes steel or destroys tissue at the site of contact

acidic anhydride A nonmetallic oxide that chemically combines with water to produce an acid

acid rain Precipitation having an approximate pH of 5.6 or less and caused by dissolved nitrogen oxides and sulfur dioxide

acrylic fiber A fiber composed of at least 85% by mass acrylonitrile units, $(-CH_2\overset{\displaystyle H}{\underset{\displaystyle CN}{C}}-)$

activation energy The minimum energy that must be supplied to a group of reactants before they begin to react chemically

activity series An arrangement of the metals in order of their decreasing ability to generate hydrogen by chemical reaction with water and acids

acute health effect Disease or impairment that manifests itself rapidly on exposure to a hazardous substance but usually subsides when the exposure ceases

acute myelogenous leukemia A cancer of the blood that develops in white blood cells, often associated with chronic exposure to benzene

addition polymer A polymer resulting from the addition of molecules, one by one, to a growing polymer chain

adsorption A physical phenomenon characterized by the adherence or occlusion of atoms, ions, or molecules of a gas or liquid to the surface of another substance

aerosol A fine mist or spray consisting of minutely sized particles, generally expelled from a nonrefillable receptacle containing contents under pressure

airborne radioactivity area For purposes of OSHA regulations, any room, enclosure, or operating area in which radioactive material exists at concentrations in excess of prescribed levels published in column 1 of Table 1 of Appendix B to 10 C.F.R. Part 20

air toxic (*see* **hazardous air pollutant**)

alcohol An organic compound each of whose molecules has a hydroxyl group $(-OH)$

alcohol, absolute Dehydrated ethanol

alcohol, denatured Ethanol to which a poisonous or noxious substance has been added to make it unfit for consumption

alcoholism A disease resulting from the misuse of ethanol

alcohol-resistant aqueous film-forming foam A foam prepared by mixing a commercially available concentrate with water, intended for use as a fire extinguisher on flammable liquid fires (abbreviated AR-AFFF)

aldehyde An organic compound each of whose molecules is composed of an alkyl or aryl group, a carbonyl group, and a hydrogen atom

aliphatic hydrocarbon A compound containing only carbon and hydrogen atoms and whose molecules are not composed of benzene or benzenelike structures

alkali metal family The elements of Group 1A on the periodic table: lithium, sodium, potassium, rubidium, cesium, and francium

alkaline Basic, as opposed to acidic; an aqueous solution or other liquid whose pH is greater than 7

alkaline earth family The elements in Group 2A on the periodic table: beryllium, magnesium, calcium, strontium, barium, and radium

alkane A hydrocarbon whose molecules are composed of only carbon–carbon single bonds $(C-C)$

alkene A hydrocarbon whose molecules are composed of at least one carbon–carbon double bond $(C=C)$

alkenyl group Any noncyclic group of atoms having the general chemical formula C_nH_{2n-1} other than the methylene group, whose formula is $-CH_2-$

alkyl/aryl chloride (*see* **chlorinated hydrocarbon**)

alkyl/aryl halide (*see* **halogenated hydrocarbon**)

alkylate Any branched-chain alkane resulting from the combination of an alkane and an alkene

alkylation The combination of an alkane and an alkene, resulting in a branched-chain alkane

alkyl group (*see* **alkyl substituent**)

alkyl substituent Any group of atoms having the general chemical formula C_nH_{2n+1}, obtained by removing a hydrogen atom from the formula for an alkane

alkyne A hydrocarbon whose molecules are composed of at least one carbon–carbon triple bond ($C \equiv C$)

allotrope A form of the same element having different physical and chemical properties

alloy A solid mixture composed of two or more metals

allyl group The designation for the group of atoms $CH_2 = CH - CH_2 -$

alpha particle A particle (α or $_2^4He$) emitted from certain radioisotopes and having the properties of a doubly ionized helium atom

alpha (α) radiation The radiation emitted from certain radioisotopes and composed of alpha particles

alternative fuel A motor fuel that replaces ordinary gasoline in whole or part

altitude sickness The inability of the human organism to extract sufficient oxygen for survival from the air at higher altitudes, where the atmospheric pressure is reduced

aluminum alkyl A compound whose molecules are composed of an aluminum atom covalently bonded to three carbon atoms, each of which is a component of an alkyl group

amalgam An alloy of mercury with one or more metals

amide An organic compound whose molecules contain the carboxamide group

amine An alkylated analog of ammonia; an organic compound derived by the substitution of one or more of the hydrogen atoms in a hydrocarbon molecule with the $-NH_2$ group of atoms

amino acid A carboxylic acid whose molecules contain the $-NH_2$ group of atoms

anemia The condition represented by a below-normal level of hemoglobin in the body

anhydride A metallic oxide or nonmetallic oxide

anoxia The absence or near absence of a supply of oxygen to the body's organs or tissues

antiknock agent A petroleum fuel additive that reduces or eliminates the noises generated within a combustion chamber when the fuel burns

aplastic anemia Bone marrow injury, often associated with exposure to benzene

aqueous Containing water as a component, usually the primary component

aqueous ammonia (*see* **household ammonia**)

aqueous film-forming foam A foam prepared by mixing a commercially available concentrate with water, intended for use as an extinguisher of fires involving flammable, water-insoluble liquids (abbreviated AFFF)

aromatic hydrocarbon A compound whose molecules are composed solely of carbon and hydrogen atoms, at least some of which are structurally similar to benzene

aryl group (*see* **aryl substituent**)

aryl substituent Typically the phenyl or benzyl group of atoms, obtained by removing a hydrogen atom from the chemical formulas of benzene and toluene, i.e., C_6H_5- and $C_6H_5CH_2-$, respectively.

asbestos The generic name of a class of naturally occurring fibrous silicates

asbestos abatement procedures Recognized methods that aim to control fiber release from asbestos-containing materials or to remove, encapsulate, repair, enclose, or encase them

asbestosis A scarring of lung tissue by asbestos fibers

asphalt (asphaltic bitumen) The residue remaining from the fractional distillation of crude petroleum, usually combined with mineral matter

asphalt kettle A vessel or container used to process, treat, hold for heating, or dispense flammable or combustible roofing materials in the form of viscous liquids

asphyxiant A gas or vapor that dilutes or displaces air and, when inhaled, causes unconsciousness or death

atactic polymer A type of polymer whose macromolecules have randomly positioned side chains along a chain of carbon atoms

atmosphere The gaseous layer that surrounds Earth

atmospheric pressure The force exerted on matter by the mass of the overlying air

atmospheric tank A tank designed to store a flammable liquid at pressures ranging from atmospheric pressure through 0.5 psi_g (3 kPa)

atom The smallest particle of an element that can be identified with that element

atomic nucleus The region at the center of the atom composed of protons and neutrons

atomic number The number of protons in an atomic nucleus; the number of electrons in an atom

atomic orbital The region in the space around an atomic nucleus in which electrons are most likely to be found

atomic weight The mass of an atom compared with the mass of the carbon-12 isotope, which is assigned a mass of exactly 12

autoignition point The minimum temperature of a liquid necessary to initiate the self-sustained combustion of its vapor in the absence of an ignition source

autopolymerization Spontaneous polymerization

available chlorine A measure of the effectiveness of commercial bleaching agents compared with elemental chlorine

aviation gasoline (*see* gasoline, aviation)

Avogadro number The number of atoms, molecules, or other units contained in a mole of a specified substance, 6.02×10^{23}

azide group The designation for the group of atoms $-N_3$

B

background radiation Cosmic radiation and inescapable low-level ionizing radiation that is emitted from naturally occurring and artificially produced radioisotopes; 300 mrem/y, or 3 mSv/y

bactericide A pesticide used to control or kill bacteria

balanced The feature of a chemical equation that has an equal number of like atoms on each side of the arrow

barometer An instrument used for determining the atmospheric pressure, usually expressed as the height of a column of mercury

base A compound whose units produce hydrated hydroxide ions, $OH^-(aq)$, when dissolved in water

base unit Any fundamental unit of measurement that describes length, mass, and other quantities

basic The property of aqueous solutions that have a pH greater than 7.

basic anhydride A metallic oxide that chemically combines with water to produce a base

basic description For purposes of DOT regulations, the sequence of the DOT identification number, proper shipping name, hazard code, subsidiary hazard code, and packing group

becquerel The SI unit of radioactivity equivalent to one disintegration per second

benzene The hydrocarbon having the chemical formula C_6H_6 with all six carbon atoms bonded one to the other in a hexagonal ring and with each carbon atom also bonded to a hydrogen atom

benzyl group The designation for the seven-carbon atom unit consisting of a benzene ring and a methylene group ($C_6H_5CH_2-$)

beta (β) radiation Negatrons or positrons

binary acid An acid composed of hydrogen and a nonmetal

biochemistry The study of the structures, functions, and reactions of the substances found in living organisms

biodiesel oil Vehicular fuel produced from vegetable products and animal fats, sometimes mixed with petroleum diesel oil

biofuel A vehicular fuel made from a domestic farm crop such as corn kernels or soybeans

biological product For purposes of DOT regulations, a virus, therapeutic serum, toxin, antitoxin, vaccine, blood, blood component or derivative, allergenic product, or analogous product used in the prevention, diagnosis, treatment, or cure of diseases in humans and animals

biological warfare agent An infectious substance that inflicts disease or causes mass casualties when disseminated throughout a population

biphenyl The hydrocarbon having the chemical formula $C_6H_5-C_6H_5$

black powder A mixture of charcoal and sulfur with potassium nitrate or sodium nitrate in prescribed amounts

blasting agent For purposes of DOT regulations, a material designated for blasting that has been subjected to prescribed tests and has been shown to have little probability of initiating an explosion or of burning and then exploding

blasting cap A small charge of a detonator designed to be embedded in dynamite and ignited by means of a spark or burning fuse

bleaching agent A substance capable of removing color, in whole or in part, from a textile

BLEVE The phenomenon in which the rapid buildup of internal pressure within a container or tank is relieved by its rapid and violent rupture, accompanied by the release of a vapor or gas to the atmosphere and propulsion of the container or its pieces; the acronym for a "boiling-liquid expanding-vapor explosion"

blood agent A chemical warfare agent that causes seizures, respiratory failure, and cardiac arrest by interfering with the blood's ability to absorb atmospheric oxygen

blood alcohol concentration The amount of alcohol in a specified volume of an individual's blood, typically measured in milligrams per deciliter (mg/dL) (abbreviated BAC)

blood pressure The constant pressure on the walls of the arteries

boiling-liquid expanding-vapor explosion (*see* BLEVE)

boiling point The temperature at which the vapor pressure of a substance equals the average atmospheric pressure

boilover The phenomenon associated with the expulsion of crude oil and the production of steam during a fire within a crude oil storage tank

booster The component of an explosive train that is actuated by the detonator and whose purpose is to trigger the detonation of the bursting charge

bottled gas (*see* liquefied petroleum gas) Broadly, any gas stored under pressure in a portable gas cylinder;

specifically, propane, butane, or a propane/butane mixture stored under pressure in a portable gas cylinder

Boyle's law The observation that at constant temperature the volume of a confined gas is inversely proportional to its absolute pressure

breeder reactor A nuclear reactor that produces fissionable uranium-233 or plutonium-239

brisance The shattering power of an explosive

British thermal unit The amount of heat required to raise the temperature of one pound of water one degree Fahrenheit (abbreviated Btu)

bromochlorofluorocarbon Any halogenated hydrocarbon whose molecules are composed solely of carbon, bromine, chlorine, and fluorine atoms (abbreviated BCFC)

bromofluorocarbon Any halogenated hydrocarbon whose molecules are composed solely of carbon, bromine, and fluorine atoms (abbreviated BFC)

BTX The acronym for benzene, toluene, and xylene

bulk oxygen system For purposes of the OSHA regulations, the assembly of equipment (storage tank, pressure regulators, safety devices, vaporizers, manifolds, and interconnecting piping) that has an oxygen storage capacity at normal temperature and pressure of more than 13,000 ft^3 (368 m^3) when connected for service, or more than 25,000 ft^3 (708 m^3) when available as an unconnected reserve

bulk packaging For purposes of DOT regulations, packaging other than watercraft including a transport vehicle or freight container in which materials are loaded with no intermediate form of containment and have (a) a maximum capacity of greater than 119 gal (450 L) for a liquid receptacle; (b) a maximum net mass greater than 882 lb (400 kg) for a solid receptacle; *or* (c) a water capacity greater than 1000 lb (454 kg) as a gas receptacle

bursting charge The main component of an explosive train whose purpose is to detonate with maximum force at one time

butane One of two alkanes having the chemical formula C_4H_{10}

C

calorie The amount of heat required to raise the temperature of one gram of water one degree Celsius

cancer A disease associated with any of the various types of malignant tumors or other abnormal growths that develop in the body

carbon black An intensely black form of carbon whose particles are submicron in diameter, used commercially in a number of products

carbon–carbon double bond Two shared pairs of electrons between two carbon atoms

carbon–carbon single bond A shared pair of electrons between two carbon atoms

carbon–carbon triple bond Three shared pairs of electrons between two carbon atoms

carbonization The process of converting an organic compound into carbon or a carbon-containing residue, typically conducted under intense temperature and pressure

carbonyl group The $\overset{\diagdown}{\underset{\diagup}{C}}{=}O$ functional group, common to both aldehydes and ketones

carboxamide group The $-\overset{\overset{\textstyle O}{\|}}{\underset{\underset{\textstyle NH_2}{\diagdown}}{C}}$ functional group, common to amides

carboxyhemoglobin The substance that forms when carbon monoxide reacts with hemoglobin

carboxyl group The $-\overset{\overset{\textstyle O}{\|}}{\underset{\underset{\textstyle OH}{\diagdown}}{C}}$ functional group, common to carboxylic acids

carboxylic acid An organic compound containing the carboxyl group (formic acid, acetic acid, etc.)

carcinogen A substance or agent that contributes to the causation of a tumor in a living organism

carcinogenesis The molecular and cellular mechanisms responsible for changing normal tissues into malignant tumors

cardiovascular Relating to or involving the heart and blood vessels

cargo tank For purposes of DOT regulations, bulk packaging that (a) is a tank intended primarily for the carriage of liquids or gases and includes appurtenances, reinforcements, fittings, and closures; (b) is permanently attached to or forms a part of a motor vehicle, or is not permanently attached to a motor vehicle but which, by reason of its size, construction, or attachment to a motor vehicle, is loaded or unloaded without being removed from the motor vehicle; (c) is not fabricated under a specification for cylinders, intermediate bulk containers, multiunit tankcar tanks, portable cars, or tankcars

carrier For purposes of DOT regulations, a person engaged in the transportation of passengers or property by land or water, as a common, contract, or private carrier, or civil aircraft

catalyst A substance that increases or decreases the rate of a chemical reaction without itself being permanently altered

catalytic cracking (*see* **cracking, catalytic**)

caustic (*see* **basic**)

ceiling limit The concentration of a substance to which an individual should not be exposed, even for an instant, to prevent the onset of ill effects or death

cellulose The substance that forms the cell walls of all plants; the major constituent of wood and cotton; a natural polymer of β-glucose

Celsius scale The temperature scale on which water freezes at 0 °C and boils at 100 °C at 1 atm

central nervous system The brain and spinal cord

cetane number The percentage of cetane ($C_{16}H_{34}$) that must be mixed with 2,2,4,4,6,8,8-heptamethylnonane to yield the same ignition performance as a sample of a diesel oil

chain reaction A multistep reaction in which a reactive intermediate formed in one step reacts during a subsequent step to generate the species needed in the first step; a self-sustaining nuclear reaction that once started, steadily provides the energy and matter necessary to continue the reaction

chalcogen family The elements in Group 6A on the periodic table: oxygen, sulfur, selenium, tellurium, and polonium

charge A characteristic of matter resulting from a deficiency or excess of electrons

Charles's law The observation that at constant pressure the volume of a confined gas is directly proportional to its absolute temperature

chemical bond The force by which the atoms of one element become attached to, or associated with, other atoms in a compound

chemical change A change that results in an alteration of the chemical identity of a substance; a chemical reaction

chemical energy The energy stored in the chemical bonds of a substance; the energy absorbed or released during a chemical reaction

chemical equation The symbolic representation of a chemical process that respectively lists the chemical formulas of its reactants and products on each side of an arrow

chemical foam A fire-extinguishing foam made from sodium bicarbonate and aluminum sulfate solutions, which react to produce a foam consisting of aluminum hydroxide and encapsulated bubbles of carbon dioxide

chemical formula A method of expressing the number of atoms or ions of specific types in a molecule or unit of matter through the use of chemical symbols for the elements and numbers or proportions of each kind of atom

chemical property A type of behavior a substance exhibits when it undergoes a chemical change

chemical symbol A notation using one or two letters to represent an element

chemical warfare agent A nerve gas, vesicant, or other substance that inflicts harm or causes mass casualties when disseminated

chemistry The natural science concerned with the properties, composition, and reactions of substances

CHEMTREC Acronym for Chemical Transportation Emergency Center; a public service hotline for firefighters, law enforcement, and other emergency responders for obtaining information and assistance for incidents involving hazardous materials

chloracne An acnelike skin affliction caused by overexposure to certain halogenated hydrocarbons, including the polychlorinated biphenyls

CHLOREP A program designed by the Chlorine Institute to aid emergency responders during incidents involving the release of chlorine into the environment during transportation mishaps or at user locations

chlorinated hydrocarbon A hydrocarbon derivative in whose molecules one or more chlorine atoms are substituted for hydrogen atoms

chlorination A chemical reaction that involves the addition of chlorine to a substance; the treatment of contaminated drinking water to kill the microorganisms that cause diseases like dysentery, cholera, typhoid, and hepatitis

chlorofluorocarbon A compound whose molecules are composed solely of carbon, chlorine, and fluorine atoms (abbreviated CFC)

chlorosilane A chlorinated derivative of silane (SiH_4)

chromium compound, hexavalent A compound containing chromium in the +6 oxidation state, the absorption of which has been shown to cause certain adverse health ailments including cancer

chronic health effect Disease or impairment resulting from exposure to a hazardous substance that manifests itself over an extended period (months to years)

cilia Fine hairlike projections such as those along the exterior of the respiratory tract that move in unison to help prevent the passage of fluids and fine particles into the lungs

cirrhosis A chronic disease of the liver characterized by inflammation, pain, and jaundice (a yellowishness of the skin)

classification code The division number of an explosive followed by its compatibility group

Clean Air Act The federal statute that empowers EPA to establish national ambient air quality standards for specified air contaminants

Clean Water Act (*see* **Federal Water Pollution Prevention and Control Act**)

coal distillation Fractional separation of the components of coal by heating the coal to vaporize its constituents, some of which may thereafter be condensed

coal gas The flammable mixture of gases and vapors that form when coal is strongly heated in the absence of air

coal gasification The chemical process that produces a mixture of carbon monoxide and hydrogen when steam is exposed to coal at high temperature and pressure

coal tar The condensed black, viscous liquid produced by heating coal in the absence of air

coal tar distillates The mixture of liquids produced by condensing the gases and vapors that evolve when coal tar is heated

coal tar pitch The residue remaining after coal is heated to approximately 662 °F (350 °C) in the absence of air

coefficient of volume expansion The change in the volume of a liquid per degree change in temperature

coke The residue that remains when coal is heated in the absence of air; *see also* **petroleum coke**

combination reaction The type of chemical reaction involving two substances that react to form a single product

combined gas law The observation that the volume of a confined gas is directly proportional to its absolute temperature and inversely proportional to its absolute pressure

combustible Easily ignitable and free burning

combustible liquid For purposes of OSHA regulations, a liquid having a flashpoint at or above 100 °F (37.8 °C); for purposes of DOT regulations, a liquid that does not meet the definition of a DOT hazard class other than 3 and has a flashpoint above 141 °F (60.5 °C) and below 200 °F (93 °C)

combustible metal Metals including magnesium, aluminum, titanium, and zinc that can readily ignite under certain conditions to produce NFPA class D fires

combustion The rapid oxidation of a material in the presence of oxygen or air

combustion, spontaneous Rapid oxidation initiated by the accumulation of heat in a material undergoing slow oxidation

commerce Trade, traffic, transportation, importation, exportation, and similar activities engaged in by shippers and carriers

common fireworks (*see* **fireworks, common**)

common name The historical name for an organic compound

common system of nomenclature The system used historically for the naming of organic compounds

compatibility group For purposes of DOT regulations, a designated alphabetical letter used to classify an explosive for control during its storage and transportation

complete combustion Combustion process involving complete oxidation

compound A substance composed of two or more elements in chemical combination

Comprehensive Environmental Response, Compensation, and Liability Act The federal statute that empowers EPA to identify and clean up sites at which hazardous substances were released into the environment (abbreviated CERCLA); called colloquially the Superfund law

compressed gas A gas confined within a vessel under pressure

concentrated acid Pertaining usually to its commercially available form that contains the greatest concentration of the substance

concentration The relative amount of a minority constituent of a mixture or solution to the majority constituent, expressed in such units as grams per liter, pounds per cubic foot, parts per million, and percentage by mass and volume

condensation The conversion of a gas or vapor into a liquid; a chemical reaction associated with the production of a polymer and a small molecule like water or ammonia

condensation polymer A polymer produced by a chemical reaction along with a small molecule like water or ammonia

condensation reaction A chemical reaction in which a polymer is formed along with a relatively small molecule like water or ammonia

condensed formula The formula of a compound in which symbols of the atoms are written next to the symbol of the atoms to which they are bonded and in which dashes are omitted or used only in a limited fashion

conduction The mechanism by which heat is transferred to the parts of a stationary material or from one material to another with which it is in contact

contaminant A foreign substance that causes a degradation in the quality of the containing medium

control temperature For purposes of DOT regulations, the temperature above which shippers and carriers are prohibited from transporting a peroxo-organic compound

convection The mechanism by which heat is transferred by the movement of the heated material itself from spot to spot

Convention on Prohibitions or Restrictions of the Use of Certain Conventional Weapons Which May Be Deemed to Be Excessively Injurious or to Have Indiscriminate Effects A convention held in Geneva, Switzerland, in December 1980, to which the United States is a signatory, consisting of four protocols that aim to restrict or prohibit during warfare the use of various conventional weapons whose effects are deemed to be excessively cruel or inhumane or that do not discriminate between the impact on civilian and military targets

conversion factor A fraction equal to 1 in which the magnitude of one unit is related in the numerator to the magnitude of another unit of the same type in the denominator

copolymer A polymer produced from two or more different monomers

corrosion Irreversible deterioration or destruction in living tissue and metals brought about by chemical action at the site of contact

corrosive material For purposes of DOT regulations, a liquid or solid that causes full-thickness destruction of human skin at the site of contact within a specified period; or a liquid that chemically reacts with steel or aluminum surfaces at a rate exceeding 0.25 in./y (6.25 mm/y) at a test temperature of 130 °F (54 °C) when measured in accordance with prescribed testing procedures

corrosivity For purposes of RCRA regulations, the characteristic of a liquid waste that possesses either of the following properties (40 C.F.R. §261.22): It is an aqueous solution and has a pH less than or equal to 2 or equal to or greater than 12.5; or it is a liquid and corrodes steel at a rate greater than 0.250 in./y (6.35 mm/y) at a test temperature of 130 °F (54 °C) when measured in accordance with prescribed testing procedures

cotton A naturally occurring fiber of vegetable origin

covalent bond A shared pair of electrons between two atoms

covalent compound A substance whose constituent atoms are bonded together mainly by covalent bonds

cracking The industrial process typically conducted at a petroleum refinery during which the covalent bonds in hydrocarbon molecules are cleaved into smaller ones

cracking, catalytic Cracking accomplished in the presence of a catalyst

cracking, thermal Cracking induced by heat

creosote oil An oily liquid obtained from the distillation of coal tar

cresol A methylated derivative of phenol

criteria air pollutants The six most commonly encountered air pollutants in the United States for which EPA has developed emission standards based on specific health criteria, namely, carbon monoxide, nitrogen dioxide, ozone, lead, sulfur dioxide, and particulate matter

criticality safety index For purposes of DOT and NRC regulations, a dimensionless number that designates the degree of control to be exercised when shippers and carriers accumulate multiple packages containing fissile material (abbreviated CSI)

critical mass The minimum mass of fissionable material that must be present before low-energy, neutron-induced fission becomes self-sustaining

critical pressure The minimum pressure that causes a gas to liquefy at its critical temperature

critical temperature The temperature above which the vapor of a liquid cannot be condensed by pressure alone

cross-linking The production of chemical bonds in multiple dimensions that is typically associated with the strength, durability, and elasticity within the macromolecules of certain polymers

crude/crude oil (*see* **petroleum, crude**)

crude petroleum (*see* **petroleum, crude**)

cryogen For purposes of DOT regulations, a refrigerated liquefied compressed gas having a boiling point colder than -130 °F $(-90$ °C) at 14.7 psi_a (101.3 kPa)

cryogenic liquid (*see* **cryogen**)

cryogenics The study of the properties of matter at extremely cold temperatures

cryogenic vessel A pressure tank, low-pressure tank, or atmospheric tank designed to contain a cryogenic fluid on which venting, insulation, refrigeration, or a combination of these is used in order to maintain the operating pressure within the design pressure and the contents in the liquid phase

cyanosis The adverse affliction resulting from exposure to cyanides and certain other substances, exhibited physically by the presence of a bluish tinge in the fingernail beds, lips, ear lobes, conjunctiva, mucous membranes, and tongue, exhibited by dizziness, headache, diarrhea, and anemia

cycloalkane A hydrocarbon whose molecules are composed of a cyclic ring of carbon–carbon single bonds

cycloalkene Any alkene whose carbon atoms are bonded to each other in a cyclic fashion

cyclonite The common name of a commercially available form of the high explosive trinitrotrimethylenetriamine

cylinder For purposes of DOT regulations, a pressure vessel having a circular cross section and designed to safely store and transport nonliquefied and liquefied compressed gases at pressures greater than 40 psi_a (275.6 kPa)

D

dangerous-cargo manifest For purposes of DOT regulations, a cargo manifest that lists the hazardous materials stowed aboard watercraft and the specific locations at which they are stowed

dangerous-when-wet material For purposes of DOT regulations, a material that by interaction with water is likely to become spontaneously flammable or to release a flammable or toxic gas or vapor at a rate greater than 28 in.^{3}/lb (1 L/kg) per hour when subjected to prescribed test procedures

decomposition reaction The type of chemical reaction involving the breakup of one substance into two or more elementary substances

deflagrating explosive (*see* **explosive, low)**

deflagration The chemical process during which a substance intensely burns instead of detonating

defoliant An herbicide that causes the leaves to drop from plants prematurely

degreaser Any organic compound used as a solvent to remove grease and oil from surfaces

degree of toxicity A relative ranking of the ability of a toxic substance to cause injury, disease, or death on exposure

dehydration The removal of water from a compound

dehydrogenation A chemical process typically conducted at a high temperature and pressure and during which molecular hydrogen is removed from a compound

denaturant A substance added to ethanol to make it unpalatable

denatured alcohol (*see* **alcohol, denatured)**

density The property of a substance that measures its compactness; the mass of a substance divided by the volume it occupies

depleted uranium Uranium and its compounds in which the concentration of uranium-235 has been reduced or eliminated by processing

depolymerization The thermal decomposition process during which a polymer produces the monomers from which it was made

dermis The layer of the skin that is located immediately below the outer layer

detonating explosive (*see* **explosive, high)**

detonation The chemical transformation of an explosive, characterized by the essentially instantaneous passage of an energy wave at a supersonic rate, the production of large volumes of gases and vapors, the evolution of energy, and a rise in pressure

detonation temperature The temperature at which an explosive detonates

detonation velocity The speed at which an energy wave moves through an explosive

detonator The component of an explosive train that detonates from exposure to heat or mechanical impact and causes the actuation of the booster

deuterium The isotope of hydrogen consisting of one proton and one neutron

deuteron The nucleus of the deuterium atom

diagnostic specimen For purposes of DOT regulations, a human or animal material, including excreta, blood and its components, tissue, and tissue fluids being transported for diagnostic or investigational purpose (but excluding live infected humans or animals), when the source human or animal has or may have a serious human or animal disease from a Risk Group 4 pathogen

diastolic The bottom number of a blood pressure reading, typically less than 85 mmHg; the pressure exerted within an individual's arteries as the heart relaxes between beats

dielectric fluid The fluid used in certain types of electrical equipment to increase their resistance to arcing and serve as a heat-transfer medium

diene An organic compound each of whose molecules has two carbon–carbon double bonds

diesel fuel (*see* **diesel oil)**

diesel oil A petroleum product commonly used as the fuel for operating diesel engines

diesel oil, regular Diesel oil having a cetane number ranging from 40 to 60

diesel oil, premium Diesel oil having a cetane number ranging from 45 to 50

diluent type A For purposes of DOT regulations, an organic liquid that does not detrimentally affect the thermal stability or increase the hazard of a specified peroxo-organic compound and has a boiling point of not less than 302 °F (150 °C)

diluent type B For purposes of DOT regulations, an organic liquid that is compatible with a specified peroxo-organic compound and has a boiling point less than 302 °F (150 °C) but at least 140 °F (60 °C), and a flashpoint greater than 41 °F (5 °C)

diluted acid Any form of an acid that has been produced by mixture with water

diol (*see* **glycol)**

dioxin The common name for 2,3,7,8-tetrachlorodibenzo-*p*-dioxin (abbreviated TCDD)

dirty bomb A clandestine explosive device charged with a radioactive or other hazardous material that is disseminated into the environment as the bomb detonates

dissociation A chemical reaction in which one substance breaks up into two or more simpler substances

distillation A physical process in which a substance or group of substances is separated from a mixture contained within a vessel by heating the mixture to a specified temperature or range of temperatures, thereby converting one or more constituents into a vapor that is subsequently condensed in another vessel

disulfide cross-linking The bonding between two sulfur atoms $(-S-S-)$ in polymer macromolecules that provides strength, durability, and elasticity in the products produced from them

division number (*see* **hazard class/division**)

DOD national stock number The 13-digit number used to identify a specific explosive for identification pursuant to DOT regulations

dome The circular fixture at the top of a railroad tank car, where the valves and pressure-relief devices are located

dose The quantity of a substance per body weight administered directly to and absorbed by an organism

dosimeter An instrument used to detect and measure radiation exposure

DOT-approved special permit For purposes of DOT regulations, a permit designated by the notation DOT-SP*** (where the asterisks represent assigned numbers) whose issuance provides carriers a means of legally transporting a hazardous material by an unconventional means for a specified period

double bond A covalent bond composed of four electrons shared between two atoms

double displacement reaction (*see* **double replacement reaction**)

double replacement reaction The type of chemical reaction in which two different compounds exchange their ions to form two new compounds

dry-chemical fire extinguisher An extinguisher containing any of several solid substances that function by removing one or more of the fire tetrahedron components from the fire scene

dry-cleaning agent A nonaqueous solvent used to remove dirt and stains from the surfaces of wearing apparel, textiles, fabrics, rugs, and other materials

dry ice Solid carbon dioxide

dry-powder fire extinguisher An extinguisher usually containing graphite powder that functions by removing heat from the scene of an NFPA class D fire

ductile The physical property of metals associated with their ability to be stretched into wires

dust Minute solid particles generated by mechanical processes such as grinding, milling, drilling, conveying, bagging, mixing, and grading that remain suspended in the air by natural forces and can ignite on exposure to an ignition source

dust explosion The combustion of dust particles suspended within a confined space

dynamite A detonating explosive containing nitroglycerin, similar organic nitrate esters, and one or more oxidizers, all of which are mixed into a stabilizing absorbent

E

ebonite A form of hard rubber used primarily to make the casings for automobile storage batteries

edema The accumulation of watery fluid in the cells, tissues, or cavities of the body

elastomer A synthetic polymer that elongates or stretches under strain but is incapable of retaining the deformation when the strain is released

electrolysis The passage of an electric current through a material, generally resulting in the formation of ions and electrons

electrolyte A substance that dissociates into ions when fused (melted) or is dissolved in solution, thereby becoming capable of conducting an electric current

electromagnetic radiation The entire range of energy that travels as waves through space

electron A fundamental particle that bears a charge of -1

electron capture A spontaneous mode of radioisotopic decay in which the nucleus captures an extranuclear electron

element A substance composed of only one kind of atom

elements, essential The elements essential for the proper functioning and survival of the human organism

elevated-temperature material For purposes of DOT regulations, a hazardous material that when offered for transportation or transported in a bulk packaging exists in one of the following: the liquid phase of matter at a temperature at or above 212 °F (100 °C); liquid phase of matter with a flashpoint equal to or greater than 100 °F (37.8 °C) that is intentionally heated and offered for transportation or transported at or above its flashpoint; *or* the solid phase of matter at a temperature equal to or greater than 464 °F (240 °C)

Emergency Planning and Community Right-to-Know Act The federal statute that empowers EPA to require a facility handling chemical substances to either submit copies of MSDSs for certain chemical substances present at the facility in amounts greater than specified threshold quantities *or* to provide a list of such substances along with their acute health hazards, chronic health hazards, fire hazards, sudden-release-of-pressure hazards, and reactivity hazards (abbreviated EPCRA)

Emergency Response Guidebook A DOT publication that provides emergency actions for implementation at hazardous materials transportation mishaps

emergency temperature For purposes of DOT regulations, the temperature at which carriers who are transporting a peroxo-organic compound are required to implement emergency measures owing to the imminent danger that the compound may thermally decompose

endocrine disrupters Compounds that mimic hormones or interfere with hormonal action when ingested into the body

endocrine glands Ductless glands, such as the adrenal, thyroid, and pituitary, that secrete fluids directly into the general circulatory system

endothermic Referring to a process that absorbs heat from its surroundings

energy The property of matter that enables it to do work

English units of measurement Units based on the yard, pound, and quart, and their fractions and multiples

enriched uranium Uranium and its compounds in which the concentration of uranium-238 has been reduced or eliminated by processing

environment The water, ambient air, and land, including the surface and subsurface strata, and the interrelationship that exists among and between them and all living things

enzyme A biological catalyst

epidemiology The branch of medical science dealing with the nature and transmission of diseases

epidermis The outer layer of the skin

epoxy resin The polymer resulting from the condensation of Bisphenol-A and epichlorohydrin

equation An expression used to symbolically represent a chemical or nuclear reaction

equation, ionic An equation that depicts only the ions that participate in an oxidation–reduction reaction

esophagus The organ that serves as the passage between the throat and the stomach

essential elements (*see* **elements, essential**)

ester An organic compound having molecules in which an alkyl or aryl group has replaced the hydrogen atom in the carboxylate group

esterification The process of producing an ester by the reaction of an organic acid with an alcohol

ether An organic compound each of whose molecules consists of an oxygen atom bonded between two alkyl and/or aryl groups

ethyl group The designation for the group of atoms CH_3CH_2-

evaporation The process by which the surface particles of a liquid escape into the vapor state

excepted package For purposes of DOT regulations, a package type in which LSA materials and SCOs are transported by exclusive-use means

EX-number A number assigned by DOT to identify an explosive that has been evaluated by prescribed methods and approved for transportation

exothermic Referring to a process that emits heat into its surroundings

explosion The sudden expansion of gases, generally accompanied by a shock wave or disruption of enclosing materials or structures

explosive For purposes of DOT regulations, a substance or article, including a device, that is designed to function by explosion, or that, by chemical reaction with itself, is able to function in a similar manner even if not designed to function by explosion

explosive article For purposes of DOT regulations, a commercially available item like a detonator or flare containing an explosive substance as a component

explosive, high An explosive that is capable of sustaining a detonation shockwave to produce a powerful blast

explosive, initiating A substance mainly used as the component of blasting caps, detonators, and primers

explosive, low A substance that usually deflagrates rather than detonates, such as black powder and igniter cord

explosive, primary An explosive that is highly susceptible to detonation by exposure to heat, mechanical shock, and friction

explosive range (*see* **flammable range**)

explosive, secondary An explosive whose susceptibility to detonation occurs only under certain circumstances such as ignition by a primary explosive

explosive substance For purposes of DOT regulations, a substance like black powder or smokeless powder contained in packaging other than an article

explosive train The sequence of charges beginning with actuation of the igniter and progressing to actuation of the bursting charge

explosophore The group of atoms bonded within the molecular or unit structure of a substance that provides its explosive nature (e.g., $-N_3$ or $-NO_2$)

extremely hazardous substance For purposes of SARA, a substance listed in Appendices A and B following 40 C.F.R. §355; a substance that can cause serious irreversible health effects from accidental releases and is most likely to induce serious acute reactions following short-term exposure

F

factor-unit method A procedure for changing a quantity expressed in one unit to a quantity of another unit by multiplying, dividing, and arithmetically canceling numbers and units

Fahrenheit scale The temperature scale on which water freezes at 32 °F and boils at 212 °F at 1 atm

family The group of elements listed in a vertical column of the periodic table all of whose members possess similar chemical properties

Federal Food, Drug, and Cosmetic Act The federal statute that empowers the Food and Drug Administration to ensure that foods are safe to eat, drugs are safe and effective, and the labeling of foods and drugs is truthful and informative

Federal Hazardous Materials Transportation Law
The federal statute that empowers DOT to regulate certain activities of shippers and carriers who offer or accept hazardous materials for intrastate, interstate, or international transportation

Federal Hazardous Substances Act The federal statute that empowers the Consumer Product Safety Commission to protect the public against unsuspecting exposure to hazardous substances contained within household products

Federal Insecticide, Fungicide, and Rodenticide Act The federal statute that empowers EPA to regulate the manufacture, use, and disposal of pesticides for agricultural, forestry, household, and other activities (abbreviated FIFRA)

Federal Water Pollution Prevention and Control Act The federal statute that empowers EPA to restore and maintain the chemical, physical, and biological integrity of the waters of the United States (abbreviated FWPPCA)

feedstock The raw material from which a new substance is manufactured at a chemical processing plant

fertilizer A substance, generally a compound containing nitrogen and phosphorus, that increases plant production when added to the soil

fetal alcohol syndrome The disorder manifested by a reduction of mental capabilities in children of women who consumed alcoholic beverages during pregnancy

fiber A polymeric substance that can be separated into threads or threadlike structures

fire A rapid, persistent chemical reaction, most commonly the combination of a combustible material with atmospheric oxygen, accompanied by the release of heat and light

fire extinguisher A substance that promotes the extinguishment of fire or the propagation of flame

fire point The lowest temperature at which a flammable liquid emits vapors at a rate sufficient to support continuous combustion

fire resistance For purposes of OSHA regulations, the ability of a material to retain its integrity for a specified time when exposed to a standard heat intensity such that it does not fail structurally and does not permit the side away from the fire to become hotter than a specified temperature

fire retardant An additive to commercial products that promotes resistance to burning

fire tetrahedron A four-sided pyramid each face of which denotes an essential fire component (fuel, oxygen, heat, and free radicals)

fire triangle A triangle each leg of which is used to designate three of the four components of a fire (fuel, oxygen, and heat)

fireworks Pyrotechnic devices intentionally designed for the purpose of producing a visible or audible effect by means of combustion, explosion, deflagration, or detonation reactions

fireworks, common Relatively small firework devices designed primarily for use by the general public

fireworks, special Relatively large firework devices designed primarily for use by experts trained in their actuation

first-degree burn A superficial injury of the skin caused by exposure to heat

fissile The ability of certain nuclei to undergo fission, during which neutrons and a relatively large amount of energy are released

fissile nuclei Uranium-233, uranium-235, plutonium-239, and plutonium-241

fission, nuclear The nuclear process during which certain nuclei are split into two lighter-mass nuclei with the simultaneous emission of one or more neutrons and a large amount of energy

flame The visible manifestation of a fire that results when minute particulate matter composed principally of incompletely burned fuel is heated to incandescence

flame retardant An approved substance incorporated into potentially flammable materials manufactured from plastics, rubber, and textiles to suppress or inhibit the initial phase of a developing fire

flammable Capable of being easily ignited, of burning intensely, or dispersing fire at a rapid rate

Flammable Fabrics Act The federal statute that empowers the Consumer Products Safety Commission to establish flammability standards for clothing textiles as well as interior furnishings, including paper, plastic, foam, and other materials used in wearing apparel and interior furnishings

flammable gas For purposes of DOT regulations, a material that is a gas at 68 °F (20 °C) or less and 14.7 psi (101.3 kPa) of pressure *and* is ignitable at 14.7 psi (101.3 kPa) when in a mixture of 13% or less by volume with air; *or* a material whose vapor has a flammable range at 14.7 psi (101.3 kPa) with air of at least 12%, regardless of the lower limit

flammable liquid For purposes of OSHA regulations and NFPA standards, any liquid having a flashpoint below 100 °F (37.8 °C), unless it is a liquid mixture having 99% or more of the volume of components with flashpoints of 100 °F (37.8 °C) or greater; for purposes of DOT regulations, either of the following: a liquid having a flashpoint of not more than 141 °F (60.5 °C); *or* a material in a liquid phase with a flashpoint at or above 100 °F (37.8 °C) that is intentionally heated and offered for transportation or transported at or above its flashpoint within bulk packaging

flammable range The set of concentrations of gases or vapors that lie between a flammable substance's lower and upper explosive limits in air

flammable solid For purposes of DOT regulations, any of the following types of materials: wetted explosives; thermally unstable materials that can undergo a strongly exothermic decomposition even without the participation of atmospheric oxygen; and readily combustible solids

flare An article containing a pyrotechnic substance designed to illuminate, identify, signal, or warn

flashover The phenomenon associated with the spread of fire from the burning area to other areas physically isolated from the initial fire source

flashpoint The minimum temperature at which the vapor of a liquid or solid ignites when exposed to sparks, flames, or other ignition sources

flowers of sulfur The finely divided powder of elemental sulfur

foam A material consisting of a mixture of small bubbles of air, water, and stabilizing agents that is designed to extinguish a fire by blanketing the fuel, excluding air, or preventing the escape of volatile vapor into the atmosphere

foam-blowing agent A low-boiling liquid mixed with polymers such as polystyrene and rigid polyurethane to produce a hard foam

forbidden explosive For purposes of DOT regulations, an explosive for which shippers and carriers have not received DOT authorization to transport; an explosive listed in the Hazardous Materials Table at 49 C.F.R. §172.101 with the word "forbidden" in column 3, signifying that the material may not be transported by any mode

forbidden material For purposes of DOT regulations, a hazardous material listed in the Hazardous Materials Table at 49 C.F.R. §172.101 with the word "forbidden" in column 3, signifying that the material may not be transported by any mode

formalin A solution of formaldehyde in water and/or methanol

formula weight The sum of the atomic weights of all the atoms in a chemical formula

fossil fuel An energy source like coal, crude oil, and natural gas, each of which was formed by the decomposition of prehistoric organisms

fractional distillation A method of separating multiple components of a mixture based on the different boiling points of its constituents, during which the vapors are collected at specified temperatures or within specified temperature ranges and subsequently condensed to liquids

fractionation (see fractional distillation)

free radical A reactive species resembling a molecule but having an unpaired electron

freezing point The temperature at which the liquid and solid states of a substance coexist at 1 atm (101.3 kPa)

freight container For purposes of DOT regulations, a reusable container having a volume of at least 640 ft^3 (18 m^3), designed and constructed to permit being lifted with its contents intact and intended primarily for containment of packages (in unit form) during transportation

Freon The trademark of the chlorofluorocarbons, hydrochlorofluorocarbons, and hydrofluorocarbons used as foam-blowing agents and refrigerants

Freon agent (see Freon)

friable Brittle or capable of readily being crumbled, as in friable asbestos

fuel A material that burns; any of several products whose combustion serves as a source of energy

fuel oil A liquid petroleum product derived in whole or part from crude oil that is typically burned for the generation of heat or power

fumigant A pesticide product that is a vapor or gas or forms a vapor or gas on application and whose method of pesticidal action is through the gaseous state

functional group An atom or group of atoms that determines many of an organic compound's characteristic chemical properties and identifies it as an alcohol, aldehyde, ester, ether, ketone, carboxylic acid, and so forth

fusible plug A piece of low-melting metal inserted as a pressure-relief device in cylinders, cargo tanks, stationary tanks, and railroad tankcars used to store or transport a compressed gas or flammable liquid

G

galvanize The process of coating a metal with a protective layer of elemental zinc

gamma (γ) radiation Electromagnetic radiation of very short wavelength emitted from some radioisotopes

gas Matter that possesses neither a definite volume nor a definite shape; for purposes of DOT regulations, a material having a vapor pressure greater than 43.5 psi$_a$ (300 kPa) at 122 °F (50 °C) or that is completely gaseous at 68 °F (20 °C) at a standard pressure of 14.7 psi$_a$ (101.3 kPa)

gas centrifugation A process utilized during uranium-enrichment to isolate $^{235}_{92}UF_6$ from $^{238}_{92}UF_6$

gasohol A blend of gasoline and ethanol used as a vehicular fuel

gasoline The complex mixture of volatile hydrocarbons commonly used as a vehicular fuel

gasoline, aviation The grade of gasoline used as aircraft fuel, having an approximate octane number equal to or greater than 100

gasoline, mid-grade The grade of gasoline used as the fuel in small motor vehicles, having an approximate octane number of 89

gasoline, premium The grade of gasoline used as the fuel in small motor vehicles, having an approximate octane number of 91

gasoline, regular The grade of gasoline used as the fuel in small motor vehicles, having an approximate octane number of 87

gasoline, unleaded Gasoline that does not contain a lead compound such as tetraethyl lead, formerly an additive that functioned as an antiknock agent

gas poisonous-by-inhalation For purposes of DOT regulations, a gas at 68 °F (20 °C) or less and a pressure of 14.7 psi_a (101.3 kPa) that is known to be so toxic to humans as to pose a health hazard during transportation; *or* in the absence of adequate data on human toxicity, is presumed to be toxic to humans because when tested on laboratory animals, it has an LC_{50} less than 5000 mL/m^3

gauge pressure The amount of pressure by which a confined gas exceeds atmospheric pressure

Geneva protocol (formally the Protocol for the Prohibition of the Use in War of Asphyxiating, Poisonous, or Other Gases, and of Bacteriological Methods of Warfare) An international agreement established by the League of Nations in 1925 that bans the use of poisonous gases against the enemy during warfare

geometrical isomerism (*see* isomerism, geometrical)

Globally Harmonized System of Classification and Labeling of Chemicals (abbreviated GHS) The United Nations system for defining and classifying hazards and communicating hazard information on labels and material safety data sheets in a common international fashion

global warming The increase in the mean temperature of Earth's atmosphere manifested by such phenomena as the melting of polar glaciers and the rise of sea levels

glycol An alcohol each of whose molecules has two hydroxyl groups; a diol

grain alcohol The common name of ethanol produced by the fermentation of a grain

gram One one-thousandth of the mass of 1 kilogram

graphite The stable allotrope of carbon at room temperature and pressure

gravity The force of attraction between two bodies

gray The unit of an absorbed dose of radiation equivalent to 1 joule per kilogram of living tissue

greenhouse effect The warming of Earth's atmosphere caused by the decrease in the rate of escape of energy from Earth's surface into outer space compared with the rate at which the Sun's radiant energy enters the atmosphere and is absorbed by Earth

greenhouse gas A substance that absorbs atmospheric energy and by so doing, affects the rate of escape of energy from Earth's surface into outer space

green salt Uranium(IV) fluoride (UF_4), an intermediate in the preparation of uranium hexafluoride

ground-level ozone (*see* ozone, ground-level)

guncotton Any of several nitrocellulose explosives made by reacting cotton with nitric and sulfuric acids

gunpowder (black) Any of several powders used in primers, fuses, and pyrotechnics and made from potassium nitrate, charcoal, and sulfur

H

half-life The time period during which an arbitrary amount of a radioisotope is transformed into half that amount

half-reaction An equation that separately depicts either oxidation or reduction

halogenated ether Any halogenated derivative of an ether

halogenated hydrocarbon A hydrocarbon derivative in whose molecules one or more halogen atoms are substituted for hydrogen atoms

halogenation A chemical reaction in which one or more halogen atoms are added to a compound

halogen family The elements in Group 7A on the periodic table: fluorine, chlorine, bromine, iodine, and astatine

halon A fire suppressant primarily composed of a substance whose molecules have one or two carbon atoms and four to six fluorine, chlorine, or bromine atoms

hazard class/division For purposes of DOT regulations, any of the nine categories of hazard assigned to a hazardous material because it complies with defining criteria

hazardous The condition of having the potential to damage a site or initiate an event that could result in injury, illness, or death

hazardous air pollutant A substance designated by EPA pursuant to enactment of the Clean Air Act and listed at 40 C.F.R. §61.01

hazardous chemical For purposes of OSHA regulations, a chemical that is a physical hazard or a health hazard and is listed at 29 C.F.R. §1910, Subpart Z

hazardous material A substance or material in any form that because of is quantity, concentration, chemical, corrosive, flammable, reactive, toxic, infectious, or radioactive characteristics, either separately or in combination with any other substance or substances, constitutes a present or potential threat to human health, safety, welfare, or the environment when improperly stored, treated, transported, disposed or used, or otherwise managed (A listing of hazardous materials is provided by DOT at 49 C.F.R. §172.101.)

Hazardous Materials Safety Permit For purposes of DOT regulations, a document required of motor carriers who transport certain hazardous materials in amounts equal to or exceeding threshold values (e.g., 3500 gal (13,248 L) of anhydrous ammonia, 3500 gal (13,248 L) of compressed methane or liquefied natural gas, or a highway route–controlled quantity of a radioactive material)

Hazardous Materials Table The compilation of hazardous materials whose transportation is regulated by DOT at 49 C.F.R. §172.101

hazardous substance (1) For purposes of CPSC, any substance or mixture of substances that (i) is toxic, (ii) is corrosive, (iii) is an irritant, (iv) is a strong sensitizer, (v) is flammable or combustible, or (vi) generates pressure through decomposition, heat, or other means if such substance or mixture of substances may cause substantial personal injury or substantial illness during or as a proximate result of any customary or reasonable foreseeable handling or use, including reasonably foreseeable ingestion by children; any radioactive substance; any toy or other article intended for use by children that presents an electrical, mechanical, or thermal hazard; any solder that has a lead content in excess of 0.2%; (2) for purposes of FWPPCA, a substance designated by EPA and listed at 40 C.F.R. §117.3; (3) for purposes of CERCLA, a substance designated by EPA and listed at 40 C.F.R. §302.4; (4) for purposes of DOT regulations, a substance or radionuclide respectively listed in Appendices A and B, 49 C.F.R. §172.101, in an amount that exceeds the reportable quantity and is transported in a single package

hazardous waste For purposes of RCRA regulations, a waste material designated by EPA and listed at 40 C.F.R. §§261.31, 261.32, 261.33(e), and 261.33(f) as well as a waste material exhibiting one or more of the hazardous waste characteristics described at 40 C.F.R. §§261.21, 261.22, 261.23, or 261.24

hazardous waste characteristic For purposes of RCRA regulations, the following properties exhibited by a hazardous waste and described at 40 C.F.R. §§261.21, 261.22, 261.23, and 261.24: ignitability, corrosivity, reactivity, and toxicity

hazard zone For purposes of DOT regulations, the four hazard levels assigned to poison gases and either of the two levels assigned to liquids that are poisonous by inhalation [49 C.F.R. §§173.116(a) and 173.133(a)]

heat The form of energy transferred from one body to another because of a temperature difference between them; energy arising from atomic or molecular motion

heat capacity The amount of heat needed to raise the temperature of either one pound of a substance one degree Fahrenheit or one gram of a substance one degree Celsius

heat of combustion The heat emitted into the surroundings when a compound is burned, yielding carbon dioxide and water vapor

heat of reaction The energy that is absorbed or evolved when a chemical reaction occurs

heatstroke The condition caused by exposure to excessive heat

hemoglobin The component of red blood cells that transports oxygen to the tissues of the body

hemotoxicant A substance that decreases the function of the blood's hemoglobin and deprives the tissues of oxygen

hepatotoxicant A substance that causes liver damage

herbicide A substance designed to control plant life

heterocyclic compound A compound whose molecules contain one or more ring atoms other than carbon

hexogen A commercial form of the explosive cyclotrimethylenetrinitramine

high blood pressure The condition in which blood is pumped through the body at a pressure that exceeds approximately 140 over 90

high explosive (see explosive, high)

high radiation area For purposes of OSHA regulations, any area accessible to personnel in which the radiation exists at such levels that a major portion of the body could receive in any 1 hour a dose in excess of 1 mSv

high-test hypochlorite A commercial oxidizer capable of producing chlorine in acidic water (abbreviated HTH)

highway route–controlled quantity For purposes of DOT regulations, the least of the following amounts: 3000 times the A_1 value for a special-form class 7 (radioactive) material; 3000 times the A_2 value for a normal-form class 7 material; or 1000 TBq (27,000 Ci) (abbreviated HRCQ)

hormone A chemical substance produced by an endocrine gland and carried through the bloodstream to target organs or a group of cells of the body, where it serves to regulate certain body functions

household ammonia An aqueous solution of ammonia; ammonium hydroxide

hydration The addition of water to a substance

hydrocarbon A compound whose molecules are composed solely of carbon and hydrogen atoms

hydrocarbon, saturated A hydrocarbon having molecules that are composed solely of $C-C$ and $C-H$ bonds

hydrocarbon, unsaturated A hydrocarbon whose molecules contain $C-H$ bonds and at least one carbon–carbon double bond or one carbon–carbon triple bond

hydrochlorofluorocarbon A compound whose molecules are composed solely of carbon, hydrogen, chlorine, and fluorine atoms (abbreviated HCFC)

hydrofluorocarbon A compound whose molecules are composed solely of carbon, hydrogen, and fluorine atoms (abbreviated HFC)

hydrogenation The addition of hydrogen to a substance, such as across a carbon–carbon double bond

hydrolysis The chemical reaction between a substance and water

hydrometer A device calibrated on its stem so the specific gravity of a liquid can be determined by reading a number at the surface of the liquid

hydrotreatment The processing of gasoline and other petroleum products with hydrogen to reduce or eliminate their nitrogenous and sulfurous content

hydroxyl group The $-OH$ functional group, as in an alcohol

hydroxyl radical The group of atoms represented as ·OH

hyperbaric oxygen therapy The treatment of various afflictions including carbon monoxide poisoning, during which the patient breathes oxygen under increased pressure inside a sealed steel chamber

hypergolic Pertaining to the ability of a substance to ignite spontaneously on contact with another substance

hypergolic liquid A liquid capable of igniting spontaneously on contact with an explosive

hypertension (*see* **high blood pressure**)

hyperthermia The excessive overheating of the body above normal body temperature

I

identification number For purposes of DOT regulations, the UN or NA number listed in column 4 of the Hazardous Materials Table

ignitability For purposes of RCRA regulations, a characteristic of any of the following wastes: a liquid, a sample of which has a flashpoint equal to or less than 140 °F (60 °C) when determined through use of a Pensky-Martens closed-cup tester; a solid capable of causing fire through friction, absorption of moisture, or spontaneous chemical changes and that burns vigorously and persistently; an oxidizer; or an ignitable compressed gas (40 C.F.R. §261.21)

igniter Any of several nitrocellulose explosives made by reacting cotton with nitric and sulfuric acids

immediately-dangerous-to-life-and-health level The atmospheric concentration of any substance that poses an immediate threat to life, causes an irreversible or delayed adverse health effect, or interferes with an individual's ability to escape during a 30-min period (abbreviated IDHL)

immiscible Incapable of combining or mixing so as to form a single phase

improvised explosive device Any explosive device that has been fabricated and designed to destroy or incapacitate personnel or vehicles (abbreviated IED)

incandescence The emission of light from a hot body

incendiary agent A substance such as elemental white phosphorus and napalm, primarily used during warfare to intentionally set fire to objects or to cause burn injury to persons through the action of flames, heat, or a combination thereof

incomplete (partial) combustion Combustion process involving incomplete oxidation

infectious substance For purposes of DOT regulations, a material known to contain or suspected of containing a pathogen

inflammation The protective response of the body's tissues to irritation or injury, characterized by redness, heat, swelling, or pain and accompanied by the loss of function

ingestion The taking in of a substance through the mouth and into the digestive system

inhalation The taking in of a substance in the form of a gas, vapor, fume, mist, or dust into the respiratory system

inhibitor A substance or mixture of substances used to arrest or retard the speed of corrosion, polymerization, or other chemical phenomenon

initial isolation distance For purposes of DOT regulations, the distance from a transportation mishap involving the release of a hazardous material to which everyone should be quickly moved in a crosswind direction

initial isolation zone For purposes of DOT regulations, the area surrounding a transportation mishap involving the release of a hazardous material in which persons may be exposed to dangerous, life-threatening concentrations of a gas or vapor that poses an inhalation hazard

initiating explosive (*see* **initiator**)

initiation The first mechanistic step of a chemical reaction during which reactive chemical species like free radicals are produced by the application of heat; the stage of cancer during which one or more healthy cells are triggered to subsequently respond to the actions of a promoter

initiator An explosive material used to detonate the main charge; the agent that triggers the formation of cancerous growths in the body

inorganic chemistry The chemistry of those substances that do not contain carbon in their composition

insecticide A substance designed to control insect life

insoluble A substance's inability to dissolve within a given solvent

inverse square law of radiation The observation that the intensity of radiation decreases as the inverse square of the distance from its source increases

ion An atom (or group of atoms bound together) with a net electric charge due to the loss or gain of electrons

ionic bond The electrostatic force of attraction between oppositely charged ions

ionic compound A compound whose atoms are bonded together mainly by ionic bonds

ionic equation (*see* **equation, ionic**)

ionizing radiation (*see* **radiation, ionizing**)

irradiation Intentional exposure to ionizing radiation, usually X rays or gamma rays, for medical treatment, sterilization or preservation of foods, and other purposes

irritant A substance capable of injuring the body's tissues by causing inflammation at the site of contact

isocyanate An organic compound whose molecules contain the $-N=C=O$ group of atoms

isomerism The phenomenon associated with compounds whose molecules have the same atoms but differ in the manner in which the atoms are structurally and/or spatially arranged

isomerism, geometrical The phenomenon associated with compounds whose molecules have the same atoms but differ in the manner in which the atoms are spatially arranged

isomerism, structural The phenomenon associated with compounds whose molecules have the same atoms but differ in the manner in which the atoms are structurally arranged

isooctane The common name given to 2,2,4-trimethylpentane within the petroleum industry

isotactic polymer A polymer whose macromolecules have side chains positioned on the same side of a chain of carbon atoms

isotope Any of a group of nuclei having the same number of protons but different number of neutrons

IUPAC names The names of organic compounds derived from use of the system adopted by the International Union of Pure and Applied Chemistry

IUPAC system of nomenclature The internationally recognized system adopted by the International Union of Pure and Applied Chemistry for the naming of organic compounds

J

joule The SI unit of energy equal to that possessed by a 2-kg mass moving at a velocity of one meter per second

K

kelvin The unit of temperature on the Kelvin scale

Kelvin temperature The absolute temperature obtained by adding 273.15 °C to a Celsius temperature

kerosene An oily liquid obtained by the fractional distillation of crude petroleum and often used as a tractor fuel, jet fuel, and heating fuel

ketone An organic compound each of whose molecules is composed of a carbonyl group bonded to alkyl and/or aryl groups

kindling point The temperature at which the combustion of a solid can proceed without further addition of heat from an external source

knocking The noises associated with the incomplete combustion of a petroleum fuel, especially in internal combustion engines

Kyoto protocol The international policy aimed at reducing the worldwide impact of global warming through curtailment of greenhouse gas emissions into the atmosphere; formally, the Kyoto Protocol to the United Nations Framework Convention on Climate Change

L

label For purposes of DOT regulations, the written, printed, and graphic matter affixed to containers of hazardous materials to rapidly communicate hazard information

lacrimator A substance that causes the eyes to involuntarily tear and close

latent health effect Disease or impairment that is manifested after an extended time period following exposure to a hazardous substance

latent heat of fusion The amount of heat required to convert a unit mass of a solid substance into a liquid

latent heat of vaporization The amount of heat required to convert a unit mass of a liquid substance into a gas

latex, natural A white liquid obtained from some species of shrubs and trees consisting in part of natural rubber suspended in water

law of conservation of mass and energy The observation that the total amount of mass and energy in the universe is constant

lead–acid storage battery The collective group of individual cells, each of which is composed of lead and lead oxide plates, immersed in a solution of sulfuric acid having a specific gravity ranging from 1.25 to 1.30

length The distance between two points

lethal concentration, 50% kill (LC$_{50}$) The concentration of a substance that when administered to laboratory animals kills half of them

lethal dose, 50% kill (LD$_{50}$) The dose of a substance that is lethal to 50 percent of the organisms tested during a specified time

leukemia Cancer of the blood-forming tissues and organs

Lewis structure A means of displaying the bonding between the atoms of a molecule by using dashes to represent a shared pair of electrons

Lewis symbol The symbol of an element together with a number of dots representing its bonding electrons

lifting power The natural ability of a confined gas that is less dense than air to rise in the atmosphere, measured as the difference between the mass of a given volume of air and that of the same volume of the gas at the same temperature and pressure

ligroin A fraction of crude petroleum; light naphtha

linen The naturally occurring fiber of the flax plant

liquefied compressed gas For purposes of DOT regulations, a gas that when packaged under pressure for transportation is partially liquid at temperatures above -58 °F (-50 °C)

liquefied natural gas Methane as a cryogenic liquid (abbreviated LNG)

liquefied petroleum gas The compressed and liquefied mixture obtained as a by-product during the refining of petroleum and consisting primarily of a mixture of propane and n-butane

liquid Matter that possesses a definite volume but lacks a definite shape; for purposes of DOT regulations, a material other than an elevated-temperature material with a melting point or initial melting point of 68 °F (20 °C) or lower at a standard pressure of 14.7 psi$_a$ (101.3 kPa)

liter The volume occupied by a cube measuring 10 cm to an edge

lower explosive limit The concentration of a gas or vapor in air below which a flame will not propagate when exposed to an ignition source (abbreviated LEL)

low explosive (*see* **explosive, low**)

low-pressure tank A tank designed to store flammable liquids at pressures above 0.5 psi$_g$ (3 kPa) but not more than 15 psi$_g$ (100 kPa)

low-specific-activity material For purposes of DOT regulations, certain radioactive material, including uranium and thorium ores and tritium-labeled water, that complies with the descriptions and activity limitations published at 49 C.F.R. §173.403 (abbreviated LSA)

lubricant (*see* **lubricating oil**)

lubricating oil Any petroleum oil capable of producing a film that coats the surfaces of moving metal parts

M

macromolecule The giant molecule of which polymers are composed, comprising an aggregation of hundreds or thousands of atoms and typically consisting of repeating chemical units linked together into chains and cross-linked into complex three-dimensional networks

macrophage A large cell that engulfs and destroys invading microorganisms

magazine A building, room, or vessel used exclusively for receiving, storing, and dispensing of explosives

main charge (*see* **bursting charge**)

malignant Tending to become progressively worse or life-threatening

malleable A physical property of metals characterized by a capability of being hammered into sheets

marine pollutant A substance denoted in Appendix B, 49 C.F.R. §172.101

marking For purposes of DOT regulations, a descriptive name, instruction, precautionary remark, weight, construction specification note, UN or NA identification number, or other information applied or otherwise placed on the outside surface of packaging for the purpose of rapidly communicating hazard information

mass The quantity of matter possessed by an object regardless of its location

mass explosive hazard For purposes of DOT regulations, a potential detonation that affects almost the entire load of explosives instantaneously

mass number The total number of protons and neutrons in a given nucleus; the atomic mass of a specific nucleus rounded off to the nearest whole number

match An item used to intentionally initiate fire when the mixture of substances contained in its heads is rubbed against a hard surface

match, safety A match designed to ignite only when its head is rubbed on a prepared surface consisting of red phosphorus and powdered glass

match, strike-anywhere A match having a "bullseye" head that consists of a white tip containing tetraphosphorus trisulfide that is actuated by rubbing it against an abrasive strip

material safety data sheet A technical bulletin prepared by manufacturers and marketers in accordance with the provisions of 29 C.F.R. §1910.1200 to provide workers with detailed information about the properties of a commercial product that contains hazardous substances (abbreviated MSDS)

matter Anything that possesses mass and occupies space

mechanism The step-by-step description (initiation, propagation, and termination) of the chemical process by which reactants are converted into products

melting The physical change from the solid to the liquid state of matter

melting point (*see* freezing point)

mercaptan An organic compound each of whose molecules is composed of an alkyl or aryl group and a thiol ($-SH$) group

mesothelioma A cancer of the membranes lining the chest and abdominal cavities resulting from exposure to asbestos fibers

meta (*m-*) Characterized by the substitution of the hydrogen atoms on the first and third carbon atoms on the benzene ring

metabolism The processes that occur in the body's cells to break down absorbed foods or other ingested substances

metal An element that conducts heat and electricity well and has high physical strength and the properties of ductility and malleability

metallic carbide An inorganic compound composed of metallic and carbide ions

metallic chromate An inorganic compound composed of metallic and chromate ions

metallic dichromate An inorganic compound composed of metallic and dichromate ions

metallic hyperoxide (*see* metallic superoxide)

metallic hypochlorite An inorganic compound composed of metallic and hypochlorite ions

metallic nitrate An inorganic compound composed of metallic and nitrate ions

metallic nitrite An inorganic compound composed of metallic and nitrite ions

metallic permanganate An inorganic compound composed of metallic and permanganate ions

metallic peroxide An inorganic compound composed of metallic and peroxide ions

metallic phosphide An inorganic compound composed of metallic and phosphide ions

metallic superoxide An inorganic compound composed of metallic and superoxide ions

metalloid (semimetal) A metal that exhibits the general physical properties of both metals and nonmetals

metastasis The formation of a malignant growth at one site in the body from cells derived from a malignancy located elsewhere in the body

meter The SI unit of length in the metric system of measurement

methane The simplest alkane; the hydrocarbon having the chemical formula CH_4; the primary constituent of natural gas

methemoglobin The substance produced when nitrogen dioxide reacts with the blood's hemoglobin

methemoglobinemia The disorder resulting from exposure to nitrogen dioxide, metallic nirtrites, metallic nitrates, and similar substances, caused by the presence of methemoglobin in the blood, and exhibited by the symptoms of cyanosis (dizziness, headache, diarrhea, and anemia)

methylene group The designation for the group of atoms $-CH_2-$

methyl group The designation for the group of atoms CH_3-

metric system of measurement The decimal system of measurement based on the meter, kilogram, and cubic meter

microbicide A chemical substance capable of destroying microorganisms

micron One-millionth of a meter

mineral An inorganic solid substance formed within planet Earth

mineral acid Any of the inorganic acids, chiefly hydrochloric acid, sulfuric acid, nitric acid, perchloric acid, phosphoric acid, and hydrofluoric acid

mineral oil transformer An electrical transformer containing mineral oil as its dielectric fluid

miscellaneous hazardous material For purposes of DOT regulations, any material that presents a hazard during transportation but does not meet the definition of any other hazard class

miscible Soluble in all proportions; capable of mixing without subsequent separation into distinct components

mixture Matter consisting of two or more substances in varying proportions that are not chemically combined

mole An Avogadro number of particles; the amount of a substance whose mass equals its molecular weight or formula weight, as relevant

molecular formula A chemical formula that indicates the number of atoms of each element in a single molecule of a substance

molecular weight The sum of the atomic weights of all the atoms in a molecular formula

molecule The smallest neutral unit of some elements and compounds, composed of two or more atoms

monomer One or more of the single substances that combine to produce a polymer

Montréal protocol (formally the Montréal Protocol on Substances That Deplete the Ozone Layer) The international agreement that phases out the worldwide production, manufacture, and use of chlorofluorocarbons and other ozone-depleting substances

muriatic acid Technical-grade hydrochloric acid

mutation A permanent change in the genetic material contained within a cell

N

napalm The mixture of aluminum compounds produced by reacting aluminum hydroxide with certain substances such as those found in coconut oil

naphtha A petroleum fraction, often further divided into light naphtha (ligroin) and heavy naphtha, and used directly as a solvent or as the feedstock for the production of certain petrochemicals

naphthalene The simplest polynuclear aromatic hydrocarbon, the molecular structure of which consists of two fused benzene rings

National Emission Standards for Hazardous Air Pollutants For purposes of EPA regulations, the standards that relate to the air emission of hazardous air pollutants that cause an increase in fatalities or a serious irreversible or incapacitating illness

National Fire Protection Association The voluntary membership organization that promotes and improves fire protection and prevention and establishes safeguards against loss of life and property by fire (abbreviated NFPA)

National Response Center The sole national point of contact for reporting hazardous substance releases into the environment within the United States and its territories

National Response Team A federal interagency group chaired by EPA and the U.S. Coast Guard responsible for distributing technical, financial, and operational information about hazardous substance releases and oil spills

natural gas The flammable gas—primarily consisting of methane—formed in nature by the decomposition of animal and plant life

natural rubber (*see* rubber, natural)

necrosis The pathologic death of one or more cells, a portion of tissue, or an organ, typically resulting from exposure to certain corrosive substances or radiation

negatron A particle (β^-) emitted from certain radioisotopes that has all the properties of an electron

neoprene A synthetic elastomer composed of *cis*-1, 4-polychloroprene

nephrotoxicant A substance that causes kidney damage

nerve gas A highly volatile liquid the vapor of which acts as a neurotoxicant when inhaled

nesting The method of securing gas cylinders upright in a tight mass using a contiguous three-point contact system wherein the cylinders within a group have a minimum of three points of contact with other cylinders, walls, or bracing

neurotoxicant A substance that adversely affects the central nervous system

neutralization A double replacement reaction between an acid and a metallic hydroxide, resulting in the production of a salt and water

neutron A fundamental particle of which matter other than protium is composed, bearing no charge

newton The unit of force in the SI system

NFPA classes of fire The five types of fire, each distinguished from the others by the nature of its fuel

NFPA classes of oxidizers Any of four groups (class 1, class 2, class 3, and class 4) into which individual oxidizers are assigned based on their ability to affect the burning rate of combustible materials or to undergo self-sustained decomposition

NFPA hazard diamond A diamond-shaped figure divided into four quadrants, each of which is generally color-coded for each of a substance's three hazards (blue for health hazard, red for fire hazard, and yellow for chemical reactivity hazard), and in which is marked a number ranging from 0 to 4 that designates the degree of the relevant hazard

NFPA hazard system The procedure of assigning one of five numbers, 0 through 4, to designate the relative degree of three hazards (health, flammability, and chemical reactivity) for a given hazardous substance

nitramine An organic compound whose molecules contain one or more groups of atoms designated as $-\overset{|}{N}-NO_2$

nitrate ester An organic compound whose molecules contain one or more groups of atoms designated as $-O-NO_2$

nitrile An organic compound whose molecules contain the $-C\equiv N$ group of atoms

nitrocellulose A flammable solid or low explosive produced by chemically treating cellulose with a mixture of nitric acid and sulfuric acid

nitro compound An organic compound whose molecules contain the $-NO_2$ group of atoms

noble gas family The elements in Group 8A on the periodic table: helium, neon, argon, krypton, xenon, and radon

nonbonding electrons An atom's electrons that do not participate in bonding with other atoms

nonbulk packaging For purposes of DOT regulations, packaging that has (a) a maximum capacity of 119 gal (450 L) or 1ess as a liquid receptacle; (b) a maximum net mass of 882 lb (400 kg) or less and a maximum capacity of 119 gal (450 L) or less as a solid receptacle; or (c) a water capacity of 1000 lb (454 kg) or less for a gas receptacle

nonflammable Incapable of being burned

nonflammable, nonpoisonous compressed gas For purposes of DOT regulations, a material or mixture that exerts in its packaging an absolute pressure of 41 psi (280 kPa) or greater at 68 °F (20 °C), or a material or mixture that does not meet the definition of a division 2.1 or 2.3 material

nonionizing radiation (*see* radiation, nonionizing)

nonliquefied compressed gas For purposes of DOT regulations, a gas that when packaged under pressure for transportation is entirely gaseous at −58 °F (−50 °C) with a critical temperature of less than or equal to −58 °F (−50 °C)

nonmetal An element that does not conduct heat and electricity well, has low physical strength, and is neither ductile nor malleable

nonmetallic oxide (*see* basic anhydride)

nonoxidizing acid An acid incapable of reacting as an oxidizing agent

non-PCB transformer For purposes of TSCA regulations, a transformer whose dielectric fluid contains polychlorinated biphenyls at a concentration equal to or less than 50 ppm

normal-form class 7 material (non-special-form class 7 material) For purposes of DOT regulations, radioactive material that does not satisfy the definition of a special-form class 7 material

***n*-propyl group** The designation for the group of atoms $CH_3CH_2CH_2-$

nuclear fission (*see* fission, nuclear)

nuclear waste A waste produced in connection with the use of one or more radioisotopes; an unwanted by-product generated during the production of a nuclear weapon

nucleus (*see* atomic nucleus)

nurse tank For purposes of DOT regulations, a cargo tank having a capacity of 3000 gal (11 m^3) or less, painted white or aluminum, never filled to capacity, and used solely for the transportation of anhydrous ammonia

O

Occupational Safety and Health Act The federal statute that empowers the Occupational Safety and Health Administration (OSHA) to protect employees from occupational illnesses and injuries caused by exposure to hazardous chemicals in the workplace

octane number A number used to represent the antiknock properties of a petroleum fuel based on a comparison with the performance of a mixture of *n*-heptane and 2,2,4-trimethylpentane

octet rule The tendency for certain nonmetallic atoms to achieve electronic stability by gaining or sharing a total of eight electrons

olefin An alkene

oleum Concentrated sulfuric acid containing additional dissolved sulfur trioxide; fuming sulfuric acid

olfactory Pertaining to the sense of smell

olfactory fatigue The temporary inability to identify the odor of an airborne substance after prolonged exposure

Omnibus Trade and Competitiveness Act The federal statute whose aim is to enhance the competitiveness of American industry throughout international markets based in part on the required use of the metric system

orange oxide Uranium(VI) oxide (UO_3), an intermediate in the production of uranium hexafluoride

organic acid (*see* carboxylic acid)

organic chemistry The chemistry of the substances that contain carbon in their chemical composition

organic hydroperoxide A derivative of hydrogen peroxide in which one hydrogen atom in the H_2O_2 molecule has been substituted with an alkyl or aryl group; an organic compound having the general chemical formula $H-O-O-R$, where R is an alkyl or aryl group

organic peroxide A derivative of hydrogen peroxide in which both hydrogen atoms in the H_2O_2 molecule have been substituted with alkyl or aryl groups; an organic compound having the general chemical formula $R-O-O-R'$, where R and R' are alkyl or aryl groups; for purposes of DOT regulations, an organic compound containing oxygen in the bivalent $-O-O-$ structure and that may be considered a derivative of hydrogen peroxide in which one or more of the hydrogen atoms have been replaced by organic radicals

organochlorine pesticide Any chlorinated hydrocarbon used exclusively for controlling, destroying, preventing, repelling, or mitigating any pest

ORM-D For purposes of DOT regulations, any "other-regulated material" whose transportation is subject to certain regulations when the material is transported domestically

ortho (*o-*) Characterized by the substitution of the hydrogen atoms on adjacent (the first and second) carbon atoms on the benzene ring

other regulated material For purposes of DOT regulations, the classification of a hazardous material that does not conform to the definition of any of the nine hazard classes but is nonetheless considered to possess a dangerous property that could pose a risk to transportation personnel and equipment during its transit (abbreviated ORM)

outage The percentage by volume by which a container of liquid falls short of being filled to the brim

outer packaging For purposes of DOT regulations, the outermost enclosure of a composite or combination packaging together with any absorbent materials, cushioning, and other components necessary to contain and protect inner receptacles or inner packages

overpack For purposes of DOT regulations, an enclosure that is used by a single consignor to provide protection or convenience in handling of a package or to consolidate two or more packages

oxidant (*see* **oxidizer** and **oxidizing agent**)

oxidation A chemical process during which a substance unites with oxygen or another oxidizing agent; a chemical process in which the oxidation number of a species increases, sometimes accompanied by the loss in a number of its electrons

oxidation number A number assigned to an atom or ion by following certain established rules to reflect its capacity for combining with other atoms or ions

oxidation–reduction reaction A chemical reaction between one or more oxidizing and reducing agents; a redox reaction

oxidizer The substance reduced in an oxidation–reduction reaction; for purposes of DOT regulations, a substance that may enhance or support the combustion of other materials, generally by yielding its oxygen

oxidizing acid An acid capable of reacting as an oxidizing agent

oxidizing agent An oxygen-rich substance capable of readily yielding some or all of its oxygen; the substance reduced during an oxidation–reduction reaction

oxyacid An acid composed of hydrogen, oxygen, and a nonmetal

oxygenate An additive that promotes the complete combustion of vehicular fuels, thereby generating lesser amounts of carbon monoxide and other pollutants

oxyhemoglobin The compound that forms when oxygen combines with the blood's hemoglobin

ozonation A chemical reaction that involves the addition of ozone to a substance; the treatment of contaminated drinking water to kill the microorganisms (including cryptosporidium) that cause waterborne diseases

ozone The allotrope of oxygen whose molecules are represented as O_3, that is, three oxygen atoms per molecule

ozone, ground-level The ozone that forms in the lower atmosphere by photochemically catalyzed reactions between volatile organic compounds and nitrogen oxides

ozone hole A thin spot in the ozone layer that has been linked to destruction of stratospheric ozone by chlorofluorocarbons and other substances

ozone layer The region of the atmosphere rich in ozone, approximately 10 to 19 mi (16 to 30 km) from Earth's surface

ozone, stratospheric The naturally occurring layer of ozone in the upper atmosphere that shields Earth's surface from the Sun's harmful ultraviolet rays

P

package For purposes of DOT regulations, the means of containment used to transport a hazardous material from one location to another

packaging For purposes of DOT regulations, the receptacle and any other components or materials necessary for the receptacle to perform its containment function and to ensure conformance with prescribed minimum requirements

packing group For purposes of DOT regulations, a grouping according to the degree of danger presented by hazardous materials (Packing Group I indicates great danger; Packing Group II, medium danger; and Packing Group III, minor danger)

para (*p-*) Characterized by the substitution of the hydrogen atoms on the first and fourth carbon atoms on the benzene ring

particulate matter Tiny solid particles typically suspended in the air or atmospheric emissions, constituting a dust, fume, mist, smoke, or smog

parts per billion The concentration of a mixture in which a certain number of units are contained within a billion of the same units (abbreviated ppb)

parts per million The concentration of a mixture in which a certain number of units are contained within a million of the same units (abbreviated ppm)

parts per trillion The concentration of a mixture in which a certain number of units are contained within a trillion of the same units (abbreviated ppt)

pascal The SI unit of pressure, equal to a force of one newton applied to an area of one square meter (symbol Pa)

passivity The property of metals like aluminum and chromium that causes them to lose their normal chemical reactivity after being treated with strong oxidizing agents

pathogen For purposes of DOT regulations, a virus or microorganism (including its viruses, plasmids, or other genetic elements, if any) or a proteinaceous infectious particle (prion) that has the potential to cause disease in humans or animals

PCB-contaminated transformer For purposes of TSCA regulations, a transformer whose dielectric fluid contains polychlorinated biphenyls at a concentration greater than 50 ppm but less than 500 ppm

PCB transformer For purposes of TSCA regulations, a transformer whose dielectric fluid contains polychlorinated biphenyls at a concentration equal to or exceeding 500 ppm

percentage The concentration of a mixture expressed as parts of a unit in 100 of the same units (symbol %)

period A horizontal row on the periodic table

periodic law The observation that the chemical properties of the elements are periodic functions of their atomic numbers

periodic table A compilation of the known elements into periods and families by increasing atomic number, so that elements with similar chemical properties are in the same column (or family)

permissible exposure limit For purposes of OSHA regulations, the time-weighted average threshold limit value of substances listed at 29 C.F.R. §1910.1000 to which workers can be exposed continuously during an 8-hr work shift without suffering ill effects (abbreviated PEL)

pesticide For purposes of FIFRA regulations, a substance or mixture of substances intended for preventing, destroying, repelling, or mitigating any pest, or intended for use as a plant regulator, defoliant, or desiccant

petrochemical A substance obtained directly or indirectly from natural gas, crude petroleum, or a petroleum fraction

petroleum (see petroleum, crude)

petroleum coke The residue remaining after asphalt is heated in the absence of air

petroleum, crude The liquid mixture consisting primarily of hydrocarbons formed in nature by the decomposition of plant life, and which has not been processed in a refinery

petroleum distillate A fraction of crude petroleum obtained by the vaporization of its components within a specific temperature range followed by its condensation

petroleum refinery A facility primarily engaged in producing on a commercial scale gasoline, kerosene, fuel oils, lubricants, and other products through fractionation, alkylation, redistillation of unfinished derivatives, cracking, blending, and other processes

pH A numerical scale from 0 to 14 used to quantify the acidity of alkalinity of a solution with neutrality indicated as 7

phase diagram A plot of the temperature versus pressure indicating the conditions at which a substance exists as a gas, liquid, and solid

phenol An organic compound whose molecules have a hydroxyl group ($-OH$) bonded directly to the benzene ring

phenolic resin Any of a group of polymeric substances obtained from the condensation of phenol and formaldehyde or chemically similar substances

phenyl group The designation for the benzene ring when it is named as a substituent on another molecule (C_6H_5-)

phlegmatizer A substance that is capable of reducing the ability of an explosive to detonate

phosphorus necrosis (see phossy jaw)

phossy jaw The disfiguring affliction caused by overexposure to phosphorus vapor

photodecomposition The breakdown of a substance by light

photon A massless packet of electromagnetic energy

photosynthesis The chemical process by which plants use carbon dioxide and water to construct cellulosic materials with the aid of sunlight and chlorophyll

phthalate An ester of o-, m-, or p-phthalic acid

physical change A change that does not result in an alteration of the chemical identity of a substance, like vaporization or sublimation

physical property A phenomenon that a substance exhibits when it undergoes a physical change, like melting and boiling

physical states of matter The three different forms in which matter is generally encountered: solid, liquid, and gas

pickling The combination of chemical reactions associated with removing surface impurities from metals by dipping them into an acid bath

placard For purposes of DOT regulations, a sign displayed by the carrier on bulk packaging, freight containers, transport vehicles, unit containment devices, or railcars to rapidly communicate hazard information concerning the hazardous material being transported

plastic explosive A high explosive that has been mixed with a gummy binder and manufactured in a flexible, hand-malleable form for an intended use at 77 °F (25 °C) such as demolition

plasticizer A substance, such as a phthalate ester, that when added in a prescribed amount to a polymeric formulation facilitates processing and provides flexibility and toughness to the end product

plastics Any of a variety of synthetic polymeric substances that can be molded and shaped

poison (see toxic substance and poisonous material)

poison gas (*see* **gas poisonous-by-inhalation**)

poisonous material For purposes of DOT regulations, a material other than a gas that is known to be so toxic to humans as to afford a hazard to health during transportation, *or that* in the absence of adequate data on human toxicity, is presumed to be toxic to humans because of data obtained from prescribed oral, dermal, and inhalation toxicity tests performed on animals; *or* is an irritating material with properties similar to those of tear gas that causes extreme irritation, especially within confined spaces

polyatomic ion An ion having more than one constituent atom

polychlorinated biphenyl Any chlorinated derivative of biphenyl, the manufacture, sale, and distribution of which have been banned in the United States; the major constituent of certain dielectric fluids formerly contained within capacitors, transformers, and other electrical equipment (abbreviated PCB)

polychlorinated dibenzofuran Any chlorinated derivative of furan, generated as an unwanted by-product during certain uncontrolled incineration, paper pulp bleaching, and chemical manufacturing operations (abbreviated PCDFs)

polychlorinated dibenzo-*p*-dioxin Any chlorinated derivative of dibenzo-*p*-dioxin, generated as an unwanted by-product during certain uncontrolled incineration, paper pulp bleaching, and chemical manufacturing operations (abbreviated PCDDs)

polyester A polymer consisting of recurring ester groups

polymer A high-molecular-weight substance produced by the linkage and cross-linkage of its multiple subunits (monomers)

polymerization The chemical reaction in which monomer molecules are linked and cross-linked into macromolecules

polynuclear aromatic hydrocarbon An aromatic hydrocarbon whose molecules have two or more mutually fused benzene or other rings (abbreviated PAH)

portable tank For purposes of DOT regulations, a closed container that is not fixed in position, has a liquid capacity of 1000 lb (454 kg) or less, and is designed to be loaded into, or on, or temporarily attached to a transport vehicle and equipped with skids, mounting, or accessories to facilitate handling of the tank by mechanical means

positron A positively charged particle (β^+) emitted by some radioisotopes but otherwise having the characteristics of an electron

preferred route (preferred highway) For purposes of DOT regulations, the designated route or highway on which a shipper or carrier is required to transport an explosive or radioactive material

premium diesel oil (*see* **diesel oil, premium**)

pressure A force applied to a unit area

pressure-relief valve A device that opens and closes at preestablished pressures on cylinders, cargo tanks, stationary tanks, and rail tankcars used to store or transport a compressed gas or flammable liquid

pressure tank (*see* **pressure vessel**)

pressure vessel A closed tank or other vessel that has been designed to store flammable liquids at pressures above 15 psi_g (100 kPa)

primary explosive (*see* **explosive, primary**)

primary hazard class (*see* **hazard class/division**)

primer (*see* **initiator**)

product A substance produced as a result of a chemical change

progression The third stage of cancer, during which tumors evolve into malignant growths in the body

promotion The second stage of cancer, during which the formation of malfunctioning cells accelerates to produce tumors

propagation The second mechanistic step of a chemical reaction during which reactant species repetitively form new species by means of multiple reactions

propane The alkane having the chemical formula C_3H_8

propellant The component of an explosive train whose purpose is to propel or provide thrust to a round of ammunition

proper shipping description For purposes of DOT regulations, the basic description of a hazardous material, its weight or volume, and the number of package units in which it is contained

proper shipping name For purposes of DOT regulations, the name in Roman print assigned to a hazardous material listed in the Hazardous Materials Table published at 49 C.F.R. §172.101

protective-action distance For purposes of DOT regulations, the minimum distance downwind from a transportation mishap involving the release of a hazardous material at which emergency responders and the public can reasonably anticipate that their health, safety, and welfare will be preserved

protective-action zone For purposes of DOT regulations, the area in which persons may become incapacitated, unable to take protective action, and/or incur serious, irreversible health effects

protein foam A film-forming foam prepared from natural protein materials and used to extinguish fires

protium The isotope of hydrogen consisting of one proton

proton A fundamental particle of which all atoms are composed, bearing a charge of $+1$

psi The unit of pressure expressed as pounds per square inch (psi_a means pounds per square inch absolute; psi_g means pounds per square inch gauge)

pulmonary Pertaining to the lungs

pulmonary agent A chemical warfare agent that damages the respiratory tract and cause pulmonary edema

pulmonary edema The excessive accumulation of fluid within the lungs

pyrophoric Capable of igniting instantaneously on exposure to air

pyrophoric material For purposes of DOT regulations, a liquid or solid that even in small quantities and without an external ignition source may ignite within 5 min after coming in contact with air when tested by prescribed procedures

R

radiant energy Energy, including heat, light, and X rays, that is transmitted by radiation as electromagnetic waves

radiation The transmission of energy in the form of electromagnetic waves from certain forms of matter; the mechanism by which heat is transferred between two materials not in contact

radiation area For purposes of OSHA regulations, an area accessible to personnel in which the ionizing radiation exists at a level in which the body could receive a dose in excess of 0.05 mSv in any one hour, or a dose in excess of 1 mSv in any 5 consecutive days

radiation, ionizing The high-energy electromagnetic radiation emitted by radioisotopes (alpha radiation, beta radiation, and gamma radiation)

radiation, nonionizing Low-energy electromagnetic radiation including microwaves and radio waves

radiation sickness The following adverse health effects resulting from short- or long-term exposure to radiation: nausea, vomiting, diarrhea, weakness, shock, skin sores, hair loss, and long-term decrease in white blood cell count

radioactive material A material containing an isotope that spontaneously emits ionizing radiation; for purposes of DOT regulations, a material containing one or more radioisotopes for which both the activity concentration and the total activity in the consignment exceed the applicable A_1 and A_2 values

radioactivity The nuclear property associated either with the spontaneous emission of alpha, beta, and/or gamma radiation, or the capture of an extranuclear electron

radioisotope An atomic nucleus that undergoes a spontaneous change by emitting a particle or by absorbing an extranuclear electron

radionuclide (see radioisotope)

Rankine temperature The absolute temperature obtained by adding 459.67 °F to a Fahrenheit temperature

rate of reaction The speed at which a chemical transformation occurs; the amount of a product formed, or reactant consumed, per unit of time

rayon The commercial fiber manufactured from regenerated cellulose

reactants Two or more substances that enter into a chemical reaction

reactivity For purposes of RCRA regulations, the characteristic of a waste that exhibits any of the following properties (40 C.F.R. §261.23): readily undergoes a violent change without detonating; reacts violently with water; forms potentially explosive mixtures with water; generates toxic gases, vapors, or fumes in a quantity sufficient to present a danger to human health or the environment; is capable of detonation or an explosive reaction if subjected to a strong initiating source or if heated under confinement; is readily capable of detonation or explosive decomposition or reaction at standard temperature and pressure; is an explosive material in hazard class 1.1, 1.2, or 1.3; or is a forbidden explosive

readily combustible solid For purposes of the DOT regulations, any of the following materials: a solid that can cause a fire through friction; a solid that burns at a rate exceeding 0.087 in./s (2.2 mm/s) when tested by prescribed means; and any metal powder that can be ignited and react over the whole length of a sample in 10 min or less when tested by prescribed procedures

redox reaction (see oxidation–reduction reaction)

reducing agent A substance that takes oxygen from another substance; the substance oxidized during an oxidation–reduction reaction

reduction A chemical process during which oxygen or another oxidizing agent is removed from a compound; a chemical process that is sometimes accompanied by the gain of hydrogen by a species; a chemical process that is sometimes accompanied by the gain of electrons

refining The combination of processes conducted at a petroleum refinery such as fractional distillation of crude oil, alkylation, cracking, and hydrotreating

refrigerant A substance that can absorb heat at one temperature and be readily compressed by a heat pump to a higher temperature and pressure, usually with a change in the state of matter, at which the absorbed heat is discharged

refrigerant gas For purposes of DOT regulations, an acceptable name for any chlorofluorocarbon or other halogenated hydrocarbon whose intended commercial use is as a refrigerant

regular diesel oil (see diesel oil, regular)

relative humidity A measurement in percent at a given temperature of the amount of water vapor present in the air compared with saturation at 100%

release Spilling, leaking, pumping, pouring, emitting, emptying, discharging, injecting, leaching, dumping, or disposing of a hazardous material or hazardous substance into the environment

rem The amount of ionizing radiation that produces the same biological effect as one röentgen of X rays or gamma rays

reportable quantity For purposes of DOT regulations, the amount listed at 40 C.F.R. §302.4 and 49 C.F.R. §172.101 of a "hazardous substance" within the meaning of DOT and CERCLA, including certain radioisotopes, the release of which triggers mandatory notification to the National Response Center (abbreviated RQ)

reporting mark A sequence of two to six identification letters assigned by the American Association of Railroads to rail carriers who operate in North America

reproductive toxicant A substance capable of producing a negative impact on a fetus on exposure of the mother to the substance

residential radon The radon that accumulates within a dwelling from the decay of naturally occurring uranium and thorium radioisotopes

Resource Conservation and Recovery Act The federal statute that empowers EPA to regulate the treatment, storage, and disposal of hazardous wastes (abbreviated RCRA)

respiration The chemical process in which the oxygen in inhaled air is conveyed from the lungs to the tissues and the cells of the body, where carbon dioxide and water are produced

respiratory toxicant A substance capable of inflaming the airways or otherwise decreasing lung function

respiratory tract The structures and organs involved in breathing, including the nose, larynx, trachea, bronchi, bronchioles, and lungs

riot-control agent A substance that rapidly produces sensory irritation or a disabling physical effect in humans and animals and disappears within a short time following termination of exposure

right-to-know law For purposes of OSHA regulations, the federal statute that addresses workers' right to know about the hazards associated with substances to which they are being exposed in the workplace

risk group For purposes of DOT regulations, a ranking of a microorganism's ability to cause injury through development of a disease as defined by criteria developed by the World Health Organization (WHO) based on the severity of the disease caused by the organism, the mode and relative ease of transmission, the degree of risk to both an individual and a community, and the reversibility of the disease through the availability of known and effective preventive agents and treatment

risk management plan For purposes of EPA regulations, the document that addresses the foreseeable risks relating to the storage, handling, and use of a hazardous air pollutant above a threshold amount, including a response plan to eliminate or mitigate its release into the environment

rodenticide A substance designed to kill rodents or to prevent them from damaging food, crops, or other materials

röentgen The amount of radiation that produces 2.1×10^9 units of charge in 1 mL of dry air at standard atmospheric pressure

room conditions A temperature of 68 °F (20 °C) and pressure of 14.7 psi (101.3 kPa)

rubber A polymer that is simultaneously elastic, airtight, water-resistant, and long-wearing

rubber, natural The polymer produced from the latex that exudes from cuts in the bark of the South American rubber tree

rubber, synthetic A polymer produced from substances other than, or in addition to, the latex from the South American rubber tree and having similar properties to those of natural rubber

rubber, styrene-butadiene The synthetic rubber produced by the polymerization of styrene and butadiene

rusting The chemical conversion of iron to ferric hydroxide

S

safety can For purposes of OSHA regulations, an approved, closed container of not more than 5-gal (18.9-L) capacity, having a flash-arresting screen, spring-closing lid and spout cover, and so designed that it will safely relieve internal pressure when subjected to fire exposure

safety match (*see* **match, safety**)

salt A compound in which the hydrogen ion from an acid has been substituted with a metallic ion

saponification The conversion of triglycerides into soaps

saturated The description of the air when it contains the maximum amount of water it can hold at a given temperature

saturated hydrocarbon (*see* **hydrocarbon, saturated**)

secondary containment A safeguarding method (such as diking around a primary containment vessel) used to prevent the unplanned release of a hazardous material into the environment

secondary explosive (*see* **explosive, secondary**)

second-degree burn An injury of the skin to a limited depth caused by exposure to heat; a partial-thickness burn

self-heating material For purposes of DOT regulations, a material that is liable to self-heat when in contact with air and without an energy supply; a material that exhibits spontaneous ignition, or if the temperature of the sample exceeds 392 °F (200 °C) during a 24-hr test period when tested by prescribed procedures

self-reactive material For purposes of DOT regulations, a material that is thermally unstable and can undergo a strongly exothermic decomposition even without the participation of atmospheric oxygen

service pressure For purposes of DOT regulations, the pressure designated in units of psi_g at 70 °F (21 °C) for each specific type of authorized compressed gas cylinder

shipping description For purposes of DOT regulations, the following minimum information for a hazardous material: proper shipping name (including the technical name, when required); hazard class or division; identification number; packing group; total quantity by mass, volume, or as otherwise appropriate; and the number and types of packages

shipping paper The shipping order, bill of lading, air bill, dangerous-cargo manifest, or similar shipping document issued by a shipper or carrier as required by DOT at 49 C.F.R. §§172.202, 172.203, and 172.204

short-term exposure limit The concentration to which workers can be exposed continuously for a short period of time without suffering irritation, chronic, or irreversible tissue damage, or narcosis of sufficient degree to increase the likelihood of accidental injury, impair self-rescue, or materially reduce work efficiency, and provided that the daily threshold limit value, time-weighted average is not exceeded (abbreviated STEL)

short-term health effect Disease or impairment resulting from single or multiple exposures to a hazardous substance over a short period, like one week

SI The abbreviation for the International System of Units (in French, le Système International d'Unités), a modernized metric system accepted as the means of measurement by the scientific community

SI base units The accepted units of measurement adopted by the International Bureau of Weights and Measures (Bureau International des Poids et Mesures, Sèvres, France) such as the meter for length, kilogram for mass, and kelvin for temperature

sievert The SI unit of dose-equivalent, equal to 100 rem

signaling smoke A type of smoke bomb whose actuation communicates prearranged information to troops or others who subsequently observe the signal from a distance

silane An organic compound whose molecules are composed of silicon and hydrogen atoms

silk A naturally occurring protein produced by the action of silkworms

single displacement reaction (*see* **single replacement reaction**)

single replacement reaction The type of chemical reaction in which one element replaces another within a compound

SI system of units The scientific standard of measurement that employs a set of units describing length, mass, time and other attributes of matter

slaking The chemical process associated with reacting calcium oxide (lime) with water to form calcium hydroxide

small-arms ammunition Shotgun, rifle, pistol, or revolver cartridges

smelter A facility that separates or refines a metal from its ore by means of a chemical change

smelting A chemical process for isolating an element from its naturally occurring ore by reacting the ore with the constituent carbon in coke

smog A mixture of air pollutants including volatile organic compounds that undergo photochemical reactions in the atmosphere to form ozone

smoke The visible plume associated with incomplete combustion and consisting of carbon particulates, ash, water droplets, and other matter

smoke bomb A device containing substances that produce smoke when actuated, generally for use in concealing military operations

smokeless powder A form of nitrocellulose

solid Matter that possesses a definite volume and a definite shape; for purposes of DOT regulations, a material that is neither a gas nor a liquid

soluble Capable of passing into a particular solvent such as water or oil to form a solution

solution A homogeneous mixture of two or more substances

solvent A substance capable of dissolving another substance to form a solution; the component of a solution that is present in the greater amount

soot The agglomeration of carbon particulates generated during the incomplete combustion of carbonaceous materials

sour crude Untreated crude oil or the fraction thereof that contains sulfurous and nitrogenous compounds

special-form class 7 (radioactive) material For purposes of DOT regulations, an indispersible, solid radioactive material that is either a single solid piece or a sealed capsule, at least one dimension of which is not less than 0.2 in. (5 mm), and in the case of the capsule, contains radioactive material that can be opened only by destroying it

specification markings For purposes of DOT regulations, the packaging identification markings including, where applicable, the name and address or symbol of the packaging manufacturer or approval agency

specific gravity The mass of a given volume of matter compared with the mass of an equal volume of water

specific heat The ratio of the heat capacity of a substance to the heat capacity of water at the same temperature

spectator ion One or more of the ions that do not participate in an oxidation–reduction reaction

sphygmomanometer A device used to measure blood pressure

spirits of nitroglycerin A medication containing a nontoxic dose of nitroglycerin often prescribed for individuals afflicted with heart and circulatory diseases

spontaneous combustion (see combustion, spontaneous)

spontaneously combustible material For purposes of DOT regulations, either a pyrophoric material or a self-heating material

standard atmospheric pressure The average value of the atmospheric pressure at sea level

stationary tank (see tank, stationary)

Stockholm Convention on Persistent Organic Pollutants (abbreviated POPs treaty) The United Nations treaty that bans or limits the use of DDT and other organochlorine pesticides

stowage The placement of materials onboard a vessel

stratospheric ozone (see ozone, stratospheric)

strike-anywhere match (see match, strike-anywhere)

strong A designation for acids and bases that are completely ionized

strong acid Any acid that predominantly forms hydrated hydrogen ions when dissolved in water

strong base Any base that predominantly forms hydrated hydroxide ions when dissolved in water

structural formula A designation in either a complete or condensed fashion of the manner by which each atom in a formula bonds to other atoms

structural isomerism (see isomerism, structural)

styrene–butadiene rubber (see rubber, styrene–butadiene)

sublimation A physical change in which a substance passes from the solid directly to the gaseous state of matter, without first liquefying

subsidiary hazard For purposes of DOT regulations, any hazard of a material (other than its primary hazard) whose hazard class numbers are listed in column 6 of the Hazardous Materials Table

substance A homogeneous material having a constant, fixed chemical composition; an element or compound

substituent A side chain or appendage of atoms on a main chain

Superfund Amendments and Reauthorization Act The federal statute that requires EPA when selecting a remedial action, to consider the standards and requirements in other environmental laws and regulations and to include citizen participation in the decision-making process (abbreviated SARA)

Superfund Law (see Comprehensive Environmental Response, Compensation, and Liability Act)

surface-contaminated object For purposes of DOT regulations, a solid object that is not itself radioactive but has radioactive material distributed on its surface (abbreviated SCO)

surfactant A substance that reduces the interfacial tension between two immiscible liquids

syndiotactic polymer A type of polymer whose macromolecules have alternately positioned side chains along a chain of carbon atoms

synergism The aspect of two substances that interact to produce an effect that is greater than the sum of the substances' individual effects of the same type

syngas or **synthesis gas (see water gas)**

synthesis reaction (see combination reaction)

synthetic rubber (see rubber, synthetic)

systemic Referring to the body as a whole

systolic The top number of a blood pressure reading, typically less than 130 mmHg; the pressure exerted within an individual's arteries as the heart is beating

T

tank For purposes of DOT regulations, a container consisting of a shell and heads that form a pressure-tight vessel having openings designed to accept pressure-tight fittings or closures but excludes any appurtenances, reinforcements, fittings, or closures

tank farm Property on which multiple stationary storage tanks have been situated

tank, stationary Packaging designed primarily to be installed in a stationary position and not intended for loading, unloading, or attachment to a transport vehicle

tear gas (see lacrimator and riot-control agent)

temperature The condition of a body that determines the transfer of heat to or from other bodies by comparison with established values for water, which is used as the standard

termination The third mechanistic step of a chemical reaction during which certain reactive species formed during the propagation step unite to produce the product

tetryl The common name of the chemical explosive 2,4,6-trinitrophenylmethylnitramine

textile A fabric that has been produced by weaving fibers

thermal cracking (*see* **cracking, thermal**)

thermite The mixture of 27% powdered aluminum and 73% iron(III) oxide

thermite reaction The chemical reaction used in some welding operations and incendiary weapons during which elemental iron is produced when elemental aluminum reduces iron(III) oxide

thermogenesis The physiological activity of microorganisms that results in the slow liberation of heat within a material and when absorbed by the material, can cause its spontaneous combustion

thermoplastic polymer A polymer that softens when heated but return to its original condition on cooling to ambient temperature

thermosetting polymer A polymer that cannot be remolded once it has solidified

third-degree burn An injury of the subsurface fat, nerves, and muscular structure, caused by exposure to heat; full-thickness burn

threshold limit value A guideline standard established by the American Conference of Governmental Industrial Hygienists (ACGIH) for airborne concentrations to which an average worker may be exposed day after day without experiencing adverse health effects (abbreviated TLV)

time-weighted average The average concentration of a substance measured over a period of exposure, distinct from a concentration obtained during an instantaneous exposure (abbreviated TWA)

TNT equivalent A comparison of the mass of an explosive substance with the mass of TNT that produces the same explosive power

tonne The metric unit of mass equivalent to 1000 kilograms

torpex The explosive material consisting of a mixture of cyclonite, trinitrotoluene, and aluminum fines

toxic A substance or material associated with its ability to cause disease, injury, or death in exposed humans and animals at relatively low concentrations

toxicant (*see* **toxic substance**)

toxicity For purposes of RCRA regulations, the characteristic of a waste, as determined by implementing prescribed test procedures, in which a representative sample of the waste is found to contain one or more of certain constituents at concentrations equal to or greater than the levels designated at 40 C.F.R. §261.24

toxic metal Any of certain metals or metalloids whose consumption, even in minute amounts, can cause illness or death

toxicology The study of the effects of toxic or poisonous substances on living organisms

toxic pollutant A substance designated by EPA under the provisions of the Federal Water Pollution Prevention and Control Act and listed at 40 C.F.R. §401.15 that after discharge and on exposure, ingestion, inhalation, or assimilation into an organism will, on the basis of information, cause death, disease, behavioral abnormalities, cancer, genetic mutations, physiological malfunctions (including malfunctions in reproduction), or physical deformations in organisms or their offspring

toxic substance A substance that can negatively affect the life processes and can cause death, temporary incapacitation, or permanent harm to humans or animals; a substance or mixture designated by EPA under the provisions of the Toxic Substances Control Act

Toxic Substances Control Act The federal statute that empowers EPA to regulate all aspects of the manufacture of toxic substances other than pesticides and drugs, including the obtaining of production and test data from industry on those toxic substances whose manufacture, processing, distribution, use, or disposal could present an unreasonable risk of injury or damage to the environment (abbreviated TSCA)

toxin A poisonous substance produced by a living organism or bacterium

transformer A device designed to transfer electrical energy between electrical circuits of different voltage

transmutation The transformation of a radioisotope to another isotope as the result of a nuclear reaction

transport index For purposes of DOT regulations, a dimensionless number that shippers and carriers assign to certain radioactive materials to designate the degree of control to be exercised during their transit

transport vehicle A motor vehicle or railcar used for the transportation of cargo by any mode

triene An organic compound each of whose molecules has three carbon–carbon double bonds

triple bond A covalent bond composed of six shared electrons between two atoms

tritium The radioisotope of hydrogen consisting of one proton and two neutrons

triton The nucleus of the tritium atom

tumor A mass of tissue that persists and grows independently of its surrounding structures and has no known physiological function

type A package For purposes of DOT regulations, a package designed to transport an A_1 or A_2 quantity of radioactive material, as appropriate

type B package For purposes of DOT regulations, a package designed to transport greater than an A_1 or A_2 quantity of radioactive material, as appropriate

U

ullage (*see* **outage**)

unsaturated hydrocarbon (*see* **hydrocarbon, unsaturated**)

unstable The characteristic of some hazardous materials to vigorously polymerize, decompose, condense, or become self-reactive when shocked or exposed to certain pressure or temperature conditions; the characteristic of radioisotopes to spontaneously disintegrate and form new isotopes

up-and-down dose The concentration in milligrams per kilogram of a substance that kills a laboratory animal within 48 hours when administered according to a prescribed procedure

upper explosive limit The concentration of a gas or vapor in the air above which a flame will not propagate when exposed to an ignition source (abbreviated UEL)

uranium-enrichment processes Isolation techniques for concentrating fissile uranium-235

V

valence electrons An atom's electrons that participate in chemical bonding to other atoms

valve A device that controls the flow or pressure of a gas within a device

vapor The gaseous form of a substance that exists either as a solid or liquid at normal ambient conditions

vapor density The mass of a vapor or gas compared with the mass of an equal volume of another gas or vapor, generally air

vaporization The process during which the molecules of a liquid escape as a gas or vapor

vapor pressure The pressure exerted within a confinement vessel by the vapor of a substance in equilibrium with its liquid; a measure of a substance's propensity to evaporate

vasodilator A substance, like nitroglycerin, exposure to which is capable of causing the blood vessels to widen

vesicant A substance used in chemical warfare to blister the skin and body tissues of the enemy

vessel Any tank, cylinder or similar device used for containment; any watercraft or other artificial contrivance used, or capable of being used, as a means of transportation on water

vinyl group The designation for the group of atoms $CH_2=CH-$

vinyl halide A halogenated derivative of an alkene in which one or more of the hydrogen atoms bonded to one or both of the carbon atoms constituting the carbon–carbon double bond have been substituted with halogen atoms; an organic compound having the formula $CH_2=CH-X$, where X is the chemical symbol of a halogen

vinyl polymer A polymer produced from one or more vinyl compounds

volatile The quality of a solid or liquid when it passes into the vapor state at a given temperature

volatile organic compounds Certain substances having a boiling point of less than approximately 392 °F (200 °C), including the constituents of gasoline vapor (VOCs); organic compounds that undergo photochemical reactions in the atmosphere to form ozone or other air pollutants

volume The capacity of matter confined within a tank or container

vulcanization The heating of natural or synthetic rubbers with elemental sulfur or certain sulfur-bearing compounds to produce disulfide cross-linking bonds within segments of their macromolecules

W

water gas The mixture of carbon monoxide and hydrogen produced by blowing steam through a bed of red-hot coke

water-reactive substance A substance that by its chemical reaction with water is likely to become spontaneously flammable, emit flammable or toxic gases, or generate sufficient heat to self-ignite or cause the ignition of nearby combustible materials

weak A designation for acids and bases that are slightly ionized

weak acid Any acid that predominantly retains its molecular identity when dissolved in water

weak base Any base that predominantly retains its unit identity when dissolved in water

weight The force of gravity acting on the mass of a particular object

wet-chemical fire extinguisher An extinguisher consisting of a substance dissolved in water that is capable of forming a soapy foam blanket when applied to burning oil

wood alcohol The common name for methanol produced by heating wood in the absence of air

wool The naturally occurring fiber obtained from the coats of sheep, llamas, and several other animals

Y

yellowcake Uranium(IV) oxide (UO_2), an intermediate in the production of uranium hexafluoride

Z

Ziegler–Natta catalyst Any of a group of compounds produced from titanium tetrachloride and an aluminum alkyl compound that is used catalytically to initiate addition polymerization

zinc dust Metallic zinc in the form of a powder

INDEX